JOURNEY TOWARDS CHRISTMAS

Convoy at Dusk

JOURNEY TOWARDS CHRISTMAS

*Official History of the 1st Ammunition Company
Second New Zealand Expeditionary Force
1939 - 45*

by

S. P. LLEWELLYN

The Naval & Military Press Ltd

Published by

The Naval & Military Press Ltd
Unit 5 Riverside, Brambleside
Bellbrook Industrial Estate
Uckfield, East Sussex
TN22 1QQ England

Tel: +44 (0)1825 749494

www.naval-military-press.com
www.nmarchive.com

In reprinting in facsimile from the original, any imperfections are inevitably reproduced and the quality may fall short of modern type and cartographic standards.

For help and encouragement in writing this history the author wishes to thank Lieutenant-Colonel W. A. T. McGuire, Majors P. E. Coutts, J. D. Fenton, R. C. Gibson, and S. A. Sampson, Captain B. J. Williams, WO I I. McBeth, WO II A. L. Salmond, Sergeants S. T. Midgley and G. McG. Mowat, Corporal G. J. McKay, and Drivers D. Falconer, G. W. Harte, and G. P. Laverick.

FOREWORD

By LIEUTENANT-GENERAL SIR BERNARD FREYBERG,
VC, GCMG, KCB, KBE, DSO, LL D.

THE publication of this history gives me an opportunity, which I welcome, of paying a tribute to the work in the Middle East and Italy of one of our units, the 1st Ammunition Company.

This book is a record of their achievements. They were a 1st Echelon unit, and were one of the few of our Division that took part in General Wavell's Desert Campaign against the Italians in 1940. From then on they fought through the whole war and finished up their great service on VE Day near Trieste. In all they were six years overseas, fifty-three months of which were spent in active operations, which included the campaigns in the Western Desert and Italy.

I am inclined to think that the New Zealand Division's greatest contribution to the war effort was during the early years in North Africa.

It has been said that 'The Western Desert was a tactician's paradise, and a Quartermaster-General's nightmare'. The campaign in North Africa was certainly a war of movement. Mobility and administration played a decisive part. A motorised force was needed. New Zealanders were ideal men for this class of warfare. They found their way across the unmapped, featureless Desert by night as well as day with uncanny skill, almost by instinct.

In this book is the story. It tells us of their formation, their work in training and in battle. It deals not only with our successes but also our disasters, such as Greece. It also tells the story of the men on leave and in the rear areas. It is a record of one of the most efficient and well trained units of the Second New Zealand Expeditionary Force.

FOREWORD

When I said good-bye to the Division in Italy I said of the New Zealand Army Service Corps, of which they are a part, that throughout the whole war they had never failed us.

I would go further and say without their resource and skill we could never have attempted 'the turning movements' in North Africa at which this Division of ours was so formidable.

I hope many will buy this book and that military students to come will study it, and glean from the pages the lessons which abound therein.

Bernard Freyberg
GOVERNOR-GENERAL

One time General Officer Commanding,
　　Second New Zealand Expeditionary Force

CONTENTS

		Page
	FOREWORD	ix
1	DESIGN FOR A UNIT	1

   Mobilisation—Training in New Zealand—Embarkation and voyage—Arrival at Port Tewfik.

2 HOW TO SEE EGYPT ON A POUND A WEEK 15

   Maadi Camp—Cairo—Advanced training—Transport is issued—Duties in Cairo Sub-area—Ferry service—A Section at El Daba—Desert training.

3 MEETING AT AMIRIYA 27

   (1) *Birth of a Happy Section.* C Section at Burnham—The Bombay incident—Arrival in Egypt.
   (2) *Working for Wavell.* A Section in the Desert—Italian prisoners—Christmas—Move to Helwan—C Section at Gebel Ruzza—Speculations.
   (3) *Diversion to a Dragon-Slaying.* B Section in England—Leave, work, and training—The Battle of Britain—Voyage to Egypt—The unit at Amiriya.

4 PICNIC BEFORE A THUNDERSTORM 51

   Piræus and Athens—First impressions—Journey north—The unit starts work—Back to Larissa—Salvage—Stukas and Messerschmitts—First casualties.

5 THE THUNDERSTORM 65

   Withdrawing troops—A bad day—Thermopylæ—Hiding from Stukas—Destruction of transport—With the 6th Brigade.

6 WITHDRAWAL FROM GREECE 86

   Anzac Day—Kea Island—Detachment at Nauplion—Back through Athens—Corinth and Tripolis—Capture at Kalamata—The last ship.

Contents

		Page
7	ISLAND INTERLUDE	103

    Arrival in Crete—Drivers as infantry—Winning the toss—Back to Egypt.

8	MURDER ON THE OLD HOOK	112

    Crown and Anchor—Reorganisation and leave—Move to the Desert—At Fuka.

9	FOX IN THE FOWL RUN	121

    Across the frontier—Chased by tanks—Travelling with Rommel—Sidi Azeiz.

10	THURSDAY, FRIDAY, AND SATURDAY	155

    Sidi Azeiz overrun—A long chase—Through the corridor to Tobruk.

11	PRISON AND THE MUSHROOM COUNTRY	172

    Wharfingers in Tobruk—Back to Fuka—Prisoners at Bardia—Christmas—A Section in Cyrenaica—Reunion at Maadi.

12	SYRIA	198

    Reorganisation—Move to Syria—Work and training—Fraternisation and the Flag—Back to Egypt.

13	WHILE SHEPHEARD'S WATCHED	221

    Carrying the 20th Battalion—Surrounded—Minqar Qaim—Stuka time—A Section returns.

14	A STUDY IN DISCOMFORT	246

    Flies—Heat—Work—Leave—The barrage begins.

15	OUT OF THE SLOUGH	263

    Battle of Alamein—Waiting to advance—Out of Egypt—Halt at Bardia.

Contents

		Page
16	JOURNEY WITH HALTS	273

Rest and training—El Agheila—On to Tripoli.

| 17 | FEEDING A CATERPILLAR | 285 |

Work at the docks.

| 18 | THE END OF THE FIRST HALF | 291 |

Medenine—Gabes—Sfax—Sousse—Back to Maadi.

| 19 | DISSECTION OF AN UNDERBELLY | 310 |

(1) *The Sangro.* Voyage to Italy—The unblown bridge—Over the Sangro—Mud and Christmas—Move to Fifth Army front.

(2) *Apollyon in the Path.* Cassino—Work—Fire at the Ammunition Point.

(3) *And So To Rome.* Isernia and San Agapito—Hove Dump—An Italian day—Forward to Rome—Narni—The next move.

| 20 | THROUGH THE VINEYARDS | 359 |

Before Florence—No. 1 Platoon with 26th Battalion—Move to the Adriatic—Winter.

| 21 | THE *MAIALE'S CASA* | 378 |

Rest and training—Albacina.

| 22 | WHITE CHRISTMAS | 387 |

Billets at Forli—Road-building under fire—The Jeep Platoon—Christmas—Back at Albacina.

| 23 | 'THY CHASE HAD A BEAST IN VIEW' | 405 |

(1) *The Rivers.* Senio—Santerno—Sillaro—Po.

(2) *Drive to a Cricket Match.* Padua and Venice—A night battle—Arrival at Ronchi—The end of the chase.

Contents

Page

24 '... AND THE REAR PARTY WILL CLEAN UP.' **444**

Unit at Villa Vicentina—Trouble with Tito—Leave unlimited—Move to Trasimene—Disbanded.

Roll of Honour   459

Honours and Awards   460

Commanding Officers   461

LIST OF ILLUSTRATIONS

CONVOY AT DUSK *New Zealand Army official (H. Paton)*	*Frontispiece*
	Facing page
AT NGARUAWAHIA, 1939 S. P. Llewellyn collection	16
ON THE *ORCADES* AT FREMANTLE J. Fenton	16
CHRISTMAS PARCELS, LIVERPOOL, 1940 Fox Photos Ltd.	17
MAADI FROM THE TURA CAVES N. Barker	17
MAIN STREET, HELWAN *New Zealand Army official (G. F. Kaye)*	32
ITALIAN PRISONERS FROM BARDIA British official	32
'BARDIA BILL' British official	33
IN THE SALT MARSHES OF BUQBUQ—A BRITISH TANK BOGGED British official	33
FORT CAPUZZO British official	64
ARRIVAL IN GREECE, PIRÆUS British official	64
NEW ZEALAND INFANTRY IN ATHENS H. G. Witters	65
KATERINE, A TYPICAL VILLAGE J. A. Carroll	65
BOMBED P. Kennedy	80
BOMBING IN LARISSA A. S. Frame collection	80
A CONVOY HALTED IN VOLOS J. A. Carroll	81
'A DIRECT HIT ON THE COOKHOUSE' S. P. Llewellyn collection	81

List of Illustrations

	Facing page
DRIVER'S WINDSCREEN REMOVED FOR VISIBILITY *S. P. Llewellyn collection*	112
TWO BOMBS IN ONE HOLE NEAR A STAFF CAR *S. P. Llewellyn collection*	112
DESTRUCTION OF WORKSHOPS' STORE LORRY *J. E. Taylor*	113
RETREAT THROUGH ATHENS—'WHEN THEY WAVED TO US' *N. Blackburn*	113
WAITING FOR NIGHTFALL, CRETE *J. E. Taylor*	128
THE PRIME MINISTER INSPECTS BEDFORDS AT MAADI *New Zealand Army official*	128
LEAVE IN TEL AVIV *S. P. Llewellyn collection*	129
RATION TRUCKS AT A SUPPLY DEPOT *New Zealand Army official*	129
THROUGH THE WIRE—THE OPENING OF THE SECOND LIBYAN CAMPAIGN *British official*	160
INSIDE THE BOUNDARY WIRE—A PLANE BURNS IN THE DISTANCE *P. E. Coutts*	160
NEW ZEALAND DIVISIONAL HEADQUARTERS AT BIR EL CHLETA *— Davey*	161
DESPATCH RIDERS—TOBRUK *P. E. Coutts*	161
A GROUP NEAR TOBRUK *O. Bracegirdle*	176
TOBRUK HARBOUR IN AUGUST 1941 *British official*	176
STUKA ATTACK ON TRANSPORT SOUTH-WEST OF GAZALA *British official*	177
DIVISIONAL AMMUNITION COMPANY DISPERSED *R. C. Gibson*	177
AMMUNITION COMPANY WORKSHOPS AT BIR EL THALATA *O. Bracegirdle*	177

List of Illustrations

Facing page

CONVOY TO SYRIA—'BEYOND THE TARMAC THE SAND WAS SOFT AND DEEP' *S. P. Llewellyn collection*	224
BAALBEK, SYRIA—'TENTS WERE PITCHED FOR LIVING IN, MESSING IN, COOKING IN' *R. C. Gibson*	224
KAPONGA BOX AFTER A RAID *R. C. Gibson*	225
BURNING LORRY, KAPONGA BOX *R. C. Gibson*	225
DESERT GRAVE *S. P. Llewellyn collection*	225
HAILSTONES IN THE DESERT, SOUTH OF ALAMEIN *R. C. Gibson*	240
DIGGING SLIT-TRENCH, KAPONGA *R. C. Gibson*	240
REST AREA NEAR COAST, BURG EL ARAB *P. E. Coutts*	241
WHEEL TRACKS, ALAMEIN *R. C. Gibson*	241
FLOODED AT FUKA *R. C. Gibson*	272
REMOVING THE RIMS FROM A 3-TON LORRY TIRE, BARDIA *R. C. Gibson*	272
A BATH AT BARDIA *R. C. Gibson*	273
LORRIES ON THE SKYLINE SOUTH OF BARDIA *R. C. Gibson*	273
DESERT FORMATION—LEFT HOOK AT EL AGHEILA *R. C. Gibson*	288
RUINS AT CYRENE *R. C. Gibson*	288
NOFILIA SIGNBOARD *C. E. Grainger*	289
TRIPOLI COOKHOUSE *R. C. Gibson*	289
TRANSPORT NEAR WADI AKARIT *O. Bracegirdle*	304

List of Illustrations

	Facing page
DISPERSAL, NEAR WADI AKARIT *O. Bracegirdle*	304
TUNISIAN BARLEY-FIELD *O. Bracegirdle*	305
THE HEIGHTS OF TAKROUNA *J. Pattle*	305
BACK TO BASE AT MAADI *New Zealand Army official (H. Paton)*	320
PASTURES AT LUCERA, ITALY *New Zealand Army official (G. Kaye)*	320
WINTER IN ITALY *R. C. Gibson*	321
MONTE MAIELLA *W. Fisk*	321
BURNT-OUT AMMUNITION DUMP, VAIRANO *R. C. Gibson*	400
'SANGRO MUD WAS NOW OUR ELEMENT' *W. Fisk*	400
UNBEATEN THAT SEASON—AMMUNITION COMPANY FOOTBALL TEAM *R. C. Gibson*	401
WATER POINT AT HOVE DUMP *New Zealand Army official (G. R. Bull)*	401
EASTER SUNDAY AT SAN AGAPITO *R. C. Gibson*	416
CASSINO UNDER SHELLFIRE *United States official*	416
FIUME PIAVE *P. E. Coutts*	417
THANKSGIVING SERVICE AT PERUGIA *P. E. Coutts*	417

LIST OF MAPS

	Facing page
Egypt 1940-41	15
Greece	51
Cyrenaica and Egypt 1941	121
Egypt 1942	221
Cyrenaica and Tripolitania	273
Tunisia 1943	291
Italy—Taranto to Rome	311
Italy—Rome to Pesaro	359
Italy—Pesaro to Trieste	405

In the biographical footnotes the occupations given are those on enlistment. The ranks are those held on discharge or at the date of death.

CHAPTER 1

DESIGN FOR A UNIT

THE MAIN BODY of the Divisional Ammunition Company went into camp on 6 October 1939, ten days after the advanced party.[1] We came from every walk of life—every walk, crawl, shuffle, and stampede. Most of us had our homes in Auckland or in the Auckland district, and after breakfast we assembled sixty-five strong at the Drill Hall, Rutland Street. The hall was cavernous and depressing and it smelt of damp mackintoshes. There was a good deal of shouting, and presently we were shepherded into a corner where there was a man with a Bible. He told us to place our hands on the Book, cut out the shoving, and repeat what he said. He read quickly from a small card and there was room on the Book for only a few hands, so most of us had to be content with gesturing towards it and moving our lips.

Soon we were marching down Queen Street in a blur of rain, a sergeant in uniform leading us. His name, we learned later, was Michael.[2] It had been raining off and on since early morning and some of us had neither mackintoshes nor overcoats. Not everyone was quite sober and our civilian clothes hung damply about us. Most of us had sugar sacks on our backs and bottles in our pockets, and as we marched, heading for the railway station, we linked arms with girls, called out to friends, and took other steps to demonstrate our amateur status.

By the time they reached Hopuhopu some members of the party were in a mood to treat everything as a gigantic joke and the sight of a group of officers in front of an endless prospect of greyish-white bell tents, the former as smart and polished as the latter were dirty and dilapidated, did nothing to damp their spirits. When the

[1] The advanced party consisted of Capt G. S. Forbes (OC), Capt O. Bracegirdle and Lt A. G. Hood (1st Composite Company, NZASC), Lt N. C. Moon (Reserve of Officers), Lt L. W. Roberts (9th Auckland Mounted Rifles). 2 Lt L. A. Radford (ammunition officer attached, 2nd Medium Battery), S-Sgt W. E. Colton, Sgt C. M. Torbet (acting CSM), acting CQMS R. F. Hood, Sgts H. Crossley, G. P. Hallam, L. D. Jones, N. K. Michael, and J. F. Seymour, and twenty-four other ranks.
[2] Lt N. K. Michael, m.i.d.; driver-mechanic; Auckland; born Fiji, 25 Dec 1913.

A

roll-call started they answered their names as loudly, cheerfully, and incorrectly as possible. 'Hallo! Hallo! Hallo!' 'That's me!' 'Right here, Colonel!' A minority—either they had served in the Territorial Army or they were already ambitious for stripes—came smartly to attention and snapped 'Sir!' Most of us, though, said 'Here' or 'Yes', and let it go at that. There were some—perhaps there were many—who had been made so miserable by recent leave-takings, the dreariness of the day, and the general beastliness of what they seemed to have let themselves in for, that when their names were called they just grunted.

On the whole, Captain W. A. T. McGuire, who was now officer commanding the company,[3] was justified in his opinion. 'You looked like a lot of tramps,' he told us long afterwards. 'My heart sank when I saw you.'

In the afternoon fifty-two others joined us, bringing our strength to about 156.

. . . .

When you volunteer for the Army you make, as it were, a pact with the Devil. You surrender not quite your immortal soul but at least your immediate hopes and ambitions, your independence and freedom, and the kindly and familiar ways of home. But the Devil is notoriously a gentleman and he grants you something in return. He frees you from the trouble of earning a living and the responsibility of thinking for yourself. Not a very good bargain perhaps, but something. The new recruit feels as though he has gone back to school, or back even further than that—back to the nursery.

Most of us, if we search our memories, will find that we were happy at Hopuhopu in spite of the continual rain, the leaky tents, the monotonous parades, the appalling food. The meals, everyone will agree, descended to a level of greasy sogginess that was seldom touched at any other time in our Army life. The food itself was

[3] Capt McGuire (1st Composite Company, NZASC) had been posted to the unit three days earlier, Capt Forbes having been transferred to the Divisional Supply Column.
Lt-Col W. A. T. McGuire, ED, m.i.d.; police officer and motor engineer; Auckland; born NZ, 22 Dec 1905; OC Div Amn Coy 3 Oct 1939-3 Oct 1941; OC NZ Base ASC 20 Oct 1941—1944 (incl AA and QMG, 6 NZ Div, 9 Mar-17 May 1943); returned to NZ 18 Sep 1944.
Maj G. S. Forbes, MBE, ED; insurance clerk; Auckland; born Christchurch, 29 Jul 1908.

fairly good except in a few instances—an issue of fish is still spoken of with respect, and from what barren and scrubby fields our potatoes were wrested we should have been interested to discover—nor were the cooks, most of whom were learners, much to blame. Crowded cookhouses and lack of equipment were the cause of the trouble. The midday meal, though, was excellent: plenty of bread, jam, cheese, and good New Zealand butter. Most of us filled up on this, and in the evenings visited the canteen, to practise patience, to develop self-assertion and, on occasions, to make a purchase.

It rained steadily at first and we spent a great deal of time in our tents. Daily our civilian clothes became damper and more disreputable and it was a relief when we were issued with serge uniforms and felt hats. By this time we had rifles and webbing as well, and the cleaning and laying-out of our kit (a performance that was subject to all the bewildering and terrifying taboos attendant on priestly ritual) presented us with an almost insoluble time problem between breakfast and company parade; but the old Army lore, much of which, no doubt, came down to us direct from the Peninsular campaigns, gained rapid currency, and after a few days it was seldom that anyone was unduly late for a parade or conspicuously badly turned-out, though the officers always found something to criticise. Once used to wearing uniform, we dropped quite swiftly into Army ways. It was easier to drill, easier to double when ordered to, and less of an affront to our native independence to stand to attention when addressing an officer, salute him, and call him 'Sir'.

Some civilian clothing still appeared in the ranks (a few figures proved too obstinate even for Company Quartermaster-Sergeant Robin Hood[4] and his assistant, neither of whom was exactly fussy about the fit of another man's coat), and 'Titch' Maybury's[5] green trilby continued for many days to be a joy to all. He wore it jauntily on ceremonial occasions as though it were freedom's flag streaming gallantly among khaki waves.

Our officers were all enthusiasts and within the limits of the little training manuals they carried about in their pockets they did their best to make things interesting for us. We were bossed about

[4] Sgt R. F. Hood, EM and clasp; department manager; born Auckland, 6 Aug 1914; injured and p.w. April 1941.
[5] Dvr C. G. Maybury; seaman; Auckland; born Brisbane, 15 Jun 1914.

and continually interfered with ('You're in the Army now!'), but we were not driven and we were seldom shouted at individually except by Staff-Sergeant Wally Colton,[6] who was now our Company Sergeant-Major. Oddly enough he was popular. It was that kind of remote popularity tinged with hero-worship—a tribute to omnipotence, perhaps—that is bestowed on any absolute monarch or successful headmaster who is not always tyrannical and oppressive. We held him in far greater awe than we did any of the officers.

No, we were not driven. Some of the NCOs, being new to authority, may have been guilty of minor pin-pricking, but on the whole the relation between the officers and NCOs and the common herd was one of friendliness tempered with caution. When was dignity endangered? At what point did cheerful obedience become servility? It was too early for people to feel quite certain of their position. Only the OC, with his kindly, unassuming smile, was at ease on Olympus, remote and awful.

The company was divided into its service components on 13 November, which gave us the nucleus of Company headquarters and Workshops Section and one complete transport section (A Section), and two days later all temporary and acting ranks were relinquished and appointments were made according to trade and other qualifications. As yet we had no transport of our own, and in spite of one or two lectures about ammunition points and the care of vehicles we had only the vaguest notion of what our real work would be. Some of us, no doubt, had schoolday memories of sumpter-mules at Hastings, of waggon trains loaded with luggage and laughing doxies at Waterloo, and of scarlet London Generals rumbling through the darkness towards Delville Wood and Bapaume; and of course we all knew that ASC stood for 'All Safe and Comfortable', but of the part played by transport in a modern war we could gather little except that you were not allowed to climb into your cab until the officer in charge of the convoy made a mounting gesture or start your engine until he made a winding one.

What else did we learn at this time? Little, one fears, that was of much practical use to us afterwards. No one showed us the

---

[6] Lt-Col W. E. Colton; Regular soldier; Wellington; born London, 22 May 1908; Director Supplies and Transport, Army HQ.

quickest way to change a spring or the best way to make a Benghazi burner. No one showed us a Bren gun or a sub-machine gun. This is not to suggest that the work of our officers and instructors was wasted. To them and to the exemplary patience with which we suffered them (at times they could be wearisome beyond words) we owed the difference between a company parade after six weeks' training and our exhibition on the day of the first roll-call. To them and to the life we were leading we owed our improved looks. Most young men are careless about getting teeth stopped and about finding boots and shoes that fit comfortably. The difference a month in the Army can make to a man's appearance and to the way he feels is quite wonderful.

Between reveille and tea-time we had scarcely a moment to ourselves. There were parades and lectures (hachures and re-entrants went round and round in our heads in company with charger guides and Section 40 of the Army Act), and there were route marches and fatigues. Each day's work ended with half an hour's physical training under Captain Bracegirdle[7] and it was the pleasantest period of all. This was when Bob Ward[8] ('Snake Gully') came into his own. Daily he provided us with one of the most warming spectacles imaginable: a fat man laughing at his own undignified convolutions. Even in those days 'Snake' was well on the way to becoming a Divisional character. On roll-calls he would answer his name with a hearty 'Hallo! Hallo! Hallo!' To him anyone below the rank of Major was 'mate'; majors and above were 'boss'. There is a story of his meeting General Freyberg later in the war. The General was wearing mufti. 'Cigarette, mate?' said 'Snake', and then, jerking his thumb towards the NAAFI[9] building, 'You work here, don't yer?' 'Snake' was no respecter of persons but he knew as much about carburettors as any man in the Division.

Fatigues. Burnt porridge to be scraped from dixies, congealed fat from baking pans. Great drums of slops to be carried gingerly towards the latrines first thing in the morning. Parades. Your

---

[7] Lt-Col O. Bracegirdle, DSO, ED, m.i.d.; clerk; Auckland; born Auckland, 14 Aug 1911; 4 Inf Bde supply officer, Jan-Apr 1940; posted to HQ NZASC (Major) 16 Jun 1941; second-in-command HQ Comd NZASC 9 Nov 1943-15 Jun 1945.

[8] Dvr R. D. Ward; mechanic; Whangarei; born Kawhia, 4 Oct 1907.

[9] Navy, Army, and Air Force Institutes.

collar biting into your neck and your jacket cutting you under the arms and Captain McGuire walking slowly between the ranks and remarking with deceptive gentleness, 'Growing a beard, soldier?' Route marches. . . .

By the middle of November we were beginning to behave and look like soldiers. 'Titch' Maybury's green hat was only a delightful memory and it was unusual for an officer to be addressed (unless, of course, by 'Snake Gully') by any title except 'Sir'. The influenza epidemic, which had filled the camp hospital and converted some of our tents into sick-bays, had abated, and so had the grey rain. The sun came out and dried the mud and the acres of wet canvas. It sparkled on burnished buttons; it hinted at far, hot lands half the world away. 'Perhaps we shall go to India,' we said. 'Or maybe Egypt.' The pessimists answered: 'No. We'll never leave New Zealand.' Usually the thought that this might be true was enough to fill us with anticipatory disappointment, but at other times, on dull evenings or on grey afternoons, some of us would feel secretly in our hearts: 'Well, I've made the gesture anyway. Hell, it would be nice back home!'

. . . .

After six weeks at Hopuhopu we moved to the newly-constructed camp at Papakura. Here we were far more comfortable. There were proper beds instead of bed-boards and straw palliasses, and we had a watertight roof over us.

The weather was fine and we made rapid progress with our training and really rapid progress in friendships. We knew now that we should be going overseas as soon as shipping was available and the knowledge made everything, friendship included, twice as important. Previously we had been at liberty to regard our relation with the Army as a military liaison—a *chaise longue* affair. Now we realised that we were committed to a proper marriage, a marriage that had every chance of being consummated on the battlefield. It was a relief to know where we stood, and our morale bounded.

There had been comparatively little skylarking at Hopuhopu—we had been too busy for one thing and for another tents make indifferent playgrounds—but at Papakura soaring spirits found an outlet in physical exuberance. Tremendous battles were fought between rival huts, with a great upsetting of beds and scattering of equipment, and nightly the sergeants had to leave their cubicles to restore order.

The days hurried by to the slap and creak of marching files and the eternally reiterated three drum beats (one—pause—two: one—pause—two) to which we were learning to subdue the rhythm of our working hours; the evenings went by to the tune of 'South of the Border' and a churning babble from the newly-opened wet canteen, the nights to the stealthy pacing of sentries and a grumble of soft snores from the long, darkened dormitories. December came and early in the month Company Quartermaster-Sergeant Robin Hood, Sergeant Athol Buckleigh,[10] and Corporal Sam Mellows[11] vanished mysteriously (they were our advanced party), and on the 14th we were placed on active service and solemnly warned that sins that had been venial once were now inexcusable. As we saw it, the drama of our situation was doubled. On the same day final leave began.

. . . . . .

Surrounded by mounds of gear—kitbags, sea kits, overcoats, packs, rifles, everything—we sat in a meadow outside Papakura Camp and waited. It was 4 January and we had been waiting since two in the afternoon. Now it was nearly six.

Company Sergeant-Major Colton stood talking to a group of officers. Without his peaked cap, tailored uniform, and Sam Browne he looked smaller, less impressive, more approachable. Like the rest of us he was wearing light khaki drill and a New Zealand felt hat. The officers seemed more approachable too. Their manner suggested a readiness to laugh, crack jokes, hand around cigarettes.

It was one of those pale, indeterminate summer evenings, neither bright nor dull, and the talk and laughter, after rising and falling in the still air hour after hour, had subsided to a low growl. Most of us were drained of energy, for the past three weeks had been extremely strenuous. First there had been final leave, then the wrench of leaving home again. New Year's Night for some of us —self-respect demanded it—had meant sneaking out of camp to

[10] Lt A. J. Buckleigh; motor mechanic; Taupo; born Palmerston North, 11 Oct 1910.
[11] Sgt S. J. Mellows; commercial traveller; Onehunga; born Dunedin, 2 Dec 1910.

take part in the celebrations. A ceremonial parade for General Freyberg had been followed by a rush of packing and a farewell parade in the Auckland Domain. The day had been very hot and we had sweated freely on the march to the railway station, but the people had cheered and cheered, and, on the whole, the majority of us had minded it a great deal less than we were prepared to admit. That evening the camp had been thrown open to the public and the final goodbyes said. Mothers, wives, sisters, and sweethearts had worn their prettiest and gayest dresses. They sat on our beds and strolled in bright groups between the severe buildings, laughing and talking. But some of the laughter seemed forced and much of the talk was rather wide of the point. Indeed the occasion was something of an emotional strain and in a way it was a relief when the time came for the visitors to go home, in silence, and with sad hearts.

That had occurred yesterday evening, and now, surrounded by our belongings, we waited for the word to move.

'Up on your feet,' ordered the sergeant-major, and soon we were marching towards Papakura station, juggling feverishly with our too-heavy loads while the camp band played an encouraging quick-step.

We were exhausted when we reached the station and more exhausted after we had battled to find seats next to our friends and card partners. There was a small crowd of women and girls on the platform and when the engine gave three snorts and started to move they waved and called out to us. As soon as we felt the movement we settled down, after the immemorial custom of soldiers at the beginning of a long train journey, to empty our water bottles and eat all our rations.

We played poker, sang 'South of the Border', and, later in the night, tried to sleep. But the carriages were cold and uncomfortable and there was room on the floor for only a few people at a time. The train rushed through the flying darkness, saying *South of the Border, South of the Border, SouthoftheBorder, Southof theBorder, SouthoftheBorder*. Some of the soldiers were very young, and their faces, streaked with smuts, were childish and weary under the dim lights.

. . . .

A night journey by rail is the perfect agency for removing

military polish and it would be interesting to equate one train-hour with its capacity for cancelling out drill periods. By the time we reached Wellington, which we did at noon on the 5th, our sergeant-major must have been consumed with a desire to set to work on us then and there. We were unshaven and crumpled and this alone encouraged us to be more than outspoken when, as we neared the wharves, NCOs were posted at the carriage doors and all the windows shut.

The train came to a halt alongside SS *Orion*. From then on events moved so quickly that by the time we had finished lunch—and an excellent lunch it was—our ship was anchored in the stream.

When we saw our quarters for the first time we jumped to the conclusion that a mistake had been made and we waited for it to be rectified with much shouting and recrimination and a vast shifting of gear. But nothing happened and we remained in our beautifully-appointed cabins, some of which contained only two berths. All were furnished with electric fires, fans, hot and cold water, white sheets, and fluffy blankets. It was a far cry from the austerities of Hopuhopu, and a slight tendency on the part of some of the officers to behave as though they had arranged the matter with the Orient Line was pardonable under the circumstances.

The *Orion* was the Commodore's ship, and at seven the next morning, followed by the *Strathaird*, the *Empress of Canada*, the *Rangitata*, and HMS *Ramillies*, she led the way out of the harbour. In Cook Strait we were joined by HMS *Leander* and the South Island contingent in the *Dunera* and the *Sobieski*. By nightfall the convoy was well out in the Tasman.

With almost empty decks the *Orion* trembled through the windy darkness, swishing and humming softly. She seemed barely to be whispering, but the loudest noises from below were quenched and smothered by that whisper. No sound, no chink of light, told of the quick traffic, the bright teeming city, enclosed and hidden. Nothing spoke of it but the currents of warm air that streamed from ventilators and tainted the salt wind with an odour of engine oil, warm bodies, hot pipes, cabbage water. Below the Plimsoll line a frilly whiteness bordered the ship's sides and her sharp forefoot divided the water into twin plumes. Otherwise she was part of the night, though once or twice she gave a sly wink. None of the ships with her made any sign. They steamed in silence, going humbly through the long darkness.

. . . .

There were 1428 troops in the *Orion* and of these 166 were members of the Ammunition Company: eight officers,[12] one warrant officer, eight sergeants, and 149 other ranks.

After a day or two we felt as though she had been our home for months. The weather was cold at first but this mattered little as we had our comfortable cabins to retire to when work was finished. The training programme continued: lectures, route marches, physical training. We practised anti-aircraft drill on the promenade deck, pausing every now and then to snatch up guns and tripods and flatten ourselves against the bulkhead as marching troops lurched past to the tune of 'Sussex by the Sea' or 'Roll out the Barrel' played by the ship's band, whose repertoire, though vigorously executed, was limited. The lectures, for the most part, took place in the first class lounge. It was furnished with deep easy chairs, and in these it was almost impossible to sit and listen for any length of time to Second-Lieutenant Radford's[13] extremely erudite talks on ammunition without dozing off. For the rest, we worked greasily in the galleys, did guard duty, spent weary hours searching for periscopes and torpedo tracks, visited the ship's cinema, gambled, and played housie-housie. ('Eyes down for the foist numbah! And here she is. Sixty-six—all the sixes.') From the canteen we bought tinned fruit, biscuits, and condensed milk, and at night we feasted in our cabins. There were wet canteens as well, and people of ordinary capacity were able to satisfy their thirst for rather less than a florin.

In the Great Australian Bight a cold wind blew and the seas tumbled. To the rattle of housie counters, the chanting of callers, the strains of 'Sussex by the Sea', and the now unnoticed hum and whisper from engines and rigging, we headed towards Fremantle, reaching it on 18 January. Those of us who were not on duty were given late leave.

[12] Several changes had been made in our establishment of officers. Our OC was now a major and Capt Bracegirdle had been posted to the 4th Infantry Brigade and Lt Hood to HQ Divisional NZASC. Newcomers were 2 Lt Torbet (who had left us to gain a commission) and 2 Lt R. C. Aitken.
[13] Capt L. A. Radford; machinist; Maeroa, Hamilton; born Hamilton, 19 Aug 1910.

On the following morning the troops in the *Orion* paraded under Colonel Crump[14] for a 14-mile route march to Perth. As we had been vaccinated a few days earlier the march was not compulsory, but the promise of leave at the end of it caused everyone who was not a cot-case to turn out. The column was to have left the wharf at eight in the morning, but there was an unfortunate delay, and by the time the march began the temperature had risen to more than 90 degrees.

The black bitumen stretched ahead of us in a tremor of heat, and Australian motorists, with mistaken kindness, drove up and down the column with iced drinks, which many of us—for the hospitality of the previous night had left burning thirsts—gulped at a draught. The result, of course, was stomach cramps. By the time we reached Perth about a third of us were in sorry case, though only ten men had dropped out. The people lining the streets gave us a tumultuous welcome, and when they understood that we had marched all the way from Fremantle they were loudly indignant and some began blackguarding our officers, which was all we needed to convince us that we had been badly used. They showed their sympathy by plying us with foaming glasses of Swan beer and by nightfall we were feeling fit enough to do the march all over again.

The convoy sailed the next day and shipboard life went on as before, only now we seemed to have more time on our hands: more time to lean over the rail (with the rather thrilling consciousness that if anyone fell overboard he would be likely to stay there); more time to watch the troubled marbles of the water and listen to their interminable swish-swishing against the ship's side, or to look up at night towards the Southern Cross, New Zealand's private constellation, and catch from the corner of an eye the sly progress of a comet, and wonder to see a whole world slipping from star to star as the flying-fish, all day long, had been slipping from wave to wave. And, after the custom of voyagers since the invention of ships, we spent long hours in exchanging preposterous information about dolphins and porpoises and the mysterious

[14] Brig S. H. Crump, CBE, DSO, m.i.d.; Regular soldier; Lower Hutt; born NZ, 25 Jan 1889; in First World War commanded NZASC in Egypt; Commander Royal Army Service Corps, 2 NZ Division, 1940-5; commanded rear party organisation in Mediterranean. 1946-7; commanded 2 NZEF (Japan) 1947; on staff of HQ BCOF and NZ representative on Disposals Board in Japan 1948-9.

phosphorescence, the dancer's spangled shawl, that trailed after us through the darkness.

Colombo was our next landfall, and on 30 January we spent an afternoon in the East. It bewildered and astonished us. The whole scene was in technicolour, with lamp-black shadows everywhere, plenty of flake white in the foreground, and a background of chrome yellow, emerald, and ultramarine. There was no haze at all and all the dimensions stood out hard and square like bricks. It was so much like something out of Hollywood (Super-Colossal! Next week at the Regal!! Better than GUNGA DIN!!!) that one half-expected to see the producer's name—Darryl F. Zanuck or David O. Selznick—blazing across Heaven.

It dazzled us and it made us thirsty, but ten rupees odd (with Japanese beer at two rupees the bottle) was insufficient for serious thirst-quenching, so most of us wandered around in the hot sun and stared and stared. It was all spread out in front of us: snake charmers—glossy bullocks—money changers—boys beautiful as Mowgli—hideous old wizards with lips and teeth scarlet from chewing areca-nut—vendors of cheap cheroots and expensive ivory or ebony elephants—Somerset Maugham Englishmen—performing mongooses—scampering rickshaws. A number of these were drawn by hot and excited Kiwis who had insisted on changing places with the rickshaw boys regardless of the heat. 'Fat' Davison[15] though, complete with cheroot, sat his rickshaw as to the manner born, and doubtless this impressive spectacle went a long way towards soothing the ruffled susceptibilities of many a sahib and memsahib.

We visited wonderful Buddhist temples, and while countless effigies of the Philosopher Prince, with crab-like arms and legs sprouting from improbable places, looked down on us from Nirvana—detached, charitable, ineffably bland—saffron-robed priests lectured us in Oxford English. Finally one would murmur, eyeing us with expectancy: 'It is the custom. . . . You pay what you like. . . .' Our initiation, done in a very delicate and priestly way, to the system of *baksheesh*.

There was no leave for us the next day and we left for Aden on 1 February, reaching it on the 8th. We were allowed ashore there

---

[15] Cpl N. A. Davison; barman-cook; Kohimarama, Auckland; born Auckland, 20 Oct 1906; wounded 27 Nov 1941; p.w. Bardia 27 Nov 1941-2 Jan 1942.

for a few hours but found little beyond narrow, goat-filled streets fit only for the swarming natives and desiccated white officials to whom Providence had abandoned them.

The Red Sea, too, was a disappointment. It was opaque and unctuous, its appearance suggesting that the Israelites had not only crossed it but had washed up in it after a particularly greasy meal. A hot wind followed us and for two days we sweated and gasped for breath in an atmosphere that seemed to be centrally-heated.

We had known for some time that Egypt was our destination, and as the end of the voyage drew near senior officers lectured us on the customs and geography of the country and medical officers gave grisly little talks from which their audiences, after squirming uncomfortably for twenty minutes, arose dedicated to continence.

The *Orion* came into the Gulf of Suez late in the afternoon on the 12th. Over us was a brassy sky, around us bare, red hills. No decent covering of farm or field or forest clothed their nakedness. It was as though part of the world's skeleton had been picked clean by vultures.

That night the *Orion* rode at anchor opposite Port Tewfik and we slept in her for the last time. We had finished the first stage of our long journey. The war, which no one really believed in yet, which no one understood yet, least of all the drivers of the Ammunition Company, had been going on for 162 days. It was generally thought that Hitler was half-beaten already—by blockade, by dissension within, by the time element. Many of us thought it would be over by Christmas.

How wise we had been to come early!

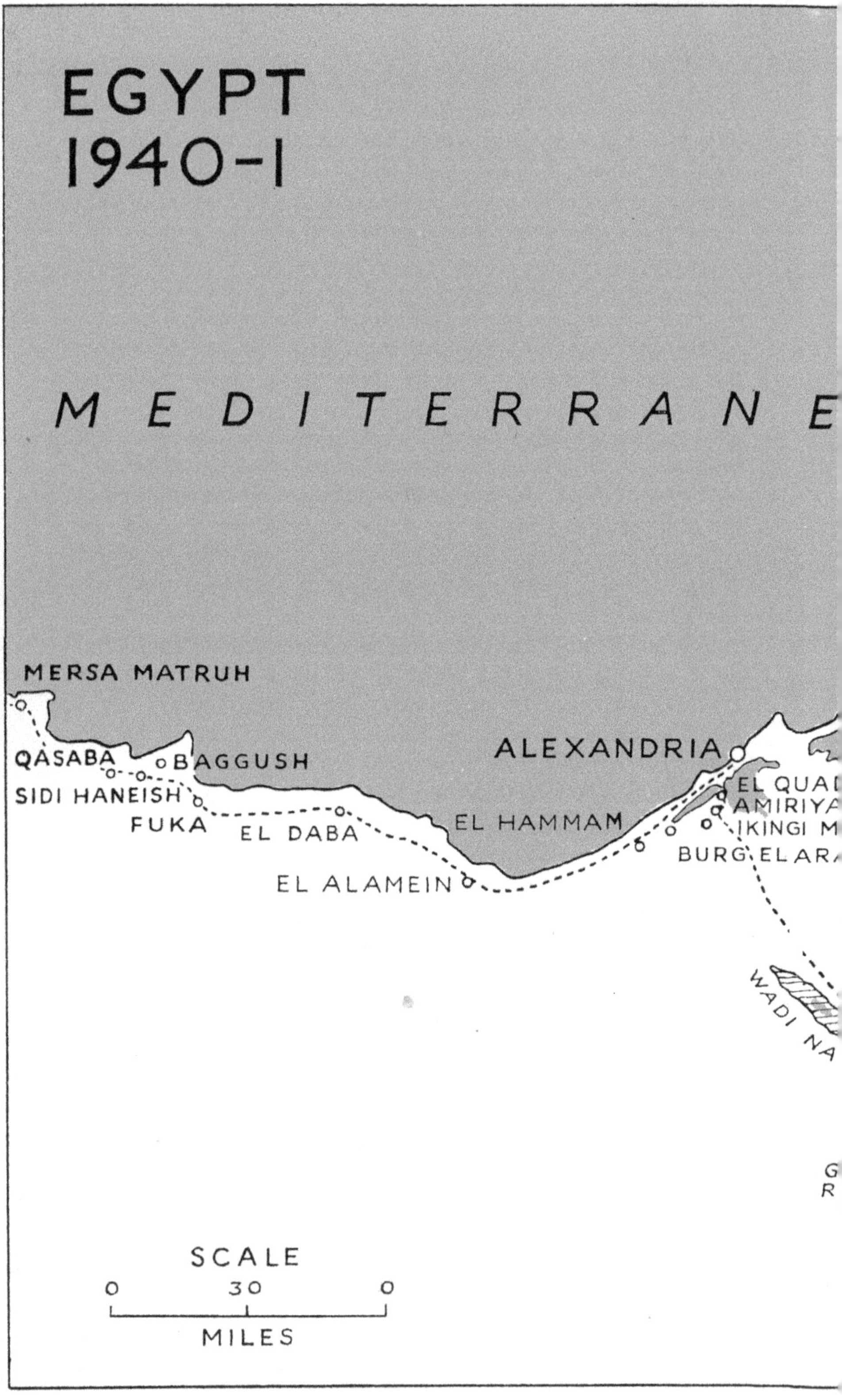

EGYPT
1940-1

MEDITERRANE

MERSA MATRUH

QASABA BAGGUSH
SIDI HANEISH
FUKA EL DABA

ALEXANDRIA
EL QUA
AMIRIYA
IKINGI M
BURG EL AR

EL HAMMAM

EL ALAMEIN

WADI NA

SCALE
0   30   0
MILES

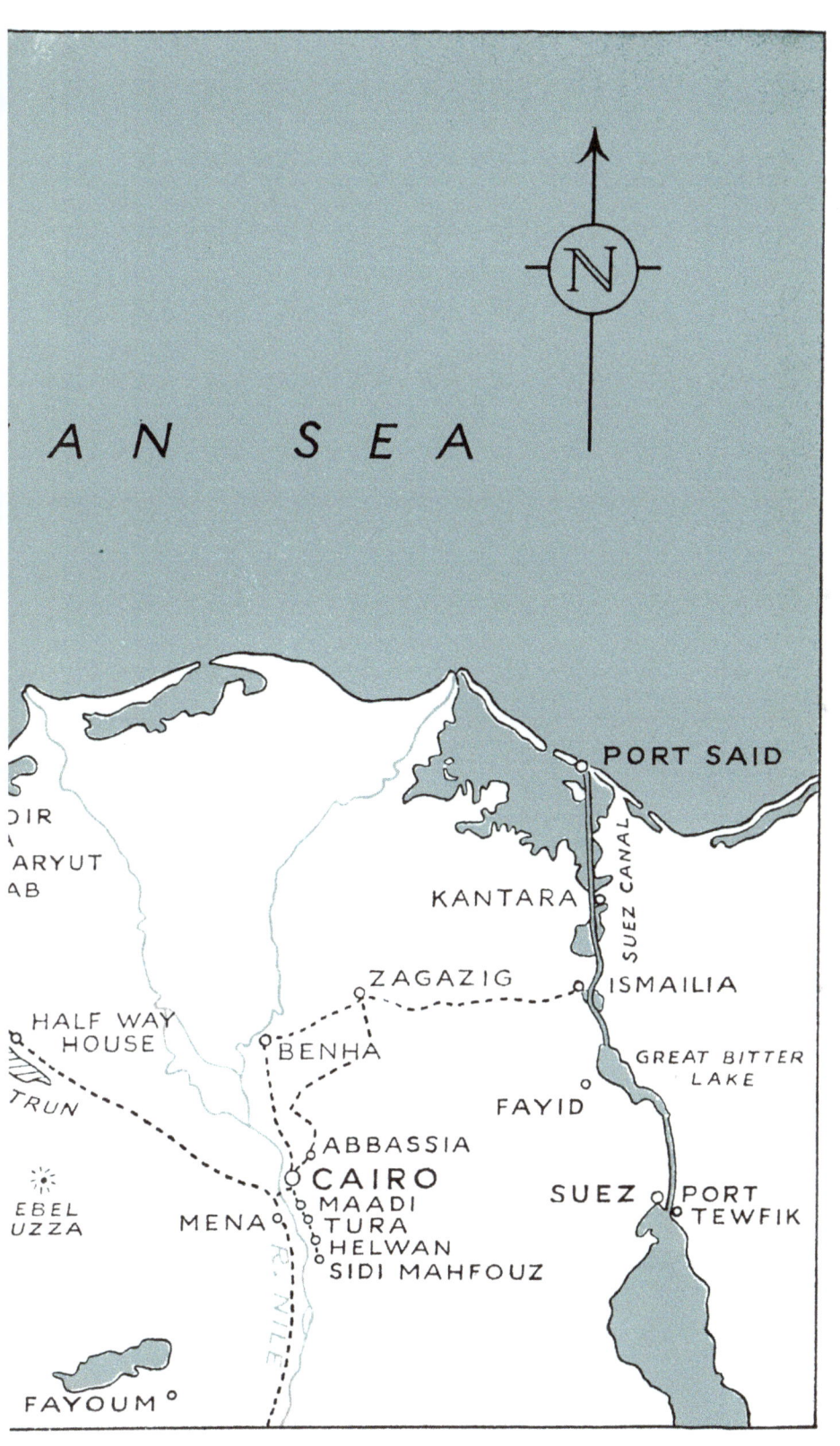

CHAPTER 2

HOW TO SEE EGYPT ON A POUND A WEEK

WE were tired after our 90-mile train journey from Port Tewfik, and by the time we had marched from Maadi siding to our lines, sorted ourselves into tent groups, rushed through the cold air in lorries to an improvised meal of M&V,[1] returned, assembled our bed-boards and made our beds, darkness had fallen. The greenish afterglow of sunset showed little except the outline of far escarpments, the towering pylons of the Marconi wireless station, the rugged contours of what was to be known later as Bludger's Hill. It was too dark to go exploring.

In places the surrounding darkness was pale with bell tents and the tops of the nearest ones glowed discreetly orange, but apart from the talk and laughter in our own lines Maadi seemed unnaturally hushed and shrouded for a great camp. The chill and silence and uncompromising solidity of the desert was strange after our long series of nights in the *Orion*'s warm, quivering, and noisy bowels.

The orderly officer made his rounds; orange faded from the tent tops; silence flowed in from the escarpments. Soon we slept, but after a while the cold crept up through our bed-boards and into our very bones.

Although it was still cold when we breakfasted the next morning, by ten o'clock we had been ordered to wear hats and shirts as a protection against the sun. The day was a holiday and we spent it in settling down and generally taking our bearings. No leave to Cairo was granted that night but as soon as it was dusk 'C. Jay' O'Brien,[2] Pat Wells,[3] and Percy Sanders[4] slipped away, returning in the small hours with news of a wonderful city of girls and music and bright lights. The beer, they said, was tolerable.

[1] Meat and vegetable ration.
[2] Dvr C. J. O'Brien; carpenter's apprentice; Cambridge; born Napier, 12 Jun 1919.
[3] Dvr F. J. Wells; labourer; Manurewa; born NZ, 5 Jun 1914; wounded 27 Jun 1942.
[4] Dvr C. P. Sanders; truck driver; Whangarei; born Dargaville, 22 Sep 1914.

After a few days the allowance of leave was quite generous and before long our knowledge of Cairo rivalled our knowledge of Auckland. Its old treasures, its languorous and heavy-lidded charm, never fresh and clean except for one instant at daybreak, never really beautiful except for one hour at sunset, made little impression on us. What we appreciated was the Stella beer and the fleshpots: whole chickens at a serving, omelets as big as chessboards, eggs by the dozen—they were small, though, and tasted as if the birds that laid them were addicted to taking snuff. The only trouble was that we were paid but 100 piastres a week; no—75, for the Major considered that we should save something against the day when longer leave would be granted.

Just to walk through Cairo, however, was an education. Its streets were pure Vanity Fair, and Sharia Wagh el Birket was the prototype of all the Broad and Flowery Paths so beloved of nursery allegorists. What pluckings at khaki sleeves! What throaty whisperings, inviting the innocent soldier (at reduced rates for a party) to witness indecent exhibitions—which, to judge from certain oblique and tasteless references to the Young and the Old Obadiah, seemed to have enjoyed some measure of popularity with an earlier generation of New Zealanders! We were a mark, too, for the vendors of pornography. After the stolen watches, the imitation jewellery, and the fountain pens (hot from the Pasha's breast pocket) had been turned down, out came the postcards and the little smudged books. The latter found ready buyers, for they were often extremely funny, but not in the sense that the authors had intended them to be. In their grubby and ill-spelt pages the expressions and sentiments of the gutter were set forth in the kind of prose style one associates with Victorian lady novelists—Marie Corelli, for instance.

But the best entertainment of all was provided by the common people of Egypt, the great army of Georges, the street-corner boys, the fellahin: in short, the Wogs (wily oriental gentlemen). Unlike the pashas, beys, and big effendis, who usually went around in a great state of moisture and fuss, as though life were a large auction sale and the best bargain was just going under the hammer, the wog took his world easily. Our attitude towards him, adopted a few days after our arrival in Egypt, never altered. It was made up of amused tolerance, intense exasperation, and the respect due to a

At Ngaruawahia, 1939

On the Orcades at Fremantle

Christmas parcels, Liverpool, 1940

Maadi from the Tura caves

person capable of expressing a whole philosophy of life in one word: *Maleesh*—never mind. In their bottomless cynicism (if ever you hear of a romantic wog give us news of him), their delight in a hard bargain, their imperturbable good nature, and their enormous splay feet—which seemed, as it were, to grin up at us, deliciously deflating human dignity, each root-like toe as comic and full of character as a Thurber drawing—we found sources of unending entertainment.

Altogether, what with resisting or surrendering to temptation, eating large and indigestible meals, drinking the mild Stella or the raging zibbib, and carrying on interminable dialogues, compounded of banter and abuse, with our friends the wogs, an evening in Cairo was not dull.

The same cannot be said for the first few weeks of our training in Maadi Camp. We had a growing conviction, common, probably, to all untried soldiers, that already we were trained to a frazzle. We began to count up the hours we had spent in route-marching and on the parade ground and to regret that we had not been able to spend them in learning, say, to play the saxophone or class wool. A sense of waste was heavy upon us.

The weather throughout February was unpredictable—it blew hot and it blew cold. Nevertheless we got through a good deal of training. On the 19th a month-long course in infantry work was started and there were special courses for NCOs, drivers, and motor-cyclists.

During March our training entered a more interesting phase. Using pool transport, we went into the desert on exercises that lasted several days at a time and gave us valuable experience in night driving and taught us a little about sleeping rough. Before that we had believed we were cold at nights! The Workshops' drivers were happier too. They had missed their tools as a man misses tobacco.

At last we were busy in what seemed to us a sensible way. There were camp driving duties, and parties were sent to Port Said and Alexandria to ferry new vehicles to Abbassia Garrison, in Cairo. This was a tremendously sought-after job, for it entailed a train journey, a night in an hotel, and the chance of a day's leave.

April was like March, but when May came we were issued with our full war establishment of vehicles, which included lorries for the two transport sections that were yet to join us—B and C.[5] There they were—brand-new Bedfords—lined up in our vehicle park. We oiled them and greased them; we checked them and polished them; we stood back and looked at them. We felt we could go a long way in those lorries.

. . . .

The wiseacres who had once predicted that we should never leave New Zealand were now busy predicting that we should never see action. Already a 'Home-before-Christmas' clique was in existence. Probably every unit of every expeditionary force in history has possessed a branch of that hardy and pathetic society that annually dissolves itself on the approach of Christmas and as regularly reassembles on Boxing Day.

Extraordinary months! On 9 April Hitler invaded Denmark and Norway and everyone was delighted and with good reason—Mr Chamberlain himself assured us that Hitler had missed the bus. It was a blow to our chances of seeing action, but still—home before Christmas! A few days later British troops landed in Norway and the sportsmen who had gambled on a long war (two years at least) began hedging. Then the British began to withdraw from Norway, but it didn't matter really, because, you see, Hitler had doubled his supply problem, and in a month a million mechanised divisions burn up a billion tons of oil, and production in Germany and Roumania—but you know the story. The Christmas Club, however, lost a few members that week. On 10 May Holland, Belgium, and Luxembourg were invaded by Germany and paratroops landed near Rotterdam. Mr Chamberlain said nothing about missing buses this time: he resigned. But there is no record of any general exodus from the Christmas Club. Winston was now in power, and the present situation, so the newspapers implied or explicitly stated, was the chance we had been waiting for. At last Hitler had been driven into the open. It would be over before we could get there.

We did take a few precautions, however. We established machine-

[5] We now had vehicles for an ammunition company planned to consist of Company headquarters (three officers and 69 ORs), three transport sections (A Section: two officers, 98 ORs, and four sub-sections with a total of 21 load-carriers; B Section: two officers, 89 ORs, and four sub-sections with a total of 19 load-carriers; C Section: ditto), and Workshops Section (one officer and 44 ORs). Total all ranks, less attachments: 399. Total vehicles, plus staff cars, motor-cycles, and Company headquarters, section headquarters, and Workshops' transport: 129.

gun posts in our part of the camp and leave was difficult to get for a day or two.

Holland stopped fighting; the Belgian Government moved to Ostend, and we looked at our new Bedfords. They were standing there in long lines, their blunt, business-like noses pointing towards the Egyptian border. We looked at them and wondered. Nearly everyone, including a few renegades from the Christmas Club, hoped that we should use them at least once.

. . . .

During the third week of May, while the Germans were driving towards the Channel ports, we left Maadi for Abbassia Garrison, moving into self-contained quarters that had been specially designed for a transport unit. For the lorries there were covered bays and a petrol pump. For us there were cool, stone barrack-rooms, a NAAFI, a shower-house, a playing pitch, and tennis courts. A tailor, a barber, and a *dhobie* had been thrown in for good measure.

We were very comfortable but we were also very busy, for with our fine quarters we had taken over the transport duties in Cairo Sub-area from the Royal Army Service Corps. In addition to supplying the Garrison with transport for its domestic needs and doing a hundred and one different jobs in and around Cairo itself, we assisted in clearing Port Said and Port Tewfik of what the newspapers would have described as an ever-increasing volume of war material. Almost every lorry was in use every day and there was only one driver for each. The A Section drivers drove their own and B Section's lorries and the forty-odd motor-cyclists in Company headquarters drove C Section's.

We worked from dawn till dusk. We delivered rations to British troops stationed in the Garrison; we ran a school bus for their children ('Hey, Mister, my daddy's a sergeant-major! Got any stamps, Mister? Got any small change?'); we carried army clerks from mess to office (this was in the days before Rommel had done away with most of the small amenities of Base life); we carted ammunition from Tura railway siding, five or six miles south of Maadi, to Tura caves, then the largest magazines in Egypt. This job meant getting up at half past three in the morning and working until dusk but we preferred it to driving for the Officer Cadet Training Unit at Abbassia. Senior British officers used to go there for refresher courses, which entailed their being driven into the desert to watch TEWTs, a pastime in which our drivers invariably declined to join, a TEWT being not, as the name suggests, a small sandpiper, but a tactical exercise without troops. Some of them were watched in extremely rough country and the care with which we negotiated it (with a trayful of senior officers bouncing at our backs) depended to a large extent on our treatment in the matter of soft drinks—these were carried in a special lorry—and the number of times we had been annoyed by unnecessary commands and gratuitous advice. A driver in his own cab likes to be treated with a little of the deference paid to a captain on his own bridge.

While the evacuation from Dunkirk was in progress we carried out a Passive Air Defence and Internal Security exercise, and from then on, whenever an alert was sounded, it was the duty of certain drivers to take their lorries at once to given points in Abbassia and the city and stand by to evacuate the wives and children of British soldiers. We had an ack-ack post in the centre of our parade ground and this was manned day and night. Shooting over empty coverts was ever an ungrateful task, and one driver, whose alertness when game was plentiful was afterwards to become a byword, was found asleep on duty, a misfortune that resulted in his being the first of us to go to a detention camp.

Others were falling asleep too, for the long trips to the ports were a severe strain with only one driver to a lorry.

The road between Cairo and Alexandria was particularly conducive to sleep. Sticky in the great heat, it stretched ahead of us like a hundred miles of black adhesive tape, mesmerizing us as a chicken is mesmerized by a chalk line. After thirty or forty miles, with the engine humming high and sweet and acting as an additional soporific, the driver would begin to nod, and his eyes, for all his efforts, would keep closing automatically as though they were being pressed shut by tiny springs. Presently he would plunge off the road into soft sand, and, jerking wide awake, swerve desperately to avoid a telegraph post. After that fright, of course, there was no danger of falling asleep, and yet, five minutes later. . . .

We were fortunate in being busy at this time. Spending hours on the parade ground practising the lying-load position would have been intolerable while the shadow of the long night, inexorable as the fulfilment of an old prophecy, was falling upon city after famous city, and Britain, watched by the whole world with pity, with wonder, or with gloating joy, was rousing herself like an old watchdog and straining cankered ears for the word that nearly everyone had believed she was too sick, too tired, too sunk in torpor, ever to hear again—the word and the drums.

No one told us, after the end of May, that we should be home before Christmas. We were promised different things: blood, toil, tears. Even the old oil story was in abeyance for a short while.

. . . . . .

Mussolini declared war on Great Britain and France on 10 June. He said they had hindered the advance of the Italian people.

One of our earliest reactions was to tumble out of bed in the small hours and into a large air-raid trench, which we manned as a defensive position. Machine guns were posted at vital points and our Boys anti-tank rifle was taken from its canvas cover. For a day or two there was quite a *Vitai Lampada* atmosphere about the Garrison, but nothing came of it—neither jammed gatlings nor dead colonels. No Fascist-inspired mobs, crying on Islam, attempted our barricades, and on the 12th the Egyptian Government broke off diplomatic relations with Italy.

None the less the war had moved immeasurably nearer. The lion was right out of his cage now, and from time to time, among the thickets of conjecture and misinformation, one glimpsed an eye or the flickering of a tawny tail. British Somaliland was invaded and a large convoy of our lorries rushed from Tura to Port Said with ammunition. For two days the drivers waited for the order to unload, but it never came. Instead there was news of an evacuation, and our next task was to take ambulances to Port Said to collect wounded. Sick and wounded soldiers were also beginning to arrive at Cairo main station and we used to meet them at night and take them to Helmieh hospital. Some were suffering from bomb wounds, others from dysentery and desert sores. They came from the Western Desert.

There the Might of Britain (to employ a euphemism popular at that time) stood guard along the Libyan border, and its supplies (which we thought of as an endless stream until the loneliness of the coast road forced itself on our attention) were brought forward by rail and by motor transport. To assist with this work the Ammunition Company inaugurated a ferry service, and from the middle of July onwards convoys left regularly for Maaten Baggush, some 170 miles east of the frontier, with mechanical transport stores.

Turning west before they reached Alexandria, the drivers travelled the rest of the way by the coast road, every inch of which was to become as familiar to them as the road between school and home. Later, the coast area was covered with tents and signposts, and parts of it resembled a used-car mart and parts a sandy slum, and the whole was cut and criss-crossed like a butcher's block. But at that time it was all virgin and we were seeing it with fresh eyes.

We saw with a shock the blueness of the Mediterranean (surely you could have dyed your shirt in it?) and how between road and sea the sand dunes were as white as sugar. On the other side the desert stretched away to the horizon, changing colour with the changing light and clouds. Rivers of vandyke brown divided continents of chrome yellow and spread evenly into far oceans of red ochre and burnt sienna. Shadows as big as armies brushed over them, travelling with the speed of trains and conquering the sunlight; and small shadows, no larger than country estates, swam singly.

We started long before dawn on these trips, rushing through sleeping Cairo and then through the cold, sweet desert. We took rations with us and our beds, and we revelled in this taste of freedom—sausages cooking over the primus, the tea stewing in the billy, bedrolls flopped out on the sand. At the end of the journey there was beer to look forward to in the Tommy mess, and perhaps —after we had unloaded the next morning—a whole day on the beach. Then home again, arriving at Abbassia in the evening with a three-day beard and a comfortable feeling of contempt for poor stay-at-homes who were ignorant of wide spaces.

The Western Desert! The magic syllables punctuated conversation in the NAAFI, clinched arguments, were spoken in sleep. Around them a cult crystallised and there was one of our number who re-oriented his whole life in relation to Maaten Baggush, even going so far as to abandon washing.

The ferry service, Cairo Sub-area duties, guard and passive air defence duties, vehicle maintenance . . . our days were full. Nevertheless, we had time for other matters besides work.

For weeks past the red flame trees bordering the playing-field had gazed down through long hot afternoons on our victorious cricket team.[6] Moon-soaked, they had seen us wandering home from late leave in Cairo, comfortably mellow and unbuttoned. Possibly they had witnessed idylls, for by now at least two of us had found time to become engaged. Certainly their slim branches, towards NAAFI closing-time, had trembled to Ted Schonau's[7] piano notes and to the voice of Tom Woodill[8] leading us in our favourite songs, the most popular of which concerned the confessions of a dying airman and the effect of Egypt on our morals and our good names.

Yes, it was all mixed and mixed. Much of it was hot desert, and mounting drowsiness, and endless adhesive tape, black and melting at noon. Much of it was long, sweaty waits outside Cairo station or in dusty, stinking streets, while one's eyes went again and again to the ever-handy buckets and barrowfuls of soft drinks—*boissons gazeuse*—and all the desire in the world centred around those cool bottles of colouring, Nile water, and animalcules. But much of it was breezy dashes down broad avenues, joyous swoopings round wide, well-cambered bends, and a streaming of scented night air beneath open windshields.

Much of it was flame trees drenched in sunlight, and a grittiness underfoot, and the yielding fusion of ball and cricket bat, and the smell, sweetish and persistent, of hot oil, hot engines, hot rubber—and flies buzzing, tirelessly important and importunate; and mosquito nets like gauzy wedding dresses misting a few sleepers in the long barrack-room; and skull-capped Abdul with white *galabieh* and red sash moving lazily along the verandahs and pausing now and then to intone mournfully and mockingly: 'Lemonë. Ver' g-o-o-d—ver' hygiene—ver' col'.'

And much of it was flame trees under the warm moon, and the NAAFI's luminous blue windows marked with whorls and

---

[6] Of 35 matches played in 1940 we won 27, drew 3, and lost 5.

[7] S-Sgt E. J. Schonau; motor mechanic; Frankton Junction; born Kent, England, 23 Jun 1913.

[8] Gnr T. H. Woodill; fellmonger; Te Papa, Auckland; born Taihape, 11 Nov 1911; wounded 4 Jul 1942.

scratches of gold light, and a blast of song, hot and outrageous, buffeting the soft darkness:

> An airman tol' me before he died
> And I don't think the bastard lied. . . .

. .

Early in September our A Section drivers—'My chickens' Lieutenant Moon[9] called them—spread their wings for the first time and set out for the Western Desert.

The date, 5 September, marks the beginning of a chapter—for the family feeling that was to hold the section together through four long years, those characteristics, that corporate personality (that soul, if you like the word) that was to distinguish A Section from its equally individual companions—aloof B Section, friendly, games-playing C Section, party-loving Workshops, Headquarters, harassed and cynical, and the still unborn Ammunition Platoon with its Cinderella complex—started from the moment when Lieutenant Moon and his chickens took flight before dawn, heading for the Western Desert to set up in business on their own.[10]

Ikingi Maryut, near the Matruh turn-off on the Alexandria road, was their immediate destination, and every mile of the journey increased their delightful sense of being on active service. This was further heightened soon afterwards by a brilliantly spectacular air raid on Alexandria. Steel helmets and slit-trenches were forgotten and they conducted themselves like spectators at a prize-fight, and with some excuse, for the sight was magnificent.

The sky above Alexandria was lit up like a Christmas tree. Tinsel and gold baubles and coloured chains scintillated in the darkness, and one white moth of a bomber, the star of Bethlehem crowning the whole tree, was caught in a cat's cradle of searchlights. It was a spectacle that our drivers were to see many times before the war ended but never again would they see it with the same shock of surprise at its beauty, its savagery, its marvellous patterns.

[9] Maj N. C. Moon, m.i.d.; commercial traveller; Takapuna, Auckland; born Auckland, 8 Feb 1912.

[10] The two officers and 70 ORs of A Section took with them only their correct establishment of transport, so the vehicles held for C Section were left without drivers. The situation was met by the attachment of one sergeant and 36 ORs from the RASC.

Lieutenant Moon, though, was not pleased. 'In future,' he said, 'you'll use your slit-trenches. And Davies,[11] I don't want any more running commentaries. Your name's not Gordon Hutter.'

Lieutenant Moon was a New Zealander to the core but he spoke with a soft drawl that suggested the Confederate States. When he abused them his chickens could almost hear the ice tinkling in the glasses of mint julep and the darkies singing in the plantations. (And when he called them his chickens, by the way, he meant very much what Long John Silver meant when he spoke of his lambs.) He had a gift for humorous exaggeration and he made full use of it. To appear on parade with the least suspicion of a side-whisker was to invite the comment that there was no room in A Section for 'any dam' John Bowles', and the driver whose vehicle raised the smallest particle of dust in the area was promptly asked what had given him the idea that he was Segrave or Sir Malcolm Campbell. His chickens cursed him and they told stories against him—'Yes,' he was supposed to have said while testing one of the vehicles, 'she certainly is runnin' mighty sick. When you get home just clean number three plug. Guess that'll fix her.'—but they laughed at his jokes and they would have followed him into any Balaclava-like situation you care to imagine if he had said to them, 'Righti-o! Get weavin'.' When he left them they gave him a silver cigarette case—and funds were low at the time.

After a week or so at Ikingi Maryut A Section moved up to El Daba, some forty miles east of Baggush, and there we can leave it for the time being.

. . . .

Towards the end of September it seemed that another transport section might be needed in the Western Desert, for since the beginning of the month the greater part of the Division had been stationed there—mostly at Baggush, where defences were being constructed. Accordingly Major McGuire decided that Lieutenant Roberts's[12] section should be acclimatised to the desert. A camp site was chosen about seven miles from Abbassia (out of reach of

[11] Dvr N. R. Davies; labourer; Paratu, Walton; born NZ. 21 May 1914.

[12] Maj L. W. Roberts, MBE, ED, m.i.d.; clerk; Auckland: born Wellington, 4 Sep 1911; appointed OC 2 NZ Amn Coy, Jun 1943; OC 1 NZ Sup Coy, Sep 1945; Regular soldier; Assistant Director Supplies and Transport, Northern Military District.

the sybaritic influence of the barracks and yet near enough for the section to continue with its Cairo Sub-area duties), and to this, on 26 September, Lieutenant Roberts and thirty-eight men moved with B Section's transport.

In spite of its voluptuous name, the Virgin's Breast (suggested to Arab geographers by the presence of two conical hills surmounted by stone cairns), the locality was sufficiently sandy and inhospitable to satisfy all requirements. As it was a toughening-up camp the drivers slept in their lorries, than which, surfeited as they were with tents and barrack-rooms, nothing could have suited them better. A wet canteen was established, and impromptu evening entertainments followed automatically. Nightly the Virgin's Breast (our drivers had a coarser name for it) vibrated to shouts and echoes robuster than any that had troubled its innocence through an eternity of passing caravans.

Time softens everything—buildings, bad liquor, memories. The bright enchanted islands—the excitements, the good times we had, the moving on crisp, sunny mornings through countries as strange to us as Cloud-Cuckoo-Land, the suppers of oyster fritters—these alone seem to stand out, whereas the dreary and bitter tasting sea in which they were pin-points only—the monotony, the homesickness, the recurrent feeling (did anyone quite escape it?) that one's individual efforts were doing rather less to win the war than those of the Man in the Moon—seemed misted over for the most part, and by mists rosy and luminous as from reflected sunsets. They *were* good times, though, and many of us can look back on those days at El Daba and the Virgin's Breast and say: 'Those were the best times I had in the Army.' Adding as a rider, of course: 'Up till then.'

CHAPTER 3

MEETING AT AMIRIYA

(1) *Birth of a Happy Section*

AT this time, while some of us were at El Daba, some at the Virgin's Breast, and some at Abbassia, C Section comes marching into our story.

The main body of the Third Echelon entered Burnham Camp in the middle of May 1940, and there all NZASC units were trained together, the idea being to give them a clear picture of how their own particular cogs would interlock with the rest of the NZASC machine. C Section's officers were Captain J. Veitch[1] and Second-Lieutenant J. D. Fenton.[2]

In no time at all the section's parade-ground work reached an extremely high standard. It could hardly have done otherwise with Sergeant-Major Bill Dillon[3]—the 'Bull'—as CSM. He was a power and a personality and his parade-ground voice possessed all the properties of an airburst, each word of command seeming to explode exactly above the centre of the centre rank and about eight feet from the ground. The 'Bull' was an example of the best type of modern Permanent Staff man—the exact antithesis of the old-fashioned sergeant-major of the waxed moustache, the face like underdone beef, and the obscene tongue. He knew exactly where he stood and where the officers stood and where everyone else stood, and he used his knowledge with wisdom and forbearance.

Sergeants-major in quest of regimental perfection can be extremely trying and the 'Bull' spared neither officers nor drivers (the former, perhaps, even less than the latter), but nearly everyone was pleased for his sake when the complete camp guard was chosen from C Section on the day of the Governor-General's visit and when the section defeated all comers from the NZASC in a drill competition.

[1] Capt J. Veitch; bus driver; born Scotland, 2 Feb 1901; transferred to 4 Res MT Coy, 3 Mar 1941; died of wounds, 3 Jun 1941.
[2] Maj J. D. Fenton, MBE, m.i.d.; foreman motor mechanic; Wellington; born Waitara, 24 Jul 1912; Regular soldier, RNZEME.
[3] Maj W. L. Dillon, m.i.d.; Regular soldier; Central Military District, Wellington; born Wellington, 11 Aug 1912.

The section was extremely fortunate in its leaders. Captain Veitch—he was called 'Scotty' always—was a man of great energy and personal charm, and the spirit of happy keenness with which he and Second-Lieutenant Fenton imbued their drivers during those early months was felt as an influence for unity long after both officers had gone their ways, one of them to a grave in Crete.

July ended and final leave came along. Then, on 27 August, C Section (91 all ranks) boarded the *Orcades* at Lyttelton. She sailed that evening and was joined in Cook Strait the next morning by the *Mauretania* and the *Empress of Japan*, which were carrying the North Island contingent. The escort consisted of HMAS *Perth* and (our drivers were glad to see) HMS *Achilles*, whose honours were fresh upon her. In those days the Battle of the River Plate was still a stirring memory.

The next morning the convoy swung west and soon the Southern Alps had faded from sight. Near the coast of Australia the *Achilles* drew a ring round the convoy, as though placing it within the protection of a magic circle, and turned back towards New Zealand. Her place was taken by HMAS *Canberra*, which had brought the *Aquitania* from Sydney.

It was cold in the Bight but the sun blazed as the ships came into Fremantle on 4 September.[4] There was leave in Fremantle and Perth and a wonderful welcome from the Australians, and on the 5th the convoy was away again, steaming under cloudless skies through seas like watered silk.

The next land was sighted on the 16th and early in the afternoon the *Orcades* anchored in Bombay harbour, which was crammed with transports, fussy steamers, and graceful dhows.

The *Orcades* tied up in the Alexandria Dock at lunch-time the next day and the drivers marched to temporary quarters in the cricket stadium. Leave was granted until a late hour. The New Zealanders spent their money on everything from expensive carved ornaments to cheap Indian babies. The latter were being sold like

---

[4] Cpl Roy Hintz, who had been admitted to the ship's hospital on the first day of the voyage with a severe abscess and had undergone an operation in Bass Strait, was put ashore at Fremantle, the intention being to return him to New Zealand. As soon as he was better, however, he took matters into his own hands, boarding an Australian troopship bound for Egypt. 'What are you doing here?' said the OC Troops. 'I got on the wrong ship,' said Roy. 'Well then,' said the OC, 'you'd better go and live in the sergeants' mess.' Roy rejoined his section on 4 January.

puppies and the line was evidently one that had proved popular among visiting troops.

On the following morning, after a long wait near the quayside, C Section went aboard the *Ormonde*. While the men were embarking they cast disgusted glances at a large pile of carcasses covered with flies. Native stevedores were dragging them through filth and dust and after they had been loaded an odour of decay permeated a great part of the ship.

Shore leave was granted that evening and early the next morning the *Ormonde* moved into the stream and dropped anchor. For lunch there was stew. It was not appetising and its smell, to put the matter mildly, was powerful. A memory of fly-covered carcasses haunted the men's minds and no one made much of a meal. And there were other reasons for dissatisfaction. The ship had been left dirty by Imperial troops, who had disembarked a few hours earlier, and quarters seemed cramped after those in the *Orcades*. To make matters worse it began to rain.

There were murmurs and more than murmurs, and while the officers were at lunch serious trouble broke out. At a quarter past one, shortly before the ship was due to sail, troops took possession of bridge and wheelhouse and told the captain that the ship would not leave port until their grievances had been redressed.

The C Section drivers, although as discontented as anyone, remained quiet. Captain Veitch and Second-Lieutenant Fenton visited their quarters and urged them (the former in no measured terms) not to side with the malcontents and to remember that the important thing was to join their comrades of the First Echelon with the least possible delay. Nearly everyone agreed and much disappointment was felt when the convoy sailed without the *Ormonde*.

That night it became imperative that the ship, which was now under arrest, should leave within a few hours if it was to catch up the convoy. Pickets were posted at all vital points at half past six the next morning and by seven the *Ormonde* was under way, catching up the rest of the convoy, which had been steaming at reduced speed, at half past three in the afternoon. The next day the ships entered the danger zone.

The Red Sea was entered early on 26 September and shortly after daylight the drivers were encouraged by the sight of a large convoy of merchantmen ploughing placidly towards the Gulf of

Aden. Even if Britannia no longer ruled every wave the old lady was still behaving as though she did. Throughout the day the coastline of Italian-owned Eritrea slid past on the port side and during the night the naval base of Massawa was left behind.

Port Tewfik was reached on the 29th. Our drivers spent the 30th on board and on 1 October they disembarked. That night they ate their evening meal at Abbassia.

    .   .   .   .   .   .

From the start the newcomers got on famously with the rest of the unit, and if they were irritated by a slight tendency on the part of some of their new friends to speak of New Zealand as though it were a country in which they had spent a happy childhood long, long ago, they did not show it. In matters such as finding one's way about Cairo and speaking Arabic they allowed the old hands to appear to advantage for a few weeks, but when it came to playing the first tentative games of the Rugby football season they could permit no patronage.

For several weeks they trained in the bull-ring, growing progressively wearier of the 'Bull's' bellow, but at last their vehicles were handed over to them. Routine duties in the Cairo Sub-area gave them plenty of driving practice under difficult conditions, and on the whole they provided their seniors in service with disappointingly few dented mudguards at which to raise a pained eyebrow. They helped, too, with the ferry service, which now extended to Mersa Matruh, some thirty miles west of Baggush.

Meanwhile the war had been changing and expanding. Over Britain the Luftwaffe had switched from day-bombing to night-bombing; German troops had virtually occupied Roumania; Mussolini had attacked Greece. But there was another side to the picture. Bombs—not many bombs and not large bombs but bombs that Goering had boasted would never fall—were causing consternation in Berlin and other cities within the Reich, and the German people had been forced to the conclusion that victory before Christmas 1940, which they had been promised by everything except a direct statement from the Fuehrer himself, was now an impossibility.

And then, on a December day, we showed that we too could attack.

(2) *Working for Wavell*

Ever since the outbreak of the Italo-Abyssinian war the Libyan frontier had been a chink in our imperial armour, and since Mussolini had sided with Hitler it had been a chink with a javelin levelled at it. And the range, since 14 September 1940, when the Italians crossed the frontier and occupied Sollum, had been pointblank.

From then on General Sir Archibald Wavell had been playing David to Graziani's Goliath.

In the circumstances our A Section drivers camped at El Daba, less than 200 miles from Sidi Barrani, towards which the Italians were advancing, had little right to the cheerful happiness they were enjoying. Every afternoon they swam in the Mediterranean, their heads dotting the blue water like corks and their common sense telling them that this was the pleasantest of all pleasant wars. In the evenings they gathered in the canteen tent to sing songs and drink bottled beer while wild dogs howled out their hearts in the surrounding desert and tame Italian pilots moved discreetly among the high clouds.

The *Regia Aeronautica*, although it was often heard and on moonlight nights and exceptionally clear days sometimes seen, was never a nuisance as far as A Section was concerned. Its behaviour justified the popular witticism: the Italians come in at 5000 feet and dive to 10,000. What they excelled in was sowing the desert with explosive fancy goods—thermos flasks, fountain pens, and other ingenious trifles.

After spending two lazy months at El Daba the section moved a few miles to a new area. Here the drivers had no sooner established themselves in comfort than they were ordered to pack up.

They moved again on the night of 4-5 December, travelling to Qasaba, between Baggush and Mersa Matruh. It must have been a day or two later, no more, that Lieutenant Roberts, who was in charge of the section at this time, Captain Moon being away on leave, came into the mess tent at lunch-time waving a slip of paper. It was a Special Order of the Day by General Wavell. He read it out, adding: 'Well, there you are, chaps. The show's on. I'm only sorry I shall be missing it.'

It was hardly Henry V haranguing his troops before Agincourt but the effect was the same. Lieutenant Roberts was not alone in being excluded from the coming show, and the few New Zealanders, the happy few whom fate had chosen to take part in it—transport drivers, engineers, signallers—would have been less than human if they had failed to compare themselves to their own advantage with gentlemen in base camps then abed.

. . . .

The campaign opened brilliantly and by the evening of the 10th Sidi Barrani was in our hands. Sollum fell on the 16th and Fort Capuzzo the next day.

At first the drivers worked between Qasaba railhead and an advanced ammunition dump a few miles farther west, and later they assisted in stocking the forward supply depots, the trips becoming longer as the campaign pursued its successful course. Occasionally they carried troops to the forward areas, but for the most part they were employed in bringing up ammunition, petrol, water, and rations. On the return trips they took salvage and prisoners to Mersa Matruh.

The Wolves of Tuscany—the Tigers of Tunis—were surrendering in their thousands and they needed no guarding. Their one fear was that they might be left in the desert and they gave trouble only when they thought the lorries were going to move without them. Then they would fly into a panic and start fighting to get aboard, yelling '*Uno momento! Uno momento!*' Sometimes it was necessary to fire a shot over their heads or, as a last resort, threaten them with freedom. Once they were in the lorries, however, the clouds melted and soon the desert would be ringing with their throaty baritones and sweet tenors. Our drivers didn't mind. There was something notable and even humorous and touching about bumping over desert tracks, or along the vile coast road, trailing clouds of Verdi and Puccini.

After Sollum had fallen the trips to the forward areas took two days, and loading and unloading usually took another day, so the drivers had little time for servicing their vehicles and less for sleep. The roads were always bad and day after day vision was reduced to only a few yards by wicked dust-storms whipped up by high winds. The lorries stumbled through a whirling ginger gloom, crashing into pot-holes, breaking springs, tearing tires.

Main street, Helwan

Italian prisoners from Bardia

'Bardia Bill'

In the salt marshes of Buqbuq—a British tank bogged

On fine days—for a while at least—the work was fascinating and the drivers were like children let loose in a toy department. The desert was dotted with wrecked diesel lorries and abandoned dugouts. Quartermasters' tents debouched boots, bales of new uniforms, knives and forks, jars of face cream, and packets of sweets. It was an *embarras de richesse*. Undiscriminating as jackdaws, they brought back mixed treasure and rubbish to their Qasaba camp: Italian groundsheets, canteens, water bottles, broken machine guns, motor-cycles. For days 'Bully' Higgins[1] was resplendent in the sky blue and gold braid of a high naval officer and the amount of useless impedimenta carted around in the ack-ack lorry drew a protest from Captain Moon: 'How the hell could you work that Goddam gun?'

It was tremendous fun for a while but from the first there had been a bad drawback. You would see the lorries break convoy, converge on some fort or system of defences, and pull up with a squeal of brakes. Then the drivers would plunge into the dugouts and reappear laden with loot, only to cast it away a second later on finding their legs and arms dusted with black fleas and their clothes hopping with them.

After the fall of Sollum and Halfaya Pass there was a lull in the campaign and during that lull came Christmas—the Christmas before which more than one optimist had promised himself he would be home. Well, what of it? Graziani's mighty army was falling to pieces in North Africa and in Albania the heroic Greeks were driving the Italians towards the sea. In Germany no oil, no warmth, no food, no hope. One hesitated to make predictions, but still, possibly by next Christmas. . . .

At Abbassia things were done in style—Christmas turkey and Christmas pudding by 'Fat' Davison, a speech by the Major—but at Qasaba it was not possible to make a great deal of the occasion, many of the drivers being on the road. However, the day did not pass unnoticed. It was honoured with beer, with extra rations, and with songs, but not, as Captain Moon had proposed, by the sacrifice of his private rooster. 'Sheriff' Davies, with guests to feed after a party, had been before him. 'I snuck out', admitted the 'Sheriff', 'and I snuck along to where that ol' rooster lived, and I screwed

[1] L-Cpl W. T. Higgins; miller; Roto-o-Rangi, Cambridge; born Blenheim, 9 May 1914.

c

that ol' rooster's neck, and I snuck him into the ol' pot.' However, under the benignant influence of the season, all was forgiven and forgotten: the theft of the rooster and Captain Moon's equally high-handed conduct in closing the bar for a week to punish his chickens for imitating Long John Silver's lambs in the matter of some looted rum.

With the arrival of the New Year Wavell struck again. Bardia, fifteen miles beyond Sollum, was the next objective, and 200 New Zealand lorries were detailed for transporting Australians to the battle zone. Accordingly, A Section embussed Australian infantry at five on New Year's morning, taking them to Buqbuq, about thirty miles east of Sollum, where they bivouacked for the night. From the direction of Bardia, on the far side of the border, came the sound of heavy gunfire and of bursting bombs. A late start was made the next day and the troops were taken to Sollum, where a halt was called at the bottom of the escarpment. 'Bardia Bill', the famous big gun of which so many stories have been told, whose capture so many units have claimed, and whose calibre has been the subject of so many arguments, sent over a few shells, some of which were considered by our drivers to have landed close enough to enable them to say they had been under fire.

Towards evening, however, under the huge shadow of the escarpment, with the battle banging away in the distance and night approaching, something inside the drivers, a little knot of anxiety and anticipation, tightened. After all, this was a real battle their passengers were going into and it was the first battle that our drivers had had anything to do with.

As it happened, that period of waiting and wondering was the most memorable part of the trip, for what followed was in the nature of anti-climax. After dark the Australians were driven up the escarpment and deposited near Fort Capuzzo, a dozen or more miles south of Bardia, and there, so far as A Section was concerned, the job ended.

Bardia fell on the morning of the 5th and Tobruk on the 22nd. By the end of the month Derna was in our hands.

As the field supply depots moved forward, the drivers, working in small convoys under their sub-section corporals, made longer and longer trips, but the novelty had worn off. What had seemed treasure once was now dirty Italian rubbish—wrecked lorries, torn groundsheets, little rifles with ridiculous folding bayonets. What had seemed high adventure was now long hours of bumping over the chunky and half-finished surface of Mussolini's preposterous Victory Avenue.

However, not all the jobs were the same. A few drivers were fortunate enough to visit Siwa oasis where the British maintained a small garrison. Here, 200 miles due south of Sidi Barrani, a surf of date palms broke against an island of mud buildings—some whitewashed and lived in, others empty and haunted. Our drivers can hardly have been unaffected by the atmosphere of the place— an atmosphere of brooding mystery and of secrets so old that only the first Gods remember them.[2] They may have noticed, too, that there were few dogs or cats about. The explanation of this was more prosaic. Siwans eat cats and dogs.

A Section's drivers did not have a monopoly of the forward areas at this time. As the fighting troops advanced so did the RASC units catered for by our ferry service, and by the beginning of February we were making round trips of from 1300 to 2000 miles and spending anything from a week to ten days on the road.

The Divisional Ammunition Company was now at Helwan Camp, some twenty miles south of Cairo, having moved there on 13 January after handing over its duties in Cairo Sub-area to the RASC. Other units of the Division had also moved to Helwan from Maadi and the Western Desert.

At Helwan we occupied large huts in No. 1 area—not that the number matters, for no area differed from another except in its relation to Shafto's cinema and the NAAFIs. All the huts were identical and around all of them was the same sand.

On the day after the move C Section was given a new job.

. . . .

When it seemed likely that the British would have to withdraw towards the Nile Delta, large quantities of petrol, oil, water, and food were buried in the desert at Gebel Ruzza, sixty miles due west of Cairo, but soon after the first victories in Libya it was decided to exhume these valuable supplies and take them to Abbassia. The task was entrusted to C Section, which, with a fatigue

[2] Between AD 20 and 1792 Siwa disappeared from history, and between then and the First World War no more than twenty outsiders are believed to have set foot in it.

party from the First Echelon, moved to Gebel Ruzza—a hill and nothing more. All around was virgin desert—miles of it, hard and ochrous, miles of it, soft, dazzlingly white, and untouched by wheel track or footprint.

When the lorries were away working and the camp was deserted except for the cooks and a mechanic from Workshops, there was a touch of fantasy about the scene, and for a moment you wondered why. Then you saw it: the tents—two big and one small—the water cart and the lorry the mechanic was tinkering with, they were the only objects of their own *size* in the whole landscape. Everything else was small like the little pebbles and the wisps of camel-thorn, or large like the hill of Gebel Ruzza, or enormous like the billowing clouds.

The supplies had been buried over a wide area and there were many dumps to be uncovered. The presence of some was proclaimed by the hard outline of a layer of packing cases and around others there was a litter of broken boards, showing that marauders had been at work, but a few had to be searched for like buried treasure.

The members of the digging party were angry and resentful at first, taking it amiss that men junior to them in service should be handling the lorries, but the weather was so fine and the air so pure (being sharp and golden like a good orange) and working in the bright sunshine stripped to the waist was such an agreeable experience that they had no choice but to enjoy themselves, and presently the various gangs were vying with one another in the amount of work they could do.

They worked all day and they worked, illicitly, at night, for it is in human nature to love digging things up—whether buried by Captain Flint or dropped carelessly by the Chaldeans. Evening after evening they would creep out of camp to dig secretly for the rum and cigarettes they believed were concealed somewhere, but they never found anything worth taking except a few cases of condensed milk and some tinned fruit.

The lorry drivers, too, were enjoying themselves. Each morning as the sunlight started to spill over the flats a loaded convoy set out for Abbassia. There were landmarks to steer by—hills and escarpments—but in the desert some demon of confusion plays draughts with landmarks and it soon became necessary to blaze a trail with empty petrol containers. That was after a convoy leader had taken a day and a half to reach Cairo, travelling 160 miles.

To everyone's disappointment the job lasted only ten days. Its importance, however, was out of all proportion to the miles covered and the tons lifted. It was the story of a party of disgruntled drivers sweating out a grievance in the sunshine and ending up by doing a good job and having a fine time. The unit was growing up and it was growing together.

. . . .

At Helwan we had neither beds nor bed-boards—only straw palliasses and a concrete floor. However, we had discovered by now that the less the Army does for you the more you are allowed to do for yourself, and we had no difficulty in making ourselves comfortable. Life was becoming daily less regimental and we were still doing enough work to keep us off the parade ground.

We worked for New Zealand units in Helwan and for several weeks many drivers were employed in carting rubble from a local quarry to a road on the outskirts of the camp. This was highly remunerative employment, for it never occurred to the Egyptian contractors, for whose benefit the transport was provided, to insult our intelligence by supposing that we should be willing to work unless it was made worth our while. Always at the end of the day a fat hand would come through the cab window and in it would be anything from 50 to 100 piastres.

Meanwhile we were still operating the ferry service, and on 26 January the unit sustained its first casualties in the Western Desert when Bob Larkin[3] stepped on an Italian thermos bomb, injuring his left foot very severely and eventually losing it. Joe Carter[4] was wounded in a leg by the same bomb but was able to return to us five weeks later.

February came, and as by now most of the vehicles had covered between 15,000 and 20,000 miles, it was decided to give them a thorough overhaul. To complete the job within a month our Workshops' drivers, helped by everyone who could be spared, slaved

[3] Dvr R. C. Larkin; truck driver; Dargaville; born Colchester, England, 27 Sep 1913; wounded 26 Jan 1941.

[4] Dvr J. Carter; mechanic fitter; Papatoetoe, Auckland; born NZ, 5 Mar 1917; wounded 26 Jan 1941.

seven days a week, and every night for two weeks they put in three hours of overtime by electric light.

February, then, was a month of hard work and long hours, but it had its lighter moments. The loss of our orderly room through fire and the consequent destruction of the greater part of our records (which we watched with complacency, believing that anything recorded about us was unlikely to be to our advantage); a Divisional regatta in which we won the assault-boat event; the finals and semi-finals of the Divisional boxing tournament, in which Frank O'Connor[5] and Jack Cave[6] made a fine showing; our victory over the 6th Field Regiment at Rugby by five points to nil—these were incidents that enlivened February, making pleasant interludes in our greasy battle with stub-axles and shell bearings.

Early in the month the return of A Section had been heralded by the arrival of the Don Rs and within a week the last group of lorries had arrived back. As they reported in they were fallen on by the mechanics, for it was now known that a move could be expected any day.

Only one question was asked: where now? Each of us possessed a clue. They hung in the huts or crowned our bedrolls, battered, some of them, already. There had been a fresh issue of sun helmets. Now that, surely, meant the Red Sea and a landing on the coast of British Somaliland. It could mean a trip round the Cape to England. Or (why not?) a landing beyond Benghazi. Greece was out of the question, for the Greeks were showing themselves quite capable of settling Mussolini's hash, and surely Germany, with her hands full in Europe and her shortage of oil, would not be crazy enough to embark on any Balkan adventure? Besides, who ever heard of sun helmets in Greece? A landing in France perhaps—but no, those damned sun helmets. . . .

March came and on the 3rd of the month 'Bull' Dillon disappeared. He was our advanced party.

The next day was our last at Helwan. Carefully we made a final check. Tanks full? Water tins full? Tucker box bulging? Hide away those Italian blankets and groundsheets and those extra spanners—might come in useful, eh?

[5] Dvr F. E. O'Connor; motor body builder; Christchurch; born Ashburton, 5 Mar 1913.
[6] Dvr J. A. Cave; farmhand; Auckland; born NZ, 23 Jul 1918.

That night the NAAFIs did a roaring trade. Bottles covered every table. Bottles covered the floor. Underfoot there was a roughness of broken glass and a sloppiness of spilled beer. Every now and again some soldier would be intolerably wounded in his deepest feelings and would strike out wildly.

The next morning was bright and chilly, very soothing to feverish brows. Long before the time to start our lorries were lined up and waiting. At last the Major's car moved slowly towards the road and we followed. Slowly we drove through Maadi, through Cairo, and along the Alexandria road. In Wadi Natrun near the Halfway House (where the drinks were ice-cold and the prices red-hot) we halted, bivouacking for the night.

The next morning we set out for Amiriya, the transit camp for the port of Alexandria.

The engines barely purred for we were travelling very slowly. Louder than the engines, louder than the tires whispering to the black bitumen, was a noise of singing. In no other way could we express our exhilaration, our confidence in the future, our delight in being on the move.

(3) *Diversion to a Dragon-Slaying*

Nothing remains but to get B Section to Amiriya. The section— the company rather[1]—had entered Papakura Camp on 12 January 1940,[2] had trained under Captain N. M. Pryde[3] and Lieutenant S. A. Sampson,[4] and on 1 May had boarded the *Aquitania*.

While the convoy was in the Indian Ocean its destination was changed from Egypt to England. Capetown was reached on 26 May;

---

[1] At that time our drivers of the Second Echelon were organised as a company, which consisted of Company headquarters, one transport section, and Workshops. This organisation continued until they joined us at Amiriya, when the transport section became B Section and the remaining drivers, most of whom were specialists, were absorbed by Company headquarters and Workshops. Their strength on entering Egypt was two officers (Maj Pryde and Capt Sampson), one warrant officer, four sergeants, and 119 ORs. Maj Pryde was posted to the Divisional Supply Column three days after he joined us.

[2] The advanced party, which had entered Papakura on 29 Dec 1939 while the First Echelon was still there, consisted of Lt P. E. Coutts, a staff-sergeant, and two drivers.

[3] Maj N. M. Pryde, MBE, ED; bank accountant; Papakura; born Waikaka Valley, Southland, 6 May 1899; served in Div Amn Coy, Nov 1939-Mar 1941; OC Div Sup Coy Mar 1941-Dec 1942; OC 2 Amn Coy Dec 1942-Jun 1943.

[4] Maj S. A. Sampson, OBE, m.i.d.; butcher; Auckland; born Auckland, 20 May 1911; OC 1 Amn Coy 26 Jan 1943-17 Apr 1944.

Freetown, capital of Sierra Leone, on 7 June. The convoy spent only a few hours here and no shore leave was granted.

Paris fell on the 14th, and two days later the convoy arrived at Greenock in the Firth of Clyde after having travelled 17,000 miles.

The troops went ashore in lighters, entraining for Aldershot the same afternoon. Their arrival, of course, was unheralded, but as they went through Scotland and over the border into England, and down through the northern counties (catching, as they rumbled through the June night, the reek of blast furnaces, and sometimes, as a wayside station flashed past, a fragrance of cottage flowers), the news of their coming went before them and at every halt they were welcomed with smiles and little bursts of applause and with sandwiches and hot drinks. They were glad, these English people, to see Australians and New Zealanders, but they were not surprised. Only Hitler was surprised. Knowing everything about the Statute of Westminster but little about the hearts of free peoples, he had expected the Empire to fly to pieces.

The Ammunition Company was taken to the village of Bourley, about four miles from Aldershot. There it went under canvas.

During their first week in England all New Zealand troops were issued with free rail warrants and granted 48 hours' disembarkation leave. Our drivers were sent off in three groups and most of them went to London.

In London there are barbed-wire entanglements and machine-gun emplacements, and the parks and the city gardens have been divested of their iron railings, which are being converted into armaments. Oxford Street and the Strand are still as crowded as ever, with hundreds of scarlet omnibuses, seemingly careless and half asleep, but as sure as cats, swooping and pouncing through the traffic. Grey pigeons, fat and pompous, stump up and down in front of Saint Paul's and the British Museum on delicate pink feet; the London sparrows are busy about Charing Cross, and the waterfowl, all the colours of the rainbow, paddle in the Serpentine. France has fallen and the swastika floats above the Eiffel Tower, but the birds of London, free in the city as in a forest, are undisturbed.

And so are the London people—the people in the scarlet omnibuses and the people who stream out into the sunlight like black ants from tube and underground, clutching their *Evening Standards*.

They are undisturbed, too, in the sleepy market towns and the villages of the South Country. Some of the drivers have gone there, taking the winding English lanes whose hedges are white with hawthorn and following them to quiet places where (close to the grey church, always so large and stately in comparison with the other houses: a stone sheepdog guarding her stone pups) the village inn—the Lion, the Three Tuns, the Death of Nelson—stands open for thirsty travellers. Here the landlord, smothering a polite belch, looks up from polishing the bar to take the order, and then, seeing the New Zealand hat, smiles broadly and refuses payment. He had a lad, or someone he knows had a lad, who was over to them parts. . . .

They are not disturbed (or if they are they are keeping it to themselves) in the big houses, walking towards which and admiring the rhododendrons New Zealanders may be seen often, a friendly old gentleman or an old lady having offered them a meal and a hot bath.

They are not worried in the village shops and the cottages (or if they are they are taking care to hide it) or in the Commercial Road and the Elephant and Castle. Gone are the days when people were eager for soft jobs, searched for funk-holes in the country, grumbled about evacuees, bought up all the tinned stuff in the neighbourhood, carped and criticised. Since France has fallen, since England has been defeated in Europe and Dunkirk has happened, there has been a new courage abroad, and a new gentleness, and a new fierceness. England is in mortal danger and her people are ready and they are waiting. David Low has published a cartoon in the *Evening Standard*. It depicts a solitary British soldier in a steel helmet. He stands on the cliffs of Dover with the Channel at his feet and shakes his fist at a ravaged and beaten continent. And the caption is this: 'Very well, alone!'

It was their finest hour.

. . . .

A week passed and the company was issued with its transport (Bedfords and a few Albions) and its second-line holding of ammunition.

During July the drivers underwent extensive training in convoy work and vehicle maintenance besides carrying out brigade transport duties and taking part in field exercises. Everywhere they went they noticed the feverish preparations that were being made to meet the invasion. Signposts had been taken down and concrete road-blocks were being erected near bridges and villages. Boys and elderly men, at all hours of the night and day, could be seen panting up and down hillsides or crouching in ditches, practising for the defence of their own corner of England. They had few rifles.

Early in August a Divisional exercise was carried out and the company established ammunition points and functioned in accordance with the rules laid down for a transport unit. Our drivers returned to Bourley on the 8th, the richer by much practical experience in operational work and by memories of the Sussex countryside. Also, they had acquired some skill at darts, England's national game.

On the night of the 27th-28th a move was made to St. Leonard's Forest near Hastings. The journey took four hours, and for the greater part of that time it was impossible to use lights as enemy aircraft were overhead. On two occasions the convoy was halted by air raids. The next day was spent in working out a scheme of protection for the bivouac area and in digging in anti-tank rifles and machine guns. On the 29th the company returned to its old area at Bourley.

Meanwhile the Battle of Britain had begun and every day it was mounting in intensity. All day long the skies over Hampshire, Sussex, and Kent were ribboned with vapour trails and speckled with white puffs as the British fighter pilots fought back and anti-aircraft guns, hidden beside rick and oast-house, in coppice and by wayside tearoom, barked and coughed. Hienkels, Dorniers, and Messerschmitts fell out of the sky, crashing in hop fields, flaming in orchards and cottage gardens. In the streets of London paper-sellers chalked up the score: 8 August, 60 planes destroyed over the Channel; 11 August, another 60; 12 August, 78 planes destroyed over the Channel; 14 August, 180 planes destroyed over Great Britain. Our own fighters, precious beyond price and for a time irreplaceable, were falling too. But at bench and assembly line the battle was also being waged. Men and women worked until they dropped, and new squadrons took the air.

August ended, and the sky over England was still ours and Germany changed her plans. If London were smashed, if the city were in flames and the Thames estuary closed, then, surely, the people would rise against their Government, mad for peace.

From their bivouac area, and in the country lanes where their convoys were held up, our drivers saw the planes come over. For a start they came in arrowhead formation with Messerschmitts above them and one large black bomber, usually, in the lead. Later they came over in tight blocks or in serried tiers, and they came from all points of the compass. Ack-ack pounded them and our fighters darted to the attack, snarling all over the sky in great circles, but many of them got through, and presently from the direction of London came the sullen rumble of bombs. Often at night the sky was pink over the city.

At first the whine of the sirens was the signal for everyone to drop what he was doing and take cover, but after a while it was decided that work should continue during alerts, unit air sentries being relied on to give adequate warning of any real danger. The company maintained two ack-ack posts.

On 6 September the unit left Bourley for Bristling Wood, near Maidstone, in Kent, the 68-mile journey, which was made at night, taking ten hours to complete on account of air raids and traffic blocks. There was something electric in the atmosphere and our drivers, as they sat in their lorries watching the white beams of the searchlights stroking the clouds and making their vast geometry over London, were conscious of a prickling excitement, half pleasurable and half fearful. Even now the barges might be leaving the French ports.[5]

The new area provided good cover, which was just as well, for concealment and camouflage had become matters of the first importance. A Dornier was shot down only half a mile from the area soon after the company arrived, and a week later a Hurricane crashed 400 yards away, the pilot escaping with minor injuries. The ack-ack posts were no longer engaging even low-flying aircraft, for orders had been issued forbidding them to open fire unless the area was directly threatened.

The weather, in the meantime, had changed. It was less warm now. Shotguns were banging away in the woods and the nuts were

[5] The move from the Aldershot area to the coast sectors in Kent and Sussex was made by the entire New Zealand contingent. It was to be held in reserve near the coast so that in the event of invasion it could launch the first counter-attack.

ripening. Leaves and bracken had turned to gold. On fine days the skies were luminous with diffused sunlight and as delicate as a robin's egg and as softly blue, with sunset a hidden pink and later a red blaze like burning London. Often, though, they were sullen and overcast and the raiders would be hidden by cloud banks or by grey vapours that scudded across the heavens like spindrift. And at times they were flat and featureless and drained of all colour, so that the raiders, tier above neat tier, had the appearance of being stationary, like verses on old vellum.

The Luftwaffe was losing heavily—on 15 September, when the battle was at its peak, 185 enemy planes were reported shot down—but still it was getting through to London. In the shopping districts there were charred ruins where great shops had stood, the names of which had been known everywhere. Buckingham Palace had been damaged and famous houses had disappeared from the West End. Rows of small villas had been wiped out in the suburbs and grass was sprouting in the slums, where for the first time in centuries fresh air had been allowed to penetrate and flowers were growing. By city wharves warehouses stood gaunt and gutted. Faces were grimmer now, for there was little sleep in London. People spent the nights in their cellars and in the tubes. Hundreds had lost their homes and many were in mourning. Gone was that strange gaiety and gladness that had come to Londoners after Dunkirk. Dogged endurance had supervened. *Very well, alone!*

In spite of her scars and of the constant danger from the skies London continued to draw our drivers like a magnet. She was to blame even more than Scotland for the fact that large numbers of New Zealanders were always absent without leave.

Under the circumstances the allowance of official leave was not niggardly. Each week three of the drivers were granted seven days' leave, and free rail travel was provided to all places as far north as Carlisle.

October arrived, and by this time autumn had become early winter. It was no weather for camping out and everyone was glad to hear that the company would be moving into winter quarters, suitable accommodation having been found at Hunton, a small village near Maidstone. Company headquarters was to occupy the village hall and the transport section an oast-house, which consisted of a two-storied building surrounded by six conical towers.

The drivers of the transport section moved to Hunton a day earlier than the rest and it was as well they did, for they had been gone only a few hours—it was the afternoon of the 3rd—when the old area was bracketed by an enemy bomber. The residue of the company did not present a large target and no damage was done.

In the days that lay ahead they were to become used to narrow escapes, for while they were at Hunton the neighbourhood received plenty of attention from the Luftwaffe. They had been there barely three days before a dive bomber dropped a stick of bombs that shook the whole village, and on the following day four high-explosive bombs fell near their billets, the nearest of them landing eighty yards from the vehicle park. There was another raid on the 13th, but, less than a fortnight later, our drivers had a taste of revenge. A Messerschmitt 109 flew over the area with three Spitfires on its tail and while circling round, losing height all the time, was engaged vigorously by the two ack-ack posts. Finally it made a forced landing about 400 yards from Section headquarters and the guard lost no time in taking the pilot into custody. He gave his name as Birk and our drivers noticed that he was decorated with the Knight's Cross and the Iron Cross. German pilots at this time were popularly supposed either to be drugged or to go into battle under the eye of the Gestapo ('My dear, in every bomber that's forced down there is always one man who has been shot through the head: he's the Gestapo man.'), and with these stories in mind our drivers gazed curiously at Herr Birk, but there was nothing in his appearance either to confirm them or to refute them. All that one could say of him was that he looked scared, as well he might do considering how people of his kind were regarded in Hunton— scared and bewildered. How did it go, that song? 'Sleep well, my kitten—we are marching against England.' Well, he could forget about marching for some time and about seeing his kitten—unless, of course, Hitler came. But Hitler was long overdue, and in England (and in Germany as well, perhaps) people were beginning to say that the Battle of Britain was already over. The first dragon, the daylight dragon, had been vanquished, and as for the second one, the dragon that flew by night breathing fire over London, burning St. Clement Danes, Westminster Hall, Our Lady of Victories, Turner's house in Cheyne Walk, and the Wren churches, that, too, was mortal.

During October bad weather interfered with the training programme but transport duties were carried out as usual. It was pleasant indeed after driving for hours through the cold rain to come home to the oast-house, which was always warm and dry. In the lower story, which was used as a mess room and cookhouse, a large open fire was kept burning night and day and its grateful warmth pervaded the sleeping quarters, which were separated from the room below by nothing except an iron grating, spread now with palliasses and bedrolls instead of with drying hops.

The New Zealand contingent moved from Maidstone to Aldershot during the first week of November. The company went to Ash, a village about four miles from the town, the drivers being billetted in a fair-sized country house called Shawfield Farm. The bad weather continued and the rest of the month was passed chiefly in a struggle to prevent the lorries from sinking below the surface of the vehicle park.

December came and brought an improvement in the weather. The mornings were bitterly cold but the ground was dry and hard and there were many of those lovely hours of December sunshine that come early in the afternoon when the last bronze leaves are eddying through the still air under milky skies, blue here, with the delicate faint blue of milk, and gold in places, but only dusted with gold, like cream. With twilight the frost seized everything and in village streets there was no warm glow from cottage window or inn door to speak of Christmas. Enemy bombers droned and hiccupped through the night sky and tighter than the grip of frost was the grip of darkness.

The company had been released from its operational role on moving from the Maidstone area, and early in the month all transport was handed in except a few lorries that were kept for domestic use. On the 11th, together with other NZASC units, the company was inspected at Mytchett, near Aldershot, by the Duke of Gloucester, Captain Pryde being in charge of the parade. That afternoon the balance of the transport was handed in.

The days flew by and it was Christmas—Christmas for German kittens whose fathers were marching against England and Christmas for the children of Ash. There was holly and mistletoe and reedy trebles touchingly out of tune singing about the shepherds and good King Wenceslas, but there were no bells. For the first time since England had known churches and the Christmas story they were silent in proud belfries and tall steeples and in little, red brick chapels. Only for the invasion would they ring out.

Christmas dinner for the whole company was held in a gaily decorated garage, part of Company headquarters' billets. The fare was excellent and until late in the afternoon everyone was merry. Then there was bad news. Sergeant Andrew Morton[6] had been killed while standing on the running-board of a lorry. He was buried on 28 December in Brockwood cemetery.

The old year ended, and early on 3 January 1941 our drivers entrained at Aldershot. For the last time they were smelling that horrible and fascinating smell—a distillation of coal-dust, egg sandwiches, and escaping steam—that would always remind them of English railways.

Half across England they travelled, reaching Newport at half past ten in the morning and boarding the *Duchess of Bedford* the same day. On the 5th the ship moved down the Severn estuary, anchoring in the Barry Roads, where she spent the 6th. She sailed at five the next morning, crossed the Irish Sea in company with three other transports, and anchored off Bangor, County Down, on the 8th.

The convoy received an addition of six transports on the 11th, and on the 12th, after sailing before dawn, was joined by eleven more off the Firth of Clyde, which brought the total number of ships to twenty-one, not counting a large naval escort.

Scotland and Ireland faded in the distance and presently the procession of ships was alone with the grey waters of the Atlantic and the circling gulls. The gulls lingered for a while and then turned towards the shore, winging their way homeward, making for their small island, their obtuse island. 'Effete,' said Germany. 'Fat,' said Italy. 'Perfidious,' said France. The charges might still stand, but there was a word to be added, and the word was 'Brave'. Every New Zealander, each in his own way, echoed it in his heart. They could say through all the years to come: 'I was there. I saw it.' They could tell how a small island, obtuse and not over brilliant in battle, had stood alone, defying dragons.

. . . .

[6] Sgt A. Morton; motor driver; born Scotland, 6 Jun 1904; accidentally killed 25 Dec 1940.

A bitter wind and green-grey tumbling seas marbled with foam. 'Hell,' says everyone, 'Must be near Iceland.' The convoy alters course and the wind mellows, smoothing the seas to long Atlantic rollers. Our drivers march round the deck wearing boots. The sea is calm now and the weather much warmer. Grinning, the orderlies swab iodine on the soldiers' arms and the doctor drives home his needle. Hotter and hotter becomes the weather. The convoy is making for Freetown. The heat in the estuary is blinding. No training is possible and the soldiers lie panting in the shade. Anti-malaria ointment has been issued and they are able to sleep on deck—a great boon. Everyone is glad when the coast of Africa fades in the distance, but it is still hot and the soldiers grumble about the food. It had seemed excellent during the first few days, as it always does on shipboard. Now they can hardly face it. Longingly they describe the kind of meals they will eat in Capetown—steaks, eggs, fish, great juicy fruit. A spicy fragrance blows from Cape Province and the *Duchess* is in harbour. Leave is granted and no one behaves outrageously so there is more leave. Old friends are greeted and the dream meals are translated into fact. Three clear days at Capetown—then to sea again. Table Mountain fades from sight and a week later the convoy is sweltering in the hottest part of the Indian Ocean. The sea is glassy and in places jet black owing to its great depth. Tropical rain falls in solid sheets and the soldiers rush into the open, bathe themselves, rinse out their underclothes. The hills behind Aden are sighted and a Blenheim bomber flies overhead as the convoy enters the Red Sea. Everyone starts to pack. At last, late in the afternoon on 3 March, fifty-seven days after leaving Newport, the *Duchess* anchors in the Gulf of Suez.

On the 5th our drivers go ashore and a train takes them to Maadi. From Maadi they get leave to Cairo. After two days they climb aboard lorries and are taken to Amiriya. The Divisional Ammunition Company is complete.

. . . .

'Old Johnny made it, eh?'

'Yeah. Got a ride up with " Cash ", his RMT brother.'

'Tough " C. Jay " missing—and " Plunger " and old " Snow ".'

'Yeah. He couldn't have made it, " C. Jay "—not with that toe of his. Not carrying this load.'

Cigarettes glowed in the darkness. Somebody cursed the cold and made reference to the peculiar effect it would have on a brass monkey. That bitter chill that creeps over the desert just before dawn was making us shiver in spite of greatcoats and a mass of equipment. We should be warm enough, though, when we started moving. Valise (with blanket-roll bound round it), haversack, kit-bag, full web equipment, rifle, ammunition, water bottle, respirator, and a hundred odds and ends—under this load it was difficult even to stand up, let alone march.

During the ten days we had spent at Amiriya we had had plenty of time in which to experiment with rolls and bundles and many of us had packed and unpacked a score of times. Leave, of course, had been out of the question and few of us had cared to slip into Alexandria and take the risk of being left behind. The arrival of B Section had broken the monotony a little but it could not be said that A Section and C Section as a whole had taken the newcomers to their hearts. They were inclined to regard us—or perhaps it was our imagination—as recruits, raw and unblooded, and their references to the Battle of Britain annoyed us. We called them the 'Glamour Boys' and 'Cook's Tourists'. Watching the lorries leave had been another distraction. Four days after our arrival at Amiriya twenty of them had left for the docks under Second-Lieutenant Fenton and on the following day the rest had left under the Major and Lieutenant Aitken.[7] The drivers in charge of them had gone too and that had necessitated finding fresh players to fill the vacancies in the poker and pontoon schools. Then there had been a sandstorm, one of the worst in our experience. Latterly we had spent most of our time in rehearsing the present scene.

Dawn drew a streak of lemon in the east and someone said: 'All right. Pick up your gear. By the centre, quick march. Keep your ranks.' Tripping over rocks, cursing our loads, which (in spite of the rehearsals) kept trying either to hamstring us or to garrotte us, we trudged across the desert in the direction of El Quadir station. It was only a short march—a couple of miles or so—but by the time it was finished the strongest of us had had enough. As soon as the train arrived the jostling and pushing that had characterised our behaviour from the moment of parading started all over again.

[7] Maj R. C. Aitken; mechanic; Wellington; born Edinburgh, Scotland, 6 Jul 1894.

Everyone was anxious to keep close to his special friends so that he could be with them on the boat. From the train stop to the wharf was a short march and we made it under the mocking glances of the Egyptians, who were far too wise and too wicked to go to war themselves. Without much delay we were got aboard HMAS *Perth*.

At a quarter past eleven in the morning the gap between the wharf and the cruiser's side started to widen. Clumsily we dressed ship, getting barked at by the Master at Arms for talking. 'Smartly there! You're in the Navy now.'

Swiftly the cruiser gathered speed, dipping through the cold and sparkling waters.

Someone asked a sailor where we were going. He looked surprised.

'Greece,' he said, 'We've been taking 'em there all week.'

Map of Greece

Fort Capuzzo

Arrival in Greece, Piraeus

New Zealand infantry in Athens

Katerine, a typical village

CHAPTER 4

PICNIC BEFORE A THUNDERSTORM

During our first six hours on Greek soil we attracted very little attention. We arrived at the port of Piraeus on 18 March on one of those hot, pale afternoons that put a circle of primrose light round the horizon. On the rim of the circle we could see the Acropolis, but not very clearly.

After we had landed and stood for some while in three ranks we were told to fall out and stay within call. Greek labourers looked up from the huge trench they were digging near the wharf and accepted cigarettes. Only one person took any real notice of us and he was the proprietor of the little wineshop at the dock gates. He responded to our presence by doubling the price of his wines and spirits. And that was all that happened. We were disappointed, for we had expected to be made a fuss of.

We were disappointed, too, in what we could see of Piraeus, and a third disappointment was the long wait at the docks. After the excitement of the past week and our dash through the Mediterranean—*Perth* had made the trip in just over twenty-four hours —it was felt as an anti-climax. Ten-ton diesel lorries had taken away one load of men, and here we were sitting on our gear on the wharf, hot and uncomfortable in pith helmets and battledress, waiting for their return.

A diversion was caused when the *Araybank*, very high in the water, sidled into the space lately vacated by the *Perth*, presenting the men ashore with a side like a red cliff, from the summit of which the drivers in charge of vehicles peered down excitedly, shouting abuse to their friends and being taunted in return with comparisons between the *Perth*'s quick journey and the eight days it had taken the *Araybank* to cross.

At last, after we had waited six hours, the ten-tonners came back and we climbed aboard. As we rumbled through Athens the shadows were bunching at the street corners and at every corner there was a crowd of people, their faces turned up to us, their eyes bright. Here was the welcome we had been waiting for. They neither cheered nor called out to us, but as each lorry went past they threw flowers and there was a burst of clapping as though we were beloved actors.

It was embarrassing but it was pleasant. It was pleasant to read, or to imagine we read, love and trust in those Greek faces, pale in the half-light—flower-like faces of children, old men's faces out of the Old Testament, women's faces that Bellini might have painted, and the many other faces equally Greek but not as clearly distinguished in that moment of sentiment: the fruitshop faces ('Yes, we have no bananas.') and the café and dining-room faces ('You lika da steak and da oyst, no?') and the blue-black jowls and greasy coiffeurs and sweating foreheads of the Levant: 'I show you the good time, my friend!'

We did not start singing until we had left Athens behind and were rushing through the leafy darkness towards Kephissia, eight miles north-east of Athens, with the cold, sweet air tasting in our throats like the aftermath of peppermints.

When we reached the camp we came down to earth with a bang. Most of the tents were only half-pitched and there was nothing to eat. Our advanced party—not his fault—had gone to the wrong camp.

We were up early the next morning—almost before the wet, white mist had cleared from the pine trees. Those of us who were not on duty (thirty lorries were employed in carting supplies and ammunition from Piraeus to various dumps near Athens) were away as soon as breakfast was finished.

Everything was a delight and a wonder—strange and yet not altogether strange, for this, too, was our heritage. Grimm was a German and Hans Andersen lived in Denmark, so half our babyhood, including Christmas trees and Santa Claus, came from Europe. The castles we saw were of the kind that Grimm's ogres emerged from; the cottages were no different from the one Hansel and Gretel lived in; the tailors sat cross-legged in their shops as they had done since fairy tales were first told; the huts of the charcoal burners and woodcutters were exactly as we had known them in childhood and the forests were the forests of our first memories.

Soon after we arrived at Kephissia everyone was given a 500-drachma note worth rather less than a pound. Such a quantity of notes had never before been seen in the neighbouring cafés and wineshops and within a few hours the whole district was swept clean of change. Messengers had to be sent to Athens to relieve the situation.

During the afternoon and evening we became acquainted with the delicious wines of the country: mavrodaphne and the ordinary crassi, as well as with Metaxa brandy, the not-so-delicious ouzo (we had known it as zibbib in Egypt and as arrack in Palestine), and a noisome liquid that tasted of turpentine and pine needles and made everyone who took too much of it allergic to fir trees for about a year afterwards.

For a little while we listened to the fruity baritones and thrilling tenors of the Greek police. (Afterwards they had to listen to us.) They wore olive-green uniforms and they seemed to be the only men of an age and condition to be called up who were not in the forward areas. They sang beautifully but with rather too much emotion, their voices reminding one of massed violas anguished among whipped cream. (Our own singing, later in the evening, was even more emotional and not nearly as good.) Their favourite chorus went to the tune of the *Woodpecker's Song* but the only word we could distinguish was Mussolini.[1] It sounded tremendously gay and gallant and defiant: it expressed the very essence of the Greek Evzones. The whole of Greece was singing it at that time, and it was still being sung more seldom, more sombrely, no less gallantly, sometimes by a group of Greek soldiers struggling home, sometimes by a single small boy, at the end of April. Doubtless the Fascist guards heard it whistled in the streets of Athens long after we had gone.

But how hopeful it sounded then! Hopeful for us and hopeful for the Greek people; hopeful for everyone who knew nothing at all about what was going on. Possibly there never had been a more hopeful convoy than the one that pulled out from Kephissia camp for northern Greece at six in the morning on 22 March 1941.

Soon we were passing through Athens, old and shabby and

[1] Jim Henderson, in *Gunner Inglorious*, suggests the following translation:
We scoff at you, Mussolini,
You and your cowardly conscripts.
Very soon you will be advancing rapidly
Backwards:
Backwards, until upon Rome itself
The Greek flag will be flying.

graceful, birthplace of half we were fighting for. The city was not wholly strange to some of us, for on the day after our arrival at Kephissia, as the Major had discovered on ordering a check parade, over sixty drivers had absented themselves from camp. With Athens behind us we came into a country of rounded hills, soft as green velvet and knotted all over with little holm-oaks: it was like tapestry. From between two of these hills, winging its way towards the convoy, came a large, black bird about the size of a young gobbler. This was too much for A Section's anti-aircraft gun crew and they started dusting it with tracer bullets as with a kind of monstrous red pepper. Two minutes later they were under arrest.

On through Thebes—pretty and full of colour as a cottage garden—through Levadhia, through Atalante, and by this time the headlamps and bonnets of the lorries were bedecked with flowers and the cabs were full of flowers too. We must have resembled a convoy of sacrificial bulls. At every halt the people came to us with wine, cakes, and eggs. ' *Kalimera,* New Zealand! ' they cried—all the rosy-cheeked children and handsome young women and shepherds and old men.

As the shadows became larger—shadows as big as sheep stations cast by tremendous hills—the halts occurred more often and it was dark when we arrived at Kamena Voula, our halting place for the night. If it had been daylight we could have looked across the water to the island of Euboea. We had covered 150 miles in a long day full of incident.

The next day we moved off at first light and travelled through fairy-tale country—pocket-handkerchief fields, patchwork-quilt farms, little laughing streams rushing from openings in the hills as from school, with, behind them and above them, terrific, towering, gold and green and purple mountains, like sleeping heroes, like sleeping gods, with huge lazy limbs sprawled carelessly everywhere and muscles like bits of the Elgin Marbles blotting out a quarter of the sky. We refuelled at Larissa railhead, camping for the night ten miles north of the town.

That night we were not so tired and some of us went to a small, tumbledown village and drank rough red wine in a little dark wineshop where Greek soldiers were singing their defiant woodpecker song. Our battledress and the badness of the wine were the only things that would have surprised Homer had he come strolling in.

Off again early the next morning and on till we reached the foot of the Olympus Pass. Then a long, winding crawl to the summit and a crawl down the other side at twelve miles an hour. Those of us who were not driving gazed wonderingly at the calm and unutterably lovely outline of the great mountain, whose top, so the ancients had said, reached Heaven, on whose grassy slopes centaurs had browsed and galloped, in whose shadowed or sunlit folds the old gods had lived human and irritable lives, feasting, quarrelling, and making love. Those of us who were driving, though, looked only at the road. Hundreds of feet below, a sheer drop from the road's edge—you could have spat down at them—were the tops of pine trees. Small and tender as asparagus tips they looked, but as menacing as massed spears.

We made camp beside the Larissa-Katerine road, a few miles west of Katerine. Our area was a tongue of land looped off from the surrounding woods and meadows by a bubbling, rock-filled stream, which could be crossed only by a narrow stone bridge. We parked our lorries beneath tall beeches. Behind us were soft hills mossy with fruit trees and old farms and browsing sheep. Beyond them was the mountain.

We had arrived at our destination and it was a lovely place.

．　　．　　．　　．　　．　　．

During the next few days we were very busy. A Section supplied a guard for ammunition at Katerine railhead, the clearing of which was the first transport job we were given. Ammunition was delivered to New Zealand units in and around Gannokhora, a village two or three miles north of Katerine, and between 26 and 30 March our lorries plied between railhead and neighbouring dumps, meeting the trains as they arrived. At that time the New Zealand Division was preparing a line that ran inland from a point on the coast fourteen miles beyond Katerine and south of the mouth of the Aliakmon River. This was part of the Aliakmon line. Farther north was a fortress line manned by Greeks—the Metaxas line.

When we were not working we were enjoying ourselves among our new surroundings. We went for long walks; we washed our bodies and our clothes in the icy stream and every evening we feasted on new-laid eggs and fresh bread, in payment for which the country people accepted empty petrol tins. There was leave to Katerine, but on our arrival the Major had held a sort of quarter sessions to deal with the accumulated charges of a fortnight, as a result of which many were confined to barracks—what a phrase to use when we were living in the very shadow of Olympus in a wood crazy with birds! It was spring, too, and the mountain's snowcaps glinted like silver in the spring sunlight, which flowed over the foothills and slanted in bright bars between the tree-trunks. Drifting mauve and white clouds blundered against the snowcaps and descended on us in fine rain, making the stream chuckle fiercely among the boulders.

On one of the clear days an enemy observation plane flew over. It hung in the air, small and silvery and innocent-looking, and we gazed up at it with intense interest and some awe. 'Don't look up, boys,' yelled Second-Lieutenant S. F. Toogood.[2] 'He'll see the whites of your eyes.' Happy laughter drowned the tiny buzzing noise, but after we had laughed ourselves out—and it took quite a time—the buzzing could be heard still: a tiny waspish whisper: 'I can see the whites of your eyes.'[3]

The enemy went away and there was nothing sinister any longer under the mountain in the clear sunshine. Spirits were high at that time; the whole green spring was only a breath away and the note was hope. Ours was the Woodpecker's attitude towards the war and we never doubted that it was shared by Generals Papagos and Wilson.[4] Probably they whistled that gay tune (or felt like whistling it) at their conferences.

We had no newspapers and no wireless sets, and when Yugoslavia signed a protocol to the Berlin-Rome-Tokio pact on 25 March it is probable that only a dozen men in the company knew what had happened. A few days later a small group of our drivers was accosted by a Greek priest in a great state of excitement. Speaking

[2] Maj S. F. Toogood, m.i.d.; theatre manager; Wellington; born Wellington, 4 Apr 1916; Ammunition officer 2 NZ Division Sep 1941-Feb 1946.
[3] The chief appointments on 2 March were: Company headquarters, Maj McGuire, Capt R. C. Gibson (posted 22 Jan 41), WO II W. L. Dillon (appointed CSM 10 Nov 40); A Section, Capt Moon, 2 Lt S. F. Toogood (attached 28 Feb 41); B Section, Capt Sampson; C Section, Capt Torbet, 2 Lt Fenton; Workshops, Lt Aitken.
The following had left us: Lt Roberts (appointed OC Base Training Depot, NZASC, Dec 40), Lt Radford (posted to HQ Command NZASC, 16 Dec 40), WO II Colton (posted to OCTU, 10 Nov 40).
[4] Commanders-in-Chief of the Greek and British forces respectively.

in Greek and pointing to a Greek newspaper, he put it into our heads that Mussolini had been assassinated and that the whole of Italy was aflame. The drivers lost no time in passing on the good news. What he had been trying to tell them, perhaps, was that the Government of Yugoslavia had been overthrown and that General Simovitch, who had replaced the pro-Nazi Prince Paul, had pledged his country to neutrality.

When we heard the news we began to talk about leave in Athens.

. . . .

On 2 April the company was ordered to unload its second-line holding of ammunition at Kilo 9 on the Larissa-Katerine road and report to Headquarters 81st Base Sub-area at Larissa.⁵ The whole company, with the exception of Workshops, which stayed where it was, moved the next day, pitching camp among smooth green hills a few miles from Larissa. That evening the Major was told that his transport would be needed for an indefinite period, and during the next five days and nights we carted ammunition, petrol, and rations from Larissa railhead to supply dumps near Servia, between forty and fifty miles north of Larissa, and to others still farther north on the far side of the Aliakmon River. These were for a mixed British, Australian, and New Zealand force—the Amyntaion detachment—that had been organised to fight a delaying action in the event of a German thrust from Yugoslavia.

We were employed in this manner when Germany invaded Greece and Yugoslavia on 6 April and swept away the last vestige of that agreeable feeling that we were guests at a Grecian picnic. We were ordered to carry arms at all times and to be on the alert for paratroops and fifth columnists. Everyone began to listen to and repeat

⁵ Second-line holding of Divisional Ammunition Company of three sections each carrying ammunition for an infantry brigade and for one-third of divisional troops units: 5400 rounds of 25-pounder ammunition for 72 guns,* 2304 rounds of 2-pounder anti-tank ammunition for 48 guns, 18,000 rounds of .55 anti-tank ammunition for 360 rifles,† 390,000 rounds of .303 ammunition for 1360 light machine guns,† 42,000 rounds of .303 ammunition for 28 medium machine guns, 7000 rounds of .5 ammunition for 28 heavy machine guns, 320,000 rounds of .303 ammunition for 8000 rifles,† 8400 rounds of .38 ammunition for 1400 pistols,† 1296 bombs (HE and smoke) for 18 three-inch mortars, 5184 bombs (HE and smoke) for 108 two-inch mortars, 1620 grenades, 1800 rounds for 300 signal pistols,† five tons of miscellaneous explosives, and 1000 active and 1000 dummy anti-tank mines. Total loads: 57 3-ton, 12 30-cwt.

* Scale per weapon: 53 rounds HE, 17 smoke, 5 AP.
† Approximately.

alarming rumours: three New Zealand machine-gunners had not been heard of since a party of men dressed as Greek soldiers had lured them into a dark wood with promises of bread and eggs: two RASC drivers had died in agony after accepting a gift of cognac from a Greek shepherd. Probably neither of these stories—and there were scores like them—had any foundation, but can you wonder that the charming old gentleman who peddled tobacco leaves in our area became suddenly a sinister figure? That behind the tinkle of sheep bells we heard the clink of Lugers? We saw a party of Greek soldiers trooping silently against the skyline, purposeful, mysterious, a led mule in the midst of them, and we wondered. . . .

We were still excited and eager—more so than ever—but the Woodpecker was glancing uneasily behind him, which was what the Germans had hoped he would be doing. André Maurois records that Jean Cocteau said to him after the disaster at Sedan: 'All you see now on the roads of France are nuns winding on their puttees'.

On the night of 8-9 April—the night before the Germans occupied Salonika—we returned to the old area at Katerine, travelling without lights and with only the slanting rain showing ahead. We had been told to pick up our second-line holding of ammunition and bring it south. The Germans were driving swiftly through Yugoslavia and Macedonia and the New Zealand Division had been ordered back to the Olympus line—not really a line at all but a series of fortified positions running from the north-west corner of Greece, near Florina, to Mount Olympus.

The roads were so greasy that they might have been smeared with lard and it was so dark in the pass that the spare drivers had to stand on the running-board and give directions. Many of us had already taken out our windscreens in spite of the cold.

The next day we took our second-line holding to an area near Dolikhe, just south of the pass on the Larissa-Katerine road. Here we unloaded and stood by to assist with the withdrawal of part of the 6th Brigade from positions forward of Katerine to a reserve position north of Dolikhe. This job was done on the 10th and on the same day we sent a convoy to Gannokhora to salvage engineers' stores.

Meanwhile the Wehrmacht was winning battles. The Greeks, who in comparison with the Germans were fighting only with the weapons of the spirit, had been overwhelmed quickly in their fortress positions near the frontier, and now the British, with too few troops to meet the attack in their hastily-prepared northern line, were withdrawing south to a shorter one. By the 11th the Amyntaion detachment, which for one had not withdrawn but had advanced, was fighting a delaying action against motorised troops and armour south-east of Florina near the Yugoslav and Albanian borders. It was outnumbered and could be expected to hold only for a little while.

We, of course, knew little or nothing of what was going on. As we saw it the deluded Germans were being allowed to hurl themselves against our impregnable positions at Servia and Olympus. Accordingly we tackled our salvage jobs with the comfortable feeling of making everything snug and shipshape enjoyed by the yachtsman as he heaves to—by Crusoe while securing his flocks.

On 11 April—by which time all New Zealand troops had withdrawn from the positions forward of Olympus with the exception of the Divisional Cavalry, one troop of the 5th Field Regiment, and a section of engineers—the Major, Captain Moon, and Second-Lieutenant Toogood took a convoy to Sphendami, nine miles north of Katerine, to salvage stores left behind by the 6th Brigade. While these were being loaded the Major received a message from the Divisional Cavalry asking him to supply transport for carting road metal. Some lorries had started for home already but the rest were put to work at once and by late evening the worst parts of the Cavalry's withdrawal route had been repaired. The lorries then went home—all, that is, except three. These, led by Major McGuire and Captain Moon and accompanied by a machine-gun party composed of A Section's ack-ack crew, set out for Aiginion, a village some seven miles north of Sphendami, where there were valuable stores that would have to be abandoned to the enemy unless they could be gathered up within a few hours. By the light of burning supply stacks the officers and drivers collected what was most valuable and by eleven o'clock that night the lorries were loaded to capacity. They headed southwards with fires bobbing behind them in the windy darkness.

The NZASC was functioning well at this time but the panzer divisions were also functioning well. There was a lot to take away and not much time. Colonel Crump, Commander NZASC, must have experienced all the sensations of a man who hurries down crowded streets with armfuls of Christmas shopping and has valuable parcels flung at him from all sides. The hour for destroying instead of salvaging was in sight.

For us it arrived on Easter Sunday, 13 April. That morning Captain Sampson took a convoy to Amyntaion near the Florina gap to pick up petrol from a dump in that neighbourhood. Unaware of the exact position of the dump he led the convoy right into Amyntaion just as a barrage was feeling its way towards that doomed village. It came marching down the hillsides by the Florina gap; it crossed Lake Petron and entered Amyntaion. Little time was lost in stopping and turning but shells were landing among the houses as the last lorries gathered speed. Early that evening—it was late afternoon when our lorries arrived—the Germans were to drive in great strength down the centre of the Veve Pass near Lake Petron and force a withdrawal.

The petrol was discovered about six miles down the Amyntaion-Servia road and as soon as the lorries were loaded our drivers turned their attention to the ration section of the dump. They were shown where the luxuries were stacked and told to help themselves. It was their first experience of what was to be known later as an 'open slather' and many of them were pink with excitement as they scrambled among the stacks, breaking open cases of tinned fruit, tinned vegetables, pickles, jam, tobacco, cigarettes. You just helped yourself. It was a childhood dream come true and for a quarter of an hour they behaved like greedy small boys. They stuffed their blouses with tins and threw cases on top of the loads of petrol.

Often, in the days that lay ahead, we were to be wet and cold, sick and sorry, tired and frightened—never hungry. Empty and half-empty tins of peas and beans and pears and raspberries were to mark the passage of our convoys. It is probably fair to say that from the day of the first 'open slather' until the day we left Greece most of us over-ate.

After our drivers had taken all they could cram into the lorries they were told to destroy what was left. It was forbidden to use fire so they spiked the petrol tins and poured kerosene over the food. While they were doing this they were ordered away by British officers who wanted the area for their guns.

The drive back to Dolikhe was unpleasant. There was mud

everywhere and the road was jammed by British armour. It was snowing too. The petrol was dumped south of the Portas Pass for the 6th Australian Division and the 4th Brigade, and it was well after midnight before our drivers got home.

They slept uneasily—' cramm'd with distressful bread '.

. . . .

After a week of rain and snow and cold winds Easter Sunday dawned bright and clear, promising fine weather. By midday our sodden hillside was drying out well. If the rain held off, we prophesied, the RAF would be able to knock hell out of the German columns.

After a big lunch of tinned meat, pickles, and fruit salad—in every section the cook's lorry was well stocked now—a convoy left to pick up the 26th Battalion at its reserve position a few miles north of our area and take it by way of Servia to a debussing point west of the nearby Portas Pass. From there the infantry would march eight miles to occupy a part of the line overlooking the village of Rymnion and the Aliakmon River.

By this time the battle for Portas Pass had begun. The Luftwaffe had opened it that morning with bombers and fighters. The Divisional Cavalry, in its position forward of Olympus, was also in action.

Those of us who stayed in camp hung out our blankets to dry in the bright sunshine, washed our clothes, kicked footballs. It was a lovely afternoon and the sunshine was spilling over the foothills by the Olympus Pass like golden syrup. At afternoon-tea time there came a hollow coughing from below the nearest of these hills, and a number of white puffs, like dabs of cotton wool, appeared on a level with the hilltops. Then we saw a flight of black Stukas come falling out of the sky. None of us had seen Stukas in action before and for a moment we wondered if they had been shot down. Then we saw black mushrooms of smoke materialising around the Bofors emplacements. After they had dropped their bombs the Stukas climbed back into the sky. Then the Messerschmitts came, diving and climbing among the woolly puffs with tremendous energy. They were followed by more Stukas, which stood on their yellow noses and fell like plummets towards the gun emplacements. A single woolly puff appeared high above them and the guns were silent. After a while the Stukas and Messerschmitts turned north and disappeared over the mountains. The smoke drifted away from where the guns had been and presently there was nothing to show, in all that sunny afternoon, that a raid had occurred.

Chattering like monkeys, the drivers came down from their vantage points. A Section's cooks went back to their interrupted task of cornering and killing a large porker that had been driven into a gully. Washing was resumed and the footballs were brought out again. It was still a lovely day. There was a sound—you could hear if you listened carefully—that suggested that other and larger footballs were being punted about beyond the mountains: Pomp! Pomp! Pomp! Pomp! Anti-aircraft guns were in action.

The drivers who had been with the 26th Battalion returned to camp. They had been dive-bombed but without damage or casualties.

Easter Monday was another lovely day.

A convoy set out to collect ammunition from a field supply depot south of Servia for delivery to Australians on the Servia front. It was not an easy trip. The earth was still ours but the Germans had already conquered the skies and there was no safety any more. Schools of Dorniers in the kind of formations you see in an aquarium swam sedately over the mountains and dropped their loads on bridges and gun positions. Slim Messerschmitts, like furious wasps, flew along the valleys and above the roads, stabbing spitefully at traffic blocks and convoys. These were our particular enemies—these and the bright-nosed, crooked-winged, black-spatted Stukas, with their swinish squeals, their mad zest, their Gadarene plunge.

The traffic was thick on the roads and it was difficult to keep any distance between the lorries. The spare driver rode behind the cab and watched the skies. As soon as he saw aircraft he banged hard on the roof. The lorry jerked to a stop and both drivers tumbled into the ditch or ran into the waist-high crops and lay quiet and quaking. As soon as the aircraft had passed the lorries moved on. The idea was to save your life without holding up the traffic more than you could help.

After one heavy raid Wally Mosen[6] was the first to get to his feet and the first to return to his lorry. A fighter-bomber, following

---

[6] Dvr W. V. Mosen; lorry driver; born Raetihi, 6 May 1916; killed in action 14 Apr 1941.

in the wake of the main wave, came up the valley strafing and Wally, who was one of the most popular drivers in B Section, was killed by an explosive bullet.

After delivering their loads the drivers headed for home, very thankful to have got rid of their ammunition. A fork in the road was now under German observation and it was being shelled, but the lorries rushed through at wide intervals and all got safely past.

The convoy was machine-gunned again and Jim Nichols[7] and Dave Forbes[8] were wounded—Jim in the forehead and Dave badly in the thigh. Both were on the same B Section lorry.

It was sundown by the time the company area was reached and everyone was very tired and shaken. It was a new area, those of us who had stayed at home having moved four miles down the road in the afternoon. We, too, had something to talk about. No sooner had we settled in than half a dozen bombers, flying close and fast like wild geese, had appeared from the east. They were quite low and there was a clicking of rifle bolts as we got ready to engage them. As soon as they were within range the rifles began to crack, sounding under the beat of engines like snapping twigs. We were too excited to feel afraid, and when the first bomb started to whistle down one or two of us had our rifles pointed straight up in the air and were almost over-balancing. The bombs came striding across the paddock, punching great brown holes in the soft ground, but the only casualties were two sheep.

That evening we told stories of our escapes and gathered in the last sunshine to examine a lorry that had been towed in with bullet holes in cab and windscreen.

In A Section's ack-ack truck, Percy Sanders and Jim Stanley[9] were loading Bren magazines. The floor of the truck was littered with shell cases.

'They stuck to their guns all day,' the drivers were saying. ' Percy worked the Bren and old Jim fired the Boys anti-tank rifle from the shoulder. They reckon he's black and blue.'

[7] Dvr J. Nichols; labourer; Matamata; born Auckland, 12 Jan 1912; wounded and p.w., Apr 1941.
[8] Dvr D. C. Forbes; general store-keeper; Hamilton; born Scotland, 20 Jan 1904: wounded and p.w., Apr 1941.
[9] Dvr F. J. Stanley; motor mechanic; born NZ, 27 Dec 1908; wounded Dec 1941.

Presently shadows covered everything—the bomb craters, the two dead sheep, the faithful Bedfords—and everyone who was not on picket duty turned in. We had a feeling that we should be wise to lay in a supply of sleep.

CHAPTER 5

THE THUNDERSTORM

SHADOWS over our drivers asleep in their Bedfords, shadows over Greece, and shadows over the cause we were fighting for. Nothing but good news for the Germans and nothing but bad news for us. Yesterday we had given up Bardia in North Africa; to-day —that dangerous and eventful Easter Monday described in the last chapter—the decision had been made to abandon the Olympus line and withdraw to one based on Thermopylae. The reason for this was that the Greeks on our left flank could not be expected to hold out much longer on their own and we could send them no help.

At a late hour that night we were visited by Colonel Crump. After he had gone there was a hurried conference of section officers and presently the sleepy drivers were shaken into wakefulness and told to get ready to move at once. A group of drivers started to strike a tent but the Major told them not to bother with it. 'You won't need that tent,' he said. The drivers looked at him for a moment, wanting to ask questions. There was something ominous in the words. They were like the small, chilling whisper from the silver reconnaissance plane: *I can see the whites of your eyes.* For days past we had been saving things that other people had left behind and now we were ourselves leaving behind things. We didn't like it.

Our orders were to take first our second-line holding and then the ammunition from a nearby field supply depot to a new area near Tyrnavos, some ten miles north-west of Larissa. These were not long trips but they were trying ones, for we were very sleepy.

Soon after midnight word arrived by Don R that all heavy non-combatant vehicles were to be through Larissa and heading south by seven the next morning. Half past six found Workshops, with its great six-wheel Thornycrofts, trying to move through the town but being unable to do so because of a heavy air raid. Eventually the drivers by-passed it by cutting across some fields and through a Cypriot camp. Then they headed towards Athens, ignorant of their destination.

Meanwhile the drivers of the section vehicles had finished clearing the field supply depot and were back in the Tyrnavos area. If they had been counting on a rest they were disappointed. With the exception of fifteen 3-ton and six 30-cwt. lorries, which had been detached on special duty under the command of Second-Lieutenant Fenton, all the load-carriers left under the Major to pick up the 25th Battalion from the Olympus sector and take it to positions covering Elasson. The 4th Brigade was stilling holding at Servia and the 5th Brigade in the Olympus Pass but the design for the withdrawal had begun to take shape.

We were on the road all through the night of the 15th-16th (stopping and starting and dozing off), and the last group of lorries did not get home until the afternoon of the next day. As they were pulling off the road into the Tyrnavos area they were machine-gunned by a single aircraft, and Stan Fisher[1] was wounded in the back while returning the fire from B Section's ack-ack truck.

The lorries that had reached home earlier in the day were already back on the road. They had been formed into a convoy to take ammunition from Dolikhe to positions covering the Vale of Tempe, south of Mount Olympus, where the New Zealand Artillery was waiting to meet a German thrust down the east coast. Soon the famous battle of the Peneios Gorge would start.

By dusk, however, all transport was in the Tyrnavos area and it looked as though we should get a night's sleep. We were dead-tired but we were happy and cheerful. The story had got around—and of course we accepted it—that the Germans were being drawn into a trap. Somewhere—at the next row of mountains probably—the exhausted panzers would run head on into the guns, the tanks, the swarms of planes. Was it then or later that we heard about the tens of thousands of Canadians that were pouring into Thrace? And the hundreds of Hurricanes at Athens that were waiting only to have their guns fitted? Yes, there was much to cheer us and as yet we were in no physical distress. The cabs of our Bedfords were comfortable and the man who was not driving could doze off for ten, twenty minutes, an hour at a time. We had plenty of tobacco and we were full of good food—hot food at that. The lorries had small inspection traps that could be opened from inside the cab

---

[1] Dvr S. A. Fisher; railway storeman, Otahuhu; Auckland; born Aoroa, Northern Wairoa, 1 Mar 1904; wounded 17 Apr 1941.

and there was always a tin of beans, sausages, or M&V wedged against the hot manifold. We were young, too, most of us—many were only boys. Hell, it was a great adventure! All we needed was a night's sleep.

No sooner had we fallen asleep, it seemed, than the NCOs were going from lorry to lorry and telling us to throw on our ammunition as quickly as possible. By a quarter past one we were heading for Volos, on the coast below Larissa, and half the transport in Greece seemed to be travelling with us. Think of a very old goods train with very loose couplings. Picture it stopping and starting twenty times in a mile and that will give you an idea of our night's progress. It was raining, too, and lorries slipped off the road continually. Sergeant Robin Hood had a motor-cycle accident and broke a leg, and a B Section lorry went over a steep bank, boxes of 25-pounder ammunition falling on 'Merry' Meredith[2] and 'Barney' George[3] who were asleep in the back. 'Merry' broke a leg and 'Barney' was injured in the leg and face.

Dawn of the 17th found us jammed nose to tail in a column that stretched as far as one could see. Fortunately it was still raining, so no aircraft came over. After we had passed through Larissa there were continual halts and at one stage of the journey we took four hours to cover two miles. By the time we had reached our destination, which was near Almiros, some fifteen miles below Volos, and uncomfortably close to an airfield, the weather had begun to mend and we lost no time in dispersing. No sooner had we done so than a flight of Stukas swept in to shoot up two aircraft that were parked nearby. After they had gone the Major searched the airfield for petrol, of which we were extremely short. He found a number of large drums of aviation spirit and from these we filled our tanks.

We ate our supper under the grey olive trees, the drivers hesitating between tinned peaches and tinned cherries, tinned pineapple and tinned pears. On a little grassy plateau a portable gramophone looted from a bombed house ground out the theme song of the campaign:

[2] Dvr R. H. Meredith; motor driver; Auckland; born Onehunga, 1 Dec 1904; injured 17 Apr 1941.
[3] Dvr C. L. George; shearer; Te Kowhai, Frankton Junction; born Dunedin, 6 Apr 1915; injured and p.w. Apr 1941; escaped 7 Mar 1945.

> He's up each morning bright and early
> To wake up all the neighbourhood.
> He brings to every boy and girlie
> His happy serenade on wood. . . .

It was still, even though we were beginning to detect in it a faint note of mockery, a delightful tune.

That night, for the first time in over eighty hours, we enjoyed a long, sound sleep.

. . . .

The detached vehicles under Second-Lieutenant Fenton had come under the direct orders of Headquarters NZASC two days earlier. On the 16th, while the three-tonners carted engineers' stores from Dolikhe to Elasson, the 30-cwts. stood by under Sergeant Buckleigh in the company's old area near Dolikhe, and that evening they set out to meet the 32nd Battery of the 7th Anti-Tank Regiment at Kokkinoplos, a hamlet perched high in the hills above the Olympus Pass. It was raining and the drivers did not know where the Germans were or whether the anti-tank gunners would be able to reach the rendezvous.

Meanwhile the three-tonners had returned to Dolikhe and Headquarters NZASC had moved south. After travelling all night with his convoy Second-Lieutenant Fenton found it at Tyrnavos, and during the morning he attended a conference about the withdrawal of the 20th Battalion from its rearguard position near Lava, some two miles south of Servia. The whole 4th Brigade Group was to be withdrawn that night—the 20th Battalion by Second-Lieutenant Fenton's detachment and a detachment from the Divisional Petrol Company.

By 9 p.m. the lorries were dispersed some distance from the embussing point, waiting for the signal to move forward to collect the troops. Stacks of supplies were burning not far away and the transport was silhouetted against the rosy light, which faded and glowed jerkily. Presently the order to move came. Leaving the fires behind, the convoy went forward into a blackness so complete that at the first halt the drivers fastened scraps of white paper to the tailboards of their lorries. As they neared the embussing point they could see where German shells were landing and sometimes the shell-bursts could be heard above the whine of the engines. The nearest one was like the sudden opening and clanging shut of a furnace door. The lorries halted, and the infantry came plodding out of the night, wet, cold, muddy, tired, hungry. They had marched many miles with full equipment but they were cheerful still. The drivers scrabbled among the gear in the backs of the lorries for tins of fruit and packets of cigarettes for their passengers and then stood by to lift up the tailboards. It was not much but it was part of the service.

When everyone was aboard the convoy set out for the Thermopylae line—the last line before Athens.

Through Elasson it went, through Tyrnavos, and, as dawn was breaking, through Larissa. Our drivers could see how thoroughly the Germans had finished what the recent earthquake had started. Alleys and courtyards were filled with rubble, houses had been sheered in half, and the dead lay among the ruins. Sections of the town were still smouldering, sending up streamers and columns of smoke, and a livid ceiling, part weather and part ruin, covered the whole of it.

Our lorries had been crawling all night—the whole of the 4th Brigade Group was on the move—but beyond Larissa the pace slowed still further. Soon they were moving only in starts and stops, and an old refugee, shambling by with his bundle, passed lorry after lorry. Presently the convoy halted for good. Stukas, it appeared, had blown the road ahead and behind, trapping a mass of transport. Aircraft came over, bombing and machine-gunning and taking no notice of the small-arms fire except to send a special squirt of bullets in the direction of any Bren-gunner who was too persistent. After they had gone several lorries were burning and others had to be pushed off the road.

A message from the head of the column seemed to suggest that the halt would be a short one, but the road was badly blocked and after a while the drivers were ordered to disperse their vehicles as best they could. The section of the column that included most of Second-Lieutenant Fenton's detachment—Sergeant Buckleigh's 30-cwts. were now close behind it—was directed on to a hillside, part of which was in crops. To reach it the drivers had to cross a stretch of meadow that was very muddy and had been badly churned up by traffic. The lorries started to shoulder their way out of the congestion, slipping and lurching over the ruined verges and sticking, some of them, and having to be towed out. And all the time, impudent and unassailable, a German reconnaissance plane was perched in the sky above them—*I can see the whites of your eyes*—sending messages to its base. The lorries ploughed through the mud and cut swathes through the bright green barley, the infantry running ahead of them, breasting the crops like bathers, seeking the cover of distant trees, desperate to get away from the road and the planes that were coming to bomb it. Farther up the road towards Larissa machine guns were firing.

Presently the whole hillside was still. The lorries were not well dispersed but the drivers had done the best they could. Their nerves were as taut as piano wire and they stood by their lorries and glowered at their nearest neighbours, each believing that it was the other fellow's fault that the transport was not better dispersed. ' Why can't the mad b—— move his b—— truck over that way? '

Soon the bombers came. They came singly and in groups. They bombed and machine-gunned the road and then they concentrated on the paddocks and hillsides. Several aircraft headed straight for where six or seven of our vehicles were parked. They flew low and there was no scream from the bombs—just a swosh-swosh-swosh that was inaudible above the hammering of the engines unless you were close to it. Then came the sheets of flame and the terrific slaps—one, two, three, four—and the sense of being smashed over the head with a rubber truncheon. Black smoke streamed over the hillside and through the murk you could see the red tracers, elongated like jelly beans and travelling, it seemed, no faster than cricket balls, wavering up towards the bombers. The bombers banked and you could see tongues of flame shooting from their wings as their machine guns fired. And then, like an afterthought, came the slap of a last bomb.

The planes had come in so low that some bombs had skidded along the soft ground without exploding.

After they had gone the drivers checked up on the damage. It might have been far worse. There had been casualties and several vehicles were on fire. One of these was parked among Sergeant Buckleigh's 30-cwts. and in the back of it cases of small-arms ammunition were burning like popcorn. A bomb had landed beside one of our 30-cwts. and the driver was muttering all the oaths he could think of. Only a few minutes before, by fitting an extra horn, he had concluded a programme of work that had made his vehicle the most complete and comfortable in the unit—in the Division. His engine would have taken him round the world and a retreat to Cape Matapan would not have emptied his larder. Now his beauty—his pretty one—was torn and blackened. It was down by the head like a bull beaten to its knees. A mixture of petrol and pineapple juice trickled from holes in its tray and he could have wept.

The enemy came over again and there were more casualties and more lorries were damaged and set on fire. While this raid was in progress Sergeant Buckleigh attended to the wounded, taking no notice of bombs and bullets. Helped by a corporal from the Divisional Petrol Company, who was afterwards awarded the Military Medal, he brought several wounded to the shelter of a small hollow. Among them was Dick Taylor[4] (C Section) who had been wounded in the left arm by an explosive bullet.

Several of the drivers did good work that day. Among them was Alf Hallmond,[5] Bren-gunner on C Section's ack-ack truck. He shot down one Dornier for certain and our drivers credited him with another. In the course of the day he burnt out two barrels and during one raid he emptied all his magazines.

During the morning something happened to Larissa. The Luftwaffe, perhaps, hit an ammunition dump—or perhaps it was the work of our engineers. At all events a pile of pearl-grey smoke, swift as a genie materialising from a bottle, built itself up, fold upon fold, layer upon billowing layer, until it was as vast as a mountain. It seemed to have the consistency of whipped cream. No one had seen anything like it before.

It was late in the afternoon before any general movement was possible on the road, and all day long the enemy passed backwards and forwards in the sky, owning it. All day long the New Zealanders crouched in culverts or bomb craters or lay hidden in the barley, watching the relentless sky or glancing longingly towards the mountains in the south, towards which the road ran straight and level across the plain. And all day long the transport stayed in the same area. It was near the village of Nikaia, six or seven miles south of Larissa.

[4] Dvr Q. R. W. Taylor; civil servant; Takapuna, Auckland; born Auckland, 10 Oct 1918; wounded 18 Apr 1941.

[5] Dvr A. J. Hallmond; labourer; Dargaville; born Dargaville, 3 Mar 1919; wounded 27 Nov 1941.

At last came the order to move. A prolonged hooting of horns brought the infantry from their hiding places at the double, and they crouched and squatted among the tangles of equipment, edging as close to the tailboards as possible.

Driving conditions were extremely bad. There was only a yard or so between each lorry and here and there a burning or disabled vehicle lay half across the road. You saw bombs bursting ahead of you and every so often a tremendous banging on the cab almost made you jump out of your skin. In one convulsive jerk you grabbed the hand brake, cut the engine, and dived into the ditch. Sometimes a Messerschmitt roared past; often it was a false alarm. In any case many of the infantry would run two or three hundred yards from the lorries—not that our drivers blamed them for this: they knew the infantry had been through far more than they had—and it might be three minutes before the column could again start to move. Progress was so slow that the passengers began to argue among themselves about the wisdom of stopping. Some were for keeping going; some were hotly against it.

'For God's sake box on, driver.'

'You're doin' all right, driver. We're not gettin' killed for these bloody jokers.'

'We'll never get any bloody place if we keep stoppin'. Get on and get it over with.'

Panicky drivers of light or unloaded vehicles—drivers of heavy vehicles too for that matter—lost their heads completely and kept blowing their horns, mad to pass everything on the road whether or not there was room to squeeze by. As soon as our drivers managed to get a little space between their lorries, these others—free lances or interlopers from convoys farther back—cut in. More than one put his vehicle over the bank. At one stage Colonel Crump drove past and shouted that all drivers were to stay in convoy and keep discipline. For a time conditions improved but soon they were as bad as ever.

The time came when the column was blocked by a burning vehicle and a huge bomb crater near a bridge. A detour had been made—was still being made, the shovels darting like tongues under the very wheels of the lorries, which jerked, stopped, screamed in anguish, jerked on—through a sodden meadow. In this several of our vehicles became bogged, and the Luftwaffe came over again.

The hold-up, however, though intensely trying to everyone's nerves, with the mountains and safety beckoning, was really a blessing. It enabled the military police to prevent any driver from turning into the meadow until the vehicle in front of him was a fair distance in the lead. Consequently when the transport got back on the road it was properly dispersed at intervals of from 50 to 100 yards. These intervals were held without much trouble until dusk.

The sun was setting as the main group of our lorries gained the foothills. Spearheads of olive shadow were thrusting across the plain of Thessaly, and the setting sun, with that strange trick it has of picking out, for no particular reason, a single farm or a field of barley, was casting pools of honey-coloured light among the mountains. Two Messerschmitts were still swooping and turning in the gorges, trying to follow the twists in the road with the object of getting in a few farewell bursts, but our drivers were happy now that the mountains sheltered them. Soon it would be dark.

After dark the column closed up until it was travelling nose-to-tail once more. Then there was another long halt and a group of drivers gathered round one of the 30-cwts. They were excited and they laughed a good deal. They ate ravenously for a little while, chopping and changing, turning from this to that—from tinned peaches to pickles, from pickles to condensed milk—but their appetites soon failed. They became solemn and spoke of Second-Lieutenant Norman Chissell[6] who had been killed at the rear of the convoy by bomb blast earlier in the day. An original member of the unit, Norman had been commissioned in March and posted to the Divisional Petrol Company.

The driver of the 30-cwt. told how Sergeant Buckleigh's detachment had collected the 32nd Battery of the 7th Anti-Tank Regiment from Kokkinoplos:

'We did the job two nights ago. They dragged us away late in the evening from a supply dump we were stuck into—beer, tobacco, good tinned stuff. We went along the main road and turned right before it goes up into the pass. We stopped in a paddock for a bit and then set off up a hell of a narrow, winding, rocky track. There was just room for a 30-cwt.—a three-tonner wouldn't have

[6] 2 Lt N. F. Chissell; garage attendant; born NZ, 25 May 1917; killed in action, 18 Apr 1941.

made the corners. We must have been the first transport to go up there since Adam was a cowboy. It was getting dark now and raining pretty steady and on the left-hand side of the road there was a sheer drop. Away in the hills you could see the guns flashing like summer lightning—at least I suppose it was the guns—but you couldn't hear them or tell whose they were.

'It took us a while to get to where we had to pick up the anti-tank jokers and when we did arrive there was a long wait. It was as black as the inside of a cow and raining hard and we had an awkward place to turn the trucks in. One of them had conked out a little way down the track, blocking it completely, so "Buck" gave orders to shove it over the cliff. We had time, though, to strip her of anything worth keeping.

'The anti-tank boys were just about finished when they arrived. They'd marched miles over the mountains and had had to destroy all their guns and transport. They didn't feel too good about it.

'We travelled south all the rest of that night and part of the next day, dropping our passengers in an area off the Larissa-Tyrnavos road. After a bit of a rest we headed back to a dump near Tyrnavos to pick up some petrol and stuff that had to be taken to Molos for the 4th Brigade.

'We got to the area, threw the load on, and still had a bit of daylight left. We were parked right by a river and it was a corker evening. We had a clean up and some of the boys got their rifles out and threw tins in the river for target practice. Some storks came over and we put a few shots round them. There was a supply dump on the other side of the road but the eyes had been picked out of it and the Greeks were lugging away what was left—bully and stuff. Some of the B Section jokers had some beer but not enough for a party.

'We had a great night's sleep and it was daylight when we woke up this morning. "Buck" came round and said we were to get cracking at once as Jerry was only one jump behind us. It was as quiet as one thing and there was hardly any stuff going past on the road. We caught up the main traffic stream just outside Larissa and got in with you jokers not long before we were bombed.'

By this time most of his listeners had wandered back to their own vehicles. With the two or three that remained the talk became general. It was agreed that fighters and anti-aircraft guns were on their way to Greece and that the Germans were certainly being drawn into a trap. The next couple of weeks might not be so good but after that—BASH! The bombing had been bad, yes, but look how few casualties there had been. And look, taking it by and large, how few vehicles had been hit! The Germans were getting it, too. Our Divisional Cavalry by the Aliakmon River, our machine-gunners at Veve—this was the story and we believed it— had caused such slaughter among drugged German infantry advancing shoulder to shoulder that many of them had vomited over their guns. But how they came on! 'A Div. Cav. joker,' said someone, 'told me that when the Jerries come to a blown bridge they drive a tank into the gap, and if that doesn't fill it they drive another one in and they keep on driving tanks in until it is filled. Then the rest of the tanks drive straight on over.'

After a while the column moved again. An endless stream of lorries was heading for the Thermopylae line, grinding, clanking, creaking, whining through the night—the drivers, dirty, unshaven, red eyed, sitting stiffly behind their steering wheels, the trays of their lorries packed with sprawling, exhausted troops or with huge, unwieldy, hastily flung on loads. The air was heavy with the reek of burnt petrol. Every now and then the miles-long column ground to a halt and then there was a long, listening silence complete except for the quick whisper of exhausts. Behind every steering wheel a cigarette glowed redly, for no one was any longer observing the rules about smoking. Then the convoy would move on again with a deep growl of lorries in low gear, which presently thinned out to a monotonous whine as the drivers shifted into third.

During the early part of the night nearly everyone drove without lights, but later, as there was no evidence that the Luftwaffe was on night duty, sections of the column, one after another, switched on their headlamps, and soon the road was twinkling and sparkling for miles. It was as though a necklace of brilliants had been flung around the dark shoulders of the hills. Sometimes a British military policeman, standing at bridge or crossroads, would call out: 'Switch off those lights, chum, or you'll get bombed'. Then a section of the necklace would vanish, only to appear again ten minutes later.

For a long time the drivers had been puzzled by a rosy glow ahead of them. It was a town burning and someone said it was

Lamia. A group of our lorries halted in the main street and the drivers were able to look around. Ordinarily they would have been sad to see Greek houses—houses from which people had waved to them and brought them presents—burning steadily, but today they had been through enough emotional experiences and if they felt anything at all it was gratitude for the pleasant warmth. A rain of soft hot ash was falling in the street and the atmosphere belonged to a drowsy afternoon in midsummer. The fire had eaten up about three-quarters of the main street and was now consuming the rest without any unnecessary fuss or noise. No one was trying to put it out: possibly everyone had fled. It was a fantastic sight. Each naked rafter wore a comb of fire and little questing flames were flickering about the charred doors and windows in search of further nourishment. In the ruins of the local cinema what was left of the grand piano glowed like a yule-log.

Leaving it behind for the Germans—it was only a tiny incident in the long Walpurgis night they had wished on the whole world—our lorries drove on into the cold darkness, which seemed to be unending and immutable. Nine miles was as much as they did in any hour and sometimes they did no more than two.

Dawn came at last and the journey ended soon afterwards. Second-Lieutenant Fenton's detachment dropped the 20th Battalion in its new positions and then set out to rejoin the unit, finding it on the coast below Molos. Sergeant Buckleigh's detachment off-loaded and remained under the command of the 4th Brigade.

. . . .

The rest of our load-carriers had also been on the road that night and the day before. On Friday the 18th at seven in the morning a Don R had arrived from Headquarters NZASC and told the Major to take us back to the Tyrnavos area. We were to help withdraw the 4th Brigade Group.

Refreshed by our sound sleep we set out in good spirits along the coast road and although we were attacked from the air no harm was done—the heavy bombing and strafing of Second-Lieutenant Fenton's and Sergeant Buckleigh's detachments was taking place on the inland road. The convoy was halted and dispersed about eight miles south-east of Larissa and the Major went ahead in his car to see how the land lay. The situation was obscure and the available reports were not reassuring. He passed through Larissa, which was being heavily bombed at the time, and carried on until he reached the Tyrnavos area, where he found the 4th Reserve Mechanical Transport Company. No instructions had been left for him.

Puzzled, he returned to the lorries and at the entrance to the dispersal area was met by a Don R who handed him a message cancelling his previous instructions and telling him to go back to Almiros to await others. The convoy started at once.

To move in or out of many roadside areas in Greece was impossible without wheel chains, and we seemed to spend half our days kneeling in the mud with broken finger nails and bleeding knuckles grappling to our chests the dead weight of those icy, slimy, accursed, indispensable chains. On the roads they were a nuisance because they worked loose and flailed the mudguards.

The drivers of a three-tonner at the tail of the convoy were removing theirs when they were attacked by three Messerschmitts. Their lorry was set on fire and destroyed, and Captain Sampson's pick-up, which they had pulled out of the mud a little earlier, had to be abandoned with four flat tires.

During the journey to Almiros the convoy was attacked several times and A Section's ack-ack crew did good work. If Percy Sanders and Jim Stanley had been using a heavier gun in Greece, a .5 instead of a .303, they would have been shooting down aircraft instead of merely opposing them. They cost the Luftwaffe some flying time (because even bullet holes have to be repaired) and probably, by keeping irresolute pilots high, they saved some lives; but a light machine gun—and this applied equally to B and C Sections' crews—was not the weapon they deserved. Percy was hitting the bombers with his Bren—you could see the tracer stubs flaking away as the bullets hit the armour-plating—but, heartbreakingly, they seldom took notice. No wonder he was seen, after one raid, to throw his gun on the ground and jump on it. Jim was using the Boys anti-tank rifle (weight 34 lb.), firing it from the shoulder until he was a mass of bruises. 'I'll get the bastard,' he used to say, savagely jamming shells into the magazine. 'He's only got to fly low and slow and I'll get the bastard.' One day the chance he had been waiting for came. He was sitting on a grassy slope with the rifle between his knees when a fighter-bomber cruised lazily down the valley. When it was level with him and about seventy yards away he heaved the rifle to his shoulder, took careful aim—Jim had smashed as many clay pigeons as any man in New Zealand—and squeezed. The rifle misfired. Jim was too upset even to swear.

However, he and Percy continued to stand by their guns, and they stood by them valiantly on 18 April on the road to Almiros. This was a town about twenty miles south-west of Volos, four from the coast, and between fifty and sixty miles by road from the Thermopylae line. No fresh orders awaited the Major when he got there, so at ten that night he set out for the Thermopylae line— an action he was told later was correct.

By daylight we were in an area near Longos, a small coast village some fifteen miles east-south-east of Molos. There was no work for us that day, so we tried to get some sleep, but sleep was almost impossible. All day long reconnaissance planes hovered noisily above our olive trees in search of dispersal areas and we strongly suspected that they were being helped by fifth columnists. We were warned to be on the lookout for a blue touring car, the driver of which was believed to be signalling to aircraft by parking it near areas that contained troops and vehicles. Only a short while before two drivers had halted just such a car, something about it having aroused suspicion. They had searched it but had found nothing to justify their detaining the driver and his passengers.

The next day was 20 April, a Sunday. It was also Adolf Hitler's 52nd birthday. Congratulatory messages were in order and the Luftwaffe decided to say it with bombs. Bombs fell on roads and bridges, bullets pruned the olive trees in the dispersal areas, and we were grateful that we were not called on to make any general move. By this time Company headquarters had shifted to an area near Atalante, some fifteen miles farther down the road, and a small convoy of our load-carriers, which had spent the night there, had to rejoin the sections at Longos. The birthday bombs fell steadily all the way but no one was hurt, the narrowest escape occurring when a piece of shrapnel the size of a flat-iron passed between two drivers and out through the back of their cab.

Workshops, which had been in an area off the Thebes-Khalkis road since the 16th, also had cause to remember the birthday. At eight in the morning three sticks of bombs fell on a nearby Australian ambulance unit. Our drivers dug slit-trenches and then went on with their work. About ten o'clock, eight aircraft (to choose the most conservative sum from a mass of hasty arithmetic) started to bomb and machine-gun the area. One bomb landed a few feet from a staff car without damaging it but shrapnel and bullets sped unerringly towards a new radiator just fitted to a load-carrier. A chin-strap broke when a steel helmet was torn from its wearer's head by blast and that was the sum of the damage—a not very impressive total for a raid in which between forty and fifty bombs had been dropped.

The raids continued and by three in the afternoon the drivers had retired in disgust to a dry creek-bed some distance from the area, leaving a sergeant, a corporal, and two volunteers to operate a report centre.

Just on dusk two Messerschmitts flew low over the area. By this time it was pitted with bomb holes, so the pilots may well have gone home to report that in addition to the damage done to Khalkis harbour during the day a scene of chaos existed beside the Thebes-Khalkis road. The truth would have disappointed them. A direct hit had been scored on the cookhouse, smashing the two olive trees between which it had been set up and strewing the neighbourhood with dixies, flour, and tinned food—this was the worst damage—and the water cart had been riddled with holes, but it would still go and the bottom half of the tank would still hold water.

Workshops inspected its area with wonder and happy pride. The drivers gathered beside the indestructible staff car to marvel at the enlargement of the original crater by a second bomb. They agreed that a greatcoat riddled by machine gun bullets had certainly been mistaken for a prone soldier and that the administrative corporal had cut a ludicrous figure while sheltering in a too-small slit-trench. They wondered how the sections had got on.

The sections had got on very well. The same euphrasia was being experienced in the Longos area and here, too, the tendency was to laugh and talk. The birthday had been going on all over eastern Greece all day.

During the afternoon the spare men and the drivers without vehicles had been given a particularly unpleasant task—the loading of the unit's transport with 3000 rounds of 25-pounder, all of which was in a single stack a few miles from the area. The Major himself superintended the operation, taking with him enough labour to load two or three lorries at a time, his idea being to get the job done quickly so that the men could return to the comparative safety of the olive trees. The lorries were held on the road two to a mile and as they were needed they were signalled forward by Dick Grant,[7] the Major's driver. His was an unenviable job and he performed it with such coolness that he was later awarded the Military Medal. Enemy aircraft were overhead nearly all the time but the Major refused to allow anyone to take cover. Consequently the stack was soon cleared and the lorries safely dispersed in the unit area.

The birthday ended when night came, and although it had been our most dangerous day so far only two men had been injured—neither seriously. They were Jack Murdoch[8] (C Section) and 'Scotty' Reid[9] (B Section).

By now the New Zealand Artillery had taken up positions in the Thermopylae line and at midnight we started delivering ammunition to the guns. While we were doing this the Major was asked for enough transport to move the 24th Battalion to a fresh position in the line. Hardly a load-carrier was in the area, but he rushed round to the various regiments and by three in the morning a convoy had been assembled and was embussing the infantry. It returned to Longos the next day, and but for the Luftwaffe, which seemed remarkably fresh after the birthday celebrations, we should have spent a quiet afternoon.

The sea was nearby and that evening many of us had a quick bathe. The water was cold but it was deliciously refreshing. When aircraft came over we stretched ourselves on the pebbles and lay still, letting the creamy surf wash over us.

Nearly everyone was in good fettle. From the downward curve of the sun you could trace the upward curve of our spirits: it was a new law of nature introduced by the Luftwaffe. We boasted about our prowess as runners, making it the subject of ridiculous comparisons ('Boy, did I move? Lovelock's a slug to me.'), and every-

[7] Dvr R. S. Grant, MM; garage attendant; Frankton Junction; born NZ, 29 Nov 1911.
[8] Dvr J. I. Murdoch; truck driver; Napier; born Kairanga, 7 Jul 1921; wounded and p.w., Apr 1941.
[9] Dvr R. T. Reid; driver; Henderson, Auckland; born Hamilton, Scotland, 21 Apr 1918; wounded 20 Apr 1941.

Bombed

Bombing in Larissa

A convoy halted in Volos

'*A direct hit on the cookhouse*' —page 79

thing at all funny was treasured and passed around. The best story, perhaps, was about a B Section driver. He was travelling along the road when he heard what he thought was a motor-cycle roaring behind him. He indicated that the road was clear and a Messerschmitt swept by at hedge height. He swears that the pilot leaned out of the cockpit and acknowledged the courtesy with a gracious wave.

Laughing and talking and calling to one another in the darkness, the drivers ate their supper and turned in. Presently the whole area was hushed. All you could hear was the footsteps of the sentries as they strolled under the olive trees, a muffled grumble of talk coming from a lorry far up the hillside, and the occasional sweet, clear pipe of a night bird. We had been told that the German paratroopers signalled to one another by imitating bird calls.

All night long, silent, impassive, stoical, in twos, in tens, in twenties, not glancing at the sentry who stood in the entrance to our area, Greek soldiers trudged south. None of them had rifles; few carried anything beyond a small bundle. They were like the chorus in a Greek tragedy. They were not men, you felt, but symbols of defeat, expressing all the pity and terror of the world we live in but feeling none of it themselves. Probably, though, they had fought with unbelievable bravery and their hearts were full of bitterness and anguish. Or were they merely dead-tired and longing for home and glad that for them the war was nearly over?

The sentry at the entrance was puzzled. He did not know that a Greek army in the Epirus had been surrounded and forced to capitulate to the Germans, nor did he know that during the day the Greek Government had asked the British to withdraw their troops from Greece.

General Wavell himself had gone to Athens to make sure of the Government's attitude. It was sensible and honourable. Greece could hold out only a few days longer. Let the British save what they could while they could.

We were told about this the next day. Our first reaction to the news, human and perfectly pardonable, was one of relief. Freedom from Stukas, a hot bath, leave in Cairo—it sounded like Heaven. Then we remembered the old men who had saluted us, the women we had seen working on the roads, and the hundreds of baby fists,

F

so cocky and confident, that had waved frantically for notice or had been held up proudly for our attention: Look, English—the sign! *Your sign!* Thumbs up! We remembered the storm of flowers and the faces in the twilight in Athens. There would, of necessity, be a ride back.

And we remembered other things. 'The English? They know only one military operation—re-embarkation.' Lord Haw-Haw had called us 'Freyberg's Circus', and how did it go, that epigram? Give the Canadians more motor-cycles and they'll break their necks; leave the Australians alone and they'll kill each other off; pay the New Zealanders an extra pound a week and they'll drink themselves to death. No libel so wounding as the one with a small core of truth. No, we were not looking forward to that second ride through Athens.

The Major told us of the withdrawal after breakfast on the morning of the 22nd and he added a word of warning. He had noticed, he said, that some of us were getting jittery. He reminded us that only one man in the unit had so far been killed, though all had had narrow escapes. It was better, he said, to go home without an arm or a leg than with broken nerves.

. . . .

After that the odd men and the drivers without vehicles stacked their kitbags in a dry ditch, guessing they had seen the last of them, though there was some talk of sending them down later. Then they climbed aboard two A Section lorries and were driven south under the command of Second-Lieutenant Toogood. Their destination was an assembly area near Daphni on the outskirts of Athens. They were attacked several times on the way down and on one occasion six A Section drivers could find nothing to shelter behind except a lone sapling whose trunk was as slender as a drainpipe. They packed round it tighter than piglets round a trough and waited for the bullets, but they never came. The Messerschmitt rose to avoid the sapling then dipped to machine-gun the drivers from the second lorry who were sheltering in a ditch by the roadside. No one was hurt.

When the party reached the Daphni area they found that Company headquarters was already there. Workshops arrived the next day.

Moved off (noted the senior NCO, Staff-Sergeant Jim Harley),[10] at 5 a.m. after making sure that everything possible was destroyed. Went through Thebes and were lucky enough to find gap on main south road through which to move our convoy. Destroyed vehicles lying on both sides of road. Had anxious time getting our heavy vehicles past some of the wrecks.

8 a.m. Were caught in Thebes Pass. Enemy planes overhead had caused a large convoy to stop. Strafing but no damage done.

9 a.m. Got going again. Orders were for us to destroy Workshops and stores waggons by running them over the steepest cliff in the pass but thought better of it and took them on.

2 p.m. Arrived Daphni. Quite a number of unit vehicles parked in olive grove. Dispersed and settled down for a meal.

4 p.m. Enemy planes over; no damage.

. . . .

The rest of the unit spent the day in the Longos area. The Luftwaffe was never very far away but good luck and good management kept our drivers safe until the friendly darkness arrived, bringing them a second night of sound sleep. The next morning, however, the enemy came over before he was expected, catching B Section off guard. Its area was heavily bombed and machine-gunned, Georgie Ireland[11] and R. V. B. Brown[12] being partly buried by a bomb and Claude Hitchon[13] and 'Kolynos' Carroll[14] wounded. Two vehicles, one an LAD,[15] were damaged.

Later in the morning we were told to keep two lorries for unit transport and destroy the rest, with the exception of eighteen from C Section, four from A Section, and three from B Section, which were to stand by to help with the withdrawal of the 6th Brigade. First we removed the petrol tanks for delivery to the New Zealand Artillery, which was short of petrol. Then we drained the oil from sumps and differentials and poured grit in them. We started the engines and pulled the throttles wide open. They hammered for a time; then they began to cough and spit. Finally they clattered

[10] Capt J. W. Harley; mechanic; Leeston; born NZ, 23 Apr 1911.
[11] Dvr G. E. Ireland; storeman; Auckland; born Auckland, 29 Jul 1917; wounded 23 Apr 1941.
[12] Dvr R. V. B. Brown; factory hand; Auckland; born Auckland, 20 Oct 1918; wounded 23 Apr 1941.
[13] L-Cpl C. B. Hitchon; taxi proprietor; Kaikohe; born Waikiora South, 13 Feb 1901; wounded 23 Apr 1941.
[14] Dvr G. M. Carroll; cook; born England, 6 Jun 1916; wounded and p.w. Apr 1941.
[15] Light Aid Detachment.

into silence. The Major went round with his tommy gun and delivered the *coup-de-grâce*. They were good lorries. They had covered thousands of miles in Egypt and the Western Desert, and in Greece most of them had done about 4000 miles. For many months they had been our jobs and our homes. Savagely we drove picks into headlamps, radiators, windscreens; we slashed tires, destroyed dynamos, batteries, distributors.

Then we sat down to refresh ourselves with tinned fruit. The twisted roots of the olive trees were now our larders, their cleft trunks our wardrobes, their branches our coat-hangers. Owning nothing except what we stood up in and could carry, and having nothing to take care of any longer except our rifles and ourselves, we felt strangely free.

Late that afternoon, with many glances at the sky, the drivers whose vehicles had been destroyed left for Daphni in the two lorries that had been kept for that purpose, Captain Moon leading them in his staff car. They arrived safely, as Workshops had done earlier in the day. The Longos area was empty now except for the transport that was standing by to shift the 6th Brigade. At eight the evening before twenty-five vehicles from the Supply Column and twenty-five from the Petrol Company had come under the Major's command for this job.

• • • • • • •

Since dawn on the 23rd the 6th Brigade, supported by the New Zealand Artillery and a number of British guns, had been holding the coast sector of the Thermopylae line. By day it had been under almost constant attack from the air and during the afternoon of the 24th our infantry and artillery engaged enemy tanks and infantry. The men were to withdraw with their guns that night, relying on speed and darkness to see them safely through the dangerous miles that lay between Molos and the 4th Brigade's covering positions at Kriekouki, a mountain pass south of Thebes.

On the evening of the 23rd the Major had been shown the embussing point in the 6th Brigade headquarters' area by the Brigade Major and told that his transport was to be there at nine o'clock the next night. Later, however, it was decided to advance the hour so as to free the narrow coast road for south-bound traffic engaged in earlier withdrawals, but because communications had broken down the Major did not get his fresh orders. (It would have made no difference if he had. Nothing—not even a motor-cycle—could move on the road in daylight.) Consequently, when the convoy did not arrive at the embussing point in the afternoon, the brigade concluded that it was lost and the 6th Field Regiment was ordered to destroy its guns to free transport for evacuating men.

Following what he thought was still the plan, the Major left the Longos area after dark with the intention of getting his convoy to the embussing point at nine o'clock, but even this was not possible. The timetable had broken down and the convoy was delayed by south-bound Artillery transport.

As the lorries drew near the danger zone—Molos had been shelled heavily a few hours earlier—flares and Very lights made everything as bright as day, and our drivers, with lorry almost touching lorry, felt certain they had been spotted. Subsequent talks with the infantry suggested that fifth columnists were responsible.

The infantry, as it happened, were also late in arriving, but at half past nine the vehicles from our unit embussed the 24th Battalion. Under the weird light, fantastic with leaping shadows and sudden gouts of darkness and sinister with small sounds, the lorries were loaded and in no time they were down on their springs with their canopies bulging like untidy parcels. Some of the men were famished and our drivers showed them where the food was kept and the cigarettes.

One by one the lorries disappeared in the darkness and went lurching along the narrow road, travelling without lights. The Major and Second-Lieutenant Butt[16] stayed behind with their lorries to pick up stragglers. They collected twenty-one and pulled out just in front of the rearguard, beating the enemy by twenty minutes and passing through Atalante ten minutes before the arrival of an enemy force that had driven down the centre of Greece.

More than a hundred miles were covered before daylight and morning found the brigade well south of the last line of defence. The 24th Battalion was hidden in an oak grove near Eleusis, a few miles north-west of Athens, with the transport camouflaged and the infantry sleeping. Re-embarkation had started already and that night the brigade was to go to a beach near Marathon.

It was Anzac Day.

[16] Capt F. G. Butt, m.i.d.; farmer; Seddon; born Blenheim, 8 Dec 1918.

CHAPTER 6

WITHDRAWAL FROM GREECE

ANZAC morning.

In the Daphni area Company Quartermaster-Sergeant Jack Williams[1] stood among piles of new shirts, new underclothes, and emergency rations. A ring of drivers hung round him like wolves but, as he had not yet been given instructions about destroying or giving away his stock, he attempted to fob them off with a few tins of sardines and some deck shoes. The moment his back was turned they snatched shirts, shorts, and underclothing, growing bolder as he grew wearier. Aircraft flew over at frequent intervals and after every alarm the heap was seen to have diminished. Soon most of the drivers were wearing new clothes.

Anzac afternoon.

In an area below Kriekouki Sergeant Buckleigh's detachment of nine 30-cwt. lorries (the tenth had been destroyed) was standing by to help with the final withdrawal of the 4th Brigade, which was to hold its present positions at all costs until the next night. The detachment was now under the command of a lieutenant from the 4th Reserve Mechanical Transport Company whose own convoy, originally of thirty lorries, had been depleted by eight. The sunshine was soft and pleasant and our drivers smoked, played cards, slept in the warm grass.

Anzac evening.

On Kea Island, fifteen miles from the Greek mainland, a party of New Zealanders left the shelter of the olive trees and the wooded gullies and made for the harbour. They had been ordered to sleep near the beaches in case a ship came. Among them were Sergeant 'Dad' Cleave,[2] Corporal Ian McBeth,[3] and some fifteen A Section drivers who had left us the day before to embark from a beach east of Athens with elements of Divisional Headquarters.

[1] Capt B. J. Williams, MC; hotel manager; Birkdale, Auckland; born Australia, 6 Jun 1905.
[2] WO II V. J. Cleave, MM; motor mechanic; Auckland; born Inglewood, 8 Aug 1910.
[3] WO I I. McBeth, m.i.d.; civil servant; Auckland; born Motueka, 1 Oct 1908.

'We were in the last landing craft to leave the beach,' said Ian McBeth. 'When she moved off we were under the impression that we were being taken to a ship but as time went by and we seemed to be heading for the open sea we began to wonder. Then the commander of the landing craft spoke to us out of the darkness. He spoke quietly in level tones. He said: "We regret that it is impossible to take you to Crete by destroyer this morning. The destroyer's departure is already overdue and we have decided to take you to Kea Island. There you will be safer from bombing and enemy action than you would be on the mainland. We intend to get in touch with a destroyer and we hope you will be picked up in the course of the next day or two. If that doesn't happen you will have to make your own way to Crete as best you can. Crete is about 150 miles due south of Kea Island."

'The voyage that followed was supremely uncomfortable. There were five or six hundred men in the landing craft and except for a small space forward she was decked over. Nearly everyone was wearing his greatcoat, web equipment, and haversack, and we were so jammed together that it was next to impossible to take anything off and difficult even to loosen a strap or unbutton an overcoat. We were standing in a mixture of water, oil, and petrol, and the fumes were so bad that smoking was quite out of the question. Soon we were wet through with sweat.

'After sailing for four and a half hours the landing craft arrived at Kea Island. The sun was well up and it was a lovely morning. We were ordered to get ashore quickly and disperse ourselves among the olive trees but to stay within call. We needed no urging, for as we were going ashore two German aircraft came skimming over the harbour a few feet above sea level. We got under cover and most of us went straight to sleep.'

They had slept soundly, many of them, even while the harbour was being bombed and low-flying aircraft were strafing the gullies and hillsides. Now, waiting by the beach, few were sleepy. They talked on into the night and no ship came.

. . . .

Sitting in the darkness and talking about ships—that was what Second-Lieutenant Fenton and the drivers of C Section's administrative vehicles were doing on Anzac evening. They were in the

Peloponnese and in sight of Nauplion harbour. They had destroyed all their lorries except one and that was being kept for New Zealand soldiers who were too sick to march. The sick men, together with some hospital orderlies, had been picked up at Daphni camp the day before and taken by way of the Corinth bridge and Argos to an area near Nauplion, a drive of between eighty and ninety miles. The convoy had been held up in Argos and severely bombed. Here twenty lost men of the Divisional Petrol Company had been taken aboard.

Late that afternoon our drivers saw a tragic sight. The *Ulster Prince*, a sizeable merchant ship, went aground in Nauplion harbour and while she was lying there helpless the Stukas came over. They circled above her, and then, with beautiful precision, one after another, they dived. The first bombs fell a little wide but the second attack was successful. The ship was hit by an oil bomb dropped by one of the leading Stukas, and the others, following behind, poured streams of incendiary bullets into her decks and superstructure. Presently she was burning from stem to stern. When the fire reached the magazine—that at least was how our drivers saw it—there was a tremendous explosion. Gear of all kinds—derricks, deck-houses, stanchions, whole and in fragments—was thrown upwards in a cloud of smoke and came pattering down into the water. Then she started to blaze in earnest.

No one was surprised when it was learned that there would be no embarkation that night.

And so, on Anzac evening, after spending all day in hiding, our drivers were waiting for the word to move. It came after dark and they marched eight miles to a quiet cove, carrying greatcoats, toilet gear, and one blanket each. When they got to the cove nothing happened, and later they were told that there was no chance of their being taken off that night. They were weary and disappointed and they dreaded another day under the Luftwaffe.

'Who goes there?' said an authoritative voice from the darkness. Our drivers were fed up and each of them waited for someone else to reply—a dangerous habit at that time. The challenge was repeated and they could make out an English officer with drawn revolver. It was time somebody answered. Replied Bill Davies,[4] wearily, lugubriously, and with perfect timing: 'Schmidt der Spy!'

[4] Dvr W. H. Davies; motor mechanic; Lower Hutt; born NZ, 12 Sep 1909.

The next day was the 26th and the sands were running out—running out in the Peloponnese, running out on the beaches near Athens, running out in the Daphni area, running out on Kea Island.

To the hungry drivers on Kea Island the 'Sheriff's' stew was beginning to smell good. Anything warm and meaty would have smelt good to them for this was their second day on the island and rations were almost exhausted.

The foundation of the stew was a chicken and a kid. The former had been given to 'Sheriff' by a village woman; the latter he had returned with after disappearing among the trees. He stirred the pot with his bayonet and thoughtfully sucked a finger. No—it was not quite ready.

Intent on their bellies our drivers were not listening for the assembly signal, but the moment it was heard—three shots in quick succession—hunger was forgotten and they hurried to where the NZASC party was forming up under a captain from the Supply Column who was on the island with 200 others from his unit.

The landing craft had moved a few miles to a safer anchorage and the troops were to march over the hills and join her. They started off in groups of twenty. It was about noon.

For a while the going was flat; then the trail started to zigzag upwards. As soon as one summit was gained another one appeared behind it. 'Dad' Cleave made his involved jokes and helped where he could and the men kept going. At last, when they were beginning to feel that everything in the world was vanity except the pleasure of stopping and stretching out at full lengh, the trail began to wind downhill. It wound through a village, and there, by the water's edge, was the landing craft. Three of our drivers were the first to reach her, which was remarkable considering that the NZASC party had been the last to start. They stripped off and fell into the sea, snatching several glorious minutes before it was realised how clearly their white bodies showed up.

The barge sailed at dusk. Early the next morning our drivers were taken aboard the anti-aircraft cruiser *Carlisle* bound for Crete.

. . . . . .

Near the Daphni area the two great Thornycrofts—the Workshops lorry with lathe and drill and work bench, and the stores lorry with pigeon-holes and bins filled with everything from cotter-pins to spare steering assemblies—had been pushed into a gully. One of them lay on its side, three of its six wheels towering high above the mechanics. The other was still upright, its nose crushed into the bank. It was the morning of the 26th.

Gear-boxes, differentials, and engines were stuffed with ammonal, the entire contents of a case about the size of a butter box being used, and fuses were laid. Later it was felt that a smaller amount of explosive would have done the job.

After lunch the drivers were told that if all went well they would embark that night, and at four in the afternoon the remaining transport, with the exception of a pick-up and a breakdown lorry, set out on its last journey, leaving Second-Lieutenant Aitken and a small party to demolish the Thornycrofts.

The fuses were lit—they were supposed to burn for a minute—and the demolition party piled into the pick-up, which stopped dead after travelling some fifty yards, its differential wedged against a rock. The explosion was expected momently and the party's efforts during the next minute may be described as frantic. As it happened there was plenty of time to spare, for the lorries were still intact when the pick-up reached the main road, which was about a mile away. By this time Greek civilians were making for the gully in search of loot, impervious to waves and shouts. The minutes dragged by and at last the lorries disintegrated with a shattering explosion, injuring or killing, our drivers felt guiltily certain, more than one civilian. A column of smoke rose in the air and a flight of Messerschmitts dived on it, their guns firing. Staying no longer, the party headed towards Athens, the driver whose job it was to clear the road of wrecked vehicles taking the breakdown lorry. When we saw him later he told us he had dragged several vehicles off the road, some of which seemed to have been placed there deliberately.

Meanwhile the main party was on its way through Athens. For most of us it was a trying experience. Flowers and cognac we did not look for. Taunts and jeers—not that we were expecting them—we could have answered with taunts and jeers. A polite acceptance of the fact of our withdrawal we could have faced with equanimity. But when they smiled at us, when they waved to us, when they held up as a gesture—the lorries were moving fast with the general stream—small presents: cakes, a white rose, a glass of wine; when they did this, the children waving, the men saluting, the women smiling, they made it hard for us. For what were we to do? Wave back? Grin all over our faces like Cheshire cats? Say over and over again our three Greek words—*kalimera, kalaspera, kalinikta?* No, it was easier to exchange waves with the two drunk Australians, their arms around girls, on a balcony. It was easier, going through Constitution Square, to taunt, as one of our drivers did, with what cruel injustice he could not be expected to know, the lonely airman seated on the café *terrasse*, elbows on an iron table: 'Having a good leave, old man?' The airman did not look up.

The hedges of the pretty road east of Athens were white with dust and in the ditches there were wrecked vehicles. We headed towards Raphena where we were to embark from D beach. After travelling for about fifteen miles we turned off the road into a grassy olive grove. We ate a hurried meal and then repacked our valises, making the final decisons about what to leave behind. We had been told we could take very little.

We were ordered to destroy Captain Moon's staff car and all but two or three of the lorries. Destroying the staff car was good fun: there was plenty of upholstery to slash, plenty of gadgets to break. The indomitable huckster who had been negotiating with a Greek business man looked on in distress. To the dozen or so civilians who had gathered to watch us we gave the back cushion, the radiator muff, a few blankets, and some tins of food. The portable gramophone—the one we had looted—went to a little boy. He was delighted with it and thanked us prettily in correct English.

He was playing it when we moved off at dusk in the remaining transport, and the 'Woodpecker's Song', which had greeted us on arrival, made a fitting requiem for the broken, deserted lorries, blurred and shadowy in the twilight, round the little boy and his gramophone.

We dismounted a few miles down the road, leaving the transport to be destroyed by engineers. Then we shuffled down to the beach in pitch darkness. We stood in a long, whispering queue and time lapsed. At last we were packed aboard a landing craft and taken, with water washing about our boots and our heads bowed beneath the steel deck, to the *Glengyle*. As soon as we had settled down in the warm, seething, and lighted hold we were given hot cocoa and large bully-beef sandwiches. Then we slept. At three on the morning of the 27th the *Glengyle* sailed for Crete.

* * *

The Greek women saw the flash of 'Tiny' Kinnaird's[5] rifle, saw the German aircraft spiralling to earth, and sprang to a perfectly natural conclusion. A second later he was being hugged and kissed by half a dozen wildly enthusiastic civilians.

Apart from that, 26 April had not been an amusing day for Second-Lieutenant Fenton and his detachment. Dawn had found them hiding beneath olive trees not far from the Nauplion beaches and from then on they had been in constant danger. Probably there was nothing personal about the strafing. The German pilots went backwards and forwards over likely dispersal areas with the thoroughness of ploughmen and our drivers were not missed. Whenever a large flight of aircraft appeared the watchers half expected to see the sky blossom with parachutes. It was known that paratroops had landed near Corinth.

The landing had occurred at breakfast-time that morning, and by then, after travelling all night, the 6th Brigade was in position between Miloi, a few miles south of Argos, and the town of Tripolis. The NZASC detachment, of course, was still with the infantry, but it was now under the command of Captain Torbet[6] (C Section), Major McGuire[7] having gone ahead on the night of the withdrawal from the Thermopylae line to report to Colonel Crump.

At that time it was the intention to withdraw the 4th Brigade across the Corinth Canal, so as soon as it was known that paratroops had landed near the bridge the 26th Battalion was ordered to send back two rifle companies to help the small mixed force that was defending it. Two of our lorries carried them.

---

[5] Dvr F. H. Kinnaird; bush hand; Ruahine, Southland; born Taumarunui, 21 Oct 1911.

[6] Maj C. M. Torbet, OBE, m.i.d.; motor engineer; Auckland; born Wanganui, 19 Dec 1909; OC 18 Tk Tptr Coy 1 Apr-6 Nov 1944.

[7] He was occupied during the 25th and 26th in locating supplies and petrol and arranging for their distribution. In the early hours of the 27th he sailed for Crete in the *Kingston*.

'We were heavily bombed and strafed on the way up,' said Reg Troughear[8] (C Section), 'and finally we were halted by a big bomb crater in the middle of the road. After we had filled it in I went back to my lorry, but I couldn't find Arthur Hearn[9] who was the other driver. I knew he was dead-tired, so I supposed he had lain down somewhere and gone to sleep. I didn't worry much because the officer in charge of us had decided not to take the other Ammunition Company lorry any farther, his idea being to keep at least one lorry in reserve in case his transport got badly knocked about. I told the drivers of this lorry, Corporal Ivan Hogg[10] and Alan Bradbury[11] (B Section) to look out for Arthur, and I went on with the rest of the convoy.

'So much time had been wasted in taking cover from aircraft that now we just carried on. We debussed the infantry and pretty soon they were in action, though they were too late to do anything about the Corinth bridge, which had been blown up before they arrived. I didn't see any paratroops but I saw plenty of enemy aircraft and I kept my head down.

'The infantry held their positions until evening and then they moved back to fresh ones—the idea, I think, was to cover the withdrawal of wounded and stragglers. Anyway it was late at night before we got back to where we had left the B Section lorry. What was left of it was still burning and there was no sign of Arthur and the others.'

Bradbury and Hogg had been told to stay where they were until a given hour and then move south. They were joined by Arthur Hearn, and later, as there was no sign of the infantry, they pulled on to the road. At once they were machine-gunned from the air, the lorry being set on fire. Alan Bradbury was killed, Ivan Hogg wounded in the hand and leg, and Arthur Hearn in the thigh.

Meanwhile the rest of the 26th Battalion had moved to positions about twenty miles north of Miloi, leaving the 25th in reserve near that town and the 24th in positions protecting Tripolis. The

[8] Cpl R. Troughear; lorry driver; Pokeno, Auckland; born Runanga, 15 Jun 1914.
[9] Dvr A. W. Hearn; lorry driver; Waihi; born Waihi, 10 Nov 1911; wounded 26 Apr 1941.
[10] Cpl I. V. Hogg; panel beater; Auckland; born Auckland, 17 Mar 1914; wounded 26 Apr 1941.
[11] Dvr A. N. Bradbury; lorry driver; born Dargaville, 24 Oct 1917; killed in action, 26 Apr 1941.

situation in the Peloponnese was getting graver every hour. German troops had crossed the Gulf of Corinth near Patras and were advancing down the west coast, threatening Tripolis and the port of Kalamata, from which the 6th Brigade was to have been the last fighting force to leave Greece. Now it was to embark from Monemvasia, far down on the south-east coast.

'These daylight moves,' said Captain Torbet, 'were made under merciless attacks from the air. We left the Miloi area at one in the afternoon with the 24th Battalion. From a hilltop near the coast we had a clear view of the Nauplion harbour. Clouds of smoke were trailing across the water and peeping through them were the masts of precious shipping.

'In Tripolis, which was being bombed as we passed through it, we were met by civic officials who handed us a sheet of typescript stating that no British troops were to stage within three miles of the town as their presence was likely to provoke air raids.

'It looked to me like fifth column work, and later I was not surprised to hear that signboards in Tripolis had been switched, so that transport, instead of going to Monemvasia, went to Kalamata.'

For No. 9 and No. 10 sub-sections, which were travelling at the end of the convoy with instructions to pick up stragglers, it might have been better had the signboards been switched earlier, for they were directed to Kalamata and had no trouble in finding their way there. They arrived at the beach that night with stragglers riding on the mudguards of the lorries and on the cab roofs, but they were not embarked.

North of the canal, too, the situation was becoming graver. Throughout the 26th, Sergeant Buckleigh's 30-cwts. had continued to stand by for the withdrawal of the 4th Brigade from its positions at Kriekouki.

'During the day,' said 'Chum' Thomas[12] (B Section), 'scraps of news came back to us. We heard that the enemy was in Thebes, then that a German column was heading towards the pass. We heard our artillery getting stuck into it and later we were told it had been driven back. Then we heard about the paratroops at

---

[12] Dvr B. A. Thomas; bricklayer; Pinner, Middlesex, England; born England, 26 Oct 1913.

Corinth and that the Germans were working round east of us, and we knew things were sticky.

'The withdrawal started when it got dark. The infantry marched to the end of the pass and we picked them up and headed towards Athens.'

. . . .

Near the beach at Nauplion Second-Lieutenant Fenton's detachment had come to the end of its second day of waiting. No paratroops had landed there, the last Messerschmitt had gone singing into the sunset, and it was dark again. Our drivers were happy in the lovely darkness and for the moment they were not worried about the morning, though they knew now that if they were not taken off that night they were unlikely to be taken off at all.

Presently they were ordered to go to the beach and they fell in at the end of a long column of Australians who were marching nine abreast. They could not believe there would be enough shipping in the harbour to hold all those Australians and themselves as well. When they got to the beach they were ordered to about turn. This put them at the head of the column, which apparently had marched past the embarkation point in the darkness. Wading, swimming, floundering, they boarded a landing craft, and as soon as it was full it put off, slapping through the salty darkness. For a while the commander of the craft believed he would have to make Crete on his own, but after heading out to sea for two miles—by now it was almost dawn—he found HMAS *Perth*, which had been embarking troops at the nearby port of Tolos.

Our drivers were given sandwiches and hot drinks and while they were enjoying these they discovered that the *Perth*'s crew, under the impression that the men ashore were in trouble, had volunteered to a man to cover their embarkation.[13]

. . . .

The next day—the 27th—was lovely—a perfect spring Sunday.

The Greeks wore their best clothes and went to church. Under the baroque spires and onion-shaped domes, pink, golden, or gleaming white, they prayed fervently for the strange soldiers, a stream

[13] That was the last planned embarkation from the Nauplion area and most of the troops who were not taken off that night were captured.

of entreaties going up to Heaven when engines were heard and ripples of dark shadow rushed over the bright domes.

The soldiers with the shabby uniforms and tired faces were still there after the service. When you spoke to them, saying '*Nike*' (victory), or gave them the thumbs-up sign, they answered in their foreign language: 'That's right, Dad. How you doin', anyway?'

'It was all pretty quiet and peaceful,' said 'Chum' Thomas, 'with people going to church and that sort of thing, but we were expecting the Germans to appear any moment. The 4th Brigade Group was dispersed between Athens and the beaches. I don't think anything interesting had happened to any of our drivers during the night journey from the Kriekouki positions. There was nothing to do, so we kept under cover and waited for orders.

'During the morning we were told to destroy our trucks. Stan Barrow[14] and I drained our oil and we were just going to put grit in the sump when the order was countermanded. By now the other boys had wrecked their trucks and started out for the beaches, but all ours needed was a fill of oil. We had mortars on board and these were wanted by one of the companies of the 20th Battalion chosen to cover the embarkation.

'We had been under attack from the air all day long and when we moved back towards Athens, heading for Markopoulon, a little village some six or seven miles west of the beaches, we were quite expecting to get it. There were about eight lorries in our convoy—our own was towing three others—and I dare say it was the last lot of British transport moving north of the Corinth Canal. Anyway, as we were coming over the brow of a hill about sixteen Messerschmitts came down on us. Our passengers tumbled into the ditch and I jumped from one side of the cab and Stan from the other. An Australian artillery officer was killed beside me and four vehicles were destroyed, ours included. The Messerschmitts stayed for some time and when the officer in command called a muster he found that eight or nine men had been killed—I think that's right—and others were casualties. There was no sign of Stan.

'Soon shells started to land near us and from then on it was rather like a bad dream. German motor-cyclists came down the road and we fired on them. I remember the tracers and the noise

---

[14] Dvr R. S. Barrow; seaman; Hamilton; born Christchurch, 24 Oct 1913; posted missing, Crete, 2 Jun 1941; escaped by boat to North Africa.

and the flash of mortars and I remember noticing that our truck was still burning. This went on for some time and then things quietened down.

'Later we were given orders to go to the beach and after a march of about five miles we went straight on to the landing craft. When I got aboard the ship I met some of the other boys and they had the surprise of their lives. Stan, who had made his way to the beach after our truck was hit, had told them I had been killed in the raid.'

The ships that sailed that night were the last to embark troops from the Athens area and in them went all Sergeant Buckleigh's drivers except two—'Chum' Arblaster[15] and Bill Dolphin[16] who had been sent out to collect supplies two days earlier and had not been seen since. That left only Captain Torbet's detachment in Greece.

. . . .

The morning of the 27th—that brilliant and interminable day—found the 6th Brigade still guarding Tripolis. It was to hold its positions until dark and then move as quickly as possible to a dispersal area near Monemvasia. As the enemy had not put in an appearance by noon the 26th Battalion started the journey in daylight. The bulk of Captain Torbet's transport stayed behind with the 25th and 24th Battalions. The latter was to be the rearguard.

All day long fighters, bombers, and little impudent Henschels, like dragonflies, brushed the tree-tops, searching for troops and vehicles. No cooking was done and our drivers moved only when the camouflage nets had to be shifted to cover the changing shadows cast by the lorries.

'Some of us,' said Captain Torbet, 'were hidden among olive trees on the slopes of a little valley, and during the day an elderly man in peasant clothes strolled down the centre of it, glancing from side to side and calling out with a strong American accent: "Come out, boys! Come on out! Don't be scared!" No one moved and he went away.'

[15] Dvr C. F. Arblaster; carpenter; Auckland; born NZ, 12 Jul 1918; p.w. Apr 1941.
[16] Dvr R. J. W. Dolphin; contractor; Auckland; born Petone, 22 Jul 1918; p.w. Apr 1941; repatriated Nov 1943.

That evening Captain Torbet heard a German radio announcer state that the Luftwaffe had been looking for the remnants of the New Zealand Divison all day but had been unable to find them owing to their clever camouflage.

One of the biggest worries was petrol. As the original intention had been to embark the brigade from Khalkis the drivers had started out with only one case of spare petrol on each lorry, and now, with a journey of about eighty miles in front of them, their reserves were dangerously low. Corporal Roy Hintz[17] had returned from a foraging expedition with six cases, and these, together with a few gallons obtained by draining the tanks of abandoned vehicles and a few more allocated to Captain Torbet from supplies requisitioned in Tripolis, enabled our lorries to move out that night with a reasonable chance of reaching Monemvasia. The total reserve—two gallons—was carried in C Section's LAD.

It was difficult to guess how much petrol the lorries would burn. They had been heavily laden when they left the Thermopylae line and since then a large number of stragglers had been picked up and many vehicles had been destroyed or had broken down, so that now the remainder were loaded far beyond the safety mark and were gulping petrol. The LAD—a 30-cwt.—was to end the journey with thirty-five men aboard.

The convoy passed through Sparta, which looked pretty and peaceful in the quiet starlight. Indeed many of the drivers recall to this day how charming it looked, and doubtless at the time, as they drove through its graceful streets, little scraps of forgotten knowledge, things they had heard in childhood about Sparta and the Spartans, stirred in more than one tired mind. And they drove on, perhaps, thinking of Sparta at war with Athens, or of the small boy and his fox, until with a start and a sudden breath-taking swerve they were back in the present and their own chapter of Greek history. Then they would remember where they were and what they were doing and determine to banish everything from their minds except the tailboard of the vehicle ahead. But thoughts would come drifting back, thoughts and dreams, and the road would fade in front of them and the unmeaning gabble of the engine change to music—the kind of music that Caliban heard on his island:

[17] Sgt R. O. Hintz; driver-mechanic; born Te Aroha, 9 Sep 1917; died in NZ, 1 Mar 1948.

' sounds and sweet airs, that give delight, and hurt not '—and mix gently with the conversation of people they had known at home. Then one of them would sleep for a few seconds—sleep soundly until an urgency of disquiet, a subconscious warning that something was terribly amiss somewhere, pulled him back on to the road to Monemvasia. At once he would wake his friend, slumped in the seat beside him drenched and drowned in sleep, so that he could have a rest from driving or at least have someone to talk to.

Before daylight brought the Luftwaffe the transport was dispersed some miles from the beaches and drivers and infantry were prepared for another day in hiding. The first wave of aircraft came over soon after dawn and sank a small ship in the harbour, and from then on the sky was seldom completely empty; but the hours dragged by and still the battalion areas had not been attacked.

The day seemed endless.

. . . . . .

And the next day—the 29th—seemed endless in a little town some miles from Kalamata, but punctually at seven in the evening, as they had said they would, the Germans arrived. They came in a British staff car, which was for two British officers, and a British lorry, which was for sixty-odd British other ranks and three New Zealanders. The Germans could congratulate themselves on a little tableau that exemplified the efficiency of the Wehrmacht in a really striking manner.

'For three nights running,' said Corporal Wally Dahl,[18] 'we had gone down to the beaches. The last was the 28th-29th and during the day and the early part of the night there had been fighting in Kalamata. In the small hours, when all hope of a planned evacuation seemed to have ended, the thousands of troops in the area were told that the senior officer present, a British brigadier, intended to surrender to the Germans at half past five in the morning. On hearing this, Les Robinson[19] and Arthur Davie[20] and I—C Section had become split up into small groups by this

[18] Sgt W. A. Dahl; tram driver; Wellington; born Wellington, 17 Jan 1905; p.w. 29 Apr 1941.
[19] Dvr L. S. H. Robinson; motor driver; born Waihi, 7 Mar 1914; p.w. 29 Apr 1941.
[20] Dvr A. C. Davie; motor driver; Walton, Auckland; born Auckland, 26 Sep 1919; p.w. 29 Apr 1941.

time—decided to put as much distance as possible between ourselves and the beaches. We marched seven or eight miles, taking our rifles with us. Towards daylight we came to an empty farmhouse, which we broke into. Here we rested for about two hours; then we heard shooting, so we pushed on, marching for five and a half hours through a swamp. Les Robinson had an infected hand and by now it was up like a balloon and he was feverish and in great pain. It was clear he couldn't be expected to carry on much longer, and as we didn't want to leave him on his own, which was what he wished, we began to think there was nothing for it but to give ourselves up. Late in the morning we came to a little town and there we were met by two British officers and a civilian. The civilian, they told us, was a German policeman. They said there were about sixty British soldiers waiting in a building in the town for the arrival of the Germans. They had sent word that they would be there at seven that evening. The officers told us we would have to surrender our rifles, so we smashed them. Then we joined the others in the building. We gave our names and addresses to a Greek woman—a Red Cross worker—and she promised to write to our families. The civilians were very decent to us, giving us soup, bread, and eggs. Then we settled down to wait for the Germans.

'They arrived punctually at seven and we were taken to a barracks near Kalamata. On the way we drove past an endless column of marching prisoners, among whom I recognised two C Section boys. Later we met the others.'

In all, seventeen of our drivers were captured in the Kalamata area.[21]

. . . .

At eleven the night before, Captain Torbet's transport had been marshalled in a dry creek-bed and about an hour later it moved off towards the beaches at Monemvasia. It crossed a bridge and this was demolished behind the last vehicle by New Zealand engineers, the flash and the explosion causing some of the drivers to think that

---

[21] They were Cpl Wally Dahl, Cpl Cam Grinter, Bill Dalton, Arthur Davie, Jack Donnelly, Ted Donnelly, Don Hourigan, Bill Johnson, Bill Leathwick, George Le Comte, 'Red' Lee, Ted Malcolm, Bob McNee, Phil Moore, Les Robinson, Jack Shaw, and Sid Spilsbury.

the convoy was being bombed. At the top of a steep cliff, about a quarter of a mile from the beach, the lorries were halted in line, the infantry forming up on the right-hand side of the road. On the other side there was a drop of 200 feet to the sea. An order was given and a dozen men collected round each vehicle. One after another the lorries were pushed over the cliff. Some smashed on rocks and others fell into deep water, while the headlamps of a few could be seen shining under the sea. It is hard to explain how they came to be switched on and why they were not broken. Some of the vehicles on the rocks started to burn. There were hundreds of vehicles below the cliff and our drivers heard later that the Navy came back and destroyed them completely by shellfire.

In groups of seventy the men marched down to the beach. A wait followed and Captain Torbet was told that about 200 men, including his own drivers,[22] might not be embarked until the next night. He was informed of the signals that would be used in that event.

Through the small hours of the morning, when the sky is darkest and the human spirit reaches its lowest ebb, the drivers waited on the beach, hope and courage running out like sand. They heard the putter of engines as landing craft went backwards and forwards in the bay and they knew there were not many troops left on the beach.

Finally, when hope was beginning to seem foolish, they were taken to the *Ajax*, the last ship to be loaded. The last boatload reached her just before four in the morning.

· · · · · · · ·

In *Ajax*, as in *Havock, Griffin, Isis, Calcutta, Vampire, Voyager, Perth, Kingston, Glengyle,* and all the other ships that had taken part in the evacuation, everything was under control. The sailors relieved our drivers of their valises and handed them up the gangway. The decks were heaped with equipment and the companion-ways were blocked with troops, but it was like stepping out of chaos into order. After days of confusion and destruction and falling back the journey had ended in a place where panic was unthinkable. You felt that if the last trump sounded and the

---

[22] One hundred and twenty NZASC all ranks were embarked from Monemvasia on the night of the 28th-29th, the majority of them under the command of Capt Torbet. There were about forty of our drivers among them.

graves started to give up their dead *Ajax* would be standing by to proceed with the evacuation of His Majesty's subjects from the four corners of the earth. The very sight of the bluejackets—their levity, their good humour, their confidence—was better than a promise from St. Michael that in the last event Hitler would not prevail. We understood now why Napoleon had failed to invade Britain, and why Hitler would fail also, and why no one would ever invade her successfully; and the reason was this: the people of Britain in time of danger, knowingly or unknowingly, thought of their island as a ship, and worked and fought her as a ship, and behind them to teach them how was the knowledge and experience and steadiness of a hundred generations of seamen. 'Stand by to repel boarders. . . .'

Soon the soldiers of the Wehrmacht would be trampling over the whole of Greece, a country lovelier and older than anything they could understand, but not over Kent and Sussex, and never would a single German soldier set foot in the *Ajax* or in any ship of the Royal Navy.

As soon as our drivers had settled down they were given hot cocoa and bully-beef sandwiches. They tried to say thank you but it was no good. ' Garn,' said the sailors. ' 'Ev another muckin' cup. My oath, Miss Weston. . . .'

. . . . . .

On 16 October 1944 the Piraeus naval radio station sent this message to the British Naval Base at Alexandria:

It is good to be with you after three years.

CHAPTER 7

ISLAND INTERLUDE

WE set off in fairly good order from Suda but after we had marched a few hundred yards the column began to go to pieces. The accumulated tiredness of days was having its effect on us and the 12-hour trip from Greece to Crete had been anything but a pleasure cruise. The *Glengyle*, carrying Company headquarters, Workshops, and some of the section drivers, had been attacked during the morning, one bomb landing near enough to buckle plates and cause interior damage, against which the Navy had claimed one plane destroyed. Bombs at sea, we had discovered, were far more terrifying than bombs ashore.

Soon men started to drop out and sit down in the cool on the low stone wall beside the lane. Some took off their boots; others opened packets of biscuits and tins of sardines. More men fell out when we came to a little wineshop and no effort was made to stop them. Presently what was left of our column, with the exception of a hardy twenty or thirty who were marching in the lead with Captain Moon, broke into small groups, each going its own pace. When they came to a spot that looked pleasant they halted and made camp. Everyone was tired but happy to be in Crete.

Just before dusk the men who had carried on for an extra mile or so were rewarded by coming to a field kitchen by the roadside. Here you could collect a cup of tea and something to eat. Soon it was dark and the flames from the cooks' fires, leaping and dancing beneath the big dixies, threw a ruddy glow on the faces of the men near them. Few talked or laughed and everyone waited patiently in the long queues as men who are already drenched wait patiently in a rainstorm. We were soaked through with tiredness.

Two searchlights fingered a silver cloud-bank and there was a distant thump but hardly anyone took cover. Hot tea first, then sleep—that was the programme.

By noon the following day all but a few of our drivers had reported at the assembly area, which was three miles west of the town of Canea. They arrived in twos and threes, dripping with sweat and caked from head to foot with white dust. On the way in they passed a clear pool in which people were laughing, splashing, and laundering their underclothes. Soon the pool was full.

The area, of course, consisted only of rows of olive trees festooned with grey blankets. Scores of little fires were burning beneath them and on every fire there was a battered tin with shaving water in it. We had the rest of the day in which to get cleaned up and that night we had another sound sleep.

. . . .

Captain Torbet's detachment went straight to Egypt and Major McGuire was given a battalion composed of the combined NZASC personnel in Crete. The rest of us were together in one area by the beginning of May but seldom were we in the same place for two nights running.

During our first four or five days on the island we did little except march, and when we were not marching we were sitting under the olive trees waiting to march or lying down under them exhausted after marching. We marched along the coast road between Canea and the Maleme airfield, casting longing glances at the dark-purple sea. We marched along narrow by-ways, kicking up the white dust with our boots—it was like flour underfoot—until the olive trees by the roadside were pale as powder-puffs. Each night we had a different olive grove for our home and a different stream for our wash-place.

We were infantry now (it exhilarated and disturbed us) and the purpose of this marching, as we saw it, was to get us fit and at the same time confuse reconnaissance pilots.

Our officers—true they were not as heavily laden as we were, their bedrolls, so it seemed, being shifted from place to place by mysterious agencies: donkeys, perhaps, were at the bottom of it—marched with us. In the lead would be Captain Moon, looking as spruce and nonchalant as though he had stepped from the white portico of the old home to give an order to his coloured overseer, and in the rear Second-Lieutenant Toogood, larding the ground like Falstaff but always ready with a word of encouragement or abuse and never too exhausted to join battle with his verbal spar-

[1] Cpl A. Falconer; truck driver; Wanganui; born Leuchars, Fifeshire, Scotland, 25 Nov 1916.

ring partner, Alan Falconer.[1] Tagging along behind, taking their time because they were encumbered with Bren guns and magazines, came the ack-ack crews. Jim Stanley still had his anti-tank rifle.

We grumbled a good deal, of course, but most of our grumbles were merely a concession to good form. We were fit, the weather was glorious, and we were more than a little pleased with ourselves. The thunder had roared, the lightning danced, and we were hardly any the worse for it. Some had been wounded, some had been taken prisoner, and some were missing, but only one had been killed—no, two; for we counted Norman Chissell.[2]

Late on 1 May, to everyone's distress, the sections were split up, though the unit remained intact. The NZASC Battalion had been absorbed into Oakes Force, which consisted of artillerymen without guns and drivers without vehicles. We were to operate under our own officers in three infantry companies.

We were reshuffled and the next day we marched to a position in the hills about a mile north of Galatas. The area Oakes Force had been given to defend extended roughly from the western outskirts of Galatas to the coast, and we were more or less in the middle of it. Facing north, we overlooked the sea and the tents of the 7th General Hospital; facing south-east, the white buildings and walled compounds of the prison of Aghya. A quarter of an hour up a winding lane took us to the village of Galatas and a good afternoon's walk to Canea. These names meant nothing to anyone at the time.

Here we settled down as infantry. We established ack-ack posts, kept paratroop watches by day, posted strong pickets at night, and at dawn sent out patrols and stood to arms. We had no shovels and no wire and not all of us had rifles. A few were without overcoats, and in the chill of the early morning when the white mists were mounting beneath the olive trees you were liable to be challenged by a figure in a grey blanket with a scarlet stripe. A head-dress of turkey feathers and the illusion would have been complete.

Some of us pretended to be bored, adopting the attitude that

---

[2] At that time we were unaware that Alan Bradbury had been killed and two sub-sections captured. Our casualties in Greece were: killed 2; wounded 12; missing 31; Total 45. The following were captured as a result of wounds, injuries, or sickness: Sgt Robin Hood, Dave Adams, Charlie Black, Alan Bush, 'Kolynos' Carroll, Dave Forbes, Barney George, Jim Nichols, Fred Wells, and Harry Wishaw.

tools and engines were our business—not this infantry nonsense. But most of us played the new game with tremendous zest.

Every morning half the unit had leave to go swimming or to visit Galatas and in the afternoon the other half was free. On fine days—and the weather was mostly fine—it was a delightful life, but we were uncomfortable when it rained. Then there was nothing for it but to crouch shivering in rough shelters made from groundsheets and green barley blades and watch the miserable thin rain—it seldom did more than drizzle—prick down like piano wire.

This open-air life made us desperately hungry and the rations, unfortunately, were slender. The cooks were using open fires—gathering wood was one of our daily jobs—and they had no gear except a few dixies and some cut-down petrol tins, but most of them went to a lot of trouble to make the bully-beef stews appetising. They added herbs and vegetables, purchased from a common fund or stolen, and Mark Brown[3] (B Section) cooked some delicious meals. The helpings were sometimes smaller than he could have wished but into the serving of them he never put less than his whole heart. From Galatas almost to the sea he could be heard calling his men to breakfast, and drowsy sentries would grip their rifles thinking the Germans had come. As the queue lengthened Mark's barrage of abuse would lift to include each new arrival—he knew everyone in the company—as he dolloped the steaming mess into a collection of old tins. Most of us had a mug or a spoon but few had a complete set of mess gear and some had nothing. Any dixies there were had to do double or treble duty.

We were hungry all the time. Grapes were all about us but they were a few weeks short of being ripe. However, we raided the local potato patches and we killed pigeons. Stuffed with onions and bread-crumbs they were delicious.

. . . .

The days slipped quickly by. Looked back on they seem all blue and gold. In spite of our many duties we were able to spend hours on the beach and hours in the little wineshops fronting the Mediterranean. For a hundred drachmae you could get as drunk as a lord, but it was more fun to stay sober and watch the Cretan fishermen at work in the calm waters between the mainland and

---

[3] Cpl J. M. Brown; cook; Rotorua; born Wanganui, 11 Jul 1900.

Theodoroi Island. Towards evening, carrying their catch—red mullet and small silver fish—they would call in for wine. Many of them wore wonderful embroidered waistcoats and a few wore gold ear-rings—like the seafaring rat in the *Wind in the Willows*, whom they rather resembled. To go with the wine there were little dishes of baked cuttle-fish, chopped egg and tomato, and—most delicious of all—rice wrapped in vine leaves and fried in oil with paprika.

Always at this hour the watch for paratroops was intensified. Throughout the lovely May evenings the watchers sat in pairs on the highest knolls in the brigade area, searching the sky and the surrounding country. The olive trees stretched away in every direction like toy soldiers, parting here and there to allow room for a vineyard, green and tender with young vines, or a square orange grove guarded by white walls. In the background, so massive that the country brushed up to them over the foothills appeared toy-like, were the White Mountains, all grey and purple and dark green until you reached the snow line. On their western slopes the sunlight lay thick and flat as though it had been put on with a paint brush—like yellow varnish. Little white boxes of houses stood out clearly among the green and gold, just a cluster of them here and there, with the biggest cluster, Galatas, close at hand.

The whole countryside looked like a wonderful old carpet that has lain for long years in a busy, sunny room. The colours, still warm and glowing, were like the ghosts of colours once unbelievably brilliant. Everything had a look of long use. Millions of hands had worn the well smooth; millions of feet had hardened the clay path. Every house and cottage—the feeling was inescapable—was so saturated with family life that the house *was* the family.

Here the pattern of life had gone on unchanged and unbroken, threaded through by the same customs, the same smells of cooking, the same unaltering round—water to be got from the well and fish from the sea, goats to be milked, grapes to be gathered in and wine pressed out in due season—since Minos was a king in Crete and Theseus slew the Minotaur and Deadalus and Icarus set out to fly to Italy. We were awed by the amount of living that had been done in one narrow island, by the tomes of history and legend that were one island's story—awed and comforted. Centuries of struggle and catastrophe and darkness, and the children still played in the streets and there was smoke in the chimneys and the girls put on their pretty dresses on Sunday. These things persisted in spite of Minotaurs and Luftwaffes.

. . . . .

On 8 May we left our rifles and ammunition for Oakes Force and marched first to a transit camp east of Galatas and then to one a mile or so south-west of Canea. We were told we should be there for a few hours but the days slipped past and we were still in Crete.

Having no arms we did no guard duties, and as we were on a moment's notice to move there was no leave. This left us unlimited time for doing what we could to make ourselves more comfortable or less hungry. Greek pedlars came round with trays of cakes and pastries and as long as our drachmae lasted we bought them, but they were not cheap. .We built shelters of rushes and groundsheets and talked interminably about our experiences in Greece. And we lay under the olive trees, content in the sunshine.

When dusk gathered we built fierce bonfires. Their life was limited to about half an hour because of the blackout, so we built them more for cheerfulness than for warmth. Sometimes a water bottle of red wine would pass from hand to hand and then there would be singing, much to the disgust of Winston Churchill, A Section's old bulldog. He would gaze into the flames with an expression incredibly wise, mournful, and disapproving, and when the concert ended, as it did invariably, to the strains of 'Ole King Farouk' (our version—not a respectful one—of the Egyptian national anthem), he would yawn with relief, showing his yellow, broken stumps.

Most of us turned in soon after dark, to lie talking for a long while or gazing up under the enormous yellow moon at a tangle of olive branches, beautiful and complicated as a rood-screen. Often we heard aircraft, and then someone would yell out, simulating panic: 'Don't look up, boys. Stop smiling, that man with the gold teeth.' The aircraft would go back to Greece, the talk die down, and a hush fall over the whole island, enabling you to sort out and separate from one another all the small noises of the night: the munching of grass from the tiny erratic tolling of a goat bell, and both from the far hiss of waves. The thin moaning that had troubled you for some time would identify itself as drunken singing. And with the increased silence (as though sound could impede scent) smells became clearer—from the hills the clean smell of thyme, from the grass a sweetness of dew, from our blankets a sour, sickish smell, and from the beaches, faint but discernible, a lovely suggestion of wet shingle, boats, lobster pots. And sleep would come.

Sometimes we were woken by the sound of strafing, and whenever the harbour was bombed the ground stirred under our ribs as though the island had coughed. Once we awoke to see a plane diving down the white beam of a searchlight, its guns firing.

They were happy and healthy days and the war situation gave us no uneasiness at all. The official news—a neighbouring British medical unit had a wireless set—was mostly bad: it told of heavy night raids on Britain; but from sources within Crete, the workings of which have never been explained satisfactorily, we were supplied with good tidings, and false and genuine news became inextricably mixed. Berlin was in ruins and Hitler had asked Churchill for three days in which to bury his dead. On being refused he had threatened to use gas and Churchill had replied: 'Go right ahead. That's just the excuse we want. We've got something that'll finish the war in three weeks.' None of us had actually heard the broadcast of this remarkable exchange but several of us knew people who had. We were sceptical, of course, but the story had its effect and we preferred it to the official news. It was better, at any rate, than Lord Haw-Haw's melodramatic gloating over what he called the Island of Doom.

. . . .

Captain G. A. Hook[4] (Supply Column) and Captain Moon (acting officer commanding Ammunition Company) stood under an olive tree and watched Father Jim Henley[5] (NZASC chaplain) spin a coin. The matter at issue was whether we should go to Egypt or stay in Crete.

Originally the intention had been to evacuate both the Supply

[4] Capt G. A. Hook; motor mechanic; Hastings; born Marton, 10 Jan 1905; p.w. 2 Jun 1941.
[5] Rev. Fr. J. F. Henley, CF; Roman Catholic priest; Eltham; born Palmerston North, 10 Sep 1903.

Column and the Ammunition Company in the *Nieuw Zeeland* but the available space had been reduced unexpectedly and it had been decided to embark either the whole of the larger unit, the Supply Column (less certain details that would have to remain in Crete to administer the detail issue depot), or the whole of the Ammunition Company plus the Supply Column's workshops and specialist personnel. But which was it to be?

Father Henley took a coin from his pocket and spun it, Captain Moon calling heads. Heads it was.

Accordingly, on 14 May, at half past one in the afternoon, we set out for Suda.

We went aboard the *Nieuw Zeeland* late in the evening and at dusk she began sidling into the stream. The gap between her side and the wharf widened like a slow yawn, and then, just as the slack water was beginning to ruffle, a man pounded on to the wharf. Without hesitating, he flung his revolver to someone on the weather deck and made a flat gangling dive. He came up gasping and spluttering and was dragged aboard at the end of a life-line. We laughed and clapped and gazed again at the lonely diminishing figures still standing on the wharf while the beautiful island, like an island in an old story, melted into the twilight behind them. The incident had been very dramatic—the revolver flying through the air, the plunge into the black water, the widening ripples, the gasping, spluttering swimmer being dragged aboard—and we knew now that we had witnessed an escape. We too, perhaps, were escaping.[6]

We knew now that the Germans would come—tomorrow or in a week's time. Captain Moon, perhaps, had known all along. Anyhow, the coin which Father Henley had tossed was in his pocket. He was keeping it as a souvenir.

· · · · · · ·

The *Nieuw Zeeland* took us safely to Port Said, which was reached at two in the afternoon on 16 May. We waited by a railway siding until ten that night, arriving in Helwan Camp at breakfast-time the next morning. On the morning after that we took part in a cere-

[6] We left behind Capt H. A. Rowe (posted to unit 27 Mar 41: attached to Supply Column 27 Apr 41), Lauris Newfield (his batman-driver), Cpl Keith Smith (sick), and Jim Winstanley and Stan Barrow (missing when we embarked).

monial parade for the Prime Minister of New Zealand, the Rt. Hon. Peter Fraser.

We had nothing in common any longer with the gipsy company that had lived in rags under the olive trees. Our island days were like a dream and it was as though we had never been away. Under the newly risen sun—it would be a scorcher later on—we stood stiffly to attention. Everything we wore, from boots to despised sun helmets, was brand-new. We were clean, shaved, regimented, and—for the moment—bored.

CHAPTER 8

MURDER ON THE OLD HOOK

CRETE was a German island and in Libya it was all to do again. In the Mediterranean the balance of sea power was against us and the Royal Air Force was at a greater disadvantage than ever. Everything was as black as night, but in the NAAFIs at Helwan no shadow fell athwart the merry-makers.

Readers of Westerns must have been irresistibly reminded of the saloons in their favourite boom towns. Pianos tinkled interminably, oceans of beer were drunk, and no one thought twice before changing a 100-piastre note. Credits were substantial after the austerities of Greece and Crete and remittances were on their way from New Zealand. 'Send her along! Get into her! Shoot her up!' That was the feeling of the hour and nowhere in the Middle East was it being translated into action with more enthusiasm than in the Helwan NAAFIs. Pandemonium reigned. South Africans and New Zealanders sang, screamed, and argued, never quite drowning, as shrieking seagulls never quite drown the surf, the steady rhythmic booming of the Crown and Anchor kings. Two at a table they sat —one to rattle the dice and put over the spiel, one to rake in the money.

'The old firm boys the old firm. Here before Greece here before Crete here before Christmas. You pick 'em and we pay 'em. We're not here to make money we're here to make friends. You gotta speculate to accumulate. A good horse never stumbles and a good sport never grumbles. The more you put down the more you pick up and a good bet to you Sport. Ten akkers on the old corner pub. Pounds crowns and browns. No bet too big no bet too small pass round the cigarettes Ted. Roll up here gentlemen to make your Palestine leave. A good bet to you Sir. Fifty on the old Sergeant-Major from our friend over there give the gentleman his change Ted. That leaves the old Mae West and the dinkie-die running for the old man. Its murder on the old hook murder on the old hook. Any more before we lift her up we can't keep these good punters waiting. She's a game of speed gentlemen and we want to keep her

Driver's windscreen removed for visibility

Two bombs in one hole near a staff car

Destruction of Workshops' store lorry —page 90

RETREAT THROUGH ATHENS
'When they waved to us' —page 90

that way. Yes we've got to lift her up and what do we see. We see the old name of the game two hooks and one crown. Just where the money lies and the old man coughs blood. Pay out on the hook and crown Ted and away to war we go again. IT'S THE OLD FIRM BOYS IT'S THE OLD FIRM. . . .'

The Crown and Anchor kings were reaping a golden harvest. Needing no equipment beyond lungs of brass, a heart the size of a pea, and the conscience of a hermit crab, they had been making a good thing out of the Army ever since the first day of the war and they intended to go on making a good thing out of it as long as the good Lord gave them health and strength. They knew all the dodges. Not for them parades and fatigues. Not for them the discomforts and limitations of the field.

For a while they were untroubled by competition and there were enough suckers to go round. Soon, however, a large number of outsiders, among whom were members of the Ammunition Company, succumbed to the lure of quick riches. They pooled their resources and set up in business on their own, and it was not long before a table shortage developed. The result was a new racket: table-broking. At auction a table would fetch anything from 15 to 30 shillings.

The authorities probably had a fair idea of what was going on, and almost nightly there were raids, but warning was given by a corps of highly-paid scouts ('OK, boys. Wrap her up. Officer of the picket.') and arrests were uncommon.

Everything went smoothly for about a month—too smoothly, for by then most of the suckers in Helwan had no money left and those who still had a few pounds were patronising the tables that offered the biggest bonuses and provided the most free beer and cigarettes. Overheads bounded as profits dwindled, and only the hermit crabs, drawing on their vast resources, were able to continue in business. The small men withdrew hurriedly, licking their financial wounds. A few of our drivers—those who had not lingered too long—had done well, and others had made money as punters, but the majority had gained nothing but experience.

The craze passed, or at least abated considerably, and remittances arrived from New Zealand to restore our finances and enable us to take advantage of the fortnight's leave due to us.

Most of us spent at least a week of it in the Holy Land. Jerusalem we found impressive but puzzling. An atmosphere of sorrow undoubtedly pervaded the place—the very stones, massive and brooding, seemed to exhale it—but an atmosphere of sanctity was not always perceptible. Money-changers were abroad in the temple, and a brisk traffic in German and Japanese hardware was being carried on in the Via Dolorosa, which was cluttered with squalid shops. The organised tours, too, though cheap and instructive, were placed so solidly on a commercial basis and the guides rattled through them with so much glibness and cheapjack assurance that it was really inevitable that an Australian soldier, a little the worse for drink, should interrupt the lay brother who was lecturing a party of us in the crypt of the Church of the Nativity at Bethlehem (possibly the most sacred spot in Christendom) by tapping him smartly on the shoulder and remarking, obviously with no wish to offend: ' You're all to hell, George. Where's them bloody bulrushes? '

Tel Aviv was different. Here there was no need for reverence, real or assumed. All you were expected to do was indulge your appetites as often and as expensively as possible. Every taste was catered for and a little unorthodoxy was not frowned on.

For our second week's leave most of us went no farther than Cairo, choosing to spend a restful holiday at the New Royal or at one of the other *pensions* established for British troops. In these it was possible—indeed, you were expected and encouraged—to lie at ease in bed during the greater part of the morning and drink American canned beer. Nothing we did or left undone could surprise the Achmeds and Mahomets. Their cynical and amused tolerance (which, we should have said, was *our* attitude towards *them*) was proof against everything.

. . . .

Thus the end of May and the first weeks of June.

Naturally we had little time in which to concern ourselves with the march of events, and perhaps it was as well, for a full knowledge of their course might have caused us to break out in a cold sweat. On a June weekend the Axis gained a decisive victory in a tank battle in the Sollum area and after that Egypt lay wide open. But for our still holding Tobruk, in which more than a division was bottled up, Rommel would probably have followed up this success. As it was, his dislike of a threatened flank stood us in good stead, though Tobruk (like Malta) was living from hand to mouth and holding only from day to day.

Not that there was nothing, or even little, on the credit side. A brilliant campaign in Abyssinia had ended in the surrender of the Duke of Aosta and his army at Amba Alagi, and in Syria (if there is anything creditable to anyone in fratricide) the struggle between French on the one hand and French and British on the other was being brought to a successful conclusion. Best of all tanks, guns, and aircraft had begun to arrive from Britain in a steady stream.

But no one could dispute that Germany held most of the cards or that it was her play. Trumps were still panzers and she could be relied on to lead trumps. On 22 June she attacked Russia.

Our first reaction towards this stupendous event was probably the common one: Britain had not so much gained an ally as lost a potential foe. It was the end of a nightmare and once again the old simple solution that had appealed to so many of us at the time of Munich seemed a possibility: let the two serpents swallow each other's tails. But the red serpent, said the military experts, was incapable of swallowing anything. They doubted if she could hold out for more than a few months. Hitler gave her a few weeks and for a while his optimism seemed justified. No matter—we had been reprieved.

·   ·   ·   ·   ·   ·   ·

Late in June we moved to the Mahfouz area, the hottest and sandiest place in Helwan. Formerly we had been on high ground; now we were in a kind of basin to which no breath of wind ever penetrated.

The move to the new area marked the division between a life of comparative indulgence and a more rigorous ordering of our days. Well, that suited us. We had enjoyed our leave and we had spent our money. Now was the moment for the hermit crabs to take their bulging pocket-books to Tel Aviv and for us to look forward. For a while there had been a tendency in our unit—it may not have been so in others—to speak of the Division as though it were finished as a fighting force and of ourselves as battered knights who would never again enter the lists. ('The old Div, Dig, she's been cut to pieces.') That was all done with now and we were as keen as ever.

The violent catharsis effected by Tel Aviv, Cairo, and the Helwan NAAFIs had left us a trifle flabby physically and spiritually, but that was changed, too, and it was not long before Major McGuire was able to march us from Helwan to Maadi in record time. We rose at dawn for a period of physical training and until midday we were busy with infantry drill, weapon training, and route marches, after which it was too hot to do anything except lie sweating on one's bed.

It was not a programme of absorbing interest and anything that presaged action—the arrival, for instance, of forty-eight reinforcements—was welcomed eagerly. 'When will we get our new vehicles?' was the most canvassed question. Most of us had not touched a steering wheel since our arrival at Helwan and we were pining to get our hands greasy again.

The lorries arrived early in August. We collected them from Abbassia and we sang as we drove them home. They were new four-wheel drive Chevrolets and they were as good as the Bedfords —better in some ways. They were easy to handle in soft sand, their cabs gave good vision, and they were well equipped with tools. We stood round them and discussed their points.

From then on we were unable to take the slightest interest in route marches and training programmes. Drivers to whom vehicles had been allotted were distinguishable from the rest by their habit of strolling through the vehicle park at odd hours as a farmer strolls down to the paddock to inspect his new Jerseys. The unlucky ones—whenever the unit was at full strength we had plenty of spare men—were hard put to it to conceal their disappointment. 'A loader's job'll do me, boy. You can have your stinking trucks. But I don't know what come over old "Dad" giving one to what's-his-name.'

A chance to test the new Chevrolets came later in the month, when first C Section, then A Section, then B Section, took part in a 36-hour desert exercise in the Mena-Fayoum area. These were only short trips but it was like old times to pull out of camp at 50-yard intervals, nod to the transport sergeant as he waved you past, and know that your tanks were full of petrol and your tucker box of rations and that your bedroll was bouncing around in the back. The lorries fulfilled all our expectations but it was not until the end of August, when A Section returned after ten days at Suez, that we realised how good they were. The section had been delivering troops and supplies to camps in the Canal zone and had covered over 33,000 miles. Complaints: one faulty generator.

September came and soon we were preparing for a move. The attendance at the morning sick parades dropped sharply and drivers who had been nursing desert sores with the assiduity of professional beggars began to pester 'Doc' Turner[1] for more effective unguents. No one wanted to stay in Helwan. We had had enough, and more than enough, of Cairo, Crown and Anchor, check parades, zibbib, and all the other ingredients of Base life. We thought nostalgically of long drives over desert tracks, meals round the cooks' lorries, bathes in the Mediterranean, evenings round our new wireless sets, cool desert nights.

We had it all worked out—a season in the desert, a race to Tripoli, back to Cairo with money in our pockets, a week's leave, and then—why not?—New Zealand. The Wehrmacht had slowed down in Russia and even the old oil story had taken on a new lease of life. If the opinion of the Oxford Institute of Statistics was worth anything the Germans would do well to go very steadily with those tanks and trucks.

On 16 September, after collecting our second-line holding of ammunition and filling our lorries with enough petrol for a satisfyingly long journey, we left Helwan. Two days later we were at Fuka, half-way between El Daba and Mersa Matruh.[2]

The new area was ideal. Headquarters and Workshops were on the beach—the drivers could jump straight out of bed into the Mediterranean—and the sections were parked near the main road. As it seemed likely that we should be there for some time we set to work to make ourselves comfortable. Those who had no lorries to live in built dugouts and each section established a wet canteen, Workshops contriving a charming little tavern at the end of the untidy straggle of dugouts and bivouacs that was, so to speak, our

[1] Cpl C. R. Turner, m.i.d.; storeman; Otahuhu; born Ti Tree Point, Hawke's Bay, 7 Jun 1900.
[2] Not all of us, though. C Section was attached to the 5th Brigade Group at this time and did not join us at Fuka until early in October, when the brigade moved to Baggush.

main street—our seafront. It looked out across the Mediterranean and in many ways it resembled those tiny taprooms that are to be found on the coasts of Devonshire and Dorset. Many were the pleasant evenings spent there by the Workshops' drivers and their visitors, and the rafters (it had rafters—yes, and settles too, cut in the sand) rang to many a good song. In units such as ours, sections, like people, come suddenly into social prominence and have their hour. The time at Fuka belonged to the Workshops' drivers. It was, you might say, their coming out. We realised all of a sudden their tremendous capacity for squeezing the last drop of enjoyment out of Army life. When there was work to be done they worked hard and to the business of enjoying themselves they brought the same keenness. Theirs were the merriest parties, the happiest homes, the liveliest adventures. At Fuka, for instance, they built themselves a raft—a crazy, delightful contraption with an old canvas bivouac for a sail—and from this they fished and bathed morning, noon, and night.

But it was not all holiday for Workshops—nor for the rest of us. We worked for the New Zealand brigades and we played our part in building up the desert supply dumps for the coming offensive. No one had told us, of course, that an offensive was in preparation, but we could read the signs: hundreds of square miles covered with stacks of ammunition, petrol, and food: camps springing up everywhere: convoys of guns and lorries. The desert railway was being pushed forward as well and we had a finger in that pie. For over seven weeks the drivers of No. 1 sub-section carted sleepers and rails for the New Zealand Railway Construction Companies, earning praise for their work.

The Luftwaffe and the *Regia Aeronautica* were not indifferent to all this activity but there was little they could do about it. The desert allows unlimited room for dispersal and the Royal Air Force of autumn 1941 was a very different proposition from that of autumn 1940. Where we had seen old Bombays and biplane Gladiators we now saw Hurricanes, Blenheims, Wellingtons, and Beaufighters, and the American Marylands, Tomahawks, Bostons, Brewsters, and Kittyhawks.

Our drivers saw little of the Luftwaffe except on one memorable occasion when an ammunition train in the Fuka railway yards was hit. Sand trickled from dugout walls and the ground quivered under us—it was as though the revolving globe had run a 'big end.'

It was our only excitement at Fuka and we made the most of it.

. . . .

The autumn days slipped past pleasantly and quietly and soon we were well into November. The desert was cold at night now and the Mediterranean was cooling too, though it was still speckled with bathers on sunny afternoons and Workshops' crazy raft still flopped up and down on it.

There had been very few changes. Major McGuire had left us for Headquarters Base NZASC and his place had been taken by Major P. E. Coutts[3] whom B Section had known as a lieutenant.[4] There had been a fuss in Company headquarters over some lost binoculars, our unit number had been changed from 24 to 69, and eight picked NCOs had been sent to OCTU.

What else had occurred? Nothing much—only a few slow processes. During the past six weeks our lorries had been run in completely and the same could be said of our unit. After the campaign in Greece new parts had been needed and old ones had required patching. Naturally there had been friction at first, but now the machine was running smoothly again and you had to have a good ear to hear a squeak or a groan. Our drivers of the 4th and 5th Reinforcements were no longer 'those new jokers': they were Dave and 'Old Baldy' and sometimes—not often—'that bastard what's-his-name'. With the unit, as with the lorries, most of the running in had been done at Fuka. A wink in the bar of the

---

[3] Maj P. E. Coutts, MBE, ED, m.i.d.; salesman; Auckland; born Auckland, 4 Dec 1903; OC 1 Amn Coy 4 Oct 1941-26 Jan 1943 and 2 Feb-6 Oct 1945; OC 18 Tk Tptr Coy Jan 1943-31 Mar 1944.

[4] The chief appointments on 9 November were: Company headquarters, Maj Coutts (posted 3 Oct 41), Capt S. A. Sampson (second-in-command), 2 Lt K. E. May (posted 28 Jun 41), Capt H. S. Jones (artillery officer attached), 2 Lt O. W. Hill (attached 24 Nov 41), WO II Dillon; A Section, Capt R. C. Gibson, 2 Lt J. M. Fitzgerald (posted 3 Sep 41); B Section, Capt D. C. Ward (posted 3 Sep 41), 2 Lt A. M. W. West-Watson (posted 18 Oct 41); C Section,* 2 Lt (T-Capt) F. G. Butt (posted 31 Mar 41), 2 Lt W. S. Duke (posted 1 Nov 41); Workshops, 2 Lt (T-Capt) A. G. Morris (posted 10 Sep 41).

The following had left us: Maj McGuire (posted to HQ Base NZASC, 3 Sep 41), Capt Moon (posted to Supply Column, 26 Aug 41), Capt Torbet (posted to Petrol Company, 25 Jun 41), Lt Aitken (posted to Base Training Depot, 28 Oct 41) and 2 Lt Toogood (posted to HQ NZASC, 3 Sep 41).

* Lt Fenton was detached from the unit at this time and on 14 November he was posted to the newly-formed 6th Reserve MT Company.

New Zealand Forces Club or a nod across the tables in the Sweet Melody, a shared grumble on a route march and then home again to your respective tents and friends—that got you somewhere. But sharing the same wave in the Mediterranean or a tin of oysters in the back of the same lorry, or driving for long hours in the same cab—that got you further.

We still cursed and grumbled, blackguarding our officers and any NCOs who were outstandingly conscientious, but we did it almost without malice and almost without meaning to. Of the squeaks of genuine distress, the hammering of round pegs in square holes, the chafing of loose parts, there remained only that irreducible minimum one finds even in hand-made machines and in hand-picked companies, and our lorries were certainly not the one nor was our unit the other. In short, we were run in. So something had happened—something quite important.

The autumn days slipped by and past our area moved new guns, new tanks, new troops, new lorries. We watched them and we knew the hour was at hand. Well, it had been a long time, but doubtless that time had been well spent. Evidently this new army they were talking about—and this new commander, General Auchinleck[5]—intended to make a bird of it. And a good thing too. Old Jerry certainly had it coming to him.

We watched the new tanks—Valentines, Crusaders, and American General Stuarts—going past in the golden dust.

'Valentines, eh? Two-pounder gun?'

'Valentines me ——. Matildas! Look at the armour.'

'Well anyway, lend us your bike, "Broady", to go down to "Cocky's".'

'Mind old Percy doesn't see you.'

. . . .

During the second week of November we were told to prepare our vehicles for large-scale desert manœuvres. 'Manœuvres me ——!' we said, echoing 'Broady' on the subject of Valentines. We knew better.

At 2.20 p.m., 13 November, we pulled on to the main road, heading west.

[5] General Sir Claude Auchinleck had succeeded General Wavell as C-in-C MEF on 5 July, and on 26 September the Eighth Army, which consisted of two main groups (30th Corps and 13th Corps), had been formed under Lieutenant-General Sir Alan Cunningham. The New Zealand Division had come under the command of 13th Corps a fortnight earlier.

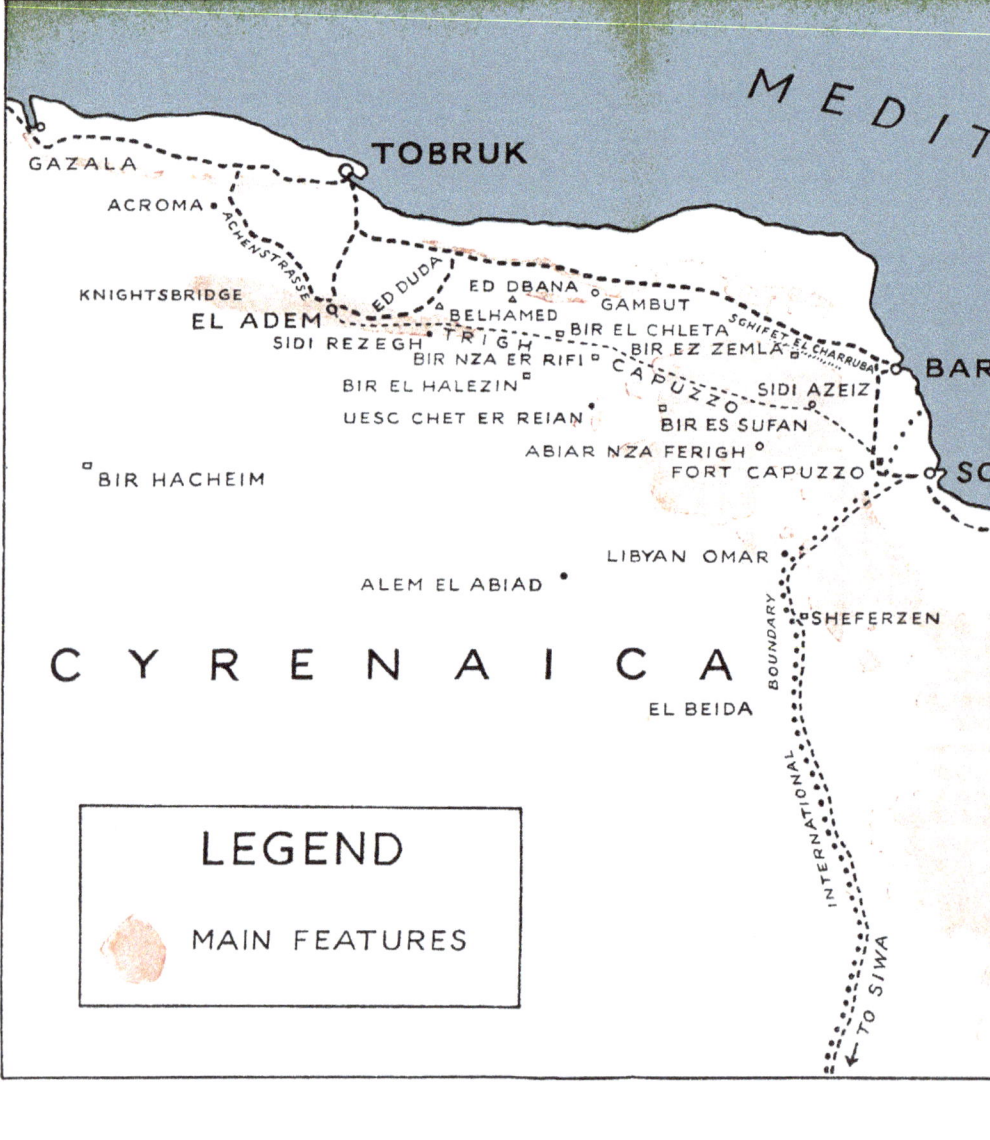

CHAPTER 9

FOX IN THE FOWL RUN

WHINING in low gear, the stream of vehicles headed for the setting sun. In the Ministries and offices the big men had done their jobs and now it was our turn—the turn of the small men. Only through our goodwill would the guns shoot, the tanks advance, the wheels go round. The columns of figures had crawled out of their pigeon-holes, the acres of maps had become solid desert, and under the uncaring sun, which was levelled on the advancing transport like a spotlight, glinting on windscreens, causing the drivers to blink and curse, the drama began to unfold.

Campaigns go well or badly; they seldom go smoothly. Already hitches had occurred. We had moved on to the road over two hours ago, and now, at 4 p.m., we were still only a few miles from where we started. Our carefully-nursed dispersal had not survived the assaults of an ambulance unit and a heavy anti-aircraft regiment, and it seemed as though the convoy would be split up hopelessly before it reached the starting point for the advance.

It was dark when we turned off the coast road to the Siwa road, and we drove without lights, following the white wraith of dust thrown up by the vehicle ahead. Every once in a while a mudguard folded up with a crunch. Finally the convoy turned into the desert, lurched another four miles over very rough ground, and halted until dawn.

A little uneasy sleep, and then, with heavy heads and stiff, cold limbs, we climbed back into the cabs and bumped off towards the west, while the chill morning, white and pale grey with a fleck of yellow in it, opened behind us like an oyster. Now we could see what the Ammunition Company looked like in desert formation. As far as the eye could see the sand was chequered with our lorries, each of which—in theory, anyway—was the moving but constant centre of 160,000 square yards of empty desert. Crawling through the colourless morning light, they resembled a regiment of monstrous snails, humped, drab, uniform. They butted their way up the sides of wadis, crunched over hard ground, ploughing fluffily through soft patches—infinitely strange and menacing and never more so than when they entered a great open expanse, flat and boundless, and moved across it with a dry rustling sound, a wisp of dust smoking behind each lorry while an angry looking sun levered itself over the horizon.

The area in which we were scheduled to spend the night of 14-15 November was reached late in the morning. We were then about thirty miles from the coast and still quite close to the Siwa road. During the afternoon our already large convoy—the 14th Light Anti-Aircraft Section, NZASC, had come under the Major's command the day before—was further augmented by the temporary attachment of the New Zealand Mobile Surgical Unit.

Throughout the 15th we travelled slowly but steadily due west, halting at dusk after travelling forty-five miles. The last three stages of the journey to the frontier were to be made at night, so we rested during the 16th, moving off at dusk in close formation. When we halted at one in the morning we had travelled twenty-five miles and were thirty-two miles due south of Buqbuq. We rested again during the 17th.

So far everything had gone fairly smoothly, but early on the night of 17-18 November the formation struck soft sand, some thirty lorries sinking axle-deep. Vehicles without four-wheel drive —some of the sections had a few Mapleleaf Chevrolets—became hopelessly bogged and this meant hard work for our breakdown lorries. Where the ground was not soft it was villainously rough and it was often necessary for the formation to narrow its front to avoid wadis or minefields. The presence of the latter was indicated by softly-glowing lanterns, but several of the drivers, confused by the darkness and knowing they were out of position, had moments of panic when they wondered if they were on the right side of the line. It was an eerie night. In ragged columns the vehicles plunged and stumbled through the desolation of stone and camel-thorn, and the heat and smell of them—you got a blast of it when a dozen or more swept crazily together to avoid an obstacle—alternated with a whisper of cold wind, forbidding as the brush of skirts when the coven gathers for its sabbat. West, in the direction of the frontier, sheets of lightning blinked and played along the horizon. North, south, and east, we were hedged in by the deep, mocking darkness.

We crossed the Egyptian frontier at dawn on the 19th, halting near El Beida, thirty-three miles south-west of Sollum. Ahead of us was the barbed-wire entanglement—never known as anything but the Wire—erected by Mussolini in 1932 to prevent the Libyan bedouin from leaving Libya and from coming back again should they manage to leave. To some of us it was a familiar landmark but others were seeing it for the first time. Six feet high and nine feet wide, it rolled away to the horizon like an enormous, hairy, reddish-brown caterpillar. It was an evil-looking thing and it had done evil. It ran from Sollum to the Great Sand Sea below Jerabub and Siwa, and to a people who in times of drought had been accustomed to move freely between Cyrenaica and Egypt in search of grazing it had meant something next door to starvation.

That night we enjoyed a sound sleep, and we spent the 20th in doing the hundred and one jobs—everything from washing a pair of socks to adjusting a noisy tappet—that accumulate in the course of a long journey.

On the following afternoon we moved through the 300-yard breach in the wire to occupy an area eleven miles north-west of El Beida and there we spent another quiet night. Many of us had begun to ask when the fun was going to start.

. . . .

It had started already.

Already advanced elements of 30th Corps, after taking Rommel by surprise and clashing with his armour, were in sight of the Sidi Rezegh escarpment, which overlooked the force investing Tobruk. Already the garrison had begun its sortie. The 13th Corps, for its part, was striking north to isolate large enemy forces in the Sollum-Bardia area.

The next morning—the 22nd—we set out for Abiar Nza Ferigh, eight miles south-west of Sidi Azeiz (itself twelve miles south-west of Bardia), which New Zealanders had captured the evening before.

After we had travelled steadily for an hour and a half the Major was told that his lorries were passing between the frontier fortress of Libyan Omar and troops of the 7th Indian Brigade who were about to attack it. If we didn't want to get shelled, he was told, we had better swing left at once. Our drivers could see the Indian infantry sitting silently in open lorries. Almost immediately, with a whine and a clap, shells began to land around us. In places, notably on our right flank, clouds of dust obscured what was going on, and new clouds were springing up every moment. It was as though the desert were being beaten like an old carpet. By this time our formation had wheeled and was heading northwest. A halt was called as soon as we were out of range and it was found that no damage had been done. Nevertheless there had been several narrow escapes and nerves were still tingling when two armoured cars came towards us across the desert. They were flying a single pennant and we had been told that our own armoured fighting vehicles would be flying two.

All was well, however, and we reached our new area, where we made haste to dig in, early in the afternoon. The New Zealand Division was now in action against Bardia and the frontier forts in the coastal sector, and our ammunition began to sell like hot cakes.

. . . .

Meanwhile the fighting around Sidi Rezegh had reached a critical stage, and the evening of the 22nd found the 6th Brigade, under the command of 30th Corps, hurrying west to lend a hand. The next day the 4th Brigade moved against Gambut, an airfield halfway between Bardia and Tobruk, the 5th Brigade staying behind to contain the enemy in the Sollum-Capuzzo-Bardia triangle. With the splitting up of the Division problems of supply at once became three times as difficult, and a Composite NZASC Company was formed under Captain L. W. Roberts (Supply Column) to serve the 6th Brigade. To this we contributed C Section under Captain Butt.

At nine on the morning of the 23rd a convoy from the new company, composed of C Section (ammunition), two sections from the Supply Column (rations), and six lorries from the 4th Reserve Mechanical Transport Company (water), left Abiar Nza Ferigh under Captain Roberts to join Headquarters 6th Brigade at Bir el Chleta, six miles south-west of Gambut. After travelling west for seven miles the convoy was met by a 6th Brigade officer who reported that enemy tanks were on Trigh Capuzzo, a rough but extremely important track running from Capuzzo to El Adem, a large airfield sixteen miles due south of Tobruk. Bir el Chleta was just north of the Trigh. Accordingly Captain Roberts decided to head west for another twelves miles before turning north.

In the usual desert formation, which by now our drivers had become adept in maintaining, the convoy pushed on, encountering here and there bad patches of soft sand in which a number of lorries gave trouble, halting all the rest. The second time this happened C Section was joined by four Royal Army Service Corps Bedfords whose excited drivers told how their convoy had been attacked and dispersed on Trigh Capuzzo.

They were in the middle of their story when five enemy armoured fighting vehicles were sighted. Some of the lorries were still bogged, so the drivers were ordered to form the transport into a defensive square and defend it with their rifles.

Fortunately the enemy sheered off without firing, and Captain Butt, grabbing a shovel, at once organised the spare drivers into a digging party. Soon every lorry was on firm ground again.

At two in the afternoon, when the convoy was due south of Gambut, Captain Roberts swung north, leading his transport through what had very recently been an Italian camp. Wine bottles and empty tins together with more offensive refuse lay thick about trenches and weapon-pits, and in one place a fire was still smouldering. On the left flank, about a mile away, unidentified armour could be seen moving west.

An hour and a half later three enemy armoured cars bore down on the convoy from the east, attempting to cut it off. The Trigh was now only five miles ahead, so Captain Roberts decided to run for it. As soon as the drivers saw the *speed-it-up* signal the transport surged forward, rocking and bumping over the rough desert. The manœuvre was successful and the armoured cars dropped away on the right flank. Just in front of Bir Nza er Rifi, which was a mile south of the Trigh and in line with Gambut, the convoy halted in deference to a mass of transport that covered the southern slope of a wadi about a mile ahead. It was taking a lot of punishment. Shells were landing among it with good effect and our C Section drivers watched from their ringside seats with growing excitement. They had decided the transport was friendly.

Meanwhile Captain Butt had gone forward in his staff car to put the matter to the test. The moment it reached the floor of the wadi the car came under machine-gun fire, and as it whipped round and came racing back a string of bullets stitched a seam in the desert a few feet from it. The position was now plain: the transport was the enemy's and the shells were from our own 25-pounders. There was no hope of crossing the Trigh and reaching Bir el Chleta.

A second later our drivers lost all interest in what was going on in the wadi. Now they were under fire themselves. They leaped for their cabs and with a throaty scream of engines the whole convoy turned and fled south while three armoured cars fired on it from one direction, two machine guns from another, and two small-calibre guns (probably anti-tank guns) from a third. The going was abominable and there were slit-trenches everywhere. Into one of these a Bedford—one of the four that had joined the convoy earlier—plunged headlong, having to be abandoned.

The pursuit was not given up until more than five miles had been covered and all the while a heavy concentration of unidentified transport had been moving north-west on the right flank.

After travelling south for eight miles, Captain Roberts turned east at Uesc Chet er Reian (through which the convoy had already passed once), carried on for another nine miles, and halted for the night at Bir es Sufan, the transport forming up in a laager just as the covered wagons did in Red Indian country.

Bir es Sufan, no doubt, meant something to Arab nomads. One could imagine—though not without an effort—their comparing it with Nza er Rifi and Uesc Chet er Reian and confessing a nostalgic preference for the former. To our drivers it was nothing—just a stretch of desert indistinguishable from any other stretch. To Captains Butt and Roberts it was a six-figure map reference (475392). Even the cistern, whose presence was suggested by the prefix *Bir*, was invisible to the uninitiated.

Some of the drivers had eaten and bedded down and others were still busy over their primuses, with canopies close drawn to prevent any light from showing, when a distant squeaking and rumbling, as of mice and volcanoes mixed, forced itself on the attention. Presently its import became unmistakable and Captain Butt sped from lorry to lorry. He said: 'Stay where you are. No noise, No lights. No smoking.'

From the monstrous racket individual sounds began to sort

themselves out: the clank of tracks, the clatter of exhausts, the chafing of metal against metal, the unruly hammering of a diesel engine. When the crescendo reached its peak our drivers could see black shapes moving past in close formation. They waited in dead silence, hardly breathing. After the column had gone by white flares of a type the Germans were using at that time shot up at intervals along its whole length. Gradually the noise died away in the distance and at last it was only a faint squeaking.

Feeling that they had experienced quite enough for one day the drivers went to bed.

. . . .

Only a few miles from where C Section's lorries were laagered but separated from them by Heaven alone knew how many wandering groups of Germans, New Zealanders, Italians, and South Africans—for by now the confusion on both sides was considerable —the Major was recording the day's events.

November 23. 0100 hours: 14 Lt A-A Section[1] ceases to be a separate entity and is marched in as an additional section of Ammunition Company. 0600: 26 lorries left for 50 FSD under 2-Lt May to load ammunition. 0900: Reported to Col Crump and was instructed to move to Uesc Chet er Reian and form ammunition point. Point to be operating at 1400 hours. 1100: Advised by HQ that enemy tanks were in the new area and unit would not move until information had been received that area was clear. 1400: Instructions received for unit to move. 1500: Convoy under 2-Lt May returns from 50 FSD. 1600: Unit moves off. After travelling seven miles convoy makes contact with rear of Divisional Headquarters. It appears to be held up while waiting for the area ahead to be cleared of enemy tanks. 1700: Unit halts and prepares evening meal. At dusk Divisional Headquarters moves on but unit stays in present area for the night. No word from C Section which left supply column lines under Capt Roberts at 0930 hours. . . .

Everyone from the Major down—from General Auchinleck down —was worried. There was reason to be. The Germans had retaken Sidi Rezegh, the sortie from Tobruk was pinned down, and worst of all the spring had gone out of our attack. It had been hammered out and for the moment we were on the defensive.

The tired sentries paced up and down beside the closly-packed

[1] 123 ORs commanded by Capt G. Fordyce.

lorries, starting nervously at distant flares, listening intently to every movement in the darkness.

. . . .

On the morning of 24 November Rommel felt the ball in his hands and he decided to make a touch-down. Collecting three armoured divisions—two German and one Italian—he made a run for the Wire. Tanks of the 7th Armoured Brigade, like plucky half-backs, went in to tackle him, but he fended them off, and in the afternoon the 21st Panzer Division under General von Ravenstein reached the frontier at Sheferzen and crossed into Egypt. Then it went merrily to work. It fanned out in columns and began shooting up transport and overrunning supply dumps, field workshops, and B Echelon areas.

Supplying the New Zealand Division with ammunition, petrol, food, and water had been difficult before; now it was a nightmare task. The 50th Field Supply Depot, from which we had been drawing ammunition, was near El Beida, on the Egyptian side of the Wire. The 5th Brigade, with its headquarters at Sidi Azeiz and large forces of the enemy all about it, was still investing the Sollum-Capuzzo-Bardia triangle. The 6th Brigade, with no safe lines of communication, was fighting in the Sidi Rezegh area, and the 4th Brigade was at Gambut. Bits and pieces of the enemy were anywhere and everywhere.

For transport units there was no safety. Every convoy was like a fleet of fishing smacks in a hostile ocean. It was now to be discovered if our drivers could really drive—if our section officers and subalterns could really read maps and lead men.

. . . .

The 24th dawned cold and bright, a pleasantly astringent Monday morning. Taking their cue from the weather, the drivers woke in good heart. Already, no doubt, steps had been taken to avoid a recurrence of the unpleasantness of yesterday.

Renewed optimism was in the air and 'Fat' Davison responded to the general feeling by giving his drivers a breakfast of fried sausages—a rare treat after a succession of bully-beef stews eked out with army biscuits.

They queued up in the chill silver and gold sunlight, avoiding the long spikes of shadow cast by the lorries and chatting happily

Waiting for nightfall, Crete

The Prime Minister inspects Bedfords at Maadi

Leave in Tel Aviv

Ration trucks at a supply depot

about the fright the armoured column had given them the night before.

'I was next in the queue,' said George Laverick,[2] 'and I was holding out my dixie and just beginning to lick my chops when two South African vehicles—a pick-up and a three-tonner—came tearing over towards us. There were about half a dozen South Africans aboard and they were in a bad way. Their clothes were in shreds and some of them were covered with blood. We particularly noticed the negro driver of one of the vehicles. His eyes were rolling in his head with fright and excitement. An officer, tattered and bloody, told Captain Butt there were enemy tanks in the neighbourhood. He said his own outfit had been cut to pieces.[3]

'A column of some sort was moving towards us from the northwest, so the cooks packed up at once and I didn't get my snarler. A few minutes later we were on the move, heading south and then east towards Abiar Nza Ferigh, which we had left the morning before.'

The rest of the unit had also started the day on an empty stomach. We were roused before dawn and the Major led us westwards across the desert, Captain H. S. Jones[4] having been sent ahead to make contact with Colonel Crump on Trigh Capuzzo. He returned some two hours later to report that he had been fired on as he was approaching what he thought was Divisional Headquarters. Soon afterwards British tank officers who were enquiring for the 6th Brigade warned the Major that enemy armour was in the neighbourhood. No one seemed to know what was what or where anything was.

In real fact the situation, at least as it affected the New Zealanders, was beginning to clear a little. The 4th Brigade was at Gambut (where advanced troops of 13th Corps had been held up for some while), and it was waiting for the word to move west and link up with the 6th Brigade, which after a hard battle was in possession of Point 175, four and a half miles east of Sidi Rezegh.

[2] Dvr G. P. Laverick; truck driver; Netherton, Paeroa; born Rangitane, Palmerston North, 14 Dec 1916; p.w. Bardia 27 Nov 1941-2 Jan 1942.
[3] On the 23rd the 5th South African Brigade had been overrun south of Sidi Rezegh and had ceased to be effective.
[4] Lt-Col H. S. Jones; company manager; New Plymouth; born Carmarthen, Wales, 21 Nov 1896.

I

The brigades linked up later in the day in readiness for the final battle for Tobruk. Facing them was the hard knot of Sidi Rezegh, Ed Duda, and Belhamed.[5]

At half past ten a Divisional military policeman told us to go to Bir el Halezin, seven miles south-west of Bir el Chleta. We moved steadily in that direction until noon when we were halted by an action between British and enemy tanks. While we were waiting for the situation to clear, Captain Ward[6] and B Section, accompanied by a Petrol Company section that had joined us early that morning after being chased by tanks, set out to serve the 6th Brigade, and twenty minutes later Captain Jones returned from Divisional Headquarters with a report that the 4th Brigade was short of ammunition. He brought with him the brigade liaison officer, who had been fired on by our own tanks, his batman sustaining severe wounds. Captain Gibson[7] and A Section left at once for the 4th Brigade and Sergeant Sam Mellows and six lorries from the former 14th Light Anti-Aircraft Section, NZASC, left about half an hour later with Bofors ammunition for the 41st Battery.

Watching from Company headquarters' area, Corporal Bev Hendrey[8] (Don R) followed the six lorries as they moved across the desert. He saw them join a fairly large formation of transport —they were just dots by now—and continue south with it. Then there was a sound of gunfire from the north and a lorry near the tail of the formation stopped and began to burn. None of those six lorries was seen again.

Later in the afternoon the Major was asked to move his transport as it was parked between German tanks and the guns of the 22nd Armoured Brigade. We moved a mile south and settled down for the night.

The light faded from apple-green to blue-grey and all over the desert groups of transport huddled close in the prescribed formation ——Custer's Last Stand. A Section's lorries were protected by the

---

[5] Gambut had fallen to the 4th Brigade the day before. Hence the shells that C Section drivers had seen landing among German transport.
[6] Capt D. C. Ward; motor driver; born NZ, 24 Apr 1905.
[7] Maj R. C. Gibson; woodwork instructor, Auckland Education Board; Auckland; born Auckland, 20 Feb 1909; OC 1 Amn Coy 17 Apr-21 Sep 1944.
[8] 2 Lt C. B. P. Hendrey; truck driver; Auckland; born Auckland, 15 Nov 1914.

4th Brigade, B Section's by the 6th, and C Section's, at Abiar Nza Ferigh, by the 5th. Sam Mellows's detachment, with the exception of Arty Meiklejohn[9] and Steve Kennedy[10] who were lying beside their burnt-out lorry almost within sight of Company headquarters, was under the protection of German panzers. The rest of the unit was near Bir el Halezin with the 22nd Armoured Brigade.

Again the sentries walked up and down beside the close-packed lorries. Nothing very alarming had occurred during the day and they spoke sagely and cheerfully of final flings, mopping-up operations, and isolated pockets of resistance—and their wireless sets did the same. True there were some disturbing stories about. Packs of panzers, it seemed, were roaming the desert and chivying and savaging unprotected convoys. And there was that business about the tanks. None of ours, a bewildered tankie had explained, fired anything heavier than a two-pound shell, whereas the lightest shell fired by the German tanks was $4\frac{1}{2}$ pounds and the heaviest 14-15 pounds.

We consoled ourselves by thinking of their horrible pink petrol. A British tank wouldn't look at it.

The blue light faded to black and later the moon rose. The burnt-out stubs of tanks and trucks stood out dark against the desert, the surface of which seemed to have been strewn with largesse—great silver pieces. They were only petrol tins, though, and empty German and Italian and British food tins.

. . . .

With eighteen men and one corpse in the back of the lorry Sam Mellows's drivers found it impossible to stretch out in comfort. They dozed for a while, then woke up aching with cold and cramps, then dozed again. Before dawn they were roused by a rumble of tanks and a confused shouting from the Germans. They shook hands in the darkness, feeling suddenly certain that the enemy was surrounded and they would soon be free.

'It started the afternoon before,' said Corporal S. T. Midgley[11] (known to everyone as 'Midge'). 'The six lorries of Sam Mel-

[9] Dvr A. B. Meiklejohn; orchardist; born NZ, 29 Oct 1918; wounded 24 Nov 1941; killed in action, 28 Nov 1941.
[10] Dvr. O. J. Kennedy; builder's labourer; Tapanui, Otago; born Tapanui, 24 Aug 1911.
[11] Sgt S. T. Midgley; traveller; Kaiapoi; born Dunedin, 12 Jun 1907.

lows's sub-section left the unit area with myself and twelve others about half an hour after Captain Gibson's convoy. We were to serve the 41st Battery, which was under the command of the 4th Brigade, and our orders were to travel four miles west and then about six north. We set out in desert formation with an interval of 150 yards between vehicles.

'After we had gone about three miles we noticed a convoy in the far distance moving towards us on our right front. Sam Mellows thought it was most probably Captain Gibson returning with A Section after failing to get through to the 4th Brigade, so he halted his vehicles. Shortly afterwards we saw six armoured cars swing across our front and drive swiftly down our left flank, keeping about half a mile out in the desert. We could now see that the formation ahead of us was an armoured column and our spirits sank when we noticed that it was accompanied by motor-cycles with side-cars.

'The six armoured cars swung left, each of them heading towards one of our lorries. Sam Mellows shouted to the drivers to stand firm. The car that had singled us out halted fifty yards away and an officer said in perfect English: "Hands up! Have you any arms?" I said: "Rifles only." The officer said: "Drive to the rear of the column." A German soldier got in beside our driver and Sam Mellows and I climbed into the back of the lorry, two guards going with us.

'Meanwhile this scene was being repeated at each of the lorries and in a matter of minutes the convoy was heading towards the German column, which had never halted, the armoured cars following behind with their machine guns trained on us. While we were catching up the column we passed within a mile or so of the company area and I could see our boys and their vehicles.

'After we had travelled a short distance we came under heavy fire from British 25-pounders and Jerry set off at full speed, altering direction from time to time. Shells were landing all round us and we heard afterwards that the lorry in which Arty Meiklejohn and Steve Kennedy were driving received a direct hit, but we could get no details. The shelling continued for about half an hour and all this time the column seemed to be travelling more or less in circles.

'At dusk we halted. After dark the column moved five miles to confuse the Royal Air Force, which had had us under observation all day long, and laagered in tight formation. It was now possible to make a rough estimate of Jerry's strength. The column appeared to consist of about sixty vehicles—load-carriers, staff cars, armoured cars, and tanks. There were plenty of motor-cycles with side-cars mounted with machine guns.

'That night we were told to unload one of the captured lorries and we were all herded into it. We were joined by six Tommies who had been captured earlier in the day and that made eighteen of us. It was pretty close quarters.

'We didn't get a feed that night. In fact the Jerries gave us nothing all the time we were with them. I don't know how we should have managed if the boys hadn't thought to grab a few tins of food while they were travelling in the backs of their lorries. Our Jerry guard—he stayed with us always and he used to sit on the camel tank, which by a mercy was full of water—was pretty generous in sharing his own rations. He was a good little joker. He came from Munich.

'An officer, an elderly man who visited us before we turned in, was quite decent too. After telling us to make ourselves as comfortable as possible he stayed for a little chat, mentioning that he knew England well and liked it. He said he used to referee soccer matches at Wembley stadium.

'As a matter of fact most of the Jerries we met treated us well. On the whole we found them a pretty free and easy crowd, much like ourselves, and there was only one incident that really got our goat. That had occurred during the afternoon. Sam and I, while we were in the back of our lorry, dug out a carton of 200 cigarettes, from which we gave the two guards a couple of packets each. They must have mentioned this outside, for during one of the halts an officer came up to us—he was a typical Prussian specimen: lean, with a cold, hard eye and a clipped, abrupt way of speaking—and demanded some of the 2000 cigarettes he understood we possessed. When we told him we had only a few packets he flew into a rage and said he could make thing very unpleasant for us if we didn't obey orders. After a bit of back-chat we gave him three or four packets and he went away muttering under his breath.

'With eighteen jokers in the back of the lorry it was hell trying to get to sleep. To make matters worse we had a most unpleasant addition to our number early in the night. An officer poked his head in and said he was going to give us a "dead comrade". Naturally we thought he meant either Arty Meiklejohn or Steve Kennedy, but what we got was a dead German with a sack over his head. No one felt like grabbing hold of him and there was a bit of hanging back before he was hoisted aboard. No one wanted him as a neighbour either and he gradually got edged forward to the front of the lorry, ending up wedged between me and the spare wheel. He was like frozen mutton.

'Before daylight we saw flares in the sky, and the Jerries started shouting and we heard tanks. We were highly delighted, believing the laager was surrounded by our own chaps. Nothing happened, though, and soon it was light enough for us to see that we had been joined by a huge enemy force—a collection of armour and lorried infantry. Rommel was there as well we heard later. We looked through peepholes we had made in the canopy and saw vehicles of all types stretching out in every direction. It was a bitter sight.

'We moved at sunrise—having first got rid of the dead Jerry, thank goodness!—with the bulk of the tanks travelling in front and a screen of armoured cars and tanks on the flanks and scout cars away out on the horizon. Royal Air Force reconnaissance planes were almost continually overhead throughout the morning and we knew that before long we should be getting it good and heavy. . . .'

By this time the fate of the six lorries was no longer a mystery. At breakfast-time Steve Kennedy had reached Company headquarters dazed with exhaustion. He said that his lorry had been destroyed and Arty Meiklejohn was lying out in the desert seriously wounded. A rescue party was formed under Captain Sampson, with Steve, although he was almost at the end of his tether, acting as guide. He told his story as they went along.

When the German column came under fire from our 25-pounders Steve's and Arty's lorry received a direct hit, both drivers being blown out of the cab. Arty's leg was almost severed by a piece of shrapnel but Steve escaped with a bad shaking. The lorry started to burn fiercely.

A German soldier pointed his revolver at Steve and told him to leave Arty and climb aboard the next lorry. Steve refused to do this and the German fired, missing him and wounding Arty in the shoulder. Ashamed of himself, or afraid of being left behind, he went away with the rest, and soon the convoy was out of sight. Our drivers stayed by their burning lorry and after a while a British tank came over to them. The crew applied a tourniquet to Arty's leg and then left, promising to send an ambulance. No ambulance arrived and they spent the night in the open. At daybreak, realising that Arty would die if no help came, Steve made his way to our lines.

After a fairly long search—Steve could give only vague directions—the burnt lorry was sighted with Arty lying beside it. He was in bad shape by now, and for the time being it was impossible to evacuate him to a field dressing station. (He was evacuated the next day and was killed on the day after that—the 28th—when New Zealand medical units were overrun between Bir el Chleta and Point 175.)

The rest of the morning passed quietly. A Section returned from serving the 4th Brigade and B Section from serving the 6th.

Early in the afternoon orders arrived for the 65th RASC Company, which had reached our lines before breakfast with the now familiar story of a chase by tanks. Their ammunition was to go to the 4th Field Regiment, their petrol and water to the supply point. It was as well that something for the guns to shoot had come forward, for C Section was now serving the 5th Brigade in the Bardia area, and until we received word that the way was open to the new replenishment point—El Beida was closed to us—nothing could be done about refilling A and B Section's lorries.

Word came at two in the afternoon and the sections set out under Captains Gibson and Ward to load ammunition at the 62nd Field Supply Depot, which was some distance from Alem el Abiad and over thirty miles south by west of Bir el Halezin.

After the convoy had travelled a dozen miles it was halted by Captain Gibson who had noticed a suspicious movement ahead. Some A Section drivers took advantage of the halt to salvage a perfectly good staff car that was sitting unattended in the desert. It was undamaged except for a few bullet holes, and a short tow started the engine. In the meantime two armoured cars had moved to a ridge on the left flank.

As soon as the convoy moved off they pounced—like a cat with a mouse. A bullet nicked Colin Cameron[12] in the thigh and a stream of bullets followed the newly-acquired staff car as it shot to the front, its driver feeling no gratification at being mistaken for an important target. Through the crackle of machine guns and the roar of engines came the heavy knock, five times repeated, of Jim Stanley's anti-tank rifle, the one he had used in Greece. He was lying on his stomach in the back of A Section's ack-ack lorry and firing over the tailboard. Before the chase was given up he managed to get away the best part of three magazines, but by that time the convoy had been driven far off its course. There was nothing for it but to return to the unit lines and report that the Germans were watching the route to the supply depot.

Our drivers got home in time for tea. The steaming bully stew and soggy rice, reassuringly familiar against the rather macabre atmosphere of the campaign, went down well after the day's excitements, a pleasant consciousness of dangers overcome adding sauce to the cooks' efforts. A and B Sections were not sorry to have had an adventure of their own, for C Section, according to all reports, had been stealing the limelight.

. . . . .

During the 24th—the day of Rommel's dash to the frontier—the Composite NZASC Company had been disbanded, C Section remaining at Abiar Nza Ferigh to serve the 5th Brigade in the Bardia area; and by the morning of the 25th seventeen lorries had been emptied. Their drivers breakfasted at dawn and set off under Captain Butt to refill at the 50th Field Supply Depot on the far side of the Wire.

After travelling south by west for eighteen miles the convoy came under fire from artillery on the right flank. Captain Butt turned it about and headed for a tank recovery section he had passed four miles back. He made his report to the officer commanding the unit but it was received sceptically and his request for an armoured escort refused. The tanks in the area, he was told, were under repair and none of them was really fit for action.

In an attempt to by-pass the danger zone he led the convoy five miles west and then turned south, only to be halted after travelling

[12] Dvr C. Cameron; butcher; born NZ, 23 Jul 1919; wounded 25 Nov 1941; killed in action (drowned) 5 Dec 1941.

a further two miles by the sight of nine or ten armoured cars on a low rise. With the early sun behind them they were dipped in pools of their own shadow and it was impossible to tell whose they were. Our drivers were inclined to think they were South African. Captain Butt, however, was taking no chances. He halted the convoy and went ahead to reconnoitre.

'I was in the staff car with Freddy Butt and his driver,' said Sergeant Bob Aro.[13] 'Freddy put the glasses on the armoured cars but what with the shadows and the haze they weren't any help. We crept nearer and stopped. We did that four times, halting finally when we were about 150 yards away. In the leading armoured car there was a joker with a black beret like our tankies wear. He was standing on the turret and waving us to come on. Then we saw someone hop down into a weapon-pit beside the car. We all saw him at the same time and Freddy yelled to his driver: "Go for your life, Jack!" Jack Girvan[14] spun the car round and as we turned I saw the joker on the turret drop behind his gun and a second later a burst of bullets ploughed up the sand alongside us. It was Jack who saved us. He did a marvellous job of driving, going flat out and flinging the car around to make it hard for the Jerries to draw a bead on us.'

Our drivers saw the staff car come tearing towards them in a spray of bullets and the lorries turned as one. As they did so five more armoured cars appeared on the right flank, and for a minute it looked as though the game was up. Bullets went between the lorries and over and under and through them. At first the convoy drew ahead, but then it struck soft sand and the armoured cars started to gain. One lorry was hit in a vital part and it clattered to a standstill, the drivers tumbling out. The enemy kept up the pursuit mile after mile, and all the time Captain Butt's staff car was like a sheepdog. Now it was in the lead indicating the course; now it was on a flank watching over a straggler. The C Section drivers were beginning to swear by him. Finally he led them back to the recovery section and only then did the armoured cars give up the chase.

[13] WO II R. G. Aro, MM; fitter and turner; Auckland; born NZ, 9 Feb 1914.
[14] L-Cpl J. F. Girvan; tractor driver; Seafield, Ashburton; born NZ, 1 Sep 1917.

Again the officer in command was asked for the loan of a light tank and this time the request was granted. However, just as the convoy was setting off, three armoured cars appeared 1300 yards north of the position. They prowled about, nosed this way and that, and then sat back on their haunches for all the world like dingoes; and like dingoes they slipped away into the haze when four I-tanks rumbled towards them. The incident, unfortunately, caused the officer commanding to change his mind. He decided that he must keep everything he had for his own protection, so Captain Butt had no alternative but to leave the convoy where it was and try to get through to Headquarters NZASC with a report.

During the next two hours the drivers amused themselves by exploring the recovery section. It consisted of a mass of vehicles —tank-transporters, lorries, and light trucks—and at the moment it was engaged in repairing British I-tanks that had come to grief in the fighting for the Omar forts, the nearest of which, Libyan Omar, was about six miles to the east. There were several light tanks in various stages of disrepair and about eight heavy ones. A number of these could be used as artillery, so the position was by no means undefended. An opportunity to defend it came early in the afternoon.

A cloud of dust appeared from the direction of Libyan Omar and rapidly grew larger. 'What can you see?' asked Tom Laverick[15] of a British sergeant, who replied: 'I can't see a lovely thing for lovely dust'. The sergeant, except that he was standing in a tank turret and was looking through field-glasses, was like Sister Anne. Even when the cloud was quite close it was impossible to see what caused it, but an ominous squeaking and rumbling told its own story. 'See anything now?' asked Tom. 'Lovely Jerries,' said the sergeant, closing the turret. A second later his gun went into action.

There was an instant—a moment suspended in time—in which the dust cloud was stabbed with orange flames and a voice could be heard shouting 'Take post!' and the Tommies could be seen manning their ruined tanks, calm through despair or through long discipline. Then the shells came over with a squeal and a short rush and the dust hid everything. On Bob Aro's orders our drivers

[15] Cpl T. D. Laverick; factory assistant; Paeroa; born Rangitane, 19 Oct 1914; wounded 27 Nov 1941.

had started their engines at the first hint of trouble and they were under way as soon as the firing started. They zigzagged among the slit-trenches while shells burst ahead of them. Twelve lorries headed west, four headed west by south, and smoke and dust separated the two groups.

Bob Aro was leading the larger group, and as soon as it was safe to do so he swung north and went straight to the Supply Column lines near Abiar Nza Ferigh—no mean feat without a compass. There he reported to an officer, but his story was laughed at, so he pushed on in the hope of finding either Captain Butt or the rest of the section. Presently he met Major Pryde, who put him on his way to the ammunition point at which the loaded lorries were still standing by. It was where he had left it that morning. Ten minutes later a staff car drove up with a short, stocky man, a tommy gun in his arms, leaning through the trapdoor in the roof. ' Freddy,' said the drivers, their hearts lightening.

Captain Butt had managed to get through to Headquarters NZASC and had been instructed to put himself and his section under the command of Captain Roberts, the Composite NZASC Company having been re-formed to serve the 5th Brigade. He had set out for the recovery section to collect the convoy and while still some distance away had seen fires break out in the area. He had closed to 1800 yards and then to 1300. By that time at least six vehicles were in flames and he was able to make out, in and about the camp and round the British I-tanks, at least fifty strange tanks and armoured cars and about 200 load-carriers. He stayed no longer.

Except for the party that had dashed off on its own—ten drivers in four lorries—C Section was complete again. Without further delay it joined the Composite Company and moved seven or eight miles north-east into the Sidi Azeiz position, where Headquarters 5th Brigade was entrenched. The company was still there when night came.

. . . .

Night came with a quickening chill, a dropping of green and pink and blue veils and a rush of shadows.

At twenty minutes to eight, after moving four miles north with the 22nd Armoured Brigade, our unit halted and made camp. The German column halted after moving five or six miles.

'It was quite dark by then,' said 'Midge'. 'The boys were in good spirits, though some of them had begun to feel a bit knocked up. We were stiff and sore and what we needed above everything else was a smoke. That was out of the question of course. A glimmer of a light would probably have been a shooting matter. The Jerries, I think, were getting a bit rattled, but you couldn't blame them for that. We'd had a taste of the Royal Air Force ourselves when thirty Blenheims came over in the afternoon. The concentration of tanks and lorries made a good target—the Jerries didn't seem to go in for dispersal on the scale we did—and columns of black smoke showed that some hits had been scored.

'Later in the afternoon, while halted, we were shelled by British 25-pounders. Several vehicles were hit and one shell landed only thirty yards away from us, spraying our lorry with stones and gravel. We were not allowed to get out and take cover and we didn't feel any too good sitting up there like Jacky. Craning from the back of the lorry, we could see the flash of the guns when they fired and after each flash we spent some lovely seconds wondering where the shells were going to land.

'Presently we moved out of range and we kept going until just on sunset, when all the tanks went forward, apparently to engage a British force. I got out of the lorry for a minute, the guard going with me. Our lorry was pointing north and the sun was going down on the left flank. Standing near me was a German officer. At that moment two 8-cwt. Chevrolets driven by Tommies came towards us out of the setting sun. When they were about twenty yards away they must have noticed my battledress and the uniforms of the two Germans, for they changed down with a flourish and one of them shouted: "What the hell goes on?" One of our boys had been watching through a peephole in the canopy and he yelled out: "Go for your bloody lives!" The two Chevrolets leapt forward, swung hard left round the tail of the lorry and hard left again round the bonnet, and were away into the setting sun, gathering speed. The officer yelled to the guard for his rifle. It was at the slung position and a second or two went by while he struggled with it. A few more were lost while the officer doubled round the lorry and by that time the target was just a diminishing cloud of dust right in the eye of the setting sun. It was hopeless and he lowered his rifle without even trying a shot. He didn't say a word.

'Taking it by and large, we felt pretty chirpy that second night. We knew we had a good chance of being rescued so long as the column stayed in the forward areas. We discussed it from this angle and that, and all the while the need for a smoke got worse and the cold struck up through the steel tray. . . .'

. . . .

Clusters of tanks, trucks, and tired men—like islands of various sizes in a cold, dark ocean—dotted the whole desert. They lay so still that it was possible for two of them to be almost cheek by jowl without knowing it. In a dozen places the leopard was lying down with the kid or with other leopards, but not in amity. The German column, for instance, as 'Midge' and his friends were to find out shortly, had chosen a British force for a bed-fellow, and it was chance alone—a dozen miles in this or the other direction—that prevented that force from being the one that was sheltering our ten lost C Section drivers and the Tommy sergeant who had driven up exclaiming: 'Blimey, chum! You bloody near shot me.' He had been mistaken for a German.

Now there were only three lorries, the fourth having broken down early in the chase, which had been very hot for a while. Rough going, however, had helped our drivers to get clear, and after putting a safe distance between themselves and the doomed recovery section they had halted and allowed the irate Tommy sergeant to come up with them. A bullet fired by Alf Hallmond had missed him by a whisker.

The drivers said to Corporal Ernie Symons[16]: 'You're the boss. What do you reckon we should do?'

There was a mass of transport on the horizon and it was decided to approach it from the flank with extreme caution. Shy as antelopes and as vulnerable, the three lorries and the Tommy sergeant's pick-up drove hesitantly across the desert, halting near a British anti-tank gun. There was a bad moment when the crew abandoned their billy of tea and sprang to action stations, but the drivers were wise enough to stay still and presently they were recognised. 'If

---

[16] Cpl A. E. Symons; farmer; Ohaupo, Waikato; born Hamilton, 11 Jul 1918.

you'd turned round, Kiwis,' they were told, 'you'd have got it proper.' Our drivers must have looked a bit white about the gills, for the Tommies handed over their billy like gentlemen. 'A nice cuppa Mike McGee,' they said. 'Get it into you, chums.'

The fugitives had found sanctuary with the B Echelon of a British armoured brigade. About 300 vehicles, including tanks and tank-transporters, were dispersed over a wide area, and the position was protected by anti-tank guns. Ernie Symons and the others were not sorry when the officer commanding the unit ordered them to stay where they were until contact could be made with a New Zealand convoy. While they were talking with the OC a scout car brought in a captured lorry and ten Germans, one of whom was an officer, a big, tow-headed Nazi who must have weighed every ounce of 16 stone. As soon as he got his feet on the ground he slung the OC a tremendous Wehrmacht salute, which was so smart as to be insulting. The Englishman just sketched the courtesy, and the honours, in point of offensiveness, were even. Meanwhile the Tommies had discovered several large tins of ham in the captured lorry. 'That,' said the big Nazi, 'is for our tea.' 'No,' said a Tommy sergeant-major. 'That's for *our* tea. For you, chummy, we've kept a nice bit of bull.' The big Nazi looked hose pipes and hunting whips but had sense enough to keep quiet.

Each of our drivers received a big helping of ham and they went to bed that night on full stomachs, to dream, perhaps, of dusk, gentle and soft as a suède glove, brushing over Auckland harbour while a liner goes past the heads lit up like a Christmas tree, and a little familiar ferry, grubby and well-loved, butts over to Devonport with all its lights in the water. Or did they dream, snoring in their three lorries, of a sea of sand, of an engine that spat and snuffled like a hairy goat, of an armoured Westphalian ham—gaining, always gaining—bestridden by a gigantic, tow-headed Nazi?

. .

Dawn in the desert comes in about five blinks. Peering through gummy eyes and a misted windscreen, you see only a lumpy pallor joggling beyond the radiator cap. You blink—and perhaps doze for five minutes—and when you open your eyes individual snail shells and spikes of camel-thorn can be seen plainly. You blink again and you recognise the lorry ahead of you as 'Mack's' by the dent in its tailboard. Again, and you can make out the staff car, louse-like, scurrying far ahead.

We had been woken by gunfire shortly before dawn—the dawn of the 26th—and the company, at the request of the 22nd Armoured Brigade, was moving back to Bir el Halezin. (Our overnight area, no doubt, was needed for a tank battle.) Captain Butt's convoy, as dawn broke, was feeling its way, engines grunting and backfiring in low gear, down a winding, precipitous track on the Sghifet el Charruba escarpment, near Bir ez Zemla, seven miles north of Sidi Azeiz, which had been left at one that morning. It was entering the 22nd Battalion's area, where an ammunition point was to be established. Some C Section drivers—those whose lorries were either empty or loaded with mines or explosives—had stayed in Sidi Azeiz, and daybreak found them huddled in their slit-trenches, rifles at the ready, and eyes straining to make out what was in front of them. For an hour past they had been listening to the groans of the wounded.

'That hour before dawn,' said George Laverick, 'was the worst part of a long, bad night. Quite early we had seen distant enemy flares and later the whole horizon was jumping with them. At the same time we could hear the growl of heavy transport going past. By midnight there were Very lights and parachute flares on all sides of us and it looked to me as though we were pretty well surrounded. However, the loaded lorries got away all right and I felt better after that. My lorry carried mines, so I didn't have to go.

'Not long before daylight a small group of transport, part of the great mass that was still moving past, got off its course or mistook us for Jerries. Anyway, it came in across the small airfield —there was one just in front of us—and when it reached the middle it was given the works. Machine-gunners opened up at point-blank range and an anti-tank gun joined the fray. This lasted about ten hectic minutes. Tommy voices could be heard calling out: they said they were British prisoners. But Germans voices could be heard too, so the officer in charge of us, suspecting a trap, refused to allow them to come in and ordered us to fire on the slightest movement. The cries from the wounded were terrible.

'When dawn came we saw a grisly sight. About thirty German and Italian soldiers were lying out there wounded and three were dead. I didn't see any wounded Tommies, but I feel certain there must have been some as about sixty prisoners had been set free and in the trays of two of the six lorries captured there were large pools of blood. There was an ambulance—occupied by a British airman and a German—that did not seem to have been hit, and there was a German staff car, the driver of which was dead.

'The day opened quietly and I was just beginning to think things were rather fun when—at about nine in the morning—a Jerry convoy came towards us, very uncertain about who we were. It came on about ten yards at a time, taking all precautions. Its inquisitiveness was soon satisfied. Our 25-pounders opened up and the way those Jerries turned and took to their scrapers reminded me of my own experiences. It was all right to see the old Jerry getting a bit of his own back.

'The next target to show itself was a crawler and a second shot from a 25-pounder sent it to its doom. The two occupants tried to escape but Kiwi machine-gunners changed their minds for them. One of them came in with his hands up. The other had no hands; he was badly burnt as well, and he died.

'Intermittent firing by our guns had been going on since early morning and it was now early afternoon. Stuff was still going past out in the desert. . . .'

Throughout the afternoon a mass of tanks and transport moved towards Bardia, passing between Sidi Azeiz and the 22nd Battalion's position at Bir ez Zemla where the ammunition point had been established. C Section's lorries were dispersed against the steep side of a wadi, at the mouth of which two sub-sections commanded by Corporal Alec Mills[17] had taken up defensive positions.

There was nothing to do except watch and wait, so Clarry Monahan[18] ('The Prune') beguiled the time by setting fire to a heap of German flares. They burnt with a beautiful blue light and everyone was extremely gratified. But not for long. Three German tanks—two Mark IVs and one Mark III—came into the wadi firing green and white recognition signals. They passed within 200 yards

[17] L-Sgt A. F. Mills; lorry driver; Te Kawa West, Auckland; born England, 21 Jan 1919.
[18] Dvr C. Monahan; builder; Hastings; born Auckland, 14 May 1905.

of Alec Mills's detachment and then stopped dead, having spotted an anti-tank gun on the right flank. The tank nearest to it opened fire and it replied in the same instant. Two more guns joined in at long range and the tanks turned and made off, one of them limping a little. Our drivers mopped their brows and implored Clarry not to do any more signalling. Then they put the billy on, for it was afternoon-tea time. In Sidi Azeiz, too, our drivers were boiling up.

For 'Midge' and his friends, passing at that moment through the gap between the two positions, there was no afternoon tea. In other circumstances Clarry could have served them from the ammunition point or George Laverick could have driven over from Sidi Azeiz with what was left in the billy, and the brew, though it might have been black and not very hot, would have at least been drinkable, for the gap was only six or seven miles wide. Germans, of course, never boil up, and in any case the column was in a hurry to reach Bardia, the perimeter of which was still five miles away.

'This was our third day with the Jerries,' said 'Midge'. 'At dawn that morning we were woken by the hell of a racket and by shouts for the driver of our lorry, Stan Wrack,[19] who was in the back with us. He was doing all the driving with a big Jerry sitting beside him. Our move the previous night had landed us next door to a mob of British armour and with first light it had spotted us and opened fire. The Jerries set off at full speed, armour-piercing shells landing among the vehicles. We could see them darting towards us, glowing red-hot in the half-light. When they hit the ground they rebounded with a flash of sparks. The British followed us for several miles, firing all the time, and as far as we could judge the Jerries lost about twenty vehicles.

'Next we were machine-gunned by fighter aircraft from a low level. At the time we were in semi-darkness, the guard having tied down the canopy at the back of the lorry. We could hear the planes swooping down on us and the rattle of their machine guns, and we cowered in the tray with our tin hats crammed over our ears. It sounded as though the planes were right on top of us and it was the hell of a feeling sitting there and not knowing what was

---

[19] L-Cpl C. S. Wrack; storeman; Whangarei; born Whangarei, 22 Dec 1918; p.w. 24 Nov 1941; evacuated from Bardia to Italy 15 Dec 1941.

J

going on outside. The raid lasted about five minutes but it seemed like five hours.

'At every halt Stan would sing out the news and the boys found this a great comfort. Stan's was a difficult job and everyone remarked that he was the right man for it. He kept cool and nursed us along over the bumps.

'During the day we were transferred to an open Jerry three-tonner, Stan coming with us. We were now near the front of the convoy.

'We had only the vaguest idea of our position but we guessed we had been making a good deal of easting since our capture and were now close to the frontier. Actually we were not far from Bardia, which we thought was in British hands. Our impression that we were with some sort of a raiding column was confirmed during the afternoon. A party of Tommy linesmen were repairing a telephone line on our right flank and they hardly bothered to look up when the huge formation rumbled past. Two of them were still working away at the top of a telephone pole when an armoured car pulled up to collect them.

'When we were a few miles from Bardia, which we could see in the distance, the German armour suddenly veered to the right, the thin-skinned vehicles carrying on towards the fortress.'

. . . . . .

It was afternoon-tea time and when the bombing was over B Section boiled up, but A Section, which had not sustained casualties, carried on towards Sidi Azeiz where there was supposed to be a field supply depot.

At two in the afternoon nineteen empty A Section lorries followed by twenty-two empty B Section lorries (plus a detachment from the 65th RASC Company) had left the unit lines at Bir el Halezin, joining Trigh Capuzzo and heading east. According to Divisional Headquarters the road to Sidi Azeiz was clear.

Shortly before three, by which time A Section was ten miles from Sidi Azeiz, Captain Gibson halted his vehicles. Behind them, a few miles down the track, a heavy air raid was in progress. 'Hell!' said the drivers. 'That's B Section getting it.' They could hear machine-gun fire and see the black smoke from the bombs.

Then the raiders headed towards A Section. The drivers started to scatter but someone sang out: 'It's all right, you jokers. They're ours.' And they were. Blenheim bombers with an escort of fighters flew over and dropped a white flare. Then bursts of bullets swept the dispersal lines. Our drivers gaped at the red, white, and blue circles, and many of them were too dumbfounded even to take cover. 'But the bastards are ours,' they cried. 'You can *see* the bastards are ours.'

There were two Very light pistols in A Section's ack-ack lorry but Jim Stanley had no idea what to do with them.

The firing stopped as suddenly as it had started and the aircraft flew away. Possibly only a warning had been intended. At all events, though the area had been raked with fire, no damage had been done. This was not the case with Captain Ward's B Section.

'We were driving along Trigh Capuzzo in column of route,' said 'Skin' Wilson,[20] 'when we saw the dust in front of us being kicked up by bullets. At first we thought we were being fired on by enemy armour but when anti-personnel bombs began to fall we realised that we were being done over from the air. Shrapnel and machine-gun bullets tore the canopies of the lorries and Tom Barlow[21]—the "Tree Man"—was wounded in the arm and chest.

'Our convoy then moved off the Trigh and got into desert formation but we were still pretty close together. Apparently we had been attacked by Blenheims with heavy fighter escort and no one could get the guts of it at all. However, hardly any damage had been done—a punctured tire for the "Tree Man" was about as far as it went.'

While the lorries were being checked a single British fighter-bomber flew over and dropped two flares. This was done to reassure them, the drivers decided, so no one was much concerned when a large Royal Army Service Corps convoy came driving through the lines of transport, making a splendid target, or when British aircraft were sighted a second time. Doubtless the mistake had been discovered and the pilots were returning to see what damage had been done. Our drivers shook their fists, but only when the planes banked to bring the sun behind them did they think of taking cover.

[20] Dvr E. D. Wilson; shop assistant; Auckland; born Te Puke, 29 Apr 1919.
[21] Dvr T. W. P. Barlow; truck driver; Auckland; born Auckland, 10 Apr 1916; wounded 26 Nov 1941.

'I stood watching the planes,' said 'Skin', 'and it was not until I saw the black blobs falling that I realised we were being bombed again. Some of us dived under lorries. Others ran for it. I got about fifty yards from my lorry and flopped down in the sand. I think there were about eighteen planes, bombers and fighters, the same ones that had attacked us before. They dropped stick after stick of little anti-personnel bombs and I didn't think I was going to come out alive.'

In the second raid Frank George[22] and Lionel Hawking[23] were wounded slightly, Doug Henderson[24] seriously, and Albert Storey[25] mortally.

'After that,' said 'Skin', 'we were told to disperse the transport properly. I could see that my lorry had been knocked about a good deal—she had three flat tires for one thing—but I got into the cab to have a go at starting her. As I did so I called out to "Yorkie" —Albert Storey—who was my cobber on the lorry. A second later I saw him struggling to his feet right at my elbow. He looked pretty bad and I had just time to hop out of the cab and grab hold of him before he slipped to the ground. He had been hit in the back by a piece of shrapnel that had passed right through the engine. Everyone thought the world of "Yorkie". He was a fine musician, and back in Petone, where he came from, he was assistant band master. His instrument was the cornet. Ray Crapp[26] bandaged him up pretty smartly but he hadn't a show.'

Six lorries belonging to the Royal Army Service Corps detachment, which had sustained one casualty, were wrecks, and four of B Section's were badly damaged. The convoy hardest hit was the one that had cut in just before the attack started. A doctor who was travelling with it attended to the wounded but there was no vehicle suitable for evacuating them to a dressing station. While waiting for one to come along, our drivers put the billy on and

[22] L-Cpl F. A. F. George; shearer; Glen Massey, Huntly; born Dunedin, 25 Nov 1913; wounded 26 Nov 1941.
[23] Dvr L. A. Hawking; dealer; Bayswater, Auckland; born Kew, England, 11 Mar 1906; wounded 26 Nov 1941.
[24] S-Sgt D. C. Henderson; clerk; born Greymouth, 15 Sep 1917; wounded 26 Nov 1941; died of wounds 14 Jul 1942.
[25] Dvr A. H. Storey; motor driver; born Yorkshire, England, 31 Jan 1905; died of wounds 26 Nov 1941.
[26] Dvr R. L. Crapp; labourer; born Whakatane, 21 Mar 1913; died in NZ, 11 Jan 1945.

then got out wheel-braces and jacks. There were many flat tires.

A Section, meanwhile, was nearing Sidi Azeiz. By now it had acquired 150 prisoners—a present from a passing artillery convoy. Three German officers were travelling in state in Jim Stanley's ack-ack lorry. The first thing they had asked for was water and then they had wanted something to eat. Two of them were polite and grateful but the third was a deplorable-looking person with small snake's eyes, thin lips, criminal forehead, and an expression in which cunning, meanness, and brutality were blended. He sat in glowering silence until Jim provoked him by saying: 'Well, the war's over for you, boy. Good thing, eh?' 'For the soldier,' he replied, 'it is better to fight.' The remark was correct enough but the tone in which it was uttered was so extraordinarily malignant that Jim picked up his Tommy gun and thoughtfully released the safety catch.

'It is not very nice, this desert warfare,' said one of the other officers in a gentlemanly sort of way, evidently wishing to dispel the unpleasant impression made by his companion. 'Myself, I do not find it very nice. These biscuits, they are very nice.' (They weren't: they were horrible.) 'You have plenty?'

He was a tall, fine-looking man with a simple face and very honest blue eyes. He was the sort of man you would be glad to get into conversation with on a railway journey. It was impossible to doubt his courage and integrity and difficult to doubt that he was inherently decent: it seemed unbelievable that a man of his stamp could have anything but contempt for the Nazi bosses. (This was in the days when people still spoke wistfully of 'honourable elements in the German officer class'.) He said he came from Bavaria. His rat-faced companion came from Munich, which seemed a most appropriate place for him to come from. One could imagine his having a high old time in the basements of the Brown House.

The remaining member of the trio was a small, plump man with a cherubic face, steel spectacles, and a really charming smile. He spoke hardly at all, contenting himself with munching biscuits, smiling his disarming smile, and bouncing up and down in the uncomfortable little seat provided by the Motley mounting. He looked like a village schoolmaster—you thought of woollen mufflers and children singing 'Silent Night'—or like one of those lovable, absent-minded German professors whom most of us had read about but none had actually seen because for ten years or more all of them had been busy, in a lovable and absent-minded way, in places like Peenemünde. Probably he was a passionate believer in the Hitler Youth, just as his gentlemanly colleague was probably an admirer of Goering's. Looking at him, however, you saw only the rosy-cheeked children, heard only the rejoicing fiddles and the crunch of snow. It was very puzzling.

Conversation with the pleasant Bavarian and the task of keeping a sharp eye on the disgusting little creature from Munich kept the ack-ack crew fully occupied until the convoy halted outside Sidi Azeiz. Twilight had fallen and the outlines of gun positions and barbed-wire entanglements were barely visible at a distance of 200 yards. Close to where Trigh Capuzzo touched the perimeter a gun and a gun-tower were burning fitfully, putting out petals of scarlet flame which bloomed and faded fantastically in the quiet dusk. No sound came from the garrison.

After halting the convoy Captain Gibson drove forward alone, tooting his horn. He stopped for a moment at the guarded entrance, then signalled the lorries forward. They went rumbling into Sidi Azeiz.

'You're lucky, boy,' a New Zealand anti-tank gunner said to Jim Stanley. 'We were watching you all the way. We were pretty near certain you were Germans and if that officer of yours hadn't come forward the way he did you'd have stopped the lot.' He pointed to where the gun and tower were glowing softly in the twilight, adding: 'That's our work.'

There was no supply depot in Sidi Azeiz, so the lorries were unable to load. The prisoners were placed under guard and given some packets of biscuits and what water could be spared, which was very little. Our drivers drew rations from the cooks' lorry—tinned sausages and tinned potatoes—and lit their primuses. Soon it was dark, and later the moon rose and went shining on through the long, cold hours.

.

It shone on Company headquarters, Workshops, and the handful of load-carriers at Bir el Halezin. It shone on the small, dirty compound in Bardia where 'Midge' and his friends, cold and supperless, were experiencing their first night as prisoners of war.

It shone between half past nine and midnight on Captain Butt and his drivers as they crept across the seven miles of desert dividing Sidi Azeiz from El Charruba escarpment, where they had spent the day with the 22nd Battalion. It showed them a dizzy pattern of Allied and enemy wheel tracks, and although they had with them the rest of the Composite NZASC Company and an escort of armoured cars they could hardly have felt more vulnerable if the convoy had been a small, lost, goods train in a siding controlled by a mad signalman. Anguished, they waited for the shining, snorting expresses to come whistling down on them.

It shone on Captain Ward's lorries, which, about midnight, were forming a laager somewhere east of Bir el Halezin. The wounded drivers, after waiting nearly five hours, had been taken in a passing British ambulance to 13th Corps' main dressing station. After that the convoy had headed for home, Captain Ward's information being that it was useless to try to reach Sidi Azeiz. It shone on Ed Duda, where, about an hour later, men from the 19th Battalion shook hands with men from the Tobruk garrison, and on Belhamed, which the 4th Brigade had captured the night before, and on Sidi Rezegh where the 6th Brigade was fighting.

It shone on the B Echelon of the British armoured brigade, one of whose six laagers sheltered the three lorries in which Ernie Symons and his nine C Section drivers were sleeping. No dreams tonight of Nazis and Westphalian hams. They had spent a pleasantly quiet afternoon tinkering with a German motor-cycle that Ernie and Alf Hallmond had brought back to camp after going out on patrol with the scout cars, and later they had taken their turn at the listening posts. Now they were sound asleep.

The aircraft came over about half past one in the morning. They circled in the moonlight and dropped several bombs, all of which fell on the same laager—the one in which the C Section lorries were parked. Two were set on fire and the third was riddled with shrapnel. Tom Laverick was wounded in the back, Bob Troughear[27] in the face and leg, Alf Hallmond in the leg, and Sid Pausina[28] in

---

[27] Cpl R. W. Troughear; carting contractor; Pokeno, Auckland; born NZ, 10 Apr 1912; wounded 27 Nov 1941.
[28] Dvr S. C. Pausina; butcher; Kaikohe; born Auckland, 4 Sep 1916; wounded 27 Nov 1941.

the thigh. The Tommies had suffered heavily as well and several of their vehicles were blazing.

Tom Laverick was stunned by blast and when he came to he was surrounded by flames.

'I opened my eyes,' he said, 'and the canopy was blazing and fire was darting all over the back of the lorry. At first I couldn't make out what had happened or where I was. A few seconds must have passed before I pulled myself together but once I did start to move I was over that tailboard like lightning. I didn't realise I had been hit until I was outside.

'At the back of the lorry there was a Tommy with his head blown off. Another Tommy had lost his leg and was screaming out and calling for his sergeant. I went over to see if I could help but there wasn't a thing I could do for him. A plane was still hovering over the burning lorries and machine-gunning.'

The driver of the third lorry, meanwhile, had climbed into his riddled cab and pressed the starter, expecting no result. To his astonishment the engine fired at once. The others scrambled aboard, the fit helping the wounded, and the lorry headed for the open desert, which seemed safety itself in comparison with that fiery neighbourhood. It was a wise move, for presently the aircraft returned, bombing and strafing.

Banners of flame streamed skywards for a long while, and afterwards, in half a dozen places, hot metal glowed cherry red.

The moon had still an hour or two left. It shone down on Sidi Azeiz, where the slit-trenches were like stencilled Ls, the gunpits like blobs of ink, and the barbed wire like faint scribbling—a dirty, blotted page. The lorries were huddled together for protection and clumps of them formed dark patches in the desert. There was resentment in the humped outlines of their canopies, in the blank stare of their stubby radiators and blind headlamps. Under the cold craziness of the moon they seemed to be on the point of trumpeting shrilly and stampeding across the desert, driven mad by all the hounding and harrying they had put up with during the past forty-eight hours.

Most of A Section's lorries were grouped roughly around an open space, which appeared at first glance to be a dumping ground for old clothes and abandoned equipment; then the moonlight,

striking a livid face or glinting on an outflung hand, showed that
it was covered with human beings, all tangled together in an effort
to keep warm. Even their uneasy stirrings, the risings and sub-
sidings in the amorphous mass, were suggestive more of the bur-
rowing of rats among rubbish than of the movements of living
people. These, though, were German prisoners—captured *Herren-
volk*—and the poor devils were as cold as any Pole and as hungry
as any Greek. Yesterday they had ranked with the world's finest
soldiers: tonight they were down and out—finished. The presence
of two Bren guns crouched over slim bipods—bird dogs inspecting
the day's bag—seemed almost superfluous.

The prisoners squirmed and muttered and some of them moaned
a little. Nothing else moved. They might have been dying men on
a dead planet. Everything looked cold and dead, and low in the
heavens, colder and more dead than anything on earth, washing
the sands with silver, emphasizing each mean and ugly detail—
the discarded tins, the rusty wire, the poor disgusting prisoners—
turning the sentries into silver statues and silvering the guns and
lorries, evil and indifferent, was the moon.

· · · · ·

The drivers lay for a minute longer—two minutes longer—in
the lovely blankets, but the voices came back: 'Right-oh, you
jokers. Pack up. Pack up. Moving in ten minutes.'

It was a little before two in the morning and Captain Roberts
had been ordered to replenish the Composite NZASC Company at
the 50th Field Supply Depot near El Beida. A Section was to go
with him, and all empty C Section lorries. The loaded ones were
to remain in Sidi Azeiz under Second-Lieutenant W. S. Duke.[29]
Counting six from the former 14th Light Anti-Aircraft Section, there
were twenty of these. The rest packed up at once.

The frozen prisoners were herded aboard—with pathetic honesty
they scuttled round trying to return old radiator muffs and bits of
canvas they had borrowed—and within ten minutes the convoy
was under way, travelling south-west. Ahead and on both flanks
German flares bobbed up and down in a manner indescribably

---

[29] Capt W. S. Duke; butcher; Dunedin; born Dunedin, 28 Jan 1913; p.w. Bardia 27 Nov 1941; evacuated from Bardia to Italy Dec 1941.

knowing and intimidating, making the outing hardly less unpleasant than a swim in a shark-infested lagoon.

The drivers who had been left behind were inclined to congratulate themselves—the No. 1 drivers anyway: the spare drivers were less certain. They had been issued with hand grenades and extra rifle ammunition and told they were infantry. They shivered in their slit-trenches, longing for dawn.

'The night was quiet,' said George Laverick, 'but Jerry put up a continual string of flares, showing that he had us taped pretty near all round. My heart was right down in my scrapers.'

Just before dawn two Hurricanes took off from the airfield. Soon it was light.

'With great sighs of relief,' said George, 'my cobbers and I made our way back to the lorries and put the billy on. When we were all set for the first mouthful—the milk was in the billy and some of the boys were getting their cups and others were fossicking around for eats—the siren sounded. The rising sun was just level with the horizon, and over a low ridge, about a thousand yards away, came German tanks, lorried infantry, and motor-cyclists, shooting and shelling.'

CHAPTER 10

THURSDAY, FRIDAY, AND SATURDAY

THE siren sounded, Bren carriers raced in from the desert, and our guns opened fire. It was 27 November.

' We were woken by the hell of a racket of shelling and machine-gunning,' said Claude Campbell,[1] ' and Harvey McCabe,[2] Don Baker,[3] and I tumbled out of bed and into the nearest slit-trenches half-dressed and only half-awake. The tracers were criss-crossing a few feet above our heads like red ribbons and shells crashed round us. Before long our lorries were burning and a load of gun-cotton and ammonal blew up.'

At the start of the attack some of our drivers were asleep; others were brewing up, folding their blankets, cleaning their teeth. Noel Orsborn[4] (14th Light Anti-Aircraft Section) was killed in the first minute of the battle while he was still asleep in the back of a lorry loaded with Bofors shells. Fortunately the whole area was scarred with Italian dugouts and British and Italian slit-trenches.

' Very soon,' said Clarry Monahan, who had taken cover behind a small stone cairn, ' the machine-gunning made it too hot for us and we retired to a small building beside which a lorry was parked. The lorry was hit and went up in a roar of flames. I began to sweat considerably for I knew it was loaded with Bofors shells. There was a mob of jokers sheltering in the building and when I suggested making a dash for an old ruin about a chain away they said: " You'll never make it." I said: " I'd as soon give it a go as wait here to be killed by those bloody Bofors." About seven of us made a dash through the machine-gun fire and all of us got away with it. The ruin was manned by three airmen with Lewis guns

---

[1] L.-Cpl H. J. C. Campbell; truck driver; Mangamuka, North Auckland; born Whangarei, 17 Dec 1914; p.w. Bardia 27 Nov 1941-2 Jan 1942.
[2] Dvr H. W. McCabe; lorry driver; Ohaupo, Waikato; born Whitianga, 5 Dec 1916; p.w. Bardia 27 Nov 1941-2 Jan 1942.
[3] Dvr D. H. Baker; contractor; Mangamuka, North Auckland; born NZ, 14 Apr 1904; p.w. Bardia 27 Nov 1941-2 Jan 1942.
[4] Dvr N. J. N. Orsborn; farm labourer; born NZ, 6 Dec 1915; killed in action, 27 Nov 1941.

and it was under heavy mortar fire. Soon afterwards the load of Bofors went up with a noise like the end of the world.'

One after another the lorries caught fire and those carrying mines or explosive blew up.

Corporal Nigel Barach[5] (B Section), and Cyril Aro[6] (C Section) were sheltering in an Italian dugout. Cyril peeped over the breastwork and saw Dick Turner[7] (C Section) crawling in the open. He was wounded in the thigh so Cyril went out under fire and dragged him to safety. While he was being bandaged an engineers' lorry caught alight about five yards from the wall of the dugout.

'We asked the engineers what was in it,' said Cyril. 'They told us it was ammonal. We asked them how much and they said "Half a dozen cases." Then they disappeared. It was impossible to move Dick, so Nigel and I decided to stay with him. About five minutes later the ammonal blew up, the blast knocking us all unconscious.'

Meanwhile the casualty list was mounting. Gil Drinnan[8] (C Section) was mortally wounded by machine-gun bullets while the battle was at its height and Charley Mann[9] (14th Light Anti-Aircraft Section) was wounded in the foot by a bullet from the same burst. He dragged Gil behind a wheel of the lorry they were sheltering under and there they were pinned to the ground by intense fire, Gil dying about half an hour later. Another C Section driver, Ben Clifford,[10] was mortally wounded by bullets and 'Fat' Davison was hit in the calf and thigh.

George Laverick had gone to ground beside the billy and the brimming mugs, and he watched the tanks close in, firing as they came.

'As soon as the leading tank was within range,' said George, 'a

[5] Cpl N. J. Barach; motor driver; Maungaturoto, North Auckland; born Auckland, 26 Jan 1908; p.w. Bardia 27 Nov 1941; evacuated to Italy 15 Dec 1941.

[6] Dvr C. Aro; driver-mechanic; Christchurch; born Auckland, 25 Aug 1915; p.w. Bardia 27 Nov 1941-2 Jan 1942.

[7] Dvr E. Turner; motor trimmer; Wellington; born NZ, 14 Sep 1909; wounded 27 Nov 1941.

[8] Dvr G. B. Drinnan; farmer; born Milton, 26 Jul 1909; died of wounds, 27 Nov 1941.

[9] Cpl C. G. Mann; grocer; Dunedin; born Dunedin, 8 Jun 1917; wounded 27 Nov 1941.

[10] Dvr J. D. B. Clifford; carpenter; born New Plymouth, 26 Nov 1903; died of wounds, 28 Nov 1941.

Bofors gun about ten yards away from me had a go at it. The Bofors fired one clip and was quiet, having received a direct hit from the tank it was engaging. A man I had known in "civvy street" came out alive but another member of the crew had the sights of the gun blown through his neck. How the others got on I am not sure.

'By this time the machine-gun fire and mortaring was terrific. I could see three tanks about 300 yards away and still coming on. I spotted a better hole farther back and started crawling towards it. I had crawled twenty-five yards and still had another twenty-five to do when Jerry spotted me and started to do a bit of peppering. It was time for me to take to the old feet and I tell you Lovelock's record had nothing on that last twenty-five yards. I landed in a good, deep hole with an 18-inch stone wall right round it. Through the cracks I could follow operations and as things progressed I am not ashamed to say that I put up a prayer or two. At one stage I noticed a little bird like a sparrow standing near me. He wasn't game to get up and fly.

'I could see things were hopeless, so I buried my diary and two or three letters, keeping only my paybook. The three tanks came within fifty yards of me, and then one of them halted and continued to let fly with all he had while the other two, firing furiously, carried on, intent on routing us out. One rumbled each side of me at a distance of about three yards, but—thank God!—I was not noticed.'

At that moment Don Baker was saying: 'Look at this.' About fifty yards away, side on to the trench in which he was sheltering with Claude and Harvey, three German Mark IV tanks, their guns shooting out flames eight feet long, were finishing off a 25-pounder.

'A few minutes later,' said George, 'things ended. I stood up and gazed about me. It was a sorry sight. The place was a mass of burning vehicles. I had seen three of our lorries disappear in black smoke, but some of the others—the ones with less sensitive loads—were still in the process of going up. Even though the firing had stopped Sidi Azeiz was still far from healthy.'

When the firing stopped Cyril Aro and Nigel Barach stood up too.

'The blast from the ammonal had knocked all of us unconscious,' said Cyril, 'and by the time we had come to and struggled out of the sand and stuff that was half burying us—the breastwork had been levelled flat—everyone was standing up with hands in the air. We helped each other out and found we were OK except for being bruised and shaken. Then we started to carry Dick Turner over to the Regimental Aid Post, where the prisoners were being herded together.'

All over the area our drivers were standing up, their hands in the air.

'The Jerry who routed us out,' said Claude Campbell, 'wore glasses and had a bit of camouflage in his tin hat. He was wildly excited and he looked grotesque, but he meant business all right. Stuff was burning all round us. One of the armoured cars that had escorted us from the 22nd's position the night before was burning like a torch, and as we watched the wireless mast started to wilt and bend slowly over.'

'I stood up,' said Clarry Monahan, 'and the white flags were flying and the German tanks came flooding in with Rommel leading them.'

It was not Rommel whom Clarry saw but a German tank commander. His tank halted near Brigadier Hargest[11] and he stood up in the turret—he was a fine-looking man with a monocle and an Afrika Korps cap—and said: 'You fought well.' Brigadier Hargest asked if his men could get their gear; the tank commander agreed.

When Clarry and another driver approached their lorries they were turned back by Germans with tommy guns. George Laverick had the same experience.

'I had about 200 yards to walk to where our lorry had been before it was blown up,' said George, 'and when I got there out pops a Jerry. He was about the biggest of the lot and why he had to pick on me I just don't know. He took one look at me and then rammed the snout of his tommy gun in the fleshy part of my back, at the same time saying something that sounded like *Ooch!* and

---

[11] Brigadier James Hargest was then in command of the 5th Brigade. He was captured with 46 officers and 650 ORs after an action lasting about an hour and a half. He escaped in March 1943 from a prison camp near Florence, eventually reaching England, and was killed in action in Normandy in August 1944.

pointing towards Rommel's tank.[12] His meaning was plain, so I didn't stop to argue the point with him. As far as I was concerned he was welcome to that part of the desert, but he wasn't satisfied and he repeated the *Ooch* performance. By this time I was under way, running with my hands in the air, which wasn't easy. A couple more pokes and a couple of *Oochs* and I was in full gallop. We kept it up between us until I reached the other prisoners at the RAP.'

It was little enough our drivers were allowed to salvage from their wrecked vehicles—a blanket, perhaps, or an overcoat. Most of them had lost everything except what they stood up in. While they were searching among charred rubbish German amateur photographers got busy with their cameras. 'Smile, please,' they said, and many of the prisoners responded automatically before realising their mistake. Other Germans were routing around for tinned food and when they discovered a tin of something particularly palatable they opened it on the spot, making a hearty meal in front of the breakfast-less prisoners. British emergency rations —slabs of rich chocolate—were in great demand, and for these a number of our drivers were searched.

With the lorries burning around them, the dead lying beside their guns, and the wounded and dying being carried to the Regimental Aid Post on stretchers, the prisoners were lined up in four ranks and counted. An inspection took place, the German guards pulling up short when they came to a New Zealander who was wearing ammunition pouches and a bayonet. 'The Jerries,' said Clarry, 'did their scones. They dragged him out of the ranks and whipped his bayonet off him, looking thoroughly disgusted. We did a big grin.'

It was their last grin for some while. After the strain, relief had brought gaiety, almost hilarity, but this had worn off now and our drivers were feeling lost and forlorn. In all fifty-four members of the Ammunition Company had been captured—forty-seven from C Section (counting former 14th Light Anti-Aircraft Section personnel) and seven from B Section.[13]

[12] George, like most of the others, had confused Rommel with the tank commander. Rommel arrived in Sidi Azeiz about half an hour after the surrender.
[13] A party from B Section had been left in charge of mines in the Abiar Nza Ferigh area and later taken to Sidi Azeiz in C Section transport.

Twenty-one lorries had been burnt or captured.

Half an hour after the battle the German transport and armour headed west, a skeleton force remaining to look after the prisoners. At last they were given the order to move, and the long column, escorted by motor-cycles with spandaus mounted on side-cars, started its 18-mile trudge to the prison compound in Bardia, those who were unable to march—among them were Charley Mann, Dick Turner, and Tom Barlow—being left behind at the aid post.

Our drivers were hungry, thirsty, and over-wrought, and the struggle to rise to their feet after the short hourly rests became increasingly painful. 'Fat' Davison, with wounds in the calf and thigh, was soon in great distress, but he struggled on gamely mile after mile until at last he was forced to sit down by the roadside. George Laverick and two others stayed with him. After a while a staff car pulled up and a German officer asked them what they were doing. They told him that 'Fat' was wounded and could go no farther. The officer said: 'Don't worry, boys. I was a prisoner in France for one day and the British treated me well. You never know your luck. You may be free tomorrow.' When the car moved off towards Bardia 'Fat' was draped across the roof—there was no room inside—like a huge fish.

The prisoners were halted near the outer defences until dark. Just on dusk a German officer drove up and shouted that there was a hot meal waiting inside the fortress. Our drivers gave him a cheer, but the announcement was made either in error or as a cruel joke, for the small walled compound into which they were eventually herded after another march contained nothing except a little water, dirty and brackish. The prisoners were parched with thirst and they rushed it, with the result that many of them failed to get a drink.

The compound was so crowded that there was hardly room to turn. This mattered little, for the night was bitterly cold and the prisoners would have had to huddle together in any event. Besides there was comfort in proximity.

They slept at last, forming under the cold moon in Bardia, as the German prisoners had formed in Sidi Azeiz, a great squalid mass of insulted and protesting flesh, a community of aches and pains, with a battle going on in each tired mind—guns firing, lorries exploding, guns firing, guns firing, on and on through the night.

Through the Wire— the opening of the Second Libyan Campaign

Inside the boundary wire—a plane burns in the distance

New Zealand Divisional Headquarters at Bir el Chleta

Despatch riders, Tobruk

The rest of the Composite NZASC Company was still free, but only, as it seemed to our drivers, by a miracle. After leaving Sidi Azeiz the convoy had travelled five miles south-west by south and then laagered for the night. It would have been senseless to go on. Enemy flares were appearing in almost every direction and some of the drivers had not slept for thirty-eight hours. They rolled themselves in their blankets beside the lorries without even taking off their boots.

At first light, with unidentified vehicles approaching from the north-east, the convoy continued on its course, swinging wide from the Wire, only to run into a semi-circle of enemy tanks, transport, and armoured cars. Turning under fire, it went back to Sidi Azeiz, the enemy pursuing. A shell whistled over Jim Stanley's head—he felt the draught of it—and landed almost between the front wheels of Les Howarth's[14] lorry but without doing damage. From the direction of Sidi Azeiz came a sound of heavy firing, and over the position there was a pall of black smoke beneath which were dabs of vermilion.

Still pursued, the convoy headed towards the 22nd Battalion at Bir ez Zemla, but was turned back once again. Captain Roberts considered next the possibility of dodging the enemy until dark and then trying to return to Sidi Azeiz, but shortage of petrol was an objection. Finally it was decided to run west towards Bir el Halezin, where the Divisional Administration Group was known to be.

Now began a breathless scramble for safety and it went on hour after hour. The going, for the most part, was terrible, and the convoy became as formless as a chariot race. Each driver chose his own path, intent only on following the leading staff car and avoiding slit-trenches. Often three or four lorries would be travelling abreast, mudguard almost touching mudguard.

What they were running from was not always clear to the drivers, but when they looked south or behind them they could usually discern little scurrying objects preceding a plume of dust, the more terrifying because they were seldom glimpsed clearly. Even when the pursuers were invisible the spectacle of between ninety

---

[14] Dvr L. J. Howarth; farmhand; Morrinsville; born Morrinsville, 4 Nov 1917.

and a hundred vehicles being thrashed along at top speed was enough to suggest that Nightmare was following with all her brood.

The man at the wheel, of course, saw only what was in front of him, and for mile after mile it was little except clumps of camelthorn round which the drifting sand had formed hard mounds anything from six inches to a foot high. They were so close together that it was impossible to steer between them and a lorry could be kept on its course only by brute strength. It was like driving over a gigantic nutmeg grater.

Hardly ever was it possible to travel in top gear, but in second and third they attained speeds they had previously thought impossible. They no longer gave a damn about their cherished engines. They prayed only that 'conrods' would not buckle like hot pokers and flying pistons mix everything into a metal omelet. But the wonderful Chevrolets never faltered. They boiled along mile after mile, screaming like circular saws. They took terrific wrenches; they were slammed up and down until it seemed next to impossible that a single spring could be unbroken; but only one Ammunition Company lorry was lost through a mechanical defect that day. It had been towed for some miles and was cast off the moment the situation became really dangerous.

A Section's anti-aircraft lorry was on the southern flank of the convoy, the muzzle of Jim's anti-tank rifle projecting over the tailboard. On several occasions he drew a bead on distant armour, but it was not until early in the afternoon that he got a chance to fire. Three midget armoured cars, which his experiences in the 1940 Libyan campaign enabled him to identify as Italian, appeared suddenly from the south and scuttled along beside the convoy at a distance of 300 yards. Jim got in several shots, the barrel of his anti-tank rifle swinging in a crazy arc between earth and sky. Some of the shots kicked up the dust a few yards away and others soared towards the Pole star, but two got home, or near enough to it to make the Italians sheer off.

After they had gone Jim threw his rifle on the deck and glowered at it. Its rubber shoulder pad had come away from the butt and he had taken some shrewd knocks. Besides, he had never quite forgiven it for misfiring in Greece. Something like a personal quarrel had developed between the two of them and twice already it had been 'given a passage'—Jim's phrase—over the tailboard, sharing the fate of a really formidable list of refractory primuses and skin-removing spanners, the majority of which had been retrieved when Jim had cooled down a little. His rages seldom lasted more than a minute or two, and now, after giving his enemy a few kicks, he began tenderly to wipe its breech.

Earlier, a number of Royal Army Service Corps vehicles had joined the convoy, and at this stage they began to run out of petrol. One after another they were abandoned, their drivers piling into a large, open Morris Commercial that was acting as a sort of lifeboat. It literally bounded across the desert, and ever and again its unfortunate occupants, wearing agonised expressions, would rise into the air in a body exactly as though they were being tossed in a blanket. It was a marvel that some of them were not thrown out.

Nor were the prisoners having a comfortable ride, but they were in good spirits. They became cockier as the day wore on, laughing and cheering whenever the convoy changed course and shouting unsolicited advice. They accused our drivers of fleeing from their own friends and perhaps they were not always wrong. Our officers were still handicapped by their inability to recognise or acknowledge friendly signals.

Time was passing and the estimated distance to Bir el Halezin had been covered, but still there was no sign of the New Zealand Division. Plainly the convoy was off course. It turned north, covered four miles, and ran into a tank battle. Then it turned east. The drivers were beginning to believe that this was one of those nightmares from which you don't wake. Finally the convoy halted. There were armoured fighting vehicles ahead of it and on both flanks. They dotted the horizon, and though it was impossible to distinguish details our drivers had the impression of being mocked and stared at by wild beasts. They could picture the bright, hot eyes, the dribble of saliva. The officers held a hurried consultation and it was decided to keep moving until the last. Petrol was short and the chase was bound to end soon one way or the other.

Like weary buffaloes the lorries swung round and charged towards the widest gap, but almost at once tanks appeared ahead of them and that was the finish. It was a relief in a way. They slowed to a trot, to a crawl, then halted. One of the staff cars went ahead and presently the convoy was waved forward. Soon the drivers were able to see the faces of the men in the tanks and armoured cars.

They were red, grinning faces, unmistakably British. Everyone began talking at once, talking and laughing with relief. It was anti-climax.

'We bin watching you blokes,' said a fat, red-headed sergeant. 'Saw you muckin' about all over muckin' desert. Had a go at heading you off twice. Suppose you thought we was muckin' Jerries. Fourth Armoured Brigade—that's us.'

The officers checked their sections and all our vehicles were accounted for except 'Dad' Cleave's and Basil Thorburn's[15] LAD. No one had seen it since eleven in the morning when three tanks had attacked the convoy from the south. A Supply Column officer and his driver had also been missed at eleven, and as their pick-up was known to have been giving trouble it was thought that 'Dad' and Basil might have been captured while helping them.

Now that the excitement was over our drivers found themselves as stiff and sore as though they had been playing the first football match of the season. In rather less than eight hours they had been chased for 100 miles over country across which you would have hesitated to drive a horse and cart.

Staying with the 4th Armoured Brigade—it was eight miles south by west of Bir el Halezin—only long enough to check their engines and distribute evenly what little petrol remained, the A Section drivers set out for the unit area, reaching it without further trouble. Captain Butt with what was left of C Section—nine lorries of the original thirty-six—arrived a quarter of an hour later, reporting that the Composite NZASC Company had been disbanded.

For the rest of us the 27th had been a quiet day. Captain Ward and B Section had returned early in the morning to report that the road to Sidi Azeiz was closed, giving the Major just time to prevent the balance of the load-carriers from trying to reach it. Then we had moved four miles north-west, our empty trays, with no comforting weight of ammunition to hold them down, rattling loosely behind us. We had halted three miles from Point 175, which, with Ed Duda, Belhamed, and Sidi Rezegh, was still in our hands, though precariously. For the rest, it was known that Rommel's armour was returning from the frontier.

There was no longer the least doubt about the gravity of the

[15] Dvr B. J. L. Thorburn; printer; Auckland; born Auckland, 1 Dec 1918.

situation and we were depressed by the emptiness of our lorries. It gave us a special feeling of failure. For the time being, through no fault of our own, we were just an incubus—dumb and idle, but eating, drinking, and burning petrol—on the shoulders of an exhausted division. It was a bitter pill for men who were accustomed to regard themselves as universal providers.

. . . . .

The sun came up next morning—the morning of the 28th—fresh and golden as a lemon, flooding the desert with clear light, easing aches and pains, making it less of a tribulation to hold ice-cold dixies at breakfast-time. It cheered everyone, even the sixty-six members of our unit who were breakfasting lightly in Bardia on cold water; even Charley Mann, Dick Turner, and Tom Barlow, who, with some sixty other sick and wounded men, had been left at Sidi Azeiz in the care of medical orderlies until someone—friend or foe—had time to collect them; even Ernie Symons's party, which, carrying its wounded in the remaining lorry, was moving slowly towards the Wire with a well-guarded convoy. Throughout the previous day its British hosts had been chased in a wide circle, Tom Laverick, who was a stretcher case, riding in a captured vehicle with the Tommy sergeant whom Alf Hallmond had done his best to kill. He had been wounded at the same time as Tom and he enlivened the trip by remarking as each shell from pursuing armour whistled overhead: 'The next one's ours, chum. It stands to reason.'

For Ernie, Tom, and the others the campaign was over and for us, too, it was drawing to a close.[16]

After breakfast on the 28th we were told to shift three miles in a south-westerly direction to allow elbow room for a tank battle, and no sooner had we begun to move than the action started. For a long while we could hear the slam of the two-pounders and the steady drilling of heavy machine guns. We moved again at 10 a.m., heading north and passing through Divisional Headquarters' area.

---

[16] All our wounded were evacuated safely to the rear. The three drivers who had been left at Sidi Azeiz, after living mainly on rice for three days and nights and being visited daily by German patrols and once by British armour, were taken on the 30th to Fort Capuzzo. Ernie Symons's party, after leaving their wounded with a New Zealand field ambulance unit on the far side of the Wire, continued in their own lorry to Fuka, where they were held until it was possible for them to rejoin us.

Our course took us close to Point 175 and over a stretch of desert that had seen much fighting. It was dotted with burnt-out tanks and vehicles—German, British, and South African. They stood in splashes of charred sand, and from the tail of each, like viscera from a squashed beetle, debouched a litter of rubbish—clothes, equipment, toilet gear, and reams of letters and photographs. Whenever the convoy halted, which it did often, our drivers would tear over to the nearest wrecks and grub frantically in these rubbish heaps. In every case the eyes had been picked out of them, but no one returned to his lorry without an armful of assorted litter that ranged from boltless Mausers to torn German overcoats.

The fascination of these lucky dips never failed. Apart from sometimes yielding a Leica camera or a Luger, they were like a peep into the enemy's private mind. Why, for instance, did German soldiers encumber themselves with cosmetics? Wherever they had been the desert was littered with pots of face cream and skin salve, bottles of sun-tan lotion and hair oil, and little tins of powder. It argued a sybaritism we had not suspected. The Italians on the other hand—reputedly delicate and effeminate—were industriously turning the desert into a midden.

The propensity of the German soldier—and of the Italian soldier too—for burdening himself with letters and photographs was easier to understand. Our soldiers were the same. Whenever a battle had occurred the state of the desert suggested that a gigantic paper-chase had been in progress concurrently with the fighting. Acres of cheap notepaper covered eastern Cyrenaica at this time, letters in German and Italian mingling in the same drifts with letters from Britain, South Africa, and New Zealand. The writing was nearly always feminine and one is tempted to add that nearly every letter might have been written by the same woman. Few of us could read German or Italian and the scribbles from Bermondsey, Capetown, and Central Otago were protected from prying eyes by a delicacy that was almost universal, but one knew instinctively what each letter said, whether it was written to *Caro Enrico, Lieber Heinrich,* or *Dear Harry.* Has the parcel arrived safely? Is the food still enough? The curtains in the sitting-room at home . . . and so on. Hardly a word, of course, about Hitler or Mussolini or Churchill or Peter Fraser. Nearly always the same letter—nearly always the same woman. Lying crumpled in the desert, mixed with torn photographs of girls, pudding-faced babies, and mournful-looking mothers—mournful behind the bright, fixed smile—they exhaled, the blue, the pink, and the mauve sheets, to an almost unbearable degree, that *weltschmerz*, that sentimental pessimism, which, though peculiarly German, is common to soldiers the world over.

Was there a hint here—it was one of the things we discussed often—that the German soldier, so like ourselves in the externals of his daily life, was subtly different from the blind automaton of legend, just as we were different, by the grace of God, from the flattering conception of the Kiwi warrior—always brave, always modest, always magnificently independent—presented by friendly journalists for the delectation of our admirers? Was it, perhaps, the whole thing, a lie and a mistake? There was no bitterness in desert warfare, no slow poison of starving civilians and smashed cottages. Rommel was a hero to both sides.

We halted at noon near Ed Dbana, three miles north of Trigh Capuzzo and eight or nine miles due east of Belhamed. Soon afterwards we came under shellfire, which continued in a desultory fashion for half an hour, damage being confined to a few shrapnel holes in our vehicles.

After tea we were warned to expect a night move. The optimists said it was the beginning of the chase to Tripoli. The pessimists said that the Division had been cut to pieces and what was left of it was going to run for the Wire. Some said we were moving to Tobruk. Most of us were too sleepy to care much.

A tank battle, like a foundry in full blast, was banging away just over the horizon, and as darkness closed in it moved nearer and soon we could see the flash of guns and watch the tracers, like rows of red stitches, curving against the night. At one stage a tank must have received a direct hit from an armour-piercing shell, for a patch of rose pink, like the instantaneous blooming of a peony, appeared suddenly in the darkness. It bloomed and faded and was gone in five seconds.

We were called soon after nine and at twenty-five minutes to ten we set off in desert formation, tacking on behind the rear of Headquarters 13th Corps half an hour later. The whole of the Divisional Administration Group—Petrol Company, Supply Column, Divisional Ordnance Workshops, and many smaller units and parts of units—was moving behind us. We went through 4th Brigade headquarters area at Belhamed and then on to Ed Duda, passing the infantry in their slit-trenches. Some of them were asleep and our drivers were desperately afraid of running over them. We were in single file because there were minefields on both sides of the track, and that was another worry. German flares were disconcertingly close and we could hear machine-gun fire. At Ed Duda, or near it, we turned north-west, and we guessed then that we were going to Tobruk. This was confirmed soon afterwards by Tommies —men of the Essex Regiment—who spoke to us from the darkness in their slow, warm voices. As long as we were driving we were strained and jumpy, but at every halt—and halts became more and more frequent—the tightness inside us unwound and billowing layers of exhaustion, like an eiderdown quilt, like an anaesthetic, came down and smothered us. Engines were switched off and sometimes it was very quiet. A cough, the squeak of a tank, the faint tapping of a machine gun, came to us close and clear like the noises that cattle make near at hand in a mist. Often a sentry would walk over to us.

'How is she, Kiwi?'

'Good. How's the war going?'

'All right, Kiwi.'

'Whose flares, d'you reckon?'

'Dunno. Must be old Jerry's.'

Silence then, and an upward jerk of the chin to avoid sleep as a swimmer avoids a wave. Sometimes the wave was too tall and it would break over its victim, drowning him quietly and suddenly —but only for a short time. He would wake up, perhaps a quarter of an hour later, to find darkness in front of him and behind him a column of silent vehicles of which he was the self-appointed leader. He would start his engine in a panic and push on as fast as he dared, praying that some sixth sense was leading him in the right direction. It was easy to make a mistake, for on both sides of the track wheel marks led confidently into the darkness, clamouring to be followed. In an agony of indecision he would make his choice and go blundering on through the night, until at last to his enormous relief the outline of a lorry appeared ahead of him. Now someone else was doing the leading. Now it was another's responsibility if half the NZASC ran into a minefield or ended up in the German lines.

Even for the Major, who was accustomed to leading large convoys under difficult conditions, it was a night of appalling anxiety. With drivers falling asleep at every stop and others following strange leaders and passing him in the darkness, he found it impossible to keep us together or watch over us. Soon after midnight he discovered his leading vehicles in a minefield, and with trampled tape all round him he directed them while they turned in the width of the track and retraced their wheel marks.

By three in the morning only a few 13th Corps vehicles were ahead of him. He was told by a British officer that the enemy was on all sides and was advised to travel on a bearing of 317 degrees, the officer adding that it was essential to get the transport clear of the corridor by daylight.

Just before dawn, by which time the bulk of our unit had passed through or gone round the whole of 13th Corps headquarters, the Major discovered that he was being followed by only half a dozen vehicles. He halted, and Corporal Bev Hendrey went back on his motor-cycle to try to find the rest. The next section of the convoy was half a mile behind, a sleeping driver heading a long line of other sleeping drivers, most of whom were ours.

We were led through the rubbish of a battlefield, among which infantry were standing-to in slit-trenches, and at a quarter past six, while it was still dark, the head of the huge column—it has been said that something like a thousand vehicles passed through the corridor that night—crossed the perimeter of Tobruk. Company headquarters was in the van and a party of our Don Rs led the way in. Behind them was the first column to enter Tobruk since the Australian 9th Division, seven and a half months earlier, had halted there on its way through Cyrenaica—had halted and fought.

'It's bloody Kiwis!' said three tired, dirty infantrymen. For thirty weeks—thirty aeons—they had been beleaguered in the world's most desolate and dangerous spot and now they were so pleased and excited they could only swear. Doubtless, under that pale dawn, they saw us as liberators. Dancing before them were visions of clean beds, Cairo food, Cairo girls, liquor. It was not for us to disillusion them or tell them of our suspicion: no one would be leaving Tobruk; people were coming in. The General commanding 13th Corps put the matter in a nutshell: 'Tobruk is relieved but not half so relieved as I am.'

By daylight nearly all our vehicles were inside the perimeter but we were not yet out of the wood. Heavy-calibre shells began feeling for the long column in which we were jammed nose to tail. They groped blindly across the desert, found the roadway, then overshot it. They tried again and this time did better. The column kept halting and it was impossible to leave the road because of minefields. When we did move it was at a maddening crawl, but our luck was in.

We got clear after what seemed an age, and the Major's logbook takes up the story:

> At 0800 hours Corps Headquarters directed us to unit area, but owing to various Corps officers giving different instructions the unit moved three times before it finally halted late in the morning. We had travelled 20 miles in 14 hours. At once the vehicles were sent to 108 AA Site to load ammunition. Fresh rations drawn and nearly empty water trailers refilled. Rations included bread. Drivers very tired, and as soon as they had completed maintenance of vehicles they were ordered to rest as much as possible.

When we had finished our jobs there was still some pale daylight left but no one wanted to have anything to do with it. The scene that surrounded us was unbelievably dismal and it was a pleasure to draw down the canopy and shut it out. Hardly a soul was abroad and the shabby lorries, with no human figures about them, looked lonely and dead-tired. Nothing can express weariness better than an army lorry. They looked not only weary and solitary, but, in some indefinable way, disgruntled. Their present surroundings, it seemed, accustomed though they were to salty and stubborn pastures, were a little too much even for them. It was as though the charred skeletons of the British and Italian lorries that littered the neighbourhood were able to remind them of mortality. It was as though they were feeling in a queer way—and an old lorry (you think with it and for it so much) is almost sentient—that no tenderness of ours, no oiling, greasing, 'four-o-sixing', could aid them against their last enemy. To just such desolate cemeteries all vehicles were bound.

And indeed, with the wind blowing thin and mean, the prospect they looked out on was so bleak, so barren, so starved and withered, that you said to yourself: ' The thing's dead. It's a bit of the world that went bad. They cut it off and threw it away.'

Our area was in the eastern part of the fortress and the ruined town with its harbour full of lost ships was hidden from view. Looking seawards, one saw only naked hills, framing here and there a fragment of the Mediterranean—hard and lustreless as blue china broken on a rubbish heap. Inland there was nothing but minefields, tangles of barbed wire, wrecked vehicles, abandoned gun positions, shell cases. The hillsides, the minefields, the area we were parked in, all were sprinkled liberally with little pieces of shrapnel. It was as though the owner of the estate, the demon who inhabited Tobruk, had gone forth to sow, and in scorn and mockery of all fertile things had sown shards.

CHAPTER 11

PRISON AND THE MUSHROOM COUNTRY

THROUGHOUT the last day of November the battle went on, the Germans fighting to close the Tobruk corridor, our forces to keep it open.

At 9 a.m. we were ordered to send eighty vehicles to serve the 4th and 6th Brigades, but the Tobruk Control Post said that a convoy of that size would not be able to get through. Finally fifteen lorries loaded with 25-pounder ammunition left at noon under Captain Gibson.

As they were moving out of the fortress they came under shell-fire but no damage was done. They moved slowly towards the Ed Duda escarpment, skirted the Essex Regiment, and travelled for a mile and a half along the *Achenstrasse*, the 40-mile road built to by-pass Tobruk. Turning left, they found the 4th and 6th Brigades a short distance south of Belhamed. The infantry were shivering in their slit-trenches, damp and frozen but still indomitably cheerful. The campaign had yielded a good harvest of loot—pistols, binoculars, cameras—and they were like children on Christmas morning.

Half the 25-pounder ammunition was delivered to the 4th Field Regiment and half to the 6th. The latter was down to its last few rounds and our drivers went straight to the gun positions and flung off their loads. Scarcely had they done so than tanks came out of the setting sun and the guns opened fire. From tray to breech, from breech to enemy tank, and all in a matter of minutes—that was how it had happened in our dreams and that was how it was happening now. It made up for any number of disappointments, frustrations, and seemingly fruitless journeys.

With 200 prisoners aboard and the guns firing around them—and not firing only but fighting, hitting out in a hot rage so that the desert shook and quivered—the lorries were hustled from the area. Presently, looking back, our drivers saw a ring of fires burning rosily in the half-light. They were supremely content, never doubting that each fire was a tank.

Doubt came the next morning—it was 1 December—when we

heard that Sidi Rezegh was again in German hands, and it deepened when we were told not to reload our lorries. We were to be used as garrison troops.

All through the day rumours of disaster, like evil birds returning to their grim roost, came back to Tobruk, and we were asked to believe, not for the first time in our Army careers, that the Division had been cut to pieces.

During the afternoon four lorries loaded with engineers' stores and mixed ammunition set out under Second-Lieutenant West-Watson[1] to serve the 14th British Brigade, which was hard pressed on the left flank of the 4th Brigade a few miles north of Belhamed. The Ed Duda route was now closed and the lorries had to travel east along the Bardia road before bearing inland. Even the infantry guides from Headquarters 14th Brigade were uncertain of the way, but at last, after being turned back by unidentified motor-cyclists and coming under mortar fire, the convoy arrived safely at the command post of the Oxfordshire and Buckinghamshire Light Infantry, for which the ammunition was intended. It was now just on dusk.

'I was told,' said Second-Lieutenant West-Watson, 'to keep the loads on wheels during the night, but as the enemy held an escarpment overlooking the post this seemed to me to be courting disaster. Accordingly I asked to be put through on the telephone to our own Divisional Headquarters so that I could get the order confirmed.

'Meanwhile the enemy had attacked and for an hour he subjected the post to heavy mortar and machine-gun fire and tried to range his mortars on the transport. Our only casualty, however, was a driver who sprained his knee while scrambling for cover.

'The enemy then broke through and drove the infantry on to the post itself. I withdrew our drivers to a nearby hollow occupied by the battalion headquarters, chose fire positions, and waited. The expected attack never came because British tanks arrived and with their support our troops were able to counter-attack.

'During the fighting that followed one lorry went forward under infantry escort with a load of mixed ammunition, and while it was away we unloaded the others in accordance with orders that had come through. As soon as the lorry returned we set out for home.'

[1] Capt A. M. W. West-Watson; branch manager; Auckland; born Carlisle, England, 31 May 1918.

While this small convoy was slipping towards Tobruk under shellfire, remnants of the New Zealand Division were heading for Egypt.

The attempt to keep open the corridor had failed and Tobruk was again isolated.

. . . .

Rumour had it that we were to be evacuated by sea and the prospect pleased hardly anyone. We had no objection to leaving Tobruk, which seemed to us not so much God-forsaken as never to have been visited by Divine Providence, but we did object strongly to the idea of abandoning our lorries. However, after being wholeheartedly accepted for a few hours, the story, on as little evidence, was as whole-heartedly rejected, and during the days that followed our employment was so strenuous that Rumour, which has an objection to busy men, wasted little time on us.

Day and night, in working parties of from thirty to a hundred, we laboured at the docks, handling petrol, ammunition, and food. Every night there was at least one air raid and often there were several. Sometimes a lone raider would make a hit-and-run attack. Sometimes four, five, or six bombers, converging on the target from different directions, would dive through the ack-ack barrage and drop their bombs in or around the harbour. Our greatest danger was from falling nosecaps and shell fragments, for the barrage was terrific—fantastic. It arched above us like the red ribs of a gigantic burning lobster pot and the noise was physically painful to listen to. There was the deep crash of the three-point-sevens, the pom-pom-pom-pom of the Bofors, the heavy stammer of the point-fives, and the light clatter, no louder in that colossal din than the tapping of a typewriter, of the Bren and Lewis guns. The technique that had saved Malta was being used. The guns were firing on fixed lines, making every square foot of the sky perilous. Only the machine guns quested at will, their red tracers, like angry bees, swarming from target to target. It was breath-taking to see aircraft fly through this inferno apparently unscathed, but they seldom stayed long enough to aim properly and damage to shipping was rare. They dropped their loads almost at random and made off at once.

Sometimes the barrage would end raggedly, with odd bangs and slams and stutterings continuing after most of the guns had ceased fire. At other times it would end as though in obedience to a conductor's baton, and then, in the immense silence that followed, you could hear the delicate patter of falling shrapnel, and occasionally the wicked swish-swish-swish of a falling nosecap or a loud clang as a fragment of shell case landed on a metal deck. The dark waters of the harbour would be ringed as by rising fish.

We worked in the ships' holds, loading the slings in which the cargo was transferred to barges and landing craft, and on the wharves, carrying an endless succession of crates, boxes, and tins to the waiting lorries, a number of which were our own.

At first, when the siren sounded, we used to go at once to the official shelters, which consisted of deep caves near the wharves, but after a while, disgusted by this waste of time, we took to ducking beneath the nearest cover, starting work again the moment the raid ended.

Unloading petrol from a motor ship was one of our first jobs. Dave Falconer[2] describes what it was like:

'When we stopped working we got as cold as hell. Our clothes were soaked in petrol and the stuff was washing about in the scuppers. This was because the petrol was in "flimsies" and in every slingful about half a dozen tins were crushed. On the first night of the job the hatches were closed on account of the danger from red-hot pieces of shrapnel, but on the second night, owing to the urgency of the work, they were left open. That night the first plane came over at nine. The beggar cut his engine and glided in unobserved—there was only one plane—and let a bomb go near our ship. He may have caught on that we were handling an important cargo, for about half an hour later we were attacked again —by the same plane we thought—and three bombs landed about a hundred yards away. We felt the blast in the hold.

'Before that we had been scuttling aft during raids to shelter under the poop but from then on we stayed in the hold. We looked at it this way: the whole ship was swimming in petrol and either she went up or she didn't.'

On 3 December we help to unload an ammunition ship—a ship whose arrival had been awaited with great anxiety, as two days

[2] Dvr D. Falconer; truck driver; born Glasgow, 30 Apr 1914.

earlier the supply of 25-pounder ammunition in Tobruk had totalled only 4000 rounds.

In some ships the winches were operated by British soldiers, often with less skill than enthusiasm. More than once our drivers were fascinated by the sight of a heavy bomb swinging nonchalantly among booms and stanchions, and on one occasion a 500-pound bomb was dumped unceremoniously on the deck a few inches from the ship's side. There it dangled, half in and half out of the sling, while the onlookers, forgetting what they had learnt about detonators, wondered if it would explode at once or wait until it toppled into the lighter below. On the petrol ship, too, the winchman was sometimes at fault, but never more so than when he jumped overboard during an air raid and swam ashore.

In spite of these anxious moments it seems safe to say that most of us enjoyed working in Tobruk harbour, though enjoyed, perhaps, is not quite the word. It was like a football game in which you would sooner not have played but are glad afterwards that you did. At all events we discovered that hard physical labour is an excellent nerve tonic. Backwards and forwards we went hour after hour, carrying boxes from the dim-lit bowels of the invasion barges to the waiting lorries, crossing and re-crossing the landing ramps, which rested on the quaysides like dropped jaws. The sweat trickled over us like warm oil and the urgent rhythm of the work was as soothing as music. How much more pleasant it was than crouching in a cold slit-trench and suffering the complete nightmare from ambulance to operating table with the passage of each plane!

Many of us, too, were affected by the poetry of the scene. All the business of the port was being carried on in the dark. Winches rattled, voices shouted, tugs fussed, lorries arrived. 'Right hand down. Bring her back on that. Keep coming. HOLD IT!' In the background were the lost ships, which, like ghosts, were the more powerfully present for being hidden. We had marvelled at them by daylight. Everywhere masts and funnels poked skywards at crazy angles. There was a small steamer with only her wheelhouse, like an old-fashioned bathing machine, showing above the surface. There was a tramp with her side torn out, so that you could see the grey water sloshing along the catwalks in her engine-room. Near the harbour's mouth there was the Italian cruiser *San Giorgio*, black with rust, and across from the wharves, on the far side of

A GROUP NEAR TOBRUK
(L. to r.) H. S. Jones, P. E. Coutts, D. C. Ward, F. G. Butt,
K. E. May, S. A. Sampson

Tobruk Harbour in August 1941

Stuka attack on transport south-west of Gazala

Divisional Ammunition Company dispersed

Ammunition Company Workshops at Bir el Thalata

the harbour, an Italian liner was aground, looking as though she had tried to rush into the hills. There were dozens of drowned ships.

Sometimes the harbour was shelled but no one took notice of that. We stopped work only for the banshee howling of the siren, which was the perfect voice for the demon that inhabited Tobruk. Then it was as though the night were being operated on without ether. While the sweat dried on us we waited for the hacking, chopping, thwacking of the many guns, and for the sky to be torn open marvellously and filled with streamers, loops, and whorls, as of scarlet and gold viscera.

No doubt a little of our contentment at that time can be attributed to the rum that was always available at the docks. We referred to it with affectionate deference as 'Tom Thumb' and agreed that it was even more potent and unpredictable in its effect than 'Uncle Joe', our name for zibbib. Most of us took it in the form of café royal, which we made with cold coffee and condensed milk and found delicious. We helped ourselves liberally to cigarettes from the NAAFI supplies we handled and it was wonderful not to have to save butts any longer. During the past fortnight some of us had been reduced to smoking cigars (Italian Army issue), than which no more powerful emetic exists, and a few of us had experimented with tea leaves.

During daylight, much to the annoyance of the night shift, the Luftwaffe concentrated on supply depots rather than on the docks, and from the unit area we saw several exciting raids. On one occasion a large bomber was brought down, and on another we were moved almost to the point of cheering by the behaviour of a tiny minesweeper, which, with bombs throwing up water spouts all round her, calmly went about on arriving at the end of her beat and steamed back on a parallel course, defending herself vigorously all the while.

At night, to the annoyance of the day shift, we were kept awake by a nearby battery of three-point-sevens and by the noise from the harbour, but if we had lost nothing except sleep during our stay at Tobruk we should have been well content. Unhappily there were casualties. Lennie De Pina[3] (A Section) was wounded in the leg by a stray piece of shrapnel, and Colin Cameron lost his life.

[3] Dvr L. De Pina; motor driver; born NZ, 2 Sep 1917; wounded Dec 1941.

The slight wound he had received earlier in the campaign had not mended and on 5 December he was evacuated from Tobruk in the merchant ship *Chakdina*, which was sunk by an aerial torpedo.

And we lost one life through sickness. Cliff Collie[4] (B Section) was taken ill in the night and died before reaching hospital. That closed our casualty list for the campaign, making it: killed, 6; wounded, 11; prisoners (at that time), 66; died of sickness, 1; total 84.

* * * * * * *

Meanwhile, in the desert outside Tobruk, defeat had become victory. The scale had been tipped by the Royal Air Force, by the skill and daring of 'Jock' columns—fast-moving, hard-hitting raiding parties named after their originator, Brigadier Jock Campbell, VC—and by the fighting qualities of the ordinary British Tommy *vis-à-vis* those of his German counterpart. By 5 December the corridor was open again, and two days later Rommel broke off the battle and began to retire westwards.

On the 7th we dumped our ammunition, and the next morning while the world was ringing with two words—Pearl Harbour—we left Tobruk, heading for the Wire with the rest of the Administration Group.

We were a sadly truncated unit. A Section, with a sub-section from Workshops, had been attached to a Composite NZASC Company newly-formed under Captain Roberts to serve the 5th Brigade; B Section, with five lorries from A Section, had stayed in Tobruk to serve elements of the 4th Brigade, and fifteen of our spare drivers with 172 others from the NZASC had stayed to do garrison duties.

The rest of us moved through the Wire at El Beida late in the afternoon and the next night found us a few miles from the Siwa road. On the 10th, at five in the afternoon, we halted in our old area at Fuka, from which we had been away a little less than a month. The vehicles came to rest in their old parking places and we jumped out to inspect our dugouts, which we found half-filled with loose sand but otherwise habitable.

As we boiled our billies that evening we were conscious of an agreeable symmetry in our affairs, a soothing pattern. The pity of

[4] Dvr C. L. Collie; barman-porter; born Sefton, 31 Oct 1914; died of sickness 4 Dec 1941.

it was that not everyone had returned to the starting point. Many were prisoners, some were in hospital, and some had gone out of the world altogether. But mixed with our melancholy was a certain smugness perfectly compatible with sincere grief. The conviction was upon us—the survivor's unwarrantable conviction—that in avoiding pitfalls into which others had fallen and returning unscathed to Fuka we had evinced judgment: had, in fact, shown merit. And our absent friends would have been the last to cavil at this: it was a feeling they would have shared.

The next day we resumed the pleasant seashore life that the campaign with Rommel had interrupted and two days later B Section returned to us. On the day after that the unit logbook contained this comfortable and laconic entry: COY AT FUKA.

But the statement was too comfortable and too laconic to be comprehensive. It took no account of A Section, which was now at Acroma, some fifteen to twenty miles west by south of Tobruk, girding its loins (it hoped) for the pursuit to Tripoli, or of the fifteen spare drivers who had been left behind in Tobruk (they were working on the wharves and being bombed nightly), or of the sixty-six members of our unit who were freezing and starving in the prison compound in Bardia.

They must have guessed that by now they had been posted as missing and struck off the unit strength, but they kept together still and they still regarded themselves as members of the Ammunition Company, though on bad days their time with the company seemed remote and clouded like childhood and fugitive like a dream. Always when they had been away from it before—on leave, on courses, in hospital—it had continued to govern and condition their lives: the parent unit. It was a train that could be caught up at the next station, and in that train above a certain seat in a certain compartment in a certain coach was a notice: RESERVED. Now it was a train that had dashed off into the darkness, none knowing how far it would travel or in what direction. At last, after a decent interval, the ticket-collector would come down the corridor, see the RESERVED sign, and take it away. Then a stranger would sit down.

The prisoners daily grew weaker and sometimes they felt dizzy and light-headed. Then nothing that had happened to them *before* and *outside*—not their homes, nor the Ammunition Company, nor the girls they remembered—had any reality beside the icy fingers of wind that tugged at the overcoats and dirty blankets beneath which they huddled close for warmth in hollows scooped out with steel helmets, the crying need for a cigarette, the grey bread and black coffee they had been given for breakfast, the macaroni, dried potatoes, dried onions, dab of sauce, small piece of bread (it was neither as filling nor as appetising as it sounds) they could look forward to for dinner. Only these things were real.

Sometimes one hardship seemed important to the exclusion of all others. Sometimes it was dysentery—many were suffering from this. Sometimes it was a smaller thing: trouble with dirty false teeth, not hearing from home, long finger-nails that broke when you tried to scratch yourself through your clothes.

Time meant nothing in the prison compound. Our drivers' watches—the ones that had not been confiscated on the first day—had gone to Italian guards in exchange for scraps of food or a few cigarettes. (One prisoner handed over a gold watch for a small black loaf.) It was enough that every morning the sun elbowed itself over the grey rim of the horizon as from a bath of filthy, freezing water, to shine dismally for a few hours or to be obscured at once by drifting veils of rain. It was enough that three times a day prison meals were served.

For the prisoners to say that they had been only two and a half weeks in Bardia meant nothing. The events of their captivity, spaced by an eternity of boredom and discomfort, stretched behind them like a perspective of telegraph posts, reaching as distant an horizon as any they had ever known. They saw, smudged at the end of the long line, the morning on which the two parties—Sergeant Mellows's drivers and the party from Sidi Azeiz—had met in prison. That was on 28 November. After the meeting they had been searched, first by German and then by Italian guards. The latter confiscated compasses, service watches (and some private watches, too), knives, razors, and anything else that took their fancy. Then about a third of the prisoners had been marched to a larger compound full of rocks and rubble, where they were joined by the rest a day or two later. Except on the north side, where it was divided from Bardia by the blank walls of houses, the new compound was bounded by a wall nine or ten feet high. The next day, to facilitate the issue of water and rations, the prisoners had been sorted into groups of thirty under their own NCOs and some old Italian mess tins and water bottles had been handed round. Then the camp sergeant-major, WO I O. A. Wahrlich[5] (5th Field Regiment), had taken the names of those who had neither blankets nor overcoats. When the blankets arrived they were filthy and there were not enough of them.

The sound of falling bombs had been comforting at first, but on 1 December nine Marylands dropped their loads near the compound and on the 2nd there was another raid, even closer this time. When that happened our drivers were filled with unreasoning bitterness. That day it rained and the camp sergeant-major wrote in his diary:

What a hell of a condition for men to live in! Some will die of exposure unless something is done soon.

And three days later: Cold again. Sick parade was attended by MO. Germans appear to have no medical supplies. Bread rather good but loaves appear to be getting smaller. Rumour that some of the officers had been taken away. Men getting a little down. Quite a number of dysentery cases. Cold at night and dust-storms as well. Saw men fighting over scraping out a dixie to get more to eat. No sign of RAF during the day but quite a bit of bombing at night.

Although the officers had been imprisoned in another compound our drivers felt more unprotected than ever when it was learned that they had been taken to Italy by submarine, Second-Lieutenant Duke with them. Now there was no authority to appeal to, however uselessly, when the Italians made unreasonable demands, as they did on the 6th. On that day Claude Campbell and others were told to clean up an area some distance from the compound, but when they got there they refused to work. They felt sure the enemy meant to use the area as a supply dump; apart from this it was littered with disgusting filth and was a favourite target of the Royal Air Force. The camp sergeant-major was sent for and he told the Italians that the order was an improper one, on which the officer of the guard flew into a passion, threatening to keep the party there without food or water until the job was done and adding that he would not hesitate to use machine guns if there was any more trouble.

[5] WO I O. A. Wahrlich, EM and clasp; lorry driver; Dunedin; born Dunedin, 11 May 1910; p.w. Bardia 27 Nov 1941-2 Jan 1942.

In face of this unanswerable argument a start was made, but the work was interrupted before long by a visit from nine Marylands. The Italians at once took to their heels and so did the working party. 'But we did not run,' said Claude Campbell, 'either as far or as fast as the Italians.'

In the midst of their miseries and discomforts our drivers were buoyed up by the belief that Bardia would fall soon. Brigadier Hargest had spoken to some of them on the first day of their captivity, telling them to be of good cheer because they would be free in a few days. They clung to this promise and they were quick to notice how on some days their guards were milder and more friendly.

Then Japan attacked Pearl Harbour and the Germans and Italians became twice as arrogant as before. Gloatingly they pointed out that New Zealand was now in the danger zone, and soon they were able to speak of the sinking of the *Repulse* and *Prince of Wales*. Our drivers did their best to discount this claim as propaganda and they repeated their own stories—Turkey was in the war on our side and the Russians were rolling up the German Army—but in their low state it was difficult to remain cheerful and confident.

They were haunted always by the fear of being sent to Italy, and fear became panic on 15 December when a hospital ship was seen approaching the harbour. Contradictory announcements were made in quick succession. Only the sick and wounded were to go. Everyone would be going. No, it was the sick and wounded plus all who had attended sick parades. These would not fill the ship so some fit men would have to go as well. A list was being made out.

There was no further change of plan and later in the day 300 prisoners, several of whom were members of the Ammunition Company, were paraded and marched down to the jetty. There they were kept waiting for about an hour, at the end of which a Medical Corps major began questioning some of the less obviously sick. In nearly every case he was given the same answer: 'I'm as fit as a fiddle. There isn't a thing wrong with me.' Whereupon he delivered a lecture on the Geneva Convention and ended by ordering all fit men back to the compound. Back they marched, happy as though the compound were paradise.

Among those who did not return—WO I Wahrlich was told later that sixty-two New Zealanders had been embarked—were Corporal Nigel Barach, Lance-Corporal Stan Wrack, Owen Blomfield,[6] Pat Dooley,[7] Bill Gamble,[8] and George Jeffrey,[9] all of whom, with the exception of Stan who had gallantly volunteered to go to Italy as a medical orderly, were sick and exhausted.

The rest could still hope, so life went on in the compound, flickering a little more feebly each day. Weakened by hunger the prisoners suffered increasingly from cold and at times they were conscious only of an empty, aching stomach to which was attached a pair of enormous numb hands and a pair of enormous numb feet. But they could still *baa* provocatively when the Italians herded them into line for roll-call; they could still discuss the gargantuan meals they would order when they reached Cairo (a whole chicken, six eggs, chips . . .); they could still, in the intervals of lying together among the rubble in inert filthy heaps, grumble a little, play poker or two-up, laugh a little. Together they would chant the schoolboy quatrain—'Cold as a frog in a frozen pool'—and someone would say: 'Put a sock in it, you jokers. Go help Fred catch a seagull.'

. .

Meanwhile Rommel was being hard pressed at Gazala, some thirty-five miles down the coast road from Tobruk. Taking part in the battle under the command of 13th Corps was the 5th Brigade, which had left the Sollum-Capuzzo area on the 9th. A Section, as part of the Composite NZASC Company serving the brigade, had gone first to El Adem, camping for a day or two on the outskirts of the great Axis airfield there. Here the section was joined by 'Dad' Cleave and Basil Thorburn who had been given up for lost (they had been with the Supply Column all the time), and here Jim Stanley was wounded in the thumb by a small explosion that occurred while he was inspecting captured equipment.

[6] Dvr O. L. Blomfield; butcher; Waitoa; born Te Kuiti, 28 Feb 1914; p.w. Bardia 27 Nov 1941; evacuated to Italy 15 Dec 1941.
[7] Dvr P. A. Dooley; farmer; Wyndham; born NZ, 28 Aug 1913; p.w. Bardia 27 Nov 1941; evacuated to Italy 15 Dec 1941.
[8] Dvr W. J. Gamble; storeman; Otahuhu; born NZ, 15 Feb 1916; p.w. Bardia 27 Nov 1941; evacuated to Italy 15 Dec 1941.
[9] Dvr G. F. Jeffrey; truck driver; Matamata; born Kakahi, 30 Jul 1913; wounded and p.w. Bardia 27 Nov 1941; evacuated to Italy 15 Dec 1941.

From El Adem, still serving the 5th Brigade, the section had gone to Acroma, twenty miles west by south of Tobruk, and from there to Gazala, from which Rommel had been driven on the 17th. On 23 December while the Eighth Army continued its advance—Derna had been occupied four days before—the 5th Brigade began the long journey to Baggush, A Section picking up the 23rd Battalion at Gazala and taking it as far as El Adem, where the drivers spent Christmas Day. For lunch they had bully beef, but it was followed by a fruit salad, the ingredients of which the cooks had been saving for a long time.

In Tobruk, in spite of an unpromising start—there was a duststorm and a heavy air raid in the morning and for some reason the NZASC party had not been included in the garrison strength for the issue of Christmas comforts—the unfortunate fifteen did tolerably. For dinner they had tinned turkey, tinned chicken, and tinned plum pudding (an issue they made to themselves), and someone produced 60 litres of bad wine.

At Fuka the fare was frugal, for it had been decided to postpone the Christmas festivities until the unit was complete. Nevertheless it was as pleasant a day as Christmas can hope to be without children, and if we enjoyed it more as a remembered happiness or as a happiness experienced vicariously—in New Zealand the red flowers were on the pohutukawa, yachts were sailing in the Waitemata, nephews, nieces, and small brothers were shrieking to high heaven, and the 5-gallon kegs were in the sinks—it was none the worse for that.

Bev Hendrey describes the day in his diary:

Most of the blokes writing home. A few of them got primed. Very quiet on the whole.

    .   .   .   .   .   .

In Bardia the prisoners fared better than they expected. A few days before, Major-General Artur Schmidt, the German commander, had paid them a visit.

'He had grey hair, pale-blue eyes, and a hard mouth—one of the old school,' said Claude Campbell. 'But his manner was friendly and when Harvey hit him up for a Christmas dinner he said: " I will promise you nodding. Your comrades outside have cut our lines of communication." He added something to the effect that we should get our Christmas dinner if the position was restored in time but we told him what he could do with it. The chaps complained about the conditions and the food and the old boy said: " This desert warfare is very uncomfortable." To which someone replied: " You said it, pal!" '

After this unpromising conversation the Christmas rations came as a pleasant surprise. Each man was given ten cigarettes, two packets of biscuits, and two or three sweets, and the usual issue of coffee and rice was augmented by an issue of bully beef (three men to a tin), cheese, jam, sugar, and cognac. From this the cooks were able to prepare two good meals, one of savoury rice, with cheese and bully, and one of sweet rice. The cognac went into the coffee. That night for the first time in twenty-seven days our drivers went to bed on full stomachs. They were comforted, too, by the feeling of having feasted from the barren fig tree. Fresh water from the rock in Horeb was hardly a greater miracle than good cheer in that flinty compound.

On Boxing Day, as in prison and workhouse the world over, it was a case of back to porridge and old clothes, but the confidence that freedom was not far away grew stronger daily. Since 16 December the prisoners had been guarded only by Italians, all Germans being needed for the front line. Each day the air raids on gun positions and supply depots became fiercer—on one occasion bombs landed only twenty-five yards from the west wall of the prison, one prisoner being killed and four wounded—and nightly the artillery bombardment was intensified.

As their eventual release was now almost a certainty few of the prisoners thought of trying to escape; indeed, as far as our drivers could discover, only one man made a serious attempt. He was Jacky O'Connor[10] (B Section). On the night of the 27th-28th, taking nothing with him, not even his greatcoat, he scaled the prison wall, hung for a moment by his finger-tips, and dropped into the darkness. He was risking a bad injury—for at the point he had chosen (it was where the latrine abutted upon the outer wall) the ground sloped steeply—but he escaped with nothing worse than a twisted ankle. Guided by the British barrage and depending on it to drown the noise he was making, he crossed a succession of

---

[10] Dvr J. S. O'Connor; painter; born NZ, 15 Feb 1915; p.w. Bardia 27 Nov 1941-2 Jan 1942; died of wounds, 14 Jul 1942.

deep gullies. He had relied on being clear of Bardia by dawn, but the going was much harder than he had expected and he was hindered by his ankle. When daylight came it found him on the side of a gorge among German infantry positions. There was nothing for it but to creep into a patch of camel-thorn and lie there all day without moving a muscle. Without water, food, or overcoat, and with so little freedom of movement that the small desert birds scuffled in the sand only a few feet from his face, he lay still for hours, hating the daylight as it has seldom been hated before and waiting for nightfall as a man waits for his girl. It came at last, late but beloved, and he moved forward cautiously until he was among the outermost German pillboxes. The escaper's supreme moment was now upon him: safety in sight and the worst dangers all about him like a thicket; on the one hand, the fun of strolling into his own lines a free man, free by his own nerve and efforts; on the other, the disaster of being dragged back to prison ignominiously or shot down like a dog.

But escapers need luck and Jacky's was dead out. He disturbed some Germans in a pillbox and began to run. There were guttural shouts from three sides of him and rifles stabbed the darkness with flashes of smoky orange. He yelled '*Kamerad!*' and the Germans came up to him.

Instead of being returned to the compound he was placed in a building outside it, and the Italian guards slyly allowed Corporal Jack Moore,[11] the commander of Group 8, to which most of C Section's drivers belonged, to continue reporting him present. When they were tired of the joke they turned nasty, threatening Jack with a diet of biscuits and water and the whole of Group 8 with the dreaded punishment of a submarine trip to Italy.

Neither threat was carried out, for by now it was plain to both prisoners and guards that their roles would shortly be reversed. On 30 December it was learned that Benghazi was in British hands, and that day each man was given ten cigarettes and there was a double issue of rice, not to mention several unexpected issues of civility.

On New Year's Eve the camp sergeant-major wrote in his diary:

At 0430 hours the attack commenced. A bombardment was

[11] Cpl J. B. Moore; carpenter; Cambridge; born Cambridge, 26 Jan 1909; p.w. Bardia 27 Nov 1941-2 Jan 1942.

kept up for more than two hours and then some tanks appeared over the horizon. Boys as excited as hell. Well, it's 1200 hours and they still seem to be cracking away and a 2-pounder has just whizzed down the gully. The idea of being a prisoner and more or less in the firing line as well is not so good. Still, here's hoping. Tons of ammo have been expended today and it can't go on like this for long. Rifle fire can be heard. Men laying odds about how long it will last. Truck arrives with water. Tanks with infantry behind can be seen approaching over the hill. Rifle fire a little closer. Brens can be heard in action.

New Year's Day was cold and bleak. Showers of rain swept over the compound and a savage wind tugged and tore at the miscellaneous rubbish from which our drivers had built shelters. The noise of the fighting was not loud but everyone knew that the end was very near. The Italian guards treated the prisoners with a solicitude that was almost tender.

January 2 (wrote the camp sergeant-major). Heavy bombardment last night. Quite a lot of shell splinters flying round our area. Garrison surrenders. . . .

At 10 a.m., after a two-hour truce, Major-General Schmidt surrendered unconditionally to General de Villiers, commander of the 2nd South African Division. The prisoners had been told earlier that their friends would be with them at 9 a.m.

In Bardia there were 1100 prisoners, gaunt, bearded, filthy, wobbly at the knees—1100 scarecrows with hearts as light as puffballs. After the first flush of joy they became 1100 craving stomachs.

'Jokers from the Div. Cav. were the first troops we saw,' said Claude Campbell, ' and as soon as they arrived they started to dish out cigarettes, tucker, and their own Christmas parcels. Don and Harvey and I got a cake. We ate it straight away and it made us crook. We were still hungry, though, and for the rest of the day, off and on, we were eating or trying to eat.'

Clarry Monahan and a friend breakfasted with German officers in a well-found mess.

'After breakfast,' said Clarry, ' we still wanted more, so we set off down a steep hill to the waterfront, where we remembered having seen a bakehouse when we marched down to the hospital ship. We were as weak as kittens but excitement kept us going. We missed out at the bakehouse, so we went over to some caves that looked as though provisions might be stored in them. Here we struck it lucky and we each filled a sandbag with Jerry pork and beans. We struggled back to the compound and dished the stuff out to our cobbers. It made them crooker than hell.'

They ate all day, many of them, with a kind of sterile lust, just for the pleasure of feeling the meat in their mouths. Dusk found them lying ill in their old places in the compound heaped with enemy blankets. The next morning, suffering from every known variety of stomach disorder, they set out for Maadi.

At Maadi they were rested and fed up (in both senses of the word) and as the weeks went by there was some danger of their forming themselves into an exclusive Bardia society. Other members of the unit who were at Base—drivers discharged from hospital —waited anxiously for our return, refraining in the meantime from complaining too loudly of cold when anyone from Bardia was present or stressing unduly that slight feeling of emptiness that assails the ordinary soldier between tea-time and bed-time.

Cold and hunger had been established once and for all as their special province—and you cannot argue with experts.

. . . .

As early as 16 December our unit had been reorganised as a reserve mechanical transport company to serve the Eighth Army, forty-seven vehicles with appropriate personnel being attached to us from other NZASC units and thirty-six from the Artillery. From then on we were under three hours' notice to move but it was not until five days after Christmas that we left Fuka for Bir el Thalata to start our new duties. A Section, which had moved from El Adem on Christmas afternoon with the 22nd Battalion aboard, was already there, and the Tobruk party rejoined us during the day, so the unit was complete again.

By this time Rommel had withdrawn to positions—temporary ones—at Agedabia at the bottom of the Benghazi bulge, and the forward units of the Eighth Army, tired now, very thin on the ground, and worried by supply problems, could do little until Benghazi port was made usable and reinforcements arrived. That was the position on 2 January when A Section ('favoured A Section,' said some, though the Major was celebrated for his impartiality) embussed the 1st Welch Regiment at Thalata and headed west, leaving the rest of us to the pleasant but not wildly exciting task of shifting miscellaneous supplies from the railhead to Tobruk.

There was a holiday feeling in the air as A Section's convoy, with the Welch sitting sedately in the backs of the lorries, slipped along the coast road, the verges of which were littered with burnt or abandoned Italian vehicles: big diesel jobs—Fiats mostly, and Lancias. Engines purred and hummed, running like a dream, and the road's tarmac surface, compared with rough desert tracks, was like a springboard. The Mediterranean, sometimes grey and tumbled but hyacinth blue and apple green on sunny days, was often at the drivers' elbows and as they drove they sang; and so they came to Derna.

Viewed from the top of the escarpment, down which the road wound in serpentine loops, the little town resembled a clutch of snow-white eggs in a green nest. Poplars, palms, and eucalyptus sprouted between the villas, which, formerly the homes of wealthy officials, were deserted now or occupied by British troops, their broken windows and chipped balconies—the Arabs had played a short, forceful, and enjoyable innings, scoring freely—looking out on trampled gardens.

The next morning a Junkers 88 bombed and machine-gunned the convoy for a few seconds, and for hour after hour a grey sky pelted it with rain, both doing something to dispel the holiday feeling. However, the green, hilly country into which the convoy had climbed on leaving Derna was lovely in our drivers' eyes after the desert, and they came happily to Giovanni Berta, a small Italian settlement that boasted a castle. Here the regiment went into bivouac for a few days.

The people of Berta, depressed and bewildered after changing their masters three times in one year, seemed never to be quite sure which salute was now in fashion. Usually they gave the wrong one.

While the regiment patrolled the surrounding country—a party of Germans was supposed to be at large—a few of our drivers empurpled themselves with some new wine discovered in a deserted factory. Once a Junkers 88 dropped bombs uncomfortably close to the transport, and always the rain fell, turning the scrubby meadows into bogs.

No one was sorry to leave Berta behind and enter the Green Mountains (Jebel el Akdar) at the top of the Benghazi bulge. Now our drivers were in true New Zealand country, and the modern village of D'Annunzio, named after the Italian poet, the hero of Fiume, might have been a large and rather beautifully designed butter factory. It was square like a factory and tall and white like a lighthouse, and among the green hills it was altogether enchanting in the flashy, Fascist manner.

Late in the afternoon they came to the end of the Green Mountains and below them was the town of Barce, between which and the coast, eleven miles away, the plain was dotted at regular intervals with little, square, box-like farmhouses: *colonizzazione*. Beyond Barce the road ran between fields and large orchards, which sparkled with raindrops and smelt lovely under a peep of late sun. At Baracca, a tiny village ten miles west of Barce and between fifty and sixty north-east of Benghazi, the journey ended. The transport was dispersed among farms and the infantry took up positions in Tocra Pass, six miles west of the village, their boots shining as brilliantly as on the day when they stepped into the lorries for the first time. Polish was something that concerned their officers deeply, and the road to Thalata must have been strewn with empty blacking tins. They could never understand why Captain Gibson worried so little about his drivers' boots or why he allowed them to wear bits of enemy uniform, protesting seriously only when they had neglected to remove insignia or badges of rank.

Section headquarters took possession of one of the little square farmhouses, which, like its fellows, contained four rooms and adjoined a sizeable outbuilding. Like its fellows it was empty. Many of the colonists had left with the retreating army; the rest— old men mostly, and women and children—had moved to the village, where they lived in pathetic squalor and discomfort. Each night a large number barricaded themselves in the church. It was not the British they were afraid of but the Senussi, who, if they were to be believed (and their panic was most convincing) had been terrorising the entire district, shooting, looting, burning, raping, and generally paying the Italians back in their own coin—or, rather, visiting on them the sins of their soldiers.

These sins (as those of us were aware who had read that delightful book *Desert Encounter* by Knud Holmboe, a copy of which was going the rounds of A Section at that time) were not light. The sect of the Senussi, which was originally based on extreme asceticism and a return to the pure teaching of the Koran, was founded more than a century and a quarter ago by Mohammed Ben Senussi, who claimed descent from the Prophet. When the Great War broke out the Senussi, by then several million strong, turned on the Italians and drove them almost out of Libya. Next, flushed with success, they moved against Egypt and were decisively beaten on Christmas Day 1915 by a small mixed force that included the 1st Battalion of the New Zealand Rifle Brigade, Sikhs, and British Yeomanry. Later they were defeated again and, after the war, when the Italians regained possession of the fertile strip along the Libyan coast, an agreement was concluded whereby the Grand Senussi, Sayyid Idris, was given rulership under the Italians of all the oases along the 29th Parallel and of Kufra to the south. In 1926. after careful preparation, the Italians attacked the oases and the remnants of the tribes fled to Kufra. The next attack came four years later, and Kufra, on which the Italians converged from three sides, was the scene of a horrible massacre. Under Graziani ('the Butcher') the survivors were ruthlessly repressed and their lot was not alleviated until Mussolini proclaimed himself the Defender of Islam and made Balbo Viceroy of Libya, ordering him to placate the tribes. His efforts, so far as our drivers could gather. had been unsuccessful.

Their sympathy, naturally—for they saw the pale, pretty children and the frightened old women and the broken old men—was with the settlers, and to give it practical expression they mustered a dozen head of cattle—miserably lean beasts that the settlers had been afraid to go into the fields to look after—and drove them into the village for distribution among the hungry. Most of them were slaughtered immediately by the non-stock-breeding members of the community.

After a day or two the drivers came to be looked on as protectors and when one of them fell ill he was taken into an Italian cottage and nursed with great tenderness.

In their snug farmhouse Captain Gibson and the drivers of Section headquarters had a pleasant time. They did little except repair the fire that burned all day and most of the night in the enormous open hearth and watch the fierce rain dash against the windows. Sometimes, catching sight of a broken toy—a child's blue go-cart—which, tidy it away how they would, reappeared continually, a death's-head at the feast, a few of them had moments of uneasiness. They wondered how homeless Italian children fared in Libya in winter. Mostly, though, they were comfortable and content.

On fine days—there were one or two—they went mushrooming, returning home with sackfuls of delicious fungi (which everyone felt almost sure *were* mushrooms) for the cooks to fry for breakfast. Duck-shooting, in spite of the scarcity of ducks and shotguns, was another popular diversion.

An order forbidding anyone to walk abroad unarmed added a spice to these expeditions, and indeed, whenever Senussi were encountered, our drivers felt the sound sense of this edict, for the tribesmen, to judge by appearances, had drifted a long way in the last century and a quarter from the pure teaching of the Koran and the principle of extreme asceticism.

Lorries went sometimes to Barce but there was little there. Once, perhaps, it had been a pretty country town; now it was indescribably shabby and bedraggled, its general appearance suggesting the aftermath of a drinking bout. It was permeated through and through by a nauseous reek of arrack, the sale of which, together with that of dusty boiled sweets and small looking-glasses, appeared to be its sole means of subsistence.

There was talk of our drivers staying with the Welch Regiment for an indefinite period and that was what both wanted, but on 10 January Colonel Crump, accompanied by our old friend Selwyn Toogood (who was a captain now), arrived with orders for the section to pack up. Goodbyes were said and meetings in Cairo arranged, our drivers little guessing that Rommel, who had just withdrawn from Agedabia to El Agheila, almost another 100 miles, would in rather less than three weeks sweep forward and engulf Benghazi, involving the Welch Regiment in fighting only less bitter and costly than that which it had seen in Crete.

The section pulled out from Baracca the next morning—and *pulled* is the operative word, for some areas were so muddy that the lorries had to be coupled together like a goods train. The return journey was diversified by a visit to Cyrene (birthplace of that Simon who was made to help carry the Cross) and Apollonia, the port of Cyrene. Above the ruins of the old city and the broken pediments and capitals of the Grotto and temple of Apollo, from which Italian archæologists had fled in haste, leaving some gear behind, grave cypresses stood sentinel, and down the precipitous hillside, haunted by spirits of grove and fountain, a cascade tumbled, spreading a green mist and a green murmur. All over the hillside were the houses of the Roman dead.

On 17 January the section joined us at Bir el Thalata.

. . . .

We left Thalata a week after A Section's return and 25 January found us at Acroma. By this time Rommel was nearing Benghazi and petrol was needed to extricate British transport. A convoy under Captain Ward was rushed to Derna, where part of the load was dumped, the rest being dropped off by the roadside on the way back. The entire trip was made in something like thirty-six hours, but at the cost of many broken springs.

By the beginning of February we were back at Thalata and for the next two or three weeks our transport was employed at the railhead in unloading trains. C Section, which had been brought up to strength partly with some of our own spare men but mostly with drivers and vehicles borrowed from other units, left for El Adem on the 15th of the month to join yet another Composite NZASC Company formed under Captain Roberts to serve the 5th Brigade, which had returned to the Western Desert from the Canal zone.

In the Thalata area we played ' 500 ', did a little desultory training, listened to our thrice-blessed wireless sets (on some days we got ' Elmer's Tune ' only five times and ' Kiss the Boys Goodbye ' only four: often it was the other way about), and always we talked and talked on and on and on and round and round and round, mostly about nothing or about private matters—rows with the Old Man, victories in love long forgotten by everyone, incidents in the past that formed the basis for great rambling chapters of autobiography: ' I've seen me go down to the boozer Saturday mornings. . . .' But it didn't matter. Nobody was compelled to listen: we knew each other so well.

Meanwhile our side was losing the war, though we were too stupid or too ill-informed or too steadfast to realise it. We were concerned only with our little setback in Cyrenaica. What was going on in the Far East was largely concealed from us by the determination of press and radio to put a brave face on matters. LITTLE ENEMY RESISTANCE TO OUR WITHDRAWAL ran a headline in a London newspaper: the story referred to the British retreat in the Malay Peninsula. Singapore, we were told, was impregnable.

At that time Germany was broadcasting a weekly programme entitled, if memory serves, ' From The Enemy To The Enemy '. It was extremely popular with the Ammunition Company, chiefly because the Master of Ceremonies made free use of Vera Lynn's best records. The general idea was that a number of jolly, laughing British prisoners were being entertained in the studio by their German hosts, the atmosphere being conveyed by bright music, tinkling teacups, and gay asides. Most of the asides were made by the MC, a light baritone of peculiar ease and charm.

' Well, I've just been having a yarn with 3456789 Bombardier Jock McGregor. Jock hails from Castle Douglas, Kirkcudbright. Jock has just bet me a fiver that Singapore will hold out until the cows come home. Well, I'm afraid we don't agree about that. My own idea is that the Japs will be in Singapore in less than a fortnight. I'm going to collect that fiver when this tiresome war's over—by the way, I see old Jock's having a bit of trouble with his raspberry jam, ha ha!—and I'm going to have a lot of fun spending it in Scotland.'

Never in our wildest nightmares did it occur to us that there might come a time when the old smoothie would be in a position to pollute Kirkcudbright, but we should have enjoyed the joke more if the Japanese had not entered Singapore on the 12th of the month and captured it just three days later.

Ah well, we reasoned, things couldn't be too desperate. Australian and New Zealand beer was coming forward from Base in comforting quantities, we were getting our mail regularly, Wickham Steed sounded hopeful, and leave had started.

We were looking forward to our own leave. Nor was it far away.

. . . .

On 23 February, leaving Captain Butt and C Section with the Composite NZASC Company, our unit turned its nose towards the coast road: first stop, Barrani. As we drew near the coast we entered a sea of poppies in which there were islands of yellow and blue spring flowers. For those of us who had not visited the mushroom country it was like finding a pound note.

The next night found us near Mersa Matruh and on the 25th we set out for El Daba.

Most of the lorries were loaded high with salvage and forty-six of them were towing other vehicles. It was a common enough sight at that time, for the campaign had been cruelly hard on transport and on everything else. At least a quarter of the vehicles heading east were towing one or more wrecks and half of them were loaded with broken or worn-out gear. It was a matter for pride that all our lorries were returning under their own power.

We spent a night at El Daba (A Section casting sentimental glances towards its old area) and a night at Amiriya, and then set out, as we thought, for Mena. The Major, however—we should have blessed him had we known—went on ahead to try to arrange for us to go straight to Maadi and for a guard to look after our vehicles that night so that everyone could have leave. He was successful and we carried straight on, grinning delightedly as we passed Mena and turned sharp left near the Great Pyramid of Cheops. Now we were bowling along a splendid boulevard, region of rich pashas and fat beys, but there stole to meet us, from the great hot heart of the city, the old mocking, familiar, fascinating stench of red pepper, caraway-seeds, stale urine, arrack, dust, sweat, coffee.

We passed the zoo, crossed the bridge over the Nile, and turned right into the Helwan road. Soon we were passing the old wharves, where feluccas were drawn up, their great wings furled, their slender tapering masts, like trout rods from Brobdingnag, bending above the water as they had bent since the world's childhood, when there was no Athens, no Rome, no London, and certainly no Berlin. Presently we swung into the Mad Mile—madder than ever it seemed after our desert solitude: more clamorous with excited *wallads* shouting their joy and derision at the sight of so much good prey returning from the unprofitable wastelands; more crowded with gesticulating fellahin trying to sell us melons and lemonade while we struggled with insoluble traffic problems—nightmare problems presented by minute donkeys from Goblin Market dragging carts with wheels twice as high as themselves, by donkeys moving like sections of hedge under shaggy masses of green-stuff, by long flat carts loaded with patient wives, shapeless in black robe and yashmak, sitting back to back (like sacks of coal except for a glint of orange or yellow gold from tooth and wrist), and by swarms of blue Fiat taxicabs, all of the same model (an old one) and each with a mechanic lying out on the off mudguard and tinkering with the carburettor—problems that might just conceivably have been solved but for the assistance of the Egyptian police, who looked warm to the point of melting in their thick black serge but were probably, in spite of their fierce excitement, quite cool.

We beat clear of it at last—that storm of donkeys, hookahs, *galabieh*s, camels, café tables, chickens, water-melons, sheep, cooking stoves, dust, trinkets, smells, and enormous bare feet—and swam, as through backwaters and calm reaches, down a long gracious avenue with the Nile again at our elbows; and so we came to the Maadi turn-off. We passed the opulent, creeper-hung villas, the carefully-tended *midans*, the Maadi Tent—still, we never doubted, unrivalled in all Egypt for ices, cakes, and salads—accelerated for the rise, got caught as usual by the bump at the railway crossing, passed the check post, entered the camp.

Under the bright desert sun Maadi Camp looked more than ever like a clean but uncomfortable palanquin on the back of an enormous, shabby, dusty camel. It was so forbiddingly neat and decorous that we took pleasure in our stained battledress, tattered shirts, bone-white boots.

News of our coming must have gone ahead of us, for hardly had we switched off our engines before the drivers from Bardia and those who had been in hospital were around us like flies. After the storm of hand-shaking, back-slapping, and all-in wrestling had subsided a little, everyone began to talk at once with no one listening. In the centre of one congratulatory group was Bruce Morice[12] whom we had seen last in Greece. Doubtless he was dying to tell us—we were dying to hear—how he and twenty others had tried to sail to Crete in a rudderless schooner, how he had lived for several weeks in a cellar in Piraeus and then for three months in the house of a Greek dentist, how at last he had been taken to

[12] Dvr J. B. Morice; truck driver; Opotiki; born Opotiki, 7 Dec 1916; posted missing, Greece, May 1941; later escaped to Egypt.

Marathon and had embarked with other fugitives in a Greek schooner, reaching Turkey and finally Smyrna where he had been taken care of by the British Consul. But he was sworn to secrecy and so was Stan Barrow who had escaped from Crete to North Africa in an open boat.

To set the seal on our contentment word went round that we would be paid and there was leave to Cairo until one in the morning. Our first thought was to get cleaned up and everyone made a bee-line for the shower-house, carrying armfuls of fresh clothes hoarded for just such an occasion. It was wonderful to stand under hot water again, though that first shower did little more than transfer the outer coating of dirt from our skins to our clean towels. Boots were polished, some pairs sucking up a whole tin of nugget, and there was a demand for needles and thread.

After we had been paid the cooks provided a meal but few of us bothered to eat it. Already many were on their way to Cairo; others were waiting for the beer bar in the NAAFI to open at six o'clock. At once, with no preliminaries or organisation, a big reunion party got under way, and it grew progressively larger and merrier until closing time. It continued in the training area and towards midnight shots were fired (enemy weapons were being tested and demonstrated) and the sky was bright with flares. The Major intervened and some arrests were made. It was a night to remember.

. . . . . .

A postscript to the campaign is provided by the last paragraph of a report submitted by the Major to Colonel Crump:

> Although some of the moves entailed long hours of driving, and, quite often, driving all night, nobody complained and all the officers, NCOs, and men did everything they were asked to do and the general standard of work and conduct was such that it is difficult to discriminate individually.

CHAPTER 12

SYRIA

THE next morning we were on the road again, threading our way through the Dead City, which was joyously alive in comparison with some of us, skirting Cairo and Heliopolis, and turning right into the Suez road. Fayid, near the Great Bitter Lake, between Ismailia and Suez, was our destination, and there we went under canvas.

During the fortnight we were at Fayid the transport was overhauled thoroughly, most of the attached personnel left us, and the unit was reinforced to a strength of 515 all ranks and reorganised to include an extra working section—'platoon', rather, to use the new term. It now consisted of Company headquarters (two officers and 25 other ranks), four transport platoons (each of two officers, 85 other ranks, and five sections of six load-carriers), Workshops platoon (one officer and 51 other ranks), and two ammunition platoons (each of one officer and 34 other ranks). A Section became No. 1 Platoon, and so on.

When the new platoon was formed everyone whose standing with authority was at all shaky was panic-stricken by the prospect of being drafted into it. With a few exceptions, however, only non-commissioned officers and key men were affected and on them devolved the not unflattering task of teaching the young idea how to shoot, the young idea being drivers from the former 14th Light Anti-Aircraft Section, NZASC (who needed no teaching), NZASC Base Transport, Base hospitals, and like sources. The platoon was commanded by Captain K. E. May,[1] under whom were Second-Lieutenant Fitzgerald[2] and Sergeants Athol Buckleigh and Morry Evans.[3]

From the start No. 4 Platoon had a personality, but not, so it seemed to many of our grave and reverend signiors, a pleasant one. As they saw it, their comfortable and well-ordered home had been invaded by a noisy and rather ill-conditioned nephew. The platoon

[1] Capt K. E. May; clerk; Wellington; born Wellington, 21 Jun 1909.
[2] Capt J. M. Fitzgerald; civil servant; Wellington; born Gore, 19 May 1917.
[3] Sgt M. G. Evans; clerk; Auckland; born Wanganui, 8 Jul 1917.

was up early in the morning, it was much underfoot during the day, and it was pining, they suspected, to lay grimy and destructive hands on the precious transport. It had none of its own yet and they foresaw an evil day when it would be permitted to play with theirs. The policy of keeping it quiet with hard training and unending guard duties had their complete approval.

Within the platoon, too, there were misgivings. Its toilsome days were ordered by Sergeant Buckleigh's whistle, which blew continually, in and out of season. The owner of the whistle, probably, was not unpopular—his nature was as immutable as his opinions and we had found him fiery, approachable, and a disrespecter of persons—but for the whistle itself the rank and file conceived an immediate and immense distaste. It blew for parades, for fatigues, for inspections, and often, so it seemed, for the sheer pleasure of blowing. In the evenings its victims gathered at the NAAFI, where they cemented old friendships and formed new ones, discussed the Whistle in all its aspects, and decided that their future was not necessarily of an unrelieved bleakness. The Whistle might break or get lost. Schemes for breaking or losing it had been formulated already. In the event, however, desperate remedies were not necessary. Buck and the Whistle were marched out to the NZASC Training Depot at Maadi, and Ted Black,[4] a man who played no musical instrument, was promoted to fill the vacancy.

The newly-formed Ammunition Platoon—there were two but we denied them their plurality, and so after a while did the unit log—left us unmoved. It was no noisy nephew to interfere with our peace or threaten our possessions. It lived, as it were, beyond the baize door. Its duties, which in Greece and during the Libyan campaign had been performed by spare drivers, consisted in the main of sorting, stacking, guarding, and issuing ammunition, and they were as dull as they sound. As its personnel was constantly in a state of flux it had no chance of developing a corporate personality, and for many months it served chiefly as a clearing house for drivers for whom there was temporarily no place in the platoons and as a convenient threat. It was a limbo to which you were consigned permanently, or a corner in which you were stood for a period, according as you were hopelessly, or only for the time

[4] L-Sgt E. Black; tramway employee; Papakura; born Scotland, 28 Feb 1901.

being, lazy, intractable, or incompetent. From the start many of its members were men of skill and conscience—they needed to be—and later all were to do difficult and occasionally dangerous work, but that time was not yet. For the present it was enough that they had no designs on our transport. No. 4 Platoon had.

Its first driving job, for which it borrowed No. 2 Platoon's lorries, was to take us to Cairo for a formal Christmas dinner at the National Hotel. We pretended to be terrified. Returning home late that night, we added to simulated terror fatigue and a degree of fretfulness. The dinner had been beautifully cooked and served but we had found it on the dainty side for our desert appetites and there had not been a great deal to drink.

The next day, 10 March, A Section—no, we must get used to calling it No. 1 Platoon—embussed troops of the 6th Brigade and set out for a far country, leaving us impatient to follow. Our appetite for the joys of Cairo, immeasurable in the desert but always easily sated when it came down to brass tacks, had been allayed by a week's leave, and what we wanted now was the open road. The wind on the heath would blow with a pleasant astringency after the debilitating atmosphere of the New Royal, the Blue Nile, the Globe, and the Pam Pam. No skull-capped pander, walking backwards in front of us, would extol its beauty and cleanliness, and no legion of filthy imps would badger us into buying worthless imitations.

Satisfaction was universal when we pulled out from Fayid behind the Supply Column on 14 March, with the sun shining brightly. At noon we crossed the canal near Ismailia by pontoon bridge, and then we were in the Sinai Desert, on which the sun blazed down with passion. The surface of the road, where it was not covered by drifting sand, was like old-fashioned liquorice, only softer and stickier, and the prospect on either side of it was exactly what you would expect to see in a desert if your knowledge of deserts was drawn solely from fiction. Former subscribers to *Chums* or to the *Boys' Own Paper* felt at home at once. The two tastefully-arranged palm trees were missing and the sailing ship floating upside down in the sky and the despairing footprints disappearing over the nearest sand dune, but the shimmering whiteness was there and the wind-ribbed surfaces and the aching distance. Beyond the tarmac the sand was so soft and deep that when a Petrol Company vehicle ran off the road the united efforts of three breakdown vehicles were necessary to recover it. After a while the landscape became less like the illustrations in the *Boys' Own Paper* and more like the kind of desert we were accustomed to—wadis and sandstone ridges—but ever and again it would flick back into pure adventure story.

Just before dusk we halted at Abu Aweigla, having travelled 165 miles. We had reached our staging area, and, wonderfully, there were shower baths.

The next morning we crossed into Palestine and drove on through country that reminded us of the illustrated Bibles of our childhood. On either side were gentle, rounded hills the colour of pea-soup— the kind of hills on which the sermons were preached and the multitudes fed—and down them, dressed as the disciples were dressed (the same kind of striped gingery blanket and head-thing), and disputing, perhaps, as the disciples disputed once, came familiar figures, padding through the grey dust. By wells we went, wells from which the water (for the homeliest, pleasantest miracle) might have been drawn for the wedding feast, and through little villages, catching glimpses of the kind of kitchen in which Martha was careful and troubled over many things.

We travelled through Beersheba, which name, starting a string of others—Abraham, Hagar, Jacob, Elijah—came droning down to us from the sleepy schoolrooms of our boyhood; through Gaza, where Samson (someone attempted to argue in all seriousness that his name was Simpson) was bound with fetters of brass and did grind in the prison-house, and so to Qastina, which said nothing to any of us. Here we spent the night in an Australian barracks with a cinema and a NAAFI. We had covered 110 miles.

The next day we passed through Masmiya, Gedera, Ramle, Lydda, and Petah Tiqva. It was like skimming through an examination paper the answers to which, tauntingly familiar, for the moment elude you, and it was a relief to pass Haifa and then Acre, history's Tobruk, with its wider selection of memories—memories of Saladin, Richard Lionheart, and Napoleon. Finally, at half past two in the afternoon, after travelling 112 miles, we halted at Ez Zib, on the coast.

Early the next morning we crossed the Syrian frontier into the Lebanon and soon we were skirting Tyre, from which for so many aeons the ships of Tarshish and the Greek and Roman galleys and the high-prowed vessels of the Phoenicians sailed the Mediterranean, loaded with precious woods, jewels, ornaments, and fabrics marvellously dyed, drawn by no other magnet than that in obedience to which the ships leave 'Frisco, Sydney, and Deptford Pool. In the early afternoon we reached Beirut, which was crowded with people of an interesting appearance, few of whom it was easy to identify with anything so innocent and prosaic as clay-pigeon shooting, clerical vacations, or honourable retirement after forty years as a lady governess. There we refuelled, and at half past ten that night, dead-tired after travelling 142 miles, which brought our total for the journey to 529, we reached our destination, which was an area twelve miles north of Baalbek.

It was not until the next morning that we were able to form an impression of our surroundings. Then we saw that we were camped in a great, wild valley. Our area, which sloped gently to the road, was covered with short, scrubby grass and outcroppings of rock. (Later, when we tried to dig cesspools and rubbish pits, we had to call in the engineers.) Facing us was a mountain which rolled away towards Baalbek, a mane of snow flung over its huge shoulders. In the foothills, tucked away in fold and crevice, were secret villages —huddles of whitish buildings that resembled the nesting places of large, untidy birds. The scene was austerely beautiful, but somehow, uncooperative. It did not look as though it had contributed greatly to the world's fruitfulness. It did not look as though it intended to. The mountain's attitude towards the meagre flocks that fossicked miserably on its lower slopes was that of a St. Bernard with fleas.

It seemed that we should be here for some time. Tents were pitched for living in, messing in, cooking in. There was talk of paths, sanitary arrangements, and regimental guards. It sounded like a lot of work of a peculiarly dull kind. Those who had vehicles could hope for transport jobs, but those whose duties were ill-defined—the drivers of No. 4 Platoon for instance—prepared for the worst. Their forebodings were fully justified.

In the shadows of the tremendous mountain, which must have frowned down on many comparable scenes in ancient times, when the Roman legionaries, under the cold surveillance of the centurions, dug fosses, erected earthworks, exercised their arms, the work went forward, the spare drivers participating fully, the rest occasionally and under protest. With the guard duties, which called nightly for one sergeant, three corporals, and forty-eight men, and daily for half that number, everyone assisted. Our camp, which our neighbours referred to as Stalag 69, was one of the best-guarded spots in the Mediterranean theatre. However, the approach of spring, an abundance of sport, a fairly liberal allowance of day-leave to Baalbek, together with an exhilarating sense of grievance, enabled us to keep happy.

. . . .

Meanwhile, at Aleppo, some thirty miles south of the Turkish border, No. 1 Platoon's drivers were enjoying themselves thoroughly. The Vannière Barracks, on the outskirts of the city, provided them with a good home, and their duties, without being burdensome, kept them busy. Under the command of the 6th Brigade they were employed in carrying supplies from Aleppo railway sidings to neighbouring depots, delivering flour to native villages, and providing transport for the infantry. Nearly every lorry was on the road nearly every day, but Captain Gibson allowed his men to make what arrangements they liked about working day on and day off and there was leave every evening for almost half the platoon.

The one fly in the ointment was that our drivers' financial resources were by no means commensurate with the amount of leisure at their disposal. Their pay—nine Syrian pounds a week—gave only an illusion of wealth. One pound would buy one bottle of beer or the first two courses of a daintily-prepared but not very substantial three-course meal. However, there were entertainments at Aleppo that were not charged for. Inter-platoon Association football matches in which an extremely catholic interpretation of the rules was permitted were contested free of charge on the fine sports ground adjoining the barracks. No charge was made for long hours of uninterrupted sunshine or for the soft Syrian evenings, through which, after collecting leave passes, our drivers set out for Aleppo.

The city was rich in historical associations, but that was a kind of riches in which few were interested. Our drivers contented themselves by observing that the old castle on the flat-topped hill resembled a dusty and battered decoration on a crushed birthday cake. All but one candle—the famous minaret—had disappeared.

They were pleased—soldiers on active service are always touchingly pleased by any evidence of what they call civilisation—to find shops, cinemas, and American bars in the main street, and they were prepared to forgive the city its many mosques, quiet courtyards, and mysterious alleys. Shuttered windows and doors heavily bolted against the intruder told stories of old Turkey, but the stories fell mostly on deaf ears, and the listeners were soon bored when sad-eyed Armenians, wearing shiny, navy-blue suits and burning to discuss the Armenian question, spoke of atrocities. Behind such doors, they said, their elderly relatives had been massacred by Turks and Kurds, and even today, behind those shuttered windows, while enlightened elements in Aleppo were wearing their blue serge and going to the pictures and speaking American, plump Turkish ladies were removing their veils to masticate confectionery, and a deadly, soft-footed dullness scented by rose water and interrupted only by indigestion was dragging on in exile, as indifferent to wars and revolutions as to Bing Crosby.

In the poor quarter of the city, where the bazaars were situated, the drivers felt at home. Here was the kind of churning, squabbling, gesticulating life with which Egypt had made them familiar. Here was a corner of the world that could risk comparison with any section of Wogdom—the riot in the quality of its confusion, the livestock in the quality of its liveliness, the vegetables in point of size (cauliflowers like medicine balls: leeks you could have played golf with), and the platters of food in the richness and variety of their colouring. It was fascinating to see a stout French officer, like someone out of *Beau Geste*, sitting down with a sheik complete with burnous and jewelled scimitar (pure Ethel M. Dell) to a meal like a Cathedral window—a window decorated with snails and purple slugs. A taste for snails, to tell the truth, was something that several of our drivers acquired at Aleppo. Frogs' legs were popular too.

And so were the wines of the country. They were many but our drivers recognised only two types: 'Woof-Woof' (white wine) and 'Purple Death' (red). Arrack could be bought on the sly but the old ruffian was generally avoided as being too unpredictable a companion for a quiet evening, and the same applied to kummel, chartreuse, curaçao, and a fiery abomination that tasted like curried lollipops.

Naturally no one drank wine or liqueurs in preference to beer. But beer, of which two brands were sold in Syria (one was quite good, the other was barely drinkable), was expensive and difficult to obtain except in the Aleppo NAAFI. This, in consequence, was our drivers' favourite resort.

Almost as popular as the beer—more popular towards the end of the week because it cost nothing—was the NAAFI orchestra, which was led by a talented Russian violinist, who claimed to have had an international reputation. Towards closing time he must often have thought nostalgically of the concert halls and conservatoires of Europe, where matters, no doubt, were arranged differently and the rule was one tune at a time. Doubtless it distressed him to have to saw his way through the Vienna Woods or tell tales of Hoffman while half his audience was deep in its deep purple dream and the other half was coming round the mountain for the twentieth time in succession and plainly intended to go on coming round it until someone put the lights out. He had one method of creating concord, however, and that never failed. He would strike up 'The Red Flag'. The Moscow front had held, the Germans had been forced to retire, and the patrons of the Aleppo NAAFI, like the rest of the world, were dazzled by Russian courage. Gladly they forsook their mountain and their deep purple dream to unite in proclaiming that the Workers' Flag was deepest red. Nor was their pleasure in the performance lessened by its spice of unconventionality. The singing of 'The Red Flag' was not at that time completely hallowed by custom, and only a year before such a demonstration would have been inconceivable in a British canteen and few would have thought it desirable.

'The Red Flag', 'The Marseillaise', 'Old King Farouk', 'God Defend New Zealand', 'God Save The King', and again in response to repeated demands ('Once more, you jokers! Give her the tit, Maestro!') 'The Red Flag', and another evening is over. Our drivers straggle out into the night, over the old bridge, across the railway tracks, and up the hill through the oats, while the stars shine down with unimaginable brilliance and fireflies, star-like themselves, swim past on tiny, earnest missions. The hill seems unaccountably steep and more than one reveller, risking the imputation that he's cast, lies down in the oats to cool off. By the railway tracks someone is howling 'Yours', and at the entrance to the barracks, grouped round the embarrassed sentry in his little box, friends of the Soviet Union are giving ' The Red Flag ' a final flutter, while others, debating across the vast breadth of an invisible rose-lit forum, assert their complete and unalterable solidarity on some point that was at issue earlier in the evening when judgments were less mellow.

' You were right, Snow. Jacobs was the name of that joker who ran the boxing down your way.'

' Yes, Jacobs. That's the name—Jacobs.'

' Bill Jacobs. Yes! That's the joker—BILL JACOBS.'

' Yes! OLD BILL JACOBS! '

In the graveyards of Aleppo dogs howl with incredible mournfulness and down by the railway station the French guard fires a few bursts from his tommy gun. The French take their guard duties seriously and at night they are best avoided. In the oats it has turned cool, and the sentry in his little box is becoming restive.

' O.K., you jokers—fair go. Better get along in, eh? Yes, she's apples. Everything good as gold. Better get along in, though.'

And off to bed they go, leaving Aleppo to the dogs, the wandering fireflies, the French patrols.

. . . . .

The reason for our presence in Syria was not, as some were inclined to suppose, that we were in need of a holiday in novel surroundings. The Division was there in the first place because Germany was expected to make a thrust either through the Caucasus or through Turkey, and in the second to foster friendly relations with the people. To accomplish the latter task we did not rely solely on our charm of manner. We came with schemes for the control of malaria and as dispensers of free medicine and distributors of free food. In all these activities the Ammunition Company took part.

For over a fortnight Corporal Owen Miles[5] and eighteen others worked under a British political officer, operating first from Homs and then from Hama, towns on the main road between Baalbek and Aleppo. For the most part they were employed in delivering wheat and dates to a tribe that claimed descent from the children of

---

[5] Cpl O. W. Miles; school teacher; born Dunedin, 16 Dec 1911; killed in action, 14 Jul 1942.

Ammon. The tribesmen boasted that they could muster 800 horsemen, and their camp, some eighty miles from the main road, was the largest in the district. To come on it suddenly at the end of the day's journey was like driving into the book of Exodus. The tents —there may have been three thousand of them—were woven from goat hair and no doubt they were identical with the ones that St. Paul made. Black they were and of no uniform design, their shape and size depending on the number of crazy flaps and wings spread round them. It was as though in the middle of that vast plain a great flock of pterodactyl, young ones and old, had been gorgonized in the act of alighting.

With the sun burning on the horizon like a blood offering and the air pungent with the smoke of fires, great flocks of sheep and goats would converge on the camp, the shepherds and the dogs leading them in (like all the shepherds in the Bible) and not, as in New Zealand, driving them. In the deepening twilight the camp fires glowed orange and ruby red—there was no nonsense about a blackout—and from the multitudinous dark life, squatting close in the tents, came a continual low murmur as of bees. Every now and then it was drowned by louder noises—the whinnying of a horse, a man's angry shout, the squalling of a hurt child—but it was there always, gentle and monotonous. So must the Israelites have murmured in the camps of Moses, of Joshua, and of Gideon.

Watching and listening, the visitor from beyond the wilderness was conscious of glimpsing the world's infancy, and the shock was considerable when a sudden spurt of flame illuminated two elegant saloon cars of an expensive French make parked nonchalantly near the Sheik's tent. It was like seeing alligators in a village duckpond or opening a Crusader's tomb and finding a cigarette lighter. And the thing became stark nonsense when a bearded patriarch, who might have been old Ammon himself, bent down to remove the ignition keys, and tested, with sandalled toe, the tire pressures.

These visits to the camp were a fascinating experience for our drivers and they were lucky in having Owen Miles with them, for he could speak good colloquial French. On arrival they would be offered minute glasses of sweet tea or thick black coffee—the smaller your glass, apparently, the warmer your welcome. Later they would be shown to a cushion-filled tent, which gave them unlimited opportunities for improvising on the Sheik-of-Araby theme. Often they were invited to dinner *en famille,* an ordeal that demanded a strong stomach and exceptionally graceful manners. The whole family—men, women, children, and dogs—would squat round an immense bowl filled with a savoury grey mess from which sheep's eyes peered glutinously. In this everyone plunged a hand, which as often as not came straight from fondling a mangy dog or a scrofulous child. As long as it was the right one—the left was an inferior member—no one minded.

The better our drivers came to know the country the more they were astonished, not by the difference between themselves and their hosts, which was vast, but by the many and often humiliating points of likeness. In incident after incident there was a homely ring. The Sheik, for instance, was open to those suspicions that sadden the lives of the best quartermasters, and when he was attending a conference in Damascus our drivers understood at once why they were asked not to dump their loads outside his tent as they had done in the past but to take them to remote parts of the camp. And they understood what was happening when they saw great wads of dates disappearing into private tents; and when grins of complicity were changed suddenly for ones of murderous rage, and sticks and stones started to fly, they understood that, too. Said someone: 'It was like those fights that spring up suddenly in the NAAFI.'

Again and again the homely ring. The chuckles of the dozen cut-throats whom our drivers taught to play '100-up', one of the more infantile card games, were echoes of their own mirth, and it was their own spirit they saw, burlesqued rudely, when the game ended at the sound of a rifle shot and the players rushed out into the night brandishing loaded Mausers. Plainly a New Zealander could become a good tribesman without much difficulty and a tribesman a good New Zealander. One man's sheep's eye was another man's toheroa.

By behaving naturally the drivers created a good impression, and when the job ended the political officer wrote to Major Coutts praising them for the tact they had shown in dealing with the natives.

・　・　・　・　・　・

Although we came to Syria with cornucopias, spilling dates, wheat, and anti-malarial unguents among tents and mud-built houses, the existence of the mailed fist—always extended in friendliness of course—was hinted at from time to time. During April the 26th Battalion, in No. 1 Platoon's transport, made a leisurely 330-mile flag-march-cum-goodwill tour through northern and eastern Syria. The trip took five and a half days and remained a dream of beehive houses (shaped like that because of the shortage of all material except stone and mud), of sudden glimpses of the Euphrates, intent on the tremendous business of sliding from the Armenian uplands to the Persian Gulf—sliding between banks that remembered the Chaldeans, the Babylonians, the Arabs of the Calipha, the Mongul invaders, and now the New Zealanders—of slowly-circling kites impervious to rifle fire, of whirring flights of sparrows in a brutal storm of machine-gun bullets, of the black tents of the barbarians and the khaki bivouacs of the children of progress, of a little tumbledown village perched on a jagged hilltop and resembling a cluster of decayed teeth, but alive, boasting a wineshop and a tiny mosque and overlooking the ruined splendour of Palmyra in which history hardly remembers who feasted and who worshipped.

Doubtless these mild displays of strength together with our affable ways did much to dispose the natives in our favour, but they were a practical people and sentiment was not allowed to interfere with business. In rifling our tool and tucker boxes, in slicing the rubber mud-flaps from our vehicles, in robbing us of canopy ropes, tail-lamps, and other external trifles, they were indefatigable. When passing through a town in convoy we drove nose to tail with a lookout posted in the back of the last vehicle. Single vehicles needed a lookout in the back all the time, otherwise they were likely to arrive at journey's end an empty husk. There was a story, almost certainly true, of six drivers who went to sleep in a tent and woke up the next morning under the clouds.

The ingenuity of the natives filled us with astonished respect, and we took to sleeping with our ammunition under our pillows like love letters and our rifles in bed beside us.

With the purely military object of the Division's stay in Syria, which, briefly, was to establish a forward outpost near the Turkish border and a main defensive position near Baalbek, our unit was concerned only indirectly. The plan of defence postulated roadblocks, strongpoints, and better communications than those that existed, so we delivered materials to the engineers and helped them to build a road by-passing Baalbek. Brigade exercises were held, and for these we provided transport, the tyros of No. 4 accompanying the elder platoons as spare drivers, which gave them experience of operational work and a rest from camp and guard duties. Their mentors, for the most part, were severe, shrinking in exaggerated distress when a gear was grated.

An indication of the variety of our duties in Syria is given by this typical entry in the unit log:

> June 1, Baalbek: All vehicles employed today—23 on 6th Brigade exercise, 8 with NZ Engineers, 9 with Political Officer at Homs, 4 attached to forward ammunition point and 4 to 27th (MG) Battalion, 2 with Field Punishment Centre, Baalbek, 6 on road construction work Baalbek by-pass, and 2 with 19 DID,[6] RASC.

Our domestic labours, which devolved on those who were unfortunate enough to be without vehicles, included the construction of paths and roads, a unit swimming pool, grounds for baseball, cricket, and hockey, a YMCA hut (in which Padre J. T. Holland[7] worked morning, noon, and night, organising lectures, card evenings, and debates), a rifle range, a small parade ground, and goodness knows what else.

And naturally, with the transport as busy as it was, Workshops was not idle. Our lorries had already covered an average of something like 10,000 miles under vile conditions, and long hauls over the Syrian hills—the Beirut-Baalbek run was a particularly severe test—were causing breakdowns.

But our Syrian interlude was not entirely filled by toil. There was one day on which we did no transport work and no unnecessary camp duties. Instead almost the entire company took part in a ceremonial parade at Baalbek at which the General presented decorations to members of the NZASC. We were pleased and surprised by the number of awards that had been won by our Corps and especially pleased that we ourselves were represented. Sergeant Bob Aro was given the Military Medal for his work with No. 3 Platoon in November 1941.

---

[6] Detail Issue Depot.

[7] Rev. J. T. Holland, CF; clergyman; Christchurch; born Newcastle-on-Tyne, England, 31 Jan 1912.

And there were holidays and pleasures. For a few there was leave to Damascus or Beirut and most of us went at least once to Baalbek, where, among the ruins of the Temple of the Sun, the Temple of Jupiter, and the Basilica of Constantine, Padre Holland made dry bones live. For No. 3 Platoon, which had returned from the Western Desert with the 5th Brigade in the middle of April, there was a belated Christmas dinner at the Palmyra Hotel, Baalbek, with Colonel Crump as guest of honour.

And there was bathing, baseball, cricket, football, boxing, wrestling, and some work that was pure pleasure: *24 April: four vehicles go to Damascus to fetch fruit and vegetables.* There were jobs that took us to Afrine, in the charming Afrine Valley, twenty-eight miles north-west of Aleppo and only a dozen from the Turkish border, across which our drivers stepped and stepped back, adding another name to their list of countries visited. A detachment from No. 1 Platoon was at Afrine for several weeks with the 22nd Battalion, living near a riverful of fish, which they caught in the scientific manner. The means—Mills bombs—justified the end, for what could be lovelier than a supper of fresh trout and eggs cooked by Dave Falconer and eaten in the cool of the evening?

And that is as good a note as any on which to end our Syrian chapter: the shadows advancing across the Afrine Valley, the Afrine trout emerging from nooks and crannies after the evening bombardment, the gentle sighing of the primus. . . .

. . . . . .

The war seemed far removed from Syria. We knew that it was still banging away in odd corners of the world—in the Far East, over England, and, rather feebly, at Gazala—and we were confident that we were still winning. Victory, we never questioned, was now only a matter of time. Some of us were anxious for just one more adventure before we went home. Others were of the opinion that they had done enough and that to tempt Providence was foolish.

The news was soothing. Malta had been awarded the George Cross and doubtless she would hold out until the end unless the Luftwaffe sank her beneath the sea by sheer weight of bombs. Tokyo had been raided by American Fortresses. Hitler had appointed himself Supreme War Lord, and over a thousand bombers had visited Cologne. Our enemies couldn't stand up to that sort of thing.

Thus the news. Little was said about drowned sailors in the Atlantic Ocean, or of convoys limping into Valetta, Murmansk, and Liverpool. We couldn't see the great hole in the heart of Exeter or the rubbish in the streets of Bath—rubbish that was once Queen Anne houses and the Assembly Rooms—or the wounds of Norwich, York, and Canterbury, England's cathedral cities.

Then Rommel moved in Cyrenaica and there was a battle at a place called Knightsbridge, about seventeen miles west of El Adem. For a while the news was good. Then it was not so good. Then it was as near to being bad as news was permitted to be at that time. By 12 June Rommel had Bir Hacheim, a desert fortress some forty miles south by east of Gazala. (Don't think of frowning walls but of slit-trenches, and barbed wire and food tins, and Free Frenchmen brave as lions.) Without this he would have felt uneasy about his lines of communication. Now everything was just right—the light, the pitch, the bowling.

On 16 June all normal work was cancelled and we were told to tune our vehicles to concert pitch. In the evening No. 1 Platoon, which had returned from Aleppo a fortnight before, marched out to the 20th Battalion. Two days later we dumped our second-line holding of ammunition and by dusk we were packed and ready to move.

We moved at twenty minutes to seven the next morning, following the Supply Column. We crossed the Syrian border early in the afternoon, swept down through Palestine, and down and down until we were running beside the Sea of Galilee, which danced and sparkled in the bright sunshine, and carried on until we reached Tulkarm. It was half past six when we got there and we had covered 197 miles. We were restless and excited, for all day long, like an extra person in the cab, mouthing and chattering, Rumour had travelled with us.

'I reckon there might be something in it.'

'I can only go by what Bill told me. He should know.'

'Anyway, I'm just telling you what the Supply Officer said. He's got a case of beer on it with his driver.'

'Hell! That would do me, eh? New Zealand!'

A driver who had left home with the 4th Reinforcements announced his intention of getting married as soon as he stepped off the boat. Basil, with the 20th Battalion, which was two days ahead of us, said that his old man had been putting aside a bottle of beer every Saturday night since January 1940. At that rate something like 130 bottles would be waiting for him.

The next day was sweltering. Our drivers opened their windscreens as wide as possible, tied back their doors to cause a draught, took off everything except their shorts. Beyond Ramle the road wound between orange groves, from which on our way to Syria laughing children had pelted us with ripe fruit, and between nicely ordered fields and neat houses. People waved to us. The women wore pretty flowered dresses and the children looked healthy and handsome. They were Jews but they seemed to have sloughed off the ancient burden of their race. After three months in Syria we had forgotten that people could look so neat, so pretty, and so clean. And our minds, with that vast complacency of which New Zealanders alone are capable, turned home.

We halted at Asluj, in the desert, at half past three in the afternoon, having travelled 128 miles. Under the cold showers in the staging area the drivers chattered like magpies. It was true all right. Embarkation lists had been made out—that was definite. The Division was going home. By nightfall the names of some of the ships were known.

It was cool when we set off at half past five the next morning and presently we ran into a thick white mist. Colder and damper than the mist was the growing conviction, born somehow, somewhere, in the night watches, that it was all lies, that the goal we were racing towards had no connection with marriages and homecomings. A few inveterate optimists were still on their way to New Zealand but the rest of us were bound for a place very different, and something told us that this time it was not going to be pleasant. All but the most careless were silent and a little gloomy.

By twenty minutes past seven we were in Egypt and we crossed the Sinai Desert without stopping. The heat was terrific. It broke over us in waves. The horizon, the black ribbon of road, the outline of the vehicle ahead—everything was dancing in it. The bitumen melted and our engines boiled. Drivers with the 24th Battalion found it necessary to mix a little oil with their petrol to prevent it from vapourising before it reached the carburettor.

After covering 123 miles we halted near the Suez Canal. New Zealand was seldom mentioned that evening. The subject was attended by too much heartache. All the talk was of what lay ahead. From the wireless news it was difficult to gather what was happening in the desert, how big the battle was, and in whose favour it was going. We were wary now of retreats that sounded like advances and advances that sounded like retreats.

Very early the next morning a large party, which included most of No. 4 Platoon, set out for Maadi to collect 120 new vehicles, fifty-seven of which were earmarked for our unit. The rest of us moved half an hour later. We travelled swiftly and smoothly and soon we were rolling through Cairo. No outward signs of crisis were apparent. There was no abatement in the tide of bootblacks, beggars, café idlers, and bull-necked pashas, nor were they in more than their customary state of excitement. As always our old sparring partners leered at us from crowded tram-cars and pointed at us from the pavements, and doubtless it was imagination that made their grimaces seem a shade more sardonic than usual, their gestures a shade more mocking and despiteful.

After we had travelled some distance along the Cairo-Alexandria road we halted for lunch. Traffic was very thick and nearly all of it was heading towards Cairo. We watched it as we munched our bully and biscuits and our hearts sank. We got on to the road again and by now the traffic was thicker than ever: a continual stream. Lorries went past loaded with troops. Nurses in open three-tonners looked strained and weary, and when we waved to them they stared straight ahead or responded only half-heartedly. There were ambulances crawling between vehicles piled high with miscellaneous equipment. One lorry was heaped with all the paraphernalia of an officers' mess, and we gave it an ironical cheer. A good many of the vehicles towed others—two, three, and sometimes four. Most depressing of all were the trailers loaded high with the wingless corpses of British fighter planes. We counted dozens of these and as each one went by the time-honoured jibe ('All our planes returned safely.') fell a little flatter. Hell, we thought, there goes our protection.

Was it now or later in the day that we heard the damnable news? It started as a chill whisper, which grew louder and louder until everyone had heard it. Tobruk has fallen. Tobruk has fallen. Rommel's in Tobruk.

Tobruk had been ours for seventeen months and at the time of our worst reverses its garrison had been a symbol of British tenacity. It was like Gibraltar, like Malta, like England herself. We couldn't believe that it was lost. We hated to believe it. We had to believe it. At every halt, above the murmur of the engines, you could hear our drivers discussing the bitter news.

We reached Amiriya at half past seven in the evening after covering 220 miles. Our transport had been behaving well, only one vehicle having dropped out. We refuelled and each lorry was loaded with enough petrol for a 400-mile journey and each man issued with two gallons of water. Early the next morning we set out for Mersa Matruh.

Traffic flowed east along the coast road throughout the day: ambulances, staff cars, tank-transporters, endlessly the dirty-yellow three-tonners, dusty, battered, with torn canopies billowing in the hot wind and loose ropes trailing behind, and again and again, sending an inward groan along the whole length of our convoy, a trailer carrying a smashed fighter, great rents in its fuselage exposing internal wounds, its red-white-and-blue rondels, like desecrated flags, showing through grease and filth. The road was not wide but often there were three columns travelling abreast, two heading east and one, our own, heading for Mersa Matruh. Some of us had only to close our eyes to be back in Greece.

It was a cheerless journey. Everything—all the vehicles that went past, all the priceless equipment—looked dirty and spoiled. And everything—the wheels, the hammering pistons, the jolting trailers—was saying over and over again: Tobruk has fallen, Tobruk has fallen, Tobruk has fallen. Lost everything in Tobruk. Stores. Transport. Twenty-five thousand men. Thirty-five thousand men. Stores. Stores. The lorries rumbled it as they swept past. They screamed it in low gear on the hills. They ticked it, quietly, persistently, throughout the many hold-ups. Tobruk has fallen, Tobruk has fallen. And our own lorries took up the refrain and repeated it all the way to Mersa Matruh.

That night our transport was dispersed at Smugglers' Cove, a few miles east of the town. We were very tired. In five days we had travelled 908 miles.

During the night, for the first time in months, German aircraft circled above us. We heard the broken beat of their engines and in Mersa Matruh ack-ack coughed and flashed.

.   .   .   .   .

The next day was 24 June.

On the beach at Smugglers' Cove—what smugglers were those, we wondered, and what was the point of smuggling anything into Matruh?—the waves folded in grave procession, curtseying and withdrawing in the bright sunlight. Beyond them the water was marvellously clear and blue except where it was stained by long streaks of amethyst or emerald. Blowing in from the sea, the breeze carried that intoxicating holiday smell of shells and seaweed and tarry row-boats. After breakfast everyone felt fine.

We put in a hard day's work on our vehicles and when evening came we had a few quiet hours to ourselves. They were delicious after the rush and bustle of the past week. We bathed, played cards, kicked a football about. It was a lovely evening.

It was a lovely evening and a new spirit was abroad. Gone was the depression of yesterday. The retreat—the withdrawal—no longer seemed disastrous and infinitely sad. It was exciting, challenging; and, anyway, worrying was of no use.

At seven the next evening the first of the new vehicles arrived loaded with ammunition (a pity this because we had already drawn our second-line holding from Mersa Matruh and now much of it would have to be returned), and by midnight sixty-two vehicles were in the area. The drivers reported that the coast road was choked with east-going transport. It was a case of every man for himself, convoy discipline being out of the question. The rest of the new vehicles arrived the next day and a detail from the Petrol Company issued them to the units concerned. After they had collected their brand-new three-ton Chevrolets and loaded them with ammunition our No. 4 Platoon drivers felt happier than they had done for weeks. No longer were they merely the boys around the place, fagging for the seniors. Their platoon was now a going concern and all they needed was a chance to make a name for themselves.

And the chance was coming—it was coming towards us from Tobruk as fast as tracks could carry it and supply lines feed it. It was coming, and we felt it as a tiny tremor in the hand and a clutch at the stomach and a restlessness in the feet.

Originally it had been the British intention to hold the frontier but lack of men and armour made this impracticable, a line not anchored firmly at both ends being an invitation to an out-flanking movement unless there was a strong mobile force in reserve. A line based on Mersa Matruh was open to the same objection (besides there was no time to organise one) so General Auchinleck decided to make a stand sixty miles from Alexandria where his right flank would be protected by the sea and his left by the Qattara Depression. When it came to naming the new line no difficulty arose. Less than a mile from the road there was a cluster of stone buildings, a water tank, and a railway station, called, simply and rather beautifully, El Alamein.

That was the position on 27 June when we pulled on to the main road shortly before midday and headed east under Captain Sampson.[8] No. 1 Platoon, though its return was expected hourly, was still with the 20th Battalion, so its ammunition was left at Smugglers' Cove with Second-Lieutenant R. K. Davis[9] and a small picket.[10] The signal to move had given no clue to our destination, the arrangement being that a guide was to meet us on the main road. The main road was ominously empty, and when they came to a deserted NAAFI our drivers were able to stop, one after another,

---

[8] He commanded the unit between 21 June and 7 August while Major Coutts was at General Headquarters, Middle East, supervising the delivery of vehicles and ammunition to the Division.

[9] Maj R. K. Davis, m.i.d.; clerk; Eureka, Waikato; born Auckland, 2 Mar 1917.

[10] The chief appointments on 27 June were: Company headquarters, Maj P. E. Coutts (with GHQ, ME), Capt S. A. Sampson, Lt O. W. Hill, WO II J. S. Bracegirdle (appointed 14 Feb 42); No. 1 Platoon, Capt R. C. Gibson, Lt T. A. Jarvie (attached 25 Dec 41); No. 2 Platoon, 2 Lt J. R. Arnold (posted 3 Mar 42), 2 Lt R. A. Borgfeldt (posted 19 Jun 42); No. 3 Platoon, Capt W. K. Jones (posted 19 Jun 42); No. 4 Platoon, Capt K. E. May, 2 Lt J. M. Fitzgerald; Workshops, 2 Lt (T/Capt) A. G. Morris; Ammunition Platoons, Lt G. P. Latimer (attached 25 Dec 41), 2 Lt R. K. Davis (posted 23 Nov 41). The following had left us: Capt D. C. Ward (posted to Base Training Depot, 1 Jun 42), Lt (T/Capt) F. G. Butt (posted to Base Training Depot, 26 Apr 42), 2 Lt W. S. Duke (p.w.), 2 Lt A. M. W. West-Watson (admitted to hospital, 11 Jun 42), WO II Dillon (posted to OCTU, 24 Feb 42).

continuing on their way the richer by cases of Canadian beer and cartons of English cigarettes. It was Greece all over again except that there were no Stukas or Messerschmitts to worry us.

After a fairly quick run, the convoy dispersed in a vacant area between Qasaba and Fuka, waited there until four in the afternoon, and then headed inland, halting an hour and a quarter later some eighteen miles from the coast road. Workshops was missing, and Captain Sampson, not knowing that the platoon had been ordered farther to the rear, was worried. Later in the evening Second-Lieutenant Davis and the picket arrived from Smugglers' Cove to report that it had been necessary to destroy No. 1 Platoon's second-line holding to prevent it from falling into enemy hands. With the help of engineers they had managed to destroy all but a few cases of small-arms ammunition. The enemy had been very close at the time.

After dark Nos. 3 and 4 Platoons—it was the latter's first important assignment—moved out to establish ammunition points, and half an hour later Company headquarters and No. 2 Platoon moved to an area south of Fuka where they spent what was left of the night. The other platoons reported the next morning with the news that it had been impossible to establish ammunition points because of the enemy's swift advance and the movements of the Division.

At half past nine Nos. 2 and 4 Platoons left under the officer commanding the 4th Reserve Mechanical Transport Company to serve the 4th and 5th Brigades (if they could be found: they were known to be heading east with the Germans after them), and an hour later, travelling behind Rear Headquarters, 2nd New Zealand Division, the rest of the unit set out for Fortress A (again don't think of a fortress), roughly in the middle of the new line, where the 6th Brigade, a centre three-quarter with the sun in his eyes and the play coming fast down the field, was standing ready. By half past eight that night the transport was dispersed in the fortress and everyone who was not on duty turned in. We were dead-tired but some of us found it hard to sleep. It was the night of 28-29 June.

After the twilight the moonlight. The full moon, with a bland and beautiful idiocy, lit everything up, making the lorries throw shadows that might have been cut out of black cardboard. By pulling the blankets over your head it was possible to ignore the moonlight, but you couldn't shut out, nor was it advisable to, the dread sound, something between a hum and a sob, of approaching bombers. They came over about nine and circled round. They sounded as though they were directly above us, but they were some distance away. When the bombs fell there was an interval of several seconds between the flash and the thud. A vehicle was set alight in the Supply Column's area and more bombers were attracted. From start to finish the raid must have lasted forty minutes, and we heard afterwards that in the Supply Column's area alone fourteen men had been killed. It was a token of what we ourselves could expect.

We were fairly busy on the 29th, very busy on the 30th. By evening we had replenished all the field regiments and No. 3 Platoon had established a Corps dump in the 6th Brigade's area. Workshops, after four days with the Supply Column, had rejoined us that morning, and all the platoons were accounted for except No. 1, which was still, we supposed, with the 20th Battalion. The unit was now operating from Qaret Somara, twenty-six miles southeast of Fortress A.

. . . .

That afternoon Rommel had arrived in front of the Alamein Line and the issue was now a straight one. They were met together, Rommel and Auchinleck, to find out which of them had the better army, and to decide, perhaps tomorrow or in a week's time, which was the stronger: the little good we believed in and were now defending or the dark, tortured spirit of our enemies. The odds were ascertainable and they were very even. We no longer thought, any of us in the Eighth Army, that fate or Providence had predestined us to victory. We understood now that it was possible for Britain to lose every battle, even the last. We should win only if the courage and endurance of our infantry and gunners, our tank crews and air crews, could equal and outlast, now in this stretch of desert, not yesterday or in a year's time, German courage and endurance, which was known to be very great. And we of the Army Service Corps—British, South Africans, Indians, Australians, and New Zealanders—knew with humility and pride that upon our faithfulness in supply the fighting man with his unappetising rations, his shells and bullets, his petrol and water, the issue depended also. It was Britain's hour, and the infantry's hour, and it was our hour, too.

And that was how matters stood on the last day of June, while the temperature rose and the flies multiplied, while Axis sympathisers in Alexandria baked cakes and formed reception committees, while Cairo waited, while Shepheard's watched.

Map of Egypt 1942

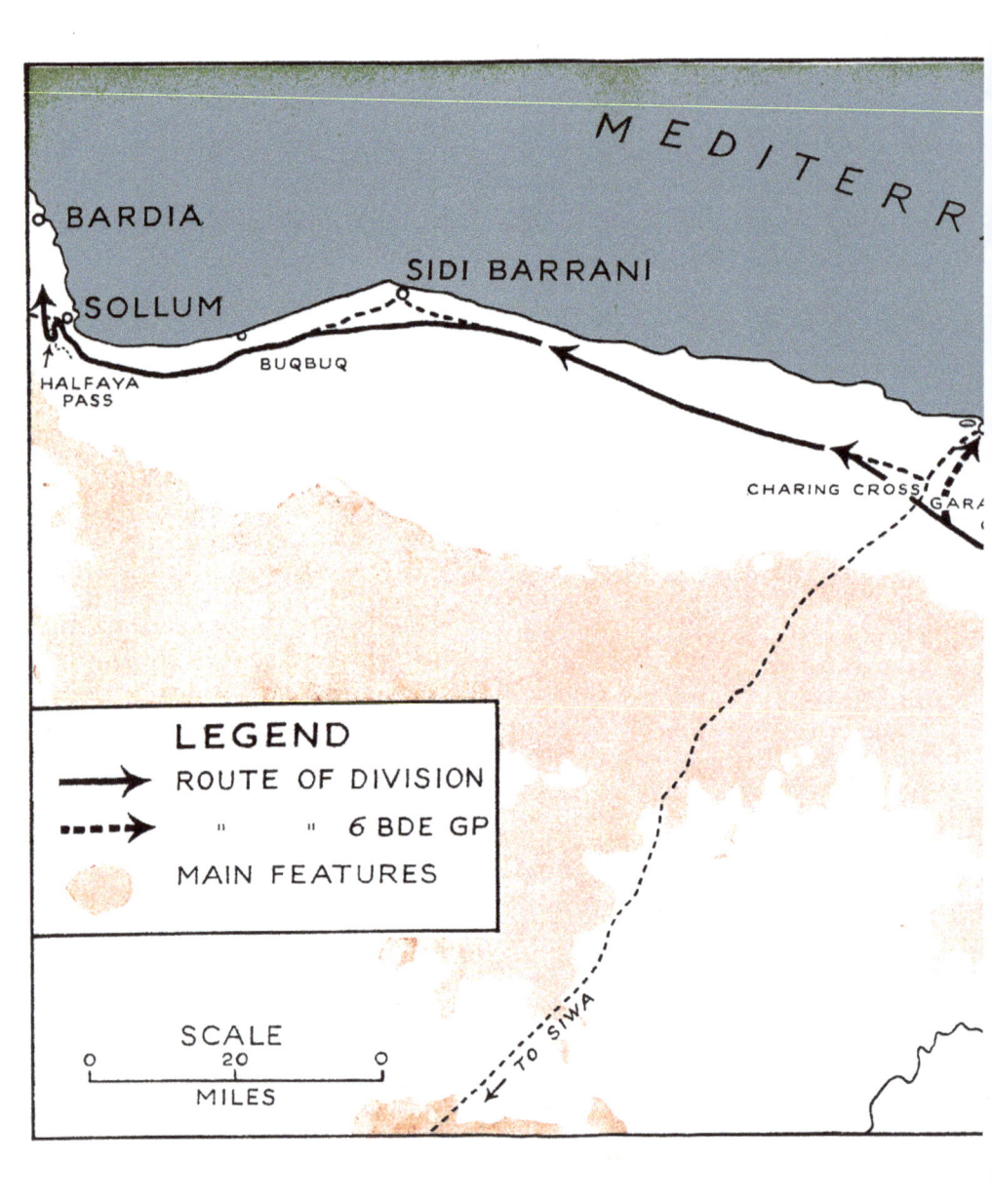

CHAPTER 13

WHILE SHEPHEARD'S WATCHED

THE sun beat down on ruined Mersa Matruh, drawing a hot reek, sweet and sickish, from crumbled masonry and dirty sand. No. 1 Platoon's drivers lay half under their lorries to get shade and a little draught. They wondered if the platoon from the 4th Reserve Mechanical Transport Company would arrive in time to relieve them, and sometimes they hoped it would but mostly they hoped not. They eschewed heroics and they had seen enough action to know how unwise it was to covet honour, but, on the other hand, here was the curtain rising on one of the great dramas of the war, and—well, they were as good as the RMT, weren't they? 'Those 20th jokers,' they told each other, 'reckon we'll do them. They're not fussy.' Our drivers were beginning to fancy themselves in a troop-carrying role.

It was 25 June and they had been with the infantry eight days—five on the journey from Syria and three in an outpost near Charing Cross, ten miles south-west of Matruh. The battalion had been relieved by Indians early that morning and now it was standing by in Matruh, waiting to move back to the Charing Cross area to cover a mine-laying party.

Two o'clock came, and as there was still no sign of the RMT the infantry boarded No. 1 Platoon's lorries and the convoy set out for Charing Cross. Detachments of the British advanced screen—Sidi Omar and Sollum had been evacuated by now—were moving east along the main road and the 20th Battalion seemed to be the only unit not in retreat. Often the road was choked by four lines of traffic, three of which, like rubbish in a chute, crashed and tumbled along in no sort of order. It was as though the desert had been tilted sideways and the British and German armies were obeying the law of gravity.

They wouldn't admit it, of course, but it was impossible for our drivers not to take a lonely pride in the situation. To be skinning her along towards Rommel, the old bull-bitch, the old Chevvy, while all the other stuff was high-tailing it for Alex—that was something. Our drivers were anxious to do well, and if for a single presumptuous moment they saw their common dust, the grey-yellow dust that spurted from beneath their wheels, mingling with the immortal spray from an armed merchantman, the slipstream from a doomed fighter, they held their peace, saying only: 'Curse those RMT slugs!'—though meaning, perhaps: Come then: let us to the task, to the battle, to the toil—' They certainly put it across us that time, those RMT jokers. They did that all right.'

The convoy passed through Charing Cross and drove along the Siwa road, halting fourteen miles south-south-west of Mersa Matruh. The infantry took up defensive positions and at 6 p.m. men from the 6th and 7th Field Companies started their job, which was to close a four-mile gap in an old minefield. Indian sappers, working from the north end of the gap, were out of sight, but our drivers could follow the New Zealand vehicles as they crawled across the perfectly flat desert. Every once in a while there was a flash and a gusty cough that told its own story. They watched breathless, and it was as though each vehicle carried a small part of them, which was felt as an ache, an absence, somewhere between heart and stomach. They did not know it, but a section from No. 4 Platoon was with the sappers.

The work was not finished until after ten, by which time four vehicles had been lost on our own mines. No one had been injured, though, and the sappers were just beginning to congratulate themselves when a 30-cwt. truck loaded with 350 mines passed over a bump and blew up, killing two men and wounding six others. For some moments a column of black smoke, much darker than the night and hooded like Death or Famine in an old print, hung in the sky, while behind it German flares blinked and spurted. The cause of the explosion was not known—then or later.

Soon afterwards the 20th Battalion withdrew through a narrow gap in the minefield and set out across country for the 4th Brigade area, ten miles south by east of Mersa Matruh, the journey, which so far as our drivers were concerned was an aimless succession of stops and starts, taking all night.

The intention was now becoming plain: our Division, less the 6th Brigade, was to fight a delaying action in front of the Alamein Line.

All reports showed that the enemy was advancing fast, and at five on the afternoon of 26 June the 4th Brigade moved ten miles south to higher ground. Two hours later the 20th Battalion was told to move north to meet an enemy formation that had broken through the Siwa road minefield.

With the sun setting, the infantry climbed into the covered lorries and our drivers closed the tailboards and sat waiting for the signal to move. There was a hum of engines, and more than twenty bombers, unidentifiable for a moment, flew out of the gold light and turned to place the darkening east behind them. They were Junkers 88s and they came in to attack at medium level. Except for the ack-ack crews, which met them with machine-gun and Bofors fire, nearly everyone was caught napping, and when the bombs began to fall some of the infantry were still in the lorries. Soon the whole area was covered by a pall of smoke through which came bursts of machine-gun fire and more bombs. Ragged lines of tracer struggled up in reply and Bofors shells blinked redly above the murk. By the time the planes left it was almost dark.

At first it was thought that the casualties had been sickeningly heavy but a check showed that four men had been killed and twelve wounded, which was less than anyone had dared to hope. None of our drivers was hurt and only two lorries were out of commission.

It was half past nine before the battalion moved, and an hour and a half later it halted and dug in. Flares were the only sign of the enemy, but later in the night our drivers heard firing close at hand. New Zealand gunners were engaging a patrol, the first to make contact with the Division. Early next morning infantry and transport moved again, halting in the brigade area in sight of an escarpment. At no time during the past twelve hours had the battalion been far from a place called Minqar Qaim, twenty-four miles south of Mersa Matruh, but our drivers didn't know where they were. They were relieved to hear that the rest of the Division was in the neighbourhood, for the hard, clean desert was very lonely and there was no pity in the clear outline of the escarpment.

While they were enjoying a cup of tea—enjoying it in spite of the brackish water from their camel tanks—a large concentration of enemy transport was seen on the horizon some miles away. Then lorried infantry and twelve tanks appeared and the battalion moved to a position below the escarpment, where it dug in as quickly as possible. The shelling became heavy just before noon.

Sheltering beside their lorries, which in some cases were parked only a stone's throw from the gun positions, which in turn were only about a hundred yards behind the infantry's slit-trenches, our drivers saw pillars of dust and smoke from our own shells form a high ragged wall along one sector of the front. Presently it started to walk towards them, halting after a while and then withdrawing.

The enemy fired back and it became dangerous to move. None the less, at one time or another during that endless afternoon, most of our drivers had to leave their slit-trenches, frenziedly hacked, scooped, chiselled in the rocky surface, to fetch ammunition, to collect wounded, and, very often, just to move their vehicles. Alan Falconer and 'Titch' Maybury, paying the price for distinguished company, had to drive out in front of the foremost slit-trenches to enable Captain C. H. Upham, VC,[1] to perch on top of their cab and direct machine-gun fire.

There were other jobs that were hardly less pleasant. When Les Howarth and Harry High,[2] of No. 1 Section, which was serving A Company, were sent to the 5th Brigade's area to pick up a load of small-arms ammunition, it took them two hours to cover two miles and they were under fire all the time. The ammunition, when they found it, was a chain in front of the firing line and one end of the stack was burning. However, they got their load.

Early in the afternoon tanks, guns, and lorried infantry started to close in on the New Zealanders, the attack developing simultaneously from north and south. The Germans tried hard to silence the 25-pounders, and shells and mortars landed all around them. Often, peeping from their slit-trenches, our drivers were unable to see the gunners for dust and smoke.

'They kept on firing though,' said George Searle,[3] 'not giving a damn. My lorry was about 300 yards from the nearest gun, but Jack McDonald's[4] was a good deal closer than that and one of his

---

[1] Capt C. H. Upham, VC and bar, m.i.d.; Government land valuer; Conway Flat, Hundalee, North Canterbury; born Christchurch, 21 Sep 1908; wounded May 1941; wounded and p.w. 15 Jul 1942.

[2] Dvr H. E. High; labourer; Timaru; born England, 23 Apr 1912; wounded May 1941.

[3] L-Cpl G. D. Searle; contractor; Waipukurau; born Owaka, 12 Aug 1914.

[4] Cpl J. McDonald; farm labourer; Eastbourne; born Makarewa, 14 Aug 1914.

CONVOY TO SYRIA. '*Beyond the tarmac the sand was soft and deep*' —page 200

BAALBEK, SYRIA '*Tents were pitched for living in, messing in, cooking in*' —page 202

Kaponga Box after a raid

Burning lorry, Kaponga Box

Desert Grave

wheels was blown off. He had to crawl out of his slit-trench to try to fix it.

'Later we watched Charlie Upham at work. We saw him standing on top of Alan Falconer's cab and directing anti-tank fire. Once, when a hidden spandau was giving trouble and our jokers couldn't find out where it was, he jumped up on the cab on purpose to draw fire. The range, though, was too great for the Brens, and finally a 3-inch mortar did the trick.

'The attack got fiercer, and a German troop-carrier, a half-track job, came tearing towards us. The anti-tank boys stopped it with a direct hit and as the old Jerries came tumbling out our Bren-gunners got stuck into them. Most of them put their hands up, but one tough Jerry was out of that carrier in a tick and down between the front wheels with his spandau going. A good boy, I should say.

'The day was hot and by this time I was as dry as a wooden god. The infantry had their tongues hanging out too, so whenever there was a lull I put the billy on in the cab, keeping the old lorry close to my trench.'

Other drivers did the same, dishing out half a mug here, a third of a mug there, or a billy-full to a platoon or a gun crew.

Meanwhile, in spite of repeated attacks, the infantry were still hanging on below the escarpment. The enemy was shooting with 210-millimetre, 105-millimetre, and 88-millimetre guns, and with captured 25-pounders. Much of the supplies and equipment lost in the last fortnight was being used against us. German soldiers with British shirts on their backs, British bully in their bellies, and Capstan cigarettes in their mouths, were driving Bedfords and Chevrolets distinguishable from ours only by a black cross painted on the door or by a piece of cloth marked with a black cross and fastened to the radiator.

As the day wore on, the area in which the 4th Brigade was fighting contracted, and at length guns, vehicles, and aid posts were crowded together much too close for safety. Here and there fires were burning, but as yet none of No. 1 Platoon's vehicles had been badly damaged, though several were out of commission with cut wiring, punctured tires, and pierced radiators, which our drivers tried to repair whenever they had a chance. When expert help was needed Sergant 'Dad' Cleave and Norm Hague[5] (the LAD crew) seemed always to be at hand.

Most of the damage, as always, was superficial. Lance-Corporal Jock Letham[6] and his driver, for instance, were pinned under their lorry for half an hour by a storm of anti-tank, mortar, and machine-gun fire, but it was hit only twice: once when an anti-tank shell carried away part of the superstructure, and again when a burst of machine-gun bullets drilled a row of holes in the tray.

Soon after half past three the sector held by the 20th and 28th Battalions was approached from one direction by twenty tanks and 200 vehicles, and from another by twenty tanks. They were engaged by the 4th Field Regiment and help was asked for from the 6th.

It was this (or an earlier threat) that caused a sudden ebb among No. 1 Platoon's headquarters' transport, which left the ack-ack lorry high and dry. Before anything could be done a pick-up squealed to a halt ten yards away and an Artillery officer got out of it, saying to 'Crassi' Cliff[7] and 'Pop' Cannell[8]: 'I want to use your bus as a screen. You chaps lie down by the wheels and I'll look after what's worrying you.' Shifting from foot to foot so that his head and shoulders were never a still target, the officer peered over the tray of the lorry and gave fire orders to his wireless operator, who was crouched in the back of the pick-up. Bullets hit the lorry and whizzed past the officer's head, but he continued to give orders and perform his clumsy dance. One bullet drilled a hole in the lorry's drive-shaft and a twist of copper-coloured metal landed in 'Crassi's' lap.

'Hi!' said 'Crassi', and the officer looked down for a second. 'How much longer do you reckon you'll be wanting us?'

'Not a lot longer,' said the officer. 'We'll get on to them any time now.'

He was a big, boyish-looking man with sandy hair and ('Crassi' thinks) spectacles.

Every now and then the wireless operator said 'Rounds fired',

---

[5] Cpl N. Hague, MM; truck driver; Cambridge; born Te Aroha, 3 Dec 1914.

[6] Sgt J. D. Letham, m.i.d.; tractor driver; Dromore, Canterbury; born Scotland, 25 Jun 1909.

[7] Dvr C. V. Cliff; station hand; Omana, North Auckland; born England, 24 May 1903; wounded 17 Jul 1942.

[8] Dvr R. W. J. Cannell; farmer; Lepperton, Taranaki; born NZ, 1 Apr 1904; wounded 27 Jun 1942.

but among the medley of sounds it was impossible to tell where the rounds were landing or from where they were coming.

At last the officer said: 'Thanks very much. We're getting out of here now. I wouldn't hang about too long if I were you. See you again, eh?'

'Like hell you will!' said 'Pop', but 'Crassi', who had no nerves, only smiled, pushing his Australian wide-awake farther back on his head and feeling for his tobacco.

The pick-up raced off and the ack-ack lorry followed, 'Pop' going through his gears in two quick rips. As soon as they were safe 'Crassi' and he found a quiet spot where they could boil up. Later they made their way back to platoon headquarters—cooks' lorry, orderly-room lorry, water cart, LAD, staff car—which was again in danger. After being cursed by several drivers (not No. 1 Platoon men) 'Pop' had to be content with a position that was far too exposed to excite jealousy. ('Pop,' said 'Crassi', 'they want you to go and park with the bloody Germans.') A vehicle still farther out in the desert came under heavy mortar fire and was set alight. Then the enemy turned his attention to the ack-ack lorry, bracketing it with bombs and finally landing one by the tailboard. 'Crassi', sheltering under the rear differential, was lifted from the ground and slammed down again unhurt. 'Pop', lying between the front wheels, was wounded in the head. Black smoke hid the lorry, which hissed air from a punctured tire and dripped water from a pierced radiator and petrol from a pierced tank.

The fire shifted to another target and 'Pop' was taken to an Advanced Dressing Station. In spite of pain and disappointment he was able to remember in which secret corner he had cached a bottle of Scotch whisky against a rainy day, and before he lost consciousness he made 'Crassi' a present of it.

'Pop' was our third casualty. The others were Ed Child[9] (No. 2 Section) and Pat Wells, whose section (No. 4) was serving D Company and had spent most of the day 400 yards in front of the 25-pounders and 100 yards behind the infantry. Ed was wounded slightly in the calf, Pat badly in the foot. 'C Jay' O'Brien, who could hardly be classed as a casualty, though he received medical

[9] L-Cpl G. E. Child; farmer; Maungakaramea, Whangarei; born Maungakaramea, 20 May 1916; wounded 27 Jun 1942.

attention, could show a paybook in two halves, a smashed fountain pen, and a spectacular graze immediately above the heart.

Not knowing how lucky the platoon had been as a whole (for the sections, of course, had stayed near their respective infantry companies all day), our drivers worried about their friends, and they were anxious, too, about the general situation. The Division, from all accounts, was encircled, 25-pounder ammunition was running low, and General Freyberg was wounded.

But dusk, thank goodness, was not far away. Dusk meant relief from tension, a cup of tea, a chance to stretch your legs. As the sun dropped towards the horizon the fire slackened and by sunset it had almost ceased. Sometimes there was a rattle of machine-gun fire from an outpost (one of those *I'm-still-here* rattles) and sometimes an armour-piercing shell, glowing red among the shadows and sending off showers of sparks when it bounced, plunged in from the surrounding gloom. Burning German transport, twinkling like camp-fires, traced a ragged circle round the beleaguered position.

Slowly, grudgingly, as though in the celestial store they were short of nights and the Divine Quartermaster was reluctant to make the issue, the darkness deepened, and presently it was as dark as it would get. Our drivers were out of their slit-trenches by now and gathered in small groups, starting at every noise. At first they were almost shy—' How d'you make out, " Pork "? '—' Not bad, eh? '—' Get a puncture? '—but soon they were talking thirteen to the dozen. Out it all poured—the bottled-up comments, suggestions, criticisms, hopes, fears, stories. It was as though an invisible butler had passed around, giving everyone just two cocktails.

' Reckon old Jerry took a hiding worse than what we did.'

' See that Jerry car get it early on? She comes bowling along the escarpment, game as one thing, and then she stops it. One of our anti-tanks, eh? Reckon they never knew what hit them.'

' Saw a two-pounder get a Jerry motor-cycle and side-car. Just blew it to nothing.'

' One Itie they reckon—Harry saw it—came into our lines on one of those motor-cycle combination things. He has his hands in the air and all he can say is " Momma! Poppa! " Riding along gripping the thing with his knees saying " Momma! Poppa! "—the poor bastard.'

'They reckon Inglis[10] took over after "Tiny" was hit. Handed over the brigade to Colonel Burrows[11]—that's our 20th joker.'

'Hell! What's the difference? They'll get the lot tonight or tomorrow. We're surrounded and we've got no 25-pounder left. It came over the German radio.'

'If anyone's in the bag it's bloody Rommel.'

'Look, boy! Rommel's just the best general. . . .'

Thus the talk, while four drivers crouch over a tin of bully and six over a tin of sausages. Round a tent, like spokes round a wheel, men lie on stretchers, blankets covering them. Every now and then one of them is picked up by two orderlies and taken inside. Sometimes an orderly glances at a man's face and pulls the blanket right over him. There is little noise—only the low murmur of voices and the hiss of primuses hidden in cabs or in the backs of lorries; only the scrape of metal against metal, the clatter of a dropped spanner, the thud of steel on rubber as groups of drivers, hot and swearing in the darkness, change tires, patch radiators, fix a drive-shaft. The ack-ack lorry is repaired and driven to a safer place. An hour later 'Dad' is able to tell Captain Gibson that all the lorries are mobile except one, which is on tow.

Captain Gibson goes round platoon headquarters, warning his men that an attempt to break out will probably be made during the night. They had better try to get some sleep.

You fetch blankets from your lorry and lie down in your slit-trench among puddles of moonlight and bars of shadow. In the distance shovels make tiny scraping noises, and 'Baldy's' pickaxe, striking hard rock, ticks steadily—like a watch under your pillow. You look up and see the thick tire, the heavy tow-bar, the ugly

---

[10] Maj-Gen L. M. Inglis, CBE, DSO and bar, MC,* m.i.d.; barrister and solicitor; in First World War commanded company in NZMG Corps; CO 27 (MG) Bn Jan-Aug 1940; commanded at various periods 4 and 6 Bdes, 9 Bde, 4 Armd Bde, and was GOC, 2 NZ Div, Jun-Aug 1942 and Jun-Jul 1943. Since 1945 has been in Germany as president of a military legal court and was recently appointed Chief Judge of the Control Commission Supreme Court in the British zone.

* First World War.

[11] Brig J. T. Burrows, DSO and bar, ED, m.i.d.; Rector, Waitaki Boys' High School, Oamaru; born Christchurch, 14 Jul 1904; CO 20 Bn 8 Dec 1941-27 Jun 1942; comd 4 Bde 27-29 Jun 1942, 5 Jul-15 Aug 1942; CO 20 Bn and Armd Regt 16 Aug 1942-12 Jul 1943; comd Adv Base 24 Jan-11 Feb 1944; 5 Bde, 29 Feb-30 Mar 1944; 6 Bde, 1 Jul-22 Aug 1944; 5 Bde, 22 Aug-4 Nov 1944.

radiator, you smell the congealing oil drained half an hour ago from the sump (remember to enter oil-change in *AB Four-One-Two* tomorrow) and are comforted. Comforted, you fall asleep.

. . . .

The orders were: ' Brigade night attack. Battalions in the following order: 19th Battalion front, 28th Battalion right rear, 20th Battalion left rear.' The intention was to break through into the open, the leading battalion's task being to clear a narrow neck of high ground to make a path for the transport. Zero hour was half an hour before midnight, and after the break-through had been made the transport would move forward to embus the infantry, the 5th Brigade following.

The transport started to form up at 11 p.m., and for ten, twenty, thirty minutes—time was difficult to reckon—all you could hear was a deep, angry growl that seemed to be coming from the earth, from the sky, from the four corners of the desert. Chequering the bone-white battlefield with swaying shadows, lurching over rocks, swerving to avoid, often unsuccessfully, shadowy slit-trenches, the transport formed up in column of route on a wide front. The growl sank to a mutter—suck of intake, rap of tappet, putt-putt of exhaust—but still you could hear nothing else.

Our drivers sat quietly in their cabs, not talking. Time passed and a battalion formed up in the open desert. The officers were speaking—you could sense it—and the men listening.

The battalion melted away and our drivers sat on, straining ears and eyes. Engines had been switched off in obedience to an order passed down from the head of the column and it was very quiet. Marvellously this great mass of transport—it seemed to be fifty or sixty yards wide and its length was impossible to judge—had attracted no fire. A flare went up and hung yellow in the sky for some seconds but no shells or bullets followed. What was wrong with Jerry? Surely he had heard the transport moving? Surely he knew what was going on? There was nothing to do except wait quietly under the moonlight, sitting tense in your cab or standing beside it and shifting from foot to foot.

. . . .

When the success signals were seen the transport moved forward through the gap and halted. One of the first shells that landed hit

No. 1 Platoon's ack-ack lorry. It was an armour-piercing shell and it struck the tow-bar side on, twisting the thick steel as though it had been tinfoil and biting a great piece out of it. From the engine came a smell of burning, and fire extinguishers were torn from the nearest vehicles. A blaze was what everyone dreaded most and our drivers would have beaten out a fire with their bare hands had that been necessary.

Mortar bombs and anti-tank shells, fired wildly and at long range, came in from the flanks, and a spandau opened up only a short distance ahead, the bullets springing from the ground and going high. Some men lay down between the lines of transport; others bustled around finding things to do, because it was easier for them to keep calm if they were busy. 'Dad' and half a dozen helpers changed the off front wheel of the ack-ack lorry in less than five minutes and Bernie Caddy[12] backed his water cart to take it in tow.

By this time the infantry had started to embus. Discipline was good though some of the men seemed almost drunk with excitement and some were wounded. For these there was no special transport and room was found for them where they were most likely to be comfortable. Everything had to be done in haste, for the weight and accuracy of the fire, though not considerable yet, was increasing momently. There was no avoidable confusion and soon the leading vehicles began to move. The whole mass surged forward.

In Thessaly, in the bright sunshine, our drivers had seen the shadows of aircraft run before them on the white road, the first warning of danger; in Libya they had been driven towards the guns like pheasants; but nothing that had happened to them before was as strange or as wild as this mad dash. It was experienced as a 'mad dash' and that is how people remember it, though in point of fact the column was moving almost sedately—probably at not more than twelve miles an hour.

The dust was so thick that at times it was difficult to see the vehicle in front, though it was never more than a few yards away and seldom more than a few feet. Bursts of light, their glare muffled by boiling dust clouds, showed where mortar bombs were bursting, but the explosions were not heard, for the long shriek

[12] Dvr A. B. Caddy; farm labourer; born Thames, 12 Jun 1914; killed in action, 15 Jul 1942.

of vehicles moving in low gear drowned everything. Bullets passed unnoticed, and afterwards our drivers found neat holes in trays and canopies. On the flanks splashes of orange and yellow showed where the infantry were making a running fight of it, firing rifles and Bren guns from behind cabs or from the backs of the lorries and drawing answering fire.

When a vehicle was knocked out or developed engine trouble or collided with another, telescoping fan and radiator, it was abandoned at once, drivers and passengers jumping on the first vehicle that slowed down.

And so it went on—not for a long time according to the clock but for ages as dreams go, and this was a kind of dream and therefore not really frightening: less frightening, our drivers were to find, than lying in a slit-trench with the lorry ten yards away and a dixie of stew cooling on the flat, ugly mudguard and the Stukas coming. There was no background of normality, no touch of everydayness, to make a nightmare out of a dream. It was Cowboy-and-Indian stuff: a picture, a story, a play—almost, in its rush and wonder, a poem.

And in the boiling dust, superimposed on the dun clouds and the wagging tailboard ahead, our drivers saw everything: all the scenes of the day, all the scenes of the night. They built them from quick phrases, chopped sentences, gabbled words; from the glimpse of a bayonet dark and bright with blood ('I tol' you I'd use 'er—I was determin' to use 'er.'); from a field dressing all one stain of blood. They saw Captain Upham standing with his hands in front of his face after throwing a grenade into the back of a lorry; and ten terrified German lads, like the ten little nigger boys, shot all together in the one big bed they were sleeping in for warmth and company; and the tracers flowing shin-high above the desert and the infantry skipping to avoid them; and a fat German officer kneeling in the back of a staff car and trying to fire his pistol. And they knew (from shouted question and answer) who was wounded and who would not be wanting, ever again, the camera safe under the seat, the zip-fastening despatch case in the tool box wrapped in rags.

After a mile and a half had been covered the head of the formation met enemy transport in a wadi and swung south to avoid it. That concluded the Cowboy-and-Indian phase. The shooting stopped as suddenly as it had started, leaving a few fires, rapidly fading in the distance, to prove that it had taken place. Soon the last fire was out of sight, but the great mass of transport, still moving in a cloud of dust, still close-packed, pressed on at the same speed. Later it sorted itself out and the drivers settled down on an eight-vehicle front and tried to dress by the centre. Once, towards morning, the head of the formation had to swing south to avoid a concentration of enemy transport, but soon it was heading east again, making for the Alamein Line. Jumbled together in the backs of the lorries the infantry slept like logs. Only the wounded were awake.

The stars paled and a streak of primrose light appeared far ahead. Light, grey and frozen, flooded the whole desert, making it possible to see the drawn, dirty faces of the men in the lorries. The tires—the thousands of tires—made a crunching noise as they turned against the hard surface.

As soon as it was light the transport moved into desert square formation and halted, while portées with anti-tank guns, and towers dragging Bofors, hurried towards the perimeter. The enemy was not far away.

The lorries were widely dispersed, and men with biscuits and lumps of bully in their hands went over to each other, meeting between the lines to exchange news. A few of our drivers, it was found, had been borrowed by the 5th Brigade, which had broken out on its own (delay in launching the attack had caused an alteration in the programme), but as far as it was known everyone was safe except Corporal 'Snow' Weir.[13] He had been missing when No. 4 Section formed up for the break-through, and Frank Humphreys,[14] his driver, after searching everywhere, had been forced to leave without him.

At seven o'clock, after an hour's halt, the journey was continued, the transport travelling in desert formation with the guns on the outside. There was no stop for lunch as the enemy was known to be following.

The day was hot and several of the wounded became feverish, and everyone, being over-tired, had a headachy, sickish feeling.

---

[13] Cpl A. McK. Weir; timber worker; born Rawene, 12 Dec 1913; p.w. Jun 1942; died while p.w., 3 Dec 1942.

[14] Dvr F. R. Humphreys; motor driver; Hamilton; born Wellington, 7 Oct 1915; wounded 4 Jul 1942.

Our drivers shared the general malaise but they were well content. It was over now and they had memories comforting to their self-esteem. 'Every man,' says Doctor Johnson, 'thinks meanly of himself for not having been a soldier.' Every soldier, he might have added, thinks meanly of himself for not having been under fire. Our drivers were proud of their unit and they would not have chosen to belong to any other, but all the same it had sometimes occurred to them—and they were sensitive on the point—that belonging to the Ammunition Company was not quite the same thing as belonging to the Black Watch. No. 1 Platoon would feel easier from now on while any spoke that fought upon St. Crispin's Day.

At nine that night—it was 28 June—the 4th Brigade reached its destination: Fortress A, Alamein Line. The wounded were evacuated to hospital, the infantry dug in, and the drivers serviced their vehicles. Then, with the guns sounding in their ears, they fell asleep.

For many days the platoon listened to the guns—our own and Rommel's. Sometimes the sound was no more than a dull rumble in the distance: often it made the air tremble. Wherever the 20th Battalion went the drivers went too, withdrawing behind the nearest cover as soon as the infantry had debussed and then standing by for the next order. There was ammunition to be fetched, hot stew to be taken to outposts, and at any moment the transport was liable to be called forward to embus the battalion and move it to another sector. It was No. 1 Platoon's testing time.

. . . .

The general position on 29 June, the day after the break-through, was that the 30th and 13th Corps were hurriedly strengthening the Alamein Line while the following units, reading from north to south, were barring the advance: 1st South African Division (Alamein Box), 18th Indian Infantry Brigade (Deir el Shein), New Zealanders (Fortress A), and 19th Indian Infantry Brigade (Fortress B).

The 29th was a quiet day for our drivers. During the morning gunfire could be heard from the direction of Fuka, where infiltration along the railway and the coast road had prevented the British from standing, and in the afternoon the 20th Battalion moved a mile or two south and dug in. The Luftwaffe was overhead all night long.

Early on the following afternoon Rommel arrived in front of the Alamein Line. The 20th Battalion was now near Deir el Munassib, twenty miles south of El Alamein, and the main attack was expected to develop during the night at a point twelve miles northeast of its position.

The attack was launched the next day (Ash Wednesday they called it in Cairo, mocking the smoke rising from army and ministerial chimneys), the first and fiercest blows being struck against the South Africans in the coast sector. The 18th Indian Infantry Brigade was heavily engaged at Deir el Shein, south of the Alamein Box, and during the afternoon tanks and infantry converged on Fortress A. When the artillery opened fire the enemy decided that the weak spot he was searching for was not there. The tanks sheered off and the infantry climbed back into their lorries. At half past six in the evening the 6th Brigade's guns opened fire on a large enemy column, which then withdrew out of range. A report from the coast sector said that the enemy was retiring west leaving burning vehicles. The last news of the day was bad: the 18th Indian Brigade at Deir el Shein, after fighting splendidly all day, had been overwhelmed, and there was now a gap in the line.

That night our drivers heard heavy bombing. The Royal Air Force was attacking panzer divisions in laager.

Rommel started 2 July by throwing the 90th Light Division, three armoured divisions, and most of his Italian infantry against the South Africans and the British 1st Armoured Division. He was trying to enlarge the gap south of the Alamein Box. The battle raged all day, and at one time he was so sure of success that he issued a communiqué in which he spoke of pursuing the defeated British towards Alexandria. The German radio said that he would lunch there on Friday.

That evening after a long, anxious day—the battalion had moved north-north-west to harry the enemy's rear and had come under shellfire—our drivers heard news of a big tank battle. The enemy was said to be retiring.

It was true. The bombing of the panzer laagers, the gallantry of the 1st Armoured Division, and the dogged resistance of the South Africans had saved the day. On 3 July Rommel tried again and the next forty-eight hours were desperately anxious. Both sides were drunk and poisoned with tiredness but they fought like lions—the Germans because they were good soldiers and the prize was in sight, the Allies because it was the last ditch and it was better to die now than to be ashamed always.

They fought in the stinking heat and through the cold darkness, and on the third morning tens of thousands of men, packed like people in a city between Alexandria and the Alamein Line, between the sea and the depression, came out of their tents, threw off dewy blankets and climbed from their slit-trenches, hoiked in the sand, scratched themselves like dogs in the sunlight, and looked around. And they said: 'She's going to be all right. I think we've done it. I reckon she's going to be all right.' They turned to their breakfasts and their tasks, forgetting their anxiety in a few hours, not remembering, after a few days, the times when she was all wrong.

. . . .

No lunch for Rommel in Alexandria—not this week. Mussolini and his white charger could go back home and stay there until they were sent for. The Eighth Army, after being beaten at Knightsbridge, after losing Tobruk, after abandoning Mersa Matruh, after failing to stand at Fuka, after seeing the new line—the last line—bend and begin to crack, was holding fast. No one could call it a great victory, but you could, if you liked, call it a miracle. Not because the Germans were much stronger than we were—they weren't—not because they had more and better weapons—they hadn't—but because somewhere on a tarmac road between Tobruk and El Daba, for ten minutes, for an hour, for a day—all who saw it will know the truth—a great army had streamed east, reeking of defeat, breathing the sour air of defeat, sick with defeat.

It was a scene best forgotten and No. 1 Platoon's drivers forgot it as soon as anyone. They were becoming used to this mobile column business and the men they carried were now their friends. The RMT could take over if it liked—but there was no tearing hurry.

And then, to spoil everything, the Stukas came.

They came in fives, in tens, in thirties, and they came every day and several times a day. They would circle above the transport and fall out of the sky one after another, screaming. And after they had gone, before the smoke had cleared or the sand settled, the ambulances would come scuttling across the desert, making for the flames and the smashed lorries. The Luftwaffe was trying to hamstring the mobile columns by knocking out their transport.

On 4 July the 4th Brigade area was raided four times. In the afternoon several lorries were damaged and two were destroyed. 'Owie' McKee,[15] sheltering in a deep slit-trench, was buried alive beside his burning lorry, and his friends dug him out with bare hands and steel helmets. He was unconscious when they lifted him into the ambulance.

Mail arrived in the evening and the last raid of the day took place while our drivers were reading letters from home by the fading light. Frank Humphreys' lorry was destroyed, Frank being badly hurt by blast when a bomb landed on the edge of his slit-trench, and George Searle's lorry, loaded with mortar bombs, sticky bombs, and hand grenades, was set on fire. After George and the others had dug out an Artillery driver who had been buried in his slit-trench they shovelled sand on the flames. Ammunition was exploding, but that did not stop Corporal Arty McDonald[16] from jumping on the tailboard and clearing it of a box of smouldering sticky bombs, which, by some miracle, was still intact. Half an hour from the start of the raid the fire was under control. The tires had gone from the back wheels and at first glance it seemed as though the lorry would have to be written off. But our drivers, knowing that transport was precious beyond price, set to work, and by ten that night, when C Company was due to move, the lorry was again mobile.

Shortly before noon the next day, while the 4th Brigade Group was shifting from east to west of the fortress, aircraft appeared out of the sun, roared down the lanes of transport, and dropped bombs from about 1000 feet. A glitter of wings, black blobs falling, an eruption of the desert, and it was all over. In the centre of the formation, where a jeep had been overturned beside a staff car, there was a bad mess, and word flew round that Brigadier J. R. Gray[17] was dead. Also among the dead, it was learned later, was

---

[15] Dvr O. R. McKee; mill worker; Auckland; born Takapuna, 19 Dec 1916; wounded 4 Jul 1942.

[16] Cpl A. G. McDonald; bricklayer; Papatoetoe; born Tonga, 24 Oct 1910.

[17] Brig J. R. Gray, ED, m.i.d.; barrister and solicitor; born Wellington, 7 Aug 1900; CO 18 Bn 5 Jan 1940-6 Nov 1941, 28 Mar 1942-29 Jun 1942; comd 4 NZ Inf Bde 29 Jun-5 Jul 1942; killed in action 5 Jul 1942.

the Brigade Major and four men attached to brigade headquarters. The 28th Battalion, which had come under the command of the brigade group the day before and was travelling on the right flank of the convoy, had lost one major, one lieutenant, and fourteen men.

After the wounded had been taken away the convoy travelled on, reaching its destination at three in the afternoon. The Luftwaffe appeared almost at once and there was a short, sharp raid. Digging was difficult and the third raid of the day caught several drivers with their slit-trenches unfinished. At 6 p.m. some sixteen planes bombed the brigade area, and the last raid, made by Stukas and Junkers 88s escorted by Messerschmitts, took place at dusk. Claude Cameron[18] was injured by blast from a 500-pound bomb and three tires were blown off Captain Gibson's staff car. Other damage was slight.

And it was like that every day. The Stukas seldom diverged from their timetable. They could be expected early in the afternoon, during the evening meal, and at sunset. Even when they failed to appear the feeling of suspended doom was almost as bad as a raid.

For a while No. 1 Platoon had a run of luck. There were no more casualties, and although several of the vehicles became like sieves none was immobilised for more than a few hours—thanks chiefly to 'Dad' Cleave and Norm Hague, both of whom earned the Military Medal for putting the job first and their lives second.

Except in an emergency, moves for which transport was needed were made only at night, and during the long, cloudless days the drivers had little to do except watch for Stukas. Most of them were very tired, but when they slept in the daytime their nerves stayed awake. Everyone developed a *listening* expression. Conversations would end suddenly and eyes go to the horizon. Ears would be cocked for the noise of engines, the thump of guns. When it was a genuine alarm Bofors started coughing in the distance, marking the course of the bombers with a mass of white puffs, thick near the centre of the target, scattered on its outskirts, like the marks on a dart-board. Our drivers seldom heard the beat of engines. As dogs in lonely districts pass on from farmstead to farmstead the chorus of warning and mad rage, drowning the step,

[18] Dvr C. L. Cameron; garage attendant; Takapuna, Auckland; born NZ, 24 May 1916; wounded 5 Jul 1942.

the sly rustle, that first caused it, so the Bofors, in area after area, barked furiously, until at last the guns with the 20th Battalion were barking them all down. 'Crassi's' spandau, a Pomeranian among wolfhounds, barked too, but you couldn't hear it.

Sometimes the Stukas dropped their bombs and went away. Often they would dive and scream among the smoke for what seemed an eternity. After they had gone our drivers would rise from their slit-trenches, stand upright in them for a few moments (as you might stand in a bath before stepping out of it) and grin shakily while they counted the fires, watched the ambulances at their grim tasks, and noted how the desert was splashed with greyish-white streaks. These were caused by small bombs that had exploded before piercing the surface.

Everyone owned that he was afraid and everyone either had a charm against fear (which sometimes worked and sometimes didn't) or was trying to find one. It helped to count the planes and watch the bombs falling. It helped to dig your slit-trench to a specification: so many yards from the lorry, so many feet deep, so many feet wide. It helped, when the planes were coming, to have a pet jingle you could let loose in your head—nonsense:

> Down where the waistline's a little longer,
> Down where the soup stain's a little stronger,
> That's where the vest begins. . . .

Or something charming from childhood:

> How many miles to Babylon?
> Three score and ten.
> Can I get there by candlelight?
> Yes—and back again.

It helped, perhaps, to pray: to pray that nothing irrevocable would happen (as yet, though people were dying around them, none of our drivers had been killed or mutilated), or to ask, with a diminution of self-respect, for the RMT to be sent.

But the RMT, the slugs, didn't come. There was no talk of the platoon's being relieved, and the Stukas went on observing their timetable.

It was hot all day, but towards evening (Stuka-time) it became cooler. The sun retracted its fierce heat and curled up in a great glowing ball, dazzling westward-gazing eyes and hiding the Stukas. After the Stukas had gone for good, the night, cool and lovely, rushed over the whole desert and flowed right to the sunset, little and distant now, like the mouth of a cave. Our drivers opened tins of pears, laughing and talking. Later—between nine and ten probably—the transport would form up in column of route, move after a wait of anything from one to four hours, and travel, most likely, all through the night, stopping and starting, while in the backs of the lorries the infantry lolled and dozed, their faces drawn and corpse-like under the moon.

All day long under the sun, under mortar bombs and shellfire, they had lain in slit-trenches, with little water and nothing to eat except biscuits and bully. But they were always cheerful—cheerful when they gave our drivers their cameras and other treasures to look after, saying 'Hang on to it if anything happens'; cheerful when you boiled up for them ('Hell! You'd think we was Father Christmas!'), cheerful and angry when they went out to die (angry with the Germans, with the authorities, with themselves for being such fools), cheerful and even happy sometimes under the hot sun, under the cold moon, under the stars.

. . . .

Ruweisat Ridge, fourteen miles south of Alamein, was the most important of a series of roughly parallel ridges cutting both the German and British positions. Whoever held it securely could sweep the coast with direct fire and operate from good cover against an exposed plateau to the south. Its capture was worth many lives.

The 4th and 5th Brigades held a ridge six miles to the south, Alam Nayil; the 5th Indian Brigade held the east end of Ruweisat and the Germans the west, which was the New Zealanders' objective. The battle was planned for the night of 14-15 July.

On the morning of the 14th the men of the 20th Battalion were in their slit-trenches, revelling in the spreading warmth after a bitter night. Greatcoats and blankets (one each) had been sent up to them the day before but had failed to keep out the cold. Since the afternoon of the 11th, when our drivers had seen them moving steadily forward into a storm of shellfire and mortar fire, there had been no extra cups of tea for them—only hard fighting and discomfort. The transport was with the B Echelon near Alam Nayil.

Shortly before noon enemy bombers escorted by fighters passed overhead on their way to bomb the Divisional replenishment area.

Hailstones in the desert, south of Alamein

Digging slit-trench, Kaponga

Rest area near coast, Burg el Arab

Wheel tracks, Alamein

Our drivers saw the black smoke, but it did not occur to them that as a result of that raid two men from their unit were dead, two dying, and six wounded.

Twelve Stukas had dived out of the sun to bomb and machine-gun No. 2 Platoon at the ammunition point. Corporal Owen Miles was killed instantly when a lorry loaded with gun-cotton, gelignite, and ammonal received a direct hit and blew up, Bob King[19] was killed by shrapnel while sheltering under a pick-up, and Doug Henderson and Jacky O'Connor were wounded mortally by shrapnel. Dave Gordon[20] was wounded in the face, and Bob Towart,[21] his mate, was wounded in the back and legs and pinned to the ground by the differential of his lorry, which was let down by the collapse of all four tires. When he had wriggled clear Dave and he began to throw cases of mortar bombs from the back of the burning lorry. Their friends had to make them stop, and after a while the lorry blew up. Second-Lieutenant Borgfeldt,[22] who had been lying beside Bob King, was wounded in the head and body, Sergeant Andy Andrew[23] (Ammunition Platoon) was hurt by blast, Joe André[24] had a compound fracture of the leg, and Len Skilton[25] was injured in the head. Two three-tonners were completely destroyed; another, which was used for carrying stores and canteen goods, had to be written off; and two vehicles, a three-tonner and a pick-up, were badly damaged. Three men who had been drawing ammunition were dead, and in all, according to the 5th Field Ambulance war diary, twenty-one deaths, had occurred in the replenishment area. Forty were wounded.

Our No. 1 Platoon drivers did not hear of the tragedy until later in the day (14 July). Headquarters 20th Battalion was bombed once during the day and the 4th Brigade area came under shellfire,

---

[19] Dvr R. King; tractor driver; born Christchurch, 25 Sep 1904; killed in action 14 Jul 1942.
[20] Dvr D. E. Gordon; labourer; Pukeatua, Te Awamutu; born Kawhia, 20 Jun 1917; wounded 14 Jul 1942.
[21] Dvr R. Towart; salesman; Christchurch; born Christchurch, 13 Feb 1908; wounded 14 Jul 1942.
[22] Lt R. A. Borgfeldt; draper; Christchurch; born Christchurch, 8 Sep 1910; wounded 14 Jul 1942.
[23] Sgt A. L. Andrew; salesman; Auckland; born Waipawa, 20 Apr 1905; wounded 14 Jul 1942.
[24] Dvr J. L. André; truck driver; Hawera; born Auckland, 11 Sep 1916; wounded 14 Jul 1942.
[25] L-Cpl L. A. K. Skilton; motor driver; Palmerston North; born Wanganui, 21 Sep 1914; wounded 14 Jul 1942.

P

but no damage was done to the transport and none of our drivers was hurt. Our own field and medium guns pounded the enemy for hours, preparing the way for the attack, and at dusk the firing quickened.

Zero hour was 11 p.m., but by then our drivers were asleep. What happened that night is not their story. They heard about it when it was over and they set out with sad hearts to collect what was left of the 20th.

They stayed where they were that night and the next day, and the news that came back to them was all bad. The battalions had gained their objectives but were in trouble with tanks. The Indians, attacking on the right, were held up. Supporting arms were unable to come forward because of tanks, and something had gone wrong about our own armour.

Our drivers thought of the infantry up there on the ridge—the tea-drinkers, the cheerful fighters, the good friends—but they had troubles of their own. The brigade area had been under shell and mortar fire since dawn, and at eleven o'clock the first bombers came over. There were ten of them and they hit No. 1 Platoon's water cart and killed Bernie Caddy in his slit-trench. Jack Voice,[26] Bernie's mate on the water cart, was wounded in the back and hurt by blast, and Lenny Hay[27] died later in the day.

It was the first time that any of the platoon's drivers had been killed violently in front of the others, and this long run of good luck had given them a sense of immunity. Now all that was shattered. They saw Bernie Caddy—dry, humorous, resourceful, very well liked—lying beside his slit-trench, unwounded, but with the life crushed out of him. They heard how Lenny Hay, asleep in the back of his lorry, had been woken by the bombs falling and had been caught before he could get to his slit-trench. He had wanted to know if any of the others were hurt. Then he had said he was hot and would like his jersey taken off. We heard afterwards from a New Zealand doctor that if courage could have saved him he would have lived.

Possessing no unusual gifts, unless a genius for friendship is a

---

[26] L-Cpl J. D. Voice; lorry driver; Invercargill; born Gore, 24 Nov 1918; wounded 15 Jul 1942.
[27] Dvr L. E. Hay; carpenter's labourer; born NZ, 16 Oct 1917; died of wounds 15 Jul 1942.

gift, possessing no unusual virtues, unless happiness and high spirits are virtues, consciously contributing nothing to the sum of the world's treasure, he was the one we could spare least. He was like Stevenson's man—an extra candle in the room. When they heard that he was dead, his friends knew then, as so many had known before them, and so many others would know later, that the war had lasted one day too long, had killed one boy too many.

The rest of the day was bad. The platoon's vehicles were parked near some abandoned German guns which drew bombs like a magnet. There was another sharp raid before lunch, and at three in the afternoon the area was raided by twenty-four bombers. Three lorries were hit—they were not ours—which brought the total for the day to six. Throughout the afternoon the area was under fire nearly all the time, but there was a lull in the bombing as soon as sixteen British fighters appeared. They patrolled the sky until shortly before sunset. After they had gone, twenty-four bombers, preceded by fighters, came out of the east. The bombers went away after dropping their loads, but the fighters stayed and were joined presently by eighteen bombers, one of which dropped twelve bombs in a row.

At dusk there was a double issue of rum and our drivers drank it gratefully, feeling the treacly fire warm and comforting all the way down to their toes, ironing out creases, steadying nerves, melting the stone under the heart. The news was shocking: Ruweisat Ridge lost, the 4th Brigade overrun, the 20th Battalion with 50 per cent casualties, the Brigadier[28] and his staff missing, and also Captain Washbourn,[29] Captain Upham, Captain Maxwell,[30] Lieutenant Moloney,[31] Second-Lieutenant Cottrell[32]—men whom our drivers had come to know in the last fortnight, know and trust.

Later that night all the transport in the area moved a short distance to Headquarters 4th Brigade. There it formed up in three lines

[28] Brig J. T. Burrows. He returned to the Division on 16 July.

[29] Capt G. W. Washbourn; bank clerk; Christchurch; born Timaru, 13 Jul 1916; p.w. 15 Jul 1942.

[30] Capt P. V. H. Maxwell, DSO; manufacturer's representative; Christchurch; born Londonderry, Ireland, 14 Feb 1906; p.w. 15 Jul 1942.

[31] Lt D. A. R. Moloney; insurance clerk; born NZ, 11 Aug 1910; died of wounds 15 Jul 1942.

[32] Capt A. I. Cottrell; solicitor; Christchurch; born Westport, 10 Feb 1907; wounded and p.w. 15 Jul 1942.

and everyone rested. German flares lit the sky but our drivers were too tired to worry. Most of them were asleep in their cabs.

The convoy moved off between three and four. Some of the lorries carried men but most were loaded only with ownerless gear, of which our drivers were the unwilling legatees. At three the next day, after a slow, roundabout journey, the brigade halted near Point 102, only a little more than ten miles east by north of Alam Nayil. The 20th Battalion was ordered to hand over all arms for delivery to the 6th Brigade, which was moving up from Amiriya. The drivers heaved sighs of relief. They had had enough.

The next morning the area was covered with neat piles of equipment, most of which was the property of dead, missing, or wounded men. From the air it must have looked like a supply dump. At all events, when the bombers appeared, which they did shortly before noon, they made straight for it. There were twelve of them and they came in at a low level, dropping bombs and strafing. Four or five bombs fell round the platoon's ack-ack lorry, which had all guns firing, and 'Crassi' Cliff dropped at his post, badly wounded in the back. Second-Lieutenant L. N. Cording,[33] who had been speaking to 'Crassi' when the raid started, was hit by a chunk of shrapnel, which almost severed his left leg. Cliff Brown[34] was unconscious and dying from blast. Ambulances arrived before the smoke had cleared and the wounded were taken to a nearby dressing station.

With heavy hearts our drivers heard that the platoon was to report to the 28th Battalion. They yielded to none in their admiration of the Maoris, but they could think of no people whose company they wanted less at that moment. The Maoris were seven miles north-west by west of Point 102 and the move was made late that afternoon. In the evening the sky was smudged with bursting shells and full of scudding aircraft.

Early the next morning the Maoris were taken to Point 102 where they were handed over to a platoon from the 6th Reserve Mechanical Transport Company. Our drivers made no pretence of feeling anything but relief. By tea-time they were back with the unit, which

---

[33] 2 Lt L. N. Cording; accountant; Wellington; born Wellington, 21 Mar 1917; wounded 17 Jul 1942.

[34] Dvr C. S. Brown; farmhand; born NZ, 20 Aug 1913; died of wounds 17 Jul 1942.

was now eighteen miles south-south-east of Alamein and sixteen miles from the coast.

We stared at the lorries as they moved into the area. Some of them were like sieves. No more would people say, with a suggestion of a sneer, that No. 1 Platoon was lucky. But the drivers were not comforted. They would rather have stayed safe and undistinguished, with Lenny, Bernie, and Cliff alive and well, and Second-Lieutenant Cording still able to play football, and Frank not dangerously ill in Helwan hospital. They would rather that no shadow had fallen on shining memories of good times at El Daba, Baracca, and Aleppo.[35]

---

[35] Our casualties between 29 June and 17 July were: killed in action, 7; wounded, 15; missing (later posted p.w.), 1.

CHAPTER 14

A STUDY IN DISCOMFORT

NEITHER side wanted a sit-down war—not with the sides as they were arranged now, anyway. Rommel had a reputation to preserve and his supply lines were too long. As for the British, they were bottled up in their own penalty area and they couldn't move without treading on one another's toes. At present, however, both armies were too weak to attack. General Auchinleck made the attempt on the night of 21-22 July, with Indians and New Zealanders striking the main blow in the centre, but it failed because our armour was not in a position to help at the critical moment. The same thing happened four nights later when he tried again with British, Australian, and South African troops. This time we lost a thousand Australians and seventy tanks were destroyed or put out of commission.

Rommel, who was overdue at Shepheard's, was equally unhappy, but there was nothing he could do about it. Until one side had more of everything than the other it would have to be a war of patrols and artillery, of ships, planes, and factories.

Minefields were extended and defences improved. Again it was the worker in Essen against the worker in Detroit, the Ruhr coalminer against the Welsh coalminer, the Neapolitan stevedore against his brother in Wapping.

. . . . . .

From an hour after sunrise until about an hour before dark we sweated all the time, even when we were in the shade. How we were able to sweat so much when we drank so little was a problem that baffled us. For a while the ration of water for each man was half a gallon a day. You got a quart (one full water bottle) to do what you liked with and the rest was pooled for the cooks. From somewhere—often from your private supply—water had to be found for your radiator. When it leaked or boiled over, no rare occurrence, there was nothing for it but to behave nobly—pour the last drop of liquid down your horse's throat, like the gentleman who brought the good news from Ghent to Aix. You got a cup of tea, not always a full one, three times a day, but the water it was made with was so brackish that it curdled the condensed milk. We were thirsty all day long and at night we had the Lemonade Dream, the hangover one, in which you swill glass after glass of something ice-cold and delicious without its doing you the least good.

Later the ration was doubled and then we were able to have an occasional wash, several drivers using one basin of water for their bodies and the same one for their socks; but the quantity was still so small that when you visited a friend for morning tea you took your water bottle with you, just as in England diners-out were careful not to forget ration cards. There was no water, of course, for washing up, but that didn't matter as dixies and frying-pans could be scoured with sand.

With chins unshaven and shirts black with sweat we resembled filthy beach-combers, but in point of fact we were cleaner and far sweeter-smelling than we should have been if water had been plentiful and the weather bitterly cold. There is a cleansing element in hot sunshine.

Worse than the lack of water were the flies.

The foul and dismaying thing about the Alamein flies was their oneness. None was separate from its fellows any more than the wave is separate from the ocean, the tentacle from the octopus. As one fly, one dark and horrible force guided by one mind, ubiquitous and immensely powerful, they addressed themselves to the one task, which was to destroy us body and soul. It was useless to kill them, for they despised death and made no attempt to avoid it. They existed only in the common will, and to weaken that we should have had to destroy countless millions of them. None the less we killed them unceasingly. We killed them singly and in detachments with fly swats, and the dead lay so thick in our lorries that we had to sweep them out several times a day. We set ingenious traps for them and they filled the traps, the living feasting ghoulishly on the dead. We slew them in mounds with our bare hands until the crunch of minute frames and the squish of microscopic viscera, felt rather than heard, became a nightmare. But what was the use? Their ranks closed at once and they went on with the all-important task of driving us out of our minds.

Although they had a common brain and a common purpose not

all were identical in appearance. About one in a thousand was larger than his fellows and of a lovely bottle-green colour. These, so the story went, and doubtless it was true, were corpse-fed. One could only suppose that the Intelligence in charge of the operation had introduced them for their moral effect, which was considerable.

Flies are attracted by any light surface, and our towels and the sun-bleached canopies of our lorries were speckled as with black confetti. Flies crave moisture, and you knew from watching your friends—and the knowledge was disproportionately humiliating and disgusting—that you too were walking around with half a hundred miniature old-men-of-the-sea clinging dourly to the back of your damp shirt. And when you shut your eyes—this is the plain truth—flies tried to open them, mad for the delectable fluid.

We couldn't always be killing them, but we had to keep on brushing them away, otherwise even breathing would have been difficult. Our arms ached from the exercise, but still they fastened on our food and accompanied it into our mouths and down our throats, scorning death when there was an advantage to be gained. They drowned themselves in our tea and in our soup. They attended us with awful relish on our most intimate occasions. They waited until our hands were full—they liked us best when we were lying beneath a lorry busy with spanner or grease gun—and then they rushed us, feet and suckers working furiously, inflicting a hundred pricks and stings.

Some of us excavated dugouts, made them fly-proof with mosquito netting, and lay grilling below ground until we could bear it no longer. Some of us wore veils, with an opening for pipe or cigarette, but there again—who wanted to wear a veil when beard and baked skin were already maddening excrescences? The best plan—the only plan—was to open the canopy of your lorry at both ends, seal the openings with netting, and turn the lorry into the hot current of air that did duty for a breeze. This made life just bearable. Unfortunately nets were scarce. Many were torn badly, some had been sold in Syria, and some had been lost. Anyway, one net was not enough for the job, so there was really no escape. The Intelligence had foreseen everything.

On 29 July while the plague was at its peak—a peak that was to be maintained effortlessly for more than two months—we were visited by a swarm of mosquitoes. For some hours we waded through a warm, whirring mist, every particle of which was able to raise a blister. This would surely have driven us mad before long, but the wind changed and the mosquitoes went away, leaving the field to the flies.

To the flies and the desert sores.

The satanic cunning ('But put forth thine hand now, and touch his bone and his flesh') was evident in these noisome ulcers. The least scratch was enough to cause them and they took rather less than a fortnight to expand round a suppurating centre to the bigness of a slice of lemon. They were irksome and humiliating rather than painful ('My flesh is clothed with worms and clods of dust; my skin is broken and become loathsome') and they took weeks, months sometimes, to heal, and when they did they left scars that remained for over a year. Nearly everyone you saw had an arm or a leg bandaged and the knuckles of both hands ringed with filthy scraps of sticking-plaster. After breakfast and before tea 'Doc' Turner's pick-up was like a lazaret, but he was impatient only when someone failed to report for a dressing.

And there were sandstorms and dust-storms. Frogs and locusts we were not troubled by, nor was there need for them. We were ready to cry 'Enough', load our lorries with Israelites, and drive them to Tel Aviv.

.   .   .   .

Until now many of us had liked the desert and some of us had loved it. But the desert had grown smaller and it had grown dirtier. In fact, robbed of spaciousness and cleanliness, it was desert no longer. It was a big sand-pit in a big slum—the kind of nightmare playground you find sometimes in great cities; sand-grey like asphalt, clumps of toddlers squalidly underfoot, rickety urchins quarrelling round swings and sweating in hot, dark clothes, the grease of poverty on everything, and the sun shining. A vast army was boxed in between the Alamein Line and the Delta. Every horizon was dotted with guns, tents, or transport, and wherever you looked there was a moving vehicle cloaked in its private dust-storm. Claustrophobia is not a disease of the desert, but many of us, cabined by the great heat, encompassed by armies, wrapped in our foul garment of flies, came near to suffering from it.

A high standard of cleanliness was expected and as far as possible enforced, but in every unit there are a few men to whom decency and order mean nothing and collectively they are formidable. Unburied tins and refuse lay among the legitimate lumber of the battlefield, adding squalor to desolation. Some men—some units—would live for a month in one area and leave it spotless. Others would stay for two hours, conduct a vast disorderly picnic, and move on, leaving behind them the kind of sour patch you find on shifting a chicken run. Near busy road junctions and the entrances to supply depots sand turned to dust—grey, clinging dust—and in this the flies multiplied. Wild animals make only a little mess, but man, divorced from his cesspits and sewers, turns all to filth.

Even under these conditions food continued to be important to us and much time was spent in grumbling about the meals. They were adequate but unpleasant, and stew—bully stew, fresh meat stew, tinned meat-and-vegetable stew, sausage stew—was nearly always the main dish. It was not a cooling diet. The daily ration of bread was grey and unpalatable and sometimes weevily, and our fresh vegetables were no longer fresh by the time they reached us. Fortunately our platoon canteens were fairly well stocked with tinned fruit at this time, so we had something good to eat. Anyway, it was no weather for gorging.

Nor was it weather for work—but it was work that saved us, giving us a sense of purpose and responsibility and something to think about beyond ourselves. The field regiments were constantly in action, and although we were now stronger by a platoon we had to borrow transport from other units to keep pace with their demands. Supplying the 25-pounders with ammunition was our greatest problem and to this No. 1 Platoon devoted all its time. The other platoons dealt in mixed loads and took turn and turn about at the ammunition point, which did not settle down in a permanent area until 17 July, when it was formed five miles south of Alam Nayil, staying there until early in August.

Whenever an unusually heavy demand for 25-pounder was on the cards, vehicles from No. 1 Platoon were sent to the point with as many loads as were likely to be needed, and this system worked fairly well. Much time would have been saved, though, if the Artillery had been able to give their orders by wireless or telephone instead of having to depend on messengers. As it was, vehicles were often standing idle at the ammunition point when they were urgently needed for other work.

The point was replenished from the unit area, the empty lorries reloading at the 86th Field Maintenance Centre, near El Hammam, some twenty-five miles north-east by north of the area, or, less often, at the Burg el Arab railhead, ten miles farther on. The round trip between the unit and the field maintenance centre was something like eighty miles and it was rough going all the way. To save time and to avoid congestion the lorries travelled singly or in small groups.

The Sun, Moon, and Star tracks, the Bottle, Boat, and Hat tracks —how well we got to know them! We drove through a hot, ochrous haze, sometimes bumping over a mile-wide pattern of ruts, sometimes ploughing between high sandbanks that the bulldozers had thrown up to make a road through the soft sand. Here the wheels, sinking a foot deep in dust before finding the buried army-track (heavy-gauge wire-netting), threw powdery bow waves, which hush-hushed under the mudguards and fell back whispering. Often there were traffic blocks, and then, while the air trembled with heat and flies in their hundreds came from nowhere, driving you from the cab, you could hear, in the uneasy silence, radiators gurgling like kettles and the grease frying on the manifolds. You swore fiercely at the delay, forgetting how urgently only five minutes ago you had craved release from the hot prison of the cab, and how since early afternoon, like a dead slave on an oar, you had been jerked and wrenched by the steering wheel, its sudden, spiteful twitches jarring you from wrist to shoulder.

With evening some magic came back to the desert. Sweat dried on face and body, the engine sweetened, and the tin of American beer hanging in a wet sock from the bracket of the driving-mirror (no money could have bought it) cooled rapidly. The lorry's spiked shadow—blunt spike of canopy, spike of cab, spike of head-lamp—ran beside you, shortening and then lengthening with each lurch. The twitch of the steering wheel was a friendly nudge now, almost a rough caress, like the touch of the dust and the warm air flowing under the windscreen.

. . . . . .

To complete this study in discomfort one would like to be able to say that we were bombed constantly, but that was not the case, though some of us were so shaken by our recent experiences that we viewed every plane with distrust and suffered vicariously when neighbouring areas were raided. We saw many Messerschmitts but they left us alone. Dog-fights took place high above us and we heard the tiny stammer of machine guns, like argument in Heaven, and watched while the aching blueness was cut by vapour trails—perfectly described arcs and segments, beautifully simple and remote: Euclid drawn large for children. Often a parachute opened like a white flower, opened and drifted to earth in a slow curve, swinging a black dot, a comma, a tiny man, while the flames crackled behind the hill and the scream still rang in our ears and the gout of black smoke, sudden as a splash of ink, drew all eyes and fingers. Ours, we wondered, or his?

Although we lived always in the shadow of the Luftwaffe we came to no harm. Individual drivers were frightened while on jobs and one day two old transport planes, Bombays, were set on fire by Messerschmitts as they were landing near the unit area, but for most of us the Luftwaffe was only a minor worry and vexation, taking its place somewhere between desert sores and curdled condensed milk. Other units were less fortunate: so much damage was done during July that we were ordered to dig bunkers for our lorries and sandbag their vitals.

Increasingly, though, the sky was filled with our own aircraft. We counted the bombers as they flew west and we counted them on their way back, and it was seldom that one was missing. The promise was coming true: 'We shall fight, with growing confidence and growing strength, in the air'.

The author of this promise visited the Eighth Army early in August. Then he went to Moscow to see Stalin, who had lost Sebastopol and Rostov. Then he returned to Egypt and paid a visit to the New Zealand sector of the Alamein Line. A reception was held and among those who attended it were men decorated in the desert. We were represented by Bob Aro.

One result of Mr. Churchill's earlier visit was a reshuffle of Generals. Lieutenant-General B. L. Montgomery was given command of the Eighth Army and two days later General Auchinleck was succeeded as Commander-in-Chief, Middle East, by General Sir Harold Alexander, famous at Dunkirk and in Burma.

We heard of these high matters and we were not encouraged. Only yesterday press and radio had been reminding us of our confidence in 'The Auk' (which was what we were supposed to call him) and saying how much he meant to us. Now they were saying the same thing about his successor, whom we were to refer to affectionately as 'Monty'. Hell!

Although it would be idle to pretend that our morale was high, at this time we were nowhere near—nowhere remotely near—breaking point. Work was still our antidote to discomfort and depression, and during the third week of August we struck a particularly busy period. At a time when we were even busier than usual it was decided to dump enough ammunition in the forward areas to last the Division three days in an emergency. While we were helping with this task a British brigade arrived in the field without any first-line ammunition, and we were told to supply half its immediate requirements.

Thus our days. As for our nights, they came with the blessedness of a recurrent miracle. They damped the flies and the dust, they hid us from the Messerschmitts, they bathed us in coolness. Then we could enjoy our tinned pineapple and pears and gather round the platoon wireless sets. Then we could sleep, dreaming of downs and lakes and the woolshed back home, dreaming until the first fly, rising earlier than any lark, stepped delicately across an eyelid, or dipped to drink, with tiny, filthy proboscis, at a desert sore.

. . . .

If you have read Kingsley's *Water Babies* you will remember with what pure delight young Tom, Grimes' the chimney-sweep's boy, tore off his sooty rags and tumbled into the stream. So we tumbled into the Mediterranean on 24 August when the unit moved to an area six or seven miles from the beach and twenty-two miles from Alamein. It was our eighth move since the beginning of July (though a month had been spent in one area), but the others had meant only more digging.

From then on we bathed daily and whether the salt water did more for body or for soul it would be hard to say. Our spirits revived in one leap, our sores started to heal, our appetites came back, and the dirt peeled off us. Only No. 2 Platoon, which had been at the ammunition point (twelve miles south by east of Alamein) since 5 August, carried on as before.

Thus marvellously recovered we heard the rumour with equanimity: Rommel was about to attack. And plainly it was more than a rumour, for men on leave were recalled to their units and everyone was ordered to be dressed and armed a quarter of an hour before sunrise as a precaution against paratroop landings.

On 30 August our fighters were unusually active, and the next day we were told to send 10,000 rounds of 25-pounder to the ammunition point. No. 1 Platoon set out with half this amount and vehicles from Nos. 3 and 4 Platoons with the rest. On their way forward our drivers passed heavy guns that were being dug in on the left of the track.

As soon as it arrived at the point No. 1 Platoon came under shellfire. This stopped before any damage was done and then there was a vicious bombing raid on transport in the next area. Our drivers looked at one another. 'The old Rommel,' they said, and they might have been speaking affectionately of a headstrong uncle who had overstepped the mark, 'he's away again.'

Rommel had said on the 30th: 'Today the Army, strengthened by new divisions, is moving into the attack for the final annihilation of the enemy.' But Alexander had said earlier, and Montgomery in a special message to the Eighth Army had repeated it: 'We will fight the enemy where we now stand; there will be NO WITHDRAWAL and NO SURRENDER.'

In the small hours of the 31st Rommel had started to lift the minefields in front of the southern sector of the line. By midday his main columns were through the minefields and then one column swung north to contain British and New Zealand troops which had been by-passed by the advance. So far the enemy had met with no opposition except from light mobile forces that had been ordered to inflict as much damage as possible before withdrawing under pressure. Rommel was using the cream of his infantry, between three and four thousand lorries, a huge number of guns, and nearly all his armour. It was the real thing.

But nothing invites disaster like a *Blitzkrieg* geared down, and on this occasion Rommel was never able to shift out of second. The story of the next three days is the story of his finding himself enclosed, and in danger of being embalmed, in a coffin-shaped salient between twenty and thirty miles deep, of his lacking room to manœuvre, of his failure to force the British armour to give battle under conditions of his own choosing, of the ceaseless bombardment of his troops, tanks, and transport from the air and by artillery, and of his finally realising that if he stayed where he was he was likely to be destroyed piecemeal.

One of the chief obstacles to an orderly withdrawal was the presence of the New Zealanders at the end of the road back. From the start our gunners had been pouring shells into the enemy's transport as it went past, and on the night of 3-4 September British and New Zealand troops launched a three-brigade attack, the object of which was to narrow Rommel's escape route and gain ground from which our artillery could inflict even greater damage on his retreating columns.

Unhappily the British units were cut to pieces by mortar and machine-gun fire, which made it impossible for the New Zealanders to consolidate their gains. A new line was formed, and it was held throughout the next day in the face of determined counterattacks, but the escape corridor remained wide open.

Pushed from behind by two British armoured brigades, Rommel made haste to use it, and by 5 September he had come to the conclusion that he had intended only ' an armoured reconnaissance in force '. By the 7th he had completed his withdrawal and the demand for ammunition was back to normal.

He had gained a little ground but he had lost a large number of admirers on both sides. In fact, a legend had been shattered.

· · · · · · ·

September the 7th was a fine, cool day, the 8th was fine but dusty, the 9th windy and wickedly dusty, the 10th—the 10th could have been the worst day of the year and we should have approved of it still, for at 9 p.m. the Division was relieved by British and Greek troops. By then a large number of us were on the beach near Burg el Arab, nearly thirty miles from the front, a smaller party was guarding the vehicles in the Swordfish area (between Amiriya and Burg el Arab and about sixteen miles inland) where most of the Divisional transport was parked, and the only platoons working were Nos. 3 and 4 at the ammunition point, which was to be kept open until the 12th. We had been paid and had been promised leave.

All but the domestic vehicles had been left in the Swordfish area, so the beach party went to bed that night under the stars, which crowded Heaven like a rash—a sort of silver chicken-pox. In their dreams our drivers heard the waves folding on the beaches and the breeze stirring the marram grass.

For a week we made holiday. Four-day leave parties (eight officers and 198 other ranks all told) left for Cairo or Alexandria on three successive mornings, and for the rest there was day-leave to Alexandria. Some of us chose to spend all our time on the beach and it would be hard to say who had the best of it.

Cairo and Alexandria were as gay and as expensive as ever. They alone in a world of worn-out playthings and depleted store cupboards had avoided the chill touch of austerity: everything was to be had at a price. All the pimps and hucksters in the Middle East, knowing they had only a few hours in which to indemnify themselves for our long absence, were at the service of the New Zealanders. In Alexandria, no doubt, there were great killings, but in Cairo it was less simple, for the genius who had organised the leave transit camp at Maadi had foreseen what would happen and had made his plans accordingly. Transit camps, as we knew them, were dusty purgatories occupied by slowly-shuffling queues of hungry and exasperated men, but this one was different. On arrival you were led to a long counter piled high with clean clothes and invited to peel off your dirty ones. Delicious meals were served at all hours and the beer bar was never closed. If you wanted a bed for the night you could have one. If you wanted to spend the night in Cairo that was all right too. You did just as you pleased, treating the camp as an hotel.

But on the beach at Burg el Arab there was bathing from sunrise to sunset and you didn't have to wear clothes. There was peace and freedom and something lovely to look at all the time. White sand, blue sky, blue sea, dark-green fig trees—the scene had a boldness and economy, a clear-cut beauty, that was breathtaking. It was so simple and at the same time so clever that a man who had never handled a brush in his life might say to himself: 'I could paint that. Couldn't miss. Couldn't go wrong.' There was plenty of beer and the meals were first-class. Indeed, they made us ask ourselves if we had been quite fair to the cooks during the past two months. That extra halfcrown a day no longer seemed a large sum when we remembered the heat, the flies, and the roar of the petrol burners, above which it was impossible to hear aircraft. 'Cook's neck' we called it, that special stiffness that came from constantly glancing skywards.

The cooks, if anyone, had earned a holiday. Well, so had we all. Between 29 June and 11 September, we had issued, among other things, 243,000 rounds of 25-pounder, 3,200,000 rounds of rifle and machine-gun ammunition, 84,000 rounds of Bofors ammunition, 345,000 rounds of tommy-gun ammunition, and 30,000 grenades. The heaviest issue of 25-pounder on any one day had been about 12,000 rounds. We had met demands for fifty-nine different commodities, which was going two better than Heinz.

On the last day of our holiday we had a second chance to see the Kiwi Concert Party, which had performed in our area on the 14th. The desert had seen wonders and absurdities before—tank battles and air raids—but nothing quite so incongruous as this. Under the blazing sun, in the middle of the wilderness, a half-circle of brown, up-turned faces (intent one moment, the next convulsed with laughter) gazed fixedly at a small, gaily-decorated stage, centre of all the beauty and merriment in the world. And all around was simmering, staring emptiness. Tanks on Saturn, battles on Mars—yes; but when a clown in a red coat and a bow tie, with yellow hair falling over his forehead, takes his quips to the moon, dances on the edge of a volcano, deluges the dead rocks with laughter, then, why then, something notable has happened.

The sweat poured from actors and audience alike and hands fluttered without ceasing, brushing away flies. But no one noticed the flies and the desert was not there—gone, vanished, along with everything else ordinary and horrible. In the suburbs of Stalingrad men were fighting like mad beasts, and all over the world, from Russia, from Germany, from Italy, from bombed Britain, the cry was going up: 'How long, O Lord? How long?' But in one bare acre of desert where Theatre had drawn its charmed circle, excluding the world's grief, none heard it. The actors worked and sweated on the small, gay stage, and the audience, forgetting all else, lived for a short hour in a country where everyone can dance and play the accordion and where no one opens his mouth except to sing golden notes or to be excruciatingly funny.

. . . . . .

Our holiday ended and on 19 September we went back to the Swordfish area where we had left the lorries. We were glad to see them again, though only a week ago we had wished them at the bottom of the sea. We went over them with spanner and grease gun, feeling in our hands that skill and knowledge, almost that tenderness, that an ostler feels when he runs his hands over a horse.

By midday on 24 September the unit was near the Divisional exercise area, thirty miles south of Burg el Arab, where the New Zealand Engineers had laid dummy minefields and dug gunpits for a mock battle—a dress rehearsal, had we known it, for the Battle of Alamein. Live ammunition was used and we established ammunition points.

We stayed in the same area until the middle of October, enjoying a thorough rest. The days were still hot and the flies were still with us, but we minded them less now. Possibly we had become used to them, like the Egyptians. One afternoon the sky clouded over and a north wind got up, exhaling a brown breath. Next the heavens opened and pelted us with hailstones as big as hens' eggs, and in a moment the desert was sprinkled with crushed ice. The stones rattled against the lorries like buckshot, cracking windscreens and drumming madly on taut canopies. On one pretext or another (there was a rifle to be rescued, a tool box to be covered) we put on our steel helmets and rushed out into the storm, loving it for its strangeness and wonder as children love snow. It stopped as suddenly as it had started and the sun shone brilliantly, melting the prize hailstones even as we unpacked our cameras. Presently there were only puddles to show where they had shone and sparkled.

The most important social event that took place while we were in this area was a party to celebrate the third anniversary of the First Echelon's entry into camp. Once enough beer had been secured the preparations were simple, all that was needed being an old canopy, four lorries from which to suspend it, and some empty jerricans for seats. Speech-making was barred, but the customary toasts were allowed, and our absent friends were honoured with more than usual solemnity, for among them (to use Henry James's phrase) were some who had achieved the extremity of personal absence. The beer worked out at about eight bottles a head, which was plenty. The first two were consumed quickly and decorously ('Later, you jokers—a man can't sing when he's stone cold.') but there was music in the third and fourth. 'Sheriff' was not there to give us the 'Pokeydoke Blues' (a lugubrious ballad about a gambler who loses his coat, his hat, and his straight-laced shoo-OOO-oos) but Basil, our crooner, sang 'The Old Rocking-chair'—in fact he sang it twice—and the rest of our star performers sang the particular songs that custom demanded of them. The fifth bottle brought new talent to light. One of the visitors sang 'Stick To Me, Bill', a genuine tear-jerker, and 'Poop-Quail' went one better with a song all about Confusion and Shame and his having only one mother. The sixth bottle, as always, had the disconcerting effect of convincing everyone whom bashfulness had prevented from singing earlier in the evening that now—this minute—was the moment for his contribution. Six singers struck up at once, each of them taking it for granted that the cries of 'Fair go, you jokers!' and 'One at a time, eh?' were meant for everyone except him.

From that point the party drifted through a golden mist to its last stage, which was reached when a small cluster of die-hards realised that all the others had wandered away to bed or to suppers of oyster and whitebait fritters, leaving them alone with the empty bottles, the upturned jerricans, the increasing chill.

'Once more, you chaps. The old "Sheriff's" song. The old "Pokeydoke Blues".'

> Ah thought Ah was er gambler—
> Ah broke every joint in town
> Until Ah met-ter gambler
> Whose name was Appledown. . . .

Gallantly they struggled on, but it was no good. Something had happened to the golden mist. All the gold had gone out of it and it was an ordinary mist now, and a mighty cold one at that.

. . . . .

The hour was approaching fast. Drivers back from leave spoke of shiploads of American Shermans, of airfields that had sprung up overnight, of pale divisions fresh from England, and of masses of guns and transport that were moving into the desert.

On 9 October seventy-two lorries loaded with 25-pounder ammunition set out under Captain May for a secret destination. The drivers had been told to say nothing to anyone about what they saw and did. The following evening found the convoy on its start line, and punctually at half past seven (timing was of the first importance) it moved west along the coast road, passing Alamein and halting two or three miles from the front. Here it was divided into groups of three, each of which was led to a gun site by an Artillery officer. The average distance from the road to the gun sites was a mile and a quarter, and an hour and thirty-five minutes was allowed for the round trip. The guns were not in the line yet—they were to be brought forward secretly two nights before the attack—so as soon as the ammunition was off-loaded all hands set to work to dig it in and camouflage it. There was no time to waste, for the transport had to be east of the start line before daylight.

Over 48,000 rounds were dumped for the ninety-six 25-pounders with the New Zealand Division (seventy-two of these were the guns of the 4th, 5th, and 6th Field Regiments), and the job was completed on the fourth night without the enemy's being aware of what was going on. On the fifth night an extra 160 rounds were taken forward for each of the New Zealand guns and 8000 rounds of Bofors ammunition were dumped for the 14th Light Anti-Aircraft Regiment.

On 16 October we moved to the Swordfish area, Captain May's detachment joining us at lunch-time. The afternoon—the ancients would have taken it for an omen—we were enveloped by one of the worst sandstorms in our experience. It was impossible to read without a light, and to walk more than a yard or two from your lorry was to get lost. At tea-time the cooks of a section of the Petrol Company, our next-door neighbour, hit on the idea of using an air-raid siren as a dinner gong, but the experiment was only partially successful, as half the men in the queue were found to be members of the Ammunition Company. The intruders, indistinguishable from one another and from anyone else under their yellow masks, had thought they were at their own cookhouse.

The next day was nearly as bad but the 18th showed an improvement. In the afternoon Captain May's convoy of seventy-six lorries set out for the Artillery waggon lines with 160 rounds a gun for the New Zealand 25-pounders, returning on the afternoon of the 19th. We moved five miles south on the 20th, and on the 21st 218 vehicles from various British and New Zealand units came under our command for the move to the Divisional assembly area, which was at Alam el Onsol, about twelve miles behind the front. At half past three that afternoon we moved towards the coast, halting in a dispersal area seven miles south of Burg el Arab.

We had a scratch meal, and shortly before half past eight the leading vehicle in our convoy passed the starting point for the night march. There were 398 vehicles under Major Coutts's command and they formed a column nearly ten miles long. On reaching the coast road we turned left, halting soon before midnight at Kilo 58, where we dispersed. The move had gone like clockwork, but most of us, conscious only that we had stopped and started a great deal, had mistaken it for another army muddle. We had enjoyed it, though. Moving under the moon in the right direction was a rare pleasure.

The next day we said goodbye to seventy-five men (drawn from Headquarters, Workshops, and the Ammunition platoons) who were to stay behind at Kilo 58 to form an administration post, and at half past seven we started the second stage of the move, reaching the Divisional assembly area before midnight.

We slept soundly that night and when we woke up the next morning it was Friday, 23 October 1942. In the course of the day the platoon commanders assembled their men and read them a personal message from General Montgomery.

When I assumed command of the Eighth Army I said that the mandate was to destroy Rommel and his Army, and that it would be done as soon as we were ready.

We are ready NOW.

The message concluded: Therefore, let every officer and man enter the battle with a stout heart, and with the determination to do his duty as long as he has breath in his body.

AND LET NO MAN SURRENDER SO LONG AS HE CAN FIGHT.

Let us pray that the Lord mighty in battle will give us the victory.

After the line had been breached, the New Zealanders, now under the command of 30th Corps for the assault, would join 10th Corps for the break-through and pursuit. There was a chance, therefore, that the supply columns might have to operate for a short time without full protection, as they had done in November. At all events we were to be prepared for anything.

What we were not told officially we learned from our grape-vine. On the northern sector, from which the 30th Corps would launch the main thrust, over 800 guns were in position. The barrage would start at twenty minutes to ten that night.

The afternoon dragged past under a sky filled with our fighters and bombers. When dusk came we made our beds so that we should be able to watch the barrage without leaving them. The nights were fairly cool now.

The moon came up unbelievably large and yellow, like a stage moon. You could have read a newspaper by it easily, and to get to sleep you had to pull the blankets over your head. We were used to turning in early because of the blackout, so when the barrage started many of us were asleep, but it woke us immediately. The horizon was on fire and it threw back a continuous hollow roar. Giants were striking matches, matches as big as pine trees, on the rough desert, and a roaring wind was blowing them out at once. The roar and the dancing flashes went on and on, and we lay in our sandy beds or stood huddled in blankets in the backs of our lorries, which were dark mouths in the silver desert, and watched and wondered.

CHAPTER 15

OUT OF THE SLOUGH

> I see you stand like greyhounds in the slips,
> Straining upon the start.

AFTER the enemy's batteries had been shelled for a quarter of an hour the barrage was switched to his front line. At 10 p.m., under a full moon, with bayonets gleaming, our infantry walked forward to the attack.

. . . .

During the next twelve days both sides fought bitterly, the British to carve a pathway for the armour, the Germans to cripple the operation then and there.

As the days went by we waited patiently in our area for the advance to begin. Eagerly we grabbed at every crumb of news that came back to us. Miteiriya Ridge, in the northern sector, was ours, but the armour was not through yet. We had driven a large salient into the enemy's defences and he was counter-attacking fiercely but unsuccessfully. The Royal Air Force had smashed a concentration of tanks and armoured vehicles. Now the Australians were fighting on the coast sector. They were doing splendidly but still the armour was not through.

Although we were seldom well-informed about what was going on we had become adept in sensing an atmosphere, and by the end of October we knew that the position was critical.

One comfort we did have. The skies, beyond a doubt, were ours. British fighters and bombers were overhead all day long, and only at night, at dawn, and at dusk, was the Luftwaffe in the least active. On the evening of the 26th a single plane dropped bombs in a minefield on the edge of our area, setting off two mines and puncturing a tire, and at dawn the next day the same plane (so we believed) almost succeeded in dropping bombs in the same holes. A driver had just lit his primus to make morning tea, earning himself considerable unpopularity. Later in the day we heard that two drivers at the administration post had been wounded by strafing, though not badly enough to warrant their leaving the field.

There was one particular plane—we swore it was always the

same one—that used to annoy us every night. It would fly round in great circles, passing over our area at intervals of about a quarter of an hour. When the moon was bright it was visible as a black lozenge, low and menacing. It dropped an occasional bomb and fired an occasional burst of bullets, but its main object, we felt certain, was to disturb our rest. In this it was eminently successful and we spent many sleepless hours in devising suitable tortures for the pilot. Once, to our joy, it was engaged by a British night-fighter, but it got away after a long chase.

A little earlier in the night, it or another plane had been using a bomb that exploded in the air with a kind of gobbling roar and a series of flashes, prompting the Petrol Company diarist to make reference to 'infernal machines'. These were butterfly bombs, a weapon the Germans had been experimenting with for some while and had lately brought to perfection. In the course of time we were to become painfully familiar with them.

The principle was as ingenious as it was simple. A large container opened in mid-air like a pod to disgorge twenty-four anti-personnel bombs, the outer casing of which split into four quarters and became a propellor, unwinding a safety device. The fuses were of several types, so some of the bombs exploded in the air, some on contact with the ground, and some at intervals throughout the night. The delayed-action type were credited with a horrible propensity for rolling into slit-trenches.

The nights passed slowly and the days, too. We were busy, but hardly as busy as we had expected to be, for although the artillery was using a great deal of ammunition there was none of that fret and delay that at other times had added so considerably to our labours. Every detail of supply had been planned with meticulous care.

On 2 November No. 2 Platoon established an ammunition point one mile west of Alamein station, and later in the day twelve vehicles from No. 4 Platoon under Lieutenant Fitzgerald opened a forward ammunition point nine miles farther on. The latter was in range of German 88s and during the day and a half of its existence it came under fire several times, two vehicles being slightly damaged by shrapnel.

Business was brisk from the moment the points opened, and no wonder, for at one the next morning every gun on the Corps' front opened up and in four and a half hours 150,000 rounds were fired.

On 3 November we were rushed off our feet. As soon as our vehicles arrived at the ammunition points Artillery vehicles backed up to them, tailboard to tailboard. In the course of the day we sold something like 10,000 rounds of 25-pounder.

At a quarter past three in the afternoon there was a call from the 4th Field Regiment for thirty-three loads of 25-pounder, and these were sent forward at once, but as it happened there was no need for them. The enemy was already broken.

Early the next morning, 10th Corps, with the New Zealand Division under its command, began the chase.

* * *

The New Zealanders, with the British 4th Light Armoured Brigade and the remnants of the British 9th Armoured Brigade, drove through the gaps in the minefields and headed south-west. Their objective was the high ground west of Fuka and their intention was to cut off the enemy's retreat. His columns were reported to be fleeing along the coast road ten-deep under a rain of bombs and bullets.

By noon the bulk of our unit was at the ammunition point near Alamein station. The area was very crowded, for the party from the administration point had rejoined us the day before and, since the beginning of the month, twelve 4th Reserve Mechanical Transport Company lorries had been attached to us to carry extra ammunition. This was because we could expect to be cut off from our supply dumps during the first stage of the advance.

The overcrowding was of little moment, for at four in the afternoon, leaving behind Headquarters and one platoon of Workshops under Lieutenant Hill,[1] we moved off in column of route behind the Major's car, heading west and a little north to hit the Boomerang track. On the way we picked up the vehicles at the forward ammunition point and the 523rd Company RASC (serving the 9th Armoured Brigade). It had been under command since 21 October.

From the New Zealand Division's headquarters the Major had received no detailed instructions about the move. He had been told to find the tail of the 6th Brigade and tag along behind it. Not knowing where the 6th Brigade was, he went forward to Divisional

[1] Maj O. W. Hill; salesman; Napier; born Auckland, 16 Jan 1917.

Headquarters for further instructions and was told to move along the Boomerang track for a few miles and then travel on a given bearing until he caught up with the tail of the brigade. On the Boomerang track we were at once enveloped in clouds of blinding dust. All the transport in the world seemed to be moving west and it was impossible to keep the convoy together.

The track and its verges had been swept by the engineers but mines were still a danger. The Germans had a trick of burying them one on top of the other with a layer of sand in between, and it sometimes happened that the bottom one was overlooked. Odd mines were buried deep enough to escape detection and these would not explode until the sand above them had been packed hard by perhaps a hundred or more vehicles. Such a one cost Les Howarth his back wheel, and shortly afterwards a wheel sailed past Major Coutts's car while a cloud of dust enveloped a 15-cwt. truck, the driver of which had been talking with the Major about mines only a few minutes before. He was injured fairly severely, and his load, part of which was General Freyberg's personal equipment, was transferred to one of our vehicles.

Most of us had long since removed the sandbags from the inside of our cabs—a protective measure adopted earlier in the year—and we felt our way forward in the greatest trepidation. We could be seen crouching over the steering wheels in stiff attitudes as though by not sitting back and making ourselves comfortable we were somehow reducing the total weight of our lorries.

Slowly and with many halts, while the dust clouds boiled about us and all we could see was the swaying canopy of the vehicle ahead and perhaps a dozen yards of deeply-rutted desert, we drove into the sunset. At six the convoy turned left off the Boomerang track and travelled for a further two miles at about a mile an hour. By then part of No. 3 Platoon and the whole of No. 4 Platoon and the 523rd Company had been cut off in the dust and darkness, so the Major decided to halt for the night, hoping that by morning it would be possible to travel at a reasonable pace. Before we bedded down we were told to be ready to move at first light. Except for the banging of a single gun, which seemed to be firing across the line of our advance, all was quiet. The moan of transport was so much part of our lives that we did not count it as a noise: it was like our own breathing.

Before dawn Captain Sampson was away to round up the missing vehicles and with the first glimmer of light we moved, halting for breakfast after travelling a few miles. Before the meal could be eaten we were off again, heading south-west. It was a beautiful morning, cold and bright, and never in the unit's history had spirits been livelier.

The desert was littered with burnt transport and smashed tanks and guns, and yet, for miles at a stretch, it was clean and almost virgin. Plainly but one battle had been fought there and that a short one. Nor was it long over. Smoke was still coiling from some of the charred wrecks, and groups of Italians, their natural gregariousness heightened by danger, were moving across the desert in tight little phalanxes, protected collectively by white sheets and individually by pathetic scraps of once-white linen and even by old newspapers. They had been disarmed, or they had thrown away their arms, and in most cases they were unguarded. Rommel, we learned later, had left five Italian divisions to their fate.

Treasure was all about us—shirts, uniforms, brand-new anti-tank rifles, berettas, and even cameras—and good eyesight paid handsome dividends. We were travelling at a fair pace now and there were few stops, but no one ignored a pistol or a camera even if it meant pulling out of convoy or simulating a petrol blockage.

Our day's march ended at six o'clock when we halted about eight miles south of Fuka. By now most of the missing vehicles had caught up with us, and the unit was complete except for Captain Morris[2] and about six Workshops' vehicles and the party that had been left behind under Lieutenant Hill. Shortly before dusk we heard gunfire, but presently all was quiet.

That night it rained.

We woke in the morning—those of us who had not been woken earlier by leaking canopies—to a prospect of puddles. We cursed bitterly as in the mind's eye we saw the enemy speeding along the coast road and our lovely Shermans floundering in a quagmire. There was still, if it cleared quickly, a hope, but it showed no signs of clearing.

After breakfast we pushed on, but much delay was caused by bogged vehicles and by hold-ups ahead of us. The large party of

[2] Capt A. G. Morris, m.i.d.; cycle and motor dealer; Ashburton; born Picton, 17 Feb 1913.

prisoners we passed at lunch-time served only as a reminder of the larger number whom the elements were helping to escape. We drove under louring skies, and during the afternoon intermittent squalls gave place to a steady downpour. At a quarter to five we halted for the night on the outskirts of a large German airfield five miles south of Sidi Haneish. Fuka was now fourteen miles behind us.

Torrential rain fell during the night.

It was still raining at dawn and showers alternated with downpours until the middle of the afternoon, by which time the desert had turned into a morass. Our officers ranged the neighbourhood in search of negotiable routes but found none, and the transport stayed where it was all day. For most of us the waiting hours were ones of extreme exertion. In every quarter of the area, especially in that occupied by the 523rd Company,[3] which had single-drive Bedfords, vehicles were in distress. Some listed at alarming angles; others, like plesiosauri aspiring from the primal ooze, pointed towards the sky, their hinder parts buried deep in wet sand; still others, travelling at a hundred yards an hour, crept agonisingly towards firmer ground while their engines screamed in anguish and their wheels ploughed parallel trenches across the desert.

During the day we were joined by Captain Morris and five Workshops' drivers. They had travelled some ten to fifteen miles along the coast road between Daba and Fuka and they were able to tell us what it was like. The verges and the road itself in some places were littered with abandoned gear and transport. Some of it was undamaged—there was a *volkswagen* that looked as though the driver had pulled up at the curb and hopped out to do some shopping—but most of it was burnt and twisted and peppered with bullet holes. For a mile and a half at one stretch there was nothing but tangled metal.

The Opel staff car that was splashing backwards and forwards through the area at a high speed with Captain Morris at the wheel had been captured after an exciting chase and some shooting. With it three German prisoners had been taken.

. . . .

Throughout 7 November the New Zealand Division and most of the flanking armour was anchored to the desert as firmly as flies

---

[3] It ceased to be under command on 11 November.

to flypaper, but the 8th, a Sunday, dawned dry and clear. The sun came out and sparkled in all the puddles and made the canopies steam. The desert was still in a foul mess but it was drying momently and we had hopes of being able to move soon. The damage, though, was done now. Rommel had been given a day's respite both from pursuit and from bombs, and his army, fleeing along the tarmac coast road, had made good use of it.

We moved at two in the afternoon, the Major's instructions being to get as far as he could towards Sidi Barrani, which was not yet in British hands.

The desert was like soggy gingerbread and vehicle after vehicle became bogged, causing much back-breaking work with shovels and the sacrifice of many German overcoats and uniforms. These were thrown under the churning wheels to give them something to grip on. Convoy discipline was in abeyance, the speed and direction of each vehicle being determined by the kind of surface it was crossing.

We skirted Sidi Haneish airfield and gazed gloatingly at the wrecked, burnt, and grounded planes. On one landing-ground we counted thirty wrecks, and were convinced, for perhaps the first time, that a great victory had been won.

As dusk gathered the going became steadily worse and when we halted shortly before nightfall we had covered only eighteen miles; the leading half of the convoy was on high ground and the rest was floundering in the mud two miles behind. Nothing could be done in darkness, so it was decided that the two halves should bivouac where they were.

Over our late tea we could speak of nothing except the great news: American and British troops had landed on the coast of French North Africa. To most of us, in the first flush of our enthusiasm, it seemed that the war was as good as over and our long journey towards Christmas as good as ended.

Bringing the convoy together was a formidable task even in daylight and it was after nine before we were ready for the road. For two hours the going was good and then we were held up while the 6th Brigade passed through a narrow gap in the minefield. Tired of waiting, the Major found an alternative gap, which led us directly to the Siwa road. The coast road was reached at half past four in the afternoon.

German and Italian equipment—objects as gross as tanks to others as small as shaving brushes—burnt, broken, and wasted, was jumbled in a black mass, among which, disgustingly mutilated sometimes and sometimes lying like men asleep, were figures in olive grey. It was certainly a notable victory.

At five we halted three or four miles west-north-west of Charing Cross, and for the first time since the start of the advance the cooks were able to give us a hot meal.

While we were enjoying it Lieutenant Hill's detachment arrived. Delayed by breakdowns and bad going, the drivers had found the journey a hard one, but at least they had seen something missed by the rest of us: the coast road between Alamein and Daba.

'The wreckage was all mixed up,' said one of them, 'with one wreck on top of another or so shoved against it that you couldn't tell which was which. There were plenty of dead Jerries and dead Ities about, and plenty of live ones too. Their arms had been taken away, and all they wanted was water and directions to the nearest pen. The Italians were much worse off than the Jerries. After the rain fell we saw them on their hands and knees in the desert scooping the stuff up from puddles. Most of them had no boots and they'd wrapped sacking round their feet, making us believe it was true what they told us: the Jerries had taken away their boots and transport and left them in the front line with only small arms.'

We were away early the next morning, passing through Sidi Barrani at noon and stopping for lunch half an hour later. It was after three before we moved again and from then on we travelled very slowly. The road was crowded with transport now and we watched the sky. During the morning some Messerschmitt 109s had swooped out of the sun and machine-gunned the tail of the convoy, damaging No. 3 Platoon's wireless set and slightly wounding two drivers with splinters from explosive bullets. At half past four we halted near Buqbuq, dispersing for the night among low mounds.

That evening there was a small delivery of New Zealand mail and parcels. Always when mail arrived the orderly-room clerks stood in the backs of their lorries and called names (as they were doing now under ragged clouds turning black and a sky almost drained of colour, the sea being black already), and then you could tell from our drivers' looks all that letters from home mean to soldiers, and hear, blowing through vanished barrack-rooms, blowing through a hundred thousand haunted squares in France, in India, in Egypt, in Africa, the old bugle call clear and cynical, mocking human needs: 'There's a letter from Lousy Lucy— there's a letter from Lousy Lou!'

The next morning we moved slowly along the coast road, dodging bomb craters and smouldering wrecks. The Sollum escarpment was ahead, and as we drew near it the press of vehicles became greater, till at last we were travelling almost nose to tail. By eleven we had reached the foot of Halfaya Pass, up which an endless stream of tanks, guns, and transport was slowly winding.

The pass had fallen during the night to a surprise attack by 110 men from the 21st Battalion ('Valiant unto Death' was the motto of the Italian division to which most of the 612 prisoners belonged) and it was quite clear now, though the occasional crack of a rifle and the duller and louder explosion of a grenade made us think that stray Axis soldiers were still being winkled out from its seamed face.

The seashore and the wide flat were dotted with the guns and transport of other New Zealand units (most of which had precedence over us), and we dispersed off the road as best we could, the ack-ack crew setting up its guns. There were one or two alarms but mercifully no heavy raids.

Six vehicles were allowed to go forward at one in the afternoon, but the rest of the convoy had to wait for another four hours and it was twilight before we started the long crawl. In many places the road had been damaged by shells or bombs and for yards at a stretch the concrete fence was down. With a sheer drop on one side of us we remembered how earlier in the day we had seen a British tank swerve off the road and go rolling down the face of the escarpment like a huge boulder. Tremendous wedge-shaped shadows, tongued like pennants, went streaming past us, merging and spreading at the bottom and covering the whole flat. Great foreheads of rock, full of madness and menace, overshadowed the slender roadway and frowned down on the labouring vehicles as though willing them to usurp control and dash like the Gadarene swine down a steep place.

By the time we reached the top our radiators were boiling and our engines running with that false sweetness that follows a long climb. Darkness had fallen before the tail of the convoy was clear of the pass and only half of us reached our destination that night. It was the junction of the Trigh Capuzzo and the Bardia road, eleven miles south by west of Bardia, which, with Sollum, Capuzzo, and Sidi Azeiz, had fallen without fighting. The rest of the convoy bivouacked at the top of the pass.

During the day—it was Armistice Day—the Battle for Egypt had been won. The Eighth Army, after advancing 278 miles in eight days, had driven the last Axis unit across the frontier.

The convoy was complete before breakfast the next morning and we moved to an area about six miles east of Sidi Azeiz, reaching it in time for lunch. The next day, travelling in desert formation, we set out for Bir el Chleta, twenty-five miles west-north-west of Sidi Azeiz. For an hour we made good time and then we halted. We got out of our cabs for a stretch and a smoke but kept near them, for we knew these halts: they might last for two hours or for two minutes. This, however, was a long halt. It lasted until 5 December.

· ·

While the New Zealand Division trained and rested near Bardia the Eighth Army continued its advance. Tobruk was entered on the 13th of the month, Derna on the 15th. (In Britain the church bells were rung for the first time since June 1940 and our Second Echelon drivers wished they had been there to hear them.) Benghazi was entered on the 20th, Agedabia on the 23rd. In twenty days the Eighth Army had advanced 778 miles and by the end of the month Rommel was behind his old line at El Agheila.

And the press said it and the radio; the fighter-bombers, homing through the dusk like thunderbolts, roared and screamed it; and we said it to ourselves comfortably over the primuses at night: 'He will not pass this way again.'

Map of Cyrenaica and Tripolitania

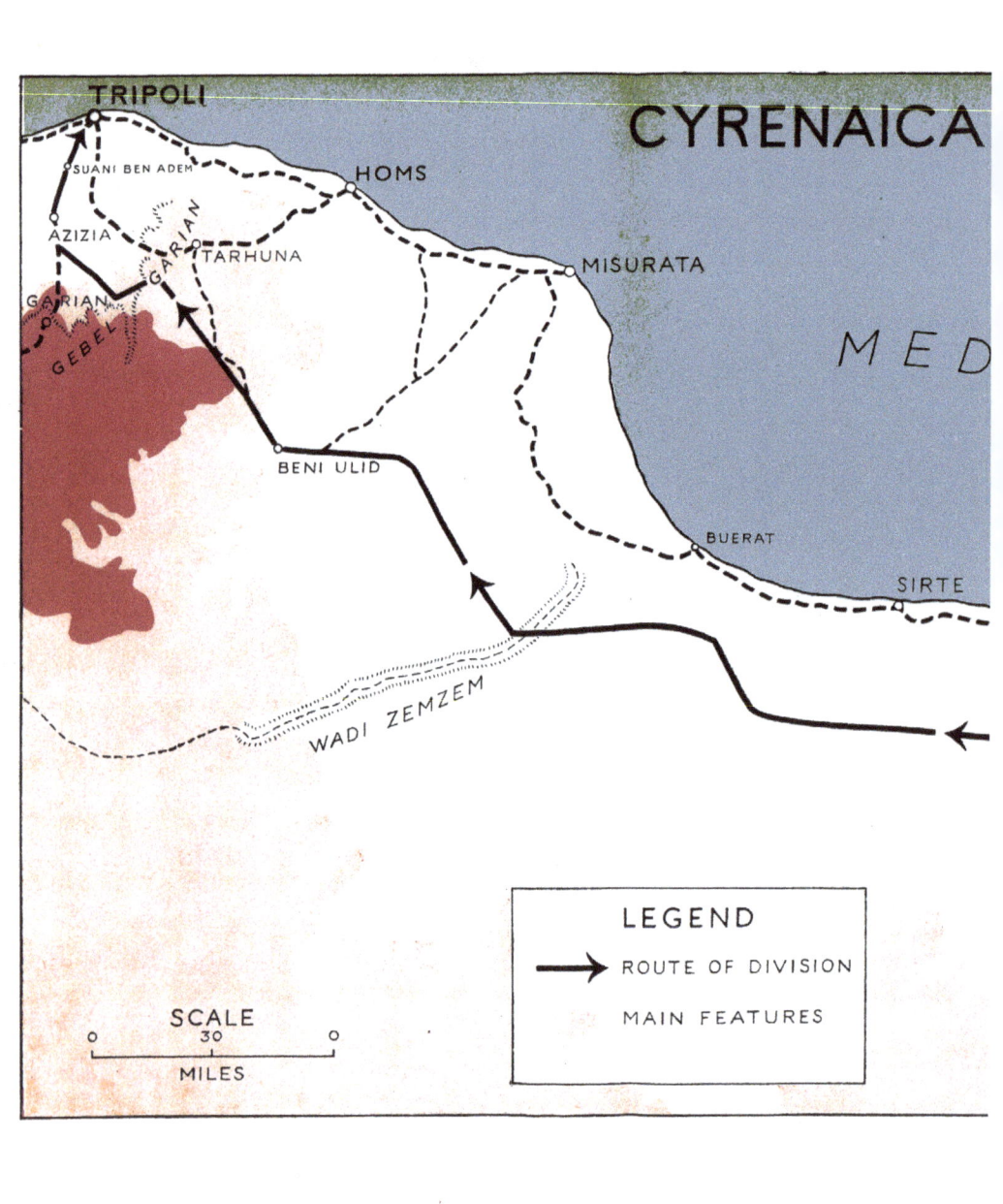

Flooded at Fuka

Removing the rims from a 3-ton lorry tire, Bardia

A bath at Bardia

Lorries on the skyline, south of Bardia

CHAPTER 16

JOURNEY WITH HALTS

WHERE we halted we stayed—one mile east of Sidi Azeiz. Morning and evening the shadows of the fighter-bombers rushed over our area, brightly while the weather was still fair and then vaguely under grey skies. A cold wind blew out of the west and it was winter. Like the ghosts of aircraft flashing over a grave the shadows went by in the desert, flicking the officers' mess, the shelters built by the cooks, the porcelain bath salvaged from Bardia by No. 2 Platoon. We had been told we should not be moving for some time and we had made ourselves comfortable.

The shadows brushed over the sand-filled trenches and the crumbling walls and the rusty wire at Sidi Azeiz, and over the blackening skeletons of the Chevrolets and the armoured cars. Survivors of the battle were in demand as guides and they would say, pointing to a heap of old iron with no paint on it: 'That was Don's lorry and that one there would be old George's.' One man went straight to where he had buried his diary a year before and a crowd gathered round while he turned the yellowing pages and tried to make out the brownish-grey writing, faint like veins.

The shadows shot past on their way to Benghazi and Agedabia and points west, and we looked up from the lecture, the route march, the parade—a training programme was in full swing—to see the avenging aircraft dance out of sight like liver spots, heading for a war only heard of now over the radio and through the newspapers. Those of us who had bad colds looked westwards through rheumy eyes, thinking: 'We'll follow soon. I'll hit the old cot straight after lunch and maybe a man'll be all right tomorrow.' But those who were sickening for jaundice shook their heavy, bursting heads and said to themselves: 'If I'm the same tomorrow she's a sick parade job. What a bastard at this stage of the piece!' There was no temporizing with yellow jaundice. Its victims experienced all the tortures of the bad sailor and the alcoholic, and if they reminded one of Mrs. Gummidge it was owing to her noisy optimism.

The weather mended towards the end of the month, but it was still very cold, especially at dawn and after sunset, and we were grateful for an issue of battle dress, winter underclothing, and extra blankets. It was perfect weather, though, for football. We managed to field a promising Rugby team, and in the first round of the Divisional knockout competition we defeated the 6th Field Regiment by 3-nil, but the Petrol Company beat us in the second round by an unconverted try after an extension of time. Inter-platoon matches were played as well.

In the evenings, except when the YMCA Mobile Cinema visited us, we played cards, cooked oyster fritters (150 bags of parcels arrived on the 26th), or listened to the platoon wireless sets. However wet or cold it was, few of us missed the news bulletins. Nowadays they were all good. The tide had turned at Stalingrad and in the Caucasus; French West Africa had come into the war on our side; French sailors had scuttled their fleet at Toulon, preventing the Germans from seizing it.

December came and Lieutenant Latimer[1] (attached) and Lieutenant Hill left us to join the 2nd New Zealand Ammunition Company,[2] which was being formed at Maadi. Suggestions from the Artillery that this was an offshoot of our unit and not an entirely different firm under an entirely different management were received coldly; nor were we pleased when we heard that we had a new name: 1st New Zealand Ammunition Company. With the war drawing to a close we regarded changes of this kind as unnecessary.

Grey day, bright blue day, day of burnished metal, whirling ginger day—they came in quick succession. Christmas, like a buoy we were for ever rounding, was in sight again, but we were destined to spend it a long way from Sidi Azeiz.

.

The El Agheila line, protected frontally by salt marshes and a system of strongpoints and minefields, and on one flank by the

[1] Capt G. P. Latimer, m.i.d.; company manager; Dunedin; born Kaitangata, 20 Mar 1910.

[2] The chief appointments on 17 December were: Company HQ, Maj P. E. Coutts, Capt S. A. Sampson, WO II I. McBeth (appointed 15 Oct 42); No. 1 Platoon, Capt R. C. Gibson, Lt A. R. Delley (posted 18 Jul 42); No. 2 Platoon, Lt J. R. Arnold, 2 Lt J. D. Todd (posted 3 Dec 42); No. 3 Platoon, Capt W. K. Jones, 2 Lt R. G. Sloan (posted 19 Jul 42); No. 4 Platoon, Capt K. E. May, 2 Lt J. M. Fitzgerald; Workshops, Lt (T/Capt) A. G. Morris; Ammunition Platoons, Lt R. K. Davis. The following had left us: Lts O. W. Hill and G. P. Latimer (posted to 2 NZ Ammunition Company, 4 Dec 42), 2 Lt R. A. Borgfeldt (admitted to hospital, 14 Jul 42), WO II Bracegirdle (posted to OCTU, 22 Sep 42).

sea and on the other by soft sand, was a formidable barrier, and Hitler had ordered Rommel to defend it to the last man. A frontal attack on its own was unlikely to succeed, but it was felt that a left hook by the New Zealanders, coincident with a frontal attack, might turn the trick.

Our unit moved on 5 December, heading for the Divisional concentration area at El Haseiat, thirty miles south-east of Agedabia and about a hundred from the El Agheila line. For three days we travelled in desert formation while the black diamonds that marked the route—they were cut out of tin and mounted on iron rods— flicked past the vehicles in the centre, and we saw, on the first day, only grey rain weeping on grey flats, then Bir Hacheim, where there had been bitter fighting in June. Here the sand had already half-buried the smashed and burnt-out tanks, guns, and lorries, as it had already buried once on the way to Siwa, too deeply for any to find, 50,000 Persians with all their arms and equipment. On the second day we followed the black diamonds across broken rocks, and on the third we passed at a good pace over flats dotted with scrub. Late in the afternoon of the fourth day we reached the concentration area. We had travelled 360 miles and none of the vehicles had given serious trouble.

On 13 December, thoroughly rested, we moved again, the black diamonds leading us south. For mile after mile there was nothing but rolling sand dunes, repeating themselves in a nightmare pattern of desolation with not even a sprig of camel-thorn to break the monotony. Just enough rain had fallen the day before to enable the leading vehicles to roll out a firm track and the going was good on the whole. We bivouacked for the night after covering sixty miles.

During the next day we touched the most southerly point of the advance, passing quite easily through Chrystal's Rift, sixty-five miles south-east of El Agheila, although, as we heard later, it had been expected to give endless trouble. For a while we were hindered by thick mists and afterwards by units ahead of us. The speed of the march was being conditioned by the ability of the petrol convoys to supply the leading vehicles and by the progress of the tanks of the Royal Scots Greys, which were under the command of the Division.

Soon after lunch we left the soft sand behind and came into

rocky country, and from then on we bumped through narrow defiles and laboured up and down steep wadis until we reached what was known as the Red Desert, where the going was fairly good. By four in the afternoon we had covered seventy-five miles, and after a halt for tea we went on through the darkness, travelling slowly on a narrow front. A burning petrol lorry spread a glow that could be seen for miles, making us wonder if all the elaborate precautions—the wireless silence, the driving away of enemy reconnaissance planes—had gone for nothing.

When we stopped for the night, after covering twenty miles since tea, we were due south of Marble Arch, which marked the Tripolitanian frontier, and well beyond the enemy's southernmost outposts. During the day the radio bulletins had said that Rommel was withdrawing from the El Agheila line.

The next day was the 15th. Following behind the 5th Brigade, we travelled slowly through the morning mist, heading in a northwesterly direction, which confirmed our opinion that the Division was making straight for the coast to try to cut off the enemy's retreat.

We were unaware of it then, but armoured cars, poked ahead of the main force like crabs' eyes, had enemy positions and troop movements under observation, and secrecy was no longer of the first importance. Wireless silence had ended at eight in the morning and although there is no record of our sending any messages we were now in a position to do so. A wireless transmission set, answer to a prayer first uttered a year ago, had been issued to us the day before.

During the afternoon and evening we travelled beside the 4th Field Regiment, halting at seven and then moving a mile and a half to clear the gun positions. We had covered ninety-five miles and were now near Merduma, more than fifty miles west of El Agheila. The sea was only eight or nine miles away.

Elements of the German 90th Light Division had been seen on the northern flank in the afternoon; the enemy was expected to try to break through from the east during the night, so we dug slittrenches and the guard was doubled. At half past six the cooks gave us a hot meal and then we turned in, dead-tired. In the distance we could hear the rumble of bulldozers and we thought: 'They're digging the guns in.'

At eleven we came under the command of the 5th Brigade. Some reports said that the 6th Brigade had managed to cut the coast road but these were untrue. The road was under fire but still open.

After breakfast the next morning, as no move seemed imminent, some of us decided to go to sleep again, but this proved impossible. Guns started to bark in the distance and soon they were joined by the 25-pounders of the 5th Field Regiment, which was still in the same position. Now we knew what was happening: Rommel's rearguard was trying to escape west.

The guns went on firing, and presently we saw a column of transport and armour on the horizon. Individual tanks and lorries (which was which it was difficult to tell though some of our drivers said they could) showed as dark, podgy shapes moving steadily among puffs of smoke.

'From Workshops' area,' said Sergeant-Major Noel Campbell,[3] 'we had a good view. We could see his transport and stuff cutting away west and our 25-pounders getting on to it. Some high explosive shells were coming back, and once we saw what looked like a German tank sneak in and have a go at a New Zealand battery. Several shells landed near our area and the closest one was only a few hundred yards away. We saw one lorry—an LAD job belonging to the Artillery—get a direct hit, and later the driver of it came over and asked us to fix him up with a new radiator. We also saw what we thought was a direct hit on one of our gun emplacements. By lunch-time everything was quiet.'

Everything was quiet, but elements of two panzer divisions and of the German 90th Light Division had escaped almost unscathed. Part had got away along the coast, and the rest—a column of thirty or forty tanks and two or three hundred vehicles—had driven swiftly and with good dispersal through a seven-mile gap between the 5th and 6th Brigades.

Next the Division tried to prevent Rommel's rearguard from escaping from Nofilia, some thirty miles along the coast. The village was strongly held, doubtless to give the German armour and transport more time to get away.

We moved late on the morning of the 17th, following the 6th

[3] Capt R. N. Campbell, m.i.d.; motor mechanic; Eureka, Waikato; born Kaitaia, 27 Jul 1915; wounded 23 Dec 1940.

Brigade and passing through winding wadis. In the afternoon No. 2 Platoon came under fire, probably from German 88-millimetre guns, but no damage was done, and at half past ten, after travelling forty-one miles, we laagered eleven miles east of Nofilia.

That night the Division had another disappointment. The 5th Brigade managed to block the coast road but not before the enemy rearguard had escaped.

At dawn we moved one and a half miles south-west to give the 6th Brigade more room, dispersing in a shallow wadi pleasantly speckled with green. We were there for three days, and on the 21st we moved to an area overlooking the sea. As soon as camp was pitched we were given permission to bathe.

The water was bright and chill and it glinted like steel. The feel of it and the news we had just heard (we were likely to stay in the area for some days) reminded us of Christmas at Fuka a year ago. Here was the same sand, the same bright and chilly water, the same coastline, and, a little farther along it, the same enemy. One Christmas of this kind was enough: two suggested the beginning of a nightmare pattern that might go on through a grey lifetime interminably repeating itself. But no—that was nonsense. What stretched in front of us was the bright thread of victory, leading to Tripoli, to Tunis, to Rome—not, of course, that we should see Rome.

. . . .

In one of its objects, to cut off Rommel's panzer army, the left hook had failed; in another it had succeeded brilliantly. Without a fight and at little expense to British arms the Germans had been forced to abandon positions loudly advertised as impregnable. The next natural line of defence was at Buerat, about 150 miles away. Rommel reached this on 26 December leaving behind delaying forces, and three days later the leading elements of the Eighth Army were in front of it.

Notwithstanding the speed of the advance—between 15 and 26 December 248 miles were covered—it is possible to think of the Eighth Army as an enormous caterpillar. When it was at full stretch, as it was when its horns (or whatever they call those things that caterpillars have) were resting on the Buerat line. it was incapable of making another lollop forward until it had drawn up its middle and hind parts, which were labelled Petrol, Ammunition, Rations, Workshops.

While this gathering together was in progress, the Division rested on the coast near Nofilia and it was here that we spent Christmas. A Corps dumping programme was started almost at once and Christmas Eve found us bringing forward supplies from the neighbourhood of El Agheila. However, thanks to the cooks, who travelled with their platoons and spent half the night building ovens, everyone had a wonderful Christmas dinner: roast pork, plum pudding, a bottle of beer, and a double issue of rum. Those who were in the unit area at Nofilia were visited by the Colonel, who congratulated us on our 'splendid work in recent campaigns and particularly on the dumping programme carried out before the offensive at El Alamein'. He said also: 'I hope that by next Christmas you will all be back in New Zealand.'

We hoped so, too, for we were neither comfortable nor over-happy at this time. Often it was bitterly cold and sometimes there were blinding sandstorms that made driving a nightmare. We used the coast road, collecting our loads at the 107th FMC,[4] near El Agheila, or the 108th FMC, near Marble Arch, where the platoons had their headquarters, and unloading either at the 109th FMC, on the coast beyond Nofilia, or at the 110th FMC, twenty-odd miles inland from Headquarters' area. The drivers with the platoons averaged 140 miles a day, and they spent their nights sometimes with Company headquarters and sometimes at Marble Arch—Arae Philaenorum to the Italians. (It was not made of marble, but it was imposing enough and even beautiful, as a design in the middle of nothing—any design—can hardly avoid being. Bas-reliefs glorifying Fascism and Mussolini covered the inside of the arch, but its long shadow fell on Douglases and Baltimores, which were using the great Axis airfield.)

The coast road was bad and the desert was worse, and round Nofilia there were mines. An NZASC driver was killed only a mile from Headquarters' area and the next afternoon a No. 4 Platoon vehicle, which was returning from the replenishment point with rations, was extensively damaged.

The New Year came and brought with it a raging sandstorm. Sand collected in deep drifts on the floors of the cabs and round

[4] Field Maintenance Centre.

the engine blocks. It stopped our noses and our ears and we could feel our lungs filling up like hour-glasses, but there was no slackening of effort, for the next move was now very near.

By the evening of 5 January 1943 our second-line holding was again on wheels. We had rations and water for eleven days and in our tanks and jerricans there was enough petrol to take us more than 400 miles. Before noon the next day we were in the Divisional concentration area south of Nofilia.

'How many miles to Tripoli?' we asked. It was three hundred as the crow flies.

'We shall be sent home,' we said, 'when we've seen Tripoli.'

The task of capturing Tripoli had been given to three divisions: the 51st (Highland) Division, the 7th Armoured Division, and the New Zealand Division (with the Royal Scots Greys under command). These composed the 30th Corps. Rommel was still occupying the Buerat line, and the plan was for the Highlanders to attack on the coast sector while the New Zealanders and the British armour, travelling inland, advanced along the enemy's right flank. If the Highlanders had no success and the armour was held up by artillery, the New Zealand Division would make another left hook.

That was the position on 10 January when we moved again, travelling west-south-west for eighty-one miles. We rested on the 11th and on the 12th travelled forty miles due west. We made hardly any progress on the 13th but by the next night we had covered a further sixty miles. We were now twenty miles east of Wadi Zemzem, which ran south-west behind the Buerat line. That night we heard the rumble of gunfire and at dawn we dispersed our vehicles.

It was now the 15th—a fine, warm day. We were on a moment's notice to move and we knew that something important was afoot. Sounds of firing came from the north-west and aircraft dropped bombs in a neighbouring area. We had an early lunch, but still there was no order to move, and the day, which had started with an air of bustle and a suggestion of great events, degenerated into a vast yawn.

The order came at last and we set off at half past three in the afternoon but travelled only two and a half miles before halting until six. Then we moved a few more miles, laagering at eight under the protection of the 28th Battalion. Carefully we collected and collated scraps of information. The Highlanders had attacked on the coast and the 7th Armoured Division on our right, but the latter had been held up by cunningly-sited guns and tanks, so the New Zealanders had been given the word *Go*. In the late afternoon the Royal Scots Greys and the 4th Field Regiment had moved round the south flank of the German line (which ran south-west from Buerat for about fifty miles, parallel with, and a little in front of, Wadi Zemzem), and shortly before dark the Shermans had gone in. Now the enemy was fleeing west again.

The next nine days remain for most of us a memory of grey dawns, and green and grey dawns, too, and dawns flecked with red like inverted sunsets, when we got up, still muddled with sleep, from our sandy blankets, and hurried to start cold engines so that we could disperse the lorries before daylight; of hasty and often unfinished meals eaten in the cab or in company with the rest of the sub-section in the back of somebody's three-tonner, with half-empty tins of bully and margarine covering every flat surface, and the billy boiling on the primus between two boxes of charges, and the water waiting to froth up in a brown cream and dribble all down the sides when the tea-leaves were thrown in; of early morning drives when the steering wheel was a circle of ice, and the gear lever a stick of ice, and your breath clouded the windscreen; of long night drives under the white ball of the moon through moon-country; and of long halts in the sunshine while the war went on ahead or seemed to have stopped altogether.

While the Highlanders advanced along the coast, clearing mines and skirting obstructions, the Armoured Division and the New Zealanders headed west through the desert, widening the distance between themselves and the northward-bending coast and hurrying along the enemy's flank. Every day the scenery was different. The afternoon of the 16th found us moving through tall scrub: a young forest you could almost call it, for the trees—bushes—were between twenty and thirty feet high. They were not much to look at and some heavy shelling had not improved them, but they were welcome after the stony wilderness and they sheltered us while we ate our tea. Glancing through their meagre branches, we saw two Messerschmitts overhead with puffs of ack-ack on their tails. When we laagered that night, again with the 28th Battalion, the moon was very brilliant and bombs could be heard rumbling in the distance. Most of us dug slit-trenches.

For the greater part of the next day—we travelled four miles before lunch and two in the afternoon—we were near a large airfield. The story was that it had been captured only that morning, but by midday convoys of lorries were passing through our lines and turning into it, and early in the afternoon great transport planes started to arrive. They landed, unloaded, and took off again.

After tea we travelled north-west until three the next morning. For most of the way the moon positively blazed, doing fantastic things to the rough, wild country. Boulders became silver nuggets and the dry river-beds looked as though they had been ploughed out of glittering metal. The flats were milky and faintly gleaming, so that they seemed like still lakes studded with misty islands and bounded by cliffs of silver. Only ourselves and the transport suffered no silver change. Dirty, unshaven faces took on a leprous tinge, and the lorries, as they lurched and staggered up steep tracks or scuttled across the open, were for all the world like beetles in a treasure-house.

The next day we came to inhabited country. Stunted, mud-coloured children and their sinister-looking parents, the women black bundles and the men a flash of ravenous teeth and a length of gesticulating black arm, pestered us for food. The children, with their cropped heads and bony knees and elbows, were hard to refuse; nor was there any need to refuse them, for most of us were carrying more army biscuits than we should ever eat. '*Biscotti! Biscotti!*' they cried, and 'Bull-biff, Johnny! Bull-biff!'—pursuing us with despairing wails long after we had passed. They fought frenziedly among themselves for handfuls of biscuits thrown from the cabs, and even when their tattered and filthy nightshirts were stuffed with treasure they continued to cry, automatically and despairingly: '*Biscotti!* Johnny! Johnny! Johnny!' (One had ceased to be George, apparently, on entering Tripolitania.) Sometimes we caught glimpses of their homes: miserable, tumbledown lean-tos and crazy boxes, half tent and half hutch, made of petrol tins, scraps of old matting, and other rubbish. We halted early in the afternoon about five miles east of Beni Ulid, which was eighty miles south-south-east of Tripoli.

We did not move again until late the next day and then only as far as the Beni Ulid road. Again we saw transport planes, Bombays, landing on a nearby airfield. That night we heard that the Germans had left Tarhuna, between fifty and sixty miles north-north-west of Beni Ulid, and were burning supplies in Tripoli.

Travelling by a road that was dusty and deeply-rutted and flanked on either side by burnt-out guns and transport, we reached Beni Ulid, an oasis and Italian outpost, the next morning. It was a pretty town of high, white-walled houses, of cave-like shops (dealing, so far as we could see, only in vegetables of the onion family and quoit-shaped loaves), and of teeming, crumbling, stinking buildings that were like nothing so much as gigantic portions of gorgonzola cheese. Bluegums grew among palm and olive trees, the grey dust, or the whitish, endlessly-trampled sand, lapping their roots. Leaving the village, we went down the steep side of a wadi, bumped for another fifteen miles over dust-filled ruts, still passing burnt-out transport, and dispersed off the road, by which time everyone, clown-like, wore a pallid mask. We were now some thirty miles south of Tarhuna.

The next day—21 January—the Division began to cross the Gebel Garian range, which separates the desert hinterland from the plain of Tripoli. Following Rear Headquarters of the Division, we moved three miles along the Beni Ulid road in the direction of Tarhuna before turning off into the desert and forming up on a fifteen-vehicle front. A long halt followed, and then we travelled parallel to the road for a short distance and stopped for the night. Darkness came down, and ahead of us, about twenty miles away, guns were in action. During the night dozens of bombers passed overhead, keeping us awake.

On the 22nd, while the leading elements of the Division debouched on to the plain of Tripoli, we moved slowly in the direction of the Gebel Garian range, passing first through deep wadis and then through extremely soft sand. We covered three or four miles in the morning and rather less in the afternoon, but after tea we moved off in column of route along a track that took us quickly to a good road, which was reached as darkness fell. We followed it for a few miles and then turned off on a rough track, which plunged us at once into the hills. These must have been the foothills of the Gebel Garian range. The going was terrible, but we were helped by brilliant moonlight, and we staggered on, twisting and turning but heading more or less in a north-westerly direction. After we had cleared the hills we were held up by units ahead of us, so we bedded down where we were. It was now midnight.

Early next morning, following elements of the 7th Armoured Division, New Zealanders entered Tripoli from the south while, almost simultaneously, troops of the 51st Division entered it from the east. Since leaving Alamein on 4 November the Eighth Army had advanced 1450 miles—roughly the distance between London and Istanbul.

For others—and we grudged them nothing, for theirs had been the greater danger—the supreme moment of marching into the fallen city to the skirl of pipes. Our own unit began to wriggle towards it with the extreme diffidence of a very young puppy approaching a dead rat.

When we moved the next morning we needed to cover only about nine miles to reach the Tripoli-Garian road, which ran due north to the city. We managed four of them. On the 25th, some time before noon, we struck the road at Kilo 65 and headed towards Tripoli in column of route. We passed Azizia, noticing what a pretty place it was, and carried on as far as Kilo 27. Here we turned left off the road into an Italian farm. Company headquarters chose an area close to the house and the platoons dispersed nearby. Compared with the interminable wadis, the oceans of stone and sand, the scene was charming. There was a little well, full of clear water. A froth of blossom clouded the peach and almond trees, so we supposed that it was spring. For many, many months all we had known of the seasons, except that they were either too hot or too cold, was about as much as prisoners in the Bastille knew of time. No wheatfields had ripened in Fortress A, no leaves had fallen at Alamein, no peach trees had budded at Marble Arch.

But now it was spring and we had reached Tripoli.

CHAPTER 17

FEEDING A CATERPILLAR

THE ugly, snub-nosed lorries, their camouflage paint subdued to a pallid fawn by dust and heat and hard service, their mudguards dented, their canopies torn and patched, but their engines still growling strongly, could be seen everywhere. They bowled along the splendid seafront and down the opulent boulevards lined with acacias and palms; they swished past offices, public buildings, and hotels (not too badly damaged), through streets of grand shops (shuttered from caution or because there was nothing in them), through parts of old Tripoli, Turkish and squalidly magnificent, and out into the surrounding country, where the tender but hectic green of the young fruit trees was interrupted by water towers and pylons, exclamatory in ferro-concrete over the marvels of Fascism, and by patches of primal desert, red sometimes and sometimes greyish yellow, the colour of an old camel, which mocked the new green varnish of Balbo's *colonizzazione*.

Of those lorries that were marked with the New Zealand fernleaf a large number were marked also with the 69 of the 1st Ammunition Company, for at that time we were doing the most work of any NZASC unit in Tripoli and perhaps of any transport unit in the Eighth Army.

When that familiar number flashed past, those who were on day-leave in Tripoli would almost wish they were working, for Tripoli was a disappointment. There was plenty to see, of course, but soldiers, apart from an earnest minority, soon weary of *seeing* things, and to interest them for long it takes more than a fine seafront, an equestrian statue of *il Duce*, some houses of disrepute, and one of Count Ciano's summer palaces. After you had seen these and had drunk tea at the Count's place, which had been taken over by the YMCA, your problem was how to fill in the time until six, when the transport left for camp.

When a thousand soldiers go on leave some seek female companionship, a larger number seek liquor, but all, including the most earnest sightseers, require an appetising and substantial meal —if possible with three eggs. Of wine and women there was a small supply in Tripoli (of indifferent quality and uncertain effect), but of food there was scarcely a vestige. To protect the civilian population General Montgomery had forbidden the Eighth Army to buy meals, though probably none could have been bought anyway, for the Axis had emptied the city of supplies and most of the inhabitants had only their meagre ration.

Consequently, long before six o'clock, our drivers were hungry and discouraged, and they would speak wistfully of returning home and putting the billy on. But returning home when he still has money to spend is anathema to the average soldier. Realising this, a handful of philanthropic New Zealanders had organised two-up schools in convenient side-streets. These gentlemen, apart from one or two unfortunates whom unfair competition at Maadi had forced into the field, were not the ones who had battened on us at Helwan and in the bar of the New Zealand Forces Club in Cairo. They, bless them, had long since retired on their savings and taken their neuroses and fallen arches back to New Zealand. The new *banditti* were different. To give them their due, they were in the game as much *pour le sport* as for profit, but that did not prevent them from doing uncommonly well. Every day a vast sum of money changed hands under their supervision and went on changing hands (in spite of the fact that two-up, unlike Crown and Anchor, is a perfectly fair game) until it reached that limited number of pairs to which all money naturally gravitates. From these it was transferred to the limited number of paybooks and there it stayed.

In quiet courtyard, then, or in shady side-street, the ring was formed, its presence being advertised by a hedge of soldiers and by the practised voice, loud but confidential, that spoke from the midst of it as from the bush on Horeb: ' I want a quid in the guts, gentlemen. I want a quid in the guts to see him go. Right, then! Are we all set? Are we all set on the side? Come in, spinner. And a good spin too. She's out the monkey, gentlemen—she's out the monkey. Right, then. I want a quid in the guts. . . .'

Black faces, lit by goggling, amethyst eyes faintly stained with saffron, formed an outer ring—not that there was anything to prevent natives, if they were men of substance, from joining the game—and Italian colonials in off-colour ducks looked on with benevolence, enacting their admiration when our drivers plunged recklessly, their sympathy when they lost.

'I want a quid in the guts, gentlemen. . . .'
'Hell, let's go back to the YM—get some more *chai*.'
'No. Let's bludge a ride home. Get back, eh, and boil up? Open a tin of tongues, eh?'

Yes, Tripoli was a disappointment. It was the kind of place where you couldn't get a good feed anywhere, and that kind of place, as everyone knows, is the New Zealander's idea of Hell.

You could, though, get a drink—not in the bars and cafés, but from a local wine factory, where the purple grape (*Vino Rosso* 1942) concluded a mad rush through a system of cylinders and metal pipes by gushing from a faucet. It was fresher than new-laid eggs and the amount you could take away—for a while at any rate—depended on the number of jerricans in your possession. It was remarkable, we found, more for its strength than for the delicacy of its flavour, especially if the cans in which we took it away had once contained high-octane petrol. It was the genuine article, though. We spoke of it as being of the port type, but it was rather more than that, for it implanted a purple stain (which wore off in the course of time) on lips and tongue, and given a few weeks it was quite capable of gnawing its way through the compressed steel of the stoutest jerrican. It was mentioned respectfully in routine orders and indeed it deserved to be. Healthy people, we found, went into a decline after drinking it, and we were forced to conclude that something had gone wrong during the period of its manufacture. A cylinder, perhaps, had been functioning incorrectly, or a pressure gauge had given a wrong reading.

We had little time, however, for carousing or for indulging our disappointment in Tripoli. The Eighth Army, that insatiable caterpillar, was waiting to be fed.

. . . . . .

On previous occasions it had been necessary to feed the caterpillar from its tail or from points (Benghazi, for instance) in the neighbourhood of its midriff; but now, thanks to our having Tripoli, the vital supplies could be injected at a point somewhere near its right shoulder—if caterpillars may be allowed to have shoulders.

The port had been badly damaged by bombs and the Germans had carried out extensive demolitions, but almost at once with the help of landing craft, it was possible to start unloading cargoes.

These had to be cleared from the wharves the moment they were landed and taken to dumps in and around Tripoli.

We started work two days after arriving at Peach Blossom Farm, and on 5 February (on the 4th 180 of us had attended a parade for Mr. Churchill) Captain May took over the duties of dock transport officer. By the 8th of the month all four transport platoons (with a vehicle availability of 120) were employed at the docks, and from then on, though detachments from other NZASC units and from the 51st Division helped us, the job was mainly our concern.

It was not practicable, of course, for the platoons to operate from Peach Blossom Farm, so they were packed as closely as possible about a square of barrack buildings some two miles from the docks. Here our drivers were very comfortable, the barracks providing many with sleeping quarters and all with a place to store their gear—the beds, boxes, and tins of which it had been necessary to strip the transport.

Captain May, controlling matters from his office with the help of a Don R and a jeep, sent the transport to the docks as it was called for. After loading, the drivers would go independently to dumps in and around Tripoli. The loading and unloading, until 11 February, was done by Highlanders. Then it was taken over by 3000 men drawn from the 5th and 6th Brigades and the Divisional Artillery. By the 14th of the month 200 vehicles, British and New Zealand, were employed at the docks under the command of our old friend and new major—Major Sampson. Major Coutts had left us on 27 January to fly to Cairo to take command of the 18th Tank Transporter Company, NZASC.

During the second week of February the transport began working in two shifts and an improvement was seen at once. On 14 February 18,471 tons were lifted, of which our unit and the detachments under our command handled 1651 tons, using 195 vehicles and travelling 10,806 miles. Between the 15th and 23rd our transport alone lifted 14,745 tons (an average of 1581 tons every 24 hours), and during the 24 hours that ended at half past five on the afternoon of the 19th we achieved a record by lifting nearly 3000 tons.

By now Company headquarters and one section of Workshops (the other was with the transport platoons) had moved to an area near Suani ben Adem, thirteen miles south by west of Tripoli. Our

Desert formation—left hook at El Agheila

Ruins at Cyrene

Nofilia signboard

Tripoli cookhouse

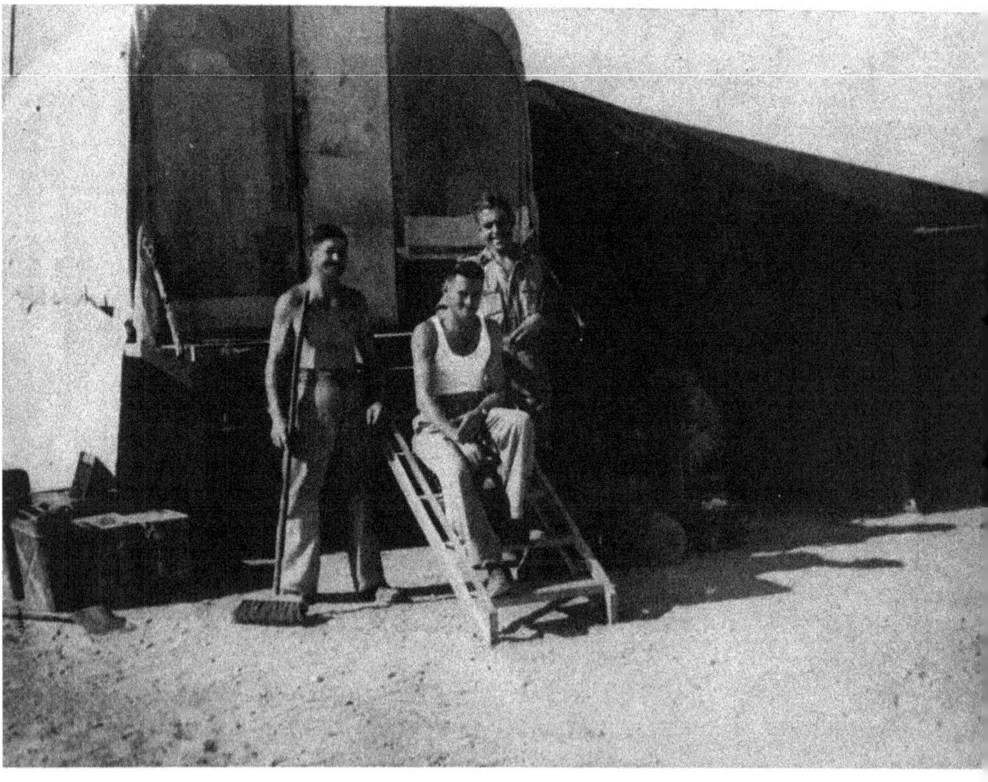

second-line holding of ammunition went with them and this had to be carefully guarded, as did everything at that time.

We had no reason to suppose that either the Italians or the natives were unfriendly towards us—the former affected to look on us as protectors and the latter as liberators—but we knew from bitter experience that both were highly acquisitive. Only once was there a suggestion of something worse and that was when a mysterious fire broke out in the barracks. A party of Workshops' drivers did their best to put it out, but the local fire-engine—a very old lorry equipped with a pump and a tank—had to be sent for. The fire, though small, burnt with unnatural fierceness, and the subsequent discovery of an artificially contrived draught seemed to suggest that it had been started maliciously, perhaps as a guide to aircraft.

The Luftwaffe, of course, was taking more than a passing interest in what was going on at the docks, but its efforts to interfere, thanks to the deadly efficiency of the anti-aircraft barrage, were unavailing. Through star-filled skies, night after night, the searchlights' tapering beams swept, steadied, and swept on, making it seem as though a phantom ship, with booms a million times bigger and busier than those working in the dark harbour below, was softly unloading stars. Then the ack-ack would start up—odd cracks first, as from a recalcitrant motor-cycle, and then a deafening acceleration of sound that reached its climax in a few seconds and stayed there. Columns of coloured balls, toppling a little at the summit, came from a hundred places, and strings of red beads from smaller and faster guns were flung all over the sky. It was hard to believe that anything could live above Tripoli, and we were not surprised when we heard that six planes had been shot down in one night. Once we saw a plane get a direct hit and fall out of the sky like a comet, lighting the whole city.

Accurate aiming was certainly out of the question, and although bombs fell near the docks and in the harbour, we never heard, while we were there, that any ships had been sunk. The harbour was full of wrecks but they belonged to Italy and Germany.

Only once were our drivers in real danger and that was on the night of 24-25 February when a solitary plane glided down to drop bombs on the docks before the ack-ack could open fire, and then came in again, flying low. Bombs showered down on our barracks,

S

wiping away the roof and the front of a garage occupied by Workshops, and blowing in windows and doors and scattering tiles around. One driver was blown five yards into an air-raid shelter, another was blown against a palm, and another woke up with a tree across the foot of his bed and a window frame round his neck. There were many cuts and bruises, but only one man was hurt badly enough to need more than first aid, and he was back with us within a few days.

The damage to the buildings mattered little, for we had done with them. The attached drivers returned to their own units and by eleven the next morning we were all together in the area near Suani ben Adem.

Vehicle maintenance, a good rest, a game of football, a picture show in the area, and it was the end of the month. On 1 March we drew six days' rations and put our second-line holding on wheels.

The caterpillar had been fed.

Map of Tunisia 1943

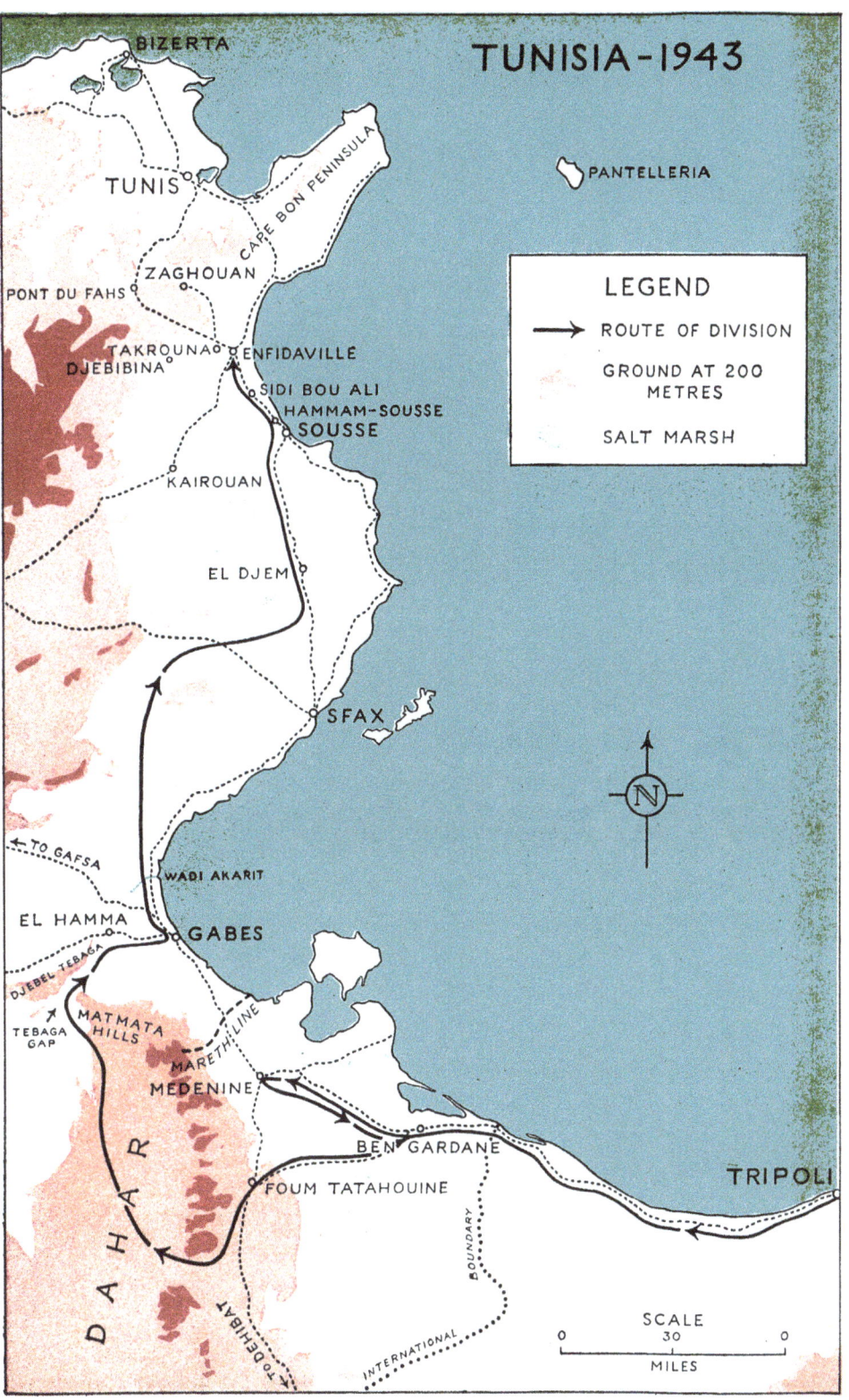

CHAPTER 18

THE END OF THE FIRST HALF

AFTER losing Tripoli, Rommel had been driven back 200 miles to the Mareth Line, once known as the 'African Maginot', which stretched for about twenty-two miles from the sea to the Matmata Hills in the west, and gave him a firm base from which to take the offensive. Soon the Americans were in trouble in western Tunisia and the Eighth Army was forced to return to the attack.

The danger then was that Rommel would break off the earlier battle, regroup quickly, and catch the Eighth Army off balance. By the end of February this seemed to be his intention, so the New Zealand Division was called forward.

The leading units moved early on 1 March, No. 1 Platoon of the Ammunition Company and ten vehicles from No. 4 Platoon travelling with the 5th Brigade. The rest of us moved at seven on the following evening and drove steadily through the night along the coast road. It was pitch dark but we were allowed to use lights.

Our first halt was for breakfast, which we ate on the Tunisian frontier. We were off within the hour and before noon we had passed through Ben Gardane, sixty miles south-east of the Mareth Line. We halted twenty miles farther on, and by half past one all the vehicles were dispersed in the new area. Sand was underfoot and wadis and bare hills surrounded us. We might have been back at El Alamein.

That evening Nos. 2, 3, and 4 Platoons moved out to open an ammunition point in the Divisional area at Medenine, fifteen miles behind the front line. Here they were awaited by No. 1 Platoon. Before them, on Rommel's right flank, were the Matmata Hills, which rose and fell like the tracings on a temperature chart—a record, it seemed, of the Tunisian fever.

The next day the fever mounted and both sides were active in the air. From the ammunition point we saw two Focke-Wulfe 109Fs, the new and alarming fighter-bombers the Germans were using, shot down in flames. The demand for ammunition was normal and we issued small quantities of all types including some 'beak' shells for the 'pheasants', the new 17-pounder anti-tank guns. (These remarkable birds, conspiratorially muffled in canvas and surrounded by a cloak-and-dagger atmosphere, had been arriving at the front for some while.) During the day Company headquarters joined the platoons at the ammunition point.

Throughout the 5th (while No. 4 Platoon started the two-day job of carting 25-pounder from the 115th FMC at Ben Gardane to the 116th FMC at Medenine) tanks, transport, and coveys of 'pheasants' flowed along the coast road towards Mareth. Plainly the crisis was at hand.

Rommel struck the next day. As the morning mist lifted columns of tanks and infantry came out of the hills. Three thrusts were launched, the most southerly against the junction of the 5th Brigade and a British brigade. None succeeded. In the afternoon a second attack on the New Zealand sector was smashed by our guns, and it was the same story all along the front. By the end of the day Rommel had had enough and that night he withdrew behind the Mareth Line, leaving fifty-two tanks and some dead and wounded.

Beaten on the ground he tried to make himself felt in the air and throughout the 7th the skies were dangerous, especially for the drivers at the ammunition point, which now made a very attractive target. (On 1 March we had been joined by a platoon from the 4th Reserve Mechanical Transport Company carrying surplus ammunition, and later by three Royal Army Service Corps detachments, which had been formed into a platoon to serve British units under the command of the Division.) During the morning fighter-bombers dropped bombs all round the ammunition point, doing no damage beyond marking a few vehicles, and that evening Workshops was machine-gunned from a low level. Two bullet holes in two bivouacs was the total damage, but in the next area the 1st New Zealand Petrol Company lost a lorry and a dump of 680 gallons of petrol.

The 8th was a quiet day in all respects and on the 9th the ammunition point closed down, the platoons joining Workshops in the unit's original area. Three days later we were told about the Division's next move.

Again it was to be a secret one. On reaching the assembly area ninety miles almost due south of Gabes, transport would move in daylight only when it was absolutely necessary and there would be no fires between dusk and dawn. Wireless silence would prevent us from employing our two wireless trucks—the one with Company headquarters and another that had joined us early in March for use at the ammunition point.

We reached the concentration area early on the 14th after travelling 120 miles, and that evening every vehicle set out for the Corps roadhead at El Dehibat, some sixty miles to the south-east. During the next twenty-four hours we were employed with other NZASC units in carting ammunition, petrol, and supplies to a field maintenance centre near the concentration area. It was dusty, windy, and cold, and we rested only when we were waiting for night to fall so that we could cover the last sixteen miles to the field maintenance centre without being observed.

During the 17th we reloaded our second-line holding and serviced our vehicles, and on the 18th we slept. After dark we moved two miles to form up behind the 5th Brigade for the coming move.

. . . .

There was talk of Gabes, a coast town twenty miles behind the Mareth Line, and of a narrow gap that was to be stormed by infantry and armour after another left hook. This gap, though we knew little or nothing about it at the time, was between the northern end of the Matmata Hills and the southern flank of the Djebel Tebaga range. It was some thirty miles south-west by west of Gabes and had once been walled up by the Romans.

We set out for the Tebaga Gap on the 19th at 7 p.m. It was a lovely night, fine and warm. Following the 5th Brigade and steering by the provosts' dim green lanterns, we travelled slowly on a nine-vehicle front, heading north through the Dahar, a rough, scrub-covered wilderness, totally uninteresting, that lay between the Matmatas and the Sahara. When we halted at half past one the next morning we had covered twenty-seven miles and were seventy miles south by west of Gabes.

The New Zealand Corps (the temporary designation of the force under General Freyberg's command for the left hook: it included the 8th Armoured Brigade, General Leclerc's Fighting French, and British medium, field, and anti-tank regiments) was now on the line reached by the Fighting French. Apart from their remarkable achievement in penetrating 1500 miles from Chad, their home territory, the French were of interest to us at that time because the day before Second-Lieutenant Todd[1] and twenty-eight of our drivers had been attached to them to carry ammunition for a flying column whose task was to protect the Corps' flank from patrols and reconnaissance parties.

After dispersing at dawn on the 20th we were told to be ready to move at half past eight that morning, the difficulty of deceiving the enemy any longer having made General Freyberg decide to push on towards the Tebaga Gap at top speed.

We did not move until after lunch—the story was that the advance had been held up by enemy guns—but throughout the afternoon, in spite of wadis, enemy minefields, and patches of soft sand, we made good progress. We stopped for tea after covering twenty-three miles, and an hour later pushed on over appalling country, travelling until half past ten at about two miles an hour. That night, for the first time since leaving the Medenine sector, we were kept awake by bombs. Flares lit up our neighbours, but we stayed in merciful darkness. We were now between twenty and twenty-five miles south by east of the gap.

The next day—the 21st—the Luftwaffe tried desperately to interfere with the advance, and while moving ten miles north-west in the afternoon our convoy was attacked twice by Messerschmitts. No damage was done.

We halted soon after dark and presently the moon rose, revealing desolate country and the outline of the Djebel Tebaga range. Here, surely, was the utmost bound of the everlasting hills spoken of in Genesis. In that timeless presence it was useless to make any distinction in period between a Roman wall and an armoured brigade. Only yesterday the horsemen had waited where we were waiting. They had ridden away, those Berbers, on their rough, strong horses and the tanks had come, and in between the mountains had scarcely found time to sigh. The echo of hooves had died away in the hills, there was a birth in Bethlehem, Titian and Shakespeare lived, Liebig discovered chloroform and Diesel invented a new kind of oil engine, and again an army stood at the Tebaga Gap.

Gunfire reverberated among the hills and there was a demand for ammunition. The night before the Eighth Army had launched a

---

[1] Lt J. D. Todd, m.i.d.; motor driver; Te Kuiti; born Waipawa, 16 Mar 1913.

frontal attack on the Mareth Line and now the New Zealand Corps was playing its part. By morning success leaned towards us for the enemy had lost Point 201.

. . . .

During the next four days the demand for ammunition increased steadily. We dumped our loads at the guns, reloaded at the field maintenance centre, and returned to the unit area—a round trip of 120 miles. A few hours' sleep and we did the same thing again. On the down trip we carried prisoners.

After moving back three and a half miles on the second day to allow the 5th Brigade to occupy our area—a report said that a panzer division was attempting an encircling movement—the unit stayed where it was, No. 1 Platoon moving forward two miles on the afternoon of the 26th to open an ammunition point.

Enemy nuisance raiders came over in daylight in twos, threes, and fours but our Bofors were too much for them. At night the Luftwaffe had things more its own way. It bumbled around dropping flares and butterfly bombs, and often we awoke in our slit-trenches to find the area lit up like a city street. Above us would be globules of yellow light, which dripped smaller globules like golden tears, and we would lie still while the wind swept the flares away or they faded with agonising slowness. We hated those butterflies with an intense hatred and it was almost a relief when the Germans dropped bombs of the good old-fashioned kind.

Then it was our turn. Throughout the night of the 25th-26th our bombers shuttled backwards and forwards, plastering the enemy's defences and filling the sky above them with clusters of golden flares. As they descended and faded others appeared over them. From where we were it seemed as though blocks of flats, unimaginably brightly lit, were being dropped by parachute. During the morning of 26 March we heard that the 1st Armoured Division had arrived. It was part of the 10th Corps, which had been switched from the Mareth front where the Eighth Army had gained and lost a bridgehead before the enemy's main position.

At half past three in the afternoon, Spitfires, Kitty-bombers, and Hurricane tank-busters came over in relays, and half an hour later the barrage started. Heavy fighting went on through the night and all the news was good. Hundreds of prisoners had been taken, many of them German. The tanks were through. They were pouring through. They were on the outskirts of El Hamma, less than twenty miles due west of Gabes. The Tebaga Gap was in our hands.

On the evening of the 27th we moved off in column of route, picking up No. 1 Platoon as we passed the ammunition point. The darkness and some patches of soft sand gave trouble but we followed the provosts' lanterns and passed safely through the gap in the minefields. We could not see the hills but we could feel them all about us. At half past two in the morning, after travelling about twelve miles, we reached the main road to El Hamma and here we halted, right in the gap, dispersing as best we could on a narrow tongue between the road and some white tapes that indicated mines. The moon was struggling over the hills, glinting on wrecked tanks and burnt-out transport. The sand round us had been scuffled up, and every yard of ground (though nothing was broken because there was nothing there to break except small stones) had that smashed and tormented look that tells of heavy and prolonged fighting. Metal embedded in the sand tinkled as we dug our slit-trenches, and the Djebel Tebagas, only four or five miles away now, scowled down on us.

That night the enemy evacuated the Mareth Line. The 30th Corps started to push towards Gabes along the coast and at dawn the advance to El Hamma continued.

It was windy in our area and blankets of brownish dust flapped disconsolately among the transport. After an early lunch No. 1 Platoon and six vehicles from No. 4 Platoon set off under Captain Gibson to join the 5th Brigade, with which a third of our second-line holding was to travel. The rest of us, thoroughly impatient now, moved at half past two, following the Divisional Reserve Group for ten miles along the road to El Hamma and dispersing before nightfall.

The next day we were told to join the 5th Brigade, which was advancing on Gabes by a route roughly parallel to the one the main body was following. By this time an out-flanking movement had caused the evacuation of El Hamma, and during the afternoon we heard that a British patrol had entered Gabes at midday.

We were off again at two in the afternoon, heading for Gabes behind the 5th Brigade. Most of us were nervous about mines, for we knew that tapes could be torn down and that even the most conscientious sweeper was fallible. During the day a Don R lightheartedly directed two No. 1 Platoon vehicles on to a minefield, on the far side of which sappers were busy with spades. They said to the drivers: 'We're burying one of our cobbers. He was talking to some infantry jokers about mines and he went into that field you've just crossed to get one. He got one all right.'

We halted at dusk and waited for almost three and a half hours with our vehicles jammed nose to tail. Then we dispersed for the night. After dark Captain Gibson's detachment pulled out with the 5th Brigade (which was again on the main axis of the advance and had been ordered to move ahead of the 6th Brigade), and the rest of us reverted to the command of the Divisional Reserve Group.

The next morning Major Sampson had trouble in finding where we were to go and we did not move until late in the afternoon. After travelling along the main axis for a mile we turned north to by-pass Gabes and follow a narrow, winding track that took us up hill and down dale and through fields of green corn. It was warm and sunny and the crops were still wet from rain that had fallen the night before. Tanks and lorries had cut swathes through them and the smell from the crushed stalks was sweet and fresh.

We reached our new area, on which Captain Gibson's detachment was converging also, shortly before dark, dispersing among pea-green hills. Four Junkers 88s were dropping bombs about a mile away and ack-ack guns were barking, but over everything there was a sort of peace.

It was that kind of evening.

. . . . .

The battle swept ahead but not out of hearing.

By the end of March the Eighth Army was in front of the enemy's next line, which was at Wadi Akarit some twenty miles up the coast from Gabes. Here the Axis made a last effort to prevent the British from linking up with the Americans. Though not as formidable as the Mareth Line the position was a strong one, its left flank being protected by the sea and its right by the now familiar salt marshes. A frontal attack was called for and while this was being planned the New Zealand Division rested.

We had day-leave to Gabes, seven miles away, and there one could bathe and listen with a mixture of envy and admiration to small children speaking their native language—French. Our welcome from the inhabitants was warm and pretty girls gave us flowers, but the Gulf of Gabes, in which we were encouraged to bathe, was cold, and as there was nothing to eat and hardly anything to drink most of us went only once to that dusty, pretty French town with the bombed jetty, the shabby, charming houses, the dilapidated green shutters, and the rusty iron balconies. It was the same at El Hamma.

In any event we were too busy for holiday-making. In four days we dumped 350 rounds of 25-pounder at each of the New Zealand gun positions—from the unit area to the field maintenance centre and back was a trip of ninety miles—and then we served a regiment of the 50th Division. That brought us to the night of 5-6 April.

In the small hours of the morning an intense barrage started and it was still going when we woke. We were under an hour's notice to move and several times we got as far as warming up our engines, but when dusk came we were still in the same area. We spent the night there, moving on the 7th behind the 6th Brigade.

The enemy was on the run now, with the armour and the leading units of the New Zealand Division in hot pursuit, but there was still a vast array of transport and fighting vehicles to move ahead of us through the narrow gaps in the minefields. By tea-time we had covered less than six miles. We were away again at a quarter past five but progress was still slow and there were enemy planes about. They had been worrying us in daylight for a week past.

We struck the Gabes-Gafsa road (Gafsa being eighty miles north-west of Gabes) at half past seven, forming up in column of route before crossing it because of minefields. We had lanterns to guide us now and the going improved. There were halts and during one of them an enemy plane dropped flares right above us. Friends could be recognised fifty yards away and the lorries stood as though in a lighted garage, casting huge shadows across the desert, but for some reason there was no attack.

We halted at half past one and bedded down for the night. Before we turned in we heard that British armoured cars had linked up with the Americans during the day.

The pursuit went on throughout the 8th but we covered only seven miles. We were now thirty miles north-north-west of Gabes and five from the coast.

The 9th, the 10th, the 11th—the next seven days in fact—stay in our minds as an interminable series of stops ('Don't wander away, chaps—they're finding out what's going on ahead.') linked together by little journeys of a mile, two miles, ten miles, all made in low gear. But little journeys add up. Following the 6th Brigade, we moved north and north-west, not always patient but happy all the time except once or twice at dusk when the Bofors barked furiously, or at night when we woke up and saw the yellow globules drifting towards us and heard the butterfly bombs gobbling in the distance.

With every mile the fields became greener and gayer. Scarlet poppies and enormous ox-eyed daisies, wild yellow pyrethrums, dandelions, mauve-coloured thistles—they spread a pinafore prettiness over the cornfields, delighting us after our arid diet of wadis. On the night of 9-10 April we dug our slit-trenches in soft turf among rabbit droppings and Kate Greenaway flowers and great clumps of furze, which later, when the flares started to fall, shrouded us in grateful shadow.

Sfax, eighty to ninety miles up the coast from Gabes, had fallen, and we travelled through wheat and barley and over rolling downland and then through olive trees—mile on mile of olive trees all so symmetrically planted that whichever way you looked you were gazing down endless avenues. We halted among them that evening —we were now twenty-two miles north-west of Sfax—and were told that we should not be moving for several days, so the next morning, after servicing the transport, we did our washing. A canvas lean-to appeared against the side of Headquarters' orderly-room lorry and courts were convened to inquire into two recent accidents.

But before the washing was dry or the findings could be promulgated we were moving through the olive trees in three columns with an air raid flashing behind us and lines of tracer showing pink in the fading light. We passed a burning lorry with a jeep on the back of it but we could hardly take our eyes from the olives. They were growing in pure white sand, smooth and tidy as a tennis-court. It lapped their roots like a clean coverlet, giving them the appearance of gnarled old paupers in a shining white hospital at evening. Their branches crackled against the canopies and our wheels ploughed tracks in the crisp sand as we pushed, like robber bands, through the strange forest.

Tall barley brushed us the next day and when we stopped for

the night we were in a cornfield. As we lay in our slit-trenches the cool green stalks bent over us, dropping beetles and fat grubs into our eyes. Being four feet high, they made our slit-trenches seem seven feet deep, which gave us an illusion of safety when the western sky started to flash and gobble.

Sousse, over seventy miles north of Sfax, had fallen, and we came by rough tracks to El Djem, thirty-six miles south of Sousse, where there was a Roman amphitheatre marvellously well preserved. Here we joined the El Djem-Sousse road, which we followed for some miles before halting and dispersing early in the afternoon.

The next morning—it was 14 April—we skimmed along the main tarmac road for half an hour, stopping two miles south of Sousse. The platoons pulled off the road to left and right and settled down among trees. It was the perfect area. Sunlight struck through the branches and fell in golden pools on patches of wild flowers and on grassy paths. Doves cooed softly and fell in succulent bundles or fluttered away to safety as a brisk volley rang out.

While the billies boiled they cooed in the woods, gently and reproachfully, in the sunlight.

. . . . .

> O! wither'd is the garland of the war.
> The soldier's pole is fall'n.

From coast to coast the enemy was facing Allied armies and behind him was the sea. His positions at Enfidaville, eighteen miles from Sousse, were about the same distance from Tunis as ours had been from Alexandria, but there the likeness ended. He had good cover, and peaks and spurs rising behind the front line provided him with observation posts.

At once we set to work to bring forward the ammunition. Before dawn on the 16th a convoy of 240 vehicles (half were ours and the rest came from the 1st New Zealand Supply Company and an RASC Company) left the unit area for a field maintenance centre near Sfax, from which 21,600 rounds of 25-pounder were to be brought forward to another centre near Sousse. The roads were good and by ten that night half the ammunition had been shifted. The greater part of what was left was diverted to our area the next day to meet a sudden demand from the Artillery.

On the 19th the unit moved forward to join No. 2 Platoon at the ammunition point, which was now a mile south of Sidi Bou Ali, twelve miles north-west of Sousse. Enfidaville was thirteen miles north-north-west of Sidi Bou Ali.

Our new area was as pleasant as the old one—the trees were as shady, the doves as abundant. Raised paths divided it into little squares of flower-filled orchard, like sunken gardens, each of which made a snug harbour for two or more lorries. Nearby, hidden by great cactus hedges, were fields of peas and beans with the pods ripening on the vines. These, we took it, went with the area. Dove and green peas—could anything be more delicious?

The ammunition point had closed on our arrival but No. 2 Platoon reopened it the next day in an area ten miles north-west of Sidi Bou Ali close to a large lake. Here the country was open and one looked across an undulating plain towards Takrouna, three or four miles west of Enfidaville, and the enemy-held hills on either side of it, pale green and primrose in the evening sunshine. Gunfire sounded hollowly among them and in places they were shrouded by a grey mist. Already the Allied general offensive was nearly sixteen hours old, and at that very moment in the hilltop village of Takrouna a small party of New Zealanders was holding a pinnacle in the face of mortars, machine guns, and grenades.

Under gathering rain clouds the hills turned to purple. Presently they were swallowed by the night but the battle went on and on, flashing and rumbling in the rain and darkness.

For a day and a night and another day the 25-pounders, jerking and smoking in the cornfields, sent over an endless stream of shells towards the hills, and endlessly our lorries shuttled backwards and forwards between the ammunition point and the unit area, the unit area and the field maintenance centre near Sousse. We carted 5152 rounds of 25-pounder on the first day and 9832 on the second.

The barrage died down and we relaxed into routine. Heavy rain had fallen, but now the weather was fine and warm and every leaf and flower was in a perfect frenzy of expansion. None of us had seen anything like it before. Here was no slow unfolding, no tentative putting forth. The flowers leapt out of their buds and were off with the first breeze—like birds. The fat pads of the cactus, obscenely glossy, morbidly succulent, seemed to be in danger of blowing up, vampire-like, and deluging everything with green blood; but already some pads had hardened and dried, horribly suggesting withered and desiccated flesh.

Plainly this spring would be over in a few hours, so we turned our attention to saving the peas and beans. After our own estates had been stripped fatigue parties were taken in lorries to a large field in the neighbourhood of the ammunition point, and here, too, the harvest was gathered in.

Picking and podding beans and peas for the community and cooking them privately in billies—with margarine they made a delicious meal even without the flesh of doves—took up a great many of our off-duty hours, and the rest were spent in bathing, for which transport to the sea was provided, and in clearing cricket and football fields. In this work a battered diesel lorry, which was in fairly good order except for a weakness of the stomach that prevented it from holding down either oil or water for any length of time and from keeping them in separate tubes when it was holding them, was of the greatest help. It puttered up and down for hours, spouting a mixture of hot oil and water and dragging a harrow made from tow chains. Like its counterpart in Workshops. another trophy of the chase, it was in demand when unofficial transport was required, and its official owner was seldom in a position to disclose its whereabouts. Often it was in Sousse or in Hammam-Sousse, three miles nearer. Not that there was much to see in either town; the former was full of bomb holes, broken glass, and vaguely deprecatory Frenchmen; the latter was a maze of small, hot alleys overhung by tall buildings and crowded with open-fronted shops in which tiny meals, compounded chiefly of garlic and olive oil. were perpetually sizzling over braziers. Or it would be fetching loaves and rock-cakes from the New Zealand Field Bakery Section. in which No. 1 Platoon had some good friends.

Sometimes our routine was interrupted by a heavy demand for ammunition or by a special job. On the morning of 27 April, for instance, 144 lorries went to the ammunition point and from there were guided in batches of sixteen to the gun positions. at each of which 400 rounds of 25-pounder were required. Dust attracted attention and there was some shelling by 88s, one of the Artillery guides being slightly wounded. In general, though, no exceptional efforts were required of us, and the days that followed, ending April and launching us into May, were lazy and pleasant.

Daily there was more heat in the sun. The flowers had faded now, peas and beans had become scarce, and the crops were the colour of new gold. The cactus leaves were powdered white with dust. Dust lay in warm drifts beside the tracks and when we returned from the ammunition point we were as white as millers.

Except for No. 4 Platoon and small detachments from Nos. 1 and 2, which were now at the ammunition point, and for Second-Lieutenant Delley[2] and ten others who were detached with the 1st New Zealand Mule Pack Company, newly formed to supply our troops in the hills, the unit was complete at this time, the drivers who had been with the French flying column having returned to us on 23 May. (Against all expectation they had few adventures to relate: there had been one air raid, one lorry had hit a mine, the French had been careless about lights.)

As always when we were together in a pleasant area and there was little work to be done the platoons were drawn close by organised games, a constant exchange of visits, and the necessity of killing time.

Time killed is time forgotten, but certain memories stand out clear and bold: evening, and Neil's old Opel Blitz and the Italian Spa from Workshops lurching towards each other across the ruts and the footballers in the back bouncing about like peas and laughing and calling out as the lorries pass . . . afternoon, and masses of khaki shirts and clean boots and a long wait for the Hon. F. Jones and people asking: 'I wonder if he'll say about going home? I wonder if the cooks are keeping the tea hot?' . . . morning, and ugly hammer-headed mules hitched to our olive trees and their masters telling us about their points and boasting of them and abusing them at the same time . . . night, and the fat thunder drops pattering on canopies and bivouacs and the warm darkness growling above the olive trees. And little things: cactus pads squashed flat by wheels, telephone wires tangled in a white hedge: the winding, crumbling road to the ammunition point and a load of salvage leaping and jangling: a Berber in a straw hat expertly fingering an army blanket: a puncture in the dust: sunshine at breakfast: beans.

[2] Capt A. R. Delley, m.i.d.; Government land valuer; Department of Agriculture, Hobart, Tasmania; born Caracas, Venezuela, 2 Sep 1916.

May the 5th. After the hot day the cool evening. Near the ammunition point the slim flamingoes settled on the lake's edge in a scarlet cloud. Gunfire no longer worried them. What they hated was being used for target practice by returning fighter planes. In the unit area there was music. A Strauss waltz played by the 5th Brigade Band stole among the olive trees, misty now with dancing midges and mosquitoes, themselves making music—a tiny, persistent shrilling. At the forward ammunition point, twelve miles west of the other point, No. 1 Platoon's drivers had finished digging their slit-trenches and some of them were making for the low hills that bounded one side of the area. From these they would get a good view of the gun flashes.

They had reached the area at lunch-time after travelling slowly along a bad road that had taken them through green and yellow hills to a wide plain. Here their lorries stood axle-deep in corn and in front of them were the wild mountains in which the fighting for Zaghouan and Pont du Fahs would take place.

The day before—the 4th—the Division had started to concentrate in the neighbourhood of Djebibina on the left flank of the Eighth Army. Its role in the coming assault was to support a French drive in the direction of Pont du Fahs, thirty miles north-west of Enfidaville, while the Americans attacked in the north, the Eighth Army in the south, and the First Army struck the main blow in the centre.

The fighting in the mountains began that night, and the next morning we were ordered to dump 163 rounds of 25-pounder at each of forty-eight gun positions. All the lorries in the unit area left for one or other of the ammunition points and at three in the afternoon they started to go forward to the guns. It was no country for motor transport, and long strings of mules loaded with ammunition and supplies for the Fighting French were plodding along the rocky, winding tracks with an air of patient disgust, which changed to anger when a gun went off near them or when one of our lorries avoided hitting them by a hair's breadth; then they snorted fiercely and erected their long ears, framing for an instant a distant foothill or a 75-millimetre gun in a gaping V-sign, which was what the French were doing, good-humouredly, with stubby fingers. The latter, in a foreign sort of way, looked extremely business-like in spite of their beards, their bizarre equipment, and the faintly exclamatory air with which they did everything, whether it was

Transport near Wadi Akarit

Dispersal, near Wadi Akarit

Tunisian barley-field

The heights of Takrouna

kicking over a motor-cycle engine (*It marches!*), making the V-sign (*Bravo, my old ones!*), or flogging a mule (*Species of an imbecile!*).

Some of the gun positions were under observation from the enemy, and as these could not be approached until after dark it was very late before many of our drivers got home. Most of them, probably, were on the road again within a few hours, for during the next three days the demand for ammunition was heavy and continuous.

When we were not working we seemed always to be congregated in or around the platoon orderly rooms, infuriating busy clerks and making them wonder why on earth it had ever occurred to them that orderly-room lorries were good places for platoon wireless sets. On Friday, 7 May, we heard that the French were in Pont du Fahs and that Bizerta and Tunis were on the point of falling, and on Saturday we heard that they had fallen the day before.

Throughout Sunday, under a hot blue sky, our lorries went backwards and forwards between the ammunition point and the unit area and the field maintenance centre, passing and repassing in a flurry of white dust; and from daylight until dark heavy traffic moving up to the forward point laid a smoke-screen across the cornfields.

The news on Monday the 10th was that large forces of the enemy, including our old friends of the 90th Light Division, were surrounded in the hills in front of the Eighth Army, British armour having swept round behind them and cut them off from the Cape Bon peninsula. During the afternoon No. 1 Platoon closed down the forward ammunition point—it was not needed now, for by this time the Division had moved back to the Enfidaville front—and relieved No. 4 Platoon in the area near the ravaged beanfields and the lake with the scarlet flamingoes.

That night a battle started in the mountains. They threw back the long echoing roar of the bombardment, and from the ammunition point our drivers could see the small foothills leaping out of the darkness as the guns flashed. The noise and the flashes went on hour after hour, and the next morning the guns were still firing. They fired throughout the day with only a few pauses, and the enemy guns answered valiantly. After dark our drivers could see

T

the yellow flash of our guns and the reddish flash of enemy shells exploding near them.

And the next morning they were still firing. It was Wednesday, 12 May, and it was the day of the races.

After lunch half the company was taken in lorries to Sidi Bou Ali where the New Zealand Mule and Donkey Turf Club (incorporated with the Mule Pack Company) was holding its first and last spring meeting. Wearing our new summer clothes (and our new summer clothes were garments in which Frankenstein's monster might have hesitated to appear in public), we milled around with five thousand other New Zealanders, shoving bundles of notes through the windows of the totalisator, losing money on Imshi (ridden by 'Sheriff' Davies) but recovering a little on Packdrill (Lieutenant Delley) and Doubtful ('Tiger' Tarrant[3]) and spending a thoroughly hot and happy afternoon. And in the distance, never quite drowned by the voice of Captain Toogood, which the public address system was diffusing over the whole field and among the olive trees, was the voice of the guns. We listened to them with quiet satisfaction, as schoolboys on the last evening of term listen to the slamming of desks. We might never hear guns again.

Von Arnim, General Officer Commanding-in-Chief, African Army Group (Rommel was safe in Europe), had been captured during the day and the enemy forces in the hills had no hope of holding out much longer, but when darkness came the guns were still firing.

Near the ammunition point, where the YMCA Mobile Cinema was screening *Something to Sing About,* a white cone of light shone steadily on the side of a three-tonner. Innumerable red pinpoints, wavering a little, glowed softly in the blue darkness, and when matches were lit sections of the audience, close-packed on rocks, on cushions, on empty jerricans, sprang out of the night haphazardly. Music and rasping dialogue competed with the rumbling of the guns, and beyond the screen, beyond the caperings of James Cagney, wild-fire flickered in the mountains. The audience took little notice of it, for the film was banal beyond belief and the business of not enjoying it occupied nearly everyone.

The film ended and the gunfire died down a little, and at a

[3] Dvr J. P. Tarrant; contractor; Pio Pio, Auckland; born Pio Pio, 1 Jul 1913.

quarter past nine it was announced by the BBC that all organised enemy resistance had ceased in Tunisia.

The guns fired fitfully through the night but by ten the next morning all was quiet. An hour and three-quarters later, Marshal Messe, Commander of the Italian First Army, surrendered unconditionally to General Freyberg. It was the 13th of May.

We were ordered to close the ammunition point, and for the last time No. 1 Platoon took the crumbling white road to Sidi Bou Ali. Heading in the same direction and riding for the most part in their own transport were thousands of prisoners. Even the drivers were prisoners in many cases. The Italians, naturally, seemed quite at home under these circumstances—they were hooting their horns in an easy civilian way and slapping along with a good deal of flash gear-changing—but the task sat oddly on the Germans. Their young, sunburnt faces were serious under the white, peaked forage caps of the Afrika Korps, and they drove slowly and conscientiously, intent on the job in hand. It seemed important to them not to scrape a mudguard.

In the backs of the lorries some of the prisoners were singing, but again steadily and gravely. They were singing, perhaps, the old battle song of the Afrika Korps:

>   With clattering trucks,
>   With engines roaring,
>   Panzers roll forward
>   In Africa.

One or two of our drivers waved to the prisoners, and many waved back, though a few were sullen.

After the surplus ammunition had been fetched from the gun sites there was no work for us. We played cricket or went swimming or stood under the olive trees near the road and watched the lorries go by. Often a lorry pulled up to allow the prisoners to get out and walk over to the ditch. It was as though a whistle had shrilled and the players were mingling on the field at half-time.

The opposing team was of enormous size, and when dusk fell the lorries were still going past our area, crushing the cactus pads deeper in the white dust by the roadside as they rolled forward in Africa.

That evening there was an issue of two bottles of beer a man. The tires hummed on the road and the mosquitoes echoed their humming in a higher key and the shadows became one shadow. And we drank our two bottles of beer slowly and appreciatively, as players, at half-time, suck lemons.

. . . .

On 15 May, at ten minutes past seven in the morning, we set out for Maadi Camp, 1864 miles away.

Engines warming up in the darkness, the shiveringly cold darkness, and moves before daylight. Overcoats shed as the sun rises, then jerseys, then shirts. Warm, dry wind blowing against bare skin and the tires on the hot bitumen making a noise like sticking-plaster being ripped off. Punctures and blowouts and plugs oiling up and bearings giving trouble.

Through Kairouan—the Prophet's barber lived there—through Gabes and the Mareth Line—a cool, green gap between the sea and the hills—through Medenine and Ben Gardane. Near Tripoli we spend the second night. The next day we dump our ammunition at a depot and then load petrol, which we are to carry for the Division. We spend two days near Tripoli and in the evenings the Kiwi Concert Party performs in our area—legs kicking under black sateen lined with white, stage a red mouth in the blue twilight: 'We're the Cancan girls from the Folies Bergères'. . . .

Jerricans full of petrol in the backs of the lorries. Jerricans shaking loose and pushing out the sides of the canopies. Vicious tugging at jerricans to unwedge them from tight rows. Jerricans being filled from 'flimsies' and the petrol spilling over the warm grass. Petrol cold and greasy on shirts and shorts.

Homs passed and Misurata. Long delay east of Misurata because the road has been washed out fifty miles ahead. Buerat passed and Sirte, and we halt near our old area at Nofilia. Marble Arch, El Agheila, and Agedabia. Benghazi, and we take aboard more petrol and spend the next day in the staging area, getting leave to town. Tocra Pass, Barce, Giovanni Berta, Derna Pass, Gazala, Tobruk, and more petrol. Bardia shining white on our left, Capuzzo, the border, Halfaya Pass, Buqbuq, Sidi Barrani, and Mersa Matruh. In front of the Lido Hotel, innocent now of rich Greeks, we splash and swim in the warm, purple water. Garawla, Qasaba, Sidi Haneish, Baggush, Fuka, El Daba, El Alamein, the Matruh turn-off, and Amiriya. The next morning—31 May—we set out on the last stage of the journey.

The road black and straight like a typewriter ribbon for miles, then the short steep climb, then round to the left, to the right, then down, then Mena House and the Pyramids, Sharia El Ihrâm, Cairo, the Nile, Maadi, the bump over the railway, NZASC Training Depot going past, the dustbowl on the far side.

The lorries halt in line, mudguard against mudguard. They sway as they halt, and one after another the hand brakes go on, croaking like frogs.

The news goes round the dustbowl like a hot wind, raising temperatures, making hearts stop for a minute and then beat faster. The lists are made out already: they will be read tomorrow; they will be read this evening; they will be read now.

Captain Gibson has papers in his hand and he begins to read:

'The following will be returning to New Zealand under the Ruapehu scheme for three months' furlough: Aicken R. B., Annan A. E., Ashton D. H. . . .'

The drivers sit quietly in the warm sand, hearing their names— the fortunate ones—but not feeling the happiness straight away, only the shock and the ache under the heart and the nearness of tears.

CHAPTER 19

DISSECTION OF AN UNDERBELLY

(1) *The Sangro*

EVERY morning a sea of pewter, burnished and dully shining. Sound of water slushing lazily in scuppers: murmur of stem and bow sighing patiently through the Mediterranean: patter of bow-spray falling on smooth swell. Already, though the sun is hardly out of the sea, the decks are alive with soldiers, most of whom suffer from a slight, an almost imperceptible, hangover, the result of sleeping between decks, the smell of oil and soft soap, the warm stickiness of the morning, the crowd in the lavatories. A rumble, a rotary impulse deep in the ship's bowels, answers an uneasiness in their own.

Breakfast in the soupy atmosphere of the troop-decks: electric lights burning: smells of porridge and of sweat and sleep: the appalling clatter of crockery: shouting and jostling of mess orderlies: dixies, warm and slippery. After breakfast, no room to move on deck because everyone has been hounded from below to leave the ship clear for inspection. Impatient waiting for 'Three Gs' to sound. Lunch, with appetites a little keener than at breakfast, and then a long, dozy afternoon, which ends with eyes and elbows aching from too much leaning on the rail and gazing seawards. Tea, but not enough of it, for appetites are ravenous now, and, after tea, cards and an examination of the day's rumours.

In wartime each troopship carries three rumours—more sometimes, but never less than three. Do they follow the ship like three albatrosses or do they sneak aboard before she leaves port? Do they crawl from deck and bulkhead like copra-beetles, come on breezes, or does the ship herself make them up, chattering through the night with doors banging, engines murmuring, heart sighing? They seldom vary in their essentials. (1) The enemy has broadcast the ship's name and her date of sailing. (2) An infectious disease has broken out. (3) Senior officers are awaiting court martial on serious charges.

These three, especially the last, go pleasantly with the cool of evening when khaki caterpillars circulate on all the decks, when destroyers fuss around laying smoke-screens, and barrage balloons (midget dirigibles that were silver earlier and are now dark like slugs) are hauled in. The distance between ships has lessened, and smoke from a dozen stacks, streaming astern in skeins, mingles in a grey net—grey shot with brown—that trails across the sea for miles. The spirit of protection and comradeship—the high, brave spirit of the convoys—is all about you. Silence then, and a light blinking quick and secret, and the drawing in, from all the corners of the sea, of the soft darkness.

. . . .

Well, it was Italy. After the doubt, the certainty. After the bustle of departure, peace and restfulness—in spite of the overcrowding, the discomfort, the indefinable malaise. After the alarms and forebodings the calm knowledge that not illness, nor the vagaries of Authority, nor the malice of fate could prevent you now from seeing the new country. Doubtless you lacked some essential article of equipment. Possibly you had missed an important inoculation. Conceivably you harboured parasites. It mattered little— they would not turn back the ship.

The bustle of departure had come at the end of a long calm, a calm starting in mid-June—in mid-June because it was not until then that our 190 ' Ruapehus ' left us after spending a fortnight in a daze of joy and alcohol—left us to miss and envy them and lose, by the process of shifting them from pocket to pocket, sending them to the laundry with our shirts, pulling them out with train tickets, innumerable scraps of paper bearing illegible addresses. After that, when the gale of long leave to Palestine, Alexandria, and Cairo had blown itself out, it was calm except for occasional squalls that spun us into Cairo for an evening's riot and sometimes into the orderly room the next morning. These occurred sporadically until our money ran out. While it lasted we were often in trouble—not bad trouble leading to courts martial but foolish trouble that caused the driver of the breakdown lorry (the unit's Black Maria) to make many trips to the Maadi Field Punishment Centre to collect people who had spent the night there. At first the provost sergeant would ask him where he came from and what he wanted. Later it was ' Huh—you again! I suppose you want T——.' T—— would be

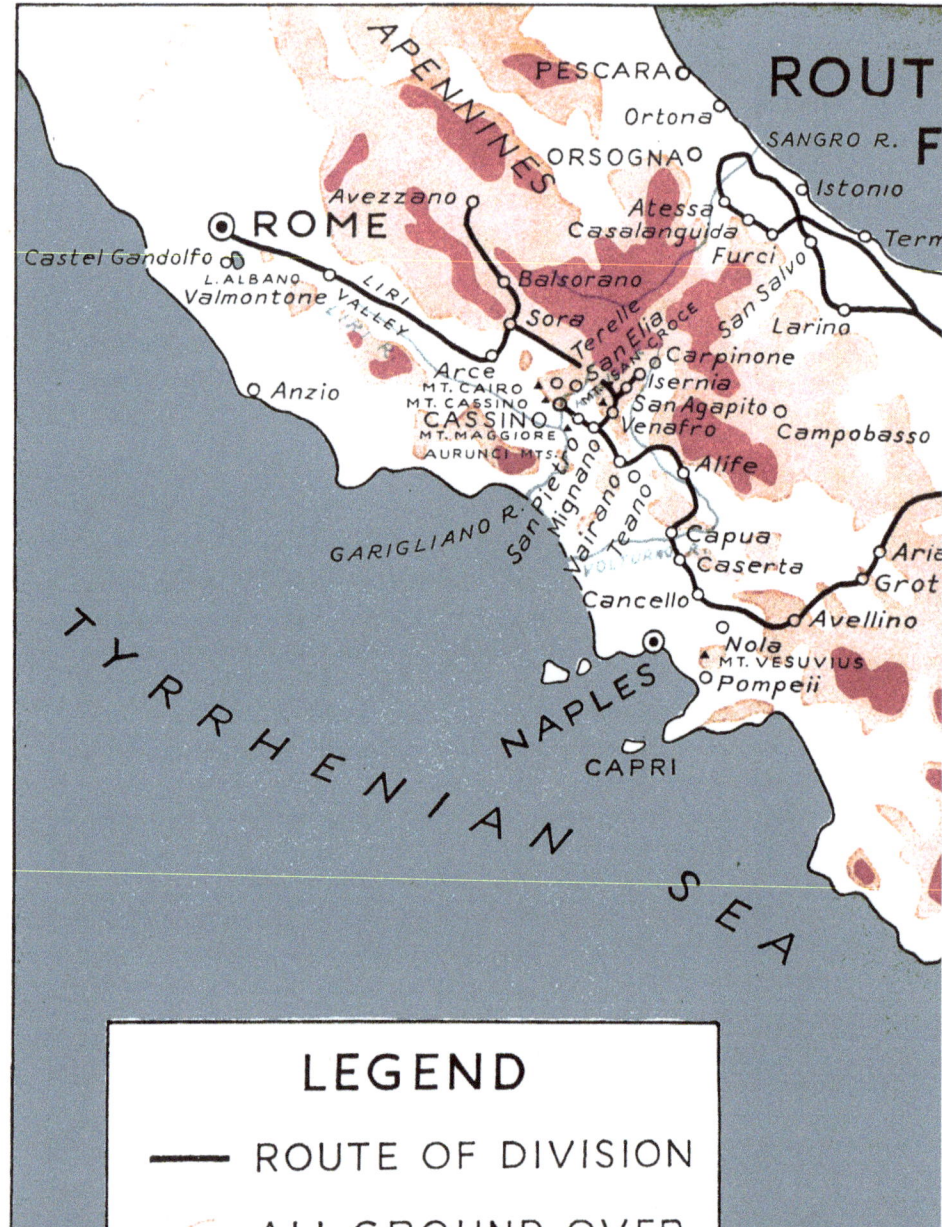

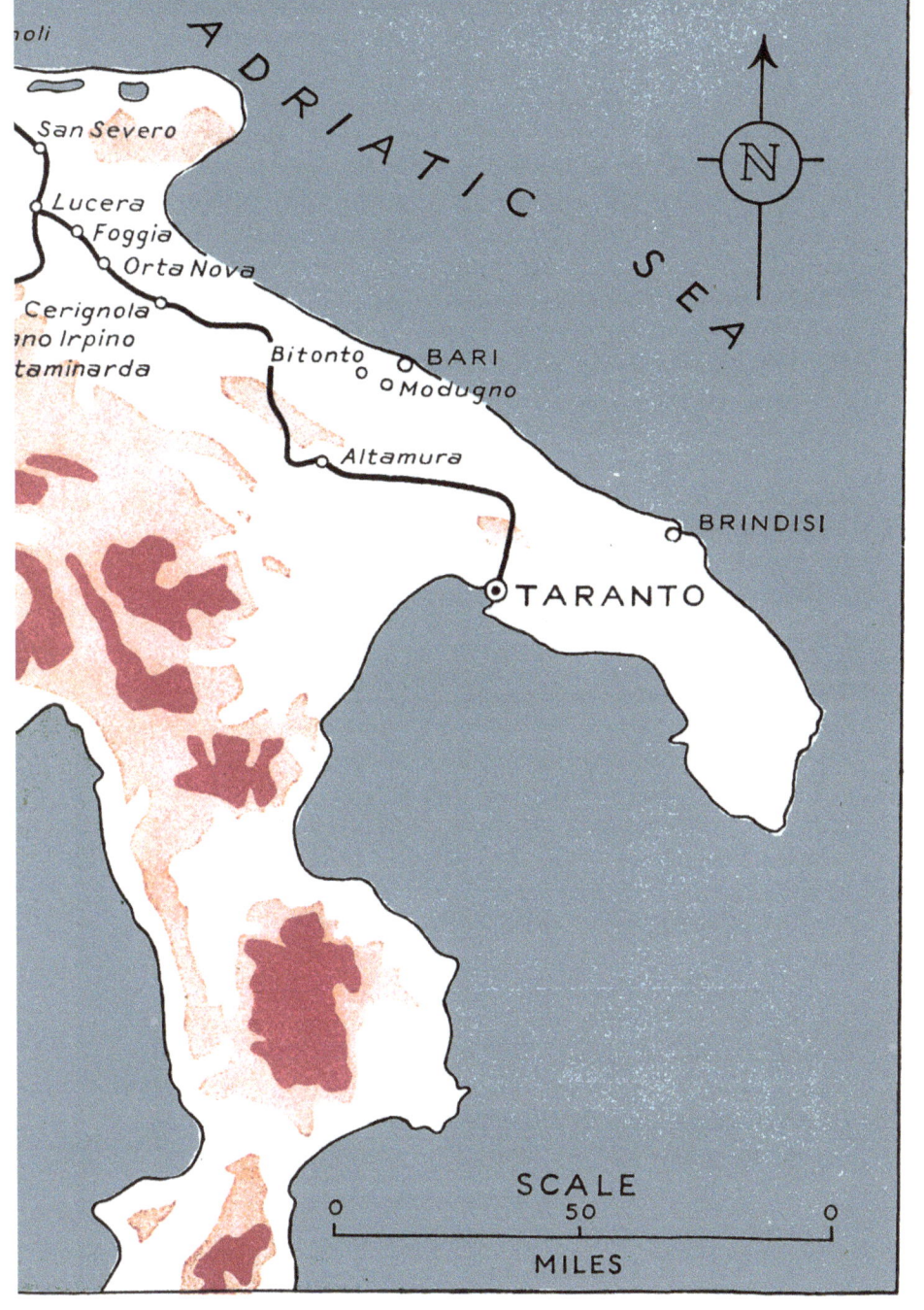

led out, scratching himself and complaining bitterly of bed-bugs.

In the second week of July 102 reinforcements were received impassively, and three days later, grumbling a little, we moved to Puttick Camp, near Mena and the pyramids, where the NZASC was concentrating. An intensive training programme began but we allowed it to interfere with our calm only a little.

Each cloudless morning was announced by the Egyptian newspaper boy with his cry of 'Very g-o-o-d news'. He had said it when France fell; he had said it, perhaps (for he was a boy only by virtue of his calling, his filthy tatters, his mocking and mischievous smile), when the German armies were surging towards Amiens in 1918, adding the improbable information that Hitler (only it would have been the Kaiser or Ludendorf then) had contracted an unmentionable disease. Cloudless morning was followed by baking day and a programme of physical training, drill, and discipline, most of which was directed at the newcomers. Each simmering evening melted into blue night, leaving us with a restlessness and longing as old as Egypt, a dark tide in the blood obedient to some star stranger and more tormented than most, a longing to make journeys—the golden journey to Samarkand, or the journey to Sicily, or the long journey home.

Daily, as the heat increased or seemed to, the programme was less arduous, and daily, as the Allies advanced in Sicily, where they had landed on 10 July, the very good news was better. On the 25th of the month Mussolini resigned, and three days later the National Fascist party was dissolved by Marshal Badoglio. On 17 August, after thirty-nine days' fighting, resistance in Sicily ended.

For a month past our transport had been moderately busy with trips to Cairo, Port Said, and Suez, and for eight No. 2 Platoon vehicles there had been a trip to Aleppo. Workshops, which had stayed at Maadi because it was easier to work there, was busy all the time. Every vehicle needed either overhauling or reconditioning.

We returned to Maadi on 27 August, and on 2 September we lost the rest of the drivers who had come overseas with the First, Second, or Third Echelons—fifty-one men. They went back to the Mena area to await their return to New Zealand under the Wakatipu scheme. The next day the Allies landed on the Italian mainland.

The segregation of the 'Wakatips', a wave of injections, the publication of the Italian armistice terms, the pervading restlessness —our peace had gone now—all convinced us that events had begun to march and that soon we should be marching too. But where? Italy? The war in Italy was nearly over. England for the Second Front? The Division had been going to England for years now and doubtless it would get there in the end—when (to borrow poetry from our newspaper boy) the apricots bloomed! Desert manœuvres? Everything pointed in that direction.

Desert manœuvres it was. That was what they told us anyway, and on 19 September Company headquarters and No. 4 Platoon moved to an area near Burg el Arab, where the Division was concentrating. The rest of the unit was to follow as soon as Workshops had finished with its transport.

The manœuvres took place without us, and what was left of September passed in a golden dream of sunshine and sea-bathing. Think what you will of the desert, of Cairo, of Egypt, but look back always with the tenderest feelings of gratitude towards the Mediterranean—our playground, our green palace from which flies and dust were rigidly excluded, our private swimming bath, our cool, our pleasant, our unfailing friend.

Nos. 2 and 3 Platoons arrived from Maadi on 26 September and by the end of the month the unit was almost complete at Burg el Arab. No. 3 Platoon had only seven vehicles and the balance of its establishment (Chevrolets) was collected from Amiriya—and a motley collection it proved to be. Most of the load-carriers—only a few were new—had neither cabs nor canopies, and although foraging parties showed much ingenuity and some lack of scruple in making these deficiencies good the drivers were still far from satisfied.

We knew now that we were going somewhere by sea and taking our vehicles with us. The unit would be divided into three flights— two of personnel and one of vehicles. The vehicle flight, which would include rather more than one driver to a lorry, would be subdivided into smaller flights, which would sail as shipping space became available. The idea was that not more than a third of any one unit should sail in one ship.

We were issued with winter clothing, with bivouacs, with two-gallon water tins, and as they handed us each article they reminded us that there would be a march of several miles at the end of the voyage. Dress rehearsals were held and they proved nothing except our inability to crawl more than a few yards under our burdens.

But where were we going? On what foreign strand were we to sink exhausted?

The answer came with October.[1] On the 3rd the first personnel flight went to Ikingi Maryut, where it was divided into two groups, both of which sailed from Alexandria on the 6th. The rest of the unit (Company headquarters and Nos. 3 and 4 Platoons) stayed at Burg el Arab until the 12th, when the second personnel flight was divided into two groups and sent to the staging area, sailing on the 18th.

The first, second, third, and fourth vehicle flights sailed on the 24th, the fifth and sixth on the 29th.

The drivers with the vehicle flights strolled casually aboard, glancing between the treads of the gangway at the filth and bobbing oranges in the black water, and dreaming happily of tool boxes stuffed with cigarettes, chocolate, and tinned fruit. The others went aboard sideways, hung with packs, valises, water bottles, and boots, stifled with overcoats and bedrolls, stuck with bivouac poles as with arrows.

And the Egyptians on the quay laughed and laughed, seeing no end to the capacity of the ruling race for making itself ridiculous and uncomfortable.

·   ·   ·

Well, it was Italy. The little ships crept slyly across the Mediterranean, leaving snail tracks in the silver water—beaten silver scaly with tiny hammer marks. The sea was calm mostly and the voyages uneventful, but not always. Lieutenant Davis's flight ran into rough weather and tanks in the forward hold broke loose,

---

[1] The chief appointments on 3 October were: Company HQ, Maj S. A. Sampson, Lt K. L. Richards (posted 2 Sep 43), WO II A. L. Salmond (appointed 10 Jan 43); No. 1 Platoon, Capt R. C. Gibson, 2 Lt R. W. Langley (posted 12 Feb 43); No. 2 Platoon, Capt R. P. Latimer (posted 2 Jun 43), Lt C. H. Haig (posted 12 Sep 43); No. 3 Platoon, Lt J. D. Todd (posted 1 Dec 42), 2 Lt G. Dykes (posted 14 Feb 43); No. 4 Platoon, Capt K. E. May, Lt A. R. Delley; Workshops, Capt A. G. Morris; Ammunition Platoon, Lt R. K. Davis, 2 Lt H. W. Boyce (posted 13 Feb 43).

The following had left us: Capt W. K. Jones (Ruapehu Draft, 4 Jun 43), Lt T. A. Jarvie (posted to Base Training Depot, 13 Jul 42), Lt R. A. Borgfeldt (wounded 14 Jul 42), Lt J. M. Fitzgerald (posted to Base Training Depot, 5 Sep 43), WO II I. McBeth (Ruapehu Draft).

threatening to smash through the ship's side. The motor vessel *Lambrook*, with Second-Lieutenant Dykes's[2] flight aboard, hit a mine when ten days out from Alexandria. A spout of water shot high in the air and came down on deck, washing an Artillery sergeant from the bridge and breaking his wrist, and soaking blankets and gear belonging to our Workshops' drivers. She limped on, with her back broken and her starboard plates rippling, to Brindisi, ten hours away.

But one by one the little ships came safely to port—those with personnel to Taranto, in the arch of the Italian boot, those with vehicles to Bari, high up on the back of the heel. The first personnel flight reached Taranto on 9 October, the second thirteen days later.

'We pulled into the wharf,' wrote Sergeant Greg Mowat,[3] who was with the second flight, ' and we could see the roofless houses, the burnt-out buildings, and the piles of rubble—a sight new to the reinforcements. Once again we piled gear on our backs, struggled up the stairway, down the gangplank, and into the new country. We left our heavy stuff on the wharf to be picked up by lorries and taken to the transit camp, which was about five miles out of town.

' The march did not seem a long one—there was so much to see. There were real stone houses with red roofs instead of wog huts made of mud and petrol tins. There were terraced vineyards, orderly and old. Real fruit grew on the trees. Only one thing reminded us of Egypt and that was the child beggars. They pestered us for cigarettes and all were mad for chocolate. Poor little bastards—they looked as if they could do with a good feed, most of them.

' When we arrived at the transit camp we found we had missed our friends of the first flight by two days—they had gone to an area near Altamura, forty or more miles north-west of Taranto. The NZASC was assembling there and we were told we should be following quite soon.

' There was leave to Taranto during the next week but no one thought it much cop. The town was dirty and damaged, the restaurants had little or no food, and the wine, though it was only

[2] Capt G. Dykes; traveller; Christchurch; born Christchurch, 26 Mar 1915.
[3] Sgt G. McG. Mowat; clerk; Wairoa; born Wairoa, 16 Dec 1914.

about sevenpence a litre, was terrible stuff. His first night in Italy old " Snow " came back to camp and all he could say was " Drunk for a bob! *Dead* drunk for a bob! "

'At the end of October Captain Gibson and fifty-three other ranks moved to Altamura and the rest followed a day or so later.'

. . .

The driver separated from his vehicle knows neither comfort nor peace of mind. He misses his home, his protection against route marches and the parade ground, the little gadgets he has contrived for his convenience. In the area near Altamura—misty mornings, bare rolling hillsides, mud and chips of stone everywhere: nothing missing, in fact, except gangs of convicts wearing broad arrows—the drivers waited disconsolately for their transport.

The first vehicles to arrive were Company headquarters' orderly-room lorry and No. 1 Platoon's cooks' lorry and water cart. The last two were needed at once by Second-Lieutenant Boyce[4] and sixty drivers who, with the help of Royal Army Service Corps transport, were forming an ammunition dump at Modugno, five or six miles south-west of Bari. That meant that the orderly-room lorry had to be turned into a taxi-cum-carrier's cart. Forgetting its usual static dignity, it dashed about collecting pay and rations and performing a hundred and one menial tasks. It even took a leave party to Bari, and it was strange indeed to be travelling for pleasure in a vehicle ordinarily so unfrivolous. An alternative means of transport was provided by the 'Altamura Express', a conveyance, ancient, crowded, and creaking, that staggered along a privately-owned narrow-gauge railway line. The engine-driver was an obliging man and for a cigarette he would stop anywhere.

Bari, with its fine harbour and magnificent buildings on the seafront, was worth a visit, and Altamura, eight miles south-west of our area, was like a town in an old story-book. Coming from a land that measures time by the extinction of the moa, we were impressed by its thirteenth-century cathedral and by a church even older. But our chief source of wonder and amusement was the Italian people. For the most part we behaved towards them as one does towards tiresome but rather attractive children; for like children they were greedy and emotional, and like children they

[4] Capt H. W. Boyce; clerk; Blenheim; born Blenheim, 20 Dec 1920.

snivelled one moment and laughed inordinately the next. Their charm, when they showed themselves charming, was child-like too.

Although the majority of us accepted the convention that all Italians had been bitter opponents of Mussolini, we remembered now and again—and the men from Bardia remembered more often —that every country gets the government it deserves. The kind the Italians had plumped for and remained complacent under for two decades had stood for bullying, boasting, and bad taste. Their vigorous apostasy, true, had a genuine ring, but it seemed not to occur to them that some deficiency in the Italian character might have contributed to their country's downfall. They had an enviable knack of dissociating themselves from the springs of their own disaster and they were not in the least put out at having been beaten at their own game by every army they had met. Possibly they were conditioned to ignominious defeat as Eskimoes are conditioned to cold and Moujiks to vodka. They shrugged off—and shrugging was something they did rather well—a quarter of a century of disaster and disrepute, and seemed to consider they had done very handsomely by everyone concerned. These were our thoughts at the time, but we might have modified them if we could have foreseen the friendliness of old men and women towards British gunners who had shelled their homes. Courage, kindliness, and patience were qualities older than Mussolini.

Anyway, it was heart-warming to find so many Sauls among the prophets. It was quite delightful to see the Wolves of Tuscany (if that was the name they had gone under in palmier days) trotting about contentedly in British battle dress dyed green—to watch the converted Blackshirt enthusiastically at work on his wall removing a DUCE from a large black VIVA and substituting a CHURCHILL, a ROOSEVELT, or a STALIN.

With this to divert us, together with camp duties and the task of familiarising our system with vast quantities of 'Purple Death', the first days of November passed pleasantly by and before long our vehicles started to reach Bari. Two flights docked on the 3rd, one on the 4th and the 5th, and two more on the 6th. On the night of the 6th-7th, while vehicles from Captain Delley's, Captain May's, and Second-Lieutenant Dykes's flights were being unloaded, there were six or seven alerts in the port area, three of which developed into raids, bombs being dropped among shipping and mines sown in the mouth of the harbour. Flares and ack-ack shells filled the sky and our drivers came on deck to enjoy the fun. Aboard the *Lambrook* enthusiasm diminished when a large piece of shrapnel made a hole in a temporary galley and another brought down the ship's barrage balloon. No damage was done to shipping and, except while the raids were actually taking place, the unloading of the vehicles was carried on beneath a smoke-screen.

By the 10th, though not quite complete—it was not that until 3 December—our unit was ready to function. We drew our second-line holding of ammunition from the new dump at Modugno and stood by for the order to move.

．　　　　　　　．

As long as Hitler held Rome and Mussolini he could claim that the Axis was still firing on all three cylinders. At present he held both. The former had been occupied on 10 September and the latter rescued from the *Carabinieri Reali* a few days afterwards. Mussolini's company Hitler could hope to enjoy for some months, but Rome—or so it seemed in the autumn of 1943—was another matter.

Naples had fallen to the American Fifth Army on 1 October, and by early November, when the Division started to join the Eighth Army on the Adriatic sector, the German line stretched across the Apennines from the north bank of the River Sangro, 130 miles north-west of Bari, to the mouth of the Garigliano River, thirty-five miles north-west of Naples. The plan was for the Eighth Army to attack between the mountains and the Adriatic, the New Zealanders' task being to cross the Sangro, cut the enemy's line, and advance quickly to threaten communications with Rome from the east.

At 8 a.m. on 12 November, Company headquarters, No. 1 Platoon, and Workshops—the other platoons were to follow—left the Altamura area for Lucera, eighty miles to the north-west as the crow flies but well over a hundred by road. Beyond Modugno. where we turned north to follow a road running roughly parallel to the coast, the country was all new. Mostly it was planted in trees and criss-crossed by low stone walls and dotted with *trulli*, stone summer-houses with roofs shaped like pudding basins, bee-hives, or ice-cream cornets. We passed through Cerignola and Orta

Nova—the kind of places you see peeping at you from the faded blues and greens of an old tapestry—and came to Foggia, twenty miles from the coast. The town had been bombed into prominence some months ago but we had not expected such complete ruin. Whole floors—great slabs of concrete and steel—lolled out at us from smashed factories like tongues. Lucera was eleven miles away and our area was five miles north-west of it.

On the 13th the 4th and 5th Field Regiments moved up to the front, and with the latter went Captain Gibson and seventeen vehicles from No. 1 Platoon and thirteen from No. 4 Platoon. Our drivers spent the night six miles north-east of the little village of Furci with the Sangro only twelve miles away. They could hear the guns. On the 14th the New Zealanders, represented by the two field regiments, and the 19th Indian Infantry Brigade (under command)—the rest of the Division was to arrive during the next ten days—took over a section of the line about thirteen miles from the sea.

Soon after two o'clock D Troop of the 5th Field Regiment fired the New Zealand Division's first round of the campaign from the neighbourhood of Casalanguida, five miles north-west of Furci. Two miles away there was a steep, winding hill, down which, in the fading light, came Captain Gibson's transport, brake-drums squeaking, rain drumming on canopies, and mud and water, the colour of weak cocoa, swishing under mudguards. Shells were landing in the valley below, and one whined over the road and hit the hillside nearby.

The enemy, obviously, was aiming at a narrow, unblown bridge at the foot of the hill, and before crossing it our drivers were warned by provosts to travel fast and at wide intervals. An artillery quad, abandoned and burning, pointed the moral. There were some anxious moments, but all got safely across, and the lorries were dispersed for the night one mile south of Casalanguida and about 500 yards from the unblown bridge. There was cover for most of them on a hillside dotted with scrub and oak but some had to be parked in the open. However, it was hoped that the gathering darkness and the grey curtain of rain would hide them from the enemy.

Platoon headquarters moved into a one-storied red house with four or five rooms, the cowshed being requisitioned by the cooks, who set to work at once. By the time tea was ready a wood fire was roaring in one of the rooms and a row of boots steamed on the stone hearth. The dixies of hot food, the loping shadows, the dancing flames lighting the drivers' faces—young, eager faces most of them—made a cheerful picture. There were cheerful noises too: the muffled thunder in the great chimney, the clink of spoons, the excited talk.

'Jerry can't be more than a mile or two away. You could hear those guns going plain as one thing.'

'The roar when she came over—like a train. Then the explosion right among us. We were tinny all right.'

The old hands affected indifference, or said (truly perhaps, for some of them had been in the field a long time) that they didn't like the look of things, but the newcomers were plainly delighted. Many of them had been bitterly disappointed at not being drafted to a fighting unit with their special friends. They felt better now. Driving the old lorry might not be so tame after all.

The flames died down, voices became sleepy, and in twos and threes, dashing through the wet darkness, the drivers went to their lorries, leaving the red house to Company headquarters and the Italians who owned it. They slept until half past one and then they were woken by mortars, grenades, and bursts of machine-gun fire.

By the unblown bridge Bren guns and spandaus were in angry argument and there was a confusion of tracer fire. Everyone was ordered to stand-to, and with hearts beating fast the drivers felt for weapons and bundled on their clothes. Some went to the trees and others crouched under the lorries. Rain was spilling out of the darkness and sheets of it slapped against the lorries and whisked spitefully beneath them. Soon everyone was wet through.

The action went on for about half an hour and then the Indian guards managed to drive off the patrol that had come down from the hills to blow the bridge. Our drivers, stiff with cold and too tired to talk, went straight to bed, the newcomers, no doubt, remembering the infantry in the rain and reflecting how pleasant it was, even if a little inglorious, to have a bed to go to.

The next morning the ammunition was dumped and the convoy left for Altamura to reload.

. . . .

On the same day—the 15th—Company headquarters and Work-

Back to Base at Maadi

Pastures at Lucera, Italy

Winter in Italy

Monte Maiella

shops moved from Lucera to an area near Larino, a small town twenty-seven miles south-east of Casalanguida, and during the next few days the platoons, helped by detachments from other units, cleared the old areas of ammunition, brought troops to the front, and established a reserve dump of 25-pounder ammunition four miles south-west of Casalanguida. On the 21st we moved to an area near there and opened an ammunition point.

These trips had been made over vile roads among mountains, the sides of which were covered with scrub oak, all scarlet and gold and orange. Yellow leaves, like largesse, drifted through the autumn air and our wheels ground them to sludge. Every hilltop supported a little village, and always from one or another of these (and at Angelus from all of them) came the sound of bells. Staggering up and down precipitous hillsides, sliding round corkscrew bends, passing everywhere reminders of Christendom's first casualty—with the Nails, the Hammer, the Spear, the Sop of Hyssop faithfully and often horribly reproduced—we drove through a mist of leaves, of bells, of rain—grey and savage sometimes, sometimes gentle and golden like autumn.

On the 25th, leaving No. 1 Platoon and Workshops to follow later, the unit moved to a windy slope about a mile from Casalanguida. Here, for the first time in weeks, we were able to give our vehicles more than the minimum attention necessary to keep them going. The bad weather, the late arrival of some of the load-carriers, the appalling hills, the shifting from area to area—these had placed drivers and transport under a severe strain and we had done well to have only eight vehicles off the road since starting work.

While we were servicing our vehicles and resting—the term is comparative—the battle of the Sangro started. On the night of the 27th-28th, under black rainclouds, the Eighth Army struck with three divisions. New Zealand infantry, on the left flank of the advance, waded through the icy Sangro and by daylight were established firmly on the north bank.

On 3 December, while the 25th Battalion was fighting desperately in the little town of Orsogna, which was blocking the entire advance —it was on a ridge eight miles from the starting point—we were told to establish an ammunition dump on the north bank of the Sangro so that it would matter less if floods or enemy action destroyed the bridges. Two days earlier we had moved a mile or so north-east to high ground, and now we had a wonderful view of the Sangro Valley and we could see our bombers at work. On the 3rd, then, forty-three lorries (fourteen from No. 1 Platoon and twenty-nine from a platoon of the 4th Reserve Mechanical Transport Company), preceded by Lieutenant Todd and Second-Lieutenants Langley[5] and Boyce, who were to supervise the dumping, set out for the Sangro with 25-pounder ammunition, heading for a Bailey bridge ten miles north-west of the area. Later they were followed by a further 117 lorries.

Rain was falling and the river was said to have risen eighteen inches in twenty minutes. It was tumbling and snarling and beyond the bridge it had overflown its banks, confronting the convoy with a formidable water hazard. Three vehicles crossed under their own power but the rest had to be winched over by a tractor. The water covered the floor-plates and whenever you stopped you could feel your lorry settling under you. The current made steering almost impossible and two vehicles came to grief. One hit a partly submerged bank, driving the fan through the radiator, and another plunged into a dip and had to be abandoned until morning. At one stage a jeep was overturned by the current.

The rain came down in torrents and everyone was soaked. Drivers floundered about waist-deep in water, and a padre, stripped off and looking the picture of muscular Christianity, did a spirited job with ropes. Finally, when all but seven lorries were across, the river became impassable and the plan of returning to the unit area had to be abandoned. Our drivers settled down for the night at the ammunition dump with the guns flashing and barking round them in the wild rain.

With its canopy drawn down, each lorry was as cosy as a lighthouse—as a cottage on the dark heath where Lear wandered mad and lost. Our drivers lay in bed warm and dry reading Auckland Weeklies and Readers' Digests by bedside lamps, a load of tinned sausages and cocoa rumbling inside them.

There was something to be said for a transport unit after all—even the newcomers admitted it.

[5] Capt R. W. Langley; clerk; Wellington; born Masterton, 14 Apr 1921.

The first attack on Orsogna ended in partial success only, and so did the second, third, and fourth. While they were being made and afterwards while the Division was engaged in what was officially known as 'offensive defence'—this period lasted until mid-January—the platoons, having completed the Sangro dump by 8 December, were employed solely in replenishing the ammunition point in the unit area from a field maintenance centre near San Salvo. It was only a dozen miles away in a straight line, but we went by a roundabout route and it took us a full day to go there and back.

This was a strange time—over us the weeping skies, under us the mountain roads, round us the ubiquitous mud. Mud was now our element. It sucked at our boots, sometimes pulling them right off; it banked up beside our labouring wheels and mounted to the differentials; it collected in pools and lay in wait for us when we stepped out of the cabs; it was thick on lorry floors and it found its way into our beds. It was so much a condition of life that we seldom thought of ourselves as dirty because we were muddy or as uncomfortable because we were conscious all the time of the feel of mud—mud wet on the handles of ammunition boxes when fingers were crushed by cold as by a vice; mud caked and dry—a woolly and unpleasant feeling—when we were warm after loading. We knew always the taste of mud, cold, mouldy, abrasive; always the look of mud, glaucous, glutinous, froggy; always the weight of mud (we were shaggy with it like old ewes), and always either the chill of mud (trouser-legs being steeped in it as in icy goulash), or the sensation of wearing (sun or primus having dried them) cardboard clothes.

But we knew other things as well—rare, bright mornings when we hurled the ammunition aboard at a mad pace to keep warm; wood-fires leaping and dancing in great hearths; the delicious drowsiness as you thawed out; the surprising friendliness of the *contadini* (the country-folk); evenings when the red wine went round in bucketfuls; birthday parties when two lorries were parked tailboard to tailboard.

On 18 December we moved four or five miles north-west to an area near the little village of Atessa. Here we took possession of several acres of the best mud. Cooks and Headquarters' drivers found homes in sheds and buildings but the rest had to live in their lorries—islands of chilly discomfort in a mud ocean. Atessa itself was a huddle of cold houses—cold and forbidding, that is, from the outside but within wonderfully warm and welcoming. Three houses—*casas* we had learnt to call them—were known as 'The First and Last', 'The Pig and Whistle', and 'The Family and Naval'.

The weather worsened towards Christmas and it was lovely after struggling with icy wheel-chains, after slipping and sliding for miles along greasy roads, after escaping a dozen times from being pushed into the ditch, to draw up beside a fire and take off sopping boots, and with new wine and old songs become progressively merrier until bedtime.

Christmas Eve dawned bleak and cold and in the afternoon it began to rain. We were very busy just then and most of us were out working, but everyone was in high spirits, looking forward to parties that night. Joyfully we flung on the last boxes of 25-pounder and turned home, where a heap of parcels, some private and some Patriotic, awaited distribution.

It came as a blow when we were told that a job impended and that anyone who drank too much could expect trouble. Possibly we ought to have cancelled our parties then and there but this was Christmas Eve. We compromised by saying that we should have a few and see how things went. No job materialised and things went very well.

Darkness came, and tiny pencillings of light, thin as a hair and invisible at ten yards—the old hands saw to that—showed where parties were in progress. Later the singing started.

. . . . . .

Nowhere is the laughter louder or the company better than in the 'Family and Naval' (No. 3 Section's house of call) where old Italian Poppa—'That the feast might be more joyous, that the time might pass more gaily, and the guests be more contented'—has fetched us his best red wine and is now making a speech, the audience applauding loudly whenever he says *Buona fortuna* (Good luck) or *Buon Natale* (Merry Christmas)—the two phrases they understand. Poppa's nut-cracker face is ready to split in half and fat Momma beams, too, and the daughters of the house, the cripple Nina and Alice whose husband is a prisoner of war, smile gently. Nicky, aged 16, has shining eyes for the soldiers.

Tiny scarlet candles burn under a picture of the Nativity and in the stable through the door sheep and great oxen sigh gustily and stamp. Pendulous shadows cast by bunches of onions and raisins nod vague approval as Poppa finishes his speech amid a chorus of *bravos* and *grazies* and then drains his tumbler. H—— rises to his feet and says with a perfectly straight face: 'Thanks, Poppa, you silly old ——. May your. . . .' What he hopes for his host is indecent and not likely to happen, but Poppa, seeing no malice in any of the faces, only shining happiness and his red wine, breaks down, wrings H——'s hand, sobs '*Grazie! Grazie! Grazie!*'

The room gives a turn and a half like a dog and by the time it has settled the two girls are singing an Italian love song in sweet, husky voices—'*Ma L'Amore No*'. The visitors bawl 'Maori Battalion', bawl 'Silent Night'. Poppa's asleep and snoring and the room gets hotter, mistier, noisier, spinning for some, for others rocking gently or floating loose in a gold cloud. Sometimes the tiny buds of light on the red candles bloom like tulips, bloom and multiply—a bank of tulips filling one whole side of the small, drunken room. Sometimes they shrink and shrink until they are swallowed up by the gloom over the cheap fretwork shelf in the corner, and then there is nothing left of those tall tulips but two imprints on the retinas, two echoes of light, two drowning motes, each smaller than a seed.

And next, or an hour later, everyone is outside in the cold and the sea of mud, lost and drunk, with the 'Family and Naval' hidden and the lorries hidden. Old George is down and muddy from head to foot and someone else is hung up on a tent-rope. Drivers stagger round in circles, roaring: 'Where's number one sub?' 'Where's Neil's lorry?' 'Where's number two?' 'WHERE'S MY BLOODY TRUCK?'

Silence and peace at last, and out of the cold darkness, welcomed only by a few hardened topers garrulous over demijohns, comes a Happy Christmas, a *Buon Natale*—the fifth since 1939.

. . .

It rained on Christmas morning and the rest of the day was dull and cold, but we enjoyed it. We enjoyed the dinner and the Canadian beer and the nuts, figs, and wine bought from regimental funds. On Boxing Day the unit diary was laconic: 'Very cold. Troops recovering from Christmas.' Work started again on the 27th.

December ended in a howling gale and we woke on New Year's morning to find the world white. Fronds and feathers were swirling from a sky dark like slate and beyond the Sangro the Apennines were blancmanges and cloud-mountains.

Not all the North Islanders had seen snow before—snow falling, anyway—and some of them were tremendously intrigued and excited. The South Islanders, of course, were quite at home in a snow kingdom and their manner was proprietary.

Much damage had been done during the night by the wild weather. Workshops was flooded and tents and bivouacs were down. Many of the roads to the Sangro were closed and after breakfast all hands were set to work with shovels. We took jerricans of vino with us and there were snow fights, but we shovelled with a will and soon the road outside our area was open again.

On 2 January we carted 8000 rounds of 25-pounder ammunition from the Istonio railhead, fourteen miles due east of our area, to the dump by the Sangro, and from then on we did the same thing every day, travelling by a roundabout route. The roads were crowded and filmed with ice and each trip took from dawn till dusk, but for all that we found time to court death and disaster on home-made toboggans. The snow lay for a week.

By the 10th we knew that the Division was to be withdrawn for a rest, and five nights later, after dumping our ammunition and removing Divisional signs, we headed north-east towards the coast with other NZASC units. It was a beautiful shining night with the road hard and frosty and a highwayman's moon overhead and the trees like lace—just the night for the Captain to stuff a brace of barkers in his skirted velvet and go riding. Our destination was secret and that gave us a feeling of adventure although we were heading only for a rest area.

We breakfasted on the coast road and then drove towards Bari until we reached San Severo, sixteen miles north-west of Foggia. Here we were joined by No. 1 Platoon, which had been to Bari for engines. The day's journey ended by the roadside three miles north by west of Lucera and the news flew round that nine vehicles from No. 2 Platoon had stopped at San Severo to load grain for Naples. When we asked our officers if that was where we were going they put us off. The next morning, however, they told us that we were on our way to join the Fifth Army and would be stationed north of Naples. The move was still a secret and all towns and villages were out of bounds.

We went south for about fifteen miles and then headed southwest across the Apennines, passing a string of places with names as pretty as girls' names—Ariano Irpino, Grottaminarda, Avellino. Gone were the barren, treeless stretches we had known round Foggia and near the coast; instead a pattern of little fields went up into the hills and mountains, stopping only where the snow started. Here the country was two months nearer summer.

The road was dry and good but we travelled slowly because other New Zealand units were on the move as well. Often we halted in crazy, charming villages that seemed to be struggling not to slip into the valleys below, and while the noses of vehicles pointed up or down at fantastic angles the villagers crowded round trying to sell oranges and apples and bad wine.

We spent the night sixteen miles east-north-east of Naples, and Vesuvius with its perpetual plume could be seen plainly. The next morning we passed through Cancello, whose great railway yards were in ruins, and Caserta, famous for its royal palace. We turned north soon afterwards, crossed the Volturno twice, and long before lunch were in the new area with the transport dispersed on dry, grassy slopes. We were now twenty-nine miles north of Naples and rather more than a mile from the little walled village of Alife. Vesuvius was hidden from us by mountains and so was another volcano destined to be famous—Monte Cassino.

． ． ． ． ． ． ．

During the next fortnight we enjoyed ourselves. The weather was warm and sunny and we played Rugby football. No. 1 Platoon did best in the inter-platoon matches and we beat the 2nd Ammunition Company 11-nil in the only extra-unit match we had time for. In the evenings, thanks to Americans of the Fifth Army who were using Alife as a rest area, we went to films and concerts. There was day-leave to Pompeii and on the way to it we caught glimpses of Naples, which was out of bounds. The little we did see was sad and shocking, reminding us of a stately mansion festering into tenements, of a lovely woman drunk and on the streets. The people who lived there—squalid children, old crones, sluttish beauties, young loafers Sydney-flash, miserable old men in the cigarette-butt industry—alternated between whining hopelessness and a sort of gamin gaiety, desperate and ferocious. It was doubtful which was the less pleasing. As in duty bound—starving people are intolerable to men who are getting three square meals a day—we voiced our indignation and disgust, but without inner conviction. Few of us really blamed them for having sunk low, for being so hungry that nothing mattered. We did what we thought we had to in our own way, and children and old people and cripples found us not uncharitable.

Poor starving Naples! She had been preyed on by the Germans —they had thought of the delightfully German trick of linking the city sewers to the water mains—and now she was the prey of Italian sharks and tigers. And all the time she was mocked by the smiling beauty of her bay and by her jewel, Capri.

The towns and villages in the surrounding district were only a little better off. When we picnicked near the ruins of Pompeii the alternative to being stared at by a hundred yearning eyes was to eat our bread and bully in the back of a vehicle with the canopy drawn down, and it was the same at Nola, fifteen miles east-north-east of Naples, where we picked up our second-line holding of ammunition.

Alife, tucked away in the country between crumbling walls, was in better case. A crowd of women and children, each with two tins —one for meat and vegetables and another for tea and sweet things —picketed our refuse pits, but they embarrassed and annoyed us only when the meal was so good or so scanty that there was little left for them. That happened seldom. The cooks gave away a good deal and in return the men dug pits and washed up and the women washed and mended. The toddlers repaid us by lisping our Christian names and hanging around the camp.

January ended, and Alife and the rest of the free world (as we liked to call it) were flooded by a warmth of optimism, making us forget disappointments on the Eighth Army front and in the

Dodecanese. Everyone, everywhere, seemed to expect great things. The hounds of spring were still snoring in their kennels, but sometimes of a sunny morning—those primrose mornings that blossom in late winter—it was as though one of them had put out a tentative paw, withdrawing it a moment later on finding the world not ready. Spring was what we longed for—spring and a surge forward to Rome, spring and the Second Front, spring and an end to this long war and to our long journey towards Christmas.

During the first week of February we heard that the Division would be in action soon. There was a word spoken, and it was Cassino.

(2) *Apollyon in the Path*

Unlike El Alamein, which from being nothing but a railway station in the desert became almost overnight nearly the whole world, Monte Cassino had enjoyed moderate fame for some centuries before it was given the freedom of every newspaper, every tavern, and every sound wave.

In the fourth century before Christ the Romans founded a colony on the banks of the Vinius (now the Rapido) and called it Cassinum. Though destroyed by Hannibal in 216 BC it grew into a town of luxury villas, and Mark Antony for one had some pleasant nights there.

In the sixth century St. Benedict came to Cassinum, changing the statue of Apollo for the Cross and the pagan temple for a church dedicated to St. Martin. That was the beginning of the Benedictine Order and of Cassino monastery.

Times were no better then than now, but with two great saints to watch over it (one the patron saint of bachelors, the other of millers) the monastery could expect great things or at least a fate different from this: to be destroyed by Lombards, sacked and burnt by Saracens, besieged and taken by Normans, smashed by earthquake, pillaged by French soldiers, and at last turned into an observation post by German officers.

Sharing house-room with chalice, altar-cloth, and illuminated missal, the Germans trained their field-glasses on American positions and spoke into their telephones. Catholic gunners volunteered to bombard Monte Cassino but for some weeks the Allies vacillated, loath to do irreparable damage to a place so venerable and so sacred.

Fifth Army troops, meanwhile, had landed at Anzio, thirty-two miles south of Rome and sixty-two miles west of Cassino, biting into the enemy's right flank and straining towards his lines of communication. The plan was to join hands with the rest of the Fifth Army and capture Rome, but success was unlikely as long as the enemy held Cassino and Monte Cassino (Monastery Hill).

Most New Zealanders know Peter McIntyre's fine painting of the scene. Monte Cairo of the almost perfect white cone is out of sight, but you get Monastery Hill (with the monastery under smoke), and you get, forward of this and a little to the right of it, Castle Hill. The once-white, once-bright town that straggled over the slopes in the foreground is represented by tooth-like stumps, for the painting was done after the bombardment.

This, then, was Cassino. Like foul Apollyon it was straddled right across the way.

. . . . . .

The Americans had done well. They had captured at heavy cost one key to Cassino, Monte Maggiore, but they had failed to make headway against the town or Monastery Hill and now they were no longer in shape to continue the battle. Therefore, on 6 February, the New Zealand Corps—a strong force that had been formed recently and which included the 4th Indian Division, British, Indian, and American artillery units, and some American armour—began to take over their sector.

A few days earlier Lieutenant Delley and one ammunition platoon had opened an ammunition point thirteen miles south-east of Cassino on the famous Route 6, highway to Cassino and Rome. A plan to establish a dump only four miles from Cassino had been abandoned and ammunition intended for this was diverted first to the point and later, when that was choked, to our unit area. On 7 February No. 1 Platoon started to establish a Corps dump a few miles from the point, but late that night was ordered to stop, the new plan being to form an artillery dump (with the code name of SPADGER) in the 6th Brigade's area. This was on Route 6 and about six miles south-west of Cassino. The next day we moved to an area nine or ten miles west by south of Alife so as to be near the main road. We called it the Vairano area, borrowing the name from the nearest village. During the next four days the transport was employed in building up Spadger dump with ammunition brought from Capua (eighteen miles north of Naples on Route 6) and from Nola, and in replenishing the point. At this time the artillery was shelling Cassino and targets south of it.

Our Indian summer was over now and the weather was horrible and the new areas seas of mud. The roads were in a bad state, too, and we were allowed to use them only at times laid down by Fifth Army Movement Control, convoys of from twenty to twenty-five vehicles being released at intervals of a quarter of an hour. This was irksome but it prevented accidents and saved time in the end.

From 12 February on we were able to draw ammunition from the Teano railhead, some six miles south-east of the Vairano area, and that halved our work by eliminating the long trip to Capua and the much longer one to Nola. There was no way of halving the rain or of doubling the size of our muddy and congested areas—they were small because a great mass of transport had to be parked close to Route 6—and there was no way of dealing with contradictory orders except by obeying them. They were contradictory, or seemed so, because the work of one insignificant transport unit had to be dovetailed with a large and ever-changing plan; but sometimes we forgot that. And remember, please, it was raining.

No matter. When the New Zealand Corps went into action against Cassino it would have all the ammunition it needed.

. . . . . .

February the 15th was a fine day. The sky was egg-shell blue and against it the summit of Monte Cairo stood out like a splash of whitewash. The monastery, 4000 feet lower but still seeming to be perched among the clouds instead of on its dun-coloured ridge, could be seen clearly from the hills behind the ammunition point. Here a group of our drivers was watching with field-glasses.

The first flight of Fortresses wheeled in the sky—very slowly it seemed—and dropped their bombs. The watchers saw great white mushrooms sprout on and around the monastery and more than a minute later they heard a sound—a hollow rumble as of thousands of tons of gravel dropping into a steel barge. Waves of heavy and medium bombers came over and the monastery was destroyed. Only its great walls were left standing.

Two nights later, while Indian troops fought in the mountains north-west of Cassino, Maori infantry crossed the Rapido south of it and advanced to the railway station. This was the New Zealand Corps' first direct assault on the Fifth Army front, and the plan was to take Cassino and Monastery Hill so that American and New Zealand armour could enter the Liri Valley, one of the gateways to Rome. All night long our artillery fired in support of the attack and by eight in the morning 17,000 rounds of 25-pounder had been cleared from the ammunition point in seventeen hours. Throughout the day the heaviest demand was for smoke shells and these were responsible for the white mist that clung to the slopes of Monte Cassino, blinding observation posts.

The Maoris fought in the railway station until four in the afternoon and then German tanks appeared. Our own tanks, in spite of heroic efforts by the engineers, had been unable to come forward because of demolitions and bomb craters, and the Maoris had no choice but to withdraw across the river. The Indians, after fighting supremely well, had failed, too.

. . . .

Apollyon was still straddled across the way and the position of the Allies at Anzio was serious.

A little grimly, and with no further thoughts of an immediate dash to Rome, we settled down to hard, slogging routine. On 19 February No. 1 Platoon moved to an area next to the ammunition point, which from then on it was responsible for replenishing, each lorry fetching two loads from the unit area every day. The rest of the transport worked between the unit area and Teano railhead.

Daily we cleared 105 loads from Teano and by the end of the month the position was very sound. Besides having met the hour-to-hour requirements of the entire Corps—with the help, of course, of the 2nd Ammunition Company and of transport attached to us —we had accumulated at the ammunition point stocks equal to our entire second-line holding.

But it was dull work and cold work. We were seldom warm and comfortable except when we were crouched over primuses or tiny charcoal braziers in the backs of our lorries with the canopy raised just enough to prevent us from being suffocated. Dry boots were the most important things in our lives—dry weather we had ceased to expect.

For some of us the monotony was broken during the second week of March by the establishment of a forward reserve dump on Route 6 about seven miles from Cassino. The nearest village was called San Pietro and the whole area was under enemy observation.

The first convoy—fifty lorries loaded with 25-pounder—went forward after dark on the 9th, the drivers taking with them enough green branches to camouflage their loads. Everything went smoothly for two nights but on the third there was a bright moon and some shelling. No ammunition was brought up on the fourth night and the transport was engaged in shifting some that was already there, neighbouring units having complained that it was too close to them. A night seldom passed without shells landing near the dump, and the party in charge of it—a detachment from No. 4 Platoon under Captain May—used to retire to the ammunition point at sunset and stay there until dawn.

On the night of the 12th-13th the weather changed. A high wind got up, scattering rain clouds, levelling tents, and playing such puckish tricks as rolling an empty hogshead all the way from Workshops to Headquarters. The morning dawned clear and sunny and the next day was even better.

The 15th was another good day, and at half past eight the bombers came over, heading for Cassino. They came over in tight formation, wave after wave—Fortresses, Liberators, Mitchells, Marauders. Our eyes ached from counting them. 'Look! More Forts! Five-six-seven. . . .' The sky sang with engines and every quarter of an hour or so we heard the long mutter—the long, collapsing mutter—of bombs. And the singing and the applause—thunder of feet and voices, far off, from some appalling stadium—went on for four hours. During that time more than 500 heavy and medium bombers of the American strategic and tactical air forces dropped over a thousand tons of bombs in an area of less than a square mile, and Cassino, which at dawn had been just another badly-battered town, was reduced to rubble. All artificial landmarks—Continental Hotel, Hotel des Roses, Botanical Gardens, Baron's Palace—disappeared. They lived on as names because one heap of masonry had to be distinguished from another but they weren't there any longer. There was nothing but stones and splintered wood and huge craters and rubbish.

This time the New Zealanders were to attack the town from the north, and the 5th Indian Brigade was to move in behind them and take Monastery Hill.

At noon, advancing behind a creeping barrage, the 6th Brigade entered the town. At first the opposition was slight, but later the Germans resisted strongly. They were paratroopers, some of the best soldiers in the world.

By evening we had Castle Hill and most of the town. Then it rained.

We lay in our lorries in our safe areas and it came down in bucketfuls. It thundered on canopies, rushed in rivulets between stacks of 25-pounder, filled petrol tins cut in half for wash-basins. It roared and gurgled and it tinkled like cracked bells, and we lay in our beds cursing it. We hoped and prayed that it would make only a little difference; but it made all the difference.

Instead of moonlight there was darkness and roaring confusion in Cassino. The Germans, who knew the town inside out, asked nothing better than this, but our men were blinded and bewildered. Engineers, struggling to bridge craters fifty to seventy feet wide to clear a path for the tanks, found their bulldozers almost useless against rubble stiff like wet concrete.

At dawn the enemy still held parts of Cassino, and although the Gurkha Rifles were on Hangman's Hill, a point below the Monastery, they were not strong enough to advance. Surprise had been lost for good and with it the chance of making a quick breakthrough with armour. Now Cassino would have to be cleared house by house.

The fighting that followed was as bitter as any in the whole war. By day the battle swayed backwards and forwards under a dark pall, by night under a waning moon. Our artillery was in action all the time, shelling gun positions and strongpoints and laying smoke-screens on Monastery Hill. The demand for ammunition was very heavy.

While our men fought in Cassino, British and Indian troops repelled counter-attacks in the surrounding heights, and the Gurkhas, isolated now on Hangman's Hill, were supplied by parachute. From the ammunition point and from No. 1 Platoon's area we watched

American Warhawks as they flew over with ammunition, water, and food. After dark we would count the gun-flashes and estimate how busy we should be the next day.

'Goin' well tonight,' we would say. 'Might stroll up tomorrow —have a couple of quickies at the Continental.'

The Continental was now as famous as the Ritz.

. . . .

Vesuvius had erupted the day before and over it hung a mass of smoke, coiled and motionless and for all the world like an enormous pearl-grey periwig. Monastery Hill, as usual, was dotted with white puffs and hazy with the smoke of battle, but at the ammunition point all was peace under the warm, golden sunshine. Washing fluttered from clothes-lines strung between stacks of ammunition, and our drivers were taking their ease after lunch. Almost the only movement in the area was round the salvage dump, backed against which were half a dozen three-ton lorries from the 18th Tank Transporter Company. The great sprawling pile of empty ammunition boxes and used shell cases should have been growing smaller daily—for our drivers were supposed never to go to Vairano or the railhead with empty lorries—but in practice the opposite was happening, the task of carting salvage, especially on cold, wet days when the stuff was awkward to handle, being one that nobody liked and some avoided, either apologetically by carting a token load or brazenly by carting no load at all.

March the 19th, however, was warm and sunny and the boxes were flying into the lorries with a brisk clatter.

Suddenly there was a quick, rustling sound and the whole area was fanned by a hot breath. Looking up from their books, their washing, their afternoon naps, our drivers saw that the salvage dump was a sheet of bright yellow flame. Their thoughts flew at once to the thousands of little blue and white bags of cordite— rejected or unused propellants—that were a feature of all our salvage dumps. Only cordite would burn as hotly and swiftly as that without exploding. Two lorries were already doomed, but others were moving off with a shriek of gears. One was unattended and just starting to burn, so Henry Blomfield[1] drove it to safety, helped to put out the flames, and went straight back to the fire.

With the first hot gust nearly everyone had made an instinctive movement towards the hills—there were thousands of pounds of high explosives in the area—but after a moment of panic non-commissioned officers and drivers rallied under Lieutenant Delley and formed human chains to clear crates of 75-millimetre ammunition from stacks near the outbreak. In the salvage dump there were many rejected 25-pounder shells—the tendency, of course, had been to load empty boxes rather than boxes with something in them —and one after another these exploded, scattering burning fragments over a wide area.

The fire had started at half past twelve, and at one o'clock Lieutenant Delley and his volunteers were joined by two teams from an American fire-fighting unit stationed at Vairano. There was now every chance of confining the fire to the salvage dump and half an hour later success seemed certain.

Then an explosion threw a burning fragment on to a stack of 75-millimetre ammunition and flames spread from crate to crate. A fire-engine was rushed to the spot, and Henry Blomfield, seizing the hose, stood within a few yards of the stack and played water on it. He was protected only by a breastwork of crated ammunition. When the flames were almost under control the water supply began to fail. It shrank from a jet to a trickle and there was nothing more to be done. The fire flared up, exploding shells and cartridges and setting alight to neighbouring stacks. From these it spread to some stacks of 105-millimetre ammunition, also wooden-crated.

By now nearly everyone had taken cover except the American firemen, Lieutenant Delley, Sergeant Bev Hendrey, Arthur Howejohns,[2] Henry Blomfield, and 'Brinny' Vedder.[3] From the first they had been moving vehicles to safety, fighting the main outbreak at close quarters, and extinguishing many lesser fires. As long as there was work for them to do they did it, undeterred by the fate of three American firemen—two were killed and one badly wounded

---

[1] L-Cpl H. C. Blomfield, m.i.d.; truck driver; Auckland; born NZ, 21 Mar 1917; wounded 3 Aug 1944.
[2] Dvr A. J. Howejohns; taxi driver; born Cardrona, Central Otago, 15 Jun 1916; died on active service, 10 May 1944.
[3] L-Cpl A. A. R. Vedder; boot repairer; Thames; born Thames, 12 Dec 1909.

—and indifferent to whizzing shrapnel and showers of ammunition boxes.[4]

The rest had been ordered to safety and they watched entranced from the shelter of ditches, stone walls, and tree trunks, and from the hills at the back of the area. Because of these, and because the only track to the main road went past the salvage dump, it had not been possible to move all the lorries to safety and some of them were taking bad knocks from shell cases and lumps of shrapnel.

Other units had fared worse or better but all were now out of danger. A platoon from the 1st Petrol Company had got clean away from an area on the far side of the main road, which was now being swept by shrapnel, and elements of the 2nd Ammunition Company, with two men hurt, had fled from an area right by the dump, leaving four tents to the flames, some ammunition, but no transport. Two men from the 18th Tank Transporter Company had been burnt in the first minute, one of them badly.

Behind the ammunition point there was a gentle slope, and here, hedged in by the hills and by a deep gully that separated them from the fire, No. 1 Platoon's domestic vehicles were parked. Lumps of metal had been landing in the gully for some time and now the fire began to creep towards it. Drivers who were sheltering there shot into the open like rabbits and made a dash for the hills. Then the flames leapt the gully, devouring tents and bivouacs on both sides of it and setting fire to a lorry that was under repair and immobile. The cooks' lorry was saved by 'Brinny' Vedder, who drove it far up the slope.

There was now more noise than any of our drivers had ever heard in battle. Armour-piercing shells, with a woof-woof-woof-woof, were trundling through the smoke and landing with a dull thud in fields and on hillsides a quarter to three-quarters of a mile away, most of them passing right over the transport and the heads of our cowering drivers. Jagged lumps of metal lopped branches from trees and smaller lumps hummed wickedly like spent bullets, while ammunition boxes and shell cases described great arcs in the sky or rose vertically to nose-dive into the flames. As a background to the larger effects burning small-arms ammunition crackled all the while. A great column of smoke, leaning over drunkenly at the top, had risen from the heart of the fire and could be seen for

[4] All were commended in a routine order for their gallantry.

V

miles, and above it, slowly expanding, there was a smoke ring through which you could have driven twelve lorries in a line. At a lower level smoke covered the whole area, making a murky twilight in which flames shone luridly. Here and there this twilight was daubed with blotches of red, orange, and violet from coloured smoke shells. The earth shook; the air trembled; the atmosphere was thin and sour.

Heads popped up and down as explosion followed explosion. There was so much to watch, and through the smoke it was still possible to catch glimpses of the Americans at their heroic task. The Brigadier[5] and Major Sampson arrived and proceeded to stroll round the area, the former missing decapitation by inches when a square something slammed past his head.

Time wore on and our drivers spoke longingly of a cup of tea. Those whose lorries were sheltered started to boil up, and others, with a contempt for death and wounds that was pardonable only under the circumstances, dashed into the danger zone to collect primuses, tea, milk, sugar. Jerseys and overcoats were in demand also, for there was no longer any warmth in the sun and many of us were wearing nothing except shorts and singlets.

By five the noise had abated and it seemed probable that the worst was over. By a quarter to six all but a few badly scared drivers were standing by their lorries and inspecting torn canopies and dented mudguards, picking up hot splinters to see if they were hot, and waiting while the billies boiled. Ever and again they ducked as the last shells exploded in the burning stacks.

By six all danger was past. It was now possible to get some idea of the damage.[6] It was far smaller than anyone would have dared to predict while the fire was at its height but even so the ammunition point presented a desolate appearance. A black mess in which embers glowed redly was spread over hundreds of square yards, and beyond this the ground was littered with shell cases, ammunition boxes, and bits of metal. Where the salvage dump had been there was a pile of hot rubbish and the smoking skeletons of two lorries. Tree trunks were burnt and blackened and still ringed with

[5] Col Crump had been promoted in September 1943.

[6] The fire destroyed, among other items, over a quarter of a million rounds of machine-gun ammunition, nearly 6000 rounds of 75-millimetre ammunition, and nearly 500 rounds of 105-millimetre ammunition—a loss of £48,000.

little circles of yellow flame, and here and there bits of canvas and of woollen underclothing smoked and glowed with the unpleasant persistence of burning string. In Headquarters' area the small, splintered corpses of eight bottles of the best Canadian beer had been laid out in a pathetic row.

To shut out this sad scene and the smell of burning, our drivers closed the backs of the lorries before settling down to supper and the task of conducting the preliminary enquiry and making a rough apportionment of praise and blame. It was agreed that the American firemen had done splendidly and that Henry Blomfield and the others had behaved with outstanding courage. It was agreed—for tolerance was in the air—that it would be proper to overlook the slight impropriety of their conduct in behaving in an outstanding manner when most people had thought it wiser to live privately and in retirement for a period. It was agreed that the Brigadier had narrowly escaped decapitation by a 4.5 ammunition box and that he had shown remarkable composure. It was agreed that the fire had been started by a cigarette butt, by Italian saboteurs, by friction; and it was agreed, vaguely and by implication (for the proposition was hard to frame) that it was to the credit of the company as a whole and to No. 1 Platoon in particular that ammunition should have exploded at the ammunition point rather than petrol at the petrol point or M&V at the supply point.

Everyone was cheerful and talkative. Everyone consumed rather more supper than he could manage comfortably (for the administrative sergeant had been truly generous with the rations) and then went happily to sleep. It had been a tiring day and a memorable one.

. . . .

The ammunition point was open again by eight the next morning and later in the day a new salvage dump was formed four miles up the road.

By now the railhead had reached Mignano, which was two and a half miles closer to the front than the ammunition point and over ten miles in advance of the unit area at Vairano—an unprecedented and highly undesirable situation. Wet weather had prevented us from moving earlier, but on the day after the fire No. 2 Platoon was able to occupy an area near the ammunition point, which it took over from No. 1 Platoon on the 25th, and on the 26th the rest of the unit moved to an area two and a half miles in advance of the point. The forward ammunition point at San Pietro had been handed over to the Artillery two days before.

Our new area, flat near the road and terraced where it sloped up into the hills, was a pleasant place, and there was room for dispersal. We were glad of this because the rear areas had lately been coming in for some shelling. The Luftwaffe seldom bothered us now and when it did it was chased all over the sky by ack-ack gunners. Nearly every American lorry mounted a .5-inch machine gun.

Smoke over Vesuvius and smoke over Cassino; but nature had shot her bolt and so had the New Zealanders. On 23 March the New Zealand Corps had been ordered to abandon the offensive for the time being and reorganise its line so that what had been gained could be held—a firm bridgehead across the Rapido, nine-tenths of the town, and Castle Hill.

After that we were not busy, the chief demand being for smoke shells and smoke canisters.

We had time to play football, to welcome back the first group of ' Ruapehus '—seventeen other ranks—and to climb the mountains at the back of our area and note how the white puffs blossomed on the slopes of Monastery Hill, the flanks of Apollyon. The ' widow-makers '—8-inch American guns—slammed their great shells through the blue sky, and ringing circles of sound, hoop upon brassy hoop, expanded and hit the mountains, smashing to pieces. Below us, on our improvised range, rifles snapped cheekily, and from the playing fields beside Route 6 a sound of cheering came to us on the stiff breeze.

And the air in the mountains was like iced soda-water and in the valleys the green leaves uncurled and there was no more winter.

(3) *And So To Rome*

The walled town of Isernia lies about twenty miles east-northeast of Cassino. It clings to the side of a steep hill, and in front of it the road divides into two white twists that sinuously embrace its old walls and meet together at the top. It is just possible to drive a jeep or a small truck through the stone archway at the bottom of the interminable main street, but only at the risk of brushing pots and pans from the open-fronted shops of the coppersmiths and of overturning trays of spring onions, small crucifixes, and coloured postcards—views of Isernia and of Monte Cassino; pictures of the Sacred Heart, of Bonzo, and of Felix: fat dog and thin cat, pale forerunners of Mickey and Minnie Mouse and older than most of the New Zealanders in Italy.

When we saw it the upper part of the town looked as though some careless giant had come strolling down the Apennines in his seven-league boots, placing a casual toe on Isernia's hilltop. This was because American pilots had paid a visit to a neighbouring bridge.

Much of Isernia, though, was only chipped, and this part was crowded beyond belief, all the squawking gesticulating life from the bombed part having been squeezed into it. Even before the disaster space must have been at a premium; for the alleys were of the narrowest, the courts and squares were minute, and there was barely room for another knick-knack in the dusty painted churches or another bulging plaster cherub in the tiny theatre. But it was very much alive, this Isernia, and though built only for foot traffic and hoof traffic, very much aware of itself as a town. In a hundred different ways it protested its patronage of the arts, the trades, the humanities, the vices. Possibly a miniature university was tucked away in one of its small courts.

About three miles to the south-west, sitting like a poor relation on another hilltop, was the little village of San Agapito. Huddled above cobblestones, the tall houses were so old that they seemed to have grown out of the hill like teeth—grey molars and crumbling bicuspids in a green gum—and the people were so miserably poor that all their best rooms, the ground-floor ones, were given over to the precious sheep, pigs, goats, and donkeys. To light them after dark many families had only the dancing open fire and tiny twists of wick in saucers of oil. The furniture was of rough, unpainted wood and more often than not fowls roosted above cupboards and in corners. For decoration there were cheap oleographs, telling with a wealth of haloes and thorns, of gold and madonna blue and vermilion, the story of the Manger and the Cross. At mealtimes the *pasta asciutta* was slapped on the bare board without benefit of plate or tablecloth. It was spread out like pastry, smeared with a tomato dressing, and sprinkled with tiny fragments of meat. Then, each from his or her different angle of approach, the members of the family forked their way steadily towards the centre, so that the big pancake became consecutively a ragged map of Australia, a map of Crete, a map of Malta, and finally vanished altogether.

The village was too insignificant to own an important bridge so not many people had been killed there—only a few men whom the Germans had taken away and shot.

Before daylight the men, women, and children of San Agapito, driving the beasts in front of them, went down by the winding, rocky lane to the orchards, potato fields, and vineyards. Some of the younger women had the downy, velvety look of a dark rose and most of them wore the traditional peasant dress, thick, pleated petticoat and bright bodice, but the old ones wore rusty black. All wore boots, mostly army boots, and all, even the ugly and old, who formed the majority (for youth flies early in San Agapito and beauty becomes a leather mask), carried themselves like queens. The children, with hardly an exception, were gay and pretty.

Spring had come to the valley, and the cherry trees were in blossom and the grass under the trees was deep green and already taller than the spring flowers. Water chuckled in the small streams and was carried through the fields by a system of channels. In the early morning birds swept through the new green leaves like bullets, scattering a spray of dew, and New Zealand lorries, going almost as fast as the birds, rushed along the white road to the ration point, raising a spray of dust.

Our platoon areas were on either side of the road, and these, as we had seen at once, had everything an area should have: green grass, trees in blossom, a stream, and—conveniently placed for relaxation and commerce—a town. True this was likely to be tyrannised over by an English Town Major or by a diligent Provost Marshal who would delight in tearing around in his jeep and interfering with the pleasures of New Zealanders, but there were villages and hamlets in the neighbourhood and on these the hand of AMGOT,[1] with its itch for controlling the sale of liquor and plastering up out-of-bounds notices, would rest but lightly. The hilltop village of San Agapito, for instance, was almost jeep-proof.

[1] Allied Military Government of Occupied Territory.

The New Zealand Corps had been disbanded on 26 March, and by 13 April the last New Zealand unit had been withdrawn from the Cassino sector. The Division, now under the command of 10th British Corps, was to assist British and Polish troops in bringing pressure against the enemy in the mountainous central sector north-east of Cassino. Our unit had moved to the Isernia area on 7 and 8 April.

During our first fortnight there we had little time for exploring the neighbourhood. Having dumped their ammunition at Mignano, the transport platoons had to make a three-day trip to a base ordnance depot at Bitonto, near Bari, to replace it. This took them into the clouds on the Apennines—eagle country where patches of snow lay like sleeping polar bears—and put them down on the familiar plains of Foggia. Two days after they got back Nos. 2 and 4 Platoons, and a detachment from No. 1 Platoon, working over three days and nights, took a paratroop brigade into the line on the Monte Croce sector (Monte Croce being fifteen miles south-west of Isernia) and brought out the 6th Brigade. Driving at night along winding mountain roads, the new drivers (it was about now that the old hands stopped calling them new) showed how little they had to learn.

When the 6th Brigade was relieved, the 5th Brigade took over the Terelle sector, near Monte Cairo. This was supplied from the Brighton dump, twelve miles west by south of Isernia, and from the Hove dump, seven or eight miles nearer the front line, the two being connected by the Inferno Track, a driver's nightmare of corkscrew bends, chasms, dizzy drops, darkness, and danger. Brighton dump was supplied from the ammunition point (now four miles south-west of the unit area) by No. 3 Platoon and a detachment from No. 1 Platoon, and from 28 April onwards NZASC convoys assembled nightly to brave the Inferno Track with ammunition, petrol, and rations.

All went well until 13 May when the Hove dump was destroyed. A shell scored a direct hit on an Artillery cookhouse and later in the day others set alight grass and scrub, the smoke showing the enemy where to aim. The dump was in a deep gully and shortly before 4 p.m. shells started to pour into it.

'After that it was lovely,' said Lance-Corporal Bill Frazier,[2]

[2] Cpl W. G. Frazier; farmer; Ashburton; born Portsmouth, 31 Jul 1914.

one of the four men from our unit in charge of the ammunition section of the dump. 'Everything started to burn—tents, bivouacs, transport, and the camouflage nets covering the ammunition stacks. Until then an Artillery officer and an other rank had been doing a nice job driving burning jeeps away from the ammunition, but now it was hopeless. Both men were wounded—the other rank badly.

'At about a quarter to six the bastards scored a direct hit on the stack of 75-millimetre ammunition just about opposite our dugout and the wooden crates started to burn. Already shells had carried away two bivvies we had rigged up alongside our dugout, so this was the finish as far as our party was concerned. We made a dash for it and got safely to a deep ravine some distance from the dump. Meanwhile the old Jerry had scored a direct hit on the petrol section and now this was going up—between two and three thousand gallons of it.

'We didn't want to hang around, so we made our way to a point on the Inferno Track where we met some machine-gunners who gave us a cup of tea and a bite to eat and some smokes. None of us had anything except what we stood up in but I was getting used to this. I lost everything when the ammunition went up on Route 6.'

That was the end of the Hove dump. From then on the Brighton dump was the terminus for all supply convoys.

. . . .

Remote from blazing dumps, shut out from the sound of guns, and with plenty of time for play, we enjoyed ourselves in our leafy valley.

They started, those golden days, with a kind of imagined click and a little whisper of wind, as though somewhere in the sky— above Carpinone perhaps—a small door had opened, not wide but just a crack. At once, with a drowsy throatiness that told of sleep-ruffled feathers and tiny yawning beaks, the first bird-calls sounded, coinciding, often, with the final despairing echoes of 'Lili Marlene' sung positively for the last time by the last returning revellers. Then the door opened wider, silently and inch by inch, and the stars paled and went out and all the birds tried over their morning songs. There was a grey moment and a green moment, and golden fingers of light rested on top of the cherry trees, and then, with a great unrolling of yellow carpets down all the western hillsides, the sun came. The birds went mad and swept through the drenched branches in clouds, and the cooks woke. They came out of their musty tents, the ones whose turn it was for early duty, and lurched into the sunshine, scratching their tousled heads and glancing grumpily at the burners. As soon as the burners were alight their hissing roar drowned everything—the bird songs and the strangled snores of the revellers.

In ones and twos the drivers who were going to Naples on day-leave climbed from the backs of their lorries and walked through the wet grass to the cookhouses for early breakfast, rattling their dixies and hoping it would be spam and beans and not (for the third time running) soya links. The leave lorry, its shadow on the sunlit grass a huge rhomboid, a huge square, a long spike, bumped over to Headquarters pursued by angry shouts from drivers who imagined they were being left behind. Not all who were having early breakfast were for Naples; some were for Campobasso and to them time was important. They gulped their soya links, shouldered their bulging haversacks, and slipped quietly along the hedge, casting sly glances at the 15-cwt. bugs in which the officers and sergeants were still snoring. They had no leave passes.

The lorry left for Naples, and from the tops of all the mountains the white mists were drawn up into Heaven like the figures in the Ascension, leaving the whole sky one stretch of blue. One after another, with a beating of shell cases and iron pipes, with a winding of sirens. the cooks called their platoons to breakfast, the drivers who responded coming with clusters of dixies and enamel mugs in both hands. We liked to lie late even in spring but we also liked breakfast, so we took it in turns to get up.

The cherry trees were dry now and the day fairly launched. Under a fire of raillery—'Four of our mosquitoes failed to return.' and 'How many d'you reckon you'll bring down today, Digs?'— the Mosquito Men shouldered their pickaxes, shovels, spray-guns, and rubber boots, adjusted the harness of their home-made flame-throwers, and set out for the creek. Presently a column of black smoke rising above the willow trees showed where they were at work. Ever since 1 May—the start of the mosquito season—they had been grubbing up boulders in the creek so as to ensure a free flow of water. burning bushes and undergrowth on its banks, and spraying ponds and puddles with a mixture of petrol and dieselene and houses and farm buildings with liquid insecticide. Using the flame-throwers was good fun but the best job of all was spraying the houses—it was also the most rewarding. At first the villagers and farm-people had been appalled by the sight of parties of soldiers advancing on them with spray-guns in their hands and flame-throwers on their backs. Old women, with tears streaming down their leathery cheeks, had called on the Holy Mother of God to protect them, and the men had gabbled and gesticulated, protesting their innocence, their poverty, their despair. There was not, there never had been, there never would be, one mosquito in the neighbourhood. ('No *zanzari! Niente—niente zanzari!*') Kindly but firmly, and with just a touch of that smugness inseparable from entering houses in the King's name, our drivers had done their duty—and lo, no beasts had sickened, no deadly poison had settled on the raisins and the Indian corn hanging in kitchen and bedroom. And then what a change! Now a visit from the Mosquito Men was like a visit from the painters and decorators—it increased the consequence of a household. Under these circumstances it was quite proper for our drivers to accept wine and eggs, make professional appointments, and allow themselves certain liberties.

The sun mounted higher and the circles of dark shade contracted about the tree trunks. Hell but she was a snorter! In the backs of some of the lorries poker and pontoon schools were in full swing and the sweat poured off the gamblers and ran down their chests in streams. Hell but she was dry work! 'What about playing my hand, "Grump", while I fix the billy?' And the driver-mechanics were saying, putting down screw-drivers and feeler-gauges and wiping their oily hands on their shorts and feeling for tobacco tins: 'Yes, Dig—what about doing the right thing? What about the old Benghazi?' The column of smoke over the willows had sunk to a grey haze, showing where the Mosquito Men were stretched out on the grass.

In hot bivouacs, in the backs of lorries tightly shut to keep out the sun, the revellers woke one after another, bathed in sweat. They stirred feverishly in their tumbled blankets, tore aside their mosquito nets (if they had bothered to use them) and said out loud: 'Off her. Definitely off her. Learnt me lesson.' With uncertain fingers they fumbled for the billy, the water can, the precious primus.

Standing outside Headquarters' orderly room, the defaulters—the ten o'clock men—were feeling the heat, too. Through some mistake (which they couldn't help feeling redounded to the discredit of Company Sergeant-Major Arthur Salmond[3]) they had been ordered to parade in battle dress. 'Ah well,' they said. '" Gibby " won't rock it in. We must be just about his first cases.' ('Gibby', of course, was Captain, now Major, Gibson. The command of the company had passed to him on 17 April when Major Sampson had left us to return to New Zealand.)[4]

But it wasn't really hot. Not hot unless you were bent over an ammunition box in the back of a stuffy lorry trying to figure out whether old 'Baldy' was sitting pat on a swinger. Not hot unless you were swollen with stale vino or wearing battle dress. Down in the bathing pool by Workshops' area—we had made it by damming the creek—no one was too hot. Here we splashed and swam, losing the last vestige of our winter pallor.

The sun was almost overhead now and it was lunch-time. The gongs sounded and the cooks dished out tinned salmon, chopped onion, and two slices of bread to each driver. There was also margarine, marmalade, and cheese.

'Three lorries to pick up 3-inch mortar from amm. point,' said Jock. 'Three of yours can go, " Goldie ". " Parky " hasn't been out this week.'

'Don't scone, Jock,' said 'Goldie'. 'Whatever you do, don't do the scone.'

The afternoon passed slowly and time itself seemed to be resting. Thunder rumbled in the distance and a ragged thunder cloud—blue-black like a Gillette razor-blade—sailed over Isernia, the skirts of its great shadow just brushing No. 2 Platoon's area.

'Seeing she's three o'clock I thought I'd bring the old mug along.'

'One load of 25-pounder and two of Mark VIIIZ for amm. point.'

'Don't do your bundle, Jock.'

[3] WO II A. L. Salmond; architect; Dunedin; born Dunedin, 23 Jan 1906.
[4] Capt May was in the same draft. Lts Sloan and Delley took charge of Nos. 1 and 4 Platoons respectively with the rank of captain.

'Goin' up the hill tonight?'

Again the gongs. There was roast beef, tinned peas, tinned potatoes, and for pudding doughnuts and treacle. For dessert a mepacrine tablet.

It was cooler after tea and footballs crashed through the branches and the teams swayed backwards and forwards on the improvised playing fields. In No. 1 Platoon's area the game was Association football without rules, the players being at liberty to come and go just as they pleased. Seldom were the sides even approximately equal.

Not all the noise, and the area was echoing with shouts and laughter, was made by the footballers. It was shower time, and George Laverick, his old felt hat on the back of his head, was standing beside No. 1 Platoon's water cart and rhythmically pumping hot water over a dozen glistening bodies, or over as many as could push their way under the single perforated jam tin that did duty as a sprinkler. Ever and again George would remove his pipe and call out in his deep, gruff voice: 'Showers on NOW! Any more for SHOWERS?'

And the shadows came together like lovers, and the sun was hidden by the trees after being tangled for one moment like a puzzle of gold wire in their topmost branches. The football ended and it was time for everyone to change into long trousers, roll down his sleeves, and smear face and hands with mosquito repellent. The footballers, flushed and sweaty, ran over to the water cart, undressing as they went and shouting to George to keep on pumping.

The mountains were jagged against the sky, and form and colour faded from the foothills, and the blue twilight came. For a wonder there were no films or ENSA shows to go to, but there were other ways of spending the evening. Already groups of drivers (and among them, no doubt, were some who had learnt their lesson as recently as that morning) were sitting or reclining on the soft grass like the guests in the *Rubaiyat*, wine glasses beside them and great jars cradled in wicker baskets. For an hour past, dressed in their 'Groppi mokka', others had been leaving for Isernia or for neighbouring villages. The four drivers who went always to San Agapito had just left. They had a tidy walk in front of them, and long before they reached the bottom of the hill the moon and the stars were shining, and the fireflies, borne on a current of warm air as on water, were swimming between the hedges and zigzagging from side to side with quick, darting movements like fish. The air was full of the sweetness of warm grass and honeysuckle and ripening fruit, and from cottages beside the lane children came running, begging *cioccolatta* and *dolci*, and getting them because of the moonlight and the fireflies and the smell of honeysuckle. It was a perfect night, peaceful and yet exciting.

From Workshops' area came the first song of the evening, 'Lili Marlene' rendered by the Salome Gang: Bill M—— (guitar), Bill S—— (guitar and songs), Joe H——, 'Snow' T——, and Dick C—— (general singing).

Below the bathing pool the fishermen paced beside the creek, smoking their battered pipes in slow content and from time to time casting a contemplative Mills bomb into the shining water.

George, who had appointed himself ARP warden (he knew more about bombing than most of us and German aircraft were overhead almost nightly), made a tour of the area, greeting each chink of light with a gruff 'What about the blackout, you jokers? Bit of a blackout man myself.'

The leave lorry returned from Naples.

Time lapsed and the nightingales sang.

In San Agapito, high above the valley, the moon was shining on one side of the main street and on all of the main square, lighting the great, broken crucifix and the great archway, beneath which three or four drivers had gathered to discuss a last litre of *vino rosso*. The stone seat struck chill through their summer clothing and they were sleepy. Only the children of San Agapito, who seemed not to need rest or warmth, were awake and alert. They listened to the rambling talk as though it were wise and beautiful beyond parallel and they could understand every word of it.

Our drivers said goodnight to the children beneath the crucifix (bare now except for the wooden hammer, one wooden nail, and a fragment of the wooden spear) and the children said: '*Buona notte, Pietre—Tubby—Giorgio. Ritornerete una seconda volta?*'

All over this corner of Italy our drivers were standing in lighted doorways and saying goodnight and thank you.

'*Buona sera, Angelo. Buona sera, Giovanni. Mille grazie.*'

'*Buona sera, Assunta.*'

'*Ritornerete. . . ?*'

The closer they got to home the louder became the noise of singing, for No. 2 Platoon was celebrating a birthday. At midnight, after which authority could hardly be expected to continue turning a deaf ear, the singing moderated; but it kept on breaking out afresh, hour after hour, as waves of beautiful feeling (beautiful solidarity, inexhaustible mirth, welling tenderness) swept over the wine-drinkers.

> The next to come was Goering's wife
> And she was anti-Nazi. . . .
> *O Trombettier, stasera non suonar.* . . .
> The Poles, the Czechs, and Germany itself. . . .

The nightingale sang too, also straining to express through his small hot throat, the inexpressible.

The party from Campobasso, sober now after a long lorry-ride, crept home, and at last, over towards Carpinone, high up and secret, a door opened. A cool breath set all the leaves dancing, and somewhere, drowsily, a bird chuckled.

.   .   .   .   .   .

Not all our days were like this. Mostly we were very busy, though it would be difficult to connect our work directly with any of the momentous events that took place during May: the full-scale attack launched by the Fifth and Eighth Armies on the night of 11-12 May when troops in the New Zealand sector made feints and the Divisional Artillery supported the Poles in an attack on Monastery Hill; the crossing of the Aurunci Mountains on the Axis right flank by French Moroccan troops (the nightmare Goums); their arrival in the Liri Valley, and the beginning of an enemy withdrawal under this threat; the resumption of the Polish attack on the Monastery; the left hook led by the 19th New Zealand Armoured Regiment and the cutting of Route 6; the final scene on the 18th when British and Polish flags flew over Monastery Hill.

These were events in which we played no outstanding part. Our job was to see that the Brighton dump never lacked ammunition, and this we did with the help of the Reserve Mechanical Transport companies, replenishing it from a field maintenance centre near Carpinone, seven miles east by north of our area.

On 23 May the Anzio force attacked from its bridgehead, linking up with the Eighth Army two days later. The next day New Zealand infantry started to advance, and by the end of the month it was plain that strangers would eat the cherries ripening in the Isernia area.

No. 2 Platoon, with detachments from No. 1 Platoon and the 2nd Ammunition Company, made the first move on 30 May, establishing a forward ammunition point near San Elia, three and a half miles north-east by east of Cassino; and on 1 June the rest of the unit moved to an area in low hills three miles south-south-east of this. The whole neighbourhood was dirty and tainted under the blue sky and the golden sunshine. Empty gunpits with all their mess spoilt the clearings in the woods and used shell cases lay thick under every hedge.

Cassino was within easy walking distance, so most of us took this opportunity of visiting it while it was still, so to speak, warm. The ruins looked moving from a distance and of course they were soaked in heroism and glory, but when you got close none of this was apparent. Then they were just dirty and insulting—like a mess on the pavement. Photographs and newspaper accounts had told us what we should see—green, stagnant pools by the Rapido, sightless houses, streets featureless as lepers, tree trunks stripped even of bark—but nothing had prepared us for the silence—the smashed stones seemed to be able to absorb sound as quicklime absorbs water—and the stench. It was not the stench of corpses, though burial parties were still busy, but of dead houses—the stifling, sweetish reek of old mortar, mice, dirty wallpaper, broken wainscots, domestic dust. Traffic rolled along Route 6 and groups of sightseers gaped at what was left of the Continental and the Hotel des Roses. Cameras clicked busily, but for all they showed afterwards they might as well have been photographing a midden.

On the day following our move to the new area the transport, helped by a platoon from the 2nd Ammunition Company, cleared the Brighton dump and the last ammunition point, and No. 1 Platoon, whose 25-pounder was more likely to be in demand than No. 2 Platoon's mixed loads, took over the ammunition point at San Elia.

When the liberation of Rome was announced we were still in the same areas—No. 1 Platoon in a green cornfield below battered San Elia, a bathing pool on its doorstep, the rest of us three miles away and getting sprayed with white dust by every passing lorry.

It was 4 June, a Sunday, and Americans had entered the city at breakfast-time that morning.

The ammunition point was closed on the 5th and the whole unit moved to an area nineteen miles north-west by north of Cassino. Here there was a shallow stream in which we could wash our clothes and our dusty bodies, and all around us were hills with toy villages perched on them. We were now in the Liri Valley, and ahead of us, fifteen miles to the north-north-west, the battle for Balsorano was ending.

The town was occupied the next morning and the New Zealanders pressed on towards Avezzano, seventeen miles north-north-west of it. Again the enemy was in full retreat and the demand for ammunition had fallen off.

That evening crowds gathered round the platoon radio sets. The news was old now—the German News Agency had announced it at 9.2 a.m., Cairo time, and an hour and a half later it had been confirmed by the Allies—but we wanted to hear it for ourselves:

> Under the command of General Eisenhower, Allied naval forces, supported by strong air forces, began landing Allied armies on the northern coast of France. . . .

At the end of the news there were recordings of scenes at the embarkation ports. We heard snatches of ragged singing—'You Are My Sunshine'—and snatches of conversation. The Tommy accents caused laughter and some of our drivers tried to imitate them, but not ill-naturedly.

In the backs of the lorries the primuses sighed and purred, and under the green tendrils of the vines the midges and mosquitoes, mixed in a grey smudge, made a noise like the highest imaginable note from a violin. And the Tommies sang 'You Are My Sunshine' —but that was yesterday and in England.

For a long time, in the darkness, the announcer went on talking about the Second Front, making it sound grave and in rather good taste. But it was chilly among the vines and only a few people were listening.

The next morning it seemed quite natural that the Second Front should have opened—natural and indeed inevitable.

. . . .

After the fall of Avezzano on 9 June the Division reverted to the command of the Eighth Army and for the time being its labours were finished. On the 13th it started to concentrate in a rest area near Arce, some fifteen miles west-north-west of Cassino. Most of the NZASC, however, was to continue working, and that suited us down to the ground. Rest areas with their unavoidable concomitants —parades and inspections—were not at all to our taste, whereas driving, oddly enough, was. A few old-timers, true, said they never wanted to touch another steering wheel as long as they lived (though when jobs at Base were advertised we did not see them rushing to the orderly room), but for most of us the sight of the lorries lined up on a sunny morning, their canopies tied down and their engines putt-putting as they warmed up, still spelt happiness.

If that was indeed so joy immeasurable lay in wait for us. We had our first taste of it on the evening of 8 June when the transport platoons, after dumping their second-line holdings in the unit area, set out for Venafro, eleven miles east of Cassino, to report to the CRASC 10th Corps. During the next week they were employed in bringing forward petrol and ammunition from Vairano to a dump forty-five miles east by south of Rome, and from Mignano to one only eight miles east of the city. The time allowed for these trips was thirty-six hours but our drivers reduced it by half a day.

Company headquarters and Workshops moved from Sora on the 13th, and the next day the transport platoons, their work with 10th Corps finished, picked up their second-line holdings and joined the rest of the unit in the new area. It was on Route 6 and three and a half miles south-west of Arce.

The surrounding country was green and beautiful but we had little time for exploring it. On the 15th the transport platoons passed to the command of the Eighth Army, and during the next three days they were employed in bringing forward ammunition and supplies from Vairano and Mignano to Valmontone, twenty-two miles east-south-east of Rome.

Since leaving Isernia we had been exposed to more new impressions than we could assimilate comfortably. We had been seeing at the loveliest time of the year some of the loveliest country in the world. It remains for most of us a beautiful blurred memory of long, slate-coloured roads dappled with sunshine and leaf patterns, but here and there a scene stands out boldly: windy weather on the 'Campagna di Roma; rain clouds like great bruises invading the blue sky and dragging their purple shadows over the new gold of

w

the cornfields; and—unforgettable—our first glimpse of Rome, all her spires gleaming, the sun going down behind her seven hills, and a voice saying: 'That round thing there—you can just see it—that's the dome of St. Peter's.'

> Last week in Babylon
> Last night in Rome. . . .

Rome! Except for one or two enterprising drivers who had taken wrong turnings none of us had seen her yet, and now, thanks to our duties with the Eighth Army, we were to live right on her doorstep. Not all of us though. The Ammunition Platoon, the Cinderella of the unit, was left to guard our ammunition in the Arce area.

We moved on the 18th—a Sunday of course—and drove along Route 6. When we were five miles from the city we turned right off the main road, and two more turnings brought us to a narrow lane on either side of which were the platoon areas. Rome, eight miles east of us, was hidden by a grey veil. Wind ruffled the grass —corn and rolling grassland was all we could see for miles—and it rained. But no one minded that.

Tonight in Rome!

. . . .

Painfully, as though searching for enemy aircraft, we twisted our necks to observe the ceiling of the Sistine Chapel. Obedient as flocks of sheep we wandered over the coloured marble of St. Peter's, catching a phrase here and there (' . . . length 669 feet . . . Michelangelo and Raphael. . . .') and all the time thinking longingly of morning tea. The Colosseum was better, for there you could at least sit down for a moment in the cool. The Torre dei Conti, the Forum of Nerva, the Forum of Augusta, the Temple of Marte Ultore ('. . . delicate Corinthian columns carefully restored. . . .')—it was educational all right and a man would be a fool to miss it, but hell it was dynamite on the old feet! Most of us wore our light sandals in Rome (for they were smarter than boots) and the protection they gave against the uncompromising stone pavements was negligible.

Standing drenched in sunlight in the Piazza Venezia, we gazed wonderingly at the Palazzo from whose small balcony the 'Bullfrog' had been accustomed to harangue the mob. (It looked—as the poor 'Bullfrog' so longed to look—clenched, massive, impervious to storms.) But what we really admired, what struck us as truly elegant, was the gold and white wedding cake built in honour of the second Victor Emmanuel. Here, we felt, they *had* something.

We admired also the shining hairdressing saloons in which flushed soldiers, faintly protesting, were being anointed with sweet oils, rubbed with unguents, assaulted with hot towels, and generally mishandled in ways that could not have commended themselves to anyone except an effeminate Latin. And we admired the shining bars and cafés whose scarlet tables and chromium-plated chairs encroached so charmingly on the busiest pavements, and we admired the shining, expensive women who looked at us with kind eyes.

For the matter of that everyone regarded us kindly, none seeming to regret that the new customers were wearing khaki instead of field grey. That we were looked on as customers pure and simple was made abundantly plain to us, and it was plain, too, that of all the new freedoms we brought the one most valued was the freedom to profiteer.

But we liked Rome. It was pleasant to feel the calm atmosphere of the buildings, their elegance, their ease, their charming disingenuousness—pleasant to walk through wide, shady streets whose creams and gentle greys were relieved by newspaper kiosks gay as bunches of flowers—and pleasant indeed to wander up the Via Nationale, passing fine shops filled with useless and expensive gewgaws, and seeing ahead, with a recrudescence of thirst, the blue folds of the New Zealand flag.

But perhaps it was pleasantest of all, once having seen the city, to stay at home, comfortably shirtless and bare-footed. Besides we were very busy at this time.

By 19 June the Eighth Army roadhead had reached Narni, forty-three miles due north of Rome, and to this we were moving ammunition from the dump we had helped to establish near our area. The round trip, which entailed crossing Rome twice, was seldom made in less than twelve hours.

What trips they were! At every bad corner Indian drivers with smiles on their kind brown faces did their best to kill us. On all the bad hills tank-transporters conditioned our speed, cutting it down to a kind of staggering crawl. Our cabs were like ovens, and British officers, tearing past red-faced and angry in 'bugs', cast scandalised glances at bare feet poking through open doors and windscreens. There were interminable traffic blocks, and when we did reach our destination, tea-less and tired, there were long waits while fussed sergeants and corporals frantically tried to discover what we were to do with our loads. Coming home at night, more hindered than helped by a single wan beam, we scuttled along in an effort not to lose touch with the bobbing tail-light of the vehicle ahead. This was tiring work, and tiring also were our daily battles with punctured tires, worn engines, worn steering assemblies.

> Man dies in full content
> Of trouble past. . . .
> So does transport.

Most of our vehicles had been on the road since 1941 and now they were slipping gently westwards. A wheeziness, a puffiness, an habitual languor, an insatiable thirst for oil—with these and with kindred ailments they were paying for the over-exertions of their youth, the excesses of their middle age, and their final folly in exposing themselves to the rigours of an Italian winter when younger lorries were either in their graves or pottering around Base in well-earned retirement. From seven in the morning until ten at night Workshops laboured to keep them rolling.

Consequently, when we did get a morning or an afternoon to ourselves, most of us preferred to stay at home, though in the evenings we often slipped away to the New Zealand flats. These, shabby-new and nominally the property of the Italians, were only a few miles from our area, and here, relaxed and perfectly at home and with no slender Corinthian columns or world-famous frescoes to reproach us, we could talk the eloquent language of chocolate and bully beef, creeping home tired and triumphant in the early hours of the morning. But not always triumphant. Sometimes there was a surly face at breakfast because she

> Would not yesternight
> Kiss him in the cock-shot light.

Hard-working days but happy ones! Once or twice they were varied by an organised picnic to lovely Lake Albano, near Castel Gandolfo, the summer residence of the Popes. This meant bathing and being able to buy armfuls of peaches at only two or three times their correct price, and a journey home along the Appian Way, and a visit to the Catacombs, where you could commune with the Christian dead or (in lighter mood) bark like a dog, stretch out in niches, humorously extinguish candles.

We returned home from one of these picnics to find that Lance-Corporal Owen Penney[5] (No. 4 Platoon), brother of No. 1 Platoon's Dick who was now in New Zealand, had been killed by the accidental explosion of a bakelite hand grenade.[6] That was the only shadow.

By the end of June the NZASC units under the command of the Eighth Army had finished clearing the Rome dump, and on 1 July they started carting ammunition to the Narni roadhead from the 21st Advanced Ammunition Depot. This was on the outskirts of a small town known to history as Antium. It was there that Coriolanus, the uncompromising patrician, sought refuge from the indignant 'plebs'. Cicero had a villa there and the Emperors Nero and Caligula were born there. Of late months the town had acquired a new title to fame under the name of Anzio.

Naturally it was in poor order, and so were most of the romantic coast villages on the Via Severiana. When we came home by this route—we used to spend the night in the unit area and go on to Narni the next day—we had on our left the Tyrrhenian Sea, blue and sparkling, on our right smashed houses. We said little ('Navy, eh? A fair sort of a towelling.'), but more than one of us, seeing for a moment through the eyes of the broken and dispossessed, thought to himself: 'They paid all right, the poor bastards. They don't owe us a thing.'

But it was no time for sentiment. All over the world full payment was being exacted both from the guilty and from the less guilty. On the Eastern Front gallant Finland was paying the last red cent (the Mannerheim Line had been broken on 18 June); in France, Germany was paying (Cherbourg had fallen on the 27th); in England, the British people were paying (for over a week now flying bombs, putt-putting through the air like motor-cycles, had been falling on their small island); and in Italy, by land and by air, items were being struck daily from Marshal Kesselring's account.

And now it was time for New Zealand to make a further payment.

[5] L-Cpl O. E. Penney; freezing works' employee; born Ohaeawai, Auckland, 27 Apr 1920; died on active service, 28 Jun 1944.

[6] Two drivers were also wounded.

We had finished the Anzio job the day before (6 July) and we were in the mood for a little relaxation. At midnight, when the move was announced, many of us were miles from the area. Worried corporals stumbled around in the dark and motor-cycles and at least one lorry were sent surreptitiously to the New Zealand flats.

In some miraculous way news of an emergency out-distanced the speeding messengers, beating them even to Rome, and when the load-carriers pulled out at three in the morning only a few drivers were unaccounted for.

Yawning and nodding, we rushed smoothly through the warm night, heading for the rest area near Arce to pick up our second-line holding and the Ammunition Platoon. The beating engines and whispering tires made their usual nonsense in our sleepy brains (anything you like: In-again-Finnigan, in-again-Finnigan, in-again-Finnigan), but it was cheerful nonsense.

The Division was going in again—hell for the fighting units perhaps, but for us (and we said it with all the apology in the world) a kind of holiday.

Route of 2nd New Zealand Division: Rome to Pesaro

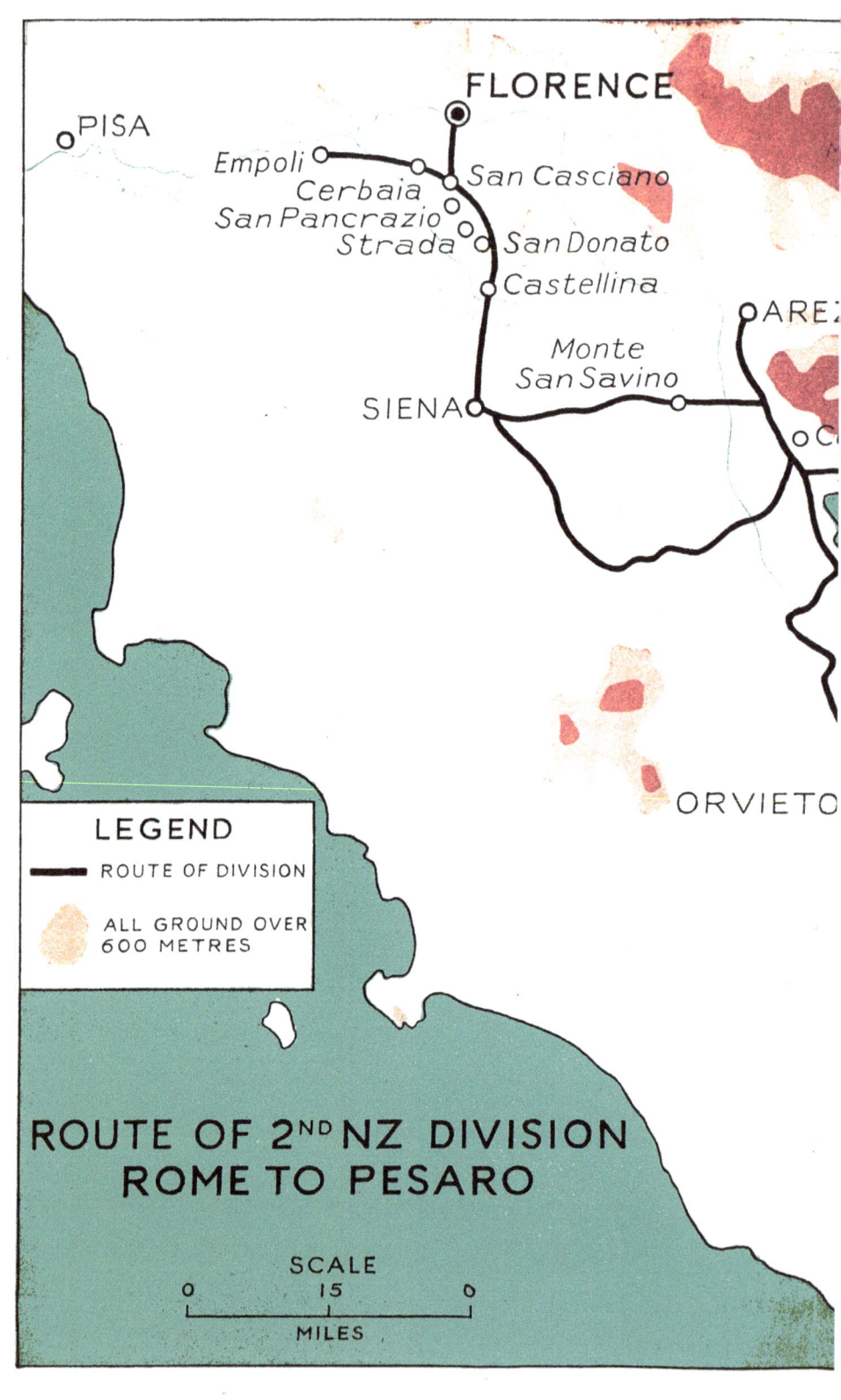

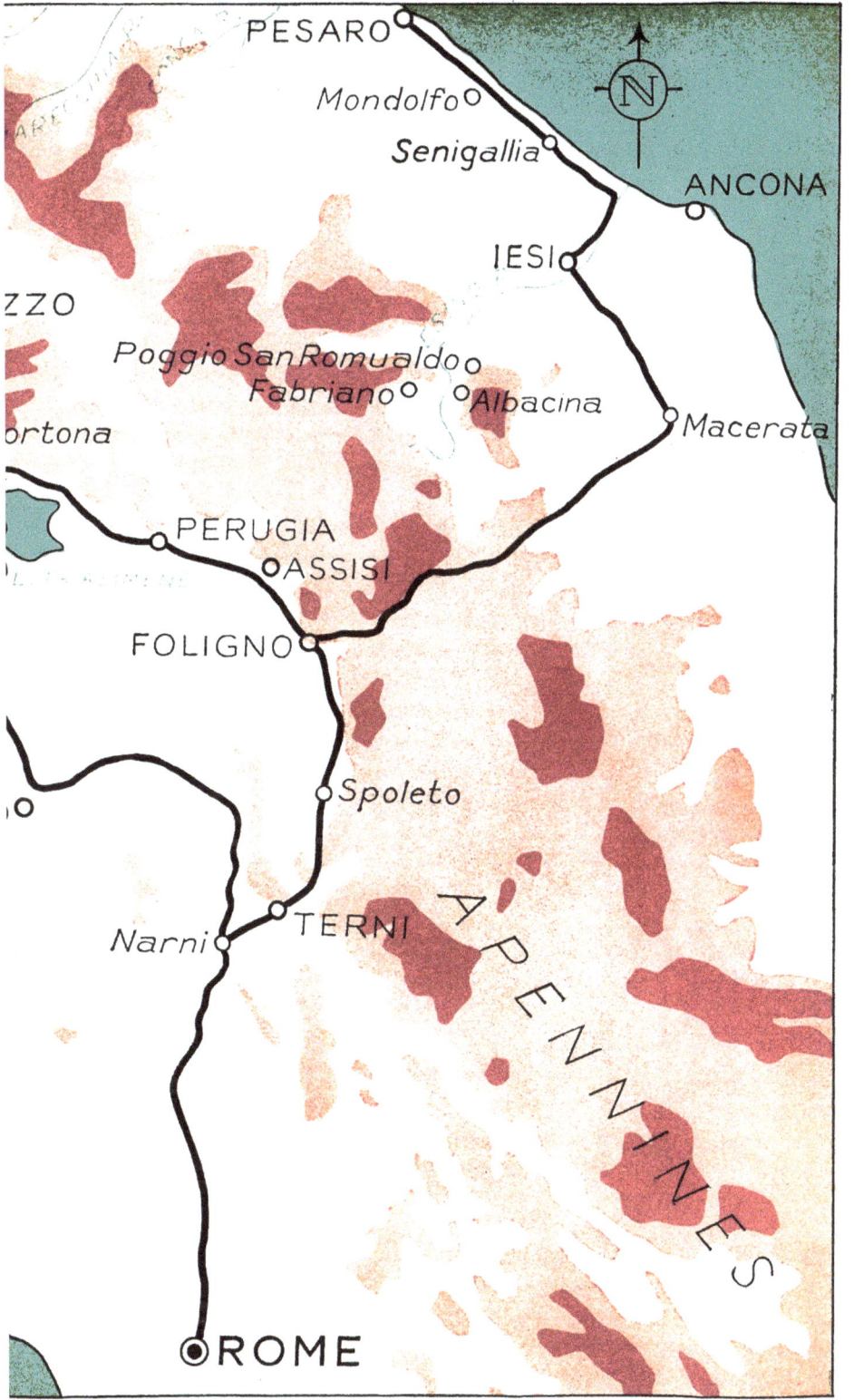

CHAPTER 20

THROUGH THE VINEYARDS

THEY were small and you knew they were not good to eat. They were dusted with silver like mistletoe and they hung in tight bunches above every ditch, tendrils half hiding them. They were delicate things to have come from such a scrawny parent (trunk twisted and branches crooked like elbows: a *doppelgänger* black and mummified) but as yet they were hard and sour—too sour to interest your wise Umbrian fox. We tried them, knowing it was a mistake, and they made our mouths dry.

Before reaching this last vineyard we had been on the move for a week, stopping a day here and a day there but making more than enough northing in the intervals to discount any benefit to our estates from the hot sunshine. Now we were eighty-six miles north-west by north of Rome and three or four miles south of Cortona. Near us was a corner of Lake Trasimene. We had been told that we should not be moving for a few days—we could rest and perhaps the grapes would ripen. This pleased us enormously, for what was the use of being in Italy in July if you were never in one place long enough to enjoy the wines of the country, make the acquaintance of your Italian neighbours, and get your washing done?

During the past week we had been able to do none of these things. First there was the journey from Arce to a staging area on the Rome-Narni road, to which Company headquarters and Workshops moved independently from Rome. Travelling north the next day ('*Neo Zealandese*,' the Italians murmured, brushing aside our clumsy attempts at deception), we passed through Terni, Spoleto, Foligno, and Perugia, and halted at last in an area near Lake Trasimene. After spending a day here we moved closer to the lake (to the relief of No. 1 Platoon's drivers between whom and a local landowner lay a little matter of some piglets), and two days later we went to the Cortona area, where No. 4 Platoon had established an ammunition point.

And here we were among the green and silver grapes. It was 13 July, and some ten miles to the north the 6th Brigade was fighting in the mountains above Arezzo. The Division, which was now under the command of 13th Corps, was to capture these so that a British armoured division and the Brigade of Guards could advance through the town and reach the River Arno north of it. The ultimate prize, of course, was Florence.

But Florence seemed likely to cost us dear, for time was what Marshal Kesselring wanted. All along the front the Germans were resisting stubbornly, while behind them, from Massa on the Gulf of Genoa to Pesaro on the Adriatic, workers from a dozen nations laboured with or without enthusiasm on the defences of the Gothic Line.

The severity of the fighting (partly because we had carried out a 300 round-per-gun dumping programme for the 5th and 6th Field Regiments) was not reflected in the demand for ammunition, which was slight during the three days that preceded the occupation of Arezzo by British armour. This happened on the 16th, and on the same day, while the 13th Corps pressed forward, the Division went into reserve.

Our sales shrank almost to vanishing point, but for four or five days we were fully employed in disposing of salvage, completing our second-line holding from a field maintenance centre twenty miles away, and bringing forward ammunition from there to another centre near Arezzo. Even so we had time to take proper notice of a very significant announcement: married men of the 4th Reinforcements were returning to New Zealand under the Taupo scheme. The news caused heart-burning and disappointment among our single Fourths and among some of our Fifths, but for many others, right down to drivers who came overseas with the 7th Reinforcements, it opened a door on hope. We gave the lucky ones (Captain Todd and seven other ranks) a rousing send-off, pledging them suitably.

After they had gone there was a feeling of flatness in the area, so No. 1 Platoon, partly to dispel this flatness and partly because 'Poodle' and 'Snow' had just returned from New Zealand furlough and it was 'Neil Mac's' birthday, gave a party. It began at sunset and went roaring on through the night, huge and formless. Captain A. E. Thodey[1] (who had taken over the platoon from Captain

---

[1] Capt A. E. Thodey; fat-stock buyer; Morrinsville; born Auckland, 16 Jul 1912.

Sloan[2] less than twenty-four hours before, the latter having been posted to the 1st Supply Company) expressed mild astonishment the next morning, but little harm had been done. The haystack round which the party had pivoted had been eviscerated and for a day gloom and despondency accompanied the platoon on its journeys and stalked unchallenged among the grape vines. By the following morning, though, cheerfulness was restored, and that was a good thing, for it looked as though the Division would be moving soon. During the day—the 20th—ten vehicles from No. 2 Platoon left to establish an ammunition point eight miles north of Siena, Siena being thirty-four miles west of Cortona, and that night the 5th Brigade began to fight its way towards Florence, which had been declared an open city. The Division had relieved French Moroccan troops near San Donato, seventeen miles north of Siena.

For two days we were employed in dumping our second-line holding in an area next to the ammunition point, but all our own concerns were overshadowed by the news from Germany. A bomb had been set to kill Hitler and something very like revolution had been attempted. For a happy moment we looked out across the heaving waters and saw a flutter of white wings and a flash of green.

The dove vanished, and on 23 July, very early in the morning while all the grapes dripped dew, we left for the ammunition point, squeezed through Siena—lovely and incomparable Siena, red rose of Tuscany, flushed with a thousand sunsets and autumn fires—and pushed on through a fog of dust. It settled on our arms and faces like warm flour and lay on the white roads in drifts six inches deep. There were no vines in the new area (which made a pleasant change); instead there were oak trees and green hillsides and the lovely litter of a forest: dead branches like antlers, great, prostrate trunks grey and lichened over, last year's acorns.

As soon as we arrived work started in earnest. The demand for ammunition mounted steadily and we were handicapped by lack of transport. On the evening of the 24th No. 1 Platoon passed to the command of the 6th Brigade, and on that and the following day Nos. 2 and 3 Platoons carted petrol for the Eighth Army.

[2] Maj R. G. Sloan; caterer; Bray, Eire; born Timaru, 15 Oct 1913.

On the evening of the 26th—it was a terrific day with the demand for 25-pounder ammunition mounting and mounting and no chance for more than a few of us to see His Majesty drive through the Divisional area—a lorry arrived at Workshops looking as though it had been picked off a bonfire. Within a few minutes word was all round the unit that No. 1 Platoon was in trouble. The drivers, however, were reassuring. The platoon had debussed the 26th Battalion near San Pancrazio, eight miles north-west by north of San Donato, and had come under shellfire while digging in on a hillside. One lorry had been set on fire and a shell had passed through the canopy of another without exploding. The rest of the damage was a punctured radiator, and now the platoon was standing by to drive the 26th Battalion to Florence.

First, though, there were obstacles to be overcome. Between the New Zealanders and Florence lay the Paula Line, based on a semi-circle of hills, and as day followed day there was no slackening in the demand for ammunition.

On the 27th we moved in independent groups to an area near San Donato. It was on high ground and from here we could see shells landing—our own and the enemy's. The next day a section from the Ammunition Platoon opened a forward point near Strada, a few miles north-west of San Donato, and soon all first-line transport was being diverted to this, for there was not a round left in the unit area. Later in the day the rest of us moved to Strada and for a short period artillery lorries stood empty in our lines. As fast as we could rush the ammunition forward from the field maintenance centres it was tail-loaded on to them. Major Gibson sent an SOS to the Brigadier and before long we were being helped by transport from the 2nd Ammunition Company, the 14th Light Anti-Aircraft Regiment, and the 7th Anti-Tank Regiment. By now we had a dumping programme of 400 rounds a gun to deal with. We worked all through the night and all through the next day, the Artillery grabbing our loads as they arrived. Between dawn and dusk we handled 34,000 rounds of 25-pounder. By the evening of the 30th—we issued 47,000 rounds of 25-pounder that day—we had enough of everything at the ammunition point to meet any predictable demands, but there was no question of easing up. The guns kept firing: 27,500 rounds, 45,000, 28,000. . . .

The New Zealanders were driving the enemy back to the last ridge of hills before the Arno and Florence and it was bitter work.

No. 1 Platoon called him Paul. A member of the 4th Reinforcements, he had joined us first in the Isernia area, and from the moment of his arrival he had gone quietly about his business. More often than not his business had taken him to Campobasso or Naples and sometimes it had entailed his being away from us for four or five days at a time. He was a powerfully built man with a heavy, purposeful face and a slow tread. Now, wearing little except a grey top hat of a kind common enough at Ascot and a dirty khaki shirt that failed to conceal his magnificent chest and splendid abdomen, he stood at the gate of his castle to receive guests—'Goldie', Des, and 'Brinny', who, halted by provosts because the road ahead was under fire, had remembered that Paul was living close by in the battalion's B Echelon area, his lorry being loaded with signallers' gear. He was said to be doing himself rather well.

He was. He showed his guests over the castle and took them to his bedroom. He drew their attention to his four-poster, falling heavily on the rich covers to demonstrate the resilience of the mattress. He threw open his wardrobe and pressed 'Goldie' to choose one of his twenty new suits.

His guests would take something, he suggested, leading them out to the lawn. He excused himself for a moment and came back with a small table, napery, crystal, and a Borgini chianti of a good year. He said he was quite satisfied with his cellar except that the former owner of the estate—a count or something—had quarters there. Ah well, the old fellow was no longer young and he had no wish to be unduly hard on him!

Presently his guests rose, mentioning that the quartermaster at battalion headquarters was probably waiting for his blankets. Paul went with them to the gate. They were to look in, he said, any time they were passing. There was always a glass of wine, a meal, a bed. When last they saw him he was moving purposefully towards the decanter, the grey topper well back on his fine but balding forehead, the dirty shirt flaring out behind with true aristocratic negligence.

The rest of the platoon was living more modestly but everyone was enjoying himself. The food was good—there was an abundance of potatoes, tomatoes, and fruit—and the job was interesting. Not that many adventures had befallen the platoon so far. After dropping the 26th Battalion near San Pancrazio it had stayed where it was until the end of the month, idle except on the 29th when it took the Maori Battalion into the line. On the 31st it moved a few miles to the 26th Battalion's B Echelon area, spending two days there. The drivers to whom new Dodges had been issued worked hard to turn them into homes, slinging bunks, fitting reading lamps, and building racks for 4.5 boxes. During the past week or so we had been issued with twenty-one Dodges to replace the worst of our Chevrolets, and a dozen of these had gone to No. 1 Platoon. They had glass cabs and were fast and smart, but they lacked four-wheel drive and were known to be hard on tires. In the event they proved far less serviceable than the ugly old Chevrolets.

The platoon's next move was made after lunch on 3 August. It went to the village of Cerbaia, where the 6th Brigade, supported by the 19th Armoured Regiment, had established a bridgehead across the Pesa River a day or two earlier. This was only seven or eight miles south-west of Florence, and from the surrounding hills our drivers could see those twinkling lights about which so much was being written at that time. The infantry was billeted in Cerbaia and the transport dispersed in the village square and along the streets leading into it.

'It was a warm, lazy afternoon,' said Corporal 'Sandy' McKay,[3] 'and the little village was curled up at the bottom of the hill asleep. There were not many civilians about, most of them having left when the fighting started, but there were plenty of soldiers—cooks sweating over burners on the pavements and Kiwis strolling in and out of shops and houses to see if the Germans had left anything behind. Most of them looked like coolies, for apart from shorts they were wearing little except wide-brimmed straw hats they had found in a neighbouring factory.

'Round about half past four, in the middle of all this peace, we heard a series of faint woomps. They were followed by low whines that rapidly became shriller and ended right among us in

---

[3] Cpl G. J. McKay; storeman; Dunedin; born Miller's Flat, Central Otago, 27 Nov 1916.

ear-splitting crumps. The shells came over in threes, sixes, sevens, and elevens. There was one batch of seventeen, four of which were duds. Between each batch there was a pause of from ten to twenty minutes. During the first couple of pauses the boys dashed out to disperse their lorries more widely and find safe places for them. One of the boys shifted his lorry from the square to a snug possie by the church and for his trouble got three punctures and a holed radiator and petrol tank. Someone else got a holed sump and Henry Blomfield's lorry was smacked as he was moving it from the main street. He was slightly wounded in the back but was able to carry on after he had been fixed up at the RAP. He was far more worried about a broken window in his new Dodge.

'After we had done what we could we all stayed under cover, crouching against walls and keeping well away from windows and doors. The shelling went on for two hours, filling the streets with choking dust. An anti-tank portée was hit and its ammunition exploded with the hell of a roar.

'As soon as it was safe to go out we found that about half our lorries had been holed by shrapnel, though none was a write-off. By a lucky chance all the new Dodges except Henry's were almost untouched. The infantry had lost two 15-cwt. bugs and sustained three casualties—very light, they told us, considering the stuff that had come over.

'Our damage was chiefly to radiators—five of these were US[4]— and tires. We had thirteen punctures all told and there were things like severed brake rods and leaking petrol tanks. This gave Jack McDonald a chance to show what a smashing LAD corporal he was. He made a trip to Workshops to get a load of spares, and then, helped by " Snow " Logan,[5] his off-sider, he worked flat out all through the night. By breakfast-time every lorry was mobile except " Bub's ".

'We loaded the infantry's gear aboard and stood by to move off at nine that morning. The story was that the battalion was to act as the vanguard of a 6th Brigade advance on Florence and everything looked pretty business-like. We had been told to roll back our canopies so that the boys could hop off quickly if we ran into any trouble. As you can imagine, we were all pretty keyed up,

[4] Un-serviceable.
[5] Dvr I. G. Logan; farmhand; Taupo; born Waipukurau, 20 Sep 1901.

and then, at about a quarter past nine, they told us the move was off.

'We were still on a moment's notice but we hung around Cerbaia until the evening of the 6th and then embussed the battalion and took it six or seven miles in a north-westerly direction to relieve an Indian outfit. They sent us back to the village to spend the night and the next day we were told we shouldn't be needed any more.'

Wearing a slightly swashbuckling air induced by their straw hats, their coloured scarves, and their adventure in Cerbaia, No. 1 Platoon's drivers joined us in an area near San Casciano (eight or nine miles south of Florence) to which we had moved piecemeal on the 3rd and 4th. During the final phase of the battle for the Paula Line, which had been broken on 3 August, we had built up a stock of 25-pounder ammunition greatly in excess of our normal establishment, and this accounted for our inability to move in one clean shift.

Our new area—but why describe it? The sun-soaked grass, the big and the little hills burdened with vines and villages—they were still with us. As for the rest: either one remembers or one doesn't. Not that there was much to remember: only the ripe fruit, golden apples of the Hesperides and pears shaped like pears and not like old tennis balls or little money-bags; only long golden days with the bees buzzing their hearts out; only short impatient nights, wine-coloured; only Italy leaping into sunlight each morning like a dolphin, leaping into the sunlight like a free spirit who could stay up all night if she wanted to, laughing defiance at that grim wheel other countries are bound on—all bound and rivetted and with no choice but to move soberly under the sun or the grey sky and be dipped back into the darkness every night. Only that and the vines and the common cabbage-whites and the common brimstones moving in the same cloud with the kind of butterflies you see in specimen cases.

For a few days we had time for these things, our sales having declined sharply after the collapse of the Paula Line. The New Zealanders engaged in mopping-up operations—No. 3 Platoon supplied them from a mine-strewn area near Cerbaia, and one of its lorries was badly damaged—were reasonable in their demands.

The official communique announcing that Florence was firmly in our hands was not issued until 22 August (the enemy's idea of an open city being rather different from ours), but the collapse of the Paula Line had decided the city's fate, and in mid-August the Division started to assemble in a rest area near Castellina, ten miles north by west of Siena. We moved back on the 14th, occupying a sheltered, dust-free area by a small stream that was very pleasant to lie in during hot afternoons. From the 17th onwards there was generous day-leave to Siena, but not for the drivers of the load-carriers. Almost at once they began carting surplus ammunition from the old area to a 13th Corps dump a dozen miles west of San Casciano, and next we were told to provide 114 lorries to help the Eighth Army move ammunition from an advanced ammunition depot at Monte San Savino, twenty miles east of Siena, to another near Iesi. Iesi was a name new to us. We drove through Perugia and Foligno, crossed the Apennines, turned north, and ended up five or six miles from the Adriatic coast. After Foligno it was new country all the way, but we saw most of it through a mist of sweat drops and white dust.

Hard on the heels of this job came a general move to the Adriatic sector, and by the end of the month the unit was complete near Iesi and all our ammunition had been brought forward. The 220-mile journey, however, proved too much for some of the lorries and Workshops' casualty ward was full again.

We had time for a quick bathe at Ancona, the famous port near Iesi, and then those insatiable 25-pounders went into action against the Gothic Line in support of the 1st Canadian Corps. The plan was for the Eighth Army to attack in the Adriatic sector, while the Fifth Army was to advance over the Apennines to Bologna. The Canadians' role was to smash the defences in a narrow corridor— part of the Gothic Line—that ran between the mountains and the sea. One end of the corridor was guarded by Pesaro, thirty-five to forty miles up the coast from Ancona, and the other, twenty miles farther on, by Rimini.

On the night of 30-31 August the load-carriers set out for the B Echelon area of the three field regiments with 107 loads of 25- pounder ammunition. The drivers travelled along the coast road for twenty-two miles, turning left just as they were beginning to wonder if the intention was to lead them slap into the battle for

Pesaro, which was flashing and banging in front of them. The convoy went eight or nine miles inland before halting and then some of the lorries were guided forward to the guns. By this time most of the enemy had withdrawn out of range.

On 2 September Pesaro was taken by Polish troops and the field regiments came out of the line. The next day, a Sunday, was a day of prayer and a special service was held for NZASC units. Five years earlier, on another Sunday, Great Britain had declared war on Germany.

Heavy thunder clouds were massing and the air was thick and clammy, and gusts of warm wind tugged at the padre's surplice, drawing attention to his khaki stockings and brown desert boots. We stood close-packed on three sides of a square, our officers out in front. On them—on the senior officers anyway—devolved the responsibility of keeping the hymns going and growling the responses, the rest of us, who could sing so lustily round a wine barrel, being given to silence or sheepish mumblings on these occasions.

The truth of the matter was that many of us had the feeling that we were there under a compulsion as much disciplinary as spiritual, and it weighed as heavily (more heavily perhaps) on those who would have attended the service in any case as on those who regarded it as a parade. And one felt, too, that the padres themselves were not always happy about the situation—that they also were circumscribed by events. Most of their sermons were little masterpieces of tact in which the Prince of Peace, the Light of the World, the Despised and Rejected of men, became an awfully decent Padre —brainy, of course, but with no side at all. It was as though, just before the service, someone (the shade of a Chaplain-General, perhaps) had tapped them on the shoulder, remarking briskly: 'Remember now! Nothing controversial—no dogma. Just general terms—simple, manly stuff. Remind the men of their homes, that always gets them.' At all events there was little or no great preaching, no scourgings from the pulpit, no cleansings of the temple, no wrestlings at Peniel.

It is only fair to say that the vast majority of us would have resented it if there had been, and one should add also that most NZASC church parades were voluntary, though it was not particularly easy to get out of them. They were conducted with no more formality than decorum required, and the Brigadier, who liked to see his units all together on a Sunday when this was possible, had a pleasantly direct manner that went down well with the rank and file. In short, the anti-clericalism in our unit (and there was not a great deal of it) was caused solely by church parades. Few could criticise the way in which the padres performed their secular duties—visiting the sick, organising libraries and recreation centres, listening with patience and sympathy to personal problems. Men like Padre Holland were an asset to any unit.

On the occasion in point—our day of prayer—the NZASC Band was in attendance, so the singing was a little lustier than usual—but as always there was something missing, though three or four hundred were gathered together. Not everyone was sorry when the fat thunder drops started to fall, and the bandsmen, grabbing their instruments, made a dash for shelter.

Later in the day the skies cleared and we beat the 1st Petrol Company 5-nil in the first of a series of NZASC Rugby games.

. .

Turkey through with the Axis, the Red Army on the Prussian border, an Allied landing between Nice and Marseilles, the German Seventh Army surrounded and smashed near Falaise, Florence liberated, Paris liberated, Roumania on the side of the angels—these were our August victories and they were notable ones. But to each and all of them the enemy made reply, as Clemenceau did in the Great War when Petain spoke of defeat: '*Je fais la guerre!*'

On the Italian front the answer was the same. Although the Canadians' first assault had taken them ten miles into the Gothic Line the enemy was fighting back stubbornly, knowing that winter was at his shoulder.

In the second week of September the Division came under the command of 1st Canadian Corps, the artillery going into action almost at once. Rimini was the next major objective, and this could be seen from the ammunition point established by No. 4 Platoon on the 11th. It was near Cattolica, a small town on the coast eleven miles below Rimini. Daily our drivers watched engagements between cruisers of the Royal Navy and the coast defences.

The rest of the unit had left Iesi on the 4th and was now in an area near Mondolfo, nineteen miles down the coast from Pesaro and a few miles inland. It was a pleasant spot, and the officers and drivers of Headquarters were able to put away their tents and take possession of a comfortable villa. The platoons bivouacked on the estate.

We were in this area for well over a week and although we were kept fairly busy we managed both to watch and to play a great deal of football. For a short time a dozen or more of our best players were released from their ordinary duties and allowed to give all their time to training and getting fit, a Divisional Rugby team being in prospect. Our next favourite pastime was bathing from the white beaches of the Adriatic.

Always the sea! If you were composing a piece of music to fit our story you would have to include, as well as the purr of primuses and the fluttering beat of choked engines on cold mornings, the soft thunder of collapsing waves, the long rasp of shingle.

On the 13th of the month the Division moved to a concentration area six or seven miles beyond Pesaro, and on that day New Zealand armour and the 22nd Motorised Battalion went into action on the coast in support of the 3rd Greek Mountain Brigade, the rest of the Division being held in reserve. Our unit moved on the 14th.

We drove along the coast road, stopping and starting with a long line of lorries, jeeps, and ambulances. At almost every turn-off Italian refugees, sitting in carts loaded with furniture and bedding, waited patiently for an opportunity to use the road. They looked as though they had been waiting for hours, lost and forlorn. Many of the carts were drawn by oxen or by horrible gaunt horses, awful spectres of reproach, and many were pushed by hand. In some cases oxen and horses, in defiance of the biblical injunction, had been yoked together, sharp horn neighbour to dribbling eye, fetlock to cleft hoof, ghastly jutting bone to shrunk crupper. In one cart a calm-faced woman was mending a torn sheet with an old sewing machine, putting saints and heroes to shame.

Pesaro, at the end of an avenue of tall trees, had been badly knocked about and the people looked at us with dull eyes. Already, though, children were at play and women were at their immemorial tasks: fetching water from wells, spreading pulped tomatoes on boards in the sun, threshing grain on the pavement outside the cottage.

Our new area was three miles inland from Pesaro and here we camped among vines, moving again on the 17th.

The coast road took us past the old castle of Gradara (which had seen the tragedy of Paolo and Francesca and looked as though it had been designed by Walt Disney for Giant Despair), past the ammunition point at Cattolica, and over the River Conca. Half a mile farther on we turned into the hills, dispersing as best we could in that part of our area not occupied by a battery of 155-millimetre guns. They fired intermittently through the night, making us nervous, but there was no counter-shelling. Not until they moved forward on the 19th were we able to spread out a little.

Our new area covered two or more hillsides, open except for some patches of bamboo, and took in a strip of flat near a by-road. On this there was a vineyard and an ugly square villa in a walled garden. The guns sounded very close and our aircraft could be heard bombing and strafing the German lines.

After dark the sky over the front was lit by sixteen searchlights, which stared all night long at nothing. The long unwavering beams bent over the countryside like lean ghosts, their heads misty blue among the clouds, their stems blue-white and astonishing as though shot from sepulchres. Some of us tried to argue that they marked the boundaries of the little neutral republic of San Marino, which we knew was quite close, but actually they were to light the battlefield for night attacks. Later we heard the term 'artificial moonlight'.

Greek and New Zealand troops entered Rimini on the morning of the 21st and the battle moved on across the Marecchia River. Our Division was now in the thick of the fighting, and the artillery, which had reverted to its normal command, was far enough forward to need a new point. No. 3 Platoon, with detachments from No. 4 and the Ammunition Platoon, opened one a mile or two south of Rimini, a dangerous district because of mines and shells.

Time and time again we were called upon to act as first-line transport and deliver our ammunition to the firing line, and often our drivers had narrow escapes. One night a shell burrowed into the road and exploded beneath a lorry, lifting it off the ground but doing little damage.

Life at Cattolica, though, consisted of more than work and occasional frights. There was leave to Rome and—after the 15th—to Florence. For the members of the 4th Reinforcements there was something better. Major Gibson and twenty-three other ranks left us to return to New Zealand on the day Rimini was entered. It was trying to rain as they drove off but they seemed not to mind. Only Paul—the Count—was reluctant to go. He had found at the last moment that he loved us, as a token of which he left most of his gear behind. He set out for New Zealand clad lightly in a pair of old shorts and nothing else, which was a pity, for the weather had broken.

Our new commanding officer was Major R. P. Latimer,[6] who had joined the unit on 5 June as a captain. Each of our majors, like the fairy god-mothers in the story, had come to us with gifts—Justice, Efficiency, Good Nature, Navigational Ability. Major Latimer brought a New Broom. There were to be changes around the place—a sergeants' mess, a newspaper, a programme of entertainment. All these ideas were excellent, but alas!—ours was a venerable institution and like most institutions of that kind rabidly conservative. For the same reason that a few old gentlemen in England, survivors of an almost extinct species, continue to call a taxi a taximeter-cab and Pall Mall 'Pell Mell', some of us still spoke of No. 1 Platoon as A Section and the unit as the Divisional Ammunition Company.

The idea of a sergeants' mess was found to be impracticable (not that anyone except the sergeants objected to it very much) but the paper was proceeded with ('Paper? Paper? There was no damned nonsense about papers when Percy was boss.'), and in due course the *Amcoy Weekly Times* made its appearance. It was neither better nor worse than other unit newspapers and it deserved a warmer reception than it got. It appeared twice and was heard of no more.

There was one innovation, however, that did survive, and this was a recreation centre organised by Padre D. V. de Candole,[7] who had been attached to us since mid-August. At first a tent was used

---

[6] Maj R. P. Latimer, m.i.d.; assistant company manager, Dunedin; born Dunedin, 10 Mar 1915; OC 1 Amn Coy 21 Sep 1944-2 Feb 1945.

[7] Rev. D. V. de Candole, CF; assistant curate, All Saints' Church, Palmerston North; born Ipswich, England, 7 Sep 1912.

for premises, but when the wet weather came the largest reception room in the ugly villa was requisitioned. Here you could write letters or read, and in the evenings, with the tea-urn bubbling in the corner, there were card parties, lectures, quiz sessions, sing-songs, and an occasional concert by the NZASC Swing Band. Only our 'Pell Mells' remembered those desert days when the voice of Vera Lynn or of Ann Shelton had issued from the bowels of the cooks' lorry and had been thought enough.

Inside the recreation room—the *Albergo* we called it—all was warmth and light and Scotch songs by Dave Falconer; outside, the bedraggled grape vines dripped miserably and lorries sank axle-deep in mud. The summer was over now and the grapes had been gathered in. If you wanted fresh fruit you ate persimmons: there was nothing else. When darkness came it was time to pull down the cover at the back of the lorry and begin searching for the primus pricker. Only our Romeos continued to go out at night, but it was no weather for love. At breakfast nowadays they seldom irritated us by looking smug and self-satisfied—like pleased cats.

Farther up the coast the infantry shivered in their weapon-pits, cursing the black mud and the grey rain. The turning of the Gothic Line had not been followed by its collapse, for the Germans were still in possession of the mountains on our left flank and from these they could dominate the battlefield. None the less the infantry pushed on doggedly, facing Panther turrets, self-propelled guns, mortars, and spandaus. The enemy fought back with skill and courage and there was truth in the subsequent verdict of a British infantryman: 'It weren't rain or bloody mountains held up advance. It were bloody Jerry sitting down behind spandau going blurp-blurp.'

By 27 September our forward troops were only 1000 yards from the Fiumicino River and the old ammunition point was too far from the front. Accordingly No. 3 Platoon returned to the unit area, and No. 4 Platoon and a section from the Ammunition Platoon opened a new point near Viserba, eight miles below the Fiumicino. Here the transport was dispersed in and around the buildings of a large linen factory with guns in action on three sides of it, the congestion being such that the ammunition had to be kept on wheels. An artillery duel took place during the night and enemy shells landed close, spattering the buildings with shrapnel. A few evenings later a plane flew over very low and dropped a stick of armour-piercing bombs 200 yards from the factory.

In spite of these drawbacks it is doubtful if the drivers at the ammunition point envied us the safety of the unit area. They were snug and dry in their large sheds whereas we were floundering in a morass with day and night made hideous by the screaming of bogged lorries. At the end of the month Headquarters forsook its sodden tents and moved into the villa, No. 1 Platoon taking over the vineyard, which, though extremely muddy, was at least accessible from the road.

Daily our transport went backwards and forwards between the unit area and a field maintenance centre near Pesaro, and between the ammunition point and the unit area, using Route 16 (misnamed the Sun Track). This was covered by a film of grease, and at times we might have been driving dodgems at a fun fair for all the control we had over our lorries. Traffic choked the road and passing vehicles flung gobbets of filth at one another. On one side of you were the foothills of the Apennines, pale and rain-swept; on the other was the Adriatic, glimpsed sometimes as a grey bosom heaving under wet silk and sometimes as a flurry of white spume. The drive through Rimini was hardly calculated to remove gloomy impressions. Much of the town had been crushed and ground into the mud and the rest seemed to have been altered by bombs primarily with a view to giving people pneumonia. Through the gaping doors and windows of once busy factories and once fashionable hotels our lorries threw sludge. Civilians, when they saw us coming, pressed tight against walls to avoid a drenching. On the coast road, just beyond the town, there were concrete gun emplacements camouflaged to look like shops or villas. One of these, with sham windows and a wicked dark slit for a huge gun, had the word *Gelata* painted above the door. *Gelata* indeed! There was no ice-cream in Rimini. None for the barefoot children with pinched faces.

Perhaps, with all this rain and all this ruin, there was an excuse for depression. The heroic failure at Arnhem—we had built on Arnhem—was only a few days old, and things on our front were going, if not badly, slowly. No, certainly not badly—not badly anywhere. It was just that everything that was happening now— the advance in Germany, the bombing of German cities into dust—

seemed, after the dramatic victories of early autumn, tame. Then we had been so certain that the Christmas towards which we were journeying and for which we had waited so long would be next Christmas. Now we were not sure.

Not that we were miserable. The word is too strong for our feeling that it was time something else happened—not victories merely but something really good, something heart-warming and exciting. German lorry-drivers had that feeling in 1940.

On 3 October something good happened for over a hundred of us. In our copies of the *NZEF Times* was a statement by Mr. Fraser. Men who had been overseas three years or more, including men of the First, Second, or Third Echelons who had returned to the Division after furlough, would be replaced progressively by men who had not yet had an opportunity to serve overseas and by those who had been overseas only a short time.

Our Fifths and our 'Coconut Bombers' (our Pacific Islanders) went around singing.

. . . .

Meanwhile the infantry had reached the River Fiumicino, which in fine weather you could wade across in gumboots. Now it was capable of drowning a tall man. Here the advance was halted and the guns on both sides began a slogging match. The Germans, who seemed to have more guns than ever, were using everything from 75 to 210-millimetres, and again shells screamed over the ammunition point or burst near it. Our drivers stayed in their linen factory and in the evening watched cinema shows in a big upstairs room.

On 11 October the rain stopped and in bright sunshine New Zealand infantry crossed the Fiumicino.

The Division had moved from the waterlogged coast sector and its new line of advance ran parallel to Route 9, the Great Emilian Way, which led straight to Bologna.

After the rain, the dust. Our drivers worked in a foul yellow fog but once off the road their wheels had only to break the surface to find mud. In the unit area we moved on piecrust.

The rain held off and our troops crossed a river, a canal, and another river. By 19 October infantry and tanks were across the Pisciatello, four or five miles beyond the Fiumicino.

A new ammunition point was needed, so No. 2 Platoon handed

over the old one to No. 3 Platoon, crossed the Rubicon (the river Uso, below the Fiumicino, is identified with the Rubicon of history), turned on to Route 9, and halted near Gambettola, twelve miles west-north-west of Rimini. The large town of Cesena, still in enemy hands, was only four or five miles away, and again our ammunition point was surrounded by guns. The next day, the 20th, the guns moved forward.

The rest of us were still at Cattolica, sunk deep in routine. It was dull, dusty, and appropriate to the yellowing year—and it was rather pleasant.

The guns moved forward, and the Brigadier called a conference of company commanders. The NZASC was to be reorganised drastically. The 6th Reserve Mechanical Transport Company, 18th Tank Transporter Company, and 1st Water Supply Section were not needed any longer and would be disbanded. The 1st Petrol Company and 1st Ammunition Company would each lose a section —their commanding officers were to decide which one.

The next day, while the skies clouded and the wind sang in the vines, we rearranged our loads to meet the new establishment, returning surplus ammunition to Pesaro. No. 2 Platoon, which was to be issued with four-tonners for carrying 25-pounder ammunition, was recalled from Gambettola, and No. 1 Platoon, which was to be left as it was (our senior platoon smiled rather smugly) took its place. At Viserba Captain Delley broke the bad news to No. 4 Platoon. It was not wanted any more. Some drivers would be used to reinforce other platoons, but the rest, irrespective of how long they had been with the unit, would go to Base.

This was an appalling prospect—Siberia and the salt mines— but there was nothing to be said or done. Theirs was the junior platoon, and the drivers, though very bitter, could not complain of injustice. They stood around and dug their toecaps into the mud; they gathered in angry groups and muttered vague threats, but there was nothing to be said or done.

During the day two New Zealand battalions advanced to the line of the River Savio above Cesena and that was as far as they went. A few hours later the Division started to move back to a rest area, leaving its sector to the Canadians.

Early on the 22nd, with needles of rain pricking through the beams of our headlamps and dawn a layer of cold fat resting on the Adriatic, Company headquarters and Nos. 2 and 3 Platoons pulled out from Cattolica. No. 4 Platoon left independently from Viserba and No. 1 Platoon stayed at Gambettola to serve the Artillery, which was to be in the line for another day or two.

We travelled down the coast road as far as the Iesi turn-off and then followed Route 76, making good time. Midday found us halted near the bottom of a deep gorge, and we got out our 'Benghazis'. We were in the Fabriano Gorge but the country was all strange to us. Our destination, they were saying, was quite close. We wondered what it would be like and were not hopeful. Rest areas were all the same—the lorries would be drawn up in lines and we should be put in tents. Certainly there would be parades and inspections, unless the Brigadier found a job for us.

The sun shone weakly, and the mountains were green at the foot and greyish yellow halfway up and purple at the top—parrot colours only more subdued, except at the bottom of the gorge where there was a wedge of sunlight, a meadow like green fire, and a flashing stream.

It was chilly and it was going to rain and we felt homeless. That was how we usually felt after leaving an area in which we had been settled comfortably for weeks; and to halt us just short of our new home (on top of an early breakfast and no wash and a long run) was the surest way of aggravating this feeling. Nor were tempers improved by the certainty of another delay while they made up their minds where they wanted the lorries parked. This was the danger period, this last mile. This was when old friends quarrelled.

'Take her easy, eh?'

'Who's driving this bloody truck?'

'Well, if you feel that way, boy. . . .'

CHAPTER 21

THE *MAIALE*'S CASA

IT was raining heavily, so the *Maiale's Casa* (which, if it means anything in English, means the Pork's House) was as good a place as another in which to hold an indignation meeting. And No. 1 Platoon's drivers were highly indignant. They were almost, but not quite, speechless with indignation, for the unbelievable had happened. They and not the drivers of No. 4 Platoon were to be sold down the river. Indignation is thirsty work and the elders of the platoon were assembled in council round a huge flagon of *vino bianco*, for which had been exchanged (it was no time for niceties of conduct) an equivalent quantity of engine oil.

The *Maiale*—the rightful tenant of the building, which she shared with an old cart, an Ariel motor-cycle, and the platoon's petrol dump—lay on her immense side and snored, presenting a soiled and impassive ham to her unwelcome guests and allowing the damp weather to draw out and accentuate her naturally powerful effluvia. Her days were numbered also. She, too, was powerless to avert fate.

Indignation or no indignation, the following morning saw the platoon's drivers preparing to evacuate their vehicles. As shopkeepers display their wares they laid out on extended tailboards any small trifles that might be expected to interest the simple villagers: German boots, biscuits, marmalade, old socks. By evening everything had been disposed of, and the next day the vehicles—the new Dodges they had worked so hard to make comfortable and keep efficient—were handed over to No. 4 Platoon. No. 3 Platoon was issued with second-hand 3-ton Dodges and No. 2 Platoon with almost new 4½-ton four-wheel-drive Dodges.

These periodic reshuffles, though we resented them on principle, were good in one way: they enabled drivers who were weary of one another to separate without fuss. If David had shared a 3-ton lorry with Jonathan, seen him morning and night, used the same soap and often the same towel, the lovely lament might have been for Saul only.

The new hands had been dealt and No. 1 Platoon was no longer in the game. Luckily there was nothing sulky or grudge-bearing in the platoon's nature. Motions of censure had been proposed and carried (as the *Maiale* could witness) and now, free from all cares and disembarrassed of their vehicles, the drivers could concentrate on making their last days with the unit as pleasant as possible. Some of them commandeered the local police station and the upper story of an old warehouse; others found private rooms.

By now 80 per cent. of our drivers were comfortably housed. Over a hundred members of Headquarters, Workshops, and the Ammunition Platoon—were settled in the only large house in the neighbourhood, the residence of a *Marchese*. He was the *padrone* of the village. We saw him sometimes—a worried little man of no great presence—when he drove up in his baby Fiat to find out if the lavatories were blocked, if his potted palms in the conservatory were being respected, and how the drive was getting on.

The villa had pink walls and green shutters and was square, pretentious, and depressing. The *Marchese*'s quarterings, surrounded by love-knots and bosomy ladies (lamentably unaphrodisiac), were painted all over the ceilings, but the general effect remained dreary. Perhaps it was because of the stone floors, the gaunt stone staircase, the chilly corridors, and the absence of all furniture except gloomy memories, sad and withering impressions, and the ghosts of a dead grandeur. There was no room in the house of which you could say with certainty: 'That was the children's room'.

Not that we criticised it as a billet. It was dry and there was glass in the windows. As for the atmosphere of the place—well, we brought our own atmosphere with us along with our blankets and 'Benghazis'. The ballroom on the first floor—few of us were affected unpleasantly by the writhing pattern of its gold and purple wallpaper—was just what the padre needed for his recreation centre.

To foregather in this room for a farewell party our former No. 1 Platoon drivers, on a showery evening, made their way towards the villa. They came up the drive between dripping laurels or let themselves in through the side door in the high garden wall. Representatives from Headquarters, Workshops, and No. 3 Platoon had been invited; No. 4 Platoon—no, we must get used to calling it No. 1 Platoon—was away from home. (Between 29 October and

7 November it was employed by the Eighth Army, its chief task being to stock a Polish field maintenance centre near Florence from depots in the Arezzo area. We were now, by the way, sixteen miles south-west of Iesi.)

The party, though tinged with sadness, was an unqualified success. When did No. 1 Platoon hold a party that was not? True, the corpulent cherubs on the ceiling had looked down on rarer wines —though the vermouth was not bad and Jock's whisky was excellent—and possibly on choicer viands—though our Field Bakery friends had done even better than usual. Perhaps they had heard wittier and wiser speeches—though none more sincere than the tribute to Sergeant Jock Letham—and perhaps lovelier singing— though members of the Salome Gang were present. What we shall not concede is the possibility of their ever hearing 'Auld Lang Syne' sung with truer sentiment. Some of us, for the shadow of a moment (but perhaps it was Jock's whisky), caught glimpses of Captain Moon and Major Gibson with glasses—not empty.

The party came to its official end and was rushed with what was left of the refreshments to the police station, where it was revived with stimulants. It died horribly in the dog-watch.

Five days later, on the first Sunday of November, while our village street was still smoking with mist after a frosty night, we said goodbye to 'Grumpy', 'Snow', 'Poodle', and George (the last of Captain Moon's chickens), to Captain Thodey and Lieutenant Langley, and to sixty-five others. Some we should see next in New Zealand. Others, and the sooner the better, would rejoin us as reinforcements.

They were driven away in the backs of the strange lorries through the mist and the sunshine and it was the end of a chapter. The lorries went bumping and lurching down the narrow, knobbly street as long, long ago other lorries—smart Bedfords fresh from Base—had bumped over the bare hills at Ikingi and down towards the salt flats. 'Righti-o, you jokers! You can spread out that a-ways. Get weavin'. . . .'

The *Maiale* came out of her *casa* to see what the fuss was about. She blinked in the spreading sunlight and was shooed home.

. . . .

The mist vanished and it was a lovely day, and so was the next and the next. Each morning our windows were blind with frost and we breakfasted in mist, stamping on the iron ground to keep warm. Then came the autumn sunshine, soft and golden like melted honey. Old women sat on stone seats in the open with their knitting and mending and girls took baskets of dirty clothes to the communal wash-place, where the water poured white and icy into an immense trough. They bared their plump brown arms, kilted their skirts, and rubbed and scrubbed in the bright sunshine, chattering like magpies. Golden leaves drifted down from the mountain and the air was sweet with wood smoke and the smell of bonfires and frosty haystacks.

We had decided we were going to like Albacina.

It was small and humble now, but many years ago, so the story went (though how much was history and how much legend we could never discover), streets and splendid buildings had stretched through the whole valley—a great and prosperous city. Then flood or pestilence or some other act of God had swept all away, leaving only a few houses on a hillside. Every old village whose origins are drowned in antiquity has a right to identify itself with lost Atlantis ('I only am escaped alone to tell thee'), but the story of Albacina was probably untrue. Not that it was contradicted by appearances. That look, common to so many Apennine villages, of having rushed into the hills to escape something dreadful was exaggerated here. A few lean and stringy houses had struggled far up the mountain, and others, less athletic, had sought safety in numbers and were huddled together like sheep on the lower slopes. Those of a full habit, such as the police station, which was prevented by its bulk either from climbing or from huddling, were left miserably at the bottom.

In this part of the village you could walk without bending forward, and it was here, in the neighbourhood of the *barbieria*, the *Maiale's casa*, and the largest of Albacina's four wineshops, that No. 2 Platoon was living. Workshops, by squeezing up a narrow and almost perpendicular lane (Via San Venanzio, if you please), had penetrated to the village square, much to the delight of the children, who, until the novelty wore off, were fascinated by everything our drivers did, whether it was cleaning their teeth, dismantling a gearbox, or making a cup of tea. Only Nos. 1 and 3 Platoons were left out in the cold. There was no room for them in the village and they had to content themselves with some soggy meadows between the villa and Route 76, but they joined us in the evenings.

The people of Albacina, tucked away in the Apennines, had seen no fighting. Soldiers therefore were less ugly in their eyes than they would have been if houses had been bombed and shelled, food and bed-clothes stolen, and gardens ground into mud. At first they were shy and cautious but after we had been in the village for a few days they began to warm to us. Perhaps it was because we were polite to the old ladies and gave all our sweets to the children and some crumbs of tobacco to the old gentlemen. Or did they believe we were poor people like themselves who would be with them not only against Nazis and Fascists but also against the larger enemies: exclusion, privilege, and bad faith?

This was the first time we had lived cheek by jowl with Italian villagers and we were able to confirm our suspicions that not all of them were dirty, cunning, and sycophantic. They treated us as guests, and as guests, for the most part, we behaved. The children called us by our first names, and their parents shared their fires with us and their macaroni and *pasta asciutta*. After that it was difficult for the young gentleman who had appropriated a sack of winter potatoes to continue to regard his act as one of pleasant daring and soldierly independence. Chickens and ducks, it was agreed tacitly, were protected birds in Albacina, and pigs were protected animals.

Was it a beautiful village? Well, it was all higgledy-piggledy, with slatternly rooftops, like the bonnets of old crones, leaning across narrow passages, and houses treading on other houses, and every single thing either back to back, face to face, or edge to edge. The church, like a patient schoolmaster surrounded by bothersome pupils, stood in the middle of all this, square and homely. Not so the *Marchese*'s villa. Sulking and exhaling damp odours, it presented the village with a gloomy pink back. In some ways it was like the *Maiale*.

And yet it was beautiful—no form but all the colour in the world and all the charm. It was warm, friendly, happy, and full of children. And more children were coming all the time.

Our duties were light while we were at Albacina. Parades occupied part of the morning and picket duties came round about once a week. The rest of the time was our own. Like most other units we were in the grip of the football fever and one of our earliest cares had been to clear a field. Inter-group matches were in full swing and so was the Freyberg Cup competition. In the latter we got through the first game successfully, beating the 1st Petrol Company 15-6, but two days later we were beaten 6-nil by the 2nd Ammunition Company. There was no disgrace in this, for our conquerors went on to reach the final round, which they lost to the 22nd Battalion after a hard game.

Golden, autumn days! The sun, shock-headed like a dandelion, drops gently towards the crossbar. The backs come down the field for the last time and just for a second, as the crowd gathers its breath, you can hear the slap of leather. Knock on! and the whistle squeaks, then shrills out loud and long: time! Laughing and shoving, the crowd moves over to the transport and the tailboards rattle down.

The autumn twilight goes swiftly and it's dark almost before tea is over. The polished stars come out one by one and lorries leave for Fabriano, the nearest large town, where there's certain to be an ENSA[1] show or a picture. The card-players wander up to the *Marchese*'s ballroom for ' 500 ', and the stay-at-homes (old George for one) climb into bed with their pipes and their Auckland Weeklies. But most of us go visiting. Hardly a soul but has a home to go to—a fire, fed sparingly with brushwood by old Momma, to sit by.

'*Cattivo*,' says Poppa, apologising for the sour red wine.

'*E buono*, Poppa! *E buono!*'

But Poppa knows better. To express its wretchedness he places stiffened fingers beneath his chin-stubble and mournfully wags his old head. If it had been good he would have grinned broadly, tilted his chin, narrowed his eyes into an expression of cunning, and screwed a stubby finger into his cheek: '*Buono! Buono!*'

Few of us have 'Sandy' McKay's mastery of the Italian language and once we have commented on the depravity of Hitler and Mussolini, the beauty of the surrounding country ('*Bella*, Poppa, *molto bella!*') and the fact that New Zealanders at home drink little wine but great quantities of beer, the topics still at our disposal are not many. This is when the children come to our

---

[1] Entertainment National Services Association.

rescue. For an hour past they have been fidgetting with the desire to show off and now they burst into song. They sing charmingly: '*Op! Op! Trotta Cavallino*'—eyes bright, small feet tapping— '*Tournerai*', '*Nel Strada del Bosco*', and our international friend the Woodpecker. Presently they push back the heavy table and start dancing—to the *apparecchio radio* if there is one—otherwise to their own music or to tunes audible only to excited children. Big sister dances with little sister. Clumsy boots shuffle on the stone floor, and shadows, black and monstrous, slip over the whitewashed walls, orange now in the soft, smoky lamplight. Roasting chestnuts crack and leap on the hearth-stone and Poppa plunges a gnarled hand into the ashes to choose a big one for each guest. '*Op! Op! Trotta Cavallino. . . .*' The room becomes stifling, faces shine with heat, and black shadows bend over walls and ceiling and brush across the sweet childlike face of the Virgin Mary; and the tiny lamp burning beneath her image glows brighter.

.   .   .   .

Around our village, like great gentle animals, lay the mountains. The big fellow who slept beside the River Esino, his flank forming one wall of the Fabriano Gorge, was Mount Pietroso. Then came Mount Cimara (that was the one, wasn't it, with the old monastery?), Mount Sella Sporta, and Mount Maliempo.

Naturally we went mountain climbing. It was climbing weather. In a jeep or on a motor-cycle you could get to the top of the big fellow in less than twenty minutes by a rough, winding track that looked like a fire-escape and consisted of corkscrew bends. On foot it was a different matter and you needed the whole afternoon. Fields and orchards struggled with you some of the way, then left you with the underbrush and the golden bushes, where the charcoal-burner, with his little cart and old, snorting donkey, worked from daylight until dark. Higher you went, with the gorge, all splashed with sunlight and great purple shadows, yawning on your left, and Albacina below you like a child's toy (a musical box, say), its bells, far off and flat and chirrupy, mixing in the frosty silence with a sound like blowflies on a sunny window, a faint buzzing sound: lorries going up the gorge. Higher and higher you went with your ears hurting from the cold and your breath coming in steaming puffs as from a kettle. Dabs of vivid pasture (slopping in places through the stone walls that tried to grapple them to the mountain) supported an odd sheep or goat, but these became fewer as you went on, and soon you were among the clouds and the boulders with the mist damp on your face.

Then, when you reached the top, the miracle happened. It was like stepping from the magic beanstalk. The track, instead of looping itself twice round a misty crag and plunging towards sea level, straightened out and led you past fields greener than life and a choppy duckpond rough as a miniature Atlantic to the enchanted village of Poggio San Romualdo, which clung, literally tooth and nail, to its emerald plateau, while the windy sunshine, blast on blast, broke over it like spray. Nothing banged or rattled in Romualdo (for everything not snugged down as on shipboard had carried away long ago) and there was little for the wind to play with except poultry feathers. Of these an unlimited quantity was provided by harassed ducks and chickens (perhaps it was their moulting season) and by fierce roosters whose chrysanthemum-like ruffs were giving them the same kind of trouble that old gentlemen have with umbrellas.

This mixture of wind and sunshine was headier than strong drink and it was a marvel that none of our drivers, tearing home to tea down the fire-escape in a borrowed jeep or on the motor-cycle from the *Maiale's Casa*, broke his neck.

Golden, autumn days. . . . A number of us had leave to Florence at this time, but later, when everything starts slipping into the mist, which shall we think of first when we smell woodsmoke, feel windy sunshine, hear bells: the Boboli gardens or a small Apennine village? Bells—what a place it was for them. On feast days and fast days our village was clangorous from dawn till dusk and on ordinary days, of which there were not many, the Angelus had to be dealt with at morning, noon, and sunset, and even the hours had to be rung in, with a few extra strokes for good measure. It was pleasant enough on a gentle autumn evening: it could be maddening at 3 a.m.

Bells in the mist muffled as from drowned ships . . . flying bell-notes going down the wind with the last leaves . . . bells merry on Sunday morning . . . bells jubilant over a white village. Yes, it snowed while we were at Albacina. We woke one day—it was Friday, 10 November—to find everything white. The funny round

Y

haystacks, each with a stout pole through its middle, looked like iced cakes; shovelfuls of snow, with soft, fat sighs, were slipping from steep rooftops; children were snowballing in the main street. Joyfully the bells clamoured.

Bells had heralded our coming and bells tolled solemnly on the morning we went away. The village came into the street to see us off. Toni the policeman was there, wearing his shabby grey uniform, his beretta, his two-day beard. Assunta was there—Assunta of the dark eyes and the modest bearing whom the 'Young Doctor' had courted so assiduously, visiting her house each evening on the pretext of teaching her English. The Monk was there, the pale young novice who had constituted himself Assunta's spiritual father and so infuriated the 'Young Doctor' by refusing him a clear field—a course of conduct that resulted in tremendous trials of patience and in late hours for all three of them. Riccardo of the baggy plus fours, the smart boy, the wide-awake boy, the boy who knew his Naples and had been around a bit and could put you in touch with the black market—he was there. And so was Vittoria, Albacina's plump beauty, who was never seen in public bare-headed. Her heavy black hair, which the partisans had cut off to punish her for loving a young Fascist, had been one of the glories of the village in the old days. The little *barbiere* was there and the old fat priest—kind, stubbly, rather dirty—and Maria and young Carlo and all the children. In fact everyone was there.

Only the *Maiale* and the *Marchese* were missing, but that was understandable.

CHAPTER 22

WHITE CHRISTMAS

V2 WEAPONS—stratosphere rockets—were falling on England. 'What,' the journalists were asking, 'has happened to Hitler?' The reunion in the Munich Bierkeller had been postponed: therefore he must be ill, mad, dead, or on his way to Japan in a giant submarine. In the east six Allied armies were attacking on a front stretching from the North Sea to the Swiss border. Through successive issues of the *Eighth Army News* the immortal but still maiden Jane, wearing brassieres and pantomime tights, fled from the odious Baloney.

On the Adriatic front the situation had changed only a little since our departure.

From Cesena Route 9 runs west-north-west in a straight line to Bologna, passing through Forli, eleven miles beyond Cesena, and then through Faenza, a further nine miles up the road. Forli had fallen on 9 November and the Eighth Army was now firmly on the line of the Lamone River just in front of Faenza, the capture of which was the New Zealanders' next task. As early as the 17th our field regiments had passed to the command of the Canadians and started to relieve British artillery near Forli, No. 2 Platoon and a section from the Ammunition Platoon opening a point for them the next day at Forlimpopoli, six or seven miles beyond Cesena on Route 9. By the 23rd the rest of us had joined them in this area.

It was bleak and windswept and consisted in the main of large, sodden fields separated from one another by narrow lanes. Beside these ran deep ditches, beyond which, further protected against trespassers by barricades of wet manure, lurked farms and cottages. Farmyards provided the only firm standing in the neighbourhood and into these we managed to cram the greater part of our transport. About 80 per cent of the drivers without vehicles found billets of a kind in lofts and storerooms, sharing them with farm implements and humid odours. The rest lived flinchingly under canvas. The skies wept, the mud slopped into our boots, and it was encouraging to learn that we should be moving to Forli as soon as that overcrowded town could accommodate us.

After delays and disappointments—No. 2 Platoon was dispossessed of a large factory on the northern outskirts of the town by the 2nd Ammunition Company, and 5th Corps denied us the Adolf Hitler barracks on the southern outskirts—we moved to Forli on 1 December.

The billeting officers of 5th Corps must have had a fellow feeling for the old woman who lived in a shoe, for the town was stuffed to bursting point at this time. The streets were crammed with transport and every building with a roof and four walls, or three walls and a bit, sheltered troops—Tommies, Canadians, Indians, New Zealanders, and even Italians. Our own area was pitted with bomb craters and heaped with rubble, but after the engineers had done a few hours' work with a bulldozer we found ourselves quite well off for room. By making use of side-streets we were able to map out a convenient circuit for the ammunition point, and we solved our parking problem by putting No. 2 Platoon in a nearby railway station and lining up the three-tonners beside the pavements like taxis in a cab-rank. Company headquarters and the officers took over a block of undamaged flats (formerly the abode of professors and other gentlemen of consequence) and Workshops moved into a school yard just across the way. By evening our domestic arrangements were complete and we were settled for the winter with a degree of comfort and even elegance that compared more than favourably with the damp and dismal sloppiness of Forlimpopoli.

Although the town made a wonderful target, the best the Germans could do was shell it at night and send over a few fighter-bombers at dusk. They would dash into the ack-ack barrage, bombing and machine-gunning, and slip away at rooftop level. Once an English sergeants' mess was hit by a bomb, but more often than not it was the civilians who suffered in health and property. On one tragic occasion a bomb landed on a crowded church, causing great loss of life.

Route 9, which we used when replenishing the ammunition point from Gambettola, was strafed fairly often, but we were lucky. On 2 December three aircraft attacked Lance-Corporal Bill Ingham's[1]

---

[1] Cpl W. O. Ingham; MM, m.i.d.; bus driver; Auckland; born Albany, Auckland, 10 Mar 1917.

section (No. 2 Platoon) and bullets danced all along the road, but only one vehicle was damaged.

The shelling was hardly more effective than the air raids and if it worried us for a night or two it was because the situation was strange. Lying in bed on a first or second floor, you felt that Forli was spread out like a race card with yourself and the hottest favourite equally a target for the impartial pin. Sometimes shells landed close enough for us to hear the tinkle of broken glass, and once all our water-cart drivers were roused in the small hours and ordered to rush their vehicles to the centre of the town where a large three-storied building had been set on fire by shrapnel.

With the capture of Faenza the shelling stopped, and that brings us to No. 2 Platoon's small but vital part in this exploit and to the first mention of No. 8 Army Jeep Platoon.

. . . . . .

By 8 December the 46th British Division was over the Lamone and had established itself firmly on the far side. The New Zealanders were on the right and their task was to cross the river and relieve the British in their newly-won positions as a preliminary to moving against Faenza. This meant two jobs for the engineers: building a Bailey bridge in the British area and making a mile-long stretch of road to close a gap in the prospective supply route.

In the small hours of the 8th No. 2 Platoon, which had been standing by with its transport stripped of canopies and canopy rails, loaded rubble from ruined buildings in Forli, and by 7 a.m. the last lorry was at the Adolf Hitler barracks, where 210 vehicles were being marshalled. The platoon moved off an hour later. The road under construction (the Lamone road) started rather more than a mile from the southern outskirts of Faenza and ended about the same distance from the new Bailey bridge (Hunter's Bridge).

The area in which the engineers were working was under direct enemy observation and the transport would not have been able to enter it except for a smoke-screen laid by mobile generators. The unloading was done by Basutos, the platoon having its narrowest escape when a Spitfire, harassed by German ack-ack fire, jettisoned its bombs. By four in the afternoon all the vehicles were back in Forli, where they reloaded.

On the 9th the platoon was held up for three hours on Route 9 by shelling, and the next day, while New Zealanders crossed the river to relieve the 46th Division, shells landed near the new road, three vehicles from a British armoured brigade being hit.

The rubble convoys were shelled and mortared again on the 11th but No. 2 Platoon's luck held. That day Canadian troops crossed the Lamone between Faenza and the coast, and our engineers, working behind a thick curtain of smoke, finished Hunter's Bridge, opening the way for the first jeep train to take supplies to the 5th Brigade in its forward positions.

The train was provided by No. 8 Army Jeep Platoon whose story opened in Company headquarters' backyard at two that afternoon under a dull sky.[2] There were five sections in the new platoon, each of which was manned by a party from an NZASC unit. We contributed the platoon commander (Second-Lieutenant R. J. Hudson-Airth[3]), the senior non-commissioned officer (Sergeant 'Sandy' Sanders[4]), and a corporal and seven drivers (No. 5 Section). The men stood about in groups, strange to one another in most cases but drawn together by a common desire for information. Little was known except that the platoon was not part of the Division though it would be attached to our Workshops for maintenance and to the 5th Brigade for operations.

They waited in the yard for three-quarters of an hour (fidgetting partly through cold and partly through excitement) and then went to a nearby Royal Army Service Corps company and took delivery of twenty jeeps with trailers and two amphibious jeeps. These were driven to an area in Forli occupied by Headquarters 5th Brigade, close to which the platoon was given billets and parking space.

[2] The chief appointments on 2 December were: Company HQ, Maj R. P. Latimer, 2 Lt M. L. O'Sullivan (posted 29 May 44), Lt L. A. Cropp (posted 28 Aug 44), WO II A. L. Salmond (posted 2 Sep 43); No. 1 Platoon, Capt A. R. Delley, 2 Lt A. K. Catran (posted 1 Oct 44); No. 2 Platoon, Capt B. J. Williams (posted 23 Aug 44), 2 Lt C. B. P. Hendrey (posted 20 Jun 44); No. 3 Platoon, Capt G. Dykes, Lt H. G. Littlejohn (posted 28 Aug 44); Workshops, Capt A. G. Morris; Ammunition Platoon, Lt H. W. Boyce, 2 Lt K. G. Miles (posted 18 Jul 44), 2 Lt R. J. Hudson-Airth (posted 20 Sep 44).

The following had left us: Capt K. E. May (posted to Advanced Base, 18 Apr 44), Capt C. H. Haig (posted to Base Training Depot, 7 Aug 44), Lt R. K. Davis (posted to 1 NZ Petrol Company, 26 Jan 44), Lt K. L. Richards (posted to 1 NZ Petrol Company, 17 Apr 44).

[3] 2 Lt R. J. Hudson-Airth; electrical salesman; Wellington; born Wellington, 13 Feb 1911.

[4] Sgt T. R. D. Sanders; farmer; Rissington, Hawke's Bay; born Darlington, England, 24 Mar 1915.

No. 8 Army Jeep Platoon—the first platoon of its kind to be manned solely by NZASC drivers—started work at seven that evening when No. 2 Section under Second-Lieutenant Hudson-Airth set out to deliver supplies to the Maori Battalion and the 23rd Battalion. The drivers returned home at breakfast-time the next morning with nothing to report except that driving conditions were bad and that they had been frozen nearly to death, once while waiting for the down-route to open and again while waiting for shelling to stop.

During the day No. 2 Platoon made its fourth trip with rubble to the Lamone road—there were no excitements—and at five in the afternoon the second jeep train moved off. Ten more jeeps with trailers had been drawn that morning and there were twenty-two vehicles in the train. It carried stores and petrol for units of the 5th Brigade, and with it went the six jeeps of No. 5 Section.

A mile and a quarter below Faenza the train turned left off Route 9 and travelled by narrow, muddy tracks to Brickworks Bridge. This spanned a tributary of the Lamone and was a mile and a quarter east of Hunter's Bridge. No traffic had crossed it yet and the approaches were so steep and slippery that three jeeps developed clutch trouble. While this was being attended to the rest of the train was halted on the far side of the bridge by heavy mortar fire. An enemy patrol was reported to be near and when the jeeps moved on every driver had his weapon handy.

Soon the Lamone road was reached. It had been built partly across open country, partly over a minor road, and partly through houses, and though it was marked by white tapes and an occasional lantern our drivers were grateful for the artificial moonlight as they crawled over its half-finished surface. Conditions worsened when the train turned on to a narrow road leading north past the 24th Battalion's forward positions, and presently a jeep slid into the ditch, blocking the rest of the train with its trailer.

'Now Jerry began mortaring the area,' said Percy Tristram.[5] 'George, Merv, "Buster", and I were together, and Jim and Murray were held up lower down the line by a second jeep that had gone over the bank. For a start we sheltered in ditches and behind our vehicles, but Jerry had the road taped and things became so

[5] Dvr P. A. Tristram; wire worker; Wellington; born Hamilton, 2 Feb 1923.

hot that we nipped over to a house occupied by some of the 24th Battalion. Here we crouched against a wall with stuff splintering all around us. It was half an hour before Jerry eased up enough to let us pull out the stuck jeep and push on. By now our boys had laid down a smoke-screen and this was a great help.

'Soon we ran into more trouble. Jim and Murray, who had caught up with us, swung out rather wide on a blocked corner and put their jeep into a ditch. Murray jumped out just as she started to go over but Jim hung on to the wheel and went all the way with the jeep. Neither of the boys was hurt and they joined each other on the road and guided the rest of the convoy past the danger spot.

'On reaching the turn-off to Hunter's Bridge Second-Lieutenant Hudson-Airth was told by our provosts that the bridge could not be used as its approaches were unsafe. The convoy was halted and a check taken, and this showed that six jeeps with trailers were missing. After that Second-Lieutenant Miles,[6] who was with us on the trip, went back with a provost corporal to help salvage the missing transport and bring out any drivers who were in trouble.

'It was now about twenty minutes since Jim and Murray had gone over the bank, and in the meantime a provost had come along and told them that Jerry's forward troops were only 250 yards away. They hadn't known this and they lost no time in doing what the provost suggested—getting the hell out of it back to the house in the 24th's lines. It was about 100 yards down the road, and here they were checked in by the picket and given a cup of tea beside a roaring fire. After they had been there five minutes a patrol came in and said it had been a nightmare getting past the corner by the ditched jeep. Jerry was right on to it with spandaus and mortars. Out in the road later the boys met Second-Lieutenant Miles who told them it was too dangerous to try to do anything about the stuck jeep and suggested they hop a ride back to Forli. Altogether six trailers and two jeeps had to be abandoned, but we got four of the trailers back some days later and one of the jeeps—less front wheels, headlamps, and so on. The rest were destroyed by the enemy.

'By this time it had been decided that what was left of the convoy should try to go forward by an old 46th Division route.

[6] Maj K. G. Miles; clerk; Christchurch; born Christchurch, 10 Jan 1921; Regular soldier.

There was an hour's wait while a bulldozer did what it could to fix the track, and it was eleven before we got to Headquarters 5th Brigade. From here the jeeps were guided to their unloading points and later the convoy reassembled. As the down-route was not due to open for another sixteen hours or so the infantry provided us with accommodation and we were able to get some sleep. We set out for home at two in the afternoon of the 13th, reaching Forli, after a long hold-up near the Lamone road, at half past five in the evening. What a trip! It had taken us a day and a night to do a job we could have done in a couple of hours under normal conditions.'

It had been a bad trip, and others of the same sort were in prospect. Against the jeep drivers were mud, cold, darkness, the nearness of the enemy, and the bad state of their transport (the previous owners had neglected it shamefully). For them were the skill of the Divisional provosts and the heroism of the engineers. On this subject our No. 2 Platoon drivers could speak with authority.

For five days now they had watched the building of the Lamone road under fire and inevitably they compared their own task, which consisted of slipping in one by one and unloading, with that of the engineers. The contrast was especially marked on the morning of the 13th when Shermans of the 4th Armoured Brigade started to move up to the Lamone under cover of a smoke-screen. While shells, mortars, and rockets from *nebelwerfers* came through and over the smoke-screen, the tanks squeaked and rattled along the new road, crushing beneath their great tracks pieces of marble mantlepiece, ornamental tiles, bits of hand-basin—things people had built and bought and lived with. Some mortar bombs landed close but none of the drivers was hurt.

By noon eighty Shermans had entered the 5th Brigade area for the attack on Faenza, and it was this move that delayed the jeep train.

The next day—the 14th—was the last one on which the rubble convoys were needed. There was heavy shellfire while some of our lorries were unloading, and two Basutos and one engineer were wounded.

At eleven that night 427 guns opened fire on the Eighth Army front and the attack started. The plan was for the 56th Division, now in the New Zealanders' old area, to simulate a crossing of the Lamone while the real attack was launched west of it by the New Zealand Division (right), the 10th Indian Division (centre), and the Polish Corps (left). Our troops were to outflank Faenza and capture Celle, a little village a mile and a quarter west of the northern outskirts of the town.

All went well, and by breakfast-time the next morning, when the down-route opened to allow the jeep train to pass through with wounded, Celle was in our hands. There was bitter fighting on the 15th, but the next day saw the Germans withdrawing from Faenza and by the morning of the 17th it was clear.

The jeep train could travel in daylight now and use Route 9 as far as the town. On the afternoon of the 18th nine jeeps under Lieutenant G. H. Littlejohn[7]—Second-Lieutenant Hudson-Airth could not be expected to command every convoy himself so our subalterns took it in turn to relieve him—travelled from Forli to Headquarters 5th Brigade in an hour and a half, which seemed quite wonderful. On the way home a shell landed beside Lieutenant Littlejohn's jeep and peppered him with shrapnel in the left side, making him the unit's first battle casualty in Italy.

By now New Zealand infantry had reached the Senio River a few miles above Faenza, and on the afternoon of the 19th six jeeps, each carrying a gun crew, helped to take a company of the 27th Machine Gun Battalion into the line between Celle and the river. That night the machine-gunners fired in support of a 6th Brigade attack, the object of which was to widen the Senio line by clearing enemy pockets from the east bank. The attack succeeded and only one strongpoint managed to hold out.

Main headquarters 5th Brigade moved to Faenza on the 20th, and on the 21st No. 8 Army Jeep Platoon was told that it was not wanted any longer as the brigade could now be supplied in the ordinary way. Second-Lieutenant Hudson-Airth called his drivers together and thanked them for what they had done; they said it had been a pleasure to work under him. The rest of the day was spent in cleaning and checking the transport before handing it back to the RASC. Some drivers were glad but most were sorry—jeeps

[7] Maj G. H. Littlejohn; student; born NZ, 24 Dec 1922; wounded 18 Dec 1944.

are pleasant things to handle and the platoon had begun to develop character.

But it was too early—and the drivers should have known this—to start grieving or rejoicing. The next morning Second-Lieutenant Hudson-Airth was warned that his platoon would probably be attached to the 4th Armoured Brigade, and on the 23rd, after half the drivers in each section had been replaced by others from the same units, Nos. 3, 4, and 5 Sections were posted respectively to the 20th, 19th, and 18th Armoured Regiments, which were then in Faenza, and Nos. 1 and 2 Sections and the administrative staff joined us in Forli.

The next day was Christmas Eve.

. . . .

It was the best Christmas we ever had in the Army. After breakfast—the cooks had been engaged half the night with more important matters and it was a sketchy meal—we paraded outside the officers' mess, with all the bells in Forli ringing their heads off, and marched through the snow to the Esperia Theatre. Here an NZASC carol service was conducted by Padre Holland. The NZASC Band was on the stage and for once in our lives we made no bones about joining in the singing. Before dismissing us the Brigadier congratulated all units on a year's good work and told us to relax and enjoy ourselves.

It was excellent advice and we took it. Each platoon had made arrangements for a sit-down dinner, and when everything was ready and the great hour arrived how gay and Christmas-like the rooms looked, their walls bright with flags and coloured streamers, their tables with oranges and silver paper and handsome chestnut and amber beer bottles!

The dinner, too, was perfect. There was roast turkey and roast chicken with stuffing, roast pork with apple sauce, mashed and roast potatoes, and cauliflower, cabbage, and green peas. Afterwards there was plum pudding with hard sauce, fruit, nuts, and chocolates. To drink there was beer, vermouth, and *vino bianco*.

Meal times were staggered to allow the Major, who had worked as hard as anyone to give us a good Christmas, to visit each mess and make the looked-for reference to the decorations, the cooking, and the year's work. No one was forgotten. Our drivers from the

Jeep Platoon, bringing a present of wine with them, had dinner with us, and those who had not been able to stay over Christmas Eve were invited to stay the night.

When no one could eat another scrap and all the toasts had been drunk there was community singing, and this was followed later by solos. Presently No. 3 Platoon opened its bar to all-comers. Wise men took a turn in the astringent air or sneaked away to lie down for an hour while the afternoon, wobbling a little, slipped into evening. Evening stayed long enough to have just three vermouths and it was night.

By now Forli was making a considerable noise—indeed, it might have been heard in Faenza if Faenza had been making less noise on her own account. Everyone was talking at the top of his voice, and talking, for the most part, confidentially—for any number of rosy partitions were dividing even the most crowded streets and rooms into little private worlds. But speech is inadequate to express deep feeling—song's the thing. Fortunately everyone was in perfect voice, and that being so everyone who could secure an audience—one was enough, fifty was perfect—burst out singing: not bawling, of course, but just letting it come, sort of smooth and easy. A great evening! One or two of the boys seemed to be getting a bit on the way and that was funny, for the stuff was quite extraordinarily easy to take—not a headache in a hogshead, not a fight or a word out of place. Definitely a good brew! It made you feel you loved everyone and it freed you from a sort of what-ye-may-call-it, so that you were able to tell your friends how you felt about them and say what you really thought about the platoon—the ol' platoon. 'Strornly easy to take. . . .

Eleven and ten got a bit joggled up and it was midnight. Bell notes boomed and tinkled in the frosty dark, and the tall tower in St Andrew's Square, the tallest tower in Forli, which had seen so many Christmasses come and go, brooded above everything—the trampled snow in the streets, the drifts piled high at the intersections, and the smooth expanses on the rooftops speckled all over with black smuts from furiously-puffing drip-burners; the Dorchester Club in the Aeronautical College and the huge, winged statue outside it (which someone, doubtless in the name of decency, had splashed with paint), and Signor Becchi's stove factory where the showers were; the Metro Theatre and the hundreds of little homes of the patient and unconsulted; the rows of lorries and tank-transporters, and the stacks of ammunition in the side-streets, and the turkey bones and parsons' noses and paper streamers; and the empty beer bottles and half-empty glasses of vermouth, and the unwashed dixies, and the noise of people singing and being sick, and all the other manifestations of a joyous and remembering spirit paying homage to the World's Birthday. *Buon Natale!*

. . . .

It is impossible to say what made this particular *Natale* so very *buono*. Perhaps it was the snow. The quantities of food, of warmth, of wine, of everything. Or perhaps we felt in our hearts that this was the last milestone of its kind in our long journey.

But did we feel that? The news was anything but satisfactory. It told of delay and loss, and even more disturbing than its content were the doubts it gave rise to, the possibilities it opened up. Field-Marshal von Rundstedt's drive in the Ardennes was now over a week old, but before it started nothing had been written in the papers or said over the air to suggest that he was capable of launching an offensive on this scale. All the talk had been of collapse and crumbling morale and a quick end. Small wonder if we were puzzled and disconcerted.

Not that we imagined they had deliberately misled us—and by They we meant the Heads, the Experts, the Very Important Persons who were for ever stepping in and out of Lockheed Lodestars with brief-cases—but we were beginning to believe, rightly or wrongly, that They ' just said things '.

Only our own front was making no demands on the public attention. The armies had settled down on either bank of the Senio, and the New Zealand battalions, with Forli and Faenza as winter bases, began working on a system that enabled each of them to spend regular periods in reserve.

The old year drew quietly to a close.

On New Year's Eve, anticipating a seasonal demand for flares, fireworks, and machine-gun ammunition, the Major posted a strong picket in the unit area, and the Jeep Platoon drivers, anticipating a demand for jeeps, took precautions also. Already they had lost one trailer and it was nothing to wake up in the morning to find a battery ground flat.

During the fighting for Faenza, and afterwards while the front was settling down, we had been fairly busy—before Christmas we had increased our surplus holding of 25-pounder ammunition from 14,000 to 18,000 rounds besides carrying out some heavy dumping programmes—but we had little to do in the first week of January and later we had even less. This was because a rationing system had been introduced, ammunition ear-marked for the Italian theatre having been diverted to Greece. Supplies of 3-inch gun, tank, mortar, and small-arms ammunition were affected and the allowance of 25-pounder ammunition per gun per day was fixed first at six and a half rounds and then at five rounds. As a result we were idle most of the time, and the day came when No. 2 Platoon's vehicles had to be sent for a short run to keep them in trim.

Only the jeep drivers were working at all regularly and their task, after 30 December, had been simplified by the withdrawal to Forli of the 18th Armoured Regiment. This enabled the five sections to take it in turn to supply the two regiments remaining in the line. Even so the drivers earned every penny they got.[8]

During late December and throughout January the 19th and 20th Armoured Regiments were employed in giving close support to the 5th and 6th Brigades and some of their Shermans were stationed in the forward defended localities. The jeep drivers had to supply these with ammunition and anything else that was needed—petrol, food, charcoal, rubble, cigarettes, beer. Also they did odd jobs—running messages, evacuating wounded, delivering mail, taking shower parties to Forli. Much of this work was done within range of enemy machine guns and some of it under direct observation. Jeeps of No. 3 Section, while serving the 20th Armoured Regiment between 23 and 30 December, came under shell, mortar, or spandau fire (and sometimes all three at once) on five trips out of six.

Nor when the day's work was done could they count on a sound sleep. Faenza was shelled nearly every night and sometimes it was bombed. There were no casualties, though, and the damage to the transport was slight.

On 30 December the jeep drivers who had been living with us

---

[8] No. 5 Section spent a week with the 18th Armoured Regiment in December and after that it took turn and turn about with No. 4 Section in serving the 19th Armoured Regiment. This was under the command of the 5th Brigade, with one squadron on a gunline, one giving close support to forward infantry, and one in reserve at Faenza.

in Forli moved to billets in the 2nd Ammunition Company's area in another part of the town where there was more room, and the next day Second-Lieutenant Hudson-Airth handed over his command to Captain A. B. Cottrell (2nd Ammunition Company).[9]

By now jeep driving was an occupation almost as rigorous as exploring the South Pole. The weather was so cold that every bomb crater was covered with two or three inches of ice, and clothes put out to dry became as stiff as buckram in a quarter of an hour and long icicles formed on them. The taps of the water carts froze solid, and as soon as you thawed them out with hot water they froze again. On 20 January 40 degrees of frost were recorded in Faenza and 36 degrees in Forli.

Other jeep drivers—Don Rs for instance—were able to improvise cabs or all-weather equipment for their protection, but the Jeep Platoon had to keep its transport cleared for action and any excresences that interfered with vision or with carrying capacity were frowned on. The drivers muffled themselves up like Eskimos but it was impossible to keep warm while driving—hands became numb after a few minutes and wherever there was an inch of bare flesh the frost bit and stung like iodine.

Driving was not only unpleasant—it was difficult and dangerous. The main roads were kept fairly clear of snow but side roads and rubble tracks were often in a terrible state. When the hard frosts came, sludge froze in solid lumps and ruts became knife-edged, cutting and tearing tires. Sometimes a slight thaw was followed by a day's drizzle, and then ice turned to mud, enabling jeep tires to pick up nails and pieces of shrapnel with the infallibility of magnets. No thaw lasted long, however, and again the jeeps would be jumping and bucketing among the ruts.

But our drivers did their job and they did it well. While the platoon was serving the 4th Armoured Brigade—in January alone 10,311 miles were covered—there was only one road accident, a minor collision in which no one was hurt.

At the end of the month Captain Cottrell relinquished his command to Second-Lieutenant G. R. Colston (2nd Ammunition Company),[10] and on the same day No. 8 Army Jeep Platoon ceased to

---

[9] Capt A. B. Cottrell, MC; carrier; Rotorua; born Rotorua, 25 Mar 1915.
[10] Lt G. R. Colston; clerk; Dunedin; born Dunedin, 5 Jul 1916. He was posted to 1 Ammunition Company on 19 February.

exist and the 8th New Zealand Jeep Platoon was formed as a component of the NZASC and placed on our war establishment. In effect it was one of our platoons.

. . . .

The rest of us had spent January very pleasantly. There was little or no work, but who wanted work when the streets were full of snow and the temperature below freezing point? Several times a week we had lunch or dinner at the NAAFI's Dorchester Club— a lengthy business this because we used to eat two meals so that we could drink two glasses of beer—and in the afternoons and evenings we joined the long queue in front of the Metro Cinema. We swept snow from our ammunition and we went for walks. We constructed drip-burners of a new and more lethal pattern and in the evenings sat round them or round Signor Becchi's stoves, taking, it may be, a glass of something—a little sweet albano or some vermouth and *vino bianco* mixed. And as we sipped we talked reverently of the progress of the Russians. By the end of January they were only ninety miles from Berlin.

Meanwhile, less dazzlingly but quite as bravely, American and British forces had bitten into the Ardennes salient, cut it in half, struck at its base, and finally driven von Rundstedt's panzers back into the Reich under a battering from the air that eclipsed even the massacre in the Falaise Gap.

And the Americans had landed on Luzon.

No wonder we were able to say goodbye to our members of the Tongariro draft (there were 112 of them and we started saying goodbye at the beginning of February) in the quiet confidence that we should all meet in New Zealand before next Christmas.[11]

On 2 February Major Latimer handed over his command to an officer who was known, if not through personal contact then through story and legend, to everyone in the unit—Major Coutts. Those who had served under him before were quick to notice that he

---

[11] Among them were Maj Latimer, Lt Catran, and 2 Lt Hudson-Airth. Sgt Denny Wells was commissioned in the field on 13 February, his request to remain with No. 2 Platoon being granted, and on the 23rd we were joined by 2 Lts W. A. Brown, R. W. W. Green, and F. M. Hill. Capt Morris, who had left us on 14 December, was replaced by 2 Lt E. G. Legge (posted 14 Nov 44), and while he was away ill Workshops was commanded first by our old friend 2 Lt Noel Campbell (detached from 1 Petrol Company) and later by another Petrol Company officer, 2 Lt K. R. Drummond.

Burnt-out ammunition dump, Vairano

'Sangro mud was now our element'
—page 323

Unbeaten that season—Ammunition Company football team

Water point at Hove Dump

had lost none of his old thoroughness. One of his first acts was to subject the transport to a frosty scrutiny. He could find little fault with it.

Saying goodbye to the Tongariro draft occupied the greater part of our nights and days for a week. The train was signalled but it didn't come, and the fact that the refreshment room was open all the time was perhaps a mixed blessing. No. 3 Platoon steadied itself long enough to hand in its Dodges and take delivery of thirty-three, six-wheel, 6-ton Macks, and our 'Tongariros' long enough to parade for the Brigadier. ('You're hanging together well, soldier,' he remarked mildly to a suffering member of No. 1 Platoon.) Three days later 105 reinforcements, mostly old members of the unit who had left in November, joined us from Base, and the next morning we cleared our throats for the last time, said 'Well . . .' for the last time, and grinned and waved with a rather overdone heartiness as the 1st Supply Company's lorries whisked our 'Tongariros' from sight. Sad though it was to lose them we gave a sigh almost of relief. Tottering on the brink of the platform and dashing in and out of the refreshment room had unnerved everyone.

A good grievance came opportunely to restore our morale—nineteen members of the disbanded Tank Transporter Company joined us to take charge of the new Macks. As few of us would admit that we were incapable of driving and maintaining, or of learning to drive and maintain, anything on wheels, their intrusion was resented bitterly, though the wisdom of employing drivers who were already experienced in the handling of heavy vehicles was not questioned. Only when we learned that no one who had a claim to stay with the unit would be sent away to make room for the newcomers were our feelings mollified.

When these excitements and distractions were over we settled down to enjoy a succession of sunny, frosty days. Business was not as slack as it had been but it was by no means brisk.

There were a few long trips during February. Early in the month Nos. 1 and 2 Platoons moved the Divisional Cavalry to an area near Fabriano, where the newly-formed 9th Brigade, of which it was to be an infantry unit, was assembling for training. The night, of course, was spent at Albacina, the villagers holding a dance. From then on the new brigade needed training ammunition from time to time and it was a pleasure to supply it. Towards the end of the month our Engineers, with the help of No. 4 Platoon transport, built several bridges across the Lamone for practice. Some of the drivers were not strange to this work, sixteen lorries from the platoon having stood by in Faenza between 19 and 21 December with bridging material for the 7th Field Company. The Jeep Platoon, meanwhile, was still serving the 4th Armoured Brigade, but the work was easier now that winter was nearly over.

The war, too, was nearing its end. Budapest had fallen after a long, bloody battle and the Allies were ashore at Iwo Jima. Turkey was in the war on our side and so was Egypt, Breslau was going to pieces brick by brick, the Russians were only an hour's drive from Berlin, troops of the American First and Ninth Armies were across the Roer, Field-Marshal Montgomery was driving towards the Rhine with more than 1500 tanks, and Germany was dying in terrible, convulsive pain. It was like cancer of the stomach on a planetary scale.

February had ended already and on 3 March we set out for Albacina, leaving detachments from No. 3 Platoon and the Ammunition Platoon to serve the Artillery until its relief was completed, No. 2 Platoon to return our 25-pounder ammunition to Ravenna and help the Engineers with bridging demonstrations, and a section of the Jeep Platoon to serve the 18th Armoured Regiment, which was to stay in the field for another ten days under the command of the Poles. The relief was being carried out by the 5th Kresowa Division.

It was still dark when we turned into Route 9 and the people of Forli were just beginning to stir.

By two in the afternoon we had reached our village.

. . . . . .

Our village hadn't changed and the people were still fond of us. A sort of frilliness had begun to cover hedges and fruit trees and the mountain was looking fine. The *Maiale* had survived Christmas and was still in her *casa*, and while we were moving into the villa the *Marchese* turned up in his baby Fiat to fuss about the electric-light bill and the plumbing. One of the toilets had been blocked during our last visit.

We were at Albacina for three and a half weeks and it was a

happy time. The load-carriers did a few jobs and Workshops laboured to bring the Jeep Platoon's transport up to the mark before the next move. Attached drivers were returned to their parent units, and by the end of the month, forty-six reinforcements having joined us on the 27th, the platoon was manned entirely by our own officers and men—Captain Boyce, Second-Lieutenant Colston, and sixty-two other ranks. Also it had been reorganised on a basis of three sections of ten jeeps and given its own administrative vehicles.

We played football, beating the 1st Petrol Company 12-nil, drawing with the 25th Battalion 9-all, and losing to the 2nd Ammunition Company 15-3. We played hockey as well, beating 1st Petrol Company 5-3, 4th Reserve Mechanical Transport Company 5-nil, and 2nd Ammunition Company 4-1, which made us winners of the NZASC competition. We brushed up our drill and as a result did not disgrace ourselves at an NZASC ceremonial parade for the GOC and Major-General H. K. Kippenberger. For the rest we did picket duties, practised marksmanship on an improvised range in the hills behind the *Marchese*'s villa, amused ourselves, kept out of trouble.

While we were doing these things the Allies crossed the Siegfried Line, raced for the Rhine bridges, captured Cologne, crossed the Rhine at Remagen between Bonn and Coblenz, swept the Germans from the Saar, crossed the Rhine in three more places and drove into the Ruhr. Meanwhile Danzig had fallen and the Russians had entered Austria. For the wretched Germans it was an agony unparalleled in history. 'What will you have? quoth God; pay for it and take it.'

Some of the days were hot but the nights were still cold, and the drivers on picket duty were grateful when the old lady who worked in the garden came down in the evening to light a fire at the gate of the *Marchese's* villa. In the evenings the trees and brushy hedges were alive with birds and full of trapped sunlight, and the air was moist with coming or past showers, and there was apt to be a rainbow, often a double one, balancing between two mountains.

The wineshops opened while it was still light, and after tea people would go into them in groups, shaking off children at the door, and order a small jug of vermouth, which was very expensive, and a large one of white wine, which was not cheap, and mix them together. Then it was pleasant, with the children calling in the street and the cool evening coming in through the window and the air not yet thick with voices and foul smoke, to take a glass. Then it was pleasant to have your say. 'They won't fix those bloody jeeps—no show in the world!' 'The old Jerry, he hangs on, eh?' 'Jane's gone off a lot lately.'

Daily there was more heat in the sun, and the tide of spring broke over the country like green surf, with a flying of green foam and a bursting of white spray and a scudding of amber mist. The birds ran shrieking through the hedges and it was Passion Week, the time of the world's ransom and of spring offensives.

Route of 2nd New Zealand Division: Pesaro to Trieste

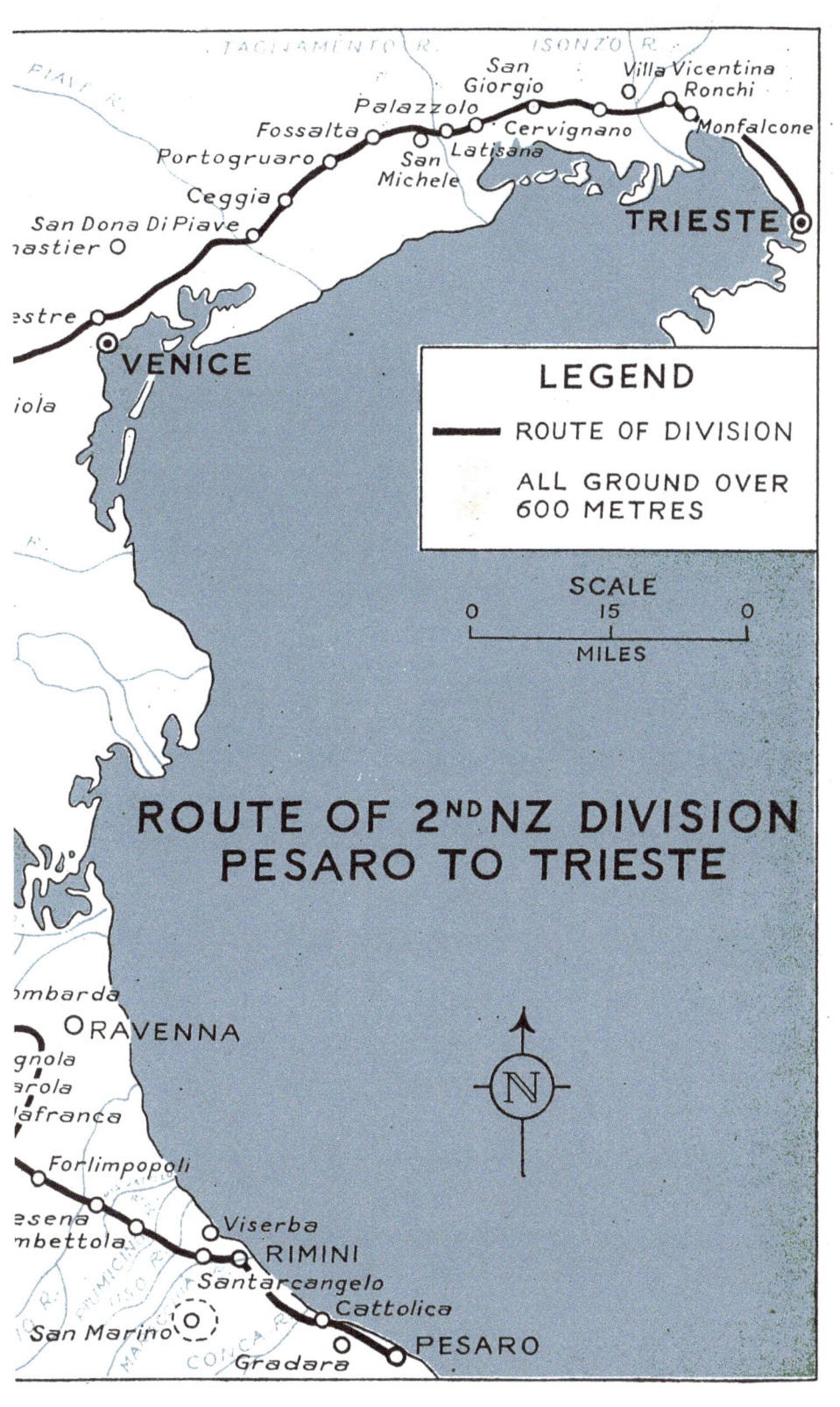

CHAPTER 23

'THY CHASE HAD A BEAST IN VIEW'

(1) *The Rivers*

WE left our adopted village on Good Friday evening with the bells ringing and the sunlight slanting across the mountain. We went first to an area a mile east of Forli, but it was unsatisfactory, and Easter Monday found Company headquarters, Workshops, and the Jeep Platoon in a large white building, once a children's clinic, on the western outskirts of the town. Nos. 2 and 3 Platoons were in paddocks a little distance away, and No. 1 Platoon, which had stayed behind to uplift the Divisional Cavalry, was still in Albacina. Its holding of 25-pounder ammunition had been brought forward by the other lorries and dumped in an area nine miles north of Forli to form the nucleus of an ammunition point. This was opened by the Ammunition Platoon later in the day.[1]

The Division was again under the command of 5th Corps, and that night the 5th and 6th Brigades relieved the 11th British Infantry Brigade on the line of the Senio, north of Faenza. The situation had changed hardly at all since the beginning of March except that the ground was now dry and firm.

During the next few days, while our infantry edged up to the south stopbank of the Senio, the start line for the coming offensive, the drivers of the load-carriers fetched large quantities of ammunition from Ravenna and Cesena, delivering some of it to the gun positions.

Those who were living with Headquarters in the children's clinic had little to do and the warm spring days passed slowly. There was none of that tense excitement that had gripped us before earlier campaigns, and the general feeling, perhaps, was one of yawning impatience for the last battle to begin so that we could finish a long, dirty job.

No one was allowed in Forli, and every evening, to pass the time away, drivers gathered at the clinic gate to exchange badinage with

---

[1] The new 6-ton Macks enabled us to carry as much as we had done before the unit was reduced by a platoon.

the *signorine* and draw conclusions from the great stream of vehicles flowing towards the river. Long after we had gone to bed the highway was loud with traffic and we would say to ourselves between sleep and waking: 'Tomorrow, perhaps'.

The day dawned like any other but by breakfast-time conjecture had crystallised into certainty. Early in the afternoon, exactly when we were expecting them, heavy bombers appeared. They came over twenty-one at a time, flying in arrowhead formation, and the roar of their engines never stopped. Sometimes they were almost transparent, like a shoal of whitebait swimming upstream, and sometimes they caught the sunlight and glittered in the clear sky like twenty-one diamonds. And with them came innumerable fighters.

Later in the day the barrage started and for a long while we listened to the big guns, thudding, thudding, thudding.

'The Two Platoon boys,' we said, 'they'll be in it.' (Four days earlier No. 2 Platoon, after leaving its ammunition with No. 3 Platoon, had passed to the command of the 5th Field Park Company and moved to a farmyard near Villafranca, five miles north of Forli. The next day it had been sent to Ravenna to load bridging.)

In the evening the YMCA Mobile Cinema showed a picture in the grounds of the children's clinic. It was a musical picture in colour and the attendance was large. In the darkness beyond the screen—ages ago it was African darkness and James Cagney— the horizon glowed and flickered, and you could hear, growling beyond Betty Grable's top notes, tanks moving forward to the Senio, night-fighters bolting through the sky, guns thundering.

It was Monday, 9 April.

. . . .

On 9 April, at half past nine in the morning, Captain Williams —this was Jack Williams who had once been our quartermaster— called his men together to tell them the story:[2]

Well, it was tonight, and five bridges were going across the Senio: Woodville high-level and low-level (100-foot and 30-foot), Raglan high-level and low-level (100-foot and 60-foot), and Sey-

---

[2] The platoon's strength had been increased by the attachment of six three-tonners from the Supply Company.

mour low-level (40-foot). The Woodville bridges were the 7th Field Company's job, and the material for the high-level one would be carried by No. 1 Bridging Train (eighteen vehicles under Captain Williams). The 5th Field Park Company would do the Raglan bridges, and the material for the high-level one would be carried by No. 2 Bridging Train (seventeen vehicles under Second-Lieutenant Denny Wells[3]).

The Eighth Army attack was a three-division job—8th Indian Division on the right, New Zealanders centre, 5th Kresowa (Polish) Division left—and the plan was to cross the Senio and get and hold a good bridgehead over the next river—the Santerno—in one operation. (Then, though Captain Williams didn't say this, the Fifth Army was to drive through the mountains to Bologna.)

The New Zealand infantry would go in at twenty minutes past seven and it was thought that the first bridging lorry would be needed an hour later. Bridging sites and assembly areas were likely to come under fire, and it was up to every man to see that he had with him his tin hat and his emergency field dressing. There would be a man with a first-aid kit in each train; if both drivers of a lorry were knocked out the second driver of the nearest lorry was to take over. If a lorry broke down it would be put on tow at once or its load transferred to an empty vehicle. Whatever happened the loads would have to be available when they were wanted. The Divisional Cavalry was providing a covering party but it was up to everybody to look after himself. Keep Bren guns, tommy guns, rifles, etc., handy. It was believed the ditches were clear of mines but it was not known for certain. Well, that was about all. . . .

Morning and afternoon passed slowly. Loads had been checked and rechecked and the moving parts of the bridging oiled and greased. The lorries were in perfect order and there was nothing more to do except think about the night's job.

Everyone knew in theory exactly what it would be like and everyone was familiar with the mechanics of bridging. While practising with the engineers during the winter our drivers had learnt to distinguish between grillage and panels, decking and skin-decking. They knew in what order the materials would be called for and

[3] Lt D. A. Wells; accountant and secretary; Gisborne; born Wellington, 21 May 1912.

how long it would take to unload them. The scene was plain in their minds.

After the bulldozer had carved a passage through the stopbank the first two lorries in the train, or the first three perhaps, would go forward to the bridge site. Swiftly but without bustle, accessories, timber, and base-plates would be off-loaded and stacked near the river's edge, and soon, with beautiful obedience, as though it were being not built but ordered into position, the bridge would begin to take shape. It would be dark and there would be little noise—only the ring of steel and the quiet orders of officers and sergeants. In the background, dim shapes in the darkness, creaking and grumbling, bulldozers would be at work on the approaches, biting, shoving, lifting. . . .

But that was the picture under a quiet sky miles from the front line. How would it be with the enemy only 1000 yards away? How would it be with flares, *nebelwerfers*, mortars, bursts from spandaus, lorries blazing? The drivers remembered what the engineers had told them: 'She'll probably be a fair bastard'. This was in Captain Williams's mind when he wrote in his diary: 'Hope none of my lads gets hurt tonight'.

The silver procession of heavy bombers was a comforting sight. No. 2 Platoon's drivers were in a better position to see it than the rest of us and by three in the afternoon they had counted 1600 aircraft. When the barrage started they heard the crack and bark of the 25-pounders and the deep, mocking guffaws from the heavy guns, and they said:

' " Sport " Williams reckons the artillery's laying on one hundred and forty thousand shells for the first barrage.'

' Hell, that's got Alamein beat! '

' They reckon it's going to be the biggest artillery show ever seen on any front of this size anywhere in the Mediterranean.'

Tea was a hurried meal and appetites were only moderate.

At twenty minutes past five, No. 1 Train, led by Captain Williams in his jeep, pulled out of the farmyard into the Hogg route, which led to an assembly point near Granarola. This was six miles northwest by north of Villafranca and about a mile and a half from the Woodville bridge sites. No. 2 Train moved out under Second-Lieutenant Wells at ten to six and went to the 5th Field Park Company's area, three miles up the Hogg route. The verges of the road were broken, the hedges battered, and the fields on either side of them white with dust.

No. 1 Train, led by six 7th Field Company vehicles carrying low-level bridging, reached the assembly point at half past six. By then the bombardment had been going on for more than three hours and the noise was deafening and almost stupefying. The countryside was alive with leaping flames and the ground beat underfoot like a pulse.

From the assembly point it was impossible to see the river, but its position was marked clearly by a great, ragged curtain of yellow smoke, among which tall columns of a darker shade aspired and dissolved. Into this dreadful pall fighter planes dived continually, and all the time, high above the tumult, like specks of soot supported by an uprush of hot air, two spotting planes hovered. As far as our drivers could see there was no enemy ack-ack fire.

When the sun was on the horizon the guns stopped firing and Wasps and Crocodiles,[4] crouched below the south stopbank, went into action with flame-throwers. Like small, dark dragons—some of our drivers had climbed a haystack and could see them—the Wasps breathed long, slim jets of fire, which arched over the Senio, faltered or seemed to falter for an instant, and then fell in gold sheets on the far stopbank. A violet haze rose over the river, darkening to purple as it mounted and glowing with the heat and fire inside it.

The artillery was silent, fighter-bombers flew up and down above the stopbanks through the smoke, and the infantry crossed the river. It was now twenty-two minutes past seven. The guns fired again, flashing more brilliantly than before, though only two minutes had passed and there was plenty of light left.

Soon German prisoners came down the road, their hands locked behind their heads and their safe-conduct passes clutched in their fingers. There were about seventeen in the first batch and they looked like men who had seen the Last Judgment. Some were smiling, twisting their mouths to express a kind of idiot good humour, but their eyes were empty and quite stupid. One or two of our drivers, curious to hear them speak, called ' *Buono sera* ', and they answered, most of them, with a sort of eager gulp: ' *Buono sera! Buono sera!* '

[4] Bren carriers and Churchill tanks equipped with flame-throwing apparatus.

They had been fine soldiers once, otherwise they would not have been where they were, but nearly all the manhood had been tortured out of them.

More prisoners went past a little while later. They came padding out of the storm, their faces and uniforms stained with dirt and battle, and the horror and strangeness of their ordeal was about them like an aura. Our drivers stared and stared.

When the barrage lifted, which it did twice in the next hour and at five-minute intervals from then on, there was an instant's lull, local and incomplete, like gaps in fog. Then, until the renewed fury drowned everything, you could hear rifles snapping like kindling wood and the rip-rip-rip of spandaus and the steady hammering of Brens.

Our drivers waited beside their lorries at the assembly point, finding difficulty in keeping still. At nine o'clock No. 1 Train was ordered to move towards the site of the Woodville high-level bridge, and it crawled forward with intervals of sixty yards between vehicles, halting on a stretch of straight road about a mile from the river. Here there was another wait.

The bombardment stopped at half past ten and it was quiet under the artificial moonlight. This started from points of dazzling whiteness and arched over the Senio, the beams broadening to spread a livid canopy over the whole battle area. Everything was blue-green—a cold, dead, ghastly colour that reminded you of an aquarium. Faces looked corpse-like, leaves and branches took on the brittle delicacy of coral—of fronds—and the whole strange landscape belonged not to this world but to one long drowned and forgotten.

As the drivers stood beside their lorries, talking in low voices and sneaking puffs from cigarettes held in cupped hands, they heard a low whistle, which rose quickly to a scream. There was a tiny instant of silence and then an explosion as a mortar bomb landed only a yard from Corporal Bill Ingham and his driver. They were lying flat and were not hurt, but their lorry was riddled with holes, and water started to pour from the radiator and petrol from the petrol tank. Bill took charge of the situation at once, telling the drivers of the next vehicle to take the wreck in tow. While the tow-chain was being fastened there was another low whistle, flash, and explosion. Mortar bombs and 88-millimetre shells began to land three or four at a time. Some drivers sheltered under the vehicles, heads against differentials. Others, forgetting about mines, lay down in the ditches, and a few took to the fields. Shells and mortar bombs, which seemed to be coming from the direction of Cotignola, a mile and a half away on the other side of the river, ranged up and down the line of vehicles, some landing beside the road, some on it. They came over with a sort of curved shriek and exploded like a box on the ear.

It was now after eleven, and the engineers' lorries at the head of the column were moving up to the low-level bridge site, which was on the left of the high-level site and only about a hundred yards from it. One of these was hit by shrapnel and the driver killed. The second driver needed help to get the vehicle moving, so Lance-Corporal Duncan McLean[5] (No. 2 Platoon), careless of his own safety, went to the rescue. For this he was later awarded the Military Medal.

By now Captain Williams had a traffic problem to deal with, the lorry at the end of the low-level train having come to grief just past the last crossroads between the bridge sites and the high-level train. With two wheels in the ditch and two in the air, it was blocking the up-route and there was no room even for a jeep to squeeze past.

Being without a winch, the three-tonner from the Supply Company next in line was unable to help, so Captain Williams shunted it down the right-hand turning at the crossroads and brought forward a No. 2 Platoon Dodge. Working under fire and in semi-darkness and with so much noise going on that they had to bellow to make themselves heard (the barrage had started again), 'Chum' Lee[6] and 'Ned' Kelly,[7] directed by Captain Williams, tried to winch the ditched lorry hind foremost on to the road. When it refused to budge Captain Williams went forward to the bridging report centre to borrow a bulldozer.

In a surprisingly short time the bulldozer came rumbling out of the night, and with shattering roars and a sort of ponderous fussiness it started to push and shove.

[5] L-Cpl D. J. McLean, MM; butcher; Balfour: born Balfour, 2 Dec 1917.
[6] L-Cpl J. G. Lee, US Bronze Star; grocer; Ohaupo, Hamilton; born Frankton, 23 Feb 1918.
[7] Dvr W. G. Kelly, m.i.d.: farmer; Te Kauwhata, Waikato; born County Westmeath, Ireland, 7 Jul 1915.

It was now that an orange glow appeared farther down the road. Above it, flames lighting its rolling belly, was an oily cloud, showing that a vehicle had caught fire. It was the one that was badly damaged and on tow. Either it had been hit again or the petrol-soaked ground had been fired by hot shrapnel; anyway it had gone up in flames with a great, gusty sigh. While Bill Ingham cast off the tow-chain other drivers got to work with fire extinguishers. The same thought was in everyone's mind: 'Hell, you could just about see this from the Alps!' The enemy certainly could see it, and after battling with the flames for ten minutes or a quarter of an hour the drivers had to take cover. Some fled to a nearby house—the Villa Agrippina of blessed memory.

The Villa Agrippina was Heaven. Thoughtful German infantry had barricaded its windows with baulks of timber and its walls were comfortingly solid. The drivers sat down on the floor in the large kitchen and never had cigarettes tasted so good. The room was full of engineers, RAP men, and tobacco smoke. At a table officers conferred and a man at a microphone spoke quietly and earnestly to someone called 'Uncle Fox-Fox'. Everyone was in high spirits and the news was good. The infantry had got across with few casualties and all was quiet at the bridge sites. The storm, though, was still raging outside: the burning lorry was still there, and the artificial moonlight, and the straight, narrow road. The idea of leaving the kitchen—the warm, cheerful, tobacco-filled kitchen—appalled everyone, and little enthusiasm was shown when Captain Williams arrived and said there was work to do.

He wanted a panel lorry to replace the one that had caused the trouble by the crossroads. Having failed to scoop it on to the road, the bulldozer had pushed it farther into the ditch (and good riddance), and the up-route was now clear in front of the burning lorry. But all the panel lorries, unfortunately, were behind it, and they could reach the crossroads only by the down-route. Getting a panel lorry on to this meant backing five lorries at the end of the train into a farmyard.

The drivers involved came out of the cosy kitchen into the storm, and when the road was clear the last panel lorry in the train, followed by three decking lorries that were needed to complete the low-level bridge, set out for the river by the down-route, Bill Ingham leading the way in Captain Williams's jeep. Captain Williams stayed behind to assess the damage to his transport.

As always on these occasions it was found to be disproportionate to the sound, the fury, and the expense of spirit. There were holes in trays, chassis, and engine cowlings. One lorry was on tow with a gash in its radiator, and the front of another was smashed. Someone had backed into it while arrangements were being made, rather hastily perhaps, to give the burning lorry more room. Two bottles of Canadian beer had been broken in a tucker box and there were six punctures, most of which had been dealt with already. 'Spieler' Sinclair[8] and Lance-Corporal Sid Bracegirdle[9] had changed a wheel under fire in less than five minutes.

Shells were still coming over but not so often now. Two heavy shells landed near the crossroads while 'Chum' Lee and 'Ned' Kelly were making themselves useful with their winch for the second time that night, the lorry in trouble being the one with that ill-omened load of panels for the low-level bridge.

It was now after midnight, and the high-level bridging was likely to be needed very soon. As the up-route was still closed to a large part of the train by the burning lorry, Captain Williams organised a fire-fighting party. When the blaze sank to a resentful smouldering he went forward to the bridge report centre to ask for a bulldozer. One was sent down immediately, and the hot mass, flaring like a stoked fire at the first touch of the blade, was shoved into the ditch.

The lorries moved forward, passing through a fringe of flame, and at one in the morning the engineers started work on the Woodville high-level bridge. As they were called for the lorries went to the bridge site in groups of three and unloaded. Then they set out independently for the Ravenna bridging dump to reload. It was as simple as that.

By the river all was orderly and quiet, though two men had been killed and several wounded. The battle had ebbed north and the work went forward smoothly and quickly under the artificial moonlight. It was no different from a rehearsal except that the far stopbank was torn and blackened and you could taste all the time

[8] Dvr A. D. Sinclair; shepherd; Tarawahi, Featherston; born Featherston, 14 Oct 1920.
[9] L-Cpl S. Bracegirdle; motor mechanic; Auckland; born Auckland, 12 Dec 1920.

(now faint, now strong, now sour, now sickly-sweet, always evil and darkly exciting) the foul breath of battlefields.

Back at the crossroads the ditched panel lorry was still making a nuisance of itself, one wheel being thrust spitefully across the road. By squeezing against the tire the 3½-ton Dodges were just able to get past, but many of them would have ended in the ditch if it had not been for Captain Williams and Bill Ingham. They gave directions under fire, and it was entirely due to their efforts that the bridging arrived on time. As a result Captain Williams was awarded an immediate Military Cross and Bill an immediate Military Medal.

By one in the morning the low-level bridge was finished, and on the far stopbank the bulldozers began to carve a passage for the long column of tanks that was waiting to move across the river.

Meanwhile, under artificial moonlight and enemy fire, work had been going forward on the other bridges—Seymour low-level, half a mile west of the Woodville bridges, and Raglan high-level and low-level, half a mile south-west of Seymour.

By half past two in the morning traffic was moving over the three low-level bridges and work on the high-level ones was going smoothly. The No. 2 Platoon vehicles in No. 2 Train came under heavy mortar and shell fire while waiting a quarter of a mile from the river with material for the Raglan high-level bridge, but none of our drivers was hurt.

The night passed and the sky lightened in the east and it was nearly six. In two hours the Woodville high-level bridge would be open. Already it was possible to cross the river and stand where the enemy had stood all through the long winter. 'Thus far,' he had said. 'No farther.'

By scorched and ravaged banks the Senio flowed sluggishly. It was narrower and meaner than our drivers had thought it would be and no sewer could possibly have looked less impressive or less worthy of a place in history. A dull gleam was on the water, and from lacerated hedge and tree, drowsily at first but louder as the dawn came and clamorously with the rising sun, birds called and answered. It was a lovely morning.

. . . . .

Raglan high-level bridge was opened three hours after Woodville high-level, and by noon the last No. 2 Platoon lorry had reloaded and was back in the farmyard off the Hogg route. After a meal our drivers serviced their vehicles, patched punctures, and examined, not without complacency, the evidence of their ordeal. Holes not noticed until now were examined and commented on.

Before dusk the platoon was dispersed in a railway yard at Granarola, and at lunch-time the next day it set out for an area two miles north-west of Cotignola. As they approached the Senio the drivers gazed gratefully at the Villa Agrippina, with interest at the burnt-out vehicle, and with solemn pride at the bridges.

That day—it was the 11th—New Zealanders crossed the Santerno and the engineers started to bridge it, our unit being represented by two members of the Jeep Platoon who were attached to the 6th Field Company at that time.

At half past one on the 12th, while the infantry struggled to hold their bridgehead, seventeen No. 2 Platoon vehicles loaded with high-level bridging set out under Captain Williams for an assembly point in the little village of San Martino, some four and a half miles north-west by west of Cotignola and less than a mile from the bridge site.

The vehicles halted in the village street, and there was a long wait while the 8th Field Company worked on the stopbank under fire. Sometimes a shell landed near the head of the bridging train but no damage was done.

San Martino was silent and shuttered but it was not deserted. Something—a warmth, an odour, an absence of desolation—suggested that behind barred doors and windows people were waiting and praying for the wave to break over them and wash away into the distance. A few of our drivers, voicing their contempt for all Fascists to show they were acting not from selfish motives but as instruments of justice, burgled some of the more likely looking houses and came out with clocks, chickens, ornaments, and mattresses. San Martino, holding its breath under the collapsing wave, kept silent.

Soon after four the first two lorries in the train were called to the bridge site and unloaded hurriedly as the front line was only a few hundred yards away. Until daybreak the lorries went forward in twos, and by then our armour was across and what was left of the train was able to move up to the river. As long as the drivers stayed below the level of the stopbank they were safe, but the bridge site itself was still dangerous because of a sniper on the left flank. He was brave but quite fanatical, and when four of his comrades appeared on the stopbank with a white flag he fired on them. It took the artillery to quieten him.

The bridge, a 100-footer, was finished by one in the afternoon, and by five the last lorry had reloaded at Ravenna and was back in the platoon area.

During the day eleven No. 2 Platoon vehicles, part of a train of seventeen under Second-Lieutenant Wells, helped the engineers build another bridge across the Santerno. It was a quiet job, for the enemy was falling back now and preparing to evacuate Massa Lombarda, a fair-sized town two miles beyond the river.

Towards evening the weather turned cooler and clouds appeared. After tea, light rain fell. Being loaded with bridging, many of the lorries were without canopies, so their drivers crouched damply under improvised shelters, hearing the rain on the roof and wondering anxiously if the weather had changed sides. Our fighter-bombers, anyway, were unaffected. A ragged wall of smoke, darker than the darkest cloud, hung over enemy positions on the far side of the Santerno. Into it plunged a succession of Mustangs, stoking the fires they had started to new fury, so that the wall rose ever higher and blacker. They came down in fives, one on the tail of the other, and it was as though some celestial card-sharp, casually dexterous, were dealing poker hands.

By morning the sky was clear and it promised to be a lovely day. Massa Lombarda had fallen at midnight, and after breakfast Second-Lieutenant Wells and a sergeant left to reconnoitre the platoon's next area, returning with the news that it was pleasant enough as a place but that the drivers must expect to see some grisly sights. The platoon moved after lunch, travelling in convoy behind the 5th Field Park Company.

Beyond the Santerno dead mules lay by the roadside and near them was a dead German. He was lying on his back in an orchard and he was pitiful in his meanness and ugliness. Death, which dignifies kings and statesmen, is unkind to soldiers.

The new area was three-quarters of a mile from Massa Lombarda (eight miles north-west of Cotignola) and the field regiments were dug in all round it. Like the last area, it was planted in fruit trees

Easter Sunday at San Agapito

Cassino under shellfire

Fiume Piave

Thanksgiving service at Perugia

—long lines of cherry, peach, and mulberry—and in the spaces between them alfalfa grew. While Italian peasants spread despairing palms over the fate of their alfalfa the lorries were parked under the trees beside long-dry drains that could be turned without much trouble into good slit-trenches. Above these drains, stretched on spars between tree trunks, was a system of parallel wires. These were for vines to cling to but they did equally well as a framework for bivouacs and canopy covers.

The work of enlarging the drains started almost at once, for no sooner had the platoon settled down than it came under shellfire. Shrapnel whizzed between the fruit trees and pinged viciously against the transport, and our drivers, stopping only to flatten out when the whistles sounded very near, dug furiously, quite altering the appearance of the area in five minutes. Shells came over throughout the afternoon and at tea-time it was learned that Divisional Headquarters had been forced to leave a neighbouring area after sustaining sixteen casualties.

That night, while the Division pressed on towards the next river —the Sillaro—our drivers were kept awake by the artillery. The 25-pounders fired all night and shells from a battery of heavy guns roared overhead like express trains going through a station.

By dawn New Zealand infantry were across the Sillaro and late in the afternoon, when the news was of bitter fighting in a small bridgehead, two trains of eight vehicles loaded with assault bridging were ordered to stand by in the platoon area. They waited all through the night and all the next day, moving at last at half past seven in the evening. At last light the whole New Zealand line was to advance behind a terrific barrage while the 6th Field Company built two low-level bridges across the river. The sites were about one hundred yards apart and about five miles west by north of Massa Lombarda.

The two trains, one under Captain Williams and one under Second-Lieutenant Wells, drove through the soft spring evening to rendezvous with guides who were waiting on a stretch of straight road some distance from the river. Here the trains halted. Away on the left among trees a rifle gave an occasional crack but no one took notice of it until bullets sang between the drivers of the leading vehicles. After that everyone avoided the skyline.

At half past eight, just as the sun was going down, both trains set out for their respective assembly areas near the stopbank.

The barrage had started now and although it was no louder than the Senio barrage (comparisons are impossible above a certain pitch) the drivers were more vividly conscious this time of the immense weight of metal screaming over them. It was like being in a high wind, and you were tempted to put your arms over your head. As night closed in the trees and houses sheltering the guns leapt out of the darkness with mechanical regularity like sky signs.

Driving along a stretch of road that ran parallel to the front, our drivers felt peculiarly exposed and naked. On their left, where the guns were firing, the whole countryside was jittering in a mad torch dance, and on their right, bathed in artificial moonlight, silver meadows and ploughed fields, newly minted, stretched away to the river. It was like driving across a stage.

After leaving the main road the trains moved slowly towards the assembly areas, taking a long time to reach them because of bad bends and the darkness. Captain Williams's train halted in open country near a farmhouse occupied by Headquarters 22nd Battalion.

The barrage was still going full blast and the sky above was in torment. In shoals, in endless processions, the shells came over, and the sound they made was almost human in its vindictiveness. They seemed to be saying, and fantastically there was a touch of Cockney in their accent: 'Yew!—Yew!—AND YEW!'

Captain Williams went from lorry to lorry, shouting: 'The show's going very well. Everything's *kapai*! The 27th boys are well across on our left and the 22nd on our right. They've got a swag of prisoners and they're doing fine.'

'"Old Sport's" as pleased as hell,' said someone. 'He thinks it's just Christmas.' The drivers grinned approvingly, grateful to Captain Williams for his habit of passing on information. Some officers treated it like racing tips.

In groups of three to indicate brigade boundaries, of ten to indicate the end of pauses in the barrage programme, red tracer shells from Bofors swam across the sky in a perfect geometrical pattern. Brilliant against Prussian blue, unhurried, travelling with the oiled smoothness of billiard balls, they sailed over the Sillaro and vanished with a blink of yellow. This happened time after time and the fascination of it never failed.

When the barrage stopped it was followed by an uneasy silence and everyone had the same thought: 'Now it's *his* turn.' Presently shells began to land between the river and the assembly area, and next there was a pattern of crumps, and then—much nearer—a whine and an explosion. Away to the left, fifty yards from the bridge site to which Second-Lieutenant Wells's train was waiting to go forward, a Sherman tank was burning rosily.

The first lorry was called for shortly before ten and the drivers found they needed all their nerve to move steadily towards the blaze. Seated in the high cab they felt they were ten feet from the ground. The tank was lying close to the floodbank with flames and smoke spouting from its open hatch as from a Benghazi burner. Its tracks were bordered with a yellow frill and the air was sickly with the stench of burning rubber. The entire bridge site was lit up, and when the lorry swung round to climb the approach to the floodbank its windscreen flashed brilliantly as though signalling to the enemy. For hours the Sherman went on burning with a horrible slow thoroughness, drawing shells, mortars, and machine-gun fire.

The bridge was completed at half past one at a cost of two casualties.

Captain Williams's drivers, meanwhile, had been forced to take cover from mortaring and shelling, most of them either in the 22nd Battalion's farmhouse or under haystacks. Drivers who had been to the bridge site said the work was going fairly smoothly and there was not much danger. The track, though, was very bad. One lorry returned with four prisoners, boys who should still have been at school. They had waded through the river to give themselves up and they were shivering with cold and excitement.

The enemy seemed to be feeling about for targets. Shells groped their way towards the farmhouse and then turned back just as they were becoming dangerous. Away on the right, where the 8th Field Company was building a low-level bridge, two men were killed and three wounded. Later, lorries from Captain Williams's train came under fairly heavy fire at their bridge site. 'Ned' Kelly and 'Chum' Lee,[10] backing to avoid a Honey tank, were blown out of their cab, escaping injury by a miracle. They found two jagged

---

[10] For his work in this campaign Dvr J. G. Lee was awarded the United States Bronze Star.

holes under the driver's seat and a hole in the petrol tank. Another lorry was holed nearby.

At the very last, when it seemed as though the job would be finished without casualties, a shell landed beside a jeep carrying Second-Lieutenant K. R. C. Rowe,[11] the liaison officer attached to No. 2 Platoon from the 5th Field Park Company. He escaped with a scratched nose but his driver was wounded in a leg and arm.

As soon as the lorries were unloaded they left independently for Lugo, four or five miles east-south-east of Massa Lombarda, which was now their replenishment point. Lumbering through the night with shafts of artificial moonlight bending over them like boomerangs and the sounds of battle dying away in the distance, our drivers felt fine. Roaring, rattling, raising great columns of dust, shaking the whole earth, the tanks were going forward to the bridges—*their* bridges.

. . . . . .

Spring with a green explosion had shattered our winter sleep and from all the coverts in the Po Valley the quarry was being flushed, but our unit was still at Forli.

With impatience, with envy, with an attempt at philosophical detachment—'Forli will do us at this stage of the piece'—Company headquarters, Workshops, and the Jeep Platoon waited at the children's clinic while the battle swept on out of sight and out of sound—across the Senio, across the Santerno, across the Sillaro. Forli was now a backwater.

No one resented this more than the Jeep Platoon drivers. They had come to look on the forward areas as their proper environment and it galled them cruelly to be left in the rear with Headquarters and Workshops. They envied No. 1 Platoon and the Ammunition Platoon their forward point near Massa Lombarda and No. 3 Platoon the day-and-night job of replenishing it, but most of all they envied the five jeep drivers who had been attached to the 6th Field Company since 6 April, and the two who had been attached to the 36th Survey Battery since the 12th. The latter returned to the unit on the 17th and were able to talk, in a way highly irritating to their colleagues, of snipers, ambushes, heavy concentrations of fire, and prisoners. ('As soon as we pulled up, three Germans

[11] Lt K. R. C. Rowe; architect; Wellington; born Wellington, 16 Dec 1910.

rushed out of the house and dived into their dugout. When we started to close in—there were about six of us—they dropped their weapons and put their hands up. One of them, a little joker aged about 15, burst into tears and held out his wallet.')

By this time—the 17th—the New Zealanders were twenty miles from their starting point. Two days earlier they had come under the command of 13th Corps and their role was no longer subsidiary to the Fifth Army's: they were to smash through to Venice and Trieste. Soon no New Zealander would be idle.

On the evening of the 17th, instead of writing in his diary 'Very quiet today and I'm off to bed early' or 'Another warm day spent in loafing', Corporal Ted Paul[12] (Workshops) was able to record:

> Tuesday—We left Forli at nine this morning and came twenty-eight miles to an area just south of Massa Lombarda, getting here at midday. The roads are terrible—narrow, full of deep potholes, and covered with a couple of inches of dust. We crossed the Senio and Santerno and found both of them disappointing.
>
> All along the road, especially near the rivers and canals, we saw signs of bombing and fighting. Huge bomb-craters, often so close together that their edges overlapped, were everywhere. Our area tonight is dirty, dusty, and stinking, and there are swarms of flies, and the smell's anything but pleasant. The big guns have been going all afternoon and the enemy has been bombed heavily. It is now half past six and planes are going over in force and the guns are still booming. The sun is fairly high but a pall of dust has risen fog-like and it threatens to hide the sun long before it sets. No. 1 Platoon and some of the Ammunition Platoon are moving forward to open an ammunition point on the far side of the Sillaro. . . .

Later the news was read. The measured voice that had warned us of disasters in Greece and Crete and Tobruk told us that American troops had reached the Czech border, that Germany proper was cut in two, that Marshal Zhukov was reported to be only twenty-eight miles from Berlin, that the Fifth Army, since launching its large-scale offensive the day before, had been making progress against fanatical German resistance and was closing in on Bologna.

In their dusty, stinking area, without a thought for the comforts of the children's clinic, the drivers made ready for bed.

[12] Cpl E. Paul; journalist; Christchurch; born Frankton Junction, 22 Sep 1907.

No. 2 Platoon moved on the 17th also. After an early lunch it left for an area near Medicina (nine or ten miles west by north of Massa Lombarda) and travelled along the main road through open country. The beast was only just out of view, as evidence of which our drivers passed two burning troop-carriers, a burnt-out Sherman, and a smouldering Honey tank. A huge dead horse, swollen and statuesque and looking for all the world as though it had tumbled from a plinth, lay on its back by the roadside, its legs stiffened in an heroic attitude. Cattle, from which civilians were ghoulishly carving steaks, blew gas among the long grass, and nearby there were German soldiers whom none had had time to bury. A dead Tommy, his shock of ginger hair pink under its powdering of dust, lay on an embankment as though asleep. Hardly a house had escaped damage, but the fields and orchards were still fairly orderly except where an acre or so had been torn up by desperate fighting. The crops, tender and green as lettuce, stood ripening under the bright sun, and everywhere Italian farmers were going about their work, ignoring the stream of transport, the crumbling houses, the intolerable stench of death.

Once a British spotting aircraft flew low over the fields and a little girl in a red dress, after looking round for cover, dropped in her tracks like a trained soldier. Presently she picked herself up, dusted her dress, and trotted off.

The new area, which was reached at three in the afternoon, was no different from the last—fruit trees and alfalfa.

On the 18th, while New Zealanders pushed on towards the Gaiana River, and during the night of the 18th-19th, while they crossed it under savage fire on a front some two and a half miles north-west of Medicina, No. 2 Platoon rested. On the 19th only four vehicles were employed, and on the 20th two trains of nine vehicles did uneventful jobs for the 7th Field Company. Early in the afternoon two companies of the 26th Battalion crossed the Idice River, the last and strongest barrier before the Po. The Germans had given it a great deal of publicity as the Genghiz Khan Line. In the evening the platoon set out for an area near Budrio, a small town six and a half miles north-west of Medicina and a mile from the river.

Every tank and lorry in Italy seemed to be rolling towards the Idice and the convoy moved through a fog of dust. The sun had gone down before the platoon was dispersed in its new area, an open field peculiarly bleak and inhospitable.

The next day was Saturday, 21 April. Poles entered Bologna at first light, and the Eighth Army, the Idice crossed and the Genghiz Khan Line broken, poured over the plains and pressed the enemy into the great bend of the River Reno between Bologna and the Po. It was a quiet day for No. 2 Platoon, only eight vehicles being employed in bridging the Idice.

At half past eight on the 22nd the platoon moved to an area seven miles north-west of Budrio and eight miles north by east of Bologna.

Beyond the Idice the countryside was different. So far as our drivers could see hardly a house was damaged. The tide of battle had flowed swiftly past, leaving little in its wake beyond an occasional splintered waggon or smashed limber. No longer were there heaps of rubble in which children and old men and women probed miserably for a piece of furniture or a twisted bicycle. Instead everything was neat and trim and the handsome villas were much as their owners had left them.

After travelling a few miles the convoy was shunted into a field beside the road and a reconnaissance party went forward. Some shells landed fairly close at lunch-time and Jimmy N—— was unfortunate enough to mistake a plate of herrings in tomato sauce for his steel helmet.

The new area was reached late in the afternoon. It followed the familiar pattern but a kind of sweetness was upon it. Summer, outdistancing the Eighth Army, had moved forward with a great leap, occupying the whole of northern Italy.

Soon after half past six a small bridging train left to join the 7th Field Company in the neighbouring village of Bentivoglio, where the Germans had blown a bridge over a canal at half past five that morning. Bentivoglio, it was easy to see, had once been charming, but now it was in a bad mess. In an attempt to block the main street the Germans had wrecked a granary, a children's home, and some buildings belonging to a military hospital.

Although Bentivoglio had had a tiring day it turned out in strength to watch our engineers at work and the bridge went across the canal to a continuous murmur of *bravos*. Plainly Bentivoglio thought it was witnessing a miracle.

Not everyone from the village was present. The report of firearms showed where partisans were happily hunting down and liquidating local Fascists in the surrounding fields. Once a jeep sped past, carrying in the back a man in Italian uniform, dying or dead. A gang of partisans, their faces stained by sun and wind as by walnut juice, and grenades hanging from their belts like clusters of fruit, came into the village square with a German sniper. He was a big, lumpy youth, pale and pimply, but he had a kind of sullen courage. One of our corporals wanted to know what would happen to him and the partisans answered with gestures and gay laughter: '*Boom-boom-boom! Finito!*' Their manner was expressive of so much innocent enjoyment that popular feeling—at any rate on the part of the New Zealanders—began to favour the prisoner in spite of the tufts of fresh grass that decorated his steel helmet. There was a chorus of disapproval—'Italian partisan bastards! Game as hell now Jerry's plucked off.' On the corporal's insistence the prisoner was handed over to our infantry, his pimply mask expressing neither relief nor gratitude. The partisans, sulking like children who have been done out of a treat, went away to find another German to play with.

On their way to Medicina to reload the drivers passed Nos. 1 and 3 Platoons, which were heading for an area a few miles from Budrio with orders to open a forward ammunition point by half past eight the next morning. It was not a pleasant night for anyone, for the Luftwaffe, forgotten for weeks, had come suddenly and disconcertingly to life. Perhaps the pilots had been ordered to use up bombs, ammunition, and petrol as an alternative to destroying them or leaving them to the enemy, or it may have been that the air squadrons in Italy had been reinforced to give a fillip to German morale. Aircraft droned backwards and forwards all night long, the sky twinkled with butterfly bombs, and the earth shook. Every so often there was the long, rattling roar of machine guns. None of our drivers was hurt, but the Supply Company suffered ten casualties in an area next to the new ammunition point.

Meanwhile a train of fifteen vehicles had been standing by under Second-Lieutenant Wells in an area near San Giorgio, a village two and a half miles west-north-west of Bentivoglio. This was to go forward to the Reno River when sent for by the 7th Field Company.

The call came at sunrise and the train moved quickly to the bridge site, which was about twenty miles north-north-east of Bologna. It was still early when our drivers got there and at once they were surrounded by excited civilians, most of whom had white flags. They had gone to bed under the New Order and woken up to find New Zealanders, the 23rd Battalion and the Maoris having crossed the river before dawn. A man who was to have been sent to a labour camp that very day grabbed his girl round the waist and did a dance of joy.

The Maoris were still at the bridge site and they were enjoying themselves. In the hurry of departure the Germans had left thirty or more vehicles near the stopbank, a roast chicken, a feast of pork and potatoes, and some beautifully groomed horses. An armourer's caravan filled with spandaus and a cooks' lorry filled with beef and bacon were giving pleasure, and so was a half-tracked vehicle that had been coaxed into life and was roaring and shrieking below the stopbank. A thunder of hooves died away in the distance as Maori huntsmen disappeared after an imaginary fox.

There was no shelling or enemy activity, but the work was slowed down by the awkwardness of the bridge site and three times extra loads of material had to be sent for. In the afternoon Field-Marshal Alexander, Lieutenant-General McCreery (Eighth Army Commander), and Lieutenant-General Freyberg, arrived. They stood on the bridge and talked while the engineers laid skin-decking and an Italian asked innocently: 'Po-leece?' Earlier he had seen a British red-cap.

The bridge was finished at four in the afternoon, by which time two regiments of tanks at the head of a mile and a half of traffic were waiting to move across the river.

In the meantime fourteen lorries had been serving the 5th Field Park Company, and the rest of No. 2 Platoon—the domestic vehicles and two load-carriers—had moved to an area near San Alberto, five miles north of San Giorgio. The New Zealand Division was still well in the lead.

By now the advance had begun to show signs of turning into a triumphal procession. On their way to San Alberto our drivers saw many partisans—here a lorry-load of men red-neckerchiefed and brandishing rifles, there a single buxom young woman with a Sten gun across her shoulders. Groups of children in party frocks held up green branches and bunches of flowers and squealed '*Ciao!*

Ciao!' In San Pietro, a town south of Alberto, people were lining the main street as though for a circus and every lorry was given a special hand-clap. Truly moved, but feeling more than a little foolish, our drivers did their best to appear gracious and at ease, and it was not their fault if they resembled performing sea lions rather more closely than they did liberators.

After tea Captain Williams called his men together in the San Alberto area and gave them the latest news. New Zealanders were racing towards the Po with British armour on their right and Americans and South Africans on their left. Ferrara, thirteen miles to the north-east and the last important bastion before the Po, was in our hands, and there was an unconfirmed report that Americans were across the river on the Fifth Army sector. The nine o'clock news added the information that Russian troops were bashing their way into Berlin from the north, east, and south. Later this long, pleasant day was brought to an end by an issue of stout and beer.

After breakfast the next morning the platoon set out for an area near Bondeno, a small town ten miles west by north of Ferrara and less than three miles south of the Po. There was a long halt on the far side of the Reno and then the lorries moved swiftly along a good road through open country. It was a lovely afternoon and it had everything—blue sky, golden sunshine, rippling cornfields, a slight breeze, careful husbandry: all the ingredients of the Georgics.

The new area—green and expansive, a spectacle of smiling plenty —was dotted with large white flags improvised from counterpanes and tablecloths. These the Italians hauled down as soon as they saw our drivers. They were friendly but their manner made it quite plain that soldiers were a visitation from Heaven like blight or frost—that they trusted their liberators with tablecloths about as far as they could see them.

. . . .

Earlier in the afternoon Second-Lieutenant Colston, eighteen men, and thirteen jeeps had passed to the command of the 6th Brigade, and eighteen men and twelve jeeps to the command of the 5th. Eight men and six jeeps were sent to the 21st Battalion, the same to the

23rd, six men and four jeeps to the 24th, seven men and four jeeps to the 25th, and five men and four jeeps to the 26th.

Very early the next morning (25 April) the 21st and 23rd Battalions crossed the Po almost without incident. Jeeps and six-pounders were ferried over in assault boats and after these came Fantails (armoured and tracked amphibious troop-carriers) and Ducks (wheeled amphibious troop-carriers). By half past three the 23rd Battalion's anti-tank platoon was on the far side of the river and with it was one of our drivers, his jeep towing a six-pounder. The driver of a second jeep would have been there too if the boat in which he was crossing had not gone aground in the mud.[13]

Soon after 5 a.m. the 7th and 8th Field Companies began building rafts for support weapons, guns, tanks, and bulldozers, and two hours later the 6th Field Company started work on a 460-foot pontoon bridge. There was no call for No. 2 Platoon's transport. It stood by all day in the Bondeno area and the drivers were not pleased.

By half past five in the afternoon the pontoon bridge was finished and ready for testing, so Lance-Corporal Sid Bracegirdle and 'Spieler' Sinclair, with their lorry under full load, drove to the bridge site, expecting to see great things. When they arrived they were disappointed. The banks of the Po were more like a contractor's yard than a battlefield and there was nothing to see except a few wrecks, the long straight line of the pontoon bridge, and an expanse of water, pale and colourless in the half-light. Gingerly, watched by an Engineer officer, they drove on to the bridge and moved slowly across it while it bent under them like a tightrope. The test was satisfactory.

They spent the night on the north bank of the Po. All night long a continuous stream of traffic flowed over the bridge and a petulant-looking moon floated in a sky like curdled milk, casting a lunar doubt on the importance of a day's history—a day's history that included the bridging of the Po, the battle for Berlin, a peasant's concern over his best counterpane, and the solemn session of the representatives of forty-seven nations at San Francisco.

[13] The jeeps towing six-pounders for the 21st Battalion were taken across by infantry.

(2) *Drive to a Cricket Match*

No. 3 Platoon's great six-tonners brushed past the bruised hedges, lolloped over potholes, loomed one after another through the dust, long, blunt, grey bonnet following dwarfed tray. The heavy tires crushed everything—spandau boxes, castaway Mausers, German overcoats, German respirators.

Our second-line holding had been doubled for the advance, and though we were helped by two Royal Army Service Corps platoons every load-carrier had to make two trips when the unit or the ammunition point moved forward.

On 18 April we left the Massa Lombarda area for one a mile and a half beyond the Sillaro.

We go forward (wrote Ted Paul), again travelling along dusty, rough roads through a badly battered piece of country. There are some great sights—houses razed, tanks shot up, masses of bomb craters. There are other sights as well and we bury three of them on arriving in the new area, which is next to a forward ammunition point established yesterday by No. 1 Platoon. There are many bodies lying in fields and under hedges near here. All have been robbed of boots and outer clothing. Many are just boys.

Later in the morning Nos. 1 and 2 Platoons and the RASC platoons brought forward ammunition from the old point, and in the afternoon a convoy of ninety-eight vehicles went to Santarcangelo, below Cesena on Route 9, to fetch 25-pounder ammunition. The roads were jammed with traffic, and by the time the lorries began to arrive back, after covering 140 miles, 45,000 rounds of 25-pounder had been sold from the ammunition point and a queue of twenty-eight artillery vehicles was waiting to be served. On the 19th, 39,000 rounds were issued, and on the 20th (while Hitler celebrated his 56th birthday and the Division crashed through the Genghiz Khan Line) 16,000 rounds. The 21st was another busy day. For the load-carriers there was the long, dusty drive to Santarcangelo, for eighteen of the jeeps the job of bringing forward small-arms ammunition from Cesena, for the Axis the loss of Bologna, and for Captain Delley and Second-Lieutenant Miles the distinction of leading the advance for a few minutes. They set out in search of a suitable area for the ammunition point, found themselves forward of the infantry, withdrew. . . .

On the 22nd the ammunition point moved to an area some two miles north-west of Budrio and the drivers of the load-carriers were again very busy. Company headquarters, Workshops, and the Jeep Platoon had a quiet day.

Clear with a cold, blustering wind (wrote Ted Paul). Two or three lots of bombers come over but on the whole there is not much doing. Late this afternoon we see two German planes and the ack-ack opens up only to close down after a few shots. It is reported that the planes carried white flags and landed on the Forli aerodrome. From all accounts many German flyers have been deserting like this during the last few days.

That night, while Nos. 1 and 3 Platoons felt their way along traffic-crowded roads to bring ammunition to the new point, tracers lit up the sky over Bologna. It was the night of the Luftwaffe's surprise appearance.

The unit moved again on the 24th.

Up soon after 5 a.m. (wrote Ted Paul), and away before six. We pass through Medicina and later through the outskirts of Bologna but we don't see much of it. The railway station and some of the suburbs are in a terrible mess. When we pull up in our new area near San Giorgio we have travelled thirty-four miles. The locals present us with eggs and *vino* and refuse payment. They take a tremendous interest in us and express surprise at our appearance, the Germans, who left here three nights ago per bullock-drawn motor trucks, having told them that the Kiwi soldier is a black man who hacks civilians about with a knife which he carries in his mouth.

Our new area—a big yard with four two-storied brick houses—is near the ammunition point, which also moved up today. We are surrounded by tall poplars, fruit trees, grape vines, and green, green pastures. It really is a beautiful district. Today's big rumour, supposed to have come from a German officer taken prisoner, is that the German armies in Italy will capitulate tomorrow.

Early the next morning, while the shadows from the poplars shortened inch by inch and the engineers worked beside the Po and the war went on unchecked, No. 1 Platoon, No. 3 Platoon, and the Ammunition Platoon went to an area near Bondeno and opened a new ammunition point. The rest of the unit followed later in the day.

We drive through beautiful country and have plenty of time in which to admire it. There is a tremendous stream of traffic on the roads—lorries travelling nose-to-tail—and we take just two and three-quarter hours to cover the first eight miles. We travel twenty-one miles and get to our new area at half past three. The people here are not over-friendly but they are interested in us. The advance is going so fast that it is impossible to pick up all the waifs and strays—especially as many houses are sheltering German soldiers dressed as civilians. Two Kiwis killed by snipers today.

We are just three miles short of the Po. The old truck is parked alongside a hayloft and the level countryside stretches away into a dusty haze.

. . . .

The field was in full cry. The scent was fresh every morning and the spoor plain. When darkness closed in the quarry could be seen disappearing among dust and shadows. *Thy chase had a beast in view.*

On 26 April the Division advanced from the Po bridgehead to the Adige River, eleven miles away, and No. 2 Platoon drove over the Po and went five miles north, halting near the village of Trecenta. That night the brigades crossed the 150-yard stretch of the Adige and with them went amphibious tanks and Fantails. Two Jeep Platoon drivers, their vehicles loaded with ammunition for B Company of the 23rd Battalion, were taken over in assault boats.

Later, a bridging train of six vehicles from No. 2 Platoon moved with the 5th Field Park Company to a staging area near the river and stood by until daylight in heavy rain. Then the lorries were called forward to the stopbank and the engineers started to build a 40-foot raft at a point five and a half miles north-north-east of Trecenta. Elsewhere a pontoon bridge was going across.

At midday the rest of the platoon moved to Crocetta, a village three miles north-north-east of Trecenta, and halted in the main street. Here the Germans had left two field guns from which they had not had time to remove even the grease and brown paper protecting the breech-blocks. Nearby, anything but brand-new, squatted between thirty and forty prisoners. Our drivers, questioning them in Italian, learned that they had given their officers the slip and hidden themselves in a house with the intention of surrendering. They were Austrian mechanics—in fact fellow tradesmen—practically workmates. Photographs of girl friends and chubby babies began to circulate and cigarettes were exchanged. Soon victors and vanquished were showing the bewildered villagers how little five years and eight months of bitter warfare weigh in the balance against a shared trade, an impulse of curiosity, and the common man's feeling that after a fight it is proper to shake hands. It was as well that no wine was available. A few litres of 'Purple Death' and they would have been hanging round each other's necks and harmonising.

That night (27-28 April) the 9th Brigade relieved the 6th Brigade and the Gurkhas relieved the 5th. The jeep drivers who had been with the infantry returned to their platoon, but not for a rest. Late that evening nine drivers and six jeeps were sent to each of the 5th and 6th Brigades, and eleven drivers and eight jeeps to the 9th. The next morning the 12th Lancers, followed by the Divisional Cavalry Battalion, the 27th Battalion, Headquarters 9th Brigade, and the 22nd Battalion, started for Venice.

No. 2 Platoon moved at 10 a.m., crossing the Adige and travelling by traffic-choked roads, muddy and slippery after the rain, to an area near Piacenza, six miles north-east of Crocetta, where it laagered beside a canal in a green meadow gay with daffodils. This the lorries at once turned into a morass.

After lunch eleven vehicles left to bridge a canal at a point three miles north-east of the area. A tank was at the bridge site and Gurkha infantry were in covering positions beside the road. By five in the afternoon the job was finished and the lorries, splashing through a heavy downpour, set out for Castel Guglielmo, six miles below the pontoon bridge, to reload.

While the rest of the platoon was eating its tea and shivering in the watery sunshine, an Italian civilian arrived in a great state of fuss to point a shaking finger at a large white building about a mile away and report that more than sixty Germans were in possession of it. Captain Williams, with an abundance of volunteers to choose from—no one was prepared to miss such an excellent opportunity of firing a shot in anger and collecting perhaps a German watch—organised three fighting patrols, and these set out somewhat incautiously across the sopping paddocks. A small armoured car, a Dingo borrowed from a neighbouring unit, went with them, and Captain Williams, revolver in hand, led the way. The plan, apparently, was to carry the building by assault. Talking heatedly about the importance of not bunching, our drivers advanced in tight little knots under an increasing barrage of Italian assurance that there was not a German within miles. They occupied the building without opposition and found it empty, the infantry having mopped up the district earlier in the afternoon. A little dashed they returned to listen to the news.

The news was quite exceptionally worth listening to. Himmler had offered unconditional surrender to Britain and America. Brescia and Bergamo, great cities at the foot of the Alps, were in our hands. Four-fifths of Berlin had fallen to Marshals Zhukov and Koniev. Graziani had been caught by partisans—so, according to reports, had Mussolini.

Darkness came down like a wet dishcloth and there was nothing to hear except drumming rain—no guns, no mortars, no sounds of war. During the day New Zealand motorised infantry had driven north-east, splitting the German defence, breaking the Venetian line, and for the first time finding that soft underbelly to which Mr. Churchill had referred hopefully long ago. By 1 a.m. on 29 April they had reached Padua, an important road centre thirty miles north-east of the Adige bridges and twenty-three miles from Venice.

At 9 a.m. on the 29th, No. 2 Platoon, following the 5th Field Park Company, plunged into the stream of traffic flowing towards Padua. The stream flowed smoothly as far as Este, where the main Padua road was reached, but at this point it started to clog, and soon everything on the road—lorries, guns, ambulances—was welded together in a solid line, down which each check was transmitted jarringly. At Monselice there was a long halt in the shadow of green hills, and from then on, while bursts of sunshine alternated with bursts of rain, the convoy crept and crawled, halting every few minutes. None the less Padua was reached before dusk.

The vehicles were parked in a side-street and the drivers told to settle down for the night. The atmosphere, though, was unsettling. In Padua, where St. Anthony worked and preached in the thirteenth century, partisans were busily proving that new machine guns sweep clean. Shots rang out and hand grenades exploded, and doubtless our drivers would have stayed quietly in their side-street had no one told them that there was a German supply dump only 400 yards away. Partisan guards, some of them very drunk, were

'THY CHASE HAD A BEAST IN VIEW' 433

turning away civilians but allowing New Zealanders to help themselves.

The inside of the building was a dipsomaniac's dream—an Aladdin's Cave of alcoholic delights. It was piled high with cases of three-star cognac, kummel, cherry brandy, *acquavitae*, and eggnog. There were some disadvantages. A hunt for Fascists was in progress close by and every so often the building would be shaken by an explosion. It was not easy to see for smoke and the air was pungent with the fumes of cordite, but this was no concern of our drivers. Their business was with the free liquor. They carried it away on bicycles and in hand-carts lease-loaned by the Italians, and by nightfall a huge number of cases had been liberated as well as a large quantity of sugar. There were no interruptions, the guards being interested only in preventing civilians from taking anything. The caretaker had been liquidated earlier in the afternoon, an incautious protest having shown him for what he was— a black-hearted, collaborating Fascist.

At eight that evening the 9th Brigade reached the Piave River at a point eighteen miles north-east of Venice, the 22nd Battalion entering San Dona on the far bank. Blown bridges made a pause necessary, and all night long tanks, guns, and lorries drove through dust and darkness to catch up with the head of the Division. Ammunition and other supplies were far in the rear but they were coming up fast. Our own unit was at Trecenta, just north of the Po, with Nos. 1 and 3 Platoons working like tigers.

Under clouded skies No. 2 Platoon left Padua the next morning and passed within six miles of Venice, which had been liberated the day before. It drove on at a fair pace through country crisscrossed with canals, their bridges unblown thanks to partisans and to the speed of the advance. Lighters and tugboats lay deserted at their moorings with ack-ack guns pointing idly at a sky no longer menacing. The platoon halted next door to the 5th Field Park Company in an area off the main Venice-Trieste road. The Piave was about five miles away.

Towards evening the skies started to drip. A chill breeze sprang up and the section corporals went round with an issue of summer clothing. Three lorries were sent to the Piave, where the 6th Field Company was building a 300-foot pontoon bridge, and a little of the liberated cognac was taken as a precaution against chills. A picket was posted and told to be particularly alert as a large force of Germans cut off by the advance was known to be in the neighbourhood.

The picket stalked glumly beneath the dripping fruit trees and the platoon slept. Tomorrow would be the first day of May.

. . . . . .

There was a sound of revelry by night.

Down the breeze came shouts, bursts of laughter, singing. 'Partisans,' said the picket, and as it was two in the morning they went to wake the relief.

The first shots were fired while the relief were pulling on their boots. Probably partisans they told each other, but as the shots sounded quite close they decided to wake the officers. Captain Williams got out of his car and listened. 'Sounds like Ities celebrating,' he said. Away on the right where Headquarters 5th Field Park Company was laagered something was burning fiercely. When another fire started, Captain Williams sent Lieutenant Rowe to find out what was going on.

The disturbance was centred round a farmyard and a long L-shaped building occupied by Headquarters 5th Field Park Company. The yard was next to the main road and through it passed a track leading to No. 2 Platoon's area, between which and the road were the closely-parked vehicles of a British FBE[1] unit. A Polish transport unit, carrying bridging for the 5th Field Park Company, was laagered nearby.

Hughie Harrison,[2] meanwhile, was making a private reconnaissance. He went through the area of the FBE unit and found some Tommies in full battle order lining a ditch. They had no idea what was going on so Hughie continued towards the Field Park Company's headquarters. On the way he was joined by a Tommy and a Pole. They reached the first vehicle in the farmyard—a YMCA van—and crept round the corner of a building. While they were doing this a man in a long greatcoat stepped from the shadows and said '*Kamerad!*' '*Tedeschi!*' yelled the Pole, and at the same instant the German bent over his Schmeizer. The

---

[1] Folding Boat Equipment.
[2] L-Cpl H. Harrison; carpenter's apprentice; Tauranga; born Te Puke, 1 May 1921.

Tommy, though, was too quick for him. He fired and the German dropped. Machine guns opened up at once and Hughie and the other two slipped away into the shadows.

He raced back to Captain Williams, and while he was telling his story there was an explosion and another fire broke out in the farmyard. Lieutenant Rowe's report also showed that the situation was serious. He had made contact with a sergeant of the 5th Field Park Company and had learnt from him that a strong German force had taken the headquarters by surprise, capturing some men and killing and wounding others. What was going on now he didn't know.

Lieutenant Rowe left to get more information and our drivers were ordered to stand-to. While they were being roused—the racket was terrific now but they were tired and the events of the past fortnight had conditioned them to night noises—bursts of tracer passed chest high between the lorries. Amazed and frightened, they prepared to defend themselves.

In the farmyard a quarter of a mile away transport and haystacks were on fire and there was a lot of noise and shouting. Tracers and explosive bullets from bredas, spandaus, and submachine guns whistled overhead, and beside these the enemy was using mortars, *panzerfaust*, and 20-millimetre guns. A continual confused shouting in German, Italian, and English made a worry of sound, like a dog-fight, but the drivers could catch a word here and there: ' *Avanti!* ' ' *Raus!* ' ' Hey, Bill! ' ' *Raus!* '

Rain fell steadily all the time, slanting in steely rods between the fruit trees and glistening against a fiery background. Flame-lit cameos, glimpsed momentarily, appeared and vanished: a figure stooping to pour petrol on and around the YMCA van; two bewildered Germans and a blue flash from a tommy gun; a group of soldiers who seemed to be wrestling among the flames.

The whistling and shouting did not stop and it was hard to tell friend from foe. In the case of the Polish drivers it was almost impossible.

Lieutenant Rowe came back with the news that the farmyard was now a scene of indescribable confusion. It and the long building were under heavy fire, the Germans having taken up positions on the floodbank of a canal on the far side of the main road. He had gone forward to the first burning haystack and had found two dead sappers beside it. A Pole had been shot in the arm while trying to sneak round the haystack. Germans were in a building on the right flank about 400 yards away and fighting was going on over a wide area.

Captain Williams decided that it was time No. 2 Platoon took part in the battle. He called his drivers together and divided them into two groups, one to defend the transport and one to go forward to the farmyard. The latter was divided into two patrols of sixteen men armed with Bren guns, tommy guns, and rifles, and these set out in open order, one advancing straight ahead under Second-Lieutenant Wells, the other swinging left under Captain Williams.

The firing and shouting had died down considerably by now but with the end of the war in sight no one was taking chances. (Our drivers, by the way, were behaving far more sensibly and professionally than they had done two days earlier in the Piacenza area.) The farmyard was reached without trouble and Captain Williams led his patrol to the back of the long L-shaped building. It seemed to be empty but when he called out 'Kiwi here!'—taking it that 'Kiwi' was a word unfamiliar to the enemy—a window opened on the top floor. He shouted 'Kiwi' twice and then fired his revolver, one driver joining in with a tommy gun and another with a Bren. An excited New Zealand voice called out: 'It's all right. You can come in. No Teds round here.'

While the building was being searched—two Germans were found but they gave no trouble—Captain Williams went into the farmyard, which was dancing and leaping in the light from five burning vehicles. Beside one of them he saw the charred body of a New Zealander. A sapper was lying badly wounded near another and he dragged him to safety. Then he shifted a jeep that was in danger of catching alight, fired at someone who failed to answer when challenged, and removed a blazing jerrican from the tailboard of a lorry loaded with petrol. It sounds simple enough, but these acts were done in the full glare of the flames when for all he knew to the contrary the neighbourhood was alive with Germans.

By now, from asking questions and listening to excited talk, our drivers were beginning to understand what had happened. A strong mixed force (in one report 500 was the number mentioned) had come shouting and singing down the main road to open fire on Headquarters 5th Field Park Company from positions behind the canal stopbank. The engineers managed to form some sort of a holding line in front of the bridging lorries—ours, their own, the Poles', and the FBE unit's—but, being hopelessly outnumbered, were unable to take offensive action. Soon the attackers swarmed into the yard and the paddocks adjoining it, setting on fire loaded vehicles with petrol, *panzerfaust*, and grenades, and starting empty ones and driving them off. (The Germans were desperately in need of transport.) Next they rushed the farm buildings and took prisoners, first killing two New Zealanders who came out of a door with their hands raised. In this action five men were killed, six wounded, and twenty-eight captured. Several vehicles were damaged, seven had been driven away, and six were on fire.

The enemy's next move was to withdraw from the farmyard to the canal stopbank and pour fire at anything and everything. Burning vehicles and haystacks made a flame-lit no-man's-land that was impassable.

After the action had lasted about an hour and a half—roughly, that is, at half past three—the whole force made off up a narrow road on the north side of the canal, taking with it its dead and wounded, its prisoners, and seventeen vehicles, ten of which belonged to a platoon of the 7th Field Company that had been ambushed and captured nearby with the loss of three men killed and fourteen wounded. Nothing was left behind except a few weapons and an ox-cart mounted with a 20-millimetre gun. The patrols from No. 2 Platoon arrived about fifteen to twenty minutes after the enemy had withdrawn.

While our drivers were helping the wounded and searching the building for stragglers, two companies of the 21st Battalion supported by tanks halted outside the farmyard. They had been sent to the rescue, but on finding all quiet they pushed on to relieve machine-gunners at the Piave bridge—the task intended for them originally. They left a Bren carrier to carry wounded.

Stray shots were still going off—those irritating and unnecessary shots that are always heard after the dust has settled and the danger is over—but the situation was completely in hand now and soon it would be light. Armed parties under Lieutenant Rowe and Second-Lieutenant Wells were posted on the canal stopbank and an enemy breda was put in working order.

In the farmhouse kitchen fourteen desperately tired German

prisoners—eight had been taken in the first building down the road—were standing in two groups, one comparatively bright and chatty, one hangdog. The latter was guarded by a party of engineers who seemed to be remembering the charred corpses and the two men who had been shot in cold blood.

A partisan arrived with a report that between a thousand and two thousand Germans were halted four kilometres down the road. Friends of his had them under observation and he wanted our drivers to borrow a few tanks and round them up. More partisans arrived and after holding a council of war the whole party went off on reconnaissance, the cock of their Sten guns and the jaunty swing of their plus fours expressing most plainly their low opinion of our platoon.

After such a night a bright sunny morning would have been appreciated but the skies were sullen. The fruit trees shivered in a cold wind and the sodden battlefield, breathing puffs and streamers of black smoke, was sordid beyond belief. Poking among a trail of broken and charred rubbish a member of the FBE unit was searching for a tin mug. He had lost everything he possessed, and he had lost his best friend as well.

．　．　．　．

It was 1 May and a wet, dull morning. Below the Alps the partisans were liberating city after famous city; on the main coast road the Division was rolling towards Trieste; in San Dona, just over the Piave, a bridging train was serving the 8th Field Company, which was improvising a 250-foot floating bridge, and the rest of No. 2 Platoon was trying vainly to keep warm and dry. In the village of Monastier, five miles north-west of the area in which the engineers had been overrun, Second-Lieutenant Colston (attached to the 9th Brigade) was witnessing the surrender of more than a thousand Germans, the force that had caused all the trouble a few hours earlier. Near the fire-station at Mestre, five miles from Venice, Corporal Ted Paul (Workshops) was bringing his diary up to date:

Yesterday—the last day of April—was one of our best days so far. We were up early and away from Trecenta by five. We took an hour to cover the first seven miles and then we crossed the Adige. Danny missed the pontoon bridge and I very nearly found out the depth of the water. As it was our front wheel went over the side and by the time we had been towed to safety the convoy was miles ahead.

We travelled forty-three miles before halting for breakfast in a little village called Mandriola, three miles below Padua. We had our meal in the lovely park of a stately old house in which Victor Emmanuel III signed the 1918 armistice. After breakfast we drove on through Padua and here it seemed as though the entire populace had turned out to welcome us. Men, women, and children—at one point they were at least ten deep—lined the streets to give us a wonderful reception, cheer us, wave flags, and whenever the convoy halted bestow the odd kiss. All along the road we got a grand hearing. It is a great sensation to be the centre of an admiring crowd of highly-delighted people even if one has done little to deserve their admiration. We were hailed as triumphant heroes yesterday. Later we passed hundreds of German prisoners, some walking, some riding in carts, all practically unescorted. They were a poor, dejected-looking lot.

We reached our destination—Mestre—at about one o'clock. We could see Venice in the distance as we turned into our area and naturally all feet were itching to pound the streets of that famous city. Even the drivers of the load-carriers, though bleary-eyed with weariness, were keen. Since the start of the campaign they have been grappling with a volume of work that can seldom have been surpassed in the history of our unit. No. 1 Platoon is already heading back south to pick up ammunition left at Ficarolo, where the Div crossed the Po.

We are a source of amazement to the people round here. They have been led to believe that the New Zealand soldier is a terrible nigger and that we have no motors, no benzine, no tires, few clothes, and little food. At tea-time last night we had an amazed audience while we ate roast beef, potatoes, green peas, pears, and custard. All the things they were told have been proved false and poor old *Tedesco* is now ' *molto cattivo, molto brutale, molto basso* '.

All day we saw partisans and they were doing their job like delighted school children. There were a good few parading up and down the streets of Mestre last night and they were armed with the wildest and weirdest collection of weapons imaginable. I doubt if some of them ever handled a rifle before as they loose off a shot on the least pretext. Ask a partisan how his rifle works and he immediately points it at something and pulls the trigger. There were bangs going on all over the place all the time and there was no feeling of security.

The news, though, was terrific—only one square mile of Berlin left to the Germans, a Russian spearhead believed to have reached Unter Den Linden, Mark Clark saying that the German armies in Italy have been virtually eliminated as a military force. Well, it was a wonderful day and I enjoyed every inch of our 70-mile journey.

May 1. We are still at Mestre and still looking towards Venice, itching to get into it. No. 3 Platoon set off for Ficarolo this morning to pick up No. 2 Platoon's holding. Most of the ammunition we brought forward during the early stages of the advance is still at Bondeno and Ficarolo and it has been decided that from now on the Company will carry only its own second-line holding of small-arms ammunition and twice its establishment of 25-pounder. The rest of the ammunition in rear dumps will be handed over to Corps.

The Jeep Platoon moved to Mestre with us and it is very pleased with itself. At present six jeeps are with the 5th Brigade, six with the 6th Brigade, two with the 9th Brigade, and three with the 6th Field Company. Four more jeeps have just set off to join the 9th Brigade.

Our jeep drivers have been sharing the experiences of the infantry to the full—mortaring, shelling, ambushes, alarms, reconnaissance, loot. They have been carting everything from Majors to margarine.

Well, it's getting on for lunch-time. Already one or two of the boys have 'snuk' away to Venice. Things are drawing to a close. . . .

· · · · · · ·

Things were drawing to a close. That evening General Freyberg shook hands with the Chief of Staff of a Yugoslav Corps that had come over the mountains from the east, and at noon the next day the war in Italy ended. Under the instrument of surrender, signed on 29 April, all land, sea, and air forces commanded by Colonel-General Heinrich von Vietinghoff, German Commander-in-Chief in the South-West and Commander-in-Chief of Army Group C, had capitulated unconditionally to Field-Marshal Alexander.

The rain had ended as well and it was a sunny day. No. 2 Platoon, travelling in convoy behind the 5th Field Park Company, was enjoying itself thoroughly. Under a blue and white sky, over roads strewn with green branches and bunches of wild flowers, beneath triumphal arches whose great letters said VIVA LA PACE E LIBERAZIONE while groups of children and young girls stood at every cottage door and farm gate chorusing '*Ciao! Ciao! Ciao!*', the convoy sped north-east, our drivers sitting up behind their steering-wheels like performing sea lions, their faces as pop-eyed with anticipation as though they were expecting juicy girls to come through the air like flying-fish.

Through Ceggia they went, through Portogruaro, Fossalta, and San Michele, over the Tagliamento, then through Latisana, Palazzolo, San Giorgio, and Cervignano. Beyond the broad Isonzo there was a change in the political sympathies of the liberated. The crown of Savoy gave place to the red star. TITO! TITO! TITO! screamed a hundred posters. The triumphal arches said VIVA IL COMITATO ESECUTIVO and VIVA LA FRATELLANZA ITALO-SLOVENA and VIVA IL MARESCIALLO TITO. The girls of Tito's army wore red stars in their forage caps and most of the men wore red neckerchiefs. Our drivers were inclined to bristle—not because they were opposed to Russia but because Tito seemed to be snapping up all the fish.

The convoy entered the town of Ronchi, drove through a sea of smiling faces, a forest of waving arms, a constellation of red stars, and down an avenue of chestnut trees, halting in a green meadow by the Ronchi railway station, eighteen miles north-west of Trieste. Everyone felt he had come to journey's end and there was also the feeling of arriving for the first cricket match of the season. Indeed, it was inescapable.

The close-cropped turf was slightly damp and the smell of mown grass, sweet and musty, was everywhere. White candles were on the chestnut trees, and the girls of Ronchi, wearing their summer frocks, stood giggling on the boundary line. High up on the right, in pearly masses rimmed and shone through by sunlight, in clumps of impenetrable blackness, thunderclouds floated. Delicate shafts of rain, as always at early cricket matches, pricked downwards and vanished, the sun triumphing. You expected to see, walking sedately across the moist turf, two umpires, white-coated, skyward-glancing, carrying stumps. . . .

•   •

New Zealanders occupied Trieste half an hour after the official end of hostilities in Italy and by the next day all the German garrison had surrendered.

History was erupting all over Europe—it was like a day out of the Apocalypse. Hitler was reported to have committed suicide—

Goebbels, too. Berlin and Hamburg, first and second cities of the Reich, were in Allied hands and Prague had been declared an open city by the new German Fuehrer, Grand-Admiral Doenitz.

At last this tremendous day ended. The shadows flowed down the hills and it was evening. At Mestre No. 1 Platoon's drivers, back from Ficarolo with ammunition, boiled water for a wash and a shave. No ammunition had been issued during the day and they could hope for a good rest. In Venice it was curfew time, and all the silver and gold and diamonds had vanished from the water and the bridges and palaces were gazing at their reflections in mirrors of ruffled jade—crinkled hoops for the reflection of bridges, trembling castles of dark green for the palaces. Drivers from the company, their faces sticky from the salt air, their eyes aching from the dazzling whiteness of the buildings, glided down the Grand Canal in a gondola, wondering if they had missed the last leave-lorry. They had visited a beer garden, the Basilica of San Marco, the Bridge of Sighs, Ponte di Rialto, the House of Desdemona, and a second beer garden. It had been a tiring afternoon.

Along a fine, broad road overlooking the sea two of our jeeps rushed towards Trieste through the pine-scented darkness with petrol for Headquarters 22nd Battalion at the Hotel Regina. Most of the jeep drivers with the 9th Brigade had been in at the death, and one of them, serving the 27th Battalion, had captured six prisoners earlier in the day. In Monfalcone, the next town beyond Ronchi, it was a night of carnival. In front of the largest hotel blazed a huge red star, and the main street was lined with laughing girls as with borders of bright flowers. A travelling fair was in town and as the chipped plaster horses, the blue swans, the scarlet gondolas, stirred smoothly by golden convoluted rods, swam round in circles, a steam organ played 'The Beer Barrel Polka' and swings tipped against the sky and the crowd danced in the side-streets. Not far away some No. 2 Platoon drivers had pacified the old people with sips of Padua kummel and were now flirting with their daughters, but not shamelessly. 'Watch yourself,' Captain Williams had said. 'This part of the world hasn't always belonged to Italy and Tito may have ideas about it. They take their politics seriously in Yugoslavia.' Tonight, though, except for some of Tito's motor-cyclists who were riding around as grimly as though the war had just started, no one was taking anything seriously. Noisy happiness was the keynote.

The darkness deepened. By the Ronchi railway station the chestnut trees were spreading their sweet English fragrance and their candles were like ghosts in the darkness; the cricket match would have been over these two hours. Now, perhaps, the grounds-man's old horse, hooves muffled in great shoes, would be dragging the heavy stone roller over tomorrow's wicket. There was quiet in the area at last and the lorries stood in pools of shadow and silence. The ladies of Ronchi, on the discovery that a paybook was missing, had been banished in disgrace, but they were still chattering in the railway yard, their faces pale ovals in the darkness and the plainest of them borrowing beauty from the event and from the hour.

From the south-east came a long column of prisoners. They sang no songs—neither the '*Horst Wessel*' nor '*Stille Nacht*'—and they gave no greetings and received none. There were thousands of them and the rhythm of their feet was like surf.

Yugoslavs were firing flares and in places the sky was crimson above Monfalcone. Away in the hills, with a sound strong and lonely—wing-tips lashing Lake Ellesmere as the swans rush into the sky—heavy machine guns were in action. But the war in Italy was over. The chase had ended and the beast was in chains.

CHAPTER 24

'... AND THE REAR PARTY WILL CLEAN UP'

ON 3 May, while the Reich Chancellery burned, while Allied forces in the west swept through Germany and took 412,000 prisoners, the 1st Ammunition Company moved from Mestre to occupy an Italian barracks at Villa Vicentina, five miles west of Ronchi and a mile from the main road. The next day it was announced that the Fifth Army and the United States Seventh Army had joined up in the Brenner Pass and that all German resistance in Holland, north-west Germany, and Denmark was at an end. On the 5th organised resistance ended in the south-western sector of the Bohemian Redoubt, where the Nazis had planned to make a last stand, and on the 6th, a Sunday, Czechoslovak flags flew in Prague for the first time in six years. In Germany prisoners flowed into the cages so fast that it was impossible to count them.

The next day, at 2.41 a.m. (French time), Colonel-General Gustav Jodl, the German Chief of Staff, signed the instrument of his country's unconditional surrender. The war in Europe had ended after lasting five years, eight months, and five days. It was announced in London that the 8th would be treated as VE Day.

. . . . . .

On VE Day the 1st Ammunition Company was at Villa Vicentina, with No. 2 Platoon close by (the 5th Field Park Company having moved from Ronchi to Vicentina on the 5th) and No. 3 Platoon on the road between Mestre and Cervignano with a load of petrol.

The long brick building occupied by Company headquarters contained many high, cool rooms, most of them empty. Their red-tiled floors, dusty and paper-strewn, were stamped with oblongs of sunlight from windows and with wedges of sunlight from half-open doors. The building was on one side of a grass square; on the others were sheds for the transport. The grass was still tall in places, but it was in process of being trampled flat and the sun was turning it into hay. It was very hot in the sheds where the lorries were parked and the dusty country lane that went past the two main entrances to the barracks was soaked in sunlight. Little familiar noises—the tinkle of a dropped spanner, the rattle of dixies, an oath and laughter—came muffled through the hot air. Girls in bright dresses cycled slowly up and down in front of the barracks and it was drowsy and quiet, except when a Yugoslav motor-cyclist, always with an air of having ridden direct from Marshal Tito's headquarters, roared by in a cloud of dust, making the dogs bark and scattering the scrawny chickens.

Only a few drivers gathered in the afternoon to listen to Mr. Churchill's speech. Many had gone to Grado for a swim—it was there that Captain Boyce, Lieutenant Hill,[1] and Second-Lieutenant Colston had taken the surrender of ninety-seven Germans two days before—and many had gone unlawfully to Ronchi, Monfalcone, or Trieste. Many were resting in their lorries, doing a little reading, a little talking, a little smoking, a little drowsing—a combination of activities that never failed to produce the very essence of boredom but was yet, in a headachy sort of way, quite pleasant. Not more than a dozen drivers were lying in the hot grass by Headquarters' wireless set.

This was the news and this was Pat Butler reading it. In England it was now possible to report the weather while the country was actually having it. There was bright sunshine in England. Admiral Doenitz had told the German people that the foundations on which the German Reich was built had gone and they must tread the road ahead with dignity, gallantry, and discipline. . . .

After the news a military band played 'A Life on the Ocean Wave'. Then Churchill spoke:

    Yesterday morning at General Eisenhower's headquarters General Jodl, representative of the German High Command, and Grand-Admiral Doenitz, designated head of the German state, signed the act of unconditional surrender of all German land, sea, and air forces in Europe to the Allied Expeditionary Force, and simultaneously to the Soviet High Command. . . .

At the end of his broadcast Mr. Churchill, shouting through all that sunlight, shouting into the dark night ahead, cried: 'Advance Britannia! Long live the cause of freedom! God save the King!'

Cease Fire was sounded by buglers of the Scots Greys and as

---

[1] Lt F. M. Hill; civil servant; Christchurch; born London, 15 Apr 1912.

our drivers got up to go they heard singing: 'Praise, my Soul, the King of Heaven'.

The hot day passed and twilight came, deepening to violet, to purple. Lights popped on in barrack-rooms and glowed softly in the backs of lorries and the King spoke:

> Today we give thanks to God for a great deliverance. Speaking from our Empire's oldest capital city, war-battered but never for one moment daunted or dismayed—speaking from London. . . .

In Villa Vicentina primuses hissed and purred. Our drivers lay on their beds in the lorries while their friends lounged against the tailboards talking and waiting for the tea water to boil. They spoke of the favourite for the Two Thousand Guineas at Newmarket next Wednesday, of Old Ted the late enemy, and of going home.

The night was warm and lovely, and there was little to drink and it was too early to go to bed. On such a night as this (thought our drivers) it would be just the job taking out the old man's V8 and sending her along and hearing the loose gravel spray up under the mudguards—yes, and watching the telephone poles flick past and seeing the rabbits and hares, silly in the dazzle, run every which way. On such a night it would be all right taking the girl into town for the dance and nipping out around supper-time for the odd rigger. Strolling home on such a night and seeing the old woolshed come up black against the hill would be just the gear—it would be all right in fact.

. . . .

With the war in Europe over we had a right to expect that the rest of our stay in Italy would be all pleasure and profit, but almost at once—almost before a member of No. 2 Platoon had time to make a present of a heavy machine-gun to a party of Yugoslavs whom he perceived to be poorly equipped—a new cloud showed over the horizon.

Tito wanted the Venezia Giulia (Trieste, the Istrian Peninsula, and some of the territory behind the peninsula) and he also wanted a part of Austria. Well, no one had expected him to want less—much less, anyway—but instead of waiting for his claims to be examined at the Peace Conference he was behaving in the best traditions of the dictators.

Soon Trieste was an armed camp with Tito's men stopping New

Zealanders in the street and demanding their leave passes. The Union Jack and Tito's red-starred tricolour floated from adjoining buildings and British and Yugoslav patrols passed one another in silence. British Honey tanks manned by Yugoslavs faced British Shermans manned by our own men, and Yugoslav guns pointed at British 25-pounders. Between Villa Vicentina and Ronchi the bridge over the Isonzo was guarded by Gurkhas and Yugoslavs. All units in Vicentina posted strong pickets at night and everyone kept his weapon handy.

On 21 May Marshal Tito started to withdraw his troops from Austria but in Trieste there was no lessening of the tension. The day before our transport had begun work at the docks, carting ammunition and Allied Military Government supplies to dumps in Udine, some forty miles north-west of Trieste, and on the 22nd Lieutenant Miles and a party from the Ammunition Platoon formed a port detachment in the city. No. 2 Platoon passed from the command of the 5th Field Park Company on 25 May, and by the end of the month our load-carriers had shifted 5800 tons and travelled 212,050 miles—only 16,950 less than the distance from the Earth to the moon. They worked day and night and there were several accidents, most of which were caused through over-tiredness.

In spite of long hours and the Tito crisis we managed to have a surprisingly good time. There was day-leave to Udine where some of us met Primo Carnera ('Old Satchel Feet'), the only Italian ever to win the world's heavyweight boxing championship. Primo, whose home was in Sequals, a town on the far side of the Tagliamento, had suffered no extraordinary privations during the past few years and was in excellent health and spirits, though he liked it to be known that he had dropped twenty pounds during the war and now weighed only 240. Extending a hand like a paddle, he said it would be a pleasure to coach any New Zealander who was interested in boxing.

For some of us there was leave to Venice where the world-famous Danieli's was now the New Zealand Forces Club. ('But,' exclaimed an English major, his horrified gaze resting on dusty, travel-stained privates, 'I spent my honeymoon there. *My honeymoon!*'). There were day trips to Klagenfurt in Austria and four-day tours of northern Italy, and at Mestre the Major opened a unit rest camp.

This, unfortunately, was found to be outside the Divisional area, and it had to close down just as it was becoming popular.

Some members of the Jeep Platoon, meanwhile, were touring Italy in their private sports cars—that was what it amounted to. A party of three travelled a thousand miles in search of a district suitable for mountaineering and ski-ing, finding one near Madonna di Campiglio, a village in the southern Dolomites.

And so May turned to June.

On that last May evening Villa Vicentina, soaked in summer and peacefulness and good sense, seemed remote from a world troubled and dangerous—remote from Damascus where French shells had started two great fires, from Westminster where a caretaker Government was in office pending the first general election in ten years, from Tito's Yugoslavs who were still tentatively building road-blocks on the far side of the Isonzo, and remote—ah, infinitely remote and separate—from *Nuova Zelanda*.

Warm rain had fallen during the afternoon and the evening was heavy with magnolias and blown roses. Headquarters and Workshops were holding a supper-dance and music beckoned from the large upper room that ran nearly the whole length of one side of the square. A sergeant from Headquarters, 'mokka'd' up in his 'Groppi', came down the stone staircase with a *signorina*. It was too dark to see what kind of a girl he had got but as he was rich and could speak passable Italian the chances were that she was lovely. The sentry at the barracks gate, seated comfortably on a low stone wall, his tommy gun between his knees, whistled as they went past.

Crunching over the gravel, drowning the music with their deep grumble, No. 3 Platoon's lorries—they occupied two sides of the square—drove one after another through the barracks gate. Elbows jutted white and sharp as the drivers struggled with their steering wheels at the corner and then the lorries gathered speed. This was the platoon's second trip to Trieste that day and none of the drivers had had more than a few hours' sleep. One driver, as he passed the sergeant and the *signorina*, flicked his ignition switch so that the engine back-fired, making the sergeant start and his girl jump into the ditch. Soon the last lorry was a red pin-point of receding tail-light and the lane was empty except for dust—a soft, sweet-scented cloud that remembered the warm rain.

With the lorries gone you could hear clearly all the noises of the dance—violins screaming, drums throbbing, saxophones wailing, girls laughing, dresses rustling. They mixed in the warm air with burnt petrol, magnolias, mosquitoes, and the soft, the rain-remembering dust.

. . . . .

For a week a fight seemed almost unavoidable, and then, on 9 June, the American, British, and Yugoslav Governments signed an agreement at Belgrade giving Field-Marshal Alexander jurisdiction over Trieste and the western half of Venezia Giulia and Tito jurisdiction over Fiume and the eastern half. On the 11th the Yugoslavs began to withdraw from the Allied zone. They marched through the streets of Trieste singing the '*Bandiera Rossa*' ('The Red Flag'), but the crowds shouted '*Viva la Liberazione! Viva i Neo Zelandesi! Viva gli Alleati!*' What they meant of course—all the pretty girls, the old women, the children—was ' Long live the landing craft and the flour and sugar! Down with looting! Down with arrests and bullets! Down with stinking politics! '

Between four and five, when No. 2 Platoon drove to the central railway station to collect 700 Italians who had arrived from prisoner-of-war camps in Germany and were to go to Mestre, the streets were full of trampled flags and flowers but almost empty of people. The crowds gathered again with the cool of evening but now it was the Communists' turn. Armed with Sten guns, rifles, and grenades, bands of Yugoslavs, with men from the Garibaldi Division and the Guarda di Popolo, marched through the main streets firing on the hated crest of Savoy and injuring civilians. *Viva la Liberazione!*

With Tito gone from Trieste the atmosphere was much more holiday-like. The transport platoons were not working so hard now and the jobs they did do were often delightful. No. 3 Platoon penetrated deep into Tito's territory to take AMG flour to Pola, near the tip of the Istrian Peninsula, and No. 2 Platoon unloaded bridging for the Royal Engineers at three points on the Trieste-Pola road.

If the period that followed was not a proud one in our history it was at least an understandable one. Discipline was relaxed—or were the authorities unaccountably blind?—and as for the moral law, of all the circumstances likely to contribute to its violation no one was wanting. We had the time—too much of it; the opportunity—no human frailty that was not catered for in Trieste; and the money—the closing stages of the campaign had been swift, arduous, but highly lucrative.

Although there was plenty of official leave many of us went away for weekends, or for a week even, just as the spirit moved us, and more than one driver kept his own establishment in Trieste. The first and last commandment of a transport unit—Your Lorry Must be Ready for the Road at All Times—was still obeyed by the majority, but there were two drivers to each vehicle and as long as the section corporal was friendly the absence of one was not noticed. And so the days slipped past, with Youth at the helm and Folly, in contravention of standing orders, at the prow.

Fortunes amassed during the advance came to an end at last but with agents of the black market prowling everywhere it would have been foolish to draw pay—foolish and rather priggish; for public opinion had removed the stigma of criminality from the sale of petrol and jeep tires and had even glamourised transactions of this kind, making them seem daring and clever. Hence, when you spoke contemptuously of the driver who dropped off a sack of AMG sugar in Trieste, or of the black-jowled gentleman in the wineshop who sprayed you with garlic and multiples of a thousand, you criticised Robin Hood.

Most of us were content merely to keep abreast of our obligations but there were some who saw in the situation a chance to provide for their old age. Their chief problem, and it worried them day and night, was how to convert lire into pound notes. They bought money orders at first and when that avenue was closed to them they bought watches, cameras, and jewellery, sometimes spending as much as 40,000 lire—the lira was worth rather more than a halfpenny—on a single article.

Others spent their money on *vino* or wasted it in ways even less rewarding. A minority stayed at home, stifling a sense of wasted opportunities.

.   .   .   .   .

The war in Europe was over and it was time to make an end. Our special function was already largely redundant, and soon the unit would be disbanded and we should be scattered to the four winds. There was no likelihood of our being reinforced and sent to the Far East as the 1st Ammunition Company, nor was that what we wanted. No, it was time to make an end.

In spirit we had broken up already. That unit consciousness, that feeling of solidarity, which for so many years had made each one of us quite certain in his own mind that the 1st New Zealand Ammunition Company was the best transport unit in the Division, had vanished some time ago. Now it was the clique that mattered—the gang, the private circle of friends. Gradual at first, the change had started when our 5th Reinforcements left us. That void in our communal life had not been filled by the replacements, many of whom came to us with loyalties older than those they owed to the Ammunition Company. A large part of No. 3 Platoon, for instance, though it worked well with the rest, was really a branch of the 18th Tank Transporter Company's Old Boys' Association. Only in No. 2 Platoon, on which, so some of us contended, the mantle of A Section had fallen, was the old spirit discernible.

And good friends were dropping out all the time. On 25 May we lost four officers and nine other ranks when members of the 6th Reinforcements (Hawea draft) left us for Bari, and on 17 June they were followed by a further thirty-four other ranks, members of the 7th Reinforcements (Waikato draft). The Sevenths were taken to Bari by No. 1 Platoon.

In the last week of June we delivered our ammunition to the 3rd Advanced Ammunition Depot, Udine, and were glad to be rid of the damned stuff. Glad, yes, but tugged at by cords of habit. Those boxes had been our constant companions, our only furniture, for Heaven knew how long. Seated on them we had played cards, stretched on them we had slept, round them we had eaten. Filled, they had been our tables and our chairs; empty, our cupboards, wardrobes, larders.

July came, and the crops stood stiff and golden in the fields and waggons piled high with maize held up our convoys in the narrow roads. The threshing machine near No. 2 Platoon's area hummed all day long and often half through the night.

There was little work to do and most of us were tired of leave—even the quota for Venice was hard to fill nowadays. We played cricket in the field behind the barracks and we bathed daily. In the evenings we drank vermouth and soda in one of Vicentina's three wineshops or took cushions with us and went to No. 2 Platoon's area for the pictures, a pleasure we shared with about a hundred village children.

There was talk of a new area for the Division near Lake Trasimene, and on 23 July, at seven in the morning, we pulled out from the barracks, heading south. It was a short convoy because Nos. 1 and 2 Platoons were staying to help shift the 6th Brigade and the Jeep Platoon had only its domestic vehicles and about six jeeps, the rest having been handed in at Cesena early in the month. There was leave to Mestre that night, to Bologna the next night, and to Fabriano and Albacina the night after that. Our village welcomed us with open arms but was sorry to learn that it might never see us again. (Dear village! Possibly it remembers us still, shaking all its bells with laughter at the thought of our execrable Italian, our inexhaustible supplies of barley sugar, our prodigious thirsts.) The journey ended the next day in an area six miles from Lake Trasimene and four miles west-south-west of Perugia, that charming hilltop town built round a corkscrew. Gentle slopes, studded with rocks and generously shaded by oak trees, went down to a small creek. Goal-posts were there already so we got out the footballs.

The oak trees spread club-shaped shadows over the rough grass and children came from nowhere to watch the game—to watch it for a while, and then, timidly at first, later with growing confidence, take part in it.

. . . . .

The Japanese ignored the Potsdam declaration. On 5 August one plane dropped one bomb and Hiroshima was destroyed.

When we heard about it our area was almost empty, the transport platoons having gone to Bari with married members of the 8th Reinforcements (Tekapo draft)—we lost three officers and twenty-five other ranks. The Jeep Platoon's transport—what was left of it—was at Madonna di Campiglio with eight drivers.

On 8 August Russia declared war on Japan and the huge Red Army in the East poured into Manchuria. An atom bomb fell next day on Nagasaki. On 10 August Japan offered to accept the

Potsdam terms if she could do so without prejudicing the prerogatives of her Emperor.

At midnight on Tuesday, 14 August, the surrender of Japan was announced in simultaneous broadcasts from London, Washington, Moscow, and Chungking. Most of us heard the news at nine the next morning from the British Forces Station in Italy and no work was done that day. Nos. 1 and 3 Platoons were back with us now but No. 2 Platoon was still at Bari. There the drivers built a bonfire and sat round it singing songs and drinking an issue of beer. In the Trasimene area, too, there was singing and drinking. The stories from Hiroshima and Nagasaki were like reports from Hell.

On 20 August the Jeep Platoon was disbanded, the drivers being distributed among the transport platoons as the Ammunition Platoon drivers had been after our second-line holding was handed in. By the end of August there were several lines of transport parked mudguard to mudguard in Workshops' area on the far side of the creek—cooks' lorries, orderly-room lorries, and staff cars and pick-ups handed in by officers who had gone home.

Our single Eighths—forty-four other ranks—joined the Tekapo draft on 10 September and the next day we lost 'Parky' Neighbours,[2] our star footballer, who was one of the thirty-nine players chosen to fly to England to train for the New Zealand Army Rugby football team. Soon afterwards Charlie Porter[3] and another driver were posted to the New Zealand Selection Camp in Austria for training and later Charlie was sent to England.

We were busy during the first three weeks of September. No. 3 Platoon carted YMCA stores from Bari to Rome and took infantry to the Divisional rest area at Mondolfo on the Adriatic coast; No. 1 Platoon took leave parties to Venice and Madonna di Campiglio, staying with them while they were there; and No. 2 Platoon, which was still at Bari, took parties to Rome, Florence, and Venice. Drivers without vehicles were employed in ferrying transport from Foligno to Trieste for UNRRA,[4] Yugoslavia.

When we were not working we played cricket on our private

[2] Cpl A. S. Neighbours; brick and pipe maker; Waimangaroa, Westport; born Westport, 23 Feb 1922.
[3] Dvr W. C. R. Porter; grocer's assistant; Palmerston North; born Wanganui, 10 Jan 1918.
[4] United Nations Relief and Rehabilitation Administration.

sports ground. Our oak trees were golden now where once they had been green, and evening by evening the shadows lengthened earlier. For most of us the journey would not be over before Christmas.

No. 2 Platoon came back from Bari on 27 September, a Thursday, having spent the night before at Albacina. This was our last contact with the village. On Friday morning the Major was told to disband Nos. 1 and 2 Platoons, and the drivers spent the day in checking their transport and returning stores to the quartermaster. That night there were farewell parties but the lush sentiment usual on these occasions was missing. We should be meeting again in the Division's next area and most of us would go home in the same ship. Nothing of value was being broken up—only an arrangement of names and numbers, only lorries and tents, only somewhere to eat, to sleep. In every important sense the unit had come to an end some time ago—on a February morning or a May evening. Just when was a matter of opinion.

On Saturday Nos. 1 and 2 Platoons were disbanded and their transport was lined up in Headquarters' area. The drivers, many of whom had sore heads, hung about looking lost and sheepish.

On Sunday morning—in the New Zealand Division everything happened on a Sunday—thirty-eight of our drivers and a party from the 4th Reserve Mechanical Transport Company set out for Trieste with Nos. 1 and 2 Platoons' transport. It was to be handed over to UNRRA, Yugoslavia. Of the drivers without vehicles eighteen were posted to No. 3 Platoon, fifty-seven to the 1st Petrol Company, forty to the 4th Reserve Mechanical Transport Company, and twenty-eight to the 1st Supply Company. Our unit consisted now of Headquarters, Workshops, and No. 3 Platoon, the bulk of which was ferrying troops to Florence.

By nightfall it was known that Headquarters and No. 3 Platoon would be disbanded on 6 October and Workshops attached to the 4th Reserve Mechanical Transport Company. Already a large part of the area was dark and quiet where once it had been lighted and noisy, and everywhere there were signs of packing. Great mounds of gear—everything from meat hooks to tommy guns— almost concealed the quartermaster's tent, but even so they were not as high as they ought to have been. They contained roughly the right number of straps, supporting, web, and helmets, steel,

Mark I, but there was a distinct shortage of boots, ankle, and an absolute dearth of blankets, woollen. Our new quartermaster-sergeant was only mildly concerned, giving it as his opinion that we couldn't be worried over trifles.

That was the last day of September, and at sunrise the next morning, though it was lovely later, there was a sharpness in the air that told of a new month.

Monday 1st (ran the unit diary). Location: Perugia. Weather: fine. Unit busy preparing to hand over vehicles to NZOC,[5] Assisi, on 6th.

('I brought this old bitch over from Egypt. She's got pistons you could tie knots in.')

Tuesday 2nd. Location: Perugia. Weather: hot morning—fair afternoon. Workshops preparing to go to 4th Reserve MT Company tomorrow night. It will be attached to that unit for UK leave scheme.

('Recovery work, eh? We'll be strung out—sections of us—all across the Riviera and France. Forte dei Marmi, San Remo, Aix. Oo! La-la!')

Wednesday 3rd. Location: Perugia. Weather: fine but windy. Unit will not move to Florence on 4th as previously instructed but will be wound up in present area on 6th when remaining personnel will be posted to 1 Supply and 1 Petrol Companies. Workshops will stay with unit until transport has been handed in. Anti-gas equipment and web gear returned to QM.

('She's all there, Ray, and you know what you can do with her.')

Thursday 4th. Location: Perugia. Weather: cold. OC returned from memorial service on Crete. Officer postings received from HQ Command.[6] OR postings received—seventy-three will go to Petrol Company, forty-two to Supply.

('Better give us your address as we're gonna be split up. Supply should be all right, though. "Bub's" there and "Hawk" and Old Harry.')

Friday 5th. Location: Perugia. Weather: fine. Transport

---

[5] New Zealand Ordnance Corps.

[6] Maj Coutts and WO II Salmond (posted to 1 Petrol Company, 9 Oct 45); Capt H. A. Wilson, Lt (T/Capt) Legge, and 2 Lt Wells (4 Reserve MT Company, 16 Oct 45); 2 Lt E. J. Stembridge* (1 Supply Company, 6 Oct 45). The following officers had left already: Capt Littlejohn (posted to 1 Supply Company, 30 Sep 45), Capt Langley (1 NZ Graves Concentration Unit, 26 Jul 45), Lt Miles (4 Reserve MT Company, 30 Sep 45), and Lt Brown (NZ Maadi Camp Composite Company, 10 Aug 45).

* Posted to unit 10 Aug 45.

lined up on football ground for checking and classing by Workshops before being handed in tomorrow. HQ domestic vehicles stripped of fittings.

('You've had your little desk, Charlie. No more charge sheets to fill in.')

Saturday 6th. Location: Perugia. Weather: fine. HQ and 3 Platoon vehicles marched out at midday for handing in to NZOC, Assisi. Drivers brought back to unit area by Petrol Company transport. Workshops marched out complete. Drivers on last ferrying detail reported in after dinner and were then marched out to Petrol Company. OC left for Florence to report to HQ Command. Rear party under Lt Wells to stay behind to clean up and await return of vehicles still on detail.

(Only two tents were left in the area. In one of them Headquarters had held a party the night before and the Major had played his accordion. Now the area was quiet and only in two places splashed with light.)

Sunday 7th. Location: Perugia. Weather: fine. One lorry from Petrol Company returned G.1098 equipment to NZOC and another took unit sports gear and wireless sets to Florence.

(These were the cricket bats we had used for the great North v. South match, the cricket balls Ray Bilkey[7] had spun so cunningly. These were the wireless sets that had given us Command Performances, Forces' Favourites, the speeches of Winston Churchill.)

Monday 8th. Location: Perugia. Weather: fine. Balance of G.1098 equipment returned to NZOC.

('*Due mila*, Pop—*due mila per tutto*. If you don't want the goods, Pop, don't handle 'em. *Due mila* finish.')

Tuesday 9th. Location: Perugia. Weather: fine. Final clean up of area and preparations completed for move of rear party to Florence tomorrow.

('*Sessanta mila*, Pop. *Sessanta mila* and they're yours. And a good bet to you, Pop. *Grazie.*')

Wednesday 10th. Location: Perugia. Weather: fine. Rear party moves out at 1100 hours to join Petrol Company in Divisional area at Florence. . . .

. . . . . .

After the rear party had moved out the Italians moved in to

[7] Cpl R. Bilkey; corset cutter; Northcote, Auckland; born Rotorua, 13 Dec 1919; wounded 20 Apr 1943.

search the ditches and rubbish pits. They found little of value and after a while they went away. Only the children stayed, twittering like birds and recalling where this cookhouse had stood and that vehicle had been parked. They threw sticks and stones into the tall walnut tree that had sheltered Headquarters' orderly-room lorry and the ripe walnuts pattered down, bouncing on the baked earth. When evening came there were still children in the area.

The shadows from the oak trees flowed down the hillside, bridging the creek and poking long fingers across the football ground. When it was quite dark and the goalposts could be seen no longer the children went home—to dream, perhaps, of the strange, friendly soldiers, the *Neo Zelandesi*, who had come, had stayed for a little while, and had moved on. And after a few months, after the weather had removed all traces of the camp and the last biscuit had been eaten and the last tin of marmalade had vanished from Momma's shelves, and the small cut foot had healed, and the bandage provided by the New Zealanders had been washed and washed until it was of no further use, they forgot. For the world was full of soldiers and they stayed for a little while and they went away.

The children forgot, yes, but not at once and not completely. Between them and the migrant soldiers there was a bridge, a bond, some fragments of a common language. They sensed, it may be, that soldiers were no different from themselves in some ways, that they, too, had a kind of innocence, and were not, in a world abounding in meanness, mean. Careless perhaps, destructive certainly, but not—not in the last resort—meriting hate and terror from children, even from burnt children in London, Naples, Rotterdam, Berlin, Hiroshima.

So the children came back for a night, two nights, three nights, to the place that remembered the soldiers and their lorries and their gear, and played until the walnut tree was deep blue in the sweet, heavy evening and the hills were purple and the stream flashed under the stars like dark silver.

ROLL OF HONOUR

KILLED IN ACTION

(includes Died of Wounds)

Dvr W. V. Mosen	14 April 1941
Dvr A. N. Bradbury	26 April 1941
Dvr A. H. Storey	26 November 1941
Dvr G. B. Drinnan	27 November 1941
Dvr N. J. N. Orsborn	27 November 1941
Dvr J. D. B. Clifford	28 November 1941
Dvr A. B. Meiklejohn	28 November 1941
Dvr C. Cameron	5 December 1941
Cpl O. W. Miles	14 July 1942
Dvr D. C. Henderson	14 July 1942
Dvr R. King	14 July 1942
Dvr J. S. O'Connor	14 July 1942
Dvr A. B. Caddy	15 July 1942
Dvr L. E. Hay	15 July 1942
Dvr C. S. Brown	17 July 1942

DIED AS PRISONER OF WAR

Cpl A. McK. Weir	3 December 1942

DIED ON ACTIVE SERVICE

Sgt A. Morton	25 December 1940
Dvr C. E. Ross	17 November 1941
Dvr C. L. Collie	4 December 1941
Dvr R. Baker	21 January 1942
Dvr W. Wood	7 March 1942
Dvr R. A. Duley	29 June 1942
Dvr D. H. Elder	29 June 1942
Dvr J. T. Meaton	24 January 1944
Dvr A. J. Howejohns	10 May 1944
L-Cpl O. E. Penney	28 June 1944
Dvr J. R. Kinross	11 March 1945

HONOURS AND AWARDS

Officer of the Order of the British Empire
Major S. A. Sampson — 5 January 1945

Member of the Order of the British Empire
Major P. E. Coutts — 7 January 1944

Military Cross
	Captain H. A. Rowe	10 April 1942
	Second-Lieut. J. R. Arnold	5 March 1943
	Captain B. J. Williams	18 October 1945

Military Medal
3519	Dvr R. S. Grant	20 September 1941
12103	Sgt R. G. Aro	27 March 1942
3438	Sgt V. J. Cleave	11 December 1942
3453	Dvr N. Hague	11 December 1942
43496	Cpl W. O. Ingham	18 October 1945
264447	L-Cpl D. J. McLean	13 December 1945

British Empire Medal
3470	Cpl W. S. Aitken	7 January 1944
25529	Cpl J. C. Connelly	5 October 1945

United States Bronze Star
477423	Dvr J. G. Lee	3 June 1945

Mentioned in Despatches
	Major P. E. Coutts	8 January 1943
	Captain S. A. Sampson	6 August 1943
	Captain A. G. Morris	3 March 1944
	Lieutenant A. R. Delley	15 February 1945
	Lieutenant J. D. Todd	8 August 1945
3545	Sgt S. H. Matthews	8 July 1941
3435	Sgt N. K. Michael	23 October 1942
21615	Dvr G. M. H. Bell	8 January 1943
3564	S-Sgt J. H. Skeates	8 January 1943
18753	WO II J. G. Pearson	6 August 1943
3579	Cpl C. R. Turner	3 March 1944
22008	Dvr D. C. Harrison	19 May 1944
21631	Dvr L. J. Moore	19 May 1944
36058	Cpl J. R. Benfield	15 February 1945
28672	Dvr H. C. Blomfield	15 February 1945

HONOURS AND AWARDS

82612	Dvr J. J. Downes	15 February 1945
17056	Sgt J. D. Letham	15 February 1945
14029	Sgt S. W. Barber	8 August 1945
16722	Sgt I. C. J. Craig	8 August 1945
46681	Cpl N. Dunn	8 August 1945
42214	WO II D. S. Finlay	8 August 1945
3595	WO I I. McBeth	8 August 1945
42080	Cpl H. K. Wallace	8 August 1945
43496	Cpl W. O. Ingham	24 November 1945
43501	Sgt L. C. H. La Roche	29 November 1945
237781	Dvr C. C. O'Hara	29 November 1945
83311	Dvr J. L. Cowan	23 May 1946
19788	Cpl W. A. Ford	23 May 1946
286496	Dvr W. G. Kelly	23 May 1946
28939	Dvr J. W. Donnelly	29 August 1946

COMMANDING OFFICERS

Major W. A. T. McGuire, ED	3 Oct 1939 - 3 Oct 1941
Major P. E. Coutts, MBE, ED	4 Oct 1941 - 26 Jan 1943
Major S. A. Sampson, OBE	26 Jan 1943 - 17 Apr 1944
Major R. C. Gibson	17 Apr 1944 - 21 Sep 1944
Major R. P. Latimer	21 Sep 1944 - 2 Feb 1945
Major P. E. Coutts, MBE, ED	2 Feb 1945 - 6 Oct 1945

www.ingramcontent.com/pod-product-compliance
Lightning Source LLC
Chambersburg PA
CBHW040319300426
44111CB00023B/2951

Adolf Bernhard Meyer, Lionel W Wiglesworth

The birds of Celebes and the neighbouring islands

Adolf Bernhard Meyer, Lionel W Wiglesworth

The birds of Celebes and the neighbouring islands

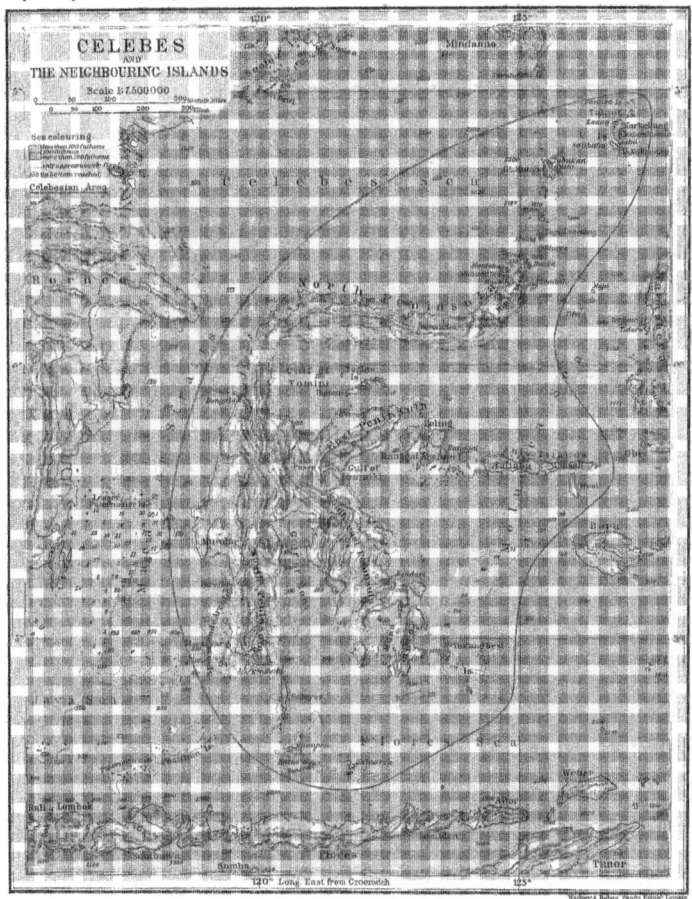

151

598.2(91.2).

THE
BIRDS OF CELEBES

AND

THE NEIGHBOURING ISLANDS.

BY

A. B. MEYER AND L. W. WIGLESWORTH.

VOLUME I.
WITH 17 PLATES (14 COLOURED) AND 7 COLOURED MAPS.

BERLIN:
R. FRIEDLÄNDER & SOHN.
1898.

PREFACE.

When we began this work six years ago, we were under the impression that it would not be premature to write "The Birds of Celebes", but the further we proceeded, the more we became aware how impossible it is at present to give a complete history of the species, as many undoubtedly still remain to be discovered in the interior and the mountains, not to speak of the islands; and in the case of others whose names are familiar to ornithologists, we encountered a great lack of knowledge touching their local distribution, their subtile variations — individual, geographical, seasonal, sexual, and developmental, — their movements or, as the case may be, migrations, with the questions bearing thereon of seasons, climate, flowering of plants, ripening of fruits and the like, their nidification and moulting, habits and economy. We, therefore, are conscious of the imperfection of our work and shall be entirely satisfied if it be useful to future workers on the Celebesian Avifauna, not doubting that a classical work could be ultimately written on the subject, of the standard of Naumann's "Vögel Deutschlands" in 12 volumes.

The principles of nomenclature are now again in process of development and, consequently, unsettled; it is impossible to meet with the approval of every one in this respect. Our endeavours to give nomenclatory expression to the minute variations of groups of individuals not yet formed into species or subspecies led us to adopt an innovation discussed on pp. 52, 53 and elsewhere in the text; we anticipate that ere long this will be superseded by something better, and only regard it ourselves as a first step towards indicating those minute complex variations of a species within itself, which occur in Nature and are of such great importance in the study of evolution.

In our synonymy an attempt at completeness has rarely been made, except in the case of endemic Celebesian birds; and, instead of an aimless repetition of the fuller synonymies in, for instance, the "Catalogue of Birds" and the "Ornitologia della Papuasia, etc.", reference is given to the authors of these works, to whom we feel deeply indebted, as indeed every ornithologist will be. As to the abbreviation of authors' names, we had proceeded too far to remodel them when the German list was issued, but we have introduced the methods recommended there at the headings of the species.

We have never used the term "Malay Archipelago", so much in vogue since Wallace, because this expression is incorrect in this sense, that not all the inhabitants of the islands comprehended are Malays. It is the "East Indian Archipelago" of which we speak, having in view the island-world from Sumatra to the Solomon Islands and from the Philippines to the Lesser Sundas.

As to localities we have been as accurate as possible, but unfortunately it is only in recent times that collectors have attached the exact locality to every specimen. Formerly "Manila" meant the whole of the Philippines, "Manado" the whole Minahassa (Northern Celebes), "Macassar" the southern part of Celebes; though, for instance, Mr. Wallace did not shoot all of his birds labelled Macassar near that town, but some at Maros and elsewhere. Of course we could not eliminate these inaccuracies. The future writer on the Birds of Celebes will happily have to deal with more exact data.

Our artist, Mr. Geisler of the Dresden Museum, wishes us to draw attention to the circumstance that the dull colours of some of our plates have been intentionally used at our request, the exact hue of the specimen painted being aimed at, sometimes at the cost of the artistic effect and clearness of tint usually seen in the English productions.

And now we have the agreeable duty of thanking all those who have lent us their welcome aid during our six years' work. In the first place Dr. P. and Dr. F. Sarasin of Basel, who placed their highly valuable and important collections entirely at our disposal; and not less the Hon. Walter Rothschild of Tring, who joined us in engaging native hunters for completing our collections and for making investigations on much new ground in our Area, as will be seen throughout the book. Mr. Nehrkorn of Riddagshausen most generously furnished us with a MS. descriptive of the eggs of Celebesian birds in his celebrated collection, and many are now described for the first time. Dr. van der Stok of Batavia and Prof. Neumayer of Hamburg helped us with our meteorological maps, as mentioned pp. *21* and *37* of the Introduction. Dr. Bowdler Sharpe of London had the great kindness to send us the proof-sheets of his part of vol. XXVI of the "Catalogue of Birds", not yet out. Mr. Veen of Kele Londej (Minahassa) and Mr. North of Sydney sent us some notes bearing upon the question of migration, quoted in our Introduction, pp. *39* and *47*. For the transmission of specimens we are much indebted to the following gentlemen: Prof. W. Blasius of Brunswick, Dr. Büttikofer of Leyden (now Rotterdam), Dr. A. Dubois of Brussels, Mr. Gurney of Norwich, Mr. Hartert of Tring, Prof. Hertwig of Munich, Prof. v. Koch of Darmstadt, Dr. v. Lorenz of Vienna, Mr. Pleske of St. Petersburg, Dr. E. P. Ramsay of Sydney, Prof. Reichenow of Berlin, and Mr. W. Schlüter of Halle; as well as for special information to Prof. de Groot of Leyden, Resident Jellesma of Manado, Prof. Kern of Leyden, Prof. Newton of Cambridge, Dr. Oustalet of Paris, Dr. Riedel of the Hague, and Count Salvadori of Turin. Inspector Lehnig of the Dresden Museum has assisted us in drawing up the lists of Geographical Distribution and of Local Avifaunas, the alphabetical Index and list of abbreviations, and has supported us in other ways; as has also Miss C. S. Fox of London by her aid in the correction of the proofs.

Should we, unhappily, have forgotten any one in rendering our thanks, we apologize for such an omission.

Finally our sincere thanks are due to the Publishers for their compliance in all our wishes as to the fashion of the book.

Royal Zoological Museum, Dresden, April 30th, 1898.

CONTENTS.

Volume I.

	Page
Preface	v
Systematic Index	x
List of plates	xvii
List of maps	xix
Alphabetical list of abbreviations of authors' names, titles of books, journals, etc.	xx
Notanda et corrigenda	xxxi

Introduction   1

Travel and Literature   2

Labillardière, 2 — Reinwardt, 2 — Salomon Müller, 3 — Quoy and Gaimard, 3 — Forsten, 3 — Léclancher, 3 — Wallace, 4 — Riedel, 4 — van Duivenbode, 4 — v. Rosenberg, 5 — Bernstein, 5 — Hoedt, 5 — Bickmore, 5 — Meyer, 5 — Conrad, 6 — Fischer, 6 — Beccari, 6 — Bruijn, 6 — van Musschenbroek, 6 — Murray, 7 — v. Bültzingslöwen, 7 — v. Faber, 7 — Teijsmann, 7 — Platen, 7 — Laglaize, 8 — Ribbe and Kühn, 8 — Guillemard, 8 — Hickson, 8 — Weber, 8 — Radde, 8 — Cursham, 9 — P. & F. Sarasin, 9 — Kükenthal, 9 — Everett, 9 — C. & E. Hose, 10 — Doherty, 10 — Waterstradt, 10 — Special List of Literature on the Birds of the Celebes, 11—16.

Seasons and Winds in the East Indian Archipelago   17

Explanation of maps, 17 — Influence of climate on the distribution of birds, 17 — The Monsoons of the East Indian Archipelago, 18 — The rainy season north of the equator, 19 — south of the equator, 20 — Change of climate at different altitudes, 21 — The seasons in Celebes, 24 — Sangi, 27 — Philippine Islands, 29 — Borneo, 29 — Sumatra, 32 — Natuna, Billiton, etc., 32 — Java, 33 — The Lesser Sunda Islands, 33 — The Halmahera Group, 34 — Buru and Ceram, Contrast of the north and south coasts, 34 — Amboina and Saparua, 34 — Banda, Kei and Aru, 35 — New Guinea, 35 — Solomon Islands, 36 — Northern Australia, 36 — Meteorological Literature, 37.

Migration in the East Indian Archipelago   38

Migration in its simplest form, 38 — Local movements, 38 — Islet nomads of the East Indies, 39 — Migration proper in the Archipelago, 40 — Table of migrants to Celebes 41, 42 — Birds at sea, 43 — Routes of different species to the East Indies, 44, 45 — Migration south of the equator, 46 — Birds which stay behind in their winter quarters, 49 — Causes of migration, 50.

Contents.   IX

Variation   53

The five phases of Variation in birds, *53* — 1. **Individual variation**: its universality, *53* — Range of individual variation, *54* — Psychological differences of individuals, *54* — Monstrosities, *55* — Albinism, melanism, etc., *55* — Dichromatism, *55* — Modifications due to foreign violence, food, etc., *56* — 2. **Geographical variation**: New species arising under geographical isolation, *57* — Increase of size in the Sangi and Talaut Islands, *58*, *59* — Decrease in Sula, *60* — Similar geographical variations of coloration, *61* — 3. **Seasonal changes**: almost unknown in the tropics, *62* — The moult, *63* — Change of coloration without a moult, *63* — 4. **Sexual differences**: classification thereof, *64* — The psychological differences of the sexes, *65* — Theories in explanation of the development of secondary sexual characters, *66—69* — 5. **Changes depending upon Age**: classification thereof, *70* — Ancestral characters, *71* — their value in questions of descent and of geographical distribution, *72* — Phylogenetic value of the tips of feathers, *73* — Hereditary effects of shelter and exposure: the tail-feathers of *Prioniturus*, *73* — of *Merops*, *76* — other cases, *77* — Effects of light, *78—79*.

Geographical Distribution   80

The contrast between the Oriental and Australasian faunas and floras, *80* — Status of geological knowledge of the East Indian Archipelago, *80* — Wallace's line: Wallace's opinions thereon, *81* — Blyth, *82* — Pijnappel, *82* — v. Pelzeln, *82* — Veth, *82* — Allen on the mammals *82* — Semper, *83* — Krümmel on the sea-depths, *83* — Martin, *83*, *85* — Wichmann, *84* — Drude, *84* — Heilprin, *84* — Marshall, *84* — Kan, *84* — Reichenow, *84* — Schuiling, *84* — Jentink on the mammals, *84* — Reclus, *85* — v. Martens on the land-shells, *85*, *87*, *88* — Warburg, *85* — Troussart, *85* — P. L. Sclater, *85* — Flower and Lydekker, *86* — van der Lith, *86* — Haeckel, *86* — Haacke, *86* — Newton, *86* — Sharpe, *86* — Guillemard, *86* — Weber on the fresh-water fishes, *87* — Beddard, *87* — Lydekker, *87* — Hedley on land-shells, *87* — Supan, *88* — Kükenthal, *88* — W. L. Sclater on the mammals *88*, — Niermeyer, *88* — The conflict of opinions, *88* — Table of the **Geographical Distribution of the Birds of the Celebesian Area**, *90—111* — Peculiar genera and species of Celebes, *112* — Table showing the estimated value of the affinities of the peculiar species of Celebes, *115—117* — Summary of the various components of the avifauna of Celebes, *118* — List of the Birds of the Sangi Islands, *120* — of the Talaut Islands, *121* — Affinities of the peculiar species of Sangi and Talaut, *122* — List of the Birds of the Peling Group, *123* — of the Sula Islands, *124* — Evidences of the former union of Peling and Sula, *125* — List of the Birds of Saleyer Island, *125* — of the Djampea Group, *126* — Affinities of the peculiar species of Saleyer and Djampea, *126* — List of the Birds of Togian Island, *126* — Peculiar species and subspecies of Celebes (Mainland), *127* — Differences of the birds of the North and South Peninsulas, *128* — Results, *130*.

SYSTEMATIC INDEX.

Order Accipitres

Family Falconidae

	Page
1. Spilornis rufipectus J. Gd.	1
— rufipectus (typical)	2
— — sulaensis (Schl.)	4
— — < sulaensis	5
2. Circus assimilis Jard. Selby	7
3. Astur griseiceps Schl.	9
4. — trivirgatus (Temm.)	11
— — rufitinctus (McClell.)	12
5. ? — tenuirostris Brügg.	13
6. Urospizias torquatus (Temm.)	15
7. Tachyspizias soloensis (Horsf.)	17
8. Spilospizias trinotatus (Bp.)	21
— — (typical)	21
— — haesitandus Hart.	23
9. Accipiter rhodogaster (Schl.)	25
10. — sulaensis (Schl.)	26
11. — virgatus (Temm.)	27
— — (typical)	27
— — affinis (Hdgs.)	28
— — manilensis (Meyen)	28
— — gularis (Temm. Schl.)	28
— — rufotibialis (Sharpe)	29
12. Spizaetus lanceolatus Temm. Schl.	32
13. Lophotriorchis kieneri (G. Sparre)	35
14. Ictinaetus malayensis (Reinw.)	38
15. Haliaetus leucogaster (Gm.)	40
16. Polioaetus humilis (Müll. Schl.)	43
— — (typical)	43
— — major n. subsp.	44
— — humilis—major	44
17. Butastur indicus (Gm.)	45
18. — liventer (Temm.)	49
19. Haliastur indus (Bodd.)	51
— indus—girrenera	54
20. Milvus migrans (Bodd.)	57
— — (typical)	60
— — melanotis (Temm. Schl.)	60
— — govinda (Sykes)	60

	Page
Milvus migrans affinis (J. Gd.)	60
21. Elanus hypoleucus J. Gd.	62
22. Pernis celebensis (Wall.)	65
23. — sp.	72
24. Baza celebensis (Schl.)	73
25. — reinwardti Müll. Schl.	75
26. Tinnunculus moluccensis Jacquin. Puch.	78
— — orientalis n. subsp.	79
— — occidentalis M. & Wg.	79
— — orientalis—occidentalis	79
27. Falco severus Horsf.	81
— — papuanus M. & Wg.	84
— — indicus n. subsp.	84
28. — peregrinus (Gerini)	85
— — (typical)	85
— — melanogenys (J. Gd.)	86
— — ernesti (Sharpe)	86
— — anatum (Bp.)	87
— — pealei Ridgw.	87

Family Pandionidae

29. Pandion haliaetus (L.)	89
— — (typical)	89
— — leucocephalus (J. Gd.)	89
— — carolinensis (Gm.)	90

Order Striges

Family Asionidae

30. Ninox ochracea (Schl.)	94
31. — scutulata (Raffl.)	95
— — (typical)	95
— — lugubris (Tickell)	95
— — affinis (Tytler)	95
— — japonica (Temm. Schl.)	95
32. Cephaloptynx punctulata (Q. G.)	100
33. Scops manadensis (Q. G.)	103
— — (typical)	103
— — albiventris (Sharpe)	105
— — rutilus (Puch.)	105
— — capnodes (Gurn.)	105

Contents.

	Page
Scops manadensis magicus (S. Müll.)	105
— — leucospilus (Gray)	106
— — morotensis (Sharpe)	106
— — brookii (Sharpe)	107
— — sibutuensis (Sharpe)	107

Family Strigidae
34. Strix flammea L.	109
— — rosenbergi (Schl.)	109
— — (typical)	111
35. — inexpectata Schl.	112
36. — candida Tick.	112

Order Psittaci
Family Loriidae
37. Eos histrio (St. Müll.)	115
— — (typical)	115
— — talautensis (M. & Wg.)	117
— — challengeri (Salvad.)	118
38. Trichoglossus ornatus (L.)	120
39. — forsteni Bp.	123
40. — meyeri Tweedd.	124
— — meyeri (typical)	124
— — bonthainensis (A. B. M.)	125
41. — flavoviridis Wall.	127

Family Cacatuidae
42. Cacatua sulphurea (Gm.)	128
— — (typical)	128
— — djampeana Hart.	130
— — parvula (Bp.)	130

Family Psittacidae
43. Prioniturus platurus (Vieill.)	133
44. — flavicans Cass.	138
45. Tanygnathus muelleri (Müll. Schl.)	140
— — (typical)	140
— — sangirensis M. & Wg.	142
— — muelleri—sangirensis	142
46. ? — luconensis (L.)	144
47. — talautensis M. & Wg.	145
48. — megalorhynchus (Bodd.)	146
— — (typical)	146
— — sumbensis (A. B. M.)	148
— — megalorhynchus—sumbaensis	148
49. Loriculus exilis Schl.	149
50. — catamene Schl.	151
51. — sclateri Wall.	153
— — (typical)	153
— — ruber M. & Wg.	154

	Page
52. Loriculus quadricolor Tweedd.	157
53. — stigmatus (Müll. Schl.)	158
54. Aprosmictus sulaensis Rchw.	170

Order Pici
Family Picidae
55. Iyngipicus temmincki (Malh.)	173
56. Microstictus fulvus (Q. G.)	175
57. — wallacei (Tweedd.)	179

Order Coccyges
Family Cuculidae
Subfamily Cuculinae
58. Hierococcyx crassirostris Tweedd.	182
59. — sparverioides (Vig.)	184
60. — fugax (Horsf.)	185
61. Cuculus canorus (L.)	187
— — (typical)	187
— — canoroides (S. Müll.)	188
62. ? — saturatus Hdgs.	191
63. Chrysococcyx malayanus (Raffl.)	194
64. — basalis (Horsf.)	195
65. Cacomantis virescens (Brügg.)	196
66. — merulinus (Scop.)	199
67. Coccystes coromandus (L.)	201
68. Surniculus musschenbroeki A. B. M.	203
69. Eudynamis melanorhyncha S. Müll.	205
70. — mindanensis (L.)	210
— — (typical)	211
— — sangirensis (W. Blas.)	211

Subfamily Centropodinae
71. Centrococcyx bengalensis (Gm.)	213
72. Pyrrhocentor celebensis (Q. G.)	221
— — (typical)	221
— — rufescens M. & Wg.	223

Subfamily Phoenicophainae
73. Phoenicophaes calorhynchus Temm.	226
— — (typical)	226
— — meridionalis M. & Wg.	227

Subfamily Scythropinae
74. Scythrops novaehollandiae Lath.	231

Order Bucerotes
Family Bucerotidae
75. Rhabdotorrhinus exaratus (Temm.)	235
76. Craniorrhinus cassidix (Temm.)	239

b*

Order Coraciae

Family Meropidae

	Page
77. Merops ornatus Lath.	248
78. — philippinus L.	253
79. Meropogon forsteni Bp.	257

Family Alcedinidae

80. Alcedo ispida L.	262
81. — moluccana (Less.)	264
82. — meninting Horsf.	266
83. Pelargopsis melanorhyncha (Temm.)	269
84. — dichrorhyncha M. & Wg.	271
85. Ceyx wallacei Sharpe.	272
86. Ceycopsis fallax (Schl.)	275
87. — sangirensis M. & Wg.	278
88. Halcyon coromanda (Lath.)	279
— — rufa (Wall.)	280
89. — pileata (Bodd.)	283
90. — sancta Vig. Horsf.	287
91. — chloris (Bodd.)	291
— — (typical)	292
92. Monachalcyon monachus (Bp.)	296
— — (typical)	297
— — intermedius Hart.	298
93. — capucinus M. & Wg.	299
94. — princeps Rchb.	300
95. Cittura cyanotis (Temm.)	303
96. — sangirensis Sharpe	305

Family Coraciidae

97. Coracias temmincki (Vieill.)	309
98. Eurystomus orientalis (L.)	312

Order Macrochires

Family Caprimulgidae

99. Caprimulgus macrurus Horsf.	317
— — (typical)	317
— — albonotatus (Tick.)	318
100. — celebensis Grant	320
101. — affinis Horsf.	321
102. Lyncornis macropterus Bp.	322

Family Cypselidae

103. Cypselus pacificus (Lath.)	327
104. Chaetura celebensis (Scl.)	329

	Page
105. Collocalia fuciphaga (Thunb.)	331
106. — esculenta (L.)	334
107. — francica (Gm.)	335
108. Macropteryx wallacei (J. Gd.)	336

Order Passeres

Family Pittidae

109. Pitta celebensis Müll. Schl.	340
110. — palliceps Brügg.	344
111. — caeruleitorques Salvad.	345
112. — inspeculata M. & Wg.	346
113. — forsteni Bp.	350
114. — sangirana (Schl.)	351
115. — cyanoptera Temm.	352
116. — irena Temm.	354
117. — virginalis Hart.	355

Family Hirundinidae

118. Hirundo rustica L.	356
119. — javanica Sparrm.	358

Family Muscicapidae

120. Muscicapa griseosticta (Swinh.)	363
121. Muscicapula westermanni Sharpe	365
122. — hyperythra (Blyth)	366
123. Siphia banyumas (Horsf.)	368
124. — djampeana Hart.	371
125. — kalaoensis Hart.	371
126. — rufigula (Wall.)	372
127. — bonthaina Hart.	373
128. Stoparola septentrionalis Bütt.	374
129. — meridionalis Bütt.	375
130. Hypothymis puella (Wall.)	376
131. — rowleyi (A. B. M.)	378
132. Rhipidura celebensis Bütt.	379
133. — teijsmanni Bütt.	380
134. Zeocephus talautensis M. & Wg.	382
135. Monarcha commutatus Brügg.	383
136. — inornatus (Garn.)	384
137. — everetti Hart.	385
138. Myiagra rufigula Wall.	386
139. Culicicapa helianthea (Wall.)	387
140. Gerygone flaveola Cab.	388
141. Pratincola caprata (L.)	390

Volume II.

Family Laniidae

142. Pachycephala sulfuriventer (Tweedd.) 394
143. — meridionalis Bütt. 396
144. — teijsmanni Bütt. 396
145. — orpheus Jard. 397
146. — grisecnota G. R. Gray 398
147. — clio Wall. 399
148. — everetti Hart. 400
149. — bonthaina M. & Wg. 401
150. — bonensis M. & Wg. 401
151. Colluricincla sangirensis Oust. 402
152. Lanius tigrinus Drapiez. 403
153. — lucionensis L. 406

Family Campophagidae

154. Graucalus bicolor (Temm.). 411
155. — leucopygius Bp. 413
156. — temmincki (S. Müll.) 415
157. — schistaceus (Sharpe) 416
158. — melanops (Lath.) 417
159. Edoliisoma morio (S. Müll.) 419
160. — salvadorii Sharpe 422
161. — talautense M. & Wg. 423
162. — emancipata Hart. 424
163. — obiense Salvad. 424
164. Lalage leucopygialis Tweedd. 425
165. — timorensis (S. Müll.) 428

Family Artamidae

166. Artamus leucogaster (Val.) 430
167. — monachus Bp. 434

Family Dicruridae

168. Dicrurus leucops Wall. 435
— — (typical) 436
— — axillaris (Salvad.) 438
169. — pectoralis Wall. 439

Family Dicaeidae

170. Dicaeum celebicum S. Müll. 441
171. — sulaense Sharpe 443
172. — sangirense Salvad. 444
173. — talautense M. & Wg. 445
174. — splendidum Bütt. 446
175. — nehrkorni W. Blas. 447
176. — hosei Sharpe 448
177. Acmonorhynchus aureolimbatus (Wall.) 449
178. — sangirensis (Salvad.) 451

Family Nectariniidae

179. Aethopyga flavostriata (Wall.) 453
180. Eudrepanis duivenbodei (Schl.) 456
181. Cyrtostomus frenatus (S. Müll.) 458
— — (typical) 458
— — saleyerensis Hart. 458
— — > saleyerensis 458
— — < saleyerensis 459
— — dissentiens (Hart.) 460
182. — teijsmanni (Bütt.) 462
183. Hermotimia auriceps (G. R. Gray) 464
184. — porphyrolaema (Wall.) 465
— — (typical) 465
— — scapulata M. & Wg. 466
185. — grayi (Wall.) 467
186. — sangirensis (A. B. M.) 469
187. — talautensis M. & Wg. 470
188. Anthreptes malaccensis (Scop.) 472
— — celebensis (Shell.) 475
— — chlorigaster (Sharpe) 477

Family Meliphagidae

189. Myzomela chloroptera Tweedd. 478
190. Melilestes celebensis M. & Wg. 481
— — (typical) 481
— — meridionalis M. & Wg. 482
191. Myza sarasinorum M. & Wg. 483

Family Zosteropidae

192. Zosterops squamiceps (Hart.) 485
193. — intermedia Wall. 486
194. — atrifrons Wall. 487
195. — subatrifrons M. & Wg. 490
196. — nehrkorni W. Blas. 490
197. — sarasinorum M. & Wg. 491
198. — anomala M. & Wg. 494
199. — babelo M. & Wg. 495

Family Timeliidae

200. Iole aurea (Tweedd.) 496
201. — longirostris (Wall.) 497
202. — platenae (W. Blas.) 498
203. Malia grata Schl. 499
— — (typical) 499
— — recondita M. & Wg. 500
204. Androphilus castaneus (Bütt.) 502
205. Cataponera turdoides Hart. 503

XIV   Contents.

	Page		Page
206. Trichostoma celebense Strickl.	504	Calornis panayensis chalybea (Horsf.)	556
207. — finschi Tweedd.	506	— — affinis (Hay)	556
208. Malacopteron affine (Blyth)	508	— — chalybes—affinis	556
		— — tytleri (Hume)	557
Family Turdidae		— — sangirensis (Salvad.)	557
209. Geocichla erythronota Scl.	509	— — panayensis—sangirensis	557
210. Merula celebensis Bütt.	510	— — altirostris (Salvad.)	558
211. Petrophila cyanus (L.)	512	— — enganensis (Salvad	558
— — (typical)	512	232. — minor (Bp.)	561
— — solitaria (P.L.S.Müll.)	512	233. — sulaensis Sharpe	561
		234. — metallica (Temm.).	562
Family Sylviidae		235. Enodes erythrophrys (Temm.)	564
212. Cisticola cursitans (Frkl.)	515	236. Acridotheres cinereus Bp.	566
213. — exilis (Vig. Horsf.)	517	237. Scissirostrum dubium (Lath.)	567
214. Phyllergates riedeli M.&Wg.	519	238. Sturnia violacea (Bodd.)	570
215. Acrocephalus orientalis (Temm.Schl.)	521	239. Basileornis celebensis G.R.Gray	572
216. Locustella fasciolata (G.R.Gray)	524	240. — galeatus A.B.M.	574
217. — ochotensis (Midd.)	526	241. Streptocitta albicollis (Vieill.)	575
218. Phylloscopus borealis (Blas.)	527	242. — torquata (Temm.)	577
219. Cryptolopha sarasinorum M.&Wg.	530	243. Charitornis albertinae Schl.	579
Family Motacillidae			
220. Motacilla flava L.	531	**Family Corvidae**	
221. — boarula L.	534	244. Corvus enca Horsf.	580
— — (typical)	534	245. Gazzola typica Bp.	584
— — melanope (Pall.)	535		
222. Anthus gustavi Swinh.	538	**Family Oriolidae**	
223. — cervinus (Pall.)	540	246. Oriolus celebensis (Tweedd.)	585
		— — (typical)	585
Family Ploceidae		— — meridionalis Hart.	586
224. Munia oryzivora (L.)	542	— — celebensis—meridionalis	586
225. — formosana Swinh.	543	247. — frontalis Wall.	589
— — (typical)	544	248. — boneratensis M.&Wg.	589
— — jagori (Marts.)	544	249. — formosus Cab.	590
— — brunneiceps (Wald.)	544	— — (typical)	590
226. — pallida Wall.	546	— — sangirensis n. subsp.	591
227. — subcastanea Hart.	548	— — formosus—sangirensis	591
228. — punctulata (L.)	548	250. — melanisticus M.&Wg.	593
— — nisoria (Temm.)	548		
229. — molucca (L.)	549	**Order Columbae**	
— — (typical)	549		
— — propinqua Sharpe	550	**Family Treronidae**	
— — molucca—propinqua	550	251. Osmotreron wallacei Salvad.	595
— — <propinqua.	551	252. — sangirensis (Brügg.)	598
— — >propinqua	551	253. — vernans (L.)	599
— — kangeanensis (Vorderm.)	551	254. Ptilopus fischeri (Brügg.)	602
		255. — meridionalis M.&Wg.	604
Family Fringillidae		256. — gularis (Q.G.)	605
230. Passer montanus (L.)	553	257. — subgularis M.&Wg.	606
		258. — melanocephalus (Forst.)	607
Family Sturnidae		259. — melanospilus (Salvad.)	608
231. Calornis panayensis (Scop.)	554	260. — chrysorrhous (Salvad.)	610
— — (typical)	555		

		Page			Page
261.	Ptilopus xanthorrhous (Salvad.)	611		Order Ralli	
262.	— temmincki (Des Murs Prév.)	613		Family Rallidae	
263.	Carpophaga concinna Wall.	615	294.	Gymnocrex rosenbergi (Schl.)	689
264.	— paulina (Bp.)	617	295.	Aramidopsis plateni (W. Blas.)	690
265.	— pulchella Tweedd.	619	296.	Hypotaenidia striata (L.)	692
266.	— intermedia M. & Wg.	619	297.	— philippensis (L.)	694
267.	— rosacea (Temm.)	620	298.	— celebensis (Q. G.)	697
268.	— pickeringi Cass.	621	299.	— sulcirostris (Wall.)	698
269.	— radiata (Q. G.)	622	300.	Rallina minahassa Wall.	699
270.	— forsteni (Bp.)	623	301.	Porzana fusca (L.)	701
271.	— poecilorrhoa Brügg.	625	302.	Limnocorax niger (Gm.)	703
272.	Myristicivora bicolor (Scop.)	627	303.	Amaurornis cinerea (Vieill.)	705
273.	— luctuosa (Temm.)	631	304.	— phoenicura (Forst.)	708
			305.	— moluccana (Wall.)	711
	Family Columbidae		306.	— isabellina (Schl.)	712
274.	Columba albigularis (Bp.)	633	307.	Gallinula frontata Wall.	713
275.	Turacoena manadensis (Q. G.)	635	308.	— chloropus (L.)	715
276.	Macropygia albicapilla Bp.	637	309.	Porphyrio calvus Vieill.	717
	— — (typical)	637	310.	— pulverulentus Temm.	721
	— — sangirensis (Salvad.)	638	311.	Fulica atra L.	722
	— albicapilla—sangirensis	638			
277.	— macassariensis (Wall.)	641		Order Limicolae	
	Family Peristeridae			Family Parridae	
278.	Turtur tigrinus (Temm. Kn.)	643	312.	Hydralector gallinaceus (Temm.)	725
279.	Geopelia striata (L.)	646			
280.	Chalcophaps indica (L.)	649		Family Glareolidae	
281.	— stephani Rchb.	653	313.	Glareola isabella Vieill.	728
282.	Phlogoenas tristigmata Bp.	654			
283.	— bimaculata Salvad.	656		Family Charadriidae	
284.	Caloenas nicobarica (L.)	657	314.	Esacus magnirostris (Vieill.)	733
			315.	Lobivanellus cinereus (Blyth)	735
			316.	? Squatarola helvetica (L.)	736
	Order Gallinae		317.	Charadrius fulvus Gm.	738
	Family Phasianidae		318.	Aegialitis vereda (J. Gd.)	741
285.	Excalfactoria chinensis (L.)	663	319.	— geoffroyi (Wagl.)	743
286.	Gallus ferrugineus (Gm.)	667	320.	— mongola (Pall.)	746
			321.	— curonica (Gm.)	749
	Family Megapodidae		322.	— peroni (Schl.)	752
287.	Megapodius cumingi Dillw.	671	323.	Strepsilas interpres (L.)	755
288.	— sangirensis Schl.	675	324.	Himantopus leucocephalus J. Gd.	757
289.	— bernsteini Schl.	676	325.	Totanus glottis (L.)	759
290.	— duperreyi Less. Garn.	677	326.	— calidris (L.)	761
291.	Megacephalon maleo (Hartl.)	678	327.	— glareola (L.)	764
			328.	Heteractitis brevipes (Vieill.)	766
	Order Turnices		329.	Actitis hypoleucos (L.)	770
	Family Turnicidae		330.	Terekia cinerea (Güld.)	773
292.	Turnix rufilatus Wall.	686	331.	Tringa acuminata (Horsf.)	776
293.	— maculosa (Temm.)	687	332.	— damascensis (Horsf.)	778
			333.	— ruficollis Pall.	780

Contents.

	Page
334. ? Calidris arenaria (L.)	782
335. Phalaropus hyperboreus (L.)	785
336. Limicola platyrhyncha (Temm.)	787
337. Gallinago megala Swinh.	789
338. Limosa novaezealandiae (G.R.Gray)	792
339. Numenius minutus J.Gd.	795
340. — variegatus (Scop.).	797
341. ?— arquatus (L.)	799
342. — cyanopus Vieill.	800

Order Ciconiiformes
Suborder Ciconiae
Family Ibidae
343. Plegadis falcinellus (L.)	803

Family Ciconiidae
344. Dissoura episcopus (Bodd.)	806

Family Plataleidae
345. Platalea sp.	809

Suborder Ardeae
Family Ardeidae
346. Phoyx manilensis (Meyen)	811
347. Ardea sumatrana Raffl.	814
348. Notophoyx picata (J.Gd.)	816
349. — novaehollandiae (Lath.)	817
350. Demiegretta sacra (Gm.)	819
351. Herodias eulophotes Swinh.	824
352. — garzetta (L.)	826
353. — alba (L.)	829
354. — intermedia (Wagl.)	832
355. Bubulcus coromandus (Bodd.)	835
356. Ardeola speciosa (Horsf.)	838
357. Nycticorax caledonicus (Gm.)	841
358. — manilensis Vig.	843
359. — griseus (L.)	845
360. Gorsachius kutteri (Cab.)	848
361. Butorides javanica (Horsf.)	851
362. Ardetta sinensis (Gm.)	854
363. — eurhythma Swinh.	856
364. — cinnamomea (Gm.)	859
365. Xanthocnus flavicollis (Lath.)	861
366. — melaenus (Salvad.)	863

Order Anseres
Family Anatidae
	Page
367. Nettopus pulchellus J.Gd.	865
368. — coromandelianus (Gm.)	866
369. Dendrocygna arcuata (Horsf.)	868
370. — guttata Schl.	870
371. Anas superciliosa Gm.	872
372. Nettion gibberifrons (S.Müll.)	874
373. Querquedula circia (L.)	879
374. Nyroca fuligula (L.)	881

Order Steganopodes
Family Fregatidae
375. Fregata minor (Gm.)	883

Family Phalacrocoracidae
376. Plotus melanogaster (Penn.)	886
377. Phalacrocorax melanoleucus (Vieill.)	888
378. — sulcirostris Brdt.	890

Family Sulidae
379. Sula leucogaster (Bodd.)	892

Order Lari
Family Laridae
380. Hydrochelidon leucoptera (Meisn.Sch.)	893
381. — hybrida (Pall.)	895
382. Sterna media Horsf.	897
383. — bergii Lcht.	899
384. — sinensis Gm.	901
385. — melanauchen Temm.	903
386. — anaestheta Scop.	906
387. Anous stolidus (L.)	908
388. Stercorarius sp.	910

Order Tubinares
Family Puffinidae
389. Puffinus cuneatus Salv.	911
390. — leucomelas (Temm.)	913

Family Diomedeidae
391. Diomedea sp.	914

Order Pygopodes
Family Podicipedidae
392. Podiceps tricolor (G.R.Gray)	915
393. — gularis J.Gd.	917

Alphabetical Index . . . . 919—962

LIST OF PLATES.

Volume I.

Plate			Pag.
I.	Spilospizias trinotatus (Bp.) imm.		21
II, III.	Spizaetus lanceolatus Temm. Schl. ad., juv.		» 32
	Pernis celebensis (Wall.) ad., juv.		» 65
IV.	Ninox ochracea (Schl.) ad. et juv.		» 94
V, VI.	Prioniturus platurus (Vieill.). Tails		» 133
VI.	Prioniturus flavicans Cass. Tails		» 138
VII.	Aprosmictus sulaensis Rchw.		» 170
VIII.	Merops ornatus Lath. Tails		» 248
	Merops philippinus L. Tails		» 253
IX.	Pelargopsis dichrorhyncha M. & Wg.		» 271
	Monachalcyon capucinus M. & Wg.		» 299
X.	Ceycopsis fallax (Schl.)		» 275
	Ceycopsis sangirensis M. & Wg. ad. et juv.		» 278
XI.	Caprimulgus celebensis Grant		» 320
	Lyncornis macropterus Bp.		» 322
XII.	Chaetura celebensis (Sclat.) fem. juv. et mas		» 329
XIII.	Muscicapula hyperythra (Blyth) mas, fem. et juv.		» 366
	Siphia banyumas (Horsf.)		» 368
XIV.	Siphia djampeana Hart.		» 371
	Siphia kalaoensis Hart.		» 371
XV.	Stoparola septentrionalis Bütt. mas, fem. et juv.		» 374
XVI.	Zeocephus talautensis M. & Wg. ad. et juv.		» 382
	Monarcha commutatus Brügg.		» 383
XVII.	Monarcha everetti Hart.		» 385
	Pachycephala teijsmanni Bütt.		» 396
	Pachycephala everetti Hart.		» 400

Volume II.

XVIII.	Pachycephala sulfuriventer (Tweedd.)		» 394
	Pachycephala bonensis M. & Wg.		» 401
XIX.	Pachycephala bonthaina M. & Wg. mas et fem.		» 401
	Cryptolopha sarasinorum M. & Wg.		» 530
XX.	Graucalus bicolor (Temm.) mas et fem.		» 411
XXI.	Graucalus leucopygius Bp. ad. et juv.		» 413
XXII.	Edoliisoma morio (S. Müll.) mas et fem.		» 419
	Edoliisoma talautense M. & Wg. mas et fem.		» 423
	Edoliisoma obiense Salvad. mas et fem.		» 424

Contents.

Plate			Page
	XXIII.	Edoliisoma salvadorii Sharpe mas et fem.	422
»	XXIV.	Dicrurus leucops Wall. ad. et juv.	» 435
		Dicrurus leucops axillaris (Salvad.)	» 438
»	XXV.	Dicaeum celebicum S. Müll. mas et fem.	» 441
		Dicaeum sangirense Salvad.	» 444
		Dicaeum nehrkorni W. Blas.	» 447
»	XXVI.	Cyrtostomus frenatus saleyerensis Hart. mas	» 458
		Cyrtostomus teijsmanni Bütt. mas et fem.	» 462
»	XXVII.	Acmonorhynchus sangirensis (Salvad.)	» 451
		Hermotimia talautensis M. & Wg. mas et fem.	» 470
»	XXVIII.	Melilestes celebensis M. & Wg.	» 481
		Myza sarasinorum M. & Wg.	» 483
»	XXIX.	Zosterops squamiceps (Hart.)	» 485
		Cataponera turdoides Hart.	» 503
»	XXX.	Zosterops subatrifrons M. & Wg.	» 490
		Zosterops anomala M. & Wg.	» 494
		Zosterops babelo M. & Wg.	» 495
»	XXXI.	Zosterops nehrkorni W. Blas.	» 490
		Zosterops sarasinorum M. & Wg.	» 491
»	XXXII.	Iole aurea (Tweedd.).	» 496
		Iole platenae (W. Blas.)	» 498
»	XXXIII.	Malia grata recondita (M. & Wg.)	» 500
»	XXXIV.	Androphilus castaneus (Bütt.)	» 502
		Phyllergates riedeli M. & Wg.	» 519
»	XXXV.	Merula celebensis Bütt. ad. et juv.	» 510
»	XXXVI.	Calornis sulaensis Sharpe	» 561
		Basilcornis galeatus A. B. M.	» 574
»	XXXVII.	Oriolus melanisticus M. & Wg.	» 593
		Ptilopus melanocephalus (Forst.)	» 607
»	XXXVIII.	Ptilopus chrysorrhous (Salvad.)	» 610
		Ptilopus xanthorrhous (Salvad.)	» 611
»	XXXIX.	Carpophaga concinna Wall.	» 615
		Carpophaga intermedia M. & Wg.	» 619
»	XL.	Macropygia albicapilla Bp. mas, fem. et pull.	» 637
»	XLI.	Megapodius cumingi Dillw.	» 671
		Megapodius sangirensis Schl.	» 675
»	XLII.	Gymnocrex rosenbergi (Schl.)	» 689
		Aramidopsis plateni (W. Blas.)	» 690
»	XLIII.	Amaurornis isabellina (Schl.)	» 712
		Gallinula frontata Wall.	» 713
»	XLIV.	Herodias eulophotes Swinh.	» 824
»	XLV.	Ardetta eurhythma Swinh.	» 856
		Drawing of Mound of Megapodius cumingi Dillw.	» 674

LIST OF MAPS.

Map I. Celebes and the Neighbouring Islands . . . . Frontispiece
        For the preparation of Map I (Celebes and the Neighbouring Islands) with the sea-depths our cartographers were directed to make use of the following publications:
            O. Krümmel: Das Relief des austral-asiatischen Mittelmeeres: Ztschr. f. wiss. Geogr. III, 1, Taf. I (1882).
            Stenfoort en Siethoff: Atlas, Ned. Bez. in Oost-Indië, 1883—1885 (Topogr. Inrichting te's Gravenhage).
            C. M. Kan: Bodengesteldheid der Eilanden en diepte der Zeeën van den indischen Archipel: Tdschr. Ned. Aardr. Gen. (2) V (Versl.), 202, Kaart IV (1888).
            id.: Kaart van den Ned.-Ind. Arch. (1:6,000,000) s. a. (after 1889).
            H. Berghaus: Atlas der Hydrographie (Berghaus' Physik. Atlas, Abth. II), Karte X (25) (1891).
        Further, the Dutch and English Admiralty Charts concerning the region round Celebes; articles in the Annalen der Hydrographie 13. Jahrg. 1885, 207, and 25. Jahrg. 1897, 353, Taf. 11; an article in the Tdschr. Ned. Aardr. Gen. (2), III (Versl.), 485: Zeediepten in den Oost-Ind. Archipel, 1886; and others.

» II. Celebes . . . . Frontispiece
        In preparing Map II (Celebes) use has been made of the maps in various Dutch and German Atlasses, of S. C. J. W. van Musschenbroek's map of the Gulf of Tomini or Gorontalo and the neighbouring territories, with its accompanying notes (Tdschr. Aardr. Gen. IV, Kaart 2, 1878, p. 93), and of the recent special maps of Drs. P. & F. Sarasin, viz.:
            1. Zeitschr. Ges. Erdk. Berlin, 1894, XXIX, Taf. 13 (Region between the Minahassa and Gorontalo, North Celebes).
            2. ib. 1895, XXX, Taf. 10 (Region between Buol and the Gulf of Tomini, North Celebes).
            3. ib. Taf. 15 (Central Celebes).
            4. ib. 1896, XXXI, Taf 2 (South-west Celebes).
            5. Verh. d. Ges. f. Erdk. Berlin, 1896, XXIII, Taf. 3 (South-east Celebes).
        Besides this Drs. P. & F. Sarasin have had the great kindness to look over our map and to express general approval of it. We have also made use of de Hollander: Handleiding Volkenk. Ned. Oost-Indië, 4. ed. 1882—4.
        The map shows only the names of the places mentioned in the text where birds have been collected.

» III. Winds and Rains: April—September . . . . Pag. 17
» IV. Winds and Rains: October—March . . . . » 17
» V. Distribution of the genus *Cacatua* . . . . » 128
» VI. Distribution of the genus *Loriculus* . . . . » 149
» VII. Distribution of the *Bucerotidae* . . . . » 242

ALPHABETICAL LIST OF ABBREVIATIONS OF AUTHORS' NAMES, TITLES OF BOOKS, JOURNALS, ETC.

Our abbreviations were made before the issue of the list used in the "Zoological Record", which is recommended by the "Deutschen Zoologische Gesellschaft" as a model for such, or we would have adopted at least a part of them, but many are far too long to be made use of in a synonymy, and others are not practical.

Portions of abbreviations here enclosed in brackets are often omitted in the text.

Abh. (Ber.) Mus. Dresden = Abhandlungen und Berichte des Kgl. Zoologischen und Anthropologisch-Ethnographischen Museums zu Dresden.
Abh. (Naturw.) Ver. Bremen = Abhandlungen herausgegeben vom naturwissenschaftlichen Vereine zu Bremen.
Abh. Senckenb. Naturf. Ges. = Abhandlungen herausgegeben von der Senckenbergischen naturforschenden Gesellschaft. Frankfurt am Main.
Acta (Nova) Acad. Leop. (Carol.) = Nova Acta Academiae Caesareae Leopoldino-Carolinae Germanicae Naturae Curiosorum. Also = Verhandlungen der Kaiserlichen Leopoldinisch-Carolinischen Deutschen Akademie der Naturforscher.
Alb., Nat. Hist. B. = E. Albin: Natural history of British Birds. 1738—40.
Am. Journ. of Sc. & Arts = American Journal of Science and Arts.
Ann. (K. K.) Nat. Hofmus. (Wien) = Annalen des K. K. Naturhistorischen Hofmuseums zu Wien.
Ann. Mus. Civ. Gen. = Annali del Museo Civico di Storia Naturale di Genova.
Ann. (& Mag.) N. H. = The Annals and Magazine of Natural History, including Zoology, Botany, and Geology.
Ann. Sc. Nat. = Annales des Sciences Naturelles. Zoologie et Paléontologie.
Arch. Nat. = Archiv für Naturgeschichte.
Atti Ac. Sc. Tor(ino). = Atti della Reale Accademia delle Scienze di Torino.
Atti Soc. It. Sc. Nat. Mil. = Atti della Società italiana di Scienze naturali. Milano.
Audubon, B. N. Am. = J. J. Audubon: The Birds of America. 1826 seq.
Auk = The Auk. Quarterly Journal of Ornithology.
Ausland = Das Ausland. Wochenschrift für Länder- und Völkerkunde.
Baird, Brew. & Ridg., Water B. N. Am. = S. F. Baird, T. M. Brewer, and R. Ridgway: The Water Birds of North America. 1884. (Memoirs of the Museum of Comparative Zoölogy at Harvard College, Vol. VII and XIII.)

Baldamus, Leben europ. Kuck. = E. Baldamus: Das Leben der europäischen Kuckucke. Nebst Beiträgen zur Lebenskunde der übrigen parasitischen Kuckucke und Stärlinge. 1892.
Bartl., (Mon.) Weaver-b. = A. D. Bartlett: A Monograph of the Weaver-birds. 1888.
Bechst., Naturg. Deutschl. = J. M. Bechstein: Gemeinnützige Naturgeschichte Deutschlands, nach allen 3 Reichen. 2. verb. Aufl. 1801—1809.
Bechst., Orn. Taschenb. = J. M. Bechstein: Ornithologisches Taschenbuch von und für Deutschland. 1802—12.
Bagbie, Malay Penin. = P. J. Begbie: The Malayan Peninsula; history, manners and customs of the inhabitants, politics, natural history etc. 1834.
Bijdr. (t. d.) Dierk. = Bijdragen tot de Dierkunde. Uitgegeven door het K. Zoologisch Genootschap "Natura artis magistra", Amsterdam.
Bijdr. taal, land, volkenk. Ned. Ind. = Bijdragen tot de Taal-, Land- en Volkenkunde van Nederlandsch Indië.
Blanf., Faun. Br. Ind(ia) B. = The Fauna of British India, incl. Ceylon and Burma. Ed. by W. T. Blanford. Birds, by E. W. Oates and W. T. Blanford. 1889—1895.
Blak., Amend. List B. Jap. = T. W. Blakiston: Amended List of the Birds of Japan, according to Geographical Distribution. 1884.
W. Blas., Braunschw. Anzeigen = W. Blasius in the "Braunschweigischen Anzeigen" (Newspaper).
W. Blas., Russ's Isis = W. Blasius in Karl Russ' "Isis": Zeitschrift für alle naturwissenschaftlichen Liebhabereien.
Blyth, B. Burmah = E. Blyth: A Catalogue of the Mammals and Birds of Burmah (Journal of the Asiatic Society of Bengal, 1875, pt. II, extra number).
Blyth, Cat. (B.) Mus. A(s). S. (B.) = E. Blyth: A Catalogue of the Birds in the Museum of the Asiatic Society. 1849.
Bodd., Tabl. Pl. Enl. = M. Boddaert: Table des planches enluminées d'histoire nat., de d'Aubenton. 1783.
Boll. Mus. Torino = Bollettino dei Musei di Zoologia ed Anatomia comparata della R. Università di Torino.

List of abbreviations.

Bonn., Tabl. Enc. Méth. = Encyclopédie méthodique, ou par ordre de matière, par une société de gens de lettres. — Histoire naturelle. — Tableau encyclopédique et méthodique: Ornithologie. Par l'abbé Bonnaterre, 1790.

Bourjot, Perr. = A. Bourjot St.-Hilaire: Histoire Naturelle des Perroquets, vol. III, 1837—38 (vols. I and II by Le Vaillant, 1801 and 1805).

Bourns & Worces., B. Menage Exped. = F. S. Bourns and D. C. Worcester: Preliminary Notes on the Birds and Mammals collected by the Menage Scientific Expedition to the Philippine Islands (Occasional Papers of the Minnesota Academy of Natural Sciences, Vol. I No. 1). 1894.

Bp., Cat. Ucc. Eur. = C. L. Bonaparte: Catalogo metodico degli Uccelli Europei. (Annali delle scienze naturali. Tom. VIII. 2e semestre.) 1842.

Bp., Comp. List B. Eur. & N. Am. = C. L. Bonaparte: A geographical and comparative list of the Birds of Europe and North America. 1838.

Bp., Consp. = C. L. Bonaparte: Conspectus generum avium. 1850—65.

Bp., Consp. Vol. Anisod. = C. L. Bonaparte: Conspectus Volucrum Anisodactylorum. 1854.

Bp., Consp. Vol. Zygod. = C. L. Bonaparte: Conspectus Volucrum Zygodactylorum. 1854.

Bp., Coup d'Oeil Ordre Pig. = C. L. Bonaparte: Coup d'Oeil sur l'Ordre des Pigeons. (Comptes Rendus hebdomadaires des séances de l'Académie des Sciences, Paris. — Articles in vols. XXXIX, 1854; XL, 1855; XLIII, 1856.)

Bp., Icon. des Pig. = C. L. Bonaparte: Iconographie des Pigeons. 1857.

Bp., Notes Orn. Coll. Delattre = C. L. Bonaparte: Notes ornithologiques sur les collections rapportées en 1853, par M. A. Delattre, et classification parallélique des Passereaux chanteurs. Paris 1854.

Brehm, Tierl. = Brehms Tierleben. Allgemeine Kunde des Tierreichs. 3. gänzlich neubearbeitete Auflage. Vögel. 1891.

Brehm, Vög. Deutschl. = C. L. Brehm: Handbuch der Naturgeschichte aller Vögel Deutschlands. 1831.

Briss., Orn. = M. J. Brisson: Ornithologia s. synopsis methodica, sist. Avium divisionem in ordines etc. 1760.

Brügg., Abh. Ver. Bremen = F. Brüggemann in Abhandlungen herausgegeben vom naturwissenschaftlichen Vereine zu Bremen.

Bütt., (Zool. Erg.) Weber's Reise (Ostind.) = J. Büttikofer in: Zoologische Ergebnisse einer Reise in Niederländisch Ost-Indien. Herausgegeben von Max Weber. 1890—97.

Buff., H. N. Ois. = G. L. Leclerc comte de Buffon, de Montbeillard (et l'abbé Bexon): Histoire naturelle des Oiseaux. 10 vols. (Small fol. ed.) 1770—86.

Bull. Ac. Imp. Mosc. = Bulletin de la Société Impériale des Naturalistes de Moscou.

Bull. Ac. Sc. Petersb. = Bulletin de l'Académie Impériale des Sciences de St. Pétersbourg.

Bull. Brit. Orn. Club = Bulletin of the British Ornithologists' Club.

Bull. Mus. Belg. = Bulletin du Musée Royal d'Histoire Naturelle de Belgique.

Bull. Mus. Comp. Zool. Cambridge = Bulletin of the Museum of Comparative Zoölogy at Harvard College, Cambridge, Mass.

Bull. of the U. S. Geol. and Geogr. Survey = Bulletin of the United States Geological and Geographical Survey of the Territories.

Bull. Soc. Philom. (Paris) = Bulletin de la Société Philomatique de Paris.

Bull. S(oc). Z(ool). Fr(ance) = Bulletin de la Société zoologique de France.

Bull. U. S. Nat. Mus. = Bulletin of the U. S. National Museum.

Buller, B. N. Zeal. = W. L. Buller: A History of the Birds of New Zealand. I. ed. 1873; 2. ed. 1888.

Butler, Foreign Finches = A. G. Butler, Foreign Finches in captivity. 1894—96.

Cab. & Hein., Mus. Hein. = J. Cabanis & F. Heine jun.: Museum Heineanum. Verzeichniss der ornithologischen Sammlung des Oberamtmanns F. Heine. 1850—63.

Calc. Journ. Nat. Hist. = Calcutta Journal of Natural History.

Campb., Ber. II. Orn. Congress Budapest = A. J. Campbell, in: Zweiter internationaler ornithologischer Congress. Budapest 1891. Haupthericht. II. Wissenschaftlicher Theil. 1892.

Cass., B. Calif. = J. Cassin: Illustrations of the Birds of California, Texas, Oregon, British and Russian America. 1853—55.

Cass., Cat. Halc. Philad. Mus. = J. Cassin: Catalogue of the Halcyonidae in the Collection of the Academy of Natural Sciences of Philadelphia. 1852.

Cass., U. S. Expl. Exp(ed). 2nd ed. = J. Cassin: United States Exploring Expedition. — Mammalogy and Ornithology. 1858.

Cat. B. = Catalogue of the Birds in the British Museum.

C. R. Congr. Int. des sc. géogr. à Paris = Congrès international des sciences géographiques tenu à Paris 1875. Compte rendu des séances. 1878—80.

C(omptes) R(end). = Comptes rendus hebdomadaires des séances de l'Académie des sciences, Paris.

Cretschm. in Rüpp. Atlas = (W. P.) E. Rüppell: Atlas zu der Reise im nördlichen Afrika. 2. Abtheilung Vögel. Bearbeitet von Ph. J. Cretschmar. 1826.

Cuv., Règne An. = G. Cuvier: Le règne animal, distribué d'après son organisation. 1829—30.

Darwin, Anim. & Plants = C. Darwin: The variation of animals and plants under domestication. 1868—69.

D'Aubent., Pl. Enl. = D'Aubenton le jeune: Planches enluminées d'histoire naturelle par Martinet, exécutées par D'Aubenton le jeune. 1765.

Daud., Tr. d'Orn. = F. M. Daudin: Traité élémentaire et complet d'Ornithologie, ou histoire naturelle des Oiseaux. 1800.

David et Oust., Ois. Chine = A. David et E. Oustalet: Les Oiseaux de la Chine. 1877.

List of abbreviations.

Des Murs & Prév., Voy. Vénus, Zool. = Des Murs et Prévost, in the Voyage autour du monde sur la frégatte "la Vénus", commandée par A du Petit-Thouars. 1845—64.

Deutsche geogr. Blätter = Deutsche geographische Blätter, herausgegeben von der Geographischen Gesellschaft in Bremen.

Diggles, Orn. Austr. = S. Diggles: The Ornithology of Australia. 1866—70.

Direct. Ind. Arch. = A. G. Findlay: A Directory for the navigation of the Indian Archipelago. 1870.

Drap., Dict. Class. H(ist). N(at). = Drapier: Dictionnaire classique des sciences naturelles etc. 1828—45.

Dresser, B. Eur. = H. E. Dresser: A history of the birds of Europe, including all the species inhabiting the Western Palaearctic Region. 1871—81, 1895—96.

Dresser, Mon. Corac. = H. E. Dresser: A Monograph of the Coraciidae. 1893.

Dresser, Monogr. Merop. = H. E. Dresser: A Monograph of the Meropidae. 1884—86.

Dumont, Dict. Sc. Nat. = C. H. F. Dumont: Dictionnaire des sciences naturelles. 1828—45.

Edw., Birds = G. Edwards: A natural history of uncommon Birds, and of some other rare and nondescribed animals, Quadrupedes, Reptiles, Fishes, Insects etc. 1743—51.

Edw., Gleam. = G. Edwards: Gleanings of natural history, exhibiting figures of Quadrupeds, Birds, Insects, Plants etc. 1758—64.

Elliot, Monogr. Bucerot. = D. G. Elliot: A monograph of the Bucerotidae, or family of the Hornbills. 1882.

Elliot, Monogr. Pitt. = D. G. Elliot: A Monograph of the Pittidae. 1893—95.

Festschr. Vers. Naturf. Braunschw. 1897 = Festschrift der Herzoglichen Technischen Hochschule bei Gelegenheit der 69. Versammlung Deutscher Naturforscher und Aerste in Braunschweig. 1897.

Festschrift zool.-bot. Ges. = Festschrift zur Feier des 25 jährigen Bestehens der k. k. zoologisch-botanischen Gesellschaft in Wien. 1876.

de Fil., Mus. Mediol. = Ph. de Filippi: Museum Mediolanum, N. 1, Animalia Vertebrata, Classis II. Aves. 1847.

Finsch, J. Mus. Godef. = O. Finsch: Zur Ornithologie der Südsee-Inseln. I. Die Vögel der Palaugruppe. II. Ueber neue und weniger gekannte Vögel von den Viti-, Samoa- und Carolinen-Inseln. (Journal des Museum Godeffroy. Heft VIII, XII.) 1875—76.

Finsch, N(eu) G(uinea) = O. Finsch: Neu-Guinea und seine Bewohner. 1865.

Finsch, Papag. = O. Finsch: Die Papageien. 1867—68.

Finsch, Vög. der Südsee = O. Finsch: Ueber Vögel der Südsee, in Mittheilungen des ornithologischen Vereins in Wien, 1884.

Finsch & Hartl., Orn. Centralpol. = O. Finsch & G. Hartlaub: Beitrag zur Fauna Centralpolynesiens. Ornithologie der Viti-, Samoa- und Tonga-Inseln. 1867.

Finsch & Hartl., Vög. O. Afr. = O. Finsch und G. Hartlaub: Die Vögel Ost-Afrikas. (C. C. von der Decken: Reisen in Ost-Afrika. 4. Bd.) 1870.

Flem., Phil. of Zool. = J. Fleming: Philosophy of zoology; or a general view of the structure, functions and classification of animals. 1822.

Forst., Descr. An. = J. R. Forsteri: Descriptiones animalium in itinere ad Maris Australis terras 1772—74 suscepto observatorum. Ed. H. Lichtenstein. 1844.

Forst(er), Zool. Ind. = J. R. Forster: Zoologia Indica. 1781.

Fraser, Zool. Typ. (Av.) = L. Fraser: Zoologia Typica, or figures of new and rare Mammals and Birds. 1841—9.

Fritsch, Vög. Eur. = A. Fritsch: Vögel Europa's. 1871.

Gadow, Vög. in Bronn's Kl. u. Ord. = H. G. Bronn's Klassen und Ordnungen des Thier-Reichs, wissenschaftlich dargestellt in Wort und Bild. 6. Band. IV. Abth. Vögel. Von E. Selenka u. H. Gadow. 1891-3.

Gätke, Vogelw. Helgol. = H. Gätke: Die Vogelwarte Helgoland. 1891.

Garnot, Voy. Coquille, Zool. Atl. = Voyage autour du monde, exécuté par ordre du roi, sur la Corvette d. S. M. la Coquille pendant les années 1822—25. Publié par L. J. Duperrey. Zoologie, par Lesson et Garnot. 1826.

Gefied. Welt = Die gefiederte Welt. Wochenschrift für Vogelliebhaber,- Züchter und -Händler. Herausg. von K. Russ.

Gerini, Orn. Meth. Dig. = Giovanni Gerini: Storia Naturale degli uccelli, trattata con metodo e adornata di figure intagliate in rame e miniata al naturale. Firenze 1767—76.

Gigl., Avif. Ital(ica) = E. H. Giglioli: Avifauna Italica. Elenco delle specie di uccelli stazionarie o di passagio in Italia. 1886.

Gigl., Avif. Ital. pt. I = E. H. Giglioli: Primo resoconto dei risultati della inchiesta ornitologica in Italia. I. Avifauna Italica. 1889.

Gigl. & Manz., Icon. Avif. Ital. = E. H. Giglioli & A. Manzella: Iconografia dell' Avifauna Italica, ovvero Tavole illustranti le specie di Uccelli che trovansi in Italia, con brevi descrizioni e note. 1882—84.

Gld., B. Eur. = John Gould: The Birds of Europe. 1832—37.

G(ou)ld., B. Asia = John Gould: The Birds of Asia. 1850—1883. [Completed by R. Bowdler Sharpe.]

Gld., B. Austr. = John Gould: The Birds of Australia. 1848—69.

Gld., B. Gr. Brit. = John Gould: The Birds of Great Britain. 1862—1873.

Gld., B. N(ew) Guinea = John Gould: The Birds of New-Guinea and the adjacent Papuan Islands, including any new species that may be discovered in Australia. 1875—88. [Completed by R. Bowdler Sharpe.]

Gld., Handb. B. Austr. = John Gould: Handbook of the Birds of Australia. 1865.

Gld., Syn. B. Austr. = John Gould: A Synopsis of the Birds of Australia and its adjacent Islands. 1837.

Globus = Globus. Illustrierte Zeitschrift für Länder- und Völkerkunde.

Gm., S(yst). N(at). = Caroli Linnaei Systema Naturae. Ed. XIII, aucta, reformata. Cura J. F. Gmelin. Lipsiae 1788—93.

List of abbreviations.   XXIII

Gould, Cent. Himal. B. = John Gould: A Century of Birds from the Himalaya Mountains. 1832.
Graafland, De Minahassa = N. Graafland: De Minahassa. Haar verleden en haar tegenwoordige toestand. 1867—69.
Grant, Handb. Game B. = W. R. Ogilvie-Grant: A Hand-book to the Game-Birds. 1895—97. In Allen's Naturalist's Library, ed. by R. B. Sharpe.
Gray, B. Trop. Is. = G. R. Gray: Catalogue of the Birds of the Tropical Islands of the Pacific Ocean, in the collection of the British Museum. 1859.
Gray, Cat. B. New Guin. = J. E. Gray & G. R. Gray: Catalogue of the Mammalia and Birds of New Guinea, in the collection of the British Museum. 1859.
Gray, Cat. Hodgs. Coll. B. = J. E. Gray, Catalogue of the specimens and drawings of Mammals, Birds, Reptiles and Fishes of Nepal and Tibet, presented by B. H. Hodgson to the British Museum. 2nd ed. London 1863.
Gray, Cruise "Curaçoa", B. = G. R. Gray: Birds in Bronchley's Jottings during the cruise of H. M. S. "Curaçoa" among the South Sea Islands in 1865. 1873.
Gray, Gen. B. = G. R. Gray: The Genera of Birds: comprising their generic characters, a notice of the habits of each genus, and an extensive list of species referred to their several genera. 1844—49.
Gray, Ill. = G. R. Gray: Hand-list of Genera and Species of Birds, distinguishing those contained in the British Museum. 1869—71.
Gray, List Acc. B. M. = G. R. Gray: List of the Specimens of Birds in the collection of the British Museum. Sec. ed. Part. I. Accipitres. 1848.
Gray, List Anseres Brit. Mus. = List of the Specimens of Birds in the collection of the British Museum. By G. R. Gray. Part. III. Gallinae, Grallae, and Anseres. 1844.
Gray, List B. Br. Mus., Columbae = List of the Specimens of Birds in the collection of the British Museum. By G. R. Gray. Part. IV. Columbae, 1856.
Gray, List (Coraciadae etc.) Fissirostr. Br. Mus. = G. R. Gray: List of the Specimens of Birds in the collection of the British Museum. Part. II. Section I. Fissirostres. 1848.
Gray, List Gall(inae) Brit. Mus. = List of the Specimens of Birds in the collection of the British Museum. By G. R. Gray. Part. III. Gallinae, Grallae, and Anseres. 1844.
Gray, List Gen. B. = G. R. Gray: List of the Genera of Birds. 1844.
Gray, Cat. gen. & subgen. B. = G. R. Gray: Catalogue of genera and subgenera of Birds in the British Museum. 1855.
Gray, List Grallae Br. Mus. = List of the Specimens of Birds in the collection of the British Museum. By G. R. Gray. Part III. Gallinae, Grallae, and Anseres. 1844.
Gray, List Psitt. B. M. = List of the Specimens of Birds in the collection of the British Museum. By G. R. Gray. Part III. Section II. Psittacidae. 1859.

Gray, Voy. Ereb. & Terror, B. = (J. Richardson & J. E. Gray) The zoology of the voyage of H. M. S. Erebus and Terror, 1839—43. 1844—45, 1875.
J. E. Gray, Zool. Misc. = J. E. Gray: The zoological miscellany. 1831.
J. E. Gray, Ill. Ind. Zool. = J. E. Gray: Illustrations of Indian zoology, consisting of coloured plates of new or hitherto unfigured Indian animals from the collection of Major-General Hardwicke. 1830—34.
Griff., An. Kingd. = G. L. C. F. D. Cuvier: The animal kingdom, described and arranged in conformity with its organization; with additional descriptions of all the species hitherto named, of many not before noticed, and other original matter, by E. Griffith, S. H. Smith, E. Pidgeon, J. E. Gray and others. 1824—33.
Güldenst., N. Comm. Petrop. XIX. = J. A. Güldenstaedt: Sex avium descriptiones: Loxia rubicilla, Tanagra melanictera, Muscicapa melanoleuca, Motacilla erythrogastra, Scolopax subarquata, Scolopax cinerea. In Novi Commentarii Academiae Imperialis Scientiarum Petropolitanae XIX 463 ff. 1774.
Guérin, Icon. Règ. Anim., Ois. = F. E. Guérin-Méneville Iconographie du règne animal de G. Cuvier, ou représentation d'après nature de l'une des espèces les plus remarquables et souvent non encore figurées de chaque genre d'animaux, pouvant servir d'atlas à tous les traités de zoologie. Oiseaux. 1829—38.
Guillemard, Australasia = F. H. H. Guillemard: Australasia. Vol. II. Malaysia and the Pacific Archipelagoes. Edited and greatly extended from A. R. Wallace's "Australasia". 1894.
Guillem. Cruise Marchesa = F. H. H. Guillemard: The Cruise of the Marchesa to Kamtschatka and New Guinea. 1886.
Gunner, in Leem. Lap. Beskr. = J. E. Gunnerus: Anmaerkninger in Knud Leems, Profe sor i det Lappiske Sprog Beskrivelse over Finmarkens Lappor 1767.
Gurney, (List) Diurn. B. of Prey = J. H. Gurney: List of the diurnal Birds of Prey, also a record of specimens preserved in the Norfolk and Norwich Museum. 1884.
Hartert, Kat. (Vög.) (Senckenb. Mus.) Frankf. M. = E. Hartert: Katalog der Vogelsammlung im Museum der Senckenbergischen naturforschenden Gesellschaft in Frankfurt a. M. 1891.
Hartert, Nov. Zool. = E. Hartert in Novitates Zoologicae. A Journal of Zoology in connection with the Tring Museum.
Hartert, Tierr(eich) = Das Tierreich. Eine Zusammenstellung und Kennzeichnung der rezenten Tierformen. Herausg. von der Deutschen Zoologischen Gesellschaft. Lief. 1. Aves: Podargidae, Caprimulgidae und Macropterygidae von E. Hartert. 1897.
Hartl., Faun. Madag. = G. Hartlaub: Ornithologischer Beitrag zur Fauna Madagascar's. 1861. (Journ. f. Orn. 1860.)
Hartlaub in Neumayer's Anleitung = G. Hartlaub in Anleitung zu wissenschaftlichen Beobachtungen auf Reisen. Herausgegeben von G. Neumayer. 1875.

List of abbreviations.

Hartl., Verzeichniss = Systematisches Verzeichniss der naturhistorischen Sammlung der Gesellschaft Museum. !. Abth. Vögel. Von G. Hartlaub. 1844.

Hartl., Vög. Madag. = G. Hartlaub: Die Vögel Madagascar's und der benachbarten Inselgruppen. 1877.

Hayes, Portr. of rare and cur. B. = W. Hayes: Portraits of rare and curious Birds, with their descriptions, from the menagery of Osterly Park. 1794.

Heine & Rchuw., Nomencl. Mus. Hein. = F. Heine & A. Reichenow: Nomenclator Musei Heineani Ornithologici. Verzeichniss der Vogelsammlung des Kgl. Oberamtmanns F. Heine. 1882—90.

Heugl., (Vög.) Orn. N. O. Afr. = M. Th. von Heuglin: Ornithologie Nordost-Afrika's, der Nilquellen- und Küsten-Gebiete des Rothen Meeres und des nördlichen Somal-Landes. 1869—74.

Hickson, Nat. in N. Celebes = S. J. Hickson: A naturalist in North Celebes. 1889.

de Hollander, Handl. Land- en Volkenk. Ned. Oost-Ind. = J. J. de Hollander: Handleiding bij de beoefening der Land- en Volkenkunde van Nederlandsch Oost-Indië, 4th ed. 1882—84.

Hombr. & Jacq., Voy. Pôle Sud = Voyage au pôle Sud et dans l'Océanie sur les corvettes L'Astrolabe et La Zélée, 1837—1840, sous le commandement de J. Dumont d'Urville. Zoologie. Par Hombron & Jacquinot. 1842—53.

Horsf., Zool. Research(es) in Java = T. Horsfield: Zoological researches in Java and the neighbouring Islands. 1824.

Horsf. & Moore, Cat. B. Mus. E. I(nd). Co. = T. Horsfield & F. Moore: A Catalogue of the Birds in the Museum of the Hon. East-India Company. 1854—58.

Hume, Rough Notes = A. O. Hume: Scrap Book, or Rough Notes on Indian Oology and Ornithology. 1869.

Hume & Marsh., Game B. Ind. = A. O. Hume and C. H. T. Marshall: The Game Birds of India, with coloured illustrations of all the known species. 1879—82.

Jacq(uinot) et Pucher(an), Vey(age) (au) Pôle Sud = H. Jacquinot & J. Pucheran: Voyage au Pôle Sud et dans l'Océanie sur les corvettes L'Astrolabe et la Zélée 1837—40. Zoologie, vol. III (Mammifères et Oiseaux). 1853.

Jard., Contr. Orn. = W. Jardine: Contributions to Ornithology. Descriptions of new or undescribed birds. 1849—52.

Jard., Nat. Libr., Orn. = The Naturalist's Library. Ornithology. Vol. IV. By W. Jardine. 1834.

Jard., & Selb(y), Ill. Orn. = W. Jardine and P. J. Selby: Illustrations of ornithology, etc. 1825—39, 1843.

(Jb.) Ver. Brük. Dresden = Jahresbericht des Vereins für Erdkunde zu Dresden.

Jb. (Ver.) Naturw. Braunschw. = Jahresbericht des Vereins für Naturwissenschaft zu Braunschweig.

Ibis = The Ibis, a Magazine of general Ornithology.

Jerd., B. Ind. = T. C. Jerdon: The Birds of India; being a Natural History of all the Birds known to inhabit Continental India. 1862—64.

Jerd., Ill. Ind. Orn. = T. C. Jerdon: Illustrations of Indian ornithology, a series of fifty coloured lithographic drawings of Indian Birds, accompanied by descriptive letterpress. 1843.

Illig(er), Prodr. = J. K. W. Illiger: Prodromus systematis mammalium et avium additis terminis zoographicis utriusque classis eorumque versione germanica. 1811.

Joest, Das Holontalo = W. Joest: Das Holontalo: Glossar und grammatische Skizze. Diss. 1883.

J. A. S. (B.) = Journal of the Asiatic Society of Bengal.

J. f. O. = Journal für Ornithologie.

J. (of the) Proc. Linn. Soc., (Zool.) = Journal of the Proceedings of the Linnean Society. Zoology.

J. R. Geogr. Soc. = The Journal of the Royal geographical Society.

J. Str. Br. R. A. S. = Journal of the Straits Branch of the Royal Asiatic Society.

Journ. L. Soc. = Journal of the Linnean Society. Zoology.

Journ. of Malacol. = The Journal of Malacology.

J(ourn.) R. Met. Soc. = Quarterly Journal of the Royal Meteorological Society, London.

Isis = Isis oder Encyklopädische Zeitung von (L.) Oken.

Isis, Dresden = Sitzungsberichte und Abhandlungen der naturwissenschaftlichen Gesellschaft "Isis" in Dresden.

Jukes, Voy. "Fly" = J. B. Jukes: Narrative of the Voyage of H. M. S. Fly in Torres Strait, New Guinea and other Islands, 1842—46. 1847.

Kaup, Classif. Säug. u.(nd) Vög. = J. J. Kaup: Classification der Säugethiere und Vögel. 1844.

Kaup, Contr. Orn. = J. J. Kaup in W. Jardine's Contributions to Ornithology. Descriptions of new or undescribed birds. 1848—52.

Kaup, Isis = J. J. Kaup in "Isis" oder Encyklopädische Zeitung von Oken.

Kaup, Natürl. Syst. = J. J. Kaup: Skizzirte Entwickluagsgeschichte und natürliches System der europ. Thierwelt. I. Vogelsäugethiere und Vögel. 1829.

Kaup, Verh. nat. hist. Ver. Hessen (Familie der Eisvögel) = J. J. Kaup: Die Familie der Eisvögel (Alcedidae). Verhandlungen des naturhistorischen Vereins für das Grossherzogthum Hessen, Vol. II (1848).

Koeler, Evol. Col. Birds (Feath.) = C. A. Keeler: Evolution of the colors of North American land birds. 1893. (Occasional papers of the California Academy of Sciences, III.)

Keulemans, Onze Vog. = J. G. Keulemans: Onze Vogels in huis en tuin. 1873.

Keys. & Blas., Wirbelth. Eur. = A. v. Keyserling & J. H. Blasius: Die Wirbelthiere Europa's. 1840.

King, Survey Int. Austr. = Ph. P. King: Narrative of a survey of the intertropical and western coasts of Australia, performed between the years 1818—1822; with an appendix, containing various subjects relating to hydrography and natural history. London 1827.

Kittl., Kupfert. (Vög.) = F. H. v. Kittlitz: Kupfertafeln zur Naturgeschichte der Vögel. 1832—33.

Koch, Baier. Zool. = K. L. Koch: System der Baierischen Zoologie. — Auch u. d. Titel: Die Säugethiere und Vögel Baierns. 1816.

List of abbreviations.

Koch, Verz. Vogelb. (aus) Cel. u. Sanghir = G. v. Koch: Verzeichniss einer Sammlung von Vogelbälgen aus Celebes und Sanghir, welche vom Grossherzoglichen Zoologischen Museum zu Darmstadt im Tausch oder gegen Baarzahlung zu erhalten sind. 1876.

Krancher's entomolog. Jahrb. = Entomologisches Jahrbuch. Herausgegeben von O. Krancher.

Krukenberg, Vergl. physiol. Studien = C. F. W. Krukenberg: Die Farbstoffe der Federn in des Verf.'s Vergl.-phys. Studien, V. Abth. u. 2. Reihe I.—III. Abth. 1881—82.

Küster, Orn. Atlas = H. C. Küster: Ornithologischer Atlas der aussereuropäischen Vögel, nach C. W. Hahn's Werke fortgesetzt. 1836—41.

Labill., Voy. à la Recherche de La Pérouse 1791—92 = J. J. Labillardière: Relation du voyage à la recherche de La Pérouse, 1791—93. [1800.] (English translation: Stockdale).

Lath., Gen. Hist. = J. Latham: A general history of Birds, being the natural history and descriptions of all the Birds (above four thousand) hitherto known or described by naturalists, with the synonymes of preceding writers; the second enlarged and improved edition, comprehending all the discoveries in ornithology subsequent to the former publication, and a general Index. 1821—26.

Lath., Gen. Syn. (Suppl.) = J. Latham: A natural history, or general synopsis of Birds. — And Supplement. 1781—1802.

Lath., Ind. Orn. = J. Latham: Index¹ Ornithologicus. 1790.

Leach, Syst. Cat. M. & B. Br. Mus. = W. E. Leach's systematic Catalogue of the specimens of the indigenous Mammalia and Birds in the British Museum. 1816. (Ed. by O. Salvin. Willughby Society. 1882.)

Lear, Ill. Parrots = E. Lear: Illustrations of the Family of Psittacidae. 1832.

Legge, B. Ceylon = W. V. Legge: A History of the Birds of Ceylon. 1880.

Less., C(ompl. (de) Buff. (Ois.) = R. P. Lesson: Complément des oeuvres de Buffon, ou histoire naturelle des animaux rares découverts par les naturalistes et les voyageurs depuis la mort de Buffon. Vol. 6. Histoire naturelle des Oiseaux. 1829.

Less., Man. d'Orn. = R. P. Lesson: Manuel d'ornithologie ou description des genres et des principales espèces d'oiseaux. 1828.

Less., Tr. d'Orn. = R. P. Lesson: Traité d'ornithologie, ou description des Oiseaux réunis dans les principales collections de France. 1831.

Less., Voy. de Bélanger = C. Bélanger: Voyage aux Indes orientales, par le Nord de l'Europe, les provinces du Caucase, la Géorgie, l'Arménie et la Perse, suivi de détails topograph., statistiq. etc. sur le Pégou, les isles de Java, de Maurice et de Bourbon; sur le Cap de Bonne-Espérance et Sainte Hélène pendant les années 1825 à 1829. Zoologie. 1831—44.

Less. & Garn., Bull. Sc. Nat. = R. P. Lesson et P. Garnot: Megapodius Duperreyi, Garn. Bulletin des Sciences Naturelles et de Géologie par le Baron de Férussac. Vol. VIII (1826).

Lesson, Voy. Coquille) Zool. = L. J. Duperrey: Voyage autour du monde, exécuté par ordre du roi, sur la corvette de S. M. la Coquille, pendant les années 1822—25 etc. Zoologie, rédigée par MM. Garnot et Lesson, etc. 1829.

Levaill., N. Hist. Guêpiers = F. Levaillant: Histoire naturelle des Promerops et des Guêpiers. 1807.

Levaill., Ois. d'Afr. = F. Levaillant: Histoire naturelle des oiseaux d'Afrique. 1799—1805.

Levaill., Ois. Parad. Rolliers = F. Levaillant: Histoire naturelle des Oiseaux de Paradis et des Rolliers, suivie de celle des Toucans et des Barbus; des Promerops et des Guêpiers, des Couroucous et des Touracos. 1803—18.

Levaill., Perr. = F. Levaillant: Histoire naturelle des Perroquets. 1801—1805.

Licht., Nomencl. Av. = H. Lichtenstein: Nomenclator Avium Musei Zoologici Berolinensis. 1854.

Licht., Verz. Doubl. Berl. Mus. = H. Lichtenstein: Verzeichniss der Doubletten des zoologischen Museums der Königl. Universität zu Berlin, nebst Beschreibungen vieler bisher unbekannten Arten von Säugethieren, Vögeln, Amphibien und Fischen. 1824.

Linn., Mant. (Plant.) = C. Linné: Mantissa plantarum altera; acced. Regni Animalis Appendix. 1771.

Linn., S. N. = C. a Linné, Systema naturae per regna tria naturae, secundum classes, ordines, genera, species, cum characteribus, differentiis, synonymis, locis. 12. ed. 1766—68.

M. et S., Verh. (Nat. Gesch.) Natuurk. Comm. = Verhandelingen over de natuurlijke geschiedenis der Nederlandsche overzeesche bezittingen, door de Leden der NatuurkundigeCommissie in Indiëen anderesehrijvers. Uitgegeven op last van den Koning door C. J. Temminck. Zoologie. (Vertebrates by Salomon Müller and H. Schlegel). 1839—44.

Madar(asz), Aquila = J. v. Madarász in the Journal "Aquila".

Madr. Journ. = Journal of Literature and Science. Published under the auspices of the Madras Literary Society.

Mag. (de) Zool. = Magasin de zoologie, d'anatomie comparée et de paléontologie. Journal destiné à faciliter aux zoologistes de tous les moyens de publier leur travaux et les espèces nouvelles ou peu connues qu'ils possèdent, par F. E. Guérin-Méneville. I. Série (1831—1838).

Malh., Picidae = A. Malherbe: Monographie des Picidées. 1861—62.

Marsh, W., Z. Vertr. Die Papag. = W. Marshall: Die Papageien (Psittaci). Zoologische Vorträge herausg. von W. M. Heft I. 1889.

Marshall, Die Spechte = W. Marshall: Die Spechte (Pici). Zoologische Vorträge herausg. von W. M. Heft 2. 1889.

Marshall, Schädelhöcker der Vög. = W. Marshall: Ueber die knöchernen Schädelhöcker der Vögel. In: Niederländisches Archiv für Zoologie. Bänd I. 1872.

Maury, Phys. Geogr. Sea = M. F. Maury: Physical Geography of the Sea. 2. ed. 1855.

Meisner & Schinz, Vög. Schweiz = F. Meisner & H. R. Schinz: Die Vögel der Schweiz; systematisch geordnet und beschrieben, mit Bemerkungen über ihre Lebensart und Aufenthalt. 1815.

Mél. Biol. Ac. Petersb. = Mélanges biologiques tirés du Bulletin de l'Académie Impériale des Sciences de St. Pétersbourg.

Mém. Mus. d'Hist. Nat. = Mémoires du Muséum d'Histoire Naturelle. Paris.

Meyen, Reise um die Erde = F. J. F. Meyen: Reise um die Erde, ausgeführt auf dem kgl. Preuss. Seehandlungs-Schiffe Prinzess Louise, commandirt von Capitän W. Wendt, in d. J. 1830—32. 1834—35.

Meyer, Ausz. Neu Guinea Reise = A. B. Meyer: Auszüge aus den auf einer Neu Guinea-Reise im Jahre 1873 geführten Tagebüchern, als Erläuterungen zu den Karten der Geelvink-Bai und des MacCluer-Golfes. 1875.

Meyer, Isis (Dresden) = A. B. Meyer in: Sitzungsberichte und Abhandlungen der naturwissenschaftlichen Gesellschaft "Isis" in Dresden.

Meyer, (Abb. v.) Vogelskel. = A. B. Meyer: Abbildungen von Vogelskeletten. 1879—1897.

Meyer & Helm, J(ahrs)b. Orn. Beob. Sachsen = A. B. Meyer & F. Helm: Jahresbericht der ornithologischen Beobachtungstationen im Königreiche Sachsen. 1885—94.

Meyer & Helm, Verz. d. Vög. Sachsens = A. B. Meyer & F. Helm: Verzeichniss der bis jetzt im Königreiche Sachsen beobachteten Vögel nebst Angaben über ihre sonstige geographische Verbreitung. [Anhang zum VI. Jahresbericht (1890) der ornith. Beobachtungstationen im Kgr. Sachsen.] 1892.

M. & Wg., Abh. Mus. Dresd. = A. B. Meyer & L. W. Wiglesworth: Abhandlungen und Berichte des Kgl. Zoologischen und Anthropologisch-Ethnographischen Museums zu Dresden.

Meyer & Wolf, Orn. Taschenb. = B. Meyer & J. Wolf: Taschenbuch der deutschen Vögelkunde oder kurze Beschreibung aller Vögel Deutschlands. 1810. 1822.

Midd., Sib. Reise = A. Th. v. Middendorff's Reise in den äussersten Norden und Osten Sibiriens. Band II. Zoologie Theil 2. Säugeth., Vögel u. Amphibien. 1851.

Milne-Edw. & Grandid., (H. N.) Ois. Madag. = A. Grandidier: Histoire physique, naturelle et politique de Madagascar. Publiée par A. Grandidier. Oiseaux par A. Milne-Edwards et A. Grandidier. 1876—81.

Mitth. Mus. Dresd. = Mittheilungen aus dem K. Zoologischen Museum zu Dresden.

M(it)th. Orn. Ver. Wien = Mittheilungen des Ornithologischen Vereins in Wien.

Mivart, Lor(iidae) = S. G. Mivart: A monograph of the Lories or Brush-tongued Parrots, composing the family Loriidae. 1896.

Mottl. & Dillw., Contr. Nat. Hist. Lab(uan) = J. Mottley and I. L. Dillwyn: Contributions to the Natural History of Labuan, and the adjacent coasts of Borneo. 1855.

Mtschr. Ver. Schutze Vogelw. = Ornithologische Monats-schrift des Deutschen Vereins zum Schutze der Vogelwelt.

Müll., S. Retz(ie) Ind. Arch(ip). = Salomon Müller: Reizen en Onderzoekingen in den Indischen Archipel, 1828—1836. 1857.

Müll., S. N. Suppl. = C. Linnés Vollständiges Natursystem nach der 12. lat. Ausgabe... Supplements- u. Register-Band über alle 6 Theile oder Classen des Thierreichs. Mit einer ausführlichen Erklärung ausgefertiget von P. L. S. Müller. 1776.

Murray, Avif. Brit. Ind. = J. A. Murray: Avifauna of British India and its Dependencies. 1887—90.

Murray, Voy. Chall., Narr. = John Murray in Report on the scientific results of the voyage of H. M. S. Challenger, 1873—76. Narrative. Vol. I, 2. 1885.

Naturaliste = Le Naturaliste. Revue illustrée des sciences naturelles.

Nature = Nature. A weekly illustrated journal of science.

Naum., Vög. Deutschl. = J. A. Naumann: Naturgeschichte der Vögel Deutschlands. Auf's Neue herausgegeben von J. F. Naumann. 1822—1860.

Naumannia = Naumannia. Archiv für die Ornithologie.

N(ed.) T(ijdschr). D(ierk). = Nederlandsch Tijdschrift voor de Dierkunde.

Nelson, Cruise "Corwin" or Report N. H. Coll. (in) Alaska = Cruise of the Revenue-Steamer "Corwin" in Alaska and the N. W. Arctic Ocean in 1881. Birds of the Bering Sea and the Arctic Ocean by E. W. Nelson. Washington 1883.

Newton, Dict. B. = A Newton: A Dictionary of Birds. Assisted by H. Gadow. With contributions from R. Lydekker, C. S. Roy and R. W. Shufeldt. 1893—96.

Newton ('s ed.), Yarrell's Brit. B. = W. Yarrell: A history of British birds. 4. ed., by A. Newton and H. Saunders. 1871—85.

Nikolski, Ile de Sakhal. et sa faune = A. Nikolski: Ile de Sakhalin et sa faune des vertébrés. St. Pétersbourg. 1889.

Nitzsch, Pterylogr. = C. L. Nitzsch: System der Pterylographie. Nach seinen handschriftlich aufbewahrten Untersuchungen verfasst von H. Burmeister 1840.

Nitzsch, Engl. Ed. = C. L. Nitzsch: Pterylography. Translated from the German (by W. S. Dallas), ed. by P. L. Sclater. 1867.

North, Nests & Eggs B. Austr. = A. J. North: Catalogue of the Nests and Eggs of Birds found breeding in Australia and Tasmania. 1889.

Notes Leyd. Mus. = Notes from the Leyden Museum.

N. T. N(ed). I(nd). = Natuurkundig Tijdschrift voor Nederlandsch Indië.

N(ouv.) Arch. du Mus. = Nouvelles Archives du Muséum d'Histoire Naturelle de Paris.

Novit. Zool. = Novitates Zoologicae. A Journal of Zoology in connection with the Tring Museum. Edited by W. Rothschild etc.

Oates, B. Brit. Burmah = E. W. Oates: A handbook to the birds of British Burmah, including those found in the adjoining State of Karennee. 1883.

List of abbreviations. XXVII

Oates, ed. Hume's Nests & Eggs = A. O. Hume: The Nests and Eggs of Indian Birds. 2. ed. by E. W. Oates. 1889—90.

Obs. meteor. de Manila de la Comp. de Jésus = Observatorio meteorológico del Ateneo municipal de Manila bajo la Dirección de los PP. de la Compañía de Jesus.

Ornis = Ornis. Internationale Zeitschrift für die gesammte Ornithologie.

Orn. Centralbl. = Ornithologisches Centralblatt. Beiblatt zum Journal für Ornithologie.

Orn. Mb. = Ornithologische Monatsberichte.

Orn. Monatsschr. = Ornithologische Monatsschrift des Deutschen Vereins zum Schutze der Vogelwelt.

Oustal., Mon. Megap. = E. Oustalet: Monographie des Oiseaux de la Famille des Mégapodes. 1880—81. In: Annales des Sciences naturelles. Zoologie.

Pall., Reis. Russ. Reichs = P. S. Pallas: Reisen durch verschiedene Provinzen des Russischen Reichs in den Jahren 1768—74. 1771—76.

Pall., Zoogr. Rosso-Asiat. = P. S. Pallas: Zoographia Rosso-Asiatica, sistens omnium animalium in extenso imperio Rossico et adjacentibus maribus observatorum recensionem, domicilia, mores et descriptiones, anatomen atque icones plurimorum. 1811.

Peale, U. S. Expl. Exped., Zool. = United States Exploring Expedition, Vol. VIII: Mammalia and Ornithology by T. R. Peale. 1848.

Pelz., Reise der Novara, Zool. or Novara Reise, Vög. = Reise der oesterreichischen Fregatte Novara um die Erde, 1857—59. Bd. I: Vögel von A. v. Pelzeln. 1865.

Penn., Ind. Zool. = T. Pennant: Indian zoology; including "Faunula Indica" etc. 1790.

Petermann's Mitth. = Petermann's Geographische Mittheilungen.

Pleske, Vög. Russ. Reichs = Th. Pleske: Ornithographia Rossica. Die Vogelfauna des Russischen Reichs. II. Sylviinae. 1891.

Pr. Acad. (N. Sc.) Philad. = Proceedings of the Academy of Natural Sciences of Philadelphia.

Pr. L. Soc. N. S. W. = Proceedings of the Linnean Society of New South Wales.

P. U. S. Nat. Mus. = Proceedings of the U. S. National Museum.

P. Z. S. = Proceedings of the Zoological Society of London.

Prév. & Knip, Pig. = C. J. Temminck: Histoire naturelle générale des Pigeons avec figures en couleurs peintes, etc. Tome II. cont. les Pigeons exotiques de Mme Knip. Le texte par F. Prévost. 1838—43.

Proc. Boston Soc. Nat. Hist. = Proceedings of the Boston Society of Natural History.

Quoy & Gaim., Voy. Astr. (Zool.) = Quoy & Gaimard in: Voyage autour du monde de la corvette L'Astrolabe exécuté 1826—29 sous le commandement de Dumont d'Urville. Zoologie. 1830—33.

Radde, Reise S.-O. Sibir. = G. Radde: Reisen im Süden von Ost-Sibirien, 1855—59. 1862—63.

Rams(ay), Tab. List (Austr. B.) = E. P. Ramsay: Tabular list of all the Australian birds at present known to the author, showing the distribution of the species over the continent of Australia and adjacent islands. 1888.

W. Rams., Tweedd. Orn. Works = The ornithological works of Arthur, IX. Marquis of Tweeddale. Ed. by R. G. W. Ramsay. 1881.

Rchb., Columbarinae or Tauben = H. G. L. Reichenbach: Die vollständigste Naturgeschichte der Tauben und taubenartigen Vögel. Columbariae. 1848.

Rchb., Pulic(ariae), Novit. = H. G. L. Reichenbach: Die vollständigste Naturgeschichte der Sumpfvögel: Novitiae ad synopsin avium. III. Rasores: Scharrvögel. I. Fulicariae: Wasserhühner. 1851.

Rchb., Handb. Picinae = H. G. L. Reichenbach: Handbuch der speciellen Ornithologie. Picinae. 1854.

Rchb., Grallat. = H. G. L. Reichenbach: Die vollständigste Naturgeschichte der Sumpfvögel: Aves Grallatores. 1851.

Reichb., Hb. sp. Orn. Alcedin. = L. Reichenbach: Handbuch der speciellen Ornithologie. Alcedineae. 1851.

Reichb., Hb. sp. Orn. Meropinae = L. Reichenbach: Handbuch der speciellen Ornithologie. Meropinae. 1852.

Reichb., Hb. Scansoriae = L. Reichenbach: Handbuch der speciellen Ornithologie. Scansoriae. 1853.

Rchb., Syn. Av. Gallinac. = L. Reichenbach: Synopsis avium. Gallinaceae. 1848.

Rchb., S. A., Natatores = H. G. L. Reichenbach: Die vollständigste Naturgeschichte der Schwimmvögel: Aves Natatores. Synopsis Avium. Vol. I. Natatores. 1845.

Rchb., Syst. Av. (Natur.) = L. Reichenbach: Avium systema naturale. Das natürliche System der Vögel. Vorläufer einer Iconographie der Arten der Vögel aller Welttheile. 1850. Also unter dem Titel: Ornithologie méthodique ou exposé des genres des oiseaux de toutes les parties du monde. Prodrome d'une iconographie des espèces des oiseaux ou synopsis avium.

Rchnw., Vogelb. = A. Reichenow: Vogelbilder aus fernen Zonen. Abbildungen und Beschreibungen der Papageien. 1878—83.

Rchnw., Consp. Psitt. = A. Reichenow: Conspectus Psittacorum. Systematische Uebersicht aller bekannten Papageienarten. In: Journal für Ornithologie. 1882.

Rchw., Vög. Deutsch O.-Afr. = Deutsch-Ostafrika. Thierwelt Ostafrikas und der Nachbargebiete. III. Bd. 2. Lf. Vögel von A. Reichenow. 1894.

Reinw., Reis. Ind. Arch. (in 1821) = C. G. C. Reinwardt's Reis naar het oostelijk gedeelte van den Indischen Archipel in het jaar 1821. Uitgegeven door W. H. de Vriese. 1858.

Rev. Zool. = Revue zoologique, par la Société Cuvierienne.

Ridgw., (B. N. Am.) Landb. N. Am. = S. F. Baird, T. M. Brewer, and R. Ridgway: A History of North American Birds, Land Birds. 1874.

Ridgw., Man. N. Am. Birds = R. Ridgway: A manual of North American Birds. 1887.

d*

List of abbreviations.

Ridgw., Sm. Rep. = R. Ridgway: in the Annual Report of the Board of Regents of the Smithsonian Institution.

Ros(enb)., Mal. Arch. = C. B. H. v. Rosenberg: Der malayische Archipel. Land und Leute. 1878.

Rosenb., Reistogt. in Gorontalo = C. B. H. v. Rosenberg: Reistogten in de afdeeling Gorontalo. 1865.

Rothsch., Av. Laysan = W. Rothschild: The Avifauna of Laysan and the neighbouring Islands; with a complete History to date of the Birds of the Hawaiian Possessions. 1893.

Rowl., Orn. Misc. = Ornithological Miscollany. Ed. by G. D. Rowley. 1875—78.

Russ, Einheim. Stubenvög. = K. Russ: Handbuch für Vogelliebhaber, -Züchter und -Händler. II. Einheimische Stubenvögel. 1873.

Russ, Fremdl. Stubenvög. = K. Russ: Die fremdländischen Stubenvögel. 1879—98.

Sail. Direct. = A. G. Findlay: A Directory for the Navigation of the Indian Archipelago etc. 1870.

Salvad., Agg. Orn. Pap. = T. Salvadori: Aggiunte alla ornitologia della Papuasia e delle Molucche. 1889—91.

Salvad., Orn. Pap. = T. Salvadori: Ornitologia della Papuasia e delle Molucche. 1880—82.

Salvad., Cat. Ucc. (dl) Borneo = Catalogo sistematico degli uccelli di Borneo di T. Salvadori con note ed osservazioni di G. Doria ed O. Beccari intorno alle specie da essi raccolte nel Ragiato di Sarawak. Annali del Museo Civico di Storia naturale di Genova. Vol. V. 1874.

Sb. Ges. natf. Freunde zu Berlin = Sitzungs-Bericht der Gesellschaft naturforschender Freunde zu Berlin.

Schinz, Abbild. Vög. = H. R. Schinz: Naturgeschichte und Abbildungen der Vögel. Nach den neuesten Systemen bearbeitet. (1831—33.) 1835—36.

Schinz, Nat. Vög. — H. R. Schinz: Naturgeschichte der Vögel. Mit kolorirten Abbildungen nach der Natur und den vorzüglichsten naturwissenschaftlichen Werken gezeichnet. Neueste (2.) umgearbeitete und sehr vermehrte Ausgabe. 1846—53.

Schl., (De) Diermt. = H. Schlegel & P. J. Witkamp: De dierentuin van het kon. zoolog. Genootschap Natura Artis Magistra. Amsterdam. 1872.

Schl., H(an)dl. (d.) Dierk. = H. Schlegel: Handleiding tot de beoefening der dierkunde. 1857—58.

Schl., Mus. P.-B. = Muséum d'Histoire Naturelle des Pays-Bas. Par H. Schlegel. Revue méthodique et critique des collections. 1862—94.

Schl., N. T. D. or Ned. Tdschr. = H. Schlegel in Nederlandsch Tijdschrift voor de Dierkunde, uitgegeven door het Koninklijk zoologisch Genootschap Natura Artis Magistra. 3. deel. 1866.

Schl., Rev. Accip). = H. Schlegel: Revue de la Collection des Oiseaux de Proie faisant partie du Musée des Pays-Bas. Accipitres. 1873. (In Mus. d'H. N. d. P.-B.)

Schl., Rev. Alcedin. = H. Schlegel: Revue de la Collection des Alcédines faisant partie du Musée des Pays-Bas. 1874. (In Mus. d'H. N. d. P.-B.)

Schl., Rev. Noctuae = H. Schlegel: Revue de la Collection des Oiseaux de Proie faisant partie du Musée des Pays-Bas. Aves Noctuae. 1873. (In Mus. d'H. N. d. P.-B.)

Schl., Revue Pitta = H. Schlegel: Revue de la Collection des Brèves (Pitta) faisant partie du Musée des Pays-Bas. 1874. (In Mus. d'H. N. d. P.-B.)

Schl., Rev. Psitt. = H. Schlegel: Revue de la Collection des Perroquets (Psittaci) faisant partie du Musée des Pays-Bas. 1874. (In Mus. d'H. N. d. P.-B.)

Schl., Valkv. (Ned. Ind.) = H. Schlegel: De vogels van Nederlandsch-Indië, beschreven en afgebeeld. — Les oiseaux des Indes Néerlandaises, décrits et figurés. Monographie 3, Valkvogels (Accipitres). 1866.

Schl., Vog. Ned. Ind. Ijsvogels (Alcedin.) = H. Schlegel: De vogels van Nederlandsch Indië, beschreven en afgebeeld. — Les oiseaux des Indes Néerlandaises, décrits et figurés. Monographie 2, Ijsvogels, (Martins-Pêcheurs). 1864.

Schl., Vog. Ned. Ind. Pitta = H. Schlegel: De vogels van Nederlandsch Indië, beschreven en afgebeeld. — Les oiseaux des Indes Néerlandaises, décrits et figurés. Monographie 1, Pitta. 1863.

Schl. & Pollen, Faun. Madag. = F. P. L. Pollen et D. C. van Dam: Recherches sur la faune de Madagascar et de ses dépendances. 2. partie: Mammifères et Oiseaux par H. Schlegel et F. P. L. Pollen. 1868.

Schmeltz, Eth. Abth. Mus. Godef. = J. D. E. Schmeltz, & R. Krause: Die Ethnographisch-Anthropologische Abtheilung des Museum Godeffroy in Hamburg. 1881.

Schrenck, Reis(e) Amur(lande). Vög. = L. v. Schrenck: Reisen und Forschungen im Amur-Lande, 1854—56. Bd. I, 2. Vögel des Amur-Landes. 1860.

Sclat., List. Vert. An. = List of the vertebrated animals now or lately living in the Gardens of the Zoological Society of London. 8. ed. 1883. By P. L. Sclater.

Sclat., Voy. Chall. B. = P. L. Sclater in Report on the scientific results of the voyage of H. M. S. Challenger, 1873—76. Zoology. Vol. II. 1881.

Scop(oli), Del. Flor. & Faun. Insubr. = J. A. Scopoli: Ornithological papers from his Deliciae florae et faunae insubricae. 1786—88. Ed. by A. Newton (Willughby Soc. 1882).

Seebohm, B. Japan. (Emp.) = H. Seebohm: The Birds of the Japanese Empire. 1890.

Seeb., Hist. Brit(ish) B. = H. Seebohm: A history of British birds with illustrations of their eggs. 1883—85.

Seeb., Distr. Charadr. = H. Seebohm: The geographical distribution of the family Charadriidae or the Plovers, Sandpipers, Snipes and their allies. 1888.

Selby, Nat. Libr., Parrots = The Naturalist's Library. Conducted by William Jardine. Ornithology Vol. VI. Parrots by P. J. Selby. 1836.

Selby, Natur. Libr., Pig. = The Naturalist's Library. Conducted by William Jardine. Ornithology Vol. V, part. III, Pigeons by P. J. Selby. 1835.

Sharpe, Mitth. Zool. Mus. Dr. = R. B. Sharpe in Mittheilungen aus dem K. Zoologischen Museum zu Dresden.

List of abbreviations.

Sharpe, Monogr. Alcedin. = R. B. Sharpe: A monograph of the Alcedinidae. 1868—71.

Sharpe, Rep. Trans. Venus Exp. B. Kerguelen = Account of the Petrological, Botanical and Zoological Collections made in Kerguelen's Land and Rodrigues during the Transit of Venus Expedition carried out in 1874—75. Birds by R. B. Sharpe. 1879.

Sharpe, Report Voy. "Alert" = Report on the zoological collections made in the Indo-Pacific Ocean during the voyage of H. M. S. "Alert", 1881—82. 1884.

Sharpe, Yarkand Mission, Aves = Scientific results of the second Yarkand Mission, based upon the collections of the late F. Stoliczka. 1878—91.

Sh. & Dresser, B. Eur(ope) = H. E. Dresser: A history of the birds of Europe, including all the species inhabiting the Western Palaearctic Region. 1871—81.

Sharpe & Wyatt, Mon. Hirund. = R. B. Sharpe & C. W. Wyatt: A monograph of the Hirundinidae, or family of Swallows. 1885—94.

Shaw, Gen. Zool. = G. Shaw: General zoology, or systematic natural history, with plates from the first authorities and most select specimens, engraved principally by Heath. Continued by Stephens. 1800—19.

Shelley, B. Egypt. = G. E. Shelley: A Handbook to the Birds of Egypt. 1872.

Shelley, Monogr. Nect. = G. E. Shelley: A monograph of the Nectariniidae, or family of Sun-birds. 1876—80.

Sitzb. Ak. Wien = Sitzungsberichte der math.-naturw. Classe der Kais. Akademie der Wissenschaften, Wien.

Sitzb. Ak. Wiss. Berlin = Sitzungsberichte der Königlich Preussischen Akademie der Wissenschaften zu Berlin.

Sitzb. Ges. Isis, Dresden = Sitzungsberichte und Abhandlungen der Naturwissenschaftlichen Gesellschaft Isis in Dresden.

Sonn(erat), Voy. N(ouv). Guin. = P. Sonnerat: Voyage à la nouvelle Guinée, dans lequel on trouve la description des lieux, des observations physiques et morales, et des détails rélatifs à l'histoire naturelle dans le règne animal et le règne végétal. 1776.

Souancé, Icon. Perr. = C. de Souancé: Iconographie des Perroquets non figurés dans les publications de Levaillant et de Bourjot Saint-Hilaire. 1857.

Sparrm., Mus. Carls. = A. Sparrman: Museum Carlsonianum, in quo novas et selectas Aves coloribus ad vivum breviqoe descriptione illustratas. 1786—89.

Steere, Coll. B. Philip. (Is.) = J. B. Steere: A List of the Birds and Mammals collected by the Steere Expedition to the Philippines. 1890.

Steph., Gen. Zool. = G. Shaw: General zoology, or systematic natural history, with plates from the first authorities and most select specimens, engraved principally by Heath. Continued by J. F. Stephens. 1800—19.

S(tr). F. = Stray Feathers. A Journal of Ornithology for India and its dependencies.

Strickl., Contr. Orn. = H. E. Strickland in W. Jardine's Contributions to Ornithology. 1848—53.

Strickl., Orn. Syn. = Ornithological Synonyms. Edited by H. E. Strickland and W. Jardine. Vol. I. Accipitres. 1855.

Studer, Reis. or Voy. Gazelle-(Reise) = Th. Studer: Die Forschungsreise S. M. S. "Gazelle" in den Jahren 1874 bis 1876. III. Theil, Zoologie und Geologie. 1889.

Sundev., Tentamen or Av. Meth. Tent. = C. J. Sundevall: Methodi naturalis avium disponendarum tentamen. Försök till fogelklassens naturenliga uppställning. 1872.

Sw(ain)s., B. W. Afr. = The Naturalist's Library. Edited by W. Jardine. Vol. XI—XII. Ornithology. Birds of Western Africa. By W. Swainson. 1837.

Sws., Zool. Illustr. = W. Swainson: Zoological illustrations, or original figures and descriptions of new, rare or otherwise interesting animals, selected chiefly from the classes of ornithology, entomology and conchology etc. 1820—23, 1829—33.

Tacz., (Faun.) Orn. Sib. Orient = L. Taczanowski: Faune ornithologique de la Sibérie Orientale. 1891—93.

T(d.) Ned. Aardr. Gen(oots.) = Tijdschrift van het Kon. Nederlandsch Aardrijkskundig Genootschap, gevestigd te Amsterdam.

Temm., Coup d'Oeil génér. Possess. Néerl. = C. J. Temminck: Coup d'oeil général sur les possessions Néerlandaises dans l'Inde Archipélagique. 1846—49.

Temm., Man. d'Orn. = C. J. Temminck: Manuel d'ornithologie, ou tableau systématique des Oiseaux, qui se trouvent en Europe, précédé d'une analyse du système général d'ornithologie, et suivi d'une table alphabétique des espèces. 2e édit. 1820, 1835, 1839, 1840.

Temm., Pig. et Gall. = C. J. Temminck: Histoire naturelle générale des Pigeons et des Gallinacées, accompagnée avec pl. anatomiques. 1813—15.

Temm., Pl. Col. = C. J. Temminck: Nouveau recueil de planches coloriées d'Oiseaux, pour servir de suite et de complément aux planches enluminées de Buffon, édit. de l'imprimerie royale 1778. Publié par C. J. Temminck et M. Langier, Baron de Chartronse; d'après les dessins de Nic. Huet fils et Prêtre. 1820—39.

Temm & Knip, Pig. = C. J. Temminck: Histoire naturelle générale des Pigeons avec figures en couleurs peintes, par Mme Knip, née Pauline de Courcelles. Le texte par C. J. Temminck. 1808—11.

Temm. & Schl., Faun. Jap. Aves = P. F. de Siebold: Fauna Japonica. Conjunctis studiis C. J. Temminck et H. Schlegel pro vertebratis atque W. de Haan pro invertebratis elaborata. Oiseaux. 1833.

Thienem., Fortpfl(anz.) Vög. = F. A. L. Thienemann: Fortpflanzungsgeschichte der gesammten Vögel. Mit 100 Tafeln Abbildungen von Vogeleiern. 1845—56.

Thunb., Act. Holm. = C. P. Thunberg in Kongliga Svenska Vetenskaps Akademiens Handlingar, Vol. XXXIII. 1772.

Tr. As. Soc. Jap. = Transactions of the Asiatic Society of Japan.

Tr. Chicago Ac. Sc. = Transactions of the Chicago Academy of Science.

List of abbreviations.

Tr. L(inn). S(oc). = Transactions of the Linnean Society of London.
Tr. Z. S. = Transactions of the Zoological Society of London.
Tr. & Pr. N. Z. Inst. = Transactions and Proceedings of the New Zealand Institute.
Tristr., Cat. (Coll.) B. = Catalogue of a Collection of Birds belonging to H. B. Tristram. 1889.
Tunstall, Orn. Brit. = M. Tunstall's Ornithologia Britannica (1771). Ed. by A. Newton. (Willughby Society 1880.)
Tweedd., Orn. Works = The ornithological works of Arthur, IX. Marquis of Tweeddale. Ed. by R.G.W. Ramsay. 1881.
Verh. D. Zool. Ges. = Verhandlungen der Deutschen Zoologischen Gesellschaft.
Verh. Ges. Erdkunde Berlin = Verhandlungen der Gesellschaft für Erdkunde zu Berlin.
V(erh). z.-b. Ges. Wien = Verhandlungen der k. k. zoologisch-botanischen Gesellschaft in Wien.
Verr. in Vinson's Voy. Madag. Annex B. = J. Verreaux in Vinson's Voyage à Madagascar. 1865.
Vieill., Analyse = L. P. Vieillot: Analyse d'une nouvelle ornithologie élémentaire. 1816. (Ed. by H. Saunders, Willughby Soc. 1883.)
Vieill., Enc. Méth. = Encyclopédie méthodique, ou par ordre de matière, par une société de gens de lettres. — Histoirenaturelle. Oiseaux, Ovipares et Serpents. (Oiseaux par R. J. E. Mauduit, revis. et augm. par Vieillot.) 1784—1820.
Vieill., Gall. Ois. = L. P. Vieillot: Galérie des Oiseaux du cabinet d'histoire naturelle du jardin du roi, ou description et figures coloriées des Oiseaux qui entrent dans la collection du muséum d'histoire naturelle de Paris. (Continuation de l'hist. natur. des Oiseaux dorés), dessinée (et lithogr.) d'après nature, par P. L. Oudart, et décrite par L. J. P. Vieillot. 1820—26.
Vieill., N. D. = L. P. Vieillot: Nouveau Dictionnaire d'Histoire Naturelle, 2e éd. (Articles contributed between 1816—19.)
Vig., Zool. J(ou)rn. = N. A. Vigors in the Zoological Journal. 1824—35.
Wagl., Mon. Psitt. = J. G. Wagler: Monographia Psittacorum. (Aus dem 1. Bde. der Denkschriften der Königl. Akademie der Wissenschaften in München 1832.) Besonders abgedruckt 1835.
Wagl., Syst. Av., = J. Wagler: Systema Avium. 1827.
Wald., Orn. Works = The ornithological works of Arthur, IX. Marquis of Tweeddale. Ed. by R. G. W. Ramsay. 1881.
Wall., Island Life = A. R. Wallace: Island Life: or, the phenomena and causes of insular faunas and floras. 1880.
Wall., Malay Archip. = A. R. Wallace: The Malay Archipelago. 1869.
Wallace, Geogr. Distr. Anim. = A. R. Wallace: The geographical distribution of Animals. With a study of the relations of living and extinct faunas as elucidating the past changes of the earth's surface. 1876.
Weber, Zool. Ergebnisse = Zoologische Ergebnisse einer Reise in Niederländisch Ost-Indien. Herausgegeben von Max Weber. 1890—97.
Whiteh(d)., Expl. (Expd.) Kini Balu = J. Whitehead: Exploration of Mount Kina Balu, North Borneo. 1893.
Wieg. Arch. = Archiv für Naturgeschichte. Gegründet von A. F. A. Wiegmann.
Wiglesw., Aves Polyn. = L. W. Wiglesworth: Aves Polynesiae. A catalogue of the birds of the Polynesian Subregion (not including the Sandwich Islands). 1891. In: Abhandlungen und Berichte des Königl. Zoologischen und Anthropologisch-Ethnographischen Museums zu Dresden. 1890/91 Nr. 6.
Wilson, B. Sandw. Is. = Scott B. Wilson: Aves Hawaiiensis. The Birds of the Sandwich Islands. Assisted by A. H. Evans. 1890—96.
Z. (Ges.) Erdk., Berlin = Zeitschrift der Gesellschaft für Erdkunde zu Berlin.
Z. wiss. Zool. = Zeitschrift für wissenschaftliche Zoologie.
Zool. Garten = Der Zoologische Garten. Zeitschrift für Beobachtung, Pflege und Zucht der Thiere.
Zool. Jahrb. Abt. f. Syst. = Zoologische Jahrbücher. Abtheilung für Systematik, Geographie und Biologie der Thiere.
Ztschr. ges. Orn. = Zeitschrift für die gesammte Ornithologie.
Ztschr. wiss. Geogr. = Zeitschrift für wissenschaftliche Geographie.

NOTANDA ET CORRIGENDA.

Page *120* in Introduction, cancel the number 42, making the total 87 instead of 88.
» 4, line 30, for Forster read Forsten.
» 4, » 31, add Saleyer (Everett *a 22*) and Togian (Meyer).
» 8, » 18, 19, alter the reference-letter *b* into *a*.
» 28, » 16, instead of chocolate-brown rufous, read chocolate-brown, without rufous.
» 53, » 18, for *H. indus* and *H. indus — girrenera*, read *H. indus* and *H. indus girrenera*.
» 55, » 26, add Togian (Meyer).
» 63, » 24, add Kalao (Everett *32*).
» 77, » 4, Hartert now calls this bird *Baza subcristata subcristata* (Nov. Zool. 1898, 47).
» 79, » 11 from below, add Djampea and Kalao (Everett *g 2*).
» 90, » *b* from below, add Kalao (Everett *d 3*).
» 91, » 33, add the Osprey has been observed nesting in India by Hume and others, according to Blandford. (Faun. Br. Ind. B. III, 1895 p. 315.)
» 95, » 25, for *Ninox lugubris affinis* read *Ninox scutulata affinis*.
» 115, » 19, for D'Aubert. read D'Aubent.
» 117, » 11, add Tagulandang and Biarro —? var. (Nat. Coll.).
» 115, » 17, 18, 20, for *c 1* read *t 1* and for *n 1* read *s L*
» 187, » 31, correct the name into *Cuculus canorus* L., without brackets.
» 188, » 7, first reference, for Tr. Z. S. read Tr. L. S.
» 191, » 21, query summer, as Platen's collecting in Mindanao was continued into the winter months.
» 217, » 14, from below, add Djampea and Kalao (Everett *w 8*, p. 176).
» 218, » 16, for *c 2* read *c 5*.
» 219, » 12, for Wellesly read Wellesley.
» 233, » 25, for *Ramphastidae* read *Rhamphastidae*.
» 239, » 13 from below, after Elliot insert Monogr. Bucerotidae.
» 248, » 16, for Cox & Hamilt. ib. read Cox and Hamilt. Pr. L. Soc. N. S. W.
» 257, title, affix an * to *Meropogon forsteni* Bp.
» 263, » 27, add Saleyer (Weber).
» 265, » 15 from below, for Kalao read Saleyer.
» 269, » 6, for Handb. read Handl.
» 285, » 14 from below, for 1885 read 1883.
» 288, » 6 from below, add Banggai (Nat Coll.).
» 294, » 6 from below, add Saleyer and Djampea (Everett *d 23*).
» 300, title, affix an * to *Monachalcyon princeps* Rchb.
» 305, » affix an * to *Ciltura sangirensis* Sharpe.
» 322, » affix an * to *Lyncornis macropterus* Bp.
» 328, line 5, for Gld., HL, read Gld., Hb.
» 329, title, before 104. *Chaetura celebensis* cancel the *.
» 335, line 6, add Kalao (Everett *13*).
» 337, » 19, add Saleyer (Everett *15*).
» 350, » 5, title, put Bp. in brackets.
» 360, » 28, add Karkellang, Talaut (Nat. Coll.).
» 389, » 15 from below, add Saleyer (Everett *14*).
» 393, » 11 from below, for *Collurincla* read *Colluricincla*.
» 400, » 18, for *S. clio* read *P. clio*.
» 407, » 6, the name *Lanius jeracopsis* is spelt *Lanius jeracopis* by de Filippi.
» 416, » 14, add Sula (Wallace).

Notanda et corrigenda.

Page 421, line 23, add Togian (Meyer *d 3*).
» 437, » 4, from below, for Luwn read Luwu.
» 440, » 22, for *carbonaria* read *carbonarius*.
» 445, title, affix an ∗ to *Dicaeum talautense* M.&Wg.
» 459, line 14, add Dongala, West Celebes (Doherty *o 1*).
» 459, » 18, erase f. 2 (♀) — this figure representing a North Celebes female (fide W. Blasius).
» 497, » 18, 19 put the names Wall., W. Blas., and Hombr. & Jacq. in brackets.
» 504, » 8 from below, title, for *Trichostoma celebensis* (Strickl.) read *Trichostoma celebense* Strickl.
» 505, » 6 for *Brachypteryx* read *Brachypteryx*.
» 513, last line, for Nat Coll. *a 25* read *a 26*.
» 525, line 6 from below, for W. Taczanowski read L. Taczanowski.
» 550, » 9, for *Munia molucca propinqua* Hart. read *Munia molucca propinqua* (Sharpe).
» 551, line 21, alter the formula *Munia molucca*>*propinqua* into *Munia molucca typica*>.
» 561, » 28, add Saleyer and Djampea (Everett *15*).
» 564, » 14, for *erythroprhys* read *erythrophrys*.
» 576, » 10, cancel the inverted commas in the name "La Pérouse".
» 605, » 24, for f. 2197 (read f. 1297).
» 606, title, affix an ∗ to *Ptilopus subgularis* M.&Wg.
» 616, » 14, add Djampea (Everett *20*.).
» 638, » 10, add Peling and Banggai (Nat. Coll.).
» 638, » 25, add Talaut—Lirung (Nat. Coll.).
» 671, last line, for *Tagegallus* read *Talegallus*.
» 676, line 18, for *bensteini* read *bernsteini*.
» 712, » 2 from below, read Tawaya, West Celebes, not West Tawaya, Celebes.
» 765, » 3 from below, add Togian (Meyer).
» 773, » 19 from below, read Petrop., not Petrov.
» 813, » 14, add Togian (Meyer *b 13*).
» 827, last line, add Sula (Wallace).
» 843, line 4, from below, read t. 153, instead of t. 155.
» 850, » 4, for *goisaki* read *goisagi*.
» 857, » 6 from below, add Karkellang, Talaut (Nat. Coll.).
» 884, » 1, for B. Kerguelen 1877 read 1879.
Plate XXVI read *teismanni*.

INTRODUCTION.

When we first planned a treatise on the Birds of Celebes, we soon found that it would be quite impossible to restrict ourselves to the mainland, as this is everywhere surrounded by larger or smaller islands which are so connected with it by their Avifaunas that they could not be left out; at the same time it proved impossible to define a natural zoological frontier between certain of these islands and the adjacent ones. Our frontispiece-map shows the limits we decided upon, viz. the inclusion of the Talaut Islands in the north, the Sula Islands in the east, and the Djampea Group in the south, though at each of these points elements from, respectively, the Philippines, the Moluccas, and the Lesser Sunda Islands are very marked. The boundary so chosen adjoins to the north the southern limit of the Philippines, as defined by Tweeddale, Worcester and Bourns, and others; to the east it coincides with Salvadori's western border, as drawn in his "Ornitologia della Papuasia e delle Molluche"; to the west it follows the eastern boundary of Borneo, as adopted in Everett's "List of the Birds of the Bornean Group", and by other writers; to the south it takes in all the islands between Celebes and the Lesser Sundas. The book may thus be said to fill up an ornithological gap, and the bounds as chosen appear also to be the most natural, except possibly (?) in the case of the Djampea Group. Moreover, the Avifauna of the adjacent groups often gives a clue to the derivation of non-Celebesian forms in Celebes; it would, therefore, be inadvisable to leave them out.

1. TRAVEL AND LITERATURE.

The naturalists and collectors who have done work among the birds of this area first deserve attention, and to the following short biographical notes concerning them we append a list of the publications on the Birds of Celebes and the neighbouring islands, based more or less directly on these travellers' results. We are afraid that our lists are not complete, either in regard to its including all the names of ornithological collectors, or all items of literature. As to the latter we have restricted ourselves, with a few exceptions, to the period after the publication of Walden's "List of the Birds known to inhabit the Island of Celebes" in the year 1872, and several papers and books, which we have not enumerated in our list, though they contain something on Celebesian Birds, will be found in the synonymy of the species, if they have not been unhappily entirely overlooked.

1793. Labillardière (Jacques Julien Houton de) 1755—1834. Frenchman. Naturalist. Accompanied Dentrecasteaux' expedition in search of la Pérouse (see: Relation du voyage par le C^{en}. Labillardière; an VIII (1800) vol. II, p. 298). The ships spent 18 days in passing through the strait between Buton and Muna, and parties landed upon both islands. There is no doubt that the "Pie de la Nouvelle Calédonie"(!), *Streptocitta albicollis* (Vieill.), was then obtained, as possibly also *Gazzola typica* Bp. Labillardière mentions some Parrots in these islands. Besides the above, he published many works and papers on botany, etc.

1821. Reinwardt (Caspar Georg Carl) 1773—1854. German. Naturalist. Sojourned from 1816—1822 in the East Indian Archipelago, visited about 1820 the Island of Saleyer, spent a few months in 1821 in North Celebes (see: Reinwardt's Reis naar het oostelijk gedeelte van den indischen Archipel in het jaar 1821 by W. H. de Vriese, Amst. 1858, pp. 503—538, Gorontalo; pp. 539—603, the Minahassa; with plates, 7 concerning Celebes, and the published Catalogues of the Leyden Museum). His ornithological collections are in the Leyden Museum, — see the list of Birds collected (125 species) in the work quoted (pp. 237—239) and the mention of 633 specimens of birds etc. sent home (p. 245). On p. 592 some birds of the Minahassa are recorded. As to ornithological papers he only wrote: "Uber die Art und den Ursprung der essbaren Vogelnester auf Java" (1838), but we do not know where this has been published. *Baza reinwardti*, occurring on islands south of Celebes and elsewhere, was named after him.

1828. Müller (Salomon). German (born at Heidelberg). Naturalist. Sojourned in the East Indian Archipelago from 1826—1837 and visited among other places South Celebes and the Island of Buton, see: Reizen en onderzoekingen in den Indischen Archipel, 1828—1836, vol. II, 1857, pp. 4—19 (on the birds of South Celebes, pp. 7—8, 64, 69—71; of Buton, pp. 12, 15, 65, 69). This work is a new and enlarged edition of a part of the "Verhandelingen over de natuurlijke geschiedenis der Nederlandsche overzeesche bezittingen, door de leden der natuurkundige Commissie en andere schrijvers", uitgegeven door C. J. Temminck, 1839—1844, fol. His ornithological collections — about 8000 specimens — are in the Leyden Museum (see: Schlegel's Catalogues; Veth: Overzicht van hetgeen gedaan is voor de Kennis der Fauna van Ned. Ind. 1879, p. 89, etc.). *Tanygnathus muelleri* from Celebes was named after him. We have nowhere been able to find the dates of birth and death of this meritorious naturalist.

1828. Quoy (Jean René Constant) and **Gaimard** (Joseph Paul) 1790—1869 and 1796—1858. Frenchmen. Naturalists. The latter took part in the expedition of the "Uranie" and "Physicienne" (1817—1820), both in that of the "Astrolabe", 1826—1829 (see: Voyage de la corvette l'Astrolabe, Paris 1830—1834, Histoire du voyage, 1833, V, 428 by Dumont d'Urville, and Zoologie, 1830, I, 165, where 10 new species of birds from Celebes are described, by Quoy and Gaimard). They only visited the Minahassa for about 5 days in the year 1828. Temminck (Coup-d'oeil gén. s. l. poss. néerl. dans l'Inde arch. 1849, III, 105—106) said about this trip: "La relâche de la corvette française l'*Astrolabe* à la factorerie de Menado, et l'excursion d'une couple de jours faite par les naturalistes français au lac de Tondano, n'ont offert, à l'une comme à l'autre expédition scientifique, q'une recolte peu nombreuse de plantes, ainsi que la capture d'un petit nombre d'animaux. Toutefois, le naturaliste a découvert dans ces acquisitions, à peu-près autant d'espèces nouvelles à faire connaître, qu'il s'est trouvé d'objets rassemblés, presque sans choix préalable; on a été non moins surpris des résultats qu'elles ont offerts à la science." Both have written many important works on Natural History, etc.

1841. Forsten (Eltio Alegondus) 1811—1843. Dutchman. Naturalist. Was elected (1836) a member of the "Natuurkundige Commissie" in the Netherlands' Indies and sojourned in North Celebes from 1841. He could not do much, however, in consequence of bad health, and died on the 2nd of January 1843 in Amboina. Nevertheless Temminck was justified in saying (Coup-d'oeil gén. s. l. poss. néerl. dans l'Inde arch. 1849, III, 106): "Les perquisitions et les travaux auxquels il lui fut possible de se livrer, nous ont valu des additions fort intéressantes à la connaissance très-superficielle qu'on avait pu acquérir jusqu'ici de cette contrée." His ornithological collections are in Leyden (see Schlegel's Catalogues, etc.). *Meropogon forsteni*, *Halcyon forsteni*, *Pitta forsteni* and *Carpophaga forsteni* from Celebes were named after him.

c. 1844. Léclancher (Charles René Auguste) 1804—1885. Frenchman. Surgeon on several warships from 1828—1844. He visited among other places Borneo and Celebes and brought home extensive collections to the Paris Museum. When with the "Favorite" from 1841—1844 he stayed at Manado in North Celebes and got two species of birds till then unknown, one of which was named after him *Dicaeum leclancheri* (see Rev. Zool. 1845, p. 93). During a former expedition of the same ship from 1830—1832 Eydoux was the doctor on board, and in 1839 he, with Gervais, described the birds then collected in the zoological part of the work on the voyage. The "Favorite" was also out from 1838—1839 with Léclancher on board, but neither this expedition, nor the one of 1841—1844, have been described, so far as we are aware.

1856. Wallace (Alfred Russell). Born 1823. Englishman. Naturalist. Amazon 1848—52, in the East Indian Archipelago 1854—62, where especially he was most successful in every respect, none of the former nor of the later naturalists there having attained anything to equal his results. He was in South Celebes from September to November 1856, and July to November 1857, in North Celebes from June to September 1859, and his Assistant, Charles Allen, collected in the Sula Islands. As is generally known, Mr. Wallace has written specially on the Avifauna of Celebes in his various important works. His separate ornithological papers concerning the Celebesian Area are: On the Ornithology of Northern Celebes, Ibis, 1860, 140; List of Birds from the Sula Islands, P. Z. S., 1862, 333; and Note on *Astur griseiceps*, Ibis, 1864, 184; but he treated of different genera and families monographically in which the Celebes birds play a great part, e. g.: On the Parrots of the Malay Region, P. Z. S., 1864; On the habits and the distribution of the genus *Pitta*, Ibis, 1864; On the Pigeons of the Malay Archipelago, Ibis, 1865; Catalogue of the Birds of Prey of the Malay Archipelago, Ibis, 1868. His ornithological collections are for the greater part in the British Museum as the "Catalogues of Birds" show, but there are to be found in many other museums and private collections specimens from his rich harvest, amounting to 8050 specimens, as he himself mentions in the preface of his "Malay Archipelago". *Prionoturus wallacei*, *Microstictus wallacei*, *Macropteryx wallacei*, *Osmotreron wallacei* and *Chalcophaps wallacei* from Celebes, as well as *Ceyx wallacei* from Sula were named after him.

c. 1856. Riedel (Johann Gerardus Friedrich). Dutchman. Born 1832 at Tondano, North Celebes, where his father was a missionary; educated in Europe. A discourse with Alexander von Humboldt in Berlin[1]) and later Mr. Wallace's presence in Celebes appear to have had much influence in awakening his interests in Natural Science. From 1853—1883 he was in the Civil Service in the East Indian Archipelago (1853—1863 in the Minahassa, 1863—1875 in Gorontalo — both in North Celebes —, 1875—1878 in Billiton, 1878—1880 in Timor, 1880—1883 in Amboina). Many papers from his pen on North Celebes are to be found in Dutch periodicals, but his chief work is: "De sluik-en kroesharigo Rassen tusschen Selebes en Papua", with many plates (1886). He made extensive ornithological and other collections everywhere, which he presented to several European Museums. His birds from Celebes are among other places at Brunswick (see: Z. f. d. gos. Orn. 1886, 81)[2], Darmstadt (see: Abh. Naturw. Ver. Bremen V, 35, 1876, and 464, 1877), Dresden (see: our work), Karlsruhe (see: J. f. O. 1883, 129), Leyden (see: Schlegel's Catalogues, etc.), Paris (to which he presented many consignments from 1864—1872), St. Petersburg (see: Z. f. d. ges. Orn. 1886, 193)[2]. *Phyllergates riedeli* and *Ardetta riedeli* from Celebes were named after him. Dr. Riedel has been living in Holland since 1883.

c. 1860. Duivenbode (Lodewijk Diederik Hendrik Alexander van Renesse van) 1832 or 1833—1881 or 1882. Dutchman (half-caste of Ternate). Planter and merchant at Manado. Son of Maarten Dirk van Renesse van Duivenbode, whom Mr. Wallace in his "Malay Archipelago" (II, p. 2) calls the King of Ternate. He sent out native hunters to make large collections of birds in the Minahassa and the neighbouring islands and presented them in part to Museums (such as the Leyden) and sold others; consequently lots of birds from "Manado" were in the European market between 1870—1880 (see, for instance, J. f. O. 1883, p. 129), and those in many collections may be traced to this source. They

---

[1]) Alexander von Humboldt asked Mr. Riedel among other things, why there are no large mammals to be found in Celebes, a question involving the whole problem which makes this island so interesting.

[2]) This collection is not from the Minahassa, as Prof. W. Blasius writes, but from Gorontalo, as we know from Dr. Riedel himself. Gorontalo does not belong to the Minahassa.

bear, however, no exact locality and date and were often mixed up with birds from other parts of the East Indian Archipelago. He appears to have been induced to collect birds from the visit of Mr. Wallace to the Minahassa in the year 1859. *Eudrepanis duivenbodei* from Sangi was named after him.

1863. Rosenberg (Karl Benjamin Hermann von) 1817—1888. German. Lived, with an interval of two years, from 1840—1871 in the East Indian Archipelago, first as soldier, then draughtsman, next in the Civil Service, finally as naturalist to the government. He travelled in North and Central Celebes from April 1863 to August 1864 and wrote concerning it: "Reistochten in de Afdeeling Gorontalo" (Amst. 1865), containing a few ornithological notes; the chapters on Celebes in his "Malayischen Archipel" (Leipsic 1878—1879) with more extensive remarks on the Avifauna (p. 270—279); and "Ein Jäger-Eldorado" (Zool. Garten 1881, 164). His determinations are, however, not throughout trustworthy, as he was more of a sportsman than of a naturalist. He also sent some hunters to the Sangi Islands in the year 1864. During his long stay in the Archipelago he made extensive ornithological collections, which now are in Darmstadt (see: Abh. Natw. Ver. Bremen, 1876, V, 35), Leyden (see Schlegel's Catalogues, etc.) and in some other Museums (e. g. Lübeck, see: J. f. O. 1877, 359; Dresden through von Schierbrand). *Strix rosenbergi* and *Gymnocorex rosenbergi* from Celebes were named after him.

1864. Bernstein (Heinrich Agathon) 1828—1865. German. Naturalist. Sojourned in the East Indian Achipelago from 1855—1865, from 1860 onwards as naturalist to the Government, and died near New Guinea. His extensive ornithological collections are in the Leyden Museum. Though he did not visit the Celebesian Area personally, except for a short stay at Macassar, he sent some of his native collectors to the Sula Islands in the year 1864 (see Schlegel's Catalogues, etc.). Bernstein's admirable ornithological papers do not concern Celebes directly. Van Musschenbroek published the diary of his last voyage to New Guinea (1864—1865) in the Bijdr. taal-, land- en volkenk. Ned. Ind. (4) VII, 1, 1883, containing much valuable information as to this lamented naturalist. *Megapodius bernsteini* from Sula was named after him.

1864. Hoedt (Dirk Samuel). Dutchman (half-caste of Amboina); secretary to the government; a passionate amateur naturalist; was nominated successor to Bernstein. He collected on Sula Besi and Sula Mangoli (1864) and on Great Sangi and Siao (1865) and forwarded his ornithological collections to the Leyden Museum (see: Schlegel's Catalogues, etc.). He died some time after 1879, but we have not been able to ascertain the year.

1865. Bickmore (Albert S.). Born 1839. American. Naturalist. In the East Indian Archipelago from 1865—1866 (see: Travels in the East Indian Archipelago, London 1868, transl. into German, 1869, and Dutch, 1873) and sojourned a short time in South Celebes (June 1865) and in North Celebes (December 1865 till January 1866). He published a list of birds collected there in the Proc. Boston Soc. Nat. Hist. as he remarks in his book, but we have not been able to find it in this Journal nor elsewhere. His ornithological collections will be in an American Museum. He now is Curator of the department of Public Instruction in the American Museum of Natural History in New York.

1870. Meyer (Adolf Bernhard). Born 1840. German. Naturalist. Travelled from 1870—1873 in the East Indian Archipelago, having been induced to go out to this part of the globe in the hope that its innumerable islands would afford the possibility of studying the variation of species in the Darwinian sense, for the publication of the "Origin of Species" had influenced his University studies (1862—70), and he selected Celebes to begin with in consequence of Wallace's brilliant speculations on the anomalous condition of its fauna, and on the scientific problems awaiting solution there. He sojourned for over a year in Celebes: November 1870, Macassar; November—July, Minahassa and the neighbouring islands; July—September, Gorontalo, Togian and Central Celebes; September—November, South Celebes; January 1873, Macassar, Gorontalo, Kema; August 1873, Macassar. His ornithological collections from there are in Dresden, Berlin, London (British Museum: Walden Collection), etc.; they amounted to about 4000 specimens. Lord Walden treated of some of them in the Trans. Zool. Soc. vol. VIII, 1872; in the Ann. & Mag. Nat. Hist. vol. VIII 1871, IX 1872, XIV 1874; Meyer himself among other places (see "Literature") in the J. f. O. 1873, 404, where he made known that he had discovered 14 new species, and 25 which had not yet been recorded from Celebes; Rowley's Orn. Misc. 1877 & 1878; Ibis 1879 (field notes); and Abbildungen von Vogelskeletten 1879—1897. *Trichoglossus meyeri*, *Cyrtostomus frenatus meyeri* from Celebes, and *Halcyon meyeri* from Togian were named after him. He has translated some of Wallace's works into German and has been in charge of the Dresden Museum since 1874.

1870. Conrad (Paul). German. Captain of a trading vessel. He collected 5 species of birds at Macassar, South Celebes, in 1870, which are probably in the Bremen Museum (see: Verh. Zool.-bot. Ges. Wien 1873, 341).

1873. Fischer (Georg). German. Army Surgeon in the Dutch Indies. Collected in Celebes and Borneo and presented his ornithological collection of 1066 specimens from the Minahassa and Sangi to the Darmstadt Museum (see: Abh. Natw. Ver. Bremen V, 1876, p. 35, and t. c. 1878, p. 538). *Ptilopus fischeri* from Celebes was named after him. In 1880—1881 he was stationed at Ternate (see: Bull. Ac. Imp. des Sc. St. Pétersb. 1884 XI, p. 109).

1873. Beccari (Odoardo). Born 1843. Italian. Naturalist. Sojourned in the East Indian Archipelago from 1865—1868 (Borneo), from 1871—1876 (Moluccas, Celebes, New Guinea), from 1878—1879 (Sumatra), and as a scientific collector takes almost equal rank with Wallace. In 1873—1874 he visited the South-eastern Peninsula of Celebes, as well as the Minahassa and Macassar, and Count Salvadori has described his ornithological collections from there, now in the Genoa Museum (see: Ann. Mus. Civ. di Stor. Nat. di Gen. 1875, VII, 641). *Aethopyga beccarii* and *Turnix beccarii* from Celebes were named after him. He lives at Radda in Chianti near Florence.

1874. Bruijn (Antonius Augustus). Dutchman. He was an officer in the Dutch Navy, but settled on Ternate as son-in-law of the great merchant M. D. van Renesse van Duivenbode (mentioned above p. 4), whose business he carried on after his death. He sent out hunters with many of his ships and sold the bird-skins collected chiefly in Paris to plumassiers, but a large and highly valuable collection was presented by him to the Genoa Museum, containing among others a series from North Celebes and Sangi (see: Ann. Mus. Civ. Gen. 1875, VII, p. 641; ib. 1876, IX, p. 50). He died about the year 1880.

1875. Musschenbroek (Samuel Cornelius Jan Willem van) 1827—1883. Dutchman. Naturalist. In the Civil Service of the Dutch Indies from 1855—1877, including a two-years' furlough in Europe. He was Resident of the Province of Manado from 1875—1876. Here he collected ornithologically, as indeed he did in all branches of Natural History wherever he was stationed (Java, Ternate), sending his collections to the Museums of the Netherlands. He presented (1879) part of his North Celebesian birds to the Dresden Museum, others to Leyden (see: Notes of the Leyden Museum 1879, I, p. 50). He published some remarks on

Introduction: Travel and Literature.   7

the birds of North Celebes (Nat. Tijdschr. Ned. Ind. 1877, XXXVI, p. 376), and amongst his other works we may mention here his large map of the Minahassa (1878, $\frac{1}{100\,000}$) and that of the Gulf of Tomini and the lands adjoining (1879). As to his sojourn in the Minahassa, see also Rowley's Orn. Misc. 1878, III, p. 115. As a gentleman of high scientific attainments he offered great help to all naturalists visiting the East Indian Archipelago. *Surniculus musschenbroeki* from Batjan, now also known from Celebes, was named after him.

1875. Murray (John). Born 1841. Scotsman (Canada). Naturalist on the "Challenger" under whose superintendence the ornithological collections were formed, and of whose note-book, and of further notes, Mr. Sclater made use in his Report (see: "The Voyage of H. M. S. Challenger" 1873—1876, Zoology vol. II, part VIII, 1880). There is, however, only one species recorded belonging to our area, viz. a Lory from Melangis[1], one of the Nanusa Islands (l. c. p. 115) to the north of Celebes. Specimens are in the British Museum. This bird was afterwards named *Eos challengeri*. Dr. Murray lives in Edinburgh as Director of the Scottish Marine Station for Scientific Research and is a member of the Fishery Board for Scotland.

1875. Bültzingslöwen (Wulf von). Born 1847. German. Sportsman. He travelled in the Minahassa in the year 1875 and brought together there a small collection of birds, which he presented to the Lübeck Museum (see: J. f. O. 1877, 359). Lives near Berlin.

1876. Faber (F. von). Dutchman. In the Civil Service of the Dutch Indies. He collected bird-skins in the Minahassa where he stayed in Amurang in the year 1876, and presented collections among others to the Dresden and Berlin Museums (as to the latter, see J. f. O. 1877, 217, and 1883, 121). Subsequently (1881) he collected ornithologically in Sumatra also. Died after 1886.

1876. Teijsmann (Johannes Elias). 1808—1882. Dutchman (of German origin). Naturalist. Lived at Buitenzorg in Java, from 1830 till his death, as Botanist and as Honorary Inspector of the Plantations. He made an official voyage in 1876 to the Moluccas and visited Sula Besi (see Nat. Tijdschr. Ned. Ind. 1877, XXXVII, p. 88); in 1877 another (besides various journeys in the Archipelago not mentioned here) to South Celebes and Saleyer (l. c. 1879, XXXVIII, p. 54), and on these occasions he collected birds among other objects. In Celebes he procured (t. c. p. 121) "893 specimens of skins of mammals, birds, etc. in 254 species". His reports also contain some ornithological notes. He visited Macassar, Pankadjene, Tjamba, Maros, Bonthain and Loka in South Celebes and Saleyer Island, but a full description of his valuable ornithological collection sent to the Leyden Museum was never given. *Rhipidura teijsmanni*, *Pachycephala teijsmanni* and *Cyrtostomus teijsmanni* were named after him (s. Notes Leyden Museum 1893, XV, pp. 167, 170, 179). In the year 1880 he had also visited the Minahassa on a short trip together with Prof. de Vriese, who died soon afterwards. (See l. c. 1861, XXIII, pp. 343—369.)

1878. Platen (Carl Constantin). Born 1843. German. Naturalist. Was a physician at Amoy, then collected, chiefly ornithologically, with a short interval in Europe in the year 1879, in 1878 in South Celebes, 1884 in Malacca, Borneo, the Moluccas and Waigiou, 1884—1886 in the Minahassa, North Celebes, 1886—1887 on Great Sangi, 1887—1892 in the Philippines, 1892—94 on Batjan. The greater part of his collections were sold to the Brunswick Museum and to Mr. Nehrkorn's Museum at Riddagshausen near Brunswick;

---

[1] We write Melangis, instead of Meangis, as the former name is on the best Dutch maps.

other specimens to different museums and collections in Germany and abroad. Dr. Platen himself wrote about his trips in Celebes and Sangi in the "Gefiederten Welt" 1879, pp. 358, 378; 1887, 193, 264. His birds from Celebes and Sangi were carefully described by Prof. W. Blasius in the Z. ges. Orn. 1885, 201 (S. Celebes, 1878); Ornis 1892, IV, 527 (Sangi, 1886); and Festschr. Vers. Naturf. Braunschw. 1897, p. 277 (N. Celebes, 1884—1886). *Aramidopsis plateni*, and *Cyrtostomus plateni* from Celebes were named after him, and *Criniger platenae* after Mrs. Platen, who has always accompanied her husband. He lives at Barth (Prussia).

c. **1880. Laglaize** (Léon). Frenchman. Collecting Naturalist. Made extensive expeditions in the East Indian Archipelago and sojourned about 1880 in Celebes and Sangi; his ornithological collections from there are in the Paris and some other museums.

1882. Ribbe (Carl) and **Kühn** (Heinrich). Born 1861 and 1860. Germans. Collecting naturalists. They made an expedition together in the East Indian Archipelago from 1882 —1885, and visited South and East Celebes in 1882—1883, North and West Celebes (on a short trip) in 1885; the island of Banggai and East Celebes for the second time were visited by Kühn alone in 1884—1885. Part of their ornithological collections from Celebes came to the Dresden Museum (see Sitzb. Ges. Isis, Dresden, 1884, Abh. 1, pp. 16, 48); for a general report on this voyage and on a second to the Bismarck Archipelago and Solomon Islands (1892—1896) by R. alone, see: Deutsche geogr. Blätter 1895, vol. 18, p. 372. Concerning Celebes Mr. Ribbe has published: "Ein Sammeltag am Wasserfall von Maros" (see Krancher's entomolog. Jahrb. 1893). He is living near Dresden; Mr. Kühn at Tual in the Kei Islands as owner of a steam saw-mill.

1883. Guillemard (Francis Henry Hill). Born 1852. Englishman. Naturalist. Visited North and South Celebes, 1883, during the "Cruise of the Marchesa to Kamtschatka and New Guinea" (1886, p. 153—215: Chapter on Celebes, where there are some notes on birds) and described his ornithological collections in the Proc. Zool. Soc. of London 1885, 542. He collected 108 species in Celebes, 3 of which had not been recorded before from there, though from elsewhere. These are mostly in the Tring Museum. Dr. Guillemard has also travelled in Lapland and in Africa, and is now living in England at Trumpington near Cambridge.

1885. Hickson (Sydney John). Born 1859. Englishman. Naturalist. Made scientific researches in North Celebes, Sangi and Talaut in 1885 to 1886 and gave some notes on the birds there (see: "A Naturalist in North Celebes", 1889, Appendix B, p. 360). His ornithological collections, which are, however, small, are in the Cambridge Museum. He now is Professor of Zoology at the Owens College, Manchester.

1888. Weber (Max Carl Wilhelm). Born 1852. German. Naturalist. Professor of Zoology at Amsterdam. During an exploring expedition in the East Indian Archipelago (1888—1889) he visited South and Central Celebes and the Island of Saleyer, and procured 97 species of birds in 234 specimens in Celebes (of which, however, only 2 species of wide distribution were as yet unknown from the island) and 14 species in 22 specimens in Saleyer. These collections are partly in the Leyden Museum, partly in the Zoological Museum of the University of Amsterdam. (See: "Zoologische Ergebnisse einer Reise in Niederländisch Ost-Indien", 1894, III, p. 269.)

1890. Radde (Gustav Ferdinand Richard). Born 1831. German. Naturalist. Accompanied two Russian Grand-dukes on a voyage to the East and visited the Island of

Buton and South-east Celebes in 1890. He has not published a report on the ornithological collections made here, unless it be in the work brought out by the Grand-dukes in Russian. A chapter of this work, from the pen of Dr. Radde, on Buton and South Celebes appeared in German in the journal "Globus" 1896, vol. 69, p. 151, wherein are some ornithological notes (s. pp. 172, 189), but the determinations are not trustworthy, and have, therefore, not been quoted in this book. Dr. Radde is Director of the Museum of Tiflis, where, or elsewhere in Russia, this collection may now be.

1892. Cursham (Charles W.). Dutchman (half-caste of Celebes). Merchant at Manado, North Celebes, who had collected birds there before 1892, which may be in some museum. Was engaged by Dr. Meyer and the Hon. W. Rothschild to form another collection and sent out native hunters from 1892—1896 to some parts of the Minahassa and the small neighbouring islands, and to the Sangi, Talaut and Banggai Groups (see: J. f. O. 1894, 237, and Abh. Ber. Mus. Dresden 1894/5 Nr. 4, Nr. 9; 1896/7 Nr. 2). In our work specimens from this source are marked: "native collectors" or "native hunters" ("nat. coll.", "nat. hunt.") and some of these skins have passed into other museums also.

1893. Sarasin (Paul and Fritz). Born 1856 and 1859. Swiss. Naturalists. The cousins Sarasin spent the years 1884—1886 in Ceylon and published the results of their investigations in the great work entitled: "Ergebnisse naturwissenschaftlicher Forschungen auf Ceylon" (1887—1893). From 1893—1896 they were in Celebes, viz. 1893—1894 in North Celebes, 1895 in central parts, 1895—1896 in the South (see: Zeitschr. Ges. Erdkunde Berlin 1894, XXIX, 351; 1895, XXX, 226, 311; 1896, XXXI, 21; and Verh. Ges. Erdkunde Berlin 1896, 337; with 5 maps). No naturalists before them have made such a thorough and many-sided exploration of the Island, and contributions of the highest importance are to be expected from their pen, for as yet they have only begun to publish some of their results. They collected in nearly every branch of Natural History and Ethnography, and we had the privilege of receiving their ornithological specimens in 9 different consignments during the time we were writing this book (see: J. f. O. 1894, 153; Abh. Ber. Mus. Dresden 1894/95, Nr. 4 and Nr. 8; 1896/97, Nr. 1), which has reached a much higher standard through their welcome aid, as may be seen on almost every page. We are, therefore, deeply indebted to Dr. P. & Dr. F. Sarasin. They obtained 207 species on the mainland of Celebes, 10 of which proved to be new to science and 12 others not yet known from Celebes. Their ornithological collections are for the most part in the Museum at Basel, where they live; they also presented some valuable specimens to the Dresden Museum. *Myza sarasinorum*, *Zosterops sarasinorum* and *Cryptolopha sarasinorum* were named after them.

1894. Kükenthal (Willy). Born 1861. German. Naturalist. When on his exploring expedition into the East Indian Archipelago from 1893—1894 spent a few weeks (in June and July 1894) in North Celebes and collected some birds there, which are now in the Senckenberg Museum at Frankfurt. Dr. Kükenthal is Professor of Zoology at Jena.

1895. Everett (Alfred Hart). Born 1848. Englishman. Naturalist. Has collected birds, etc., beginning about 1870, in the East Indian Archipelago, viz. in Borneo (see his "List of the Birds of the Bornean Group of Islands": J. Str. Br. R. As. Soc. 1889, Nr. XX), the Philippines (see Ibis 1872, and Proc. Zool. Soc. 1877—1879), Natuna, Savu, Lombok, Timor, etc. (see Novit. Zool. 1893 and 1896), and visited South Celebes and the islands to the south in 1895 (see Novit. Zool. 1896, pp. 69, 148, 256; 1897, 170). His ornithological collections are chiefly in the British and Tring Museums, but many of his duplicates are to be found elsewhere, for instance, in the Dresden Museum. *Androphilus everetti* from South Celebes,

Pachycephala everetti and *Monarcha everetti* from the Island of Djampea were named after him. He has been during most of this time in the Sarawak Government Service, and now resides in Labuan, when not engaged in zoological collecting.

1895. Hose (Charles and Ernest). Born 1863 and 1872. Englishman. Visited the Minahassa, on a collecting trip in 1895. Charles, the elder brother, had already explored the mountainous parts of North Borneo and made extensive collections there; he has written about the mammals and birds of Borneo (1893). He is in the service of the Rajah of Sarawak, as Resident of the Province of Baram. *Dicaeum hosei* from Celebes was named after him. Ernest Hose also lives in Sarawak, when not engaged in collecting work.

1896. Doherty (William). American from Cincinnati. Naturalist. He has travelled extensively for many years in Europe, Persia, India and the East Indian Archipelago and visited South Celebes in 1887 and 1891, South and West Celebes (Palos Bay) in 1896, Talaut in 1892 and Sula in 1898. His ornithological collections are chiefly in the Tring Museum (see: Nov. Zool. 1896, 153). *Pitta dohertyi* from Sula was named after him, this species, however, could not be treated of in our book, as it was not described until after we had finished. He is highly distinguished as a linguist and entomologist and has written a great deal on Lepidoptera from India. He is still travelling in the East.

1897. Waterstradt (Johannes). Born 1869. Dane. Collecting naturalist. Has made extensive expeditions in Ceylon, Malacca and the East Indian Archipelago (chiefly Borneo) since 1888 and sent his Bornean hunters to the Talaut Islands in 1897. The ornithological collections of this expedition are in the Tring Museum (see: Nov. Zool. 1898, 88). He is still collecting, and is settled for the moment in Labuan.

Special List of Literature on the Birds of Celebes.[1])

1850. Gould J. The Birds of Asia. 1850—1883. 7 vols. folio (completed by R. B. Sharpe).
1857. Schlegel H. Handleiding tot de beoefening der dierkunde. Vol. I 1857. 530 pages. 8 plates in folio.
1860. Wallace A. R. On the ornithology of Northern Celebes. Ibis p. 140—7.
1862. Schlegel H. Muséum d'Histoire Naturelle des Pays-Bas. Revue méthodique et critique des collections. 1862—1880. 7 vols. and 1 vol. Index by F. A. Jentink.
1862. Wallace A. R. List of Birds from the Sula Islands (east of Celebes), with descriptions of the New Species. P. Z. S. p. 333—46. Plates XXXVIII—XL.
1863. Schlegel H. De Vogels van Nederlandsch Indië (Pitta, Ijsvogels, Valkvogels). 1863—1866. 3 parts 4°. 185 pages, 50 plates.
1864. Wallace A. R. Note on Astur griseiceps, Schlegel. Ibis p. 184, plate V.
1865. Finsch O. Neu-Guinea und seine Bewohner. 185 pages. (On p. 154—85 list of Birds including Celebes.)
1866. Schlegel H. Observations Zoologiques I. Ned. Tijdschr. Dierk. III p. 181—213.
1868. Bickmore A. S. Travels in the East Indian Archipelago.
1868. Sharpe R. B. A monograph of the Alcedinidae. 1868—1871. 4°. LXXXII+304 pages, 120 plates.
1869. Wallace A. R. The Malay Archipelago. 2 vols. XXIV + 1002 pages. (German edition by A. B. Meyer, in the same year.)
1871. Meyer A. B. Brief über Merops Forsteni von Celebes. J. f. O. p. 231—2 (translated in Gould's Birds of Asia vol. I pl. 39 1873).
1871. Walden A. On a new species of Trichoglossus (T. Meyeri) from Celebes. Ann. Mag. Nat. Hist. (4) VIII p. 281—2.
1872. Cabanis J. [Note on Oriolus formosus n. sp.] J. f. O. p. 392—3.
1872. Walden A. Description of a supposed new species of Cuckoo from Celebes. Ann. Mag. Nat. Hist. (4) IX p. 305—6.
1872. Walden A. On some supposed new species of Birds from Celebes and the Togian Islands. Ann. Mag. Nat. Hist. (4) IX p. 398—401.

1872. Walden A. A List of the Birds known to inhabit the Island of Celebes. Tr. Z. S. VIII p. 23—108, Plates III—X. 4to.
1872. Walden A. Appendix to a List of Birds known to inhabit the Island of Celebes. Tr. Z. S. VIII p. 109—98, Plates XI—XIII.
1873. Cabanis J. [Note on Gerygone flaveola n. sp.] J. f. O. p. 157—8.
1873. Finsch O. und P. Conrad. Ueber eine Vogelsammlung aus Ostasien. Verh. Zool.-bot. Ges. Wien p. 341—61 (5 species from Celebes).
1873. Meyer A. B. Notiz über die Vögel von Celébes. J. f. O. p. 404—5.
1873. Pelzeln A. v. [Liste von Vögeln grösstentheils aus Celebes.] Verh. Zool.-bot. Ges. Wien p. 10 of sep. copy.
1873. Sharpe R. B. On three new species of Birds. P. Z. S. p. 625—6 (1 new species from Celebes).
1874. Walden A. Descriptions of some new species of Birds. Ann. Mag. Nat. Hist. (4) XIV p. 156—8 (1 new species from Togian).

---

[1]) Where not otherwise specialized the book is octavo.

1874. Sharpe, Soebohm, Sclater, Gadow, Hartert, Grant, Hargitt, Salvadori, Salvin, Saunders: Catalogue of the Birds in the British Museum. 27 volumes. 1874—1898.
1875. Gould J. The Birds of New Guinea and the adjacent Papuan Islands. 1875—1888. 5 vols. folio (completed by R. B. Sharpe).
1875. Gurney J. H. Notes on a "Catalogue of the Accipitres in the British Museum". Ibis p. 87 et seq. (These notes continue till 1882 and contain remarks on many of the Birds of Prey of Celebes.)
1875. Meyer A. B. [Ueber Coryllis.] Gef. Welt IV p. 229—30.
1875. Meyer A. B. Ornithologische Mittheilungen I. Mitth. Zool. Mus. Dresden vol. I p. 19. 4to.
1875. Murray J. Voyage of H. M. S. Challenger. Zoology vol. II part VIII p. 115. 4to.
1875. Pelzeln A. v. Africa-Indien. Verh. Zool.-bot. Ges. Wien p. 33—62.
1875. Salvadori T. Intorno a due collezioni di uccelli di Celebes inviate al Museo civico di Genova dal Dr. O. Beccari e dal Sig. A. A. Bruijn. Note. Ann. Mus. Civ. di Stor. Nat. di Genova VII p. 641—81, Tav. XVIII.
1876. Brüggemann F. Beiträge zur Ornithologie von Celebes und Sangir. Abh. Naturw. Ver. Bremen V p. 35—102, Taf. III—IV.
1876. Koch G. v. Verzeichniss einer Sammlung von Vogelbälgen aus Celebes und Sanghir, welche vom Grossherz. Zoologischen Museum zu Darmstadt im Tausche oder gegen Baarzahlung zu erhalten sind.
1876. Pelzeln A. v. Ueber eine . . . Sendung von Vogelbälgen. Verh. Zool.-bot. Ges. Wien p. 716—20, Taf. XIII.
1876. Salvadori T. Intorno a due piccole collezioni di Uccelli, l'una di Pottà (Isole Sanghir) e l' altra di Tifore e di Batang ketcil, inviate dal Signor A. A. Bruijn al Museo Civico di Genova, nota. Ann. Mus. Civ. Genova IX p. 50—65.
1876. Salvadori T. [Letter on Celebes Birds.] Ibis p. 385—6.
1876. Shelley G. E. A monograph of the Nectariniidae or family of Sun-Birds. 1876—1880. CVIII + 393 pages, 121 plates. 4to.
1876. Wallace A. R. The Geographical Distribution of Animals. 2 vols. XXXII + 1108 pagg. (German edition by A. B. Meyer, in the same year.)
1877. Brüggemann F. Nachträgliche Notizen zur Ornithologie von Celebes. Abh. Naturw. Ver. Bremen V p. 464—6.
1877. Lenz H. Mittheilungen über malayische Vögel. J. f. O. p. 359—82.
1877. van Musschenbroek S. C. J. W. Jets over de Fauna van Noord-Celebes en zijne naaste omgeving. Nat. T. Ned. Ind. XXXVI p. 376—84.
1877. Reichenow A. [Note on Birds from Celebes.] J. f. O. p. 217—8.
1877. Rowley G. D. [and A. B. Meyer]. Broderipus formosus (Cab.). Orn. Misc. II p. 227 —9, plate LVI. 4to.
1877. Rowley G. D. [and A. B. Meyer]. On a few species belonging to the genus Loriculus. Orn. Misc. II p. 231—54, plates LVII—LX. 4to.
1877. Rowley G. D. [and A. B. Meyer]. On the Genus Pitta. Orn. Misc. II p. 259—69, 321—3, plates LXII, LXIV—V. 4to.
1877. Salvadori T. Intorno alle specie di Nettarinie della Papuasia, delle Molucche e del gruppo di Celebes. Note. Atti d. R. Acad. Sc. Torino XII p. 299—321.
1877. Teijsmann J. E. [Report on a journey to the Moluccas with the mention of some birds from Sula Besi.] Nat. T. Ned. Ind. XXXVII p. 75—148.
1878. Fischer G. Bemerkungen über zweifelhafte celebensische Vögel. Abh. Naturw. Ver. Bremen V p. 538.
1878. Meyer A. B. Description of two species of Birds from the Malay Archipelago. Rowley's Orn. Misc. III p. 163—4. 4to.

1878. Reichenow A. Vogelbilder aus fernen Zonen. Abbildungen und Beschreibungen der Papageien. 1878—1883. Folio. 33 plates and letter-press.
1878. v. Rosenberg H. Der Malayische Archipel. Land und Leute. 615 pages, 2 maps.
1878. Rowley G. D. [and A. B. Meyer]. Domicella coccinea (Latham). Orn. Misc. III p. 123—9, plate XCVIII. 4to.
1878. Rowley G. D. [and A. B. Meyer]. On the genus Cittura. Orn. Misc. III, p. 131—43, plates XCIX—CII. 4to.
1878. Salvadori T. Descrizione di trentuna specie nuove di uccelli della sotto regione papuana, e note intorno ad altre poco conosciute. Ann. Mus. Civ. Genova XII p. 317—47.
1878. Salvadori T. Descrizione di tre nuove specie di uccelli e note intorno ad altre poco conosciute delle isole Sanghir. Atti d. R. Accad. d. Sc. Torino XIII p. 1184—9.
1878. Sharpe R. B. On the collections of Birds made by Dr. Meyer during his expedition to New Guinea and some neighbouring Islands. Mitth. Zool. Mus. Dresden III p. 349—72, plates XXVIII—XXX. 4to.
1879. Meyer A. B. Field Notes on the Birds of Celebes. Ibis p. 43—70, 125—47.
1879. Meyer A. B. [On Pitta Forsteni] in Gould's Birds of New Guinea vol. IV, letter-press to plate 30. Fol.
1879. Meyer A. B. Abbildungen von Vogelskeletten. 2 vols. 1879—1897. XXXVIII + 192 pages, 242 plates. 4to.
1879. Platen C. C. Reiseskizzen aus Süd-Celebes. Gef. Welt p. 358—60, 378—81.
1879. Schlegel H. On Strix inexpectata. Notes Leyden Mus. I p. 50—2.
1879. Schlegel H. On an undescribed species of Ardea. Notes Leyden Mus. I p. 113—4.
1879. Teijsmann J. E. [Report on a journey to South Celebes with the mention of some birds of South Celebes and Saleyer.] Nat. T. Ned. Ind. XXXVIII p. 54—125.
1880. Frenzel A. [and A. B. Meyer]. Ueber Flederemauspapageien (Gattung Coryllis). Monatsschr. D. Ver. Schutze Vogelwelt p. 8—28, with plate.
1880. Legge W. V. A History of the Birds of Ceylon. XLVI + 1237 pages, 34 plates. 4to.
1880. Meyer A. B. De Vogels van Celebes. Een Handboek voor de Ingezetenen van het Eiland. Prospectus in Dutch and Malay, 4 pages, with plate (5 sp. of King-fishers fig.). 4to. (Never published on account of insufficient subscriptions.)
1880. Meyer A. B. [Letter on Streptocitta torquata and caledonica.] Ibis p. 249 and 373.
1880. Oustalet E. Monographie des oiseaux de la Famille des Mégapodidés. Ann. Sc. nat. Paris 6. sér. X, Art. Nr. 5. 60 pages, pl. 20—23; XI, Art. Nr. 2. 182 pages, pl. 2—3.
1880. Salvadori T. Ornitologia della Papuasia e delle Molucche. 3 vol. and 3 App. 1881—91. XLIII + 2119 pages. 4to.
1880. Schlegel H. On an undescribed Bird of the Timalia-Group. Malia grata. Notes Leyden Mus. II p. 165—7.
1880. Wallace A. R. Island Life: or, the phenomena and causes of Insular Faunas and Floras. XVII + 526 pages.
1881. Blasius W. Ueber eine Sendung von Vögeln aus Nord-Celebes. Braunschweigische Anzeigen Nr. 247 (newspaper!) and Gef. Welt p. 534.
1881. Meyer A. B. [Letter on Gymnophaps poecilorrhoa and Ptilopus Fischeri.] Ibis p. 169—70.
1881. Meyer A. B. Ueber Vögel von einigen der südöstlichen Inseln des malayischen Archipels. Verh. Zool.-bot. Ges. Wien p. 759—74.

1881. Oustalet E. Notes d'Ornithologie. (2° série.) Bull. Soc. philomath. Paris 7° sér. vol. V 18 pages (1 new species from Sangi).
1881. v. Rosenberg H. Ein Jäger-Eldorado. Zool. Garten p. 164—8.
1882. Elliot D. G. A monograph of the Bucerotidae. XXXII pages, 59 plates with letter-press. Folio.
1882. Blasius W. Ueber neue und zweifelhafte Vögel von Celebes. (Vorarbeiten zu einer Vogelfauna der Insel.) J. f. O. p. 113—62.
1883. Sharpe R. B. Notes on some species of Birds of the Family Dicaeidae. P. Z. S. p. 578—80 (1 new species from Sula).
1884. Dresser H. E. A monograph of the Meropidae. 1884—1886. XX + 144 pages, 34 plates. Folio.
1884. Meyer A. B. Ueber neue und ungenügend bekannte Vögel, Nester und Eier aus dem Ostindischen Archipel im K. Zoologischen Museum zu Dresden. Sitzber. u. Abh. der naturw. Ges. Isis Dresden. Abh. 1. 64 pages.
1884. Salvadori T. Remarks on the 8. and 9. Vol. of the "Catalogue of Birds". Ibis p. 322—9 (among others some remarks on Celebes Birds).
1885. Blasius W. Beiträge zur Kenntniss der Vogelfauna von Celebes. I. Vögel von Süd-Celebes, 1878 gesammelt von Herrn Dr. Platen bei Mangkassar und im District Tjamba. (Mit vier colorirten Tafeln.) Z. für d. ges. Orn. p. 201—328, Taf. XI—XIV.
1885. Guillemard F. H. H. Report on the Collection of Birds obtained during the Voyage of the Yacht "Marchesa" — Part IV. Celebes. P. Z. S. p. 542—61.
1885. Sharpe R. B. and C. W. Wyatt. A monograph of the Hirundinidae, or family of Swallows. 2 vols. LXXV + 673 pages, 129 plates. 4to.
1886. Blasius W. Ueber Vögel von Celebes. Braunschweigische Anzeigen Nr. 52. (Newspaper!)
1886. Blasius W. Beiträge zur Kenntniss der Vogelfauna von Celebes. II. Vögel von Nord-Celebes, 1866 und 1867 gesammelt in der Minahassa und 1868 dem Herz. Naturhist. Museum in Braunschweig geschenkt von Herrn Resident J. G. F. Riedel, damals in Gorontalo. Z. für d. ges. Orn. p. 81—179.
1886. Blasius W. Beiträge zur Kenntniss der Vogelfauna von Celebes. III. Vögel von Nord-Minahassa, gesammelt in der Minahassa und zu verschiedenen Zeiten (hauptsächlich 1869 und 1876) dem Zool. Museum der Kais. Akad. der Wiss. zu St. Petersburg geschenkt von Herrn Resident J. G. F. Riedel, damals in Gorontalo. Z. für d. ges. Orn. p. 193—210.
1886. Guillemard F. H. H. The Cruise of the Marchesa to Kamtschatka and New Guinea. 2 vols. LIV + 683 pages.
1887. Blasius W. [Notiz über Sangi-Vögel.] Braunschweigische Anzeigen Nr. 75 p. 695. (Newspaper!)
1887. Meyer A. B. [Ueber Coryllis catamene Schl.] Gef. Welt XVI p. 264—5. 4to.
1887. Platen C. C. Ornithologische Skizzen aus der Minahassa. Gef. Welt p. 193—4, 205—6, 217—9, 230—1. 4to.
1887. Platen C. C. Der Fledermauspapagei von Sangir (Coryllis catamene, Schl.). Gef. Welt p. 263—4. 4to.
1888. Blasius W. [Notiz über Sangi-Vögel.] Braunschweigische Anzeigen Nr. 9 p. 86 (Newspaper!) and Russ' Isis p. 78. 4to.
1888. Blasius W. Die Vögel von Gross-Sanghir (mit bes. Berücksichtigung der in den Jahren 1886 und 1887 von Herrn Dr. Platen und dessen Gemahlin bei Manganitu auf Gross-Sanghir ausgeführten orn. Forschungen) nebst einem Anhange über die Vögel von Siao. Mit 2 Tafeln. Ornis IV p. 527—646, Taf. III—IV.

1889. Hickson S. J. A Naturalist in North Celebes. With Maps and Illustrations. 392 pages.
1890. Meyer A. B. Brush-Turkeys on the Smaller Islands North of Celebes. Nature vol. XLI p. 514—5. 4to.
1892. Büttikofer J. A Review of the Genus Rhipidura. Notes Leyden Mus. XV p. 65 —110 (2 new species p. 79 and 80).
1892. Büttikofer J. On Merula javanica and its nearest allies. Notes Leyden Mus. XV p. 107—10 (1 new species p. 109).
1892. Meyer A. B. [Letter on the Togian Islands.] Ibis p. 178—80.
1893. Büttikofer J. On two new species of Pachycephala from South Celebes. Notes Leyden Mus. XV p. 167—8.
1893. Büttikofer J. On two new species of the genus *Stoparola* from Celebes. Notes Leyden Mus. XV p. 169—70.
1893. Büttikofer J. On two new species of Birds from South Celebes. Notes Leyden Mus. XV p. 179—81.
1893. Büttikofer J. On two new species of Birds from Java and Celebes. Notes Leyden Mus. XV p. 260—1.
1893. Dresser H. E. A monograph of the Coraciidae. XX + 111 pages, 27 plates. Folio.
1893. Elliot D. G. A monograph of the Pittidae. 1893—1895. XXVII pages, 51 plates and letter-press. Folio.
1893. Meyer A. B. und L. W. Wiglesworth. Leucotreron fischeri meridionalis n. subsp. Orn. Monatsber. I p. 12—3.
1894. Büttikofer J. Ornithologische Sammlungen aus Celebes, Saleyer und Flores. Weber's Zool. Erg. III p. 269—306, Taf. XVII—XVIII. 4to.
1894. Grant W. R. O. On the Birds of the Philippine Islands. II. Ibis p. 519 (1 new species from Celebes).
1894. Meyer A. B. Neue Vögel aus dem Ostindischen Archipel. Abh. u. Ber. Mus. Dresden 1894/5, Nr. 2. 4 pages, with plate. 4to.
1894. Meyer A. B. und L. W. Wiglesworth. Neue Vögel von Celebes. Abh. u. Ber. Mus. Dresden 1894/5 Nr. 4. 3 pages. 4to.
1894. Meyer A. B. und L. W. Wiglesworth. Beschreibung einiger neuen Vögel der Celebes-Region. J. f. O. XLII p. 113—6.
1894. Meyer A. B. und L. W. Wiglesworth. Ueber eine erste Sammlung von Vögeln von den Talaut Inseln. J. f. O. XLII p. 237—53, Taf. III.
1894. Sarasin P. und F. Reiseberichte aus Celebes. I. Ueberlandreise von Menado nach Gorontalo. II. Erforschung des Bone-Flusses. Z. Ges. Erdk. Berlin XXIX p. 351—401 (with notes on birds).
1895. Büttikofer J. A Revision of the genus Turdinus and genera allied to it. Notes Leyden Mus. XVII p. 65—106.
1895. Sarasin P. und F. Reiseberichte aus Celebes. III. Von Buol nach dem Golf von Tomini. Z. Ges. Erdk. Berlin XXX p. 226—34. IV. Reise durch Central-Celebes vom Golf von Boni nach dem Golf von Tomini. Ibid p. 311—52 (with notes on birds).
1895. Meyer A. B. und L. W. Wiglesworth. Bericht über die von den Herren P. und F. Sarasin in Nord-Celebes gesammelten Vögel. Abh. u. Ber. Mus. Dresden 1894/5 Nr. 8. 20 pages. 4to.
1895. Meyer A. B. und L. W. Wiglesworth. Eine zweite Sammlung von Vögeln von den Talaut Inseln. Abh. u. Ber. Mus. Dresden 1894/5 Nr. 9. 9 pages. 4to.
1896. Blasius W. Vögel von Pontianak (West-Borneo) und anderen Gegenden des indo-malayischen Gebietes. Mitth. Geogr. Ges. Lübeck 2. Reihe Heft X p. 90 —145 (p. 124—5 on Celebes birds).

1896. Hartert E. Preliminary descriptions of some new Birds from the mountains of Southern Celebes. Nov. Zool. III p. 69—71.
1896. Hartert E. On ornithological collections made by Mr. Alfred Everett in Celebes and the Islands south of it. Nov. Zool. III p. 148—83.
1896. Hartert E. A few additions to former Notes. Nov. Zool. III p. 255—6.
1896. Meyer A. B. und L. W. Wiglesworth. Bericht über die 5. bis 7. Vogelsammlung der Herren Dr. P. und Dr. F. Sarasin aus Celebes. Abh. u. Ber. Mus. Dresden 1896/7 Nr. 1. 17 pages. 4to.
1896. Meyer A. B. und L. W. Wiglesworth. Eine Vogelsammlung von Nordost-Celebes und den Inseln Peling und Banggai. Mit einer Karte. Abh. u. Ber. Mus. Dresden 1896/7 Nr. 2. 20 pages. 4to.
1896. Mivart St. G. A monograph of the Lories. I.III + 193 pages, 51 plates. 4to.
1896. Radde G. Besuch auf Buton und Süd-Celebes. Globus, Band 69 p. 151—5, 171—74. 4to.
1896. Sarasin F. Durchquerung von Südost-Celebes. Verh. Ges. Erdk. Berlin p. 339—57 (with notes on birds).
1897. Hartert E. Das Tierreich. 1. Lief. Aves; Podargidae, Caprimulgidae und Macropterygidae. 98 pages.
1897. Hartert E. On some necessary and some desirable changes of names . . . Nov. Zool. IV p. 11.
1897. Hartert E. Mr. William Doherty's Bird-Collections from Celebes. Nov. Zool. IV p. 153—66.
1897. Hartert E. Descriptions of . . . one new subspecies from Djampea. Nov. Zool. IV. p. 172.
1897. Sharpe R. B. [Dicaeum hosii n. sp.] Ibis p. 449.
1897. Blasius W. Neuer Beitrag zur Kenntniss der Vogelfauna von Celebes. (Nach Sammlungen des Herrn Dr. C. Platen von Rurukan in der Minahassa, Nord-Celebes. Mit einer Farbentafel. Festschrift der herz. Techn. Hochschule bei Gelegenheit der 69. Vers. D. Ntf. u. Aerzte in Braunschweig p. 275—395. (Too late for our letter-press.)
1898. Hartert E. List of a collection of Birds from the Island of Lirung or Salibabu, the largest of the Talaut Group. Nov. Zool. V p. 88—91. (Too late for our letter-press.)
1898. Rothschild W. [2 new species from Sula Mangoli.] Bull. Br. Orn. Club Nr. LI p. XXXIII—IV. (Too late for our letter-press.)
1898. Hartert E. [On Surniculus musschenbroeki from Macassar]. Nov. Zool. V p. 119, note. (Too late for our letter-press.)
1898. Hartert E. List of a collection of Birds made in the Sula Islands by William Doherty. Nov. Zool. V Nr. 2. (Too late for our letter-press.)

2. SEASONS AND WINDS IN THE EAST INDIAN ARCHIPELAGO.

See Maps III and IV.

The red colour on the maps denotes "*fine season*", the blue colour "*rainy season*", while purple distinguishes districts where it would be incorrect to term the period either wet or fine. The arrows fly with the wind. The short arrows distributed over the Archipelago show the persistence of the wind in the given direction (those grouped 5 mm apart denoting 50 per cent of the winds as coming more or less directly from the quarter indicated, those 4 mm apart denoting 60 per cent, and so on). The long arrows, *chiefly* outside the Archipelago, demonstrate the prevailing direction of the wind during the six months concerned, without reference to percentage.

It should be observed that our sketch-maps are more or less arbitrary and hypothetical, for data have not been accurately recorded from all parts, and others are hidden in papers in periodicals or in special works, and were not consulted, as lying too far from the aims of this book. We have been obliged to satisfy ourselves with an approximately correct picture of the winds and rains of the Archipelago, and this remark refers both to the arrows on our maps and to the colours. The general results and general points of view of our reasoning will not be greatly affected, even if some one should prove that there are faults here and there in our maps, which we are very ready to concede.

Among the many causes which effect the dispersal and the distribution of birds (cf. A. R. Wallace, Geogr. Distr. Anim. vol. I, chap. III, 1876), winds and seasons play an important part — the winds directly by carrying birds involuntarily to new lands, or in offering barriers to their wandering across certain zones; the seasons indirectly by their influence upon the abundance or scarcity of food, which forms the strongest of several motives for migration and local movements. In temperate and cold climes the alternation of the seasons, summer and winter, is, as it were, accompanied by a flow and ebb of vital energy in the vegetable kingdom expressed in the sprouting of foliage and the fall of the leaf of deciduous trees, and, at the approach of the cold season, insects, such as feed upon leaves and flowers, etc., hibernate or perish with the disappearance of their food, seeds and grain are buried or hidden under snow, molluscs and batrachians hibernate; consequently, most insectivorous, granivorous and other birds betake themselves at this season to warmer latitudes, the majority invading the countries and islands of the tropics, where specimens often fall to the gun or blow-pipe of the collector, sometimes to be named as new species or subspecies by learned ornithologists at home. Amongst the endemic birds of tropical countries, such as the East Indian Islands, periodic wanderings across the sea are very rare (though not unknown amongst certain Pigeons); the birds do not as a rule shift their quarters in an extensive manner, for they have not far to go to find the needed food, since one side of an island often has its dry season when the other has its wet one, and the highlands in general an entirely different climate from that of the districts near the sea-shore or the plains. Birds, therefore, need not cross the sea to find new feeding-grounds; it is probable however that local movements, depending upon the ripening of certain fruits or blossoming of flowers (see, for instance in the text, Meyer's observations on some small Parrots in Celebes, pp. 122, 150, 159) are common, and the food-supply seems to be regulated by the season. The seasons are determined in the East Indies by the monsoons, the monsoons themselves by the position of the sun over the greater masses of land.

Monsoons of the East Indian Archipelago. — In consequence of the superior power of the sun about the equator, the heated atmosphere there rises, and an indraught of the cooler air from the north and the south flows in to supply its place, taking the form of N.E. and S.E., instead of due North and South, winds, owing to the rotation of the earth. These are the Trade-winds, which blow with a general regularity from year's end to year's end over most of the Pacific and over parts of the other great Oceans. The return of the rising equatorial air through higher strata towards the north and south, its meeting in the upper atmosphere with high N.E. and S.E. Trade-winds blowing from the Poles, their stopping one another, piling up and descending to the globe about lat. 30°—35° N. and S., giving rise to a high barometer and zones of calms (cf. Maury, Phys. Geogr. Sea, 14th ed. 1869 p. 80), the starting from these belts of the true Trade-winds and of low winds returning towards the Poles, are questions upon which the meteorologist may be consulted, but which need not further concern the weather-chart of the Archipelago. Here, there are four principal winds, two north of the equator and two south of it, which blow alternately throughout the year generally speaking without much interruption, except at the periods of the shifting of the winds, the N.E. and S.E., which blow each for about half a year, when they are displaced by other winds, the true Monsoons.

The South-west Monsoon. — When the sun in the Northern Hemisphere draws towards the Tropic of Cancer (our summer), the plains and table-lands of Asia become greatly heated[1]), and the N.E. Trade-wind, instead of continuing

---

[1]) "L'été de Pekin, qui est sous le quarantième degré de latitude, donne une moyenne de chaleurs égale à celle du Caire (30° lat.), et son hiver, une moyenne de froids égale de celle d'Upsal (60° lat.)!" (David, N. Arch. du Mus. 2ᵉ sér., 1885, VIII, 5).

to blow towards the parts of the Indian Ocean and islands about the equator, now no longer the hottest quarter, stops, and, turning back, commences to move towards the heated Continent as the S.W. Monsoon. This wind is general from about April to October in the Indian Ocean and the East Indian Archipelago north of the equator, sometimes extending as far east as the Marianne Islands and as far north as the south of Japan.

The South-east Monsoon. — During the time that the sun continues north of the equator and the South-west Monsoon is blowing, the S.E. Trade-wind in general has free course in the Southern Tropics, and from April to October is the wind of the East Indian Archipelago south of the equator, where it is commonly spoken of as the South-east Monsoon. The general direction of it and of the S.W. Monsoon of the Northern East Indies is shown on Map III.

The North-east Monsoon. — When the sun, at the autumnal equinox, passes into the southern hemisphere, the N.E. Trade-wind reasserts itself in the parts where it has had to give way to the S.W. Monsoon, and, displacing that wind, it blows from some time in October, to April. It is then the prevailing wind of the Archipelago north of the equator.

The North-west Monsoon. — At this time of the year, in the zone of calms about the equator between the N.E. and S.E. Trade-winds, a westerly monsoon, unaccountable to meteorologists at the time of Maury's celebrated work (1869), sets off and blows in a narrow, curved belt across nearly the whole width of the great oceans; much, one may suppose, as the return current flows back towards the buttresses of the bridge under which a swift river passes. Out of this belt there arises the N.W. Monsoon of Australia and the East Indies south of the equator. Corresponding with the S.W. Monsoon of the north of the equator, the N.W. Monsoon evidently is originated by the heating of the interior of Australia and New Guinea during the southern summer, and it is the prevailing wind throughout the Archipelago from the equator southwards during the period October to April. See Map IV.

The Rains: north of the equator. — Between April and October the S.W. Monsoon, arising in the Indian Ocean about the equator or to the south of it near Sumatra, reaches the northern half of that island saturated with moisture and produces the rainy season there and along the west coast of the Malay Peninsula. The mountain-ranges running through the middle of that peninsula probably hold back the clouds, for, as the wind passes over the east coast, it is the fine season there, though there are occasional showers. In the Gulf of Siam the wind again commences to take up moisture; here, during the cruise of H. M. S. "Saracen" between 1855—58 a rough sea was experienced at this time of year, and, on the opposite coast "strong breezes with much rain and occasionally a fresh gale" (Direct. Ind. Arch. 1870 p. 17). Cambodia and the neighbouring parts of Siam have then their wet season. Further east the wind, passing over the middle of Sumatra and gathering vapour afresh in the south part of the

China Sea, brings the season of most rain to Borneo north of the equator, though the climate of that country is moist at all times of the year. The Sooloo Islands and the southern and western parts of the Philippines have now their rainy season. See Map III.

During the other half-year, October to April, the N.E. Monsoon is operative and reverses the work of the S.W. Monsoon. Laden with moisture taken up in the Pacific it deposits much of it on the northern and eastern parts of the Philippines, and, on arriving in Borneo north of the equator, does not bring with it so much rain as the S.W. Monsoon. In the Siamese Peninsula, Annam, facing the wind, now has plentiful rains, but on the opposite side Siam and Cambodia have their finest time of year: "at this time the sky is frequently unclouded for a week together", but the wind again becomes saturated in passing over the Gulf of Siam, and on the opposite coast along the eastern shore of the Malay Peninsula "the weather is wet and stormy".

The Rains: south of the equator. — In the parts of the Archipelago lying south of the equator the S.E. Trade-wind — here called the S.E. Monsoon — is the prevailing wind from April till October. Blowing from out of the arid deserts of Australia, it leaves the north coast of that country hot and dry, and has not time to take up much vapour before it reaches Timor and the chain of islands stretching between there and Java. Consequently these islands have now their dry season, and the vegetation of Timor — which country is probably the driest of all — is said by Wallace to have an aspect strikingly similar to the Australian. Before reaching the west end of Java the S.E. Monsoon, having passed over a wide stretch of ocean, has gathered moisture, and this part of the island now receives ample quantities of rain, though not in such abundance as is the case there during the returning N.W. Monsoon, which is the bad season, and the vegetation here is consequently most luxuriant. The same holds good even in a still greater degree for the west coast of Sumatra south of the equator, where there is still less difference between the two seasons. During this Monsoon it is also the fine season in South Borneo and in almost the whole of Celebes; but further east this is now no longer the case. The winds that reach the shores of the Gulf of Tomaiki in East Celebes, the southern coasts of Buru and Ceram, the Islands of Kei and Aru, the S. W. Coast of New Guinea, etc., pass from the South Pacific Ocean either through the Torres Straits or over the Cape York Peninsula across the Gulf of Carpentaria; thus, there are no broad lands in the way to receive their moisture until they reach the above-mentioned territories, which now have their rainy season. The high mountain-chains of New Guinea and the ranges which intersect the islands of Ceram and Buru serve to retain the clouds brought up by this Monsoon, and here again occurs the phenomenon of the rainy season on the south side of an island and the fine season on the north.

These conditions are reversed during the counter Monsoon, the N.W. Monsoon, which predominates from October to April in the Archipelago south of the equator, and the lands, which had their dry season before, now get their wet one, and those, which then had their rainy season, now have their fine one. Though easy to be understood, this alternation of the seasons is often not a little striking. See Map IV.

Professor van der Stok of Batavia recognises 4 different types of Monsoons in the East India Archipelago, about which he has most obligingly sent us (in lit.) the following particulars.

"There are various types of Monsoons in the Indian Archipelago:

"First, the perfectly regular, the S. E. Trade-wind, or the S. E. Monsoon blowing out of Australia, which prevails in the southern parts of the Archipelago from April till October, characterized by dry weather (instances: Java, Bima in the Lesser Sunda Islands, Macassar, S. Celebes, and Banjermassin, S. Borneo); while from October to April the heavily saturated West Monsoon is in force.

"A second type is found, as your map[1] also shows accurately, in South and Middle Sumatra, especially on the west coast, and in the middle portion of Borneo, where all through the year a tolerably equal quantity of rain descends (instances: Padang, Siboga, Sintang, Singapore). For these conditions also your map is suitable.

"It is otherwise with the third type, of which examples in North Sumatra and North Borneo are recorded. Here it is not possible to divide the year into two halves in such a manner that the one contains the dry, the other the wet season. (See below p. *29* under Borneo.)

"Finally a fourth type is found in the eastern parts of the Archipelago where, as in Amboina, Saparua, etc., the Monsoon-periods — at least on the south side of the Island — are exactly reversed [as compared with the first type], as your map also shows. As a whole Celebes also lies under the southern Monsoon-division, for in North Celebes there is still to be found only a trace of the February minimum whereby North Sumatra and North Borneo are characterized".

Change of climate at different altitudes. It is probably always the case that the highlands of tropical islands have a very different climate from the coasts and plains. In April, 1871, when the fine season had begun at Manado, Meyer could not start for the mountains of the Minahassa, as the rainy season was still going on there. The following table shows this:

	Average of years	April mm	May mm	June mm	July mm
Masarang (highlands)	15	256	249	228	135
Manado (coast)	17	205	167	179	125

In Java where meteorology has been much more thoroughly studied than elsewhere in the Archipelago, Dr. J. J. de Hollander (Handl. Land- en

---

[1] An original MS. map, since revised and modified.

Volkenk, Ned. Oost-Ind. 1882, 4th ed. vol. I pp. 178—181) describes four different climates peculiar to different altitudes — climates having not only differences of temperature, but also of winds and rains. As the existence of such have great influence upon bird-life we translate his remarks:

"The climatic conditions of Java are very varied, and especially dependent upon the altitude, as well as upon the Monsoons. In respect of this the surface of Java may be divided into four zones or belts: the First, or Hot Zone extending from the sea-level up to 2000 feet; the Second or Temperate Zone from 2000 to 4500 feet; the Third or Cool Zone from 4500 to 7500 feet; and the Fourth or Cold Zone from 7500 to 10 000 feet and upwards.

"In the First or Hot Zone the mean temperature is 29.7° C. (85.5° F., 23.8° R.) on the strand and 24.2° C. (75.5° F., 19.3° R.) on the upper boundary. At Batavia the greatest heat is experienced in April, the least in January; the nights and mornings are, however, coolest from June to August. In this Zone the atmosphere is very damp. This dampness naturally increases as one descends from a higher level to the strand, so that the atmosphere at Batavia contains a mean amount of 84 per cent of vapour, in other words, a cubic metre of air holds in suspension 20.25 grammes of water-vapour, whereas at the sea-level the air would be completely saturated with 26.39 grammes of vapour to the cubic metre. The damp is greatest in the months of January and February, and least in August. Near the sea-shore the air is filled with pernicious vapours developed in great quantities by the heat from the morasses, where many plant and animal remains are always rotting, ... these exhalations, however, do not appear to rise to a height of more than 900 ft. The Monsoons operate very regularly. The rainy Monsoon prevails continuously from November to March, the dry Monsoon from May to September or October; the shifting of the winds takes place in April and October or November. The most rain falls in December, January, February and March; and, although at this time brighter and rainless days sometimes occur, the sky is usually heavily clouded over, and the rain comes down in copious streams, sometimes for days in succession, causing great floods. ... In the East Monsoon the dryness is most marked in July and August, when — save for the daily alternating land- and sea-breeze — hardly any wind is perceptible. The moisture in the atmosphere is then deposited as heavy dew, to be taken up again with the warmth of the sun in the morning, forming itself in the upper air into clouds which are driven landwards by the sea-breeze (felt up to 2500 ft. above sea-level) and become heaped up in the Second Zone. They sometimes disburden themselves in thunderstorms in the afternoon, especially in mountainous districts, such as Buitenzorg, where storms accompanied with heavy showers are of almost daily occurrence.

"In the Second or Temperate Zone the mean temperature is 23.6° C. (74.5° F., 18.9° R.) at the lower and 18.7° C. (65.7° F., 15° R.) at the upper boundary, with a very marked difference between the warmth of the day and of the night, especially on the table-lands. The warmth of the day itself is also subject to more or less variation, according as the moisture out of the First Zone rises sooner or later and in greater or less quantity, which, forming into clouds, intercepts the rays of the sun. Owing to the lesser heat and to the ground also being less damp, less vapour rises here than in the First Zone and consequently the atmosphere in general contains less moisture, the mean quantity of vapour being 15·7 grammes to the cubic metre, while 21·15 grm. at the lower and 16·88 grm. at the upper boundary are necessary for saturation. The degree of moisture also varies much in different localities; it is much greater over the wet rice-fields ('sawahs') and dense woods than over stretches of grass or 'alang-alang', or over plantations of shrubs (tea, coffee). Also the masses of mist driven up from lower territories by the sea-breeze produce great differences. These mists condense here more quickly than in the warmer temperature of the First Zone, and often very heavy storms and showers suddenly result. As to this division it should be remarked that in the west part of Java the atmosphere is much damper than in the eastern portions of the island. The West Monsoon in the highest parts of this Zone is already felt in less force, and consequently the difference between the seasons (the wet and the dry) is here much less marked than in the lower districts, and even when the West Monsoon is in full force in the First Zone, the East Monsoon (the Trade-wind) often blows here for days in succession.

"In the Third or Cool Zone the mean temperature is 18·7° C. (65·7° F., 14·9° R.) at the lower and 13° C. (55·4° F., 10·4° R.) at the upper boundary. The difference of temperature between day and night is here much less marked than in the First and Second Zones; the plateau of Mt. Diëng (6300 ft.) presents an exception to this rule, the difference here being so great that the dew on bright nights sometimes freezes into rime. In this Zone the air, which, in consequence of the diminished warmth cannot carry so great a quantity of water, is entirely saturated with water-vapour (16·88 grm. to the cubic metre at the lower and 11·60 grm. at the upper boundary). The mists rising from the lower regions condense here to such an extent that this Zone might literally be called the Zone of Clouds. They sometimes begin to form as early as nine o'clock in the morning, especially on declivities covered with forest; from 11 or 12 till 2 or 3 o'clock everything is covered with thick fog, which discharges itself — often simultaneously in different places — in storms of thunder and rain, after which alone the sun makes its way again through the clouds. But when the clouds are not broken up in this manner, so thick a fog covers everything for the rest of the day that it is impossible to distinguish an object at twenty-five paces, and it is not till after sun-down that the fog settles on the earth as dew. This, however, is more particularly applicable to the lower parts of this Zone, where the clouds gather most thickly; they seldom ascend to the upper parts, and then in less quantity, in consequence of which the showers there are rarer and less heavy. The influence of the West Monsoon is here almost entirely imperceptible. The S.E. wind in general blows continuously, though, when the rainy season is going on in the lower regions, it may be replaced for a few days only by the West wind or by complete calm, the latter being the condition which nearly always reigns at night. The rain also is not heavier or more continuous here during the time of the West Monsoon than in the other period, but falls in tolerably equal quantity almost daily throughout the year.

"In the Fourth or Cold Zone (7500 to 10000 ft. and upwards) the mean temperature is $13°$ C. ($55.4°$ F., $10.4°$ R.) at the lower boundary, and $8°$ C. ($46.4°$ F., $6.4°$R.) at a height of 10000 ft. above the sea-level. The difference in warmth of day and night is usually not very great; this is to be ascribed to the comparatively small extent of solid ground for the sun to play upon. On the highest mountain-tops, in places where there is no shelter from bushes and other objects, the temperature sometimes descends to the freezing-point, so that water, removed from the ground which contains warmth, receives a coating of ice in the open air and the grass is covered with rime. The moisture of the atmosphere is much less here than in the lower Zones, not exceeding 11.60 grm. at the lower limit and 8.70 grm. at 10000 ft. The air is consequently more rarified, purer, more transparent and fine; sound does not travel well, and breathing is more difficult. The few mists that rise up so high fail to form into clouds; rain is consequently very rare, and then only occurs as a fine drizzle. During great calms, however, it happens that the mists from the lower districts ascend right up into this Zone and then, becoming at once solidified by the ice-cold atmosphere, fall as hail. An East wind prevails uninterruptedly on these heights, though usually falling to a calm at night. Only very rarely, when the West wind is blowing strongly in the lower regions and is driven up the declivities of the mountains, it makes itself felt in the undermost parts of this Zone, bringing mists and fogs with it. At 10000 ft. a West wind is unknown. Obviously there can be no question of a rainy season in this Zone."

Although we have not particulars of the climatic variation at different altitudes in other parts of the Archipelago, it is not to be expected that the high mountains south of the equator will present any great differences from those of Java; those found north of the equator will be affected by other winds.

It is now proposed to examine the different parts of the Archipelago in greater detail.

Celebes. — The greater part of the country, being south of the equator, is under the influence of the Monsoons of the southern hemisphere. Over the Northern Peninsula, which lies just north of the equator, the S.E. Monsoon of April to October seems to be deflected by the S.W. Monsoon of the north of the equator, and from October to April the N.E. Trade-wind of the north similarly deflected by the N.W. Monsoon of the south; in consequence of this fairly due South and North winds respectively figure rather prominently here on Dr. van der Stok's monthly charts[1]). The fine season over most of the island is during the S.E. Monsoon between April and the beginning of November, the rainy season from November till March, when North-west or North winds are predominant. To this rule there are many exceptions, sometimes due to location, sometimes to shelter from the high mountains.

Touching the Minahassa, Graafland writes (De Minahassa 1867 I, 1): "The changing of the Monsoons here takes place almost imperceptibly. One passes over from the East to the West Monsoon without noticing it otherwise than by the more or less plentiful showers and thunderstorms; and even this is not regular. There are years in which the West Monsoon brings so little rain that poor rice-harvests are the sensible result, while there are again other years when too much rain causes the rotting of the crop. The West or rainy season is calculated to be from the middle of October to the middle of April, but this is not at all certain". Dr. Riedel writes (in lit.) that during the N.W. Monsoon the sea is rough on the north coast of the Peninsula, which faces the wind, while on the south coast the wind is less heavy and blows out to sea. The plantations are harvested everywhere at the same time, and the rice is sown in October—November. There is, however, as Dr. van der Stok's tables show, a marked difference in rainfall between the north and south coasts of the Peninsula; when the N.W. Monsoon is blowing[1]) Manado and Kwandang on the north receive two or three times as much rain as Kema and Gorontalo some 20 miles distant on the south. The interior of the country is mountainous, and, as is clear from Dr. de Hollander's remarks on Java, the N.W. Monsoon is a superficial, somewhat shallow wind, and it is doubtless held back and deprived of its moisture to a great extent by the hills. During the N.W. Monsoon the shipping is carried on at Kema, while during the S.E. Monsoon everything in this way is done at the more important settlement of Manado. Meyer arrived at Manado in November, 1870, having been misinformed by Mr. Wallace that October is the beginning of the fine season for this region. Travellers should go there in April and the following months, though on the south coast of the Minahassa, at Kema for instance, the weather is much better and even fine in the rainy season. September is the driest month of the year.

The western portion of the N. Peninsula seems to be exposed at most times of the year to N. and N.W. and S.W. winds blowing out of the Celebes Sea, and Tontoli at the N.W. angle of the Peninsula cannot be said to have a rainy season, but has a tolerably equal rainfall throughout the year.

The following tables, extracted from the "Regenwaarnemingen in Nederlandsch-Indië", 1895, show the differences in rainfall and rainy days at different places in the N. Peninsula:

---

[1]) See: van der Stok: Wind and Weather, Currents, Tides and Tidal Streams in the East Indian Archipelago (Batavia, 1897. Broad folio).
[2]) The "Directory for the Indian Archipelago" 1870 p. 22, states that on that part of the island situated N. of the equator the N.E. Monsoon in October replaces the S.W., wrongly adding that it makes the fine season.

Average monthly rainfall and number of rainy days in the North Peninsula.

	Situation	Altitude in metres	Number of years	Rain	January	February	March	April	May	June	July	August	September	October	November	December	Total
Manado	N. Coast	4	17	mm	478	337	271	205	167	179	125	121	82	125	295	418	2713
				Rainy days	22	18	17	15	15	16	10	12	8	11	16	21	181
Masarang	Interior	900	15	mm	235	215	224	256	249	228	135	121	108	182	239	265	2448
				Rainy days	19	16	16	19	19	18	13	12	8	14	17	17	188
Kema	S. Coast	—	15	mm	145	139	149	160	124	156	89	88	60	75	160	150	1495
				Rainy days	12	11	11	14	10	13	9	9	5	7	13	11	125
Kelo-Londej	Interior	892	8	mm	278	152	268	362	287	244	194	194	107	185	283	223	2777
				Rainy days	23	16	21	23	22	22	18	19	10	15	19	19	227
Kwandang	N. Coast	—	15	mm	234	247	181	167	198	191	127	144	81	170	263	280	2283
				Rainy days	13	12	10	10	13	11	8	9	6	10	13	14	129
Gorontalo	S. Coast	—	15	mm	103	86	95	138	106	145	88	118	49	70	117	119	1234
				Rainy days	9	7	8	10	10	11	8	9	4	6	9	11	102
Limbotto	near S. Coast	—	15	mm	137	97	132	179	129	161	88	90	47	79	141	188	1468
				Rainy days	10	8	11	12	11	9	7	9	4	8	12	13	112
Tontoli	N. W. Coast	2	15	mm	263	200	178	130	197	310	243	251	185	226	177	246	2696
				Rainy days	14	12	11	10	12	17	13	15	11	16	12	14	157

In the South Peninsula the rainy and dry seasons are generally much more strongly contrasted than in N. Celebes. On the west side of this Peninsula the N.W. Monsoon brings great quantities of wet and the S.E. Monsoon for some months very fine weather. On the opposite east coast the converse of this is the rule. This is well shown by the rainfall at Balang Nipa which lies on the east coast of the Peninsula in about the same latitude as Macassar some 60 miles distant on the west coast: May, June and July are among the fine months at Macassar, while great quantities of rain fall at this time at Balang Nipa; but December, January, February and March are fine months at Balang Nipa, during which Macassar receives deluges of rain. Bonthain on the south coast of this Peninsula is sheltered from the N.W. Monsoon by the great Bonthain mass of mountains, and its seasons correspond with those of Balang Nipa, except that much less rain falls. Dr. van der Stok has most obligingly sent us tables showing the direction of the winds at Bonthain in the years 1886, 1887 and 1888. These are chiefly westerly from December to April, veering from S.W. to W.N.W. and generally changing a point or two in the course of each day; from May till the end of November the general direction is east, N.E. to S.S.E., with similar changes during the course of the day.

The following are the tables in the "Regenwaarnemingen in Ned. Ind." for the South Peninsula of Celebes (see next page).

One of the only two injurious winds known in the Dutch East Indies is found on the west coast of Celebes between Maros and Mandar and called the "Barubu". It blows yearly during the months of July, August and the beginning of September from the E. N. E. and extends about a geographical mile seawards. It causes a difficulty in breathing, dries up the lips and the throat, bringing about inflammation of the eyes and often long-lasting fevers (de Hollander, 1882, I, 86). The botanist Teijsmann experienced this wind in the South Peninsula at Pankadjene, Tjampa (6th September) and Bantimurang (26th September). He describes it as a wind which covers everything with fine dust, as very unpleasant, and at sea often very dangerous (N. T. Ned. Ind. 1879, 60, 78). A similar obnoxious wind is the "Anging bolo" of Bima, Sumbawa.

The temperature of Celebes is not high, seldom exceeding 32° C. (26° R., 90° F.). The tables show that August and September are the driest months at nearly all spots where the rainfall has been observed. On the whole, as a glance at our maps III and IV will show, Celebes has the same seasons as the islands lying south of the equator, as indeed should be the case from its geographical position; but at a few spots both in North and South Celebes traces of the minimum of rainfall in February, which is characteristic of N. Borneo and N. Sumatra, may be noticed.

Sangi. — The rainy season seems to set in after October. Dr. Platen, writing from Great Sangi in January, 1887, speaks of having been confined to the house for weeks by ceaseless rain falling in the N. W. Monsoon. (Gefied. Welt, 1887, 263).

Average monthly rainfall and number of rainy days in the South Peninsula.

	Situation	Altitude in metres	Number of years	Rain	January	February	March	April	May	June	July	August	September	October	November	December	Total
Segeri	near W. Coast	3	15	mm	719	463	398	244	216	137	93	28	53	125	387	730	3593
				Rainy days	23	18	16	15	13	10	7	3	3	7	15	21	151
Pankadjene	near W. Coast	3	17	mm	876	548	420	235	174	165	80	22	35	110	328	760	3753
				Rainy days	23	21	17	14	11	9	6	3	3	7	16	22	152
Tjamba	Interior	300	15	mm	516	346	261	207	177	168	124	53	31	82	220	378	2565
				Rainy days	21	18	17	16	16	14	10	5	3	7	14	20	161
Macassar	W. Coast	2	17	mm	744	544	422	130	101	120	51	13	18	47	189	663	3042
				Rainy days	25	21	19	12	9	9	6	3	2	5	12	23	146
Alloe	S. Coast	3	14	mm	457	402	242	95	90	97	50	12	12	54	165	452	2137
				Rainy days	17	16	13	8	8	8	4	2	1	3	10	18	108
Bonthain	S. Coast	1	15	mm	145	97	110	125	179	197	159	53	31	41	75	93	1286
				Rainy days	14	11	12	12	14	14	10	6	2	4	8	13	120
Salayer	Salayer Id., W. Coast	2	15	mm	239	120	159	198	164	167	90	10	16	31	148	240	1572
				Rainy days	12	10	10	13	11	11	6	2	1	3	8	13	100
Kadjang	E. Coast	8	15	mm	147	163	168	261	541	395	312	76	46	51	98	175	2453
				Rainy days	14	14	13	16	19	17	14	5	3	4	6	13	138
Bikeroe	Interior	245	14	mm	321	276	306	391	407	18	372	122	58	60	186	249	3102
				Rainy days	17	18	18	16	20	18	16	9	4	5	11	18	170
Balang Nipa	E. Coast	3	17	mm	199	118	130	299	161	441	382	134	52	105	81	119	2451
				Rainy days	12	11	12	17	20	20	17	11	5	6	7	13	151

Sooloo Archipelago and Celebes Sea. — According to the Sailing Directory, easterly winds are experienced here from October till May; from May onwards westerly winds are accompanied by "rains, squalls and tempests", especially in July and August (p. 22). From van der Stok's charts the direction of the more prevalent winds is seen to be southerly from June till October, and northerly or north-easterly from November till April.

Philippine Islands. — The windward shores of the islands are in general the wet ones; thus, on the north and east coasts the most abundant rainfall takes place in the half-year from October till March when the N. E. Tradewind is in force, while the territories to the South and West, which lie exposed to the S. W. Monsoon, have their rainy season during the other half-year. At Manila the winds vary much in the course of each day, owing to land- and sea-breezes. In July, August and September S. W. winds prevail, in November and December N. E. winds, becoming more S. E. in March and April. The town, being on the west coast, has its rainy season during the summer months of the northern hemisphere (cf. Obs. meteor. de Manila de la Comp. de Jésus, 1870—73, 1876). The following observations were obtained by Meyer from Iloilo on the Island of Panay. The S. W. Monsoon commences in June, exceptionally in May, but it does not then blow uninterruptedly. The rain falls in varying quantities, being sometimes heavy, at other times not so; and it may last — though this is the exception — for a month continuously; fourteen days in succession may be taken as the rule, judging from four or five occasions when this was experienced. June—September generally afford disturbed weather; most of August fine. The N. E. Monsoon commences about the middle of October, and there is occasional rain from October to December, in January only very little; from the end of January till March it is dry. In April and May, when the Monsoons are changing, there-are occasional thunderstorms.

In Palawan the Monsoons are liable to much interruption, and there appears to be no rainy season in particular (Sail. Direct. 1870 p. 23).

Borneo. — "Owing to the great extent of alluvial ground with which Borneo is nearly everywhere surrounded, to the numerous water-courses irrigating the land in every direction, and to the vapours arising from the dense forest with which most parts of the island are covered the atmosphere is always damp." Land- and sea-breezes are felt far inland. The temperature is more moderate than might be expected; at Pontianak in the west almost on the equator, at sunrise 24·5° C. (19·5° R., 76° F.), at midday not more than 33·3° C. (26·7° R., 92° F.), mean 27·8° C. (22·2° R., 82° F.). As the island is divided by the equator, the northern and southern halves are subjected to the influence of different monsoons.

South of the equator the character of the season is well determined by the particular monsoon in force. The S. E. Monsoon prevails, roughly speaking, in the time of our summer, and the finest months of the year, as is shown by the reports sent in to Dr. van der Stok from six stations in Central and S. E. Borneo, are July, August and September. June is also a fine month, though somewhat more rain falls than in October, when the returning N. W. Monsoon, the true rainy wind, can hardly have commenced to blow home. The wettest months of the year are from November to March, inclusive, when the N. W. Monsoon is in force. It has been stated that while Borneo south of the equator is having its rainy season under this wind, the parts of the island north of the equator, over which the N. E. Monsoon is blowing, are fine; but this is not altogether true. The great western projection of the island upon which Sarawak (near its north coast), Sinkawang (west coast), Pontianak (south coast) are situated, receives its heaviest amount of wet at the same time as South Borneo. This is also the case at Sintang far inland a little north of the equator. The N. E. Trade-wind does not blow in force right down to the equator, and, when the N. W. Monsoon is going on in the islands south of the equator, an indraught of the deflected N. E. wind must needs take place. In the South China Sea in the islands of Bintang, etc. this deflected wind is felt for half the year as a prevailing North wind (see, p. *32*), and, no doubt, as such, or even as a N. W. wind, it passes over Borneo a little north of the equator, bringing with it great quantities of moisture from the China Sea.

In Northern Borneo the seasons vary much according to locality. Dr. de Hollander speaks of the S. W. (April—October) as the rainy monsoon, and the N. E. (October—April) as the fine one, and so also the Sailing Directory of the Indian Archipelago; but there is much to take exception at in these statements. In an article on the climate of British North Borneo by Mr. Robert H. Scott (Journ. R. Met. Soc. 1889, pp. 206—219), to which Dr. van der Stok has kindly called our attention, it is stated that "the true wet season occurs at Sandakan (on the N. E. coast) in the N. E. Monsoon, and includes the months of November, December and January, and generally part of October or February or both . . . The true dry season immediately follows this true wet season, and includes March and April, and generally the whole of May and part of February . . . This true dry season is followed by a period of moderate rainfall, commencing generally about June. The first month or six weeks of this period almost deserves to be called a second wet season, and the rest of the period up to the commencement of the true wet season might be described as the second dry season". With these variations, the actual figures taken at Sandakan and Kudat show that by far the larger half of the total rainfall is deposited during the N. E. Monsoon, which is not to be wondered at, seeing that these places then present a windward shore to the Monsoon passing over the Sooloo Sea.

At Labuan, the average for 11 years shows that the first four months of the year — the closing ones of the N. E. Monsoon — are the driest.

The following tables are taken from those of Dr. van der Stok (Regenwaarnemingen in Ned. Ind. 1895, 416, 417), and Scott, *l. c.* (the inches of the latter converted to mm).

Introduction: Seasons and Winds.

Average monthly rainfall in Borneo.

Locality	Situation	Nr. of years	January	February	March	April	May	June	July	August	September	October	November	December	Total
South of the equator															
Muarah Teweh	114°43′ E × 0°55′ S	12	287	295	312	321	249	207	148	161	159	223	313	296	2971
Buntok	114°30′ E × 1°15′ S	15	308	272	345	311	263	163	108	123	118	186	331	354	2882
Ammuntai	115°10′ E × 2°15′ S	17	310	287	363	225	210	154	113	85	79	124	250	325	2525
Barabei	115°16′ E × 2°17′ S	16	275	305	300	258	241	205	111	127	88	160	289	350	2709
Pengaron	115°15′ E × 3°15′ S	17	334	289	345	217	242	178	111	109	80	149	242	327	2623
Banjermassing	114°35′ E × 3°19′ S	17	315	303	313	221	173	170	112	113	87	144	221	315	2457
Pontianak	109°20′ E × 0°1′ S	17	262	179	250	230	248	234	171	237	206	447	398	369	3231
North of the equator															
Sinkawang	109°00′ E × 0°55′ N	16	379	211	212	233	213	204	166	301	195	432	438	494	3478
Sintang	111°32′ E × 0°7′ N	17	347	253	343	314	292	272	229	327	261	403	362	364	3797
Kuching, Sarawak	110°8′ E × 1°28′ N	5	690	610	256	255	231	222	121	225	198	252	345	653	4058
Labuan	115°20′ E × 5°30′ N	11	221	132	184	213	376	353	302	385	275	371	355	352	3519
Limbuak	117°5′ E × 7°15′ N	1	285	79	610	188	218	138	183	562	117	457	361	450	3648
Kudat	116°52′ E × 6°54′ N	3	407	212	326	58	111	126	174	143	126	242	364	603	2942
Sandakan	118°12′ E × 5°49′ N	10	496	235	194	109	143	221	145	188	258	244	398	514	3145

As a general conclusion it may be stated that Borneo south of the equator has its fine season during the months of our summer and its wet one in those of our winter. A little north of the equator the conditions are much the same, the differences being that the rainfall is more copious, and that the wet and fine seasons commence a little earlier in the year. August, when the S.W. Monsoon of Asia is in full force, appears to be a very wet month here, as it is also at Labuan, Mempakol, Gaya on the N.W. coast of the island, and at Banguay Island off the north point of Borneo. On the N.W. coast the figures from four stations show a strongly marked minimum rainfall in February; the fine period appears to last for about three months only, January—April. On the North and N.E. coast the true wet season takes place in the N.E. Monsoon and the true dry season about the period of the shifting of the winds, February—May.

Sumatra. — Dr. van der Stok shows that very varying conditions prevail upon the different coasts of this great island.

In the Malacca Straits land- and sea-breezes are general, neutralizing the effect of the Monsoons. The wettest months along this part of the coast of Sumatra are from October to December, the rains being apparently brought up by the N.E. Monsoon out of the China Sea and the Straits. The dry months are February and March, and also June and July, there being here, as in North Borneo, a second rainy and fine period.

At the northern end of Sumatra the S.W. Monsoon, is much more marked, being felt from May till October, and bringing the rains with it from the Indian Ocean. February and March, when the E. Monsoon is blowing, are the finest months.

Along the N.W. coast down to the equator it is hard to speak of any rainy — or one might better say of any fine — season. South of the equator down to the Straits of Sunda the wettest months are from September to December, and the driest May to August, with a reduced rainfall in February.

In South and S.E. Sumatra the Monsoons are well marked, the dry season being produced by the S.E. Monsoon from April or May till September, the wet accompanying the northerly and westerly winds which prevail from November to March.

The Bintang, Lingga, Karimon, Timbulan, Anambas, Natuna, and Serasan Islands: groups of small islands in the South China Sea between Malacca and Bangka and Borneo. — These islands receive the rains of both Monsoons and are very wet almost the whole year, the greatest number of fine days occurring in January, February and March. The climate is not considered unhealthy.

Banka, Billiton and the Straits of Karimata and Gaspar. — Here the climatic conditions are very changeable, varying on the land at different altitudes. The greatest amount of wet is brought up by the N.W. Monsoon in November, December and January; there is a minimum of rainfall in February, but the driest months are July, August and September. In Banka the temperature on the coast is given as $21°-24°$ C. ($17°-19°$ R., $70°-75°$ F.) during the night and morning, rising sometimes to 32°—35° C. at midday. In the interior it is less hot, the nights being even cold and damp. In Billiton the damp atmosphere is sometimes very oppressive, although the temperature in the morning and evening is 22°—23.5° C., and rarely more than 29° C. at midday; the nights very cool (de Hollander, t. c, 812, 828; van der Stok, Wind and Weather).

Java. — The chief meteorologic conditions of this island as recorded by de Hollander have been already given (pp. *22—24*). Bad weather is encountered during the N.W. Monsoon (October to April); the fine season accompanies the S. E. Monsoon in the months of our summer. On the south coast much wet is also brought up by the S.E. Monsoon from the Indian Ocean, particularly towards the western parts of this coast, but the true rainy season here as elsewhere is during the N.W. Monsoon. It has been said that bad weather marks the shifting of the Monsoons; there set in "wild storms from the W. and N.W."; "storms of wind and rain beneath a clouded sky alternate with severe gales and heavy winds" (Jansen in Maury's Phys. Geogr. Sea, 14th ed., 380); but the extensive observations of van der Stok lead him to the opposite conclusion — that "the condition of the sea is at its best when the Monsoons turn, i. e. in March, April, and November" (op. cit. p. 57).

The Lesser Sunda Islands (Bali, Lombok, Sumbawa, Flores, Sumba, Timor, Rotti, Timorlaut, and the intermediate smaller islands). — The wet and the dry seasons are here very strongly contrasted, especially in Timor. In this island hardly a drop of rain falls during the five months June—October, while an abundance comes down in December, January and February. The rivers are said to be then overstreaming, but during the S. E. Monsoon many are dried up, and the thermometer then rises to 52° C. in the sun and 35° C. in the shade. A similar drought in summer is found at least on the north coast of the other islands (Bali, Sumbawa). Flores is subject to manifold and sudden changes in the atmosphere, making it very unhealthy. An injurious wind, the "Anging bolo", occurs, as mentioned above (p. *27*), in Sumbawa where the climate is considered unhealthy. The Lesser Sunda Islands as a whole receive far less rain than Java, Celebes or the Moluccas, and only about one-half or one-third the amount of that which falls in Borneo, and ornithologists should not neglect to make studies of possible climatic variation among allied species of birds in these regions, such as have been made on certain birds in North America by Allen (see, below p. *58*). It may be that the climate has had something to do with "Wallace's Line" as far as it goes (see, below pp. *81—89*), for not all animals and plants can exist indifferently in a wet climate like that of Borneo and a dry one like that of the Lesser Sundas. A general similarity between Timor and Australia has been noticed, and it should not be forgotten that the S.E. Monsoon, which is productive of the drought in the Lesser Sundas blows out of the arid deserts of Australia, and it may bring many things directly with it, just as the returning N.W. Monsoon may carry to Australia any thing that is capable of sustaining a voyage through the air.

The Halmahera Group. These islands lying under or near the equator are chiefly under the influence of northerly winds from December to April and southerly ones from June to October, and as they are therefore not sheltered by any great land-masses they receive much rain with both winds. According to Dr. van der Stok there is "a principal maximum in June and July and a secondary maximum in January", but the rainfall seems to be chiefly determined by the position of the place concerned — whether it lies on the windward or lee shore of the island with sheltering hills behind. The climate is stated to be healthy. In Ternate the thermometer seldom rises higher than 30° C. (23.5° R., 85° F.); in Halmahera the mean temperature on the coast is about 30° C. at midday and 23° C. at night.

Buru and Ceram. — In consequence of the high mountain-chains which intersect these islands from west to east, a wall is presented to the alternating N. W. and S. E. Monsoons, so that, when the former is blowing, the northern or windward sides of the islands have their rainy season, while the southern sides, being sheltered by the mountains, are fine. This wind dominates from December to March. From May or June till October, when the S. E. Monsoon is blowing, the previous state of things is reversed; the southern parts now get their rainy and the northern their fine season.

In the "Jahresb. des Ver. f. Erdk. zu Dresden", 1892, 159, 160, Mr. C. Ribbe writes: "The climate of Ceram is one of the heathiest and most agreeable experienced by me in the Indies; from my tables of the temperature I find that the greatest degree of heat at Illu was $32\frac{1}{1}$° C. in the shade, the lowest 20° C. ... By shifting quarters according to the time of year, it is possible to live in a perpetual spring, for the great heat, as also the wearisome rainy season can be avoided: you build your hut now on the south and then on the north coast of the island. ... It would not be safe, however, to conclude that these weather conditions recur from year to year with mechanical regularity; on the other hand frequent exceptions to the rule take place, from which the travelling naturalist may have to suffer severely".

The rainfall has been studied at two places in Ceram — at Wahaai on the north coast and at Amahei nearly opposite on the south coast, spots which are separated by mountains from 6000—10,000 feet high. The following monthly averages for 15 years show the contrast in the seasons at the two spots.

	Jan.	Febr.	Mch.	Apr.	May	June	July	Aug.	Sept.	Oct.	Nov.	Dec.	Total
Wahaai	280	476	293	234	140	106	113	94	93	107	110	212	2258
Amahei	127	101	121	191	274	424	537	538	241	174	101	109	2938

The figures for February on the one hand and for July and August on the other are particularly instructive. The respective Monsoons are then especially well developed.

Amboina and Saparua. These islands lying under the shelter of the south coast of Ceram, have their rainy season during the months of our summer, when the S. E. Monsoon is blowing, and a great abundance of rain then falls. The north coast of Amboina receives less rain than the south, but the seasons are not so strongly contrasted as in Ceram.

Banda Islands. "The East Monsoon brings rain and storm with it." April, May and June are the wettest months, and August to November somewhat the driest.

Kei Islands. — The fine season seems to be restricted to three months of the year, viz. August, September and October.

Aru Islands. — A plentiful and generally even amount of rain fell here during the year 1890. Dr. de Hollander writes: "In the West Monsoon, which lasts from the beginning of December to April, the day-temperature is very high, but the nights, when a heavy dew usually falls, are on the contrary cold and raw; during the East Monsoon from the middle of April till the middle of November the heat of the day is less oppressive and the nights less cold. The most rain falls at the turning of the Monsoons; at this time fever occasionally occurs."

New Guinea. — In Dutch New Guinea, as in other islands which are intersected by high mountains, the north coast has its fine season while the rainy season is on the south. Some observations at Doreh and Andei in the Geelvink Bay gave the prevailing winds as east — generally due east — from May to November, which is the dry season, and westerly during the other half-year — the rainy season. Stormy or violent winds were rare and occurred most frequently in the E. Monsoon (Meyer, Ausz. Neu Guinea Reise 1875, p. 20). Van der Stok records the winds at Mansinam, western shore of the Geelvink Bay, as N. E. from May to September, the drier period, and W. from November to April, with a maximum rainfall in February. De Hollander records the temperature as rising to as much as $35.5°$ C. $(28.5°$ R., $96°$ F.) during the E. Monsoon.

On the west coast the "Sailing Directory of the East Indian Archipelago" states that the S. E. Monsoon blows from April to November, bringing great quantities of rain chiefly from June to September; the N. W. Monsoon prevails from October to May, when the weather is fine and calm.

Further down the north and north-east coast, some interesting figures, on a parallel with others already given, have been sent in from three settlements in Kaiser Wilhelms-Land: Finschhafen, Constantinhafen and Hatzfeldhafen. Finschhafen[1]) lies at the head of the Finisterre Peninsula, south of the lofty Finisterre Mts. which intersect the peninsula near its north coast: Constantinhafen is on the other side of the range at the head of Astrolabe Bay, and Hatzfeldhafen[1]) higher up the coast. Owing to its position, Finschhafen, together with the adjacent north coast of the Huon Gulf, presents a windward shore to the

---

[1]) Finschhafen and Hatzfeldhafen are now abandoned.

S. E. Monsoon of June—September, while Constantinhafen lies sheltered from wind and rain by the mountains, which also, with some other nearer hills, protect Hatzfeldhafen. During the N. W. Monsoon the case is reversed, Finschhafen is now sheltered by the Finisterre Mts., and the other two exposed to the wind. Consequently Finschhafen has its fine season with this, the N. W. Monsoon, and its rainy season with the S. E. Monsoon, while at Constantin- and Hatzfeldhafen exactly the opposite takes place. The total rainfall for the year being given as 100, the following shows the percentage of rain when the different Monsoons are in force:

	Finschhafen.	Constantinhafen.	Hatzfeldhafen.
Dec.—April 1886/87	17%	58%	62%
Dec.—April 1887/88	18%	69%	68%
June—Sept. 1886	58%	—	11%
June—Sept. 1887	62%	18%	17%

These results held good for the subsequent reports in 1888 and 1890. (Nachrichten über Kaiser Wilhelms-Land, 1888, 160).

In South-eastern New Guinea Capt. Moresby states that "the N.W. Monsoon blows from November till March, accompanied with occasional westerly gales, and with fine-weather intervals". D'Albertis says that a S.E. wind blows at Yule Island for 8 months and confirms Moresby's remarks in stating that the rainy months are November—February (N. G. 1880, I, 402). The same traveller found that heavy rains fall in the valley of the Fly River from December to April. In the dry and cooler season May—August the max. heat was 29.5^0 C., and during the hotter months rarely rose to 35^0 C.

Solomon Islands. — This group is subject to variable winds, violent squalls, and heavy rainfall. The N.E. Monsoon from the end of November to the end of March is considered to be the rainy season. Heavy gales from the west and north-west are not infrequent at this period. The S.E. Trade-wind from April to November seems to blow in fits and starts, interrupted by calms, variable winds, and often heavy squalls and much rain. The temperature varies little: 75^0 F. at night to 90^0—95^0 F. at noon (Leeper, J. R. Met. Soc. 1885, 309—313).

Northern Australia. — Here the N.W. Monsoon brings the rains, and the S.E. Monsoon, blowing out of the interior, is of course very dry.

This rough sketch of the winds and rainfall in the East Indian Archipelago may, it is hoped, prove to be not without use in the study of the climatic variation of birds, of their local movements, their nidification and moulting. The winds and rains and temperature are also factors which should be taken into consideration in questions concerning the geographical distribution of birds.

The following publications have been consulted in preparing this chapter:

Findlay: A Directory for the navigation of the Indian Archipelago (London, 1870).
  id.: A Directory for the navigation of the South Pacific Ocean (London, 1877).
de Hollander: Handleidung bij de beoefening der Land- en Volkenkunde van Nederlandsch Oost-Indië, 4th ed. 2 vols. (Breda, 1884).
Leeper: Journ. Roy. Meteor. Soc. 1885, 309—313.
Maury: Physical geography of the Sea, 14th ed. (London, 1869).
Meyer: Auszüge aus den auf einer Neu Guinea-Reise im Jahre 1873 geführten Tagebüchern (Dresden 1875).
Nachrichten über Kaiser Wilhelms-Land (Berlin, 1887—90).
Neumayer: Segelhandbuch für den Indischen Ocean, with Atlas (Hamburg, 1891).
  id.: Segelhandbuch für den Stillen Ocean, with Atlas (Hamburg, 1897).
Observatorio meteorológico del Ateneo municipal de Manila (Manila, 1876—78).
Scott: Journ. Roy. Meteor. Soc. 1889, XV, Nr. 72, October.
van der Stok: Regenwaarnemingen in Nederlandsch-Indië. 27. Jahrgang, 1895 (Batavia, 1896).
  id.: Wind and Weather, Currents, Tides and Tidal Streams in the East Indian Archipelago (Batavia, 1897).

Excerpts from other writings will be found in the text.

We are also indebted to Dr. van der Stok of Batavia for examining an original draught of our maps III and IV and for kindly pointing out and describing an alteration which was neccessary. We had, too, the advantage of Prof. Neumayer's kind advice.

3. MIGRATION IN THE EAST INDIAN ARCHIPELAGO.

Migration in its simplest form. — As Mr. Whitlock points out in his recent critique on Gätke's theories (see: "The Migration of Birds" 1897), the first stage in the history of migration is probably seen in the daily journeys of certain species to their feeding grounds and their return to their roosting places in the evening. In England this may be particularly well observed in the Rooks and their comrades, the Jackdaws; they are very methodical in their daily visits to certain fields, though their movements are by no means governed with the regularity and punctuality of a pendulum. This sort of thing will of course be found in all animals, human or avian or other, which sleep in one spot and dine in another. One of the most striking cases which we have seen from the East Indies is that recorded by Dr. Hagen of the large Hornbill, *Cranorrhinus corrugatus*, in E. Sumatra. These ill-flying birds feed by the sea and return in the evening to their roosting places inland, making use of certain trees about every kilometer of their way as travellers' rests. "The resting places are fixed spots, and, if they are not scared, the birds may be expected with tolerable certainty every evening at the appointed time" (see: text p. 244).

Local movements. — A further development of the principal of migration is seen where species do not spend the hours of a day but remain some weeks or months in one locality, and then depart elsewhere. These movements are probably common in the tropics and depend upon the abundance of certain foods at these periods. Thus Meyer observed that the small Parrots, *Loriculus exilis* and *stigmatus*, *Trichoglossus ornatus* and *meyeri*, visited Manado in great numbers at March or, respectively, April and May, the cause being apparently the flowering or fruit-bearing of certain trees; and the Sarasins found *Munia pallida* abundant at Macassar during the rice-harvest (June, July), but they personally did not see them afterwards as late as September. There is much to be learnt still about movements of this kind in Celebes, but it would be easy to multiply instances in the Indian countries and elsewhere. Dr. E. P. Ramsay (see: "Ornis" 1885) terms the movements of all Australian birds "nomadic", but there are also some species there which are as true migrants as those of the temperate parts of the northern hemisphere (see, below, p. *48*).

Local movements are the more to be expected among the birds of Celebes and other parts of the Archipelago in consequence of the great differences in the season often to be found in spots a few miles apart. The east side of the Southern Peninsula, for instance, has its wet season when the west side has its dry one, and when the west coast is deluged with rain, the east coast has fine weather, though the two districts are only some 30—60 miles apart. In a similar manner the north and south coasts of Ceram are contrasted, and traces of the same condition are seen on the north and south coasts of the North Peninsula of Celebes. The climate of the mountains is also strongly contrasted with that of the plains. We have been at some pains to obtain particulars of the movements of the birds from gentlemen resident in the East Indies, but have not received any data suited to the requirements of the present work, except some notes from Mr. H. Veen of Kele Londej, a place situated at a height of about 3000 ft. in the Minahassa. Mr. Veen has observed in the case of a few species ("Sonsoliat", "Tegi", "Tangkuitj" = *Hypothymis puella*, "Keresow") that their movements are affected by the abundance or scarcity of food; they generally arrive in groups and start again after 3—4 weeks. He adds: "The blossoming and the ripening of the fruits goes on the whole year, according to the altitude, e. g. the Lansap *(Lansium domesticum)*; near the coast this fruit is ripe towards the end of December and in January; a little higher up in February and March; still higher from April to August. The Durian and the Mango are ripe near the coast as early as December and January, but at Langowan (in the hills) in April—June. The Coffee-tree is in blossom in Kele Londej the whole year round, but at its fullest from November to February, and it begins partially to ripen in May. At Langowan this occurs about a fortnight earlier, and at Tondano a fortnight earlier still. Last year (1893) coffee was gathered every month, though in various quantities. And this though Kele Londej is only about 400 ft. higher than Langowan, and Langowan only about 200 ft. higher than Tondano".

Other phaenological observations from Celebes than these, except stray ones, have not reached us, though we possess such from Middle Luzon, which, however, would not afford a crucial test for Celebes.

Islet nomads. — A curious phase of migration is displayed by several Pigeons of the East Indian Archipelago. The species occurring in the Celebesian area and displaying the characteristic in question are *Caloenas nicobarica*, *Myristicivora bicolor*, *Carpophaga concinna* and *C. pickeringi*. These birds — at least the two first and better known species — are highly gregarious; they repair to breed on certain fixed islets and during the rest of the year seem to wander from one small island to another within their range, only occurring exceptionally on the neighbouring mainlands. In this manner the *Myristicivora* and the *Caloenas* have a range of three or four thousand miles, and that of *Carpophaga concinna* is also large. The Nutmeg Pigeon, *Myristicivora*, is often excessively numerous; we read of the lofty trees of a small island being simply covered with thousands of these white birds, of their fairly swarming at times on other islets, of great flocks literally hanging in clusters on the trees of the little island of Manado tua. So, too, the Nicobar Pigeon, *Caloenas*, is described as swarming by thousands upon its almost inaccessible breeding islet of Batty Malve in the Nicobars. It is obvious at once that these small islets cannot and do not support such a population of Pigeons for long, consequently the birds are repeatedly on the move, flying over the sea in search of fresh feeding grounds. Instances of their being seen in the act of crossing the sea are given in the text (p. 629, 659). It appears that they visit their breeding islands seasonally, but everything has still to be learnt about the periodicity, if any, of their visits to the other islands of their ranges. The four Pigeons in question have no near affinity with one another; the Nicobar Pigeon is the sole representative of a subfamily; the other three have more or less near affinities with mainland species.

As is well known, certain sea-birds, such as the Gannets, Albatroses, and some Terns, resort to particular rocks or other islets to breed. In their case, as in that of the Pigeons, protection of their brood from animals destructive to their eggs and young may well have been the original motive for the adoption of these habits, yet with the Pigeons it remains strange that they avoid the mainland after their young have been safely reared. (For further remarks hereon, see pp. 616, 629, 659—661).

For the sake of the general reader, who may be apt to suppose that narrow straits of the sea offer no barrier to the geographical distribution of tropical species, it may be mentioned that, so far from this being the case, there is reason to believe that resident species never, or only very exceptionally, cross the sea; were it otherwise the species would not be found with such restricted ranges as is actually the case.

Migration proper in the East Indian Archipelago. — The following is a list of most of the more prominent migratory birds of Celebes. A few species, well known to be migrants, offered difficulties which have led to their being omitted, while a large number of other species have been left out because their migrations are as yet hidden in such obscurity that it would probably be misleading to attempt to trace them categorically. In the case of the species given it has often been impossible, in the absence of positive data, to avoid speculation in the use of the signs for summer and winter haunts, but it will generally be found that too little has been said, rather than too much.

○: signifies "summer visitor". (It would generally be safe to assume that the species breeds in the localities so marked, but in not one-half of them have the nest and eggs yet been discovered.)

×: signifies "winter visitor", or "passes through on migration".

†: signifies "rare", or "a straggler".

Introduction: Migration.

Localities	*Pachycephala schlegeli* (Horsf.)	*Accipiter virgatus gularis* (T. & S.)	*Butastur indicus* (Gm.)	*Ninox scutulata japonica* (T. & S.)	*Hierococcyx sparverioides* (Vig.)	*Cuculus canorus canorinus* (S. Müll.)	*Cacomantis coromandus* (L.)	*Merops ornatus* Lath.	*Merops philippinus* L.	*Halcyon pileata* (Bodd.)	*Halcyon sancta* Vig. Horst.	*Caprimulgus periphus* (Lath.)	*Pitta cyanoptera* Temm.	*Hirundo rustica* L.	*Muscicapa griseisticta* (Swinh.)	*Lanius tigrinus* Drapiez	*Lanius lucionensis* L.	*Petrophila cyanus solitaria* (P. L. S. Müll.)	*Acrocephalus orientalis* (Temm. Schl.)	*Locustella fasciolata* (G. R. Gray)	*Locustella ochotensis* (Midd.)	*Phylloscopus borealis* (Blas.)	*Motacilla flava* L.	*Motacilla boarula melanope* (Pall.)	*Anthus gustavi* Swinh.	*Anthus cervinus* (Pall.)	*Sturnia violacea* (Bodd.)
Northern Europe	–	–	–	–	–	–	–	–	–	–	–	–	–	–	–	–	–	–	–	–	–	–	O	–	O	O	–
Centr. & S. Europe	–	–	–	–	–	–	–	–	–	–	–	–	–	–	–	–	–	–	–	–	–	–	†	O	–	O	–
South-west Asia	–	–	–	–	–	–	–	–	–	–	–	–	–	–	–	–	–	–	–	–	–	–	×O	–	–	×	–
Africa	–	–	–	–	–	–	–	–	–	–	–	–	–	–	–	–	–	–	–	–	–	–	×	–	–	×	–
Madagascar	–	–	–	–	–	–	–	–	–	–	–	–	–	–	–	–	–	–	–	–	–	–	–	–	–	–	–
Siberia, West	–	–	–	–	–	–	–	–	–	–	–	–	–	–	–	–	–	–	–	–	–	–	O	–	O	O	–
Siberia, Central	–	–	–	–	–	–	–	–	–	–	–	–	–	O	–	–	O	–	–	–	–	O	O	–	O	O	–
Siberia, N.E.	–	–	–	–	–	–	O	–	–	–	–	–	–	O	–	–	O		–	–	O	O	–	–	O	O	–
Arctic Regions	–	–	–	–	–	–	–	–	–	–	–	–	–	–	–	–	–		–	–	–	–	O	–	–	–	–
Bering Islands	–	–	–	–	–	–	–	–	–	–	–	–	–	–	–	–	–	breeding in the North	–	–	O	–	O	–	–	×	–
Alaska	–	–	–	–	–	–	–	–	–	–	–	–	–	–	–	–	–		–	–	–	–	O	–	–	×	–
Kamtschatka	–	–	†	–	–	–	–	–	–	–	–	–	–	–	–	–	O		–	O	O	O	O	–	O	O	–
S.E. Sib., Amurland	–	O	O	O	–	–	–	–	–	–	O	–	–	O	–	O	–		–	–	×O	O	×O	O	O	×O	–
Mongolia	–	–	O	–	–	–	–	–	–	–	–	–	–	O	–	–	O		–	O	–	×O	×O	O	–	–	–
Manchuria	–	O	O	–	–	–	–	–	–	–	–	–	–	–	–	–	–		–	–	–	–	–	–	–	–	–
Corea	–	O	O	–	–	–	–	O	–	–	O	–	–	O	–	–	O		†	O	O	O	–	×O	–	O	–
Japan	–	O	O	O	O	O	–	–	–	O	–	–	–	O	–	O	–	the South,	†	O	O	O	×	×	×	–	O
China, North	O	O	×	×O	O	O	–	O	–	O	–	–	O	–	O	–	–		×O	O	×	×O	O	×O	×	×	–
China, South	O	×	×	×O	–	O	O	–	O	O	O	–	–	O	–	O	–	South, wintering in	×	×	×O	O	×	×	×	×	–
Formosa	–	×	–	–	–	–	–	–	–	–	–	–	–	O	–	–	–		×	×	–	×	×	×	×	×	–
Philippines	×	×	×O	×	×	×	–	×O	×	–	–	–	–	×	×	–	×		×	×	×	×	×	×	×	×	×
Borneo	×	×	×	×	×	×	–	×	×	×	–	×O	–	×	–	–	×		×	×	×	×	×	×	×	×	×
Talaut Islands	×	–	×	×	–	×	–	–	×	–	–	–	–	×	–	–	×		–	×	–	×	×	–	–	–	–
Sangi Islands	×	–	×	×	–	×	–	–	×	×	–	–	–	×	–	–	×		×	–	–	×	×	–	–	–	–
Celebes, North	×	×	×	×	×	×	–	–	×	×	–	×	×	×	†	×	×		×	×	–	†	×	×	–	–	–
Celebes, South	–	–	×	–	×	–	–	–	×	–	–	–	–	×	–	–	–		–	–	–	–	×	×	–	–	–
Moluccas	–	–	×	–	×	–	×	–	×	–	–	–	–	×	–	†	×		×	×	–	–	×	×	–	–	×
Papuasia	–	–	×	–	×	–	×O	–	×	–	–	–	–	×	–	–	–		–	–	–	–	–	–	–	–	–
India	–	–	–	O	×O	×	–	–	×O	×	–	×	–	–	–	†	–		–	†	–	–	–	–	×O	–	×	–
Ceylon	–	–	–	–	–	×	–	–	×	†	–	–	–	–	–	†	–		–	–	–	–	–	–	×	–	–
Islands, B. of Bengal	×	–	–	–	×	–	–	–	×	–	–	–	O	–	–	–	–		–	–	×	–	–	×	×	–	×	–
Further India	†	×	–	–	×	×O	–	–	×O	×	–	×	O	–	–	†	–		–	×	–	–	×	×	×	–	×	–
Malay Peninsula	×	×	×	–	×	–	×	–	×	–	–	×	×	–	–	×	–		×	×	×	–	×	×O	×	–	–
Sumatra	×	×	–	–	×	–	–	–	×	–	–	×	×	–	–	×	–		×	†	–	×	×	×	×	–	–
Java	×	×	×	–	×	–	×	–	×	–	–	×	–	–	–	×	–		–	×	–	–	×	×	×	×	–	–
Lesser Sunda Is.	–	×	–	–	–	–	–	–	×	–	–	×	–	–	–	×	–		–	×	–	–	×	×	×	×	–	–
Australia	–	–	–	–	×	–	–	O	–	–	O	×	–	–	–	–	–		–	–	–	–	–	–	–	–	–
Polynesia	–	–	–	–	–	–	–	–	–	–	–	†	–	–	–	–	–		–	–	–	–	–	–	–	–	–
New Zealand	–	–	–	–	–	–	–	–	–	–	–	–	–	–	–	–	–		–	–	–	–	–	–	–	–	–
North America	–	–	–	–	–	–	–	–	–	–	–	–	–	–	–	–	–		–	–	–	–	–	–	–	×	–

Introduction: Migration.

Localities	Fulica atra L.	Glareola isabella Vieill.	Lobivanellus cinereus (Blyth)	Squatarola helvetica (L.)	Charadrius fulvus Gm.	Aegialitis veredus (J. Gd.)	Aegialitis geoffroyi (Wagl.)	Aegialitis mongola (Pall.)	Aegialitis curonica (Gm.)	Strepsilas interpres (L.)	Totanus glottis (L.)	Totanus calidris (L.)	Totanus glareola (L.)	Helodromas ochropus (Vieill.)	Actitis hypoleucos (L.)	Terekia cinerea (Güld.)	Tringa acuminata (Horsf.)	Tringa damacensis (Horsf.)	Tringa subarquata (Güld.)	Phalaropus hyperboreus (L.)	Limicola platyrhyncha (Temm.)	Gallinago megala Swinh.	Limosa novaezealandiae (G. R. Gray)	Numenius minutus J. Gd.	Numenius variegatus (Scop.)	Numenius cyanopus Vieill.	Ardetta cinnamomea Swinh.
Northern Europe	O	–	–	O	–	–	–	–	O	O	O	O	O	–	O	O	–	–	–	O	O	–	–	–	–	–	–
Centr. & S. Europe	×O	–	–	×	†	–	–	–	×O	O	×	O×	O×	–	O×	†	†	–	–	×	×	–	–	–	–	–	–
South-west Asia	×O	–	–	×	–	–	()	O	O	×	×	×O	×	–	O	×	–	–	–	×	×	–	–	–	–	–	–
Africa	–	×	–	×	–	–	×	×	×O	×	×	×	×	–	×	×	–	–	–	×	×	–	–	–	–	–	–
Madagascar	–	–	–	×	–	–	×O	–	–	×	–	–	–	–	×	×	–	–	–	–	×	–	–	–	–	–	–
Siberia, West	–	–	–	–	–	–	O	–	–	–	–	O	O	–	O	O	–	–	–	–	–	–	–	–	–	–	–
Siberia, Central	–	–	–	–	O	O	O	–	–	–	–	O	O	–	O	O	–	–	–	O	–	–	–	–	–	–	–
Siberia, N.E.	–	–	O×	×	–	–	O	–	–	–	O	O	O	–	O	O	O	–	–	O	–	–	×	–	O	–	–
Arctic Regions	–	–	O	O	–	–	–	–	–	O	O	–	–	–	O	×	O	–	–	–	–	–	–	–	–	–	–
Bering Islands	–	–	×	O	–	–	O	–	–	×	–	O	×	O	×	×	×O	×	O	–	–	–	×	–	–	×O	–
Alaska	–	–	×	O	–	–	–	–	–	–	–	–	–	–	×	–	–	–	–	–	O	–	–	–	–	†	–
Kamtschatka	–	–	O×	O	–	–	O	–	–	–	–	–	–	–	×	O	O	O	–	–	–	–	×	–	O	–	–
S.E. Sib., Amurland	O	–	×	O	–	–	–	–	–	–	×O	–	O	×	O	×	×	–	–	O	×	O	×	O	×	O	O
Mongolia	–	–	O	–	×O	O	–	O	O	–	–	–	–	×	×	×	–	–	×	×	O	–	O	–	–	–	–
Manchuria	O	–	×	×	–	–	–	–	–	–	O	–	–	–	–	–	–	–	–	–	–	–	–	–	–	–	–
Corea	O	–	O	×	–	–	–	–	–	×	×	–	O	–	×	×O	×	–	×	O	–	×	–	–	×	O×	–
Japan	O	–	O	×	×	–	†	×	×O	×	×	O	×	×	O	×	×	×	O	×	×	×	×	×	†	×	–
China, North	×O	–	×O	×	×	O	–	×	×O	×	×	×	×	×	O	×	×	×	×	×	×	×	×	×	×	×	O
China, South	×	–	×	×	×	–	×O	×	×	×	–	×	O	×	×	×	–	×	×	×	–	×	×	×	×	×	–
Formosa	–	–	×	×	×	×	×O	×	×	×	×	–	×	O	×	×	–	×	×	×	–	×	×	×	×	×	×
Philippines	×	–	×	×	×	–	×	×	×	×	×	–	×	–	×	×	×	–	×	×	–	×	–	×	×	×	×
Borneo	–	×	–	×	×	–	×	×	×	×	–	×	×	×	×	×	–	×	×	×	–	–	–	×	–	×	×
Talaut Islands	–	–	–	×	×	–	×	×	–	–	–	×	×	–	–	×	–	–	–	×	–	–	–	–	–	–	–
Sangi Islands	–	–	–	–	×	–	×	×	–	–	–	×	×	–	×	×	–	–	–	×	–	–	–	×	–	–	–
Celebes, North	×	×	†	†	×	–	×	×	×	–	×	×	×	×	×	×	–	×	–	×	–	×	–	×	×	×	×
Celebes, South	–	×	–	×	×	×	×	×	×	–	×	×	×	–	–	×	–	–	–	×	–	–	–	×	×	×	–
Moluccas	–	×	–	×	×	×	×	×	–	–	–	×	×	–	×	×	–	–	–	×	–	–	–	×	×	×	–
Papuasia	–	×	–	×	×	×	×	×	×	–	–	×	×	–	×	×	–	–	–	×	–	–	–	×	×	×	–
India	×O	–	×	×	–	×	×	×O	×	×	×	–	×O	×	†	–	×	–	×	–	–	–	–	–	–	–	–
Ceylon	–	–	×	×	–	×	×	×	×	×	×	–	×	†	–	×	–	†	–	–	–	–	–	–	–	–	–
Islands, B. of Bengal	–	–	–	×	–	×	×	×O	×	×	×	–	×	–	×	–	×	–	×	–	–	–	–	–	–	–	–
Further India	×O	–	×	×	×	–	×	×	×	×	×	–	×	–	×	–	×	–	×	–	–	–	–	–	–	–	–
Malay Peninsula	×O	–	–	×	–	×	×	×	×	×	×	–	×	–	×	–	×	–	–	–	–	–	–	–	–	–	–
Sumatra	–	–	–	×	×	×	×	×	–	×	×	–	×	–	×	–	×	–	–	–	–	–	–	×	–	–	–
Java	×	×	–	×	×	×	×	×	×	×	×	–	×	×	×	×	×	–	–	–	–	–	–	×	×	†	×
Lesser Sunda Is.	–	×	–	×	×	–	×	×	–	×	×	–	×	–	–	×	–	–	–	×	–	–	–	×	×	×	–
Australia	–	O	–	×	×	×	×	×	–	–	×	–	×	–	×	–	–	–	×	–	–	–	–	×	×	×	–
Polynesia	–	–	–	×O	–	×	–	–	–	–	–	–	–	–	×	–	×	–	×	–	–	×	–	×	×	×	–
New Zealand	–	–	–	×O	–	–	–	–	–	–	–	–	–	–	×	–	–	–	–	–	–	–	–	×	–	†	–
North America	–	–	×	O	–	–	–	–	–	–	–	–	–	–	–	–	–	–	–	O×	–	–	–	–	–	–	–

During our studies it has become abundantly evident to us that ornithologists are not generally aware that migration goes on in the East Indies to the great extent it does. Naturalists travelling in the Archipelago have rarely had a word to say on this subject, for the reason that they themselves are continually on the move from place to place and are consequently unable to say whether the birds they see are visitors or stationary. Among the residents or travellers of longer standing it would be difficult to mention the names of more than two or three competent observers; of these Mr. Everett has as yet published nothing on migration to our knowledge beyond a few terse notes on individual species, but there are a number of interesting observations extant from the pen of Mr. Whitehead. We quote from this gentleman the following passages relating to the S. E. parts of Palawan: "When the great rush of birds from the far north takes place these plains have a much more lively aspect; hundreds of Wagtails, Pipits, Snipes, and other small Waders are continually flying up on your approach . . . Towards the middle of September, after we had collected all the resident species within our reach, the sea-coast, with its rocky points and estuaries, was by far the most attractive hunting-ground; for about that time the great winter migration from the north reaches the coasts and forests of Palawan. Most Waders passed between 5 and 6. P. M., all in one direction, S.W.; if a small flock settled and was disturbed, never did the birds return, but still hurried on their southward course. By continuing this line of flight they would touch Balabac, and then turn due south down to the coast of Borneo, where some remain for the winter, but most seem to travel further still. When the wind was blowing gales from the S.W., bringing up heavy clouds loaded with rain, then was the liveliest time for moving; on calm, or even moderate days, it was seldom worth while to visit the coast. All these great travellers were as fat as butter, and in no state for a bird-collector" (Ibis 1890 p. 40). Out of a total of 157 species in Palawan 49 are migrants.

Birds at sea. — Nearly all naturalists when at sea in these regions have been visited by birds of migratory species for a temporary rest on board ship. When sailing ships were common this was possibly a more usual occurrence than in these days of steamers. The following passage from F. J. F. Meyen's "Reise um die Erde", 1835 II, 195, is of interest. When crossing the China Sea between Macao, Canton River, and Manila, "besides the ordinary sea-birds a number of various small land-birds were seen, which, as we quickly perceived, were migrating and resting upon our ship. We obtained on this occasion the *Lanius phoenicurus* (Pall.) (= *L. lucionensis*), the *Hirundo domestica* Pall, and the *Motacilla flava*, all of them birds which spend the summer months in Southern Siberia . . . As it appeared all these birds were directing their course towards the Moluccas; the Swallows came along in such numbers that we caught eight of them in the evening. One day the head of a palm-tree drifted by our ship . . . Large birds were resting on this swimming island as it came towards us, and a host of fishes swam in front of it, while innumerable crabs sat upon it and sunned themselves".

In the special articles on the species records of individuals captured or observed at sea are quoted, such are: *Tachyspizias soloensis* (China Sea, near the Pescadores, May); *Accipiter virgatus* (China Sea, near the Natuna Is., 14th November); *Merops philippinus* (Indian Ocean, November); *Halcyon sancta* (Pacific Ocean, 300 sea-miles S.E. of the Louisiade group); *Hirundo rustica* (several records from the seas between the Moluccas and S. China); *Acrocephalus orientalis* (40 miles N. of the Loochoo Islands); *Motacilla flava* (Indian Ocean and China Sea); *Tringa ruficollis* (China Sea, May 14th); *Hypotaenidia philippensis* (Pacific Ocean far at sea east of Australia); *Myristicivora bicolor* and *Caloenas nicobarica* (Seas of the East Indian Archipelago).

Routes. — The above tables tend to prove that each species has its own route or routes of migration; nevertheless the species allow of a certain amount of co-ordination, as follows.

From Northern Europe and Siberia to the East Indian Archipelago. — So far as species occurring in Celebes are concerned, migration reaches its highest development in the Arctic Chiffchaff, *Phylloscopus borealis* (J. H. Blas.) and the Siberian Pipit, *Anthus gustavi* Swinh. The former is known to breed as far west as Northern Norway (Collett), the latter as far as the Petchora (Seebohm and Harvie-Brown); they also breed in various parts of Siberia. In the winter there are no observations to show that individuals stop short of S.E. Asia and the East Indies; on the other hand they are observed to pass through China on migration, the Pipit holding to the countries washed by or near the Pacific Ocean and not visiting the Indian countries, while the Chiffchaff occurs both in the territories invaded by the Pipit and in Further India and the neighbouring islands as well. So far as can be judged the general direction of the two species in autumn is from west to east across Siberia and then south or south-south-east. They are in singular contradiction to Gätke's picturesque theory of an east to west flight at this season, and *Anthus gustavi* should be contrasted with *Anthus richardi* V., which is supposed to migrate in the opposite direction, and is often referred to for proof by Gätke.

From Northern Europe to Africa, and from Siberia to the East Indies or further. — Many species migrate thus, but, as a rule, appreciable racial differences may be seen in the western and eastern individuals, with intermediate forms from other localities between them. Such are: *Hirundo rustica*, *Cuculus canorus*, *Petrophila cyanus*, *Motacilla flava*, *Motacilla boarula*, *Anthus cervinus*, *Totanus calidris*. Other forms are commonly separated as distinct species in the West and East; such are *Charadrius pluvialis* and *fulvus*, *Acrocephalus turdoides* and *orientalis*, *Tringa minuta* and *ruficollis*, *Limosa lapponica* and *novaezealandiae*, *Numenius phaeopus* and *variegatus*. Others do not differ in the West and East, or at least no prominent racial differences in them have as yet been insisted upon; such are: *Squatarola helvetica*, *Aegialitis curonica*, *Strepsilas interpres*, *Totanus glottis*, *T. glareola*, *Actitis hypoleucos*, *Terekia cinerea*, *Phalaropus hyperboreus*, etc. It may be that these last commingle more freely in their breeding grounds than the others, and have not yet adopted routes of migration of an equally unvarying character.

From China, or China and S. E. Siberia, to Further India, the Philippines and Sunda Islands as far as North Celebes. — A large number of visitors to Celebes are distributed in summer and winter, respectively, as above. Celebes appears to be reached by individuals which have travelled over the Philippines, and not by birds coming from Borneo and Java. This is shown by the fact that these species occur in the Northern Peninsula, but much more rarely, if at all, in other parts of Celebes. Migrants descending south through the Philippines and across the Celebes Sea are confronted by a lofty barrier 400 miles long formed by this peninsula, and the majority of individuals do not pass over it into South Celebes or into the Moluccas. If the migrants came from Borneo or Java they would reach the west and south coasts first, and their presence on the northern coast could only be accounted for as an aimless progression towards the north-east, of which there is no evidence. The simpler explanation may be assumed with much confidence to be the correct one. The winter visitors, therefore, from China and the North to the Northern Peninsula are: *Tachyspizias soloensis*, *Accipiter virgatus gularis*, *Hierococcyx sparverioides*, *Coccystes coromandus*, *Halcyon pileata*, *Pitta cyanoptera*, *Lanius lucionensis*[1]), *Petrophila cyanus solitaria*[1]), *Locustella ochotensis*, *Lobivanellus cinereus*, *Tringa damascensis*, *Ardetta eurhythma*.

There is no species occurring as a migrant in the Philippines which has been found in South Celebes and not in the North Peninsula, except *Limicola platyrhyncha*, which is as yet known from Celebes by one specimen only from the South-central part of the island.

Species with the above summer and winter distribution, but which pass on further into South Celebes, the Moluccas and Papuasia are: *Butastur indicus*, *Ninox scutulata japonica*, *Cuculus canorus canoroides*, *Acrocephalus orientalis*, *Gallinago megala*, *Numenius minutus*. To these should be added most of the Waders, and the two Wagtails, the Pipit, *Anthus cervinus*, the Arctic Chiffchaff, *Phylloscopus borealis*, etc., which, however, have a higher northern breeding range.

From S. E. Siberia and China to Further India and the Sunda Islands, avoiding the Philippines. — This route is pursued by *Lanius tigrinus*, in striking contrast to its compatriot *Lanius lucionensis*, which visits the Philippines in abundance, and the Indian countries and Southern Sunda Islands only sparingly (See pp. 405, 408—410). Most of the individuals of *Pitta cyanoptera* seem to take a route like that of *L. tigrinus*.

West-Pacific migration. — There are several species occurring in Celebes which in winter visit the countries in the more immediate neighbourhood of the Pacific, not crossing to the Malay Peninsula, Sumatra, Java and the islands in

---

1) Has occurred in the Moluccas.

the Bay of Bengal. *Sturnia violacea*, for instance, is not known from the continent of Asia, but breeds in Japan, from where it seems to fly directly in autumn to the Philippines, Borneo, N. Celebes and the Moluccas. *Muscicapa griseosticta* and *Locustella fasciolata* breed in North China or N. E. Asia and occur in winter in the East Indies which are washed by the Pacific, as far as New Guinea and. the Moluccas, respectively. *Heteractitis brevipes* wanders in winter down the West Pacific coasts from unknown breeding grounds in the high North as far south as Australia. A remarkable traveller across the Pacific is seen in *Limosa novaezealandiae;* this bird has been found breeding in Alaska and Arctic Siberia, and it visits New Zealand in great numbers in winter, the majority of individuals apparently flying directly across the Pacific without making use of the East Indies as a resting-place, for the number of examples recorded from there is comparatively very small (see p. 794). In the case of this species, as also in that of the Pacific Cuckoo, *Urodynamis taitiensis* (Sparrm.), it is perhaps erroneous to attempt to avoid the assumption of a "sense of direction". Still ocular remembrance of the sparsely scattered atolls and high islands of the Pacific on which the birds may land or pass over, together with the positions of the sun and stars at certain hours, and the direction of the roll of the waves should not be left out of account as a means by which they may alter and regulate their course over hundreds of miles of trackless ocean (see Möbius: "Ein Beitr. z. Frage üb. d. Orientirung der wandernden Vögel": Ausland 1882 p. 648).

Migration south of the equator. — Professor Newton remarks (D. B. 555): "If the relative proportion of land to water in the southern hemisphere were at all such as it is in the northern we should no doubt find the birds of southern continents beginning to press upon the tropical and equatorial regions of the globe at the season when they were thronged with emigrants from the north . . . but we know almost nothing of the migration of birds in the other hemisphere". In this comment — very true, apparently, in regard to the comparatively small amount of migration south of the equator — Prof. Newton has almost overlooked a great point of difference in northern and southern migration, namely that the birds of the South proceed towards the equator in the time of our summer and leave the tropics again for their breeding quarters about the time that the equatorial countries are invaded by the migrants from the North. The southern territories which call for consideration in this book (the East Indies and Australia) have furnished very few thorough-going migrants; it is certain that the number is very small compared with that of the northern hemisphere, but there is also a regrettable lack here of competent observers and of published observations. One or two passages relating to migration across the Torres Straits seem to show that among the islands here it is possible that a second Heligoland may be found some day. Remarks to this effect are made in Moseley's 'Naturalist on the "Challenger"', 1879, p. 364: "Most of the birds of Cape York are constantly migrating, and the resident official at Somerset told me that the constant change from month to month of the birds seen about his place was most astonishing. The Torres Straits Islands serve as resting places for the birds crossing from New Guinea; Booby Island is evidently thus used, and the number of its land-birds is thus to be accounted for. This island corresponds thus in this respect with such an island as Heligoland. . . . It is the last place in the world, as viewed from the sea, with clouds of Boobies hovering over it, from which one would expect two new land-birds [a Dove and a Rail described by Gould] to hail. Our officers laughed at the notion of there being quails or anything to shoot upon it. . . . On August 13th, 1841, the officers of the 'Beagle' shot 145 quails, 18 pigeons, 12 rails of two species, and 3 pigeons ["doves" in Stokes' 'Discoveries in Australia' 1846 II, 330]." The contents of the game-book of the "Beagle" when among the islands of the Arafura Sea are large and interesting, though unscientific; an extract is given by Stokes, l. c. The transit of the Australian Bee-eater, *Merops ornatus*, across the Torres Straits has been remarked upon by two observers (see p. 250). This bird takes some weeks to travel from here to its breeding grounds in New South Wales.

Touching migration in Australia Mr. A. J. North (*in lit.* 7. VIII. 1894) has most obligingly furnished us with the following: "There is nothing published in the Proceedings of any society beyond a paper contributed by Dr. E. P. Ramsay to the Ornithological Congress held . . . in Vienna about twelve years ago . . . As there pointed out by Dr. Ramsay we have, comparatively speaking, but very few migratory birds in Australia, but a great number of nomadic species which shift from one point of the country to another, many of them appearing regularly every season in the spring to breed and returning north or west directly the cold weather sets in. From my notes which I have kept for the past twenty years I will give you the date or time of arrival and departure of the species asked for . . .

"*Chaetura caudacuta*. This bird arrives in New South Wales during the hottest months of the year. I have noted them as early as December, but they usually arrive in January and depart again about the middle of April. I have never seen them resting, they pass the whole of the day on the wing.

"*Cypselus pacificus*. Arrives and departs at the same time as the preceding species, with which it is more often than not seen in company. It is not, however, so numerous as *C. caudacuta*.

"*Eurystomus pacificus* and *Scythrops novaehollandiae*. These species arrive in Northern Queensland about the end of August, but their appearance is influenced greatly by the season; sometimes it is at the end of September, and in 1892 it was as late as the 12th of October when *Eurystomus* arrived. They leave again on the approach of cold weather about the end of April. In New South Wales *Eurystomus* arrives usually about the middle of September and departs again early in April. I saw young birds nearly fledged taken from their nesting place in the hollow limb of a tree near Newcastle on the 3rd of October, 1893; this was very early for New South Wales.

"*Halcyon sancta*. The main body of these birds arrive in N. S. W. during August and the early part of September, breed and depart again by the end of March; I have, however, noted straggling pairs during the intervening winter months. In Northern Queensland this bird is a resident species.
"*Circus assimilis* (*C. jardinii* Gould). This bird arrives in N. S. W. during the month of August and is frequently met with in the small clumps of pine (*Callitris*) growing out on the plains in the inland portions of the colony, and in which it is often found breeding during September and October. It takes its departure again about the end of January or middle of February."

The paper of Dr. E. P. Ramsay's referred to by Mr. North will be found in the "Ornis" 1885, pp. 581—584. The author remarks: "One thing with respect to our Australian fauna must be carefully borne in mind, strictly speaking we have no migration among our birds"; and again: "The term 'migratory' as understood by European naturalists, does not apply to any Australian species, the term I propose for these is 'nomadic'". Dr. Ramsay considers that Australian birds wander from place to place in quest of fresh feeding grounds and suitable breeding quarters; when they have reared their young, they retire to another part, sometimes only 10 or 20 miles away. "The *Hirundinidae*, our species of *Gallinago*, *Rhynchaea*, *Merops*, *Artamus* and some *Rallidae*, make the closest approach to a migration here as they sometimes wander from the north to the south of Australia". We know nothing about the endemic species of Australia, but, as regards others which occur in Celebes, Dr. Ramsay's statements are certainly more or less erroneous and misleading. Two to four natives of Australia occur only as migrants in Celebes, in the same manner as a number of natives of the northern hemisphere reach the island and proceed as far as Australia in the south. Evidences of remarkable migrations across the Torres Straits have been given above. Of birds breeding in Australia *Merops ornatus* (though a few probably stay out of the general migration on occasions) is known to be a migrant to Celebes and the Moluccas; *Glareola isabella* and *Halcyon sancta* are almost equally certainly such; the Australian form of *Eurystomus orientalis* has been recorded from Celebes and probably correctly, though we confess to an inability to distinguish the supposed races of this bird. *Circus assimilis* and *Scythrops novaehollandiae* seem to be residents in Celebes, though it would be a matter for no surprise if their numbers ultimately proved to be affected by migration to and from Australia. Some of the greatest migrants of the North travel to Australia, among them being the following visitors to Celebes: *Cypselus pacificus*, *Squatarola helvetica*, *Charadrius fulvus*, *Aegialitis vereda*, *A. geoffroyi*, *A. mongola*, *Strepsilas interpres*, *Totanus glottis* and *glareola*, *Heteractitis brevipes*, *Actitis hypoleucos*, *Terekia cinerea*, *Tringa acuminata*, *Tringa ruficollis*, *Limosa novaezealandiae*, *Numenius minutus*, *N. variegatus* and *cyanopus*. There are also several other species about which particulars as to migration would be very welcome, such as *Turnix maculosa*; the Herons, *Herodias eulophotes*, *H. alba*, *H. garzetta*, *H. intermedia*, *Notophoyx novaehollandiae*, *N. picata* and *Nycticorax caledonicus*; and the Rails, *Hypotaenidia philippensis*, *Amaurornis cinerea* and *moluccana*. Dr. Ramsay's term "nomadic" for wandering Australian birds denotes an initiatory form of migration not of a regular half-yearly character, but, as is shown by the above species, a development of the migratory habits is sometimes found as high as in many species of the northern hemisphere.

The return-migration. — Birds do not appear always to return in spring by the route pursued in autumn, often, apparently, being rare or absent in one season in districts through which they pass in abundance in the other season.

Birds which stay behind in their winter quarters. — It seems to be a very common occurrence for a few individuals to stay throughout the year in the winter resorts of the species. Among Celebesian species which have been found in the island after their fellows have departed to their breeding localities in the North, or, respectively, the South, may be mentioned: *Merops ornatus*, *Motacilla flava*, *Charadrius fulvus*, *Aegialitis geoffroyi*, *A. mongola*, *A. curonica*, *Totanus glottis*, *T. glareola*, *Heteractitis brevipes*, *Tringa acuminata*, *T. ruficollis*, *Numenius variegatus*. To these *Motacilla boarula* and *Anthus gustavi* should probably be added; the first has been sent to us from Manado tua as late as May, the Pipit as late as May 26th. Other instances of migrants killed in the Moluccas and elsewhere at dates when the main body of their species is absent will be found in the text. During his travels in Polynesia Dr. Finsch found many Waders of several species on the Marshall and Gilbert atolls of the Central Pacific in the summer months; and similar observations will be found in "Stray Feathers" and elsewhere.

It is unprofitable to speculate on the cause of this violation of the general rule, as the possible explanations are many and the individual judgment is prone to select that which conforms best to its own prejudices. Several reasons may indeed work together in inducing these birds to stay behind. The following is likely to escape general observation, viz. the time of shedding the remiges varies in individuals[1]) and a bird with its powers of flight thus diminished at the spring migration might well hesitate at attempting the journey. Or an accident — a broken wing or leg — may delay an individual, and as observers know, one or more sympathetic companions will be likely to remain by the injured one with a devotion equal to that of man himself. In other cases, as, for instance, the individuals on the atolls in the middle of the Pacific, it may more plausibly be supposed that the birds had lost their way. At other times it looks as if the birds remain behind simply from choice. But, though in Natural History it is almost always impossible to assert that a certain this, and this only, is the true explanation, it is happily sometimes possible to show that some other is an untrue explanation. Thus, it has been asked in Baird,

---

[1]) See pp. 739, 744, 747, 761, 762, 765, 768, 772, 798.

Brewer and Ridgway's "Water Birds of North America" 1884 I, 123, whether those birds which spend the summer in the winter haunts of their species are not old, effete and barren individuals. This suggestion is negatived by the fact that among such visitors to Celebes five species have been known to breed in some part, or parts, of their winter range, viz. *Charadrius fulvus, Aegialitis geoffroyi, A. mongola, Strepsilas interpres* and *Limosa novaezealandiae*.

Causes of migration. — It may be tritely said that birds migrate in autumn to feed, in spring to breed! At the approach of winter most birds must of necessity proceed towards the tropics, or starve, owing to the disappearance of their food through death or hibernation, or through concealment under snow or ice. In spring the temperate and arctic regions produce an abundance of food and, it may be presumed, offer safer and easier conditions for the propagation of the species than is found in the tropics; the birds then repair to their native haunts. Naturalists, who seek for physiological conditions to account for the actions of the subject, may find a stimulus for the spring migration in the annual development of the reproductive systems, while the approaching autumnal moult of the remiges may sometimes serve as a warning to species that it is time to accomplish their flight towards the equator, for many of them leave long before the cold sets in. Waders killed in their winter quarters in the Celebesian area in the late autumn or winter months are generally found to be moulting some of their remiges. Possibly, however, the chief motive for the spring migration is to be found in the love of home so strongly developed in birds, for without this it is conceivable that they would attempt to establish themselves in the tropics for breeding purposes. Instances of the marvellous regularity with which individuals return to the old breeding haunt after a journey of hundreds, or more often thousands, of miles have excited the admiration of all field-observers. That the young birds also sometimes follow their parents to their birth-place is shown by a case given on p. 48 (text). But it seems to be the case — which is not so generally known — that birds display a very similar adhesiveness in the choice of their winter quarters. An instance of enormous numbers of Wagtails and Swallows returning two years in succession to roost in a coffee garden in Ceylon, as observed by Mr. S. Bligh, will be found on p. 537; and Davison states that a large number of migratory *Collocaliae*, which had taken to roosting upon a certain spot about a yard square against the roof of a shed in the Andamans, disappeared when the building was pulled down, only to come again and occupy the same spot on a new shed, which had been put up on the identical site (p. 332). There are no rookeries near Dresden, but every year about the beginning of November great numbers of Rooks and Jackdaws pass over the neighbourhood for many days in succession[1]), and some spend the winter there.

---

[1]) The flocks fly high, so that it is sometimes difficult to detect them with the naked eye. The mode of progress is very slow in a S. or S.W. direction, perhaps at the rate of 12 miles an hour, and conducted with much cawing and calling and circling in the air, as if none wish to have the responsibility of leading the way.

Two or three individuals often make their appearance in the daytime in the Bismarckplatz in the town each year, though they are not always driven there by cold and hunger, and they are most likely the same birds each season. There is an autumnal migration route of Wild Geese (*Anser segetum*) at Grossenhain near Dresden, the birds making their appearance during their transit to the South every year on the fish-ponds at this place, but only quite exceptionally on the fish-ponds at Moritzburg some 10 miles to the S.E. Mr. Schwarze of the Dresden Museum informs us that Kingfishers did not occur in summer at his native village in Saxony, but were to be found every winter, when the fish-pond was frozen over, at the inlet or trap where water was let in, and where there were plenty of small fish. The Grey Crow, *Corvus cornix* L., visits the Eastern counties of England in great numbers every autumn; these birds do not proceed far inland, being unknown in the Western counties; and such is almost equally the case in the Southern Midlands, for instance, in Northern Buckinghamshire. Yet these parts of England appear to be quite as capable of sustaining them as the Eastern counties. Mr. W. Eagle Clarke's "Digest of the Reports on Migration" should also be read in this connection.

If the birds remember their final and intermediate halting-places so well, it appears that they must find their way in migrating by means of such familiar land-marks and stations. It is well known that "homing" Pigeons are lost if turned out in a strange country, and these birds are trained for long flights by breaking the journey into a number of stages and thus gradually lengthening the familiar landscape. In the same way it can well be understood how such a bird as *Phylloscopus borealis*, which migrates from Norway across Siberia to the East Indies, may originally have been a native of East Siberia, but extended its breeding range for a mile or two at intervals, the advanced individuals flying back over the known track to their comrades in autumn; while the young generation gathered experience from the older travellers, and thus the ancient traditional route was increased in extent.

Nevertheless it should be pointed out that Professor Newton, who was one of the first to examine the "homing" faculty of Pigeons in the hope that it might afford a clue to explaining how migrating birds find their way, has since been led to abandon it, holding that ocular memory as the guiding medium is disproved by three facts; first, that migrating birds fly over the open sea, sometimes thus traversing as much as a thousand miles before they can reach land; secondly, that much migration is done in darkness at night, sometimes at high elevations; thirdly, that "among migrants the young and old always journey apart and most generally by different routes". These reasons may well be considered beyond all objection by many ornithologists, but by others not so. A bird may take its direction across the vast open expanses of the Pacific, as has been already remarked, by means of the familiar lay of the land at starting, the rise and setting of sun and stars, and, as Prof. Möbius has suggested, by the direction of the roll of the waves; also by ascending to a height of 10 000 feet (which seems to be within what has actually been observed — Dict. B. p. 563) a hill of 2000 feet could be seen at a distance of 190 English miles, and similar heights in the land the bird is leaving would disappear at this distance behind.[1]) This of course could only take place under suitable atmospheric conditions, but, should such conditions be thought impossible, it may be mentioned that Meyer when in Celebes was able from the top of Mt. Klabat (over 6000 feet) to see the islands of Maju and Tifore in the Molucca Straits, a distance of about 90 English miles, and the guide, the Hukum kadua of Ayermadidi, who had ascended the mountain four times, stated that Ternate could sometimes be seen, a distance of about 170 miles. The hills of Tifore are only 300—400 feet high, but the Peak of Ternate reaches 5300 feet. The effects of refraction may still further increase the length of view. It would be of interest to know whether the birds ever attempt their great journeys across the Pacific at night. As to nocturnal migration it should be remembered that nights are seldom of a "pitch" dark character, when we are not aware that birds[2]) are heard migrating, and it is not unreasonable to suppose that they keep in view the outlines of the country over which they wander. As to the young birds migrating alone, it may be that a small percentage of adults among them may have been overlooked by those who have made these statements; like many others we do not entirely trust the evidence. But if it be accepted, such cases seem to point to "inherited experience", the young bird recognising the right way by innate knowledge, in the same manner as a newly hatched chick knows that grains of rice are good to eat. If birds have a "sense of direction", why do they take such indirect routes to their destinations?

---

[1]) Prof. Pattenhausen of the Royal Technical School in Dresden has most obligingly given us the following formula for making this computation: Granted the atmosphere as perfectly transparent and, further, that the rays of light travel directly onward from the surface of the sea, the equation is:

$$s = 3 \cdot 8 \, (\sqrt{h} + \sqrt{h'})$$

when h = the height of the point of view in metres,
h' = the height of the object seen in metres,
and s = the radius of view in kilometres.

[2]) Diurnal birds almost certainly have a long and clear range of vision, although it may not be fair to cite the Vulture, Kite, and Kestrel in proof. At night they seem to experience much the same difficulty as ourselves. Those who have practised netting along hedgerows with a beater on one side and a double-handed clap-net on the other are well aware that it is next to no use attempting to catch birds except on dark nights, and even then the great majority see the net against the sky and avoid it. But the manner in which a scared party flutter in some neighbouring hedge to which they have flown, is in part due to the difficulty they have in finding a perch in the dark, and shows that they do not see distinctly.

4. VARIATION.

The phases of Variation, or Modification of Structure and Plumage displayed among the birds of Celebes may be conveniently considered under the following headings:
1. Individual Variation; the differences peculiar to the individual.
2. Geographical Variation; as shown by local races, subspecies, or species.
3. Seasonal Changes; such as peculiar summer and winter plumages in birds.
4. Sexual Differences; the secondary sexual characters.
5. Changes depending upon Age; the development and decadence of the individual.

1. Individual Variation.

The assumption that no two individuals are ever exactly alike seems to be completely justified by facts. No one, probably, is so fully aware of this, as the zoologist, who is called upon to make the closest possible comparison of large series of individuals of the same race. In the course of writing the present book, for instance, which is chiefly based upon a study of the external coverings, and bills, legs and feet of Celebesian birds, with occasional reference to their skeletons — some thousands of specimens have been examined, yet to the best of our knowledge no two of them were exactly alike; moreover in the text several thousand measurements of parts will be found, yet we believe that hardly any two cases occur in which four terms (wing, tail, tarsus, bill) are the same. There are some very close observers of Nature to whom a knowledge of this infinite diversity of form is perceptible; who, as children, are conscious of the peculiarities of individual Sparrows or the differences in blades of grass; others, and amongst them men of learning, have never had their eyes opened to the fact, and assert that exactly the opposite is the case. Thus the idea of a uniformity of the individuals of a species is encouraged by the latter, with its consequence, that species were evolved per saltum. This position is partly the result of a system of nomenclature which no longer meets the needs of our time. There are of course species — groups of individuals possessing some character (or characters) never so found in any other group, — but each individual has its own peculiarities, and an ideal system of nomenclature can only be attained when each specimen is compared with the type — the first-described individual — of its species, and a formula representing the difference appended to the binomial name. Such a method is however impossible at present for practical reasons; there is no means of measuring the differences in question. Another reason why individual variation finds so little mention from those who have most experience of it is that for a long time a strong effort has been made to enforce the recognition of subspecies; the effect of this has been to give undue prominence to geographical variation and to divert attention from individual and other phases of variation, which form exceedingly difficult factors in the study of local races, and are for that very reason disregarded by careless or ignorant writers.

Range of individual variation. — Compared with some other animal forms, for instance, certain Beetles, individual variation in birds (other than domesticated races) keeps within somewhat narrow limits. The maximum as regards variation of plumage is seen in the male Ruff *(Machetes pugnax)*. The extinct Solitaire *(Pezophaps)* of Rodriguez was also extremely variable; its osteological remains show that "the variability of colour he (Leguat) had noticed in the females — some fair, some brown — was paralleled by the marvellous variability displayed by almost every bone of the skeleton" (Newton, D. B., 890). Other familiar examples of high individual variability among birds are furnished by the Honey Buzzard *(Pernis apivorus)*, the Common Partridge *(Perdix cinerea)*, and the Common Crossbill *(Loxia curvirostra)*. The minimum of individual variation is perhaps found among certain highly local species of Pigeons, such as those of the genera *Ptilopus* and *Osmotreron*; also many Sunbirds, Kingfishers, Parrots, etc., seem to be very stable in this respect. The differences separating many species of Pigeons are so small that it creates a feeling of surprise to find that large series of specimens do not present every sort of intermediate form and other modification. Under domestication, on the other hand, the Blue Rock Dove has displayed a most remarkable variability. A high degree of individual variability is sometimes found among species of birds in which the sexes are dissimilar (e. g. *Machetes*, *Macropygia*, *Gallus)*, but the sexes are dissimilar also in *Ptilopus*, *Osmotreron*, the Sunbirds, and other stable forms, so that this factor evidently has no direct influence in the matter.

Some species vary much in the measurements of various parts of the body, but little in coloration (e. g. *Streptocitta*, *Amaurornis moluccana*); others vary much both in size of parts and in coloration (e. g. *Halcyon chloris*, *Xanthocnus flavicollis*). Highly specialized features are generally very variable: for instance, the long tails of *Streptocitta*, *Phoenicophaes* and *Fregata*, the long hind claws of *Centrococcyx* and *Hydralector*, the long bills of *Limosa* and *Numenius*, the long tarsus of *Himantopus*; but age also plays an important role in this connection.

Psychological differences of individuals. — Persons who have closely observed or have kept wild species in confinement have often noticed differences of temperament in individuals; some are bold and fierce, others more gentle; some more clever, others more stupid; some trustful, others shy; and so on. A marked "individuality" in birds may sometimes be noticed in their choice of their feeding grounds and nesting spots.

Monstrosities. — Cases of exceptional individual variation, infraction of the rule of bilateral symmetry[1]) etc., have not fallen under our observation among Celebesian birds. There is a tame Duck with webless toes in the Sarasin Collection from North Celebes.

Albinism, Melanism, etc. — Among genera occurring in Celebes albinism, partial or complete, seems to be most frequent in the Coucals, *Centropus* and *Centrococcyx*. The Hornbill. *Cranorrhinus* sometimes displays white spots on the tail, but this may be a partial reversion to a form with a white band across the rectrices, as seen in some allied Hornbills. Cases of albinism are so common in the Heron, *Demiegretta sacra*, that this species may be rightly termed dimorphic. A further advance of albinism is seen in species which are now always white, such as certain Herons and Swans, for it seems certain that these birds were at one time coloured species.

According to the observations of Mr. K. G. Henke (see, besides, Z. ges. Orn. 1886 III, 268), albinism, when partial, does not conform to the rule of bilateral symmetry.

Partial melanism occurs in a highly variable degree in the large Talaut Oriole, *Oriolus melanisticus*, which appears to be developing into a species with a black upper surface. The Bittern, *Xanthocnus melaenus* (Salvad.) may ultimately prove to represent a case of frequent melanism in *X. flavicollis*. Permanently black species among Celebesian birds are *Ictinaetus malayensis* (when adult), *Surniculus musschenbroeki*, *Eudynamis* (males), *Rhabdotorrhinus exaratus* and *Cranorrhinus cassidix* (females), *Dicrurus*, *Corvus*, *Limnocorax*.

Examples of individual xanthochroism in *Trichoglossus ornatus* have been mentioned by Brüggemann, Meyer, and Guillemard among Celebesian birds (see text, p. 121). The species of *Cacatua* and *Myristicivora* are probably permanently xanthochroistic forms. The sulphur tint of the plumage of the former in life[2]) is due to the absence of the pigment fuscin, the colour parrot-green being caused by the yellow pigment psittaco-fulvin lying on the fuscin (see, Meyer, Sitzb. Ak. Wiss. Berlin 1882, 518); and the similar tint in the Nutmeg Pigeons is probably caused in the same manner, very possibly by the same pigment.

Dichromatism. — The phenomenon of dimorphism seems to be classifiable under Individual Variation, although there are cases where it appears to mark the commencement of the evolution of a new species. The best illustration of

---

[1] See below p. 68.
[2] It fades through exposure to the light in course of time in skins and stuffed examples, leaving the plumage white.

dichromatism among Celebesian birds is afforded by the Heron, *Demiegretta sacra*; this bird is ordinarily slate-coloured, but a pure white form is very frequent, and in some parts of its range, as in the Andamans, white individuals number some 20 per cent of the species. The two forms are known to breed together, and pure white young ones, as well as the usual dark ones, are known from the nest; piebald specimens are also not uncommon. It may well be, as has been suggested by Dr. Stejneger, that *Demiegretta sacra* will in the end become a pure white species, like the allied Herons of the genus *Herodias*, one of *Ardea*, and *Bubulcus* (when not breeding). In other parts of the world three more Herons are known among which white individuals are of very frequent occurrence (see text, p. 822).

The genus *Spilornis* is supposed by some ornithologists to be dichromatic when young, but further proof of this is wanting (see p. 3). *Ardetta eurhythma* perhaps makes an approach to dichromatism, since the male is sometimes known to breed in its immature dress (the ordinary female dress), and a female is occasionally found in the male plumage. Other cases of dichromatism have not been found among Celebesian birds, and the phenomenon is indeed always rare in ornithology.

Modifications of the individual due to foreign violence, such as injuries to feathers and parts of the body, disease, effects of shelter and exposure, of food etc. cannot be discussed here, as leading too far. Only a remark or two. The heredity of oft-repeated external action on feathers is discussed below, pp. *73—79*. Remarkable effects may sometimes be produced by food, such as the conversion of Canaries or white Fowls into red ones by feeding them with Cayenne pepper. Isabelline-coloured Pigeons fed with crumbs coloured with Methyleosine were turned into birds of a permanent red tint, and green Australian Parrakeets (*Mellopsittacus*) supplied with millet coloured with Methylviolet were converted into blue birds, the yellow forehead becoming white or dirty white (Sauermann, Mitth. Ornith. Vereins Wien 1890, pp. 92—94). Parrots (*Sittace*) in Brasil are made to change colour into yellow by plucking out feathers and inoculating the wounds with a frog's or toad's blood or with the milky secretion from its skin; when the new feathers grow the colour is changed. The common Amazonian green Parrot (*Chrysotis*), if fed with the fat of large Siluroid fishes, becomes beautifully variegated with red and yellow feathers. In Halmahera *Lorius garrulus* is said to be influenced in a similar manner. (After v. Martius and Wallace, communicated by Meyer: Sitzb. Ak. Wiss. Berlin 1882, 521.)

Thus we imagine that a bird flying to a neighbouring island and finding there another sort of food, may, if settled there, acquire some new character in coloration; e. g. *Loriculus stigmatus* flying over to Banggai becomes *sclateri ruber*, to Togian Island *quadricolor*, etc. etc., these forms only differing slightly. Such alterations may occur per saltum or at least quickly, not requiring generations. A fruit from Batavia planted at Manado does not always remain the same, as is well known there; wood-ruff transplanted from the woods loses its aroma. Examples could be added by scores. Though we cannot explain these alterations mechanically, facts remain; so it is with insignificant variations in the colours of birds in new localities, if isolated. How easily colour may be influenced is shown by the following case of one of the Musophagidae, *Corythaix albocristatus*, first made known by Dr. Chenu (Encyclopédie D'Hist. Nat. Oiseaux 2ᵐᵉ partie 1855 p. 55): "Une particularité remarquable, dont nous devons la communication à Jules et Édouard Verreaux, si bons observateurs, c'est que les douze ou quatorze pennes alaires, qui sont d'un si beau pourpre violâtre, perdent cette couleur chez les individus vivants, lorsqu'elles ont été mouillées par la pluie: si, dans cet état, on vient à les toucher ou à les frotter avec les doigts, ceux-ci se trouvent aussitôt rougis par la couleur pourprée qui a déteint sur eux; et, en séchant, ces mêmes plumes reprennent leur éclat primitif. Sur la dépouille de l'Oiseaux, aucun effect semblable ne se produit. Ce fait nous paraît unique dans la classe des Oiseaux."[1]) Though this be exceptional, no doubt chemical or mechanical alterations of colour occur elsewhere, be they dependent on food, light or other external influences, touching which we know next to nothing at present. Alteration of colour in individuals, gone astray to isolated localities, leads us to geographical variation, which should next be treated of.

2. Geographical Variation.

Although it is conceivable, and indeed likely, that a new species may sometimes owe its origin to dimorphism, a condition which may be ultimately due to the successful multiplication of a single case of exceptional individual variation, it is nevertheless far more certain that the great majority of the peculiar forms of Celebes and the neighbouring islands are what are termed geographical species or local races, which have developed their distinctive characters while geographically isolated from one another. In the Celebesian area there are about 150 species of this description now known, not to speak of a large number of partially formed races. The latter are in many respects the most interesting, as they show species in the first stages of their differentiation, and their study holds out the best hope of solving the problem of the origin of species — or at least of the majority of species. The differences seen are often very small, but of a very palpable description, as, for instance, the broader black border to the secondaries of *Eos histrio* in Sangi, the narrower border in Talaut; the darker grey of the head of *Phoenicophaes calorhynchus* in North Celebes, the lighter grey in the South, and so on. These differences may be due to an inherent tendency in the individuals in question to evolve

---

[1]) Compare Schlegel: J. f. O. 1858, 381; A. Bogdanow: C. R. Ac. Sc. Paris 1857 XLV, 311 and 1862 LIV, 660; Brehm Tierl. 3. ed. 1891 V, 138; Krukenberg: "Die Farbstoffe der Federn" in his Vergl.-Physiol. Stud. 3. Abt. 1881, 75.

in a certain direction (as a complete ceasing from all variation, even under unaltered conditions, is inconceivable in the course of the propagation of organic forms), or they may be caused by local influences. For some cases the former assumption appears unavoidable; for other cases there is satisfactory evidence of the effect of local influences, though the exact nature of these latter is almost always uncertain; as a rule, probably, both causes operate together, but it very rarely happens that an opinion either way is permissible at present. In illustration of independent development in the same direction it may be mentioned that the genus *Loriculus* has produced two species with very similar red crowns in Ceylon and in the Southern Philippines; and a corresponding distribution of markings is seen in the plumage in the two forms of Celebes (*L. stigmatus*) and Sooloo (*L. bonapartei*), which are not closely related to one another (see p. 163 et seq.). Local influences are sufficiently indicated when a number of species are found to vary in the same place in a corresponding manner; for instance, two phases of modification have been detected by Mr. Allen among North American birds and are recapitulated by Professor Newton (Dict. B. 1896, p. 1005) as follows: there is "a general increase of intensity toward the south and development of dark markings at the expense of the light intervening spaces, so that of brightly-coloured species southern individuals are the most brightly coloured, and some tints, which to the northward cannot be called brilliant, become vivid in a lower latitude. In respect of longitude Variation occurs with like regularity, the differences appearing to hold a direct relationship to the humidity of the climate. Thus on the dry plains of the middle and western parts of the continent birds have a pallid complexion, while on the Pacific slope they resume nearly the tints of the eastern form, though further to the northward, in the rainy belt that extends along the coast of British Columbia, they acquire a depth of colour far in excess of that which they display on the Atlantic border."[1]) In such cases the direct influence of climate upon the colours appears to be proved.

The following instances of correlated geographical variation are noticeable among the birds of the Celebesian Province.

Increase of size in the Sangi and Talaut Islands. — The local species or races of Sangi and Talaut having their nearest affinities with species on the mainland of Celebes are:

---

[1] This quotation, which we have reprinted, is only from a résumé of Mr. Allen's original memoir: "On the Mammals and Winter Birds of East Florida" (Bull. Mus. Comp. Zool. Cambridge, 1870—1, II, pp. 161 —450, plates IV—VIII) which should be consulted here, chiefly pp. 239 sq., where he says: "Causes of Climatic Variation. — ... The southward increase in depth of color and in iridescence in birds specifically identical coincides also with the general increase in brilliancy of color in birds, taken as a whole, in the lower latitudes (as well as in insects and animals generally), the maximum being reached in the tropics. — The longitudinal Variation, or the westward increase in color, seems to be also coincident with the increased humidity to the westward, the darker representatives of any species occurring where the annual rainfall is greatest and the palest where it is least ..."

Introduction: Variation.

Sangi	Talaut
1. *Tanygnathus muelleri sangirensis* M.Wg.	1. *T. muelleri—sangirensis*
2. *Ceycopsis sangirensis* M.&Wg.	
3. *Cittura sangirensis* Sharpe	
4. *Hypothymis rowleyi*[1]) (Meyer)	
	2. *Dicaeum talautense* M.&Wg.
5. *Acmonorhynchus sangirensis* (Salv.)	
	3. *Hermotimia talautensis* M.&Wg.
6. *Anthreptes malaccensis chlorigaster* (Sh.)	
7. *Zosterops nehrkorni* W. Blas.	
8. *Calornis panayensis sangirensis* (Salv.)	4. *Calornis panayensis sangirensis* (Salv.)
9. *Osmotreron sangirensis* (Brügg.)	
10. *Ptilopus xanthorrhous* (Salv.)	5. *Ptilopus xanthorrhous* (Salv.)
11. *Macropygia albicapilla sangirensis* (S.)	
12. *Megapodius sangirensis* Schl.	6. *Megapodius sangirensis* Schl.

To the Talaut list the following species having their nearest affinities in the Philippines and elsewhere should be added: 7. *Tanygnathus talautensis* M.&Wg.; 8. *Zosterops babelo* M.&Wg.; 9. *Oriolus melanisticus* M.&Wg.; 10. *Carpophaga intermedia* M.&Wg.; and to the Sangi list should be added *Oriolus formosus* Cab., allied to Philippine and Sula forms.

There are about 17 species in Sangi with near affinities to forms on the mainland of Celebes, but which have developed more or less appreciable local differences. Of these 17, no fewer than 12 named in the above list have increased in size in Sangi, while two others, *Prioniturus platurus* and *Dicaeum sangirense*, are probably also a trifle larger than their relatives on the mainland. *Dicrurus leucops*, *Hermotimia sangirensis* and *Pitta caeruleitorques* have not increased in size, but they are at the same time not smaller than their Celebesian allies. In no case have Celebesian species decreased in size in Sangi.[2])

The imperfectly explored Talaut Islands are at present known to possess 8 peculiar species allied to forms belonging to Celebes or the Philippines, not counting 2 with Moluccan affinities; and to these eight should be added 4 Celebesian forms which Talaut possesses in common with Sangi. Ten of these twelve species display a marked increase of size in Talaut, and the other two show no reduction.

The converse supposition that the large forms of Sangi and Talaut represent the original size and that the races of Celebes and the Philippines are those which have undergone alteration, viz. reduction in size, is not plausible. The islands seem most obviously to have been colonised chiefly from the mainlands

---

[1]) Nearest affinities uncertain.

[2]) From the island of Siao belonging to the Sangi group a single very small specimen of an Owl like *Scops menadensis* has been described by Schlegel as *Scops siaoensis*. Its distinction is not admitted in this work (see p. 104).

of Celebes and the Philippines, and not Celebes and the Philippines from them; their volcanic or coral character (see de Hollander, Land- en Volkenkunde Ned. Oost. Ind. 1884, II, 234) and the absence of heavy, ill-flying birds and of peculiar generic types speak for the recent upheaval and colonisation of these islands[1]). It might happen that a species or two on the mainlands subsequently became smaller, but there is no reason to assume that the twenty species which remained in Celebes and the Philippines all became small hereafter, while those which had peopled the islands all maintained the original sizes of the species. On the other hand there is reason to anticipate that the individuals of these twenty species which had emigrated to the new islands would undergo alteration of some kind or other, for the conditions of existence are not precisely the same there. It appears, therefore, quite safe to assume that **Celebesian and Philippine birds develop as a rule into larger races in Sangi and Talaut**[2]).

As is usual in such cases it is not difficult to find more than one explanation why these things should be so, but not easy to decide which explanation is the true one. The most plausible suggestion is that the dangers are fewer and the struggle for partners perhaps more severe in Sangi and Talaut than on the mainland. Hawks and Falcons which prey upon birds seem to be very rare in these islands; so far not one bird-eater, strictly speaking, has been killed in them, for *Tachyspizias soloensis*, the most dangerous to small birds, is not only a migrant but feeds to a great extent on insects. Moreover there are no monkeys in Sangi and Talaut, and other enemies, which could be dangerous to breeding birds, their nestlings and eggs, are rarer than in Celebes and the Philippines. The chief competition therefore that goes on would appear to be among the birds themselves, and the largest and strongest will be more likely to secure nesting quarters and partners than the smaller and weaker[3]).

Decrease in size in Sula. — Two Birds-of-prey and two Pigeons display a slight reduction in dimensions in the Sula Islands, viz. *Spilornis rufipectus, Accipiter sulaensis, Turacoena manadensis* and *Macropygia albicapilla.* Sula seems to resemble Celebes in its Birds-of-prey.

Differences in size in the North and South Peninsulas of Celebes. — As a rule the birds agree in their dimensions in these districts, but where there is a difference it seems to be in the direction of an increase in size in South Celebes. *Pachycephala meridionalis* and *Stoparola meridionalis* are much larger than the

---

[1]) Prof. Hickson noticed evidences of recent slight elevation in Talaut (Nat. in N. Celebes 1889, 131, 137,.

[2]) The case affords a good illustration of our postulate, that colonists become more changed than stayers-at-home (see, p. 162).

[3]) It may be added that many ornithologists are of opinion that the males of most species of birds are more numerous than the females, and Dr. Platen certainly collected many more males than females in the Sangi Islands (see W. Blasius, Ornis 1888, pp. 527—646.; but it may well be that the superior plumage of the males leads to their being shot and skinned more frequently, and it is preferable not to introduce this doubtful element into the argument.

representative species, *P. septentrionalis* and *S. septentrionalis*, of the North; also *Carpophaga paulina* shows a slight increase in size in the South, and a very small increase is noticeable there in *Streptocitta albicollis* and in *Phoenicophaes calorhynchus*.

Similar geographical variations of coloration. — The Lories of the genus *Trichoglossus* range from Australia as far as Celebes and consist of two groups, *Trichoglossus* proper of Australia, the Lesser Sundas, Papuasia, the Moluccas and Celebes, and *Psitteuteles* of Australia, the Lesser Sunda Islands and Celebes. Count Salvadori (1891) recognises 16 species, all of which have a yellow (in two cases red) band across the base of the remiges, except in the Celebesian area, where there are three species, *T. ornatus* and *P. meyeri* in Celebes, and *P. flavoviridis* in Sula, which have no yellow band. We have, however, found small evidences of a yellow band in two or three immature specimens out of 17 examples of *T. ornatus* and in three young specimens and one female in a smaller series of *P. meyeri* — a significant indication that these species are derived from birds which possessed the band, such as are found inhabiting the countries to the east and south of Celebes to-day. Why the birds have lost the band in the Celebesian area it appears useless to speculate.

Pitta forsteni of Celebes wants the usual white wing-band.

The Serpent-harrier, *Spilornis rufipectus*, and the Sparrow-hawk, *Accipiter rhodogaster*, of Celebes, are represented in the Sula Islands by two closely allied, but slightly smaller forms (*S. rufipectus sulaensis* and *A. sulaensis*). Both of these have undergone a similar modification of the wing, viz. the bars on the under surface of the remiges have become narrower in Sula, or have increased in width in Celebes, as the case may be.

The Cuckoos of the genus *Eudynamis*, which range from the Himalayas to Australia, have pale bills, and the Kingfishers of the genus *Pelargopsis*, ranging from India to the Sula Islands, have red bills, except in Celebes, where both the Cuckoo and the Kingfisher have the bill black, while the bill of the latter is varied with black and red in the neighbouring Banggai Archipelago.

Out of the ten known geographical species of the Talaut Islands three display melanotic influences or, at least, a darkening of their tints; these are *Oriolus melanisticus*, *Dicaeum talautense*, and *Pitta inspeculata*. The Lory, *Eos histrio talautensis*, has, however, slightly less black on the wings than the *typical Eos histrio* of Sangi.

The above cases are included under the heading Geographical Variation, because their peculiarities of coloration seem most probably to be connected with some unknown local influences; there are in the Celebesian area, however, other cases of similar variation, the cause of which seems to be in no way connected with the locality. Such are *Pernis celebensis* and *Spizaetus lanceolatus* which are similar, adult to adult, and young to young, as are in the same way also *Spilospizias trinotatus* and *Accipiter rhodogaster*; while *Muscicapula westermanni*,

Lalage leucopygialis and *Graucalus bicolor* correspond in coloration to a considerable extent, male with male, and female with female. These cases call for consideration later on.

3. Seasonal Changes.

The modifications which birds undergo at certain periods of the year seem to depend sometimes upon climatic, sometimes upon sexual conditions. The breeding season however is regulated by climatic conditions, the young being brought forth at a period when food is abundant; consequently climate should be regarded as promoting all periodic variation. Climate alters the appearance of the surface of the earth — causes it to be clothed with a luxuriant vegetation or covered with snow and ice, now bringing forth an abundance, and then removing the supply of food — and organisms are modified to suit these conditions. In the tropics, as, for instance, in Celebes, where a contrasted summer and winter does not exist, but only a fine and a rainy season, strongly marked periodic changes in the plumage of the birds are rarely seen. More than 160 peculiar species are now known from the Celebesian area, and seasonal changes are not known to occur in a single one of them, though sexual differences are common enough. A few tropical or subtropical Herons (*Ardeola*, *Herodias*, *Bubulcus*), a *Cisticola*, and perhaps one or two others which are resident in Celebes differ when in nuptial and simple plumage, but, in order to see seasonal variation in full evidence, it is necessary to look to the northern temperate and arctic regions. Here, as is well known, most remarkable contrasts of summer and winter plumage are abundantly represented; as, for instance, the varied dress of the Ptarmigan (*Lagopus mutus*) in summer, its snow-white plumage in winter; the black under surface of the Golden Plover (*Charadrius*) in summer, the whitish of these parts in winter. Many northern forms visit Celebes in winter, often in an attire very different from that in which they breed in the North; amongst them may be mentioned the Eastern Golden Plover (*C. fulvus*), and the Grey Plover (*Squatarola*) which undergo a similar seasonal change; the Stints and Godwits which are suffused with rufous in summer; the Glossy Ibis (*Plegadis*) which has the under parts chestnut in summer, earthy brown in winter; the Phalaropes; certain Terns (*Hydrochelidon*), etc. These changes are not of a sexual nature, as the sexes differ little or not at all in coloration, and both are subjected to the same seasonal changes; but in many — probably in most — cases where there are any secondary sexual differences these characters are intensified in the breeding season and new markings are sometimes added in the male sex (e. g. the ruff of *Machetes*, the black facial markings in some species of *Aegialitis*, the long tail-feathers of *Vidua*). It may, however, also happen that the sexes are less similar in the winter season than when breeding; this seems to be the case to a slight extent with *Anthus cervinus*.

The moult. — In the temperate and cold regions of the northern hemisphere it is generally admitted that a complete moult takes place in birds in autumn after the breeding season; many species moult again in spring, and some a third time in summer. The principal time for moulting in Celebes, Sangi, and Talaut seems to be from July to Dece[m]ber, when the birds probably undergo a complete post-nuptial moult. It is questionable whether signs of a spring moult can be found; on the other hand some species may be found moulting during most months of the year. For instance, specimens in the Dresden Museum of *Heteractitis brevipes* are moulting in January, April, July, August, November; of *Actitis hypoleucos* in January, March, July, November.

Some of the Waders, autumn visitors from the North to Celebes (*Aegialitis geoffroyi*, *Heteractitis*), seem to moult first on the under parts, then the remiges, and finally on the upper surface. No regular order in moulting is pursued in the Black Sunbirds (*Hermotimia*), among which the transition from young to adult male dress can be particularly well observed. The characteristic metallic subgular stripe of the Celebesian species makes its appearance first, but the rest of the plumage is developed without any such regular sequence, and there is a specimen of *Hermotimia talautensis* in the Dresden Museum (C 15377) in almost complete adult dress except on the forehead and crown, while a second of the same species (C 13847) has the plumage of the adult on the forehead and most of the crown, but the young dress on most of the other parts. This proves that the transition from the young to the adult dress does not take place in perfect phylogenetic order; that is, the adult characters are not necessarily developed in the young male bird in exactly the same sequence as that in which they were acquired in the evolution of the race (see, also, immature male, pp. 469, 471).

Besides their feathers some birds are known to shed certain corneous appendages or coverings on their bills; for instance, the white Pelican of America has a horny knob on the culmen during the breeding season, but which falls off when that period is over; and the Puffin (*Fratercula*) moults the horny sheath of its bill and the outgrowths over the eyes (Newton, D. B., pp. 599, 600). It is possible that a similar moult of the ribbed plates at the base of the bill of the Celebesian Hornbill, *Cranorrhinus cassidix*, takes place. It is believed by the natives to add one rib-plate each year; and, though this notion is certainly wrong, it is possible that a shedding of the plates has been observed.

Change of coloration without a moult. — In a recent number of the "Auk" (1897, April) Dr. Chadbourne has furnished what seems to be the first really conclusive evidence that a change of colour may take place in the perfect feather, this being caused by a redistribution of the pigments already present in the shaft and barbs. The observations were made on the male Bobolink, *Dolichonyx oryzivorus* (L.), but there can be now no doubt that the principle is general among birds. The difficulties of making observations are great, and no certain evidence has been adduced from Celebesian birds, but Prof. W. Blasius holds that *Centrococcyx bengalensis* when passing into adult dress is subjected to certain colour changes without moulting (see, also, p. 215 of our text).

The changes in colour of certain corneous or epidermal parts, such as the bills and legs of certain Herons (*Herodias*, *Bubulcus*) in the breeding and winter seasons, may perhaps be placed in the same category as changes of colour in the feathers without a moult (see p. 838).

4. Sexual differences.

In relation to sex it is convenient to gather birds into three groups:

1. **Male more highly developed than the female.** Examples: *Paradiseidae*, *Trochilidae*, *Cinnyridae*, many *Phasaniidae*, many *Anatidae*, etc., etc.

2. **Sexes alike.** Examples: *Pittidae*, *Artamidae*, many *Ploceidae*, many *Alcedinidae* and *Cuculidae*, most *Ardeidae* and *Laridae*, etc., etc.

3. **Female more highly developed than the male.** Examples: *Turnicidae*, *Phalaropus*, *Limosa*, *Hydralector*, *Centrococcyx*, *Rhynchaea*, *Eudromias*, *Casuarius*, *Dromaeus*, the *Crypturi* and others.

To these may be added doubtfully:

?4. **Sexes developed on independent lines of evolution.** *Eudynamis*, *Monachalcyon*, *Cittura*. The sexes either differ in coloration from the nest or after the first plumage: nevertheless there is some reason to think that the adult female represents an earlier stage in the evolution of the race, and that the species concerned should, therefore, be placed in the first group. Thus the female of the Black-billed Koel, *Eudynamis melanorhyncha*, resembles another Cuckoo, *Centrococcyx bengalensis*, when the latter is in first plumage; *Monachalcyon monachus*, especially the female, bears resemblance to *Halcyon hombroni* of the Philippines; the female of *Cittura cyanota* is more like both sexes of *C. sangirensis* than is the male.

It is not to be understood that these groups are always sharply characterized and easily distinguishable; on the other hand, gradual transitions from one group to another are found: from such contrasts of the sexes as are seen in *Paradisea*, *Gallus* and *Cinnyris* in which the male far surpasses the female in adornment, to *Tanygnathus* and *Zosterops* in which the female is hardly inferior to the male, to *Pitta* and *Myristicivora* in which there is nothing to the human eye to choose between the two sexes, to *Limosa* and *Phalaropus* in which the female becomes rather the finer bird, and so on to *Turnix* in which she is much superior to her partner. It also happens at times that the male is the more advanced in one respect and his mate in another; thus, among the Birds-of-prey the male generally has the more highly developed plumage, but the female is of larger size.

The psychological differences of the sexes. — The rule found among mammals — that the male is more active and wars and works for the sake of the female, while the female is more passive and gentle and devotes herself more to the care of the young — holds good also for large numbers of birds, but in many others the sexes seem to be much alike in temperament and to share duties, while in some species the rule is more or less completely subverted, the male undertaking the "female duties", and the female assuming the usual role of the male. The fact is important, as it shows that there are no mental peculiarities originally bound up with the primary fact of sex. It appears, moreover, that these psychological conditions often (but not always) accompany the three conditions of development of plumage and structure mentioned above; namely, when the male is more highly developed than the female, he is noisy, combative and extravagant of display in his courtship, while the female builds the nest or most of it, incubates the eggs, and takes the chief or sole care of the young; when the sexes are alike, the males are less quarrelsome in the breeding season, less demonstrative in their courtship, and share the work of incubating the eggs and rearing the young; when the female is the more highly developed, she is noisy, pugnacious with other females and courts the male, leaving him to do most or all of the work of hatching the eggs and caring for the young. Thus the highly coloured males of the *Trochilidae*, many *Anatidae* and *Gallinae* seem not to concern themselves for the brood to which the plain-looking female devotes herself most faithfully, whereas the large and handsome female *Turnix* roams about and calls and fights other females, leaving the smaller and plainer male to attend chiefly to the incubation of the eggs and the welfare of the chicks, though indeed she does most of the nest-building and assists a little in hatching the eggs (Krohn, Gefied. Welt, 1894, 190). The female of one of the Emus which is larger than the male and wears a slight top-knot has been observed in captivity not only to leave the entire work of incubation to the male, but apparently to use her utmost endeavours to destroy her young when hatched (Darwin, Descent of Man 1871, II, p. 205).

Among species the sexes of which are much alike in appearance and which share the duties of incubation may be mentioned the Tits, *Paridae*; many Warblers, *Sylviidae*; some Larks, *Alaudidae*; some Buntings, *Emberizidae*; certain Finches, *Fringillidae*; Woodpeckers, *Picidae*, and others; while in other cases the male feeds the brooding female and sometimes relieves her in sitting for a short time (cf. e. g. Naumann's Vögel Deutschlands, 1824, IV, 93 and in many other places). But it is by no means always the case that the finer one sex in birds is in comparison with the other, so much the more he (or she) will abandon the nest, eggs and young to the humbler consort, and that the more similar they are in appearance, the more evenly will they share duties. As instances to the contrary may be mentioned the male Ostrich, which, though the finer bird, broods on the eggs of his wives at night; the females of the Birds-of-prey, which are usually superior to the males in point of size (though not in coloration), fulfil the usual maternal duties properly; the males of the Swallow, Goldfinch and Hoopoe which, though very like their mates, are said to take no share in the incubation of the eggs; while the male Reed Bunting and Blackcap, which are more highly developed than their mates, nevertheless help a little in hatching the eggs. There appears, therefore, to be no hard and fast law of correlation in the evolution of higher organic development and of mental functions of the "masculine" type; in other words, the structural differences and the psychological differences of the sexes seem to have been developed independently.

Theories in explanation of the development of secondary sexual characters. — Several have been advanced:

1. Darwin (Descent of Man, 1871, pp. 38—238, Birds) accounts for the superiority of the male by reason of the choice by the female of the male which pleases her best (sexual selection), and

2. partly by the survival of the fittest in combat.

3. Wallace (Darwinism, 1889, 289, et seq.) believes that the secondary sexual differences have risen to a higher development in one sex owing to a prepotency of vitality or growth-power, and some evidence is adduced tending to make it plausible that the accessory plumes of the males are developed over centres of high muscular or nervous activity.

4. Wallace (t. c. p. 277, Darwin, t. c. p. 166) adds the complementary theory that the need for protective coloration in the brooding female has prevented her by natural selection from acquiring many of the bright colours and showy ornaments of the male.

5. Stolzmann (P. Z. S. 1885, 421, et seq.) bases a theory on the assumption that among birds the males are more numerous than the females. Any development of colour or markings which is disadvantageous to the males, by rendering them more conspicuous and more easily destructible to foes (whether predaceous animals or males of their own species), will be advantageous for the species, because the superfluous males are parasites devouring food which would be useful to the breeding birds, persecutors of the brooding females, destroyers of the eggs, etc. It is argued that natural selection will favour the preservation of those females which produce male offspring handicapped with such peculiarities of structure, plumage, temperament, etc., as are likely to bring about the destruction of these males.

6. Beddard (Animal Coloration, 1892, p. 282 *et antea*) finds that "the secondary sexual characters of animals are dependent upon the germ glands themselves; and that the sexual diversity of animals is also associated with differences of disposition and habit".

Touching Darwin's theories it is obvious that natural selection in the "law of battle" affords a simple explanation of the development of certain offensive and defensive organs, greater size, strength, activity and courage.

Mr. Wallace's theory of the development of protective coloration in the female is also easily to be understood on the ground of the elimination of the disadvantageously coloured birds of this sex, yet the explanation does not seem to apply to the majority of cases, in many of which the female is like the male, and in others she is only a little less bright or wants some special marking and appears then to represent a lower stage in the history of the race, as the immature male is often like her. Many males assist in incubation. The female of the Cuckoo, *Eudynamis*, which lays its eggs in Crows' nests does not appear to be protectively coloured, but the male, being black, might be thought to be so.

Darwin's theory of sexual selection has been much contested of late years. The author cites cases of certain female birds in captivity mating by preference with certain males and avoiding others; though allowance must perhaps be made for this in nature, there is now a strong opinion in favour of the view of a passive role being generally played by the female, the male expelling his rivals and making the female yield to him.

There is much to be said for Mr. Wallace's view of an excess of vitality or growth-force in the male as the cause of the development of superfluous decorative plumes, etc., though a localization of such growths in the skin "over centres of high nervous or muscular activity" is not tenable. For instance, the second primary of the male *Macrodipteryx*, an African Nightjar, is developed into an enormous racket-feather capable of erection; three long racket-feathers sprout from each side of the head of the male of the Paradise-bird, *Parotia*, one very long one in *Pteridophora*, etc., etc. As the principal muscular and nervous centres are not different in birds, such a great diversity in the location of the accessory growths could not arise from this cause. Why does the male *Paradisea* have its ornamentation chiefly on the side of the breast, and another Bird of Paradise, *Lophorhina*, on the occipital region and jugulum?

Mr. Wallace's theory appears to include "the normal development of colour due to the complex chemical and structural changes ever going on in the organism" (Darwinism, p. 288), for the sex which possesses the most growth-force will be the first to undergo these necessary modifications. It is probable that a great number of sexual differences owe their origin to this developmental law. (See *Loriculus*, antea p. 57).

Mr. Stolzmann portrays the two sexes as naturally inimical to one another's well-being. The males above a certain number are useless parasites, they diminish the food-supply and persecute the females; ill-fed females produce an excess of male offspring, and the female for her own preservation produces males which are disastrously equipped for the struggle for existence. We are unable to grasp the argument, if indeed it is a valid one, for it appears to us that the handicapped males will be the first to perish, and the males which will perpetuate the species will be the sons of females which produce the best-equipped offspring. Their qualities being inherited, these males will somewhat counteract the tendency on the part of certain females to produce inferior males, and the latter females will be less likely to survive than their sisters. As their inferior male offspring will not be allowed to perpetuate itself it would appear that such females will have to breed with the finer males and their harmful effect on that sex will be done away with in course of generations, because the male qualities are transmitted in part to the female offspring as well as to the male.

Mr. Beddard's opinion that sexual dimorphism is mainly dependent upon the reproductive organs is based upon such rare cases as that of a hermaphrodite Chaffinch (p. 262) in which one side of the bird was found to be like the male in coloration, the other side like the female, with the generative systems correspondingly divided[1]). As bearing upon the same matter may be cited the circumstance that old females which have lost their fertilily sometimes assume the male dress.[2]) Such facts appear to be very instructive, perhaps proving that the sexual glands themselves through nervous influences determine the coloration of the integument, intricate questions, which we are not prepared to discuss here. But when Mr. Beddard suggests that the differences in disposition and habit of very many males and females are dependent upon the sexual germ glands themselves he appears to be contradicted by reasons given *antea*, p. *65*.

Exceptional cases. An examination of exceptions often throws more light upon a matter than is afforded by the contemplation of the rule. The female of the Coucal, *Centrococcyx bengalensis*, is much larger and stronger than the male, though of similar coloration; it utters remarkable cries and is not known to take any share in incubating the eggs; the male is small, silent, and it broods on the eggs. Moreover the male possesses only one testicle, the left one being entirely wanting. The conditions have been fully described by their discoverer, Bernstein, in the Natuurk. Tijdschr. van Ned. Ind. 1860, pp. 27—49, pl. I; and mentioned J. f. O. 1859, 185; 1860, 269. (See also, *subtus*, pp. 219, 220). Apparently both Darwin, Wallace, and Beddard might claim this case as supporting their different theories. Darwin, though he seems to have overlooked the fact, anticipated the possibility of such a condition: "If we might assume that the males . . . have lost some of that ardour which is usual to their sex, so that they no longer search eagerly for the females . . . then it is not improbable that the females would have been led to court the males, instead of being courted by them" (Descent, p. 207). For Wallace's view it might be claimed that the structural deficiency of the male points to a lower status of vitality, sufficient to account for its smaller size and quiet habits. In accordance

---

[1]) The Chaffinch quoted by Prof. Beddard was described by Prof. Weber (Zool. Anzeiger 1890, 508). Compare Prof. Cabanis' descriptions of such differently coloured halves in *Pyrrhula vulgaris* and *Colaptes mexicanus* (J. f. O. 1874, 344); of v. Rosenberg's of a Chaffinch with two anterior halves of the body, the one in coloration a male, the other a female (M. O. Ver. Wien 1884 VIII, 87 & plate) and of Kleinschmidt's of a bilateral-asymmetrically coloured specimen of the Common Kingfisher (Abh. u. Ber. Zool. Mus. Dresden 1898/9 Nr. 2 p. 73, plate III). Also the remarks of Prof. Brandt on *Arrhenoïdia lateralis* (Z. wiss. Zool. 1889 XLVIII, 107) should be consulted. Lorenz asserted that he had seen a similar case in *Tetrao tetrix* (see: Tichomirow: "On Hermaphroditism in Birds" — written in Russian — 1887 p. 21 note), but we doubt this.
[2]) "Hahnenfedrigkeit". Comp. Meyer: Auer-, Rackel- und Birkwild 1887 p. 33, and Abh. Ber. Mus. Dresden 1894/5 Nr. 3, as well as Brandt's paper quoted in note 1.

with Beddard's theory, the reduction in size and the quiet disposition of the male bird should be due to the partial atrophy of the male organs. Beddard's theory here seems to contain the most probable explanation, and it would be well to make observations on other *Centropodinae*. It still remains to be demonstrated how such an aberration has arisen, which is perhaps comparable to the development of only the left ovary[1]) in the female of all birds.

In the case of *Turnix nigricollis*, in which the male does most of the work of incubating the eggs, the large and strikingly coloured female is supposed from observations by Mr. Krohn (Gefied. Welt, 1894, 190) to be given to polyandry. Darwin cites Jerdon's remark that the females are "much more commonly met with than the males". It is difficult to reconcile this statement with the former supposition, but these cases are mentioned here as showing that the contrast in size, habits, etc., may sometimes, as in *Centrococcyx*, be accompanied, and perhaps determined, by a deficiency of reproductive energy or capacity, or sometimes, as perhaps in *Turnix*, by an excess of reproductive power.

In addition to the 6 theories of the origin of secondary sexual characters mentioned above at least two more can be indicated.

7. Secondary sexual characters as "recognition markings". Mr. Wallace (Darwinism, 1889, p. 217 et seq., and in other works) seems to have been the first to define conspicuous markings and patches of colour as useful means by which individuals of a species may at once recognise others of their own kind. He applies his theory to species and further on (p. 284) to the sexes.

8. Development of accessory sexual characters owing to external violence or excessive physiological employment of the parts in question. Use promotes the development of a part in the individual, disuse its atrophy. In the next section of this chapter reason is also given for the opinion that mutilations of feathers — and hence of other parts — if repeated for generations are inherited.

In the present case out of 8 theories of the origin of secondary sexual characters it may well be that 6 have been actually operative in Nature, working alone or more likely in different combinations and degrees. These causes are:

1. The differences of the reproductive organs (Theory 6).
2. Higher development owing to a prepotency of growth-force (Theory 3).
3. Survival of the fittest in combat (Theory 2).
4. The stimulation of parts to a higher development by use and external violence or irritation (Theory 8).
5. Development of recognition-characters by natural selection and preferential mating of males and females which can distinguish one another (Theory 7).
6.[2]) Protective coloration for the sex which broods on the eggs (Theory 4).

---

[1]) A rudimentary right ovary is usually present (Gadow, Vög. in Bronn's Kl. u. Ord. 1891, p. 842).
[2]) As to Darwin's theory of sexual selection authors are disagreed.

5. Changes depending upon Age.

The modifications of plumage and structure displayed during the life-time of the individual, the phenomena of its development and decadence, may fitly be placed at the end of this chapter, as one form or other of the four preceding phases of variation — sexual, seasonal, geographical, and (if perpetually recurrent) individual variation — is often repeated during the growth of the young towards maturity.

Classification of the developmental phases. — Charles Darwin (Descent of Man, p. 187) gives six "classes of cases or rules under which the differences and resemblances, between the plumage of the young and the old, of both sexes or of one sex alone, may be grouped". Keeler (Evol. Col. Feath. 1893, p. 213) adds two classes more. All eight of them have representatives among Celebesian birds, and they allow of re-grouping according to the phase of variation which exerts a predominant influence in each case.

Sexual influences predominate in four classes:
1. Male more highly developed than female: young like female (*Loriculus*, *Cinnyris*, etc.).
2. Female more highly developed than male: young like male (*Turnix*).
3. Male like female: young like the parents (Many *Psittaci*, *Columbae*, etc.).
4. Male unlike[1]) female: young male like adult male, young female like adult female (*Monachalcyon*, *Cittura*)[2]).

The influence of seasonal variation appears to be prepotent for the following:

5. Male like female; young like the adults in winter plumage (*Bubulcus*), or like them in summer plumage (*Alca*), or intermediate between summer and winter plumage (*Charadrius*).

The influence of some previous condition in the history of the race (hereditary geographical or individual modification) is sometimes satisfactorily, more often doubtfully, displayed under the following conditions:

6. Male like female: young different from both (*Munia*, *Larus*, *Ardea*, etc.).
7. Male unlike female: young different from both (*Siphia*, *Chalcophaps*, *Eudynamis*,[3]) etc.).
8. Male unlike female: young ones different, and differing sexually from one another (*Graucalus bicolor*).

---

[1]) Probably a higher development: see *antea*, p. 64.
[2]) The condition — male unlike female: young male like female, young female like male — is not known.
[3]) In *Eudynamis* the coloration of the young is supposed to be protective (see Whitehead, Ibis 1888, p. 410; and Expl. Kina Balu 1893, p. 145).

9. Male unlike female: young female like adult female, young male peculiar (*Microstictus* partly; *Dryobates* and *Xenopicus*: Keeler, p. 224).
It seems to be a very true remark of Darwin's that these several classes graduate into one another.

Ancestral characters. — At the present time much interest turns on the difficult question of the manifestation of the past history of the race occasionally to be read in the plumage of the young or in the less highly developed sex. Among Celebesian birds the following are some of the more interesting and undeniable examples of ancestral indications in the young.

The Kingfishers of the Oriental genus *Pelargopsis* have the lower back and rump blue, except in the Celebesian area, where *Pelargopsis melanorhyncha* and *P. dichrorhyncha* have these parts buff. The young of *P. melanorhyncha* is known to have the parts in question blue — proof that the species was once so coloured (pp. 269, 270 of text).

The Lories of the subgenera *Trichoglossus* typical and *Psitteuteles* have a yellow (or red) band across the under side of the wing, except in *Trichoglossus ornatus* and *Psitteuteles meyeri* of Celebes, and *P. flavoviridis* of Sula, which have the wing uniform below. Traces of yellow, where the band should be, are often seen in young individuals (occasionally in an apparently adult female) of *P. meyeri*, and now and then in the young of *T. ornatus*, proving that these two species once possessed the wing-band (p. 126 of text).

The Stork, *Dissoura episcopus*, has no contour-feathers, but only down, on the sides of the head and on the neck, though it is not to be doubted that it once had these parts feathered. The young has the sides of the head feathered, and some feathers of blackish brown are produced on the neck, but they soon fall out. These feathers indicate what the species was like at some period of the past (pp. 807, 808 of text).

The Parrots of the genus *Prioniturus* have the two middle tail-feathers furnished with long projecting rackets. Young birds before the first moult display attenuated projecting ends or half-formed rackets (see pl. V, figs. 1, 2), showing, according to the argument pursued below, p. 74, an earlier stage in the formation of these growths.

The Tree Duck, *Dendrocygna guttata*, has round white spots on the flanks; in the young these spots take the form of stripes similar to those of *D. arcuata* at all ages; a proof that the round spots are a recent development (p. 872). The Blackbird, *Merula celebensis*, when young is spotted like a Thrush (see pl. XXXV).

The little slate-and-vinous Hawks, *Accipiter rhodogaster* and *sulaensis* and *Spilospizias trinotatus* are totally different when young, resembling Kestrels (*Tinnunculus*); and the Pigeon, *Chalcophaps*, in first plumage has no resemblance to the adult (an unusual circumstance among Pigeons), but has the coloration of the Pigeon-genus *Macropygia*. It appears hardly possible to doubt that these are ancestral indications (pp. 25, 26, 650, 652 of text).

The Kingfisher *Melidora* of New Guinea, a curious form with a hooked bill, is held by Sharpe to be the lowest type of the family, and *Cittura* of Celebes and Sangi has the nearest affinities with it, but wants the maxillary hook. When quite young *Cittura* has this hook (p. 307). Though not a feather-character, this point is of equal significance.

It has been already remarked that, when the two sexes are not alike, one (usually the female) seems to show a lower development than the other. It is probable that such females preserve more ancestral features than the males, which have acquired more new features than have the females; yet direct proof of this is hard to find. Among Celebesian birds, a female of *Psitteuteles meyeri* displays, as mentioned above, a trace of the yellow ancestral wing-band; and the rackets of the females of *Prioniturus* are seldom so long as in the males. Indirect proof of the phylogenetic value of the female plumage is furnished when the young of both sexes are like the mother, for such facts as those given above render it pretty certain that the young tend to display ancestral characters. It sometimes happens that the mother and young of one species resemble the adults of both sexes of another species less highly developed than the male of the first.

These considerations place in the hands of the student of geographical distribution an important and (to ourselves) new means of proof in tracing the land of origin of particular species or genera — provided that our supposition be admitted that emigrants, cut off from their native country, are more likely to get altered than the stayers-at-home.[1]) In this manner it has proved possible to trace the genus *Loriculus* (of which over 20 geographical species are known between India and New Guinea) as having originated in Asia, and to construct a genealogical tree of two main branches showing the descent of the species from the Asiatic *L. vernalis* or its ancestor, this species being supposed to have extended its range in process of time across the Archipelago, undergoing some new modification with each change of habitat, viz. with each new isolation. The more eastern forms now throw back by their females and young to more western forms, and in this manner the two branches of the genus finally converge upon a form like *L. vernalis*. The case is fully discussed, pp. 160—169 of text, Map VI.

On similar grounds it is possible to trace the origin of the Blue-and-rufous Flycatcher of Kalao Island to Celebes. The sexes are slightly different, and the male of *Siphia kalaoensis* is the most specialized member of its group; its female is like the male from Djampea Island, *S. djampeana*; the female from Djampea is like the male from Saleyer Island and Celebes, *S. banyumas*, which is thus indicated as having emigrated first to Djampea and later from there to Kalao.

In the same manner the blue back of the young of *Pelargopsis melanorhyncha*

---

[1]) For proof see variation in Sangi and Talaut, antea, p. 58.

of Celebes may be held to prove the derivation of that species from the blue-backed forms of the Oriental Region, and the indications of a wing-band in the young *Trichoglossus* and *Psitteuteles* of Celebes to demonstrate their descent from the banded species of the Australian Region.

In applying this argument one is apt, however, to stumble on such difficulties as the following. Müller's Green Parrot, *Tanygnathus muelleri*, of the Celebesian area occasionally displays blue on the head when young, suggesting its derivation from the blue-headed *T. luconensis* of the Philippines. But the young *T. luconensis* has the head green, which might be taken as an indication of its descent from the green-headed *T. muelleri*. Is *T. muelleri* descended from *T. luconensis*, and *T. luconensis* from a pre-existing green-headed Parrot, or is the coloration of the head of the young simply due to some chemical condition imposed upon it by the respective parents?

Mr. Keeler (Evolution Colors Birds 1893, p. 178) has suggested, without producing any real proof, that a different colour at the basis of a feather may have a phylogenetic value and denote what the colour of the bird at this spot was at some period of the past. On the contrary our own observations have persuaded us that a different basal colour sometimes shows what colour the feather is going to become. The adult male of the eastern form of the Blue Rock Thrush, *Petrophila cyanus*, has the breast and abdomen chestnut; the immature bird has the feathers of these parts terminally fringed with whitish, next to which is a subterminal bar of dusky, below this usually a little blue, and then a large area above the extreme base chestnut — the colour which the bird will become. Also the jugulum, head and upper parts of the adult are blue, but in the young this blue occupies the basal part of the feathers. Not the base, but the tip of the feather may sometimes have a phylogenetic worth. Evidence of this is shown by the buff-backed Kingfisher of Celebes, *Pelargopsis melanorhyncha*, the young of which by its pale blue back throws back to the other members of the genus, all of which (except another Celebesian form) have blue backs. Now the blue in the young *P. melanorhyncha* is confined to the tips of the feathers; below this they are buff, though there is usually also a faint buff fringe round each feather. In the young of this species the tendency to change into a form with a buff back does not set in in force until the tips of the feathers have already been developed; these tips present the point wherein it agrees with the rest of the genus — apparently therefore a character of long standing, while the buff at the base betrays the character which will soon be assumed.

Hereditary effects of shelter and exposure. — It is proposed here to show some evidence drawn from Celebesian birds that modifications of shape or colour of feathers as caused by the ever-repeated action of mechanical attrition, or by the action of light, are ultimately transmitted to offspring.

The racket-tail-feathers of *Prioniturus*. The two middle tail-feathers are prolonged much beyond the others, and in adult birds the overreaching portion of these two rectrices is converted into a bare shaft tipped with a spatule of ordinary web (see pl. VI, figs. 1, 4, pl. V, fig. 5). The question of the formation of these racket-feathers has been broached by several writers, especially by Prof. W. Blasius (Ztschr. ges. Orn. 1885, pp. 212—219, figs.). Dr. Finsch remarked (Papag. 1868, II, 401) that the bareness of the shafts was manifestly due to the attrition of the barbs of the feathers; Meyer showed (Ibis 1879, 49) that this view, as of a direct mutilation of the individual, is incorrect, since many specimens were shot by him in which the racket-feathers were growing, and the bare rachis lay upon the surface of the other feathers protected from foreign contact. Prof. W. Blasius has expressed the opinion that the shafts do not grow out naked from the first, but become bare later, owing perhaps to a physiological casting-off of the webs.

The specimens in the Dresden Museum prove that the webs are neither rubbed off, nor bitten off as in the case of the Motmots (see Salvin, P. Z. S. 1873, p. 433). Two specimens figured on plate VI, figs. 2, 3, display the growing racket as found underneath the upper tail-coverts (here removed to show the conditions); the shaft is already webless even where it is still enclosed in the corneous husk or follicle out of which the young feather has grown, and where it could of course be neither rubbed nor bitten. On removing a third younger sprouting racket (♂ ad. *P. platurus*) by the root and taking off the epidermal husk (pl. V, fig. 4), it was found that the web (rami) is present on either side of the shaft, but some of the rami appear not to be attached at all but to run, soldered together, parallel to the shaft almost to its root; other rami have become individually broken off or have fallen off from the shaft, and it was easy to see that, as the feather grew longer, all would have fallen from the shaft. In a growing racket with the shaft 35 mm cut out of the tail of an adult male bird it was not possible to detect any signs of barbs with certainty. Plate VI, fig. 3 displays 44 mm of a growing shaft (♂ ad. *P. flavicans*), which would reach a length of 67 mm (judging from the length of the other perfect racket); this shaft was found to be bare down to its point of attachment by the side of the oil-gland; near the base alone some corneous matter of uncertain determination, but perhaps feather-material, was adhering to it.

These investigations tend to prove that no web at all is produced with long-shafted rackets, but rackets of a lower stage of development have imperfect or unattached webs which fall off before the racket is fully exposed.

The inquiry as to how the middle tail-feathers originally began to be lengthened and narrowed and finally formed into long rackets may be answered by a hypothesis which, if it is a correct explanation of the facts, may be not without weight in its bearings upon theories of heredity.

It is easy to obtain a practical demonstration as to how racket-feathers may be formed by holding a feather by the barrel and scraping the webs with a knife; a bare stem with a spatule at the tip then quickly forms itself, the yielding pliancy of the tip making it difficult to remove the web from this point without cutting off the end of the feather altogether. In Nature any feather of sufficient stiffness, prolonged so as to stand out beyond the other feathers, will be liable to such a process as this, attrition against the twigs of trees, the walls of their nesting holes etc., supplying the place of the knife. Assumed that the two middle tail-feathers of *Prioniturus* were originally a little longer than the rest[1]), the ends, if sufficiently prolonged, are liable to attrition; and a narrowing of the tips, such as is now seen in the young birds (pl. V, fig. 1), will result. The friction at the ends of the feathers causes irritation to the roots; an increased supply of blood ensues there, with the result of an increased size of these feathers. These longer feathers are more liable to attrition, and half-formed rackets (pl. V, fig. 3) take shape; the increased irritation and consequent lengthening of the feather results in the production of other stages (pl. V, fig. 4), up to the most advanced development of the present time (pl. VI, fig. 1). Yet the striking features shown in the plates were not obtained in one generation, as has been proved; on the other hand this appears to have been a process of ages, more and more advanced results being obtained in successive generations and transmitted by heredity. The simplest stages of this formation are displayed by young birds in first plumage which in respect of the tail probably resemble the first ancestors of the genus (pl. V, fig. 1); the second moult, when the webs are often quite absent on the shafts of the rackets, which are about half the full length in old birds, seems to show a later period in the history of the race; while the highest development of these feathers, as seen in old birds (especially males) of the present day, is probably the most recent stage in the evolution of the genus.

The following are the arguments in proof that these rackets are the inherited effects of attrition:

1. It has been shown that such can easily be formed artificially by scraping, the size of the spatule depending upon the stiffness of the feather.

2. Where the shafts are not exposed to attrition they are not bare. It is only on the projecting part of the middle tail-feathers that the shafts are bare; and as far as the ends of the lateral feathers, by which the middle ones are protected from attrition, they are fully webbed. If the bareness were due to something else, it might be expected that the naked shaft would not in every species[2]) arise just at this point of the tail, but sometimes much higher up, or sometimes much lower down[3]).

3. Rackets do not occur on unexposed feathers sheltered from attrition.

---

[1]) A very common condition in birds.
[2]) We have examined *P. luconensis, cyaneiceps, suluensis, discurus, flavicans, platurus*, and the plate of *verticalis*.
[3]) Genera in which the racket-feathers are longer and consequently heavier (e. g. *Bhringa*) usually have larger spatules and the attenuation of the webs on the shaft continued towards the base of the tail — a result of friction upon the other feathers.

4. Rackets are present in birds having no affinity with one another, and in the most varied positions on the wing, tail, or head, where a sufficiently stiff feather projects so as to be liable to attrition. Thus they are found on projecting feathers on the sides of the head in the Paradise-bird, *Parotia*, on the projecting second primary of the Nightjar, *Macrodipteryx*, on the overreaching tail-feathers of *Prioniturus*, of the Indian Drongos, *Bhringa* and *Dissemurus* (the web of the racket on the outside only), of the Kingfishers of the genus *Tanysiptera*, etc.

5. Remains of the web are often to be found on the shaft of the racket (pl. V, fig. 5).

6. There appears to be no other means of accounting for the origin of these racket-feathers. They are not sexual characters, nor is it conceivable that they are useful and hence developed by natural selection. The theory of "recognition markings" fails, because they are not present in the young and because they are present and very similar in different species living in the same localities (e. g. *Prioniturus platurus* and *flavicans*).

7. The Motmots of America have the curious habit of forming rackets artificially on the lengthened middle tail-feathers by biting or tearing off the web behind the tip. The result appears now to be partially inherited, since a very pronounced narrowing of the web here is seen in young birds (see Salvin, P. Z. S. 1873, pp. 431, 432 with figures). The habit of tearing away the web also appears to be inherited, for young birds reared by hand began to tear away the webs of the middle tail-feathers when these had reached their full length (see Cherrie, Auk 1892, 323).

As an argument against the loss of the webs through attrition during the individual development, it has been pointed out that when a narrow fringe of web is found on one side of the shaft, it is almost always on the outside that this occurs, where it is said that it would be most likely to get rubbed (Meyer and W. Blasius, ll. cc.). Due weight should, however, be given to the following considerations: first, birds rarely spread out their tails except in flight, and in the position of rest one middle tail-feather lies over the other so that little of the latter is seen, and the inner web of the one racket would receive a good deal of the pressure and friction put upon the outer web of the other; and second, the webs on the inside would be liable to get crossed, interlocked, sawed and broken by one another.

The attenuated tail-feathers of *Merops*. The two middle tail-feathers of all the species of Bee-eaters of the genus *Merops* are prolonged beyond the others when the bird is adult; the tip is not furnished with a spatule as in *Prioniturus*, but attenuated for its terminal projecting portion and for a little distance on the non-projecting part (see plate VIII, fig. 1). These attenuated ends are not formed by attrition at the sides during the lifetime of the individual, as is shown by young feathers sprouting out of the follicles thus perfectly developed (see pl. VIII, figs. 2, 3). Yet the argument for attrition continued during generations without number applies equally well here. The habits of *Merops* are very different from those of *Prioniturus*; the Parrot breeds in holes in trees, but the Bee-eater forms a burrow, like the hole of a mouse or rat, for a depth of one to three metres in a bank of sand or earth. The friction caused by the sand, against which the terminal portion of the feather is chiefly brushed, seems sufficient to account for this peculiar shape. If a feather of ordinary shape be taken, and rubbed and drawn between two sheets of sand-paper, a ragged similitude of a *Merops*-rectrix may be obtained.[1]

Other cases. If once the theory that the racket-tail-feathers of *Prioniturus* are the inherited results of attrition is admitted, a principle is arrived at by which a host of other cases are capable of explanation. Among feather-formations may be mentioned: the bifid tips of the remiges of *Merops* and *Hirundo*, explicable by the habit of these birds of supporting themselves on their wings when commencing their nests (for *Merops*, see p. 252, note), the oscillation of the body forcing the webs apart at the tips of the feathers and so forming a little notch, just as is done by rubbing the tip of a feather on blotting-paper or by knocking the tip gently with the finger; the stiff, tapering tails of Woodpeckers and *Nasiterna*, stimulated to strong growth and worn down to shape by the habit of using them as a prop in climbing; the curiously attenuated first primaries of many Hornbills and Pigeons, so shaped by the friction caused in flight to these reduced quills which lie under the other remiges, against which they vibrate and by which they are rubbed; the narrowing of the outer webs of the lateral tail-feathers of all birds and the gradual increase in width of these webs from one feather to another until on the two middle feathers they are of approximately equal width, the middle feathers being protected by the lateral ones from the friction of objects against which the tail is repeatedly getting brushed, the lateral feathers being exposed to this attrition, — most of all the outermost pair in which the outer web is narrowest. Also in the narrowing of the outer webs of the remiges, though feathers of this shape are apparently essential to flight, mechanical attrition, caused by the rush of air in flight, may have worked together with natural selection in determining their shape. The friction may have acted as a stimulus to the lengthening of these feathers which are far larger and stronger than contour-feathers. Other parts may be modified in the same manner as feathers by the inherited effects of wear and tear: such as the bills of *Anastomus*, *Esacus* and *Demiegretta*, worn away so as no more to close properly by the rough shells and crustaceans upon which the birds feed; the bill of adult Hornbills not meeting for a space where the bird lays hold of objects in climbing and feeding and even swings from them on occasion suspended by its bill (see, Legge, B. Ceylon 1880, p. 274); the skin of the head of the Cock, drawn out into a comb and with the formative feather-

---

[1] See also Meyer's remarks on and figures of the two lengthened middle tail-feathers of *Paradisea minor* etc. in Abh. Ber. Mus. Dresden 1898/9 Nr. 2 p. 44 plate II.

papillae destroyed by the beaks of antagonists; the face of the adult Rook from which the feathers fall at the base of the bill as a result of dirt and wear for generations; the head and face of the adult Moleo, naked owing to ages of attrition from the sand in which it burrows; and so on. Examples drawn from man and other animals could be given. The principle is of importance, as a cause of, or directive stimulus to, variation; it should therefore not be accepted without criticism. For some cases the principle of natural selection affords an explanation (e. g. the remiges), but for others the argument furnished thereby can hardly be made to commend itself to impartial judgment (e. g. the rectrices), and for others again this principle appears to fail completely (e. g. the rackets of *Prioniturus*, the comb of the Cock[1]).

Effect of light. In course of time most colours in mounted specimens and skins of birds fade with exposure to the light. Among Celebesian birds the effect is particularly well seen in the buff of the Nutmeg Pigeons and the wash of salmon-colour on the under parts of the Moleo, which soon fade in exposed skins, leaving the respective parts white. Nor does light seem to be operative solely upon the dead.

Where the wing rests upon the body. — In nearly all birds a change of colour takes place on the under side of the remiges where they rest upon the body with the wing closed, so that this part differs from the distal ends and more external parts of these feathers. Sometimes merely a slight change of gloss is seen, but all stages of difference may be found from this up to the most marked contrasts. Among Celebesian birds some of the most striking examples are: the Cuckoo-shrike, *Graucalus bicolor*, with the remiges white below where they rest upon the body, black on the other portions: the Parrots, *Prioniturus Loriculus*, with the remiges below verditer-blue against the body and partly where they cover one another, black elsewhere; the Roller, *Coracias temmincki* — remiges blue against the body, black changing with the light to bronze on the free parts[2]); the Flycatcher, *Zeocephus*, with the said parts ferruginous and blackish respectively; and so on. A tendency to blackness is generally seen on the distal ends and external portions of these feathers.

Where the tail-feathers are concealed by the upper tail-coverts. — A change of colour in the shafts and webs of the rectrices is generally seen on their concealed bases, very commonly a tendency to paleness or white, suggesting a loss of pigment. The most striking examples occurring in Celebes are the Cuckoo-shrike, *Graucalus bicolor*; the two Nutmeg Pigeons, *Myristicivora bicolor* and *luctuosa*; the Pratincole, *Glareola isabella* (as also *G. orientalis*). In these birds all that part of the tail which is concealed by the upper tail-coverts is white, and all

---

[1]) As shown by Stolzmann (P. Z. S. 1885, 430) this may bring disaster to the wearer of it, but the author attempts to explain this by natural selection.
[2]) On the upper surface of the remiges these colours are reversed, being blue above where they are bronze below, and bronze above where they are blue below, but the lines of demarcation do not exactly coincide.

the terminal exposed part is black, the division of the colours being sharply conterminable with the tips of the longest upper tail-coverts.

The concealed bases of the contour-feathers. — Here again a difference of pigmentation or of gloss, or of both, is seen, the bases being usually of paler or duller hues than the tips. Thus the bases of the contour-feathers in *Corvus enca* are white, the terminal portions glossy black; in the Parrot, *Aprosmictus sulaensis*, the bases are grey or greenish, the exposed terminal portions bright blue on the mantle; and so on.

Apart from the phylogenetic value of the different parts of a feather, there is convincing evidence that light must be cited as an important agent affecting the distribution of the pigments of a feather, either through physiological stimulation, or direct action, or both. No better test case could be found than the male of the Celebesian Cuckoo-shrike, *Graucalus bicolor*. Seen from above with its wings closed it is a black bird, for, though the rump and upper tail-coverts are white, these parts are then probably concealed by the wings. The under surface is white. The wings and tail are black; yet that part of the tail which lies hidden beneath the upper tail-coverts is white, and so is the wing below where it rests upon the sides of the body; also the black contour-feathers of the upper parts are white on their concealed bases. It may be said that wherever the feathers are exposed to the sun they are black; where they are in shadow or concealment they are white. It is preferable to attempt no explanation of these facts here, but it seems permissible to suggest that the case is similar to that of *Prioniturus*, the difference being that the inherited effects of attrition are assigned as the cause of the formation of the racket-feathers of the Parrot, whereas the action of light is regarded as having in the course of generations in some way brought about the distribution of the pigments in *Graucalus*.

Direct evidence of the action of sunlight upon plumage is afforded by the following statement. As Dr. Russ writes: "The Goldfinch when kept caged in a dark place often becomes black, and even in a light room the bright colours after moulting often appear fainter and more impure, but this can be prevented if the Goldfinch is placed as much as possible in the open air and sun" (Einheim. Stubenvög. 1873, II, 265).

The soft, glossless plumage of nocturnal birds, viz. Goatsuckers and Owls, also calls for consideration in this connection, as well as examples already suggested, and innumerable other ones in which the action of light, or the want of it, appears less obvious.

5. GEOGRAPHICAL DISTRIBUTION.

In the intermediate seas between the Euro-asiatic and the Australian continents there is stretched out the largest and most numerously membered archipelago of the earth, with a fauna and flora derived partly from the West and partly from the East. Where do we now find the frontier of these two faunas and floras, which contrast so strongly with one another in their extreme forms; or do they pass into one another so insensibly that a sharply defined frontier cannot be traced? It would be very premature to attempt a sketch of the geological history of this region of the earth in view of the quite insufficient knowledge available concerning the living and, especially, concerning the extinct fauna and flora of the Archipelago, for we are acquainted with only a small fragment of the latter (the extinct); we must content ourselves with an attempt to answer the above interesting question with the aid of the rather better established data of the present time [1]), and, in accordance with the character of our book,

---

[1]) The status of geological knowledge as to Celebes is very defective as yet, though some valuable work has been done recently and more is to be expected from Drs. P. & F. Sarasin. In this state of things we have found it preferable to abstain entirely from discussing the past history of the island, but give some of its literature, where references to further geological, palaeontological and mineralogical papers are to be found:

1883. K. Martin: Wiss. Aufg., welche der geologischen Erforschung des Indischen Archipels gestellt sind. (Lecture. Leyden, Brill). Id.: Die wichtigsten Daten unserer geologischen Kenntniss vom Niederl. Ost-ind. Arch.: Bijdr. taal- land- en volkenk. van Ned. Ind., uitg. ter gelegenheid van het 6. intern. Congres der Orientalisten te Leiden. Land- en volkenkunde. 1883, 17 (Celebes p. 23) with summary of literature.
1888. C. M. Kan: Bodengesteldheid der eilanden en diepte der Zeeën van den ind. Arch.: Tdschr. Ned. aardr. Gen. (2) V (Versl.), 202, Kaart IV.
1890. K. Martin: Die Kei Inseln und ihr Verhältniss zur australisch-asiatischen Grenzlinie, zugleich ein Beitrag zur Geologie von Timor und Celebes: Tdschr. Ned. aardr. Gen. (2) VII, 241. Id.: Zur Geologie von Celebes: l. c. 1891 VIII, 180.
1890. A. Wichmann: Bericht über eine ... Reise etc. II. Celebes. l. c. (2) VII, 921 Tab. II; 1892 IX, 258. Id.: Die Binnenseen von Celebes: Petermann's Mitth. 1893, 215 Taf. 16. Id.: Petrogr. Studien über den Ind. Arch. I. Leucitgesteine von der Insel Celebes: Nat. Tschr. Ned. Ind. LIII, 315, with plate. 1895 II. Zur Geol. der Insel Saleyer: l. c. LIV, 236 pl. V.
1894. W. F. van Vliet jr.: De verticale ligging en de geologische bouw van Celebes: Tdschr. gesch., land en volkenk. 9. Jaarg., 257 (with summary of literature).
1895. I. W. Retgens: ... Gesteenten van Celebes: Jaarb. Mijnwezen in Ned. Oost-Ind. 24 Jaarg., 124.
1896. Encyclopaedie van Ned. Indië (by van der Lith and others) I, 317 (s. a.). A general report on the geological formation of Celebes.
1896. K. Martin: Zur Frage nach der Entstehung des ost- und westindischen Archipels: Geogr. Zeitschr. II, 376.

It is nearly the same with the flora of the Archipelago. Though relatively much is already known, we are very far from a thorough knowledge, which would enable us to draw trustworthy conclusions. We, therefore, likewise abstain from touching these questions, which are discussed in the works of Grisebach (Vegetation der Erde 1872, Ges. Abh. u. kl. Schr. zur Pflanzengeographie 1880), Engler (Versuch einer Entwicklungsgeschichte der Pflanzenwelt, insbesondere der Florengebiete seit der Tertiärperiode 1882), Drude (Die

restrict ourselves to the ornithological facts at our disposal, taking the Avifauna of Celebes as the basis for this purpose. Conclusions which may be drawn from ornithological facts alone must, however, be weighed very carefully, as birds have their own modes of dispersal. We shall then see in how far these conclusions differ from those arrived at by other means.

Wallace's line.

As is generally known Mr. Wallace drew a line to the west of Celebes by which the Archipelago was divided into two widely differing halves. This division was welcomed with much approbation on account of the fascinating speculations of its inventor, though these speculations were more suggestive than substantially founded upon and backed by facts, some of which were not taken into consideration, and others were not available with our defective knowledge of 20 or 30 years ago, nor indeed are they available to-day.

Mr. Wallace has, however, in the course of his later studies modified his views in some respects. At first, as in the "Malay Archipelago" (1869) and in the "Geographical Distribution of Animals" (1876) — not to mention earlier writings[1]) — the line passes between Bali and Lombok, through the Macassar Strait west of Celebes, turning to the east between Mindanao and Halmahera; while he adds in "Island Life" (1880, 431) "that the present land of Celebes has never (in Tertiary times) been united to the Asiatic continent, but has received its population of Asiatic forms by migration across narrow straits and intervening islands". He draws in the latter work (p. 434) the following conclusion: "We have in this island a fragment of the great eastern continent which has preserved to us, perhaps from Miocene times, some remnants of its ancient animal forms"; and (p. 509): "I now look upon Celebes as an outlying portion of the great Asiatic continent of Miocene times, which either by submergence or some other cause had lost the greater portion of its animal inhabitants and since then has remained more or less completely isolated from every other land. It has thus preserved a fragment of a very ancient fauna along with a number of later types which have reached it from surrounding islands by the ordinary means of dispersal". He further says in his "Australasia" (ed. by Dr. Guillemard 1894, p. 287): "The peculiarities of the animal life of Celebes may be best explained by supposing it to be an outlying portion of that Miocene continent, which became detached from it, and has since never been actually joined to any Asiatic or Australian land. It has thus preserved to us some descendants of ancient types, and these have become intermingled with such immigrants from both east and west as were enabled to establish

---

Florenreiche der Erde, Petermann's Erg. Heft Nr. 74 1884, Atlas der Pflanzenverbreitung 1887, Handbuch der Pflanzengeographie 1890), Warburg (Die Flora des asiatischen Monsungebietes: Verh. Ges. Deutscher Naturf. Allg. Theil 1890, especially concerning South Celebes), etc.

[1]) These earlier writings are to be found in the Ibis 1859, 450; J. of the Proc. Linn. Soc., Zool., 1860, IV, 172; P. Z. S. 1863, 481; J. R. Geogr. Soc. 1863, XXXIII, 217; Edinburgh Philos. Journ., new ser., 1864, XIX, Nr. 1, etc.

themselves in competition with the ancient inhabitants. To the naturalist, therefore, Celebes is an island of extreme interest. It cannot be said to belong either to the eastern or the western divisions of the archipelago, but to stand almost exactly midway between them; the relic of a more ancient land, and dating from a period perhaps anterior to the separate existence of any of the islands."

If we now glance over the scientific literature on "Wallace's line", as Huxley baptised it (P. Z. S. 1868, 313), it should be understood that we do not pretend to give an exhaustive extract, but only quote such writings as have been within easy reach or which have appeared sufficiently characteristic. There are also heaps of other books and papers in which Wallace's line is mentioned.

E. Blyth, in 1871 (Nature III, 428), recognizes the line. He has a Celebesian Sub-region of the Melanesian Region and it comprises: Celebes, Lombok, Sumbawa, Flores, Wetter, Timor and Sandalwood Island.

J. Pijnappel, in 1872 ("Enkele aanmerkingen op Wallace's Insulinde": Bijdr. taal, l. en vk. Ned. Ind. 3. ser. VII, 159), made some serious objections and is of opinion, that as Geography, Anthropology, Ethnography and Botany are opposed to the line, Zoology alone cannot uphold it; the less so, as it sometimes requires the most hazardous hypotheses as to geological convulsions, upheavals and submergences in order to explain the occurrence of a single mammal.

A. v. Pelzeln in a paper entitled "Africa-Indien", published 1875 (see: Verh. z.-b. Ges. Wien p. 33), adopted the line; he considered Celebes as belonging to the Australian Region and enunciated as peculiar bird-genera (p. 48): *Monachalcyon*, *Cittura*, *Ceycopsis*, *Artamides*, *Gazzola*, *Streptocitta*, *Scissirostrum*, *Enodes*, *Basileornis*, *Prioniturus* and *Megacephalon*. He takes as identical (p. 47) *Scops manadensis* from Celebes and Madagascar, and *Ortygometra flavirostris* from Celebes and Africa, and mentions eighteen species which are common to the Ethiopian and the Indo-Malayan Region. In 1876 he confirmed his general conclusion in a paper on the Malayan mammalia (see: Festschr. z.-b. Ges. p. 53).

P. J. Veth, in 1875, gave a lecture on the line before the International Geographical Congress in Paris ("Observations sur les lignes de Wallace": C. R. Congr. Int. des sc. géogr. à Paris, 1878, 305), and treated the matter with the acumen usual to him. He said that it rests on an inadequate basis hydrographically, that the flora was not taken into consideration, that Wallace only referred to mammals, birds, some insects and land-shells instead of to the whole fauna; that it is, therefore, zoologically insufficiently proved, and that it is not evinced by the facts of anthropology (see, also, l. c., p. 276 and Veth's translation of Wallace's paper: "Over de physische Geogr. van den Ind. Arch.", with notes, Zalt-Bommel, 1865).

J. A. Allen, 1878 ("The Geographical Distribution of the Mammalia considered in relation to the principal ontological regions of the earth and the laws that govern the distribution of animal life": Bull. of the U. S. Geol. and Geogr.

Survey, vol. IV, 363—377), is one of the few earlier zoologists who do not recognise Wallace's line. His Indian Region, being part of the Indo-African Realm, has for its eastern frontier a line drawn west of the Moluccas and Aru. He says (p. 358): "I fail to see any good reason for assigning Celebes and all the smaller Sunda Islands to the Papuan Province, as Mr. Wallace and others have done, but abundant evidence that such is not their real affinity." And p. 364: "The Australian Realm will be here restricted so as so embrace none of the islands situated to the westward of the Moluccas."[1]) His Insular or Malayan Province forms part of the Indian Region; it includes all the Sunda Islands, the Philippines and Celebes. His Papuan Province (p. 367) takes in the Molucca and Aru Islands to the west, but he considers the Molucca Group (p. 364) to be a transitional link between the Indo-African and the Australian Realm, faunistically more loosely allied to the latter than to the former.

K. Semper, 1880, in his work: "Die natürlichen Existenzbedingungen der Thiere" (II, 136), discussed the problem fully. Though he found that facts do not speak everywhere in favour of Wallace's line, he was nevertheless inclined to adopt it in a general way; he explained the differences of the faunas to the east and west not, however, by former land-connections, but by the sea-currents transporting the animals, a hypothesis which, as far as we are aware, has not been accepted elsewhere.

O. Krümmel, in 1882, published (see: Ztschr. wiss. Geogr. III, 1, Taf. I) an important map: "Tiefenkarte des australasiatischen Mittelmeeres", on which he drew the line, but remarked (p. 2) that the depths of the Macassar Straits are quite insufficiently known and (p. 3) that in the Straits of Lombok only one sounding very near the coast of Bali, which was broken off at 50 fathoms, serves as a basis for the assertion that a deep gap in the chain of islands exists here! He further mentioned (p. 5) that there are no soundings whatever known from the three large gulfs of Celebes.

K. Martin, in a lecture on the "Wissenschaftlichen Aufgaben, welche der geologischen Erforschung des Indischen Archipels gestellt sind", held in Leyden in the year 1883, considered the line entirely erroneous. In his opinion (p. 28) the continental frontier between Asia and Australia is approximately identical with the chain of volcanoes in the Archipelago. The same author says in a paper: "Die wichtigsten Daten unserer geologischen Kenntniss vom niederländisch Ost-indischen Archipel" (see: Bijdr. taal-, land- en volkenkunde Ned. Ind. uitg. ter gelegenheid van het 6. intern. Congress der Orientalisten te Leiden, Land- en Volkenkunde. 1883, 27): "As far as our knowledge of to-day goes, Wallace's line is geologically unjustifiable. ... Nothing hinders us from drawing

---

[1]) Previously (Bull. Mus. Comp. Zool. Cambridge, 1870—71, II, p. 381) Mr. Allen had uttered the following opinion: "The Australian Realm embracing Australia, New Zealand, New Guinea and their dependent islands, including those to the eastward [?] as far as Timor and Celebes, is zoologically as distinct ..." This is not at all clear to us, but as this prominent writer later (see above) was quite intelligible, it is not necessary to discuss his former intimation.

the frontier to the north-west of Timor; the sea-depths would allow it just as well, and in this case at least a separation of geognostically different regions would be attained." (A. Wichmann, however, appears to reckon Timor, etc., to the former Asiatic continent, see: Samml. des geol. Reichs-Mus. in Leiden, I. Ser. Bd. II, p. 201, 1887.)

O. Drude, 1884 ("Die Florenreiche der Erde": Erg. Heft Nr. 74 zu Petermann's Mitth. p. 62 b), acknowledges the line as a floristic frontier, to which opinion he still adheres in 1890 ("Handbuch der Pflanzengeographie", p. 150 and Map I; see also: "Atlas der Pflanzenverbreitung" Berghaus' Phys. Atlas. 5, No. 1, 1887).

A. Heilprin, 1887 ("The Geographical and Geological Distribution of Animals" p. 107, and map), adopts an Austro-malaysian Transition Region, which is bounded to the west by Wallace's line, to the east and south by New Guinea and Australia, to the north by Mindanao.

W. Marshall, 1887 ("Atlas der Tierverbreitung": Berghaus' Phys. Atlas. VI. Abth. p. I ͣ, Map III), simply adopts the line.

C. M. Kan, in a paper published in 1888 on the "Bodengesteldheid der eilanden en diepte der Zeeën van den ind. Archip." (see: T. Ned. Aard. Gen. 2. ser. vol. V. Meer uitgebreide art. p. 219, with map IV) does not recognize the line for reasons indicated in the title of his paper.

A. Reichenow, in 1888 ("Die Begrenzung zoogeographischer Regionen vom ornithologischen Standpunkt": Zool. Jahrb. Abt. f. Syst. III, 699), recognizes an Eastern Zone with an Eastern Temperate Region, an Ethiopian Region and a Malayan Region as far as Wallace's line (see also map XXVI), and a Southern Zone which extends to the west up to Wallace's line; he, therefore takes Celebes as non-Oriental.

R. Schuiling, in a special dissertation: "De grenslijn van Wallace eene continentale grens", 1888 (T. Ned. Aard. Gen. 2. ser. V, p. 523), came to the conclusion (p. 548), that Geology, Zoology and Oceanography teach: "Celebes belongs to Asia".

F. A. Jentink, in 1889 (l. c. VI, 244), showed (p. 246), that we are very far from such an adequate knowledge of the mammals of Bali and Lombok as to justify Wallace's affirmation (Island Life 1880 p. 4): "Bali and Lombok differ far more from each other in their birds and quadrupeds than do England and Japan," neither was Wallace justified in basing an argument on 16 land-mammals as the ascertained number from Celebes, because as early as 1878 21 were already known and this large island has not been at all thoroughly investigated yet. In 1888 there were already 26 land-mammals and 19 bats extant from there, a number which is probably still far from the true total. There was therefore no good reason whatever for drawing important conclusions with such scanty knowledge.[1])

---

[1]) We can only point to Mr. Whitehead's recent discoveries of mammals on the high mountains of the Philippines (see: Ann. Mag. N. H. 1895, 6. ser. vol. XVI, 160), in the conviction that such an experienced

E. Reclus, 1889 (Nouv. Géogr. univ., vol. XIV, 209) sticks to the line and says: "De tous les côtés elle [Celebes] apparait isolée; c'est une terre dont l'isolement complet est un fait géologique datant des ages les plus reculés"(!).

E. v. Martens published in 1889 his "Tagebuch-Notizen" from Banda, Timor and Flores (see: Z. d. Ges. f. Erdk. zu Berlin, vol. 24, p. 83) and concludes (p. 104) that Timor, Celebes, the Philippines, and the islands east of Java represent the region of intermixture of the eastern and western animal worlds and may just as well belong to neither as to both. "Nearly every zoological genus presents a different frontier, a sharply defined common frontier does not exist in nature, nor here."

O. Warburg remarked in 1890 ("Die Flora des asiatischen Monsungebietes": Verh. Ges. Deutsch. Naturf. Bremen, Allg. Theil, p. 15 of sep. copy) that important as Wallace's line is for understanding the evolution of the floras in detail, the character as a whole was not altered by the separation; the greatest part of the present flora would have already transmigrated before the separation, thus certainly long before Miocene times.

K. Martin in a paper of 1890: "Die Kei-Inseln und ihr Verhältniss zur Australisch-Asiatischen Grenzlinie" (Tdschr. Kon. Ned. Aardr. Gen. 2. ser. VII, 273) says: "To the west of Great Key and to the north-west of Timor lies a natural and geognostically well-founded line of separation between the islands dismembered from the Asiatic and Australian continents." He adds, however, that it is not to be expected that on the continental borders the present faunistic and floristic character of single islands should have a direct connection with the geological line of separation or be congruent with it, because peripheric parts of the continental masses are at times connected or separated.

E. L. Trouessart, in 1890 ("La Géographie Zoologique", pp. 89, 131, 243 and map p. 9), simply accepts Wallace's line.

E. v. Martens showed in 1891 ("Landschnecken des Indischen Arch.": Weber's Zool. Erg. II, 263) that the land-shells do not allow of a sharp line being drawn between Celebes and Borneo, though they differ considerably, for North Celebes cannot be separated from the Philippines, and the differences between Java on the one side, and Flores and Timor on the other, are less conspicuous. The region to the east of Celebes does not offer any uniformity with this island and cannot be regarded as constituting a unit with it.

P. L. Sclater, who, as is well known, first divided the earth into six ornithological regions now widely adopted for animals in general ("On the general Geographical Distribution of the Members of the Class Aves": J. Proc. Linn. Soc. 1858 II, 130—145)[1] in 1891 recognised the line ("On Recent Advances in our knowledge of the Geographical Distribution of birds": Ibis, p. 515), though he gave Celebes a special heading (p. 530) and says (p. 533) that Celebes is "a

collector as he would gain a similar harvest on the high mountains of Celebes, which rise to nearly 10 000 feet, and from where next to nothing is as yet known.

[1] See, also, his lecture on "The Geogr. Distribution of Mammals" Sc. Lect. f. the people (6) 1874 p. 80.

debatable land between the Oriental and Australian Regions, but more properly attributable to the former".

W. H. Flower & R. Lydekker, in 1891 (Introd. to the Study of Mammals p. 102), regard Celebes as the typical representative of the Austro-Malayan transitional region or sub-region, but they do not define it and do not recur to Wallace's line (except on p. 97).

P. A. van der Lith, 1893, gave (Nederl. Ost-Indië, 2. ed. I, 11) a sketch of the facts and views concerning the line, but was inclined not to adopt it. A very readable résumé is to be found in J. F. van Bemmelen's book: "Uit Indië", 1895, p. 146 et seq.

E. Haeckel in 1893 ("Zur Phylogenie der australischen Fauna" in Semon's Zool. Forschungsreisen, I p. V) adopts the line without entering critically into this difficult and complicated question, though he presents us with the following astonishing affirmation: "An keinem anderen Punkte unserer Erde stehen zwei benachbarte Thiergebiete in so auffallendem Gegensatze, als auf der schmalen Grenze zwischen der indo-malayischen und austral-malayischen Region. Überschreiten wir die schmale Meerenge am Südende dieser Grenze, die tiefe Lombok Strasse, so treten wir mit einem Male aus der Gegenwart in das mesozoische Zeitalter [!]. Obgleich die beiden Nachbar-Inseln Bali und Lombok nur wenige Meilen entfernt und im Allgemeinen denselben klimatischen Bedingungen unterworfen sind, erscheint dennoch die charakteristische Landesfauna derselben gänzlich verschieden; und noch mehr gilt das, wenn wir die Mangkassar Strasse überschreiten und von dem indischen Borneo nach dem australischen Celebes übersetzen. Der durchgreifende Gegensatz ihrer Vogel- und Säugethier-Welt ist so gross, dass er zu den schlagendsten chorologischen Argumenten des Transformismus gerechnet werden muss." (!)

W. Haacke, 1893, simply adopts the line (Schöpfung der Thierwelt" p. 238).

A. Newton, also in 1893, ("A Dictionary of Birds", p. 317—363, and Map facing p. 1), likewise accepts the line. Concerning Celebes he says: "To the Papuan Region may be assigned, though with doubt, the wonderful island of Celebes, presenting perhaps more anomalies than any other in the world, and yet anomalies which, by the use of strictly scientific inference (as Mr. Wallace has shewn us), may possibly tell a story that sounds so romantic and yet will satisfy those who judge it more severely".

R. B. Sharpe, likewise in 1893 ("On the Zoo-Geographical Areas of the World, illustrating the Distribution of Birds": Natural Science III, 100 and map', applies the Wallace-line as western frontier to his Australian Region; he recognizes a Celebean Sub-Region, which, with the exception of the Sula Islands, coincides with our Celebesian area, and a Moluccan Sub-Region, comprising everything between Lombok, a line east of Celebes, New Guinea and Aru to the east, and Australia to the south.

F. H. H. Guillemard, when editing Mr. Wallace's "Australasia" in 1894, likewise appears to have neglected literature when he says (p. 347 — a passage not in the first edition of 1879, p. 419, and we are not aware whether it is in subsequent ones): "We thus have the Sunda Chain divided distinctly and definitely into an Asiatic and an Australian portion, the dividing line coinciding with the deep-sea channel existing between Bali and Lombok. This boundary is now universally known as 'Wallace's line'".

M. Weber, 1894, in his important paper: "Die Süsswasser-Fische des Indischen Archipels, nebst Bemerkungen über den Ursprung der Fauna von Celebes" (Zool. Ergebnisse III, 468) came to the result that Celebes has no Australian, but a highly impoverished Indian character in its fish-fauna, and remarked as to the general problem (p. 473): "The unhappy line of Wallace, which he himself has not formally retained for Celebes, has worked its way deeply into the brains of numerous zoologists as something fascinatingly simple. Text-books which touch upon zoogeography and get rid of the subject in a few words maintain their hold on this classical frontier. And thus the Australian Fauna of Celebes lives notwithstanding various protests." Prof. Weber concludes (p. 476): "The original line of Wallace separates groups of islands, of which the western (Borneo, Sumatra and Java) received, on account of their size, but chiefly in consequence of their longer connection with the Indian continent, a rich Oriental fauna and, therefore, have developed specific forms of Indian character. Of the eastern, Celebes was first separated from the Indian continent and remained cut off. In consequence, it retained single older forms, which developed independently. — Consisting in earlier times of single smaller islands, its fauna has remained poor."

F. E. Beddard, in 1895 ("A Text-Book of Zoogeography"), recognises Wallace's line (p. 103 and frontispiece-map) as a frontier between the Oriental and the Australian Regions (p. 103 and 113), though (p. 113) he says that Celebes "probably" belongs to the latter, but (p. 106) treats of it under the heading of the Malayan Sub-region of the former.

R. Lydekker in 1896 ("A Geographical History of Mammals", p. 45 and map), adopts Heilprin's Transition Region (see above) as an Austro-Malayan Region and as one of four Regions of the Notogaeic Realm (p. 27): "Poverty, and an admixture of Australian and Malayan types, with a very marked preponderance of the latter, are the leading features in its mammalian fauna". He says, however, that from the living mammalian fauna one might be inclined to place the whole area within the limits of the Oriental Region. He evidently hesitates in giving Celebes a fixed position, the more so as "there is absolutely no palaeontological evidence to help us in regard to past history".

C. Hedley ("Mollusca of the Oriental region": Journ. of Malacol. IV, 53) showed, 1895, that the line between Bali and Lombok has no value for the Mollusca, as the land-shells of these two islands do not differ essentially.

Likewise E. v. Martens showed in 1896 (Sb. Ges. natf. Freunde zu Berlin p. 157), that of 10 land-shells from Lombok 3 are geographically neutral, 4 are assigned to the great Sunda Islands, 3 to the eastern islands, that, therefore, no sharp frontier exists for land-shells between Bali and Lombok.

The same eminent conchologist said in 1897 ("Süss- und Brackwasser-Mol. des Indischen Archipels": Weber's Zool. Ergebnisse IV, 298): "The frontier between Bali and Lombok is for the fresh-water Mollusca quite imaginary, as long as we know next to nothing of the species living on these islands;" and he proved further (p. 297) that the fresh-water Mollusca from South Celebes are most closely allied to those of Java and Flores, those of North Celebes most closely to those of the Philippines, which is not consistent with Wallace's line, but with the geographical position.

A. Supan, 1896 ("Grundzüge der physischen Erdkunde", p. 557, and Maps XIX, XX) sticks to the "celebrated" line.

W. Kükenthal, in 1896 (Abh. Senckenb. Naturf. Ges. XXII, 130), abnegates Wallace's line.

W. L. Sclater, 1896, treating of the Mammals of his Celebesian Subregion ("The Geography of Mammals": Geogr. Journ. VIII, 388 with Maps), finds that the Australian element in the mammalian fauna of Celebes does not in any way require the supposition of an ancient land-connection with that Region, but that the greater amount of Oriental forms suggests such a former connection with Asia; he, therefore, annexes the Celebesian Subregion to the Oriental and not to the Australian Region. In the beginning of this important paper (l. c. 1894, III, p. 97, with Map, and IV, p. 35, with Map) Mr. Sclater draws Wallace's line to the east of Celebes and between Bali and Lombok (see, also, l. c VIII, p. 378) and takes this as the frontier between the Australian and the Oriental Region, reckoning the Sula Islands to the former, Celebes to the latter, as "on the whole the evidence of the mammals, at any rate, serves to connect it more closely with the Oriental Region" (see l. c. IV p. 36).

F. J. Niermeyer, finally, in 1897 ("De Geschiedenis van de lijn van Wallace": Tijdschr. Kon. Ned. Aard. Gen. 2. ser. XIV, 758), has given a very readable historical sketch. He rightly censures zoologists, botanists, and geographers for often writing on the problem without having consulted Wallace himself, or the manifold literature extant on this subject, and still advocating a frontier which specialists have long since abandoned. He shows in detail how Wallace himself has altered his opinion from 1860 to 1863, 1869, 1876, and 1880, and what Weber's merits are in promoting knowledge on this question.

On going over these different opinions on Wallace's line it will be seen that they are fairly equally divided, though they must be weighed and not counted, many writers on the general subject not plunging deeply into the problem, but uncritically following the authority of this eminent naturalist. One must also take into account that errors, when once they have crept into books, disappear from them with great difficulty. On the other hand also, some specialists of the highest standing acquiesce in the line, partly disregarding the circumstance that Wallace himself has to a certain degree altered his views; whereas others of the same rank encounter insuperable obstacles in adopting such a frontier between the Oriental and Australian Regions. There can be no doubt that in our present state of knowledge it is premature to define the problem for solution, however interesting and suggestive it may be, and that it is, therefore, waste of time to speculate on it with the help of an up-and-down system for the islands and continents, just as required. It is characteristic of an inadequate hypothesis that it is always in need of a new one which should sustain it, and as geology and palaeontology are as yet powerless to guide us, we must restrict ourselves to zoology, though we know that here also our knowledge is defective in a high degree. Let us see, however, what the ornithology of Celebes in its present state teaches, and whether our results agree at least with those arrived at by others.

What are the characteristic elements of the Celebesian Avifauna and where did they originate? This is the only question we put, and which we will try to answer — always bearing in mind that our ornithological knowledge of Celebes, especially of the centre and high mountains, is imperfect —, leaving all further speculations to the naturalist of the future.

The following table of the Geographical Distribution of the species treated of in this book will facilitate the answer to our question. It will be observed that the Celebesian Area is flanked to the left by the Nearctic, Ethiopian, Palaearctic and Oriental Regions, to the right by the Australian and Neotropical Regions, the generally adopted Sclaterian division having been accepted for convenience's sake, though we are aware that that of Prof. Newton (D. B. 1893, p. 315 et seq.) is an improvement upon it. (His main divisions are the New Zealand, Australian, Neotropical, Holarctic, Ethiopian and Indian Regions; uniting under the Holarctic the Nearctic and Palaearctic, and separating the former Australasian Region into an Australian and a New Zealand Region.) The affinities of the Celebesian Avifauna make it preferable to break up the Oriental Region into several parts, inserting between them Japan — a section of the Palaearctic Region, as follows: Indian Province, Chinese Province, Japan, the Malay Peninsula, Sumatra, Java, and Borneo. For similar reasons we have divided the Australian Region into Papuasia, Australia, Polynesia, and New Zealand. As to the middle parts of the East Indian Archipelago we advocate, as will be seen later on, the recognition of a broad Transition-Zone, comprising four areas — a Philippine, a Celebesian, a Lesser Sundan, and a Moluccan, — although the three first display a preponderance of Asiatic elements, while the Moluccas correspond naturally to their geographical position between Sula and Papuasia.

GEOGRAPHICAL DISTRIBUTION OF THE

Number	Name of species	Page in the text	Nearctic Region	Ethiopian Region	Palaearctic Region	Indian Province	Chinese Province	Japan	Malay Peninsula	Sumatra	Java	Borneo	Philippine Islands	Lesser Sunda Islands
1	Spilornis rufipectus J. Gd. (typical)	2	—	—	—	—	—	—	—	—	—	—	—	
	— rufipectus sulaensis (Schl.)	4	—	—	—	—	—	—	—	—	—	—	—	
	— rufipectus<sulaensis	5	—	—	—	—	—	—	—	—	—	—	—	
2	Circus assimilis Jard. Selby	7	—	—	—	—	—	—	—	—	—	—	—	
3	Astur griseiceps Schl.	9	—	—	—	—	—	—	—	—	—	—	—	
4	Astur trivirgatus (Temm.) (typical)	11	—	—	—	*	*	—	—	*	*	*	*	
	— trivirgatus rufitinctus (McClell.)	12	—	—	—	*	—	—	*	—	—	—	—	
5	? Astur tenuirostris Brügg.	13	—	—	—	—	—	—	—	—	—	—	—	
6	Urospizias torquatus (Temm.)	15	—	—	—	—	—	—	—	—	*?	—	—	
7	Tachyspizias soloensis (Horsf.)	17	—	—	—	*	*	—	*	*	*	*	*	
8	Spilospizias trinotatus (Bp.) (typical)	21	—	—	—	—	—	—	—	—	—	—	—	
	— trinotatus haesitandus Hart.	23	—	—	—	—	—	—	—	—	—	—	—	
9	Accipiter rhodogaster (Schl.)	25	—	—	—	—	—	—	—	—	—	—	—	
10	Accipiter sulaensis (Schl.)	26	—	—	—	—	—	—	—	—	—	—	—	
11	Accipiter virgatus (Temm.) (typical)	27	—	—	—	*	—	—	*	*	*	—	—	
	— virgatus affinis (Hdgs.)	28	—	—	—	*	*	—	—	—	—	—	—	
	— virgatus manilensis (Meyen)	28	—	—	—	—	—	—	—	—	—	—	*	
	— virgatus gularis (Temm. Schl.)	28	—	—	*	—	*	*	*	*	*	*	*	
	— virgatus rufotibialis (Sharpe)	29	—	—	—	—	—	—	—	—	—	*	—	
12	Spizaetus lanceolatus Temm. Schl.	32	—	—	—	—	—	—	—	—	—	—	—	
13	Lophotriorchis kieneri (G. Sparre)	35	—	—	—	*	—	—	*	*	*	*	*	
14	Ictinaetus malayensis (Reinw.)	38	—	—	—	*	—	—	*	*	*	*	*	
15	Haliaetus leucogaster (Gm.)	40	—	—	—	*	*	—	*	*	*	*	*	
16	Polioaetus humilis (Müll. Schl.) (typical)	43	—	—	—	*	—	—	*	—	—	*	—	
	— humilis major n. subsp.	44	—	—	—	*	—	—	—	—	—	—	—	
	— humilis—major	44	—	—	—	*	—	—	—	—	—	—	—	
17	Butastur indicus (Gm.)	45	—	—	*	*	*	*	*	—	—	*	*	
18	Butastur liventer (Temm.)	49	—	—	—	*	*	—	—	—	*	*?	—	
19	Haliastur indus (Bodd.)	51	—	—	—	*	—	—	*	*	*	*	*	
20	Milvus migrans (Bodd.) (typical)	60	—	*	*	*	—	—	—	—	—	—	—	
	— migrans melanotis (Temm. Schl.)	60	—	—	*	*	*	*	—	—	—	—	—	
	— migrans govinda (Sykes)	60	—	—	—	*	—	—	—	—	—	—	—	
	— — affinis (J. Gd.) & m.—affinis	60	—	—	—	*	*	—	*	*	—	—	*	
21	Elanus hypoleucus J. Gd.	62	—	—	—	—	—	—	*	*	*	*	*	
22	Pernis celebensis (Wall.)	65	—	—	—	—	—	—	—	—	—	—	—	
23	Pernis sp.	72	—	—	—	—	—	—	—	—	—	—	—	
24	Baza celebensis Schl.	73	—	—	—	—	—	—	—	—	—	—	—	
25	Baza reinwardti Müll. Schl.	75	—	—	—	—	—	—	—	—	—	—	—	
26	Tinnunculus moluccensis orient. n. subsp.	79	—	—	—	—	—	—	—	—	—	—	—	
	— moluccensis occidentalis M.&Wg.	79	—	—	—	—	—	—	—	—	*	*	—	
	— molucc. orientalis—occidentalis	79	—	—	—	—	—	—	—	—	—	—	—	
27	Falco severus papuanus M.&Wg.	84	—	—	—	—	—	—	—	—	—	—	—	

Introduction: Geographical Distribution.

SPECIES OF THE CELEBESIAN AREA.

Talaut Islands	Sangi Islands	North Peninsula	West Celebes	East Peninsula	Togian Islands	Central Celebes	South Peninsula	S.E. Peninsula and Buton	Saleyer Island	Djampea Group	Peling Group	Sula Islands	Moluccas	Papuasia	Australia	Polynesia	New Zealand	Neotropical Region	Name of species
*?	*	—	*	*	*	*	*	*	—	—	—	—	—	—	—	—	—	—	Spilornis rufipectus (typ.)
—	—	—	—	—	—	—	—	—	—	—	*	*	—	—	—	—	—	—	— rufipectus sulaensis.
—	—	—	—	—	—	—	—	—	—	—	—	—	—	—	—	—	—	—	— — ⟨sulaensis.
—	*	—	—	*	*	—	—	—	—	—	—	—	—	—	*	—	—	—	Circus assimilis.
—	*	*	—	—	*	*	—	—	—	—	—	—	—	—	—	—	—	—	Astur griseiceps.
—	*	—	—	—	—	—	—	—	—	—	—	—	—	—	—	—	—	—	Astur trivirgatus.
—	—	—	—	—	—	—	—	—	—	—	—	—	—	—	—	—	—	—	— — rufitinctus.
—	*?	—	—	—	—	—	—	—	—	—	—	—	—	—	—	—	—	—	? Astur tenuirostris.
—	—	—	—	—	—	—	*	—	—	—	—	—	*	*	—	—	—	—	Urospizias torquatus.
*	*	*	—	—	—	—	—	—	—	—	*	*	*	—	—	—	—	—	Tachyspizias soloensis.
—	*	*	—	—	—	—	—	—	—	—	—	—	—	—	—	—	—	—	Spilospizias trinotat. (typ.).
—	—	—	—	—	—	—	—	—	—	—	—	—	—	—	—	—	—	—	— — haesitandus.
—	*	*	—	—	—	*	—	—	—	—	—	—	—	—	—	—	—	—	Accipiter rhodogaster.
—	—	—	—	—	—	—	—	—	—	—	*	*	—	—	—	—	—	—	Accipiter sulaensis.
—	*	—	—	—	—	—	—	*	—	—	—	—	—	—	—	—	—	—	Accipiter virgatus (typ.).
—	—	—	—	—	—	—	—	—	—	—	—	—	—	—	—	—	—	—	— virgatus affinis.
—	—	—	—	—	—	—	—	—	—	—	—	—	—	—	—	—	—	—	— virgatus manilensis.
—	*	—	—	—	—	—	—	—	—	—	—	—	—	—	—	—	—	—	— virgatus gularis.
—	—	—	—	—	—	—	—	—	—	—	—	—	—	—	—	—	—	—	— virgatus rufotibialis.
—	*	—	—	*	*	—	—	—	—	—	—	*	—	—	—	—	—	—	Spizaetus lanceolatus.
—	*	—	—	—	—	—	—	—	—	—	—	—	*?	—	—	—	—	—	Lophotriorchis kieneri.
—	*	—	*	—	*	—	*	—	—	—	—	*?	*	—	—	—	—	—	Ictinaetus malayensis.
—	*	—	—	—	*	—	—	*	—	—	—	*	*	*	*?	*?	—	—	Haliaetus leucogaster.
—	—	—	—	—	—	—	—	—	—	—	—	—	—	—	—	—	—	—	Polioaetus humilis (typ.).
—	—	—	—	—	—	—	—	—	—	—	—	—	—	—	—	—	—	—	— humilis major n. subsp.
—	—	—	—	—	—	—	—	—	—	—	—	—	—	—	—	—	—	—	— humilis—major.
*	*	*	—	—	—	*	*	—	—	—	—	*?	*	—	—	—	—	—	Butastur indicus.
—	—	—	—	—	—	—	—	—	—	—	—	—	—	—	—	—	—	—	Butastur liventer.
*	*	*	—	—	*	—	*	—	—	—	*	*	*	*	—	—	—	—	Haliastur indus.
—	—	—	—	—	—	—	—	—	—	—	—	—	—	—	—	—	—	—	Milvus migrans (typical).
—	—	—	—	—	—	—	—	—	—	—	—	—	—	—	—	—	—	—	— migrans melanotis.
—	—	—	—	—	—	—	—	—	—	—	—	—	—	—	—	*	—	—	— migrans govinda.
—	*	—	—	—	—	*	*	—	—	*	—	*	—	—	—	—	—	—	— migrans affinis, etc.
—	*	—	—	—	—	*	*	—	—	*	—	*	—	—	—	—	—	—	Elanus hypoleucus.
—	—	—	—	—	—	—	—	—	—	*	—	*	—	—	—	—	—	—	Pernis celebensis.
—	*	—	—	*	*	—	—	—	*	—	*	*	—	—	—	—	—	—	Pernis sp.
—	—	—	—	—	—	—	—	—	—	*	—	—	—	—	—	—	—	—	Baza celebensis.
—	*	—	—	—	—	—	—	*	—	—	*	*	—	—	—	—	—	—	Baza reinwardti.
—	—	—	—	—	—	—	—	—	*	—	—	*?	—	—	—	—	—	—	Tinnunculus mol. orient.
—	*	—	—	—	*	—	—	—	*	—	*	—	—	—	—	—	—	—	— moluccensis occident.
—	—	—	—	—	—	—	—	—	—	—	—	—	—	—	—	—	—	—	— — orient.—occident.
—	—	—	—	—	—	—	—	—	—	—	—	—	—	—	—	—	—	—	Falco severus papuanus.

Introduction: Geographical Distribution.

Number	Name of species	Page in the text	Nearctic Region	Ethiopian Region	Palaearctic Region	Indian Province	Chinese Province	Japan	Malay Peninsula	Sumatra	Java	Borneo	Philippine Islands	Lesser Sunda Islands
27	Falco severus indicus n. subsp.	84	—	—	—	*	—	—	—	—	—	—	—	—
	— severus papuanus—indicus	84	—	—	—	—	—	—	—	—	—	—	—	—
28	Falco peregrinus (Gerini) (typical)	85	*?	*	*	*	—	—	*	—	*	*	*	*
	— peregrinus melanogenys (J. Gd.)	86	—	—	—	—	*?	—	—	*	*	*	*	*
	— peregrinus ernesti (Sharpe)	86	—	—	—	—	—	—	—	—	—	*	*	*
	— peregrinus anatum (Bp)	87	*	—	*?	—	—	—	—	—	—	—	—	—
	— peregrinus pealei Ridgw.	87	*	—	—	—	—	—	—	—	—	—	—	—
29	Pandion haliaetus (L.) (typical)	89	—	—	—	—	—	—	—	—	—	—	—	—
	— haliaetus leucocephalus (J. Gd.)	89	—	*	*	*	*	—	—	—	*	—	*	—
	— haliaetus carolinensis (Gm.)	90	*	—	—	—	—	—	—	—	—	—	*	*
30	Ninox ochracea (Schl.)	94	—	—	—	—	—	—	—	—	—	—	—	—
31	Ninox scutulata (Raffl.) (typical)	95	—	—	*	—	—	*	*	*	*	—	—	—
	— scutulata lugubris (Tickell)	95	—	—	*	—	—	—	—	—	—	—	—	—
	— scutulata affinis (Tytler)	95	—	—	*	—	—	—	—	—	—	—	—	—
	— scutulata japonica (Temm. Schl.)	95	—	—	*	—	*	*	—	—	—	*	*	*
32	Cephaloptynx punctulata (Q. G.)	100	—	—	—	—	—	—	—	—	—	—	—	—
33	Scops manadensis (Q. G.) (typical)	103	—	—	—	—	—	—	—	—	—	—	—	—
	— manadensis albiventris (Sharpe)	105	—	—	—	—	—	—	—	—	—	—	—	—
	— manadensis rutilus (Pucher.)	105	—	*	—	—	—	—	—	—	—	—	—	—
	— manadensis capnodes (Gurney)	105	—	*	—	—	—	—	—	—	—	—	—	—
	— manadensis magicus (S. Müll.)	105	—	—	—	—	—	—	—	—	—	—	—	—
	— manadensis leucospilus (Gray)	106	—	—	—	—	—	—	—	—	—	—	—	—
?	— manadensis morotensis (Sharpe)	106	—	—	—	—	—	—	—	—	—	—	—	—
	— manadensis brookii (Sharpe)	107	—	—	—	—	—	—	—	—	—	*	—	—
	— manadensis sibutuensis (Sharpe)	107	—	—	—	—	—	—	—	—	—	—	*	—
34	Strix flammea rosenbergi (Schl.)	109	—	—	—	—	—	—	—	—	—	—	—	—
	— flammea L. (typical) [1]	111	—	—	—	—	—	—	—	—	—	—	—	—
35	Strix inexpectata Schl.	112	—	—	—	—	—	—	—	—	—	—	—	—
36	Strix candida Tick.	112	—	—	*	*	—	—	—	—	—	—	*	—
37	Eos histrio (St. Müll.) (typical)	115	—	—	—	—	—	—	—	—	—	—	—	—
	— histrio talautensis M. & Wg.	117	—	—	—	—	—	—	—	—	—	—	—	—
	— histrio challengeri (Salvad.)	118	—	—	—	—	—	—	—	—	—	—	—	—
38	Trichoglossus ornatus (L.)	120	—	—	—	—	—	—	—	—	—	—	—	—
39	Trichoglossus forsteni Bp. (typical)	123	—	—	—	—	—	—	—	—	—	—	—	*
	— forsteni djampeanus Hart.	124	—	—	—	—	—	—	—	—	—	—	—	—
40	Trichoglossus meyeri Tweedd. (typical)	124	—	—	—	—	—	—	—	—	—	—	—	—
	— meyeri bonthainensis (A. B. M.)	125	—	—	—	—	—	—	—	—	—	—	—	—
41	Trichoglossus flavoviridis Wall.	127	—	—	—	—	—	—	—	—	—	—	—	—
42	Cacatua sulphurea (Gm.) (typical)	128	—	—	—	—	—	—	—	—	—	—	—	—
	— sulphurea djampoana Hart.	130	—	—	—	—	—	—	—	—	—	—	—	—
	— sulphurea parvula (Bp.)	130	—	—	—	—	—	—	—	—	—	—	—	*
43	Prioniturus platurus (Vieill.)	133	—	—	—	—	—	—	—	—	—	—	—	—
44	Prioniturus flavicans Cass.	138	—	—	—	—	—	—	—	—	—	—	—	—
45	Tanygnathus muelleri (Müll. Schl.) (typ.)	140	—	—	—	—	—	—	—	—	—	—	—	—

[1] Distribution not clearly defined.

Introduction: Geographical Distribution. 93

Talaut Islands	Sangi Islands	North Peninsula	West Celebes	East Peninsula	Togian Islands	Central Celebes	South Peninsula	S. E. Peninsula (and Buton)	Saleyer Island	Djampea Group	Paling Group	Sula Islands	Moluccas	Papuasia	Australia	Polynesia	New Zealand	Neotropical Region	Name of species
—	—	—	—	—	—	—	—	—	—	—	—	—	—	—	—	—	—	—	Falco severus indicus.
—	*	—	—	—	*	—	—	—	—	—	—	—	*	—	—	—	—	—	— — papuan.—indicus.
—	*?	—	—	—	—	—	—	—	—	—	—	—	*	—	—	—	—	—	Falco peregrinus (typ.).
—	*	—	—	—	—	—	—	—	—	—	—	—	—	*	*	*	—	—	— — melanogenys.
—	—	—	—	—	—	—	—	—	—	—	—	—	—	—	—	*	—	—	— — ernesti.
—	—	—	—	—	—	—	—	—	—	—	—	—	—	—	—	—	—	*	— — anatum.
—	—	—	—	—	—	—	—	—	—	—	—	—	—	—	—	—	—	—	— — pealei.
—	*	*	—	—	—	—	—	—	—	—	—	—	—	—	—	—	—	—	Pandion haliaetus (typ.).
*	*	*	—	—	—	—	—	—	*	*	—	*	*	*	*	*	—	—	— — leucocephalus.
—	—	—	—	—	—	—	—	—	—	—	—	—	—	—	—	—	—	*	— — carolinensis.
—	*	—	*	—	—	—	—	—	—	—	—	—	—	—	—	—	—	—	Ninox ochracea.
—	*	—	—	—	—	—	—	—	—	—	—	—	—	—	—	—	—	—	Ninox scutulata (typical).
—	—	—	—	—	—	—	—	—	—	—	—	—	—	—	—	—	—	—	— — lugubris.
—	—	—	—	—	—	—	—	—	—	—	—	—	—	—	—	—	—	—	— — affinis.
—	—	—	—	—	—	—	—	—	—	—	—	—	—	—	—	—	—	—	— — japonica.
—	*	*	—	—	—	*	—	—	*	—	*	*	—	—	—	—	—	—	Cephaloptynx punctulata.
*	*	—	—	—	—	*	*	—	—	—	—	—	—	—	—	—	—	—	Scops manadensis (typ.).
—	—	—	—	—	—	—	—	—	—	—	—	—	—	—	—	—	—	—	— — albiventris.
—	—	—	—	—	—	—	—	—	—	—	—	—	—	—	—	—	—	—	— — rutilus.
—	—	—	—	—	—	—	—	—	—	—	—	—	—	—	—	—	—	—	— — capnodes.
—	—	—	—	—	—	—	—	—	—	—	—	—	*	—	—	—	—	—	— — magicus.
—	—	—	—	—	—	—	—	—	—	—	—	—	*	—	—	—	—	—	— — leucospilus.
—	—	—	—	—	—	—	—	—	—	—	—	—	?	—	—	—	—	—	? — — morotensis.
—	—	—	—	—	—	—	—	—	—	—	—	—	—	—	—	—	—	—	— — brookii.
—	—	—	—	—	—	—	—	—	—	—	—	—	—	—	—	—	—	—	— — sibutuensis.
*	*	—	—	—	—	*	—	—	—	—	—	—	—	—	—	—	—	—	Strix flammea rosenbergi.
—	*	—	—	—	—	—	—	—	*	—	—	—	—	—	—	—	—	—	— flammea (typical)[1].
—	—	—	—	—	—	—	—	—	—	—	—	—	—	—	—	—	—	—	Strix inexpectata.
—	—	—	—	—	*	—	—	—	—	—	—	—	—	—	*	*	—	—	Strix candida.
*	*	—	—	—	—	—	—	—	—	—	—	—	—	—	—	—	—	—	Eos histrio (typical).
*[?]	—	—	—	—	—	—	—	—	—	—	—	—	—	—	—	—	—	—	— histrio talautensis.
—	—	—	—	—	—	—	—	—	—	—	—	—	—	—	—	—	—	—	— histrio challengeri.
—	*	*	*	*	—	*	*	—	*	—	—	—	—	—	—	—	—	—	Trichoglossus ornatus.
—	—	—	—	—	—	—	—	—	—	—	—	—	—	—	—	—	—	—	Trich. forsteni (typical).
—	—	*	—	—	—	—	—	*	—	—	—	—	—	—	—	—	—	—	— forsteni djampeanus.
—	—	—	—	—	*	—	—	—	—	—	—	—	—	—	—	—	—	—	Trich. meyeri (typical).
—	—	—	—	—	—	—	—	—	—	—	—	—	—	—	—	—	—	—	— meyeri bonthainensis.
—	—	—	—	—	—	—	—	—	—	—	*	—	—	—	—	—	—	—	Trich. flavoviridis.
—	*	*	—	—	*	*	*	—	—	*	—	—	—	—	—	—	—	—	Cacatua sulphurea (typ.).
—	—	—	—	—	—	—	—	—	—	—	—	—	—	—	—	—	—	—	— sulphurea djampeana.
—	—	—	—	—	—	—	—	—	—	—	—	—	—	—	—	—	—	—	— sulphurea parvula.
*	*	*	—	—	*	*	*	—	—	*	—	*?	—	—	—	—	—	—	Prionitorus platurus.
—	*?	*	—	*	—	*	—	—	—	—	—	—	—	—	—	—	—	—	Prionitorus flavicans.
—	*	—	*	*	*	*	*	—	—	*	*	—	—	—	—	—	—	—	Tanygnath. muelleri (typ.).

[1] Lit. Nanusa Islands.

Number	Name of species	Page in the text	Nearctic Region	Ethiopian Region	Palaearctic Region	Indian Province	Chinese Province	Japan	Malay Peninsula	Sumatra	Java	Borneo	Philippine Islands	Lesser Sunda Islands
45	Tanygnathus muelleri sangirensis M.&Wg.	142	—	—	—	—	—	—	—	—	—	—	—	—
	— muelleri—sangirensis.	142	—	—	—	—	—	—	—	—	—	—	—	—
46	? Tanygnathus luconensis (L.). . . .	144	—	—	—	—	—	—	—	—	—	—	*	—
47	Tanygnathus talautensis M. & Wg. .	145	—	—	—	—	—	—	—	—	—	—	—	—
48	Tanygnath. megalorhynchus (Bodd.) (typ.)	146	—	—	—	—	—	—	—	—	—	—	—	—
	— megalorhynchus sumbensis (Meyer)	148	—	—	—	—	—	—	—	—	—	—	—	*
	— megalorhynchus — sumbensis . . .	148	—	—	—	—	—	—	—	—	—	—	—	—
49	Loriculus exilis Schl.	149	—	—	—	—	—	—	—	—	—	—	—	—
50	Loriculus catamene Schl.	151	—	—	—	—	—	—	—	—	—	—	—	—
51	Loriculus sclateri Wall. (typical) . .	153	—	—	—	—	—	—	—	—	—	—	—	—
	— sclateri ruber M. & Wg.	154	—	—	—	—	—	—	—	—	—	—	—	—
52	Loriculus quadricolor Tweedd. . .	157	—	—	—	—	—	—	—	—	—	—	—	—
53	Loriculus stigmatus (Müll. Schl.) . .	158	—	—	—	—	—	—	—	—	—	—	—	—
54	Aprosmictus sulaensis (Rchw.) . . .	170	—	—	—	—	—	—	—	—	—	—	—	—
55	Iyngipicus temmincki (Malh.)	173	—	—	—	—	—	—	—	—	—	—	—	—
56	Microstictus fulvus (Q. G.)	175	—	—	—	—	—	—	—	—	—	—	—	—
57	Microstictus wallacei (Tweedd.) . .	179	—	—	—	—	—	—	—	—	—	—	—	—
58	Hierococcyx crassirostris (Tweedd.) .	182	—	—	*	*	—	—	—	—	—	—	—	—
59	Hierococcyx sparverioides (Vig.) . . .	184	—	—	*	*	*	—	*	—	*	*	*	—
60	Hierococcyx fugax (Horsf.)	185	—	—	*	*	*	*	*	*	*	*	*	—
61	Cuculus canorus (L.) (typical)	187	—	*	*	*	—	—	—	—	—	—	—	—
	— canorus canoroides (S. Müll.) . .	188	—	—	*	*	—	*	*	—	*	*	*	*
62	? Cuculus saturatus Hdgs.	191	—	—	*?	*	*	*?	—	*?	*	*	*?	*?
63	Chrysococcyx malayanus (Raffl.) . . .	194	—	—	—	—	—	—	*	*	*	*	*	—
64	Chrysococcyx basalis (Horsf.)	195	—	—	—	—	—	—	—	—	*	—	—	*
65	Cacomantis virescens (Brügg.)	196	—	—	—	—	—	—	*	—	*	—	—	*
66	Cacomantis merulinus (Scop.)	199	—	—	—	*	*	—	*	*	*	*	*	*
67	Coccystes coromandus (L.)	201	—	—	—	*	*	—	*	*	*	*	*	—
68	Surniculus musschenbroeki A.B.M. .	203	—	—	—	—	—	—	—	—	—	—	—	—
69	Eudynamis melanorhyncha S. Müll. .	205	—	—	—	—	—	—	—	—	—	—	—	—
70	Eudynamis mindanensis (L.) (typical)	210	—	—	—	—	—	—	—	—	—	—	*	—
	— mindanensis sangirensis (W.Blas.)	211	—	—	—	—	—	—	—	—	—	—	—	—
71	Centrococcyx bengalensis (Gm.) . . .	213	—	—	—	*	*	—	*	*	*	*	*	—
72	Pyrrhocentor celebensis (Q. G.) (typical)	221	—	—	—	—	—	—	—	—	—	—	—	—
	— celebensis rufescens M. & Wg.)	223	—	—	—	—	—	—	—	—	—	—	—	—
73	Phoenicophaes calorhynch. (Temm.)(typ.)	226	—	—	—	—	—	—	—	—	—	—	—	—
	— calorhynchus meridionalis M.&Wg.	227	—	—	—	—	—	—	—	—	—	—	—	—
74	Scythrops novaehollandiae Lath. . .	231	—	—	—	—	—	—	—	—	—	—	—	*
75	Rhabdotorrhinus exaratus (Temm.) .	235	—	—	—	—	—	—	—	—	—	—	—	—
76	Cranorrhinus cassidix (Temm.) . .	239	—	—	—	—	—	—	—	—	—	—	—	—
77	Merops ornatus Lath.	248	—	—	—	—	—	—	—	—	*?	*?	—	*
78	Merops philippinus L.	253	—	—	—	*	*	—	*	*	*	*	—	*
79	Micropogon forsteni Bp.	257	—	—	—	—	—	—	—	—	—	—	—	—
80	Alcedo ispida L.	262	—	*	*	*	*	*	*	—	*	*	*	*

Introduction: Geographical Distribution.

lebesian Area														Name of species
Celebes														
East Peninsula	Togian Islands	Central Celebes	South Peninsula	S. E. Peninsula (and Butou)	Saleyer Island	Djampea Group	Paling Group	Sula Islands	Moluccas	Papuasia	Australia	Polynesia	New Zealand	Neotropical Region	
—	—	—	—	—	—	—	—	—	—	—	—	—	—	—	Tanygn. muell. sangirensis.
—	—	—	—	—	—	—	—	—	—	—	—	—	—	—	— muelleri—sangirensis.
—	—	—	—	—	—	—	—	—	—	—	—	—	—	—	?Tanygnath. luconensis.
—	—	—	—	—	—	—	—	—	—	—	—	—	—	—	Tanygnath. talautensis.
—	—	—	—	—	—	—	—	—	*	*	—	—	—	—	T. megalorhynchus (typ.).
—	—	—	—	—	—	—	—	—	—	—	—	—	—	—	— — sumbensis.
—	—	—	—	—	*	—	—	—	—	—	—	—	—	—	— megalorh.—sumb.
—	—	—	—	—	—	—	—	—	—	—	—	—	—	—	Loriculus exilis.
—	—	—	—	—	—	—	—	*	—	—	—	—	—	—	Loriculus catamene.
—	—	—	—	—	—	—	*	—	—	—	—	—	—	—	Loriculus sclateri.
—	—	—	—	—	—	—	—	—	—	—	—	—	—	—	— sclateri ruber.
—	—	—	—	—	—	—	*	*	—	—	—	—	—	—	Loriculus quadricolor.
*	—	*	*	—	—	—	*	*	—	—	—	—	—	—	Loriculus stigmatus.
—	—	—	—	—	—	—	—	—	—	—	—	—	—	—	Aprosmictus sulaensis.
—	*	—	*	—	—	—	—	—	—	—	—	—	—	—	Iyngipicus temmincki.
—	—	—	—	—	—	—	—	—	—	—	—	—	—	—	Microstictus fulvus.
*	—	*	*	*?	—	—	—	—	—	—	—	—	—	—	Microstictus wallacei.
—	—	*	*	—	—	—	—	—	—	—	—	—	—	—	Hierococcyx crassirostris.
—	—	—	—	—	—	—	—	—	—	—	—	—	—	—	Hierococc. sparverioides.
—	—	—	—	—	—	—	—	—	—	—	—	—	—	—	Hierococc. fugax.
—	—	—	—	—	—	—	—	—	—	—	—	—	—	—	Cuculus canorus (typ.).
—	—	*	—	*	*	—	—	*	*?	*?	*?	*	*?	—	— canorus canoroides.
—	—	*?	—	—	—	—	—	—	—	—	—	—	—	—	? Cuculus saturatus.
—	*	*	—	—	—	—	—	—	*?	*	*	—	—	—	Chrysococcyx malayanus.
—	—	*	—	—	—	—	—	—	*	*	—	—	—	—	Chrysococcyx basalis.
*	—	—	*	—	—	*	—	—	—	—	—	—	—	—	Cacomantis virescens.
—	*?	—	*	—	—	—	—	—	—	—	—	—	—	—	Cacomantis merulinus.
—	—	—	*	—	—	—	—	—	—	—	—	—	—	—	Coccystes coromandus.
—	—	—	*	—	—	—	—	*	—	—	—	—	—	—	Surnicul. musschenbroeki.
*	*	—	*	—	—	*	*	—	—	—	—	—	—	—	Eudynam. melanorhyncha.
—	—	—	—	—	—	—	—	—	—	—	—	—	—	—	Eud. mindanensis (typical).
—	—	—	—	—	—	—	—	—	—	—	—	—	—	—	— — sangirensis.
—	—	*	*	*	—	*	—	—	*	—	—	—	—	—	Centrococcyx bengalensis
*	*	—	*?	—	—	—	—	—	—	—	—	—	—	—	Pyrrhocentor celeb. (typ.).
*	—	—	*	—	—	—	—	—	—	—	—	—	—	—	— celebensis rufescens.
*	*	—	*	*?	—	—	—	—	—	—	—	—	—	—	Phoenicoph. calorh. (typ.).
—	—	*	*	—	—	—	—	—	—	—	—	—	—	—	— — meridionalis.
—	—	—	—	—	—	—	—	*	*	*	—	—	—	—	Scythrops novaehollands.
*	—	*	*	—	—	—	—	—	—	—	—	—	—	—	Rhabdotorrhinus exarat.
*	*	*	*	—	—	—	—	—	—	—	—	—	—	—	Cranorrhinus cassidix.
—	*	—	—	—	—	—	*	*	*	*	—	—	—	—	Merops ornatus.
—	—	*	*	—	—	—	—	—	—	—	—	—	—	—	Merops philippinus.
—	—	—	—	—	—	—	—	—	—	—	—	—	—	—	Meropogon forsteni.
—	—	—	—	—	*	—	—	*	—	—	—	—	—	—	Alcedo ispida.

Number	Name of species	Page in the text	Nearctic Region	Ethiopian Region	Palaearctic Region	Indian Province	Chinese Province	Japan	Malay Peninsula	Sumatra	Java	Borneo	Philippine Islands	Lesser Sunda Islands
81	Alcedo moluccana (Less.)	264	—	—	—	—	—	—	—	—	—	—	—	—
82	Alcedo meninting Horsf.	266	—	—	—	—	—	—	—	—	—	—	—	—
83	Pelargopsis melanorhyncha (Temm.)	269	—	—	—	*	*	—	*	*	*	*	*	*
84	Pelargopsis dichrorhyncha M.&Wg.	271	—	—	—	—	—	—	—	—	—	—	—	—
85	Ceyx wallacei Sharpe	272	—	—	—	—	—	—	—	—	—	—	—	—
86	Ceycopsis fallax (Schl.)	275	—	—	—	—	—	—	—	—	—	—	—	—
87	Ceycopsis sangirensis M.&Wg.	278	—	—	—	—	—	—	—	—	—	—	—	—
88	Halcyon coromanda rufa (Wall.)	280	—	—	—	—	—	—	—	—	—	—	—	—
89	Halcyon pileata (Bodd.)	283	—	—	*	*	*	—	*	*	—	*	—	—
90	Halcyon sancta Vig. Horsf.	287	—	—	—	—	—	—	—	*	*	*	—	*
91	Halcyon chloris (Bodd.) (typical)¹)	292	—	—	—	—	—	—	—	—	—	—	—	—
92	Monachalcyon monachus (Bp.) (typical)	297	—	—	—	—	—	—	—	—	—	—	—	—
—	monachus intermedius Hart.	298	—	—	—	—	—	—	—	—	—	—	—	—
93	Monachalcyon capucinus M.&Wg.	299	—	—	—	—	—	—	—	—	—	—	—	—
94	Monachalcyon princeps Rchb.	300	—	—	—	—	—	—	—	—	—	—	—	—
95	Cittura cyanotis (Temm.)	303	—	—	—	—	—	—	—	—	—	—	—	—
96	Cittura sangirensis Sharpe	305	—	—	—	—	—	—	—	—	—	—	—	—
97	Coracias temmincki (Vieill.)	309	—	—	—	—	—	—	—	—	—	—	—	—
98	Eurystomus orientalis (L.)	312	—	—	*	*	*	—	*	*	*	*	*	*
99	Caprimulgus macrurus Horsf. (typical)	317	—	—	—	—	—	—	*	*	*	*	*	*
—	macrurus albonotatus (Tick.)	318	—	—	—	*	—	—	—	—	—	—	—	—
—	macrurus—albonotatus	318	—	—	—	*	*	—	—	—	—	—	—	—
100	Caprimulgus celebensis Grant	320	—	—	—	—	—	—	—	—	—	—	—	—
101	Caprimulgus affinis Horsf.	321	—	—	—	—	—	—	—	*	*	*	—	—
102	Lyncornis macropterus Bp.	322	—	—	—	—	—	—	—	—	—	—	—	—
103	Cypselus pacificus (Lath.)	327	—	—	*	*	*	*	*	—	—	—	—	—
104	Chaetura celebensis (Scl.)	329	—	—	—	—	—	—	—	—	—	—	*	—
105	Collocalia fuciphaga (Thunb.)	331	—	—	—	*	*	—	*	*	*	*	—	—
106	Collocalia esculenta (L.)	334	—	—	—	—	—	—	—	—	—	—	—	—
107	Collocalia francica (Gm.)	335	—	*	—	*	—	—	*	—	—	—	*	—
108	Macropteryx wallacei (J. Gd.)	336	—	—	—	—	—	—	—	—	—	—	—	—
109	Pitta celebensis Müll. Schl.	340	—	—	—	—	—	—	—	—	—	—	—	—
110	Pitta palliceps Brügg.	344	—	—	—	—	—	—	—	—	—	—	—	—
111	Pitta caeruleitorques Salvad.	345	—	—	—	—	—	—	—	—	—	—	—	—
112	Pitta inspeculata M.&Wg.	346	—	—	—	—	—	—	—	—	—	—	—	—
113	Pitta forsteni Bp.	350	—	—	—	—	—	—	—	—	—	—	—	—
114	Pitta sangirana (Schl.)	351	—	—	—	—	—	—	—	—	—	—	—	—
115	Pitta cyanoptera Temm.	352	—	*	*	*	—	—	*	*	*	*	?	?
116	Pitta irena Temm.	354	—	—	—	—	—	—	—	—	—	—	—	*
117	Pitta virginalis Hart.	355	—	—	—	—	—	—	—	—	—	—	—	—
118	Hirundo rustica L. (races)¹)	357	—	—	—	—	—	—	—	—	—	—	—	—
119	Hirundo javanica Sparrm.	358	—	—	—	*	—	—	*	*	*	*	*	*
120	Muscicapa griseosticta (Swinh.)	363	—	—	—	—	*	—	—	—	—	—	—	*

¹) Distribution of races not capable of exact definition.

Introduction: Geographical Distribution

Tukan Islands	Sangi Islands	North Peninsula	West Celebes	East Peninsula	Togian Islands	Central Celebes	South Peninsula	S.E. Peninsula (and Buton)	Saleyer Island	Djampea Group	Paling Group	Sula Islands	Moluccas	Papuasia	Australia	Polynesia	New Zealand	Neotropical Region	Name of species
—	*	—	*	—	—	*	—	*	—	*	*	*	*	—	—	—	—	—	Alcedo moluccana.
—	*	*	—	*	—	*	*	*	—	—	—	*	—	—	—	—	—	—	Alcedo meninting.
—	*	*	*	—	*	*	—	*	—	—	—	*	—	—	—	—	—	—	Pelargops. melanorhyncha
—	—	—	—	—	—	—	—	—	—	*	—	*	—	—	—	—	—	—	Pelargops. dichrorhyncha.
—	—	—	—	—	—	—	*	—	—	—	—	—	—	—	—	—	—	—	Ceyx wallacei.
—	*	*	—	—	—	*	—	—	—	—	—	—	—	—	—	—	—	—	Ceycopsis fallax.
*	*	—	—	—	*	—	*	—	—	—	*	*	—	—	—	—	—	—	Ceycopsis sangirensis.
*	*	—	—	*	—	*	—	—	—	—	—	—	—	—	—	—	—	—	Halcyon coromanda rufa.
—	*	—	—	—	—	—	—	—	—	—	—	—	—	—	—	—	—	—	Halcyon pileata.
*	*	—	*	—	*	—	*	*	*	*	*	*	—	*	*	*	—	—	Halcyon sancta.
*	*	—	*	*	*	*	*	*	*	*	—	—	—	—	—	—	—	—	Halcyon chloris (typ.)[1].
—	*	—	—	—	—	—	—	—	—	—	—	—	—	—	—	—	—	—	Monachalc. monach. (typ.).
—	—	*	—	—	—	—	—	—	—	—	—	—	—	—	—	—	—	—	— — intermedius.
—	—	—	*	—	*	—	—	—	—	—	—	—	—	—	—	—	—	—	Monachalcyon capucinus.
—	*	—	—	—	—	—	—	—	—	—	—	—	—	—	—	—	—	—	Monachalcyon princeps.
—	*	*	*	—	—	—	*	—	—	—	—	—	—	—	—	—	—	—	Cittura cyanotis.
*	*	—	*	—	—	*	—	*	—	—	—	—	—	—	—	—	—	—	Cittura sangirensis.
*	*	—	—	—	*	—	*	—	*	*	*	*	—	—	—	—	—	*	Coracias temmincki.
—	—	—	—	—	—	—	—	—	—	—	—	—	—	—	—	—	—	—	Eurystomus orientalis.
—	*	—	—	—	—	*	—	—	—	—	—	—	*	*	*	—	—	—	Caprimulg.macrurus(typ.).
—	—	—	—	—	—	—	—	—	—	—	—	—	—	—	—	—	—	—	— albonotatus.
—	—	—	—	—	—	—	—	—	—	—	—	—	—	—	—	—	—	—	— macrurus—albonotat.
—	*	—	*	—	—	*	—	—	—	—	—	—	—	—	—	—	—	—	Caprimulgus celebensis.
—	*	—	—	—	—	—	—	—	—	—	—	—	—	—	—	—	—	—	Caprimulgus affinis.
*	—	—	—	—	—	—	—	—	—	—	—	—	*	*	—	—	—	—	Lyncornis macropterus.
—	*	—	—	—	—	—	—	—	—	—	—	—	—	—	—	—	—	—	Cypselus pacificus.
—	*	—	—	—	*	—	—	—	—	*	*	*	—	*	—	—	—	—	Chaetura celebensis.
—	—	—	—	—	—	*	—	—	—	*	*	*	*	—	—	—	—	—	Collocalia fuciphaga.
—	—	—	—	—	—	—	—	—	—	*	*	*	*	—	—	—	—	—	Collocalia esculenta.
—	*	—	*	*	*	*	—	*	—	—	*	*	—	—	—	—	—	—	Collocalia francica.
—	—	—	—	—	—	—	—	—	—	—	—	—	—	—	—	—	—	—	Macropteryx wallacei.
—	*	*	*	*	*	—	—	—	*	—	—	—	—	—	—	—	—	—	Pitta celebensis.
*	—	—	—	—	—	—	—	—	—	—	—	—	—	—	—	—	—	—	Pitta pallicops.
—	—	—	—	—	—	—	—	—	—	—	—	—	—	—	—	—	—	—	Pitta caeruleitorques.
—	*	—	—	—	—	—	—	—	—	—	—	—	—	—	—	—	—	—	Pitta inspeculata.
—	*	—	—	—	—	—	—	—	—	—	—	—	—	—	—	—	—	—	Pitta forsteni.
*	—	—	—	—	—	—	—	—	—	—	—	—	—	—	—	—	—	—	Pitta sangirana.
—	*	—	—	—	—	—	—	—	—	—	*	*	—	—	—	—	—	—	Pitta cyanoptera.
*	*	—	—	—	—	—	—	—	—	—	—	—	—	—	—	—	—	—	Pitta irena.
—	—	—	—	—	—	—	*	—	—	—	—	—	—	—	—	—	—	—	Pitta virginalis.
*	*	—	*	*	—	*	—	—	—	—	—	*	*	*	—	—	—	—	Hirundo rustica L.(races)[1]
*	*	—	—	—	—	*	—	—	—	—	—	—	—	—	—	—	—	—	Hirundo javanica.
—	—	—	—	—	—	—	—	—	—	—	—	—	—	—	—	—	—	—	Muscicapa griseosticta.

98 Introduction: Geographical Distribution.

Number	Name of species	Page in the text	Nearctic Region	Ethiopian Region	Palaearctic Region	Indian Province	Chinese Province	Japan	Malay Peninsula	Sumatra	Java	Borneo	Philippine Islands	Lesser Sunda Islands
121	Muscicapula westermanni Sharpe	365	—	—	*	—	—	—	—	—	—	—	—	*
122	Muscicapula hyperythra (Blyth)	366	—	—	*	—	—	—	*	*	*	*	*	*
123	Siphia banyumas (Horsf.)	368	—	—	—	—	—	—	*?	*?	*	—	*?	*?
124	Siphia djampeana Hart.	371	—	—	—	—	—	—	—	—	—	—	—	—
125	Siphia kalaoensis Hart.	371	—	—	—	—	—	—	—	—	—	—	—	—
126	Siphia rufigula (Wall.)	372	—	—	—	—	—	—	—	—	—	—	—	—
127	Siphia bonthaina Hart.	373	—	—	—	—	—	—	—	—	—	—	—	—
128	Stoparola septentrionalis Bütt.	374	—	—	—	—	—	—	—	—	—	—	—	—
129	Stoparola meridionalis Bütt.	375	—	—	—	—	—	—	—	—	—	—	—	—
130	Hypothymis puella (Wall.)	376	—	—	—	—	—	—	—	—	—	—	—	—
131	Hypothymis rowleyi (A.B.M.)	378	—	—	—	—	—	—	—	—	—	—	—	—
132	Rhipidura celebensis Bütt.	379	—	—	—	—	—	—	—	—	—	—	—	—
133	Rhipidura teijsmanni Bütt.	380	—	—	—	—	—	—	—	—	—	—	—	—
134	Zeocephus talautensis M.&Wg.	382	—	—	—	—	—	—	—	—	—	—	—	—
135	Monarcha commutatus Brügg.	383	—	—	—	—	—	—	—	—	—	—	—	—
136	Monarcha inornatus (Garn.)	384	—	—	—	—	—	—	—	—	—	—	—	—
137	Monarcha everetti Hart.	385	—	—	—	—	—	—	—	—	—	—	—	*
138	Myiagra rufigula Wall.	386	—	—	—	—	—	—	—	—	—	—	—	—
139	Culicicapa helianthea (Wall.)	387	—	—	—	—	—	—	—	—	—	—	*	—
140	Gerygone flaveola Cab.	388	—	—	—	—	—	—	—	—	—	—	—	—
141	Pratincola caprata (L.)	390	—	—	*	*	—	—	—	—	*	*?	*	*
142	Pachycephala sulfuriventer (Tweedd.)	394	—	—	—	—	—	—	—	—	—	—	—	—
143	Pachycephala meridionalis Bütt.	396	—	—	—	—	—	—	—	—	—	—	—	—
144	Pachycephala teijsmanni Bütt.	396	—	—	—	—	—	—	—	—	—	—	—	—
145	Pachycephala orpheus Jard.	397	—	—	—	—	—	—	—	—	—	—	—	*
146	Pachycephala griseonota G.R.Gray	398	—	—	—	—	—	—	—	—	—	—	—	—
147	Pachycephala clio Wall.	399	—	—	—	—	—	—	—	—	—	—	—	—
148	Pachycephala everetti Hart.	400	—	—	—	—	—	—	—	—	—	—	—	—
149	Pachycephala bonthaina M.&Wg.	401	—	—	—	—	—	—	—	—	—	—	—	—
150	Pachycephala bonensis M.&Wg.	401	—	—	—	—	—	—	—	—	—	—	—	—
151	Colluricincla sangirensis (Oust.)	402	—	—	—	—	—	—	—	—	—	—	—	—
152	Lanius tigrinus Drapiez	403	—	—	*	*	*	*	*	*	*	*	*	—
153	Lanius lucionensis L.	406	—	—	*	*	*	—	*	*	*	*	*	*
154	Graucalus bicolor (Temm.)	411	—	—	—	—	—	—	—	—	—	—	—	—
155	Graucalus leucopygius Bp.	413	—	—	—	—	—	—	—	—	—	—	—	—
156	Graucalus temmincki (S. Müll.)	415	—	—	—	—	—	—	—	—	—	—	—	—
157	Graucalus schistaceus (Sharpe)	416	—	—	—	—	—	—	—	—	—	—	—	—
158	Graucalus melanops (Lath.)	417	—	—	—	—	—	—	—	—	—	—	—	*
159	Edoliisoma morio (S. Müll.) (typical)	419	—	—	—	—	—	—	—	—	—	—	—	—
	— morio septentrionalis M.&Wg.	420	—	—	—	—	—	—	—	—	—	—	—	—
	— morio—septentrionalis M.&Wg.	421	—	—	—	—	—	—	—	—	—	—	—	—
160	Edoliisoma salvadorii Sharpe	422	—	—	—	—	—	—	—	—	—	—	—	—
161	Edoliisoma talautense M.&Wg.	423	—	—	—	—	—	—	—	—	—	—	—	—
162	Edoliisoma emancipata Hart.	424	—	—	—	—	—	—	—	—	—	—	—	—
163	Edoliisoma obiense Salvad.	424	—	—	—	—	—	—	—	—	—	—	—	—
164	Lalage leucopygialis Tweedd.	425	—	—	—	—	—	—	—	—	—	—	—	—

Introduction: Geographical Distribution.

	Celebesian Area															
		Celebes														
Talaut Islands	Sangi Islands	North Peninsula	West Celebes	East Peninsula	Togian Islands	Central Celebes	South Peninsula	S. E. Peninsula (and Peeleng)	Saleyer Island	Djampea Group	Peling Group	Sula Islands	Moluccas	Papuasia	Australia	Polynesia	New Zealand	Neotropical Region	Name of species
—	—	—	—	—	—	—	*	—	—	—	?	—	—	—	—	—	—	—	Muscicapula westermanni.
—	*	—	—	—	—	*	*	—	—	—	—	—	—	—	—	—	—	—	Muscicapula hyperythra.
—	*	*	—	—	—	*	*	—	*	—	—	—	—	—	—	—	—	—	Siphia banyumas.
—	—	—	—	—	—	—	—	—	—	*	—	—	—	—	—	—	—	—	Siphia djampeana.
—	—	—	—	—	—	—	—	—	—	*	—	—	—	—	—	—	—	—	Siphia kalaoensis.
—	*	—	—	—	—	*	*	—	—	—	—	—	—	—	—	—	—	—	Siphia rufigula.
—	—	—	—	—	—	—	*	—	—	—	—	—	—	—	—	—	—	—	Siphia bonthaina.
—	*	—	—	—	—	—	*	—	—	—	—	—	—	—	—	—	—	—	Stoparola septentrionalis.
—	—	—	—	—	—	—	—	—	—	—	—	—	—	—	—	—	—	—	Stoparola meridionalis.
—	*	*	*	*	*	*	*	*	—	—	—	*	*	—	—	—	—	—	Hypothymis puella.
—	—	—	—	—	—	—	—	—	—	—	—	—	—	—	—	—	—	—	Hypothymis rowleyi.
—	—	—	—	—	—	*?	—	—	—	*	—	—	—	—	—	—	—	—	Rhipidura celebensis.
—	—	—	—	—	—	*	—	—	—	—	—	—	—	—	—	—	—	—	Rhipidura teijsmanni.
*	—	—	—	—	—	—	—	—	—	—	—	—	—	—	—	—	—	—	Zeocephus talautensis.
*	*?	—	—	—	—	—	—	—	—	—	—	—	—	—	—	—	—	—	Monarcha commutatus.
—	—	—	—	—	—	—	—	*	*	*	*	*	—	—	—	—	—	—	Monarcha inornatus.
—	—	—	—	—	—	—	—	—	—	—	—	—	—	—	—	—	—	—	Monarcha everetti.
—	—	—	—	—	—	—	—	—	—	—	—	—	—	—	—	—	—	—	Myiagra rufigula.
—	*	—	—	—	*	*	*	—	—	*	—	—	—	—	—	—	—	—	Culicicapa helianthea.
—	—	—	—	—	—	—	—	—	—	—	—	—	—	—	—	—	—	—	Gerygone flaveola.
—	*	*	—	*	*	*	*	—	—	—	—	—	—	—	—	—	—	—	Pratincola caprata.
—	*	—	—	*	*	—	*	—	—	—	—	—	—	—	—	—	—	—	Pachycephala sulfurivont.
—	—	—	—	—	—	*	—	—	—	—	—	—	—	—	—	—	—	—	Pachycephala meridional.
—	—	—	—	—	—	*?	*	—	—	—	—	—	—	—	—	—	—	—	Pachycephala teijsmanni.
—	—	—	—	—	—	—	*	—	—	—	—	—	—	—	—	—	—	—	Pachycephala orpheus.
—	—	—	—	—	—	—	—	—	—	—	*	*	—	—	—	—	—	—	Pachycephala griseonota.
—	—	—	—	—	—	—	—	—	—	*	*	*	—	—	—	—	—	—	Pachycephala clio.
—	—	—	—	—	—	—	*	—	—	—	—	—	—	—	—	—	—	—	Pachycephala everetti.
—	—	—	—	—	*	—	—	—	—	—	—	—	—	—	—	—	—	—	Pachycephala bonthaina.
*	*	—	—	—	—	—	—	—	—	—	—	—	—	—	—	—	—	—	Pachycephala honensis.
—	—	—	—	—	—	—	—	—	—	—	—	—	—	—	—	—	—	—	Colluricincla sangirensis.
*	*	—	—	—	—	—	—	—	—	—	—	—	*	—	—	—	—	—	Lanius tigrinus.
*	*	—	—	—	—	—	—	—	—	—	—	—	—	—	—	—	—	—	Lanius lucionensis.
*?	*	*	*	*	—	—	*	—	—	—	—	—	—	—	—	—	—	—	Graucalus bicolor.
—	*	*	*	—	—	—	—	—	—	—	—	—	—	—	—	—	—	—	Graucalus leucopygius.
—	—	—	—	—	—	—	—	—	—	—	—	*	*	—	—	—	—	—	Graucalus temmincki.
—	—	—	—	—	—	—	—	—	—	—	*	*	*	*	*	—	*	—	Graucalus schistaceus.
—	—	—	—	*	*	—	—	—	—	—	—	—	—	—	—	—	—	—	Graucalus melanops.
—	—	—	—	—	—	—	—	—	—	—	—	—	—	—	—	—	—	—	Edoliisoma morio (typ.).
—	*	*	*	*?	—	—	—	—	—	—	—	—	—	—	—	—	—	—	— morio septentrionalis.
—	—	—	—	—	—	—	—	—	—	—	—	—	—	—	—	—	—	—	— morio—septentrion.
*	—	—	—	—	—	—	—	—	—	—	—	—	—	—	—	—	—	—	Edoliisoma salvadorii.
—	—	—	—	*	—	—	—	—	—	—	—	—	—	—	—	—	—	—	Edoliisoma talautense.
—	—	—	—	—	—	—	—	—	—	—	—	—	—	—	—	—	—	—	Edoliisoma emancipata.
—	—	—	—	—	—	—	—	—	—	*	—	—	—	—	—	—	—	—	Edoliisoma obiense.
—	*	*	*	—	—	*	*	—	—	—	*	*	*	—	—	—	—	—	Lalage leucopygialis.

13*

100 Introduction: Geographical Distribution.

Number	Name of species	Page in the text	Nearctic Region	Ethiopian Region	Palaearctic Region	Indian Province	Chinese Province	Japan	Malay Peninsula	Sumatra	Java	Borneo	Philippine Islands	Lesser Sunda Islands
165	Lalage timorensis (S. Müll.)	428	—	—	—	—	—	—	—	—	—	—	*?	*
166	Artamus leucogaster (Val.)	430	—	—	*	—	—	—	*	*	*	*	*	
167	Artamus monachus Bp.	434	—	—	—	—	—	—	—	—	—	—	—	—
168	Dicrurus leucops Wall. (typical)	436	—	—	—	—	—	—	—	—	—	—	—	—
	— leucops—axillaris	437	—	—	—	—	—	—	—	—	—	—	—	—
	— leucops axillaris (Salvad.)	438	—	—	—	—	—	—	—	—	—	—	—	—
169	Dicrurus pectoralis Wall.	439	—	—	—	—	—	—	—	—	—	—	—	—
170	Dicaeum celebicum S. Müll.	441	—	—	—	—	—	—	—	—	—	—	—	—
171	Dicaeum sulaense Sharpe	443	—	—	—	—	—	—	—	—	—	—	—	—
172	Dicaeum sangirense Salvad.	444	—	—	—	—	—	—	—	—	—	—	—	—
173	Dicaeum talautense M. & Wg.	445	—	—	—	—	—	—	—	—	—	—	—	—
174	Dicaeum splendidum Bütt.	446	—	—	—	—	—	—	—	—	—	—	—	—
175	Dicaeum nehrkorni W. Blas.	447	—	—	—	—	—	—	—	—	—	—	—	—
176	Dicaeum hosei Sharpe	448	—	—	—	—	—	—	—	—	—	—	—	—
177	Acmonorhynchus aureolimbatus (Wall.)	449	—	—	—	—	—	—	—	—	—	—	—	—
178	Acmonorhynchus sangirensis (Salvad.)	451	—	—	—	—	—	—	—	—	—	—	—	—
179	Aethopyga flavostriata (Wall.)	453	—	—	—	—	—	—	—	—	—	—	—	—
180	Eudrepanis duivenbodei (Schl.)	456	—	—	—	—	—	—	—	—	—	—	—	—
181	Cyrtostomus frenatus (S. Müll.) (typical)	458	—	—	—	—	—	—	—	—	—	—	—	—
	— frenatus saleyerensis Hart.	458	—	—	—	—	—	—	—	—	—	—	—	—
	— frenatus > saleyerensis	458	—	—	—	—	—	—	—	—	—	—	—	—
	— frenatus < saleyerensis	459	—	—	—	—	—	—	—	—	—	—	—	—
	— frenatus dissontiens (Hart.)	460	—	—	—	—	—	—	—	—	—	—	—	—
182	Cyrtostomus teijsmanni (Bütt.)	462	—	—	—	—	—	—	—	—	—	—	—	—
183	Hermotimia auriceps (G. R. Gray)	464	—	—	—	—	—	—	—	—	—	—	—	—
184	Hermotimia porphyrolaema (Wall.) (typ.)	465	—	—	—	—	—	—	—	—	—	—	—	—
	— porphyrolaema scapulata M. & Wg.	466	—	—	—	—	—	—	—	—	—	—	—	—
185	Hermotimia grayi (Wall.)	467	—	—	—	—	—	—	—	—	—	—	—	—
186	Hermotimia sangirensis (A. B. M.)	469	—	—	—	—	—	—	—	—	—	—	—	—
187	Hermotimia talautensis M. & Wg.	470	—	—	—	—	—	—	—	—	—	—	—	—
188	Anthreptes malaccensis (Scop.) (typical)	472	—	—	*	—	—	—	*	*	*	*		*?
	— malaccensis celebensis (Shell.)	475	—	—	—	—	—	—	—	—	—	—	—	—
	— malaccensis chlorigaster (Sharpe)	477	—	—	—	—	—	—	—	—	—	—	*	—
189	Myzomela chloroptera Tweedd.	478	—	—	—	—	—	—	—	—	—	—	—	—
190	Melilestes celebensis M. & Wg. (typical)	481	—	—	—	—	—	—	—	—	—	—	—	—
	— celebensis meridionalis M. & Wg.	482	—	—	—	—	—	—	—	—	—	—	—	—
191	Myza sarasinorum M. & Wg.	483	—	—	—	—	—	—	—	—	—	—	—	—
192	Zosterops squamiceps (Hart.)	485	—	—	—	—	—	—	—	—	—	—	—	—
193	Zosterops intermedia (Wall.)	486	—	—	—	—	—	—	—	—	—	—	—	*
194	Zosterops atrifrons Wall.	487	—	—	—	—	—	—	—	—	—	—	—	—
195	Zosterops subatrifrons M. & Wg.	490	—	—	—	—	—	—	—	—	—	—	—	—
196	Zosterops nehrkorni W. Blas.	490	—	—	—	—	—	—	—	—	—	—	—	—
197	Zosterops sarasinorum M. & Wg.	491	—	—	—	—	—	—	—	—	—	—	—	—
198	Zosterops anomala M. & Wg.	494	—	—	—	—	—	—	—	—	—	—	—	—
199	Zosterops babelo M. & Wg.	495	—	—	—	—	—	—	—	—	—	—	—	—
200	Iole aurea (Tweedd.)	496	—	—	—	—	—	—	—	—	—	—	—	—

Introduction: Geographical Distribution. 101

Celebesian Area														
Celebes														
West Celebes	East Peninsula	Togian Islands	Central Celebes	South Peninsula	S. E. Peninsula (and Buton)	Saleyer Island	Djampea Group	Peling Group	Sula Islands	Moluccas	Papuasia	Australia	Polynesia	New Zealand	Neotropical Region	Name of species
---	---	---	---	---	---	---	---	---	---	---	---	---	---	---	---	---
—	—	—	—	*	—	*	*	—	—	—	—	—	—	—	—	Lalage timorensis.
—	*	*	—	*	—	*	*	*	—	*	*	*	—	—	- -	Artamus leucogaster.
—	—	—	—	*	—	—	—	*	*	—	—	—	—	—	—	Artamus monachus.
*	*	*	*	*	*	*	—	—	—	—	—	—	—	—	—	Dicrurus leucops (typ.).
—	—	—	—	—	—	—	—	—	—	—	—	—	—	—	—	— leucops—axillaris.
—	—	—	—	—	—	—	—	—	—	—	—	—	—	—	—	— leucops axillaris.
—	—	—	—	—	—	—	*	*	*	—	—	—	—	—	—	Dicrurus pectoralis.
*	—	*	—	*	—	—	—	*	*	—	—	—	—	—	—	Dicaeum celebicum.
—	—	—	—	—	—	—	—	—	—	—	—	—	—	—	—	Dicaeum sulaense.
—	—	—	—	—	—	—	—	—	—	—	—	—	—	—	—	Dicaeum sangirense.
—	—	—	—	—	—	—	—	—	—	—	—	—	—	—	—	Dicaeum talautense.
—	—	—	*?	—	*	*	—	—	—	—	—	—	—	—	—	Dicaeum splendidum.
—	—	—	*	—	—	—	—	—	—	—	—	—	—	—	—	Dicaeum nehrkorni.
—	—	—	—	—	—	—	—	—	—	—	—	—	—	—	—	Dicaeum hosei.
*	—	—	—	*	*	—	—	—	—	—	—	—	—	—	—	Acmonorh. aureolimbatus.
—	—	—	—	—	—	—	—	—	—	—	—	—	—	—	—	Acmonorh. sangirensis.
*	—	—	*	*	*	—	—	—	—	—	—	—	—	—	—	Aethopyga flavostriata.
—	—	—	—	—	—	—	—	—	—	—	—	—	—	—	—	Eudrepanis duivenbodei.
—	—	—	—	—	—	—	*	*	*	*	*	—	—	—	—	Cyrtostomus frenatus(typ.)
—	—	—	—	—	*	—	—	—	—	—	—	—	—	—	—	— frenatus saleyerensis.
*	*	*	—	*	—	*	—	—	—	—	—	—	—	—	—	— — >saleyerensis.
—	—	—	—	*	—	—	—	—	—	—	—	—	—	—	—	— — <saleyerensis.
—	—	—	—	—	—	—	—	—	—	—	—	—	—	—	—	— frenatus dissentiens.
—	—	—	*?	—	*	—	*	*	*	*	—	—	—	—	-	Cyrtostomus teijsmanni.
*	*	—	*	*	*	—	—	—	—	—	—	—	—	—	—	Hermotimia auriceps.
—	*	*?	—	—	*?	—	—	—	—	—	—	—	—	—	—	H. porphyrolaema (typ.).
—	—	—	—	—	—	—	—	—	—	—	—	—	—	—	—	— — scapulata.
—	—	—	—	—	—	—	—	—	—	—	—	—	—	—	—	Hermotimia grayi.
—	—	—	—	—	—	—	—	—	—	—	—	—	—	—	—	Hermotimia sangirensis.
—	—	—	—	—	—	—	—	—	—	—	—	—	—	—	—	Hermotimia talautensis.
*	*	*	—	*	*	—	*	*	—	—	—	—	—	—	—	Anthrept.malaccens.(typ.).
—	—	—	—	—	—	—	—	—	—	—	—	—	—	—	—	— — celebensis.
—	—	—	—	—	—	—	—	—	—	—	—	—	—	—	—	— — chlorigaster.
—	—	—	*	—	*	*	—	—	—	—	—	—	—	—	—	Myzomela chloroptera.
—	—	—	*	—	—	—	—	—	—	—	—	—	—	—	—	Melilestes celebensis(typ.).
—	—	—	—	—	—	—	—	—	—	—	—	—	—	—	—	— — meridionalis.
—	—	—	*	—	—	—	—	—	—	—	—	—	—	—	—	Myza sarasinorum.
*	—	*	*	—	*	—	—	*	—	—	—	—	—	—	—	Zosterops squamiceps.
*	—	*	—	—	—	—	—	—	—	—	—	—	—	—	—	Zosterops intermedia.
—	—	—	—	—	—	—	—	—	—	—	—	—	—	—	—	Zosterops atrifrons.
—	—	—	—	—	—	*	*	—	—	—	—	—	—	—	—	Zosterops subatrifrons.
—	—	—	—	—	—	—	—	—	—	—	—	—	—	—	—	Zosterops nehrkorni.
—	—	*	*	*?	—	—	—	—	—	—	—	—	—	—	—	Zosterops sarasinorum.
—	—	—	—	—	—	—	—	—	—	—	—	—	—	—	—	Zosterops anomala.
—	—	—	—	—	—	—	—	—	—	—	—	—	—	—	—	Zosterops babelo.
—	*	—	—	—	—	—	—	—	—	—	—	—	—	—	—	Iole aurea.

Number	Name of species	Page in the text	Nearctic Region	Ethiopian Region	Palaearctic Region	Indian Province	Chinese Province	Japan	Malay Peninsula	Sumatra	Java	Borneo	Philippine Islands	Lesser Sunda Islands
201	Iole longirostris (Wall.)	497	—	—	—	—	—	—	—	—	—	—	—	—
202	Iole platenae (W. Blas.)	498	—	—	—	—	—	—	—	—	—	—	—	—
203	Malia grata Schl. (typical)	499	—	—	—	—	—	—	—	—	—	—	—	—
	— grata recondita (M.&Wg.)	500	—	—	—	—	—	—	—	—	—	—	—	—
204	Androphilus castaneus (Bütt.)	502	—	—	—	—	—	—	—	—	—	—	—	—
205	Cataponera turdoides Hart.	503	—	—	—	—	—	—	—	—	—	—	—	—
206	Trichostoma celebensis (Strickl.)	504	—	—	—	—	—	—	—	—	—	—	—	—
207	Trichostoma finschi Tweedd.	506	—	—	—	—	—	—	—	—	—	—	—	—
208	Malacopteron affine (Blyth)	508	—	—	—	—	—	—	*	*	*	*	—	—
209	Geocichla erythronota Scl.	509	—	—	—	—	—	—	—	—	—	—	—	—
210	Merula celebensis Bütt.	510	—	—	—	—	—	—	—	—	—	—	—	—
211	Petrophila cyanus (L.) (typical)	512	—	*	*	*	—	—	—	—	—	—	—	—
	— cyanus solitaria (P.L.S.Müll.)	512	—	—	*	*	*	*	*	—	*	*	*	*
212	Cisticola cursitans (Frkl.)	515	—	*	*	*	*	*	*	*	*	—	*	*
213	Cisticola exilis (Vig. Horsf.)	517	—	—	—	*	*	—	*	—	*	—	*	*
214	Phyllergates riedeli M.&Wg.	519	—	—	—	—	—	—	—	—	—	—	—	—
215	Acrocephalus orientalis (Temm. Schl.)	521	—	—	*	*	*	*	*	*	*	*	*	*
216	Locustella fasciolata (G.R.Gray)	524	—	—	*	—	*	*	—	—	—	—	*	*
217	Locustella ochotensis (Midd.)	526	—	—	*	—	*	—	—	—	—	—	*	*
218	Phylloscopus borealis (Blas.)	527	*	—	*	*	*	*	*	*	*	*	*	*
219	Cryptolopha sarasinorum M.&Wg.	530	—	—	—	—	—	—	—	—	—	—	—	—
220	Motacilla flava L.	531	*	*	*	*	*	—	*	*	*	*	*	*
221	Motacilla boarula L. (typical)	534	—	*	*	—	—	—	—	—	—	—	—	—
	— boarula melanope (Pall.)	535	—	—	*	*	*	*	*	*	—	*	*	*
222	Anthus gustavi Swinh.	538	—	—	*	—	*	*	—	—	—	—	*	—
223	Anthus cervinus (Pall.)	540	*	*	*	*	*	—	—	—	—	—	*	—
224	Munia oryzivora (L.)	542	—	—	*	—	*	*	*	*	*	*	*	*
225	Munia formosana Swinh. (typical)	544	—	—	—	—	*	—	—	—	—	—	—	—
	— formosana jagori (Marts.)	544	—	—	—	—	—	—	—	—	—	—	—	*
	— formosana brunneiceps (Wald.)	544	—	—	—	—	—	—	—	—	—	—	*	—
226	Munia pallida Wall.	546	—	—	—	—	—	—	—	—	—	—	—	*
227	Munia subcastanea Hart.	548	—	—	—	—	—	—	—	—	—	—	—	—
228	Munia punctulata nisoria (Temm.)¹)	548	—	—	—	—	—	—	*	*	*	—	—	—
229	Munia molucca (L.) (typical)	549	—	—	—	—	—	—	—	—	—	—	—	—
	— molucca propinqua Sharpe	550	—	—	—	—	—	—	—	—	—	—	—	—
	— molucca—propinqua	550	—	—	—	—	—	—	—	—	—	—	—	—
	— molucca<propinqua	551	—	—	—	—	—	—	—	—	—	—	—	—
	— molucca typica>	551	—	—	—	—	—	—	—	—	—	—	—	—
	— molucca kangeanensis (Vordm.)²)	551	—	—	—	—	—	—	—	—	—	—	—	—
230	Passer montanus (L.)	553	—	*	*	*	*	*	*	*	*	*	—	*
231	Calornis panayensis (Scop.) (typical)	555	—	—	—	—	—	—	*	*	*	*	—	*
	— panayensis chalybea (Horsf.)	556	—	—	—	—	—	—	—	—	—	—	—	—
	— panayensis affinis (Hay)	556	—	—	—	*	—	—	—	—	—	—	—	—
	— panayensis chalybea—affinis	556	—	—	—	—	—	—	—	—	—	—	—	—
	— panayensis tytleri (Hume)	557	—	—	—	—	—	—	—	—	—	—	—	—

¹) Five races recognised by Sharpe. ²) Kangeang.

Introduction: Geographical Distribution.

Talaut Islands	Sangi Islands	North Peninsula	West Celebes	East Peninsula	Togian Islands	Central Celebes	South Peninsula	S. E. Peninsula (and Buton)	Saleyer Island	Djampea Group	Peling Group	Sula Islands	Moluccas	Papuasia	Australia	Polynesia	New Zealand	Neotropical Region	Name of species
—	—	—	—	—	—	—	—	—	—	—	*	*	—	—	—	—	—	—	Iole longirostris.
—	*	—	—	—	—	*	—	*?	—	—	—	—	—	—	—	—	—	—	Iole platenae.
—	*	—	—	—	*	—	—	—	—	—	—	—	—	—	—	—	—	—	Malia grata (typical).
—	—	*	—	—	*	—	—	—	—	—	—	—	—	—	—	—	—	—	— grata recondita.
—	—	*	*	—	—	*	—	—	—	—	—	—	—	—	—	—	—	—	Androphilus castaneus.
—	—	—	*	—	—	*	—	—	—	—	—	—	—	—	—	—	—	—	Cataponera turdoides.
—	—	*	*	—	—	*	—	—	—	—	—	—	—	—	—	—	—	—	Trichostoma celebensis.
—	—	—	*	—	—	—	—	—	—	—	—	—	—	—	—	—	—	—	Trichostoma finschi.
—	*	*	—	—	—	*	—	—	—	—	—	—	—	—	—	—	—	—	Malacopteron affine.
—	—	—	—	—	—	*	—	—	—	—	—	—	—	—	—	—	—	—	Geocichla erythronota.
*	*	*	—	—	—	—	—	—	—	—	—	—	*	—	—	—	—	—	Merula celebensis.
—	—	—	—	—	—	—	—	—	—	—	—	—	—	—	—	—	—	—	Petrophila cyanus (typ.).
—	*	*	—	*	*	*	—	—	—	—	*	—	—	—	—	—	—	—	— cyanus solitaria.
—	*	*	—	—	*	—	—	—	—	—	*	—	*	*	*	—	—	—	Cisticola cursitans.
—	—	—	—	—	*	—	—	—	—	—	—	—	*	—	—	—	—	—	Cisticola exilis.
*	*	—	—	—	—	—	—	—	—	—	—	—	*	—	—	—	—	—	Phyllergates riedeli.
—	*	—	—	—	*	*	—	—	—	—	*	—	*	—	—	—	—	—	Acrocephalus orientalis.
—	*	*	—	*	—	*	—	*	*	—	*	—	*	—	—	—	—	—	Locustella fasciolata.
—	—	—	—	—	—	—	—	—	—	—	—	—	—	—	—	—	—	—	Locustella ochotensis.
—	*	—	*	—	—	*	—	*	—	*	*	—	*	—	—	—	—	—	Phylloscopus borealis.
—	—	—	—	—	—	—	—	—	—	—	—	—	—	—	—	—	—	—	Cryptolopha sarasinorum.
—	*	—	—	—	—	—	*	—	—	—	*	*	—	—	—	—	—	—	Motacilla flava.
—	*	—	—	—	—	*	—	—	—	—	*	—	—	—	—	—	—	—	Motacilla boarula (typ.).
—	*	—	—	—	*	—	—	—	—	—	—	—	*	—	—	—	—	—	— boarula melanope.
—	—	—	—	—	*	—	—	—	—	—	—	—	—	—	—	—	—	—	Anthus gustavi.
—	*	*	*	*	—	*	—	—	—	—	—	—	—	—	—	—	—	—	Anthus cervinus.
—	—	*	—	—	—	—	—	—	—	—	—	—	*	—	—	—	—	—	Munia oryzivora.
—	—	—	—	—	—	—	—	—	—	—	—	—	—	—	—	—	—	—	Munia formosana (typical).
—	—	—	—	—	—	—	—	—	—	—	—	—	—	—	—	—	—	—	— — jagori.
—	*	—	—	*	*	*	—	—	—	—	—	—	—	—	—	—	—	—	— — brunneiceps.
—	*	—	—	—	—	—	—	—	—	—	—	—	—	—	—	—	—	—	Munia pallida.
—	—	—	—	*	*	—	—	—	—	—	*	*	—	—	—	—	—	—	Munia subcastanea.
—	—	—	—	—	—	—	—	—	—	—	—	—	—	—	—	—	—	—	Mun. punctulata nisoria[1].
—	—	—	—	—	—	—	—	—	—	—	—	—	—	—	—	—	—	—	Munia molucca (typ.).
—	*	—	—	—	—	*	*	—	—	—	*	*	—	—	—	—	—	—	— propinqua.
—	—	—	—	—	—	—	*	*	—	—	—	—	—	—	—	—	—	—	— molucca—propinqua.
—	—	—	—	—	—	—	—	—	—	—	—	—	—	—	—	—	—	—	— <propinqua.
—	—	—	—	—	—	—	—	—	—	—	—	—	—	—	—	—	—	—	— molucca typica>.
—	—	—	—	—	—	—	—	—	—	—	—	—	—	—	—	—	—	—	— kangeanensis[2].
—	*	—	—	—	—	—	*	—	—	—	—	—	—	—	—	—	—	—	Passer montanus.
—	—	—	—	—	—	—	—	—	—	—	—	—	—	—	—	—	—	—	Chlornis panayensis (typ.).
—	—	—	—	—	—	—	—	—	—	—	—	—	—	—	—	—	—	—	— panayensis chalybea.
—	—	—	—	—	—	—	—	—	—	—	—	—	—	—	—	—	—	—	— — affinis.
—	—	—	—	—	—	—	—	—	—	—	—	—	—	—	—	—	—	—	— — chalybea—affinis.
—	—	—	—	—	—	—	—	—	—	—	—	—	—	—	—	—	—	—	— — tytleri.

Introduction: Geographical Distribution.

Number	Name of species	Page in the text	Nearctic Region	Ethiopian Region	Palaearctic Region	Indian Province	Chinese Province	Japan	Malay Peninsula	Sumatra	Java	Borneo	Philippine Islands	Lesser Sunda Islands
231	Calornis panayensis sangirensis (Salvad.)	557	—	—	—	—	—	—	—	—	—	—	—	—
	— panayensis—sangirensis	557	—	—	—	—	—	—	—	—	—	—	—	—
	— panayensis altirostris (Salvad.)[1]	558	—	—	—	—	—	—	—	—	—	—	—	—
	— panayensis enganensis (Salvad.)[2]	558	—	—	—	—	—	—	—	—	—	—	—	—
232	Caloruis minor (Bp.)	561	—	—	—	—	—	—	—	—	—	—	*	—
233	Calornis sulaensis Sharpe.	561	—	—	—	—	—	—	—	—	—	—	—	—
234	Calornis metallica (Temm.)	562	—	—	—	—	—	—	—	—	—	—	—	—
235	Enodes erythrophrys (Temm.)	564	—	—	—	—	—	—	—	—	—	—	—	—
236	Acridotheres cinereus Bp.	566	—	—	—	—	—	—	—	—	—	—	—	—
237	Scissirostrum dubium (Lath.)	567	—	—	—	—	—	—	—	—	—	—	—	—
238	Sturnia violacea (Bodd.)	570	—	—	—	—	*	—	—	—	—	*	*	—
239	Basilcornis celebensis G.R.Gray	572	—	—	—	—	—	—	—	—	—	—	—	—
240	Basileornis galeatus A.B.M.	574	—	—	—	—	—	—	—	—	—	—	—	—
241	Streptocitta albicollis (Vieill.)	575	—	—	—	—	—	—	—	—	—	—	—	—
242	Streptocitta torquata (Temm.)	577	—	—	—	—	—	—	—	—	—	—	—	—
243	Charitornis albertinae Schl.	579	—	—	—	—	—	—	—	—	—	—	—	—
244	Corvus enca (Horsf.) (Celebesian races)	581	—	—	—	—	—	—	—	—	—	—	—	—
245	Gazzola typica Bp.	584	—	—	—	—	—	—	—	—	—	—	—	—
246	Oriolus celebensis (Tweedd.) (typical)	585	—	—	—	—	—	—	—	—	—	—	—	—
	— celebensis meridionalis Hart.	586	—	—	—	—	—	—	—	—	—	—	—	—
	— celebensis—meridionalis	586	—	—	—	—	—	—	—	—	—	—	—	—
247	Oriolus frontalis Wall.	589	—	—	—	—	—	—	—	—	—	—	—	—
248	Oriolus boneratensis M.&Wg.	589	—	—	—	—	—	—	—	—	—	—	—	—
249	Oriolus formosus Cab.	590	—	—	—	—	—	—	—	—	—	—	—	—
250	Oriolus melanisticus M.&Wg.	593	—	—	—	—	—	—	—	—	—	—	—	—
251	Osmotreron wallacei (Salvad.) (typical)	595	—	—	—	—	—	—	—	—	—	—	—	—
	— wallacei pallidior Hart.	597	—	—	—	—	—	—	—	—	—	—	—	—
252	Osmotreron sangirensis (Brügg.)	598	—	—	—	—	—	—	—	—	—	—	—	—
253	Osmotreron vernans (L.)	599	—	—	*	*	—	—	*	*	*	*	*	*
254	Ptilopus fischeri (Brügg.)	602	—	—	—	—	—	—	—	—	—	—	—	—
255	Ptilopus meridionalis M.&Wg.	604	—	—	—	—	—	—	—	—	—	—	—	—
256	Ptilopus gularis (Q.G.)	605	—	—	—	—	—	—	—	—	—	—	—	—
257	Ptilopus subgularis M.&Wg.	606	—	—	—	—	—	—	—	—	—	—	—	—
258	Ptilopus melanocephalus (Forst.)	607	—	—	—	—	—	—	—	—	—	—	*	—
259	Ptilopus melanospilus (Salvad.)	608	—	—	—	—	—	—	—	—	—	—	—	—
260	Ptilopus chrysorrhous (Salvad.)	610	—	—	—	—	—	—	—	—	—	—	—	—
261	Ptilopus xanthorrhous (Salvad.)	611	—	—	—	—	—	—	—	—	—	—	—	—
262	Ptilopus temmincki (Des Murs & Prév.)	613	—	—	—	—	—	—	—	—	—	—	*	—
263	Carpophaga concinna Wall.[3]	615	—	—	—	—	—	—	—	—	—	—	—	—
264	Carpophaga paulina (Bp.)	617	—	—	—	—	—	—	—	—	—	—	—	—
265	Carpophaga pulchella Tweedd.	619	—	—	—	—	—	—	—	—	—	—	—	—
266	Carpophaga intermedia M.&Wg.	619	—	—	—	—	—	—	—	—	—	—	—	—
267	Carpophaga rosacea (Temm.)	620	—	—	—	—	—	—	—	—	—	—	—	*
268	Carpophaga pickeringi Cass.[3]	621	—	—	—	—	—	—	—	—	—	—	*	*

[1] Nias. [2] Engano. [3] Islet nomad.

Introduction: Geographical Distribution.

Talaut Islands	Sangi Islands	Celebesian Area — Celebes — North Peninsula	West Celebes	East Peninsula	Togian Islands	Central Celebes	South Peninsula	S. E. Peninsula and Buton	Saleyer Island	Djampea Group	Peling Group	Sula Islands	Moluccas	Papuasia	Australia	Polynesia	New Zealand	Neotropical Region	Name of species
—	*	—	—	—	—	—	—	—	—	—	—	—	—	—	—	—	—	—	Calornis panayens. sangir.
—	*	—	—	—	—	—	—	—	—	—	—	—	—	—	—	—	—	—	— panayensis – sangir.
—	—	—	—	—	—	—	—	—	—	—	—	—	—	—	—	—	—	—	— — altirostris.
—	—	—	—	—	—	—	—	—	—	—	—	—	—	—	—	—	—	—	— — enganensis.
—	—	—	—	—	*	—	*	*	—	—	*	—	—	—	—	—	—	—	Calornis minor.
—	—	—	*	—	—	—	—	—	—	—	*	*	*	*	—	—	—	—	Calornis sulaensis.
—	—	—	—	—	—	—	—	—	—	—	*	—	—	—	—	—	—	—	Calornis metallica.
—	—	*	—	—	—	*	—	—	—	—	—	—	—	—	—	—	—	—	Enodes erythrophrys.
—	—	—	—	—	*	—	—	—	*	—	—	—	—	—	—	—	—	—	Acridotheres cinereus.
—	*	*	*	*	—	*	—	—	—	*	—	—	—	—	—	—	—	—	Scissirostrum dubium.
—	*	—	*	—	—	*	—	—	—	—	—	—	*	—	—	—	—	—	Sturnia violacea.
—	—	—	—	—	—	—	—	*	*	—	—	—	—	—	—	—	—	—	Basileornis celebensis.
—	—	—	—	—	—	*	*	—	—	—	—	—	—	—	—	—	—	—	Basileornis galeatus.
—	*	*	*	*	*	—	—	—	—	—	—	*	—	—	—	—	—	—	Streptocitta albicollis.
—	*	—	*	*	—	*	*	—	—	—	—	*	—	—	—	—	—	—	Streptocitta torquata.
—	—	—	—	—	—	—	—	—	—	—	—	*	—	—	—	—	—	—	Charitornis albertinae.
—	—	—	—	—	—	—	—	—	—	—	—	—	—	—	—	—	—	—	Corvus enca (Cel. races).
—	*	*	—	*	—	—	*?	—	—	—	—	—	—	—	—	—	—	—	Gazzola typica.
—	—	—	—	—	—	—	—	—	—	—	—	—	—	—	—	—	—	—	Oriolus celebensis (typ.).
—	—	*	—	*	*	—	—	—	—	—	—	—	—	—	—	—	—	—	— — meridionalis.
—	—	—	—	—	—	—	—	—	—	—	—	—	—	—	—	—	—	—	— celebensis—meridion.
—	—	—	—	—	—	—	—	—	—	*	*	—	—	—	—	—	—	—	Oriolus frontalis.
—	*	—	—	—	—	—	—	*	—	—	—	—	—	—	—	—	—	—	Oriolus boneratensis.
—	—	—	—	—	—	—	—	—	—	—	—	—	—	—	—	—	—	—	Oriolus formosus.
*	—	—	—	—	—	—	—	—	—	—	—	—	—	—	—	—	—	—	Oriolus melanisticus.
—	*	*	*	*	—	*	—	*	—	*	*	—	—	—	—	—	—	—	Osmotreron wallacei (typ.).
—	—	—	—	—	—	—	—	—	*	—	—	—	—	—	—	—	—	—	— wallacei pallidior.
—	*	—	—	—	—	—	—	—	—	—	—	—	—	—	—	—	—	—	Osmotreron sangirensis.
—	*	*	—	*	*	—	—	—	—	—	—	—	—	—	—	—	—	—	Osmotreron vernans.
—	*	—	—	—	—	—	—	—	—	—	—	—	—	—	—	—	—	—	Ptilopus fischeri.
—	—	—	—	—	*	—	—	—	—	—	—	—	—	—	—	—	—	—	Ptilopus meridionalis.
—	*	*	*	—	—	*	—	—	—	—	—	—	—	—	—	—	—	—	Ptilopus gularis.
—	—	—	—	—	—	—	—	—	—	—	—	—	—	—	—	—	—	—	Ptilopus subgularis.
—	—	—	—	—	—	*	*	—	*	—	—	—	—	—	—	—	—	—	Ptilopus melanocephalus.
—	*	*	*	—	*	*	—	—	—	—	—	—	—	—	—	—	—	—	Ptilopus melanospilus.
—	—	—	—	—	—	—	—	—	*	*	*	—	—	—	—	—	—	—	Ptilopus chrysorrhous.
—	*	—	—	—	—	—	—	*	—	—	—	—	—	—	—	—	—	—	Ptilopus xanthorrhous.
—	—	—	—	—	—	—	—	—	—	—	*	*	—	—	—	—	—	—	Ptilopus temmincki.
—	*	*	*	—	*	*	—	—	*	*	—	—	—	—	—	—	—	—	Carpophaga concinna.
—	—	—	*	—	—	—	—	—	—	—	—	—	—	—	—	—	—	—	Carpophaga paulina.
—	—	—	—	—	—	—	—	—	—	—	—	—	—	—	—	—	—	—	Carpophaga pulchella.
—	—	—	—	—	—	—	—	—	—	—	—	—	—	—	—	—	—	—	Carpophaga intermedia.
—	*	—	—	—	*	—	—	*	—	—	*	*	—	—	—	—	—	—	Carpophaga rosacea.
—	—	—	—	—	—	—	—	—	—	—	—	—	—	—	—	—	—	—	Carpophaga pickeringi.

Number	Name of species	Page in the text	Nearctic Region	Ethiopian Region	Palaearctic Region	Indian Province	Chinese Province	Japan	Malay Peninsula	Sumatra	Java	Borneo	Philippine Islands	Lesser Sunda Islands
269	Carpophaga radiata (Q. G.)	622	—	—	—	—	—	—	—	—	—	—	—	—
270	Carpophaga forsteni (Bp.)	623	—	—	—	—	—	—	—	—	—	—	—	—
271	Carpophaga poecilorrhoa Brügg.	625	—	—	—	—	—	—	—	—	—	—	—	—
272	Myristicivora bicolor (Scop.)[1]	627	—	—	—	*	*	—	*	—	*	*	*	*
273	Myristicivora luctuosa (Temm.)	631	—	—	—	—	—	—	—	—	—	—	—	—
274	Columba albigularis (Bp.)	633	—	—	—	—	—	—	—	—	—	—	—	—
275	Turacoena manadensis (Q. G.)	635	—	—	—	—	—	—	—	—	—	—	—	—
276	Macropygia albicapilla Bp. (typical)	637	—	—	—	—	—	—	—	—	—	—	—	—
	— albicapilla sangirensis (Salvad.)	638	—	—	—	—	—	—	—	—	—	—	—	—
	— albicapilla—sangirensis	638	—	—	—	—	—	—	—	—	—	—	—	—
277	Macropygia macassariensis (Wall.)	641	—	—	—	—	—	—	—	—	—	—	—	—
278	Turtur tigrinus (Temm. Kn.)	643	—	—	—	—	*	—	*	*	*	*	*	*
279	Geopelia striata (L.)	646	—	*	—	*	*	—	*	*	*	*	*	*
280	Chalcophaps indica (L.)	649	—	—	—	*	*	—	*	*	*	*	*	*
281	Chalcophaps stephani Rchb.	653	—	—	—	—	—	—	—	—	—	—	—	—
282	Phlogoenas tristigmata Bp.	654	—	—	—	—	—	—	—	—	—	—	—	—
283	Phlogoenas bimaculata Salvad.	656	—	—	—	—	—	—	—	—	—	—	—	—
284	Caloenas nicobarica (L.)[1]	657	—	—	*	—	—	—	—	*	*	*	*	—
285	Excalfactoria chinensis (L.)[2]	663	—	—	*	*	—	*	*	—	*	*	*	*
286	Gallus ferrugineus (Gm.)	667	—	—	*	*	—	*	*	—	*	*	—	*
287	Megapodius cumingi Dillw.	671	—	—	—	—	—	—	—	—	—	—	*	—
288	Megapodius sangirensis Schl.	675	—	—	—	—	—	—	—	—	—	—	—	—
289	Megapodius bernsteini Schl.	676	—	—	—	—	—	—	—	—	—	—	—	—
290	Megapodius duperreyi Less. Garn.	677	—	—	—	—	—	—	—	—	—	—	—	*
291	Megacephalon maleo (Hartl.)	678	—	—	—	—	—	—	—	—	—	—	—	—
292	Turnix rufilatus Wall.	686	—	—	—	—	—	—	—	—	—	—	—	—
293	Turnix maculosa (Temm.)	687	—	—	—	—	—	—	—	—	—	—	—	—
294	Gymnocrex rosenbergi (Schl.)	689	—	—	—	—	—	—	—	—	—	—	—	—
295	Aramidopsis plateni (W. Blas.)	690	—	—	—	—	—	—	—	—	—	—	—	—
296	Hypotaenidia striata (L.)	692	—	—	*	*	—	*	*	*	*	*	*	—
297	Hypotaenidia philippensis (L.)[4]	694	—	—	—	—	—	—	—	—	—	—	—	—
298	Hypotaenidia celebensis (Q. G.)	697	—	—	—	—	—	—	—	—	—	—	—	—
299	Hypotaenidia sulcirostris (Wall.)	698	—	—	—	—	—	—	—	—	—	—	—	—
300	Rallina minahassa Wall.	699	—	—	—	—	—	—	—	—	—	—	—	—
301	Porzana fusca (L.)	701	—	—	*	*	*	*	*	*	*	*	*	—
302	Limnocorax niger (Gm.)	703	—	*	—	—	—	—	—	—	—	—	—	—
303	Amaurornis cinerea (Vieill.)	705	—	—	—	—	—	—	—	—	—	—	—	—
304	Amaurornis phoenicura (Forst.)	708	—	—	*	*	—	*	*	*	*	*	*	*
305	Amaurornis moluccana (Wall.)	711	—	—	—	—	—	—	—	—	—	—	—	—
306	Amaurornis isabellina (Schl.)	712	—	—	—	—	—	—	—	—	—	—	—	—
307	Gallinula frontata Wall.	713	—	—	—	—	—	—	—	—	—	—	—	—
308	Gallinula chloropus (L.)	715	—	*	*	*	*	—	—	—	—	*	*	*
309	Porphyrio calvus Vieill.	717	—	—	—	—	—	—	—	—	—	*	*	—

[1] Islet nomad.   [2] Introd. into Mauritius and ? Bourbon.   [3] Introduced.   [4] Mauritius.

Introduction: Geographical Distribution.

Talaut Islands	Sangi Islands	North Peninsula	West Celebes	East Peninsula	Togian Islands	Central Celebes	South Peninsula	S. E. Peninsula (and Buton)	Saleyer Island	Djampea Group	Pelling Group	Sula Islands	Moluccas	Papuasia	Australia	Polynesia	New Zealand	Neotropical Region	Name of species
—	—	*	—	—	—	—	*	—	—	—	—	—	—	—	—	—	—	—	Carpophaga radiata.
—	*	*	—	—	*	*	—	—	—	—	—	—	—	—	—	—	—	—	Carpophaga forsteni.
*	*	*	—	—	—	—	—	—	—	—	—	—	—	—	—	—	—	—	Carpophaga poecilorrhoa.
*	*	*	*	—	*	*	—	—	*	*	*	*	*	*	—	—	—	—	Myristicivora bicolor [1].
*	*	*	*	*	*	—	—	—	—	*	*	*	*	—	—	—	—	—	Myristicivora luctuosa.
*	*	—	—	—	—	—	—	—	*	—	—	—	—	—	—	—	—	—	Columba albigularis.
—	*	*	—	*	*	*	*	—	—	*	*	—	—	—	—	—	—	—	Turacoena manadensis.
—	*	—	*	*	*	*	*	—	—	*	*	—	—	—	—	—	—	—	Macropygia albicap. (typ.).
—	—	—	—	—	—	—	—	—	—	—	—	—	—	—	—	—	—	—	— sangirensis.
—	*	—	—	—	—	—	—	—	—	—	—	—	—	—	—	—	—	—	— albicapilla—sangir.
—	—	—	—	—	—	—	—	*	*	—	—	—	—	—	—	—	—	—	Macrop. macassariensis.
*	*	*	—	*	*	*	*	*	—	—	—	—	*	*	—	—	—	—	Turtur tigrinus.
*	*	*	*	*	*	*	*	*	—	*	*	*	*	*	*	—	—	—	Geopelia striata.
—	*	*	*	—	*	*	*	—	—	—	—	—	*	*	—	—	—	—	Chalcophaps indica.
—	*	*	—	—	—	*	*	—	—	—	—	—	—	—	—	—	—	—	Chalcophaps stephani.
—	—	—	—	—	—	*	—	—	—	—	—	—	—	—	—	—	—	—	Phlogoenas tristigmata.
—	—	—	—	—	—	—	*	—	—	—	—	—	—	—	—	—	—	—	Phlogoenas bimaculata.
—	*	*	—	—	—	—	—	—	—	—	—	—	*	*	—	—	—	—	Caloenas nicobarica [1].
—	—	—	—	—	—	—	—	—	—	—	—	—	—	—	*	—	*	—	Excalfactoria chinensis [2].
*	*	—	—	*	*	*	*	—	—	—	—	—	—	—	—	—	—	—	Gallus ferrugineus.
—	—	*	—	—	—	—	—	—	—	—	—	—	—	—	—	—	—	—	Megapodius cumingi.
*	*	*	—	—	—	—	—	—	—	—	—	—	—	—	—	—	—	—	Megapodius sangirensis.
—	—	—	—	—	—	—	—	—	—	—	—	*	—	—	—	—	—	—	Megapodius bernsteini.
—	—	—	—	—	—	—	*	*	—	—	—	—	—	*	*	—	—	—	Megapodius duperreyi.
*[3]	—	*	—	—	—	—	—	—	—	—	—	—	—	—	—	—	—	—	Megacephalon maleo.
—	*	*	—	—	*	—	*	—	—	—	—	—	—	—	—	—	—	—	Turnix rufilatus.
—	—	—	—	—	*	—	*	—	—	—	—	—	—	—	*	*	—	—	Turnix maculosa.
—	—	*	—	—	—	—	—	—	—	—	—	—	—	—	—	—	—	—	Gymnocrex rosenbergi.
—	—	*	—	—	—	*	—	—	—	—	—	—	—	—	—	—	—	—	Aramidopsis plateni.
—	—	*	—	—	—	—	*	—	—	—	—	—	*	*	*	*	*	—	Hypotaenidia striata.
—	—	*	—	—	—	—	*	—	—	—	—	—	—	—	—	—	—	—	Hypotaenidia philippens. [4]
—	—	—	—	—	—	—	—	—	—	—	—	*	—	—	—	—	—	—	Hypotaenidia celebensis.
—	*	*	—	—	—	*	—	—	—	—	—	—	—	—	—	—	—	—	Rallina minahassa.
—	—	*	—	—	—	—	—	—	—	—	—	—	—	—	—	—	—	—	Porzana fusca.
—	—	*	—	—	—	—	*	*	—	—	—	—	—	—	—	—	—	—	Limnocorax niger.
*	*	*	—	—	*	—	*	*	—	—	—	—	*	*	*	*	—	—	Amaurornis cinerea.
*	*	*	—	—	*	*	*	—	—	—	—	—	*	*	—	—	—	—	Amaurornis phoenicura.
—	—	*	—	—	—	—	—	—	—	—	—	—	*	*	—	—	—	—	Amaurornis moluccana.
—	—	*	—	—	—	—	*	—	—	—	—	—	—	—	*	—	—	—	Amaurornis isabellina.
—	*	*	—	—	—	—	—	—	—	—	—	—	*	*	—	—	—	—	Gallinula frontata.
—	*	*	—	—	*	—	*	—	—	—	—	—	—	—	—	—	—	—	Gallinula chloropus.
—	—	—	—	—	—	—	—	—	—	—	—	—	*	*	*	*	*	—	Porphyrio calvus.

14*

Introduction: Geographical Distribution.

Number	Name of species	Page in the text	Nearctic Region	Ethiopian Region	Palaearctic Region	Indian Province	Chinese Province	Japan	Malay Peninsula	Sumatra	Java	Borneo	Philippine Islands	Lesser Sunda Islands
310	Porphyrio pulverulentus Temm.	721	—	—	—	—	—	—	—	—	—	—	*	
311	Fulica atra L.	722	—	*	*	*	*	*	—	*	*	*?	*	
312	Hydralector gallinaceus (Temm.)	725	—	—	—	—	—	—	—	—	—	—	—	
313	Glareola isabella Vieill.¹)	728	—	—	—	—	—	—	—	—	*	*	—	
314	Esacus magnirostris (Vieill.)	733	—	—	—	*	—	—	—	*	*	*	*	
315	Lobivanellus cinereus (Blyth)	735	—	—	*	*	*	*	—	—	—	—	—	
316	? Squatarola helvetica (L.)	736	*	*	*	*	*	*	*	—	*	*	*	
317	Charadrius fulvus Gm.	738	*	*	*	*	*	*	*	*	*	*	*	
318	Aegialitis vereda (J. Gd.)	741	—	—	*	*	*	—	—	—	*	—	*	
319	Aegialitis geoffroyi (Wagl.)	743	—	*	*	*	*	*	*	*	*	*	*	
320	Aegialitis mongola (Pall.)	746	*	*	*	*	*	*	*	*	—	*	*	
321	Aegialitis curonica (Gm.).	749	*?	*	*	*	*	*	*	—	*	*	*	
322	Aegialitis peroni (Schl.)	752	—	—	—	—	—	—	—	—	—	—	*	
323	Strepsilas interpres (L.)	755	*	*	*	*	*	*	*	*	*	*	*	
324	Himantopus leucocephalus J.Gd.	757	—	—	—	—	—	—	—	—	*	*	*	
325	Totanus glottis (L.)	759	*	*	*	*	*	*	*	*	*	*	*	
326	Totanus calidris (L.)	761	—	*	*	*	*	*	*	*	*	*	—	
327	Totanus glareola (L.)	764	—	*	*	*	*	*	—	*	*	*	*	
328	Heteractitis brevipes (Vieill.)	766	—	—	*	—	*	*	—	—	—	*	*	
329	Actitis hypoleucos (L.)	770	—	*	*	*	*	*	*	*	*	*	*	
330	Terekia cinerea (Güld.)	773	—	*	*	*	*	*	*	—	*	*	*	
331	Tringa acuminata (Horsf.)	776	*	—	*	*	*	*	—	—	*	*	—	
332	Tringa damascensis (Horsf.)	778	—	*	*	*?	*	*	—	—	*	*	—	
333	Tringa ruficollis Pall.	780	—	—	*	*?	*	*	*	—	—	*	*	
334	? Calidris arenaria (L.)	782	*	*	*	*	*	*	—	—	—	*	*	
335	Phalaropus hyperboreus (L.)	785	*	*	*	*	*	*	—	—	—	—	—	
336	Limicola platyrhyncha (Temm.)	787	—	*	*	*	*	*	—	—	*	—	*	
337	Gallinago megala Swinh.	789	—	—	*	*	*	*	—	—	—	—	—	
338	Limosa novaezealandiae (G.R.Gray)	792	*	—	*	—	*	*	—	—	*	*	*	
339	Numenius minutus J.Gd.	795	—	—	*	—	*	*	—	—	*	*	—	
340	Numenius variegatus (Scop.)	797	—	—	*	—	*	*	*	*	*	*	*	
341	? Numenius arquatus (L.)	799	—	*	*	*	*	*	*	*	*	*	*	
342	Numenius cyanopus Vieill.	800	*	—	*	—	*	*	—	—	—	*	*	
343	Plegadis falcinellus (L.)	803	*	*	*	*	*	—	—	*	*	*	—	
344	Dissoura episcopus (Bodd.)	806	—	—	—	*	—	—	*	*	*	*	*	
345	Platalea sp.	809												
346	Phoyx manilensis (Meyen)	811	—	—	*	*	—	—	*	*	*	*	*	
347	Ardea sumatrana Raffl.	814	—	—	*	*	—	—	*	*	*	*	*	
348	Notophoyx picata (J.Gd.)	816	—	—	—	—	—	—	—	—	—	—	—	
349	Notophoyx novaehollandiae (Lath.)	817	—	—	—	—	—	—	—	—	—	*?	—	
350	Demiegretta sacra (Gm.)	819	—	—	*	*	*	*	*	*	*	*	*	
351	Herodias eulophotes Swinh.	824	—	—	—	*	*	*	—	—	—	—	—	
352	Herodias garzetta (L.)	826	—	*	*	*	*	*	*	*	*	*	*	
353	Herodias alba (L.)²)	829	—	*	*	*	*	*	*	—	*	*	—	

¹) Add Dilliton.   ²) Varies geographically.

Introduction: Geographical Distribution.

Talaut Islands	Sangi Islands	North Peninsula	West Celebes	East Peninsula	Togian Islands	Central Celebes	South Celebes	S. E. Peninsula (and Buton)	Saleyer Island	Djampea Group	Peling Group	Sula Islands	Moluccas	Papuasia	Australia	Polynesia	New Zealand	Neotropical Region	Name of species
*	—	—	—	—	—	—	—	—	—	—	—	—	—	—	—	—	—	—	Porphyrio pulverulentus.
—	*	—	—	—	—	—	—	—	—	—	—	—	*?	*	*?	—	—	—	Fulica atra.
—	*	—	—	—	*	—	—	—	—	—	—	—	*	*	*	—	—	—	Hydralector gallinaceus.
—	*	—	—	—	—	—	—	—	—	*	—	—	*	*	*	*	—	—	Glareola isabella[1]).
—	*	—	—	—	—	—	—	*	—	—	—	*	*	*	*	—	—	—	Esacus magnirostris.
—	*	—	—	—	—	—	—	—	—	—	—	—	*	*	*	—	—	—	Lobivanellus cinereus.
—	*?	—	—	—	—	—	—	—	—	—	—	—	*	*	*	*	—	*	? Squatarola helvetica.
—	*	—	—	*	*	—	*	—	—	—	—	—	*	*	*	*	*	—	Charadrius fulvus.
—	*	—	—	—	*	—	—	—	—	—	—	—	*	*	*	—	—	—	Aegialitis vereda.
*	*	—	—	—	*	—	—	—	—	*	—	—	*	*	*	*	—	—	Aegialitis geoffroyi.
—	*	—	—	—	—	*	—	—	—	—	—	—	*	*	*	—	—	—	Aegialitis mongola.
—	*	—	—	—	*	—	—	—	—	—	—	—	—	—	—	—	—	—	Aegialitis curonica.
—	*	—	—	—	*	—	—	—	—	—	—	—	—	—	—	—	—	—	Aegialitis peroni.
—	*	—	—	—	—	*	*	—	—	—	—	—	*	*	*	*	*	*	Strepsilas interpres.
—	*	—	—	*	*	*	*	—	—	—	—	—	*	*	*	—	*	—	Himantopus leucocephal.
—	*	—	—	—	*	—	*	—	—	—	—	—	—	*	*	—	—	—	Totanus glottis.
—	*	—	*	—	*	—	—	—	—	—	—	—	*	—	—	—	—	—	Totanus calidris.
—	*	—	—	*	*	—	—	—	—	—	—	—	*	*	*	—	—	—	Totanus glareola.
*	*	—	—	*	*	—	*	—	—	—	—	—	*	*	*	*	—	—	Heteractitis brevipes.
—	*	—	—	—	*	—	—	—	—	—	—	—	*	—	*	—	—	—	Actitis hypoleucos.
—	*	—	—	—	—	—	—	—	—	—	—	—	*	*	*	*	—	—	Terekia cinerea.
—	*	—	—	—	—	—	—	—	—	—	—	—	*	*	*	—	—	—	Tringa acuminata.
*	*	—	—	—	—	—	*	—	—	—	—	—	*	*	*	—	—	—	Tringa damascensis.
—	*?	—	—	—	—	—	—	—	—	—	—	—	*	*	—	—	—	*	Tringa ruficollis.
—	*	—	—	—	—	—	—	—	—	—	—	—	*	*	—	—	—	—	? Calidris arenaria.
—	*	—	—	—	*	—	—	—	—	—	—	—	—	—	—	*	—	—	Phalaropus hyperboreus.
—	*	—	—	—	—	*	—	*	—	—	—	—	*	*	*	—	—	—	Limicola platyrhyncha.
*	*	—	—	—	—	—	—	—	—	—	—	—	*	*	*	*	—	—	Gallinago megala.
—	*	—	—	—	—	—	—	—	—	—	—	—	*	*	*	—	—	—	Limosa novaezealandiae.
—	*	—	—	—	—	*	*	—	—	—	*	—	*	*	*	—	—	—	Numenius minutus.
*	*?	—	—	—	—	—	—	—	—	—	—	—	*	*	*	—	—	—	Numenius variegatus.
—	*	—	—	—	—	—	—	—	—	—	—	—	*	*	*	—	*	—	? Numenius arquatus.
—	*	—	—	—	*	—	—	—	—	—	—	—	—	—	—	—	—	—	Numenius cyanopus.
—	*	—	—	*	*	—	—	—	—	—	—	—	—	—	—	—	—	—	Plegadis falcinellus.
—	*	—	—	—	—	—	—	—	—	—	—	—	—	—	—	—	—	.	Dissoura episcopus.
—	*	—	*	*	*	*	—	—	—	—	—	—	*	*	*	—	—	—	Platalea sp.
—	*	—	—	—	—	*	—	—	—	—	—	—	*	*	*	—	—	—	Phoyx manilensis.
—	*	—	—	—	—	—	—	—	—	—	—	—	—	—	—	—	—	—	Ardea sumatrana.
*	*	—	—	—	—	—	*	—	—	—	—	—	*	*	*	*	*	—	Notophoyx picata.
—	*	—	—	—	—	—	*	—	—	—	—	—	*	*?	*?	—	—	—	Notoph. novaehollandiae.
*	*	—	*	—	*	—	—	—	—	—	—	—	*	*	*	—	—	—	Demiegretta sacra.
*	*	—	—	—	—	—	—	—	—	*	*	—	*	*	*	—	*?	—	Herodias eulophotes.
																			Herodias garzetta.
																			Herodias alba[2]).

Introduction: Geographical Distribution.

Number	Name of species	Page in the text	Nearctic Region	Ethiopian Region	Palaearctic Region	Indian Province	Chinese Province	Japan	Malay Peninsula	Sumatra	Java	Borneo	Philippine Islands	Lesser Sunda Islands
354	Herodias intermedia (Wagl.)	832	—	*	—	*	*	—	*	*	*	*	*	*
355	Bubulcus coromandus (Bodd.)	835	—	*	*	*	*	*	*	*	*	*	*	*
356	Ardeola speciosa (Horsf.)	838	—	—	—	*	*	—	*	*	*	*	*	*
357	Nycticorax caledonicus (Gm.)	841	—	—	—	—	—	—	—	—	*	*	*	*
358	Nycticorax manilensis Vig.	843	—	—	—	—	—	—	—	—	—	—	*	—
359	Nycticorax grisens (L.)	845	*	*	*	*	*	*	—	*	*	*	*	—
360	Gorsachius kutteri (Cab.)	848	—	—	—	—	—	—	—	—	—	—	*	—
361	Butorides javanica (Horsf.)¹)	851	—	*	*	*	*	*	—	*	*	*	*	*
362	Ardetta sinensis (Gm.)²)	854	—	—	*	*	*	*	—	*	*	*	*	*
363	Ardetta eurhythma Swinh.	856	—	—	*	—	*	*	—	—	—	—	—	—
364	Ardetta cinnamomea (Gm.)	859	—	—	—	*	*	—	*	*	*	*	*	*
365	Xanthocnus flavicollis (Lath.)³)	861	—	—	—	—	—	—	—	—	—	—	—	—
366	Xanthocnus melaenus (Salvad.)	863	—	—	—	—	—	—	—	—	—	—	—	—
367	Nettopus pulchellus J.Gd.	865	—	—	—	—	—	—	—	—	—	—	—	—
368	Nettopus coromandelianus (Gm.)	866	—	—	—	*	*	—	*	*	*	*	*	—
369	Dendrocycna arcuata (Horsf.)	868	—	—	—	—	—	—	—	*	*	*	*	*
370	Dendrocycna guttata Schl.	870	—	—	—	—	—	—	—	—	—	*	*	*
371	Anas superciliosa Gm.	872	—	—	—	—	—	—	*?	*	—	—	—	*
372	Nettion gibberifrons (S. Müll.)	874	—	—	—	—	—	—	—	*	*	*	—	*
373	Querquedula circia (L.)	879	—	—	*	*	*	*	—	*	*	*	*	*
374	Nyroca fuligula (L.)	881	—	*	*	*	*	*	—	—	—	*	*	—
375	Fregata minor (Gm.)⁴)	883												
376	Plotus melanogaster (Penn.)⁵)	886	—	—	—	*	*	—	*	*	*	*	*	—
377	Phalacrocorax melanoleucus (Vieill.)	888	—	—	—	—	—	—	—	—	—	—	—	*
378	Phalacrocorax sulcirostris Brdt.	890	—	—	—	—	—	—	—	—	—	*	—	—
379	Sula leucogaster (Bodd.)⁶)	892												
380	Hydrochelidon leucoptera (Meisn. Sch.)	893	—	*	*	*	*	—	—	—	*	*	*	—
381	Hydrochelidon hybrida (Pall.)	895	—	*	*	*	*	—	—	—	*	*	*	—
382	Sterna media Horsf.	897	—	*	*	—	—	—	*	—	*	—	—	*
383	Sterna bergii Lcht.	899	—	*	—	*	*	*	—	*	*	*	*	*
384	Sterna sinensis Gm.	901	—	—	*	*	*	*	—	*	*	*	*	*?
385	Sterna melanauchen Temm.⁷)	903	—	—	—	*	*	*	—	*	*	*	*	*
386	Sterna anaestheta Scop.	905	—	—	—	*	*	*	—	*	*	*	*	*
387	Anous stolidus (L.)⁸)	908	*	*	—	—	—	—	—	*	*	*	*	*
388	Stercorarius sp.	910												
389	Puffinus cuneatus Salv.	911												
390	Puffinus leucomelas (Temm.)	913	—	—	—	—	*	*	—	—	—	*	*	—
391	Diomedea sp.	914												
392	Podiceps tricolor (G. R. Gray)	915	—	—	—	—	—	—	—	—	—	—	—	*
393	Podiceps gularis J. Gd.	917	—	—	—	—	—	—	—	—	*	*?	—	

¹) With racial differences. ²) Add Seychelles. ³) Racial differences not perfectly understood. ⁴) Indian and Pacific Oceans.

Introduction: Geographical Distribution.

		Celebesian Area																	
					Celebes														
Talaut Islands	Sangi Islands	North Peninsula	West Celebes	East Peninsula	Togian Islands	Central Celebes	South Peninsula	S. E. Peninsula (and Buton)	Saleyer Island	Djampea Group	Peling Group	Sula Islands	Moluccas	Papuasia	Australia	Polynesia	New Zealand	Neotropical Region	Name of species
—	*	*	—	—	—	*	—	—	—	—	—	—	*	*	*	—	—	—	Herodias intermedia.
*	*	*	—	—	—	*	*	*	—	—	—	—	*	—	—	—	—	—	Bubulcus coromandus.
—	*	—	—	—	—	—	—	—	—	—	—	—	—	—	—	—	—	—	Ardeola speciosa.
*	*	—	—	—	—	—	—	—	—	*	*	—	*	*	*	*	*	—	Nycticorax caledonicus.
—	*	—	—	—	—	—	—	—	—	—	—	—	—	—	*	—	—	—	Nycticorax manilensis.
—	*	—	—	—	—	—	—	—	—	—	—	—	—	—	—	*	—	—	Nycticorax griseus.
*	—	—	—	—	—	—	—	—	—	—	—	—	—	—	—	—	—	—	Gorsachius kutteri.
—	*	—	—	—	—	*	*	*	—	—	—	—	*	*	*	*	—	—	Butorides javanica [1].
—	*	—	—	—	—	—	—	—	—	—	—	—	*	*	*	*	—	—	Ardetta sinensis [2].
—	*	—	—	—	—	—	—	—	—	—	—	—	—	—	—	—	—	—	Ardetta eurhythma.
—	*	—	—	—	—	*	—	—	—	—	—	—	—	—	—	—	—	—	Ardetta cinnamomea.
*	*	—	—	—	—	*	—	—	—	—	—	—	—	—	—	—	—	—	Xanthocnus flavicollis [3].
*	*	—	—	—	—	—	—	—	—	—	—	—	—	—	—	—	—	—	Xanthocnus melaenus.
—	*	—	—	—	—	—	—	—	—	—	—	—	*	*	*	—	—	—	Nettopus pulchellus.
—	*	—	—	—	*	*	—	—	—	—	—	—	—	—	—	—	—	—	Nettop. coromandelianus.
—	*	*	—	*	*	—	—	—	—	—	—	—	*?	*	*	*	—	—	Dendrocycna arcuata.
—	*	—	*	—	—	—	—	—	—	—	—	—	*	*	—	—	—	—	Dendrocycna guttata.
—	*	—	—	*	*	—	—	—	—	—	—	—	*	*	*	*	*	—	Anas superciliosa.
—	*	—	—	*	*	—	*	—	—	—	—	—	—	*	*	*	*	—	Nettion gibberifrons.
—	*	—	—	—	—	—	—	—	—	—	—	—	*	—	—	—	—	—	Querquedula circia.
—	—	—	—	—	—	—	—	—	—	—	—	—	—	—	*	—	—	—	Nyroca fuligula.
—	*	—	—	—	*	—	—	—	—	—	—	—	—	—	—	—	—	—	Fregata minor.
—	*	—	—	—	*	—	—	—	—	—	—	—	—	—	—	—	—	—	Plotus melanogaster.
—	*	—	—	—	—	—	—	*	—	—	—	—	*	*	*	*	*?	—	Phalacroc. melanoleucus.
—	*	—	—	—	—	—	—	—	—	—	—	—	*	*	*	—	—	—	Phalacroc. sulcirostris.
*	*	—	—	—	*	—	—	—	—	—	—	—	*?	—	—	—	—	—	Sula leucogaster.
*	*	—	—	—	*	—	—	—	—	—	—	—	—	—	—	—	*	—	Hydrochelidon leucoptera.
—	*	—	*	—	*	—	—	—	—	—	—	—	*	*	*	—	—	—	Hydrochelidon hybrida.
—	*	—	*	—	*	—	—	—	—	—	—	—	—	*	*	—	—	—	Sterna media.
—	*	—	—	—	*	—	—	—	—	—	—	—	*	*	*	*	—	—	Sterna bergii.
—	*	—	—	—	—	—	—	—	—	—	—	—	—	*	*	—	—	—	Sterna sinensis.
—	*	—	—	—	—	—	—	—	—	—	—	—	*	*	*	*	—	—	Sterna melanauchen.
*	*	—	—	—	—	—	*	—	*	—	—	—	*	*	*	*	—	—	Sterna anaestheta.
—	—	—	—	—	—	—	—	—	—	—	—	—	—	—	—	—	—	—	Anous stolidus.
—	—	—	—	—	—	—	—	—	—	—	—	—	—	—	—	—	—	—	Stercorarius sp.
—	*	—	—	—	—	—	—	—	—	—	—	—	—	—	—	*	—	—	Puffinus cuneatus.
—	*	—	—	—	—	—	—	—	—	—	—	—	*	*	*	—	—	—	Puffinus leucomelas.
—	—	—	—	—	—	—	—	—	—	—	—	—	—	—	—	—	—	—	Diomedea sp.
—	*	—	—	—	—	—	—	—	—	—	—	—	*	*	—	—	—	—	Podiceps tricolor.
*	*	—	—	—	—	—	—	—	—	—	—	—	—	*	*	—	—	—	Podiceps gularis.

[5] Add Madagascar.   [6] The tropical and subtropical seas of the globe.   [7] Islands of the Indian Ocean.   [8] Tropical seas.

The Island of Celebes is now known to possess 15 peculiar genera of birds, some of which, indeed, range into other parts of the Celebesian area. These forms are of unequal values, that is to say some of them are very distinct from genera existing in other parts of the world, while others hardly possess any structural difference to warrant their distinction. We have, therefore, divided the Celebesian genera into three classes, according to their taxonomic value.

The following four are I. Class genera:

Megacephalon   *Cittura*
Streptocitta   *Scissirostrum*.

The Moleo, *Megacephalon*, has affinities with *Talegallus* and *Aepypodius* of New Guinea.

Streptocitta, a bird very similar in appearance to a Magpie, seems to occupy an intermediate position between the *Sturnidae* and *Corvidae*. It is probably most nearly allied to *Basileornis*, a Celebesian genus occurring on the mainland and known also by distinct species from Banggai and Ceram; the latter genus seems to have its nearest affinities with *Melanopyrrhus* of New Guinea. Hence *Streptocitta* does not seem to be of Oriental origin, but of Australasian (pp. 576, 573 of text).

Cittura, a low type of Kingfisher, has certain affinities with the still more primitive *Melidora* of New Guinea, but its direct descent from such a form is improbable, and its land of origin is hidden in doubt (p. 307 of text).

Scissirostrum, a Grosbeak-like Starling, has been variously treated by systematists, and in our opinion it has as near affinities with the Oriental *Acridotheres* as with anything, but its curious beak undoubtedly entitles it to an isolated position among the *Sturnidae* (p. 569 of text).

The following are II. Class genera:

Rhabdotorrhinus   *Malia*
Meropogon   *Cataponera*
Ceycopsis   *Enodes*
Myza   *Aramidopsis*.

The small Hornbill, *Rhabdotorrhinus*, is apparently most nearly related to the Philippine genus, *Penelopides* (p. 237 of text).

Forsten's Bee-eater, *Meropogon*, finds its nearest allies in *Nyctiornis* of the Oriental Region — not including the Philippines (pp. 259, 260 of text).

The little Kingfisher, *Ceycopsis*, seems to be intermediate between the red-backed section of *Ceyx* which ranges from India to the Philippines, and the blue-backed section of that genus found from the Philippines to the Solomon Islands, but it differs from both in possessing a minute inner toe, which is quite obliterated in all members of the genus *Ceyx*. We believe that indications may be found showing that the red-backed section of *Ceyx* is more ancient than the blue (pp. 274, 277 of text).

Introduction: Geographical Distribution. 113

The Honey-sucker, *Myza*, belongs to the purely Australasian family of the *Meliphagidae*, a family which cannot, however, be sharply distinguished from the *Nectariniidae* of the Ethiopian, Oriental and Australian Regions. The nearest affinities of *Myza* are doubtful (p. 483 of text).

The Bulbul, *Malia*, appears to belong to Oates' *Crateropodinae*, a subfamily of the *Timeliidae*. The geographical limits of the group have not been defined, but it is very strongly represented in the Oriental Region and very sparingly in Papuasia. The nearest affinities of *Malia* are uncertain (p. 501 of text).

Cataponera, a Babbler, is related to genera of the Oriental Region (p. 504 of text).

Enodes, an aberrant Starling, is somewhat intermediate between the Indo-Australian *Calornis* and the Oriental *Acridotheres* (p. 565 of text).

Platen's Rail, *Aramidopsis*, is most closely allied to *Aramides* of South America (p. 691 of text).

The remaining three genera are what we have termed III. Class and are not of higher taxonomic value than 10 strongly characterized species. They are:

Spilospizias   *Charitornis*   *Gazzola*.

The ten species estimated to be of equal value with III. Class genera are:

1. *Microstictus fulvus* (Q. G.) or *wallacei* (Tw.)
2. *Hierococcyx crassirostris* (Tweedd.)
3. *Cranorrhinus cassidix* (Temm.)
4. *Monachalcyon monachus* (Bp.), or *princeps* R.
5. *Pachycephala bonensis* M. & Wg.
6. *Melilestes celebensis* M. & Wg.
7. *Carpophaga poecilorrhoa* Brügg.
8. *Phlogoenas tristigmata* Bp.
9. *Gymnocrex rosenbergi* (Schl.)
10. *Amaurornis isabellina* (Schl.)

The above ten, equal to III. Class genera, we propose to term I. Class species. Their differences from their nearest allies (not counting local forms of them in the Celebesian area itself) are about as great as those between *Corvus monedula* L. and *Corvus corone* L., or *Buteo vulgaris* Leach and *Aquila chrysaetus* (L.).

Then follow 22 II. Class species, differing as markedly from their nearest allies outside Celebes as, for instance, *Turdus musicus* L. from *Turdus merula* L. They are:

1. *Astur griseiceps* Schl.
2. *Accipiter rhodogaster* (Schl.)
3. *Trichoglossus ornatus* (L.)
4. — *meyeri* (Tweedd.)
5. *Prioniturus platurus* (Vieill.)
6. *Loriculus exilis* Schl.
7. — *stigmatus* (Müll. Schl.)
8. *Iyngipicus temmincki* (Malh.)
9. *Phoenicophaes calorhynchus* Temm.
10. *Coracias temmincki* (Vieill.)
11. *Hypothymis puella* (Wall.)
12. *Graucalus bicolor* (Temm.)
13. — *temmincki* (S. Müll.)
14. *Dicaeum nehrkorni* W. Blas.
15. *Acmonorhynchus aureolimbatus* (Wall.)
16. *Zosterops squamiceps* (Hart.)

17. *Zosterops anomala* M. & Wg. 20. *Ptilopus fischeri* Brügg.
18. *Trichostoma celebense* Strickl. 21. — *gularis* (Q. G.)
19. *Basileornis celebensis* G. R. Gray 22. *Carpophaga forsteni* (Bp.)

There are about 45 species in Celebes which may be relegated to the III. Class, differing from one another about as much as *Corvus corone* L. from *Corvus cornix* L., and for the most part geographical species. They need not be tabulated again here, but in the following table these III. Class species have been taken as the unit in computing the relationship of the avifauna of Celebes with those of the neighbouring countries. We have adopted the following scale:

    One III. Class species, or V.(!) Class genus = 1
    One II. Class species, or IV. Class genus = 2
    One I. Class species, or III. Class genus = 4
    One . . . . II. Class genus = 8
    One . . . . . . I. Class genus = 16

Species of lower value than the III. Class, and first class subspecies are valued at ·50; less pronounced races at ·25. We have not troubled to make finer estimations than these last[1]).

In attaching a value in this manner to species or genera, room for error due to the "personal equation" must be allowed for, but the method is obviously better than the usual addition and substraction of genera and species as if they were, respectively, units of equal worth. It is very evident, for instance, that *Megacephalon* suggests much more about ancient Celebes than does *Gazzola*, but it is another question whether the value of the former (four times as great as the latter) is correctly estimated.

---

[1]) Besides the peculiar species of Celebes and the neighbouring islands, there are about 200 other Celebesian species which are either migratory or extend their range beyond the bounds of the area. They may be tabulated as follows, with the premonition that there is great uncertainty as to whether some species should be termed migratory or not:

    Species occurring both in the Oriental and Australasian Regions (often migratory)   c. 60
    Asiatic migrants . . . . c. 64
    Australian migrants . . . . c. 5
    Asiatic non-migrants . . . . c. 38
    Australian non-migrants . . . . c. 34

It is sufficiently obvious that these species have nothing to say about the former distribution of land and water in the East Indies, and they should not be taken into consideration in questions of Geographical Distribution.

Introduction: Geographical Distribution.

Estimated value of the affinities of the peculiar species of Celebes.[1])

Doubtful	Indo-Australian	Oriental	Bornean	Javan	Philippine	Name of species	Lesser Sundan	Moluccan	Papuasian	Australian	Australasian
—	—	—	1	—	—	1. *Spilornis rufipectus* J. Gd.	—	—	—	—	—
—	2	—	—	—	—	2. *Astur griseiceps* Schl.	—	—	—	—	—
4	—	—	—	—	—	3. *Spilospizias trinotatus* (Bp.)	—	—	—	—	—
2	—	—	—	—	—	4. *Accipiter rhodogaster* (Schl.) .	—	—	—	—	—
—	—	—	—	—	1	5. *Spizactus lanceolatus* (Bp.) . .	—	—	—	—	—
—	1	—	—	—	—	6. *Pernis celebensis* (Schl.) . .	—	—	—	—	—
—	—	—	·5	—	·5	7. *Baza celebensis* Schl. . . .	—	—	—	—	—
—	1	—	—	—	—	8. *Ninox ochracea* (Schl.).	—	—	—	—	—
—	—	—	—	—	—	9. *Cephaloptynx punctulata* (Q. G.) . . .	—	—	1	—	—
—	·5	—	—	—	—	10. *Scops manadensis* (Q. G.) (typical) . . .	—	—	—	—	—
1	—	—	—	—	—	11. *Strix inexpectata* (Schl.)	—	—	—	—	—
—	—	—	—	—	—	12. *Trichoglossus ornatus* (L.)	—	1	1	—	—
—	—	—	—	—	—	13. *Trichoglossus meyeri* Tweedd.	2	—	—	—	—
—	—	—	—	—	—	14. *Cacatua sulphurea* (Gm.) (typical) . . .	·5	—	—	—	—
—	—	—	—	—	2	15. *Prioniturus platurus* (Vieill.).	—	—	—	—	—
—	—	—	—	—	1	16. *Prioniturus flavicans* Cass.	—	—	—	—	—
—	—	—	—	—	1	17. *Tanygnathus muelleri* (Müll. Schl.) . .	—	—	—	—	—
—	—	—	—	—	—	18. *Loriculus exilis* Schl.	2	—	—	—	—
—	2	—	—	—	—	19. *Loriculus stigmatus* (Müll. Schl.) . .	—	—	—	—	—
—	—	—	—	—	2	20. *Iyngipicus temmincki* (Malh.)	—	—	—	—	—
—	—	—	—	—	4	21. *Microstictus fulvus* (Q. G.)	—	—	—	—	—
—	—	4	—	—	—	22. *Hierococcyx crassirostris* Tweedd. .	—	—	—	—	—
—	—	—	—	—	—	23. *Cacomantis virescens* (Brügg.)	—	·5	—	—	—
—	1	—	—	—	—	24. *Eudynamis melanorhyncha* S. Müll. . .	—	—	—	—	—
—	—	—	—	1[2])	—	25. *Pyrrhocentor celebensis* (Q. G.) . . .	—	—	—	—	—
—	—	—	—	2	—	26. *Phoenicophaes calorhynchus* Temm. . .	—	—	—	—	—
—	—	—	—	—	8	27. *Rhabdotorrhinus exaratus* (Temm.) . .	—	—	—	—	—
—	—	—	—	—	4	28. *Cranorrhinus cassidix* (Temm.)	—	—	—	—	—
—	—	8	—	—	—	29. *Meropogon forsteni* Bp.	—	—	—	—	—
—	—	—	·5	—	·5	30. *Pelargopsis melanorhyncha* (Temm.) . .	—	—	—	—	—
—	8	—	—	—	—	31. *Ceycopsis fallax* (Schl.)	—	—	—	—	—
—	—	·5	—	—	—	32. *Halcyon coromanda rufa* (Wall.) . . .	—	—	—	—	—
—	—	—	—	—	2	33. *Monachalcyon monachus* (Bp.)	2	—	—	—	—
—	—	—	—	—	·5	34. *Monachalcyon princeps* Rchb.	·5	—	—	—	—
16	—	—	—	—	—	35. *Cittura cyanotis* (Temm.)[3]).	—	—	—	—	—

[1]) Species occurring on the mainland, sometimes ranging into the islands of the area. Where two or more geographical races are found in the Celebesian area, the first described only is reckoned.
[2]) Literally Kangean.   [3]) Probably Australasian.

116 Introduction: Geographical Distribution.

Doubtful	Indo-Australian	Oriental	Bornean	Javan	Philippine	Name of species	Lesser Sundan	Moluccan	Papuasian	Australian	Australasian
—	—	2	—	—	—	36. *Coracias temmincki* (Vieill.)[1]	—	—	—	—	—
—	—	—	—	—	1	37. *Lyncornis macropterus* Bp.	—	—	—	—	—
—	—	—	—	—	1	38. *Caprimulgus celebensis* Grant	—	—	—	—	—
—	—	1	—	—	—	39. *Macropteryx wallacei* (J. Gd.)	—	—	—	—	—
—	—	—	—	—	.5	40. *Pitta celebensis* Müll. Schl.	—	.25	.25	—	—
—	—	—	.25	—	.25	41. *Pitta forsteni* Bp.	—	.5	—	—	—
—	—	—	—	—	1	42. *Siphia rufigula* (Wall.)	—	—	—	—	—
—	—	1	—	—	—	43. *Stoparola septentrionalis* Bütt.	—	—	—	—	—
—	—	2	—	—	—	44. *Hypothymis puella* (Wall.)	—	—	—	—	—
—	—	—	—	—	—	45. *Rhipidura teijsmanni* Bütt.	—	—	—	1	—
—	—	—	.5	—	.5	46. *Gerygone flaveola* Cab.	—	—	—	—	—
—	—	—	—	—	1	47. *Pachycephala sulfuriventer* (Tweedd.)	—	—	—	—	—
—	4	—	—	—	—	48. *Pachycephala bonensis* M. & Wg.	—	—	—	—	—
—	2	—	—	—	—	49. *Graucalus bicolor* (Temm.)	—	—	—	—	—
—	1	—	—	—	—	50. *Graucalus leucopygius* Bp.	—	—	—	—	—
—	—	—	—	—	—	51. *Graucalus temmincki* (S. Müll.)	—	2	—	—	—
—	—	—	—	—	—	52. *Edoliisoma morio* (S. Müll.)	—	1	—	—	—
—	—	1	—	—	—	53. *Lalage leucopygialis* Tweedd.	—	—	—	—	—
—	—	—	—	—	—	54. *Artamus monachus* Bp.	—	1	—	—	—
.5	—	—	—	—	—	55. *Dicrurus leucops* Wall.	—	—	—	—	—
—	—	—	1	—	—	56. *Dicaeum celebicum* S. Müll.	—	—	—	—	—
—	—	1	—	—	—	57. *Dicaeum nehrkorni* W. Blas.	—	1	—	—	—
—	—	1	—	—	—	58. *Dicaeum hosei* Sharpe	—	—	—	—	—
—	—	—	—	—	—	59. *Acmonorhynchus aureolimbatus* (Wall.)	2	—	—	—	—
—	—	1	—	—	—	60. *Aethopyga flavostriata* (Wall.)	—	—	—	—	—
—	—	—	—	—	—	61. *Cyrtostomus frenatus* (S.Müll.)(loc. races)	—	—	—	—	.5
—	—	—	—	—	—	62. *Hermotimia porphyrolaema* (Wall.)	—	1	—	—	—
—	—	.5	—	—	—	63. *Anthreptes malacc. celebensis* (Shelley)	—	—	—	—	—
—	—	—	—	—	—	64. *Myzomela chloroptera* Tweedd.	—	—	1	—	—
—	—	—	—	—	—	65. *Melilestes celebensis* M. & Wg.	—	4	—	—	—
—	—	—	—	—	—	66. *Myza sarasinorum* M. & Wg.	—	—	—	—	8
—	—	—	1	1	—	67. *Zosterops squamiceps* (Hart.)	—	—	—	—	—
—	—	—	—	—	—	68. *Zosterops atrifrons* Wall.	—	1	—	—	—
—	—	—	—	—	—	69. *Zosterops sarasinorum* M. & Wg.	.5	—	—	—	—
—	2	—	—	—	—	70. *Zosterops anomala* M. & Wg.[2]	—	—	—	—	—
8	—	—	—	—	—	71. *Malia grata* Schl[3]	—	—	—	—	—
—	—	—	1	—	—	72. *Androphilus castaneus* (Bütt.)	—	—	—	—	—
—	—	8	—	—	—	73. *Cataponera turdoides* Hart.	—	—	—	—	—
—	—	2	—	—	—	74. *Trichostoma celebense* Strickl.	—	—	—	—	—
—	—	—	—	—	—	75. *Geocichla erythronota* Scl.	1	—	—	—	—

[1] Partly African.   [2] Chiefly Oriental.   [3] Probably Oriental.

Introduction: Geographical Distribution. 117

Doubtful	Indo-Australian	Oriental	Bornean	Javan	Philippine	Name of species	Lesser Sundan	Moluccan	Papuasian	Australian	Australasian
—	—	—	—	1	—	76. *Merula celebensis* Bütt.	—	—	—	—	—
—	—	1	—	—	—	77. *Phyllergates riedeli* M. & Wg.	—	—	—	—	—
—	—	—	—	·5	—	78. *Cryptolopha sarasinorum* M. & Wg. .	·5	—	—	—	—
—	—	—	—	—	·25	79. *Munia formosana brunneiceps* (Wald.)	—	—	—	—	—
—	—	—	—	—	—	80. *Munia subcastanea* Hart.	1	—	—	—	—
—	—	—	—	—	—	81. *Munia molucca—propinqua* . .	·25	—	—	—	—
—	8	—	—	—	—	82. *Enodes erythrophrys* (Temm.) .	—	—	—	—	—
—	—	—	—	1	—	83. *Acridotheres cinereus* Bp.	—	—	—	—	—
16	—	—	—	—	—	84. *Scissirostrum dubium* (Lath.)[1] . . .	—	—	—	—	—
—	—	—	—	—	—	85. *Basilornis celebensis* G. R. Gray . .	—	2	—	—	—
16	—	—	—	—	—	86. *Streptocitta albicollis* (Vieill.)[2]. . . .	—	—	—	—	—
4	—	—	—	—	—	87. *Gazzola typica* Bp.	—	—	—	—	—
—	—	·5	—	—	—	88. *Oriolus celebensis* (Tweedd.). . . .	—	—	—	—	—
—	—	—	—	·5	—	89. *Osmotreron wallacei* Salvad.	—	—	—	—	—
—	—	—	—	—	2	90. *Ptilopus fischeri* Brügg.	—	—	—	—	—
—	—	—	—	—	2	91. *Ptilopus gularis* (Q. G.).	—	—	—	—	—
—	—	—	—	·25	—	92. *Ptilopus melanospilus* (Salvad.) . . .	·25	—	—	—	—
—	—	—	—	—	·5	93. *Carpophaga paulina* (Bp.)	—	—	—	—	—
—	—	—	—	—	1	94. *Carpophaga radiata* (Q. G.)	—	—	—	—	—
—	—	—	—	—	2	95. *Carpophaga forsteni* (Bp.)	—	—	—	—	—
4	—	—	—	—	—	96. *Carpophaga poecilorrhoa* Brügg.[3] . .	—	—	—	—	—
—	1	—	—	—	—	97. *Myristicivora luctuosa* (Temm.) .	—	—	—	—	—
—	—	—	—	—	—	98. *Turacoena manadensis* (Q. G.) . .	1	—	—	—	—
—	—	—	—	—	—	99. *Macropygia albicapilla* Bp.	—	·5	—	—	—
—	—	—	—	—	—	100. *Macropygia macassariensis* (Wall.) . .	·5	—	—	—	—
—	—	—	—	—	2	101. *Phlogoenas tristigmata* Bp.	—	—	2	—	—
—	—	—	—	—	—	102. *Megacephalon maleo* (Hartl.). . . .	—	—	16	—	—
—	—	—	—	—	—	103. *Turnix rufilatus* Wall.	1	—	—	—	—
—	—	—	—	—	—	104. *Gymnocrex rosenbergi* (Schl.)	—	2	2	—	—
8	—	—	—	—	—	105. *Aramidopsis plateni* (W. Blas.)[4]. . .	—	—	1	—	—
—	—	—	—	—	—	106. *Hypotaenidia celebensis* (Q. G.) . . .	—	—	—	—	—
—	—	—	—	—	1	107. *Rallina minahassa* Wall.	—	—	—	—	—
—	—	—	—	—	—	108. *Amaurornis isabellina* (Schl.) . . .	—	—	—	—	2
79	31	37·5	5·75	7·25	44·5		15	7·25	33·75	2	10·5

[1] Probably Oriental. [2] Probably Australasian. [3] Probably Oriental. [4] S. American.

The above totals for each area added together give the sum of 273·5. The percentage of "Doubtful" components in the Celebes avifauna is then:

$$\frac{79 \times 100}{273 \cdot 5}, \text{ or } 29 \text{ per cent.}$$

In the same manner the percentage of the other components is obtained.
The total number of peculiar forms in Celebes is 108. Dividing the value of a species where the nearest affinities with these peculiar forms are found in two or more of the neighbouring areas, the shares are:

Doubtful	10	forms
Indo-Australian	12	"
Oriental (incl. Bornean and Javan)	29·25	"
Philippine	23·75	"
Lesser Sundan	13	"
Australasian	20	"
	108	

The table then affords the following percentages:

	Number of forms	Comparative value
Forms of doubtful affinities	9·3 per cent	29 per cent
Forms of Indo-Australian affinities	11·1 per cent	11·3 per cent
Forms of Oriental affinities (incl. also strictly Bornean and Javan)	27·0 per cent	18·46 per cent
Forms of Philippine affinities	22·0 per cent	16·27 per cent
Forms of Lesser Sundan affinities	12·0 per cent	5·5 per cent
Forms of Moluccan affinities	5·0 per cent	2·6 per cent
Forms of Papuasian and Australian affinities	13·5 per cent	16·94 per cent
	99·9	100·07

Thus, the forms of doubtful affinities in Celebes have the highest value, viz. a little over 3 each;
  those of Indo-Australian and of Australasian affinities have the value 1
  those of the Oriental Region about ·7
  those of the Philippines about ·74
  those of the Lesser Sunda Islands only about ·5.

Consequently, the "Doubtful" elements seem to be the oldest, and the Lesser Sundan generally the most recent in Celebes.

Including the Philippines in the Oriental Region, and comparing the area thus formed with the Australian Region, Celebes shows itself to be decidedly Oriental.

  Oriental components . . 49 per cent, value 35·73 per cent
  Australasian[1]) components 18·5 per cent, value 19·54 per cent

---

[1]) The Lesser Sundan forms are not added, being shared about equally between the Australian and Oriental Regions.

That is to say, one-half of the peculiar birds of Celebes have their nearest affinities in the Oriental Region, and one-fifth only in the Australian Region; but the Australasian forms seem to be on an average rather more strongly differentiated than the Oriental forms.

While the Philippines display nearly as many points of affinity with Celebes as do all the other parts of the Oriental Region taken together, they also show fewer points of dissimilarity — i. e. a far smaller number of genera not found in Celebes.

A comparison of all the genera occurring in the Philippines, Borneo, Java, and Papuasia and the Moluccas with those of the Celebesian area gives the following results[1]):

	Present in the Celebesian area	Absent in the Celebesian area
Genera occurring in the Philippines (not including Palawan)	154	68
Genera occurring in the Bornean Group	150	147
Genera occurring in Java	135	125
Genera occurring in Papuasia and the Moluccas	136	169

The figures show that, as regards genera, the Celebesian area agrees much better with the Philippines than with the neighbouring countries.[2])

We now turn to examine the various parts of the Celebesian area itself, giving lists of the birds of the different groups of islands, with a few remarks on their general affinities and derivation. The relations of the Northern and Southern Peninsulas of Celebes to one another are similarly discussed.

---

[1]) The numbers are taken respectively from Prof. D. C. Worcester's "Contributions to Philippine Ornithology" (Pr. U. S. Nat. Mus. 1898 XX, pp. 551—564), Mr. Everett's "List of the Birds of the Bornean Group of Islands" (J. Str. Br. R. A. S. 1889, pp. 96—212), Dr. Vorderman's "List of the Birds of Java" (Nat. Tdschr. Ned. Ind. 1885, XLIV, pp. 189—207), and Count Salvadori's "Orn. Papuasia e Moluce." 1880—82 vols. I—III.

[2]) It would be suggestive to compare the number of the endemic genera and species of Celebes with those of other islands of the earth, say Borneo, Mindanao, Java, Timor, New Guinea, New Zealand, Madagascar, etc., etc., but such data are not yet readily available. It would, however, be worth while to draw up such lists, as we possess them to a greater or less extent for plants. We mention for instance that endemic genera of *Phanerogamae* are known from Fiji 13, Ceylon 21, New Zealand 22, Sandwich Islands 32, the Mascarenes 34, New Caledonia 38, Japan 48, Madagascar 91, and that from New Guinea about 35 are as yet known (see O. Warburg, Bot. Jahrb. 1891 XIII, 231).

List of the Birds of the Sangi Islands.

Name of species	Great Sangi	Siao	Tagulandang Ruang & Biaro	Name of species	Great Sangi	Siao	Tagulandang Ruang & Biaro
1. *Spilornis rufipectus* J. Gd. (?typical)	—	*	—	*46. *Eudrepanis duivenbodei* (Schl.)	*	—	—
2. *Tachyspizias soloensis* (Horsf.)	*	*	—	*47. *Hermotimia sangirensis* (A. B. M.)	*	*	*
3. *Butastur indicus* (Gm.)	*	*	—	48. *Anthreptes malaccensis chlorigaster* (Sharpe)	*	*	—
4. *Haliastur indus* (Bodd.)	*	*	*	*49. *Zosterops nehrkorni* W. Blas.	*	—	—
5. *Pandion haliaetus* (L.) (typical)	*	—	—	*50. *Iole platenae* (W. Blas.)	*	—	—
6. *Pandion haliaetus leucocephalus* (J. Gd.)	*	*	—	51. *Petrophila cyanus solitaria* (P. L. S. Müll.)	*	—	—
7. *Ninox scutulata japonica* (Temm. Schl.)	*	*	*	52. *Locustella fasciolata* (G. R. Gray)	*	—	—
8. *Scops manadensis* (Q. G.) (typical)	*	*	—	53. *Phylloscopus borealis* (Blas.)	*	—	—
9. *Strix flammea rosenbergi* (Schl.)	*	—	—	54. *Munia molucca — propinqua*	*	*	*
*10. *Eos histrio* (St. Müll.) (typical)	*	*	*	55. *Calornis panayensis sangirensis* (Salvad.)	*	*	—
11. *Prioniturus platurus* (Vieill.)	*	*	—	* *Calornis panayensis — sangirensis*	—	—	*
12. *Prioniturus flavicans* Cass.	?	*	—	*56. *Oriolus formosus* Cab. (typical)	—	*	—
*13. *Tanygnathus muelleri sangirensis* M. & Wg.	*	—	—	*57. *Oriolus formosus sangirensis* M. & Wg.	—	*	—
14. ? *Tanygnathus luconensis* (L.)	*	—	—	* *Oriolus formosus — sangirensis*	—	—	*
15. *Tanygnathus megalorhynchus* (Bodd.) (typ.)	*	*	*	*58. *Osmotreron sangirensis* (Brügg.)	*	*	*
*16. *Loriculus catamene* Schl.	*	—	—	59. *Ptilopus xanthorrhous* (Salvad.)	*	*	*
17. *Cuculus canorus canoroides* (S. Müll.)	*	—	—	60. *Carpophaga concinna* Wall.	*	*	*
18. *Eudynamis mindanensis sangirensis* (Blas.)	*	*	*	61. *Myristicivora bicolor* (Scop.)	*	*	*
19. *Centrococcyx bengalensis* (Gm.)	*	*	*	62. *Columba albicularis* (Bp.)	—	—	*
20. *Merops ornatus* Lath.	*	*	—	*63. *Macropygia albicapilla sangirensis* (Salvad.)	*	*	—
21. *Alcedo ispida* L.	*	*	—	* *Macropygia albicapilla — sangirensis*	—	—	*
22. *Alcedo moluccana* (Less.)	*	*	—	64. *Turtur tigrinus* (Temm. Kn.)	—	*	*
*23. *Ceycopsis sangirensis* M. & Wg.	*	—	—	65. *Chalcophaps indica* (L.)	*	—	—
24. *Halcyon coromanda rufa* (Wall.)	*	*	—	66. *Caloenas nicobarica* (L.)	—	*	—
25. *Halcyon sancta* Vig. Horsf.	*	*	—	67. *Gallus ferrugineus* (Gm.)	—	*	*
26. *Halcyon chloris* (Bodd.) (typical)	*	*	*	68. *Megapodius sangirensis* Schl.	*	*	*
*27. *Cittura sangirensis* Sharpe	*	—	—	69. *Megacephalon maleo* (Hartl.)	—	*	—
28. *Eurystomus orientalis* (L.)	*	*	*	70. *Amaurornis phoenicura* (Forst.)	—	—	*
29. *Cypselus pacificus* (Lath.)	—	*	—	71. *Amaurornis moluccana* (Wall.)	—	*	—
*30. *Pitta polliceps* Brügg.	*	—	—	72. *Aegialitis geoffroyi* (Wagl.)	*	*	—
*31. *Pitta caeruleitorques* Salvad.	*	—	—	73. *Heteractitis brevipes* (Vieill.)	*	*	—
*32. *Pitta sangirana* (Schl.)	*	—	—	74. *Actitis hypoleucos* (L.)	*	*	—
33. *Pitta irena* Temm.	—	*	—	75. *Tringa rufescollis* Pall.	*	*	—
34. *Hirundo rustica gutturalis* (Scop.) and *H. rustica — gutturalis*	*	*	—	76. *Numenius variegatus* (Scop.)	*	*	—
35. *Hirundo javanica* Sparrm.	*	*	—	77. *Ardea sumatrana* Raffl.	*	*	—
*36. *Hypothymis rowleyi* (A. B. M.)	*	—	—	78. *Demiegretta sacra* (Gm.)	*	*	—
*37. *Monarcha commutatus* Brügg.	*	*	—	79. *Herodias gorzetta* (L.)	*	*	—
*38. *Culicicincla sangirensis* (Oust.)	*	—	—	80. *Bubulcus coromandus* (Bodd.)	*	*	—
39. *Lanius luconensis* L.	—	—	*	81. *Nycticorax caledonicus* (Gm.)	—	*	—
40. *Graucalus leucopygius* Bp.	*	—	—	82. *Xanthoenas flavicollis* (Lath.)	*	—	—
*41. *Edoliisoma salvadorii* Sharpe	*	*	*	83. *Xanthoenas melanous* (Salvad.)	—	—	*
*42. *Dicrurus leucops — axillaris*	—	*	*	84. *Sula leucogaster* (Bodd.)	*	*	—
*43. *Dicrurus leucops axillaris* (Salvad.)	*	—	—	85. *Hydrochelidon leucoptera* (Meisn. Sch.)	*	—	—
*44. *Dicaeum sangirense* Salvad.	*	—	—	86. *Sterna anaestheta* Scop.	*	—	—
*45. *Aemorhynchus sangirensis* (Salvad.)	*	—	—	87. *Anous stolidus* (L.)	*?	—	—
				88. *Podiceps gularis* J. Gd.	*	—	—

* signifies autochthonous.

List of the Birds of the Talaut Islands.[1])

Name of species	Karkellang	Kahruang	Salibabu	Name of species	Karkellang	Kahruang	Salibabu
1. *Tachyspizias soloensis* (Horsf.)	*	—	—	35. *Motacilla flava* L.	—	*	—
2. *Butastur indicus* (Gm.)	*	—	*	*36. *Munia molucca typica* >[2])	*	—	*
3. *Haliastur indus—girrenera*	—	*	*	37. *Calornis panayensis sangirensis* (Salvad.)	*	*	*
4. *Pandion haliaetus* (L.)	—	—	*	*38. *Oriolus melanisticus* M. & Wg.	*	*	*
Pandion haliaetus leucocephalus (J. Gd.)	*	—	—	39. *Ptilopus xanthorrhous* (Salvad.)	*	*	*
5. *Ninox scutulata japonica* (Temm. Schl.)	—	*	*	40. *Carpophaga concinna* Wall.	*	*	*
*6. *Eos histrio talautensis* (M. & Wg.)	*	*	*	*41. *Carpophaga intermedia* M. & Wg.	*	*	*
*7. *Prioniturus platurus talautensis* Hart.	*	—	*	42. *Carpophaga pickeringi* Cass.	—	*	*
*8. *Tanygnathus muelleri—sangirensis*	—	*	*	43. *Myristicivora bicolor* (Scop.)	*	*	*
*9. *Tanygnathus talautensis* M. & Wg.	*	*	*	44. *Chalcophaps indica* (L.)	—	—	*
10. *Tanygnathus megalorhynchus* (Bodd.) (typ.)	*	*	*	45. *Macropygia albicapilla sangirensis* (Salvad.)	—	—	*
11. *Cuculus canorus canoroides* (S. Müll.)	*	*	*	46. *Megapodius sangirensis* Schl.	*	*	*
12. *Eudynamis mindanensis sangirensis* (W. Bl.)	*	*	*	47. *Amaurornis cinerea* (Vieill.)	*	*	*
13. *Centrococcyx bengalensis* (Gm.)	*	*	*	48. *Amaurornis phoenicura* (Forst.)	—	*	*
14. *Scythrops novaehollandiae* Lath.	—	—	*	49. *Porphyrio pulverulentus* Temm.	*	—	—
15. *Merops ornatus* Lath.	*	—	—	50. *Charadrius fulvus* Gm.	*	*	—
16. *Alcedo ispida* L.	*	—	*	51. *Aegialitis vereda* (J. Gd.)	*	—	—
17. *Alcedo moluccana* (Less.)	—	—	*	52. *Aegialitis geoffroyi* (Wagl.)	—	*	—
18. *Halcyon coromanda rufa* (Wall.)	*	—	*	53. *Totanus glareola* (L.)	*	—	—
19. *Halcyon sancta* Vig. Horsf.	*	*	—	54. *Heteractitis brevipes* (Vieill.)	—	*	—
20. *Halcyon chloris* (Bodd.) (typical)	*	*	*	55. *Actitis hypoleucos* (L.)	*	*	*
21. *Eurystomus orientalis* (L.)	*	*	*	56. *Limosa novaezealandiae* (G. R. Gray)	—	*	—
*22. *Pitta inspeculata* M. & Wg.	*	*	*	57. *Numenius variegatus* (Scop.)	—	*	—
23. *Hirundo javanica* Sparrm.[2])	*	—	—	58. *Demiegretta sacra* (Gm.)	—	*	—
24. *Muscicapa griseosticta* (Swinh.)	*	—	—	59. *Herodias garzetta* (L.)	—	*	—
*25. *Zeocephus talautensis* M. & Wg.	*	*	*	60. *Herodias alba* (L.)	—	*	—
26. *Monarcha inornatus* (Garn.)	*	—	*	61. *Bubulcus coromandus* (Bodd.)	—	*	—
27. *Lanius lucionensis* L.	*	*	—	62. *Gorsachius kutteri* (Oab.)	*	—	—
*28. *Edoliisoma talautense* M. & Wg.	*	*	*	63. *Ardetta eurhythma* Swinh.[2])	—	*	—
*29. *Dicaeum talautense* M. & Wg.	*	—	*	64. *Xanthocnus flavicollis* (Lath.)	—	*	*
*30. *Hermatimnis talautensis* M. & Wg.	*	*	*	65. *Dendrocygna guttata* Schl.	*	*	—
*31. *Zosterops babelo* M. & Wg.	*	*	*	66. *Dendrocygna arcuata* Horsf.[2])	*	—	—
32. *Petrophila cyanus solitaria* (P. L. S. Müll.)	—	*	*	67. *Sterna bergii* Licht.	—	*	—
33. *Locustella fasciolata* (G. R. Gray)	*	—	*	68. *Podiceps gularis* J. Gd.	—	*	—
34. *Phylloscopus borealis* (Blas.)	*	*	—				

[1]) A small collection of forty species obtained by Mr. Waterstradt's Bornean hunters on Salibabu has recently been recorded by Mr. Hartert (Novit. Zool. 1898, pp. 85—91). Of these 21 were new to Salibabu, and 3, *Pandion haliaetus*, *Alcedo moluccana*, and *Munia molucca*, had not yet been recorded from the group, though the last is present in the Dresden Museum from Karkellang, as well as a small form of *Pandion*. *Trichoglossus ornatus* (L.), *Iyngipicus temmincki* (Malh.), and *Lyncornis macropterus* Bp., are included in Mr. Hartert's list, but omitted in the above, as we have little doubt that they were obtained in Celebes. For the same reason Hartert omits *Oriolus celebensis* and *Dicrurus leucops*.

[2]) *Hirundo javanica*, *Ardetta eurhythma*, and *Dendrocygna arcuata* were included in a third collection from Talaut, received in 1897.

[3]) The formula *Munia molucca typica* > is more accurate than the *Munia molucca* > *propinqua* used in the text, p. 551, the Talaut birds being, as one might say, more typical than the type!

Introduction: Geographical Distribution.

The nearest affinities of the peculiar species of the Sangi Islands.

	Celebes	Philippines	Borneo Group	Moluccas and Papuasia
1. *Eos histrio* (S. Müll.) (typical)	—	—	—	*
2. *Tanygnathus muelleri sangirensis* M. & Wg.	*	—	—	—
3. *Loriculus catamene* Schl.	—	—	—	*
4. *Ceycopsis sangirensis* M. & Wg.	*	—	—	—
5. *Cittura sangirensis* Sharpe	*	—	—	—
6. *Pitta palliceps* Brügg.	*	—	—	—
7. *Pitta caeruleitorques* Salvad.	—	*	—	—
8. *Pitta sangirana* (Schl.)	—	*	*	—
9. *Monarcha commutatus* Brügg.	—	—	—	*
10. *Hypothymis rowleyi* (A. B. M.)	*	—	*	—
11. *Colluricincla sangirensis* (Oust.)	—	—	—	*
12. *Edoliisoma salvadorii* Sharpe	—	—	—	*
13. *Dicrurus leucops axillaris* (Salvad.)	*	—	—	—
14. *Dicaeum sangirense* Salvad.	*	—	—	—
15. *Aemonorhynchus sangirensis* (Salvad.)	*	—	—	—
16. *Eudrepanis duivenbodei* (Schl.)	—	*	—	—
17. *Hermotimia sangirensis* (A. B. M.)	*	—	—	—
18. *Zosterops nehrkorni* W. Blas.	*	—	—	*
19. *Iole platenae* (W. Blas.)	Toglan Sula	—	—	—
20. *Oriolus formosus* Cab.	—	*	—	—
21. *Osmotreron sangirensis* (Brügg.)	*	—	—	—
	12	4	2	6

The nearest affinities of the peculiar species of the Talaut Islands.

	Celebes	Philippines	Borneo Group	Moluccas and Papuasia
1. *Eos histrio talautensis* M. & Wg.	—	—	—	*
2. *Tanygnathus muelleri sangirensis* M. & Wg.	*	—	—	—
3. *Tanygnathus talautensis* M. & Wg.	—	*	—	—
4. *Prioniturus platurus talautensis* Hart.	*	—	—	—
5. *Pitta inspeculata* M. & Wg.	—	—	—	*
6. *Zeocephus talautensis* M. & Wg.	—	*	—	—
7. *Edoliisoma talautense* M. & Wg.	—	—	—	*
8. *Dicaeum talautense* M. & Wg.	*	—	—	—
9. *Hermotimia talautensis* M. & Wg.	*	—	—	—
10. *Zosterops babelo* M. & Wg.	—	*	—	—
11. *Munia molucca typica* >	—	—	—	*
12. *Oriolus melanisticus* M. & Wg.	—	*	—	—
13. *Carpophaga intermedia* M. & Wg.	—	*	*	—
	4	5	1	4

Six of the Talaut forms (numbers 1, 2, 7, 8, 9, 12) have still nearer affinities in Sangi, and there are five other species or subspecies which are "identical" in Sangi and Talaut. One of these latter (*Macropygia albicapilla sangirensis*) has its nearest affinities in Celebes, one (*Eudynamis mindanensis sangirensis*) in the Philippines, and three (*Calornis panayensis sangirensis, Ptilopus xanthorrhous, Megapodius sangirensis*) in Celebes and the Philippines alike. Consequently it might be claimed that the Philippines are known at present to have 9 forms in Talaut, and Celebes only 8 forms; but as the avifauna of Sangi presents the strongest agreement with Talaut, and Sangi belongs to Celebes, it is convenient to include Talaut with Sangi in the Celebesian area.

The peculiar birds of the Sangi and Talaut Islands seem to be of comparatively recent origin; there is not a form among them which can be termed an ancient type. There is not a single peculiar genus, and all, or almost all, the endemic species are geographical races of forms in the lands lying near at hand to north, south, east or west. Moreover, ill-flying birds, such as the *Bucerotidae* and *Phoenicophainae*, are absent, or at least not known as yet. Everything points to the recent colonisation of these islands, and their highly volcanic or coral character and the deep sea around them are suggestive of their recent upheaval[1]).

As has been pointed out elsewhere, almost all of the peculiar species of Sangi and Talaut have increased in size (see p. 58).

List of the Birds of the Peling Group.

Name of species	Peling	Banggai	Name of species	Peling	Banggai
*1. *Spilornis rufipectus* < *sulaensis*	*	*	*13. *Loriculus solateri ruber* M. & Wg.	*	*
2. *Accipiter sulaensis* (Schl.)	*	*	14. *Aprosmictus sulaensis* Rchw.	*	—
3. *Haliaetus leucogaster* (Gm.)	—	*	15. *Cacomantis virescens* (Brügg.)	—	*
4. *Polioaetus humilis* (Müll. Schl.) (typical)	*	—	16. *Eudynamis melanorhyncha* S. Müll.	*	—
5. *Haliastur indus—girrenera*	*	—	17. *Merops ornatus* Lath.	*	—
6. *Pernis celebensis* (Wall.)	*	—	18. *Alcedo moluccana* (Less.)	*	—
7. *Baza celebensis* Schl.	*	—	19. *Alcedo meninting* Horsf.	*	—
8. *Tinnunculus moluco. orientalis—occidentalis*	*	—	*20. *Pelargopsis dichrorhyncha* M. & Wg.	*	*
9. *Pandion haliaetus leucocephalus* J. Gd.	—	*	21. *Halcyon coromanda rufa* (Wall.)	*	—
10. *Trichoglossus ornatus* (L.)	*	*	22. *Halcyon chloris* (Bodd.) (typical)	*	*
11. *Prioniturus platurus* (Vieill.)	*	*	23. *Halcyon sancta* V. & H.	—	*
12. *Tanygnathus muelleri* (Müll. Schl.) (typical)	*	*	24. *Eurystomus orientalis* (L.)	*	*

[1]) Such a change need call for no surprise; Worcester shows (Pr. U. S. Nat. Mus. 1898, 581) that such has evidently been the case with Siquijor, an island, with an area of about 90 sq. miles, to the north of Mindanao. "There is a tradition among the natives to the effect that the island has been thrown up from beneath the sea within a comparatively short time, and there is abundant geological evidence that this tradition is founded on fact. Every stone cracked open by the hammer shows evident signs of its coral origin. The tops of the highest hills, which rise a thousand feet above sea level, are strewn with the shells of the very same mollusks which to-day live along the shores. The hills themselves are mere masses of coral rag, to which a few trees cling with difficulty, as the soil washes down into the valleys almost as fast as it is formed. The fresh-water streams are without fish." Our native collectors sent 16 species in a small collection from Ruang, a volcano rising out of the sea close to Tagulandang in the Sangi Islands, and it is pretty certain that these species must have settled there since the eruptions of 1870 and 1871, which destroyed the vegetation (see p. 634 of text).

Name of species	Peling	Banggai	Name of species	Peling	Banggai
25. *Collocalia esculenta* (L.)	*	*	44. *Cisticola cursitans* (Frkl.)	*	—
26. *Macropteryx wallacei* (J.Gd.)	*	*	45. *Cisticola exilis* (Vig. Horsf.)	*	—
27. *Hypothymis puella blasii* Hart.	*	*	46. *Munia molucca—propinqua*	*	—
28. *Monarcha inornatus* (Garn.)	*	*	47. *Calornis sulaensis* Sharpe	*	*
29. *Culicicapa helianthea* (Wall.)	—	*	48. *Scissirostrum dubium* (Lath.)	*	*
30. *Pachycephala olio* Wall.	*	*	*49. *Basileornis galeatus* A.B.M.	—	*
31. *Graucalus schistaceus* (Sharpe)	—	*	50. *Oriolus frontalis* Wall.	*	*
32. *Graucalus melanops* (Lath.)	*	—	51. *Osmotreron wallacei* (Salvad.) (typical)	*	*
33. *Edoliisoma obiense* Salvad.	*	*	*52. *Ptilopus rubgularis* M. & Wg.	*	*
34. *Lalage leucopygialis* Tweedd.	*	*	53. *Ptilopus chrysorrhous* (Salvad.)	*	*
35. *Artamus leucogaster* (Val.)	*	*	54. *Carpophaga paulina* (Bp.)	*	*
36. *Artamus monachus* Bp.	—	*	55. *Myristicivora luctuosa* (Temm.)	*	*
37. *Dicrurus pectoralis* Wall.	*	*	56. *Columba albigularis* (Bp.)	—	*
38. *Dicaeum sulaense* Sharpe	—	*	57. *Turacoena manadensis* (Q. G.)	*	*
39. *Cyrtostomus frenatus* (S. Müll.)	*	*	58. *Macropygia albicapilla* Bp.	*	*
40. *Hermotimia auriceps* (G. R. Gray)	*	*	59. *Chalcophaps indica* (L.)	*	—
41. *Anthreptes malaccensis celebensis* (Shell.)	*	*	60. *Glareola isabella* Vieill.	*	—
*42. *Zosterops subatrifrons* M. & Wg.	*	—	61. *Aegialitis geoffroyi* (Wagl.)	*	—
43. *Iole longirostris* (Wall.)	*	*	62. *Nycticorax caledonicus* (Gm.)	—	*

List of the Birds of the Sula Islands.

*1. *Spilornis rufipectus sulaensis* (Schl.)
2. *Tachyspizias soloensis* (Horsf.)
3. *Accipiter sulaensis* (Schl.)
4. *Spizaetus lanceolatus* Temm. Schl.
5. ? *Ictinaetus malayensis* (Reinw.)
6. *Haliastur indus—girrenera*
7. *Baza celebensis* Schl.
*8. *Pisorhina sulaensis* Hart. †
9. *Ninox scutulata japonica* (Temm. Schl.)
*10. *Trichoglossus flavoviridis* Wall.
11. *Tanygnathus muelleri* (Müll. Schl.) (typical)
*12. *Loriculus sclateri* Wall. (typ.) (also from Mangoli †)
13. *Aprosmictus sulaensis* Rchw.
14. *Eudynamis melanorhyncha* S. Müll. (= *fascialis* Wall.)
15. *Cuculus canorus canoroides* (S. Müll.) †
16. *Cacomantis virescens* Brügg. †
17. *Merops ornatus* Lath.
18. *Alcedo moluccana* (Less.)
*19. *Pelargopsis melanorhyncha eutreptorhyncha* Hart.†
*20. *Ceyx wallacei* Sharpe
21. *Halcyon coromanda rufa* (Wall.)
22. *Halcyon sancta* Vig. Horsf.
23. *Halcyon chloris* (Bodd.)
24. *Eurystomus orientalis* (L.)
25. *Macropteryx wallacei* (J.Gd.)
26. *Pitta erena* Temm.

*27. *Pitta dohertyi* Rothsch. (Bull. B. O. C. 1898, p. XXXIII)
28. *Hirundo javanica* Sparrm.
*29. *Hypothymis puella blasii* Hart. †
30. *Monarcha inornatus* (Garn.)
31. *Pachycephala grisenota* G. R. Gray (= *lineolata* Wall.)
32. *Pachycephala olio* Wall.
*33. *Rhinomyias colonus* Hart. †
34. *Graucalus melanops* (Lath.) †
35. *Graucalus schistaceus* (Sharpe)
36. *Graucalus temmincki* (S. Müll.)¹)
37. *Edoliisoma obiense* Salvad.
38. *Lalage leucopygialis* Wald. †
39. *Artamus monachus* Bp.
40. *Dicrurus pectoralis* Wall.
41. *Dicaeum sulaense* Sharpe
42. *Cyrtostomus frenatus* (S. Müll.)
43. *Hermotimia auriceps* (G. R. Gray)
44. *Anthreptes malaccensis celebensis* (Shell.)
45. *Zosterops subatrifrons* M. & Wg. (Sula Mangoli: Doherty) †
46. *Iole longirostris* (Wall.)
47. *Phyllscopus borealis* (Blas.) †
48. *Munia molucca—propinqua*
49. *Calornis sulaensis* Sharpe
50. *Calornis metallica* (Temm.)

† Mr. Hartert has kindly sent us (April, 1898) an early copy of his "List of a Collection of Birds made in the Sula Islands by William Doherty" (Nov. Zool. vol. V, Nr. 2, May 1898); all species marked † are new additions.
¹) Accidentally omitted in the distribution of the species in the text.

51. *Baxileornis galeatus* A. B. M. †
*52. *Charitornis albertinae* Schl.
53. *Corvus enca* Horsf. (Celebesian race)
54. *Oriolus frontalis* Wall.
55. *Osmotreron wallacei* Salvad. (typical.
56. *Ptilopus chrysorrhous* (Salvad.)
*57. *Ptilopus mangoliensis* Rothsch. (Bull. B. O. C. 1898, p. XXXIV)
58. *Carpophaga paulina* (Bp.)
59. *Myristicivora luctuosa* (Temm.)
60. *Columba albigularis* (Bp.) †
61. *Turacoena manadensis* (Q. G.)
62. *Macropygia albicapilla* Bp. (typical)
63. *Chalcophaps indica* (L.)
*64. *Megapodius bernsteini* Schl.
*65. *Hypotaenidia sulcirostris* (Wall.)
66. *Rallina minahassa* Wall.
67. *Amaurornis moluccana* (Wall.) †
68. *Eseacus magnirostris* (Vieill.)
69. *Numenius variegatus* (Scop.)
70. *Herodias alba* (L.)
71. *Herodias garzetta* (L.) ¹)
72. *Querquedula circia* (L. †

Peling and Sula. The Peling or Banggai Archipelago and the Sula Group seem to have formed in comparatively recent times one large island. Although the island of Peling lies only about 12 miles from the coast of East Celebes, the majority of the characteristic Celebesian genera (*Microstictus, Pyrrhocentor, Phoenicophaes, Cranorrhinus, Rhabdotorrhinus, Monachalcyon, Cittura, Ceycopsis, Meropogon, Coracias, Myza, Malia, Cataponera, Enodes, Streptocitta, Megacephalon, Aramidopsis*) were not included in the only collection yet made upon the island, while the peculiar species or subspecies are generally the same as Sula forms. Fifteen species are identical with Sula forms, or are local races thereof, but only four agree with Celebes forms not known to occur in Sula; ten Sula species were not sent from Peling and Banggai (though some of them are pretty sure to occur there); while twenty-two Celebes species not known from Sula were contained in the above collection from the neighbouring mainland of Celebes, but not from Peling or Banggai (see, Abh. Mus. Dresd. 1896, Nr. 2, pp. 1—6).

List of the Birds of Saleyer Island.

1. *Spilornis rufipectus* J. Gd. ¹)
2. *Haliaetus leucogaster* (Gm.)
*3. *Pernis sp.*
4. *Cuculus canorus canoroides* (S. Müll.)
5. *Alcedo moluccana* (Less.) ¹)
6. *Alcedo ispida* L. ¹)
7. *Halcyon chloris* (Bodd.) ¹)
8. *Caprimulgus macrurus* Horsf. (typical)
9. *Macropteryx wallacei* (J. Gd.) ¹)
10. *Siphia banyumas* (Horsf.)
11. *Culicicapa helianthea* (Wall.)
12. *Gerygone flaveola* Cab. ¹)
13. *Pratincola caprata* (L.)
*14. *Pachycephala teijsmanni* Bütt.
15. *Pachycephala orpheus* Jard.
16. *Lalage timorensis* (S. Müll.)
17. *Artamus leucogaster* (Val.)
18. *Dicrurus leucops* Wall. (typical)
19. *Dicaeum splendidum* Bütt.
*20. *Cyrtostomus frenatus saleyrensis* Hart.
21. *Myzomela chloroptera* Tweedd.
22. *Zosterops intermedia* Wall.
23. ? *Malia grata* Schl. (typical)
24. *Phylloscopus borealis* (Blas.)
25. *Motacilla flava* L.
26. *Munia molucca* < *propinqua*
27. *Calornis minor* (Bp.) ¹)
28. *Osmotreron wallacei* Salvad. (typical)
29. *Ptilopus melanocephalus* (Forst.)
30. *Macropygia macassariensis* (Wall.)
31. *Geopelia striata* (L.)
32. *Megapodius duperreyi* Less. Garn.
33. *Amaurornis phoeniura* (Forst.)
34. *Charadrius fulvus* Gm.
35. *Strepsilas interpres* (L.)
36. *Himantopus leucocephalus* J. Gd.
37. *Totanus glottis* (L.)
38. *Totanus calidris* (L.)

¹) Accidentally omitted in the distribution of the species in the text.

39. *Actitis hypoleucos* (L.)
40. *Tringa ruficollis* Pall.
41. *Gallinago megala* Swinh.
42. *Bubulcus coromandus* (Bodd.)
43. *Butorides javanica* (Horsf.)
44. *Nettion gibberifrons* (S. Müll.)
45. *Sterna anaestheta* Scop.

List of the Birds of the Djampea Group.

1. *Urospizias torquatus* (Temm.)
2. *Accipiter virgatus gularis* (Temm. Schl.)
3. *Elanus hypoleucus* J.Gd.¹)
4. *Baza reinwardti* Müll. Schl.
5. *Tinnunculus moluccensis occidentalis* M. & Wg.¹)
6. *Pandion haliaetus leucocephalus* (J. Gd.)¹)
7. *Ninox scutulata japonica* (Temm. Schl.)
8. *Strix flammea* L. (typical)
*9. *Trichoglossus forsteni djampeanus* Hart.
*10. *Cacatua sulphurea djampeana* Hart.
*11. *Tanygnathus megalorhynchus — sumbensis*
12. *Cuculus canorus canoroides* (S. Müll.)
13. *Centrococcyx bengalensis* (Gm.)¹)
14. *Alcedo moluccana* (Less.)
15. *Halcyon chloris* (Bodd.)¹)
16. *Eurystomus orientalis* (L.)¹)
17. *Caprimulgus macrurus* Horsf. (typical)
18. *Collocalia francica* (Gm.)
19. *Collocalia esculenta* (L.)¹)
*20. *Pitta virginalis* Hart.
*21. *Siphia djampeana* Hart.
*22. *Siphia kalaoensis* Hart.
(*?)23. *Rhipidura celebensis* Bütt.
24. *Monarcha inornatus* (Garn.)
*25. *Monarcha everetti* Hart.
26. *Myiagra rufigula* Wall.
*27. *Pachycephala everetti* Hart.
*28. *Edoliisoma emancipata* Hart.
29. *Lalage timorensis* (S. Müll.)
30. *Artamus leucogaster* (Val.)
31. *Dicaeum splendidum* Bütt.
(*?)32. *Cyrtostomus teijsmanni* Bütt.
33. *Myzomela chloroptera* Tweedd.
34. *Zosterops intermedia* Wall.
35. *Phylloscopus borealis* (Blas.)
36. *Motacilla flava* L.
37. *Anthus gustavi* Swinh.
38. *Munia molucca < propinqua*
39. *Calornis minor* (Bp.)¹)
*40. *Oriolus boneratensis* M. & Wg.
*41. *Osmotreron wallacei pallidior* Hart.
42. *Ptilopus melanocephalus* (Forst.)
43. *Carpophaga rosacea* (Temm.)
44. *Carpophaga concinna* Wall.¹)
45. *Myristicivora bicolor* (Scop.)
46. *Macropygia macassariensis* (Wall.)
47. *Megapodius duperreyi* Less. Garn.
48. *Amaurornis cinerea* (Vieill.)
49. *Esacus magnirostris* (Vieill.)
50. *Nycticorax caledonicus* (Gm.)
51. *Phalacrocorax melanoleucus* (Vieill.)

Saleyer and the Djampea Group. It has been shown by Büttikofer and Hartert that these islands have many points of affinity with Timor and the other Lesser Sunda Islands, as well as with Celebes. On counting out the respective forms, and allowing for the nearest affinities of the peculiar species, it appears that there are 9 Celebesian forms and 14 Lesser Sundan forms in these islands. Of the nine Celebesian forms four are not known from the Djampea Group, but only from Saleyer, as likewise two or three of the Lesser Sundan forms, so that it appears that the Djampea Group has much stronger relations with the islands to the south than with Celebes.

List of the Birds of Togian Island.

1. *Spilornis rufipectus* J.Gd.¹)
2. *Haliastur indus — girrenera*¹)
3. *Trichoglossus ornatus* (L.)
4. *Prioniturus platurus* (Vieill.)
5. *Prioniturus flavicans* Cass.
6. *Tanygnathus muelleri* (Müll. Schl.) (typical)
*7. *Loriculus quadricolor* Tweedd.
8. *Microstictus fulvus* (Q. G.)
9. ? *Cacomantis merulinus* (Scop.)
10. *Eudynamis melanorhyncha* S. Müll.

---

¹) Accidentally omitted in the distribution of the species in the text.

11. *Pyrrhocentor celebensis* (Q. G.) (typical)
12. *Phoenicophaes calorhynchus* Temm. (typical)
13. *Cranorrhinus cassidix* (Temm.)
14. *Merops ornatus* Lath.¹)
15. *Alcedo meninting* Horsf.
16. *Pelargopsis melanorhyncha* (Temm.¹
17. *Halcyon coromanda rufa* (Wall.)
18. *Halcyon sancta* Vig. Horsf.
19. *Halcyon chloris* (Bodd.) (typical)
20. *Macropteryx wallacei* (J. Gd.)
21. *Pitta celebensis* Müll. Schl.
22. *Hirundo rustica gutturalis* (Scop.) and *H. rustica — gutturalis*
23. *Hirundo javanica* Sparrm.
24. *Hypothymis puella* (Müll. Schl.)
25. *Graucalus bicolor* (Temm.)
26. *Graucalus leucopygius* Bp.
27. *Edoliisoma morio* (S. Müll.)¹)
28. *Artamus leucogaster* (Val.)
29. *Dicrurus leucops* Wall. (typical)
30. *Dicaeum celebicum* S. Müll.
31. *Cyrtostomus frenatus > sulyercuris*
32. ? *Hermotimia porphyrolaema scapulata* M. & Wg.
33. *Anthreptes malaccensis celebensis* (Shell.)

*34. *Iole aurea* (Tweedd.)
35. *Cisticola exilis* (Vig. Horsf.)
36. *Munia formosana brunneiceps* (Wald.)
37. *Calornis panayensis* (Scop.) (typical)
38. *Scissirostrum dubium* (Lath.)
39. *Streptocitta torquata* (Temm.)
40. *Corvus enca* Horsf. (Celebesian race)
41. *Oriolus celebensis* (Tweedd.) (typical)
42. *Osmotreron wallacei* Salvad. (typical)
43. *Ptilopus melanospilus* (Salvad.)
*44. *Carpophaga pulchella* Tweedd.
45. *Turacoena manadensis* (Q. G.)
46. *Macropygia albicapilla* (Bp.) (typical)
47. *Chalcophaps indica* (L.)
48. *Gallus ferrugineus* (Gm.)
49. *Megapodius cumingi* Dillw.
50. *Charadrius fulvus* (Gm.)
51. *Totanus glareola* (L.)¹)
52. *Actitis hypoleucos* (L.)
53. *Dissoura episcopus* (Bodd.)
54. *Phoyx manillensis* (Meyen)¹)
55. *Herodias garzetta* (L.)
56. *Hydrochelidon hybrida* (Pall.)
57. *Sterna media* Horsf.¹)

Only two species, *Loriculus quadricolor* and *Iole aurea*, are known to be peculiar to this group of islands in the Gulf of Tomini; for a third form, *Carpophaga pulchella*, which has been separated, is probably the same as *C. paulina*.

Peculiar species and subspecies of Celebes (Mainland).

1. *Spilornis rufipectus* J. Gd. (typical)
2. *Astur griseiceps* Schl.
3. ? *Astur tenuirostris* Brügg.
4. *Spilospizias trinotatus* (Bp.) (typical)
   Spilospizias trinotatus haesitandus Hart.
5. *Accipiter rhodogaster* Schl.
6. *Ninox ochracea* (Schl.)
7. *Cephaloptynx punctulata* (Q. G.)
8. *Strix inexpectata* Schl.
9. *Trichoglossus meyeri* Tweedd. (typical)
   Trichoglossus meyeri bonthainensis (A. B. M.)
10. *Cacatua sulphurea* (Gm.) (typical)
11. *Loriculus exilis* Schl.
12. *Loriculus stigmatus* (Müll. Schl.)
13. *Iyngipicus temmincki* (Malh.)
14. *Microstictus fulvus* (Q. G.) ǂ
15. *Microstictus wallacei* (Tweedd.)
16. *Hierococcyx crassirostris* Tweedd.
17. *Pyrrhocentor celebensis* (Q. G.) (typical) ǂ
    Pyrrhocentor celebensis rufescens M. & Wg.
18. *Phoenicophaes calorhynchus* Temm. (typical) ǂ
    Phoenicophaes calorhynchus meridionalis M. & Wg.
19. *Rhabdotorrhinus exaratus* (Temm.)

20. *Cranorrhinus cassidix* (Temm.) ǂ
21. *Meropogon forsteni* Bp.
22. *Ceycopsis fallax* (Schl.)
23. *Monachalcyon monachus* (Bp.) (typical)
    Monachalcyon monachus intermedius Hart.
24. *Monachalcyon capucinus* M. & Wg.
25. *Monachalcyon princeps* Rchb.
26. *Cittura cyanotis* (Temm.)
27. *Coracias temmincki* (Vieill.)
28. *Caprimulgus celebensis* Grant
29. *Lyncornis macropterus* Bp.
30. *Pitta forsteni* Bp.
31. *Pitta celebensis* Müll. Schl. ǂ
32. *Siphia rufigula* (Wall.)
33. *Siphia bonthaina* Hart.
34. *Stoparola septentrionalis* Bütt.
35. *Stoparola meridionalis* Bütt.
36. *Rhipidura teijsmanni* Bütt.
37. *Gerygone flaveola* Cab.
38. *Pachycephala sulfuriventer* (Tweedd.)
39. *Pachycephala meridionalis* Bütt.
40. *Pachycephala bonthaina* M. & Wg.
41. *Pachycephala boweri* M. & Wg.

¹) Accidentally omitted in the distribution of the species in the text.
ǂ Occurs also on Togian Island.

42. *Graucalus bicolor* (Temm.) †
43. *Graucalus leucopygius* Bp. †
44. *Edoliisoma morio* (S. Müll.) (typical)
    Edoliisoma morio septentrionalis M. & Wg.
    Edoliisoma morio — septentrionalis M. & Wg.
45. *Dicaeum celebicum* (S. Müll.) †
46. *Dicaeum nehrkorni* W. Blas.
47. *Dicaeum hosei* Sharpe
48. *Acmonorhynchus aureolimbatus* (Wall.)
49. *Anthopyga flavostriata* (Wall.)
50. *Cyrtostomus frenatus* > *saleyerensis* †
    Cyrtostomus frenatus < *saleyerensis*
    Cyrtostomus frenatus dissentiens (Hart.)
51. *Hermotimia porphyrolaema* (Wall.) (typical)
    Hermotimia porphyrolaema scapulata M. & Wg.
52. *Hermotimia grayi* (Wall.)
53. *Melilestes celebensis* M. & Wg. (typical)
    Melilestes celebensis meridionalis M. & Wg.
54. *Myza sarasinorum* M. & Wg.
55. *Zosterops squamiceps* (Hart.)
56. *Zosterops atrifrons* Wall.
57. *Zosterops sarasinorum* M. & Wg.
58. *Zosterops anomala* M. & Wg.
59. *Malia grata* Schl. (typical)
    Malia grata recondita M. & Wg.
60. *Androphilus castaneus* (Bütt.)
61. *Cataponera turdoides* Hart.
62. *Trichostoma celebensis* (Strickl.)
63. *Trichostoma finschi* Tweedd.
64. *Geocichla erythronota* Sol.
65. *Merula celebensis* Bütt.
66. *Phyllergates riedeli* M. & Wg.
67. *Cryptolopha sarasinorum* M. & Wg.
68. *Munia subcastanea* Hart.
69. *Enodes erythrophrys* (Temm.)
70. *Acridotheres cinereus* Bp.
71. *Basileornis celebensis* G. R. Gray.
72. *Streptocitta albicollis* (Vieill.)
73. *Streptocitta torquata* (Temm.) †
74. *Gazzola typica* Bp.
75. *Oriolus celebensis* (Tweedd.) (typical) †
    Oriolus celebensis meridionalis Hart.
    Oriolus celebensis — meridionalis
76. *Ptilopus fischeri* (Brügg.)
77. *Ptilopus meridionalis* M. & Wg.
78. *Ptilopus gularis* (Q. G.)
79. *Ptilopus melanospilus* (Salvad.)
80. *Carpophaga radiata* (Q. G.)
81. *Carpophaga forsteni* (Bp.)
82. *Carpophaga poecilorrhoa* Brügg.
83. *Phlogoenas tristigmata* Bp.
84. *Phlogoenas bimaculata* Salvad.
85. *Megacephalon maleo* (Hartl.)
86. *Turnix rufilatus* Wall.
87. *Gymnocrex rosenbergi* (Schl.)
88. *Aramidopsis plateni* (W. Blas.)
89. *Hypotaenidia celebensis* (Q. G.)
90. *Amaurornis isabellina* (Schl.)

Contrast between North and South Celebes. Almost all links between the Lesser Sunda Islands and Celebes occur in the Southern Peninsula, but not always in the North. This is shown by *Butastur liventer*, *Chrysococcyx malayanus* and *basalis*, *Muscicapula westermanni*, *Lalage timorensis*, *Zosterops intermedia*, *Munia pallida* and *M. punctulata nisoria*, *Calornis minor*, *Macropygia macassariensis*, *Geopelia striata*, which are not known from the North. Up to the present *Loriculus exilis*, though allied to *L. flosculus* of Flores, is only known from the North, but we anticipate its discovery, or of a race of it, in the South.

On the other hand there are as well as *L. exilis* Schl. several peculiar species which occur in the North and not in the South, for instance, *Ninox ochracea* (Schl.), *Strix inexpectata* Schl., *Meropogon forsteni* Bp., *Lyncornis macropterus* Bp., *Pitta forsteni* Bp., *Myza sarasinorum* M. & Wg., *Enodes erythrophrys* (Temm.), *Carpophaga poecilorrhoa* Brügg., *Gymnocrex rosenbergi* (Schl.), *Aramidopsis plateni* W. Blas., *Megacephalon maleo* (Hartl.) and others, besides such migratory species as only touch the North. It must, however, be taken into consideration that it is more than probable that at least some of these species will still be found in the South, which is much less thoroughly explored than the North, and the Centre is almost unknown. Conclusions, therefore, cannot be drawn from these data.

That there are a number of forms differing specifically or subspecifically in the North and South Peninsulas will appear from the pages of our book. These representative forms are:

North Peninsula	South Peninsula
Spilospizias trinotatus (typical)	S. trinotatus haesitandus
Trichoglossus meyeri (typical)	T. meyeri bonthainensis
Microstictus fulvus	M. wallacei
Phoenicophaes calorhynchus (typical)	P. calorhynchus meridionalis
Monachalcyon monachus	M. capucinus
Stoparola septentrionalis	S. meridionalis
Pachycephala sulfuriventer	P. meridionalis
Pachycephala bonensis	P. bonthaina
Edoliisoma morio septentrionalis	E. morio (typical)
Cyrtostomus frenatus > saleyerensis	C. frenatus < saleyerensis
Hermotimia grayi	H. porphyrolaema
Melilestes celebensis (typical)	M. celebensis meridionalis
Malia grata recondita	M. grata (typical)
Trichostoma celebense	T. finschi
Streptocitta torquata	S. albicollis
Oriolus celebensis (typical)	O. celebensis meridionalis
Ptilopus fischeri	P. meridionalis
Phlogoenas tristigmata	P. bimaculata

When it is remembered that the distance from the extreme ends of the North and South Peninsulas is between 800 and 900 miles and that the interior is in most parts very mountainous, the difference in the birds of the North and South need not cause surprise, since isolation, one of the essential conditions for the origin of a new species or subspecies, can occur here very readily.

There are differences of other kinds in other classes of animals in the North and South, but as the fauna of Celebes is so insufficiently known, the cases cannot yet be grouped together from a more general stand-point. We may mention, however, that Prof. v. Martens showed (in Weber's Zool. Ergebn. 1891 II, 259), that of the 64 land-shells known from Celebes only 2 are doubtless identical in the North and South, while 23 occur only in the North and 21 only in the South-west, etc. Among the Land Planarians collected by Dr. P. & Dr. F. Sarasin (Verh. D. Zool. Ges. 1897, 114), Prof v. Graff found that in North Celebes the Oriental, in South Celebes the Australian character prevails. It may be added that Prof. Wichmann (Tijdsch. K. Nederl. Aardr. Gen. 2. ser. 1890 VII, 978, and Petermann's Mitth. 1893, 281) surmises that during the second half of the Tertiary age single parts of South Celebes were raised as islands above the surface of the sea and only later, when the whole of it was upheaved, became united with Central Celebes as one land. Whether certain differences in the fauna of North and South Celebes may be explained hereby, we leave over to future decision. That, for instance, the Moleo of the North does not occur in the South is no zoological proof of former geological conditions, for as a rule animals have a restricted distribution.

The result of our study of the birds of Celebes, as well as of those of the countries around, is that by its Avifauna Celebes has far stronger connections with the Philippines than with any of the other neighbouring lands, and that the relation of its birds with the Oriental Region is more than twice as strong as with the Australian Region.

The line between Celebes and Borneo, though not that between Bali and Lombok, no doubt represents a conspicuous faunistic frontier, which remains unaltered even if the oldest continental frontier in earlier times was more to the east, but this line between Celebes and Borneo has not the fundamental significance which is still attributed to it by many writers. Even to-day the broad strait is nearly bridged over by shallows between South Celebes and Borneo (see map I). The line is not the western frontier of the Australian Region. The origin of the Celebesian Avifauna is principally an Asiatic one, but Celebes as a whole, or as a group of islands, was separated early from the continent, or never was intimately connected with it; its Avifauna, therefore, remained poor and must be pronounced an impoverished Asiatic one, but in consequence of isolation, peculiar forms were developed. The Papuan elements in it can be simply explained in view of the geographical position by the dispersal of birds through flight. This agrees very well with the results arrived at by Prof. Weber and Prof. v. Martens and others (see above pp. *85*, *87* et seq.).

The special faunas of Celebes, however, and of all islands of the East Indian Archipelago are far from worked out and we shall not live to see this. It will be the labour of a century and more. The future, therefore, only can decide, whether the ornithological facts as at present known teach us correctly that Celebes belongs to the Oriental Region and not to the Australian, and that it is most appropriate and safe to adopt a Transition-Zone between these two Regions, comprising a Celebesian Area, besides severally a Philippine, a Moluccan and a Lesser Sundan Area, of which the Celebesian has been treated of as to its Avifauna in our present essay.

After all we have not been able to discover anything very extraordinary about the birds of the island of Celebes. Its most striking feature is not that it has so many highly peculiar forms, but so extremely few. It has nothing among its birds to compare with a Dodo, or a Kiwi; it has not even a single peculiar avian family; only a few well marked peculiar genera, a large number of well characterized species belonging to genera not peculiar to the island, a still larger number of less well characterized species, local races or "subspecies", others which only very close observers believe they can discriminate; while the rest are by common consent termed absolutely identical with the individuals of their kind in the neighbouring lands. Islands like New Caledonia and Fiji have in proportion to their size quite as much that is peculiar about them, as has Celebes. The chief interest in the latter depends upon its intermediate position between Asia and Australia, the faunas of which are so vastly different.

SYSTEMATIC PART.

ORDER ACCIPITRES.

The diurnal Birds of Prey: Hawks, Eagles, Falcons, the Osprey, Vultures; distinguishable by the rigid, hooked, and sharply pointed bill; powerful, hooked claws; three toes in front and the fourth behind (except in the Osprey, in which the outer toe is reversible); usually of great powers of flight; 11 primaries; plumage often varied in coloration, but of sober tints: brown, black, white, grey, rufous, and purplish being found, but pure yellow, blue, red, bright green, and metallic tints are wanting. The *Accipitres* usually build a nest of sticks; the eggs are white or whitish in ground-colour, in many genera very handsomely varied with markings of rufous or brown; the young are hatched helpless and clothed with down.

FAMILY FALCONIDAE.

Containing all the Birds of Prey, except the Osprey (which is distinguished from them by its having the outer toe reversible and pterylologically by its having no aftershaft to the feathers) and the Vultures (which have the head and neck bare, or clothed in down).

GENUS SPILORNIS G. R. Gray.

Birds of Prey of medium size (about as large as a Raven), stout and compact in form, wings moderate; bill not denticulate; a broad nuchal crest; tarsi naked (except the upper fourth anteriorly), reticulated with hexagonal scales; toes short; under parts marked with transverse spots or bars of white; food: chiefly reptiles and amphibians; number of eggs laid: one or two. The genus contains about 12 species of a local and stationary character, distributed from the Himalayas and South China to the Andaman Islands, Celebes and Sula.

* 1. SPILORNIS RUFIPECTUS J. Gd.[1])

Russet-breasted Serpent-harrier.

Under this specific name we include two well-pronounced geographical races, *1. the typical Spilornis rufipectus* of the mainland of Celebes, *2. Spilornis rufipectus*

---

[1]) The abbreviation of the author's name at the head of each article is taken from the Berlin "Liste der Autoren" (1896), though the form adopted does not always meet with our approval.

sulaensis of the Sula group, and *3.* the individuals inhabiting the Peling group, which are intermediate between these two races. For the treatment of this, and of similar cases, the following method of nomenclature may be adopted without prejudice to ornithology.

+ 1. The typical Spilornis rufipectus.

a. **Spilornis rufipectus** *(1)* Gould, P. Z. S. 1857, 222; *(II)* id., B. Asia, I pl. IX (1860); *(3)* Wall., P. Z. S. 1862, 338, pt.; *(4)* id., Ibis 1868, 16, 21; *(5)* Wald., Tr. Z. S. 1872, VIII, 35; *(6)* Sharpe, Cat. B. I, 1874, 291; *(7)* Salvad., Ann. Mus. Civ. Gen. 1875, VII, 643; *(8)* Brügg., Abh. Ver. Bremen 1876, V, 46; *(9)* Gurney, Ibis 1878, 96, 102; *(10)* W. Blas., J. f. O. 1883, 135; *(11)* Gurney, List Diurn. B. of Prey 1884, 17; *(12)* W. Blas., J. f. O. 1885, 403; *(13)* id., Ztschr. ges. Orn. 1885, 222; *(14)* Guillem., P. Z. S. 1885, 544; *(15)* Hickson, Nat. in N. Celebes 1889, 89; *(XVI)* Meyer, Vogelskel. II 1892, 27, t. CLVII; *(17)* Bütt., Zool. Erg. Webers Reise 1893 III, 271; *(18)* Sharpe, Ibis 1893, 552; *(19)* M. & Wg., Abh. Mus. Dresd. 1895 no. 8, p. 3; *(20)* iid., ib. 1896 no. 1, p. 7; *(21)* iid., ib. 1896, no. 2, p. 7; *(22)* Hartert, Nov. Zool. 1896, 161; *(23)* id., ib. 1897, 159.

b. **Circaetus bacha celebensis** *(1)* Schl., Mus. P.-B. Buteones 1862, 27; *(2)* Rosenb., Malay. Archip. 1878, 271.

c. **Circaetus rufipectus** *(1)* Schl., Valkv. Ned. Ind. 1866, 37, 72, pl. 23, figs. 1—3; *(2)* Gray, HL. 1869, I, 15; *(3)* Schl., Rev. Accip. 1873, 114.

"**Kokodschi**", Tjamba, S. Celebes, Platen *a 13.*
"**Berna**" (albescent young), Tjamba, Platen *a 13.*
"**Buliba-mohengo**", Gorontalo Distr., N. Celebes, v. Rosenberg *b 2.*
"**Kiokkiok**", near Manado, Nat. Coll. in Dresd. Mus.
"**Boina**", Balante, E. Celebes, Nat. Coll.
"**Sikep utang besar**", Lembeh Id., Nat. Coll.

Figures and descriptions. Gould *II. 1*; Schlegel *c I*; Meyer *XVI* (skeleton); Sharpe *6.*

Diagnosis of race. Wing relatively longer than in *S. rufipectus sulaensis* (see, table of measurements), remiges below greyish white, broadly tipped and barred with blackish. These bars coalesce on the secondaries and base of primaries, enclosing spots of white, mottled with brown.

Old female. Upper surface dark brown, glossed with purple; sides and top of head, crest, and throat black, ear-coverts washed with grey; hind neck dusky, the margins of the feathers here and there, and on the crest, fulvous brown; secondaries and some of the upper tail-coverts tipped with white; carpal edge spotted white, tipped with whitish, and crossed with four broad black bands — the basal one rather indistinct; breast mummy-brown; remaining under-parts — including under wing- and tail-coverts — darker, the lower breast spotted, the sides, abdomen, thighs and under tail-coverts closely barred with white; under side of wing broadly barred and spotted with white (Manado, Nr. 6682).

Younger female. Like the above, but the brown of the upper plumage paler and duller without so much purple gloss; hind neck pale brown, without (or with only a few) yellow-brown margins here and on the crest, the black feathers of which are more or less broken up with yellowish white; tail crossed with three black bands, the basal one indistinct; breast much paler and more of a dark wood-brown tone (S. Celebes: Platen — C 10707).

Male. Similar to the female, but the white bars of the under parts more sharply defined and extending further towards the breast (whereas in the female they form sooner into disconnected spots); the brown bars on the under tail-coverts narrower (♂ vix ad. S. Celebes — Platen, C 10708; 3 ♂♂ ad. & vix ad. Sarasin Coll., N. & S. Celebes).

"Iris gold-yellow; periocular skin and cere green-yellow; bill blue-black; feet lemon- (or gold-) yellow" in both sexes (Platen *a 13*).

Female in albescent immature plumage. Head, crest and neck fulvous white with dark brown shaft-streaks; the upper parts display a varied plumage of sepia and fulvous brown, the feathers in general having dark centres and pale bases and margins; primaries and secondaries tipped with white; tail pale brown above, white below, and crossed with 5 to 6 indistinct dark bands, tip white; whole of under surface buffy white, streaked from the breast downwards with dark brown, which often spreads out in a washy manner in lighter brown over much of the feather. In this specimen the cross-barred feathers of maturity are sprouting at the flanks (S. Celebes — Nr. 6683).

A male in albescent plumage recently obtained by the Drs. Sarasin at Kema, Aug. 5th, 1893, corresponds with the above description of the female; ear-coverts and subocular region black; under surface purer white, with fewer and smaller brown streaks, here and on the upper surface not showing a general wash of rufous apparent on comparison in the other specimen. "Iris yellow; legs and feet yellowish grey; bill black, at the base blue" (Sarasin). A second male is much more rufo-fulvous in general tint than the other (Macassar, 12. IX. 95: Sarasin Coll.).

A specimen in the Leyden Museum (N. Celebes — Faber, 1883) is half in albescent plumage, half in adult. Wings and tail as in albescent specimens; under wing-coverts white, some tipped with rufous brown; back, breast, abdomen and thighs much as in ♀ ad.

First plumage. The full-fledged young of this species is not known, but in the cases of *Spilornis bacha* (Java), *S. cheela* (India) and *S. spilogaster* (Ceylon), young birds of each in the first stage of dress have been described or figured (Schlegel, Valkv. N. I. pl. 22, f. 3; Sharpe, Cat. B. I, 287; Legge, B. Ceylon 1880, 62; Bernstein, J. f. O. 1860, 425), from which it will be seen that the first plumage often — perhaps always — much resembles that of full maturity. Whether young birds always assume this mature-looking dress on first leaving the nest, and then lose it and put on the immature albescent plumage, and finally recover the adult type of coloration, or, whether the members of the genus are dimorphous when young — both mature-plumaged and albescent individuals existing from the nest, — are questions upon which opinion is divided, and facts, unfortunately, are as yet insufficient to allow of their being answered. The albescent type of immature plumage probably occurs in all species of the genus. Specimens in this dress are figured by Schlegel in the cases of *S. bacha*, *S. rufipectus* and *S. sulaensis* (b I; c I; Schl., Valkv. pl. 22, f. 3); similar inmature birds of *S. davisoni* Hume, *S. rutherfordi* Swinh. and *S. cheela* Lath. have been described (Hume, Str. F. II, 148; Bingham, ib. IX, 144; Oates, B. Brit. Burmah, II, 194; Sharpe, Cat. B. I, 287), and there is a ♀ specimen of *S. holospilus* Vigors from Mindanao in this plumage in the Dresden Museum (Nr. 13822). Gurney considered this to be the second plumage *(8)*, Schlegel and Colonel Legge express the opinion that it is a more or less frequent variation of dress assumed in the nest itself. Longitudinal streaks on the feathers of birds appear to represent a more original, less highly differentiated plumage than do cross-bars; this is shown by the fact that almost all birds of prey, which when adult acquire a cross-barred under-side, have this region streaked or drop-marked when immature.

If Gurney's view, that the pale, streaked specimens of *Spilornis* are in the second plumage, be correct, the curious case would be seen of a species regularly reverting from a higher stage of dress to a lower one and, subsequently, re-acquiring the more highly differentiated coloration.

As pointing to the probability that both albescent, streaked individuals and also dark, spotted ones exist from the nest in the case of *Spilornis*, Colonel Legge points out that of the Booted Eagle *(Nisaetus pennatus)* both dark and light young ones have been taken out of the same nest; but the case is not strictly a parallel to that of *Spilornis*, inasmuch as *N. pennatus* has two different phases of adult dress, a light and a dark one, and, when dimorphous pairs of young ones have been found, they are said to be sprung from a light male and dark female, or vice-versa (B. Ceylon, 62).

The albescent plumage is found in both sexes.

Skeleton. Length of cranium . . . . . . 77.5 mm | Length of fibula . . . . 84.0 mm
Greatest breadth of cranium 42.0 » | » » tarso-metatarsus . . . 76.0 »
Length of humerus . . . . . 97.0 » | » » sternum . . . . 61.0 »
» » ulna . . . . 110.0 » | Greatest breadth of sternum . . . 35.5 »
» » radius . . . . 105.0 » | Height of crista sterni . . . . . 13.5 »
» » manus . . . . 87.0 » | Length of pelvis . . . . 72.0 »
» » femur . . . . 65.0 » | Greatest breadth of pelvis . . . 31.5 »
» » tibia . . . . 105.5 »
(Siao, Sangi in Mus. Berol. XVII.)

Nidification. Unknown.

Distribution. Celebes. South Peninsula (Wallace *a 1*, Guillemard *a 14*, Platen *a 13*, Weber *a 17, etc.*); Central Celebes — Luwu Distr. (Weber *a 17*, P. & F. Sarasin *a 20*); S. E. Peninsula — Kendari (Beccari *a 7*); E. Peninsula (Nat. Coll. *a 21*); N. Peninsula (Forster *b 1*, v. Duivenbode *c 3, etc.*); Talissi Id. (Hickson *a 15*); Lembeh Id. (Nat. Coll. in Dresd. Mus.); Siao — known only from skeleton (Meyer *a XVI*).

+2. Spilornis rufipectus sulaensis (Schl.).

d. Circaetus sulaensis (1) Schl., Valkv. Ned. Ind. 1866, 38, 72, pl. 23, figs. 4—6; *(2)* Gray, HL. 1869 I, 15.

e. Spilornis sulaensis (1) Wall. Ibis 1868, 16; *(2)* Sharpe, Cat. B. 1874, I, 292; *(3)* Gurney, Ibis 1878, 102; *(4)* id. Diurn. B. of Prey 1884, 17; *(5)* Sharpe, Ibis 1893, 552.

f. Circaetus rufipectus sulaensis (1) Schl., Mus. P.-B. Rev. Accip. 1873, 114.

Figures and descriptions. Schlegel *d I*; Sharpe *e 2*.

Diagnosis. Wing relatively shorter than in the *typical S. rufipectus* (see table); under-side of quills greyish white, passing into blackish at the distal ends, and crossed by three or four well-marked bars of blackish, much narrower than those of the typical form. These bars do not coalesce on the basal half of the quills in the same manner as in that form, but pass separately across the wing.

Distribution. Sula Islands, Sula Besi and Sula Mangoli (Allen *d 1*, Bernstein & Hoedt *c I, e 1*).

In the Leyden Museum are seven specimens — 3 ♂ ad., 3 ♀ ad. and 1 ♀ juv. albescent — from Sula. The males have the breast paler; on the lower breast and abdomen the white bars are broader and the brown ones narrower, and the barring on the lower breast is better defined than in the other sex; under tail-coverts white (in one specimen slightly barred towards the tip). The females of Sula have the under tail-coverts barred and in regard to the barring of the under surface would appear to resemble the males of Celebes, but there are only one or two males with the sex satisfactorily ascertained in the Leyden Museum for comparison. The female of Celebes is more spotted below.

3. Spilornis rufipectus < sulaensis.[1])

g. **Spilornis sulaensis** *(1)* M. & Wg., Abh. Mus. Dresden 1896 no. 2, p. 7.

"Alaji Kabut", Peling; "Alaji", Banggai, Nat. Coll.

Diagnosis. Intermediate between the *typical S. rufipectus* and *S. rufipectus sulaensis*, but on the whole more like the latter race.

Distribution. Peling group between Sula and E. Celebes: — Peling and Banggai (Nat. Coll.).

Measurements.	Wing	Tail	Tarsus	Culmen from cere
(Nr. 6682) [♀] North Celebes	370-5	250	79	28
(Nr. 2190) [♀] North Celebes	340-4	250	74	26
(*C* 846) [♀] N. Celebes, March 1871 (Meyer).	360	240	75	—
(*C* 10844) [♀] N. Celebes, Aug.—Sept. 92 (Nat. Coll.). .	344	230	74	27
(*C* 10845) [♀] N. Celebes, Aug.—Sept. 92 (Nat. Coll.). .	353	236	74	27.5
(Sarasin Coll.) ♂, N. Celebes, 20. Oct. 93 (P. & F. S.) .	340	230	78	25.5
(Sarasin Coll.) ♂, N. Celebes, 6. Oct. 94 (P. & F. S.) .	342	242	78	27
(Sarasin Coll.) ♂ juv., N. Celebes, 5. Aug. 93 (P. & F. S.)	337	230	73	27
(*C* 14247) [♂] Lembeh Id. March 95 (Nat. Coll.). . . .	337	215	73	25
(*C* 10708) ♂, South Celebes, 6. May 78 (Platen) . . .	330	215	75	28
(*C* 10707) ♂, South Celebes, 12. March 78 (Platen) . .	325	220	73	27.5
(Schlüter Coll.) ♀, South Celebes (Platen)	344	229	72	—
(Sarasin Coll.) ♂, South Celebes, 21. Sept. 95 (P. & F. S.)	322	223	74	29
(Sarasin Coll.) ♂ juv., S. Celebes, 12. Sept. 95 (P. & F. S.)	332	232	70	26
(Sarasin Coll.) ♀, Centr. C., Palopo, 21. Jan. 95 (P. & F. S.)	328	—	72	28
(*C* 14295) [♀], East Celebes, May-Aug. 95 (Nat. Coll.). .	363	236	78	—
(Norwich Mus.) ad. Macassar (Wallace). .	Communicated by Mr. Gurney:	345	—	71	—
(Norwich Mus.) ♂ ad. Macassar (Wallace)		345	—	71	—
(Norwich Mus.) ♂ ad. Macassar (Wallace)		338	—	69	—
(Norwich Mus.) ad. Macassar (Wallace) .		345	—	76	—
(Norwich Mus.) imm. N. Celebes (Meyer) .		345	—	71	—
(Norwich Mus.) vix ad. N. Celebes (Whitely)		355	—	76	—
5 adults, Peling Id., V—VIII. 95 (Nat. Coll.)	312-38	—	—	—
4 adults, 2 juv. Banggai Id., VI. 95 (Nat. Coll.). . . .	305-25	—	—	—
(Norwich Mus.) ♂, Sula Islands	314	221	73	27
(Norwich Mus.) ♀ juv., Sula Islands	305	220	72	26

[1]) As will be found more fully explained under the heading: *Haliastur indus*, a long connecting hyphen (*s. g. S. rufipectus — sulaensis*) is used in this work to indicate the connecting forms between subspecies; or, where it may be safely said that such individuals have stronger leanings to one subspecies than to the other, the sign > or <, respectively (*S. rufipectus > sulaensis*) indicating that the affinities are more with the typical form, or (*S. rufipectus < sulaensis*) with the latter form.

The fine series of this Serpent-harrier before us displays great variation in tint; on the breast, for instance, from pale russet to dark Vandyck-brown. The darkest examples are probably the oldest. The specimens from North Celebes are, with one exception, darker than those of the southern peninsula, and Mr. Gurney writes that the specimen from Kema (N. Celebes) in the Norwich Museum "is decidedly darker than our four from Macassar, especially on the breast". Two in the Leyden Museum from Pare-Pare near Macassar do not appear to differ from those of the north, and Mr. Hartert remarks *(a 22)* of some fresh specimens from the southern peninsula in the Tring Museum that the breast is much paler in some examples, darker in others. None of the southern examples before us appear to be old birds. The birds from Peling and Banggai vary in the same way, — one is Vandyck-brown on the breast, most of the others much paler.

Outside the Celebes Province a very near ally of *S. rufipectus* is found in Dr. Sharpe's recently described *Spilornis raja* (Bull. B. O. C. 1893, no. X; Ibis 1893, 552, 569), a specimen of which was sent by Mr. Edward Bartlett from Kuching, Sarawak. This is said to be most like the Sula race, but differs in having the white bars of the breast, abdomen and axillaries strikingly narrower. From Sharpe's measurements it would appear to have the wing of the Sula form (viz. 309 mm), but a shorter tail (178 mm).

The food of Serpent-harriers consists chiefly of frogs, snakes and other Reptilia. The Indian species lay one or two eggs; most usually only one.

The genus *Spilornis* is an important one in questions of geographical distribution. Owing to the small number of eggs laid, the species do not suffer from overcrowding; and there appear to be no causes for its ever shifting its quarters. The genus is a purely Indian one, and is not found further east than the Sula Islands; and the close connection of these islands with Celebes is shown by the fact of their possessing the same species, which also — if identical — occurs on Siao in the Sangi Islands.

Attention might here be called to the close similarity of the plumage of *Circus assimilis* Jard. & Selby to certain species of this genus, especially *Spilornis holospilus* Vig. of the Philippines.[1]) Mimicry is here out of the question, as *C. assimilis* does not occur in the Philippines, and the similarity must be taken as pointing to kinship of the two genera (see below under *Circus assimilis* and *Pernis celebensis*).

GENUS CIRCUS Lac.

The Harriers are of slender form, with very long wings, tail, and tarsus, and somewhat short and not powerful toes. A more or less well developed facial ruff extending from behind the ear-coverts across the throat:

---

[1]) Cf. also *Spilornis panayensis* Steere (List B. Philip. Is. 1890 p. 7).

bill somewhat weak, with a blunt festoon: tarsi naked (except at the top anteriorly), clad in front with transverse shields, elsewhere with small reticulate scales. Food: amphibians, reptiles, small mammals, etc. Eggs 2—4 in number. About 18 species, migratory and stationary, distributed over the greater part of the world.

+ 2. CIRCUS ASSIMILIS Jard. Selby.
Allied Harrier.

Circus assimilis *(1)* Jard. & Selby, Ill. Orn. 1826, I, pl. 51 type examd.[1]); *(2)* Schl., Mus. P.-B. Circi, 1862, 9; *(III)* id., Valkv. 1866, 29, 66, pl. 20, f. 2, 3; *(4)* Walden, Tr. Z. S. 1872, VIII, 37; *(5)* Sharpe, Cat. B. 1874, I, 63; *(6)* Gurney, Ibis, 1875, 225; *(7)* id., Diurn. B. of Prey, 1884, 23; *(8)* W. Blas., Ztschr. Ges. Orn. 1885, 205, 234; *(9)* North, Nests & Eggs B. Austr. 1889, 1, pl. II, f. 4 (egg); *(10)* Büttik., Webers Reise in Ost-Ind. 1893 III, 272; *(11)* M. & Wg., Abh. Mus. Dresden 1896 no. 1, p. 7; *(12)* Hartert, Nov. Zool. 1896, 163.

a. **Circus jardinii** Gld., P. Z. S. 1837, 141; *(1)* id., B. Austr. 1848, I, pl. 27; *(2)* S. Müll., Reizen Ind. Arch. 1857, II, 8; *(3)* Gld., Handb. B. Austr. 1865, I, 60; *(4)* Schl., Rev. Acc. 1873, 50.

b. **Spilocircus jardinii** *(1)* Kaup, Isis, 1847, 102.

c. **Strigiceps jardinii** *(1)* Bp., Consp. 1850, I, 34.

"Bokan buri", S. Celebes, Platen 8.

For further references see Sharpe *5*.

Figures and descriptions. Jardine & Selby *I*; Gould *a I*, *a 3*; Schlegel *III*; North *9* (egg); Kaup *b I*; Sharpe *5*; W. Blas. *8*.

Male, nearly adult. General colour above brownish ash, darker on head; forehead, ear-coverts and crown with rufous margins to the feathers; secondaries pure ashy, banded with dark brown — indistinctly on the inner web; wing-coverts, scapulars and upper tail-coverts marked with short bars or large spots of white, which are more indistinct and ashy on exposed parts of the plumage; shoulder rufous; tail above ashy, below white, crossed with seven bars of blackish and terminally margined with white; under surface — including under wing- and tail-coverts and thighs — cinnamon-rufous, lighter on the thighs and abdomen, and spangled all over with white spots arranged two and two at short intervals on the opposite webs of the feathers. "Iris sulphur-yellow; cere and bill bluish grey (cere pale yellow — Wallace); tip of bill black; feet citron-yellow" (Platen). Nr. 6735, Tjamba, May).

Old. Crown of head, cheeks and ear-coverts tawny-rufous, with blackish mesial streaks to the feathers (♂, Lake Posso, 14. Feb. 95, P. & F. Sarasin).

Female. Like the male, but larger.

Young. Above brown with fulvous margins to the feathers; upper tail-coverts white washed with rufous and having dark brown centres; tail sepia-brown tipped with

---

[1]) The type of *Circus assimilis* J. & S. in the British Museum is immature and not normal, differing from all other specimens there of this species in the coloration of the wings and tail. The tail is nearly uniform brownish ashy with a rufous wash at its sides, marked with 3 or 4 imperfect bars of brown towards the base, followed by a clear space, with an imperfect terminal bar. Upper tail-coverts white, a few of the longer ones with a bar of brown towards the tip.

tawny-buff and crossed with six bands of black; below pale tawny-buff, lightest on abdomen and thighs, and streaked with dark brown on breast and under tail-coverts (ex Sharpe).

Eggs. 2 or 3, white, with a bluish green tinge on the inner surface; 51—52 × 38—39 mm (Australia — A. J. North *9*). Uniform white, a little smaller than those of *C. rufus* of Europe: 49—51 × 39—39 mm: Nehrkorn, MS.).

Nest. Flat, of small sticks and twigs, lined with green leaves; usually placed among the thick branches of a low tree (Australia).

Breeding-time in Australia. Sept.—Nov.

Measurements.

	Wing	Tail	Tarsus	Culmen from cere
a. (Sarasin Coll.) ♂ ad. Lake Posso, Central Celebes, 14. II. 95 (P. & F. S.)	375	235	84	20
b. (C 10709) ♂ vix ad. S. Celebes, 16. IV. 78 (Platen)	390	260	93	19
c. (C 10710) ♀ vix ad. S. Celebes, 25. V. 78 (Platen)	425	275	98	20.5
d. (Nr. 6735) ♂, S. Celebes, 6. V. 78 (Platen)	385	250	93	—
e. (C 11293) [♀] New South Wales	435	280	101	21
f. (C 11092) [♀] juv. Australia	437	280	107	22

Distribution. N. Celebes—Minahassa (Riedel *b 4*), Gorontalo District (Forsten *2*, Rosenberg *b 4*), Central Celebes (P. & F. Sarasin *11*), S. Celebes (S. Müller *2*, *b 2*, Weber *10*, Platen *8*, Everett *12*); Australia—apparently throughout (Ramsay *9*); Tasmania (Norwich Mus. *0*).

This well-marked Harrier is remarkable for its distribution, occurring, as it does, in Celebes and Australia and, so far as is yet known, on none of the intervening islands. To the north and west of Celebes, the Chinese *Circus spilonotus* Kaup is found in the Philippines and Borneo, and, according to Mr. Everett and Mr. Whitehead (J. Straits Br. R. A. S. 1889, 190; Ibis, 1890, 43), this species is "a regular winter migrant" to Borneo and Palawan. East of Celebes, a Harrier (*C. spilothorax*, Salvad. & D'Alb.) has been discovered at Yule Island in the Papuan Gulf, South New Guinea, corresponding closely in coloration with *C. maillardi* Verr. (*C. macrocelis* N̦ewton) of Madagascar and Réunion and with *C. wolfi* Gurney and *C. gouldi* Bp. of New Caledonia and Australia, respectively (Salvad., Orn. Pap. I, 71); but neither this species, nor *C. spilonotus*, have anything to do with *C. assimilis*.

The similarity of the adult plumage of this Harrier to that of certain species of *Spilornis* is worthy of notice. The type of coloration may be ancient. It should be remarked that Gurney, whose arrangement of the *Falconidae* we follow, places the subfamily *Circinae* much nearer to the *Circaetinae* (containing *Spilornis*) than has been the custom with most other authors (*7*).

C. assimilis is, we believe, a stationary species in Celebes, though we have only been able to collect the following few dates of occurrence there ranging from February to November:

Birds of Celebes: Falconidae.

a. Macassar	♀ juv.	March	(2)
b. Tjamba	♂ nearly ad.	April 16th	(7, C 10709)
c. Tjamba	♂ nearly ad.	May 6th	(Nr. 6735)
d. Tjamba	♀ nearly ad.	May 25th	(7, C 10710)
e. Gorontalo	♂ ad.	June 8th	(a 4)
f. Gorontalo	♂ ad.	July 16th	(a 4)
g. Ayer-Pannas	♀	August 18th	(a 4)
h. Bone	♀ ad.	Nov. 23rd	(a 4)
i. Bone	♀	Nov. 23rd	(a 4)
j. Lake Posso	♂ ad.	Febr. 14th	(11)

Mr. Everett *(12)* obtained six specimens between Sept. 16 and November.

If stationary, differences might be expected to have arisen between Celebesian and Australian birds; nevertheless the numerous specimens in the British and Leyden Museums appear to be quite similar.

When at Macassar, Salomon Müller observed that this Harrier "flies like the Harriers of the northern quarters of the globe, haunts open fields and meadows and settles in the same manner by preference silently and unnoticed upon the ground, or on a stone, or a low pole, or the like, where, for its own security and for watching for its prey, it gets a wide view" *(a 2)*.

GENUS ASTUR Lac.

The Goshawks range in size from medium to small (Kestrel-size); stout, compact birds; wings rather short, reaching to about the middle of the tail; tomia with a blunt festoon; tarsi stout, feathered on their upper half (except posteriorly), the lower half clad with transverse shields anteriorly, elsewhere irregularly scaled; toes stout, the middle one not exceptionally prolonged. Food: any land-vertebrates weaker than the bird itself. Eggs 2—4. The genus is found in most parts of the five continents and Australia, but authors are not agreed as to its systematic limits. There are migratory and stationary forms.

┬ * 3. ASTUR GRISEICEPS Schl.
Grey-headed Goshawk.

a. **Falco griseiceps** [Temm. in Mus. Leyden].

Astur griseiceps *(1)* Schl., Mus. P.-B. Astures, 1862, 23; *(II)* Wall., Ibis, 1864, 184, pl. V; *(3)* Finsch, Neu-Guinea 1865, 155; *(IV)* Schl., Valkv. 1866, 19, 58, pl. 11, f. 1, 2; *(5)* Wall., Ibis, 1868, 6, 20; Gray, HL. 1869, I, 30; *(6)* Schl., Rev. Acc. 1873, 66; *(7)* Sharpe, Cat. B. 1874, I, 106; *(8)* Brüggem., Abh. Ver. Bremen 1876, V, 43; *(9)* M. & Wg., Abh. Mus. Dresden 1895 Nr. 8, p. 3; *(10)* iid., ib. 1896, Nr. 1, pp. 4, 7; *(11)* Hart., Nov. Zool. 1897, 165.

b. **Lophospiza griseiceps** *(1)* Kaup, P. Z. S. 1867, 178; *(2)* Walden, Tr. Z. S. 1872, VIII, 33; *(3)* Lenz, J. f. O. 1877, 365; *(4)* Meyer, Ibis 1879, 55; *(5)* W. Blas., J. f. O. 1883, 134; *(6)* id., Ztschr. ges. Orn. 1885, 221.

c. **Lophospizias griseiceps** *(1)* Gurney, Ibis 1875, 355; *(2)* Salvad., Orn. Pap. 1880, I, 67; Gurney, Diurn. B. of Prey 1884, 29.

"**Rurunbalu**", Tjamba, S. Celebes, Platen b 4.
"**Sikep burik sedang**", juv. Manado, Nat. Coll.
Figures and descriptions. Wall. *II*; Schl. *IV, 1, 6*; Sharpe 7; Brüggem. 8; W. Blas. b 6.

Adult male. Head, neck, cheeks and ear-coverts blue-grey, shafts of the feathers dark; rest of upper surface, including primaries, warm (Prout's) brown with a purple gloss in certain lights; upper tail-coverts washed with slaty; tail crossed with four bands of blackish, the basal one concealed and imperfect; chin and throat white, bounded on either side by a malar streak of slaty black and marked down the middle with a stripe of the same colour; underparts white, uniform on the lower abdomen and under tail-coverts, but marked on breast, sides and abdomen with broad longitudinal streaks of blackish brown, and closely barred on thighs and flanks with blackish; under wing-coverts white; quills below white, at distal ends brown, and imperfectly crossed with about four narrow bars of dusky (♂, Rurukan, N. Celebes: April 1894: P. & F. Sarasin).

Adult female. Much like the male, but larger; the head browner grey, upper surface paler brown; the white of the under surface more tinged with buff, and the longitudinal stripes thereon warm dark brown (Prout's); the dusky bars on the thighs and flanks broader; five bars (instead of four) generally distinguishable on the tail — the two basal ones more or less obliterated and concealed; under wing-coverts white, with a few brown spots on them — sometimes also seen in the male (♀ ad., Kema, N. Celebes, 30th Aug. 1893: P. & F. Sarasin).

"Iris gold-yellow; feet yellow; cere greenish yellow; bill grey-blue" (♀ — Sarasin).

Young. Above warm dark brown, with whitish terminal edges to the feathers, the middle and lateral parts of the feathers more buff-brown; head above blacker brown, with whitish tips; face and under surface white, sometimes almost unstriped with brown, usually striped with dark brown, chiefly along the mesial part of the feathers of the breast, and stained (as it appears) with a weak solution of the same brown chiefly on the face and breast; the thighs with dark transverse spots and bars; 5—6 dark bars on the tail, basally indistinct (♀ juv., Lake Matanna, S. E. Central Celebes, 27. II. 96: P. & F. Sarasin; and other examples).

"Iris yellow; cere greenish; legs green-yellow, on the shins dark" (P. & F. S.).

Measurements.	Wing	Tail	Tarsus	Culmen from cere
a. (Nr. 6701) ad. Minahassa, N. Cel. (v. Musschenbroek)	206	180	56	—
b. (Nr. 6703) [♂] ad. Gorontalo, N. Cel. (Riedel) . . .	183	157	49	—
c. (Sarasin Coll.) ♀ ad. Kema, N. Cel., 30. VIII. 93 .	200	160	54	18
d. (Sarasin Coll.) ♀ ad. Tomohon, N. Cel., 2. VIII. 94	204	170	—	17.5
e. (Sarasin Coll.) ♂ ad. Rurukan, N. Cel., IV. 94 .	176	142	46	15
f. (Sarasin Coll.) ♂ ad. Lake Limbotto, N.Cel., 18.III.95	181	141	—	16.5
g. (Sarasin Coll.) ♂ ad. Lake Towuti, S.E. Cel., 3.III. 96	182	142	—	14.5
h. (Sarasin Coll.) "♂" [? ♀] j. Mapane, C. Cel.,3. III. 95	200	168	—	17.5
i. (Sarasin Coll.) ♂ j. Lake Matanna, C. Cel., 27. II. 96	195	161	—	17
j. (c 10847) juv. Minahassa, VIII-IX. 92 (Nat. Coll.) . .	203	164	54	18
k. (c 13454) juv. Minahassa, 15. VIII. 94 (Nat. Coll.) . .	200	165	—	16.5

In the case of the tarsus, on account of the difficulty of measuring this part in many birds of prey, the figures given should be taken only as approximate.

Distribution. Celebes. S. Peninsula (Wallace *5*, Platen *b 4*), S. E. Central Celebes, Lakes Matanna and Towuti (P. & F. Sarasin); Dongala, W. Celebes (Doherty *11*); N. Central Celebes, Mapane (P. & F. Sarasin); N. Peninsula, Minahassa (Wallace *5*; etc.), Gorontalo Distr. (Forsten *1*, etc. *6*, *8*).

Ceram was also put down as a locality for this species in 1865 by Dr. Finsch *(3)*, but it is now evident that this is an error.

This small Goshawk was discovered about fifty years ago by Forsten. Its range is not known to extend beyond the mainland of the island, where it seems to be not uncommon, though specimens are not abundant in European Museums. Professor W. Blasius *(b 3)* was able to examine 21 skins collected by van Duivenbode's hunters in North Celebes. A nice series of three adult males, two adult females, and two young specimens were obtained in North and Central Celebes by the cousins Sarasin, and these, combined with the specimens in the Dresden Museum, enable us to say with safety that the species does not vary locally to any appreciable degree, if at all, and also to point out the difference between the sexes.

Nothing has been recorded of the habits of this Goshawk. It is most nearly related to the next species, *Astur trivirgatus* (Temm.) which differs by having the breast rufous-brown, or striped with rufous-brown, as against the long blackish or blackish brown stripes in the present bird.

4. ASTUR TRIVIRGATUS (Temm.).

Crested Goshawk.

This species is composed of the two following subspecies:

✢ 1. The typical Astur trivirgatus (Temm.)

a. Falco trivirgatus *(1)* Temm., Pl. Col. 1824, no. 303.

b. Astur trivirgatus Cuv., Règne An. 1829, pt. I, 332; *(1)* Jerd., B. Ind. 1862, I, 47 *(2)* Schl., Mus. P.-B. Astures, 1862, 22; *(III)* id., Valkv. 1866, 18, 57, pl. 10; *(4)* Swinh., Ibis, 1866, 395; Wall., Ibis 1868, 6; Schl., Rev. Acc. 1873, 65; *(5)* Sharpe, Cat. B. I 1874, 105 pt.; *(6)* Salvad. Ucc. di Borneo, 1874, 17; *(7)* Sharpe, Ibis, 1876, 32; *(8)* Hume, Str. F. 1879, VIII, 43; *(9)* Legge, B. Ceylon, 1880, 20; *(10)* Nicholson, Ibis, 1882, 52; *(11)* W. Blas., Verh. z. b. Ges. Wien, 1883, 21; *(12)* Oates, B. Brit. Burmah, 1883, II, 177 pt.; *(13)* Guillem., P. Z. S. 1885, 545; *(14)* Sharpe, Ibis, 1888, 195; *(15)* W. Blas., Ornis, 1888, 303; *(16)* Sharpe, Ibis, 1889, 67; id., ib. 1890, 274, pt.; *(17)* Everett, J. Str. Br. R. A. S. 1889, 180; *(18)* Steere, Coll. B. Philip. Is. 1890, 7; *(19)* Oates, Hume's Nests & Eggs, 1890, III, 119, pt.; *(20)* Vorderman, N. T. Ned. Ind. 1890, 379; *(21)* Hose, Ibis, 1893, 418; *(22)* Hartert, Ornis 1891, 122; *(23)* Bourns & Worces., B. Menage Exped. 1894, 32; *(24)* Everett, Ibis 1895, 31.

12 Birds of Celebes: Falconidae.

c. **Lophospiza trivirgatus** Kaup, Contr. Orn. 1850, 65; *(1)* David et Oust., Ois. Chine, 1877, 22; *(2)* Bourdillon, Str. F. 1880, IX, 299; *(3)* W. Blas., J. f. O. 1890, 144; *(4)* Hartert, Nov. Zool. 1895, 476.
d. **Lophospiza indica** *(1)* Hume (nec Hodgs.) Str. F. 1875, III, 24.
e. **Lophospizias trivirgatus** *(1)* Gurney, Ibis, 1875, 355; *(2)* id., Str. F. 1877, V, 502; *(3)* id., Diurn. B. of Prey, 1884, 29; *(4)* Blanf., Faun. Br. India B. 1895 III, 401.
f. **Lophospiza trivirgata** *(1)* Hume, Str. F. 1877, 8, 124.
g. **Astur (Lophospizia) trivirgatus** *(1)* Hartert, J. f. O. 1889, 375.

Figures and descriptions. Temminck *a 1*; Schlegel *III*; Sharpe, *5, 16*; Gurney *d 1, d 2*; Hume, *e 1*; Legge *3*; Oates *12*; W. Blasius *11*; Blanford *e 4*; d'Alton, Skel. d. Raubv. 1838, Taf. VI figs. *n, o* (Skull); Selenka, Bronn's Kl. u. Ord. VI. Abth. IV, pl. III fig. 6 (wing-bones).

Young. Above warm dark (Prout's) brown, the feathers furnished with small rufescent tips; head and crest blackish; ear-coverts and neck more rufescent; upper tail-coverts crossed with a blackish band and broadly tipped with white; tail — not fully grown — crossed with three black bands; below white, on the flanks and thighs barred and elsewhere marked with streaks and drops of brown; a dark streak down the middle of the throat (Java — Nr. 5692).

Adult. The upper plumage — brown in the young — becomes slaty grey with maturity, palest on head, browner on wings; tail crossed with four dark bands; throat as in the young; chest tawny-rufous; rest of under surface white, barred with rufous and brown, except on the under tail-coverts which are uniform. Bill black; iris orange-yellow; feet yellow.

"Iris light brown; bill black; tarsus yellow" *(a 13)*.

Measurements.

	Wing	Tail	Tarsus	Culmen
(*a* Nr.-5692) juv. Java	180	130	56	22
(Sharpe *3*) ♂ ad. (Java?)	198	160	55	—
(Gurney *5*) ♀ ad. Java	228	—	58	—

Distribution. Indian Peninsula (Hume *f 1*, Gurney *e 1*); Ceylon (Hume *f 1*, Legge *b 9*); Pegu (Oates *b 12*); Formosa (Swinh. *b 4*); Philippines (Gurney *e 1*, Platen *c 3*, Steere *b 18*); Palawan (Whitehead *b 14*, Platen *b 15*); Borneo (S. Müll. *b 2*, etc. *b 6, b 17*); Sumatra (S. Müll. *b 2*, Forbes *b 10*); Java (Boie, Bernst. *b 2*); North Celebes (Guillem. *b 13*).

2. Astur trivirgatus rufitinctus (McClell.).

h. **Spizaetus rufitinctus** *(1)* McClell., P. Z. S. 1839, 153.
i. **Astur trivirgatus** pt. *(1)* Sharpe, Cat. B. 1874, I, 105; *(2)* Oates, B. Brit. Burmah, 1883, II, 177 pt.; *(3)* Sharpe, Ibis, 1890, 274 pt.
j. **Lophospizias indicus** (? Hodgs.), *(1)* Gurney, Ibis, 1875, 355.
k. **Lophospiza indica** *(1)* Hume, Str. F. 1877, V, 8.
l. **Lophospiza rufitinctus** *(1)* Hume, Str. F. 1877, 124; *(2)* Hume & Davison, ib. 1878, VI, 7.
m. **Lophospizias rufitinctus** *(1)* Gurney, Str. F. 1877, V, 502.
n. **Astur rufitinctus** *(1)* Hume, Str. F. 1879, VIII, 152.
o. **Lophospizias trivirgatus subsp. rufitinctus** *(1)* Gurney. Diurn. B. of Prey, 1884, 29.

Birds of Celebes: Falconidae.   13

Diagnosis. Like the typical *A. trivirgatus*, but larger and more rufescent. Apparently a highland race. (See: Gurney *j 1*, Hume *k 1*, *l 1*).

Distribution. Himalayas from Nepaul to Assam, Cachar, Sylhet, Tipperah (Hume *k 1*, *l 1*); Nagpore, Bengal (Gurney *j 1*); Tenasserim (Davison *l 2*); Malay Peninsula (Blyth, Hume *n 1*).

Dr. Sharpe includes in the smaller race a young specimen obtained by Mr. Wallace in the Malay Peninsula *(b 5)*; and it is possible — judging from the length of wing — that Swinhoe's specimen from Formosa, though placed by Gurney in the smaller race, approaches more nearly to the larger Himalayan birds *(b 4, e 1)*.

The right of this species to be included among the birds of Celebes rests upon a young specimen obtained at Likoupang, North Celebes, by Dr. Guillemard during the cruise of the "Marchesa".

In India two, or rather three, nests of this species have been found, from which it would appear that only two eggs are laid *(b 19)*. It is not surprising, therefore, to find a sufficient number of dates of occurrence in Borneo to show that the species is a resident there *(b 16, b 7, b 11)*. Its occurrence in Celebes is of interest, but, in consideration of the fact that this species is represented there already in a near relation, *A. griseiceps* Schl., it is more probable that it is a straggler from the north or west than a resident.

It differs from *A. griseiceps* by its brownish-rufous breast, or where the breast is white with stripes and spots thereon, by the brown-rufous tint of these markings, as against the blackish, or very dark brown, colour of the lengthy stripes of *A. griseiceps*; the claws also seem to be longer in the latter species.

The Crested Goshawk was found by Captain Legge to be very partial to lizards, while other authors note that it preys fiercely upon fowls, pigeons and other birds *(b 9)*.

For further references to literature on this species see: Sharpe *b 5*, Legge *b 9*, Oates *b 12*.

+ 5. ? ASTUR TENUIROSTRIS Brügg.

Astur tenuirostris *(1)* Brüggem., Abh. Ver. Bremen 1876, V, 43.
a. **Urospiza iogaster** *(1)* W. Blas. (? nec S. Müll.), J. f. O. 1883, 151.
b. **Urospizias hiogaster** pt.? *(1)* Gurney, Diurn. B. of Prey 1884, 36.

Brüggemann described an example from Celebes from v. Rosenberg, without exact locality, as follows:

Young ♂. Bill somewhat thin, elongated; the ridge scarcely curved downwards at the base, almost straight, and not bent markedly till near the point; the projecting hook only slightly developed; edge of upper mandible along the gape straight, somewhat turned inwards about the middle, without any signs of a tooth, and with only a trace of a flat hollow before the point; feet and their covering as in *A. iogaster*: tarsi somewhat slender, feathered in front for about a third of their length, at the base with overhanging feathers, behind with a double row of rather small, and for the most part multiangular, scales; toes middle-sized; claws relatively weak.

Entire upper surface uniform dark brown, passing into dark grey on the front and sides of the head; bases of the feathers white; lesser wing-coverts, tertiaries and points of the secondaries and rectrices furnished with rust-red edges; tail uniform dark brown, in certain lights with faint traces of narrow cross-bars [separated by intervals of about 10 mm]; inner webs of the quills and tail-feathers with moderately broad indistinct cross-bars of darker brown, more apparent near the rust-red margins and on the under side; under wing-coverts reddish white; outer quills below ashy grey, inner ones pale rust-reddish; tail-feathers below whitish grey, reddish at the edge of the inner web; body below white, with a rusty yellowish wash on the abdomen and thighs; feathers of the breast with dark brown shafts and with a rhombic, sharp-pointed, brown spot before the end; towards the belly these spots are arrow-shaped, and pass into rather broad and somewhat washy cross-bars on the sides of the body and thighs; lower abdomen, anal region, under tail-coverts and feathering of the tarsi unspotted. Wing 182 mm (when adult perhaps ca. 200), tail 115; culmen from forehead 23; height of bill at base 14.5, at front edge of cere 12; tarsus 53; middle toe 28 (Brüggemann).

Through the kindness of Professor von Koch we have been able to examine Brüggemann's type, a young specimen most nearly resembling, among other Celebean species, the young of *Astur trivirgatus* and *griseiceps*. From these species it differs by its more slender tarsus and toes, especially in this respect from *A. griseiceps*, and the character of the barring on the wings below and tail is quite different from both.

Professor W. Blasius *(a 1)* thinks it possible that this specimen is a young one of *A. hiogaster* or the young of a new species of Celebes, the adult of which has not yet been found. Jacquinot and Pucheran have already indicated Macassar as a haunt of *A. hiogaster*, but neither Walden, nor Salvadori, nor Blasius believe in the correctness of the locality noted for the specimen in question. It is not possible to form a positive opinion about it. If *A. tenuirostris* is the same as *Urospizias hiogaster* the following is the synonymy:

→ **Urospizias hiogaster** (S. Müll.)

a. **Falco hiogaster** S. Müll., Verh. Nat. Geschied. 1839—44, 110.
b. **Epervier océanien** *(1)* Hombr. & Jacq., Voy. Pole Sud, Atl. pl. 2, f. 1.
c. **Accipiter hiogaster** Gray, Gen. B. 1845, I, 29.
d. **Nisus iogaster** *(1)* Schl., Mus. P.-B. Astures 1862, 43; *(II)* id., Valkv. 1866, 27, 65, pl. 18, f. 1, 2, 3; *(3)* id., Rev. Acc. 1873, 89.
e. **Erythrospiza iogaster** *(1)* Kaup, P. Z. S. 1867, 173.
f. **Erythrospiza iogastra** *(1)* Wald., Tr. Z. S. 1872, VIII, 34.
g. **Astur hiogaster** *(1)* Sharpe, Cat. B. 1874, I, 104; *(2)* id., Mitth. Mus. Dresd. 1878, 353, 355.

Urospizias hiogaster *(1)* Gurney, Ibis 1875, 365; id., Diurn. B. of Prey 1884, 36.
h. **Astur tenuirostris** *(1)* Brüggem., Abh. Ver. Bremen 1876, 43.
j. **Urospizias iogaster** *(1)* Salvad., Orn. Pap. 1880, I, 47; *(2)* W. Blas. & Nehrk., Verh. z.-b. Ges. Wien 1882, 413; *(3)* Salvad., Agg. Orn. Pap. 1889, 18.
k. **Urospiza iogaster** *(1)* W. Blas., J. f. O. 1883, 151.
For further synonyms and references see Salvadori *j 1*.
Figures and descriptions. Hombron & Jacquinot *b I*; Schlegel *d II*, *d 1*, *d 3*; Kaup *e 1*; Sharpe *g 1*, *g 2*; Salvadori *j 1*; W. Blasius & Nehrkorn *j 2*.

The absence of any exact locality for Brüggemann's *Astur tenuirostris* (the type-specimen now bears no original label at all) renders it impossible to decide at present whether *A. tenuirostris* is a new Celebesian species, or *A. hiogaster* in young plumage, or whether the bird is really an inhabitant of Celebes, or has been recorded from there owing to an erroneous label. Salvadori (O. P. I, 48) does not mention Celebes as a locality for *hiogaster*, and therefore does not acknowledge Jacquinot and Pucheran or express an opinion on *tenuirostris* Brüggem. It is a priori not altogether likely that *hiogaster* of Ceram and Amboina should occur in Celebes, especially as an ally of *hiogaster*, *A. pallidiceps* Salvad., belongs to the intermediate island of Buru, and it always remains surprising that no other traveller obtained the bird in Celebes. v. Rosenberg himself sent only one specimen of *hiogaster* from Amboina to the Leyden Museum.

GENUS UROSPIZIAS Kaup amended by Gurney.

Hawks of rather small size, in structure hardly differing from *Astur*; tail long, crossed with 8—24 bars (Kaup); tarsus rather long, the upper fourth feathered in front, toes moderate. About 20 local species, inhabiting Australia, Papuasia, the Moluccas, the Lesser Sunda Islands to Lombok, Polynesia as far as Fiji and the Mariannes.

+ 6. UROSPIZIAS TORQUATUS (Temm.).
Rufous-collared Hawk.

a. **Falco torquatus** "Cuv." *(1)* Temm., Pl. Col. Nrs. 43 ad., 93 juv. (1823).
b. **Nisus torquatus** *(1)* Schl., Mus. P.-B. Astures 1862, 39; *(II)* id., Valkv. Ned. Ind. 1866, 25, 63, pl. 17, figs. 1—5; *(3)* id., Rev. Accip. 1873, 91.
c. **Astur cruentus** *(1)* Gould, P. Z. S. 1842, 113; *(II)* id., B. Austr. 1848, I, pl. 18?; *(3)* id., HB. B. Austr. 1865, I, 43.
d. **Astur torquatus** *(1)* Sharpe, Cat. B. I, 1874, 125; *(2)* Hartert, Nov. Zool. 1896, 177.
Urospizias torquatus *(1)* Gurney, Ibis 1875, 365; *(2)* Salvad., Orn. Pap. I, 1880, 60, 549; *(3)* id., Ibis 1881, 606; *(4)* Gurney, Ibis 1881, 263—66; *(5)* id., Diurn. B. of Prey 1884, 37; *(6)* Salvad., Agg. Orn. Pap. 1889, 20; *(7)* Gurney jr., Ibis 1893, 349; *(8)* Salvad., Ann. Mus. Civ. Gen. 1896, XXXVI, 60.
For further synonymy and references cf. Salvadori *2*, *6*.

Figures and descriptions. Temminck *a 1*; Schlegel *b II*; ? Gould *c II*; Sharpe *d 1*; Salvadori *2*.

Adult. Above brownish slate-grey, paler on ear-coverts; a broad collar of rufous on hind-neck and upper mantle; fore-neck and breast paler rufous, almost uniform on jugulum, but taking the form of narrow pale rufous bars on a whitish ground on breast, sides, abdomen and under wing-coverts; under tail-coverts whiter, thighs almost without bars; tail like the upper parts, below whitish grey, crossed with 10—12 almost obliterated dusky bars: wing 256 mm; tail 200; tarsus 61; middle toe with claw ca. 45; culmen from cere 20 (ad. Dammar Id., near Timor: Riedel — Nr. 669b).

Measurements. ♂ adult wing 195 mm, ♀ adult wing 230 mm (Hartert *d 2*).

Young. "Above brown; the feathers white at the base, some of this white being shown on the hind-neck; all feathers margined with rusty rufous; shoulders deep rufous. Underside white; chin, throat, and breast longitudinally striped with brown; abdomen with more rounded pale rufous spots. Thighs entirely rufous, the feathers with paler edges. Quills more distinctly barred than in adult birds" (Hartert *d 2*).

Distribution. Java?, Sumbawa, Flores, Samao, Timor, Babbar, N. & W. Australia, New Guinea, Waigiou, Batanta (cf. Schlegel *b 1, b 3*; Salvadori *2, 6*). In the Celebesian Province — Djampea and Kalao Islands (Everett *d 2*).

In the above treatment of this species we have followed Count Salvadori, though it should be pointed out that the late J. H. Gurney, J. H. Gurney jun., and Dr. E. P. Ramsay hold a different opinion in regard to its distinctness in Australia and New Guinea. Temminck originally recorded its distribution as N. Australia, Timor, and the Moluccas. Several specimens were recently obtained by Mr. Everett in Djampea and Kalao between Flores and Celebes, and recorded by Mr. Hartert, hence the inclusion of the species in the Celebes list.

In Mus. P.-B. Astures, 1862, 42, Schlegel recorded under *Nisus cruentus* a ♂ ad. (Nr. 10) said to have been collected in Celebes by Reinwardt. Later (Rev. Acc. 1873, 88) it was included by him with 67 other specimens from various localities under the title *N. rufitorques*, but the locality, Celebes, was then cancelled and stated to be erroneous. This specimen is included by Count Salvadori in the synonymy of *A. griseigularis* of the Halmahera group.

An *Astur torquatus* labelled "?Celebes" is recorded by Mr. Hartert in his Katalog der Vög. Frankf. Mus. 1891, 180. As in the case of a specimen of *A. hiogaster* in the Dresden Museum similarly labelled "?Celebes", no authority is given for the locality, and both in all probability are incorrect.

GENUS TACHYSPIZIAS Kaup amended by Gurney.

A small migratory Hawk, in structure very like *Astur*; wing rather long and pointed, 3rd quill longest, from carpus to the tip of the secondaries only three-fifths of the length of the wing; bill moderate, tomia furnished with a large festoon or almost-semicircular tooth, cere very large, convex and bloated-looking; tarsi slender, upper fourth feathered anteriorly. Food:

insects; birds. Only one species, ranging from North China and Tenasserim
east as far as Waigiou in Papuasia.

7. TACHYSPIZIAS SOLOENSIS (Horsf.).
Horsfield's Short-toed Hawk.

a. **Falco soloensis** Horsf., Tr. L. S. 1821, XIII, 137.
b. **Falco cuculoides** *(1)* Temm., Pl. Col. 1823, Nrs. 110, 129 (juv. et ad.).
c. **Astur soloensis** Less., Man. d'Orn. 1828, I, 94; *(1)* Sharpe, Cat. B. 1874, I, 114, pl. IV,
f. 1.; *(2)* Hume & Davison, Str. F. 1878, VI, 8; *(3)* Bingham, Str. F. 1880, IX,
143; *(4)* Oates, B. Brit. Burmah 1883, II, 180; *(5)* Guillem, P. Z. S. 1885, 544;
(6) Everett, J. Str. Br. R. A. S. 1889, 180; *(VI bis)* Murray, Avif. Brit. Ind.
1889, 20, fig.; *(7)* Sharpe, Ibis 1889, 68; 1890, 274; *(8)* Steere, Coll. B. Philip.
1890, 7, 5; *(9)* Styan, Ibis 1891, 488; *(10)* De la Touche, Ibis 1892, 410; *(11)* Blanf.,
Faun. Br. India, B. III, 1895, 400.
d. **Nisus minutus** Less.; *(1)* Pucher., Rev. Zool. 1850, 210.
e. **Nisus soloensis** Loss., Tr. d'Orn. 1831, 60; *(1)* Schl., Mus. P.-B. Astures, 1862, 44; *(II)*
id., Valkv. 1866, 28, 66, pl. 19, f. 4—6; *(3)* id., Rev. Acc. 1873, 97.
f. **Tachyspiza soloensis** *(1)* Kaup, Classif. Säug. u. Vög. 1844, 117; *(2)* id., P. Z. S. 1867,
169, 172; *(3)* Walden, Tr. Z. S. 1872, VIII, 110; *(4)* Meyer, Ibis 1879, 56.
g. **Micronisus soloensis** *(1)* Gray, List Acc. B. M. 1848, 75; *(2)* Swinh., P. Z. S. 1863,
261; *(3)* id., P. Z. S. 1871, 342; *(4)* Vorderman, N. T. Ned. Ind. 1891, 205; *(5)* id.,
ib. 1895, 318.
h. **Micronisus badius** Swinh. (nec Gm.), Ibis 1860, 359.
j. **Accipiter virgatus** Swinh. (nec Temm.), Ibis 1861, 264.
k. **Astur cuculoides** *(1)* Sharpe, Cat. B. 1874, I, 115, pl. IV, f. 2; *(2)* David & Oust.
Ois. Chine 1877, 24; *(3)* Styan, Ibis 1891, 488; *(4)* id. Ibis 1893, 333.
Tachyspizias soloensis *(1)* Gurney, Ibis 1875, 365, 366; *(2)* Salvad, Orn. Pap. 1880, I,
65; *(3)* Gurney, Diurn. B. of Prey, 1884, 32; *(4)* Pleske, Bull. Ac. Sc. Petersb.
1884, 113; *(4a)* W. Blas., Ornis 1888, 544; *(5)* Salvad., Agg. Orn. Pap. 1889, 21;
(6) M. & Wg., Abh. Mus. Dresden 1895 no. 8, p. 4; *(7)* iid., ib. 1895 Nr. 9, p. 1.
l. **Tachyspizias cuculoides** *(1)* Gurney, Ibis 1875, 366; *(2)* id. (? subsp.), Diurn. B. of Prey
1884, 32.

"Meo", Karkellang, Talaut Is., Nat. Coll.
"Sikep abu-abu mera", Minahassa, iid.
For further references see Salvadori 2.

Figures and descriptions. Temminck *b 1*; Schlegel *e II, e 1, e 3*; Sharpe *c I, k I*; Murray
c VI bis; Kaup *f 1, f 2*; Pucheran *d 1*; Gurney *l*; Hume & Davison *2*; Bingham
c 3; Salvadori *2*; Oates *c 4*; Vorderman *g 4*; Blanford *c 11*.

Adult *"Astur cuculoides"*. Entire upper surface slate-grey with dark slate margins to the
feathers, the unexposed basal part of the feathers white; wing-coverts and quills
blacker; ear-coverts smoke-grey; breast pale vinaceous-cinnamon; abdomen and
sides varied with grey which tends to form into bars; remaining under parts
buffy white, the feathers on chin and throat having dark shaft-streaks; inner webs
of secondaries and basal parts of the inner webs of primaries below pure
white; distal ends of the latter black; tail below whitish, crossed with about six
imperfect narrow dark bands. [Celebes: C 10480.)

18   Birds of Celebes: Falconidae.

A second specimen, moulting (male), is similar to that described, and a few brown feathers of immaturity among the scapulars and secondaries reveal the fact that the pale vinaceous under surface of this form (known as *T. cuculoides*) is not a result of age, but is assumed as the young dress is cast off. "Iris dark brown; cere orange; feet gold yellow" (♂, Kema, N. Celebes, 3. Oct. 93; P. & F. Sarasin).

Adult in more rufous plumage: "*Astur soloensis*". Similar to the first-described bird, but darker slaty above; the jugulum and breast rufous, obscurely barred, the bars becoming more distinct on the flanks (Main, N. Celebes, 28. Feb. 1894: Nat. Coll. C 13239).

A young specimen (moulting), assuming adult plumage of the rufous form, has the head, neck and mantle, and single feathers on the scapulars and lower back, dark slate; the wings (many remiges lost), tail, and other upper parts in the brown plumage of the young; the breast (adult) deep rufous (Manado tua Id., end of May, 1894: Nat. Coll. C 13355).

Young. Very different from the adult: above brown (instead of slaty), darkest on head; below white, broadly streaked on the breast with rufous, and barred on the abdomen, flanks and thighs with paler rufous.

A young specimen — ♀, Mindoro — kindly lent to us by Mr. Nehrkorn resembles Schlegel's figure *(e II)*, but the streaks on the breast are larger and broader; the upper surface more uniform; ear-coverts without any grey wash; head dark clove-brown — almost black.

Eggs. Similar to those of *Accipiter nisus*, white with a few dark brown convolutions and blots at the small end: 39 × 30 mm (Nehrkorn in litt.).

Measurements.	Wing	Tail	Tarsus	Culmen of cere
a. (C 10480) ad. Celebes.	198	130	45	13
b. (Nr. 6726) ? ad. Sangi.	198	132	44	13
c. (C 13753) ad. Talaut Is., Nov. 1894 Nat. Coll..	190	130	42	12.5
d. (C 13239) ad. Minahassa, Feb. 94 iid.	199	133	43	13.5
e. (C 13358) vix ad. Manado tua Id., May 94 iid.	—	133	44	13.5
f. (Sarasin Coll.) ♂ vix ad. Minahassa, 3. Oct. 93 (P. & F. S.)	190	126	43	12

Distribution. North China, Amoy, Foochow, Tientsin, Pescadores, etc. (Fortune *g 1*, Swinhoe *g 2*, *g 3*); Tenasserim (Davison *c 2*, Bingham *e 3*); Nicobar Islands; Malacca; Sumatra; Java; Philippines — Mindoro (Mus. Nehrkorn; Salvad. 2; Schl. *e 1*, *e 3*); Palawan (Platen); Borneo (Mottley 2, Ussher, Whitehead *c 6*, *c 7*); Talaut Islands — Karkellang (Nat. Coll.); Sangi (v. Rosenb., Hoedt *e 3*; Meyer); Sino (Hoedt *e 3*); Celebes — Northern Peninsula (Forsten *e 1*, Rosenb. *c 3*, Meyer *f 3*, *f 4*, Guillemard *c 5*); Sula Islands — Sula Besi (Bernst., Hoedt *e 3*); Halmahera; Ternate; Batchian; Morty; Gagie; ? New Guinea (Salvad. *2*); Waigiou (Platen *5*).

Some strong grounds have been given for separating *A. cuculoides* Temm. from *A. soloensis* Horsf. as a distinct species. Dr. R. B. Sharpe *(k I)* distinguishes *A. cuculoides* chiefly by its pale vinous under surface mixed with ashy, and pure white under wing-coverts, from *A. soloensis* in which these parts are vinous-chestnut and buffy white, respectively. The late J. H. Gurney *(1)* adds from information supplied by Swinhoe, in whose cabinet he was able to examine specimens, that the two forms lay eggs which differ remarkably, and that they have differently coloured irides. Nevertheless Gurney, at the time, was doubtful about their specific distinctness, and, in his "List of the Diurn. B. of Prey in Norwich Mus." p. 32, he only admits *A. cuculoides* as a questionably valid subspecies. It should be remembered, that both occur together in the same localities — China, Java (Dresd. Mus.), Celebes, and, that at one time Swinhoe was not quite sure of the correct nomenclature for *A. soloensis*, *A. virgatus* and *A. badius*, but on one or two occasions misemployed these names, whence it appears not unjustifiable to suppose that the eggs in the Swinhoe collection, referred to by Gurney as those of *A. soloensis* and *A. cuculoides*, were in reality those of *A. soloensis* and *A. virgatus*, especially as *A. cuculoides* is not mentioned by Swinhoe except in the synonymy of *A. soloensis* (g 2) and we are not aware that any fourth species, corresponding to it, was ever spoken of by him.

After examining the series of these forms (nearly 70 in all) in the British and Leyden Museums (where is the type of *T. cuculoides*), in addition to six at Dresden, we at first were of opinion that the pale, more uniform plumage of *T. cuculoides* represents the old *T. soloensis*, especially the old male, though there is one very pale specimen at Leyden marked ♀. There is a male in the British Museum, assuming the adult dress, which is of as dark a rufous on the breast as the female, viz. specimen "*i* ♂ ad. Pescadores (Swinhoe)" of Dr. Sharpe's Catalogue (*c I*) which retains some immature plumage; other specimens, mostly males, if sexed, in the Leyden Museum afford transitions between such and the type of *cuculoides*. But this supposition — that the pale vinous specimens are always old — is controverted by a specimen (♂) in the Sarasin Collection described above; it is in the pale plumage of *T. cuculoides*, but a few brown feathers not yet moulted show that it is only a second year's bird. The only conditions possible seem to be, therefore, either that *T. soloensis* and *cuculoides* are two distinct species, as Sharpe supposed, or that they form one species which varies a good deal in the intensity of the rufous on the breast and slaty on the back. The presence of intermediate specimens, the perfect agreement of the two forms in structure, and their occurrence in the same localities, are arguments against their being two species; we believe the other explanation to be the correct one.

Very little has been recorded of the habits of *A. soloensis*. Kaup, examining the peculiar formation of the bill, believed himself justified in pronouncing that its food, especially when it has young, would prove to consist only of insects (*f 2*); in Celebes v. Rosenberg found it to be "ein Hauptinsectenvertilger" (Malay. Archip. 1878, 271). We cannot put much stress upon either of these two statements. The specimen obtained by Zelebor during the visit of the "Novara" to Kar Nicobar was shot, however, while it was unsuccessfully chasing an *Oriolus macrourus* (Novara-Reise, Vög. p. 12).

When a species is found to range far over a group of islands and yet develop no differences of coloration in the various localities, it is usually safe to assume that it is migratory. In the present case, in the absence of any direct statements to this effect, the dates, which have been recorded by various authors, tend to show that, while the species breeds in China (Swinhoe), it descends in the cold season with the N. E. monsoon into the Malay Peninsula and the East Indian Archipelago and stays till the S. W. monsoon begins, — that is to say from September-October to March. We make this statement with some hesitation, inasmuch as it is certain that some specimens remain behind (though perhaps not south of the equator), a specimen in the Leyden Museum having been obtained in Morty Island by Bernstein on August 1st, 1861.

On the other hand, all the dates but one, which we have been able to find, point to its being a migratory species. Davison and Bingham speak of it as a rare straggler in Tenasserim, and a specimen given to Swinhoe by Captain Ebert flew on board ship off the Pescadores early in May, facts which also point to migration.

1. ♀ ad. Amoy, China, April 1867 (Swinhoe) Sharpe, c I.
2. ♀ ad. Amoy, China, April 1867 (Swinhoe) Sharpe, c I.
3. ♂ Ningpo, 22. May 1874 (Swinhoe) in Brit. Mus.
4. ? N. China, 15. April 1884 (Fortune Coll.) in Brit. Mus.
5. ♀ Amoy, 20. April 1867 (Swinhoe) in Brit. Mus.
6. ♀ Amoy, Sept. 1866 (Swinhoe) in Brit. Mus.
7. ♂ ad. Pescadores, April 1866 (Swinhoe) Sharpe c I.
8. ? Pescadores, May 1866 (Ebert) Swinhoe (? same as Nr. 3).
9. ♀ Thoungyeen, Tenasserim, 12. April 1880 (Bingham) in Brit. Mus.
10. ♂ juv. Tenasserim, 1. Oct. 1875 (Hough) in Brit. Mus.
11. ♀ juv. Nicobar Is., Dec. 1873 (Wimberley) in Brit. Mus.
12. ? Nicobar Is., 25. Feb. (Zelebor) Pelz, Novara, Vög. 12.
13. ♀ juv. Nicobar Is., 31. Jan. 1874 (Wimberley) in Brit. Mus.
14. ♀ juv. Salangore, 11. Nov. 1879 (Davison) in Brit. Mus.
15. — juv. Malacca, Feb. 1880 (Davison) in Brit. Mus.
16. ♂ ad. Mindoro, 20. April 1890 (Platen) in Mus. Nehrk.
17. ♀ juv. Mindoro, 9. Sept. 1890 (Platen) in Mus. Nehrk.
18. — ad. Labuan, Dec. 1889 (Everett) in Brit. Mus.
19. — juv. Labuan, Dec. 1889 (Everett) in Brit. Mus.
20. — juv. Labuan, Feb. 1883 (Everett) in Brit. Mus.
21. ? Kini Balu, Borneo, 26. Feb. 1887 (Whitehead) Everett, c 6.
22. — ad. Talaut Is., Nov. 1894 (Nat. Coll.)
23. ♀ ad. Sangi, 3. Jan. 1866 (Hoedt) Schl. e 3.
24. ♀ juv. Sangi, 24. Oct. 1864 (Rosenb.) Schl. e 3.
25. ♂ juv. Sangi, 3. Nov. 1864 (Rosenb.) Schl. e 3.
26. ♂ ad. Sangi, Oct. 1865 (Hoedt) Schl. e 3.
27. ♂ ad. Sangi, Oct. 1865 (Hoedt) Schl. e 3.
28. ♀ ad. Sangi, Oct. 1865 (Hoedt) Schl. e 3.
29. ♂ juv. Celebes, 22. Sept. 1863 (Rosenb.) Schl. e 3.
30. ? Celebes, Mch. 1871 (Meyer) f 4.
31. — imm. Manado tua Id., May 1894 (Nat. Coll.).
32. — ad. N. Celebes, Feb. 1894 (Nat. Coll.).
33. ♂ vix ad. N. Celebes, 3. Oct. 1893 (P. & F. Sarasin).

34. ♀ ad. Sula,	Feb. 1864 (Bernst.)	Schl. *e 3*.
35. ♀ juv. Sula,	Dec. 1864 (Bernst.)	Schl. *e 3*.
36. ♂ ad. Morty,	1. Aug. 1861 (Bernst.)	Schl. *e 3*.
37. ♂ ad. Morty,	11. Dec. 1861 (Bernst.)	Schl. *e 3*.
38. ♂ — Ternate	28. Dec. 1879 (Fischer)	Pleske, *4*.
39. ? Ternate,	Dec. 1874 (Bruijn)	Salvad. *2*.
40. ♀ ad. Gagie,	14. Nov. 1864 (Bernst.)	Schl. *e 3*.

Dr. Steere *(c 8)* obtained the species in Mindanao in October or December. The apparent absence from South Celebes, though many specimens have been obtained in the north during our winter, is also indicative of migration.

GENUS SPILOSPIZIAS Salvad.

A small weak-winged Hawk, peculiar to Celebes, and easily recognisable there by the three rows of large, round spots of white on the tail. Point of wing very blunt, the 3rd and 6th primaries only about 4 mm shorter than the 4th and 5th, the secondaries relatively long, the distance from the carpus to the tip of the inner ones being about five-sixths the length of the wing; tail rounded and slightly decurved; tomin with a large festoon; cere small, short, and dark in colour; toes much as in *Tachyspizias*, but the anterior claws blunter and straighter; food unknown. Only one species, evidently strictly stationary, as it differs racially in North and South Celebes.

⋌ * 8. SPILOSPIZIAS TRINOTATUS (Bp.).
White-spot Hawk.
Plate I.

It is not surprising to find that this weak-winged Celebesian Hawk differs racially to some extent in the Northern and the Southern Peninsulas of the island. The two races are:

⋌ 1. The typical Spilospizias trinotatus.

a. **Accipiter trinotatus** *(1)* Bp., Consp. 1850, I, 33 [ex Temm., MS.]; Strickl., Orn. Syn. 1855, 115; *(2)* Wall., Ibis 1868, 8 pt.; Gray, HL. 1869, I, 34.

b. **Astur trinotatus** Bp., Rev. Zool. 1850, 490; *(1)* Sharpe, Cat. B. 1874, I, 101, pt.; *(2)* Brüggem., Abh. Ver. Bremen 1876, V, 44; *(3)* Guillem., P. Z. S. 1885, 544.

c. **Sparvius trinotatus** Bp., Rev. Zool. 1854, 538.

d. **Nisus trinotatus** *(1)* Schl., Mus. P.-B. Astures, 1862, 45; *(II)* id., Valkv. 1866, 27, 65, pl. 19, f. 1—3; *(3)* id., Rev. Acc. 1873, 90.

e. **Erythrospiza trinotata** *(1)* Kaup, P. Z. S. 1867, 172; Walden, Tr. Z. S. 1872, VIII, 33; *(2)* Meyer, Ibis 1879, 55; *(3)* Hickson, Nat. in N. Celebes, 1889, 87.

f. **Erythrospizias trinotatus** *(1)* Gurney, Ibis 1875, 364; *(2)* id., Diurn. B. of Prey 1884, 32.

g. **Spilospiza trinotata** *(1)* Salvad., Ann. Mus. Civ. Gen. 1876, 643; W. Blas., J. f. O. 1883, 134.

h. **Spilospizias trinotatus** *(1)* Salvad., Orn. Pap. I, 1880, 47; *(2)* M. & Wg., Abh. Mus. Dresd. 1895, Nr. 8, p. 4; *(3)* Hart., Nov. Zool. 1897.

"**Sikip-batta-batta**", Minahassa, Meyer *e 2*.
"**Sikep abu-abu**", Manado, Guillem. *b 3*.
"**Sikep werreng selak**", Alfurous, Minahassa (Nat. Coll.).

Figures and descriptions. Schlegel *d II, d 1, d 3*; Bonaparte *a 1*; Kaup *e 1*; Sharpe *b 1*; Gurney *f 1*; Brüggem. *b 3*,

Adult male. Above slate-grey, blacker on the upper tail-coverts; quills above blackish, notched on the inner webs towards the base of the feathers with white; wing below and under wing-coverts white, free end of wing below grey; tail black, marked on the inner webs of the feathers with three rows of large white spots, intervals of about 25 mm separating the rows; on the outer webs of the feathers almost obliterated brownish spots corresponding to the white ones are just discernible; tail narrowly tipped with white, mainly confined to the inner webs; ear-coverts smoke-grey, paler on chin and throat; below rufous vinaceous-cinnamon, paling into whitish on thighs, flanks and under tail-coverts (♂, Tomohon, Minahassa, 7. IV. 94: P. & F. Sarasin); bill and cere above the nostrils black; cheeks and orbits orange-yellow; feet deep orange-yellow; iris chrome-yellow (Wall. *2*).

Young. Head and neck dark brown with ferruginous borders to the feathers; entire upper surface dark cinnamon-rufous, tail varied with black — the rufous predominating on the outer webs, the black on the inner — and marked as in the adult with three rows of large white spots on the inner webs, and tipped on the inner webs with white; under surface fulvous white, marked with long stripes of blackish brown on the breast and sides, and with a narrow dark streak down the middle of the throat; under wing-coverts and base of wing fulvous white; quills below reddish, marked on inner webs towards the bases of the feathers with short bars of blackish not extending completely across the web. (Manado, Nr. 6733).

Changing plumage. Another specimen is in a stage midway between the above described young plumage and that of the adult: middle parts of upper surface slaty grey as in the adult, with some cinnamon-rufous tipped feathers about the occiput and mantle; first primary and one or two of the secondaries cinnamon-rufous, the rest dark slaty; shoulders and scapulars cinnamon-rufous varied with slaty; breast vinaceous-cinnamon varied with a few broad streaks of dark brown, particularly on the left side of the specimen (Tondano, Aug.—Sept. 1892: Nat. Coll. — C 10790).

A very similar example in process of assuming the adult dress is in the Sarasin Collection. (See plate). "Iris, malar skin, and feet yellow; bill black" (♂ imm., Kema, 31. July 1893).

Measurements.	Wing	Tail	Tarsus	Culmen from cere
a. (Nr. 8150) ♂ Manado	170	148	55	—
b. (Nr. 6731) ad. Manado	164	146	52	14
c. (Nr. 6732) ad. Manado	180	148	56	16
d. (c 10848) ad. Manado, Aug.—Sept. 92 (Nat. Coll.)	165	145	55	15
e. (c 10849) ad. Manado, Aug.—Sept. 92 (Nat. Coll.)	164	150	51	15
f. (c 10790) vix ad. Tondano, Aug.—Sept. 92 (Nat. Coll.)	172	150	52	14.5
g. (Nr. 6733) juv. Manado	160	138	49	13.5
h. (c 10850) juv. Manado, Aug.—Sept. 92 (Nat. Coll.)	164	144	51	15
i. (Sarasin Coll.) ♂ ad. Tomohon, N. Cel. 7. April 94	155	130	53	12.5
j. (Sarasin Coll.) ♀ ad. Masarang, N. Cel. 9. Oct. 94	166	141	54	14
k. (Sarasin Coll.) ♂ vix ad. Kema, N. Cel. 31. July 93	169	140	51	15
l. (Sarasin Coll.) ♂ juv. Kema, N. Cel. 13. Sept. 93	173	136	53	—

Distribution. Northern peninsula of Celebes: Minahassa (Wallace *a 2, b 1*, Rosenberg *d 3*, Meyer *e 2*, Beccari *g 1*, etc.), Gorontalo District (Forsten *d 1*), Talissi Island (Hickson *e 3*); Tawaya and Dongola, W. Celebes (Doherty *h 3*).

2. Spilospizias trinotatus haesitandus Hartert.

i. Accipiter trinotatus pt. *(1)* Wall., Ibis 1868, 8 (Macassar).
i. Astur trinotatus *(1)* Sharpe, Cat. B. 1874, I, 101, pt. (Macassar); *(2)* Büttik., Zool. Erg. Webers Reise 1893, III, 271 (Pare-Pare).
Spilospizias trinotatus haesitandus *(1)* Hartert, Nov. Zool. 1896, III, 162, *(2)* id., ib. 1897.

Diagnosis. Differs from the typical S. trinotatus of the N. Peninsula in having the abdomen paler, white for its greater part, the vent and under tail-coverts pure white, the thighs pure white or with a very faint rosy shade on their upper part only (Hartert *1*). In the typical form these parts are whitish, washed with pink buff.

Measurements.

	Wing	Tail	Tarsus	Culmen from cere
a. (C 15443) ♀ ad. (cotype) Indrulaman, S. Col. Sept. 95 (Everett)	175	143	52	16

Distribution. Southern Peninsula, Celebes: Macassar (Wallace *i 1, j 1*), Pare-Pare (Weber *j 2*), Peak of Bonthain (Everett *l*), Doherty *k 2*.

The differences between this form and the northern birds are slight and apparently not always pronounced.

This little Celebesian Hawk, which Mr. Wallace speaks of as one of the most beautiful Hawks of the East, much resembles *Accipiter minullus* of South Africa in the unusual markings of the tail, but differs from it and other species in so far that it must be placed in a genus or, better, subgenus for itself. In 1867 Kaup made the genus *Erythrospiza* for this species and *A. hiogaster* and *griseigularis*, but later the two latter species were very properly removed by Gurney and *E. trinotatus* made the sole representative of the genus, the name of which he amended to *Erythrospizias*. Count Salvadori, almost simultaneously, pointed out that *Erythrospiza* had been employed already by Bonaparte for a Loxiine genus; and he advanced the name *Spilospiza*[1]) in its stead (*g 1*). The original description of *Erythrospiza* as applying to *S. trinotatus* is imperfect and misleading; an analysis of the original description of the two genera *Erythrospiza* and *Teraspiza* (= *Accipiter*) shows that the only tangible means mentioned for distinguishing between the two forms consists in the statement that in *Teraspiza* the inner vanes of the first four primaries are emarginated, while in *Erythrospiza* the emarginations extend to the fifth; but even this does not hold good for *E. trinotatus* in which the emarginations are usually only slightly cut out on the second, third and fourth, and not at all on the fifth primary. *S. trinotatus* in reality differs markedly from *Teraspiza* — taking *T. rhodogaster* for comparison — in the rounded point of the wing, there being little difference

---

1) We prefer *Spilospizias* to *Spilospiza*, as Gurney did for other accipitrine genera with names ending in *spiza*, because σπίζα means any small chirping bird, whereas σπιζίας is a falcon or sparrow-hawk. (σπίλος = spot.)

in length between the 3rd, 4th, 5th and 6th primaries; in the blunt, weak ends of these feathers, significant of quiet flight; in the rounded tail; in the relatively much shorter toes, especially the middle one; whereas *T. rhodogaster*, which cannot be separated from *Accipiter* (as likewise the other members of *Teraspiza*), has the point of the wing and the tips of the individual feathers sharpened, hard and strong; the 6th (about 12 mm) and succeeding primaries much shortened; the tail square; the toes — especially the middle one — lengthened. Doubtless the two birds differ much in habits.

From *Urospizias hiogaster* and *griseigularis* which were included in the same genus by Kaup, but removed by Gurney, *S. trinotatus* differs chiefly in its short, rounded and weak wings, its less massive beak, and in its curiously marked and weak tail-feathers. *Spilospizias* stands as a genus on a somewhat better footing than *Tachyspizias*, *Urospizias*, *Scelospizias* and others into which the great genus *Accipiter* with its eighty or ninety species has been split up. Whether it be advisable to recognize these small divisions at all is very doubtful, but if once the subdivision of *Accipiter* into *Accipiter* and *Astur* is made, the other subdivisions appear necessary. The student of general ornithology would do well, and be sufficiently within the bounds of accuracy in speaking of this great body of Hawks broadly under the generic name *Accipiter*. Although *Spilospizias* is an interesting Celebesian form, it cannot be looked upon as of equal importance with *Meropogon*, *Basileornis*, *Scissirostrum*, etc.

On the island of Talissi just off the Northeastern extremity of the Minahassa Dr. Hickson speaks of this Hawk as "a bird often seen on the orange trees and the higher branches of the trees in the mangrove swamps". Nothing is known with certainty about its breeding habits, but it is probable that it makes its nest in mountain districts, where the young birds remain until they have assumed the adult plumage. From December to March 1870-71 young birds were not met with by Meyer in the plains near Manado, and the natives asserted that they did not occur there; on the other hand, in June one or more young specimens were shot at a height of 2000 ft. near Kakas on Lake Tondano (e 2). The Dresden Museum has, however, since received 3 young examples from the neighbourhood of Manado.

GENUS ACCIPITER Briss.

The Sparrow-hawks are of small size, distinguishable from *Astur*, etc. by their long slender toes, the middle one being especially prolonged, overreaching the others by the entire length of the claw or more; tarsus slender, rather long, the upper fourth feathered anteriorly; tomia festooned; wing rather short, secondaries reaching about three-fourths the length of the wing. Preys chiefly upon birds. Eggs 3—6. About 23 species, migratory and stationary, of almost cosmopolitan distribution.

⚦ * 9. ACCIPITER RHODOGASTER (Schl.).

Vinous-breasted Sparrow-hawk.

a. **Nisus virgatus rhodogaster** *(1)* Schl., Mus. P.-B. Astures, 1862, 32.
Accipiter rhodogaster *(1)* Gurney, Ibis 1863, 450; *(2)* Wall., ib. 1868, 7; Gray, Hl. 1869, I, 33; *(3)* Sharpe, Cat. B. 1874, I, 145; *(4)* Gurney, Ibis 1875, 471, 484; *(5)* Brüggem., Abh. Ver. Bremen 1876, 45; *(6)* Gurney, Ibis 1882, 452; *(7)* id., Diurn. B. of Prey 1884, 39; *(8)* Guillem., P. Z. S. 1885, 545; *(9)* W. Blas., Ztschr. ges. Orn. 1886, 85; *(11)* M. & Wg., Abh. Mus. Dresden 1895, Nr. 8 p. 4; *(12)* iid., ib. 1896, Nr. 1, p. 4; *(13)* Hartert, Nov. Zool. 1896, 162; *(14)* id., ib. 1897.
b. **Nisus rhodogaster** *(1)* Schl., Valkv. 1866, 21, 60, pl. 12, f. 5, 6; *(2)* id., Rev. Acc. 1873, 76.
c. **Teraspiza rhodogaster** *(1)* Kaup, P. Z. S. 1867, 171.
d. **Teraspiza rhodogastra** *(1)* Walden, Tr. Z. S. 1872, VIII, 33, 109, pl. IX (♀ juv.); *(2)* Meyer, Ibis 1879, 55; *(3)* W. Blas., J. f. O. 1883, 134.

"**Sikep werreng kokie**", Alfurous, near Tondano, N. Celebes (Nat. Coll.).

Figures and descriptions. Schl., *b 1*, *b 2*; Walden *d 1*; Kaup *c 1*; Sharpe *3*; Gurney *4*, *6*.

Adult ♀. Above dark slate-grey, paler on the ear-coverts and sides of neck; wing-coverts blackish; quills and tail above browner, obscurely crossed with dark bars, these bars are very distinct below on a pale greyish brown ground, 7 or 8 in number on outermost rectrice, 4 or 5 on the middle feathers, and about 9 on the longest primary; under wing-coverts fulvous white mottled with grey; chin and throat greyish white; breast, sides and upper abdomen rufous-vinaceous-cinnamon; lower abdomen and thighs pale grey, slightly varied with rufous; crissum and under tail-coverts white. (♂ ad. Tomohon, N. Cel. 10. April 1894: Sarasin Coll.) "Bill black; cere dusky yellow; feet yellow; iris bright chrome-yellow" (Wallace).

The female is very much larger than the male; the upper parts are blackish slate, the under wing-coverts buff-white with blackish spots thereon, the bars on the tail plainer and more decided; in other respects like the male. ([♀] Manado: Meyer Nr. 6722).

Young. Kestrel-like. Above bright hazel-rufous, the quills and tail crossed with four bars, and the wing- and tail-coverts and back marked with large heart-shaped spots of black; head and neck dark brown with fine rufous margins to the feathers; below buffy white, finely streaked about the sides of the face and down the centre and sides of the throat with blackish, and broadly streaked with blackish brown on the remaining under parts (with the exception of the under tail-coverts); under wing-coverts deeper buff, the lowest row spotted with brown, which is present as a few fine streaks on some of the others; secondaries crossed with five dark bars, primaries with about seven. (N. Celebes: v. Musschenbroek Nr. 6721).

Measurements.	Wing	Tail	Tarsus	Bill from cere
a. (Nr. 6723) [♀] ad. N. Celebes (Riedel)	211	156	63	16
b. (Nr. 6722) [♀] ad. Manado, N. Cel. (Meyer)	212	157	60	16
c. (Sarasin Coll.) [♀] juv. Kema, N. Cel. end Sept. 93	201	151	57	15
d. (C 10789) [♂] ad. Manado, VIII—IX. 92 (Nat. Coll.)	167	125	53	12
e. (Sarasin Coll.) ♂ ad. Mt. Soputan, N. Cel. 16. IV. 95	166	120	51	11.5
f. (Sarasin Coll.) ♂ ad. Tomohon, N. Cel. 10. IV. 94	161	116	51	11.5

(Measurements continued.)

	Wing	Tail	Tarsus	Bill from cere
g. (Sarasin Coll.) ♂ vix ad. E. of Peak of Bonthain, c. 900 m. S. Cel. 25. X. 95	170	125	54	12.5
h. (Sarasin Coll.) ♂ juv. Tomohon, N. Cel. 3. X. 94 .	168	129	50	12.5
i. (Nr. 6721) [♂] juv. N. Celebes (v. Musschenbr.) . .	161	122	51.5	13

Distribution. Celebes: Minahassa (v. Rosenberg *b 2*, v. Duivenbode *b 2*, *d 3*, etc.), Gorontalo Distr. (Forsten *a 1*, Rosenberg *b 2*, Riedel *9*), Southern peninsula — Macassar (Wallace *2, 3*, Everett *13*), Mt. Bonthain (P. & F. Sarasin, Everett *13*), Dongala, W. Celebes (Doherty *14*).

Nothing is known of the habits of this Hawk. It was made the type of Kaup's genus *Teraspiza*, but the distinguishing characters given are very slight and the differences not appreciable at first sight. How insufficient they are may be gathered from the fact that even Kaup failed to see that the allied *A. sulaensis*, of which he had a specimen, belonged to the same genus, and placed it in another (*Uraspiza* Kaup), though Schlegel would not allow it later even subspecific distinction from *A. rhodogaster* (*c 1*, *b 2*). It is distinguishable from that slightly differentiated geographical species by the bars on the wing below being nearly twice as broad, and in the male they are noticeable down to the tip of the wing, whereas in the male of *A. sulaensis* the bars are quite obliterated on the free end of the wing for about 4 cm; the Sula form also seems to be smaller. It is interesting to find that a closely corresponding process of wing-differentiation and of size has occurred in the Celebesian and Sulan forms of *Spilornis* (cf. antea, pp. 2, 4). *Accipiter sulaensis* differs further by its vinous (not grey) ear-coverts.

The resemblance of this Sparrow-hawk to the White-spot Hawk, *Spilospizias trinotatus*, is very perfect, though the curious white spots on the tail of the latter, and the structural differences pointed out in our article (p. 24), at once serve to distinguish them. What is more remarkable is that the young, which are totally different from the parents, are closely similar in both species. Moreover, the young, especially at least that of *A. rhodogaster* (by reason of the plain bars on the wings and tail), are just like Kestrels! The young of birds by their first dress often seem to "throw back" to a distant ancestral form, but the indications generally seem as if obscured by superimposed influences inherited from the parents, and, when not so obscured, the question comes in, whether the similarity really has the meaning which one is prompted to attach to it.

+ * 10. ACCIPITER SULAENSIS (Schl.).
Sula Vinous-breasted Sparrow-hawk.

a. **Nisus sulaensis** *(1)* Schl., Valkv. 1866, 26, 64, pl. 16, f. 3, 4.
b. **Uraspiza sulaensis** *(1)* Kaup, P. Z. S. 1867, 176.
Accipiter sulaensis *(1)* Wall., Ibis, 1868, 10, 20; Gray, HL. 1869, I, 34; *(2)* Sharpe, Cat. B. 1874, I, 146; *(3)* Gurney, Ibis, 1875, 484; *(4)* M. & Wg., Abh. Mus. Dresden 1896, Nr. 2, p. 7.

c. **Nisus rhodogaster** *(1)* Schl., Rev. Acc. 1873, 76 pt. (Sula).
d. **Urospizias sulaensis** *(1)* Salvad., Orn. Pap. 1880, I, 65.
e. **Accipiter sulaensis** subsp. Gurney, Diurn. B. of Prey 1884, 39.
"Alaji sasoko", Banggai Id., Nat. Coll.

Figures and descriptions. Schl. *a 1, c 1*; Kaup *b 1*; Sharpe *2*.

Adult male. Similar to the adult male of *A. rhodogaster*, but the cheeks vinaceous-cinnamon with a grey shade (instead of dark grey), chin and throat whiter vinaceous, the dark bars on the wing below much narrower and absent on the distal 40 mm of the wing, no bars distinguishable on the outermost tail-feather, the others also more indistinct ([♂] ad. Banggai Id., V—VIII. 95, Nat. Coll. — C 14627).

Female. Very much larger than the male. Differs from the female of *A. rhodogaster* by having the bars on the wing and tail much narrower, those on the outermost tail-feather partly obliterated; cheeks vinaceous grey ([♀] ad. Peling Id. V—VIII. 95: Nat. Coll. — C 14502).

Immature. The under parts as in the adult male; above with the rufous, Kestrel-like plumage of the young, varied with the slate-grey feathers of the adult ([♂] Banggai, V—VIII. 95: Nat. Coll. — C 14626).

Measurements.	Wing	Tail	Tarsus	Culmen from core
a. (C 14502) [♀] ad. Peling.	185	135	56	15
b. (C 14627) [♂] ad. Banggai	160	118	48	13
c. (C 14626) [♂] imm. Banggai	—	115	49	13

Distribution. Sula and Peling groups: Sula Besi (Bernstein *a 1, c 1, b 1*); Banggai (Nat. Coll.), Peling (Nat. Coll.).

Schlegel did not admit the validity of this local species in 1873 *(c 2)*, but it was again upheld by Sharpe *(2)*, and after examining the type, we agree with the latter in holding the two forms distinct. It is a very rare species in collections. Gurney *(3)* once remarked that the type appeared to be the only specimen extant in any collection; there was, however, a second in Darmstadt *(b 1)*; five are now known. The young one described by Kaup at the same time as that of this species is probably either *A. rubricollis* Wall. or *A. hiogaster* S. Müll., as is pointed out by Schlegel *(c 1)* and Salvadori *(d 1)*. *A. sulaensis* varies geographically in Sula in the same manner as *Spilornis rufipectus sulaensis*.

11. ACCIPITER VIRGATUS (Temm.).

Jungle Sparrow-hawk.

This species apparently consists of the following subspecies:

1. The typical Accipiter virgatus.

a. **Falco virgatus** Temm., Pl. Col. 1823, Nr. 109 (Java).
See also: Schl., Valkv. Ned. Ind. 1866, 20, 59, pl. 12, f. 1—4; Sharpe, Str. F. 1879,

VIII, 440; Gurney, t. c. 443; Hume, Str. F. 1880, IX, 231; Legge, B. Ceylon, 1880, 26; Oates, B. Brit. Burmah, 1883, II, 182; Gurney, Diurn. B. of Prey, 1884, 39, 164—177; Blanford, Faun. Br. Ind., B. III 1895, 404; Grant, Ibis 1896, 104—109.

Diagnosis. Size small, breast in adult male rufous, in female browner, upper surface slaty (after Gurney).

Distribution: Java (type), Sumatra (Mus. Leyd.), Malay Peninsula, Burmah, India and the Himalayas (Hume), Ceylon (Legge).

2. Accipiter virgatus affinis (Hdgs.).

b. **Accipiter affinis** Hodgs. in Gray's Zool. Misc. 1844, 81.
c. **Accipiter virgatus** *(1)* Sharpe, Str. F. 1879, VIII, 440—442 pt. (larger race of inner Himalayas); *(2)* Gurney, t. c. 443; *(3)* Hartert; Nov. Zool. 1894, 482.
d. **Accipiter gularis** Hume (nec. T. & S.), Str. F. 1880, IX, 231; Oates, B. Brit. Burmah, 1883, II, 183, pt.
e. **Accipiter virgatus affinis** Gurney, Diurn. B. of Prey 1884, 39, 168—173.

Diagnosis. Size large, breast chocolate-brown rufous (after Gurney).

Distribution. Himalayas from Sikkim to Mussorie (Hume); Formosa! (Gurney—in migration?).

(3?). Accipiter virgatus manilensis (Meyen).

f. **Accipiter manillensis** *(1)* Meyen, Nova Acta Acad. Leop. 1834, XVI, Suppl. p. 69, pl. IX; *(2)* Grant, Ibis 1894, 503; *(3)* Grant, Ibis 1895, 438; *(4)* Bourns & Worces., B. Menage Exped. 1894, 32.
g. **Accipiter stevensoni** Tweed., P. Z. S. 1878, 938, pl. LVII.
h. **Accipiter virgatus manillensis** Gurney, Diurn. B. of Prey 1884, 40, 173—177.

Diagnosis. Size small, breast of adult female as rufous as in the male, above somewhat browner than the typical form.

Distribution. Philippine Islands: — Luzon (Meyen, W. Ramsay, Whitehead *f 2)*, Mindanao (Everett).

+ 4. Accipiter virgatus gularis (T. & S.).

i. **Astur (Nisus) gularis** *(1)* Temm. & Schl., Faun. Jap. Aves, 1845, 5, pl. II.
j. **Accipiter nisoides** *(1)* Blyth, J. A. S. B. 1847, XVI, 727, *(2)* id., Ibis 1865, 28; 1866, 240; 1870, 158; *(3)* Gurney, Diurn. B. of Prey 1884, 40, 165—177; *(4)* id., Ibis 1887, 362; *(5)* Gigl. & Salvad., P. Z. S. 1887, 581; *(6)* De La Touche, Ibis 1892, 485.
k. **Accipiter stevensoni** *(1)* Gurney, Ibis 1863, 447, pl. XI; *(2)* Swinhoe, ib. 1874, 430; *(3)* Sharpe, Cat. B. 1874, I, 152 (note); *(4)* Gurney, Ibis 1875, 482; *(5)* Sharpe, Str. F. 1879, VIII, 442; *(6)* Gurney, Ibis 1880, 217; *(7)* Hume, Str. F. 1880, 109, 231; *(8)* Legge, B. Ceylon 1880, 29; *(9)* Oates, B. Brit. Burmah 1883, II, 183.
l. **Accipiter virgatus** *(1)* Sharpe, Cat. B. 1874, I, 150, partim; *(2)* David & Oust., Ois. Chine 1877, 26; *(3)* Styan, Ibis 1887, 233; *(4)* Sharpe, ib. 1889, 68; *(5)* Everett, J. Str. Br. R. A. S. 1889, 180; *(6)* Styan, Ibis, 1891, 326, 488; *(7)* Hose, Ibis 1893, 418; *(8)* Everett, Ibis 1895, 38; (?9) De La Touche, ib. 337.
m. **Accipiter gularis** *(1)* Gurney, Ibis, 1875, 481—484, pt.; *(2)* id. Str. F. 1879, VIII, 443; *(3)* Seebohm, B. Japan. Emp. 1890, 205; *(4)* id., Ibis 1893, 52; *(5)* Grant, Ibis

1896, 104; *(6)* Hartert, Nov. Zool. 1896, 177; *(7)* Blanf., Fauna Br. Ind. B. III, 1895, 405.
n. ?**Teraspiza virgatus** *(1)* Wald., Tr. Z. S. 1875, IX, 141.
o. ?**Accipiter stephensoni** *(1)* Kutter, J. f. O. 1883, 294.

Diagnosis. The barring on the under parts of the female continued up to the throat, while in the females of the *A. virgatus* group the chest is, on the whole, longitudinally marked and blotched, or nearly uniform in colour, in contrast to the barred breast and under parts; the ashy black line down the middle of the white throat very narrow, in *A. virgatus* much wider and more strongly marked. In *A. gularis* the fourth primary quill is considerably longer than the fifth, in *A. virgatus* only slightly longer. (Grant *m 5*). Sharpe says *A. gularis* has a shorter middle toe (c. 28 mm as against 30.5).

Distribution. Siberia and Mantchuria (David *l 2*); Corea (Prince of Savoy *j 5*); North China (David *l 2*, etc.); South China (Styan *l 3, l 6*, De La Touche *j 6*); Japan (Gurney *j 3*, Seeb. *m 3*); Philippines? (Meyer *n 1*, Gurney *j 3*); Cochin China (Gurney *j 3*); Natuna Islands (Conrad *m 2*); Malacca (Wallace *k 5*); Sumatra (Gurney *j 3*); Java (Gurney *j 3*); Timor (Wallace *k 5*); Palawan (Everett *l 3*); Borneo (Everett, Whitehead, Mottley *l 5, l 4*); North Celebes (Riedel in Petersb. Mus.); Djampea (Everett, fide Hartert *m 6*).

5. Accipiter virgatus rufotibialis (Sharpe).

p. **Accipiter rufotibialis** Sharpe, Ibis 1887, 437; id., ib. 1889, 68, pl. II; id., ib. 1890, 274; Everett, J. Str. Br. R. A. S. 1889, 181; Hose, Ibis 1893, 418.

Diagnosis. Under tail-coverts uniform chestnut, thighs chestnut.
Distribution. Mt. Kini Balu, Borneo (Whitehead).

The above subdivision of this troublesome and badly understood species is based chiefly upon the conclusions of Gurney, Hume, Sharpe and O. Grant, but it can only, we think, be regarded as a provisional one. In his elaborate and careful article on *A. virgatus* (*l 3*), Gurney distinguished three forms of this species, viz: the typical one, *A. virgatus*, a large race of the Himalayas (*A. affinis* subsp.) and a third, differing in coloration, from the Philippines (*A. manilensis* subsp.). Gurney distinguishes these from the typical form as follows:

"The larger form *(affinis)* chiefly differs, as regards coloration, from the typical *A. virgatus* (the range of which is decidedly more southern, though both races inhabit the most northerly parts of India) in the bright rufous, which usually characterizes the under surface of the old males of *A. virgatus*, being replaced in those of the northern race by a non-rufous chocolate brown". "The nearly allied hawk of the Philippine Islands (*A. manilensis* subsp.) is remarkable for having the rufous colouring of the breast as strongly developed in the adult female as in the male, which I believe is never the case in the typical *A. virgatus*; and, in addition to this, it also differs from *A. virgatus* in the somewhat browner and less slaty tint of the upper surface, and in the dark gular stripe being, in some adult specimens, much less distinctly marked."

A fourth form, which we have termed *A. virgatus gularis* (T. & S.), is treated by Gurney as a distinct species — *A. nisoides* Blyth[1]). The means of distinguishing this form from the typical *A. virgatus* are not pointed out; Dr. R. B. Sharpe, however, considers it[2]) "a paler breasted form of *A. virgatus* with a shorter middle toe (28 mm as against 30.5 mm circa), the female being barred underneath with brown, but not with rufous. It ranges from China down the Malay Peninsula, to Java and Timor, visiting these localities in winter" *(k 5)*. Hume speaks of it as wanting the central gular stripe, but this does not always hold good, — at least for Chinese birds *(k 7)*. Mr. Ogilvie Grant *(m 5)* has quite recently re-investigated this species, arriving at much the same result as Mr. Gurney, and holding *A. gularis* for a good species. Whether treated as a species or a subspecies does not lessen the difficulty of determining it, and Grant's careful analysis seems to be applicable only to the old female. Mr. Hartert *(m 6)*, with Gurney's and Grant's results before him, found much difficulty in determining Everett's 3 examples from Djampea; the wing-formula and the markings on the breast did not correspond with Grant's diagnosis. When the migration of the subspecies is taken into consideration, it is probable rather that they were the *typical A. virgatus* than the East Asiatic form, *gularis*, which in conformity with many other migrants from the north might be expected not to travel so far south.

Dr. Sharpe's *A. rufotibialis* may be distinguished from the others by its red thighs.

Facts, as at present understood, appear to point to the circumstance that *A. virgatus* in the warmer parts of its distribution is stationary and has become modified there into several local races; in winter, however, these localities are invaded by quantities of individuals migrating southwards from China, Japan and probably elsewhere, where they have developed some racial differences, and which now become intermingled with the stationary races, and this circumstance makes a satisfactory understanding of the species hardly possible at the present day. In the spring the northern birds, apparently, separate themselves from their southern relations and proceed again to their breeding quarters in China and Japan. Had facts relating to these points been more plentiful when Gurney wrote, it is probable that he would have had reason to modify his views; facts, however, are still scanty. In North China, near Pekin, Abbé David says that the species arrives in spring in great numbers *(l 2)*; about the lower Yangtse Basin Mr. F. W. Styan marks it as a non-breeding species, which passes in migration, and mentions a pair of specimens taken at sea, between Shanghai and Nagasaki on 6th May *(l 6)*; in South China, Foochow, Mr. De La Touche notes it as occurring in spring and autumn *(j 6)*; in a note sent with a specimen from Kini Balu, Mr. Whitehead speaks of it as being evidently a migrant

---

1) Blyth's name was published in 1847, Temminck and Schlegel's Raptores of the Fauna Japonica, Aves, in 1845 (see Seebohm, B. Japan. Empire, 1890, 3); consequently these authors have the priority.
2) Under the name *A. stevensoni*.

to that mountain of North Borneo *(l 4)*; Gurney mentions a specimen captured at sea off the Natuna Islands on 14th November (not "Nantura" as Gurney writes, see *l 1*), and we cannot agree with him that it had "accidentally" wandered so far to the southward *(m 2)*; the same author elsewhere expresses the opinion that it may occur as "only a winter visitor to Java" (P. Z. S. 1878, 938), and Dr. Sharpe takes the same view. But, since other species, such as *Tachyspizias soloensis*, *Butastur indicus* and *Ninox scutulata japonica*, migrating southwards from China and Japan, appear to fix their winter limits in Borneo and North Celebes, going little south of the equator, it is desirable to ascertain whether the birds, which invade Java in our winter, are not wanderers from India and Tenasserim, making a movement similar to that which appears to take place with *Butastur liventer* of the same regions.

More inquiry should also be made as to whether the Philippine Islands are really inhabited by a stationary and distinguishable race, inasmuch as the few specimens known from those quarters appear to have been obtained during the time of the southern migration from China. Luzon, January; Guimaras, March (Meyer *n 1*); Mindanao, Dec. — April (Koch *o 1*); Mindanao — 4 specimens, April (Everett *j 3*); Luzon, 2 sometime between Jan. 1st and April 3rd (Whitehead *f 2*), 1 in winter, Luzon (Whitehead *f 3*).

A single specimen of *A. virgatus gularis* obtained in North Celebes (Gorontalo) by Dr. Riedel has been most kindly lent to us by Mr. Pleske; it is now preserved in the St. Petersburg Museum. It is a new addition to the avifauna of Celebes. Djampea Island, from where Mr. Hartert records three specimens, also falls within the Celebesian Province as defined in this book.

Appended is a general description of the species and of the above specimen from Celebes.

Figures and descriptions. Temminck & Schlegel *i 1*; Gurney *k 1, j 3, k 4, m 1, m 2*; Sharpe *l 1, k 5*; Hume *k 7*; David & Oust. *l 2*; Grant *m 5*; Blanford *m 7*.

Adult (general description). Above blackish slaty, ear-coverts and sides of neck greyer; tail ashy grey crossed with three or four bars of blackish; throat, abdomen and under tail-coverts white, the former marked with a dark mesial streak; rest of under surface vinous chestnut; under wing-coverts whitish marked with brown; wing below ashy crossed with blackish bars.

Measurements (from Grant *m 5* converted to mm).	Wing	Tail	Tarsus
Typical *A. virgatus* ♀	183	145	51
A. virgatus affinis ♀	203-221	168-180	55-56
A. virgatus gularis ♀	183-190	127-142	50-53
A. virgatus gularis ♂	157-167	114-119	41-47
A. virgatus manilensis ♀	175-178	137-140	52-53
A. virgatus manilensis ♂	152-155	122-124	51
A. virgatus rufotibialis	149-150	114-117	46

Young. A. virgatus gularis. Above a somewhat greyish dark brown, the nape a little varied with white, the hind neck with rufous; wing-coverts, secondaries and feathers of back terminally margined with rufous; tail hair-brown, crossed with four dark bars, the outer feathers with about eight; throat and under tail-coverts white, the former with a slight trace of a gular streak; remaining under surface white, marked with streaks of rufous brown on the breast, with heart-shaped spots and bars on the abdomen, sides, thighs and under wing-coverts (Celebes — Riedel, St. Petersburg Museum).

Wing 170 mm; tail 122; tarsus 45; middle toe 27 (without claw), culmen from cere 9. The young of *Accipiter rhodogaster* readily distinguishes itself from the young of *A. virgatus* by the hazel-rufous of its upper plumage, barred on the wings and tail and spotted on the back with black; also the middle toe (32 mm) and the beak (13 mm) are longer.

GENUS SPIZAETUS Vieill.

The Hawk-eagles are powerful birds of large-medium size (larger than a Raven); with or without a crest; wing somewhat short, secondaries long, from carpus to the tip of the secondaries about five-sixths of the length of the wing; tomia festooned; tarsus completely feathered down to the base of the toes; toes somewhat short; claws large, strongly hooked, the hinder one nearly half as long again as the toe-joint. The species of the genus are stationary, about ten in number, found in the Oriental Region as far as Celebes, Africa, Central and South America. They prey upon large birds, mammals, lizards, and lay one or two eggs.

+*12. SPIZAETUS LANCEOLATUS Temm. & Schl.
Celebesian Hawk-eagle.
Plates II and III.

Spizaetus lanceolatus *(1)* Temm. & Schl. (Celebes tantum), Faun. Jap. Aves 1845, 8; *(2)* Bp., Consp. 1850, I, 29 partim; *(3)* Wall., Ibis 1868, 13; *(4)* Pelz., Verh. z.-b. Ges. Wien 1872, 426 (Aru!); *(5)* Sharpe, Cat. B. 1874, I, 270; W. Blas., J. f. O. 1883, 134; *(6)* id., Ztschr. ges. Orn. 1886, 195; *(7)* Büttik., Zool. Erg. Webers Reise in Ost-Ind. 1893, III, 272; *(8)* M. & Wg., Abh. Mus. Dresden 1895, Nr. 8, p. 4; *(9)* iid., ib. 1896, Nr. 1, p. 7.

b. **Spizaetus cirratus** part. *(1)* Schl., Mus. P.-B. Astures 1862, 9 (Celebes); *(II)* id., Valkv. Ned. Ind. 1866, 16, 55, pl. 7, f. 2 (Celebes) et f. 3 (Sula); *(3)* id., Rev. Accip. 1873, 57 (Celebes, Sula); Rosenb., Malay. Arch. 1878, 271.

c. **Spizaetus fasciolatus** "Temm." (err.) Schl., Astures 1862, 9; id., Valkv. 1866, 53; id., Rev. Accip. 1873, 58.

d. **Limnaetus lanceolatus** Gray, HL. 1869, I, 13, pt.; *(1)* Wald., Tr. Z. S. 1872, VIII, 34, 110; Gurney, Ibis 1877, 424; *(2)* Meyer, Ibis 1879, 56; *(3)* Salvad., Orn. Pap. 1880, I, 5; Gurney, Diurn. B. of Prey 1884, 49; Guillem., P. Z. S. 1885, 545.

"Bakatoa", Tjamba, S. Celebes (Platen in Dresd. Mus.).
"Koheba burik", Banka Id., N. Cel. (Nat. Coll.).
"Kiokkiok"[1]), Manado (Nat. Coll. in Mus. Dresd.).

Figures and descriptions. Schlegel *b II, b 3*; Walden *d 1*; Sharpe *5*; W. Blasius *6*.

Adult male. Head and neck blackish brown, the bases of the feathers white, the margins and sub-basal part of the feathers of the neck and sides of the head broadly washed with cinnamon, giving a streaked appearance to these parts; back and wings sepia glossed with purple, darkest on the mantle and shoulders; lower back paler; upper tail-coverts narrowly tipped with whitish; tail whitish brown, slightly mottled and washed with darker shades and crossed with four bands of blackish brown, a broad space (about 45 mm) separating the two endmost ones; throat white, with a black stripe down the middle[2]) of it and at the sides (sub-malar stripe); breast and sides pale brownish rufous marked with broad drop-shaped spots of black and, towards the abdomen, irregularly with white ones; remaining under parts white, closely barred with brown, of a rufous tinge on the abdomen and blacker as well as more closely barred on the legs; under wing-coverts white, thickly spotted and barred with dark brown; quills below white varied with grey, brown towards the distal ends, and with more or less perfect remains of about five dark cross-bars, which are also perceptible on the upper side on the inner webs. "Iris gold-yellow; core bluegrey; bill black; feet citron-yellow. Length 550 mm; expanse 1160." (♂, Tjamba, S. Celebes: Platen — Nr. 6670).

Adult female measured by Dr. R. B. Sharpe has: wing 376; tail 274; tarsus 82.5 (? 92.5); culmen 42.

A second specimen, from the neighbourhood of Manado (Nat. Coll.) is darker on the under surface than the above, especially on the abdomen. Head black; the black moustachial and throat stripes broad and well-developed (C 10846).

Female. Much larger than the male.

Young. Head and neck white, narrowly streaked on the crown and more broadly on the neck with dark brown; all other upper parts dark brown with a purplish gloss as in the adult, the unexposed inner webs of the secondaries white; upper tail-coverts tipped with white, the basal part of the feathers varied with white and barred with brown; tail much as in the adult, crossed with three broad bars; under surface white, almost uniform on chin and throat, somewhat sparsely marked with drop — shaped spots of dark brown on the breast, a few broad but ill-defined markings of washy rufous on abdomen, indistinct bars of brownish rufous on the abdomen and under tail-coverts, and close bars of dark brown — narrower and less dark than in the adult — on the flanks and thighs; under wing-coverts white, spotted somewhat sparsely with dark brown; quills and tail below much as in the adult. (N. Celebes — Riedel in St. Petersburg Mus.)

Another specimen, still younger, much resembles the above, but has still more white about it; sides of wing more mottled with white; head and neck more narrowly streaked with brown; under surface white, marked with a few streaks of brown on the breast, thighs and flanks narrowly barred with rufous or brown, very indistinct on the inner side of the thighs; some brownish bars just discernible on

---

[1]) A name given for *Spilornis rufipectus* also.
[2]) Prof. W. Blasius *(6)* has expressed the opinion that the black stripe down the middle of the throat mentioned by Bonaparte *(2)* does not really exist in this species. We find it to be always present in the adult and nearly adult, as shown by four before us, and Schlegel *(b II)* has described it in three in the Leyden Museum. It is, however, absent in the very differently plumaged young bird (see plate).

the under tail-coverts; under wing-coverts still more scantily spotted with brown than in the above. In almost every detail the description of the young *Pernis celebensis* would apply equally well to this bird (N. Celebes — Dresd. Mus.).

Immature. Intermediate between the young and adult: head whitish, broadly streaked with black, breast marked with guttate streaks of blackish brown; under parts mottled with pale brown and whitish, with feathers regularly barred with dark brown intermixed; flanks and thighs narrowly barrred as in the adult; upper parts dark brown, tail with five bands, the basal one indistinct (Banka Id., N. Celebes, 17. May 1893: Nat. Coll. — C 12235).

Two young specimens in the Leyden Museum have the head all white.

	Wing	Tail	Tarsus	Mid. toe without claw	Culmen from cere
a. (Nr. 6670) ♂ ad. S. Celebes (Platen) . . .	355	258	91	45	30
b. (C 12235) [♂] imm. Banka Id., N.Cel. 17. V. 93 (Nat. Coll.)	375	265	—	50	31
c. (Sarasin Coll.) ♂ vix ad. Lake Posso, 17.II. 95 (P. & F. S.)	358	265	91	—	29.5
d. (C 14246) [♂]juv. Lembeh Id. 28.II. 95 (Nat. Coll.)	355	260	—	—	31
e. (Mus. Petersb.) [♂] juv. N. Celebes (Riedel) .	370	270	91	49	31
f. (C 10846) [♀] ad. near Manado VIII—IX. 92 (Nat. Coll.)	410	290	97	53	35
g. (Sarasin Coll.) ♀ juv. Rurukan Forest, N. Cel. 2. VII. 94 (P. & F. S.)	395	276	—	—	34.5
h. Nr. 14077 [♀] juv. N. Celebes (Riedel) . . .	406	285	90	54	34

Distribution. Celebes and Sula. Gorontalo Distr. (Forsten *b 1, b 3*); Minahassa (Wallace *5*, v. Rosenberg *b 3*; etc.); Banka Island off the Minahassa (Nat. Coll. in Dresd. Mus.); Lembeh Island (Nat. Coll. ib.); Central Celebes — Lake Posso (P. & F. Sarasin *9*); South Celebes — Macassar (Wallace *5*), Tjamba Distr. (Platen in Dresd. Mus.), Maros (Weber *7*); Sula Islands — Sula Besi (Bernstein *b II, b 3*).

Aru, recorded as the land of origin of this species by von Pelzeln *(4)*, may be safely ascribed to erroneous labelling, as v. Pelzeln himself subsequently concluded (see: Salvadori *d 3*).

Nothing is known of the habits of this Hawk-eagle. The allied *S. limnaetus* of Java preys on rails, waterfowl, ducks, chickens, robbing also the nests of other birds (Schl. Astures, 7). Colonel Legge found the favourite food of *S. ceylonensis* (Gm.) to be a large lizard, *Calotes*. It also devours Squirrels and other small mammals (B. Ceylon, 59). The allied species, *S. cirrhatus* (Gm.) and *S. ceylonensis*, lay as a rule only one egg; *S. limnaetus* one or two eggs (Legge, l. c.; Oates, Hume's Nests and Eggs, 1890, III, 147, 149). Consequently, among the Hawk-eagles, there is no overcrowding; the birds are stationary, and a number of local races have been built up.

Touching the remarkable similarity of the plumage of this species to *Pernis celebensis*, see the descriptions of the latter and plates).

The form of *Spizaetus* most nearly allied to *S. lanceolatus* is perhaps *S. philippensis* Gurney. A figure of this species given by Walden (Tr. Z. S. 1875 IX, pl. XXIV) displays a bird very like the adult Celebesian Hawk-eagle, but differing by having a crest, and the cross-bars on the under surface commencing first on the flanks, instead of extending on to the lower breast and abdomen. The *Spizaetus limnaetus* (or *caligatus* Raffl. as it is called by Gurney), ranging from India to Java, occurs, as Gurney remarks (Ibis 1877, 425), "under two very distinct phases of plumage, if, indeed both be really referable to one species" (see, also, Whitehead, Ibis 1889, 71), one black-brown all over, the other particoloured. *S. lanceolatus*, when adult, seems to be readily distinguishable by its barred legs from the latter.

GENUS LOPHOTRIORCHIS Sharpe.

A crested Hawk-eagle of medium size, differing from *Spizaetus* in having a much longer, more pointed wing, the secondaries falling short of the tip by one-third of the length of the wing (instead of one-sixth); the middle toe much longer, being three-fourths the length of the tarsus, instead of about one-half that length as in *Spizaetus*. Evidently a form of more sustained flying-powers. Preys on birds and small mammals. Probably not strictly stationary. One species ranging from the Himalayas to Celebes; a second in Columbia.

13. LOPHOTRIORCHIS KIENERI (G. Sparre).

Rufous-bellied Hawk-eagle.

a. **Astur Kieneri** *(1)* G. S.[1]), Mag. Zool. 1835, pl. 35 (ad.).
b. **Spizaetus cristatellus** *(1)* Jard. & Selby, Ill. Orn. pl. 66.
c. **Spizaetus albogularis** Tickell apud Blyth, J. A. S. B. 1842, XI, 456.
d. **Limnaetus kieneri** Strickl., Ann. N. H. 1844, XIII, 33; *(1)* Jerd., B. of Ind. 1862, I, 74; Legge, Str. F. 1875, 198; *(2)* Gurney, Ibis 1877, 433; *(3)* Salvad., Orn. Pap. 1880, I, 5.
e. **Nisaetus kieneri** *(1)* Jerd., Ill. Ind. Orn. 1847, p. 5, pl. I.
f. **Spizaetus kieneri** Gray, Gen. B. 1845, I, 14; Bp. Consp. 1850, I, 29; Blyth, J. A. S. B. 1850, 335; Horsf. & Moore, Cat. B. Mus. E. Ind. Co. 1854, I, 34; Strickl., Orn. Syn. 1855, I, 71, *(1)* Schl., Mus. P.-B. Astures, 1862, 11; *(2)* Gurney, Gld. B. of Asia 1863, pt. XV; *(3)* Wall., Ibis 1868, 14; Gray, HL. I, 13; *(4)* Hume, Rough Notes 1869, 201, 216; *(5)* Schl., Rev. Acc. 1873, 58; *(6)* Hume, S. F. 1873, I, 310; *(7)* Salvad., Ucc. Borneo, 1874, 16; *(8)* Reichnw., J. f. O. 1877, 218; *(9)* W. Blas, ib. 1883, 121, 122; *(10)* Hartert, ib. 1889, 374.
g. **Spizaetus kieneri** Blyth, Cat. B. Mus. As. S. B. 1849, 26.
Lophotriorchis kieneri *(1)* Sharpe, Cat. B. 1874, I, 255, 458; *(2)* Legge, B. Ceylon, 1880, 43—46; *(3)* Hume, Str. F. 1880, IX, 273, 274, 277; *(4)* Gurney, Diurn. B. of Prey 1884, 50; *(5)* Sharpe, Ibis 1889, 71; id., ib. 1890, 274; *(6)* Everett, J. Str. Br. R. A. S. 1889, 182; *(7)* Steere, List B. Philip. Is. 1890, 7; *(8)* Salvad., Ann. Mus. Civ. Gen. 1891, XII, 41; *(9)* Hose, Ibis 1893, 419; *(10)* Blanf., Faun. Br.

---

[1]) Dr. R. B. Sharpe and others ascribe these initials to Geoffroy St. Hilaire; Count Salvadori and Schlegel to Gervais; Colonel Legge shows good reason for believing them to belong to G. Sparre.

Ind. B. III 1895, 345; *(11)* Grant, Ibis 1895, 438; *(12)* Hartert, Nov. Zool. 1896, 575.

Figures and descriptions. G. S. *a 1*; Jardine & Selby *b 1*; Jerdon *d 1*; Schlegel *f 1*; Gurney *f 2*; Sharpe *1*; Hume *f 6, 3*; Legge *2*, Blanford *10*.

Adult. Above blackish brown, 6 or 7 narrow indistinct bars on tail; occipital crest about 60 mm long; chin, throat and breast white, passing into ferruginous on the remaining under parts, which, together with the sides of the breast, are marked with dark shaft-streaks; tail below greyish; quills below white, greyer towards tip; under wing-coverts with the greater series tipped with white and edged with fulvous.

Measurements.

	Wing	Tail	Tarsus	Culmen from cere
♂	360—394	208—228	68.5—76	25.5—28
♀	432—444	254—317	76	30.5 (from Legge *2*).

Immature. Crown and sides of head and neck tawny-brown, becoming more fulvous on neck; the bases of the feathers white (producing a mottled appearance on the crown in which the tawny-brown feathers of a more advanced stage of dress are sprouting); short occipital crest (1.6 in.) brown as head; upper parts Prout's Brown, becoming drab-colour mixed with whitish on the greater wing- and upper tail-coverts; secondaries and the three inmost primaries tipped with whitish; tail brown as back, crossed with five narrow equidistant bars of black of fairly equal width and separated by spaces of about 25 mm; all under parts, including legs, pure white; the flanks and outer side of the thighs with some feathers of brownish intermixed. Wing 362 mm; tail 230 mm; tarsus 71 mm; middle toe 53 mm. Celebes: v. Muschenbr. Nr. 6671).

Distribution. Himalayas (Sparre *a 1*; Inglis *2*, etc.); India (Jerdon *d 1*, Tickell *2*); Ceylon (Bligh, Legge *2*); Malacca and Singapore (Hume *3*); Sumatra (Hartert *f 10*, Modigliani *8*); Java (Gurney *f 2*); Borneo (Wallace *f 3*; Whitehead, Fischer, *5, 6*); Philippines, Luzon (Gevers *f 1, f 5*, Whitehead 11); Mindanao, Panay, Marinduque (Steere *7*); Celebes (Faber *f 8, f 9*, van Musschenbroek in Mus. Dresd.); ? Batchian (Norwich Mus. *d 2, d 3*); Satonda, Lesser Sunda Is. (Doherty *12*).

The first mention of the occurrence of this species in Celebes was made by Prof. Reichenow in reference to a specimen brought from that island by v. Faber, and later, in answer to a communication from Prof. W. Blasius, Dr. Reichenow confirmed his first note *(f 9)*. A second specimen in the Dresden Museum, labelled Manado by van Musschenbroek, is in immature plumage, and corresponds fairly well with one described by Hume, Str. F. IX, 274. We have described it above.

Probably because the immature plumage of this species remained for a long time unknown, there has been much misconception touching the specimen in the Leyden Museum from Luzon referred by Schlegel to this species. Gurney *(f 2)* included this specimen under his *Spizaetus philippinensis*, overlooking the much longer tarsi and tail of that species; and other authorities have followed him. Consequently the Philippine Islands have been generally struck out of the range of this species. There is no reason to doubt the correctness of Schlegel's identification of the specimen from the Philippines, and in the

Leyden Museum we have seen just such another immature bird from the Philippines as that described above from Celebes.

The specimen in the Norwich Museum stated to have come from Batchian bears no collector's name on the label. It was purchased of Bouvier of Paris, and most probably the locality indicated is quite correct, but, until more evidence is forthcoming, it must be looked upon as of uncertain origin (see Salvadori *d 3*).

The genus *Lophotriorchis*, founded by Dr. Sharpe for this species and *Sp. isidori* of South America and to which *L. lucani* Sharpe & Bouvier of S. W. Africa has since been added[1]), occupies a position about midway between *Nisaetus* and *Spizaetus*, resembling the former by its long wings, feet and tarsi, and the latter by its bill and immature plumage. The specimen in the Dresden Museum agrees in many respects of coloration with *Spizaetus alboniger* Blyth, juv., from Malacca and Borneo, but the following scheme shows how widely they differ structurally:

L. kieneri juv. wing $\frac{14.25}{2.80}$ mid. toe $\frac{2.10}{2.80}$
tarsus tarsus

S. alboniger juv. wing $\frac{11.00}{2.625}$ mid. toe $\frac{1.375}{2.625}$
tarsus tarsus

It will be found that, in proportion to the tarsus, the wing of *Lophotriorchis kieneri* is 2½ inches longer than that of *Spizaetus alboniger*, and the middle toe 0.64 inches longer.

L. kieneri is a rare species, and little is known of its habits. Colonel Legge points out that it is a hill-haunting bird. Its food consists of birds and small mammals; Mr. Wallace's Sarawak example was killed while devouring a pigeon — a fact which points to its fine flying powers, which have been remarked upon by Legge and others.

From its rarity Mr. Whitehead *(5)* believes this to be a migratory species in Borneo, visiting the country during the N. E. monsoon, towards the end of which (March 20th, 1887) his specimen from Kini Balu was obtained. Mr. Hume's example from Singapore and Count Salvadori's from Sumatra were likewise shot during the N. E. monsoon, viz: Jan. 19th 1880, and Feb. 3rd 1891, respectively. Dr. Steere found it in Mindanao, Panay and Marinduque in Oct.— Dec., January and May, the last being a very late date for a migratory species; but until more evidence is forthcoming of course no conclusion as to migration can be drawn.

GENUS ICTINAETUS Jerd.

This Eagle is well characterized by its foot: the inner toe, including the claw, is much longer than the middle one; the claws very little curved, that of the hallux and of the inner toe very long, exceeding the length of the

---

[1]) Gurney considered this species to be *Nisaetus spilogaster* juv. (Diurn. B. Prey, 52, note).

toes; tarsi completely feathered; wings very long; edge of upper mandible without any tooth or festoon. A single species, ranging from the Himalayas to the Moluccas; possibly migratory. It carries off birds' nests, for which purpose its foot is adapted, and devours the eggs or young; feeds also on reptiles, mammals.

+ 14. ICTINAETUS MALAYENSIS (Reinw.).
Black Eagle.

a. **Falco malayensis** *(1)* Reinw. in Temm., Pl. Col. Nr. 117 (1824).
b. **Aquila malayensis** Vig., Zool. Jrn. 1824, I, 337; *(1)* Schl., Mus. P.-B. Aquilae 1862, 11; *(II)* id. Valkv. 1866, 8, 49, pl. 3, f. 1, 2; *(3)* id., Rev. Acc. 1873, 116; *(4)* Rosenb., Malay. Arch. 1878, 271.
c. **Aquila (Heteropus) pernigra** Hodgs., J. A. S. B. 1836, V, 227.
d. **Nisaetus ovivorus** Jerd., Madr. Journ. 1844, XIII, 158.
e. **Neopus perniger** Hodgs. in Gray's Zool. Misc. 1844, 81.
f. **Ictinaetus perniger** Blyth, Ann. N. H. 1844, XIII, 114.
g. **Onychaetus malayana** Kaup, Classif. Säug. u. Vög. 1844, 120.
h. **Neopus malayensis** Gray, Cat. Hodgs. Coll. B. 1846, 42; *(1)* Beavan, P. S. Z. 1868, 396, pl. XXXIV; *(2)* Sharpe, Cat. B. 1874, I, 257; *(3)* Brüggem., Abh. Ver. Bremen 1876, V, 45; *(4)* Gurney, Ibis 1877, 423; *(5)* Legge, B. Ceylon 1880, 47; *(6)* Salvad., Orn. Pap. 1880, I, 6; *(7)* Oates, B. Brit. Burmah 1883, II, 190; *(8)* Davison, Str. F. 1883, X, 335; *(9)* Sharpe, P. Z. S. 1888, 268; *(10)* id., Ibis 1889, 71; id., ib. 1890, 274; Everett, J. Str. Br. R. A. S. 1889, 181; *(11)* Oates, Hume's Nests and Eggs 1890, III, 145; *(12)* Salvad., Orn. Pap. Agg. 1889, 11; *(13)* Hartert, J. f. O. 1889, 376; *(14)* Sharpe, Ibis 1893, 563; *(15)* M. & Wg., Abh. Mus. Dresd. 1896, Nr. 1, p. 4; *(16)* Büttik., Notes Leyden Mus. 1896, XVIII, 195.

Ictinaetus malayensis *(1)* Blanf., Faun. Brit. India B. III, 1895, 347.

"Oopo", Celebes (v. Rosenb. *b 4*).

For full synonymy and references see Salvadori *6*.

Figures and descriptions. Temminck *a 1*; Schlegel *b II*; Beavan *h 1*; Sharpe *h 2*; Gurney *h 4*; Legge *h 5*; Salvadori *h 6*; Oates *h 7, h 10* (egg); Blanford *1*.

Adult. Entire plumage brownish black; tail marked on the upper surface with about 7 indistinct and broken bars of brownish grey, which are whitish and distinct on the under surface (Halmahera Nr. 2945). Iris brown; bill, gape and feet yellow; tip of bill black.

Young. Above brownish black; greater wing-coverts and secondaries tipped, the tail-coverts barred, with white; the feathers of the back, lesser, and middle wing-coverts with a small whitish spot at the tip of each; tail black, indistinctly crossed with numerous bars of grey-brown; feathers of head and neck broadly margined and tipped with whitish; face, throat and under surface pale tawny, with broad sooty brown margins to the feathers of the breast and sides; wings and tail below marbled with broken bars of whitish (♂ juv. Loka, S. Celebes, 6. X. 95: P. & F. Sarasin).

Measurements.	Wing	Tail	Tarsus	Bill from cere
a. (Sarasin Coll.) ♂ juv. Loka, S. Cel. 6. X. 95 . . .	530	294	73	26
b. (Sarasin Coll.) ♀ [?] juv. Rurukan, N. Cel. 20. XI. 94	538	285	71	30
Small ♂ adult (from Legge *h 5*).	520	305	81	31
Large ♀ adult (from Legge *h 5*) . .	635	375	97	35

Distribution. India, Ceylon, Burmah, Tenasserim, Malacca, Sumatra, Java, Borneo, Celebes, Ternate, Halmahera (fide Legge *h 5*, Salvadori *h 6*), Nias (Nieuwenhuisen & v. Rosenberg *h 16*. In Celebes: — Minahassa (v. Duivenbode *b 3*, Fischer *h 3*, P. & F. Sarasin *h 15*); S. Peninsula, Loka (P. & F. Sarasin); (?) Sula (apud G. R. Gray, HL. 1869, I, 11).

Though the locality Sula may very probably be correct, G. R. Gray, as Salvadori has remarked, has never shown his grounds for this indication.

The remarkable shape of the foot of this Aquiline species is connected with something very unusual in its habits. "It subsists", writes Colonel Legge. "as far as can be observed, entirely by bird-nesting, and is not content with the eggs and young birds which its keen sight espies among the branches of the forest trees, but seizes the nest in its talons, and decamps with it, and often examines the contents as it sails lazily along. Furthermore, Mr. Bligh informs me that he once found the best part of a bird's nest in the stomach of one of these Eagles which he shot in the Central Province! . . . . . The long inner claws of this bird seem especially adapted for the work of carrying off loose and fragile masses such as the nests of birds". Their length and straightness also, it may be added, would enable the bird to let fall a nest, when it has done with it, without difficulty and risk of entanglement, — no small consideration, when it is remembered into what a panic of alarm and rage a wild animal is thrown at finding itself clung to by an object from which it tries in vain to free itself. It also occasionally carries off large birds, but this may be of rare occurrence; a rat in the stomach is noted by Beavan *(h 1)*; a rat, a bird's egg and a snake's egg by Mr. Wray *(h 9)*; a snake by Davison *(h 8)*. Temminck states probably from information of Reinwardt that it eats insects, as well as birds and reptiles. In India Jerdon remarks that "doves, and perhaps some other birds breed at all times in the year; and it may, perhaps, obtain eggs or nestlings at all seasons, by shifting its quarters and varying the elevations, if not, it probably may eat reptiles; but of this I cannot speak from observation".

This Eagle is rare in Celebes and in the other parts of the Malay Archipelago where it has been found, and we suspect, it occurs only in migration, or as a somewhat frequent straggler. It was never met with by Mr. Wallace in any island; Meyer never saw it in Celebes, and the only examples recorded from there prior to 1894 were two, both immature — one in the Leyden *(b 3)* and the other in the Darmstadt Museum.

In November 1894, a third example, also young, was obtained by the cousins Sarasin in the same province as the previous specimens, the Minahassa, and in October, 1895, a young example from the extreme south of the island was added to their collection.

When von Rosenberg in his "Malayischen Archipel", p. 271, speaks of it as being "nicht sehr häufig", there we do not doubt that he must have mistaken some other large bird of prey for it; especially as there is no specimen obtained by him in the Leyden Museum. No doubt the movements of this bird will be found to depend mainly upon the time of production of its favourite diet, the eggs and young of other birds; and this varies in different localities. Mr. Whitehead has obtained it in Borneo in June. Other notices are: N. Celebes, November; S. Celebes, October; Ternate, March (Bernstein *b 3*) and December (Bruijn *h 6*); Halmahera, December (Bruijn *h 6*); Borneo, October and February (Doria and Beccari, Salvad. Ucc. Borneo 1874, 5, Whitehead *h 10*); Malay Peninsula, Sept.—Oct. (Wray *h 9*). In the winter months also it has been noticed to be more plentiful in the Himalayas at an elevation of seven or eight thousand feet than it is during the hot season (Legge *h 5*), a curious reversion of the usual rule of migration, namely, that of going to a warmer climate during the cold season, and showing that food is of greater consequence to birds than is the weather.

GENUS HALIAETUS Sav.

The Sea-eagles are of large size; wing very long, tail rounded, "scarcely exceeding the closed wings" (Legge); tarsus stout, the upper third feathered in front, the lower two-thirds covered in front with transverse shields; toes moderate; claws semicircularly hooked, grooved below, soles very rugose, furnished with sharp horny points (adapted for holding slippery booty like fish); bill large, basally — where covered by the cere — straight, tomia with a slight festoon. Feeds on mammals, reptiles, fish, carrion. Lays two eggs. Six species, migratory and stationary, distributed over most of the tropical and temperate sea-coasts of the world, except those of South America.

15. HALIAETUS LEUCOGASTER (Gm.).

White-bellied Sea-eagle.

a. **Falco leucogaster** Gm., S. N. 1788, I, 257 (ex Latham).
b. ? **Falco blagrus** Daud., Tr. d'Orn. 1800, II, 70 (ex Levaill.).
c. **Falco oceanica** *(1)* Temm., Pl. Col. 1823, Nr. 49.
Haliaetus leucogaster Vig., Zool. Journ. 1824, I, 336; *(1)* Gld., Syn. B. Austr. 1838, pl. 37, f. 1; *(2)* Schl., Mus. P.-B. Aquilae 1862, 14; *(III)* id., Valkvogels 1866, 9, 50, pl. 4, f. 1, 2; *(4)* Finsch & Hartl., Orn. Centralpol. 1867, 1; *(5)* Schl., Rev. Accip. 1873, 117; *(6)* Sharpe, Cat. B. 1874, I, 307; *(7)* Legge, B. Ceylon 1880, 68; *(8)* Vidal, Str. F. 1880, 32; *(9)* Gurney, Ibis 1882, 235; *(10)* Oates, B. Brit. Burmah 1883, II, 199; *(11)* Gurney, Diurn. B. of Prey 1884, 59; *(12)* Vorderman, N. Tijd. Ned. Ind. XLII, 3 (sep. copy); *(12bis)* Guillem., P. Z. S. 1885, 545; *(13)*

Styan, Ibis 1888, 232; *(14)* Sharpe, Ibis 1888, 195; *(15)* id., ib. 1889, 73; 1890 274; *(16)* Hartert, J. f. O. 1889, 376; *(17)* Everett, J. Str. Br. R. A. S. 1889, 183; *(18)* North, Nests and Eggs B. Austr. 1889, 7, pl. I, f. 2 (egg); *(19)* Steere, List B. Philip. 1890; 7; *(20)* Oates, ed. Hume's Nests and Eggs Ind. B. 1890, III, 164; *(21)* Hose, Ibis 1893, 419; *(22)* Büttik., Zool. Ergebn. Weber's Reise in Ost-Ind. 1893, III, 272, 286, 290; *(23)* Studer, Reis. Gazelle 1889, III, 198; *(24)* Hagen, Td. Ned. Aard. Genoots. 1890 *(2)* VII, 130; *(24 bis)* Hickson, Nat. in N. Celebes 1889, 35; *(25)* Bourns & Worces., B. Menage Exped. 1894, 33; *(26)* Vorderm., N. T. Ned. Ind. 1895, 318; *(27)* Blanf., Faun. Br. Ind. B. III 1895, 318; *(28)* M. & Wg., Abh. Mus. Dresden 1895, Nr. 8 p. 4; *(29)* iid., ib. 1896, Nr. 2, p. 7.

d. **Haliaetus sphenurus** *(1)* Gld., Syn. B. Austr. 1838, pl. 37, f. 2.

e. **Ichthyaetus leucogaster** *(I)* Gld., B. Austr. 1848, I, pl. 3; *(II)* Diggles, Orn. Austr. 1866, pt. 5.

f. **Cuncuma leucogaster** Gray, Cat. Acc. Br. Mus. 1848, 24; *(1)* Hume, Str. F. 1876, IV, 422, 461, 462; *(1a)* Meyer, Ibis 1879, 56; *(2)* Salvad., Orn. Pap. 1880, I, 7—11, 548, et Agg. 1889, 11; *(3)* W. Blas., P. Z. S. 1882, 698; *(4)* id., Ornis 1888, 304; *(5)* Salvad., Orn. Pap. Agg. 1889, 11; *(6)* Hartert, Nov. Zool. 1895, 476; *(7)* Salvad., Ann. Mus. Civ. Gen. 1896, XXXVI, 58.

"**Kadjawali**", Main, Minahassa, Nat. Coll.
"**Koheba laut**", Lembeh Id., iid.
"**Bunia**", Tonkean, E. Celebes, iid.
"**Kuwajang**", Banggai Id., iid.

For further synonymy and references see Salvadori *f 2, f 5*.

Figures and descriptions. Temminck *e I*; Gould, *I, d I, e I*; Schlegel III; Diggles *e II*; North *18* (egg); Finsch & Hartlaub *4*; Schlegel *5*; Sharpe *6*; Hume *f 1*; Salvadori *f 2*; Legge *7*; W. Blasius *f 3*; Vorderman *12*; Blanford *27*; etc.

Adult. Head, neck and all the under parts white; upper surface of body ashy grey; primaries and base of tail slaty-blackish; end of tail white for the terminal 90 mm (ca.), (E. Celebes, Nat. Coll. — C 14355). "Bill black, base and cere lead colour; feet very pale yellow (dirty white — Whitehead *15*); iris olive brown" (Wallace).

Young. Above brown, more whitish tawny on the region of the head and neck, darker on the back, all the feathers edged with paler brown, fulvous or whitish; below rufous brown and tawny, varied with whitish; tail whitish, terminal third blackish brown (juv. Minahassa, 9. II. 94, Nat. Coll. — C. 13236).

The following are taken from extreme measurements made by Legge, Sharpe and W. Blasius *(6, 7, f 3)* and three adults from Celebes in the Dresden and Sarasin Collections.

Measurements.	Wing	Tail	Tarsus	Culmen with cere
Small ♂	535	240	87	50
Large ♀	610	315	107	55

Eggs. Generally 2, white; seen against the light blackish green; often smeared with brown or yellowish — probably from the fæces of the parent bird. Average measurements: 71.4 × 52.6 mm (Oates *20*, North *18*).

Breeding season. West coast of India and Ceylon, October—December *(8, f 1, 7)*; mouth of the Hoogly and Nicobar Islands, end of January *(20)*; North Borneo, a nest with nearly fledged young ones on April 12th — hence the eggs were laid at the end of February or early in March *(15)*; New South Wales, first eggs laid in July *(18)*.

Nest. A large mass of sticks, lined with finer ones, situated in a great tree or — where this is wanting — on a rock *(18, 20)*.

Distribution. India and South China to Ceylon, Laccadive, Andaman and Nicobar Islands; Siam, Cochin China, Philippines; throughout the East Indian Archipelago to New Guinea, the Admiralty and Solomon Islands; Australia; Tasmania; ? Tonga Islands. [For exact localities and authorities, cf. Salvadori, *f 2*; and add Cochin China (Moreau *9*), Palawan (Platen *f 4*, Whitehead *14*), Natuna Is. (Hose *f 6*), Noordwachter Id. (Vorderman *26*), Dana Id. (Studer *23*)].

In the Celebesian area: — (Forsten. *2*, Wallace), Minahassa (v. Duivenbode *5*, Meyer *f 1a*, etc.), Lembeh Id. (Guillemard *12bis*, Hickson *24bis*, Nat. Coll.), E. peninsula (Nat. Coll. *29*), S. peninsula (Weber *22*); Saleyer Id. (Weber *20*); Banggai Id. (Nat. Coll. *29*).

The original authority for the Tonga Islands appears to be Lesson (Man. d'Orn, 1828, I, 85). No reliable confirmation of the species having occurred there has been produced since this date.

Until more evidence is forthcoming it is advisable to exclude the Cape of Good Hope from the range of this species. "Le Blagre" of Levaillant, commonly identified with it, was indicated as an African species, and there is a specimen in the British Museum from the late Jules Verreaux with the locality Cape of Good Hope indicated; but, since the country has become better known, no confirmation of these statements has come to hand. (See Sharpe *6*, Salvad. *f 2*, Legge *7*).

Schlegel *(5)* points out that this Sea-eagle, like *Haliastur indus*, often has black shaft-streaks in the white adult plumage, and also points out that in their first plumage the two species are similar. It is not the case, however, that specimens with dark shaft-streaks are peculiar to any special part of the range of the species.

The White-bellied Sea-eagle is a bird of the sea-coast, though sometimes found to a distance of 60 miles or so up the mouths of the larger rivers of Burmah (Oates), and Colonel Legge notes the fact of its breeding at some of the large inland lakes of Ceylon. Its food consists chiefly of sea-snakes, fish, cuttle-fish and other marine animals. In Java Mr. de Bocarmé *(2)* remarks that it preys on dead carcases thrown up on the strand at the mouths of rivers: and at Pigeon Island off the W. coast of India Hume shot one while bearing in its claws the stomach and liver of a Goat, and found the remains of a sheep's head, amongst the bones of sea snakes, etc., beneath the nest of another; but it does not appear to attack mammals when alive, though it has been known to carry off wounded birds.

The normal number of eggs is two. The time of laying varies much, even in districts not far from one another, as is pointed out above.

H. leucogaster has no very close allies. It is easily distinguishable from other birds of prey in Celebes by its large size, and the adult by its striking coloration.

GENUS POLIOAETUS Kaup.

In structure this form is very like *Haliaetus*, though smaller, but it may be distinguished by its claws, which are rounded below, and by its shorter, more rounded wings, which when closed fall considerably short of the tip of the tail (Legge), and in which the 4th and 5th quills, instead of the 3rd, are the longest; the tibial plumes are short, not overreaching the upper third of the tarsus. There are two species, inhabiting the Indian Region from the Himalayas to Celebes; they prey chiefly upon fresh-water fish, and lay 3 eggs.

16. POLIOAETUS HUMILIS (Müll. Schl.).

Lesser Fishing-eagle.

Of this species there are two pronounced races, which are said to blend in Cachar.

+ 1. The typical Polioaetus humilis (Müll. Schl.)

a. **Falco humilis** *(1)* M. et S., Verh. Nat. Gesch. Natuurk. Comm., Aves, 1839—44, 47, pl. 6.
b. **Ichthyaetus nanus** Blyth, J. A. S. B. 1842, XI, 202; 1843, XII, 304 (Singapore).
c. **Pandion humilis** Kaup, Classif. Säug. u. Vög. 1844, 122; *(1)* Schl., Valkvogels 1866, 13, 53, pl. 5 f. 3.
d. **Polioaetus humilis** Kaup, Contr. Orn. 1850, 73; *(1)* Wall., Ibis 1868, 14; *(2)* Sharpe, Cat. B. 1874, I, 454; *(3)* Salvad., Ucc. Borneo 1874, 6; *(4)* Gurney, Ibis 1878, 455—458; *(5)* Hume, Str. F. 1880, IX, 244; *(6)* Oates, B. Brit. Burmah 1883, II, 223; *(7)* W. Blas., Ztschr. ges. Orn. 1885, 222; *(8)* Everett, J. Str. Br. R. A. S. 1889, 183; *(9)* Hartert, Nov. Zool. 1895, 476.
e. **Haliaetus humilis** Schl., Mus. P.-B. Aquilae, 1862, 18.
f. **Pontoaetus humilis** *(1)* Blyth, Ibis 1863, 22.
g. **Polioaetus humilis subsp.** Gurney, Diurn. B. of Prey 1884, 60.
h. **Polioaetus plumbeus humilis** *(1)* M. & Wg., Abh. Mus. Dresd. 1896, Nr. 2, p. 7.

"**Kuajan pupusi**", Peling Id., Nat. Coll.

For further references see Sharpe *d 2*, Salvadori *d 3*, Oates *d 6*.

Figures and descriptions. Müller & Schlegel *a 1*; Schlegel *c 1*; Sharpe *d 2*; Gurney *d 4*; Hume *d 5*; Oates *d 6*; W. Blasius *d 7*.

Diagnosis. Size smaller; bill much smaller; foot and tarsus markedly slenderer, though not shorter (Hume *d 5*); the dark band across the end of the tail a little more distinct (Gurney *d 4*).

Adult. Abdomen, thighs and under tail-coverts white; all other parts ashy brown, more chocolate colour on the back and wings; throat and cheeks slightly streaked with whitish; terminal fourth of tail indistinct dark brown, tip whitish (ad. Peling Id., V—VIII, 95: Nat. Coll., C 14497). "Bill and cere dusky lead-colour; feet pale bluish white; iris light yellow" (Wallace *d 1*).

Young male. Above brown, the feathers margined with fulvous brown; on lower back and rump varied with white; tail-feathers brown, white at the base and on the inner web, and indistinctly barred with blackish brown; forehead, sides of head, hind neck and entire under surface white, the feathers margined on the first-named parts, on the breast and greater under wing-coverts with very pale ashy brown (from Sharpe).

Measurements.	Wing	Tail	Tarsus	Exposed Culmen
♂ juv. (Sharpe *d 2*)	350	193	72	
♀ ad. (W. Blas. *d 7*)	415	223	74	40
(C 14497) ad. Peling Id.	415	215	76	40

Distribution. Cachar, India (Inglis *d 5*); Assam (Brit. Mus. *d 2*); Burmah (Hume *d 6*); Malay Peninsula (Blyth *f 1*); Singapore (Strickl. *d 3*); Bunguran (Hose *d 9*); Borneo (Doria & Beccari, Everett, Platen *d 8*); South Celebes — Macassar (Wallace *d 2*); Tjamba (Platen *d 7*); Peling Island (Nat. Coll.).

2. Polioaetus humilis major n. subsp.[1]

Diagnosis. Size larger (wing, generally, 430—460 mm, but sometimes 483 mm — Blanford), bill much larger, foot and tarsus stouter (Hume).

Distribution. "The sub-Himalayan ranges and submontane tracts (occasionally in the cold season straying some distance into the plains) from the borders of Afghanistan to Suddya in Assam" (Hume, Str. F. 1877 V, 130).

3. Polioaetus humilis — major.

Diagnosis. Intermediate.

Distribution. The birds of Cachar are said to be intermediate between the typical *P. humilis* and *P. humilis major*. (Blanford, Faun. Br. Ind., B. III 1895, 372.)

This species, like the allied *P. ichthyaetus* Horsf., probably lives almost entirely upon fish, caught chiefly in inland waters. In N. E. Cachar Mr. Inglis noticed that it was more often to be found in the neighbourhood of rivers than jheels. In the Malay Archipelago it appears to have been very seldom met with. Only two specimens, as Prof. W. Blasius points out, had hitherto been recorded from Celebes, both of them from the Macassar Peninsula, but another has now been obtained in Peling Island by the native hunters working for the Dresden Museum. One specimen from Borneo in the Norwich Museum was mentioned by Gurney, Ibis 1878, 458, and not more than three others from there, we believe, were known to Mr. Everett at the time of compiling his List (1889). In Java it does not yet appear to have been noticed.

The large race of the Himalayas is stated to build a great nest in trees, the same pair returning yearly and adding to the structure. The usual number of eggs is three (Oates, Hume's Nests and Eggs, III, 169).

The allied *P. ichthyaetus* (Horsf.) is, as Gurney points out, a larger bird, having, when adult, the tail white[2].

---
[1] This form has often been called *P. plumbeus* Hodgs., which is a nomen nudum, and of uncertain application.
[2] **Polioaetus ichthyaetus** (Horsf.).
In giving the distribution of this species Cat. B. I, 453, Dr. Sharpe notes Celebes among other localities,

GENUS BUTASTUR Hdgs.

The Buzzard-eagles are of small-medium size (about as big as a Crow); wing long, primaries overreaching the secondaries by about one-third of the wing-length; tail moderate, slightly rounded; tarsus naked (except for the upper fourth in front), covered with polygonal scales, the largest in front; toes rather short; cere large, occupying two-fifths of the exposed culmen, bill not powerful for a bird of prey. Preys on reptiles, amphibians, insects. Four species, among them migratory and probably stationary forms, found from India and S. E. Siberia to New Guinea; also N. E. Africa.

+ 17. BUTASTUR INDICUS (Gm.).
Grey-faced Buzzard-eagle.

a. **Javan Hawk** Lath., Gen. Syn. Suppl. 1787, 32.
b. **Falco indicus** Gm., S. N. 1788, I, 264.
c. **Falco poliogenys** (1) Temm., Pl. Col. 1825, Nr. 325.
d. **Buteo poliogenys** Less., Man. d'Orn. 1828, I, 103; (1) Temm. & Schl., Faun. Jap. Aves 1845, 21, pl. VII B (juv.); (2) Schl., Mus. P.-B. Buteones, 1862, 22; (III) id., Valk-vogels, 1866, 33, 70, pl. 21, f. 2, 3; (4) id., Rev. Acc. 1873, 111.
e. **Poliornis poliogenys** Kaup, Classif. Säug. u. Vög. 1844, 122.
f. **Poliornis barbatus** (Eyton).
Butastur indicus (1) Sharpe, Cat. B. I, 1874, 297; (2) Walden, Tr. Z. S. 1875, IX, 143; (3) id., P. Z. S. 1877, 689, 757; (4) id., ib. 1878, 612; (5) David & Oust., Ois. Chine, 18; (6) Hume & Davison, Str. F. 1878, VI, 19—21, 497; (7) Salvad., Orn. Pap. 1880, I, 14, et Agg. 1889, 12; (8) Kelham, Ibis 1881, 365; (9) Gurney, ib. 1882, 235; (10) Kutter, J. f. O. 1883, 295; (11) Oates, B. Brit. Burmah 1883, II, 197; (11bis) Gurney, Diurn. B. of Prey 1884, 73; (12) Pleske, Bull. Ac. Petersb. 1884, XII, 111; (13) Guillem., P. Z. S. 1885, 253, 545; (14) W. Blas., Ornis 1888, 541; (15) Sharpe, Ibis 1888, 195; (16) W. Blas., Ornis 1888, 304; (17) Sharpe, Ibis 1889, 72; 1890, 274; (18) Everett, J. Straits Br. R. A. S. 1889, 183; (18bis) Salvad., Orn. Pap. Agg. 1889, 12; (19) Tristr., Cat. Coll. B. 1889, 63; (20) Whitehead, Ibis 1899, 42; (21) Steere, Coll. B. Philip. Is. 1890, 7, 5; Seebohm, B. Japan 1890, 196; (22) Styan, Ibis 1891, 488; (23) Seebohm, Ibis 1893, 52; (24) Hose, t. c. 418; (25) Tacz., Faun. Orn. Sib. Orient. I, 1891, 69; (26) Stejn., P. U. S. Nat. Mus. 1893, 624; (27) Grant, Ibis 1894, 503; (28) Bourns & Worces., B. Menage Exped. 1894, 33; (29) Everett, Ibis 1895, 32; (30) Grant, t. c. 438; (31) Blanf., Faun. Br. Ind. B. III, 1895, 365; (32) M. & Wg., Abh. Mus. Dresden 1895 Nr. 8, p. 4; (33) id., ib. Nr. 9, p. 1; (34) iid., ib. 1896 Nr. 1, p. 4; (XXXV) Meyer, Abb. v. Vogelskeletten 1897, Taf. CCXI.

---

but we do not know upon what grounds this statement was made. As this learned ornithologist omits Celebes from the range of P. *ichthyaetus* in his tables showing the distribution of Bornean birds in The Ibis 1890, 274, we conclude that he has since found reason to believe himself to have been in error, especially as no such bird has been recorded from Celebes of late years. When Hume (Str. F. V, 129), Legge (B. Ceylon, 74), Oates (B. Br. Burmah, II, 222), and Blanford (Faun. Br. Ind. B. III 1895, 370) also make mention of Celebes as the most eastern bounds of its distribution, it would appear that their statements are founded upon that of Dr. Sharpe.

"**Sikep sedang**", Minahassa, Nat. Coll.
"**Dandape**", Great Sangi, iid.
"**Tagi**", Talaut Islands, iid.
For synonymy and further references see Salvadori 7, *18bis*.
Figures and descriptions. Temminck *c I*; Temm. & Schl. *d I*; Schlegel *d III, d 2*; Sharpe *I*; Hume & Davison *6*; Salvadori *7*; Oates *11*; Taczanowski *25*; Blanford *31*; Meyer *XXXV*, etc.
Adult. Above dark grey-brown, shafts blackish, bases of the feathers drab; head above and nape blacker, with rufous edges; wing-coverts hazel-brown, a few whitish terminal spots thereon; secondaries grey-brown, tips whitish, primaries and primary coverts blacker, the unexposed parts rufous-hazel, crossed with narrow partially obliterated bars; upper tail-coverts broadly tipped and barred with white; tail crossed with about four blackish bands, tip whitish; cheeks and ear-coverts drab, with black shaft-lines, submalar stripe and one down the middle of the throat blackish; chin and throat laterally white; under parts white, crossed with brace-shaped bars of rufous-brown, chest nearly uniform rufous-brown; under tail-coverts buff-white; under wing-coverts buff-white, with sparing hastate spots of rufous; wing-below, where it rests upon the body, white, the bars nearly obliterated, distal ends of remiges blackish (♂ ad. Tomohon, N. Celebes, 9. IX. 94: P. & F. Sarasin.
"Iris golden yellow; bill chrome-yellow at base, rest black; cere chrome; legs dull chrome-yellow; claws black" (Everett *13*).
Female. Similar to the male, but larger.
Young. Differs from the adult in having the under parts white, streaked with rufous-brown (not barred); the dark stripe down the middle of the throat much reduced or absent; cheeks and ear-coverts blackish brown (not drab); head above white, saturated with rufous on occiput and nape, and streaked with blackish brown; feathers of the upper parts with pale terminal edges, white on some of the wing-coverts (♂ juv. Rurukan, N. Cel., 22. XI. 94: P. & F. Sarasin).
"Iris dark yellow; bill black, cere yellow; legs and feet yellow" (P. & F. S.).
Measurements (extremes from small male to large female). Wing 306—345 mm; tail 190—213; tarsus 58—62.
Eggs. Not positively known.
Distribution. Japan, Ussuriland *(26)*, China, Formosa, Cochin China, Tenasserim, Malacca, the Philippines, including Palawan *(4, 15, 16)*, Balabac *(20)* and Sooloo *(13)*, Borneo, Labuan, Java, Celebes — Northern Peninsula (Rosenberg, Duivenbode, Riedel *d 4*, Guillemard *13*, etc.), Siao (Hoedt *d 4*), Great Sangi (Rosenberg, Hoedt *d 4*, Platen *14*, Nat. Coll.), Talaut Islands — Salibabu and Karkellang (Nat. Coll.), Ternate, Halmahera, Morty, Salawatti, Waigiou (cf. Salvadori *7*, Oates *11*, Kelham *8*, W. Blasius *14*, Kutter *10*].

Count Salvadori includes Burmah (Hume) in his list of localities, but only Tenasserim is noted for it in Mr. Oates' and Mr. Blanford's more recent works.

It is possible to show that this species occurs in the East India Archipelago only as a migrant from China, Ussuriland and Japan during the N. E. monsoon, the winter in the latter countries. Such Mr. Whitehead considered it undoubtedly to be in Borneo *(17)*, and Mr. Everett *(18)* states that "it appears in Labuan and Northern Borneo in September and remains through the winter. It is quite the most abundant of the migratory as *Haliastur intermedius* is of the resident birds of prey in those parts of the island". Mr. Whitehead also marks it as a migrant in Palawan *(20)*. Abbé David states that it breeds in the mountains near Peking, though it appears not to be plentiful in China *(5)*; further south, it passes through the Lower Yangtse country, as Mr. F. W. Styan writes, "on migration in March and April. A good number travel together, and remain a week or so among the hills on their way; they seem to avoid the plains" *(22)*. Apparently the species is resident, or some remain to breed, in the Philippines, an egg, which appears to belong to this species, having been obtained in Mindanao by Schadenberg and Koch *(10)*. The following recorded dates afford fairly conclusive evidence that this Hawk descends into the East India Archipelago at the beginning of the cold in the northern hemisphere; whither they take their departure again with the returning S. W. monsoon.

	Locality	Specimens	Time of year	Reference
	Luzon	5 or more	Nov., Dec., Feb., April, July	*2, 19, 21*
	Panay	1 » »	January	*21*
	Guimaras	2 » »	March, December	*2, 21*
	Negros	1 » »	February	*21*
Philippines	Cebu	2 » »	March	*3, 21*
	Siquijor	1 » »	February	*21*
	Mindanao	3 » »	October, December, March	*10, 21*, Dresd. Mus.
	Basilan	1 » »	November	*21*
	Cuyo	1 » »	December	*2*
	Palawan	1	November	*4*
	Sooloo Is.	1	December	*13*
Borneo		9	October to March	*17*
Malacca		1	February	*8*
Tenasserim		2	March	*6*
Talaut Is.		8	October, November	Dresd. Mus.
Sangi Is.		7	October to January	*14*
Great Sangi		1	December	Dresd. Mus.
N. Celebes		4	October, February	*d 4*
N. Celebes		1	December	Dresd. Mus.
N. Celebes		6	October, November	P. & F. Sarasin
Ternate		3	November to February	*7, 12*
Halmahera		3	January, April 1st	*d 4*
Morty		1	January	*d 4*
Salawatti		1	November	*d 4*

From the above list of about seventy specimens, furnishing all the recorded dates which we have been able to find for specimens obtained in these islands, it will be seen that every one was obtained between October and April, except Dr. Steere's notice from Luzon, the northernmost island of the Philippines, which furnished one or more in July. Its prey consists probably almost exclusively of reptiles, amphibia and insects which become scarce or quite disappear in the northern parts of its range on the approach of winter; hence the motive for the autumnal migration.

To explain the returning migration in spring other motives than hunger must be looked for; it is evident that no diminution of food takes place in warm climates at this time of the year, but rather the opposite. It is, however, possible to argue that the influx of northern migrants overstocks these regions, rendering breeding there more inconvenient than in more northern climes, which now also offer a plentiful food supply. This reason goes hand in hand with another one, the remembrance of, and attachment for, the old breeding quarters. Many birds will not forsake their nesting spot, as almost every field-naturalist knows from experience, even after their eggs or young have been taken and nest destroyed; and a pair of the same species — probably, as a rule, the same identical pair[1]) — return year by year to the same spot, even to the same tree or bush.

The Grey-faced Buzzard-eagle has as yet been recorded only from the Northern Peninsula of Celebes, — Gorontalo and the Minahassa. So, likewise, *Tachyspizias soloensis* and *Ninox scutulata japonica* have been recorded only from North Celebes. These are also migratory species, which descend in the winter from China apparently, and reaching North Celebes via the Philippine Islands

---

[1]) Some curious cases, showing that it need not follow that it is always the identical pair of parent birds, which returns to the same spot, are recorded by Seebohm in the case of the Merlin *(Falco aesalon)*. A pair of these birds was found breeding on a certain bank and were killed; nevertheless for several succeeding years other pairs of these birds always made their appearance and tried to rear their young on the same bank, though one or both birds were killed off every season (British Birds 1883, I, 35). The only way, as Seebohm thinks, by which these facts may be accounted for, is on the supposition that Merlins have certain recognised breeding haunts, which are seized upon by a pair out of a party in migration, provided the former owners are not already in possession. This may be the case; but it appears to us more probable that the birds, which attempted to nest at the spot, were the young which had been reared there some years earlier and had been expelled from their native haunt by the their parents (or possibly by the more courageous members of the household), the usual occurrence among birds of prey and probably among other birds.

An interesting case in this connection has been communicated to us from personal observation by Mr. W. Schnuse of Dresden. A pair of Storks had nested for a few years on the top of a disused tall chimney in the village of Walterniemburg, Province of Saxony. In 1894, three young ones were reared, and the birds left for the winter. Next spring, two Storks, evidently the parents, returned and took possession of the old nest. A day or two afterwards three other Storks arrived and tried to occupy the nest, and tremendous fights took place every day for at least a week for its possession. The three (apparently the young of the former year), which roosted at night on an old disused nest on the parsonage about 200 yards off, at length relinquished their efforts. It is a pity that no one seems to have taken the trouble to amass facts of this description.

Since the above remarks were made, Prof. Newton's thoughtful considerations on the question have appeared (Dict. B. pt II, 1893, 555), and may well be inserted here. "When we consider the return movement which takes place some six months later, doubt may be entertained whether scarcity of food can be urged as its sole or sufficient cause, and perhaps it would be safest not to come to any decision on this point. On one side it may be urged that the more equatorial regions which in winter are crowded with emigrants from the north, though well fitted for the resort of so great a population at that season are deficient in certain necessaries for the nursery. Nor does it seem too violent an assumption to suppose that even if such necessaries are not absolutely wanting, yet that the regions in question would not supply sufficient food for both parents and offspring — the latter being, at the lowest computation, twice as numerous as the former — unless the numbers of both were diminished by the casualities of travel. But on the other hand we must remember what has been advanced in regard to the pertinacity with which Birds return to their accustomed breeding-places, and the force of this passionate fondness for the old home cannot but be taken into account, even if we do not allow that in it lies the whole stimulus to undertake the perilous voyage".

are content for the most part to pass the cold season there, without going further south.

The next species, *Butastur liventer* — distinguishable by its rufous hazel tail and yellow bill — is perhaps also migratory, but this species has only been found in South Celebes. It is not known from China and the Philippines; its northernmost known quarters are Pegu, where it breeds, and hence its wandering probably takes place by way of Malacca and the chain of islands from Sumatra eastwards, so reaching South Celebes before any other part of the island. So, also, *Polioaetus humilis*, likewise a Burmese and Indian species not known in China or the Philippines, has only been obtained in South Celebes (in two examples) and at Peling Island. See also *Lanius tigrinus* and *lucionensis*.

18. BUTASTUR LIVENTER (Temm.).
Rufous-tailed Buzzard-eagle.

a. **Falco liventer** *(1)* Temm., Pl. Col. 1527, Nr. 438.
b. **Buteo liventer** Cuv.; *(1)* Schl., Mus. P.-B. Buteones 1862, 21; *(II)* id., Valkvogels 1866, 33, 69, pl. 21, f. 1; *(3)* id., Rev. Acc. 1873, 111.
c. **Buteo pallidus** Less., Tr. d'Orn. 1831, 82.
d. **Astur liventer** Gray, List. Acc. B. M. 1844, 34.
e. **Poliornis liventer** Kaup; *(1)* Walden, Tr. Z. S. 1872, VIII, 37; *(2)* Hume, Str. F. 1873, I, 318; *(3)* Salvad., Cat. Ucc. Borneo 1874, 9; *(4)* Hume, Str. F. 1875, III, 31; *(5)* Meyer, Ibis 1879, 56.
f. **Circaetus liventer** (G. Müller) *(1)* Kaup, Isis 1847, 267.
Butastur liventer *(1)* Sharpe, Cat. B. 1874, I, 296; *(2)* Hume & Davison, Str. F. 1878, VI, 21; *(3)* Oates, ib. 1882, X, 180; *(4)* id., B. Brit. Burmah 1883, II, 196; Gurney, Diurn. B. of Prey 1884, 73; *(5)* W. Blas., Ztschr. ges. Orn. 1885, 207, 208, 233; *(6)* Everett, J. Straits Br. R. A. S. 1889, 183; *(7)* Oates, Hume's Nests & Eggs 1890, III, 161; *(8)* Hartert, Kat. Vög. Frankf. M. 1891, 176; *(9)* Vorderman, N. T. Ned. Ind. 1892, LI, 375; *(10)* Büttik., Zool. Ergebn. Webers Reise Ost-Ind. 1893, III, 271; *(11)* Blanf., Faun. Br. Ind. B. III 1895, 364; *(12)* M. & Wg., Abh. Mus. Dresden 1896, Nr. 1, p. 7; *(13)* Hartert, Nov. Zool. 1896, 162.

"Buetje", Tjamba, S. Celebes (Platen *5*).
For further references see Sharpe *1*; Salvadori *e 3*; Oates *4*.
Figures and descriptions. Temminck *a 1*; Schlegel *b II*; Kaup *f 1*; Sharpe *1*; Hume *e 4*; Oates *3, 4, 7*; W. Blas. *5*; Vorderman *9*; Blanford 11.
Adult. Above greyish brown, more rufous brown on upper tail-coverts and scapulars; shaft-streaks dark; primary coverts and remiges hazel-rufous, the inner webs of the latter crossed with about six narrow dark bars, the outer webs for the terminal half ashy brown; tail hazel-rufous crossed with from two to seven narrow bars — the basal ones being sometimes absent; throat whitish; under surface drab-brown like the face, but mottled on the abdomen with broken cross-bars of white; thighs, under tail-coverts and under side of wing white, the dark bars of the quills being rather faint.

"Iris light yellow; eyelids, cere, feet and bill yellow, tip of latter black" (♀, S. Celebes: Platen — Nr. 6738 and others).

Three specimens in the Dresden Museum from Celebes — all of them at the end of the moult — and one in the Sarasin Collection are in this plumage.

Female. Like the male; a little larger perhaps.

Young. Similar to the adult, but the head above brown, with blackish brown shaft-streaks, the wing-coverts mottled with whitish, the grey-brown breast marked to some extent with whitish transverse spots (♂, 11. Dec. 1894, Macassar: P. & F. Sarasin.

A young specimen of three months' age described by Mr. Oates *(3)* had the bars on the tail very clearly defined.

Measurements.	Wing	Tail	Tarsus	Bill from cere
a. (6738) ♀, S. Celebes (Platen)	305	155	63	21
b. (C 11205) ♀, S. Celebes, 6. III. 78 (Platen)	290	162	66	20.5
c. (Sarasin Coll.) ♀, Macassar, 14. VII. 95 (P. & F. S.)	285	150	65	20
d. (C 11204) ♂, S. Celebes, 8. II. 78 (Platen)	285	150	65	20
e. (Sarasin Coll.) ♂ juv., Macassar, 11. XII. 94 (P. & F. S.)	276	146	63	19

Eggs. 2; white, very smooth and compact; seen against the light — clear dark green. Size 46 × 37 mm ca. (Oates *7*).

Nest. Of small sticks, situated in a tree.

Breeding season in Burmah. March *(7)*.

Distribution. Pegu (Oates etc. *e 2*, *e 4*, *4*, *7*); Tenasserim (Davison *2*); Siam (Gurney *e 1*); Java (Reinwardt *b 1*); Timor (S. Müller *b 1*); ? Borneo (*b 1*, *e 3*, *6*); South Celebes, — Macassar (S. Müller *b 1*, Wallace *1*, Meyer, Platen *5*, etc.); Pare-Pare (Weber *10*); Luwu, Gulf of Boni (Weber *10*); ? Moluccas (Frankf. Mus. *8*).

Mr. Everett declines to admit this species into his list of the birds of Borneo for want of sufficient evidence, nor have we seen any reference to actual specimens from that island. If the locality be correctly indicated, there is a specimen in the Frankfurt a. M. Museum from the Moluccas — apparently the only one as yet recorded east of Celebes. For the present it should be treated as erroneous; no collector is mentioned.

There is some reason to suspect that this Hawk will prove to be a partially migratory species in the East Indian Archipelago. In lower Burmah it is abundant, and breeds there, laying 2 eggs in March. In Java a note of Mr. de Bocarmé states that it is generally seen in December and January and appears not to nest in that island. It is fairly abundant in marshes and damp meadows; preys upon frogs, grasshoppers and small birds, also snakes and crabs (*b 1*, *4*). One specimen from Java in the Leyden Museum is dated May. Five specimens reported upon by Prof. W. Blasius *(5)* and another one in the Dresden Museum were shot by Dr. Platen in February, March and April 1878; and one was obtained near Macassar by Meyer in October, 1871[1]).

Prof. Weber obtained four in the southern Peninsula, one in October, one in February, and the other two apparently prior to October *(10)*.

---

[1]) This specimen, which is in the Berlin Museum, was erroneously entered in the Catalogue there as having come from Manado, although it is labelled "Macassar, Oct. 1871". This mistake has led to the locality Manado being recorded for the species in Ibis 1879, 58, which should now be cancelled and Macassar substituted in its place.

Mr. Everett, who arrived at Macassar on September 16th, 1894 *(13)*, sent specimens to the Tring Museum and reported it to be common. Drs. Sarasin, as shown above, obtained a young male in December, 1894, but a female in worn plumage in July, 1895. It is thus clear that the species occurs in Celebes all the year round, though it may be more plentiful in the months of the northern winter.

This species appears to be an inhabitant of the plains, and its presence in the southern Peninsula up to Luwu at the head of the Gulf of Boni is an interesting fact when contrasted with the occurrence of its fellow-species *B. indicus* in the north of Celebes, while the latter is absent in the south. Whether *B. liventer* visits S. Celebes on migration or not, there is no doubt that it came into the island from the south-west, while *B. indicus* comes to N. Celebes every autumn from the north-west.

Buteo desertorum Vieill.

A single specimen in the Leyden Museum labelled North Celebes (v. Faber, 1883). According to Dresser (Birds of Europe V, 458) the range of this species is over South-East Europe, Africa and India — the Neilgherries and Himalayas (Hume, Rough Notes, 270); in winter it has occurred as far south as Ceylon and Thatone in Tenasserim (VII.—X. orn JB. Sachs. 1886, p. 2), so that its occurrence in Celebes is well within the bounds of possibility. Faber collected only in Celebes and Sumatra.

GENUS HALIASTUR Selby.

The Brahminy Kite is of medium size, wings long, reaching nearly to the tip of the tail (Legge); tail moderate, rounded; tarsus short, about one-eighth the length of the wing, upper third feathered, bare part with transverse shields in front; toes short, claws somewhat small and not much bent. Devours animal food of almost any description. A single, stationary species, occurring from India to Australia. Lays 2 eggs.

+ 19. **HALIASTUR INDUS** (Bodd.).
Brahminy Kite.

This species ranges in great abundance from the Indian countries and Ceylon, throughout the East India Archipelago as far as New Guinea and the Solomon Islands, to Australia.

As it extends its range from India eastwards and southwards, it diminishes in size and the dark shaft-streaks on the white plumage of the head, neck and breast gradually diminish in breadth, until in New Guinea and the neighbouring islands they quite disappear. In India and Burmah the stripes are broad and take in some of the web of the feather on either side of the shaft; in Ceylon, Siam, Malacca and Sumatra the stripes are generally narrower, a further decrease is seen in specimens from Java, Sumbawa, Sumba, Flores, Borneo and the Philippines, where the streaks appear to be confined to the shafts, which are very distinct and black, and Javan specimens were named intermedius by the late J. H. Gurney; a little more East a still further decrease in the dark shaft-streaks is to be seen in specimens from Sangi, Celebes, Sula, Halmahera, Obi major, Amboina, Buru, Timor, and North Australia, where the dark shaft lines are, as a rule, thin and scanty, or are quite absent, leaving the head, neck and breast pure white; in New Guinea and the surrounding islands, New Ireland, Solomon Islands and Aru, these parts are, apparently, in general, pure white, and this most eastern form of the species has received the name *girrenera* from Vieillot. But the species is so plentiful throughout its range, and the transition from the broad-striped specimens of India to the white ones of New Guinea is so gradual and regular that it is impossible to draw any distinct lines of demarkation between specimens from neighbouring localities; and any geographical sections of the whole which have been called by distinct names are, by definition, only subspecies, or even subsubspecies.

In how far is it practically profitable to distinguish local sections of this species by name, and what form of nomenclature is convenient? According to the most prominent present-day definition of ornithologists, species are well-differentiated groups of individuals which are not connected with one another by means of a series of intermediate intergrading forms, whereas subspecies are interconnected by such a series of intermediate forms. Just as a genus is always known under a single name, so also a species should always be known under a binomial, and subspecies under trinomials. In the present case of *Haliastur indus*, it seems advisable to recognise at least the two extreme forms of the species, as subspecies, viz: the typical race of India with the broadest stripes, and that of New Guinea without any stripes. Unhappily, modern rules of nomenclature make no provision for distinguishing as a subspecies the local race first described, and we are compelled to speak of both the species as a whole, and of the typical subspecies, as *Haliastur indus* (Bodd.). Granting that it be worth the trouble to recognise subspecies at all by name, it is certainly adventageous that one and all should be spoken of under trinomials in order to distinguish them from species, of which they only represent a part. This might readily be done by strictly confining the original binomial to the species, as at present defined, and then adding the word typicus to the typical subspecies first described, but it is not yet advisable to employ this method, inasmuch as — until some rule of nomenclature be made to prevent this result — the name of the writer, who appended the word typicus to the originally described race, would be likely to be quoted by any one writing down the trinomial hereafter and the name of the true author overlooked. In referring to such subspecies we are, therefore, compelled to attach the vernacular term "the typical", to the scientific name of the species, instead of adding a classical trinomial[1]).

Haliastur indus arouses another question in nomenclature. If subspecies be distinguished by a trinomial, under what form of nomenclature are the forms which interconnect them to be spoken of? In the present case, if the two extreme races only of *H. indus* are treated of as subspecies, the main body of the species is made up of intermediate forms. Specimens from Java, which stand in characters about midway between those of India and New Guinea, have been named *intermedius* by Gurney; and those of Celebes which are again intermediate between the latter and those of New Guinea were named *H. indus* var. *ambiguus* by the late Dr. Brüggemann, but the last named, like those intermediate between *intermedius* and the typical *indus*, cannot be regarded as more than subsubspecies; and if such a subdivision ever come into use four names must be employed for it. In such a case it is certainly much easier to remember the facts than the names which are intended to call them to mind, and the nomenclature employed would defeat its own purpose. Therefore we prefer to give no long names, but to designate the forms intermediate between the typical *H. indus* and *H. indus — girrenera* in the following manner: *Haliastur indus — girrenera*, with the remark that a long hyphen connecting the names of two subspecies serves to designate the intermediate forms which connect them[2]). It does not appear to be advisable to recognise *H. indus intermedius* as a third subspecies, since this is not aberrant, but distinctly intermediate between the two extremes of variation in *H. indus*.

Perhaps in future — when the want becomes sufficiently pressing to necessitate the step — a somewhat considerable change in the nomenclature of the present day may be effected as follows: species as at present defined will remain under their original binomials; subspecies under trinomials; but the degree of relationship between the interconnecting forms to these subspecies will be displayed by the use of numbers — somewhat after the manner of chemical formulae. Thus, in the case of *Haliastur indus* — taking four degrees of relationship into consideration — the typical subspecies will be *Haliastur indus typicus*, that of New Guinea *H. indus girrenera*; that of Celebes, which may be supposed to have three times as strong a connection with *girrenera* as with *typicus*, will be represented as $H.\ indus_1\ girrenera_3$; that of Java being just about midway in characters as $H.\ indus_2\ girrenera_2$; that of Malacca as $H.\ indus_3\ girrenera_1$. This method could be carried to any degree of refinement, and

---

[1]) Since this article was written, Dr. Jordan and Mr. Hartert have published some very similar remarks in the "Novitates Zoologicae" and in "The Ibis". Mr. Hartert, however, did not hesitate to append the classical trinomial typicus to the typical form, but he has since decided that a doubling of the specific name (e. g. *Acredula caudata caudata*) is better (Zool. Anzeiger 1897 XX, 41—47). It must be left to a general agreement, which mode of expression be adopted, the one being nearly as good or bad as the other.

[2]) We are not so sanguine as to believe, that our brother ornithologists will adopt our innovation of nomenclature, but we trust that future "rules of nomenclature" will also take into consideration cases like this, and make some proposition which can be generally adopted.

is certainly less complex than the use of a quadrinomial such as *Haliastur indus girrenera ambiguus*.

Better suited to the ornithological needs of the present day, so long as mathematical accuracy which may be expressed in numbers is generally impossible, are the signs $>$ and $<$, as explained in a subsequent article. We use them as follows:

Haliastur indus $>$ *girrenera* means that the specimen so indicated is more like the typical *indus* than *girrenera* (viz. ordinary birds from Ceylon, Sumatra etc.);

Haliastur indus $<$ *girrenera*, more like *girrenera* than *indus* (Celebes etc.);

Haliastur indus $=$ *girrenera*, sharing equally the characters of *indus* and *girrenera*, intermediate (Java, etc.).

Haliastur indus — girrenera.

a. **Haliastur indus** partim *(1)* Schl., Mus. P.-B. Aquilae 1862, 19; *(II)* id. Valkv. 1866, pl. 4, fig. 4; *(3)* id., Rev. Acc. 1873, 119—123; *(4)* Salvad., Ucc. di Borneo 1874, 12; *(5)* Lenz, J. f. O. 1877, 366; *(6)* Legge, B. Ceylon 1880, 76; *(7)* A. Müller, J. f. O. 1882, 428; *(8)* Hartert, J. f. O., 1889, 405; *(8bis)* Hagen, Td. Ned. Aard. Genoots. 1890 (2), VII, 130; *(9)* Vorderman, Notes Leyden Mus. 1891, 122; *(10)* M. & Wg., J. f. O. 1894, 238; *(11)* iid., Abh. Mus. Dresd. 1895, Nr. 8, p. 4; Nr. 9, p. 2; *(12)* iid., ib. 1896, Nr. 2, p. 7.
b. **Haliastur intermedius** (Gurney), subsp.; *(1)* Sharpe, Cat. B. 1874, I, 314; *(2)* Gurney, Ibis 1878, 460—466; *(3)* id., Diurn. B. of Prey 1884, 79.
c. **Haliastur intermedius** Gurney; *(1)* Walden, Tr. Z. S. 1875, IX, 142; *(2)* Salvad., Orn. Pap. 1880, I, 19; *(3)* Meyer, Verh. z.-b. Ges. Wien 1881, 760; *(4)* Guillem., P. Z. S. 1885, 253; *(5)* id., t. c. 502; *(6)* id. t. c. 562; *(7)* Sharpe, Ibis 1889, 74; *(8)* Everett, J. Str. Br. R. A. S. 1889, 183; *(9)* Steere, Coll. B. Philip. 1890, 7; *(10)* Salvad., Ann. Mus. Civ. Gen. 1891, 41; *(11)* Sharpe, Ibis 1894, 244, 258; *(12)* Grant, t. c. 407; *(13)* Bourns & Worces., B. Menage Exped. 1894, 33; *(14)* Vorderman, N. T. Ned. Ind. 1895, 329, 333; *(15)* Grant, Ibis 1895, 438; *(16)* Everett, Nov. Zool. 1896, 597.
d. **Haliastur indus** var. **ambiguus** *(1)* Brüggem., Abh. Ver. Bremen 1876, 45; *(2)* Büttik., Zool. Ergebn. Webers Reise Ost-Ind. 1893, III, 271.
e. **Haliastur leucosternus** Walden; *(1)* Meyer, Ibis 1879, 56; *(2)* Joest, Das Holontalo 1883, 105.
f. **Haliastur girrenera** var. **ambigua** *(1)* Salvad., Orn. Pap. 1880, I, 17; *(2)* W. Blas., J. f. O. 1883, 135; *(3)* id., Ztschr. ges. Orn. 1885, 227; *(4)* id., Ornis 1888, 542.
g. **Haliastur girrenera** *(1)* Meyer, Isis 1884, 9.

"Asiare" or "Kasihare", adult; "Jamba", young, Talaut Is., Nat. Coll.
"Kasiahe", Siao, Tagulandang and Ruang, "Kasiaheng", Gt. Sangi, iid.
"Koheba mera", Banka Id., N. Cel., iid.
"Koheba dada puti" i. e. Bird of Prey with white breast, Malay vernacular of Celebes, Meyer e 1.
"Bulia", Gorontalo, Joest e 2.
"Tjangeh", Tjamba, Maros, S. Celebes, Platen *f 3.*
"Alaij pau", Peling Island, Nat. Coll.

For further references see Sharpe *b 1*, *c 7*.
Figures and descriptions. Schlegel *a II*, *a 1*, *a 3*; Sharpe *b 1*; Brüggemann *d 1*; Lenz *a 5*; Gurney *b 2* (specimens described from Sumatra, S. Borneo, Flores, Macassar, Togian Islands, Amboina, N. E. Australia); A. Müller *a 7*; Meyer *g 1*; W. Blasius *f 3*, *f 4*.
Adult. Head, neck, breast and upper part of abdomen white, marked to a greater or less extent with dark shaft-streaks, which are broad — but narrower than in the typical *H. indus* — in the western parts of its range, and gradually become fine and scanty in the eastern parts; primaries black, passing into cinnamon-chestnut towards the bases of the feathers; all other parts cinnamon-chestnut, whitish at the tip and on the under side of the tail. "Bill bluish white; feet citron-yellow" (Platen *f 3*). Iris variable — probably with age: dark chocolate, warm chocolate, olive-brown, brown, light brown, dull yellow, reddish yellow (Gurney *b 2*, Sharpe *c 7*, W. Blasius *f 4*).
Young. The white of the adult replaced by cinnamon, whole under surface of this colour.
Measurements after Legge, Salvadori and Gurney. Wing 345—406 mm, tail 170—241, tarsus 46—57, bill from core 28—31. The smallest measurements are generally those of the most eastern parts.
Distribution of intermediate specimens. Ceylon (Legge *a 6*); Malacca (Sharpe *b 1*); Salanga (A. Müller *a 7*); Sumatra (Gurney *b 2*; Salvad. *c 10*); Bangka (Schl. *a 3*); Billiton (Vorderman *a 0*); Java (Schl. *a 3*, Sharpe *b 1*); Lombok (Vorderman *c 14*); Sumbawa (Guillem. *c 4*); Sumba (Meyer *c 3*); Flores (Gurney *b 2*); Timor (Wall. *b 1*); Philippines (Schl. *a 3*, Meyer *c 1*, Steere *c 0*); Sooloo Islands (Guillem. *c 4*); Borneo (Schl. *a 3*, Everett *c 8*); Talaut Is. (Nat. Coll. *a 10*, *a 11*); Great Sangi (Platen *f 4*, Nat. Coll.); Siao (Hoedt *a 8*, Meyer *g 1*, Nat. Coll.); Tagulandang and Ruang (Nat. Coll.); Celebes — North (Forsten, Rosenb. *a 1*, *a 3*, Meyer *e 1*, *etc.*), — S. W. Peninsula, Macassar (Wallace *b 1*); Maros and Tjamba (Platen *f 3*), Tete Adji (Weber *d 2*); Peling (Nat. Coll.); Sula Besi (Hoedt *a 3*); Buru (Wallace *b 1*); Amboina (Schl. *a 3*); Obi Major (Guillem. *c 4*); Halmahera (Meyer in Dresd. Mus.); Australia (Gurney *b 2*, Ramsay).

In the west and east extremes of the range given above normal specimens of the typical *H. indus* and of *H. indus girrenera* respectively occur intermingled with others that are not normal. Thus, in Malacca, Mr. Hume records a specimen with the dark streaks as strongly developed as in any Indian specimen (Str. F. IX, 1880, 120); Colonel Legge, on the other hand, regards the Malaccan birds as further removed from the Indian than are the Cingalese ones, which he considers intermediate between the two. It is very probable that in Malacca, and elsewhere, an influx of specimens from the north takes place in winter, for, as is pointed out by Gurney, Abbé David notes a migratory movement among the birds of Tchékiang and Kiangsi: "il disparait de ces provinces pendant l'hiver et se retire dans la Cochinchine" (David and Oust. Ois. Chine, 15), but it is not certain to what race the Chinese birds belong. In Halmahera, Amboina, Celebes, etc. pure white plumage on the head, neck and breast is common; from Flores, also, such a bird is mentioned by Schlegel (*a 3*) and we have recently received an almost typical *girrenera* from Talaut (*a 10*). The most logical way of treating such specimens is to record them under the name of the normal subspecies, *H. indus girrenera* or the typical *indus*, as the case may be; while those which deviate from these subspecies in the same localities should bear the sign, *H. indus—girrenera*.

As regards animal food, *H. indus* seems to be almost omnivorous. Fowls, fish, lizards, molluscs, crustaceans, large orthoptera, refuse of ships, are noticed by Gould, Schlegel, Jerdon, Legge etc.; and Pryer observed it catching bats and Edible-nest Swifts on the wing in front of the great *Collocalia* caves of British North Borneo (P. Z. S. 1884, 536). Probably, its feeding habits vary somewhat in different localities; in Ceylon, for instance, although Mr. Layard has known it to seize a fowl, "this must be a rare occurrence" (Legge), whereas the more narrowly striped specimens of Java carry off domestic fowls daily (Bocarmé), and in Borneo Mr. Whitehead found it to be a great robber of chickens. The species builds in trees, and throughout its dominions — from India to Australia — the ordinary number of eggs laid at a sitting is two. North and south of the equator the breeding seasons are different, as would appear from the following:

North of equator.

Locality	Season	Eggs	Authority
India	Jan.—April	2 (or 3)	Oates, Nests and Eggs Ind. B. III, 170.
Burmah	Dec.—Feb.	2—3	Oates, B. Brit. Burnah.
North Ceylon	Dec.—Feb.		Parker, Ibis 1883, 193.
South Ceylon	Feb.—Mch.	2	Legge, B. Ceylon, 76.
Labuan	Dec. 1873	1	Sharpe, P. Z. S. 1879, 323.

South of equator.

Java	May—June	2	Schl., Valkr. 10, 51.
Australia (Port Essington)	July—Aug.	2	Gld., B. Austr., text, pl. 4.

It is, of course, not to be supposed that any abrupt difference in the breeding seasons is to be found N. & S. of the equator. In Tenasserim lat. 16° N., for instance, Colonel Bingham noticed a pair breeding, "but the nest, when examined on the 4th April, was still unfinished", this being only about a month earlier than in the case with the species in Java. In such countries as lie in the more immediate neighbourhood of the equator and which experience no very great difference of temperature at any time of the year, it may ultimately prove that the breeding-seasons of the birds are mainly influenced by the character of the monsoons — wet or fine; Legge states that in Ceylon (B. Ceyl. p. XXIII) birds breed during the rains, the times varying on opposite sides of the island. The most important condition probably is that the food supply be at its maximum when the young are hatched.

It will be noticed in the above list that an egg obtained at Labuan was taken in December, while in Java the species lays in May and June, that is nearly half a year later; at Labuan December is the final month of the rainy season; May in Java is at the beginning of the fine season.

In almost all parts of its range *H. indus* exists in great abundance. Thus, it is spoken of as a very common species at Macassar (v. Rosenb.), in N. Celebes

(Meyer), in Ceylon (Legge), "extremely common in Burmah" (Oates), "excessively common in Tenasserim" (Davison), "everywhere abundant on the strand in Ceram" (Ribbe); "the most plentiful resident bird of prey in Borneo" (Everett), "the commonest bird of prey near Batavia" (Vorderman), "the commonest and most impudent bird of prey" in Sumatra (Hagen). Owing to its great numerical abundance, to the expulsion of the young by the parents, the search for partners and for suitable breeding territories well supplied with food, a continuous crowding out of the surplus population of this species doubtless takes place, and the movement is as nothing to a bird possessing such fine power of flight. It is, however, naturally a stationary bird, having no call for migration by reason of its omnivorous habits; consequently, it does not proceed farther, when forced to make a move, than to the nearest suitable locality; but this serves to keep up a continuous intermixture of individuals from neighbouring quarters, so as to prevent its becoming marked off into distinct smaller species. Why, in spreading from the continent of Asia to the islands of the East, *H. indus* has gradually lost the black shaft-streaks on the white plumage of the head, neck, breast and abdomen, so acquiring a pure white on these parts (or — viceversa — has acquired black shaft streaks in extending its range from New Guinea to India [see p. 64 *antea*¹) are questions worthy of consideration, but the explanation of which must be left to future time, if the development of this species as a whole has not by then undergone changes, be it that it has split into sharply separated forms or that it has settled down into one.

GENUS MILVUS Cuv.

The Kites are medium-sized birds of prey and well characterized by their long, forked tail. Wing very long; tarsus and toes short, upper third of the tarsus feathered, lower part in front transverse-scutate, claws not strongly hooked; bill rather weak, culmen narrow, compressed, tomia slightly festooned. Preys on small vertebrates of any description; also, carrion. Three to six species found in the Old World, including Australia; migratory and stationary; laying 2—3 eggs.

20. **MILVUS MIGRANS** (Bodd.).
Black Kite.

The following references bear upon what may be called the Indian Kitequestion:

Homeyer, J. f. O. 1868, 252; Anderson, P. Z. S. 1872, 79; Hume, Str. F. 1873, 160; Brooks, Ibis 1874, 461; Sharpe, Cat. B. I, 1874, 323—326; Anderson, P. Z. S. 1875, 25; Brooks, Str. F. 1875, 229; Hume, l. c. note; Hume & Davison, Str. F. 1878, VI, 23; Oates, Str. F. 1878, VII, 44; Gurney, Ibis 1879, 76—82; Brooks, t. c. 282—284; Scully, Str. F. 1879, VIII, 227—229;

Brooks, t. c. 466, 467; Hume, l. c. note; Legge, B. Ceylon, 82; Scully, Str. F. 1881, X, 95; Oates, S. F. 1882, X, 181; id., B. Brit. Burmah 1883, II, 202—204; Seebohm, British B. I, 1883, 80; Gurney, Diurn. B. of Prey 1884, 80, 81; Brooks, Ibis 1884, 238; Menzbier, t. c. 311, 312; W. Blas., Ztschr. ges. Orn. 1885, 230—232; Brooks, Ibis 1885, 385, 386; Tristr., Cat. Coll. B. 1889, 63, 64; Zaroudnoi, Bull. Ac. Imp. Mosc. 1889, 755; Oates, Hume's Nests and Eggs 1890, III, 173—177; Blanford, Faun. Brit. Ind. B. III 1895, 374—378.

The controversy referred to above, after having lasted over a period of a quarter of a century and produced written matter enough to fill an octavo volume, remains almost as far as ever from a satisfactory settlement. It still needs years of close and accurate observations, agreeing, moreover, better in themselves than is at present the case, before any decided opinion upon the Indian Kite-question can be formed. The points in dispute may be stated as follows: on the one side Mr. Hume and others maintain that there are three forms of Black Kite in India — a big one (*M. melanotis* T. & S.), a middle-sized one (*M. govinda* Sykes) and a small one (*M. affinis* Gld.); the other side, which may be identified with Mr. Brooks' name, affirms that the middle-sized and the small Kite of Hume are one and the same, and that the big Kite — and not this species — is the true *Milvus govinda* Sykes.

Taking first into consideration the question of transferring the name, *M. govinda*, to the large Kite, we believe that no sound and sufficient reason can be urged to necessitate this step. The only strong argument in its favour is, that one of Sykes' types in the possession of Mr. Brooks is undoubtedly a specimen of the large Kite. But the weight of this argument is counterpoised by the fact that Sykes was not aware that there were two races of Kite in India, and that at least one other specimen said to be a type is in preservation elsewhere (see Legge l. c.) and belongs to the smaller, commoner, race. This in itself would be sufficient reason for declining to make an inconvenient change in nomenclature, but, beyond this, though the original description applies equally well to either form as regards coloration, the accompanying measurement of the tail (11 in.) betoken that it was made from a small specimen of the smaller Kite. Mr. Brooks urges that the body — 26 in., the only other measurement given — is that of the large Kite, but there is reason to believe that these measurements were not made from a freshly killed specimen, but from a skin, in which case — though that of the tail holds good — that of the body is, of course, of no value whatever as a discriminative character[1]. That the description was made from a skin is proved first from the fact, as Mr. Hume has pointed out, that Sykes' paper was written in England; and, secondly, that no naturalist would be likely to take the measurement of the tail alone of a specimen before skinning (which can be done equally well afterwards), and neglect to note down the sex, colours of soft parts and expanse, which cannot be ascertained at any

---

[1] Of two skins of *Loriculus stigmatus* before us one measures 123 mm and the other 187 mm, or half as long again as the former, though in the flesh no doubt they were about equal in size.

other time. Also, the habits of *M. govinda* as described by Sykes appear to be those of the smaller Kite (cf. Legge). Thus, as regards nomenclature, the smaller Kite has the stronger claim to the name *govinda*, under which it is most commonly known.

Next, touching the question of the occurrence of two forms of small Kite in India, or of only one, it would be impossible — without a great quantity of material and without a much more accurate knowledge of the habits of these birds in India and elsewhere than can be obtained from the imperfect and, sometimes, conflictory statements of field naturalists — to offer any solution of the matter in dispute; nevertheless we would throw into the discussion a fresh element, which may not be without effect upon the knotty points concerned.

It should be borne in mind that *M. govinda* and the large *melanotis* are only worthy of subspecific separation from *M. migrans* of the western Palaearctic Region. In India the two forms possibly do not intergrade; after most accounts they differ in habits; the large Kite is a hill-race breeding in the Himalayas, and, though *M. govinda* breeds there as well, it would seem possible that this form generally lays somewhat later than the other, judging from the statement in "Hume's Nests and Eggs" 1890, III, 173, 176, that *M. govinda* usually lays in the Himalayas in April and May, and *M. melanotis* from January to the beginning of May. The differences between *M. govinda* and a supposed smaller race, occurring in India and generally spoken of as *M. affinis*, are by no means so well made out by the supporters of this view. It is inconceivable that a race, hardly separable from *M. govinda* at the best, should occur as an independant breeding bird intermingled with the latter, and it still remains to be shown whether a smaller race migrates from Burmah or elsewhere at certain seasons¹), or whether the slight differences of coloration commonly found in larger and smaller Indian specimens are not the natural accompaniment of a slight individual increase or diminution of size. In the cases of *govinda* and *affinis* any movements taking place will be likely to be of a very local character, depending upon the abundance of food; true migrations are hardly to be expected, as the bill of fare of Kites is a somewhat broad one, and in warm countries, like those under consideration, some kind or other of food palatable to it must always be present. There is evidently a difference at least of size between the race of Australia and that of Burmah, which is spoken of by Mr. Oates and others as *affinis*. As is pointed out below, facts tend to show that the Burmese specimens lay a somewhat larger egg. The Australian specimens in the British Museum are also on the whole somewhat more rufous than Indian or Tenasserim ones. Also Legge points out that his Cingalese specimens of *M. govinda* differed as adults from a Macassar example of *affinis* in the less rufous coloration of the head, hind-neck, and lesser wing-coverts, and, in youth, in the less rufescent character of the upper-surface tippings; also there are differences between the

---

¹) See Anderson, P. Z. S. 1872, 79.

Cingalese *Milvus* and the typical Indian *M. govinda*, — facts pointing to the stationary habits of these birds. *M. govinda*, however, as well as the typical *migrans*, visits Afghanistan in migration (St. John, Ibis 1889, 153).

Thus, we conclude, that there are neither two nor three species of Kite in India, but two (or three) subspecies, one of which proceeds to further racial variation in other localities — Ceylon, Celebes, Australia.

For the present it seems best to treat of *M. migrans* as consisting of four subspecies — *Milvus aegyptius* being accounted distinct — viz.:

1. The typical Milvus migrans (Bodd.).

Diagnosis. Size medium (♀ wing c. 430 mm); the edges of the feathers of crown and nape whitish instead of light brown or rufous; abdomen more distinctly ferruginous; usually little or no mottling or banding on the basal portion of the quills in adults. (Blanford, Faun. Br. Ind., B. III 1895, 378.)

Distribution. Europe, N. W. Africa, Asia east as far as Tomsk on the Obi, Turkestan, Afghanistan.

2. Milvus migrans melanotis (Temm. & Schl.).

Diagnosis. Size large (♀ wing 485—546 mm); feathers of head edged with tawny or rufous; usually a conspicuous white patch on wing below at base of quills; abdomen and under tail-coverts usually much paler than in the next form.

Distribution. East Siberia, Japan, India, Burmah; Borneo (Everett).

3. Milvus migrans govinda (Sykes).

Diagnosis. Size medium (♀ wing 430—495 mm, Blanford); head as in *M. m. melanotis*; plumage browner than in *M. m. affinis*.

Distribution. Afghanistan, India, Ceylon, gradually blending with the next form from Burmah to New Guinea.

4. Milvus migrans affinis (Gould).

Diagnosis. Size small (wing about 400 mm, Australia), plumage more rufous than in *M. m. govinda*.

Distribution. Australia, passing into the preceeding form in the localities between Australia and India.

These intermediate birds will be contained under the formula *M. migrans govinda — affinis*, should a special term for them be thought necessary. For the present we prefer to take the birds ranging from Australia to Burmah as one subspecies:

+ Milvus migrans affinis (Gld.).

a. **Milvus affinis** *(1)* Gld., P. Z. S. 1837, 140; *(II)* id., B. Austr. 1848, I, pl. 21; *(3)* Schl., Mus. P.-B. Milvi 1862, 3; *(IV)* Diggles, Orn. Austr. 1866, pt. I; *(V)* Schl., Valkv. 1866, 30, 67, pl. 20, f. 1; *(6)* Wall., Ibis 1868, 16; *(7)* Sharpe, Cat. B. 1874, I, 323; *(8)* Hume & Davison, Str. F. 1878, VI, 23; *(8 bis)* Gurney, Ibis 1879, 76—82; *(8 ter)* Hume, Str. F. 1879, VIII, 45; *(9)* Legge, B. Ceylon 1880, 82; *(10)* Salvad., Orn. Pap. 1880, I, 21; Oates, Str. F. 1882, X, 181(?); *(11)* id., B.

Brit. Burmah 1853, II, 202(?); *(12)* Brooks, Ibis 1884, 238(?); id., ib. 1885, 385(?); *(13)* W. Blas., Ztschr. ges. Orn. 1885, 230—232; *(XIV)* North, Nests & Eggs Austr. B. 1889, 10, pl. IV, f. 5, 6 (eggs); *(15)* Oates, Hume's Nests & Eggs 1890, III, 176(?); *(16)* Büttikofer, Notes Leyd. Mus. 1892, 197; *(17)* Newton, Dict. B. 1893, 491; *(18)* Büttik., Webers Reise in Ost-Ind. 1893, III, 271; *(20)* Vorderman, N. T. Ned. Ind. 1895, LIV, 329, 332.
b. **Milvus migrans** pt. *(1)* Schl., Rev. Acc. 1873, 126.
c. **Milvus govinda? affinis** *(1)* Gurney, Diurn. B. of Prey 1884, 81.
d. **Milvus migrans affinis** *(1)* Hartert, Nov. Zool. 1896, 598.
"**Latjana**", S. W. Celebes, Platen *a 13*.

For further references see Salvadori *a 10*.

Figures and descriptions. Gould *a II*; Schlegel *a V, a 3*; Diggles *a IV*; North *a XIV*; Sharpe *a 7*; Legge *a 9*; Salvadori *a 10*; Oates *a 11*; W. Blasius *a 13*; Eyton, Osteol. Av. 1867—75, 3, 7 (sternum).

Adult. Head and neck broccoli brown, marked with dark shaft-streaks; chin, upper throat and ear-coverts whitish with dark shaft-streaks; rest of upper surface sepia brown, the feathers margined (on the wing-coverts strongly washed) with the colour of the head; tail — in young specimens plainly, in old specimens indistinctly — crossed with numerous dusky bars; under surface bistre brown with dark shaft-streaks, darkest on the under wing-coverts and lightest on the under tail-coverts and thighs, longest under wing-coverts grey-brown, with dark bars; primaries black, washed and mottled with pale brown towards the base of the wing, whitish at the concealed base of the outer quills (♀, S. Celebes 26. I. 78: Platen, C 11110, and others).
"Iris brown; cere light yellow; bill black; feet citron-yellow" (Platen *a 13*).

Younger birds have a more mottled appearance than those fully adult, owing to the pale margins of the feathers being more distinct and the banding of the tail strongly expressed.

Measurements.	Wing	Tail	Tarsus	Bill from cere
a. (6750) ♂, S. Celebes (Platen)	420	315	54	—
b. (Sarasin Coll.) ♂, Macassar, 13. VII. 95 (P. & F. S.)	405	260	54	24
c. (C 11110) ♀, Macassar, 26. I. 78 (Platen) . . .	437	310	56	—
d. (C 11109) ♀, S. Celebes, 5. V. 78 (Platen)	410	265	55	—
e. (Sarasin Coll.) ♀, Macassar, 13. VII. 95 (P. & F. S.)	400	260	53	24

Eggs. Australia; 3; dull white ground-colour, with reddish irregular spots and dots: mean of two measured 45.5 mm × 37.8 (North *a XIV*). Burmah; usually 3; egg-lining bright green; shell tolerably smooth and glossless, dull white, marked and blotched with rust colour, bright in the majority, but pale in a few. Average of 12 eggs: 53 × 41.4 mm (Oates *15*).

It would thus appear that the Australian bird usually lays a considerably smaller egg.

Nesting season in Pegu from the 3rd week in January to the end of March. Nest composed of twigs, etc. situated in a tree.

Distribution. Australia (Gould *a 1*, etc.); Yule Island (D'Alb. *a 10*); Duke of York Id. (Ramsay *a 10*); Timor (S. Müller *a 3*, Wallace *a 6*); Sumba (ten Kate *a 16*); S.W. Celebes — Macassar (S. Müller *a V*, Bernst. *b 1*, Wallace *a 6*, etc.), Tjamba (Platen *a 13*), Tete Adji (Weber *a 18*); N. Celebes — Manado (Faber in Leyden

Mus.); Lombok (Vorderman *a 20*, Everett *d 1*); Sumatra (S. Müll. *a 3*); Singapore (Davison *a 8*); Malacca (Davison *a 8*, Hume *a 8ter*); Tenasserim.(Davison *a 8*); Cochin China (Gurney *a 8bis*); Burmah (Oates *a 11*).

For some reason the Black Kite is very local, rather than rare, in the East Indian Archipelago. It has never been met with in the majority of the islands. In Celebes, notwithstanding the amount of collecting done in the Minahassa and Gorontalo, it had not been found there until 1883, when v. Faber sent a specimen from Manado to the Leyden Museum. In the southern end of the south-western Peninsula, as in Timor, S. Müller notes that it is very common about the mouths of the rivers *(a V)*. Whether it is a migrant from Australia here, or a resident, there is insufficient evidence to show, but the latter view is the more probable one, since it appears probable that Australian birds are a little smaller, and more rufous. The food of this species consists of any such birds, small mammals, reptiles, fish, etc. as it can catch and kill; also, carrion.

GENUS ELANUS Sav.

The Black-shouldered Kites are of small-medium size and easily distinguishable by their coloration — bluish grey above, with most of the wing-coverts black and white under parts; also by the long, pointed wing, with the 2nd primary longest; tarsus with the upper two-thirds in front feathered, the lower part covered with small polygonal scales; tail long, very slightly forked; bill compressed in front, the nostrils oval, protected by the loral bristles. Preys upon insects; also small land-vertebrates. Four species, two at least of doubtful validity, found in the warmer parts of both the Old World and America. Eggs 3—5.

+ 21. ELANUS HYPOLEUCUS J. Gd.
East Indian Black-shouldered Kite.

a. Falco melanopterus Horsf. (nec Daud.), Tr. L. S. 1822, XIII, 137.
Elanus hypoleucus (1) Gould, P. Z. S. 1859, 127; *(2)* id., B. Asia vol I, pl. 12 (1860); *(3)* Sclat., P. Z. S. 1863, 207; *(4)* Wall., Ibis 1868, 17; *(5)* Gray, HL. 1869, I, 28; *(6)* Wald., Tr. Z. S. 1872, VIII, 36; *(7)* Schl., Rev. Acc. 1873, 130; *(8)* Sharpe, Cat. B. 1874, I, 338; *(9)* Salvad., Ucc. Borneo 1874, 12; *(10)* Wald., Tr. Z. S. 1875, IX, 142, 249; *(11)* id., P. Z. S. 1877, 757; *(12)* id., ib. 1878, 939; *(13)* id., ib. 1879, 69; *(14)* Meyer, Ibis 1879, 56; *(15)* Salvad., Ann. Mus. Civ. Gen. 1879, 173; *(16)* Sharpe, Ibis 1879, 236; *(17)* id., P. Z. S. 1879, 314; *(18)* Gurney, Ibis 1879, 333; *(19)* Nicholson, ib. 1881, 140; *(20)* id., ib. 1882, 67; *(21)* Vorderman, N. T. Ned. Ind. 1882, XLI, 3; *(22)* Gurney, Diurn. B. of Prey 1884, 84; *(23)* Guillem., P. Z. S. 1885, 253; *(24)* Everett, J. Str. Branch R. A. S. 1889, 184; Tristr., Cat. Coll. B. 1889, 64; *(25)* Steere, List Coll. B. Philip. Is. 1890, 7; *(26)* Hartert, J. f. O. 1891, 299; *(27)* Salvad., Ann. Mus. Civ. Gen. 1891, 42; *(28)* Büttik., Zool. Ergebn. Webers Reise in Ost-Ind. 1893, III, 271;

(29) Sharpe, Ibis 1894, 244, 258; *(30)* Bourns & Worces., B. Menage Exp. 1894, 33; *(31)* M. & Wg., Abh. Mns. Dresd. 1896 Nr. 1, p. 7; *(32)* Hartert, Nov. Zool. 1896, 177.
b. **Elanus intermedius** *(1)* Schl., Mus. P.-B. Milvi 1862, 7; *(II)* id., Valkvog. Ned. Ind. 1866, 31, 68, pl. 24, f. 2, 3; Gray, HL. 1869, I, 28.
c. **Elanus melanopterus** *(1)* v. Martens (nec Loach), J. f. O. 1866, 9.

Figures and descriptions. Gould *1*, *II*; Schlegel *b II*; Sharpe *8*; Gurney *18*; Vorderman *21*.

Male adult. All under parts, forehead, sides of head and neck pure white; superciliary streak black; upper wing-coverts (except the longest) black; upper surface, including the two middle tail-feathers, ashy grey, darkest about the interscapulary region; end of quills dusky, below greyer and becoming pure white where the wing rests upon the sides of the body; the five outer pairs of tail-feathers white (S. Celebes, January: Meyer, Nr. 1680). "Feet citron-yellow; claws black; cere yellow; under the eyes yellowish; iris fiery red" (Meyer *14*).

Measurements. Wing 300 mm; tail 137; tarsus 38; bill from cere 19.5.

Female. An adult ♀ in the Sarasin Collection answers to the above description of the male: size — wing 310, tail 140, tarsus 39, bill from cere 20 mm (♀, coast between Macassar and Bonthain, 1. X. 95: P. & F. S.).

Young (assuming adult plumage). Above grey-brown, with white margins to the feathers, varied with new feathers of grey (maturity); wing-coverts black and under surface white as in the adult (♂, Lake Posso, Central Celebes, 19. II. 95: P. & F. Sarasin. Iris light yellowish brown (Everett).

Distribution. Celebes — Macassar (Wallace *4*, *8*, Meyer *14*), Luwu, Gulf of Boni (Wobor *28*), Lake Posso, Centr. Cel. (P. & F. Sarasin), North Celebes (Forsten *b 1*), Manado (Faber in Loyd. Mus.); Java (S. Müller *b 1*, Forbes *19*, *20*, etc.); Sumatra (Boccari *15*, Modigliani *27*); Borneo (Schwaner *b 1*, Mottley *3*, Treachor *16*); Sooloo Islands (Burbidge *17*, *23*, Everett *29*); Philippines — Luzon (Jagor *10*, Steere *25*), Cebu, Zamboanga, Basilan (Everett *11*, *12*, *13*), Mindoro, Negros, Guimaras (Steere *25*, *26*), Calamianes (Bourns & Worcester *30*).

The adult specimen from Celebes described above does not show a trace of black upon the tips of the under primary coverts, apparent in Gould's figure *(II)* of the type from the same locality, but it is present to a slight extent in the Sarasins' adult example, while our four Javan specimens all show it more or less, as does also Schlegel's figure of one from this locality *(b II)*. Perhaps it is lost with age, as one of our Javan specimens has only a trace of it.

The Dresden Museum possesses one adult specimen and the Leyden Museum five from the West Coast of Sumatra, which have the terminal third of all the primary coverts and the under side of all the primaries (except a very small extent about the base of the shafts) black. These specimens differ from *E. axillaris* (Lath.) of Australia in having the two central tail-feathers, not uniform dirty white like the other tail-feathers, but of the dark colour of the upper surface, which is also altogether of a darker grey than in the Australian form.

From *E. coeruleus* they are distinguishable by the presence of this black on the under primary coverts and by the black under surface of the primaries. From *E. hypoleucus* they differ in the same respects, but the constrast of colour of the under surface of the primaries is, of course, greater than is seen when the specimens are compared with *E. coeruleus*. Count Salvadori identifies his Sumatran specimens with *E. hypoleucus (15, 27)*; it is impossible to do this with those birds (Nr. 6740, Dresd. Mus.) which we propose to call provisionally *E. coeruleus—axillaris*, an intermediate form.

As specimens of *E. axillaris* have been reported from Java (Gould *II* and Gurney *18*), these perhaps belong to the same intermediate form, which may in fact link together *E. coeruleus* of Europe — India and *E. axillaris* of Australia.

Among the 17 examples in the Leyden Museum there is one from Java which has the wings nearly as black below as the Sumatran birds (♂; Voy. Diard, Cat. Nr. 6); two others from Java and one from Manado (Faber) have the primaries below blackish with the basal part white; another from Banjermassing, Borneo, is a true *hypoleucus*. These form, in fact, transitions between the Sumatran birds and six others from Java and Celebes, which have the under wing-coverts and under surface of quills nearly all white and are the true *hypoleucus*.

It thus appears probable enough that we have not to do with several species, but with a set of subspecies of one form, and that *E. hypoleucus* is very closely related to, and probably only subspecifically distinct from, *E. coeruleus* (Desf.) known from S. E. Europe, Africa, India to the Malay Peninsula (Hume, Str. F. VIII, 45).

The form *hypoleucus* is a rare, or very local, bird in Celebes, as also, apparently in Borneo: no specimens have been included in the large collections made of late years in North Celebes, from which locality the single specimen of Forsten in the Leyden Museum and that more recently obtained by Faber appear to be the only ones on record; nor was it represented in the collection formed by Dr. Platen near Macassar in 1878.

Its food consists largely of lizards, though doubtless, like *E. coeruleus* in Ceylon, it devours coleoptera and rats and mice as well (Legge, B. Ceylon, 87). In India the latter species, like *E. axillaris* Lath. of Australia, lays three or four eggs; but, considering its rarity in Celebes, it is probable that *E. hypoleucus* breeds more slowly than these, if it is stationary there.

Gould *(II)* remarks that there is not a more distinct and better defined group of Hawks than those forming the genus *Elanus*, which occurs both in the Old and New Worlds. Its occurrence in two species in Australia, and its absence, so far as is known, in Papuasia and the Moluccas is worthy of note.

GENUS PERNIS Cuv.

The Honey-buzzards are of medium size and easily recognised by their feathered lores; the bill is weak, tapering, not strongly hooked, the culmen rounded, tomia not festooned, the cere very large, occupying two-fifths of the exposed culmen, the nostrils oblique, covered posteriorly by an operculum of the cere-skin; wings and tail long, the tail slightly rounded; tarsus short, reticulate, the upper half feathered in front; toes rather long. Feeds chiefly on the larvae of *Hymenoptera*. 3 or 4 species, migratory and stationary, found in Europe, Africa and Asia as far as Celebes. Eggs 1—2.

† * 22. PERNIS CELEBENSIS (Wall.).
Celebesian Honey-buzzard.
Plates II and III.

a. **Pernis cristatus** part. *(1)* Schl., Valkv. Ned. Ind. 1866, 39, 73, pl. 26, f. 4; Gray, HL. 1869, I, 26; *(2)* Schl., Rev. Acc. 1873, 132, Nr. 16 only.
b. **Pernis cristatus** var. **celebensis** *(1)* Wall., Ibis 1868, 17.
c. **Pernis ptilorhyncha** *(1)* Wald. (nec Temm.), Tr. Z. S. 1872, VIII, 36.
Pernis celebensis *(1)* Wald., t. c. p. 111; *(2)* Sharpe, Cat. B. 1874, I, 349; *(3)* Tweedd., Ibis 1877, 287; *(4)* Meyer, Ibis 1879, 56; *(5)* Gurney, Ibis 1880, 216; *(6)* Hume, Str. F. IX, 1880, 448; *(7)* Gurney, t. c. 1881, 448; *(8)* W. Blas., J. f. O. 1883, 114, 126, 135; *(9)* Gurney, Diurn. B. of Prey 1884, 87; *(10)* Guillem., P. Z. S. 1885, 545; *(11)* Newton, Dict. B. 1893, 427; *(12)* M. & Wg., Abh. Mus. Dresd. 1895, Nr. 8, p. 4; *(13)* iid., ib. 1896 Nr. 2, p. 7.
d. **Baza celebensis** Rosenb. (nec Schl.), Malay. Archip. 1878, 271.
"**Goheba**" v. Schierbrand in Mus. Dresd.
"**Koheba**", Manado, Guillem. *10.*
Figures and descriptions. Schlegel *a 1, a 2*; Walden *1, 3*; Sharpe *2*; Gurney *5, 7, 10.*
Adult. Above Prout's brown with a purplish gloss; head above black; feathers of the neck broadly margined with rufous; tail whity brown crossed with about five bands of black and mottled in the interspaces with wavy lines of brown, a space of about 57 mm separating the terminal and subterminal bands; extreme end of tail whitish; lores and region around the eye smoke-brown; chin and throat white, marked with black streaks gathered into black stripes down the middle and at the sides of the throat; breast light russet striped with black; remaining under parts, including under wing-coverts and thighs, white closely barred with sepia-brown; under side of quills whitish with about seven bars of sepia-brown, broken up at the base. "Iris light yellowish brown ("dark brown" — M.); length in the flesh 620 mm" (Ribbe and Kühn, Maros, S. Celebes, Nr. 3920). Another specimen (Nr. 5756 Manado) closely resembles that described.

Young (first plumage). Back, wings and upper tail-coverts sepia, varied with drab-brown and white on the greater wing-coverts and scapulars; some feathers of dark Prout's brown present in the interscapular region; tail much as in the adult; head and neck white, the latter streaked with Prout's brown, many feathers of the same colour sprouting on the occiput; in front of eyes smoky grey; under parts white, uniform on chin, throat, abdomen and under tail-coverts, narrowly streaked with dark brown on the breast, barred with paler brown on the thighs, barred with rufous on the axillaries, and very faintly with pale rufous on the anal region and under tail-coverts. At the flanks the specimen has obtained some of the feathers of maturity barred with dark brown and white. (N. Celebes Nr. 14011).

Two other young specimens in the Leyden Museum, apparently rather older than that described above, have the head clothed with feathers with white bases, and black centres passing into light brown at the margins, giving a streaked appearance; feathers about eye brownish slaty; the rest much as in the above, but the white under surface more abundantly streaked with brown (N. Celebes — Faber in Leyd. Mus. Nr. 45, 35).

Another specimen, a little older, has the thighs, belly, axillaries and under tail-coverts well barred with brown; under wing-coverts with sagittate spots of brown showing a tendency to form into bars; breast buffy white, streaked with sepia-brown (Minahassa, v. Musschenbr. in Leyd. Mus.).

The specimen in the Leyden Museum figured by Schlegel *(a 1)* is not quite adult, though more advanced than the immature bird described by Gurney *(6)*; under wing-coverts white, touched up with a few small streaks of dark sepia or little bars of faint brownish or rufous; the bars — especially the white ones — on belly, flanks and under tail-coverts much broader than in adult (♀, Bone, Rosenb. in Leyd. Mus.).

Measurements.	Wing	Tail	Tarsus	Mid. toe with claw	Culmen from core
a. (3920) ad. S. Celebes (Ribbe & Kühn) . . .	370	270	48	—	—
b. (5756) ad. Manado, N. Cel.	370	250	46	—	22
c. (C 14500) ad. Peling Id. V—VIII. 95 (Nat. Coll.)	370	245	50	70	25
d. (Sarasin Coll.) ♀ ad. Kema, N. Cel. 5. X. 93	375	260	—	—	25
e. (14011) juv. N. Celebes . .	350	247	50	—	—

Distribution. Celebes, Peling. — Bone in Gorontalo (v. Rosenberg *a 2*), Minahassa (Meyer *1, 4*, Guillemàrd *10*; etc.), S. Peninsula, Maros (Ribbe & Kühn); Peling Id. (Nat. Coll.).

This rare Honey-buzzard was known until recently only from the northern Peninsula of Celebes, but, as is shown above, there is one in the Dresden Museum from South Celebes and one from Peling Island. There is a fine series of eleven in the Leyden Museum.

The similarity of the adult plumage of this Honey-buzzard to that of the adult Celebesian Hawk-eagle, *Spizaetus lanceolatus*, was noticed by Schlegel when figuring the species *(a 1)*, and has since been commented upon by other writers. The resemblance in coloration may, indeed, be spoken of as perfect (see our plate), as remarkable as that of *Xenopirostris polleni* and *Tylas edwardsi* in Madagascar, where the plumage of one bird is, as Prof. Newton says, counterfeited in the other, "feather for feather" (Dict. B. p. 574). Gurney, who knew more about birds of prey than anybody, termed the case of the Celebesian *Pernis* and *Spizaetus* as extraordinary. The young in first plumage of the Honey-buzzard, now described and figured for the first time, is entirely different in coloration from its parents, but just like the Hawk-eagle at the same age! Mr. Wallace has cited a corresponding case seen in two American Hawks, a Sparrow-hawk and an insect-eating *Harpagus* (Contrib. Nat. Select. 107; and Newton, Dict. B. 574) — young like young, and adult like adult. On an earlier page of this book (p. 26), we have drawn attention to another case of exactly the same nature — that of the Hawks *Spilospizias trinotatus* and

Accipiter rhodogaster. The similarity of the adult Honey-buzzard and Hawk-eagle of Celebes naturally did not escape Mr. Wallace, and he remarks (b. 1) that it must be ascribed to the influence of some local peculiarity or to mimicry; the same explanation of mimicry is advanced by him for the American *Harpagus* and *Accipiter*.

As to the Honey-buzzard and Hawk-eagle, the first suggestion — that of some local peculiarity — seems to us to be insufficient of itself alone to account for the production of a striking correspondence in the characteristic markings of two birds of different habits of life; though soil, climate, food, and the coloration of surrounding objects are doubtless not without effect upon the coloration of a bird's plumage.

Touching the argument for mimicry it is possible to urge that either species may be benefited by resembling the other: (a) that the weaker Honey-buzzard, through assuming the dress of the Hawk-eagle, would be avoided by any competitors for the same class of food, suffer less from the persecution of Crows and other birds, and so enjoy a freer field of action when searching for its food and attending to its nest; (b) that the Hawk-eagle, from its likeness to the harmless, insect-eating Honey-buzzard, would be at an advantage in approaching the Rails, Waterfowl and other large birds and small mammals upon which it preys.

Apart from anything else, it must be objected to the first argument that the Honey-buzzard apparently stands alone among the higher animals as a devourer of wasps' nests and their contents, and other destroyers of wasps' grubs are only to be found among small insects (such as *Stylops*, which deposits its eggs in the bodies of *Hymenoptera*), which — it need not be said — could not be afraid of it in the dress of a Hawk-eagle; also, birds of prey are hated and persecuted by other birds just on account of their predatory habits, and it might be well for *Pernis celebensis* in this respect if it resembled some more harmless bird. As it is, there is reason to suppose that the Honey-buzzard is not much molested; in his "Birds of Ceylon" p. 92, Colonel Legge mentions a specimen of the allied *Pernis ptilonorhynchus* which was shot in the fort at Trincomalie while "associating with Crows, and flying round the barrack-room at the dinner hour in company with them, on the look-out for scraps thrown out from the verandahs". The European Honey-buzzard, however, according to Naumann and Dresser, is much given to robbing the nests of other birds, and is vigorously mobbed in consequence.

The second argument which we have set up — that the Hawk-eagle in the disguise of the *Pernis* would be at an advantage in approaching its prey — is stronger; but there are reasons which go far to destroy it. In order that the animals upon which it preys should be rendered indifferent to the appearance of the *Spizaetus*, it is necessary that the harmless *Pernis* should be a very familiar bird and the Hawk-eagle rare. This is not the case. In four of the largest collections of Celebesian birds of prey — the Leyden[1]), British, and Norwich Museums, and the Duivenbode-collection reported upon by Prof. W. Blasius — *Pernis celebensis* numbered in all 11 skins, *Sp. lanceolatus* 22. There are before us in Dresden 5 of the former and 7 of the latter.

Supposing — which is probably not the case, since birds are better field-ornithologists than most naturalists — that the birds etc., which suffer from the frequent depredations of the Hawk-eagle, do not at once discriminate its differences from the Honey-buzzard, they will avoid both species with dread; but it is more likely that the differences of size, structure, flight and general bearing cause the dangerous species to be detected, possibly even before its peculiar coloration is made out.

Moreover, in the Philipines, as will be noticed presently, a *Spizaetus*, *S. philippensis*, has been found closely resembling the Celebesian species, though the bird is unsatisfactorily known, but in these islands *Pernis celebensis* is represented by the highly variable *P. ptilonorhynchus*. Also in Malacca and Sumatra a *Pernis* has been described as showing much similarity to *P. celebensis*, but in those quarters the Hawk-eagle is *S. limnaetus*, which seems to be a very variable species and can by no means be said to match the Honey-buzzard, except that, perhaps, the latter possesses the same quality of great variability in common with it.

Furthermore, a type of plumage, bearing much resemblance to that of the adult *Pernis* and *Spizaetus* of Celebes, is seen in a South American Buzzard, *Buteo melanoleucus* Vieill. *Spizaetus mauduyti* of South America also bears some resemblance to the *Pernis* and *Spizaetus* of Celebes.

If we were to seek for cases of close resemblance among birds of different genera not occurring in the same locality, we are confident that many might be found. To mention two: *Spilornis holospilus* of the Philippines is very like *Circus assimilis* J. & S. of Celebes and Australia in its coloration, and *Ninox punctulata* of Celebes to *Glaucidium jardini* Bp. of S. America. Here the theory of mimicry is, of course, worthless. Touching the albescent plumage of the young, so completely different from the adult dress, those who maintain the theory of mimicry might perhaps here advance the suggestion that the causes, whatever they may be, by which one species is advantaged by resembling the other were the same in past ages as today; consequently, when the one was albescent, the other must needs become albescent too, and when the first assumed its present adult plumage, the other must needs follow it in this variation; while phylogenetic reasons here as elsewhere come in to account for the repetition in immature birds of what had been gone through by their ancestors.

As containing an alternative explanation to that of Mr. Wallace, the phylogenetic relationship of the two forms, *Pernis celebensis* and *Spizaetus lanceo-*

---

[1]) Since the publication of Schlegel's "Revue", the number of specimens of *P. celebensis* has been increased to 11.

latus should be considered. That the plumage of birds sometimes betrays their descent is about as certain as any thing zoological can be. For instance, the Kingfishers of the genus *Pelargopsis* have blue rumps, but the Celebesian species has the rump buff; we have received the latter (one specimen) when not yet adult, and the rump is strongly washed with blue. The greatest caution is, however, required in applying this line of argument, for the plumage of the young seems also to be affected by all sorts of other causes not at all well understood. There is reason to suppose that the plumage displayed by the young of *Pernis celebensis* and *Sp. lanceolatus* — a white under surface with dark streaks — not only represents that of an ancestral stock common to both these forms, but also more or less accurately that of a vast number of other birds of prey. Almost all *Falconidae*, which acquire a transverse-barred pattern of under plumage when adult (*Astur, Accipiter, Falco peregrinus* etc.), are in the first dress, assumed on leaving the nest, brown above and white with dark longitudinal streaks below. The opposite case — of a cross-barred young one developing into a streaked adult bird — would, we believe, be an unheard-of phenomenon, though a temporary reversion of the kind sometimes apparently takes place, in *Spilornis* (see p. 3); as a rule striped birds of prey, like the Falcons, have young much resembling the females. If these facts have any meaning at all, they must signify that all these birds are descended from streaked races of Hawks of past ages, similar in coloration, perhaps, to the Buzzards and Falcons of the present day. Consequently there is, at all events, no reason to suppose that the *Pernes* and *Spizaeti* have been developed from a more remote stock of *Falconidae* than that which is now represented by their first plumage.

We take it that the young plumage, shown on our plate, is very ancient in type, though how closely similar it may be to the plumage of the ancestor of *Pernis* and *Spizaetus*, or how much changed by later influences, we have no opinion. From this common ancestor the present full dress of the *Pernis* and *Spizaetus* of Celebes may have been evolved under one of three conditions: 1. the structural differences were first developed and then the evolution of the plumage proceeded similarly in both and the birds grew alike, or 2. the evolution of the plumage proceeded on similar lines while the structural changes were going on; or 3. the present adult plumage is ancient and had been acquired before the structural differentiation commenced.

It is difficult to form an opinion as to which of these three explanations is the more probable; the first may be true or false; there seem to be no methods for testing it, proving, or disproving it. If we look at Plate III of the young of today, and imagine we are viewing the birds as they were when adult at some period of the past, after the structural change had taken place, we may well wonder why these two birds, structurally so different, should undergo a great but identical change of coloration. We cannot imagine that two birds may have undergone such a structural change and have assumed such different habits while the causes of the evolution of the coloration have undergone no change. One feels compelled to seek some natural selective agent to explain why the coloration did not develop in very different directions in the two birds, instead of in an identical direction.

The second hypothesis — that the change of structure and the evolution of the coloration proceeded quietly together through perhaps a long period of time, — is more acceptable because it makes a less abrupt and violent appeal to the imagination, but there seems to be nothing further in its favour.

The third hypothesis — that the present adult dress had been acquired before ever the structural differentiation into *Pernis* and *Spizaetus* (not to speak of other genera) had taken place — admits of discussion from several points of view.

It is certain (A) that small structural changes may take place without causing any marked changes of the adult coloration: witness many of the Kingfishers, where the type of coloration of the four-toed *Alcedo* may be found in certain species of the three-toed *Ceyx* (the more recent form in respect of the foot) and others, or certain Flycatchers like *Siphia*, *Muscicapula* (pls. XIII—XIV) and others, with one or two types of coloration running through closely-allied genera, or the Flowerpeckers *Acmonorhynchus* and *Pachyglossa*, much alike in coloration, but the former (the more recent form in respect of the wing) has nine primaries, the latter the customary ten.

(B) The hypothesis involves the assumption that the land of origin of the genera *Pernis* and *Spizaetus* was ancient Celebes. There is, we believe, a form of prejudice, against which we ourselves have to struggle, which pre-supposes that ancient Celebes (whatever sort of island, or islands, or adjunct of Asia it may have been) has simply been a receiver of birds from other parts, a land for colonisation, which has kept to itself the bird-forms it received. In the main this view is very likely right, but it is probable that in some cases Celebes may have sent out bird-colonists of its own.

The genus *Pernis* is a link between Celebes and Asia, not being found in the Australian Region; only one aberrant *Spizaetus*, *S. gurneyi*, is found east of Celebes, and the two genera may, of course, have become differentiated structurally from one another in Celebes as well as anywhere else. There is something to be said for this view. In the Great Sunda Islands and the Philippines as far as India, both the Honey-buzzard *(Pernis ptilonorhynchus)* and the Hawk-eagle *(Sp. limnaetus)*, instead of possessing — as seems to be the case in Celebes — a fixed adult dress, agree in being wonderfully variable. Now, it appears that long ages in a given locality, with the same climate, food, dangers, predilections and regular routine of habits, and, perhaps, other more hidden influences, lead to the assumption of a more or less fixed, stable coloration in birds, as is shown by the species of all tropical islands sufficiently remote from one another¹); but, when some of these species by reason of overcrowding, a storm, or other cause, are transported to a fresh locality where the conditions of existence are changed, the original, fixed coloration becomes disturbed — if there was originally a useful purpose in it, that purpose is now removed, — variability has free play for the time, and the amount of variation is made greater and more complicated by the interbreeding of the changing individuals. On these grounds the Celebesian Hawk-eagle and Honey-buzzard should be regarded as ancient forms, the variable birds of Borneo, etc. which have not yet settled down into a fixed type of plumage of their own, as of more recent origin. Leaving out of the discussion *Pernis tweeddalei* of Malacca which is only known in three or four specimens, the only remaining species of *Pernis* is our own *Pernis apivorus* — very variable and therefore, also recent. *Spizaetus*, however, has developed into a number of local species elsewhere than Celebes viz. the Indian Region, Africa, S. America. Let it be remembered, that Celebes, as is shown by some of its mammals as well as by some of its birds, is an ancient land, zoologically speaking, and that neither *Pernis* nor *Spizaetus* (except in the aberrant *S. gurneyi*) pass into the still more ancient Australian Region, and it will be seen that *P. celebensis* and *S. lanceolatus* have some claim to be regarded as the most ancient members of their genera.

(C) One of the most convincing arguments made use of by Charles Darwin in tracing the descent of domestic Pigeons from the Blue Rock Pigeon *(Columba livia)* was drawn from the occasional reversion, partial or nearly complete, of the former to the plumage of the Blue Rock. If the variable *Spizaetus limnaetus* and *Pernis ptilonorhynchus* are descended from specimens of *S. lanceolatus* and *P. celebensis* which have flown across from Celebes to Borneo, or the Philippines and further, similar cases of reversion should be found. Now, in his Valkvogels Ned. Ind. Schlegel portrays nine specimens of the variable *Spizaetus limnaetus* (called by him *S. cirratus*) and six of *Pernis ptilonorhynchus* (here called *cristatus*: pls. VI-VIII and XXV, XXVI), and among them some fairly good matches to the two forms of Celebes may be seen. About four specimens of a Honey-buzzard called *Pernis tweeddalei*, Hume, have been described from Malacca and Sumatra; it bears much resemblance to *Pernis celebensis*, and it seems very doubtful if they are more than individual variations of the variable *P. ptilonorhynchus*, since that bird lives not only on both sides of, but also in the same localities as the supposed distinct form (cf. Hume *6*, and Str. F. 1887, X pl., p. 513; Schlegel *a I*, and Mus. P.-B. Pernes 1862, 2; Schl. & S. Müll., Verh. Natuurk. Comm. 1839—44, 49, pl.; Kelham, Ibis 1881, 369; Gurney *5, 7*). The little known *Spizaetus philippensis*, which seems to be much like the Celebesian Hawk-eagle (see Walden, Tr. Z. S. 1875, IX, pl. XXIV), is not unlikely to prove

---

¹) As an exception the extinct Solitaire *(Pezophaps)* of Rodriguez must be mentioned (see: Newton, Dict. B. "Solitaire"). This large bird was most variable in structure and colour. It used evidently to get much knocked about in fighting. The male Ruff, another great fighter, is probable the most variable bird known. See remarks in Introduction on inherited mutilations of feathers, etc.

to be an individual variety of the variable *S. limnaetus*; Prof. W. Blasius finds that bird in Mindanao (J. f. O. 1890, 144). It is possible that all these forms are reversions to the two original forms of Celebes.

To these reasons may be added, (D) that both *Pernes* and *Spizaeti* lay only one or two eggs at a sitting and, thus, they are not likely to get crowded out of a large island, but might exist there with the conditions of reproduction and destruction at a balance for any required length of time.

In how far these remarks are well, or ill, grounded can, of course, only be ascertained by a much more thorough knowledge and study of the two genera, *Pernis* and *Spizaetus*, than has been possible to us; but the case serves to show — and we believe it is only one in very many — that, where some striking similarity between two members of distinct genera occurs, this may be looked upon as an ancestral character retained, rather than as something acquired by the natural selection of varieties of one species, which obtained some advantage by resembling the other. Whoever has examined young and old specimens of these birds will find it hard to listen to the supposition, that the parallelism between the two forms is a fortuitous coincidence, that they are of independent development, but in our present state of ignorance any one who will is perfectly free to make this suggestion and hold to it.

The label on a specimen in the Dresden Museum, stating that the stomach contained wasps and their larvae, shows that *P. celebensis*, like the other species of the genus, feeds mainly upon hymenoptera, its peculiarly feathered face protecting it against their stings. The allied *P. ptilonorhynchus* is found as a rule to lay two eggs, but one egg at a sitting is not uncommon (Oates, Hume's Nests and Eggs, 1890, III, 181).

23. PERNIS SP.

In the Novitates Zoologicae 1896, 177, Mr. Hartert describes a Honey-buzzard obtained by Mr. Everett in Saleyer in November, 1895, but which he was unable to identify with certainty. "A large bird (female), remiges moulting, wing 440 mm. The whole underside is buff or ochraceous buff, some of the feathers (older ones) paler, others (the new ones) darker and brighter. The throat is surrounded by an irregular black band, the feathers of the lower throat and upper breast have narrow deep brown shaft-lines, but all the breast, abdomen, flanks, scapulars [axillaries?] and under wing-coverts are uniform without a trace of bars or bands. Upper side dark brown as in most *Pernes*, not differing from many specimens of *P. ptilonorhynchus*."

Mr. Hartert believes the specimen to belong to the latter species, and not to *P. celebensis*. It is manifestly immature, but from the description evidently differs from the young *P. celebensis* by its larger size, and the absence of bars on the flanks, thighs and under tail-coverts. *P. ptilonorhynchus* (Temm.) is a highly variable species, as is pointed out in the preceding article, but it may always be distinguished from *P. celebensis* by its large size, and the adult — usually at all events — has a long nuchal crest.

GENUS BAZA Hdgs.

Birds of small-medium size, with a nuchal crest, long and nearly square tail, and moderately long wings; especially characterized by the edge of the upper mandible which is furnished with two sharp points or teeth[1]; tarsus short, upper half feathered in front, the rest covered with polygonal scales, much larger in front than behind. Feeds on insects. Lays 3 eggs. About 11 species of a local character, found from India to Australia, and in parts of Africa.

✢ * 24. BAZA CELEBENSIS Schl.
Celebesian Baza.

a. **Falco (Lophotes) reinwardti**, partim, *(1)* S. Müll. & Schl., Verh. Nat. Comm. 1839—44, 37, 38 (young ♀ only, from Tondano, N. Celebes: Forsten — fide Schl.).
b. **Baza reinwardtii**, partim, *(1)* Schl., Mus. P.-B. Pernes 1862, 5, Nr. 5 only.
c. **Baza magnirostris** *(1)* Wall. (nec Gray), P. Z. S. 1862, 337; *(II)* Schl., Valkv. 1866, 40, 75, pl. 28, f. 4 only (Celebes, Sula); *(3)* Wall., Ibis 1868, 18; *(4)* Wald., Tr. Z. S. 1872, VIII, 36, pt. (Celebes, Sula); *(5)* Meyer, Ibis 1879, 56.
d. **Baza reinwardti** Finsch (nec M. & S.), New Guinea 1865, 154 (Celebes).
Baza celebensis *(1)* Schl., Rev. Acc. 1873, 135; *(2)* M. & Wg., Abh. Mus. Dresden 1896, Nr. 1, p. 7; *(3)* iid., ib. Nr. 2, p. 7; *(4)* Hartert, Nov. Zool. 1896, 162.
e. **Baza erythrothorax** *(1)* Sharpe, P. Z. S. 1873, 625; *(II)* id., Cat. B. 1874, I, 357, pl. X, f. 2; *(3)* Brüggem., Abh. Ver. Bremen 1876, 46; *(4)* Gurney, Ibis 1880, 462, 469; *(5)* W. Blas., J. f. O. 1883, 114, 135; *(6)* Gurney, Diurn. B. of Prey 1884, 90.
Figures and descriptions. Schlegel c *II, 1*; Sharpe *e 1, e II*; Brüggemann *e 3*; Gurney *e 4*.

Adult male. Above dark brown, blackish on mantle and carpal region; head and crest black; face and ear-coverts dark grey; tail brown, crossed with three bands of darker brown — the endmost one much the broadest — and broadly tipped with white; throat and chest russet-brown; a black streak down the middle of the throat; rest of under surface broadly banded with darker brown and white, the markings more rufous on thighs and under wing-coverts; under tail-coverts whitish, blotched with rufous (♂, Enrekang, S. W. Central Celebes, 7. VIII. 95, P. & F. Sarasin).

Female. Differs from the adult male in having the feathers of the head and neck black broadly margined with rufous, crest tipped with rufous; face and ear-coverts rufous like throat and breast (not dark grey); the bands on the under parts rufous like the breast, only a little browner on flanks; upper parts glossy brown, not so black as in the male (♀, Enrekang, 7. VIII. 95, P. & F. S.).

We at first took this specimen and a similar one before us for the young, but the presence of some old worn feathers in the wings shows that it is at least in its second year.

"Bill lead-colour, black above; feet white; iris yellow" (Wallace).

Young. Like the female, but the banding on the under surface ill-defined.

---

[1] "In the young bird, the tooth is often single" (Blanford).

Measurements.

	Wing	Tail	Tarsus	Culmen from care
a. (C 10479) [♀], Celebes	295	195	35	22.5
b. (Sarasin Coll.) ♀, S. E. Central Celebes, 7. VIII. 95	323	202	37	24
c. (Sarasin Coll.) ♂, S. E. Central Celebes, 7. VIII. 95	300	190	37	24.5
d. (C 14625) ad. [♂], Banggai Id., V.-VIII. 95 (Nat. Coll.)	290	187	36	26

The specimen from Banggai is like the adult male from S. E. Central Celebes in the Sarasin Collection. As a rule the species of Sula do not differ from those of Banggai and Peling. It is possible, however, that when more specimens from Sula are to hand they will be found to present some points of difference, as is suggested by the two examples in the Leyden Museum.

The following notes were made there:

    a. ♂, Sula Besi (Bernstein),
    b. ♂, Sula Mangoli (Bernstein).

Similar examples with grey cheeks and ear-coverts, dark slaty head with black crest; throat-streak black, more developed in one (that figured by Schlegel, Valkv. pl. 28, fig. 4) than in the other; chest uniform light chestnut-rufous; rest of under surface banded with the same colour (somewhat browner at sides) and white:

    c. ♀ [?], North Celebes (Forsten),
    d. ♂, Manado (v. Duivenb.),
    e. —, Manado (v. Musschenbr.).

Three similar examples, corresponding with a and b in having grey cheeks and ear-coverts and dark slaty head with black crest; but the bars on the under surface are not of the same tint as the chest, being browner (hair-brown). Probably the adult plumage, but in one the first primary on each side is not fully grown:

    f. —, Manado (v. Musschenbr.),
    g. —, Minahassa (v. Musschenbr.).

Two immature (or female) specimens, with cheeks and ear-coverts rufous (or rufous brown); crown of head rufous with black centres and tips to the feathers; crest black; breast rufous; under surface banded with the same colour, one specimen agreeing in this respect with the Sula birds, though those appear to be adult; in the other specimen (probably the younger) the rufous colour predominates on the under surface, the white cross-bands being very ill-determined.

Adult males from Sula may, therefore, prove to differ from adult males from the Minahassa by having the under surface banded with rufous like the chest, instead of with hair-brown.

Distribution. Celebes to Sula. — Minahassa (Forsten *a 1*, *b 1*, *1*, v. Duivenbode *1*, Meyer *c δ*, etc.); S. E. Central Celebes (P. & F. Sarasin *2*); South Peninsula (Wallace *c II*); Banggai Island (Nat. Coll.); Sula Besi and Sula Mangoli (Allen *c 1*, Bernstein *1*, Hoedt *1*).

Under the original description of *Baza reinwardti (a 1)* there were included a female specimen obtained near Lake Tondano, North Celebes, by Forsten, a female from Pontianak, Borneo, and three males — at first erroneously stated to have come from Manado, but which were afterwards found to have been obtained in Amboina and Ternate (Schl., Rev. Acc. 1873, 133, Nos. 1, 2 and 42). These last three specimens belong to a very distinct species and should be recognised as the true *Baza reinwardti* as is done by modern authors, since the plate given (under *a 1)* represents two of them and the first of the descriptions is made from one of them. The two specimens from Celebes and Borneo were then identified by Schlegel and others with the closely allied *B. magnirostris* Gray of the Philippines[1]) *(c II)*; finally the Celebes and Sula Islands' form was marked off as a distinct species and named almost simultaneously *B. celebensis* by Schlegel and *B. erythrothorax* by Sharpe (6, 7). The description of Schlegel was published in July, 1873; that of Dr. Sharpe received at a meeting of the Zoological Society of London in June of the same year; but, as such papers are not published in the Proceedings till October, the priority rests with Schlegel. In the Catalogue of the Birds of Prey in the British Museum, 1874, the specimen from Borneo was overlooked except that Schlegel's figure of it was included in the synonymy of the Celebesian species. For a long time it remained the only example on record from Borneo. It was identified by Salvadori (Cat. Ucc. Borneo 1874, 11) with *Baza jerdoni* Blyth of Malacca, but was named as distinct *B. borneensis* by Brüggemann *(e 3)*. Ultimately Dr. Sharpe received 3 specimens from the Baram District of Borneo shot by Mr. Hose, and the distinctness of the Bornean birds as a species is upheld by him, though their identity with the Sumatran form is regarded as possible (Ibis 1893, 554—557). The Sumatran bird, according to Sharpe, is the same as *B. jerdoni* of Malacca and *B. incognita* Hume of Tenasserim and Native Sikkim. Mr. Blanford (Faun. Br. Ind. B. 1895 III, 411), who remarks that only one specimen is known from Tenasserim and one from Sikkim, agrees with Dr. Sharpe in uniting the birds from Sikkim to Sumatra as one species.

The Celebesian *Baza* is interesting as a link between Celebes and Sula, it is also of interest from its being very closely allied to the Philippine, Bornean and Sumatran species[2]), but not so with *B. reinwardti* of the Moluccas and Papuasia from Buru to New Guinea. It is a rather rare bird in collections; probably, like its congeners, it inhabits thick forest and preys entirely upon insects, as Davison observed to be the case with *Baza lophotes* (Str. F. VI, 1878, 24). Of the breeding of the *Bazae* next to nothing is known. In Australia Mr. Ramsay procured a nest of *B. subcristatus* containing three eggs, and others have been obtained there also. Quite recently Mr. Blanford has described the nest and 3 eggs of *B. lophotes* from Tenasserim (B. Br. Ind. 1895 III, 410; North, Nests and Eggs B. Austr. 1889, 15).

25. BAZA REINWARDTI Müll. Schl.

Reinwardt's Baza.

a. **Falco (Lophotes) Reinwardtii**, part. *(I)* S. Müll. & Schl., Verh. Natuurk Comm. 1839—44, Aves, 35, pl. 5, fig. 2.

---

[1]) The type is labelled "Juno, Island of Manilla, South" by Cuming (Sharpe, Ibis 1893, 555).
[2]) The genus has not yet been discovered in Java.

Baza reinwardti *(1)* Schl., Mus. P.-B. Pernes 1862, 5, Nrs. 1, 2, 4, 7; *(II)* id., Valkvogels Ned. Ind. 1866, 40, 77, pl. 27, figs. 1—3; *(3)* id., Rev. Accip. 1873, 133, pt., *(4)* Sharpe, Cat. B. I, 1874, 358; *(5)* Salvad., Orn. Pap. I, 1880, 26, 549; *(6)* W. Blas., J. f. O. 1883, 115, 124, 131; *(7)* Salvad., Orn. Pap. Agg. 1889, 12; *(8)* Gurney, Ibis 1893, 339; *(9)* Meyer, Abh. Mus. Dresd. 1893, Nr. 3, pp. 4, 5; *(10)* Hartert, Nov. Zool. 1896, 177, 247, 598; *(11)* Salvad., Ann. Mus. Civ. Gen. 1896, XXXVI, 59.
b. **Baza stenozona** *(1)* Gray, P. Z. S. 1859, 169, 189.

For further synonymy and references cf. Salvadori 5, 7, 11.
Figures and descriptions. Müller & Schlegel *a I*, Schlegel *II*; Sharpe *4*; Salvadori *5*.
Adult. Above slate-grey, carpal region, upper tail-coverts, and a short crest blacker, secondaries with a subterminal band of blackish, primaries crossed with 4 or 5 bands, basally obliterated; scapulars and innermost remiges brown; terminal third of tail black, basal part slate-grey, crossed with three nearly obliterated black bars; face, throat, and chest grey, paler than on head; remaining under parts white, passing into rufous cinnamon on middle of abdomen, thighs, and under tail-coverts barred with brownish grey on breast, sides and abdomen; under wing-coverts whitish, rufous cinnamon about the middle (ad. Ceram: Dr. Riedel — Nr. 6743).

"Iris yellow; cere, mandible, basal half of maxilla light plumbeous; apical half of maxilla jet-black; feet white, claws brown" (Everett *10*).

Young (just about able to fly). Similar to the adult, but the throat white, with black shaft-lines; breast cinnamon, the middle portion of the feathers grey; the bars on the under parts brown and somewhat narrower; the feathers of the upper parts browner with pale edgings; tail broadly tipped with grey. The crest is already present (juv. Amboina: Dr. Riedel — Nr. 6745).

Immature (apparently assuming adult plumage). Browner above than the adult described from Ceram; throat whitish, chest greyer, under parts stained with rufous cinnamon and barred on breast, sides, abdomen, and axillaries with rufous brown (not dark brownish grey); the terminal black band on the tail only one-fifth of the entire length (Bonerate, 1895: P. & F. Sarasin).

Measurements.	Wing	Tail	Tarsus	Bill from cere
a. (Sarasin Coll.) vix ad. Bonerate	320	203	37	22
b. (6743) ad. Ceram (Riedel)	287	168	34	21
c. (6744) ♀ ad. Buru (Riedel)	318	175	—	20
d. (2963) ad. Andei, N. Guinea (Meyer)	292	180	—	—
e. (2964) juv. Andei, N. Guinea (Meyer)	300	175	—	—
f. (5800) ad. loc. incert. (v. Schierbr.)	316	180	—	21
g. (5801) juv. loc. incert. (v. Schierbr.)	270	170	—	20

Distribution. Moluccas — Amboina, Ceram, Buru (cf. Salvad. *5*), Papuasia — Waigiou, Salawatti, Mysol, Misori, New Guinea, Kei, Aru (cf. Salvad. *5*, *7*), Timor (Wallace, v. Rosenb. *5*), Lombok (Everett *10*), Djampea Id. (Everett *10*), Bonerate Id. (P. & F. Sarasin).

Reinwardt's Baza has to be included in the Celebesian list in virtue of two specimens from Djampea and Bonerate between Celebes and Flores, the first collected by Mr. Everett, the other presented to the Messrs. Sarasin when in Celebes. The latter specimen is remarkable for the length of its tail and the narrowness of the black terminal portion, and for the rufous brown bars of the under surface. It is not fully adult. It differs from *Baza rufa* by the colour of the chest — all grey. By its head, throat, and chest it resembles *B. timorlaoensis* Meyer, but the banding below is different. Mr. Hartert has most kindly sent us particulars about the Djampea specimen: "after comparing it with the large series in the British Museum I had not the slightest hesitation in referring it to *B. reinwardti*. In my Djampea skin the bars below are also red-brown with a grey shade or sometimes even a distinct grey band along the upper margin, but we have one from Waigiou in which these bands are of an even purer red-brown. The terminal band on the tail is broader than in some of our New Guinea specimens, narrower than in others. The wing of our bird is 322 mm. Altogether my bird has the darkest under tail-coverts of all in the Tring Museum. At present I can see no reason for separating it even subspecifically from *B. reinwardti*, but it would be valuable to get skins from Timor, Flores etc., to see whether perhaps the 'Austro-Malayan' birds differ from those of the Papuan Islands".

In the Solomon Islands — *Baza gurneyi* Rams., in New Britain — *B. bismarcki* Sharpe, and in Timorlaut — *B. timorlaoensis* Meyer, the adult of which is still unknown, have been marked off as distinct from *B. reinwardti*. *B. subcristata* Gould of Australia is also nearly allied, and *B. rufa* Schl. of the Halmahera-group is likely to cause trouble. The last seems to be distinguishable by its having the chest grey washed with rufous, and the under parts more deeply stained with that colour, as well as having the bands thereon red-brown.

As is pointed out under *B. celebensis* the habitat of *B. reinwardti* was originally stated to be Celebes, and the error was not discovered till Schlegel wrote his "Valkvogels" in 1866. This has given rise to much error and mislabelling. Dr. Finsch in his list of birds in "New Guinea" 1865, 154 indicated Celebes as a locality of the species, but was obviously misled by previous writings and the wrongly labelled specimens in the Leyden Museum. No further proof of the occurrence of the species in Celebes can be drawn from three examples of unknown origin — but included among a lot of Celebesian birdskins — examined by Prof. W. Blasius (*6*). There is no certain evidence that this species has ever occurred in Celebes.

Baza rufa Schl.

A specimen of this Moluccan species in the Dresden Museum, marked as having come from Celebes (Nr. 2197) is mentioned by Meyer, Z. ges. Orn. 1884, 272. An immature specimen of the same species is similarly labelled "Celebes" (Nr. 2196). Both were purchased prior to 1874 of Frank of Amsterdam, and bear the name of no collector, but only labels affixed by that dealer, determining them simply as "*B. reinwardtii*, Celebes". This indicates that they were bought of Frank who got his determinations from the Leyden Museum before the distinctness of *B. rufa* was made out, that is to say before 1866, when *B. reinwardti*, owing to the original error of Temminck and Schlegel, was believed to be a Celebesian species. The conclusion to be drawn is, that the specimens were determined and the locality indicated on what were at the time sufficiently good hypothetical grounds, but which must now be regarded as almost certainly erroneous.

GENUS TINNUNCULUS Vieill.

"The Kestrels differ from the true Falcons by having a shorter bill and a much smaller and weaker foot, the middle toe without the claw being only $^2/_3$ to $^3/_4$ the length of the tarsus. The tail is longer and the feathers graduated, the outer rectrices 1 to $1^1/_2$ inches shorter than the middle pair, and the wing is shorter. The sexes differ, and the females and young have the upper parts banded black and rufous" (Blanford). They prey upon mice, insects, birds, frogs, etc., seeking their booty on the ground, hovering over it, and pitching down upon it from above. Lay 4—6 eggs. About 15 species, some migratory, others stationary, scattered over nearly the whole world.

26. TINNUNCULUS MOLUCCENSIS Jacquin. Puch.

Moluccan Kestrel.

a. **Falco tinnunculus** *(1)* S. Müll. (nec. L.), Verh. Nat. Geschied. Natuurk. Comm. 1839—44, 87 (Celebes), 209 (E. Indies); *(2)* id., Reiz. in Ind. Archip. 1857, II, 8 (Celebes).
b. **Cresserelle des Moluques** *(1)* Hombr. & Jacq., Voy. Pôle Sud, Atl. 1843, pl. I, f. 2· (Amboina); *(2)* T. & S., Faun. Jap. Aves 1845, 3 (E. Ind.).
Tinnunculus moluccensis "Schl.", Bp., Consp. 1850, I, 27 (nomen nudum); Jacquinot et Pucheran, Voyage au Pôle Sud, Zoologic, 1853, vol. III, p. 47; *(1)* Wall., Ibis 1868, 5; *(2)* Meyer, ib. 1879, 55; *(3)* Salvad., Orn. Pap. 1880, I, 37, et Agg. 1889, I, 14; *(4)* Gurney, Ibis 1881, 469; *(5)* Meyer, Verh. z.-b. Ges. Wien 1881, 760; *(6)* W. Blas., P. Z. S. 1882, 700; *(7)* Sclat., ib. 1883, 51, 194, 200; *(8)* Meyer, Isis 1884, 9; *(9)* Gurney, Diurn. B. of Prey 1884, 97; *(10)* Pleske, Bull. Acad. Petersb. 1884, 112; *(11)* W. Blas., Ztschr. ges. Orn. 1885, 220; *(12)* Guillem., P. Z. S. 1885, 562; Vorderman, N. T. Ned. Ind. 1886, 7 (sep. copy); Tristr., Cat. Coll. B. 1889, 67; *(13)* Salvad., Agg. Orn. Pap. 1889, I, 14; *(14)* M. & Wg., Abh. Mus. Dresd. 1895, Nr. 8, p. 4.
c. **Falco moluccensis** Schl., Naumannia 1855, 253; *(1)* id., Mus. P.-B. Falcones 1862, 28; *(II)* id., Valkv. 1866, 6, 47, pl. 1, f. 3, 4, 5; *(3)* id., Rev. Acc. 1873, 42; *(4)* Rosenb., Malay. Arch. 1878, 271; *(5)* Vorderman, N. T. Ned. Ind. 1884, 189.
d. **Cerchneis moluccensis** *(1)* Sharpe, P. Z. S. 1874, 583; *(2)* id., Cat. B. 1874, I, 430; id., Mitth. Zool. Mus. Dresden 1878, III, 357; H. O. Forbes, P. Z. S. 1884, 431; *(3)* Vorderm., N. T. Ned. Ind. 1895, LIV, 331.
For further synonymy and references cf. Salvadori *3, 13*.
Figures and descriptions. Hombron & Jacquinot *b I*; Schlegel *c II, c 1, c 3*; Sharpe *d 1, d 2*; Salvadori *3*; Gurney *4*; W. Blasius *6, 11*.

This species ranges from Timorlaut and the Moluccas as far as Borneo, but it varies locally within this area, the extremes of difference being found in Halmahera on the one side, and in Celebes and the Lesser Sunda Islands on the other, while some of the Moluccas, such as Buru and apparently Amboina (which furnished the type of the species) are inhabited by birds which seem to be intermediate. A little consideration will make it clear that there is no good purpose served by applying trinomials to the forms intermediate between two races — such names are not only useless and cumbersome, but also misleading since they bring with them the idea: here the species has developed into a pronounced race! The only places where trinomials may be employed without damage are where the extreme racial variations are found. In the case of the present species there is no typical subspecies. The extremes and means, as at present understood, may be identified in the following manner:

1. Tinnunculus moluccensis orientalis n. subsp.

e. Falco moluccensis (I) Schl., Valkv. Ned. Ind. 1866, 6, 47, pl. I, f. 3 (Halmahera).
Diagnosis. Under surface darker; cheeks fulvous-brown; under wing-coverts pale fulvescent brown, spotted with black.
Distribution. Halmahera-group: Halmahera, Morty, Ternate, ? Kaiva, March, Tidore, Batchian.

2. Tinnunculus moluccensis occidentalis M. & Wg.

f. Falco moluccensis (I) Schl., Valkv. Ned. Ind. 1866, 6, 48, pl. 1, figs. 4, 5.
g. Tinnunculus moluccensis occidentalis (1) M. & Wg., Abh. Mus. Dresd. 1896, Nr. 2, p. 8; *(2)* Hartert, Nov. Zool. 1896, 162, 178, 589, 597; *(3)* id. ib. 1897, 159.
"Sikep burik mera", Minahassa, Nat. Coll.
"Sikep tarak take", Tondano, N. Celebes, Nat. Coll. (the young of *Spilospizias trinotatus* and of *Accipiter rhodogaster* are confused with the present bird under this name).
"Oopo", Gorontalo, Joest, Holontalo 1883, p. 106.
"Zere zere", Tjambu, S. Peninsula, Platen 11.
Diagnosis. Under surface paler; cheeks albescent-grey; under wing-coverts white, sparingly marked with black spots.
Distribution. Celebes, Borneo (Schwaner), Java (Whitehead *g 2*), Lombok (Vorderman *d 3*, Everett *g 2*), Sumba (Riedel *5*, Doherty *g 2*), Flores, Letti (?), Timor, Timorlaut (?).

3. T. moluccensis orientalis — occidentalis.

Diagnosis. Connecting links.
Distribution. Amboina, Buru, Ceram, Goram, Peling.
General description:
Adult female. Above russet, the feathers of the head streaked with black, those of the back and wings plentifully marked with cross-bars and triangular spots of black; quills blackish with light margins, inner webs barred with russet; tail and upper tail-coverts smoke-grey, the former crossed with a broad subterminal band of black and marked with 1—9 narrow, more or less obsolete, bars; tip whitish; under surface brownish fawn-colour, throat, thighs and under tail-coverts paler, breast, sides and

abdomen marked with drop- and arrow-head spots of black; under wing-coverts white or ochraceous buff spotted like the under surface.

Adult male. Like the female, but both above and below less plentifully spotted and banded with black; tail clear smoke-grey, crossed only by the broad subterminal black band.

"Iris brown; bill greyish blue; cere and skin round the eyes yellow; feet deep citron-yellow, claws black" *(M. 2)*.

Young. Appears to differ little from the female.

Measurements. Wing 210—240 mm; tail 160; tarsus 43—44; culmen from cere 15—16.

Attention was called as long ago as 1866 by Schlegel to local deviations of coloration, when it was pointed out that the supposed specimen from Borneo and two from Timor and Flores respectively agreed with those from Celebes, while the more strongly coloured race of Halmahera was also represented in the Leyden Museum in specimens from the neighbouring islands of Morotai, Ternate, Mareh, Tidore and Batchian.

In a good series intermediate specimens may occur in localities indicated for the pronounced races. For instance, we have one from the Minahassa, N. Celebes, which by its dark tawny brown tints approaches three from Halmahera, but the under wing-coverts are lighter.

A male from Timorlaut (Nr. 6693) appears to have a somewhat greyer head and the spots on the back smaller than in any others, but, as the specimen is a bad one, it is difficult to form a good judgment of its characters. This may be a hidden subspecies. Schlegel *(c II)* mentions that the Flores specimen also has smaller and fewer spots on the upperside.

In Borneo the dark Japanese form of the Common Kestrel *(Tinnunculus alaudarius)* occurs in winter (Everett, List B. Borneo, in J. Str. Branch R. A. S. 1889, 186), and the same species — probably also the same local form of it — has been obtained near Luzon at sea (Finsch & Conrad, Verh. Z. B. Ges. Wien, 1873 June 4). The statement, therefore, of Horsfield (Tr. Z. S. 1821, XIII, 135) that *F. tinnunculus* belongs also to Java is likely enough to be true, although his note has always been taken as referring to *T. moluccensis*.

The food of the Moluccan, like that of the Common Kestrel is pretty varied. In Celebes Salomon Müller noted that it fed on "mice, small lizards and birds, grasshoppers etc."; v. Rosenberg speaks of it as "ein Hauptinsektenjäger"; Meyer observed that it preyed on little birds. It is possible for a bird with so varied a diet as this to remain stationary in a tropical island. As the numerous dates recorded by Schlegel and others show, the species occurs all the year round in Celebes, where it is extremely plentiful, and from a note of Dr. Fischer, recorded by Mr. Pleske, it is present also throughout the year in Ternate. Consequently, it is not surprising that in these two neighbouring localities distinguishable subspecies have arisen.

GENUS FALCO L.

The true Falcons range in size from small to medium; the wing is long and very pointed, the secondaries short, not reaching as far as the middle of the wing; bill well hooked, furnished with a tooth; tarsus short, reticulate; middle toe as long or longer than the tarsus; tail moderate, square to rounded in shape. They prey chiefly on birds, catching them on the wing. About 30 species distributed over nearly the whole world. Eggs 3—5.

+ 27. FALCO SEVERUS Horsf.
Chestnut Hobby.

Falco severus (1) Horsf., Tr. Z. S. 1821, XIII, 135; (2) Schl., Mus. P.-B. Falcones 1862, 23; (III) id., Valkv. Ned. Ind. 1866, 4, 45, pl. 2, f. 2, 3; (3bis) id., Rev. Acc. 1873, 39; (4) Sharpe, Cat. B. 1874, I, 397; (5) Brüggem., Abh. Ver. Bremen 1876, V, 454; (6) Bourdillon, Str. F. 1876, 354; (7) Hume, Str. F. 1879, VIII, 43, 81; (8) Sclat., P. Z. S. 1880, 65; (9) Legge, B. Ceylon 1880, 110; Reid, Str. F. 1881, X, 4; (10) H. O. Forbes, Ibis 1881, 140; (11) Oates, B. Brit. Burmah 1883, II, 216; Davison, Str. F. 1883, X, 333; (12) Everett, P. Z. S. 1889, 225; (13) id., J. Str. Br. R. A. S. 1889, 186; Tristr., Cat. Coll. B. 1889, 66; (14) Stoere, List Coll. B. Philip. Is. 1890, 8; (15) Sharpe, Ibis, 1893, 562; (16) Madarasz, Aquila 1894, 88; (17) Bourns & Worces., B. Menage Exped. 1894, 33; (18) Grant, Ibis 1895, 39; (19) Blanf., Faun. Br. Ind. B. 1895 III, 423; (20) Hartert, Nov. Zool. 1896, 256, 533.

a. Falco aldrovandi "Reinw." (1) Temm., Pl. Col. 1823, Nr. 128.
b. Falco guttatus Gray, Ann. N. H. 1843, XI, 371.
c. Falco rufipedioides Hodgs., Calc. Journ. Nat. Hist. 1844, IV, 283.
d. Hypotriorchis severus (1) Gray, Gen. B. 1844, I, 20; (2) id., List Accip. Br. Mus. 1848, 53; (3) Jerdon, B. India 1862, I, 34; (4) Blyth, Ibis 1863, 9; (5) Hume, Rough Notes 1869, 1, 87; (5bis) Wald., Ibis 1872, 98; (6) Holdsw., P. Z. S. 1872, 410; (6bis) Wald., Tr. Z. S. 1875, IX, 139; (7) Salvad., Ann. Mus. Civ. Gen. 1875, 643, 750; (7bis) Wald., P. Z. S. 1878, 937; (8) Salvad., Orn. Pap. 1880, I, 33; 1882, III, 507; and Agg. 1889, 14; (9) Gurney, Ibis 1882, 131, 153; id., Diurn. B. of Prey 1884, 102; (10) Ploske, Bull. Acad. Sc. Petersb. 1884, XII, 112; (11) Finsch & Meyer, Z. ges. Orn. 1886, 2; (12) W. Blas., Ornis 1888, 303; id., Ibis 1888, 373; Tristr., Cat. Coll. B. 1889, 66; (13) Whitehead, Ibis 1890, 43; (14) W. Blas., J. f. O. 1890, 138.
e. Falco frontatus (1) Schl., Valkv. 1866, pl. 2, f. 5.

For further synonymy and references see Salvad. d 8.

Figures and descriptions. Temminck a 1; Schlegel III, e I; Sharpe 4; Legge 9; Salvadori d 8; Oates 11; Finsch & Meyer d 11; Blanford 19.

Adult. Head, neck, cheeks and ear-coverts black; quills and lesser wing-coverts also black; rest of upper surface slate-colour, palest on the rump, shafts of the feathers black; chin and throat white, slightly washed with rufous; rest of under surface dark cinnamon-rufous, almost unmarked with black; under wing-coverts like the under surface, but marked with black about the carpal joint; quills and tail below brownish grey, obscurely barred with cinnamon-grey towards their bases

82   Birds of Celebes: Falconidae.

(Lotta, Minahassa, 23. June 1894: Nat. Coll. — C 12237). An adult ♂ (Macassar, Wallace in the British Museum) has the bars present on the basal half of the wings, but much obliterated; tail, not barred, above blackish slate, below shining dusky; under surface of body deep cinnamon-chestnut. This description appears to apply to another specimen — from the bars being perceptible on its tail apparently a female — shot at Lotta N. Celebes 29. May 1893 by our native collectors, now in the Tring Museum.

Young. Above resembling the adult; below the dark cinnamon-ground colour is plentifully marked with streaks and spots of black, which take the form of sagittate marks and bars on the under wing-coverts; the bars on wings and tail below more distinct than in adults from the same locality (Java, C 10553, S. E. New Guinea, Nr. 8726).

A young specimen from Celebes (Tweedd. Coll. in Brit. Mus.) does not correspond as regards the tail with the above adult ♂ from Macassar, having it barred with cinnamon as in a young one from Sikkim.

Measurements.	Wing	Tail	Tarsus	Culmen from cere
a. (C 10568) ♂ Calcutta	246	120	35	15.5
b. (C 10552) Java	220	108	32	13
c. (C 10553) juv. Java . . .	217	100	31	12.5
d. (8727) ♀ S. E. New Guinea. . . .	247	112	34	15
e. (8726) ♂ juv. „ „	211	98	29	13
f. (C 12237) [♀] N. Celebes . . .	245	117	34	15.5
g. (Tring Mus.) [♀] N. Celebes	235	109	32	14

Distribution. Himalayas (Legge 9, Salvad. d 8); Bengal (Blyth d 4); Ceylon (9, d 8); ? Burmah (Oates 11); Philippines (Cuming d 2); Mindanao (Everott d 7*bis*), Negros (Layard 6*bis*), Mindoro (Platen in Mus. Nehrkorn), Cebu, Siquijor, Tawi-Tawi, Calamianes, Romblon, Sibuyan (Bourns & Worcester 17), Palawan (Platen d 12, Whitehead d 13, Steere 14), Sooloo Islands (Platen d 14); Borneo (Fischer 5, Everett 15); Tenasserim and Malacca (d 8), Java (d 8, Forbes 10); South Celebes Macassar (Wallace 4, Beccari d 7), North Celebes, Manado (v. Duivenbode 9*bis*, Nat. Coll.); Ternate (Fischer d 10); Halmahera (d 8); Ceram (d 8); Salawatti (d 8); Jobi (d 8); S. & S. E. New Guinea (d 8, d 11); New Britain (Brown 8).

The range of *F. severus* appears to be checked and bounded by the occurrence of two rival species of Hobby, the Palaearctic *F. subbuteo* L. on the one hand, and the Australasian *F. lunulatus* Lath. on the other. The ranges of both of these species overlap that of *F. severus*, and that of *F. lunulatus*, in particular, encroaches far into it; *F. severus* and *F. subbuteo* are found together in India, where the former, according to Colonel Radcliffe "is local, while *F. subbuteo* is migratory"; and, to the south, both *F. severus* and *F. lunulatus* have been obtained in New Britain *(d 9)* and also in Ceram and Ternate.

Whether *F. severus* is a migratory species in the East Indian Archipelago, there is no sufficient evidence to show; probably it is not such. In the cold season, however, as was noted by Blyth, "it visits the plains of Bengal, where it is somewhat rare"; in Ceylon it can only be regarded as a straggler according to Colonel Legge, who believes that it finds its way to that island during the season of migration. There is no evidence of a similar movement in the East

Indian Islands. More evidence is, however, to be desired, since the closely similar *F. lunulatus* occurs also among the same islands; and in Palawan Whitehead regards *F. severus* as a migratory bird *(d 13)*. It is much to be regretted — as has been already remarked — that the dates when specimens are shot are so rarely recorded; sufficient of them would enable ornithologists to draw conclusions as to the stationary or migratory habits of species. In the present case we have been able to find only the following:

Locality	Specimens	Months	Season	Reference
Java	3	June	Fine	2, 3bis, 10
Negros	1	February	?	d 5bis
Mindanao	1	April	?	d 7bis
Mindoro	1	October	?	Platen
Palawan	3 or more	Summer and Autumn	?	d 12, d 13, 14
N. Celebes	1	May	Fine	Nat. Coll.
N. Celebes	1	June	Fine	Nat. Coll.
S. Celebes	1	January	Rainy	d 7
Finschhafen, New Guinea	1	March	Fine	Mus. Dr.

It will be seen that specimens have been obtained in one or other part of the Archipelago in summer and winter, fine season and wet; hence it is improbable that the species is migratory in the strict sense of the term, and this condition is what is wanted before local differences can be insisted upon. The three New Guinea specimens in the Dresden Museum, and four from New Guinea, Salawatti and New Britain in the British Museum, are of a blacker slaty above than is usual in Indian specimens and in two before us from Java. In the coloration of the under surface the New Guinea form agrees with the Indian one, but the under side of the wing and tail are nearly uniform dusky brown, only a few bars on the inner webs being present at the base of the primaries, while on the tail the bars are almost, or quite, completely obliterated. Indian specimens have the inner web of the quills and tail barred with cinnamon for almost their entire length, but the barring on the under side of the tail is less distinct in the adult male than in the adult female or in the young. In the adult male the tail above is almost uniform dark ashy, with a subterminal black bar; in the adult female it is distinctly barred with black throughout its length. Young specimens from India have the tail above blackish brown, not barred; below barred with cinnamon. Javan specimens, young and old, appear to run midway between those of India and New Guinea, judging from the two in the Dresden Museum. As to Celebes, the adult male in the British Museum is not to be separated from New Guinea specimens; and two since received from our native collectors are apparently similar ones; the young specimen from Celebes in the British Museum corresponds best, however, with a young bird from Sikkim as regards the markings on the tail. Javan and Celebesian birds must therefore be marked as intermediate between the geographical extreme developments of the species. Birds from the most eastern bounds of the range of the species have departed most widely in coloration in a certain direction, and we think it advisable to distinguish the two extremes — the races of S. E. New Guinea and of India especially — as subspecies:

1. Falco severus papuanus M. & Wg.
Abh. Mus. Dresd. 1893 Nr. 3, p. 6.

Diagnosis. Adult: above darker; tail nearly black; below darker; remiges and tail below unbanded.

Young: wings and tail below obscurely banded, under wing-coverts covered with a network of black.

Distribution. New Guinea.

2. Falco severus indicus n. subsp.

Diagnosis. Adult: above paler slaty; tail brownish slaty; wings and tail below barred on the inner webs with pale cinnamon.

Distribution. India (Calcutta: Henderson — C 10568).

Intermediate forms are known to us from Java (whence the type of the species) and Celebes, and it is probable that the larger part of the East Indies are inhabited by such.

The legs and feet of *Falco severus* are stouter and stronger, and the middle toe and wings relatively shorter than in *F. subbuteo* L., the type of Boie's genus, *Hypotriorchis*. *F. severus* thus forms a link between the most typical Falcons and *Hypotriorchis*, which genus was separated from them chiefly on the ground of its possessing a long, thin, middle toe, and slender, somewhat lengthy tarsus, but these slight structural peculiarities lose all generic worth when it is seen that they hold good only for the extreme forms.

In India the food of this species is said to consist chiefly of small birds (Legge *9*), but it probably preys largely upon insects as well. Mr. de Bocarmé found that the species in Java fed chiefly on Orthoptera, the remains of which, such as the wings and legs, are to be seen in abundance on the rocks in craters to which the bird resorts *(2)*. The specimen obtained in Ceylon by Mr. Bligh was shot while hawking dragon-flies.

Colonel Legge considers *F. severus* a mountain species; it was found breeding in the Himalayas in 1860 by Col. Radcliffe, who got young birds unable to fly; Mr. Thompson gives evidence in Hume's "Rough Notes" as to its breeding in Kumaon and the Ghurlwal; Mr. Bourdillon is of opinion that it nests in Travancore in South India — all more or less mountainous provinces, and Mr. de Bocarmé found it among the volcanoes of Java as pointed out above. Nevertheless — as has already been remarked — Blyth observes that it visits the plains of Bengal in the cold season and, as it has been also shot at Macassar, it frequents the plains in Celebes too.

In almost all parts of its range *F. severus* appears to be a rare species. In Celebes it certainly is so. It was not included in the large collections sent from Celebes by Dr. Fischer to the Darmstadt Museum, nor in those of

v. Duivenbode, Riedel and Platen investigated by Prof. W. Blasius; neither was it obtained there by Meyer during his stay of about twelve months, nor by Dr. Guillemard during the cruise of the "Marchesa", nor by the older travellers. Only three naturalists were more successful, two or more specimens were shot at Macassar by Mr. Wallace, from whom the British Museum received its two specimens (? the same); another was obtained at the same place by Beccari. There is also a specimen from Celebes in the Norwich Museum, as we learn from Mr. Gurney. The only record of its occurrence in North Celebes appeared to consist in the single specimen in the Leyden Museum, until two specimens from Lotta near Manado were sent to us in 1894. Possibly its rarity in collections has something to do with the difficulty of invading its mountain haunts.

† 28. FALCO PEREGRINUS (Gerini).
Peregrine Falcon.

The Peregrine Falcon has developed some considerable differences of coloration in certain areas of its range and has been divided into five species or subspecies, as follows:

1. The typical F. peregrinus.

a. **Accipiter Falco peregrinus** Briss., Orn. 1760, I, 321.
b. **Accipiter peregrinus** Gerini, Orn. Meth. Dig. 1767, I, 55, pl. 23, 24 (see Seebohm c 7).
c. **Falco peregrinus** Tunstall, Orn. Brit. 1771, I; *(1)* Dresser, B. Europe VI, 31, pl. 372 (1876); *(2)* Legge, B. Ceylon 1880, 101; *(3)* Salvad., Orn. Pap. 1880, I, 31; *(4)* Scully, Ibis 1881, 416; *(5)* Gurney, Ibis 1882, 292—304, 438; *(6)* Oates, B. Brit. Burmah 1883, II, 214; *(7)* Seebohm, Brit. B. 1883, I, 23, pl. 3 (egg); *(8)* Gurney, Diurn. B. of Prey 1884, 106; *(9)* Styan, Ibis 1887, 223; *(10)* Sharpe, Ibis 1888, 195; *(11)* W. Blas., Ornis 1888, 303; *(12)* Whitehead, Ibis 1890, 43; *(13)* Wiglesw., Aves Polyn. 1891, 1; *(14)* Styan, Ibis 1891, 489.
d. **Falco communis** Gm., 1788; *(1)* Schl., Valkv. 1866, II, 44, pl. 1, f. 1; *(2)* id., Rev. Acc. 1873, 32 pt.; *(3)* Sharpe, Cat. B. 1874, I, 376; *(4)* Rosenb., Malay. Archip. 1878, 271; *(5)* Sharpe, P. Z. S. 1881, 790; *(6)* id., Ibis 1889, 75; *(7)* Everett, J. Str. Br. R. A. S. 1889, 186; *(7 bis)* Steere, List Coll. B. Philip. Is. 1890, 7; *(8)* Sharpe, Ibis 1890, 274; *(9)* id., Bull. Brit. Orn. Club 1892, II, Nov.; *(10)* Ribbe, Ver. Erdk. Dresden XXII, 1892, 168; *(11)* Sharpe, Ibis 1893, 116, 559; *(12)* Tacz., Faun. Orn. Sib. Orient. 1891, I, 77.

For synonymy see Sharpe *d 3*; Legge *c 2*; Salvad. *c 3*; Oates *c 6*; Seebohm *c 7*.
Figures and descriptions. Schlegel *d 1*; Dresser *c 1*; Seebohm *c 7*; Sharpe *d 3*; Gurney *c 5*; etc., etc.
Osteology. Giebel, Z. ges. Naturw. 1858 V/VI, 43 (hyoid); Magnus, Knoch. 1868 pl. XVII, III, fig. D (sternum); id., Vogelkopf. 1870, III, 11 (lacrymal); Milne-Edw., Ois. foss. 1867—71, pl. 179—181 (fragments); Eyton, Osteol. Av. 1867—75, pl. IIIA (skel.), pl. IIA, f. 3 (metatars.); pl. VIA, f. 5 (pal. bones), pl. VA, f. 4 (pelvis), pl. VIIA, f. 2 (coracoids, etc.), pl. IV, f. 3 (base cran.); Parker, Tr. L. S. 1875 *(2)* I. pl. 25, ff. 12, 13 (vomer); Fürbringer, Unters. 1888 III, 6, 103, 104, IV, 23, 131, VI, 83 (clavicula, etc.).

Adult. Head, neck and primaries dusky brown; rest of upper plumage slaty grey, each feather crossed with several bars of dusky brown; below white washed with buff, under side of wings, sides, abdomen, thighs and under tail-coverts closely barred with brown, on the breast a few drop-shaped brown spots; hinder part of cheeks usually white.

Young. Above brown, all the feathers with pale margins; below white, marked from below the throat downwards with drops and streaks of brown, which tend to form into bars on the flanks.

	Wing	Tarsus	Middle toe
Small ♂ (Sumba)	312	43	44.5
Large ♀ (Spain)	375	53	56

(Gurney *c 5*).

Distribution. Europe, Africa, Asia, (?) America (see Legge *c 2*); Philippines (Salvad. *e 3*, Steere *d 7*ᵇⁱˢ), Palawan (Whitehead *c 10*, Platen *c 11*); Borneo (*c 3*, *d 7*, Everett *d 9*, etc.); Bangka (*c 3*); Sumatra (*c 3*); Java (*c 3*); Sumba (Mus. Norw. *c 5*); (?) N. Celebes (Rosenb. *d 4*); Ternate (*c 3*); Ceram (*c 3*, *d 10*).

2. Falco peregrinus melanogenys (Gld.).

e. **Falco melanogenys** *(1)* Gld., P. Z. S. 1837, 139; *(II)* id., Syn. B. Austr. 1838, pl. 40, fig. 2; *(III)* id., B. Austr. 1848, I, pl. 8; *(4)* Sharpe, Cat. B. 1874, I, 385; *(5)* Gurney, Ibis 1882, 302—304; *(6)* W. Blas., "Braunschw. Anzeiger", 1886 March 3; *(7)* Salvad., Orn. Pap. Agg. 1889, I, 14; *(8)* North, Nests & Eggs B. Austr. 1889, 16, pl. III, f. 4 (egg); *(9)* Styan, Ibis 1891, 489; *(10)* Hartert, Nov. Zool. 1896, 597.

f. **Falco communis** *(1)* Schl., Valkv. 1866, 2, 44, pl. 1, f. 2 (tantum).

g. **Falco peregrinus** *(1)* Meyer, V. z.-b. Ges. Wien 1881, 760.

g¹. **Falco melanogenys** subsp. Gurney, Diurn. B. of Prey 1884, 107.

For synonymy see Sharpe *e 4*; Salvad. *e 7*.

Figures and descriptions. Gould *e 1*, *e II*, *e III*; Schlegel *f 1*; North *e 8*; Sharpe *e 4*; Gurney *e 5*.

Diagnosis. Differs from the typical *peregrinus* in having nape and sides of head black; the black transverse bars on the under surface of adult specimens narrower, more regular and, generally, closer together; the outer and inner toes apparently a little longer in relation to the middle toe; size a little smaller.

Measurements.	Wing	Tarsus	Middle toe
Small ♂ (Australia)	296	39	48
Large ♀ (Philippines)	348	51	56

(Gurney *e 5*).

Distribution. Australia; Tasmania; New Caledonia; New Hebrides; Fiji Islands, — Viti Levu; New Guinea, Aru, New Britain, Jobi, Java, Sumatra, Borneo, Philippines (fide Salvad. *e 7*); Sumba (Riedel *g 1*); North Celebes (Platen *e 6*); Lombok — seen only (Everett *e 10*). It has also been recorded from China (*e 9*, Rickett, Ibis 1894, 223), but the bird seems to have been *F. peregrinator* (cf. Grant *i 3*).

3. Falco peregrinus ernesti (Sharpe).

h. **Falco melanogenys** *(1)* Gurney (nec Gld.), Ibis 1882, 302, 304.

i. **Falco ernesti** *(1)* Sharpe, Ibis 1894, 545; *(2)* Grant, Ibis 1895, 438; *(3)* id., Ibis 1896, 529.

j. **Falco atriceps** *(1)* Eagle Clarke (nec Hume), Ibis 1895, 529.

Descriptions. Under the above references.

Diagnosis. Blacker than *F. p. melanogenys*; under surface shaded with cinereous; under wing-coverts and axillaries black, barred with small white lines (Sharpe *i 1*).
Distribution. Borneo (Pretyman *h 1*, Hose *i 1*), Luzon (Whitehead *i 2*), Negros (Keay *j 1*, Whitehead *i 3*), Malicollo, New Hebrides (Wykeham Perry *i 2*). This is supposed by Sharpe to be the resident form in the East India Islands.

4. Falco peregrinus anatum (Bp.).

k. **Falco anatum** Bp., Comp. List B. Eur. & N. Am. 1838, 4.
l. **Falco peregrinus anatum** *(1)* Ridgw., Man. N. Am. Birds 1887, 247.
Adult. "Chest usually immaculate" (Ridgw. *j 1*).
Young. "More deeply colored, with ground color of lower parts frequently deep ochraceous" (Ridgw. *j 1*).
Distribution. "Whole of America south as far, at least, as Chili; eastern Asia?" (Ridgw. ib.).

5. Falco peregrinus pealei Ridgw..

m. **Falco polyagrus** pt. Cass., B. Calif. 1853, pl. 16 (hinder figure only), (fide Gurney, Ridgw.).
n. **Falco peregrinus pealei** *(1)* Ridgw., Landb. N. Am. 1874, III, 137; *(2)* Gurney, Ibis 1882, 297, 298; Ridgw., l. c. note; *(3)* Gurney, Diurn. B. of Prey 1884, 106; *(4)* Ridgw., Man. N. Am. Birds 1887, 247.
Adult. "Top of head deep slaty, or plumbeous slate, uniform with back; chest heavily spotted with blackish, and dusky bars of remaining under parts very broad" (Ridgw. *n 4*).
Young. "With lower parts sooty black, streaked with pale buffy, or buffy white; the feathers of upper parts without rusty margins" (Ridgw. *n 4*).
Distribution. "Aleutian Islands, west to Commander Islands and south along Pacific coast to Oregon" (Ridgw. ib.).

Of these five subspecies the three first named only concern the ornithology of Celebes.

Taking first the typical *F. peregrinus* into consideration it appears that the only authority for the occurrence of this species in Celebes is v. Rosenberg, who remarks that it is rare there, but that he got a fine specimen in Kema in the North. As, however, he adopts the nomenclature of Schlegel, who did not discriminate between *F. peregrinus* and *melanogenys* and had not recognised *F. ernesti*, it is not possible to say to which race v. Rosenberg's specimen belonged; and, if it is still extant, we do not know in what museum or collection it is to be found. *F. peregrinus* has been obtained in islands lying on every side of Celebes, — the Philippines, Borneo, Sumba, Ceram, Batjan and Ternate, so that it is impossible that it could fail to occur in Celebes as well, though satisfactory evidence on this point is wanting. In Ceram Mr. Ribbe found *F. communis* [? *melanogenys* or *ernesti* in part] and *moluccensis* very plentiful both in the forest as well as on the coast and mountains. There is considerable reason for believing that this race is only a winter visitor to the East Indian Archipelago. It is stated to be a migrant in India, though a good many specimens remain there throughout the year *(c 2)*, and Mr. Oates is inclined to regard it as a resident in Burmah. The southward migration through Gilgit and the Hindoo Koosh has been observed by Major Scully to take place in October and the return in April *(c 4)*. About the lower Yangtse Mr. F. W. Styan speaks of it as a common resident species, but an influx of specimens from the North takes place in autumn. Further south in China it is a common winter visitor, at Foochow, Mr. De la Touche found it abundant from October to spring. In North Borneo and Labuan birds of this race, according to Mr. Everett, appear in the N. E. monsoon, "and are doubtless regular winter migrants, probably from China" *(d 7)*. Mr. Whitehead, also, reports it to be a winter migrant to Palawan.

The Australian Peregrine, *melanogenys*, is less well understood. It is known to breed in Australia laying two or three eggs in September — October, the Australian spring *(e8)*. At this time of year a specimen has been obtained in Sumatra by Beccari, and Bocarmé's notes show that a form of Peregrine Falcon breeds in Java, making its nest in trees in the mountain forests; but, as Schlegel, who published these remarks, did not distinguish the different races of *F. peregrinus* by name, it is not certain to which race — if to only one — Bocarmé's observations refer. Dr. Sharpe believes that Java and Borneo are inhabited by a peculiar local race of Peregrines (*ernesti*); Seebohm considered that *F. melanogenys* breeds in Sumatra and Java as well as Australia *(c7)*. In the lower Yangtse Basin, Mr. F. W. Styan has recorded a pair of *F. melanogenys* as breeding there, but the dark form in South China has been identified by Mr. H. H. Slater with the Indian *F. peregrinator* Sundev. *(j 1, i 3)*. *Falco ernesti* was observed breeding in Negros by Mr. Whitehead. Any conclusion as to a possible migration from Australia during the winter of that country — the months of our summer — cannot be made, owing to the absence in literature of actual observations and of dates when specimens were shot. The bird is either decidedly rare in the archipelago, or more difficult to obtain than the generality of birds of prey.

That *F. melanogenys* occurs very rarely in Celebes — possibly as a straggler — is evident from the fact that the only specimen of which there is any record from this island, is one which Prof. W. Blasius laid before a meeting of the Naturhistorischen Verein at Brunswick in 1886, as reported in a newspaper account of the sitting *(e 6)*. It is possible, however, that v. Rosenberg's note, stating that he found *F. communis* not common in the island *(d 4)* may concern this form of Peregrine.

A closer examination of the specimen from Sumba sent by Dr. Riedel to the Dresden Museum has served to show us that the bird belongs to the southern race, and not to the northern one with which it was at first identified. Therefore both subspecies occur on Sumba *(c 5)*.

FAMILY PANDIONIDAE.

The family contains only one species, the Osprey, peculiar by its having the outer toe reversible, and no aftershaft to the feathers. The tarsus is naked (except at the top anteriorly) and covered with very rugose small scales. The claws very long, and curved almost to a semicircle; the toes furnished below with pads covered with small, strong, sharp points or spines, adapted for holding the fish upon which it preys. Size large medium; wing very long, extending beyond the tip of the tail. Migratory and stationary. Eggs 2—3. Almost cosmopolitan in range, but varies locally to some extent.

GENUS PANDION Sav.

Description as for the family.

+ 29. PANDION HALIAETUS (L.).
Osprey.

The genus *Pandion* contains only one species, *P. haliaetus*, which is almost cosmopolitan in its range. In the Old World, in America, and in Australia some differences of size and coloration have arisen, which makes it advisable to adopt the three subspecies already distinguished.

1. The typical Pandion haliaetus.

a. **Falco haliaetus** Linn. 1766; *(1)* Naum., Vög. Deutschl. 1822, I, 241, pl. 16.
b. **Pandion haliaetus** Less., Man. d'Orn. 1828, I, 86; *(1)* Gld., B. Eur. 1837, pl. 12; *(2)* Schl., Mus. P.-B. Aquilae 1862, 22 pt.; *(III)* id., Valkv. 1866, 12, 52, pt., pl. 3, f. 3; *(4)* Newton, Yarrell's Brit. B. 1871, I, 34; *(5)* Schl., Rev. Acc. 1873, 123 pt.; *(6)* Sharpe, Cat. B. 1874, I, 449; *(7)* Salvad., Cat. Ucc. Borneo 1874, 7; *(VIII)* Dresser, B. Europe VI, pls. p. 139 (1876); *(9)* Legge, B. Ceylon 1880, 122; *(10)* Gurney, Ibis 1882, 594—598; *(11)* Soebohm, Brit. B. 1883, I, 56, pl. 3 (egg); *(12)* Oates, B. Brit. Burmah 1883, II, 220; *(13)* W. Blas., J. f. O. 1883, 118; *(14,* Gurney, Diurn. R. of Prey 1884, 112; *(15)* W. Blas., Ornis 1888, 539; *(16)* Sharpe, Ibis 1889, 75; *(17)* Everett, J. Str. Branch R. A. S. 1889, 187; *(18)* Whitehead, Ibis 1890, 43; *(19)* Sharpe, t. c. 274; *(20)* Styan, Ibis 1891, 486; *(21)* De La Touche, Ibis 1892, 483; *(22)* Meyer & Helm, Jhrsb. Orn. Sachsen 1892, 79; *(23)* Tacz., Faun. Orn. Sib. Orient. 1, 1891, 52; *(24)* M. & Wg., Abh. Mus. Dresd. 1896 Nr. 2, p. 8.
For synonymy and further references see Sharpe b 6.

2. Pandion haliaetus leucocephalus (Gld.).

c. **Pandion leucocephalus** *(1)* Gld., P. Z. S. 1837, 138; *(II)* id., Syn. B. Austr. 1838, pt. III, pl. 6; *(III)* id., B. Austr. 1848, I, pl. 6; *(4)* Sharpe, Cat. B. 1874, I, 451 (sbsp.); *(5)* Meyer, Ibis 1870, 56; *(6)* Salvad., Orn. Pap. 1880, I, 11 and Agg. 1889, 11;

(7) Gurney, Ibis 1882, 595—598; *(8)* id., Diurn. B. of Prey 1884, 113 (subsp.?); *(9)* Meyer, Isis 1884, 6; *(10)* North, Nests & Eggs B. Austr. 1889, 23, pl. 5, f. 1, 2 (eggs); *(11)* Steere, List Coll. B. Philip. Is. 1890, 8; *(12)* Meyer, Abh. Mus. Dresden 1893, 3, p. 4.

d. **Pandion haliaetus leucocephalus** *(1)* Brüggem., Abh. Ver. Bremen 1876, V, 45; *(2)* Wiglesw., Aves Polyn. 1891, 1; *(3)* Hartert, Nov. Zool. 1896, 178.

For synonymy and further references see Salvadori *e 6*.

3. Pandion haliaetus carolinensis (Gm.).

e. **Pandion carolinensis** (Gm. 1788); *(1)* Audubon, B. N. Am. pl. 81 (1831); *(2)* Gurney, Ibis 1882, 595—598; *(3)* id., Diurn. B. of Prey 1884, 113 (subsp.?).

f. **Pandion haliaetus carolinensis** Ridgw., B. N. Am. 1874, III, 182; *(1)* id., Man. N. Am. Birds, 1887, 254.

Key to distinguish the 3 subspecies:

A. Below purer white than in the other races, breast often immaculate; "shafts of the tail-feathers continuously white":. . . . . *P. haliaetus carolinensis.*
   Distribution. "Temperate and tropical America in general, north to Hudson's Bay and Alaska" (Ridgw. *f 1*).

B. Breast always more or less varied with brown:
   a. Larger; wing from 437 (♂)—528 mm (♀) (ex Gurney *b 10*); crown of head usually more strongly marked with brown streaks than in *leucocephalus*: . . . . *The typical P. haliaetus.*
   Figures and descriptions. Naum. *a I*; Gld. *b I*; Dresser *b VIII*; Sharpe *b 6*; Legge *b 9*; Gurney *b 10*; Seebohm *b 11*; etc. etc.
   Osteology. Selenka, Bronn's Kl. u. Ord., pl. XVI, f. 7 (sternum); Milné-Edw., Ois. Foss. 1867—71, pl. 179; Eyton, Ost. Av. 1867—75, pl. III,' f. 10 (frons), pl. V, f. 5 (pelvis); Blanchard, Ost. Ois. 1859, pl. II, 7, 8; Fürbringer, Unters. 1888, I, 20; III, 102; VI, 81 (shoulder); Giebel, Z. ges. Naturw. 1858, V/VI, 49.
   Distribution. Europe; Africa; Asia, as far south as Java (♂ with wing 450 mm — Vorderman), Borneo *(b 16, b 17)*, Palawan *(b 18)*, and the Sangi Islands *(b 15)*; Celebes (in Dresden Mus.).

   b. Smaller; wing 385—493 mm; crown of head sometimes nearly, or quite, pure white: . . . . *P. haliaetus leucocephalus.*
   Figures and descriptions. Gould *c 1, c II, c III*; Schl. *b III*(?); Gurney *c 7*; North *c 10*. Eyton, Ost. Av. 1867—75, pl. I (skel.), pl. II, f. 1 (metatarsus), pl. III, f. 2 (sternum).
   Distribution. New Zealand *(b 11)*; Australia and New Caledonia throughout the intervening islands to Celebes and Sangi (Salvad. *c 6*); Banggai Id. (Nat. Coll.); Talaut Is. (Nat. Coll.); Bawean Islands (Schl. *b 2*); S. E. Borneo *(b 2)*; Philippine Islands — Marinduque (Steere *c 11*); Pelew Islands (Mus. Hamburg *d 2*).

The species *P. haliaetus* may be generally described as follows:

Adult. Head and neck white, with dark brown centres to the feathers present in greater or less abundance on the crown and hind neck; quills black; all other parts of the upper surface dull dark brown, the margins of the feathers somewhat paler;

ear-coverts blackish brown continued into a streak down the sides of the neck; chin and throat white, often marked with brown shaft-streaks; chest brown, sometimes pure white; under wing-coverts nearest the body white, the outermost ones brown with white margins; remaining under parts of body — except the axillaries, which are spotted with brown — pure white.

Iris yellow; cere blue; feet blue or bluish white.

Young. Above darker, richer brown than the adult, all the feathers of the back, upper tail- and wing-coverts terminally margined with white or fulvous; crown of head and neck much more saturated with blackish brown than in the adult; numerous cross-bars — imperceptible in the adult — plainly apparent on the tail of the young.

Distribution of the species. Almost cosmopolitan. Absent in the southern countries of S. America and in the islands of the Pacific Ocean, except two or three (cf. Dresser *b VIII*, Newton *b 4*, Legge *b 9*, Salvadori *c 6*).

The American form of the Osprey, *P. carolinensis* does not further concern this work; both of the other two forms, the typical Osprey of the Old World and the smaller Australian bird in our opinion occur in the Celebesian area.

Schlegel and Sharpe include Celebes within the range of the *typical* form (*b 5*, *b 19*); Brüggemann, Gurney, Meyer, Blasius and Salvadori within that of the Australian form (*d 1*, *b 10*, *c 5*, *b 13*, *c 6*) and Schlegel and Blasius (*b 15*) identify the Sangi Islands as a locality for the *typical P. h.*, Gurney and Salvadori as a locality for *P. h. leucocephalus*, — so, also, Meyer (*c 9*). These conflicting opinions will be answered in the future in one of two manners — either one or both forms will be found to be migratory visitors to Celebes and Sangi; or, the birds inhabiting these islands will be found to present an interconnected, intermediate race between the northern and southern Ospreys. If migratory, then the northern race may be expected to occur in Celebes and Sangi in the months of our winter and the southern race during the months of the Australian winter, i. e. in our summer. The northern Osprey is known to migrate south in autumn, passing the winter south of about 45° N. lat. (*b 22*); in Ceylon it is only present during the cool season (*b 9*), and this also appears to be the case in Burmah (*b 12*) and possibly throughout all countries south of the Himalayas. No record of its nesting in India was known to Mr. Oates in 1889. In Borneo three specimens of the northern Osprey have been obtained in December by Mr. Whitehead, who believes it is a migrant there — i. e. in Northern Borneo (*b 16*), and it is also marked by him as migratory in Palawan (*b 18*). Another was obtained at Sarawak by Beccari on March 3rd, 1867 (*b 7*). A fifth mentioned by Schlegel (*b 2*) was killed at Pagattan at the S. E. extremity of the island, and has the wing only 420 mm ("15¹/₂ French inches") long; thus evidently — though a male — belonging to a smaller race. For the Sangi Islands and Celebes the few specimens, which have been properly dated, run as follows:

92   Birds of Celebes: Pandionidae.

Sangi

a. ♂	wing 420 mm		. . . . .	9. Aug. 1886 (Platen 15)
b. ♂	» 495 »		. . . .	20. Jan. 1887 (Platen 15)
c. ♀	» 485 »		. . . .	20. Jan. 1887 (Platen 15)
d. ♀			. . . .	4. Aug. 1865 (Rosenb. b 5)
e. ♀	wing 437—470 mm		. . .	3. Nov. 1866 (Rosenb. b 5)
f. ♀			. . . .	24. Jan. 1866 (Hoedt b 5)

Talaut

g. 1 wing 435 mm . . . . Autumn 1896 (Nat. Coll.)

North Celebes

h. ♂	» 422 (or 429) mm . . .	. . . .	10. Oct. 1863 (Rosenb. b 5)
i. 1	» 430 mm	. . . .	May 1871 (Meyer, Dresd. Mus.)
j. 1	» 450—470 mm (2 or more specimens)	.	March 1871 (Meyer c 5)
k. 1	» 450 mm .	. . . . . .	12. May 1893 (Nat. Coll., Dr. M.)
l. 1	» 420 » . . .	. . . .	24. Apr. 1893 (Nat. Coll., Dr. M.)
m. 1	» 470 » .	. . . . . .	11. Dec. 1894 (Nat. Coll., Dr. M.)
n. 1	» 470 »	. . . . .	27. Feb. 1894 (Nat. Coll., Dr. M.)
o. 1	» 450 » .	. . . . . .	March 1895 (Nat. Coll., Dr. M.)
p. 1	» 430 » . . .	. . . .	March 1895 (Nat. Coll., Dr. M.)
q. 1	» 426 » .	. . . . . .	Feb. 1894 (Nat. Coll., Dr. M.)

East Celebes and Banggai Id.

r. 2	» 430 and 470 mm .	. .	May—Aug. 1895 (Nat. Coll., Dr. M.)
s. 1	» 430 mm, Banggai Id. .	. . .	May—Aug. 1895 (Nat. Coll., Dr. M.)

Prof. W. Blasius identifies specimen a with b and c as belonging to the northern race; but its size would seem to indicate that it belongs to the smaller one. Our own specimens of the larger race have winter dates. The other two were shot in our winter. On the whole the above facts seem to indicate that the larger form is a migrant, and the smaller a resident in Celebes and Sangi.

A change of disposition in this species, as recorded by Mr. C. Ribbe in Ceram, also renders it very probable that the southern form is resident in the Archipelago: "One of the most importunate birds of prey, which I met with in the Indies, was *Pandion haliaetus*; this species, elsewhere, in consequence of sharp persecution, a very shy bird, was bold and impudent on the coast of Ceram, where I was able to observe it daily. If the fishermen do not look out, it will steal the fish in their boats; when fishing with the line is being done, it will watch for the moment when the fish is drawn out, and pitch down upon it. Once, on a larger fishing expedition in which I joined, it happened that an Osprey paid for its boldness with its life, for it was beaten to death with sticks by the boatmen". (Jb. Ver. Erdk. Dresden, XXII 1892, 168).

The northern Osprey lays at the end of April or beginning of May *(b 11)*; in Australia, the smaller form breeds from July to October *(c 10)*. A northern bird, staying behind in the East Indian Archipelago, might thus be prevented from interbreeding with the smaller race. The food of this species consists entirely of fish, caught by plunging into the water.

? Gyps sp.

In the Nat. Tdschr. Ned. Ind. 1876, XXXVI, 381, van Musschenbroek mentions his having encountered a Vulture at Kwandang, North Celebes, without his being able to identify the species. He suggests *Gyps indicus*.

ORDER STRIGES.

The Owls are nocturnal or crepuscular in habits and are best characterized by their eyes, which are very large and not placed laterally on the sides of head as in most other birds, but directed forwards, each eye being surrounded by a disk of feathers radiating from it. The four toes are placed two in front, one behind, and the outer one laterally on the foot; the last capable of being completely turned backwards. The plumage is very soft and fluffy, the most prevalent tints being brown — rufous brown or yellow-brown, — grey, and white; and the patterns are often completely broken up, chequered and vermiculated. The round, white eggs are usually laid in cavities in trees or rocks. The larger species prey upon vertebrates of various kinds, some on fish, many of the small Owls on insects. In size the Owls vary from that of a Lark to that of an Eagle.

FAMILY ASIONIDAE.

The sternum furnished with a manubrial process; the furcula, not attached to the keel of the sternum, often consists of but two stylets which do not even meet one another, the posterior margin of the sternum with two pairs of projections, one pair on each side, with corresponding fissures between them; the tarsus furnished with a bony ring or loop bridging the channel holding the common extensor tendon of the toes (Newton, Dict. B. 672). Basal and second joints of middle toe subequal in length (Blanford).

GENUS NINOX Hdgs.

Owls of rather small size, facial disk wanting, causing a very Accipitrine appearance; bill moderate, the nostrils formed by semi-tubular swellings of the cere placed well forward; wings long; tail long, more than 3 times the length of the tarsus; toes bristly, rather long, the soles covered with rough horny points and laterally fenced with stiff bristles; tarsus feathered to an extent varying in different species. They prey upon insects, small mammals, birds, lizards. Eggs 2—4. About 30—40 species, one at least of them migratory, distributed from India and China, to Australia, New Zealand and Madagascar.

+ * 30. NINOX OCHRACEA (Schl.).

Ochraceous Hawk-owl.

Plate IV.

a. **Noctua ochracea** *(1)* Schl., Ned. Tdschr. Dierk. 1866, III, 183; *(2)* id., Rev. Noctuae 1873, 22; *(3)* Rosenb., Malay. Archip. 1878, 271, 583.
b. **Athene ochracea** Wall., Ibis 1868, 23; Gray, HL. 1869, I, 41; Wald., Tr. Z. S. 1872, VIII, 38.
Ninox ochracea *(1)* Sharpe, Ibis 1875, 258; *(II)* id., Cat. B. 1875, II, 167, pl. XI, fig. 2; *(3)* M. & Wg., Abh. Mus. Dresd. 1896, Nr. 2, p. 8.
"**Totosik**", Lotta, Minahassa, Nat. Coll.
"**Keketi**", Tonkean, E. Celebes, Nat. Coll.

Figures and descriptions. Sharpe *II*; Schlegel *a 1, a 2*; Rosenberg *a 3* (ex Schl.).

Adult. Above dark brown with a chestnut wash, more dusky on the head; the underlying scapulars with large patches of white, chiefly on the outer webs of the feathers; greater and middle wing-coverts marked on the outer webs with a few spots of white, often concealed; quills brown, outer webs washed with rufous, the secondaries notched somewhat deeply on the inner and slightly on the outer webs with white, the remains of which are perceptible on the outer edge of the primaries; tail brown like the quills, marked on the inner webs (except on the middle pair) with notches of white, deepest towards the base of the feathers — these white notches mark the spaces between obscure bars, of which about seven may be made out on the middle feathers; — bristle-plumes of the forehead and lores whitish with black shafts, which are continued into fine hair-like points reaching considerably beyond tip of bill; above eye, and malar region, whitish; under surface tawny raw-umber, darkest on the chest and gradually becoming lighter, until on the under tail-coverts it is buff; under wing-coverts light cinnamon marked with brown near the edge of the wing, the greater series and inner webs of quills below notched with white (Lotta, North Celebes, 4. Dec. 1894: Nat. Coll. — C 13865).

"Iris somewhat dark brown, bill brownish white" *(a 1)*.

Measurements. Wing 184 mm; tail 108; tarsus 27; middle toe without claw 22.5; culmen from cere 14.

A specimen in the Brit. Mus. measures: wing 200 mm; tail 127; tars. 25 (Sharpe).

Young. Differs from the adult in having the under surface white or whitish; throat and jugulum dark brown; upper parts dark brown, without any chestnut hue (juv. Main, N. Celebes, Feb. 1894: Nat. Coll. — C 13240).

Measurements.	Wing	Tail	Tarsus	Bill from cere
a. (10439) ad. Gorontalo (Riedel)	184	108	27	14
b. (C 13865) ad. Minahassa, 4. XII. 94 (Nat. Coll.) . . .	189	101	26.5	12
c. (C 14357) ad. E. Celebes, V.-VIII. 95 (Nat. Coll.) . .	192	100	27.5	13
d. (C 13240) juv. Minahassa, Feb. 94 (Nat. Coll.) . . .	178	85	27	12

Distribution. North Celebes: Gorontalo District — Negri-lama (Rosenb. *a 1*) and Gorontalo (Riedel in Mus. Dresden); Minahassa (v. Duivenb. *a 2*, Nat. Coll.); Tonkean, E. Cel.

This little Owl is a somewhat rare species in collections. It was discovered by von Rosenberg at Negri-lama on the coast of the Gulf of Tomini in 1863.

There are now examples in the Dresden, Leyden, British, Tring, and apparently, Vienna Museums. It is nearly allied to the Little Owl of the Philippines, *N. philippensis* Bp., which differs from it in having the upper wing-coverts spotted, the tail shorter, etc. Another species, showing much similarity in tone of coloration, but having the under surface marked with cross-bars instead of uniform or with indistinct stripes, is the *Ninox hantu* of Buru, figured by Dr. Sharpe on the same plate with *N. ochracea*.

31. NINOX SCUTULATA (Raffl.).
Brown Hawk-owl.

This species apparently consists of four races. We have not been able to investigate those of the Indian countries, but accept the views thereon of Schlegel and Dr. Sharpe, with the corrections of Mr. Hume and more recently of Mr. Blanford (Bull. B. O. C. 1894 XVIII, p. XLII, Ibis 1894, 526; Faun. Br. Ind. B. III 1895, 309). The races may be diagnosed as follows:

1. The typical Ninox scutulata.

Diagnosis. Size medium, wing 197—210 mm; 4th primary scarcely longer than the 3rd and 5th, which are about equal in length (Schlegel *a I*).
Distribution. India and Ceylon, south-eastwards as far as Java and Borneo.

2. Ninox scutulata lugubris (Tickell).

Diagnosis. "Rather paler, with a greyer head" (Blanford).
Distribution. "Found in India and Burmah generally, chiefly in the less damp parts of the country" (Blanford). The typical form, on the contrary, is believed by Mr. Blanford to belong to those parts where the rainfall is heavier — Malabar, Ceylon, parts of Burmah, etc. (Faun. Br. Ind. B. III 1895, 311).

3. Ninox lugubris affinis (Tytler).

Diagnosis. A small race: wing 168—193 mm (Blanford).
Distribution. Andaman Islands.

4. Ninox scutulata japonica (T. & Schl.).

Diagnosis. Size large, wing c. 218—235 mm; 3rd and 4th primaries about equally long, or the third a trifle the longest, the 5th (according to Schlegel and W. Blasius) much shorter and nearly equal to the 2nd. (In our specimens with three exposed bands on the tail this seems to hold good for the 5th primary, but in three of our four with more rufous tails and four exposed bands, the 5th is only about 10 mm shorter than the 3rd, instead of about 20 mm shorter, like the 2nd; in the fourth example the fifth quill is missing. Prof. W. Blasius measurements *(i 3)* show however, that the wing-formula is subject to great variation, the 5th quill being in one specimen (a female) only 85 mm shorter than the longest, in another specimen 21 mm

96 Birds of Celebes: Asionidae.

shorter, both examples being from the same locality, Great Sangi, and the specimens before us also, most indeed moulting, seem to have no great stability in the quill-lengths. We have noticed this in other Owls.)

Distribution. Japan *(a 1, d 1)*; Loo-choo Islands *(g 4, h 8)*; S. E. Siberia and China from the Amoor southwards *(g 1, h 1, h 9, h 10, i 1)*; migrating in winter to the Philippine Islands (Platen *i 2*, Steere *j 1*, Whitehead *h 7*); Borneo *(c 2 g 5)*; Talaut Is. — Kabruang and Salibabu (Nat. Coll.); Sangi Islands (Rosenb., Hoedt *e 1*, Meyer *h 4*, Platen *i 1*); Tagulandang (Nat. Coll); — Celebes, N. Peninsula (Rosenb. *e 1*, Meyer *c 3*), S. Peninsula, Bonthain (Weber *h 12*); Sula Mangoli (Hoedt *e 1*); Kalao (Everett *l 4*); Ternate (Rosenb. *e 1*, Fischer *h 5*); Flores (Wallace *f 1*).

The following synonymy belongs to the last-named subspecies:

Ninox scutulata japonica (T. & Schl.).

a. **Strix scutulata japonica** *(1)* Temm. & Schl., Faun. Jap. Aves 1845, 28, pl. 9 B.
b. **Athene japonica** Bp., Consp. 1851, I, 41.
c. **Ninox japonicus** *(1)* Bp., Rev. Zool. 1854, 543; *(2)* Wald., Tr. Z. S. 1872, VIII, 40; *(3)* Meyer, Ibis 1879, 57; *(4)* Sharpe, Ibis 1889, 80; id., ib. 1890, 274; *(5)* Campbell, ib. 1892, 243.
d. **Noctua hirsuta japonica** *(1)* Schl., Mus. P.-B. Striges 1862, 25.
e. **Noctua hirsuta**, partim, *(1)* Schl., Rev. Noctuae 1873, 23 (ex China, Japan, Celebes, Sula, Sangi, Ternate).
f. **Athene florensis** *(1)* Wall., P. Z. S. 1863, 488; id., Ibis 1868, 23.
g. **Ninox japonica** Sharpe, Ibis 1875, 258; *(1)* David & Oust., Ois. Chine 1877, 36; *(2)* Blakis. & Pryer, Ibis 1878, 246; *(3)* Taczan., J. f. O. 1881, 179; *(3*bis*)* Gurney, P. Z. S. 1878, 940; *(4)* Stejneger, Pr. U. S. Nat. Mus. 1887, 401; *(5)* Everett, J. Str. Br. R. A. S. 1889, 179; *(6)* Bourns & Worces., B. Menage Exped. 1894, 33; *(7)* Grant, Ibis 1896, 111, 463, 531.
h. **Ninox scutulata** *(1)* Sharpe, Cat. B. 1875, II, 156—167 partim; *(2)* id. (nec Raffl.?), Ibis 1877, 4; *(3)* Salvad., Orn. Pap. 1880, I, 80, and Agg. 1889, 23; *(3*bis*)* W. Blas., Verh. z.-b. Ges. Wien 1883, 22; *(4)* Meyer, Isis 1884, 14; *(5)* Pleske, Bull. Ac. Petersb. 1884, 114; *(6)* Hartert, J. f. O. 1889, 373(?); *(7)* Whitehead, Ibis 1890, 44; *(8)* Seebohm, B. Japan Emp. 1890, 187; *(9)* Styan, Ibis 1891, 326, 486; *(10)* De La Touche, Ibis 1892, 482; *(11)* Seebohm, Ibis 1893, 52; *(12)* Büttik., Zool. Erg. Weber's Reise in Ost-Ind. 1893, III, 272; *(13)* Hartert, Nov. Zool. 1895, 476.
i. **Ninox macroptera** *(1)* W. Blas., Ornis 1888, 545—555; *(2)* id., J. f. O. 1890, 145.
j. **Ninox lugubris** *(1)* Tweedd., P. Z. S. 1878, 940; Gurney, Ibis 1884, 170; *(2)* Steere (nec Tickell), List. Coll. B. Philip. Is. 1890, 8.
k. **Ninox hirsuta japonica** *(1)* Tacz., Faun. Orn. Sib. Orient. 1881, I, 131.
l. **Ninox scutulata japonica** *(1)* M. & Wg., J. f. O. 1894, 239; *(2)* iid., Abh. Mus. Dresd. 1895, Nr. 8, p. 5; *(3)* iid., ib. Nr. 9, p. 2; *(4)* Hartert, Nov. Zool. 1896, 177.
m. ? **Ninox scutulata lugubris** *(1)* Eagle Clarke, Ibis 1895, 476.
"**Mamejo**", Talaut Islands, Nat. Coll.
"**Mawututu**", Tagulandang Id., iid.
Figures and descriptions. Temm. & Schl. *a 1*; Schlegel *d 1*; David & Oust. *g 1*; Stejneger *g 4*; Meyer *h 4*; W. Blas. *i 1*; Sharpe *h 1* pt.
Adult. Above Vandyke-brown, washed on the head with greyish, and with rufous on the back, sides of neck, throat and wings; wing-coverts unspotted, the outer ones dark purplish brown, the inner ones like back, scapulars with concealed white spots; outer

web of primaries light cinnamon-rufous with two or three partially obliterated yellowish bars; tail reddish brown crossed with five dusky bars and terminally margined with buff; under parts white with long oval light chestnut-brown spots; under side of wing banded on the inner webs of the feathers with yellowish isabelline, except on the free ends of the primaries.

"Iris golden yellow; bill dark lead-grey; feet light yellow" (Platen *i 1*).

Some examples have the head greyer than others, and in the former the chestnut stripes on the under surface seem to be a little broader. A greater difference is seen in the tail. Two before us from Celebes and two from Talaut, generally greyer and more broadly striped below, have the tail greyer brown above, crossed with three exposed black bands and two bands concealed by the upper tail-coverts. Two from Talaut, one from Sangi and one from Tagulandang are (in two or three cases) less grey above and less broadly striped, displaying more white below; these have the tail more rufous brown, crossed with four exposed black bands narrower than in the other birds, with three or two concealed under the upper tail-coverts. Two of the greyer birds with three exposed bands in the Sarasin Collection are females, but we are inclined to regard the difference as one of age rather than of sex, the greyer birds being probably older. See, also, on sexual differences: W. Blasius *(i 8)*.

Measurements.	Wing	Tail	Tarsus	Culmen from cere
a. (8257) imm.? Great Sangi (Meyer)	235	132	—	16
b. (C 13056) imm.? Talaut Is. Nov. 93 (Nat. Coll.)	226	120	—	17
c. (C 13055) ad. Talaut Is. Oct. 93 (Nat. Coll.)	224	120	27	15
d. (C 13755) ad. Talaut Is. Nov. 94 (Nat. Coll.)	228	120	—	15
e. (C 15258) imm.? Talaut Is. Oct. 95 (Nat. Coll.)	233	127	29	17
f. (C 13455) imm.? Tagulandang, Aug. 94 (Nat. Coll.)	230	136	—	17
g. (Sarasin Coll.) ♀ ad. Rurukan, N. Cel. 9. XI. 94	225	115	25	13
h. (Sarasin Coll.) ♀ ad. Kema. N. Cel. 24. X. 93	218	118	25	14.5

Viewed as a species composed of the above four races *N. scutulata* ranges from the Himalayas, the Amoor River and Japan (except, perhaps, Jesso — *h 8*) southwards throughout India and China to Ceylon, the Andaman and Nicobar Islands, Flores, Celebes, Ternate.

This species varies considerably, and naturalists — starting with the assumption that it is non-migratory — have conferred specific names on specimens killed in a dozen different parts of its range. The result of Dr. Sharpe's valuable, but, from the nature of the difficulties, by no means conclusive, researches *(h 1)* was that there are only three forms of this bird — two light and one dark. Mr. Blanford, as shown above, finds a fourth race in the Andamans. Mr. Hume has recorded the light race of the Andamans from Cachar east of the Brahmapootra, and it seems open to inquiry whether the Andamans are not visited in winter by migrant individuals from India, as well as having a stationary race of their own.

A single specimen of *N. scutulata lugubris*, the type of *N. madagascariensis* Bp., is said to have been obtained in Madagascar. Dr. Sharpe expressed the opinion that the locality was erroneously indicated; Dr. Hartlaub, in his "Vögeln Madagaskars" 1877, says there can be no question as to the correctness of the locality; Mr. Grandidier, however, takes an opposite view and considers it certainly wrong (Ois. Madag. 1879, I, 126, note). Although we incline to the latter opinion, it need, nevertheless, create no great surprise that a lost straggler, migrating southwards from the Himalayas in winter during the N. E. monsoon, should be aided by it across the Indian Ocean to Madagascar; just as individuals of migrant North American birds occasionally lose their way and succeed in reaching the coast of England.

There remains the dark form of Dr. Sharpe, which, following Schlegel *(d 1)* and Prof. W. Blasius *(i 1)*, we have split into two differing notably in the formula of the primaries (see *supra*), a larger Eastern or Japano-Chinese race and a smaller Western or Indian one. Whether these divisions will prove sufficient, or not, can only be ascertained by means of prolonged investigation and a much more complete series of specimens and facts than the Dresden Museum and the various ornithological periodicals and works up to the present afford; the Eastern race, *N. s. japonica*, may, however, be shown on good grounds to be migratory; while the Western one may be a migratory, or a stationary bird, or the number of stationary birds, slightly differing from all others (f. i. *N. borneensis* [Bp.] Sharpe, Ibis 1889, 80), are perhaps increased at a certain season of the year by migratory ones. The correct answer to these difficulties is not likely to influence Celebes, which is visited by *N. s. japonica*, judging by the length of the wing; but it is to be hoped that some careful ornithologist will see fit to take the matter in hand.

The remarkable denticulate toes of this species and its claws as sharp as pins are suggestive of a class of food preyed upon that is difficult to hold. Colonel Legge says that it feeds almost exclusively on beetles, moths and grasshoppers. It captures insects on the wing (B. Ceylon, 148). The stomach of a specimen killed in Borneo by Mr. Everett was "distended with beetles. chiefly *Buprestidae*", another had swallowed a small gecko-lizard *(h 2)*. The Drs. Sarasin found in the stomach of one killed in N. Celebes 9 recognisable mole-crickets, several cockroaches. With the disappearance of this food at the approach of winter in Amoor Land, China and Japan, the Owls in these regions must of necessity migrate southwards; and, judging from analogies in other parts of the world, it is probable that the birds from the Amoor, in which Prof. W. Blasius and Mr. Taczanowski point out some differences *(i 1, g 3)*, do not proceed so far south as the more southern ones of China and Japan. In Japan it has not yet been recorded with certainty from the northernmost island at all, but Seebohm states that it is "not uncommon in summer" near Yokohama and Nagasaki. It is just at this season that it appears to be entirely absent in the East Indian Archipelago, where, with the exception of the Philippines, we have only been able to find a single specimen dated between the end of April and the end of September. On the other hand all the dates recorded of specimens killed south of the Philippines — over two dozen in number — are in our winter months. As these specimens may be rather important we give a list of them.

Locality	Date	Reference
1. Borneo	April 9. 1896	Sharpe *c 4*
2. Talaut	Nov. 1893	Dresden Mus. (supra)
3. Talaut	Oct. 1893	Dresden Mus. (supra)
4,5. Talaut (2 sp.)	Nov. 1894	Dresden & Tring Mus.
6. Talaut	Oct. 1885	Dresden Mus.
7. Sangi	Oct. 30. 1864	Schlegel *e 1*
8. Sangi	Nov. 23. 1865	Schlegel *e 1*
9. Sangi	Dec. 5. 1865	Schlegel *e 1*
10. Sangi	Jan. 17. 1866	Schlegel *e 1*
11. Sangi	Dec. 11. 1886	W. Blasius *i 1*
12. Sangi	Jan. 20. 1887	W. Blasius *i 1*
13. Sangi	Febr. 1. 1887	W. Blasius *i 1*
14. Sangi	Febr. 3. 1887	W. Blasius *i 1*
15. Sangi	Dec. 18. 1886	W. Blasius *i 1*
16. Siao	Oct. 26. 1865	Schlegel *e 1*
17. Tagulandang (south of Sangi)	Aug. 1894	Dresd. Mus.
18. 19. N. Celebes (2 specimens)	Sept. 24. 1863	Schlegel *e 1*
20. N. Celebes	Spring 1871	Meyer
21. N. Celebes	Oct. 24. 1893	P. & F. Sarasin
22. N. Celebes	Nov. 9. 1894	P. & F. Sarasin
23. S. Celebes	Winter 1888—89	Weber *h 12*
24. Sula Mangoli	Nov. 30. 1864	Schlegel *e 1*
25. Ternate	Nov. 16. 1879	Pleske *h 5*.

The following relate to the western race (the typical *N. scutulata*, or *borneensis*):

Borneo (3 specimens)	Oct. 1885	Sharpe *c 4*
Borneo	March 1886	Sharpe *c 4*
Borneo	March 1875	Sharpe *h 2*
Borneo (2 specimens)	Oct. 1881	W. Blasius *h 3bis*
Borneo	Jan. 25. 1891	Salv.(A.M.C.G. 1891,42).

Except that Mr. Whitehead — to whom ornithologists are indebted for many useful observations on migration in Borneo — marks *N. scutulata* as a migratory visitor to Palawan *(h 7)*, in addition to supplying dates pointing to the same condition in Borneo, where Dr. Sharpe had already expressed the opinion that it was a migratory bird (P. Z. S. 1879, 325), there appears to exist no statement based on direct observations of the migration of this species in the East Indian Archipelago, where observations on migration have been as yet generally neglected by all travelling naturalists. In China and Japan this is happily not so much the case: Mr. Campbell speaks of it as a summer visitor to Corea *(c 5)*; Mr. Styan as a breeding summer visitant to the Lower Yangtse Basin *(h 9)*; Mr. De La Touche as a species "not uncommon in May" at Foochow, and "rather common at Swatow in April", *(h 10)*, thus, presumably, for the mos part a passing migrant in South China; in Japan, as Seebohm says, it is rather common in summer. At this time of year as stated above, we have been able to find hardly any record of its having been killed in the East Indian Archipelago (except in the Philippines); like *Butastur indicus* and *Tachyspizias soloensis* it now breeds in China and Japan; only to appear again, like those species, in the Archipelago at the approach of the northern winter.

From the circumstances of this migration and the correspondence of our Sangi and Talaut birds with Prof. W. Blasius' Sangi examples we cannot admit the right to specific or subspecific rank of *N. macroptera*, which is named as doubtfully entitled to it in his important article on this species *(i 1)*.

In N. Celebes and Sangi it appears to be only a visitor, at least the above dates do not permit of any other interpretation than that it breeds from Corea and Japan to Central China in our summer, passes as a migrant through Southern China in spring and autumn and sojourns in N. Borneo, Celebes, Sangi, Sula, Ternate, and perhaps elsewhere during our winter. On the more northerly Philippines a few may even remain to breed, or these are individuals, left behind during the seasonal wandering to the customary breeding grounds of the species.

GENUS CEPHALOPTYNX Kaup.

This genus differs from *Ninox* in having the nostrils placed in two pyriform sack-like swellings of the cere, situated much more basally on the bill (the distance from the nasofrontal suture to the anterior edge of the nostrils being about ⅓ the length of the entire bill from the nasofrontal suture to tip, whereas in *Ninox* it is approximately ½ this length), nostrils hidden by bristle-feathers; culmen round, not compressed; wing somewhat short, secondaries extending about ⅝ of the length; tail short and weak, less than 3 times the length of the tarsus; tarsus entirely feathered, toes shorter and stouter, and claws larger than in *Ninox*, the soles less rugose and not fenced laterally with bristles. One species in Celebes and one in the Solomon Islands.

† * 32. CEPHALOPTYNX PUNCTULATA (Q. G.).

Spotted Hawk-owl.

a. **Noctua punctulata** *(1)* Quoy & Gaim., Voy. Astr. Zool. 1830, I, 165, Atl. pl. 1, f. 1; *(2)* Schl., Mus. P.-B. Striges 1862, 29; *(3)* id. Rev. Noctuae 1873, 29.

b. **Athene punctulata** Gray, Gen. B. 1845, I, 45; *(1)* Wall., Ibis 1868, 22; Wald., Tr. Z. S. 1872, VIII, 38; Brüggem., Abh. Ver. Bremen 1876, V, 47; Meyer, Ibis 1879, 57; W. Blas., J. f. O. 1883, 135.

Cephaloptynx punctulata *(1)* Kaup, Tr. Z. S. IV, 209; *(II)* Meyer, Abb. Vogelskel. 1897, pl. CCXII.

c. **Spiloglaux punctulata** Bp., Rev. Zool. 1854, 544.

d. **Ninox punctulata** *(1)* Sharpe, Ibis 1875, 259; *(2)* id., Cat. B. 1875, II, 182; *(3)* Sclater, P. Z. S. 1877, 108; *(4)* Salvad., Orn. Pap. 1880, I, 87; *(5)* Meyer, Ibis 1882, 234; *(6)* Guillem., P. Z. S. 1885, 546; *(7)* Sharpe, P. Z. S. 1888, 184; Lister, t. c. 527; *(8)* Salvad., Orn. Pap., Agg. 1889, I, 24; *(9)* M. & Wg., Abh. Mus. Dresd. 1895, Nr. 8, p. 4; *(10)* Hartert, Nov. Zool. 1896, 161; *(11)* id., ib. 1897.

For further references see Sharpe *d 2*.

"**Totosik**", Manado (Nat. Coll.).

Figures and descriptions. Quoy & Gaimard *a 1*; Schlegel *a 2*; Sharpe *d 2*.

Adult. Above Prout's brown, with a reddish tinge in it, sprinkled all over with minute light spots, which take the form of short bars on either web of the tail-feathers and on the outer webs of the quills; above the eyes from forehead to ear-coverts a stripe of white; a broad band of white extending from below the ear-coverts on the sides of neck, on throat and chin; chest brown like the back and spotted or barred with whitish; remaining under surface more rufous marked with broken bars and spots of whitish; thighs, under tail-coverts, under wing-coverts and some barring at the base of the quills light buffy (Gorontalo — Nrs. 2203, 2204). "Iris brown" (Guillem. *d ♂*).

Other specimens from the Minahassa are apparently older adults and differ from the two above described from Gorontalo in having the upper surface darker and without the reddish tinge, the spots fewer and much less conspicuous; the abdomen and the middle part of the body below white, like that described by Dr. Sharpe. "Iris brown; bill pale yellow-grey; feet yellowish" (P. & F. Sarasin).

Young. Differs from the adult in being dark brown below; obscurely spotted with whitish: thighs brown, indistinctly barred, hinder side and crissum buff; throat dark brown, submalar region and chin whitish; upper surface dark chocolate-brown, unspotted on mantle and back, spotted as in the adult on the wing-coverts, remiges, rectrices, and occiput; rest of head almost unspotted: "iris dark sepia" (♂ juv. Tomohon, N. Cel. 26. IV. 94: P. & F. Sarasin).

Measurements.	Wing	Tail	Tarsus	Culmen from cere
a. (2204) ad. Gorontalo 1871 (Meyer)	160	75	32	18
b. (2203) ad. Gorontalo 1871 (Meyer)	155	68	33	—
c. (C 10793) vix ad. Tondano, VIII—IX. 92 (Nat. Coll.)	170	76	—	20
d. (C 10792) ad. Tondano, VIII—IX. 92 (Nat. Coll.)	173	77	—	18
e. (C 10851) ad. Manado, VIII—IX. 92 (Nat. Coll.)	170	77	33	—
f. (Sarasin Coll.) ♂ ad. Tomohon, 26. III. 94 (P. & F. S.)	157	76	—	18.5
g. (Sarasin Coll.) ♀ ad. Kema, 13. IX. 93 (P. & F. S.)	165	77	—	19
h. (Sarasin Coll.) ♀ ad. Tomohon (P. & F. S.) . . .	157	72	—	—
i. (Sarasin Coll.) ♀ ad. Tomohon, 16. IV. 94 (P. & F. S.)	166	75	—	18
j. (Sarasin Coll.) ♀ ad. Tomohon, 28. III. 94 (P. & F. S.)	167	—	—	17.5
k. (Sarasin Coll.) ♂ juv. nr. Tomohon, 26. IV. 94 (P. & F. S.)	165	71	—	17

Distribution. Celebes: South Peninsula, Macassar (Wallace *b 1*, Everett *d 10*), Indrulaman (Everett *d 10*); foot of Mt. Bonthain (Doherty *d 11*); N. Peninsula, Minahassa (Forsten *a 2*, v. Duivenbode *a 3*, etc.); Gorontalo (Meyer).

This species — peculiar, so far as is yet known, to the Island of Celebes — has a near ally in *Ninox (Cephaloptynx) granti* Sharpe of Guadalcanar in the Solomon Islands, which has the mantle unspotted, the under surface much more distinctly barred, and the fore part of the thighs brown, by which means, as Dr. R. B. Sharpe points out, it may be distinguished from the Celebesian species *(6, 7)*. Another species probably closely allied is the *Ninox odiosa* Sclat. of New Britain, which differs, according to Sclater and Salvadori, from *C. punctulata* in being considerably larger, in having the brown of the upper surface paler and duller and the spots much larger, and in its wanting the line of white or fulvous down the middle of the under surface *(3, 4)*. A young specimen in down in the Dresden Mus. (Nr. 13051) shows the same distinguishing characters, besides it has the whole under surface broadly barred. Although the genus *Ninox* is found from India and Japan throughout the intervening countries as far as Australia and New Zealand, and also in Madagascar, it contains many aberrant forms, one of which is the Celebesian *N. punctulata*, which served as type for Kaup's genus *Cephaloptynx* and now finds its nearest allies in New Britain and the Solomon Islands. It is quite possible that other nearly related species may yet be found in the intervening countries as well as in the great islands to the west of Celebes, but in any case the occurrence of an Owl in an out-of-the-way part of the world has no real zoo-geographical worth. Owls, in consequence of their nocturnal habits, of their thick, fluffy plumage and lightness of weight, buoyant flight, and their powerlessness to bear up against a strong head-wind, are more liable than most species to distribution by wind and storms. This may explain the occurrence of *Strix flammea* and of *Otus brachyotus* on certain remote islands of the Pacific, as well as their being distributed over the greater part of the world besides.

Nothing is known of the food or nidification of *C. punctulata*.

Attention may be drawn to the resemblance of the plumage of this species to certain neotropical members of the genus *Glaucidium* — especially *Glaucidium jardini* Bp. of New Granada — but which is much smaller.

GENUS SCOPS Sav.

The Scops Owls are of small size, furnished with small "ear"-tufts; bill rather weak, cere long (about $^2/_5$ the entire length of the bill), laterally swollen to enclose the nostrils; tarsus feathered; toes generally naked (sometimes feathered above), not bristly; wings rather long; tail moderate, rounded; plumage highly variegated and vermiculated. Feeds on insects and small mammals. Eggs 4—6. Some 25 species, inhabiting the tropical and temperate parts of all the continents, but absent in Australia.

33. SCOPS MANADENSIS (Q. G.).

Manado Scops Owl.

The little Eared Owl of Celebes, the *typical S. manadensis*[1]), as the first described of a number of local races not worthy of separation as distinct species, should combine under its specific title, *S. manadensis* itself and some eight other subspecies known up to the present as occurring in Celebes, Sangi, the Moluccas, Borneo, Sooloo, Flores and Madagascar.

1. The typical Scops manadensis.

a. **Strix manadensis** *(1)* Quoy & Gaim., Voy. Astrol. Ois. 1830, I, 170, Atl. pl. 2, f. 2.
b. **Otus manadensis** *(1)* Temm. & Schl., Faun. Jap. Aves 1845, 26.
c. **Ephialtes menadensis** *(1)* Gray, Gen. B. 1845, I, 38; *(2)* Wall., Ibis 1868, 25 pt. [Celebes only); *(3)* Wald., Tr. Z. S. 1872, VIII, 40 pt.; *(4)* Meyer, Ibis 1879, 57; *(5)* W. Blas., J. f. O. 1883, 135.
d. **Pisorhina menadensis** *(1)* Kaup, Isis 1848, 769.
e. **Scops menadensis** *(1)* Bp., Consp. 1850, I, 47; *(2)* Schl., Mus. P.-B.'Oti 1862, 20; *(3)* id., Rev. Noctuae 1873, 12; *(IV)* Sharpe, Cat. B. 1875, II, 76, pl. VIII, fig. 2; *(5)* Brüggem., Abh. Ver. Bremen 1876, V, 48; *(6)* Salvad., Ann. Mus. Civ. Gen. 1876, 644; *(7)* Hartl., Vög. Madag. 1877, 46; *(8)* Meyer, Isis 1884, 13; *(9)* Guillem., P. Z. S. 1885, 545; *(10)* W. Blas., Ornis 1888, 544; *(XI)* Meyer, Abb. Vogelskelet. 1892, pl. CLXI, pt. XVI, 31; *(12)* M. & Wg., Abh. Mus. Dresd. 1895 Nr. 8, p. 5; *(13)* iid., ib. 1896 Nr. 1, p. 7.
f. **Megascops menadensis**. Kaup, Tr. Z. S. IV, 230.
g. (?) **Scops sinoensis** *(1)* Schl., Rev. Noctuae 1873, 13.
"**Hot Hot**", Manado, N. Celebes, Nat. Coll.
For further references see Sharpe e *IV*.
Figures and descriptions. Quoy & Gaimard *a I*; Sharpe *e IV*; Meyer *e 8, e XI* (skeleton); Schlegel *e 3, g 1*; Brüggemann *e 5*; W. Blasius *e 10*.
Adult. General colour above, of head, neck, back and wing-coverts, mummy brown, passing into a pale dull or drab-brown on quills, upper tail-coverts and tail, the whole thickly and minutely variegated with spots of tawny olive or rufous and with black, most strongly expressed as streaks about the shafts and broken up into unformed cross-bars and fine vermiculations on the webs of the feathers; the exposed outer webs of the scapulars white or buffy white tipped with black; superciliary region whitish; breast cinnamon variegated with white and with black vermiculations and central streaks; sides, abdomen and under tail-coverts like the breast, but much more mottled with unformed bars and spots of white; tarsi — feathered to near the toes — wood-brown speckled with black; primaries marked on the outer webs with pale cinnamon bars of about 4 mm wide and separated by intervals about twice as broad of brownish (N. Celebes and Sangi, Nr. 8253, 8254, etc.).
Variation. As in some other forms of *Scops* Owl, browner and more rufous individuals of this race occur, the latter looking like the former saturated in a rufous dye. A fine series of thirteen in the Sarasin Collection shows that this difference is not due to

---

[1]) This word has been variously spelt *manadensis, menadensis, minadensis,* but, not only is *manadensis* the original spelling of Quoy & Gaimard, but also the correct orthography of the town, Manado.

sex, nor to locality, also that it is not to be ascribed to dimorphism, since the series shows transitions from greyer, dull brown birds to strong rufous. Young and nestling specimens — of which there are four in the Sarasin Collection and one in the Dresden Museum — are all of the grey-brown type, and it is therefore extremely probable that the intense rufous colour is simply a sign of old age.

Young. A young one clothed in nestling down, except for the wings, tail and some feathers of the breast, has the bars on the outer webs of the quills white; the upper plumage more broccoli brown and less rufous, than in the adult; the feathers on the breast show that this part is more strongly mottled with white and that the under side is altogether lighter than in old birds. In this it resembles a young one of *S. magicus* from Ceram and another of *S. leucospila* from Halmahera — the latter marked ♂ — in the Dresden Museum, which look as though they might have been haunting a flour-mill, and bear about the same degree of resemblance to their dark rufous parents that a miller does to his fellow men. Wing 152 mm (Nr. 7755 Manado).

A young one in down is whitish brown above and below, barred with dusky brown (♂, Tomohon, 13. VI. 94, P. & F. S.).

Two, a little older, are assuming the mature-looking variegated brown plumage on the upper parts; the under surface is whitish, somewhat sparingly marked with irregular dusky bars (♂, ♀, Macassar, 15. IX. 95, P. & F. S.).

Measurements (15 specimens not including young ones). Wing 144—157, average about 150 mm; tail c. 80; tarsus c. 25—28; bill from nostril c. 10.5—12 (N. & S. Celebes and Gt. Sangi — Mus. Dresden and Sarasin Coll.).

Skeleton.	*S. manadensis*	*S. rutilus* (Madagascar)
Greatest breadth of cranium	31.0 mm	34.0 mm
Length of humerus	44.5 »	51 »
Length of ulna	47.5 »	57 »
Length of metacarpus	24.5 »	25 »
Length of principal digit	16.8 »	19 »
Length of femur	32 »	35 »
Length of tibia	46.5 »	54 »
Length of tarso-metatarsus	25.0 »	30 »
Greatest breadth of sternum	17.5 »	21 »
Height of crista sterni	5.2 »	8 »
Length of coracoid	20.5 »	22 »

(ex Meyer e *XI* and Milne-Edw. & Grandid., Hist. Nat. Madag. 1876, XIII, pl. 40 A, and texte 1879, XII, 134.)

Distribution. Celebes: — Minahassa (Q. & G. *a I*, Forsten *e 2*, Meyer *c 4*, etc.); Macassar (Wallace *c 2*, *e IV*, P. & F. Sarasin); Kandari, S. E. Peninsula (Beccari *e 6*); — Siao (v. Duivenbode *e 3*); Great Sangi (Meyer *e 8*, Platen *e 10*).

The specimen from Siao in the Leyden Museum, the only one as yet known from that island, is smaller than any specimen as yet recorded from Celebes and Sangi, viz: wing 126 (4 in. 9 lines French), measured straight with compasses, Schlegel's usual method, as against 144 mm (measured over the wing) of the smallest specimen in the Dresden Museum, being 18 mm smaller. *S. manadensis*, however, varies in size and it occurs in Great Sangi, between which island and Celebes Siao lies, so that it is rather more probable than not, that the type of *S. sioensis* comes within the limits of variation of size characteristic to the species. A greater amount of variation in size is shown amongst specimens of *S. leucospila* of Batchian and Halmahera, the wing of the type (Batchian) measures (159 mm) those of three specimens from Halmahera in the Dresden Museum measure 192 mm, 191 and 179 (young).

2. Scops manadensis albiventris (Sharpe).

h. **Scops manadensis** Wall. & Wald. *c 2, c 3,* pt.
i. **Scops albiventris** *(1)* Sharpe, Cat. B. 1875, II, 78, pl. VIII, fig. 1.
Diagnosis. Differs from the *typical S. manadensis* in having a distinct wash of grey on the upper surface; cheeks greyish white; belly white with very scanty cross markings and lines (Sharpe ib.).¹)
Distribution. Flores (Wallace).

3. Scops manadensis rutilus (Pucher.).

j. **Scops rutilus** *(1)* Pucher., Rev. Zool. 1849, 29; *(2)* Sharpe, Cat. B. 1875, II, 80; *(3)* Hartl., Vög. Madag. 1877, 44.
k. **Scops manadensis** Hartl. (nec Q. & G.), Faun. Madag. 23; Grandid., Rev. Zool. 1867, 255, 321; Milne-Edw. et Grandid., H. N. Ois. Madag. 1876, pl. XL, XLA (skelet.), texte 1879, 133.
l. **Scops madagascariensis** Grandid., Rev. Zool. 1867, 85.
For further synonymy and references see Sharpe *j 2*; Hartl. *j 3*.
Diagnosis. Differs from the *typical S. manadensis* in having the inner lining of the quills nearly uniform, with only a few broad bars of yellowish white near the base of the feathers; whereas in the *typical S. manadensis* the whole wing is narrowly barred with fulvous for its whole extent; upper surface of the latter more stellated, the spots being more yellowish in appearance (Sharpe *e IV*).
Observation. We do not think that Dr. Sharpe's diagnosis will be found to answer for all cases; the under side of the remiges displays much variation, it is the paler (in our opinion younger) birds in which the wing below is most markedly barred; one or two of the rufous specimens have no bars thereon at all except at the base, and there is a freckled appearance towards the tip. For some osteological differences see Meyer under *e XI,* and table *supra* p. 104.
Distribution. Madagascar.

4. Scops manadensis capnodes (Gurney).

Scops capnodes J. H. Gurney, of Anjouan Island in the Comoro Group has been recently separated as a species from *S. rutilus* of Madagascar (Ibis 1889, 104). Plumage much darker, also less mingled with white, the pale portion of the scapulars (white in six, light rufous in a seventh Madagascar specimen, compared) much less extended, absent in one specimen; wing a little longer; lower portion of tarsus bare.
Observation. The remaining three races of *S. manadensis* may be usually distinguished from the foregoing by their larger size: Wing 165—195 as against 130—165 mm, but individuals sometimes overstep these limits.

5. Scops manadensis magicus (S. Müll.).

m. **Strix magica** S. Müll., Verh. Nat. Geschied. Natuurk. Comm. 1839—44, 110 (Amboina).
n. **Otus magicus** *(1)* Temm. & Schl., Faun. Jap. Aves 1845, 25 (Amboina, Celebes).
o. **Scops magicus** *(1)* Bp., Consp. 1850, I, 46; *(2)* Schl., Mus. P.-B. Oti 1862, 22; *(III)* Sharpe, Cat. B. 1875, II, 70, pl. V; *(4)* Brüggem., Abh. Ver. Bremen 1876, V, 47; *(5)* Salvad., Orn. Pap. 1880, I, 73, 76; Agg. 1889, I, 22.

¹) Can this be a young individual?

p. **Ephialtes leucospila** *(1)* Wall., Ibis 1868, 25, 27 partim (Celebes).
For synonymy see Salvad., ib.
Diagnosis. General plumage of a yellow tint, both above and below; ruff and ear-coverts ochraceous buff; no white, but yellowish, on the scapulars; no perceptible collar round hind neck; entire upper surface very strongly banded with fulvous [Sharpe *m I*].
Observation. According to Salvadori's key, specimens, agreeing with the following two subspecies in having the upper surface darker than the lower, occur, but such may be distinguished from these other subspecies by the markings of the under surface being scantier, but broader.
Distribution. Amboina; Ceram *(m 2)*.
Observation. There are two specimens of a *Scops* in the Leyden Museum, mentioned by Schlegel as having the "teinte générale d'un roux foncé très ardent", which are stated to have been obtained at Gorontalo by Forsten: a third in the Darmstadt Museum, also from Celebes, was collected by Dr. Riedel probably at Gorontalo likewise. Both Schlegel and Brüggemann agree in separating these from *S. manadensis*, and in recording two forms of *Scops* Owl from Celebes. Dr. Sharpe only allows the island one race, and that the typical *S. manadensis*. Brüggemann found his Celebesian *S. magicus* perfectly identical with a Halmahera specimen; Schlegel his two with one from Ternate. On the other hand we now consider these rufous birds from Celebes to be simply old examples of the local race (see *supra*).

6. Scops manadensis leucospilus (Gray).

q. **Scops leucospilus** (Gray), *(1)* Sharpe, Cat. B. 1875, II, 72, pl. VI; *(2)* Salvad., Orn. Pap. 1880, I, 72, 74; *(3)* Pleske, Bull. Ac. Petersb. 1884, 522.
r. **Scops bouruensis** *(1)* Sharpe, Cat. B. 1875, II, 73, pl. VII, f. 2 (fide Salvad.).
Diagnosis. Differs from *S. m. magicus* in having the markings of the under surface more finely diffused and numerous; and from *S. m. morotensis* in being paler and having the scapulars spotted with white, not rufescent white (Salvad. ib.).
Distribution. Batchian, Halmahera; Ternate; Buru *(q 2, q 3)*.

7. (?) Scops manadensis morotensis (Sharpe).

s. **Scops morotensis** *(1)* Sharpe, Cat. B. 1875, II, 75, pl. VII, f. 1; *(2)* Salvad., O. Pap. I, 72, 76; Agg. 1889, I, 22.
Diagnosis. Like *S. m. leucospilus*, but darker; the spots on scapulars rufescent white (Salvad. ib.).
Distribution. Morty; Ternate *(s 2)*.
Observation. Mr. Pleske considers on good grounds that this form is identical with *leucospilus*. Although it is not advisable to unite them until more material from Morty Island is forthcoming, it appears to us very probable that what is called *morotensis* are only old specimens of *leucospilus*. As three specimens in the Dresden Museum of the *typical manadensis, magicus,* and *leucospilus* with more or less remains of nestling down show, as does also a specimen figured by Dr. Sharpe (pl. VI), the young of these owls have a remarkably pale, blanched, appearance; hence and from specimens before us we conclude that the more saturated, rufous, coloration is a sign of greater age. Of the four specimens of *morotensis* in the British Museum the three labelled "Molucca Islands" may, of course, have come from Morty, or from one of the neighbouring islands, which have furnished *leucospilus*. Three specimens of the

latter in the Dresden Museum from Halmahera afford interesting gradations of plumage; the first (♂ juv.) is one of the blanched birds above mentioned with remains of nestling down, the second (♂) is a normal *leucospilus* with white on the scapulars, the third — marked ♀ — is altogether darker and more rufous and has the outside of the scapulars marked with pale rufous as in the normal *morotensis*! The second stands midway between this and the first.

8. Scops manadensis brookii (Sharpe).

t. **Scops brookii** *(1)* Sharpe, Bull. Brit. Orn. Club Nr. II, 1892, p. IV; id., Ibis 1893, 117; *(II)* Hose, t. c. 417, pl. XI.

Diagnosis. "Differs from *S. bouruensis* and all its allies in having the triple band on the head and hind neck white instead of ochraceous, the pattern being the usual one of the group, viz., a white occipital spot; a second, larger one, on the nape; and a third on the hind neck forming a broad cervical collar (Hose t *III*). The broad band on the side of the crown, extending to the ear-tufts, is also white. The tibial joint has a large patch of chestnut barred with black".

Distribution. Borneo.

9. Scops manadensis sibutuensis (Sharpe).

u. **Scops sibutuensis** *(1)* Sharpe, Ibis 1894, 244.

Diagnosis. Differs from *S. manadensis* "in having all the markings of the upper surface very fine and not all over as in that species. The quills have also more bars in the Celebesian bird than in the species from Sibutu".

Distribution. Sibutu, Sooloo Islands.

Observation. Both the characters indicated by Dr. Sharpe as discriminative for this form are subject to great variation in the *typical S. manadensis*.

To these subspecies should perhaps be added *S. mantananensis* Sharpe of Borneo (Ibis 1893, 117, 559; 1894, 244).

The little Eared Owl, *S. manadensis*, presents a case of the faunistic relationship of Celebes and the surrounding islands with Madagascar. For a long time *S. rutilus* Pucher. of that island was regarded as identical with *S. manadensis* of Celebes and Flores as has been already pointed out, and we do not believe that it is always possible to distinguish them, though Dr. Sharpe considers them distinguishable by means of the different aspect of the inner lining of the quills. In six specimens in the Dresden Museum the number of spots on the outer web of the first primary ranges between six and eight. The Flores bird *S. albiventris* Sharpe, is also separated from the typical Celebes form of *S. manadensis*; but one before us does not answer to Dr. Sharpe's diagnosis except that the belly is whiter than in almost any from Celebes. The breast of this example is more tawny brown and less boldly streaked with black.

The *typical Scops manadensis* is highly variable. It would be an error to suppose that the fact that *S. rutilus* is not more than subspecifically distinct from *S. manadensis* lends aid to the theory of a former land-connection between Celebes and Madagascar in bygone ages: such aid must be sought for in allied

genera — not in the almost perfect identity of two species, which points to
another cause. *S. rutilus* is possibly a recent immigrant to Madagascar — one
that was carried out to sea by a storm at night from anywhere within its range
and, somewhat aided by the strong and continous S. E. tradewind, flying, as
all birds do, at an angle with the wind (though it must of course travel much
quicker) had the good fortune to reach as the nearest land Madagascar. (See
for the winds in the Indian Ocean, which very well explain such a case, the
Atlas of the Indian Ocean, published by the "Deutsche Seewarte" 1891 pl. 21.)
According to Gätke ("Vogelwarte Helgoland" 1891, 72) a bird may even fly
a geographical mile in a minute (!?), the distance from Celebes or the small Sunda
Islands to Madagascar, therefore, would require between 17 and 18 hours (70 degrees of 15 geographical miles each). As a bird will scarcely be able to remain on the wing for such a time without rest, it could only in exceedingly
rare cases reach Madagascar from its eastern quarters, through there are various
intermediate islands of small size. There are no other cases known of such a
similarity between birds of the East Indian Archipelago and Madagascar. Owls,
for reasons already given, appear to be especially liable to distribution in this way.

Nothing is known of the habits of *S. manadensis*. The Madagascar race
preys on mice, small birds, lizards and large insects; a similar diet, with the
addition of bats, is described by Colonel Legge for *S. minutus* of Ceylon
(B. Ceylon, 144).

Scops mantis (T. & Schl.).

This species is noted by v. Rosenberg as occurring in Celebes (Malay.
Archip. 1878, 271), but, as Prof. W. Blasius remarks, it is obviously taken
from Gray's Hand-List (I, p. 46, Nr. 477) without any special new proof of
its presence there (cf. W. Blasius, J. f. O. 1883, 125). Rosenberg has often
rendered his statements untrustworthy by a similar proceeding, as is now
generally known.

[1] As an excellent parallel to the occurrence of *Scops manadensis* in Madagascar may be cited Prof.
König's discovery of *Glaucidium siju* Cab. of Cuba in the Canary Islands (J. f. O. 1890, 336—340), a single
specimen of which was killed on Adeje, August 22nd, 1889 by Don Ramon Gomez, who informed Prof.
König that he met with it after strong winds in the S.W. part of the island at a height of 1074 foot.
Prof. König remarks: "Altogether it looks as if African continental forms are entirely absent in the Canaries,
while more and more of such from the New World are always being discovered. And does it not seem manifest that this most striking immigration of American types is to be traced to the air-currents which pass from
America across the sea? ... The prevailing winds ... which are rarely or hardly ever directed from the
African Continent towards the island-group, blow on the other hand (probably indeed with some constancy)
from the New World across the great stretches of the Atlantic Ocean and continue as far as the Old."

FAMILY STRIGIDAE.

The sternum is without a manubrium; the furcula formed by the clavicles is perfect, there is only one notch on either side in the hinder margin of the sternum; the ring on the upper part of the tarsus in front is present[1]), but reduced to a cartilaginous loop; the second joint of the middle toe is considerably longer than the basal joint, the claw thereof is pectinated; the facial disk is very perfect, the radiating feathers on either side meeting to form a sort of ridge or "hog-mane" over the forehead and base of the culmen; the disk is surrounded by a well-defined ruff of stiff feathers.

GENUS STRIX L.

Now that *Photodilus* has been removed from the Barn Owl section in virtue of its not having the clavicles united and of its having two processes on the hinder margin of the sternum (cf. Newton, Dict. B. 673), the genus *Strix* is the only one of the *Strigidae* and may be recognised by the characters given for the family. There are five species. The range of the genus is almost cosmopolitan. 3—4 white eggs are laid. The birds prey chiefly upon small mammals.

34. STRIX FLAMMEA L.
Barn Owl.

Although the Barn Owl, *Strix flammea*, and its races have been most ably discussed by Dr. Sharpe in the Catalogue of Birds, II, 291—303 and in Rowley's Ornithological Miscellany, I, 269—298; II, 1—21, it would be hardly possible to divide this cosmopolitan species into subspecies which would answer to all the peculiar conditions of the case. This work must be left to the ornithologist of the future, supposing he fulfil the ideas concerning the necessity of trinomial nomenclature with all its consequences. In Celebes and Sangi *Strix flammea* presents a great differentiation, and it was named as a distinct species by Schlegel after its discoverer, Rosenberg. It is not easy, however, to point to any difference between it and the Javan form of *S. flammea (javanica* Gm.); though some, as a rule, are to be found.

+ **1. Strix flammea rosenbergi** (Schl.).

a. **Strix flammea** sp. *(1)* S. Müll.. Verh. Nat. Comm. 1839—44, 87; *(2)* id., Reizen 1858, II, 8.
b. **Strix rosenbergi** *(1)* Schl., Ned. Tijdschr. Dierk 1866, III, 181; *(2)* Wall., Ibis 1866, 26, 27; *(3)* Wald., Tr. Z. S. 1872, VIII, 41; *(4)* Schl., Rev. Noctuae 1873, 16 (Celebes tantum); *(5)* Wald., Tr. Z. S. 1875, IX, 146; *(6)* Brüggem., Abh. Ver. Bremen 1876, V, 48; *(7)* Schl., Notes Leyd. Mus. 1878, I, 50; *(8)* Rosenb., Malay. Archip. 1878, 271, 583; *(9)* Meyer, Ibis 1879, 57; *(10)* W. Blas., J. f. O. 1883, 135; *(11)* Moyer, Isis 1884, 6, 14; *(12)* Guillem., P. Z. S. 1885, 546; *(13)* W. Blas., Ztschr. ges. Orn. 1885, 235; *(14)* id., Ornis 1888, 556; *(15)* Büttik., Zool. Erg. Weber's Reise Ost-Ind. 1893, III, 272; *(16)* Hartert, Nov. Zool. 1896, 161.
c. **Strix flammea** *(1)* Sharpe, Cat. B. 1875, II, 293; id., Rowl. Orn. Misc. 1866, I, 297; II, 14.
d. **Strix flammea rosenbergi** *(1)* M. & Wg., Abh. Mus. Dresd. 1895, Nr. 8, p. 5.

---
[1]) Has been supposed to be absent (Newton, Dict. B. 1894, 672).

"**Ngiong**", Minahassa, Nat. Coll.
"**Wada-Watonga**", N. Celebes, Gorontalo ?, Rosenb. *b 8*.
"**Karin**", Tjamba, S. Celebes, Platen *b 13*.

Descriptions. Schlegel *b 1*, *b 4*; Sharpe *c 1*; Brüggemann *b 6*; Rosenberg *b 8*; W. Blasius *b 13*, *b 14*.

Adult. Above drab-brown, each feather finely vermiculated with white and having a spot of white in the blackish middle-ground of the end of the feather; shoulders and wing-coverts tawny buff, coarsely vermiculated towards the ends of the feathers with dusky brown and white and tipped with spots of white as on the back; quills and tail-feathers above paler than the back, this colour is gathered into five distinguishable dusky bars on the tail and into several imperfect ones on the quills, the intermediate spaces coarsely speckled and vermiculated with drab-brown, dull tawny and whitish; facial disk greyish white, in front of eyes dark brown; facial frill varied with feathers of raw umber with dark centres, of pure white and of white tipped with dark brown or rufous, the last-named being prevalent under the chin; breast buff, the feathers furnished with a small terminal spot, and some with broken terminal margins, of dusky; the remaining under-parts and under wing-coverts ochraceous buff, each feather marked with one or more small spots of brown (N. Cel. C 10417).

"Iris dark brown; bill whitish; feet greyish yellow" (Platen *5*).

Variation. This Owl varies to a considerable extent. The general tint above of the specimen described is the brownest among a number of specimens, and the bars on the are the least decided. Other specimens are much greyer above, ranging from a tail moderately darker grey to much blacker grey; four exposed bars on the tail and one concealed under the coverts are sharply defined; the white spots are more or less numerous. The ruff round the face is usually varied with white feathers, sometimes tawny with umber brown tips to the feathers; the under parts are darker or paler ochraceous buff, the blackish spots variable in number and size.

Two specimens from Batavia in Java (Nr. 5909 and 2202) have the under surface more barred and the spots on the back more conspicuous. These differences are slight, when seen in the actual specimens, and not sufficient in themselves alone to justify even the subspecific separation of the Celebesian Owl from the Javan one (*Strix javanica* Gm.), but we do not unite them as it remains to be seen what a greater quantity of material will prove.

Measurements.	Wing	Tail	Tarsus	Bill from cere	Mid. toe without claw
a. (C 10417) — North Celebes	360	150	82	28	50
b. (8266) — North Celebes . . . . „	340	150	77	28	43
c. (8267) — Great Sangi . . .	336	155	75	26	41
d. (C 10852) — North Celebes	340	150	78	27	—
e. (C 13355) — North Celebes, May 94 (C.) . .	355	140	—	28	—
f. (C 13864) — North Celebes, Dec. 94 (C.) . .	345	145	—	26	—
g. (Sarasin Coll.) ♂, Kema, N. Cel., 1. XI. 93 .	335	153	—	28	—
h. (Sarasin Coll.) ♀, Tomohon, N. Cel., 28. III. 94	348	156	—	29	—
i. (Sarasin Coll.) ♂, Kema, Nov. 93 . .	340	150	—	28	—
j. (Sarasin Coll.) ♂, Tomohon, 31. III. 94 .	355	150	—	29	—
k. (Sarasin Coll.) ♀, Maros, S. Cel., 22. I. 96 . .	335	150	—	—	—
For comparison: *S. javanica* Gm.					
(5909) — Batavia (v. Schierbrand)	360	160	81	26.5	49
(2202) — Batavia	343	150	73	26.5	—

Distribution. Celebes, Sangi. — Great Sangi (Meyer *b 11*, Platen *b 14*). Celebes, N. Peninsula: Minahassa (v. Duivenbode *b 4*, *b 10*, Meyer *b 9*, etc.); Gorontalo Distr. (v. Rosenberg *b 1*); S. Peninsula: Macassar (S. Müller *a 1*, *a 2*, Wallace *b 2*, Weber *b 15*, Everett *b 16*); Tjamba Distr. (Platen *b 13*), Maros (P. & F. S.); Indrulaman (Everett *b 16*).

Schlegel has identified a young female from the Philippines with *S. rosenbergi*, but the only Barn Owl known from there at present is *Strix candida* Tickell, which is perfectly distinct, and Schlegel's indication therefore requires further investigation.

Prof. Blasius (*b 13*) calls attention to the great difference between this form and the typical *Strix flammea*. "It does not appear to me to be correct to account this form, as Sharpe has done (*c 1*), simply a variety of *Strix flammea*. The stouter bill, tarsus and toes, the bristly feathering of the lower half of the tarsus, the rusty brownish upper surface of the tail crossed with four bands with vermiculate markings between them, the darker quills without any clear cross-banding, the darker coloration of the plumage in general and the peculiar character of the spots with which it is marked, which are very accurately described by Brüggemann (*b 6*), — all this gives the bird a thoroughly different character".

Wallace (s. Mal. Arch.) also has called especial attention to the much greater size and strength of this species as compared with *Strix javanica*. We would point out, however, that one of the specimens from Batavia in the Dresden Museum is as large as our biggest Celebesian ones (see p. 110).

The earliest notice of this Barn Owl appears to be one of Sal. Müller, who speaks of a species of Barn Owl as resorting to old temples (!), caves, etc. near Macassar (*a 1*, *a 2*). Mr. Wallace obtained it at this place in 1856 in bamboo thickets. In the stomach of one of their specimens the Drs. Sarasin found a Rat. These animals abound in Celebes.

✝ 2. The typical Strix flammea.

e. **Strix flammea** (*1*) Hartert, Nov. Zool. 1896, 177.

Diagnosis. Size much smaller; bill, tarsus, toes and claws much weaker; remiges light cinnamon-rufous, with about 4 well-defined vermiculate bars of grey and dusky; dusky without grey on the inner webs; upper parts much paler, the white spots smaller and more elongated.

Wing c. 270 mm; tail 115; tarsus 57; middle toe without cl. 30; bill fr. c. 18.

Distribution. Not clearly defined. In the Celebesian Province: Kalao Island between Flores and Celebes (Everett *e 1*).

Quite recently Mr. Hartert has recorded "typical *Strix flammea*" from the area included in this work. It may possibly be a migrant there from the Indian countries, but it is more likely to prove a resident. It is, as shown above, very different from *Strix rosenbergi*. *S. candida* is about the same size, but has a very long slender tarsus, bare for its lower third, and much longer, stouter toes.

For fuller particulars about *S. flammea*, see Sharpe (*c 1*) and the principal works on European and Indian ornithology.

⚁ * 35. STRIX INEXPECTATA Schl.
Unexpected Owl.

Strix inexpectata Schl., Notes Leyden Mus. I, 1878, 50; J. f. O. 1879, 426; W. Blas., J. f. O. 1883, 124.

Description. Differs as follows from *S. rosenbergi*. Wing short, the point of it overreaching the secondaries by only 45—55 mm (20—24 lines French) as against 95—115 mm in that species; the secondaries crossed with 7, the first primary with 8 and the others with 9 narrower black bars; under surface of the wing darker (silvery grey), the bars indistinct and quite obliterated on the first primary; tail crossed with nine bars; tarsi covered with downy feathers down to the toes, which are somewhat shorter and much more slender than in *S. rosenbergi*; upper parts very bright rufous, unmixed with grey and marked with smaller white spots; below deep tawny rufous as in the darkest specimens of *S. rosenbergi*; face darker, as in *S. castanops* and *S. novaehollandiae* ("light chestnut, the feathers in front of the eye black, the loral plumes also slightly tipped with black" — Sharpe, description of *S. castanops*, Cat. B. II, 305). (Ex Schlegel.)

Measurements. Wing 258 mm; point of wing 45—55; tail 122; tarsus 68; middle toe 36 (converting the French lines of Schlegel to millimeters).

Distribution. North Celebes.

This species is only known by one specimen in the Leyden Museum, obtained by van Musschenbroek in the Minahassa. A very good species, the nine bars on the tail and the tarsi feathered down to the toes easily distinguishing it from the other Owls occurring in the Celebesian area.

⚂ 36. STRIX CANDIDA Tick.
Grass Owl.

Strix candida *(1)* Tickell, J. A. S. B. II, 1833, 572; *(II)* Jerd., Ill. Ind. Orn. 1847, pl. 30; *(3)* id., B. Ind. I, 1862, 118; *(IV)* Gld., B. Asia pl. 18 (1872); *(5)* Sharpe, Cat. B. II, 1875, 308; *(6)* David & Oust., Ois. Chine 1877, 46; *(7)* Hume, Str. F. VI, 1878, 27; *(8)* Sharpe, P. Z. S. 1882, 335; *(9)* Davison, Str. F. X, 1883, 341; *(10)* Oates, B. Brit. Burmah 1883, II, 168; *(11)* E. P. Ramsay, Tab. List Austr. B. 1888, 2; *(12)* North, Nests & Eggs Austr. B. 1880, 24, pl. VI, fig. 5; *(13)* Tristr., Cat. Coll. B. 1889, 68; *(14)* Oates, ed. Hume's Nests & Eggs 1890, III, 95; *(15)* Steere, List Coll. B. Philipp. 1890, 8; *(16)* Wiglesw., Av. Polyn. 1891, 4; *(17)* Büttik., Zool. Erg. Webers Reise Ost-Ind. 1893, III, 272; *(18)* Rickett & H. Slater, Ibis 1894, 222; *(19)* Bourns & Worces., B. Menage Exped. 1894, 33.

a. **Strix longimembris** *(1)* Jerd., Madras Journ. 1839, X, 86.
b. **Scelostrix candida** Kaup; *(1)* Wald., Tr. Z. S. 1875, IX, 145.
c. **Strix amauronota** *(1)* Cab., J. f. O. 1866, 9; 1872, 316.
d. **Strix pithecops** *(1)* Swinh., Ibis 1866, 396, 397.
e. **Strix walleri** *(1)* Diggles, Orn. Austr. pt VII, pl. (1866).
f. **Strix oustaleti** *(1)* Hartl., P. Z. S. 1879, 295; *(2)* Sharpe, ib. 1882, 335.

For further references see Sharpe *5*, *f 2*; Oates *10*.

Figures and descriptions. Jerdon *II*; Gould *IV*; Diggles *e I*; Sharpe *5*; David & Oust. *6*; Davison *9*; Oates *10*, *14* (egg); North *12* (egg).

Adult. Much like *S. flammea* in coloration (see description of *S. flammea rosenbergi*, *antea*), but easily distinguishable by its much more slender tarsi (girth of tarsus a little above the foot 23 mm, as against 34 mm in *rosenbergi*), lower third bare; under surface white, washed with buff on the chest, narrowly fringed with brown on the frill, and sparsely marked with small brown spots on the lower parts (Cebu, Oct. 1879 — Nr. 8269). Bill and cere pinky white; legs and feet bluish brown; irides deep brown; claws horny, tinged with bluish (Davison *9*).

Nestling. The young of *S. candida* is tawny, of *S. flammea* white (Gld. *IV*). Covered with long filamentous down of a dull orange-buff colour; the feathers of the upper parts, as far as developed, of a dark brown colour, spotted near the end of the shaft of the feathers with white, the basal parts of the dorsal plumes bright orange (Sharpe *5*).

Measurements. ♂ ad. Wing 335 mm; tail 135; tarsus 84; bill from gape 48; total length in the flesh 376; expanse 1155; weight 14 oz. (Davison *9*).

Egg. India — 4 or 5. Pure white, with very little gloss, and of a more elongated oval than those of *S. flammea*: 42 × 32.5 mm (Hume *14*); Australia — white, but have a slight bluish tinge; oval, rather swollen about the centre: 43 × 32 (North *12*).

Nest. Little or none, at most a little grass scattered and smoothed down in the midst of heavy grass-jungle; always on the ground. — India (Tickell *14*).

Breeding season. October—December: India (Hume *14*).

Distribution. India (Jerdon *3*, *IV*, etc.); Tenasserim — Tonghoo (Lloyd *7*, *10*); S. China (Rickett *18*); Formosa (Swinh. d *1*, *6*); Philippines — Luzon (Jagor c *1*, Steere *15*); Cebu (Meyer, Mus. Dresd., b *1*); Siquijor (Steere *15*); Calamianes (Bourns & Worcester *19*); Celebes — Luwu, Gulf of Boni (Weber *17*); Australia — Queensland and New South Wales (Diggles e *1*, Ramsay *11*); Fiji Islands — Viti Levu (Kleinschm. f *1*, *16*, Parr *8*).

The right of this species, the Grass Owl of Indian naturalists, to be included in the Celebes list rests upon a single specimen obtained by Prof. Max Weber in the Luwu District at the head of Gulf of Boni in February, 1889 (Büttik. *17*). This discovery helps to fill up a gap in the anomalous distribution of the species; and it may also be expected to turn up in other localities, where there is plenty of jungle-grass, between India and Australia. Hitherto its occurrences in widely isolated localities have led to an increase in its synonymy; as Sharpe remarks, "on a Philippine specimen being discovered, it was named *S. amauronota* by Prof. Cabanis; and in the same year Mr. Swinhoe found it in Formosa and called it *S. pithecops*. Shortly after, it turned up in Queensland, only to be named *Strix walleri* by Mr. Diggles; and now its last appearance, in the Fiji Islands, has gained it the additional cognomen of *S. oustaleti*" (*f* *2*). Dr. Sharpe believed these unexpected occurrences to be accounted for by its migrations, but it is now known to be a breeding species in Australia as well as in India. The distribution of Owls offers many peculiarities, which set all zoo-geographical boundaries at naught. Its possible causes — the nocturnal habits of these birds, the lightness of their plumage, and their liability to be carried by winds — are mentioned more fully elsewhere (p. 108).

Though very like *S. flammea* in general coloration, *S. candida* differs much in habits, living, as Jerdon says, almost exclusively in long grass, for which its lengthy tarsi are well adapted, and its nest is made on the ground. Gould *(IV)* remarks upon the difference in coloration of the young of the two species, those of the Grass-Owl being tawny, of the Barn Owl white.

ORDER PSITTACI.

A Parrot may be recognised among other birds by its large hooked bill, covered at the base by a cere, by its toes placed two in front and two behind, the fourth one being reversed, — these two characters being indeed shared by the Owls; the foot is used on occasion as a hand in which it raises its food towards its bill; the bill, the upper mandible of which is moveable, is made use of in climbing, which it does much, though heavily; its food is generally of a vegetable character, fruits, seeds, nuts, etc., but some species eat insects also. The colours are generally bright, and all tints are found, including pure red, blue, and yellow, but green (especially Parrot-green) is the commonest colour. As to size, the Parrots number amongst them some of the smallest birds, with a body scarcely larger than a Humble Bee, while the largest attain the dimensions of a Raven or small Eagle. The tongue is thick and fleshy, or sometimes brushy or fringed; as is well known some forms are clever at imitating human speech.

The Parrots lay white eggs, 2—10 in number (Finsch), depositing them in holes chiefly in trees; the young are hatched helpless and naked. For further particulars: cf. Finsch, Papageien 1867, I, pp. 1—238; Salvadori, Cat. B. XX Psittaci 1891, pp. 1, 2.

FAMILY LORIIDAE.

"Bill much compressed, generally longer than deep, not notched, and smooth; culmen rounded and narrow; lower mandible rather long with the gonys narrow, straight, and obliquely slanting upwards, not flattened in front and with no keel-like ridge; upper mandible with no file-like surface on the under surface of the hook. Tongue brushy. Cere broader over the culmen and gradually becoming narrower along the sides of the bill. Tail graduated or rounded, sometimes even, rarely longer than the wings, generally shorter. Wing acute, with the three first quills generally the longest. Range: Australian Region (except New Zealand, but including Polynesia)" — Salvadori, Cat. B. XX, 11.

GENUS EOS Wagl.

Lories of about the size of a Turtle-dove (Salvadori), with a rather long — about $^3/_4$ the length of the wing — strongly-graduated tail, the outermost rectrice being about 4 cm shorter than the middle ones; wings moderately long, the three outermost primaries terminally narrowed, 2nd and 3rd the longest, secondaries short, the outermost reaching about half the wing-length; bill orange-red; plumage chiefly red, black and blue; wings chiefly red, or red and black (except in *E. fuscata*) by which *Eos* differs from *Lorius*, in which the wings are green. Salvadori recognises 11 species; they range from Talaut through the Moluccas to New Guinea and the Solomon Islands, occurring also in the Tenimber Islands and in Ponapé, Caroline Islands.

✝ * 37. EOS HISTRIO (St. Müll.).
Red-and-blue Lory.

Under this species-name we include three subspecies, viz: — *the typical*[1]) *E. histrio* (Müll): Sangi; *E. histrio talautensis* M. & Wg.: Talaut; *E. histrio challengeri* Salvad.: Melangis.

1. The typical Eos histrio (Müll.).

a. **Psittaca indica coccinea** Briss., Orn. 1760, IV, 376, pl. 25, f. 2.
b. **"Perruche des Indes Orientales"** *(1)* D'Aubert., Pl. Enl. 143.
c. **Psittacus histrio** Müll., S. N. Suppl. 1776, 76.
d. **"Lori Perruche violet et rouge"** Buff.; *(1)* Levaill., Perr. 1801—5, pl. 53.
e. **Psittacus indicus** Gm., S. N. 1788, I, 318.
f. **Psittacus coccineus** Lath., 1790; *(1)* Shaw, Gen. Zool. 1811, VIII, 461, pl. 68.
g. **Lorius coccineus** Steph.; *(1)* Schl., Mus. P.-B. Psittaci 1864, 128; *(2)* id., Rev. Psitt. 1874, 58.
h. **Eos indica** *(1)* Wagl., Mon. Psitt. 1832, 557; *(2)* Wall., P. Z. S. 1864, 290, 295; *(3)* Garrod, P. Z. S. 1873, 465, 634; 1874, 587.
i. **Eos coccinea** Sclat., P. Z. S. 1860, 227; Salvad., Orn. Pap. I, 1880, 268.
j. **Domicella coccinea** *(1)* Finsch, Papag. 1868, II, 800 (nec Talaut); *(II)* Rowley, Orn. Misc. 1878, III, 123, pl. 98; *(3)* Meyer, l. c. remarks; Meyer, Ibis 1879, 55; *(4)* Platen, Gefied. Welt 1887, 263.
k. **Eos histrio** Gray; *(1)* W. Blas., Ornis 1888, 563; *(2)* Salvad., Cat. B. 1891, XX, 21 (Sangi); *(III)* Mivart, Loriidae 1896, 23, pl. VII, figs. 1, 3.
l. **Lorius histrio** Koch; *(1)* Brüggem., Abh. Ver. Bremen 1876, V, 41, 101; Fischer, ib. 1878, 538, note.
m. **Domicella histrio** *(1)* Rchnw., J. f. O. 1881, 167; Consp. Psitt. 103; *(II)* id., Vogelb. 1883, t. XXXI, f. 1.
"**Sumpihi**", Siao and Great Sangi, Nat. Coll. in Dresden Mus.
For further synonymy and references see Salvadori *k 2*.

---

[1]) The Sangi bird, as the first known from an exact locality, has the best right to this title. Levaillant's plate *(d I)* is not sufficiently accurate to represent the differences between it and the Talaut form.

Figures and descriptions. Rowley *j II*; Mivart *k III*; Levaillant *d I*; Shaw *f I*; Rchnw. *m II*; Wagl. *h I*; Finsch *j I*; Brüggem. *l I*; Reichenow *m I*; W. Blasius *k I*; Salvadori *k 2*.

Adult. Poppy red; vertex (but not forehead and occiput), ear-coverts, hind neck, mantle, shoulders, breast and long tibial plumes bright hyacinth-blue; scapulars, tips of greater wing-coverts and of quills together with the exposed outer webs and tips of primaries black; numerous feathers on the red abdomen and under tail-coverts tipped with blue or black; tail dusky washed with violet, the inner webs red (Great Sangi — Nr. 1898).

"Iris orange-yellow; bill orange-red; feet dark grey" (Platen *k 1*).

In the young in certain parts of the plumage the blue is replaced by red, and the red by blue. Two young "males" described by Brüggemann have the whole head above and nape violet-blue, with some red tipped feathers on forehead and nape; mantle carmine-red changing into violet; the black of the back and wings more extended than in adult, the small upper wing-coverts tipped with black; the red feathers of the under surface irregularly tipped with violet, the colour of the breast band; rump dark carmine red (brownish red in adult). Similar young males are mentioned by Brüggemann (*l 1*) and Prof. W. Blasius (*k 1*). Iris light brown; bill orange-yellow; feet grey (Platen *k 1*).

An immature bird from Siao (5. VII. 93) has the entire head above and neck poppy-red varied with violet feathers on the occiput and neck; under surface from breast downwards violet, washed with red, most strongly on the sides and middle of body where it is almost as much red as blue; red of rump rather darker than that of head. This specimen appears to be intermediate, passing into adult plumage (C 12588).

Variation of the young. It is in this species very puzzling. Of six specimens before us three (2 Talaut, 1 Tagulandang) have the crown, occiput and nape blue (in one mixed with red), the under parts in the Tagulandang bird and one of the others red, mixed with blue or dusky, in the third blue, mixed with a little red. The mixture of colours is not caused by entire feathers of different tints, but the individual feathers are particoloured. These specimens we take for birds in first plumage.

A second stage is that shown by the young example from Siao described above, with the under surface blue mixed with red, but the crown, occiput and nape red like the forehead (instead of blue). Some (old?) blue feathers are intermingled on the crown and nape, and there are a few blue feathers sprouting on the vertex, as well as red ones here and elsewhere on the head.

A young specimen, slightly more advanced towards maturity, has the under parts (except the thighs and flanks) almost uniform red, the blue breast-band only just commenced on the sides of the breast, the entire head above red, except for a few blue-tipped feathers (apparently old ones) on the vertex, and some new blue feathers are sprouting there (C 15296).

In the next stage there is a narrow breast-band and a narrow blue band on the vertex (C 13436).

These specimens therefore seem to prove the following:

First, the young has the head above, except the forehead, blue; the under parts varied with red and dusky blue. Next, the entire head and under parts change by moulting into red. Finally, the blue breast-band and vertex is acquired; at first it is narrow, afterwards broader.

Measurements.	Wing	Tail	Tarsus	Bill from cere
a. (Nr. 1898) ad. Gt. Sangi	175	135	20	23
b. (Nr. 3329) ad. Sangi Is.	161	124	19	20
c. (Nr. 3328) ad. Sangi Is.	178	136	20	24
d. (Nr. 3327) ad. Sangi Is.	154	134	18	20
e. (Nr. 3833) ad. Siao Is.	161	120	19	21
f. (C 12588) juv. Siao Is.	152	105	19	20

Four additional adult specimens from Siao (June—July 1893) have wing 164 to 165 mm, tail ca. 135; five other adults from Great Sangi (July) have wing 159—172.

Distribution. Great Sangi (Forsten *g 1*, Wallace *h 2*, Meyer *j 3*, Fischer *k 1*, Platen *j 4*), Siao (Hoedt, v. Duidenbode *g 2*, Meyer *j 3, 2*).

† **2. Eos histrio talautensis** (M. & Wg.).

n. **Domicella coccinea** *(1)* Finsch, Papag. 1868, II, 800 (ex Talaut).
o. **Eos indica** *(1)* Hickson, Nat. in N. Celebes 1889, 155.
p. **Eos histrio** *(1)* Salvad., Cat. B. XX, 1891, 21 (Saha).
q. **Eos histrio talautensis** *(1)* M. & Wg., J. f. O. 1894, 240; *(2)* iid., Abh. Mus. Dresd. 1895, Nr. 9, p. 3.
r. **Eos histrio variety talautensis** *(1)* Mivart, Loriidae 1896, 24A.

"Sampiri", Talaut, Nat. Coll.

Diagnosis. As shown elsewhere *(q 1)*, the Talaut bird differs from the typical form of Sangi as follows:

Talaut	Sangi
More red on the wings.	More black on the wings.
Secondaries red, with a black terminal edging 2—5 mm wide, narrowest in old individuals.	Secondaries red with a black terminal edging, 7—12 mm wide, narrowest in old individuals.
Wing-coverts in the adult almost uniform red, a black tip only on isolated feathers of the greater or middle series.	The greater wing-coverts tipped with black, forming a band across the wing; a second band is generally formed by black tips to the midde coverts.
The 1. primary narrowly edged on its basal half, the 3. primary broadly, with red.	The first 3 primaries externally black or the 1. only very narrowly edged with red, the 4. as broadly edged with it as the 3. in the Talaut birds.

Measurements (9 adults from Kabruang, Nov. 2nd—5th, 2 adults from Salibabu, Oct. 29th, and 6 adults from Karkellang, autumn 1896). Wing 163—171; tail 117—136; culmen from cere 19.5—21 mm. One immature: wing 157, tail 110.

Nearly all of these specimens are moulting, but the majority still retain their old primary quills. The moult probably lasts from September to December.

Distribution. Talaut Is. — Kabruang, Salibabu and Karkellang (Nat. Coll. in Dresd. and Tring Museums).

Observation. The first mention of an *Eos* in the Talaut Islands was made by von Rosenberg in a communication to Dr. Finsch; again Dr. Platen (Gefied. Welt 1887, 263) speaks of its existence in Talaut, remarking that this is the habitat of the birds seen in captivity in Sangi and Manado. Dr. Hickson saw the bird in plenty in

Talaut, and the good series obtained by our native hunters confirms the opinion of its abundance there.

In his beautiful Monograph of the Lories Dr. Mivart shows that he has misgivings as to the validity of the Talaut race, inasmuch as he supposes that the differences described as racial fall within the scope of individual variation of the form of Sangi. Probably extremes meet — otherwise the Talaut bird would come under the definition of species, not subspecies — but we have not yet found such extremes. 15 adults from Talaut and 5 adults from the Sangi Islands before us can be correctly sorted by a glance at the wing-coverts and tips of the secondaries, without looking at the labels, and these are only about one-half of those examined by us.

3. Eos histrio challengeri (Salvad.).

s. **Eos indica** *(1)* Sclat., P. Z. S. 1878; 578; id., Voy. Chall. B. 1881, 115; *(3)* Murray, Voy. Chall., Narr. 1885, I, 2, 669.

t. **Eos challengeri** *(1)* Salvad., Cat. B. XX, 1891, 22 (type examd.); *(II)* Mivart, Loriidae 1896, 25, pl. VII, f. 2.

Diagnosis. Very much like *E. histrio*, but much smaller, and with the blue colour on the breast less extended and more or less mingled with red (Salvad. *o 1*). Feet black; bill orange; eyes red, or light brown in the male (Murray *n 1*). Wing 152; tail 102, tarsus 17 mm.

Distribution. Melangis, Nanusa Islands (Murray *o 1*).

Observation. When the "Challenger" was dredging off the Nanusa Islands on 10th Febr. 1875, some natives came off in a boat from the southern island bringing with them four specimens of this parrot. Three of these are now in the British Museum, one of which we have examined. They appear to be the only birds yet known from these islands, and we question whether they are adult.

The Red-winged Lory, *Eos histrio*, affords an interesting case of differentiation into races within short distances, and it affords food for considering whether such variations may have occurred per saltum or by imperceptible gradations, and what may have been the cause. Such simple examples may contain perhaps a more ready answer to these difficult questions than the more complicated ones often attacked by students of evolution. It should be borne in mind, however, that "per saltum" and "gradually" are really vague terms; what one person may call a "jump", another may call a "gradation".

The genus *Eos*, comprising Red-winged Lories as distinguished from the typical *Lorius* in which the wings are green, ranges, according to Professor Reichenow and Count Salvadori, from the Solomon Islands through Papuasia to the Moluccas, occurring also in the Tenimber group and on the islands lying between North Celebes and Mindanao. The latter authority includes within the range of *Eos* the remote island of Ponapé or Puynipet in the West Carolines, the habitat of *Eos rubiginosa* (Bp.), which occurs on the island, as Dr. Finsch observed, in great numbers, doing much damage to the plantations of the colonists. The occurrence of *Eos* in localities north and east of Celebes, and its absence in this country as an indigenous species is somewhat remarkable; in the neighbourhood of Manado specimens of *Eos histrio* have indeed been obtained, but these, according to Meyer's experience, as a rule show signs of having recently been in captivity in their worn tail-feathers, etc., and are to be regarded as specimens brought over from Sangi (or, as Dr. Platen affirms, Talaut) in the boats of the natives, and which have escaped.

At Manado "there seldom arrives a boat without bringing some living birds or the like" (Meyer *j 3*); and the locality Halmahera once stated to be the home of this species was most likely recorded on the ground of examples having been bought or of escaped birds shot there. The species as Mr. Wallace remarks is also brought over to Ternate, and Meyer saw a pair even in Cebu in the Philippines (*j 3*).

To a similar cause, perhaps, is due the labelling of an example of *Eos riciniata* and two of *Lorius garrulus* in the British Museum as from Celebes, species belonging to the Halmahera group as shown by Count Salvadori (Cat. B. XX, 29, 41).

Eos histrio was first recorded from its true habitat, the Sangi Islands, by Forsten; later it was again obtained there by Mr. Wallace, and in the year 1871 in numerous examples by Meyer, and later again by Dr. Fischer. Writing in 1887 Dr. Platen remarks that it is by no means a common bird in Great Sangi, but that, in consequence of the ever-widening extent of the cocoa-nut plantations, it has retired more and more into the mountainous interior of the country, and the caged examples both here and those brought to Manado are derived from the neighbouring Talaut Islands (*j 4*); but it is perhaps not impossible that the time of year of Dr. Platen's visit — the rainy season — may have had something to do with their scarcity near that part of the coast where he collected. The Talaut Islands, as mentioned above, were first noted as a locality for *Eos histrio* by von Rosenberg. Here the species was recently found in great abundance by Dr. Hickson, who confirms Dr. Platen's statement in remarking that it is comparatively rare in the Sangi Islands. An observation of much interest to ornithologists is recorded by Dr. Hickson in his account of his visit to the Saha Islands (Saka of the Dutch maps), two small islands, the one about three-quarters of a mile in diameter, the other about half as large, lying three or four miles, apparently, from the coast of Salibabu, one of the larger islands of the Talaut group. "My attention was called to these islands by a flock of lories, consisting of many hundred individuals, which flew from the main island to the larger of them as the sun was setting on the previous evening". This is one of the rare instances on record of a local species in these parts of the tropics voluntarily crossing a stretch of sea between one island and another, though indeed the birds probably only flew to their roosting place.

The circumstance, that wider excursions as a rule do not appear to occur frequently in these islands, tends to offer a contradiction to the opinion of writers of pessimistic views that the struggle for existence is distressingly severe. At all events the fact that birds of many families in the tropics become differentiated into peculiar species in islands lying within sight of one another, in many cases being protected against overcrowding by laying only one or two eggs, does not lend any support to such a view.

According to the assurances of the best native hunter formerly in Dr. Meyer's service, *Eos histrio* in Sangi, like *Eos riciniata* (Bechst.) of Ternate, lays either two eggs or only one; in the latter case the young one produced is said to be bigger *(j 3)*. The allied *E. rubiginosa* of Ponapé nests in holes in trees, laying only one egg (Finsch, J. f. O. 1880, 284).

Eos histrio, according to Mr. Wallace (Geogr. Distr. I, 420), obliges us to place the Sangi Islands with the Moluccas instead of with Celebes; and this strange connection seems to be warranted by some other facts!

GENUS TRICHOGLOSSUS Vig. Horsf.

The species of this genus vary in size from that of a Lark to that of a Turtle-dove, the upper surface chiefly parrot-green; remiges below black or shining dusky grey; crossed with a band of yellow or red (not seen in the species which inhabit Celebes and Sula, except as an occasional remnant); tail strongly graduated; breast usually barred. About 16 species, ranging from Australia to New Caledonia and the New Hebrides, west through the Papuan Islands and the Amboina-group (but absent in the Halmahera-group), to Sula, Celebes, Timor.

+ * 38. TRICHOGLOSSUS ORNATUS (L.).
Ornate Lory.

a. **Psittacus ornatus** *(1)* Linn., S. N. 1766, I, 143 (ex Seba, Briss.); *(II)* Shaw, Gen. Zool. 1811, VIII, 416, pl. 60; *(3)* M. & S., Verh. Natuurk. Comm. 1839-44, 90, 182; *(4)* S. Müll., Reizen Ind. Archip. 1858, pl. II, 12; Reinw., Reis. Ind. Archip. in 1821, 1858, 592.

b. **La Perruche Lori** *(1)* Levaill., Perr. 1801, pl. 52.

Trichoglossus ornatus *(1)* Gray, Gen. B. 1846, II, 411; *(1bis)* S. Müll., Reiz. Ind. Arch. 1858, II, 69; *(2)* Rosenb., J. f. O. 1862, 60 (Sula); *(3)* id., N. T. Ned. Ind. 1862, XXV, 139 (Sula); *(4)* Wall., P. Z. S. 1864, 291, 295; *(5)* Schl., Mus. P.-B. Psittaci 1864, 112; *(6)* Finsch, Papag. 1868, II, 842; *(7)* Finsch & Conrad, V. z.-b. Ges. Wien 1873, 2, 15; *(8)* Schl., Mus. P.-B. Rev. Psitt. 1874, 49; *(9)* Salvad., Ann. Mus. Civ. Gen. 1875, VII, 645; *(10)* Brügg., Abh. Ver. Bremen 1876, V, 42; *(11)* Rosenb., Malay. Archip. 1878, 274; *(12)* Meyer, Ibis 1879, 53, 145; *(13)* Salvad., Orn. Pap. 1880, I, 299; *(14)* Rchnw., J. f. O. 1881, 160; id., Consp. Psitt. 1881, 96; *(XV)* id., Vogelb. 1878-83, pl. VIII, fig. 7; *(16)* W. Blas., Ztschr. ges. Orn. 1885, 220; *(17)* Guillem., P. Z. S. 1885, 544; id., Cruise "Marchesa" 1886, 180; *(XVIII)* Meyer, Vogel-Skelett. 1884, pl. LXX; Platen, Gefied. Welt 1887, 230; *(19)* Salvad., Cat. B. 1891, XX, 61; *(20)* Büttik., Z. Erg. Weber's Reise Ost.-Ind. 1893, III, 273; *(21)* M. & Wg., Abh. Mus. Dresd. 1895 Nr. 8, p. 5; *(22)* iid., ib. 1896 Nr. 2, p. 8; *(23)* Everett, Nov. Zool. 1896, 161; *(XXIV)* Mivart, Loriidae 1896, 119, pl. XXXIX; *(25)* Hartert, Nov. Zool. 1897, 159, 165.

"**Parkitje**" (from the Dutch), Minahassa, Meyer *12*.
"**Kerut**", inland name, Minahassa, Meyer *12*.
"**Ulolito**", Gorontalo, Rosenb. *6*, *11*; Joest.
"**Dorra**", Tjamba, S. Celebes, Platen *16*.
"**Burong nuri**", Bonthain, S. Celebes, Ribbe & Kühn in Mus. Dresd.
"**Koloje**", Tonkean and Balante, E. Celebes, Nat. Coll.
"**Kaleki**", Peling and Banggai, Nat. Coll.
For synonymy and further references see Salvad. *19*.
Figures and descriptions. Mivart *XXIV*; Shaw *a II*; Levaillant *b I*; Reichenow. *XV*, *14*; Meyer *XVIII* (skeleton); Finsch *6*; Brüggemann *10*; W. Blasius *16*; Salvadori *19*.

Adult male. Head above and ear-coverts hyacinth-blue; occiput poppy-red with dusky blue tips to the feathers; on side of neck from behind the ear-coverts a band of gamboge-yellow; all other upper parts parrot-green, some large, concealed, subterminal yellow spots on the feathers of the mantle, terminal fourth of the tail-feathers — except the two middle ones — yellow-red below where concealed by the under tail-coverts, etc.; cheeks and chin poppy-red; throat and breast poppy-red with broad terminal margins of dusky blue, giving a barred appearance; rest of under surface parrot-green becoming yellow-green on the lower abdomen and under tail-coverts, and chiefly gamboge-yellow on the sides; metacarpal edge parrot-green, under wing-coverts yellow; basal part of tail-feathers red (C 1028, Limbotto, N. Celebes, Aug.).

"Iris orange; bill light coral-red; cere dark brown; feet greyish green" (Platen *16*). Iris blackish blue and black; bill yellow-orange and red with black; feet greyish green and greenish (Ribbe & Kühn).

Female. Like the male, probably not smaller, the smallest wing-measurement (122 mm) of 14 examples in the Dresden Museum being that of a ♂.

Young. Red feathers of occiput nearly absent; parrot-green of abdomen much varied with yellow; dark bars on chest narrower than in adult; some feathers on cheeks margined with yellow (Minahassa, Nr. 3999).

Measurements. The wings of 17 specimens vary between 119—134 mm. Measurements of the smallest and the largest specimens; tail 75—84; tarsus 16; culmen from cere 19—20.5 (C 1031, Nr. 1947).

Skeleton.

Length of cranium	43 mm	Length of tibia	42 mm
Greatest breadth	23 »	Length of tarso-metatarsus	15 »
Length of humerus	27 »	Length of sternum	43 »
Length of ulna	29 »	Greatest breadth of sternum	22.5 »
Length of radius	27 »	Height of crista sterni	14 »
Length of manus	38 »	Length of pelvis	43 »
Length of femur	29 »	Greatest width of pelvis	22 »

Egg. White, 25 × 17 mm (Meyer *12*).
Breeding season. Near Manado, February *(12)*. As Finsch remarks, the *Trichoglossidae* nest in holes in trees, laying 2—4 eggs.
Variation. Cases of xanthochroism in this Parrot are apparently not rare. Specimens displaying a completely or partially yellow plumage are mentioned by Brüggemann *(11)*, Meyer *(12)* and Guillemard *(17)*. (As to xanthochroism in parrots see Meyer, Sitzb. Preuss. Akad. Wiss. 1882, 517.)
Distribution. Celebes, Togian Islands, Buton. North Peninsula — Minahassa (Forsten *5*, Wallace *19*, etc.); Gorontalo District (Rosenb. *8*, Meyer *12*); Banka and Mantehage

Island (Nat. Coll.); Togian Islands (Meyer *12*); E. Celebes — Balante and Tonkean (Nat. Coll.); S. E. Peninsula — Kandari (Beccari *9*); Buton (S. Müller *3*, *4*, *5*); West Celebes (Doherty *25*); South Peninsula—Bonthain (Weber *20*, Ribbe & Kühn in Dresd. Mus.); Macassar (Bernst. *8*, Wall. *10*); Boni (Mus. Leyd. *5*); Tjamba (Platen *10*); Sula Islands — Peling and Banggai (Nat. Coll.).

Von Rosenberg first recorded the Sula Islands as a locality for this species on information of one of the native chiefs of the islands *(3)*, but there was no evidence in proof of the correctness of this till specimens were obtained by our native hunters in Banggai and Peling. The hunter Kamis Birahi, who accompanied Bernstein, Wallace, v. Rosenberg and Meyer on some of their journeys, and who visited Sula, informed Meyer that *T. ornatus* did not occur in Sula.

It has not been recorded from Sangi, but Meyer received a specimen from Siao not differing in plumage from those of Celebes (MS. note), though this may have been from captivity.

As Dr. Finsch remarks Celebes forms the north-westernmost boundary of the genus *Trichoglossus* (the Wedge-tailed Lories) viewed as composed of Prof. Reichenow's subgeneric groups *Glossopsittacus*, *Charmosyna*, *Oreopsittacus*, *Neopsittacus* and *Trichoglossus*. Two of Prof. Reichenow's groups are further subdivided by Count Salvadori (Cat. B. XX, 1891): — *Charmosyna* into *Hypocharmosyna*, *Charmosynopsis*, *Charmosyna*; and *Neopsittacus* into *Psitteuteles*, *Ptilosclera* and *Neopsittacus*, but the last named genus, according to Salvadori, does not belong to the *Loriidae* at all, but to the *Cyclopsittacidae*. The Blue-headed group of Wedge-tailed Loris, to which *T. ornatus* belongs, ranges, as Count Salvadori shows, from Celebes throughout most of the intervening islands to New Caledonia and the New Hebrides, occurring also in the Lesser Sunda Islands and Australia. Curiously enough it is absent, so far as is yet known, in the Halmahera group, just as it is in Sangi.

The Blue-headed Lory of Celebes is a very distinct species, differing from all its fellows in having the cheeks, chin and throat red instead of blue, and the quills below uniform shining brownish smoke-grey, instead of having only their distal part of this colour and their basal part orange-yellow or red. Its nearest allies appear to be *T. massena* Bp. of S. E. New Guinea, New Britain and the neighbouring islands as far as the New Hebrides, and *T. cyanogrammus* Wagl. of Western Papuasia as far as Buru.

Although confined to Celebes, *T. ornatus*, like some of the *Trichoglossi* of Australia, is by no means a strictly sedentary bird in the island, where it is the commonest parrot. Meyer found it at all times and everywhere in the Minahassa from January till July; at Limbotto in August; near Gorontalo in September; on the Togian Islands in August; in South Celebes in October and November. About the end of March, 1871, it suddenly appeared in crowds in the neighbourhood of Manado. The *Trichoglossi* feed largely on the dews of flowers, using the tongue in a licking manner (Finsch, Papag. II, 816); fruit also would appear to form a large part of their sustenance. *T. ornatus* feeds, "according to the season, on all possible fruits; in captivity they prefer bananas above everything, but also like rice" (Meyer *12*). In S. Celebes Ribbe & Kühn found in its stomach the fruits and the seeds of trees. The local movements of the species are most likely regulated by the time of ripening of fruit, or of flowering of certain trees. Gould observed, as Dr. Finsch points out (l. c. 815), that certain *Trichoglossi* are more or less birds of passage: "a few species seem to make periodically settled wanderings from the South, where they breed, to the North, when they gather themselves into countless flocks, which, hasten through the air with rushing speed like a cloud in regular evolutions and accompanied by deafening cries"; but, as the author adds, nothing is known about similar migrations, if such there be, among the insular species.

"*T. ornatus* smells, as all the allied parrots do, very agreeably of hyacinths" (Meyer *12*). It is kept as a pet by the natives of Celebes, but also used as an article of food.

+ 39. TRICHOGLOSSUS FORSTENI Bp.

Forsten's Lory.

Trichoglossus forsteni *(1)* Bp., Consp. Av. 1850, I, 3 (ex Temm. in Mus. Lugd.); *(2)* Finsch, Papag. II, 1868, 826; *(3)* Rchnw., J. f. O. 1881, 157 (Consp. Psitt. 23) syn. emend.; *(4)* Guillem., P. Z. S. 1885, 502; *(5)* Salvad., Cat. B. XX, 1891, 51; *(VI)* Mivart, Loriidae 1896, 93, pl. XXIX; *(7)* Hartert, Nov. Zool. 1896, 176, 572.

a. **Trichoglossus immarginatus** *(1)* Blyth, J. A. S. B. 1858, XXVII, 279.

For further synonymy and references see Salvadori *5*.
Figures and descriptions. Mivart *VI*; Finsch *2*; Reichenow *3*; Salvadori *5*.
Adult male. Above parrot-green; entire head, face, throat and mantle purple blackish blue, the forehead and cheeks striated with brighter blue, the feathers of the mantle with concealed red spots; around neck (not throat) a collar of greenish yellow; breast, sides, and under wing-coverts vermilion; abdomen blue-black; flanks, thighs, and under wing-coverts greenish yellow, barred with parrot-green; remiges below black, a band or patch of yellow across the basal part of the inner webs: "bill red, yellow at tip; feet olive-green" (Guillemard *4*). Wing 141 mm; tail 110; tarsus 17; bill from cere 18.5 (♂ nat. coll.; Djampea Id., Dec. 1895: Everett, C 14861).
Sex. The sexes are closely similar in coloration (Finsch *2*).
Distribution. Sumbawa (Forsten *2*, Guillemard *4*); Djampea between Flores and Celebes (Everett *7*).

This Lory, previously known only from Sumbawa, was found recently to be common on Djampea Island by Mr. Everett, whose specimens were compared with the type of the species by Mr. Hartert who remarked that they were perfectly identical, but later found some slight differences (see below). It is one of a group of eleven known species, with the head and cheeks blue or dusky (or for the most part of these colours) and having a yellow — in one case a red — band across the base of the remiges; they are found from Australia to

Timor[1]) and Sumbawa in one direction, and from New Caledonia to the Moluccas in another. The present bird seems to be related to *T. haematodes* (L.) of Timor and to *T. novaehollandiae* (Gm.) of Australia, the former differing by its yellow breast and green occiput, the latter by its much brighter blue tints. *T. mitchelli* G. R. Gray, of which Prof. Mivart gives a good figure, appears to stand still closer to Forsten's Lory, but it evidently is easily distinguishable by having the head and middle of the abdomen dusky green, instead of purplish blue-black. The distribution of this form is as yet unknown.

Mr. Hartert *(7)* says that two males received by him from Sumbawa "have not such a broad blue patch behind the pale greenish band on the hind neck as all those from Djampea have"[2]).

+ * 40. TRICHOGLOSSUS MEYERI Tweedd.
Meyer's Lory.

This species is known in two forms in the northern and southern Peninsulas of Celebes, respectively:

1. The typical Trichoglossus meyeri.

a. **Trichoglossus flavoviridis** part. *(1)* Wall., P. Z. S. 1862, 337; 1864, 295 (Manado).
b. **Trichoglossus meyeri** *(1)* Wald., Ann. & Mag. N. H. 1871, VIII, 281; *(II)* id., Tr. Z. S. 1872, VIII, 32, pl. IV; *(3)* Schl., Rev. Psitt. 1874, 50; *(4)* Salvad., Ann. Mus. Civ. Gen. 1875, VII, 646; *(5)* Brüggem., Abh. Ver. Bremen 1876, V, 42; *(6)* Meyer, Ibis 1879, 54; *(7)* Rchnw., J. f. O. 1881, 156; Consp. Psitt. 1881, 92; *(VIII)* id., Vogelb. 1878-83, t. XI, fig. 3; *(IX)* Meyer, Vogel-Skelett. 1881-82, t. XXIV; *(10)* W. Blas., J. f. O. 1883, 125, 134; *(11)* Guillem., P. Z. S. 1885, 544; *(12)* Platen, Gefied. Welt 1887, 206, 230; *(13)* M. & Wg., Abh. Mus. Dresd. 1895 Nr. 8, p. 5; *(14)* iid., ib. 1896 Nr. 1, p. 7.
c. **Psitteuteles meyeri** *(1)* Salvad., Cat. B. 1891, XX, 63; *(II)* Mivart, Loriidae 1896, 125, pl. XLI.
"**Parkitje lolaro**" (Mangrove Paroquet), Manado, Meyer *b 6*; Guillem. *b 11*.
Figures and descriptions. Walden *b II*, *b 1*; Mivart *c II*; Reichenow *b VIII*, *b 7*; Meyer *b IX* (skeleton); Salvadori *b 4*, *c 1*; Brüggem. *b 5*.
Adult male. Parrot-green; feathers of mantle with yellow bases; head above golden-olive, yellower on forehead; ear-coverts bright yellow; feathers of lores, cheeks and chin barred with ochraceous and tipped with dull green; below yellow with a light greenish wash, each feather broadly bordered with parrot-green, abdomen and under tail-coverts strongly washed with green; under wing-coverts light yellow-green, at the metacarpal edge grass-green; quills below uniform shining brownish smoke-grey; tail below deep olive-yellow (♂ Rurukan, 3. IV. 04: P. & F. S.). "Iris cherry-red; feet greyish blue; bill orange-red" (Meyer *b 1*).

---

1) In recording the habitat of the allied *T. haematodes* (L.) Count Salvadori mentions only Timor. Other writers add Wetter (Schl., Rev. Psitt. 1874, 48), Sanao (Rchnw., Consp. Psitt. 1881, 98) and Sumba (Meyer, Verh. z.-b. Ges. Wien 1881, 762; Hartert, Nov. Zool. 1896, 586), but these localities appear to have been intentionally omitted in the Catalogue of Birds, in anticipation of the occurrence of local differences of plumage in these islands. This caution appears to be fully justified; the two specimens from Sumba in the Dresden Museum differ from two others of unknown origin in being much larger, and in having the under wing-coverts yellow instead of flame scarlet.

2) Since this was written Mr. Hartert has named the Djampea birds *T. forsteni djampeanus* (Nov. Zool. 1897, 172).

Female. Differs from the male in having the head above much darker and duller — viz. brownish green with a suggestion of purplish therein in certain lights. (♀ Rurukan, 3. IV. 94, P. & F. S.).

Younger. Crown of head more yellow-green, scarcely yellower on the forehead; yellow of under surface greener; inner margin of quills yellowish; some of the same colour spread out on the inner webs of some of the inner primaries. A trace of green barring near tip of tail (Nr. 1746). Bill dark.

Distribution. North Celebes — Manado (Wallace *a 1*, Meyer *b 1*, v. Duivenb. *b 3*, etc.); Rurukan — 3000 ft. (Platen *b 12*, P. & F. S.); (?)Gorontalo (Brit. Mus. *c 1*); Central Celebes between Lake Posso and Mapane (P. & F. S. *b 14*).

✝ 2. Trichoglossus meyeri bonthainensis (Meyer).

d. **Trichoglossus meyeri** var. **bonthainensis** *(1)* Meyer, Isis 1884, 16; *(2)* Salvad., Cat. B. 1891, XX, 64 note; *(III)* Mivart, Loriidae 1896, 127, pl. XLI, figs. 2, 3.

Figures and descriptions. Mivart, Meyer, Salvadori ll. cc.

Diagnosis. Key to the two subspecies:
1. The subterminal bars on the feathers of lores and cheeks broader and ochraceous in colour; the green bars of the under surface narrower. *The typical T. meyeri.*
   Distr. North Celebes.
2. The subterminal bars on feathers of lores and cheeks narrower, dull apple-green; the green bars of the under surface broader; the yellow bases of feathers of mantle less developed . . . . *T. meyeri bonthainensis.*
   Distr. Mt. Bonthain, S. Celebes. (Ribbe & Kühn.)

Measurements.	Wing	Tail	Tarsus	Culmen from cere
a. (Nr. 1745) Manado	108	67	13	14.5
b. (Nr. 1747) Manado	103	63	12	14.5
c. (Nr. 1746) imm. Celebes	101	57	12.5	13.5
d. (Sarasin Coll.) ♂ ad. Rurukan, 3. IV. 94	102	63	13	15.5
e. (Sarasin Coll.) ♀ ad. Rurukan, 3. IV. 94 .	101	58	12	14
f. (Sarasin Coll.) ♂ juv. Rurukan, 4. X. 94 .	102	61	—	14
g. (Sarasin Coll.) juv. Centr. Celebes, 23. II. 95 . . .	101	62	11.5	13.5
h. (Nr. 6007) Bonthain, S. Celebes (type of *bonthainensis*)	106	68	13	15.5

The type-specimen of *bonthainensis* remains up to the present the only one on record from any part of South Celebes.

The group of small Green Wedge-tailed Lories of the subgenus *Psitteuteles*, according to Count Salvadori, is found in one species in Australia, one in the Timor group, one in Celebes and one in Sula. It is curiously absent in the Moluccas and Papuasia, as also in Sangi. In the latter islands, the genus *Eos*, which is absent in Celebes, occurs, but *Trichoglossus*, which is represented in two species in Celebes is not found there at all, or at least it has not yet been discovered there. Perhaps these facts have a certain relation to one another, it may be suggested that one species ousts the other from this or that locality by seizing the food they both eat. In Amboina and elsewhere nearer to the eastern centre of development of the parrot tribe, however, both a species of *Eos* and of *Lorius* and a Blue-headed *Trichoglossus* occur together.

Count Salvadori's excellent keys for the determination of the Green Wedge-tailed Lories of the subgenus *Psitteuteles* and the Blue-headed ones of the subgenus *Trichoglossus* awake the attention to an interesting fact. In Celebes the members of both subgenera have the underside of the quills uniform shining brownish smoke-grey; in all parts of Australasia where they occur both subgenera have the basal part of the quills crossed by a band of bright colour — yellow usually, but in two cases — one in each subgenus — red. The circumstance is, of course, suggestive of a common cause, but a difficulty to the maintaining of this opinion is presented by the fact that the *Trichoglossus*, which has the red band, does not occur in the same locality as the *Psitteuteles* with the red band, but one *(T. rosenbergi* Schl.*)* on Mysore Island in Geelvink Bay, New Guinea, and the other *(Ps. chlorolepidotus)* in Australia. We are, however, from the following circumstance able to conclude at least on good hypothetical grounds that the Celebesian forms are of a more recent development than the Australasian ones: three young specimens and one female of *T. meyeri* described above differ from the old ones amongst other points in having some amount of yellow on the inner webs of some of the inner remiges, in one case very obviously the remains of a wing-band; in two or three immature specimens of *T. ornatus* also (Nrs. 3999, 6580) we find a trace of the yellow band on one or two feathers of the secondaries, but not in any adult specimens, nor in other specimens not quite adult. This appearance indicates that *T. meyeri* and *ornatus* are sprung from two forms which possessed a yellow wing band [1]. There is thus reason to suppose that the *Trichoglossi* of Celebes are a development more recent than those of Australia. For this reason no one can very well advance the supposition that the Australasian forms came from Celebes, but on the other hand it will generally be felt that this is some sort of evidence that Celebes was originally colonised by *Trichoglossi* from Australasia, and then for some unknown reason they lost the wing-band in Celebes.

Trichoglossus meyeri finds its nearest ally in *T. flavoviridis* Wall. of the Sula Islands, which shares with it the peculiarity of having no coloured band on the under side of the wing. As already mentioned Count Salvadori's work shows that this group of Wedge-tailed Lories does not occur further East; and the single species inhabiting the Lesser Sunda Islands, *T. euteles* Temm. (a form which will perhaps have to be broken up into several subspecies) differs markedly from those of Celebes and Sula. The first notice of *T. meyeri* was made by Mr. Wallace, who obtained a specimen in Manado, but the skin was destroyed and the author identified the bird doubtfully with *flavoviridis* of Sula *(a 1)*. In 1866 a specimen was sent to the Leyden Museum by van Duivenbode, but apparently not recognised as new by Schlegel; five years afterwards numerous specimens were obtained in the Minahassa and recognised as new by Dr. Meyer;

---

[1] Young specimens of *Charmosyna pulchella* Gray also have a very plain yellow band at the base of the secondaries and inner primaries, which disappears in the adult.

it was named *T. meyeri* by Lord Walden. Wallace believed the species to be rare in Celebes, in consequence of the competition of the abundant *T. ornatus*, which does not occur in Sula; like *T. ornatus*, however, it makes considerable local movements, depending no doubt on the time of flowering or fruit-bearing of certain trees, and at the end of April, 1871, *T. meyeri* appeared in the neighbourhood of Manado in flocks (as did also about the same time *T. ornatus*, *Loriculus stigmatus* and *L. exilis*), and could always be procured till the 18th of May, after which date it became scarce. Dr. Platen appears to have found both *T. ornatus* and *meyeri* plentiful at Rurukan (3000 ft.). Of the latter he writes that it is never kept in captivity like *T. ornatus*, though both will feed on broken biscuit and powdered canary seed. *T. meyeri* is of a much quieter and gentler nature than *ornatus*, perhaps the cause of its not being cared for as a pet *(b 12)*. A tame female of *T. meyeri* in Dr. Platen's possession laid an egg in November. The breeding season of the species in freedom is unknown.

✝ * 41. TRICHOGLOSSUS FLAVOVIRIDIS Wall.
Yellow-green Lory.

Trichoglossus flavoviridis *(1)* Wall. P. Z. S. 1862, 337, pl. XXXIX; id., 1864, 295, 292 pt.; *(2)* Finsch, Papag. 1868, II, 849; *(3)* Wald., Ann. N. H. 1871, VIII, 281; *(4)* Schl., Rev. Psitt. 1874, 49; *(5)* Rchnw., J. f. O. 1881, 156 (Consp. Psitt. 1881, 92); W. Blas., J. f. O. 1883, 125.

a. **Psitteuteles flavoviridis** *(1)* Salv., Cat.B.XX, 1891, 63; *(II)* Mivart, Lor. 1896, 123, pl. XL.
For further references see Salvadori *a 1*.
Figures and descriptions. Wallace *I*; Mivart *a II*; Finsch *2*; Reichenow *5*; Salvad. *a 1*.
Adult. Green; head olive-yellow; nape with a dusky collar; face, cheeks and chin dusky green, each feather margined with yellow; face, neck, breast, and upper part of the belly bright yellow, each feather narrowly margined with dark green, producing a regular scaly appearance; belly, vent and under tail-coverts yellowish green with green edges to the feathers; the interscapulary feathers with concealed yellow bands; quills dusky black underneath; tail-feathers beneath ochre-yellow: bill orange-red; orbits bare, yellow; feet lead-colour; iris orange. Total length 216 mm; wing 122; tail 81; bill 18; tarsus 13 (Salvad. *a 1*).
Distribution. Sula Islands (Allen *I*) — Sula Besi (Hoedt *3*); Sula Mangoli (Bernst. *3*).

This species is the only member of the *Loriidae* known to occur in the Sula Islands, being very nearly related to *T. meyeri* of Celebes, from which it differs in having the head yellow, the chin and a nuchal collar dusky, and the yellow of the breast without any greenish tint. It was first obtained by Wallace's assistant, Allen, who appears to have found the species abundant. The islands in which he collected were the southern and eastern ones of the group, i. e. Sula Mangoli and Sula Besi. Ten specimens from these islands were subsequently sent by Hoedt and Bernstein to the Leyden Museum in 1864. As already noticed, the occurence of closely allied species in Sula and Celebes, and the non-occurrence of the subgenus *Psitteuteles*, so far as is known, in the Moluccas a little further east, is of interest in questions of geographical distribution.

FAMILY CACATUIDAE.

"Sternum complete; orbital ring completely ossified, with a process bridging the temporal fossa (Garrod, P. Z. S. 1874, 594); nostrils open in a cere not much swollen, generally naked but sometimes feathered; bill very deep, deeper than long, with the upper mandible generally much compressed; hook of the upper mandible nearly perpendicular, except in *Licmetis*, and with a file-like surface underneath; tarsus short; head always crested; as a rule only the left carotid present" (Salvadori, Cat. B. XX, 101). Coloration uniform.

Five genera found in Australia, two thereof passing into Papuasia, and one of the latter, *Cacatua*, further to the Philippines, Celebes, and the Lesser Sunda Islands.

GENUS CACATUA Vieill. ex Briss.

Map V.

The species of this genus are of rather large size — from that of a Jackdaw to a Raven almost — and are well characterized by their white or rosy — (in one case deep rose-coloured) plumage, their crests, and perpendicularly-hooked bills. Fuscin, a colour which, when overlaid with psittacofulvin, gives parrot-green, is absent in the Cockatoos (Krukenberg). The genus *Cacatua* is composed of 15 species, which, as shown in the following article, fall into three natural groups, with a distribution from Australia to Lombok, Celebes, the Philippines, Moluccas, the Papuan Islands to the Solomon group.

+42. CACATUA SULPHUREA (Gm.).

Sulphurous Cockatoo.

According to Mr. Hartert (Nov. Zool. 1897, 165) this form embraces three subspecies. They are:

1. The typical Cacatua sulphurea.[1]

a. **Psittacus sulphureus** *(1)* Gm., S. N. 1788, I, 330 (ex Briss. and Edwards); *(2)* S. Müll., Verh. Natuurk. Comm. 1830—44, 90, 182; id., Reizen Ind. Arch. 1857, 12; *(3)* Russ, Fremdl. Stubenvög. 1881, IV, 655—658.
b. **Psittacus cristatus** *(1)* Labill., Voy. à la Recherche de la Pérouse 1791—92, II, 301.
c. **Cacatua sulphurea** *(1)* Vieill., N. D. 1817, XVII, 10; *(2)* Schl., Mus. P.-B. Psitt. 1864, 137, pt.; *(3)* Fraser, P. Z. S. 1865, 227; *(3bis)* Schl., Ned. Tdschr. Dierk. 1866, III, 319, 321; *(4)* Garrod, P. Z. S. 1873, 460, fig. 6 (carotids), 461, 465, *(5)* id., ib. 1874, 587, 588, 591, 595; *(6)* Schl., Rev. Psitt. 1874, 66; *(7)* Salvad., Ann. Mus. Civ. Gen. 1875, VII, 644; *(8)* Brüggem., Abh. Ver. Bremen, 1876, V, 37; *(9)* Rosenb.,

---

[1] Many authors spell this word "*sulfurea*", but this is not the original spelling of Gmelin, nor indeed the more correct.

Mal. Archip. 1878, 275; *(10)* Meyer, Ibis 1879, 44, 145; *(11)* Krukenberg, Vergl. physiol. Studien 1882, II, 2. Abth., 29, 33, 35; *(XII)* Meyer, Vogelskelette 1881—82, t. XVIII; *(13)* W. Blas., Ztschr. ges. Orn. 1886, 195; *(14)* W. Marsh., Z. Vortr. Die Papag. 1889, 31; *(15)* Salvad., Cat. B. XX, 1891, 121; *(16)* Büttik., Z. Erg. Weber's Reise Ost-Ind. 1893, III, 273; *(17)* M. & Wg., Abh. Mus. Dresd. 1895, Nr. 8, p. 5; *(18)* iid., ib. 1896, Nr. 1, p. 7; *(19)* Hartert, Nov. Zool. 1897, 164.

d. **Plyctolophus sulphureus** *(I)* Lear, Ill. Parrots 1832, pl. 4; *(II)* Selby, Nat. Libr., Parrots 1836, 129, pl. 14; *(3)* Nitzsch, Pterylogr. 1840, 146; id., Engl. Ed. 1867, 102.

e. **Psittacus sulphureus minor vel moluccensis** *(I)* Bourjot, Perr. 1837—38, pl. 80.

f. **Plictolophus sulfureus** *(1)* Finsch, Papag. I, 1867, 296—300, pt., II, 1868, 942; *(II)* Rchnw., Vogelbilder, 1878—83, t. IV, f. 1.

g. **Plissolophus cristatus** *(1)* Rchnw. (nec. L.), J. f. O. 1881, 29 (Consp. Psitt. 1881, 29).

"**Katella**" or "**Catala**", Gorontalo Dist., Rosenb. *d 1*, 9.
"**Gatalla**" and "**Cacatua puti**" (malay), Meyer *10*.

For further synonymy and references see Salvad. *15*.

Figures and descriptions. Lear *d I*; Selby *d II*; Bourjot *e I*; Reichenow *f II*; Meyer *c XII* (skeleton); S. Müller *a 2*; Schlegel *c 2*; Garrod *c 4*, *c 5* (anat.); Finsch *f 1*; Salvadori *c 15*.

Adult ♂. White, with a light wash of sulphur-yellow, especially strong on the under surface, under side of wings and of tail; crest (85 mm), ear-coverts and basal part of the remaining feathers of head deep sulphur-yellow. (The mantle and wings above of this and of 8 other specimens are nearly white without any yellow wash to speak of, but this hue seems to fade through exposure, as it may be found at the bases of the feathers and on the lower back and upper tail-coverts under cover of the wings. — Kwandang, North coast of Celebes, — Nr. 3536). Iris of adult red (Meyer *10*).

Female. Like the male, but with a smaller bill.

Young. The yellow of the plumage much less intense than in the old ones; iris dark — either black or brown (Meyer *10*).

According to Finsch the nestlings of the white Cockatoos are clothed in long white down (Papag. I, 272); the same author mentions a young one of the Flores species, which is very closely related to the Celebesian one, in which the yellow feathers of the ear-coverts were in process of development (l. c. 297)[1].

Measurements.	Wing	Tail	Tarsus	Culmen from core
a. (Nr. 3536) ♂ Kwandang, N. Cel.	236	110	19.5	39
b. (Nr. 3535) ♂ Kwandang, N. Cel.	235	116	20	40
c. (Nr. 3537) ♀ Paguatt, Tomini Bay	232	114	19	35
d. (C 1331) ♀ Paguatt, Tomini Bay	235	116	—	34.5
e. (C 1329) ♀ Paguatt, Tomini Bay	230	114	—	35.2
f. (C 1325) ♂ Paguatt, Tomini Bay	243	122	—	37
g. (C 1330) ♀ Kwandang, N. Cel.	221	110	—	35
h. (C 6590) ? Gorontalo Distr.	230	110	—	38.5
i. (Nr. 3538) ? Gorontalo Distr.	243	113	—	40.5
j. (Sarasin Coll.) ♀ Buol, N. Cel., 15. VIII. 94	237	115	—	34
k. (Sarasin Coll.) ♂ Maroneng, Gulf of Mandar, 18. VIII. 95	234	110	—	39

[1] The entire series of this genus in the British Museum are, without exception, marked by Count Salvadori as adult!

Distribution. Celebes and Buton — Tomini (Forsten *c 2*), Paguatt, Tomini Gulf (Rosenb. *c 6*, Meyer *c 10*), Posso, Tomini Gulf (Meyer *c 10*), Kandari, S. E. Celebes (Beccari *c 7*), Buton (Labillardière *b 1*, S. Müller *a 2*), Maros, S. Celebes (Weber *c 16*), Mandalli, W. Coast (Meyer *c 10*), Maroneng (P. & F. Sarasin), Dongala and Tawaya, W. Coast (Doherty *c 20*), Kwandang, N. Coast, and the small islands in Kwandang Bay (Meyer *c 10*, Rosenb. *c 6*, *c 9*), Buol (P. & F. Sarasin).

2. C. sulphurea djampeana Hartert.

Nov. Zool. 1897, 164.

h. **Cacatua sulphurea** *(1)* Hart., Nov. Zool. 1896, 176.

Diagnosis. Exactly like the *typical C. sulphurea* of Celebes, but with a smaller beak measured straight from the outer margin of the cere, where maxilla and mandible meet, to the tip — in two examples 24 mm, in one only 23.5, as against 27 mm in females from Celebes (Hartert l. c.).

Distribution. Djampea Island (Everett *h 1*).

3. C. sulphurea parvula (Bp.).

For synonymy, etc. cf. Salvadori, Cat. B. XX. 1891, 120.

Diagnosis. Similar to the *typical C. sulphurea* of Celebes, but the ear-coverts paler and much less yellow (Hartert).

Distribution. Lombok, Sumbawa, Flores, Semao, Timor.

Observation. We cannot regard Mr. Hartert's division as conclusive; if one were only to investigate closely enough further racial peculiarities would no doubt be found.

The distribution of *Cacatua sulphurea* in Celebes is a remarkably interrupted one. In the Minahassa it is unknown as a wild bird, as it is also at Macassar and Maros in the South; this also appears to be the case at Gorontalo, though it is to be met with on the coast of the gulf at Paguatt and Posso, as also twenty miles from Gorontalo at Kwandang on the north coast, and again further west at Buol. At Posso, on the south shore of the Gulf of Tomini, Meyer first met with it in large flocks; but on the same coast further east at Todjo it was not then to be found, neither does it appear to occur on the Togian Islands[1]. Labillardière in one of the early French voyages was the first to notice it at Buton Island; afterwards Salomon Müller saw it in large numbers there: "Before nearly all the houses we saw Cockatoos and other species of parrots fastened by a small double ring of buffalo-horn or cocoa-nut on one foot to sticks, etc."; *Psittacus sulphureus* was most abundant, in much smaller numbers *Ps. ornatus* and *Ps. setarius (P. platurus)*. A good reason for this peculiar distribution is not easy to find, but the data known are scanty and insufficient and may partly rest on observations of local movements and if more facts were to hand, a different sketch of distribution of the species might perhaps be drawn up.

The distribution of the genus *Cacatua* is likewise somewhat anomalous. The

---

[1] The species is included in the list of Togian birds collected by Meyer in 1871 *(10)*, but this indication is a mistake.

fifteen species comprising the genus, as recognised by Count Salvadori, are placed by Prof. Reichenow in his Conspectus Psittacorum in two subgenera, — the characters of which were first clearly shown by Dr. Sclater in 1864 (P. Z. S. 1864, 188) — recognisable by possessing a long, recurved crest of narrow feathers, or a crest of broadened feathers generally decurved, respectively. Each of these subgenera consists of two sections, the characters of which are well pointed out by Reichenow.

Dr. Finsch makes a different division of the genus *Cacatua* in his key to the species, gathering the species into two groups distinguishable by having 1. the nostrils and cere naked and the bill black, or 2. the nostrils and cere feathered and the bill light.

To our mind the genus shows itself to be composed of 3 natural groups:

1. Bill large, black; cere and nostrils naked; crest long, of narrow feathers, curving upwards at the extremity:

    C. galerita (Lath.): Australia, Tasmania.
    C. triton (Temm.): New Guinea and the islands close by; Aru.
    C. citrinocristata (Fraser): Sumba.
    C. parvula (Bp.): Timor, Semao, Flores, Sumbawa, Lombok.
    C. sulphurea (Gm.): Celebes; Buton.

2. Bill comparatively small, white or horn-colour or yellowish; cere and nostrils feathered; crest of short broad feathers, decurved — in one case *(C. leadbeateri)* lengthened and recurved:

    C. sanguinea Gld.: N. Australia.
    C. goffini (Finsch): Timorlaut.
    C. ducorpsi J. & P.: Solomon Is.
    C. gymnopsis Sclat.: Australia.
    C. haematuropygia (P. L. S. Müll.): Philippine and Sulu Islands.
    C. roseicapilla Vieill.: Australia.
    C. leadbeateri (Vig.): S. Australia.

3. Bill large, black; cere and nostrils naked; crest of very long broadened feathers, more or less decurved:

    C. alba (P. L. S. Müll.): Halmahera Group.
    C. ophthalmica Sclat.: New Britain.
    C. moluccensis (Gm.): Ceram, Amboina.

To these points it may be added that in *C. galerita* of Group 1, and in *C. leadbeateri* of Group 2, the oil-gland has been found to be present; in *C. alba* of Group 3 it is generally absent, or reduced to a small membranous mamilla (Garrod 5, Nitzsch *b 3*). It is stated by Garrod that the oil-gland has not been found in *C. sulphurea*, but, according to Nitzsch, it is present.

It has been stated that Meckel found a right carotid artery of reduced calibre in *Cacatua sulphurea* (*4*, *5*); in a specimen of that species dissected by Garrod "the left only was present, as in *C. cristata* (= *alba*), *C. leadbeateri* and *C. galerita*" (*5*). In all other Parrots examined by Garrod the two carotids were present; and that author remarks that the genus *Cacatua*, like the *Passeres* and many others, has lost its right carotid, and in this respect has departed most from the ancestral type.

Judging from the fact that three of the five genera of *Cacatuidae* recognised by Count Salvadori are confined to Australia, and that the other two (though *Microglossus* is perhaps rather Papuan than Australian) also occur there, and that five of the fifteen species of *Cacatua* proper are peculiar to Australia and the other ten distributed interruptedly between Australia and the Philippines, it is evident that unless proof to the contrary is found Australia should be looked upon as the chief region of development of the family. Elsewhere no two species of the genus *Cacatua* are known in the same locality. It would appear that, by reason of competition, no island is able to harbour more than one species of a section of the genus *Cacatua*; and the suggestion, that one species has crowded out another of another section affords perhaps the most reasonable ground to account for their anomalous distribution.

Gould says that in Australia *C. sanguinea* and *C. galerita* are often seen in company. This, of course, primarily denotes that their food is the same, and that there is plenty of it; if there were not sufficient for both, one species would soon become impatient of the other's company.

As Krukenberg discovered, the yellow colour of the crest of *C. sulphurea* and of some other Cockatoos is due to the presence of pure psittacofulvin, the same pigment, which when laid over a darker ground pigment (fuscin) — not present in the Cockatoos, — gives the colour known as parrot-green. The yellow pigment was also found by him in the apparently white feathers of *C. sulphurea*, a circumstance which led him to suppose that the young feathers are yellow and lose the yellow effect as they become full grown, by reason of the distribution of the pigment over a wider area, and of the action of light (*11*); but this is not the case as is shown by sprouting feathers in skins before us. Light, however, certainly exercises a blanching effect upon them. The nestling of other species is clothed in long white down, as Finsch states: that of *C. sulphurea* is unknown. Prof. Marshall expresses the opinion that the delicate red and yellow tones of the plumage of Cockatoos is due to the fine epithelial dust cast off by the powder-down feathers — which are highly developed in Parrots — acting like a powdery pigment on the contour feathers (*14*). But this is not the case in *C. sulphurea*, where the powder-down feathers are white and the dust cast off of the same colour. The white Pigeons of the genus *Myristicivora* are saturated with a similar yellow, which soon fades after death on exposure to the light, leaving the birds white. Such is also the case with the salmon-colour with which the white breast of the Moleo is tinted.

Nothing is known of the breeding of *C. sulphurea* in freedom, but in captivity cases are on record of its laying two or three eggs at a sitting (Fraser *3*, Russ *a 3*).

FAMILY PSITTACIDAE.

"It is extremely difficult to define the *Psittacidae* by positive characters. They are separated from the *Stringopidae* by having a complete sternum, from the *Nestoridae*, *Loriidae*, and *Cyclopsittacidae* by the file-like surface on the palatine portion of the hook of the bill, and from the *Cacatuidae* by the absence of a crest (except in *Nymphicus*)" — Salvadori, Cat. B. XX, 137. The family ranges almost all over the warmer parts of the world.

Count Salvadori recognises 6 subfamilies of the *Psittacidae*, one of which, the *Palaeornithinae*, is represented by several genera in the Celebesian area. In this subfamily the tail is soft, sometimes very long and graduated, sometimes short and square, or wedge-shaped; the furcula is present (except in the genera *Agapornis* and *Psittacula*), the left carotid is normal, like the right one running in the hypoapophysial canal; orbital ring always incomplete; sides of the head feathered, if naked, only immediately round the eyes; bill moderate, or very strong, deeper than long (in *Loriculus* long and thin), smooth, mostly red, sometimes black, or yellow; sexes generally different. Australian, Oriental and Ethiopian Regions — cf. Salvadori, t. c. pp. 137, 386.

GENUS PRIONITURUS Wagl.

The Racket-tailed Parrots are easily recognisable when adult by the two middle tail-feathers, the shafts of which are prolonged beyond the others and tipped with a spatule of ordinary feather-construction; tail — except for the two middle feathers — square. The general colour is yellowish parrot-green; the under surface of the remiges where they rest upon the body verditer-blue, elsewhere black or blackish. Nine species and races have been described; they inhabit the Philippines, the Celebesian Province, and, possibly, Buru.

* 43. PRIONITURUS PLATURUS (Vieill.).
Blue-billed Racket Parrot.
Plates V and VI.

a. **Psittacus platurus** *(1)* "Temm.", Vieill., N. Dict. N.H. 1817, XXV, 314 (Nouvelle Calédonie!); *(2)* Russ, Fremdl. Stubenvög. 1881, 491.
b. **Psittacus setarius** *(1)* Temm., Pl. Col. 1824, 15 (Timor!); *(2)* Müll. & Schl., Verh. Natuurk. Comm. 1839—44, 90, 107, 182 (Buton); S. Müll., Reiz. Ind. Archip. 1857, II, 12.

Prionīturus platurus *(1)* Wagl., Mon. Psitt. 1832, 523; *(2)* Wall., Ibis 1860, 141; *(3)* Wald., Tr. Z. S. 1872, VIII, 32; *(4)* Salvad., Ann. Mus. Civ. Gen. 1875, VII, 645; *(5)* Brüggem., Abh. Ver. Bremen 1876, V, 39; *(6)* Lenz, J. f. O. 1877, 362; *(7)* Meyer, Ibis 1879, 49, 145; *(8)* Rchnw., J. f. O. 1881, 255 (Consp. Psitt. 1881, 143); *(IX)* id., Vogelbild. 1878—83, t. XXVII, f. 5; Meyer, Isis 1884, 6; *(X)* id., Vogelskel. 1884, t. LXVII; *(11)* Guillem., P. Z. S. 1885, 543; *(12)* W. Blas., Ztschr. ges. Orn. 1885, 212—219, figs.; *(13)* Platen, Gefied. Welt 1887, 219, 231; *(14)* W. Blas., Ornis 1888, 559; *(15)* Salvad., Cat. B. XX, 1891, 415; *(16)* M. & Wg., Abh. Mus. Dresd. 1895 Nr. 8, p. 5; *(17)* iid., ib. Nr. 9, p. 2; *(18)* iid., ib. 1896 Nr. 2, p. 8; *(19)* Grant, Ibis 1895, 466; *(20)* Hartert, Nov. Zool. 1896, 160.

c. **Psittacus spatuliger**, mas. *(I)* Bourjot, Perr. 1837—38, pl. 53.
d. **Prioniturus wallacei** Gld., MS.; *(1)* Gray, List Psitt. B. M. 1859, 18; *(2)* Schl., Dierent. 1864, 70.
e. **Prioniturus setarius** *(1)* Sclat., P. Z. S. 1860, 223, 226; *(II)* Gld., B. Asia, VI, pl. 12 (1862); *(3)* Wall., P. Z. S. 1864, 284, 293.
f. **Eclectus platurus** *(1)* Schl., Mus. P.-B. Psitt. 1864, 45; *(2)* id., Rev. Psitt. 1874, 22.
g. **Pionias platurus** *(1)* Finsch, Papag. 1868, II, 395.

"**Kring-kring**", Minahassa, Rosenb. *g 1.*
"**Ili-ili**", Gorontalo, Rosenb. *g 1.*
"**Kulli-kulli**" or "**Kelik-kelik**" (Nat. Coll.), Minahassa, Meyer 7.
"**Cacatua birotti**", Malay name, N. Celebes, Meyer 7; Guillem. *11.*
"**Bawan buzar**", Tjamba, S. Celebes, Platen *12.*
"**Kelean**", Tonkean, E. Celebes, Nat. Coll.
"**Tulik**", Peling and Banggai, Nat. Coll.
"**Urili**", Talaut Islands, Nat. Coll.
"**Kulili**", Siao, Nat. Coll.

For further synonymy and references see *Salvadori 15.*

Figures and descriptions. Gould *e 2;* Temminck *b 1;* Bourjot *c 1;* Reichenow *IX;* Meyer *X* (skeleton); Sclater *e 1;* Finsch *g 1;* Salvadori *4, 15;* Brüggemann *5;* W. Blasius *12.*

Adult male. Head, neck and under surface yellowish parrot-green, the yellow predominant on the under tail-coverts; on crown a spot of poppy-red; occiput lilac; feathers of lower neck above interscapular region tipped with deep orange, forming a collar; wing-coverts greyish pea-green, the lesser series lilac; rest of upper surface dull parrot-green, almost grey-green on mantle and scapulars, tertiaries margined with yellow, lower back touched up with lilac; tail-feathers except the 2 middle ones dark blue for the terminal 2,5 cm; the two middle tail-feathers green, the shafts produced naked for 40—55 mm beyond the rest of the tail, but having both webs left at the tip, forming a small pear-shaped spatule dark blue, green towards its base; quills and tail below verditer blue, changing to beryl-green according to the light, the former black near the shafts, which are themselves black (Minahassa — Nr. 1942). Iris dark brown; feet greyish blue, claws grey; bill bluish grey or whitish (N. Celebes — Meyer 7): iris brown; cere black; bill horn-grey black; feet lead-grey (Tjamba — Platen *12).*

Adult female. Above entirely dark parrot-green; below yellow-green, inclining to orange on the tips of the under tail-coverts; tail as in the male, but the bare shafts (21 mm) of the rackets shorter (Nr. 3983).

Young male (of 2nd year?). Wants the red spot on the crown, the yellow-green of the hind-neck and orange collar of the adult male; the two middle tail-feathers only about

15 mm longer than the others, the webs of the projecting portion much narrowed, but not formed into rackets. This specimen is an abnormal one, apparently; on the left side of the head, the lores, superciliary region, ear-coverts and cheeks are lilac, like the occiput, on the right side the same parts are parrot-green; some of the lilac colour also appears on the left side of the lower neck, but not on the right (Manado — Nr. 6089). In a nearly adult specimen (Nr. 6088) a little lilac appears on the right ear-coverts, but not on the left.

Young (of 1st year ?). Like female, but the two middle tail-feathers only projecting as much narrowed tips about 10 mm beyond the others; the blue terminal portion of the other tail-feathers less broad than in the adult; rump, upper tail-coverts and the basal part of the tail-feathers on the right side strongly washed with greyish blue (N. Celebes — Nr. 6090). Feet blue-grey (Platen *12*).

These 2 young dresses may perhaps point to a tendency in young birds to develop blue tints.

Variation. We have paid much attention to the geographical variation of this species, with the usual result in such cases — that of discovering how little we really know about it. The bird certainly varies locally to a considerable extent, but its division into subspecies should be left to the future and not undertaken by the uninitiated, among whom we include ourselves. In certain localities it is possible to point to some differences:

1. Mainland of Celebes.

In adult males the occipital patch is lilac or light grey-blue, the broad green cervical collar is strongly stained with orange on the mantle, the racket-feathers (differing greatly with age) are long-shafted and comparatively large-headed; lesser wing-coverts at the carpal bend washed with the lilac of the occiput. Measurements: wing 170—192 mm[1]); tarsus 19; tail (without the rackets) 80—100; culmen from core 21—23.

2. Peling and Banggai Islands.

Adult males from these islands have the patch on the occiput and carpal bend of a somewhat brighter and clearer blue (less grey) than on the mainland of Celebes; the lower half of the cervical collar is deep orange; the rackets are small, the shafts thereof comparatively short, the size is small: wing in 5 adult males 167—176 mm.

3. Eastern Peninsula, Celebes.

Two adult males from here take a somewhat intermediate position between the Peling form and that of N. & S. Celebes, though on the whole being more like the latter. Their size is rather small: wing 166, 171 mm.

4. Lembeh Island.

Two adult males from this island, close to the coast of N. Celebes, have the occipital patch and carpal bend somewhat brighter blue than in our mainland birds, and the collar above the mantle purer orange. The orange part of the collar is less bright than in the Peling form, the back is greener and less grey, the rackets are larger. Wing 176, 186 mm.

---

[1]) This represents the extremes of size in 24 specimens. The females do not appear to be smaller than the males.

5. Talaut Islands.

Adult males from this group have the patch on the occiput and carpal bend light bright blue (Cambridge or flax-flower blue — Ridgway IX, 14), purer even than in the Peling form, and this tint is a good deal more extended on the lesser wing-coverts near the carpus; cervical collar above the mantle deep orange; size rather large: wing in 14 examples of different sex and ages 177—194 mm.

6. Sangi.

The locality is recorded by Brüggemann *(5)* alone; he describes the collar as purer green, with the yellow part over the mantle brighter than in birds from Celebes, the size large: wing ca. 190 mm.

7. Siao.

Schlegel *(f 2)* considered the collar above the mantle to be brighter and the spot on the crown broader[1] than in Celebesian males. There are two males in the Leyden Museum from Siao, examined by one of us; of these one certainly has a lighter and brighter collar than males from Celebes, the other — perhaps not fully coloured — is like the latter. Three young specimens from Siao in the Dresden Museum measure: wing 172—187 mm.

Skeleton.

Length of cranium	50.0 mm		Length of tibia	51.0 mm
Greatest breadth of cranium	28.0 »		Length of tarso-metatarsus	19.0 »
Length of humerus	42.0 »		Length of sternum	48.0 »
Length of ulna	51.0 »		Greatest breadth of sternum	27.5 »
Length of radius	50.0 »		Height of crista sterni	16.0 »
Length of manus	54.0 »		Length of pelvis	54.0 »
Length of femur	36.0 »		Greatest breadth of pelvis	28.0 »

Distribution. Celebes and the neighbouring islands: Minahassa (Wallace *2, e 3, 15*, Meyer *7*, etc.); Gorontalo District (Rosenb. *f 2*, v. Duivenb. *f 2*, Meyer *7*); Togian Is. (Meyer *7*); Lembeh Islands (Nat. Coll.); Kandari, S. E. Celebes (Beccari *4*); Buton S. Müll. *b 2, f 1*); S. Peninsula (Wallace *2, e 3, 15*, Platen *12*, etc.). — Siao, Sangi Is. (v. Duivenb. *f 2*, Nat. Coll.); (?) Sangi Islands (Fischer *5, 14*); Talaut — Karkellang Id. (Nat. Coll.); (?) Buru (Hoedt *f 2*); Peling and Banggai Is. (Nat. Coll.).

Schlegel records a young female of this species killed by Hoedt in the Bay of Bara on the North-east coast of Buru; we are unable to form an opinion as to this occurrence.

The Racket-tailed Parrots of the genus *Prioniturus* range from the Philippines, including Palawan and Sooloo to Celebes and possibly Buru. Nine distinct forms have been described, two or three of them being local races and only worthy of subspecific distinction. *P. platurus* may fairly be regarded as the most highly differentiated of the group. The adult male excels its compatriot in Celebes, *P. flavicans*, by having two patches of colour, a red and a lilac one on its green head, whereas the male *P. flavicans* has a blue head

---

[1] Varies with age, and the shape thereof depends much upon the manner of preparation of the skin.

above with a red patch thereon. *P. verticalis* Sharpe (Ibis 1894, 248, pl. VI), discovered by Everett in the Sooloo Islands, and *P. montanus* Grant (Ibis 1895, 466), discovered recently by Whitehead in the mountains of Luzon, are allies of *P. flavicans*. We take *P. platurus* for a highly-differentiated form of *P. luconensis* of Luzon, and *P. flavicans* as an offshoot of the other Philippine form represented by *P. discurus*. The young birds are all green, the adult *P. luconensis* is also all green, and it is therefore, we believe, of a more ancestral character than the adult male *P. platurus*.

The young of this species resemble the young of *flavicans* in coloration, but they may be readily distinguished by certain differences between the two forms pointed out by Sclater *(e 1)*, Brüggemann *(5)* and W. Blasius *(12)*:

	Prioniturus platurus	*Prioniturus flavicans*
Under tail-coverts:	Elongated; in adult as long as the rectrices, in young a little shorter.	Comparatively short.
Cere round nostrils:	Naked	Feathered.
Under bill: . . .	With a strong indentation on either side near the end, the cutting edge at the end projecting upwards.	The cutting edge much more even.
Colour of bill: .	Horn-greyish blue	Horn-white. (Platen).

Prof. W. Blasius adds that *P. platurus* has a shorter but stouter upper bill, as well as its being of a different colour, but this statement we are unable to confirm from the specimens in the Dresden Museum, in which the upper mandible varies much in size. *P. platurus* is altogether rather the smaller bird.

The Racket-tail-feathers of *Prioniturus, Merops*, etc. present facts of interest to students of evolution, and they will be found discussed in our Introduction. (See figures thereof plates V, VI and VIII.)

According to Meyer *P. platurus* is solely an inhabitant of the low-lands and *P. flavicans* of the mountains of Celebes; but this view must be modified since Dr. Platen obtained both species in Rurukan at over 3000' (Coll. Nehrk.) and the Drs. Sarasin both at Tomohon, *P. platurus* in April, *P. flavicans* at the end of May. They may, however, shift their ranges at different seasons.

"The bird flies much during the night, and can often be heard crying on the wing over one's head. It feeds in the night on the fruits of gardens and fields, and is fond of Indian corn, rice, and fruits like 'lansa' (*Lansium domesticum* Jack.), 'pakawa' (?), etc. During the daytime it is seldom to be met with in the plantations, but is to be seen flying very high and crying loud, seldom alone. It makes its nest in hollow trees. On trees it does not move much, but sits quietly. If one is shot down from a group the others do not stir, but lie, concealed by their green plumage, between the leaves, just as I have noticed in the case of other Parrots. The natives of the Minahassa assert that, if the 'kulli-kulli' is taken by surprise in the rice-fields, it becomes confused, or terrified, falls down, and then can easily be caught" (Meyer 7, and in Gould's Bds. of

N. Guinea pt. VI, 1878). Specimens in captivity in Europe were unknown to Dr. Russ *(a 2)*, and Dr. Platen failed to obtain living specimens *(13)*. S. Müller saw it in captivity in Buton *(b 2)* and Meyer once had a tame specimen at Manado, but it appeared to be very unhappy in its cage *(7)*.

* 44. PRIONITURUS FLAVICANS Cass.
White-billed Racket Parrot.
Plate VI.

Prioniturus flavicans *(1)* Cass., Pr. Acad. N. Sc. Philad. 1853, VI, 73; *(2)* Sclat., P. Z. S. 1860, 229, 226; *(III)* Gould, B. Asia VI, pl. 13 (1862); *(4)* Wall., P. Z. S. 1864, 284, 293; *(5)* Brüggem., Abh. Ver. Bremen 1876, V, 40; *(6)* Lenz, J. f. O. 1877, 363; *(7)* Meyer, Ibis 1879, 51, 145; *(8)* Rchnw., J. f. O. 1881, 255 (Consp. Psitt. 143); *(IX)* id., Vogelb. 1878—83, t. XXVII, f. 4; *(10)* Guillem., P. Z. S. 1885, 543; *(11)* W. Blas., Ztschr. ges. Orn. 1885, 212, 213—218; id., ib. 1886, 83; *(12)* Platen, Gefied. Welt 1887, 219, 231; *(13)* W. Blas., Ornis 1888, 560; *(14)* Salvad., Cat. B. XX, 1891, 416; *(15)* Sharpe, Ibis 1894, 248; *(16)* Grant, Ibis 1895, 466; *(17)* M. & Wg., Abh. Mus. Dresd. 1895, Nr. 8, p. 5.

a. **Prioniturus discurus** *(1)* Wall. (nec. V.), Ibis 1860, 141.
b. **Eclectus flavicans** *(1)* Schl., Mus. P.-B. Psitt. 1864, 45; *(2)* id., Rev. Psitt. 1874, 23.
c. **Pionias flavicans** *(1)* Finsch, Papag. 1868, II, 399.

"**Kulli-Kulli**" and "**Cacutua birotti**" (Malay), Minahassa: — not distinguished by the natives from *P. platurus* (Meyer 7).

For further synonymy and references see Salvadori *14*.

Figures and descriptions. Gould *III*; Reichenow *IX*, *8*; Sclater *2*; Brüggemann *5*; Lenz *6*; Schlegel *b 1*; Finsch *c 1*; W. Blasius *11*; Salvadori *14*.

Adult male. Parrot-green, darkest on the wings, lighter on the sides of head and face; breast ochraceous green washed with golden, deeper and brighter on the hind neck, forming a sort of collar; pileum coeruleam blue; in the middle of it on the crown a large patch of poppy-red; abdomen, flanks and under tail-coverts yellower green than the back; under wing-coverts bright green; the two middle tail-feathers above green, bearing long rackets, with black spatules; the other tail-feathers green, black at the ends and on the inner webs; tail below glossy verditer blue, changing with the light to a brighter blue; wing below also verditer blue on the inner webs, black at the ends and near the shafts of the feathers; first primary edged with bluish green on the outer web above (C 11008 — Minahassa, Aug.—Sept.). "Iris grey-brown; bill horn-white; feet blue-grey" (Platen in Mus. Nehrkorn, Nr. 892).

Adult female. Like the male, but without the red spot on the crown, and the blue of the pileum less extended; the racket-feathers equally long (length of naked shaft with spatule 86 mm). (Minahassa, Aug.—Sept. — C 11009). Feet ash-grey (Platen in Mus. Nehrk.).

Young male (in second year?). Like the female, but the blue of the head less extended; one or two red feathers on the crown; the racket-feathers not yet properly formed, only 25 mm longer than the other rectrices, and the shaft narrowly webbed (Nr. 6085).

Younger. Only a little blue on head, the blue-tipped feathers with a few green ones intermixed (Nr. 6086).

First plumage. The young ones are quite green (Meyer 7).
Measurements. Wing 183—195 mm (14 specimens); tail 95—104 (without the rackets); tarsus 19; culmen from cere 22½—25.
Distribution. North Celebes, the Togian Islands and (?) the Sangi Islands. — Minahassa (Forsten 2, b 1, Wallace a 1, Meyer 7, etc., Platen 12); Banka Id. (Nat. Coll. in Dresd. Mus.); Lembeh Id. (Nat. Coll.); Gorontalo District — Limbotto (Meyer 7); Gorontalo (Riedel in Dresd. Mus.); Togian Islands (Meyer 7); (?) Sangi Islands (Brüggem. 5).
Observation. Brüggemann (5) received a single specimen, marked "Sangir ♂", amongst 31 specimens from Dr. Fischer and Dr. Riedel. It presented some slight differences of coloration, though, as the author remarked, these might have been due to immaturity. Further evidence should be obtained before the Sangi Islands be admitted as a habitat of *P. flavicans* or of a local form of it. It would be a matter of no surprise if these islands or at least some of them should be found to harbour races of both *P. flavicans* and *platurus*.

That *P. flavicans* on the mainland has only been found in the northern Peninsula up to the present is remarkable.

This Racket-tailed Parrot is, to our mind, by no means so highly differentiated a species as *P. platurus*, and the nearest affinities of it and its near allies, *P. verticalis* Sharpe (Bull. B. O. C. 1893, X, Nov. 28) of the Sooloo Islands (similarly marked, but differing in the shades of red and blue on head and greenish on under surface) and *P. montanus* Grant of Luzon are not difficult to find. The female and immature male of *P. flavicans* differ from the adult *P. discurus* (Vieill.) — both male and female — of the Philippines in little, except that the Celebesian form is a good deal larger, the blue of the crown less pale, and the underside of the quills and tail paler and greener. The adult male of *Prioniturus flavicans* has the addition of a poppy-red patch on the crown of the head; it is therefore most likely that *P. flavicans* is derived from a form similar to *P. discurus* — or, rather, its ancestor — the adult male of which in Celebes has become larger and obtained the addition of a red patch on the head. This view appears to us more plausible than the converse hypothesis, that *P. discurus* is descended from *P. flavicans*, and that the adult male of it does not attain to its full development in the Philippines.

The genus *Prioniturus* has a very restricted range over the Philippines and the Celebes group exclusive as far as is known of Sula, though it may occur on Buru and in this case no doubt also on Sula.

The known forms from the Philippines, except perhaps *P. verticalis* and *montanus*, have a simpler, less specialized appearance than the two Celebesian ones. Over the greater part of the former occurs *P. discurus* (V.) and on Luzon *P. luconensis* Steere and *montanus* also, whereas Mindoro, Palawan and Sooloo have forms and species of their own: *P. mindorensis* Steere, *P. cyaniceps* Sharpe, *P. suluensis* W. Blas. and *P. verticalis* Sharpe, the two last on different islands of the Sooloo group. A better knowledge will perhaps prove that there are further differences amongst the specimens from the different islands of the Philippine group.

GENUS TANYGNATHUS Wagl.

This genus is always distinguishable from *Prioniturus* by its strongly graduated, almost wedge-shaped tail, as well as by its larger size and larger, broader bill. The colour is chiefly yellowish parrot-green, relieved on some of the upper parts by blue, occasionally by yellow and black, but no red, except on the bill, where it is found as a rule, though sometimes in the male alone. Wing below dusky, more shiny where it rests upon the body. There are four well distinguished species, of which some seven additional local forms have been described. Range: from the Philippines to Celebes and Sumba, through the Moluccas as far as N.W. New Guinea and the Tenimber Islands.

* 45. TANYGNATHUS MUELLERI (Müll. Schl.).
Celebesian Green Parrot.

This Parrot varies geographically in size. In Sangi, following a general rule there, the birds are very large, always exceeding the dimensions of those of the mainland of Celebes. In Talaut they are somewhat smaller, small individuals from there being occasionally, though very rarely, equalled or exceeded in size by large ones of Celebes, while large individuals are similar to Sangi birds. We prefer to group and label the forms as follows:

+ 1. The typical Tanygnathus muelleri.

a. ? **Psittacus sumatranus** *(1)* Raffl., Tr. Z. S. 1822, XIII, 281.
b. **Psittacus muelleri** Temm. in Mus. Leyd.; *(1)* M. et S., Verh. Natuurk. Comm. 1839—44, 108, 182 (Buton); S. Müll., Reiz. Ind. Archip. 1858, II, 69, 70; *(2)* Russ, Fremdl. Stubenvög. 1881, 452.
c. **Tanygnathus sumatranus** *(1)* Souancé, Icon. Perr. 1857, pl. 46; *(2)* Brüggem., Abh. Ver. Bremen 1876, V, 38.
d. **Tanygnathus muelleri** *(1)* Bp., Consp. Av. 1850, I, 5; *(II)* Souancé, Icon. Perr. 1857, pl. 45; *(3)* Wall., P. Z. S. 1864, 286, 294; *(4)* Wald., Tr. Z. S. 1872, VIII, 31 (pt. excl. Samar, Sangi); *(5)* Garrod, P. Z. S. 1874, 587; *(6)* Salvad., Ann. Mus. Civ. Gen. 1875, VII, 644; *(7)* Meyer, Ibis 1879, 47, 145; *(8)* Salvad., Orn. Pap. 1880, I, 135; *(9)* Krukenb., Vergl. physiol. Studien 1882, 213; *(10)* Rchnw., J. f. O. 1881, 245; Consp. Psitt. 1881, 133; *(XI)* id., Vogelb. 1878—83, t. XXVII, f. 9; *(XII)* Meyer, Vogelskel. 1883, t. XLV; *(13)* Guillem., P. Z. S. 1885, 542; *(14)* W. Blas., Ztschr. ges. Orn. 1885, 209; *(15)* id., Ornis, 1888, IV, 556; *(16)* Hickson, Nat. in N. Celebes 1889, 59, 86 (Talisse); *(17)* Salvad., Cat. B. XX, 1891, 430; *(18)* Büttik., Zool. Erg. Weber's Reise Ost-Ind. 1893, III, 273; *(19)* M. & Wg., Abh. Mus. Dresden 1895, Nr. 8, p. 5; *(20)* iid., ib. 1896, Nr. 2, p. 8; *(21)* Hartert, Nov. Zool. 1896, 160.
e. ? **Psittacus leucorhynchus** *(1)* Reinw., Reiz. Ind. Archip. in 1821, 1858, 592 (Tondano).
f. **Tanygnathus albirostris** *(1)* Wall., P. Z. S. 1862, 336 (Celebes, Sula); 1864, 286, 294; *(2)* Sclat., P. Z. S. 1868, 262; 1871, 494; *(3)* Rchnw., J. f. O. 1881, 246; Consp. Psitt. 1881, 134; *(4)* Dallwitz, J. f. O. 1885, 103; *(5)* Guillem., P. Z. S. 1885, 543.

g. **Electus muelleri** *(1)* Schl., Mus. P.-B. Psitt. 1864, 48; *(2)* id., N. T. D. 1865, III, 185, pt.; *(3)* Finsch, Papag. 1868, II, 357; *(4)* id. et Conrad, Verh. z.-b. Ges. Wien 1873, 2, 14 (sep. copy); *(5)* Schl., Rev. Psitt. 1874, 25; *(6)* Lenz, J. f. O. 1877, 361; *(8)* Joest, Das Holontalo 1883, 106.

"**Olia**", Gorontalo, Joest *g 8*.
"**Cacatua idiu**", Malay name, Meyer 7.
"**Danga**", Tjamba, S. Celebes, Platen *14*.
"**Kaleak**", Natives near Manado, Nat. Coll. in Mus. Dresd.
"**Keja**", "**Kejak**" or "**Keak**", E. Celebes, Peling, and Banggai, Nat. Coll.

For further synonymy and references see Salvadori *d 17*.

Figures and descriptions. Souancé *d II, c I*; Reichenow *d XI*; Meyer *d XII* (skeleton); Finsch *g 8*; Brüggemann *c 2*; Salvadori *d 6, d 17*; W. Blasius *d 14*.

Adult male. Head, neck, mantle, scapulars, upper tail-coverts, and entire under parts yellow-green, brighter on the head; lower back and rump bright turquoise-blue; wings parrot-green, the lesser coverts and some of the inner median coverts margined broadly with blue, darker than that of the rump, the other wing-coverts and tertiaries margined with yellow-green; tail above parrot-green, yellowish at the tip; tail below golden olivaceous, shafts below white, above blackish; quills below shining brownish mouse-grey (Tjamba, S. Celebes, April 1878 — C 10423). "Iris light yellow; cere brownish; bill coral-red, point lighter; feet grey-brown (Platen). Another old male (Mus. Nehrkorn) labelled by Dr. Platen, Rurukan, has: "bill sealing-wax-red (lackrot); cere grey-black; feet yellow-brown."

Adult female. Like the male, but duller and greener; mantle and scapulars parrot-green like the wings; little or no trace of blue tips to the lesser wing-coverts (Limbotto — C 1904). "Iris light yellow; bill horn-white; cere grey-black; feet yellow-brown" ($\female$ Rurukan, Platen in Mus. Nehrkorn).

The female may be at once distinguished from the old male by its green mantle and scapulars, and white bill, though in some females (perhaps very old ones) this may assume a reddish colour.

Young male. Head duller darker green than in the female, otherwise like it; lower back and rump duller blue than in the old male. "Iris light yellow; cere brownish; bill light red; feet brown-grey" (Tjamba, 25th April, 1878 — Platen C 11206).

Measurements. Wing (34 specimens) 196—212 mm; tail 121, 137 (largest- and shortest-winged specimens); culmen from cere 30—34½; tarsus 19—21. Females as a rule are a trifle smaller than old males, and the bill is smaller.

Skeleton.

Length of cranium	62.0 mm	Length of tibia	56.0 mm
Greatest breadth of cranium	34.4 »	Length of tarso-metatarsus	19.0 »
Length of humerus	46.6 »	Length of sternum	50.2 »
Length of ulna	55.5 »	Greatest breadth of sternum	32.2 »
Length of radius	52.0 »	Height of crista sterni	15.7 »
Length of manus	59.0 »	Length of pelvis	60.7 »
Length of femur	39.6 »	Greatest breadth of pelvis	28.7 »

Distribution. Buton (S. Müller *b 1*); S. W. Peninsula — Macassar (Wall. *d 3, d 17*, Conrad *g 4*), from Maros to Tanette (Meyer *d 7*, Weber *d 18*), Tjamba (Platen *d 14*); N. Peninsula — Minahassa (Forsten *g 1*, Wall. *d 3, d 17*, Rosenb. *g 5, etc.*), Gorontalo District (Rosenb. *g 5*, Meyer *d 7*), Lembeh Id. (Meyer), Talissi Id. (Guillem. *d 13*, Hickson *d 16*); Banka, Manado tua, and Mantehage Is. (Nat. Coll.); Togian Islands (Meyer *d 7*); S. shore of Gulf of Tomini — Posso and Todjo (Meyer *d 7*);

East Celebes (Nat. Coll.); Peling and Banggai (Nat. Coll.); Sula Islands — Sula Besi and Sula Mangoli (Allen *f 1, d 17*, Bernst. *g 5*, Hoedt *g 5*).

Observation. We are unable to find that this subspecies differs in North and South Celebes and Buton, and the Islands Banka, Manado tua and Mantehage off the Minahassa, the size is equal in all parts of the island; there is considerable difference in coloration between the old male and the female or young male, and old male specimens vary amongst themselves in the same locality (Manado), having the blue tint of the rump lighter or darker, etc.

† 2. Tanygnathus muelleri sangirensis M. & Wg.

h. **Eclectus muelleri**, partim, *(1)* Schl., N. T. D. 1865, III, 185.
i. **Tanygnathus muelleri**, partim, *(1)* Wald., Tr. Z. S. 1872, VIII, 31; *(2)* Salvad., Ann. Mus. Civ. Gen. 1876, IX, 53; *(3)* Rchnw., J. f. O. 1881, 245 (Consp. Psitt. 133); *(4)* Meyer, Isis 1884, 6; *(5)* W. Blas., Ornis 1888, 556; *(6)* Hickson, Nat. in N. Celebes 1889, 155; *(7)* Salvad., Cat. B. XX 1891, 431 (Sangi).
j. **Tanygnathus muelleri sangirensis** *(1)* M. & Wg., J. f. O. 1894, 113.

"**Kakatua**", Great Sangi, Nat. Coll. in Dresd. Mus.

Diagnosis. Like the *typical T. muelleri*, but much larger: wing 226—235 mm as against 205, the average size in Celebes.

Measurements.	Wing	Tail	Tarsus	Culmen from cere
a. (C 12696) ad. [♂] Gr. Sangi, 21. VII. 93	226	143	23	33
b. (C 1184) imm. ♂ Gr. Sangi	235	145	24	35.5
c. (C 12605) [♀] Gr. Sangi 25. VII. 93	231	137	—	31
d. (Mus. Tring) [♀] Gr. Sangi 14. VII. 93	227	130	—	29.5
e. (C 1180) juv. Gr. Sangi	230	134	22	31

Distribution. Great Sangi (Meyer, Nat. Coll. and Leyd. Mus.).

3. Tanygnathus muelleri — sangirensis.

k. **Tanygnathus muelleri sangirensis** *(1)* M. & Wg., J. f. O. 1894, 239; *(2)* lid., Abh. Mus. Dresd. 1895 Nr. 9, p. 2.

"**Areaā**" or "**Area uwawi**", Talaut, Nat. Coll.

Diagnosis. Intermediate between the Celebes and Sangi forms.

Measurements. Wing (19 examples) 210—227 mm; tail c. 140 mm; culmen from cere c. 32 mm.

Distribution. Talaut Islands: Kabruang and Karkellang (Nat. Coll.).

Observation. There is, as a rule, more blue on the wing-coverts about the bend of the wing in Talaut and Sangi birds than in those of Celebes, but this character varies much in different individuals and with age.

It is curious that no form of *T. muelleri* has been found on Siao or the other islands between Celebes and Sangi. It can hardly be that the islands are too small to support two forms of *Tanygnathus*, *T. megalorhynchus* being there already, for Kabruang and Karkellang in the Talaut group are inhabited by three species of this genus, and the former island seems to be but little larger than Siao.

Müller's Green Parrot has two very close allies in *T. everetti*, Tweedd. of Samar, Panay and Mindanao, and *T. burbidgei* Sharpe of the Sooloo Islands.

These three species form a group, which stands nearer to *T. luconensis* (L.) of the Philippines than to the other members of the genus. *T. luconensis* differs chiefly by having the head above blue when adult. At first sight it appears as if a certain interest attaches to the fact that now and then young birds of *T. muelleri* show traces of a blue head; we have found this in a specimen from Sangi, in one from East Celebes, in one from Banggai. It may be a hint that the blue-headed form represents an earlier stage in the development of *T. muelleri*. But the young of the blue-headed *T. luconensis* has the head green and the bill red like the adult; it therefore seems to suggest that a form like *T. muelleri* was an early stage in the development of *T. luconensis!* It is just possible that two stages in the phylogenetic history of the race is betrayed by the young of the two species, but it seems quite unsafe as yet even to attempt to interpret the meaning of the facts.

The question of the significance of the red or white bill in this species — points which have led to its being split into two — must, we think, undoubtedly be answered according to the conclusion of Prof. W. Blasius — a conclusion at which indeed we had independently arrived from over 60 specimens examined — that the white-billed individuals are females or young males, the red-billed individuals are old males *(d 14)*. Occasionally females, as Brüggemann says, acquire a reddish bill *(c 2)*. In 1871, in order to solve the question whether the white- and red-billed Parrots were one and the same species or different ones, Meyer procured a very large series of specimens and came to the conclusion that the bill of the young bird is white, and gradually assumes a red colour as the bird grows older, and deep red with age; but it was not then known — as, indeed, it seems to have escaped ornithologists generally up to the present — that the female closely resembles the young. Dr. Hickson arrived at a somewhat different conclusion: "My boys and I shot a great many of these birds, partly to settle this vexed question and partly for food, and I found without exception that those with scarlet bills were males and those with white bills were females" *(d 16)*. We do not know whether Dr. Hickson was aware that the young male is like the female in coloration, but his observations taken together with those of Meyer and others of Dr. Platen made on the spot *(b 2)*, and of the Drs. Sarasin, and with the fact that in Zoological Gardens two or more cases are known of examples, which arrived with a white bill, but acquired a red bill subsequently (see Finsch *g 3*, W. Blasius *d 14*, Salvadori *d 17*), all serve to prove that females and young males have white bills (sometimes reddish), and old males red bills[1]).

Little is known of the habits of this Parrot. A living one, which was in

---
[1]) A series of 37 specimens, collected in Aug.-Sept. 1892 for the Dresden Museum (though not sexed) vary as follows: b with the yellow-green scapulars and mantle of old males have the bill very deep crimson; 2 or 3 others with the yellow-green colour less strongly developed (younger males) have the bill red; 2 others, getting yellowish green on the shoulders, have the bill white (immature males); the remainder have generally a parrot-green back and white bill (females and young males).

Dr. Meyer's possession at Manado, fed on rice and bananas, and was generally unintelligent, idle, quiet, or grumbling *(d 7)*. In North Celebes the species is very common, and Dr. Hickson speaks of their resorting to the lower branches of the trees towards sunset, and keeping up a constant chattering noise until past midnight. Does it feed in the night like *Prioniturus platurus*? In South Celebes (Tjamba) it appears to be much rarer, as Dr. Platen's graphic description shows: "When one is on the watch in the woods of South Celebes, or passing through them as noiselessly as possible, Müller's Parrot, anxiously avoiding open places and keeping by preference in the darkness of the wood, may often be seen, usually alone, more rarely in twos or threes, as it flies past with its heavy flight and disappears apparently without leaving a trace in the crown of a thick-leaved tree. On remaining quiet and keeping the spot in sight, it is noticed, often not until after a lapse of half an hour or more, that the birds, which during this time have been sitting perfectly motionless and gazing at us, at last become lively and go about their daily duties, without however letting a sound be heard. In freedom they make the same quiet and serious impression as in captivity, never excite the attention of the hunter, but on the other hand contrive through motionless behaviour to be so deceptive that the eye passes them over without notice, even when such a bird is sitting openly on a bough. A single time my Malay hunter informed me as something remarkable, that he had seen a flight of six examples of this Parrot. As to its breeding, I could unfortunately learn nothing, for the nesting places are all to be found amongst inaccessible steep cliffs, and besides the natives show no interest whatever in capturing and rearing such birds" *(b 2)*.

†46. ? **TANYGNATHUS LUCONENSIS** (L.).

Blue-headed Green Parrot.

a. **Psittacus luconensis** Linn., S. N. 1766, I, 146 (ex Brisson).
b. **Eclectus luconensis** *(1)* Finsch, Papag. 1868, II, 362.
c. **Tanygnathus luzonensis** *(1)* Brüggem., Abh. Ver. Bremen, 1876, V, 38; *(2)* Koch, Verz. Vogelb. aus Cel. u. Sanghir, Febr. 1876, 1; *(3)* Salvad., Ann. Mus. Civ. Gen. 1876, IX, 53.
Tanygnathus luconensis *(1)* Sclat., P. Z. S. 1871, 479; *(2)* Meyer, Isis 1884, 6; *(3)* Salvad., Cat. B. XX, 1891, 424.
d. **Tanygnathus luzionensis** *(1)* W. Blas., Ornis 1888, IV, 559.
For synonymy and references see Salvadori *3*.
Adult male. Parrot-green, yellower below and on hind neck, mantle and upper tail-coverts; crown and occiput light cerulean-blue; feathers of the lower back tipped with paler (turquoise-) blue; wing-coverts blue, broadly bordered with dark golden ochraceous, the greater series green, with terminal part blue and light green and yellow-brown margins, the primary coverts without these margins, the least series (bend of wing) black, tipped with blue and golden ochraceous; wings below shining dusky; tail below yellow shaded with olive (Bataan, Luzon — Nr. 3993). Wing 190, tail 124; tarsus 19; culmen from cere 31 mm.

Young male. Like the adult, but with only a trace of blue on the head; the mantle parrot-green; the back darker blue, which colour is much more extended; the wing-coverts dark parrot-green, — darkest at the edge of the wing, — bordered with dark golden-ochraceous (Luzon — Nr. 3988).

Distribution. "Philippine Islands including Palawan and Mantanani (small island to the North-west of Borneo) and also the Sooloo Islands" (Salvad. c 3); Sangi (Fischer 1).

Observation. This species is included in the avifauna of the Celebesian province in virtue of six specimens in the Darmstadt Museum labelled by Dr. Fischer as coming from the Sangi Islands — most likely Great Sangi — and recorded by Brüggemann. Like Prof. W. Blasius we see no reason why the indication Sangi should be considered as possibly erroneous, though confirmatory evidence of the occurrence of this widely-spread species, or a local race of it, in Great Sangi — where it was not obtained by Hoedt, v. Rosenberg, van Duivenbode, Meyer, Bruijn, nor Platen — would be welcome.

This species varies a good deal in size, specimens from the Sooloo Islands and Talaut being the largest yet recorded. They also vary in coloration: the two specimens described above from Luzon — especially the young one — have the blue on the back well developed; in three from Cebu (adult and young) this part is parrot-green without any blue; one from Palawan (♂ juv.) has only a slight trace of blue; in two from Mindanao (♂, ♀, immature) it is better developed, though not so strongly as in the Luzon specimens, from which they further differ slightly in other points. Brüggemann's Sangi specimens have "the hind part of head scarcely washed with bluish; rump sea-blue; lesser wing-coverts blackish green with light green borders, the greater ones dark green bordered with greenish yellow. Wing 187—195; tail 119—127 mm. From the above series (of six), which contains different degrees of age, it may be seen that the blue on the head increases in intensity and extent with age, while the same colour on the rump disappears more and more, and at last (through attrition of the feathers [!]) becomes quite lost" (1).

The Sangi birds seem from the description to correspond with those of Talaut, as far as the intense blue of the head and pure green back of fully adult birds is concerned.

Only a large series of specimens from all the different islands can prove whether these differences of coloration are bound to the locality.

⌁ * 47. TANYGNATHUS TALAUTENSIS M. & Wg.

Talautese Blue-headed Green Parrot.

a. **Tanygnathus luzonensis** (nec Linn.) *(1)* M. & Wg., J. f. O. 1894, 239.
Tanygnathus talautensis *(1)* M. & Wg., Abh. Mus. Dresden 1895, Nr. 9, p. 2.
"Area rusipang", Karkellang, Nat. Coll.

Diagnosis. Similar to *T. luconensis* but larger, the head above and ear-coverts cerulean-blue, sharply cut off from the olivaceous yellow of the neck, becoming green on forehead and loral region; malar region washed with blue (ad. Karkellang — C 13766, type of species; and others).

Young. The occiput only washed with blue; the wing-coverts green, scarcely any blue showing, the green-yellow edgings lighter than in the adult, the carpal region green, not black (Karkellang — C 15265).

Measurements. (20 specimens not including young ones) wing 202—222 mm, average 210—215; tail 115—142; bill from core c. 33.5—39; tarsus c. 20.

Distribution. Talaut Islands — Kabruang and Karkellang (Nat. Coll.).
Observation. This species is a large and handsome local race of the Blue-headed Green Parrot, *T. luconensis*, of the Philippines from which it is easily distinguishable, according to our experience by its much larger size and fine blue head. Philippine specimens measured by us have the wing 174—189 mm, but Salvadori (Cat. B. XX, 426) records the wing as being 190—201 mm. Specimens from Manila and Sooloo indeed evidently attain to the dimensions of average *T. talautensis*, as the wing of one from Sooloo is given by Salvadori as 211 mm, tail 132 mm, but we prefer not to make *T. talautensis* a subspecies, with these forms as the necessary connecting links, since we do not believe that Philippine birds ever acquire such a blue head as those of Talaut. Possibly the Sangi birds mentioned in the preceding article will be found to affect this question.

+ 48. TANYGNATHUS MEGALORHYNCHUS (Bodd.).
Big-billed Green Parrot.

Under several forms this bird ranges from Talaut to the Moluccas, Sumba, Timorlaut, and New Guinea. In the Southern Moluccas and in Timorlaut, respectively, the birds have been struck off as species, *T. affinis* Wall. and *T. subaffinis* Sclat. The rest have been called *T. megalorhynchus* by Salvadori, but the Sumba birds were separated as a subspecies by Meyer. Those from Djampea seem to be intermediate. There seem to be other local variations, as well as much individual variation, but the question demands greater study than we feel disposed to devote to it, and the following method of nomenclature will answer our purpose.

1. The typical Tanygnathus megalorhynchus.

a. **Psittacus megalorhynchos** *(1)* Bodd., Tabl. Pl. Enl. 1783, 45.
b. **Perroquet à bec couleur de sang**, *(1)* Levaill., Perr. 1805, pl. 83.
c. **Psittacus macrorhynchus** *(1)* Gm., S. N. 1788, I, 338; *(II)* Shaw, Gen. Zool. 1811, VIII, 2, p. 530, pl. 79; *(3)* Reinw., Reis. n. Ind. Archip. in 1821, 1858, 592.
d. **Tanygnathus megalorhynchus** *(1)* Wall., P. Z. S. 1864, 285, 294; *(2)* Meyer, J. f. O. 1873, 405; *(3)* Brüggem., Abh. Ver. Bremen 1876, V, 37; *(4)* Salvad., Ann. Mus. Civ. Gen. 1876, IX, 52; *(5)* Meyer, Rowl. Orn. Misc. 1878, III, 127; *(6)* id., Ibis 1879, 48; *(7)* Salvad., Orn. Pap. 1880, I, 129; Agg. I, 1889, 30; *(VIII)* Rchnw., Vogelb. 1878—83, t. XI, f. 6; *(9)* Meyer, Verh. z.-b. Ges. Wien 1881, 762; Isis 1884, 6; *(10)* W. Blas., Ornis 1888, IV, 557; *(11)* Hickson, Nat. in N. Celebes 1889, 158; *(12)* Salvad., Cat. B. XX, 1891, 426.
e. **Eclectus megalorhynchus** *(1)* Schl., N. T. D. 1866, III, 184; *(2)* Finsch, Papag. II, 1868, 351; *(3)* Schl., Rev. Psitt. 1874, 23 pt.
"**Kalea**", Sangi Islands, v. Rosenb. *e 2*.
"**Karea**", Gt. Sangi and Siao, Nat. Coll. in Dresd. Mus.
For further synonymy and references see Salvadori *d 12*.
Figures and descriptions. Levaillant *b 1*; Shaw *c II*; Reichenow *d VIII*; Finsch *e 2*; Salvadori *d 7, d 12*; W. Blasius *d 10*.
Male. Head, neck, and sides of breast yellow parrot-green; sides and under wing-coverts deep yellow, the greater series sulphur-yellow with dusky bases; rest of

under surface of body greenish yellow; wings above dark blue, the lesser coverts, the smaller scapulars adjoining them, and some of the middle and greater coverts furthest from the edge of the wing, black; some of the lesser coverts and the scapulars broadly tipped with blue, the middle and greater coverts broadly bordered with deep yellow, the terminal part of some of the primaries green, the tertiaries green bordered with yellow-green, the longer scapulars green with a black spot surrounded by blue about the end of the shaft; quills below shining brownish smoke-grey; lower back and rump turquoise-blue; upper tail-coverts yellow apple-green; tail parrot-green, yellowish at tip; below golden, shaded with olive (Tabukan, Sangi, Nr. 13268).

"Iris light yellow; bill sealing-wax red; feet grey-greenish" (Platen *d 10*).

Female. Has the bill smaller than the male (Salvad. *d 12*); feet grey-green *(d 10)*.

Young. Has the scapulars and upper wing-coverts not so black and more greenish, and the yellow edges of the same paler (Salvad. *d 12*).

Measurements.

	Wing	Tail	Tarsus	Culmen from cere
a. (Nr. 13268) ♂ Sangi	255	147	23	50
b. (C 12693) ad. Gr. Sangi, 23. VII. 93	266	165	—	51
c. (C 12692) imm. Gr. Sangi, 18. VII. 93	248	152	—	49.5
d. (C 12694) ad. Gr. Sangi, 17. VII. 93	248	153	—	46
e. (Mus. Tring) ad. Gr. Sangi, 26. VII. 93 .	255	160	—	46
f. (Mus. Tring) ad. Gr. Sangi, 25. VII. 93 . . .	244	142	—	45.5
g. (C 12589) ad. Siao, 6. VII. 93 . . .	255	160	—	50
h. (C 12590) juv. Siao, 8. VII. 93	247	158	—	45
i. (Mus. Tring) ad. Siao, 4. VII. 93	255	160	—	44.5
k. (Mus. Tring) ad. Siao, 5. VII. 93 . . .	239	146	—	44
l. (Mus. Tring) ad. Siao, 7. VII. 93	244	158	—	48
m. (C 13431) ad. Tagulandang, 1. VIII. 94	262	162	—	51
n. (C 13432) juv. Tagulandang, 7. VIII. 94	244	150	—	42
o. (C 13435) ad. Gunungapi, Tagulandang, 25. VIII. 94.	256	150	—	51
p. (C 13434) ad. Gunungapi, Tagulandang, 27. VIII. 94.	257	160	—	49
q. (C 13433) ad. Biarro, 4. IX. 94	252	150	—	50
r. (Mus. Tring) [♂] Mantehage, IV. 93 . .	250	150	20	46
s. (C 12265) [♀] Mantehage, 27. IV. 93 . .	250	145	24	43
t. (Nr. 13266) ♂ Manado, Celebes	245	145	23	54
u. (Nr. 13267) (♀ ?) Manado, Celebes	236	143	—	46.5
v. (Nr. 1949) (♀ ?) New Guinea	245	149	21	43
w. (Nr. 3436) (♀ ?) Moluccas	237	145	20.5	45

8 specimens from Talaut (Kabruang and Karkellang) measure: wing 229—259; tail 130—155; bill from cere 43—53 mm.

Distribution. North Celebes — (?)Manado (v. Musschenbroek *d 9*), (?)Tondano (Reinwardt *c 3*); Mantehage or Mantrau Id. near Manado (Meyer *d 5*, *d 6*, Nat. Coll.); Biarro and Tagulandang (Nat. Coll.); Sangi Islands — Siao (Hoedt *e 3*, Nat. Coll.); Great Sangi (Wall. *d 1*, Rosenb. *e 3*, Hoedt *e 3*, Meyer, Fischer *d 3*, Bruijn *d 4*, Platen *d 10*, Nat. Coll.); Talaut Islands — Saha (Hickson *d 11*); Karkellang and Kabruang (Nat. Coll.); Halmahera Group — Halmahera, Obi, Moor, Tidore, Motir, Ternate, Batchian, Makian, Morotai, Weeda; Western Papuan Islands — Sorong, Mysol, Salvatti, Batanta, Waigiou, Guebeh; W. coast of N. New Guinea (Salvad. *d 7*, *d 12*).

2. Tanygnathus megalorhynchus sumbensis (Meyer).

f. **Tanygnathus megalorhynchus** var. **sumbensis** *(1)* Meyer, Vorh. z.-b. Ges. Wien 1881, XXI, 762.
g. **Tanygnathus megalorhynchus** *(1)* Salvadori, Cat. B. XX, 1891, 428 footnote.
h. **Tanygnathus megalorhynchus sumbensis** *(1)* Hartert, Nov. Zool. 1896, 588.
Diagnosis. Under wing-coverts greenish yellow, instead of deep yellow; under surface greener than in the typical form.
Distribution. Sumba (Riedel *f 1*, Doherty *h 1*).

3. Tanygnathus megalorhynchus — sumbensis.

i. **Tanygnathus megalorhynchus** *(1)* Hartert, Nov. Zool. 1896, 176.
Diagnosis. Intermediate (C 15857—58).
Distribution. Djampea Island (Everett).
Observation. Mr. Hartert (*h 1*) considers these specimens more like the typical form, we more like *sumbensis*; probably an intermediate position will not be far wrong.

The first notice of the occurence of this Parrot in Celebes was made in 1821 by Reinwardt, who simply mentions that *Psittacus ornatus*, *Ps. macrorhynchos* (i. e. *megalorhynchus*) and *Ps. leucorhynchus* were the three species of Parrot met with by him at Lake Tondano, but he may have perhaps misnamed the red-billed *muelleri*: *macrorhynchos*. In 1871 it was found by Meyer on Mantehage, a small island north of Manado about 8½ miles off the coast, — not in the Togian Islands as stated by Count Salvadori *(d 12)*. Again in 1879 two specimens were sent to the Dresden Museum from van Musschenbroek with the locality "Menado" attached in that gentleman's writing. In the Sangi Islands the species abounds *(d 5)*; it is also, according to Dr. Hickson, not uncommon in the Talaut Islands. As was remarked by Meyer in 1879, *T. megalorhynchus* appears to be a species which is extending its range.

The Sumba birds have all greenish yellow under wing-coverts, instead of deep yellow, by which means the two forms may most readily be distinguished; also the under surface is greener — not so decidedly yellow-green, as in specimens of the typical form. Seen from above the two forms are not to be distinguished.

T. megalorhynchus with its allies, *T. affinis* (Ceram group) and *subaffinis* (Timorlaut) forms a very well marked section of the genus *Tanygnathus*, with great bloated-looking red bills, inhabiting the Moluccas, some of the Papuan Islands, etc. *T. megalorhynchus* with its blue wings and strongly contrasted colours on the wing-coverts appears to be the most highly differentiated form of the three. The connection of this section with the other members of the genus appears to be through *T. affinis* of Buru etc. and *T. luconensis* of the Philippines. *T. luconensis*, though much smaller, has like the *megalorhynchus* group its bill red in both sexes and in the young (upper mandible red, lower orange in adult, according to Salvadori), and the coloration of the wings shows much similarity to *T. affinis*; the head of the adult *T. luconensis* is, however, blue or greyish lilac. In immature specimens of *T. affinis* (Nrs. 3437, 3438), which are assuming the green head of maturity, considerable traces of blue may be seen on the crown. This appearence perhaps gives a hint that *T. luconensis* represents an earlier stage of development. The genus *Tanygnathus* has its chief centre in the Celebes-Philippine province.

GENUS LORICULUS Blyth.
Map VI.

The Lorikeets are of small size — from that of a Blue Tit to that of a Lark —, the general colour yellowish parrot-green, with the rump and upper tail-coverts always red (accept in *L. tener*); the tail is short, half the length of the wing, or less; the first three primaries longest and almost subequal; tail below and wing below (where it rests upon the body) verditer-blue; bill rather long, and not much hooked, longer than deep; the reversed toe nearly equal to the middle (third) one in length. 23 species and subspecies have been described, with a range from India as far as New Guinea and the New Britain group.

+ * 49. LORICULUS EXILIS Schl.
Green Lorikeet.

Loriculus exilis *(1)* Schl., Ned. Tdschr. Dierk. 1866, III, 185; *(2)* Wald., Tr. Z. S. 1872, VIII, 32; *(3)* Schl., Rev. Psitt. 1874, 60; *(IV)* Rowl., Orn. Misc. 1877, II, 243, pl. LIX; *(5)* Meyer, op. cit. 233, 244—247; *(6)* Rosenb., Malay. Archip. 1878, 274, 593; *(7)* Meyer, Ibis 1879, 52; *(VIII)* id., Vogelskel. 1882, t. XXII; *(9)* W. Blas., J. f. O. 1883, 134; *(10)* Guillem., P. Z. S. 1885, 544; *(11)* Salvad., Cat. B. XX, 1891, 521; *(12)* M. & Wg., Abh. Mus. Dresden 1896 Nr. 1, p. 4.
a. **Coryllis exilis** *(1)* Finsch, Papag. 1868, II, 729, pl. 5; *(1bis)* Frenzel, Mtschr. Ver. Schutze Vogelw. 1880, 8, 11, 22; *(II)* Rchnw., Vogelb. 1878—83, t. XV, f. 6; *(3)* id., J. f. O. 1881, 226 (Consp. Psitt. 114); *(4)* Platen, Gefied. Welt 1887, 206, 230.
b. **Psittacus exilis** *(1)* Russ, Fremdl. Stubenvög. 1881, III, 817.
"**Tintis kitjil**" (little Lorikeet), Malay name, Manado, Meyer 7.
Figures and descriptions. Rowley *IV*; Rchnw. *a II*; Meyer *VIII* (skeleton) *5*; Schlegel *1*; Finsch *a 1*; Guillemard *10*; Salvadori *11*.
Adult. Parrot-green, with a yellowish wash, the wings darker and duller; rump and upper tail-coverts, which reach to the end of the tail, poppy-red, the bases of the feathers yellow; tail above green, like the wings, the feathers — except the two middle ones — tipped with greenish yellow; under surface yellower green than the upper; on the middle of the throat a lengthened spot of poppy-red, the parts surrounding it washed with verditer-blue or beryl-green; under side of wings and tail bright verditer-blue, the outer webs and tips of the primaries and the inner webs near the shafts black, the least series of under wing-coverts green, the middle series blue washed with green (Manado, Nr. 14030). "♂: Eyes yellow; feet orange-yellow somewhat reddish; bill coral-red" (Meyer 7, Platen *a 4*). "♀: iris brownish" (Platen *a 4*).
Young. Without the red spot in the throat (Finsch): bill brownish yellow; feet and cere

yellowish; iris light brown (Platen *a 4*); another example ♂ juv.: bill orange-red, cere and feet yellow-brown, iris orange (Platen, Nehrkorn in lit.).

According to Dr. Guillemard *(10)* the female is without the red spot on the throat, though the two sexes of species were at first (after a specimen with a red throat-spot determined as ♀ by von Rosenberg) stated to be identical by Schlegel and Finsch. The five specimens in the Dresden Museum have all a red spot on the throat, but the sex is not marked, so that they do not help to decide what is the truth of the matter. A young male, sent by Platen to Mr. Nehrkorn, with an orange-red bill has no spot. Dr. Guillemard's birds may have been young females.

A specimen marked ♀ in the Sarasin Collection has a small spot of red on the throat.

Measurements.

	Wing	Tail	Tarsus	Culmen from cere
a. (Nr. 14030) Manado	69	31	8	7.5
b. (Nr. 1935) Manado	67	34	8.5	—
c. (Nr. 1742) Manado	67	31	8	7.5
d. (Nr. 14029) Manado	66	—	8.5	7.5
e. (C 1100) Manado	65	32	8	—
f. (Sarasin Coll.) ♀, Rurukan, 12. IV. 95.	69	30	8.5	7

Skeleton.

	mas.	fem.		mas.	fem.
Length of cranium	23.0	24.0 mm	Length of tibia	23.2	23.0 mm
Greatest breadth of do.	13.8	14.3 »	Length of tarso-metat.	7.8	7.7 »
Length of humerus	14.8	14.8 »	Length of sternum	21.5	22.0 »
Length of ulna	17.5	17.6 »	Greatest breadth of do.	12.7	12.4 »
Length of radius	16.2	15.9 »	Height of crista sterni	8.0	8.0 »
Length of manus	21.4	21.7 »	Length of pelvis	21.7	22.8 »
Length of femur	16.1	16.0 »	Greatest breadth of do.	11.4	11.5 »

Distribution. North Celebes — Tulabulo and Paguatt, Gorontalo Province (Rosenberg *1, 3, 6*); Manado, Minahassa (Meyer *2, 5, 7*, Guillem. *10*); Rurukan, Minahassa (Platen *a 4*, P. & F. Sarasin).

This minute Parrot ranks with *L. aurantiifrons* Schl. of Mysol and New Guinea as the smallest species of the genus *Loriculus*. It was first discovered by von Rosenberg, who obtained five specimens in a garden at Tulabulo, N. E. of Gorontalo in 1864. In 1871 it was found again by Meyer at Manado; in the month of March only a single pair was met with, but in May it suddenly appeared in large flocks, frequenting the mangrove bushes near the sea-shore, and about a hundred specimens were obtained by the natives with blowpipes. The contents of the stomachs of specimens examined appeared to be composed, as far as Meyer could make out, of the juices of flowers; but doubtless it eats fruits as well, as von Rosenberg states *(1)*, since no animal can live upon honey alone, from the fact that it contains no albumen (Marshall, Papag. 1889, 21). Some tame specimens belonging to Dr. Platen were fed on a soft milk food, consisting of finely powdered biscuit. flavoured with bananas or sugar water. The Ceylonese species *L. indicus* feeds on the juices of both fruits and flowers (Legge, 182). At Rurukan at a height of over 3000 ft. Dr. Platen appears to have found *L. exilis* in plenty, and he mentions that both this bird and *L. stigmatus* breed twice a year, viz. in February and August, and always prefer the sugar palms, the dead lower leaf-stems of which offer them convenient nesting holes *(a 4)*. Like other members of the genus, *L. exilis* is of a very affectionate disposition, as was shown by Dr. Platen's tame specimens at Rurukan.

The nearest known ally of *L. exilis* is *L. flosculus* Wall. of Flores, which differs from the Celebes form, as Count Salvadori shows, in being larger (wing 76 mm), in having the nape tinged with orange, the tips of the tail-feathers yellowish, stained with red, and in wanting the area of verditer-green surrounding the red spot on the throat. The presence of this wash of verditer or berylgreen on the throat of *L. exilis* suggests a relationship with *L. vernalis* of India, a species which appears to have more ancestral, or fewer recent, characters than other members of the genus; and on the same part of the throat a bluish tint, much like that in *L. exilis*, is found in the male of *L. vernalis* and is, sometimes also apparent in the female; but in this species the red spot on the throat is not developed.

The genus is more specially considered under the heading, *L. stigmatus*.

In respect of its bill *L. exilis* is the most strongly differentiated form of its genus. The under mandible appears at first sight deformed, the edge of it is hollowed out at the sides, the terminal part is then sharply curved up and lengthened, fitting into the hollowed-out upper bill like an incisor tooth, which prevents the bill from shutting, and at each side a semicircular hole is formed. This construction of the bill is perhaps connected with some peculiarity in its feeding; the same formation is seen on a less pronounced scale in the blackbilled *L. amabilis* of the Moluccas and *L. aurantiifrons* of New Guinea and in some of the other black-billed species, also in the orange-billed *L. pusillus* Gray of Java and in *L. flosculus* Wall. of Flores. The resemblance is strongest in the last case. The orange-billed species have in general the under mandible with a straight cutting edge.

+ * 50. LORICULUS CATAMENE Schl.
Sangi Lorikeet.

a. **Loriculus amabilis** (part.) Wald., P. Z. S. 1871, 333 (Sanghir); id., Tr. Z. S. 1872, VIII, 26 (Tweedd. Orn. Works 1881, 131).
Loriculus catamene *(1)* Schl., Ned. Tdschr. Dierk. 1871, IV, 7; *(2)* id., Rev. Psitt. 1874, 62; *(III)* Rowl., Orn. Misc. 1877, II, 236, pl. LVII (♂ juv., ♀); *(4)* Meyer, t. c. 233, 237; *(5)* id., Gefied. Welt 1887, 264; *(6)* W. Blas., Ornis 1888, IV, 560; *(7)* Salvad., Cat. B. XX, 1891, 537.

b. **Coryllis catamenia** *(1)* Rchnw., J. f. O. 1881, 230 (Consp. Psitt. 118); id., Vogelb. Nachtr. 1883, Nr. 52.

c. **Psittacus catamene** Russ, Fremdl. Stubenvög. 1881, 805.

d. **Coryllis catamene** *(1)* Platen, Gefied. Welt 1887, 263.

"**Lusint**", Great Sangi, Platen *d 1*.
"**Lunsihi**", Great Sangi. Nat. Coll.

Figures and descriptions. Rowley *III*; Schlegel *1*; Reichenow *b 1*; Platen *d 1*; W. Blasius *6*; Salvadori *7*.

Adult male. Bright parrot-green, lighter and washed with yellowish on the under surface; sinciput (a short cap), lower back, rump and upper tail-coverts (which extend beyond the tip of the tail) poppy-red; a spot of the same colour on the middle of the throat; under tail-coverts — longer than the tail — scarlet, the extreme edges of the feathers yellow-green; the exposed outer webs and ends of quills above green, the inner webs black; the inner webs, as far as they rest upon the sides of the body, bright verditer-blue, the external part of the feathers black; under wing-coverts varied with yellow-green and verditer-blue (Great Sangi: — Mus. Nehrkorn, Nr. 1299). "Iris yellow or orange-red; bill black; cere brownish yellow; feet yellow-orange or orange" (Platen *6*).

Adult female. Like the male, but the head all green, without the cap of poppy-red; the under tail-coverts yellow-green varied with scarlet intermixed (Great Sangi: — Mus. Nehrk., Nr. 1300). "Iris brown or light brown; bill black; cere brownish yellow; feet yellow-orange or orange" (Platen *6*).

Young male. Like the adult female; not a trace of red on the head; the under tail-coverts approaching those of the adult male in intensity of colour (W. Blasius *6*). Iris light brown; bill black.

Young female. In two young females the red throat-spot is much less developed than in the adult of both sexes; the upper tail-coverts red, the ends of them reaching far short of the tip of the tail; the under tail-coverts green-yellowish with rather broad reddish tips: in one with the shortest upper tail-coverts the bill is yellowish; in the other dark brown. Iris brown (W. Blasius *6*).

Measurements.	Wing	Tail	Tarsus	Culmen from cere
a. (Mus. Nehrk. 1299) ♂, Great Sangi	82	45	9	10
b. (Mus. Nehrk. 1300) ♀, Great Sangi	84	42	9	10
c. (Nr. 1740) ♀, Great Sangi	80	41	9	—
d. (C 12699) [♂] ad. Great Sangi, 15. VII. 93	85	43	9.5	11
e. (Mus. Tring) [♀] ad. Great Sangi, 17. VII. 93	84	40	—	10
f. (C 12698) imm. Great Sangi, 17. VII. 93	83	38	—	10

Distribution. Great Sangi Island (Hoedt *1, III*, Meyer *4*, Platen *d 1, 6*, Nat. Coll.).

This little Parrot is only known from the principal island of the Sangi Group, where it was first discovered in 1864 by Hoedt, who apparently only obtained one specimen, the type, an adult male. A few more were procured by Meyer in the year 1871. Dr. Platen, during his residence in the island in 1886—87, obtained a fair number of specimens, forwarding to Mr. Nehrkorn, to whom we are indebted for the loan of the two described, as many as thirteen carefully labelled examples, which are discussed by Prof. W. Blasius (*6*): but the species according to Dr. Platen is not particularly plentiful. The nearest known ally of *L. catamene* may be seen in *L. amabilis* Wall. of Halmahera and Batchian, which corresponds with it in all the principal details of coloration and in having the upper tail-coverts in the adult bird longer than the tail. The

Sangi Lorikeet may be distinguished by its larger size and also by its scarlet under tail-coverts. These are, however, almost entirely yellow-green in the young, much as in *L. amabilis* both young and old; this species has the under tail-coverts of a uniform green. This might appear to indicate that *L. amabilis* represents an earlier race, which developed scarlet under tail-coverts in Sangi. But *L. amabilis* has also developed some red colouring in its plumage, and, indeed, in a place where it is not found in *L. catamene*, viz: on the metacarpal edge and in the females at the bases of the feathers of the forehead. Therefore, whoever would argue that one species is directly descended from the other, must admit that the red has become lost in one place and developed in another. It is possible that a species with red on both the metacarpus and under tail-coverts remains to be found, since there is a female specimen (apparently a cage-bird) in the British Museum, labelled Halmahera and doubtfully referred to *L. amabilis* by Count Salvadori, which has red on both of these parts. The question can be raised whether the male of the form from which *L. amabilis* developed, had a red cap or not. The fact that the young males of *L. stigmatus, quadricolor, sclateri,* and — in all probability — *amabilis,* acquire the red carpal edge and *L. catamene* the red under tail-coverts, at an earlier age than the red cap, which is never produced in *L. sclateri*, render the opinion plausible that the ancestral form of *L. amabilis* and *catamene* was without a red cap, and so was more like the *L. flosculus* and *exilis* of to-day. These species are marked as of an earlier development in our genealogical tree of the genus (p. 163), a position which receives further confirmation from the fact that the young female of *L. catamene* is known to have the under tail-coverts nearly all green, a spot of red on the throat and a yellowish bill (W. Blasius *6*), so corresponding in nearly every detail of coloration with the mature *L. exilis* and *flosculus.*

* 51. LORICULUS SCLATERI Wall.

Sula Lorikeet.

In Sula the birds of this species have a more orange, in Peling and Banggai a redder back, and the following nomenclature may be employed for these two races:

1. The typical Loriculus sclateri.

a. **Loriculus sclateri** *(1)* Wall., P. Z. S. 1862, 336; pl. XXXVIII; 1864, 267, 294; *(2)* Schl., Mus. P.-B. Psitt. 1864, 132; *(3)* id., Ned. Tdschr. Drk. 1866, III, 186; 1871, IV, 8; *(4)* Wald., Tr. Z. S. 1872, VIII, 32 (Orn. Works 1881, 136); *(5)* Schl., Rev. Psitt. 1874, 61; *(6)* Meyer, Rowl. Orn. Misc. 1877, II, 233, 251; id., Ibis 1879, 52; *(7)* Rosenb., Malay. Archip. 1878, 274; *(8)* Sclat., List Vert. An. 1883, 326; *(9)* Tristr., Cat. Coll. B. 1889, 76; *(10)* Salvad., Cat. B. XX, 1891, 533; *(11)* M. & Wg., Abh. Mus. Dresd. 1896, Nr. 2, p. 9.

b. **Coryllis sclateri** *(1)* Finsch, Papag. 1868, II, 697; *(1ᵇⁱˢ)* Frenzel, Mtsschr. Ver. Schtz.

Vogelw. 1880, 12, 23; *(2)* Rchnw., J. f. O. 1881, 230 (Consp. Psitt. 118); *(3)* id., Vogelb. Nachtr. 1883, Nr. 51.

c. "**Loriculus wallacei**" G. R. Gray" (!), *(1)* Wald., Ann. & Mag. N. H. 1872, IX, 398, 399; id., Orn. Works 1881, 125.

Figure and descriptions. Wallace *a 1*; Finsch *b 1*; Salvad. *a 10*.

Adult. Parrot-green, quills and tail darker, under surface lighter; mantle and back golden-orange, many feathers tipped with scarlet-vermillion; rump and upper tail-coverts bright dark poppy-red; a patch on the throat and the metacarpal edge of the same colour; under wing-coverts yellowish green washed with blue, the longer ones verditer-blue; tail below, inner webs of secondaries, and inner webs of primaries, as far as they rest upon the sides of the body, bright verditer-blue (Sula Besi — Teijsman 1877 — C 14319).

"Bill black; cere and base of the upper mandible yellow; feet yellow; iris yellow" (Wall. *a 1*).

Young. Green colour duller; the patch on the back is dull orange, throat-spot small, red; the carpal edge greenish yellow (Sula Besi — Teijsmann — C 14518).

Measurements.	Wing	Tail	Tarsus	Culmen from cere
a. (Nr. 1999) ad. ? Sula Islands	90	40	10	13
b. (Mus. Petersb.) ? Sula Islands	87	—	11	14
c. (C 14319) ad. Sula Besi	91	37	11	12
d. (C 14518) juv. Sula Besi	89	—	—	12.5

Distribution. Sula Besi in the Sula Islands (Allen *a 1*, Bernstein *a 5*, Hoedt *a 5*, Teijsman); (?) Celebes — Bone and Negri lama, Gorontalo Province (Rosenberg *a 3*, *a 5*, *a 7*).

Observation. Sula Besi is, up to the present, the only island in the Sula group from which this form is known. There are now 25 specimens in the Leyden Museum, many of which were obtained by Teijsman in 1877. Its apparent absence from Sula Mangoli is surprising.

+ 2. Loriculus sclateri ruber M. & Wg.
Abh. Mus. Dresd. 1896, Nr. 2, p. 9.

"**Sinsin**" or "**Sinsing**" of the natives.

Diagnosis. Like the typical form, but the mantle deep scarlet, and the feathers of the forehead red, except at the tips; the red of the rump scarlet like the mantle.

Size. Wing ca. 90 mm; tail 40; tarsus 12.5; bill from cere 11—12.

Distribution. Peling and Banggai (Nat. Coll.).

Observation. The young (and perhaps the females) are without red on the forehead and have but little red or orange on the mantle, but the green seems to be brighter than in the young from Sula Besi.

Mr. Büttikofer had the great kindness to compare this form with the 25 specimens from Sula Besi in the Leyden Museum and found that it had more red on the forehead and mantle than the bulk of the specimens from that island; one specimen, however, which he sent us for comparison was almost as red on the mantle as the new race, though the tint thereof was more orange; a second specimen had red on the forehead, though less extensive than in adults from Peling and Banggai. There are a number of small intermediate islands between these and Sula, and it appears pretty certain that these will furnish birds which intergrade with both races.

L. sclateri is one of a little group of three local forms — *L. stigmatus* (S. Müll.) of Celebes, *L. quadricolor* (Wald.) of the Togian Islands and *L. sclateri* (Wall.) of Sula, Peling and Banggai; it is, therefore, perplexing to find two properly labelled specimens (with sex and dates when obtained) in the Leyden Museum stated to have been killed at Bone and Negri lama in North Celebes by von Rosenberg, and in the case of the second specimen on the very day when a specimen of *stigmatus* was obtained by him there *(5)*. The record of *L. sclateri* in Celebes is very probably due to specimens escaped from confinement, for the *Loriculi* are much valued by the natives as pets and transported from place to place. If not, we can only conclude, in view of the highly "local" character of *L. sclateri* and its allies, that the specimens in some way got mislabelled.

The nearest known ally of *L. sclateri* is *L. quadricolor* of the Togian Islands, from which the former differs chiefly in that the adult male has the head green, like that of the female, whereas the adult male of *L. quadricolor*, as also of *L. stigmatus, catamene* and *amabilis*, has the additional ornament of a red cap. The commencement (or a relic, as the case may be) of this red cap is indeed seen on the forehead of *L. sclateri ruber*, and sometimes to a small extent in the *typical* form. Mr. Wallace has indicated *(a I)* the base of the upper mandible as yellow, but this does not seem to be the case in adults, though the young have much of the bill yellow.

The yellow supposed to be at the base of the bill, and the absence of the red cap appear at first sight to entitle *L. sclateri* to be regarded as an earlier stage of development from which *L. quadricolor* and *stigmatus*, and thence, perhaps, also *L. amabilis* and *catamene* may have been differentiated. But much may be urged against such an assumption: it seems much more probable that *sclateri, quadricolor* and *stigmatus* were severally differentiated in Sula, Togian and Celebes from a form formerly spread over all three localities, and that in Togian and in Celebes and not in Sula the males acquired a red cap, in all cases undergoing the variation independently of one another. That such analogous variation may take place appears to be undeniable; thus *L. indicus* (Gm.) of Ceylon, a local form of the wide-spread *L. vernalis* of India to the Malay Peninsula, and *L. apicalis* Souancé of the southern Philippine Islands have both developed a very similar red crown, quite independently as far as can be seen.

In the genus *Loriculus* certain groups of species occur, and one such is what we call the *stigmatus*-group, consisting of *L. stigmatus, quadricolor* and *sclateri*, another is that formed by *L. catamene* and *amabilis*, another is presented by the birds inhabiting the Philippines. The *amabilis*-group has affinities with *L. exilis* of Celebes, as has been shown, by the shape of the bill and the tail-coverts prolonged beyond the tip of the tail. *L. exilis* has nothing to do with the Philippine species, but as is remarked elsewhere, with the more southern forms, *L. flosculus* of Flores and *L. pusillus* of Java, which share to a great extent its peculiarly shaped bill and lengthened tail-coverts. As to the *stigmatus*-group it may not be apparent at first sight whether it links on to the *amabilis*-group and thence to *L. exilis* and the forms of Flores and Java, or on to the Philippine species. *L. bonapartei* Souancé of the Sooloo Islands comes most into question among the latter birds, for, though its allies have yellow bills, this form sometimes has a black bill (in the adult, as we believe), while the bills of others were indicated by Dr. Guillemard as "red", "brown-black", "brownish", "very dark yellow" and by Everett (Ibis 1893, 249) as "dull orange, clouded and tipped with black". These, we believe, are younger birds, since young specimens of *L. stigmatus*, *catamene* and *amabilis* are known to have yellowish bills, and we do not share the opinion, which Count Salvadori expresses (Ibis 1891, 48—51), that the labels of Dr. Guillemard's specimens are not correct. *L. bonapartei* has "a rather long bill, and the exact style of plumage of the red-billed *L. apicalis*. The blue cheeks of the female also betray its real affinities with *L. apicalis* and the other Philippine species".

The adult males of *L. amabilis* and of *L. bonapartei* both agree with the adult males of the *stigmatus*-group in having the bill black, a red patch on the throat and a red cap (absent in *L. sclateri*). But the key to the answer of the question, whether the *stigmatus*-group has its closest affinities with *L. amabilis* or with *L. bonapartei*, is to be found in the females. The females of the *stigmatus*-group have a red patch on the throat, no red on the crown and no blue on the sides of the face; the females of the *amabilis*-group likewise have a red patch on the throat, no red on the crown or blue on the face; the female of *L. bonapartei*, on the other hand, has a red crown and blue cheeks, but no red patch on the throat. There is thus an almost perfect parallelism in the coloration of both male and female of the *stigmatus*- and *amabilis*-groups, sex for sex, and this is further borne out by some points of similarity in the general pattern of plumage, such as the abruptly marked-off red cap of the males (absent in *sclateri*) and the red metacarpal edge (absent only in *L. catamene*); the bill also agrees best in these two groups. The evolution of the sexes of *L. bonapartei* is not on a parallel with that of the *stigmatus*-group; the males of the two groups agree, but the females do not. The fact should not be overlooked that the adult males of birds usually differ from the females in having some character, or characters, superadded to those of the female, as, for instance, the possession of an additional patch of colour, a crest, spurs, a song; in such cases the young of both sexes very frequently resemble the female, and the female here may be reasonably regarded as having retained the dress of an earlier stage in the history of the race, whereas the dress of the male has been further developed. This rule is well illustrated by the genus *Loriculus*, in which the immature males much resemble the old females, the quite young birds having a still simpler dress. We therefore look to the females of the *stigmatus*- and *amabilis*-groups for indications of the relations of these species and infer that they are sprung from a race with a red spot on the throat, but with no red cap. Such a race is seen in *L. exilis* of Celebes and in *L. flosculus* of Flores, and these species — though, no doubt, somewhat changed — may perhaps be regarded as more ancient types.

⊣ * 52. LORICULUS QUADRICOLOR Tweedd.
Togian Lorikeet.

Loriculus quadricolor *(1)* Wald., Ann. & Mag. N. H. 1872, IX, 398; id., Tr. Z. S. 1872, VIII, 109 (Orn. Works, 1881, 125, 207); Meyer, J. f. O. 1873, 404; *(2)* id., Rowl. Orn. Misc. 1877, II, 233, 251, 252; *(3)* id., Ibis 1879, 52, 145, 146; *(IV)* Salvad., Cat. B. XX, 1891, 534.

a. **Coryllis quadricolor** *(1)* Rchnw., J. f. O. 1881, 231 (Consp. Psitt. 119); id., Vogelbild. 1883, 55.

Figures and descriptions. Salvadori *IV*; Walden *I*; Reichenow *a l*.

Adult male. Like the adult *L. sclateri*, but with the addition of a scarlet sinciput; the interscapulars and back golden orange, without the orange-red tinge in the middle; the rump and upper tail-coverts of a darker red (ex Salvadori and Walden).

Adult female. Like the male, but without the scarlet sinciput; the feathers of the forehead probably with scarlet bases.

Young. Like the female, but with the metacarpal edge greenish yellow, instead of red.

A specimen in the Dresden Museum marked young male evidently corresponds well with one so labelled in the British Museum (specimen *c* of the Catalogue), but which Count Salvadori wrongly believes to be a nearly adult female (the bases of the feathers on the forehead being red) as in very many cases the young males correspond with the old females. This is the case with its nearest ally *L. stigmatus*.

Measurement. (Nr. 6031) ♂ juv. Togian Id. Wing 90, tail 41, tarsus 10, culmen from cere 11 mm.

Distribution. Togian Island, Togian group in the Gulf of Tomini (Meyer *1*, *3*).

Only six specimens of this Parrot were collected by Meyer, three of which are in the British and one in the Dresden and the other two in the Berlin Museum. They were shot near the village of Togian on the chief island of the group in the Gulf of Tomini in the month of August, 1871. *L. quadricolor* is intermediate in coloration between *L. stigmatus* of Celebes and *L. sclateri* of Sula. The Banggai Peninsula, the eastern limb of the mainland of Celebes, jutting out between the Togian and Sula Islands, was pointed to by Meyer in 1877 as preserving the answer to the following problem: "It would be a very interesting point to ascertain which species lives there; for if it be *L. stigmatus*, the two allied forms *L. sclateri* and *L. quadricolor* will prove to be insular forms derived from a parent stock, *L. stigmatus*, both changed in a somewhat similar manner, perhaps through the same (say 'insular') conditions, but not quite in the same manner. If, on the other hand, the species which inhabits Banggai should prove to be not *L. stigmatus*, it would certainly be of interest to know whether it is *L. sclateri* or *L. quadricolor*, or (as is possible) a form which is intermediate between these two" *(2)*.

The collections from the Eastern Peninsula made by native hunters in 1895 for the Dresden Museum proved that *L. stigmatus* occurs there, and that therefore *L. quadricolor* and *sclateri* are insular forms, though perhaps not derived from *L. stigmatus* as it is at the present day.

* 53. LORICULUS STIGMATUS (Müll. Schl.).
Celebes Lorikeet.

a. **Psittacus (Psittacula) stigmatus** *(1)* Müll. & Schl., Verh. Natuurk. Comm. 1839—43, 182.
Loriculus stigmatus *(1)* Bp., Rev. Zool. 1854, 155; *(2)* Wall., P. Z, S. 1864, 287, 294; *(3)* Schl., Dierentuin, 1864, 70, fig.; *(4)* id., Mus. P.-B. Psitt. 1864, 131; *(5)* Wald., Tr. Z. S. 1872, VIII, 32; *(6)* Schl., Rev. Psitt. 1874, 60; *(7)* Salvad., Ann. Mus. Civ. Gen. 1875, VII, 645; *(8)* Brüggem., Abh. Ver. Bremen 1876, V, 41; *(IX)* Rowley, Orn. Misc. 1877, II, 254, pl. LX; *(10)* Meyer, t. c. remarks 233, 234, 250; *(11)* Lenz, J. f. O. 1877, 363; *(12)* Meyer, Ibis 1879, 51; *(13)* Guillem., P. Z. S. 1885, 543; *(14)* W. Blas., Ztschr. ges. Orn. 1885, 218; *(15)* Salvad., Cat. B. XX, 1891, 535; *(XVI)* Meyer, Vogelskel. 1892, pt. 18, 44, t. CLXX; *(17)* Büttik, Z. Erg. Weber's Reise Ost-Ind. 1893, III, 273; *(18)* M. & Wg., Abh. Mus. Dresden 1895, Nr. 8, p. 6; *(19)* iid., ib. 1896, Nr. 1, p. 7; *(20)* iid., ib. 1896, Nr. 2, p. 9; *(21)* Hartert, Nov. Zool. 1896, 160; *(22)* id., ib. 1897, 165.

b. **Nanodes stigmatus** *(1)* S. Müll., Reiz. Ind. Archip. 1858, pt. II, 60, 71.

c. **Coryllis stigmata** *(1)* Finsch, Papag. 1868, II, 694; *(1bis)* Frenzel, Mtsschr. Ver. Schtz. Vogelw. 1880, I, 10; *(II)* Rchnw., Vogelbild. 1878—83, t. XV, f. 7; *(3)* id., J. f. O. 1881, 231 (Consp. Psitt. 119); *(4)* Platen, Gefied. Welt 1887, 206, 230.

"Tindito", Gorontalo, v. Rosenb. *c 1*; Joest.
"Tintis (Meyer *12*) or "Tientis" (Guillem. *13*), Manado.
"Tintis", Tonkean, E. Celebes, Nat. Coll.
"Bawan kidjili", Tjamba, S. Celebes, Platen *14*.

For further synonymy and references see Salvadori *15*.

Figures and descriptions. Rowley *IX*; Reichenow *c II*; Meyer *XVI* (skeleton); Schlegel *3*; Finsch *c 1*; Brüggemann *8*; Meyer *10*; Lenz *11*; W. Blasius *14*; Salvadori *15*.

Adult male. Parrot-green, yellower green on the under surface, the interscapulary region with a slight tinge of orange; sinciput, a patch on the throat and the metacarpal edge bright poppy-red; rump and upper tail-coverts crimson, much darker than the cap, the basal part of the feathers yellowish green; under wing-coverts yellowish green, the greater series verditer-blue, quills and tail below verditer-blue, the ends and exposed outer part of the quills black (Nr. 1931 — North Celebes).

"Iris light yellow; bill black; cere and feet orange" (Platen *14*).

Adult female and immature male. Like the adult male, but without the red sinciput (Nr. 1932). In the old female and young male the feathers of the forehead sidewards as far as the neighbourhood of the eyes possess a red basal half, showing through plainly, so as to give the forehead a reddish colour (W. Blas. *14*). Iris of female brown (Platen *c 4*).

First plumage. Like the adult female, but without the red bases to the feathers of the forehead; throat without any red patch, a yellowish space occupying the spot where it appears later; carpal edge greenish yellow (Nr. 1933 — N. Celebes). A young one about a fortnight old was quite green, except its light yellow shoulder-edges; under parts lighter green; nape light orange-tinged; the red of the upper tail-coverts already perfect; bill yellow; feet yellowish brown (Meyer *10*).

Eggs. "The eggs collected by Dr. Platen at Rurukan, the most highly situated village in the Minahassa, are almost spherical and extremely thin-shelled. Measurements: 19×17 mm" (Nehrkorn MS.). Colour white.

Nest and breeding season. "Breeds twice a year, viz: in February and August, and always prefers the sugar-palm, the lower withered leaf-stems of which afford it convenient nest-holes; *Coryllis stigmatus* lines the nest-holes some centimeters high with shreds of leaves" (Platen *c 4*).

Measurements. Wing (14 specimens, N. Celebes) 93—100 mm; tail 40—45; tarsus 10—11.5; culmen from core 11—11.5. The longest- and shortest-winged specimens (Nrs. 1931, 1741) are both adult males. A young yellow-billed specimen (Nr. 1933) is one of the largest, wing 98 mm.

Skeleton.

Length of cranium	33.8 mm	Length of tarso-metatarsus	11.5 mm
Greatest breadth of cranium	18.3 »	Length of digitus III	19.0 »
Length of humerus	22.4 »	Length of sternum	33.5 »
Length of ulna	25,6 »	Greatest breadth of sternum	20.0 »
Length of radius	23.4 »	Height of crista sterni	11.8 »
Length of manus	30.0 »	Length of coracoideum	18.8 »
Length of metacarpus	16.8 »	Length of scapula	24.0 »
Length of digitus princip.	13.3 »	Length of clavicula	12.0 »
Length of femur	22.5 »	Length of pelvis	30.5 »
Length of tibia	31.5 »	Greatest breadth of pelvis	18.0 »
Length of fibula	14.0 »		(Meyer *XVI*).

Distribution. Celebes, Minahassa (Forsten *a 1, 4*, Wallace *2, 15*, Rosenberg *6*, etc.); Lembeh, Mantehage and Banka Islands (Nat. Coll.); Gorontalo Province (Forsten *a 1, 4*, Rosenb. *6*, Meyer *10, 12*); Paguatt, N. coast of Gulf of Tomini (Rosenb. *6*); Posso, S. coast of Gulf of Tomini (Meyer *10, 12*); E. Celebes — Tonkean (Nat. Coll.) Macassar, S. Celebes (Wallace *2, 15*); Tjamba, S. Celebes (Platen *14*); Luwu (Weber *17*); Tawaya and Dongala, W: Celebes (Doherty *22*).

L. stigmatus has been found in most parts of the Northern Peninsula of Celebes as far as its base at Posso, and in the Eastern Peninsula; it has also been recorded from two or three spots in the South-western Peninsula, though Meyer remarked that it was less plentiful here than in the north *(10)*. In all probability, therefore, it exists in Central Celebes as well. It was not obtained in the south-eastern part at Kandari by Beccari, nor was it observed by the earlier travellers in Buton Island.

This Lorikeet is a common bird in North Celebes, ranging from the sea-coast high into the hills. Dr. Platen found it breeding at the village of Rurukan (over 3000 ft.). Meyer observed that it lived singly or in pairs, not in flocks; it is — like *Loriculus exilis*, *Trichoglossus ornatus* and *T. meyeri*, — not a strictly stationary bird, but in the habit of making local movements, regulated, no doubt, by the time of ripening of certain fruits. In the beginning of March, 1871, it was especially plentiful near Manado. It feeds on soft fruits, such as bananas and the like, and is therefore to be found in the plantations near the villages *(10)*. In captivity Dr. Platen found it ate readily canary-seed and biscuit, which *L. exilis* rejected until the food was bruised and made up into a sort of pap with bananas or sugar-water. A tame bird of Meyer's when at Manado was very fond of tea.

The *Loriculi* according to Dr. Frenzel (see Mtsschr. des D. Vereins zum Schutze d. Vogelwelt 1880, No. 1, 15) are much superior to the other dwarf Parrots in intelligence. The above-mentioned example was taken very young and was in the possession of Meyer and his wife for a long time in Celebes, and evinced a great affection for the latter. "It followed my wife everywhere in the house, and did not rest till it was near her, then, without help, climbed up from the ground to her shoulder or her head. It loved best to take food from her mouth, and licked up tea from her lips. When we took our tea in the afternoon, and the little bird only heard the rattling of the cups, it became much excited, and did not rest until its cage was opened and it could come near the table; it then took the tea out of a small basin or a spoon. When in its cage, and my wife passed by, it clapped its wings till she let it out. When we left Manado for a fortnight on a boat-tour, ... a neighbour reported that it had been melancholy all the time. When it saw my wife again, it became much excited; this I observed myself. Being placed in the same cage with the smaller *Loriculus exilis*, it always bit it; they could not remain together. But with the larger *Trichoglossus meyeri* it became anxious; nevertheless it attacked the bird as much as possible" (Meyer *10*). During a second and longer absence of Meyer and Mrs. Meyer from Manado, the bird, though properly cared for, died; it was reported to have cried incessantly, and at last was found dead. "The man in whose care it was, had seen and knew exactly how we had treated the bird; it therefore did not die on account of wrong or improper food, etc. I will not decide whether the explanation (of our neighbour) that it died from sorrow was the right one or not; at all events it was an amiable lovely bird for whose death we mourned. ... But the case serves to show that even these small Parrots are very sagacious and attractive creatures" *(10)*. Although far exceeding the Dwarf Parrots *(Psittacula)* in trustfulness and affection for their owner, they do not possess the same inseparable attachment as these for one another; on the other hand, they are quarrelsome when together, and Dr. Frenzel mentions a case of a young male of *L. galgulus* belonging to Dr. Russ which bit an old male to death, while three in his own possession would not let one another eat in peace, so that separation was necessary, or two of them would have been starved (op. cit. p. 16).

The genus Loriculus.
Map VI.

In 1877 Meyer *(10)* gave a key showing the geographical distribution of the genus *Loriculus*. The accompanying key is based upon this, modified according to Salvadori's treatment of the genus and to recent discoveries.

The Geographical Distribution of the genus Loriculus.

Species	India and Burmah	Ceylon	Andamans	Malay Peninsula	Sumatra	Borneo	Java	Flores	Luzon	Negros, Panay	Samar, Leyte	Cebu	Mindoro	Siquijor	S. Philippines	Sooloo	Sangi	Celebes	Togian	Sula	Peling, Banggai	Halmahera, Batchian	Mysol	New Guinea	Fergusson Id.	Duke of York Id.
1. L. vernalis	*		*	*																						
2. L. indicus		*																								
3. L. galgulus[1]				*	*	*																				
4. L. pusillus							*																			
5. L. flosculus								*																		
6. L. philippensis									*																	
7. L. regulus										*																
8. L. worcesteri											*															
9. L. chrysonotus												*														
10. L. mindorensis													*													
11. L. siquijorensis														*												
12. L. apicalis															*											
13. L. bonapartei																*										
14. L. catamene																	*									
15. L. exilis																		*								
16. L. stigmatus																		*								
17. L. quadricolor																			*							
18. L. sclateri																				*						
19. —— subsp. ruber																					*					
20. L. amabilis																						*				
21. L. aurantiifrons																							*	*		
22. —— subsp. meeki																									*	
23. L. tener																										*

In a key showing the geographical distribution of the *Loriculi*, like the above, it is not possible to arrange all the species according to their nearest affinities, since the genus is composed of two main branches, each comprising about half of the species, but it is sufficient to show that the birds are spread almost without interruption from India throughout the East India Archipelago to New Guinea (South-east) and the Duke of York Island in the New Britain Group, continually presenting new forms as one proceeds from point to point across this area, and it is pretty certain that still more species are to be discovered here.

Did the genus arise in the Asiatic Continent and spread its range southeastward to the Papuan Subregion, or did it proceed from New Guinea, where

---

[1] Frenzel (1ᵇᵒ) mentions a pair with blackish wings, which he supposes to be the sign of a local race, a supposition highly probable in consideration of the distribution of the species.

parrots abound, to India, or did it range in earlier times, when the whole was one mass of land, throughout the then Eastern Continent and was more or less differentiated, split into species, after the formation of the Archipelago?[1])

The conclusion seems unavoidable — and this remark is made after carrying the matter in mind for a period of about five years — that *Loriculus* (in some cases at least) extended its range by flight across the sea.

In addition to this we must state, as necessary to the following argument, that emigrants get altered more than stayers-at-home. It may in a certain sense be compared to the axiom of Euclid: "If unequals be added to equals, the wholes are unequal", the unequal condition in the zoological case being the new conditions of existence for the colonists.

Further, it may be stated that there is much reason to assume that the plumage of young birds is often ancestral in character. In proof of which it could hardly be possible to point to a better case than the racket tail-feathers of *Prioniturus*, which in the young clearly show an initiatory stage of development (see plates with figures and remarks thereon in our Introduction), and other good cases among Celebesian birds are the blue back in the immature *Pelargopsis melanorhynchus*, the traces of a yellow wing-band in the young *Trichoglossus meyeri*, the Kestrel-like plumage of the young of the Hawks *Accipiter rhodogaster* and *Spilospizias trinotatus*, the identical young plumage of the Hawk-eagle *Spizaetus lanceolatus* and the Honey-buzzard, *Pernis celebensis*. In many cases among birds the female shows a minor differentiation than the male, for instance, again, *Prioniturus*; very rarely a higher one *(Turnix)*. As Gadow remarks (Newton's Dict. B. 1893 p. 100): "Instances, too well known to be repeated here, show clearly how the changes of bygone ages of the ancestors are recapitulated in the yearly moult of the growing individual until with maturity its present stage of perfection is reached". The very utmost caution should, however, be used in making use of this principle, for all sorts of disturbing side-influences forbid its acceptance as thorough-going: sometimes, for instance, as in some of the Kingfishers, the young display no differences from their parents, sometimes the differences seen in them seem to be of a special protective or other character rather than ancestral, sometimes the mother's influence preponderates though her characters may be more recent than those of the father (for instance, in the genus *Edoliisoma*), sometimes, when the male seems to be the less differentiated, the young take after him *(Turnix)*, sometimes the male young one is like the father, the female young one like the mother (for instance, *Monachalcyon*, *Cittura*, *Eclectus*). It is excessively difficult to form an opinion as to whether the dress of a young bird is ancestral in character or not; one feels justified in doing so only when the young throws back very plainly to some form which is existing to-day.

---

[1]) We do not take into consideration the possibility, that the genus spread out east and west from one of the central island groups, as there appear to be no indications for such a supposition.

This appears to occur in many cases in the genus *Loriculus* and affords a reason for the construction of the accompanying genealogical tree:

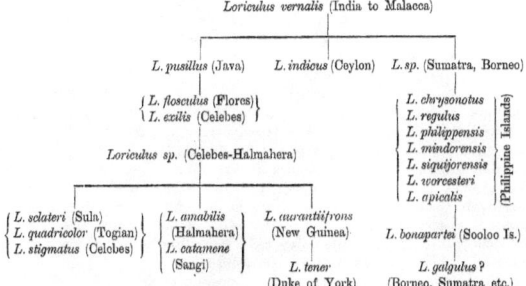

Loriculus vernalis (India to Malacca)

A glance at the above genealogical tree will show that *L. vernalis*, the "Stammform", to which we hypothetically trace the different species of the genus, is supposed to have given rise to two main branches, as well as to a brief offshoot, *L. indicus* of Ceylon, which has not been differentiated further. Hitherto, by several authors, the genus has been divided for practical purposes — as an aid to the determination of the species — into two groups, characterized by the possession of an orange-coloured or a black bill; but it would be an error to suppose that on this difference alone the genus can be divided into two natural groups. In the young of the black-billed *L. catamene, stigmatus, sclateri*, and *amabilis* the bill is known to be yellowish or whitish at first, in the black-billed *L. galgulus* it is dull yellow, shaded with dusky (Davison), and is stated by Guillemard to be "red", "very dark yellow" etc. in specimens of the black-billed *L. bonapartei*, a species in which we suspect the quite black bill to be present only in old birds. Furthermore, the black-billed *L. bonapartei* and probably *L. galgulus* are really as far removed as possible from the other black-billed species, and the former at all events should be placed at the extreme end of the other branch of the genealogical tree, notwithstanding the fact that there is a considerable resemblance between the adult males of *L. bonapartei* and *L. stigmatus*. The ontogenetic reasons on which this division of the genus rests are the following:

In addition to having a general colour of parrot-green, all the *Loriculi*, with one exception, possess a red rump and upper tail-coverts, a character assumed with the first plumage[1]), and which may, therefore, be regarded as of

---
[1]) The young *L. galgulus* has only the edges of the feathers red (Salvadori).

long standing. The exception is *L. tener* of Duke of York Island in the New Britain Group, in which the rump and upper tail-coverts are greenish yellow. The differences of plumage, which show that the genus is composed of two main branches originating in a form like *L. vernalis*, or even a form like the female of *L. vernalis*, which most likely represents a lower stage than the male adult, are the following:

The throat-spot. All the members of the left or, as we will call it, the southern branch of the genealogical tree possess a spot or patch of colour on the throat in one or both sexes, and in the higher forms even in the young. In the adult male of *L. vernalis* it is bluish (nearly or quite absent in the female); in *L. pusillus* of Java it is yellow (in the female, as in the male — according to Finsch; paler — according to Salvadori; absent — according to Reichenow), in *L. flosculus* and *exilis* it is red (whether present or absent in the female there is insufficient evidence to say); in the seven remaining species it is present as a red spot like that of *L. flosculus* in both male and female, and makes its appearance at an early age in the young, in the first plumage being sometimes, if not always, yellowish in colour.

In the species on the right or northern branch of the "tree" the throat-spot is always absent in the young and in the females; in the adult males alone it is developed as a final addition to their coloration in the form of an extensive patch of red on the throat.

In the male *L. galgulus* Dr. Frenzel observed from living examples that "first the blue spot on the crown shows itself, next the orange-coloured, triangular spot on the nape makes its appearance, and lastly the beautiful scarlet on the breast (and throat) appears; the change of coloration goes on very slowly and the bird is not in full dress till the third year" *(c 1 bis)*. But in the higher members of the southern branch the colour on the throat is among the first characters — not the last — to make its appearance. Since young specimens of *L. stigmatus, amabilis, sclateri*, and *catamene* with yellowish bills display it to some extent, it may prove to be present as yellow if not red in these birds from the time they leave the nest.

Thus, in the southern branch the throat-spot appears to be a somewhat ancient character, in the northern branch it appears to be the most recent acquisition of the adult male.

The coloration of the crown. The two groups again present a contrast in regard to the coloration of the crown. In the members of the left or southern branch, the females of all the species may broadly be said never to possess any special colouring on the head (in the females of *L. stigmatus, quadricolor, sclateri, amabilis* and *aurantiifrons* the green feathers of the forehead have reddish bases). In the adult males of the latter species (except *sclateri*) and in *L. catamene* a sharply defined red sinciput is produced (in *aurantiifrons* yellow); in the simpler species, *L. exilis, flosculus* (as far as known), *pusillus* and *vernalis* no sign of a red crown is to be seen in either sex, neither does it occur in the immature males which gain it when adult; these resemble the females when young. It is the latest character to be assumed in the ontogenetic development of the male individual as it gradually acquires the markings of the perfect bird, and it would, therefore, appear that the red sinciput is the latest acquisition in the phylogenetic development of this branch of the species.

In the members of the right or northern branch of the tree the process is strongly contrasted with this. Not only do the males possess red on the crown (except *L. galgulus* in which there is a blue spot on the crown, but we have found some traces of red in a young specimen), but also the females (except, again, *L. galgulus*) have red crowns — sometimes confined to the fore part of the head, sometimes extended to the nape, — and the immature males generally — probably always — resemble the females. Only the very young birds are quite green on the head, so bearing much likeness to *L. vernalis*, female. In this group, therefore, one may conclude that the red crown is a character of rather long standing.

Ontogenetic and phylogenetic development. As may be gathered from the above, the development of the plumage of the two branches of the *Loriculi* starting in common with a first plumage of green with a red rump and upper tail-coverts, afterwards proceeds on two quite different lines. It is known that in *L. amabilis* and *stigmatus*, two most highly specialized species of the left branch of the genealogical tree, a yellowish or whitish beak and a yellowish spot on the throat is found in the quite young bird (in *stigmatus* first "light yellow and then orange-yellow", according to Brüggemann 8). A nowexisting species with a yellow patch on the throat and a yellow bill is the simply coloured *L. pusillus* of Java, which therefore may be the not-much-changed ancestor of *L. amabilis* and *stigmatus* and its fellows. *L. amabilis* and *stigmatus* in the next stage of their immature development become like *L. flosculus* and *exilis*, in that the yellow spot changes into red, but the bill now or soon after loses its pale colour and becomes black, and the carpal edge and the bases of the feathers of the forehead show a red tinge; so that, though the young bird has still much in common with *L. exilis*, the parallelism in coloration is not perfect. The females do not develop their coloration beyond this stage, which is also the dress of the young males before they acquire the red sinciput, and the same characters are found in the females and immature males of four species — *L. stigmatus*, *quadricolor*, *sclateri* (the male of which never acquires a complete red cap) and *amabilis*). These considerations afford some ground for the supposition that these four species are descended from a common ancestor a good deal resembling the adult males of *L. flosculus* and *exilis* in coloration. The young female of *L. catamene* corresponds still more closely with *L. exilis* and *flosculus*.

The ontogeny of the plumage of the species of the northern branch of *Loriculus* proceeds by the young birds (*L. regulus*, *L. philippensis*) first of all assuming a plumage not unlike that of the immature, or female, bird of *L. vernalis*. This young form in both sexes (as is known in, at least, four cases[1]) develops into a bird like the adult female of its species, viz., a bird with a red crown of greater or less extent, and blue cheeks and chin. As the male becomes adult, it loses the blue on the cheeks and chin, and the red on the head generally undergoes some change in extent and tint, and the extensive patch of red on the breast appears. All the females have a general likeness among themselves, and an immature male of one species before us, *L. apicalis* (Nr. 6012) corresponds remarkably well with the female of another, *L. chrysonotus* (Nr. 1739). There is, therefore, reason to suppose that the Philippine species, including *L bonapartei* of Sooloo, are sprung from a common ancestor much like their females.

L. galgulus of Borneo, Bangka, Sumatra and the south part of Malacca is a very aberrant and puzzling form, differing in the adult male from the males of this branch in having a blue spot on the crown and no red there, and in the female from the females in wanting the blue on the face and red on the crown, yet it appears to belong to this branch of the *Loriculi* from the fact that the female has no red spot on the throat and that the male, like the Philippine males, develops a large patch of red here as a final adornment. We place it provisionally not far from *L. bonapartei*, in virtue of its black bill and that a young specimen from captivity and showing traces of xanthochroism before us offers one or two red feathers, tipped with greenish yellow, on the crown: but we doubt if further research will prove this position to be its true one.

The grounds given for regarding the *Loriculi* as consisting of two main branches — a left and a right one as shown in the above genealogical tree — may be briefly restated as follows:

Left (southern) branch:
1. The females of the more specialized black-billed forms have a spot of colour on the throat.
2. The males of the black-billed species assume the throat-spot at a very early age.
3. The females are without a red crown.
4. In the black-billed adult males only, a red sinciput makes its appearance.
5. The ontogeny of the plumage of the more highly specialized species tends to show that they sprang from forms like the simpler ones — green, with red rump and tail-coverts, a yellow or red spot on the throat, but no red on the crown.

Right (northern) branch:
1. The females have no spot on the throat.
2. Young males have no spot on the throat; a large patch of colour here and on the chest is the last thing to make its appearance in the male plumage.
3. The females possess a red crown.
4. The immature males have a red crown like that of the females.
5. The development of the Philippine races appears to betoken that they come immediately from a form like their females — green, with red rump and tail-coverts, no spot on the throat, but with a red crown and blue chin and cheeks.

[1] *L. regulus, philippensis, siquijorensis* (fide Steere), *apicalis*.

According to the above methods of reasoning, therefore, the genus *Loriculus* consists of two branches and one aberrant species of uncertain position, *L. galgulus*. The two branches by their young throw back to a form resembling *L. vernalis* of the Indian countries, and the more highly-differentiated species of the southern branch by their females and immature males throw back to the simpler species of their branch, which seem to stand nearest to *L. vernalis*. *L. vernalis* should, therefore, stand nearest to the ancestral type. From the postulate, that colonists get most changed, the Indian countries should, therefore, be the original habitat of the ancestral type, which became more and more changed as it proceeded to the Philippines on the one hand, and to Java, Celebes, Halmahera, New Guinea, on the other.

The alternative explanations should now be examined. It may be argued in the first place that the reverse of all the above process has taken place, — that the complex species of New Guinea, Halmahera, and Celebes were the original forms and extended their range westward, gradually losing their specializations of plumage until they became simplified into the modest-looking *L. vernalis* of India. There is the more reason for advancing this argument, as a retrograde development, or at least a change from more complex differentiations of structure and colour to simpler ones, may certainly take place. Species may have a culminating point set to their evolution beyond which they cannot develop, and having attained this they turn again into less specialized forms till they die out, — not from external causes, but because their vitality is exhausted and their career is over. As it is with the individual, so it may be with the species, the genus, the family, and so on. The very centre of the parrots in the Old World is at the present time (and may always have been) Australia and the eastern part of the East Indian Archipelago; Wallace looks upon the Oriental region as the land of origin of the *Psittacidae*, but this is far from being proved in any way, and Reichenow, for instance, offers reasons for making the hypothesis of the eastern origin just as plausible. At all events one has in either case to deal with mere probabilities, as proofs will be wanting till palaeontological facts allow men to abstain from speculations. (Compare Fürbringer's remarks in his fundamental "Untersuchungen" pp. 1116, 1287 and 1293). If the "Stammform" of *Loriculus* originated in New Guinea or the neighbourhood and spread from there westward, the ontogenetic characters, which are treated above as remnants of the phylogenetic development of the genus, must have another meaning, which we cannot explain. The argument that *Loriculus* spread its range from Papuasia to India, though it has this in its favour that the genus issued from the countries which are the richer in Parrots, fails, however, to account for the existence of the branch found in the Philippines and Ceylon, in which the ontogenetic development differs from that of the southern branch to which the Papuan species belong.

The third alternative explanation is that *Loriculus* ranged from India to

Papuasia when the whole was one undivided mass of land, and developed into distinct species when it became isolated by the breaking up of this ancient continent into the present archipelago. The family of the *Psittacidae* is in all probability an old one, existing at least from Miocene times. Unfortunately our knowledge of fossil *Psittacidae* is utterly defective as yet. Only *Psittacus verreauxi* M.-E. of the Lower Miocene of Allier proves that in a milder climate Europe was then inhabited by parrots. Extinct species or genera from other quarters (the Mascarenes, Seychelles, Comoros, and West Indies — from Rodriguez probably allied to *Eclectus*) are more or less subfossil only. As Prof. Fürbringer says (p. 1574):

"The palaeogeography of birds, as based on real finds, still stands on utterly weak legs: only single spots here and there of the earth have as yet been investigated, and discoveries, available for the phylogeny of the bird-class, are restricted to very few localities in Europe and North America. However the geography of living forms allows of some conclusions. Every ornithologist knows the brilliant reasonings of Wallace, the genial conclusions, attained by this author as to the early history of the distribution of birds. Whom do they not captivate, whom do they not induce to make some steps on the same route? Their formal value cannot be too highly appreciated. But sober investigation knows, that nearly everywhere only probabilities come into question here, nay, to a great extent even possibilities only, which may be controverted by other, not less justified suppositions. Very many families may have originated in the oriental region, but more cannot be said now-a-days. Our conclusions as to the distribution of land and water and as to other separating barriers offer some probability only for the caenozoic era; as to the condition before the Eocene we can only make suppositions. And at this period all suborders and gentes of the lower and the far greater number of those of the higher birds very probably were already defined. If someone were to assert that there was a nearly universal distribution of all the various chief-types at this period when the climate was everywhere fairly uniform and explain their later localisation in one region or the other by secondary extinction in the others — no one could offer sure counter-proofs. The possible decisions pro and contra, based upon the present geographical and morphological conditions, have nothing more than a greater or less degree of probability." Further p. 1572: "That in the beginning of the Eocene higher birds were already represented, proves, that the beginnings of the specialization of the *subordines* and *gentes* in view at least already took place at the end of the cretaceous period. It is even very possible that this occurred still earlier, for no one can assert that merely with these poor finds in our hands we already possess the typical forms of those times; the upper cretaceous, especially in its marine deposits, probably contains a great variety of such types as yet undiscovered." And p. 1109:

"Already the Eocene shows us forms which appear to be built after the specialized type of the now-living bird-divisions . . . The variety of Eocene birds is extremely large and authorizes us to suppose that, what has been found till now from this period, only represents a very small part of the bird-life, already very richly developed in those times. In the Miocene the similarity with the living fauna is further increased ... and in some cases one can refer the fossil forms even to living genera ... This similarity with the present time increases much more in the pliocene and quaternary times, when not only living genera but also recent species are represented ... During tertiary times the present geographical distribution was regulated. Eocene and Miocene with their tropical and subtropical climate still offer in our latitudes birds, which now only live in the tropics or their nearest neighbourhood ... At the same time towards the end of the cretacious period and in the course of the Eocene and Miocene the continental separations and connections took place (e. g. the separation of Australia from the oriental region at the end of the cretacious period ...), which ... were of predominant influence on the distribution of birds."

The foregoing deductions show that there is no reason why the genus *Loriculus* could not have existed from miocene times; by the argument it is assumed that it existed when New Guinea was terrestially joined on to Asia, and ranged over this continent now broken up. The line of reasoning may then be as follows: There exists in all organic beings a tendency to develop in a certain direction, a tendency which is better able to assert itself under isolation. We have already been induced to draw the conclusion that the appearance of a similar red crown in *L. indicus* of Ceylon and *L. apicalis* of the Southern Philippines, and the production of two considerably similar — but not nearly related — birds in the adult males of *L. stigmatus* and *L. bonapartei*, may be ascribed to a tendency to develop red on certain parts. Thus, under long separation *L. stigmatus* of Celebes may have become in virtue of this tendency from a beginning like *L. vernalis* first like *L. pusillus* of Java and have been changing ever since with time in a definite direction; *L. pusillus*, more recently isolated in Java, may have only reached its first simple stage of differentiation, but what *L. stigmatus* now is, or the like, *L. pusillus* may once become, so far at least as local influences allow.

Such or a similar palaeogeographical explanation of the distribution of *Loriculus* meets, however, with many difficulties, which we abstain from detailing, as we here are on too hypothetical grounds. We have only wished to bring before the reader some proofs of our endeavours in searching for the truth, which perhaps future ornithologists studying this case will attain, where we end more or less with a query, though we are inclined to look on our first explanation of the genesis of the species of the genus *Loriculus* as the most reasonable one. We bear, however, in mind, that the facts of geographical distribution and of variation of species are far too complicated to be mastered at the present time; much more knowledge than that of the present day is required to find the clue which will be satisfactory to the critical mind and the impartial judgment of the naturalist.

GENUS APROSMICTUS J. Gd.

This handsome genus is easily distinguished from all the other Parrots occurring in the Celebesian area by its long tail, as long or longer than the wings and strongly-graduated, the outermost feathers not reaching much more than half the length of the tail; fourth primary longer than the first; the reversed toe with the claw about as long as the middle toe without the claw; bill red or mostly red; principal colours red, blue, and green. Eight species ranging from Australia and Papuasia to the Moluccas, Sula and Peling.

† * 54. APROSMICTUS SULAENSIS Rchw.
Red-and-blue Parrakeet.
Plate VII.

a. **Platycercus dorsalis** var. *(1)* Wall, P. Z. S. 1862, 335, 337; 1864, 282, 293 (Sula).
b. **Platycercus dorsalis** *(1)* Schl., Ned. Tdschr. Drk. 1866, 184 (Sula); *(2)* id., Rev. Psitt. 1874, 38 (Sula).
c. **Platycercus amboinensis** *(1)* Finsch, Papag. 1868, II, 250 (Sula).
Aprosmictus sulaensis *(1)* Rchnw., J. f. O. 1881, 128 (Consp. Psitt. 64); *(2)* id., Vogelb. 1878-83 Nachtr. Nr. 16; *(3)* Salvad.; Orn. Pap. 1882, III, 515; *(4)* id., Cat. B. XX, 1891, 492; *(5)* M. & Wg., Abh. Mus. Dresden 1896 Nr. 2, p. 8.
"**Kakas**", Peling, Nat. Coll.

Adult. Entire head, neck, and under parts deep poppy-red, the under tail-coverts with broad mesial streaks of dusky blue; carpus, mantle, back and tail dark blue, brightest on the rump, duskier and more violet according to the light on the tail; the tail-feathers below black, all except the two middle ones tipped with red; wings dark parrot-green, a band of this colour passing across the blue of the upper mantle; metacarpal edge and under wing-coverts blue, slightly mixed with green; remiges below shining dusky. Bill reddish; feet blackish (ad. Peling, V—VIII, 1895, C 14758).

Younger. Differs from the adult in having the cervix and upper mantle green like the wings, the bill varied with reddish and dusky (Peling, C 14514).

Measurements. (3 examples in the Dresden and 2 in the Tring Museum from Peling) wing 182—190 mm; tail ca. 190—195; tarsus ca. 20; bill from cere ca. 20.

Distribution. Sula and Peling. Sula Islands — (Allen *a 1*); Sula Mangoli (Bernstein and Hoedt *b 1, b 2*); Peling (Nat. Coll. in Dresd. & Tring Mus.).

Observation. This, and three closely-allied Parrots, may be distinguished as follows:
   a. Interscapular region blue in adult.
     a'. Tail below black.
         A. dorsalis (Quoy & Gaim.): Distrib. — N. W. New Guinea and the surrounding western Papuan Islands (Salvad.).
     b'. Two outer tail-feathers broadly edged with red.
         A. amboinensis (L.): Distrib. — Amboina and Ceram (Salvad.).

b. With a green space on the mantle, or interscapulary region green with blue margins in adult.

  a″. Larger (wing 215—230), the outermost tail-feathers broadly, the next two pairs narrowly edged with red.
  A. buruensis Salvad.: Distrib. — Buru.

  b″. Smaller (wing 175); the tail-feathers not edged with rosy, only tipped therewith.
  A. sulaensis Rchw.: Distrib. — Sula Islands and Peling.

The Sula form of this Parrot was first discovered, either in Sula Mangoli or Sula Besi or in both islands, by Mr. Wallace's assistant, Allen, and Wallace observed that the Sula birds possessed a reddish bill, whereas in those of New Guinea it is black with the base of the upper mandible red. Schlegel's specimens from Sula Mangoli, from the collections of Bernstein and Hoedt, did not confirm the validity of the differences pointed out, the three first sent to the Leyden Museum all had black bills with some red at the base of the upper mandible. The perfect red bill seems to be a sign of maturity.

The discovery of this handsome species on Peling Island, which is separated from Celebes by a strait of less than 20 kilometers, and its absence on the mainland (so far as is known), is a curious circumstance. It is one of those facts which have led us to consider as probable the opinion that the Peling and Sula groups are the remains of one large island, which has been broken up in comparatively recent times. The genus *Aprosmictus* finds its maximum development in New Guinea, but the eight species recognised by Count Salvadori range from Eastern Australia to Peling.

The Buru form, which probably stands nearest to that of Peling and Sula, is larger and has a broad red border on the inner web of the outer tail-feathers.

Compared with *Tanygnathus*, *Prioniturus* and *Loriculus*, *Aprosmictus* occupies a somewhat remote position; it differs chiefly by its shortened reversed toe, by its long, graduated, broad-feathered, blue tail, and by its general coloration of red and blue.

ORDER PICI.

The Woodpeckers form a part of the heterogeneous group of birds combined under the term *Picariae*, perching-birds which are neither Birds-of-Prey, Parrots, Passeres, nor Pigeons. They include the Woodpeckers, Cuckoos, Hornbills, Toucans, Trogons, Rollers, Swifts, Humming-birds, Kingfishers, Bee-eaters, and others. Some of these birds contrast as strongly with others as do the long-acknowledged Orders; for instance, the Woodpeckers differ from the Swifts, the Hornbills from the Humming-birds as greatly as do the *Accipitres* from the Parrots or the *Passeres* from the Pigeons. In the "Catalogue of Birds", the sections of the *Picariae* are titled Suborders; in the "Fauna of British India, Birds" vol. III, Mr. Blanford prefers to treat them as Orders. We have employed the latter term for this work. The Pici are represented by the single

FAMILY PICIDAE.

Woodpeckers in life attract attention chiefly by their habit of clinging and climbing on the stems and branches of trees, pecking away the bark and wood, often making thereby a loud hammering, in search of the insects and their larvae on which they feed, or in forming the cavities in which they may eventually breed. The eggs laid therein are glossy white; the young in first plumage are known sometimes to resemble the male, when the sexes differ in adult plumage. The birds have no song, properly so called, though in the breeding-season they may utter peculiar notes. The well-known "drumming" made at this season by some species is caused by the bird striking a dead branch and then suffering the vibrating wood to beat against the tip of the bill. The flight is weak and undulating. In size the Woodpeckers vary from the dimensions of a Wren to those of a Crow; the most common colours are green, black, white, ochraceous, red, and yellow; pure blue is, we believe, absent.

A Woodpecker has a zygodactyle foot — the outer (fourth) toe being reversed, a character which at once distinguishes it from certain *Passeres*, such as the Nuthatches, Tree-creepers and Tits, which share to some extent its habits; the claws are curved to a semicircle (circa), are very narrow and deep. The bill is strong, straight, or nearly so, often sharpened laterally at the tip to a chisel-like point; the "tongue is very long, worm-like, provided with a pointed horny barbed tip, capable of great protrusion, the hyoid cornua extending backwards over the skull (except in *Sphyropicus* and *Xenopicus*)" (Hargitt, Cat. B. XVIII, 2); the occipital foramen appears to be placed well under the skull; the skin of the neck fits closely, so that in skinning the skull often cannot be passed through it; the keel of the sternum is low, the manubrium is furnished posteriorly with two processes on either side. There are twelve tail-feathers, which in the true Woodpeckers are stiffened and spiny, and used to support the body in clinging and climbing; in the Piculets *(Picumninae)* and Wrynecks *(Iynginae)* the tail-feathers are soft and ordinary, and these forms are distinguished as Subfamilies. For further particulars see Marshall, "Spechte", 1889; Hargitt l. c., Blanford l. c., and others.

GENUS IYNGIPICUS Bp.

The Pygmy Woodpeckers are about the size of a Sparrow; the bill about as long as the head tapering to a point; the wing rather long, the first primary very small, about $^1/_4$ the length of the second, the 3rd, 4th and 5th are the longest; the reversed fourth toe is longer than the third toe; the tail is short, about $^1/_2$ the length of the wing, the lateral rectrices normal and not much stiffened, the middle feathers and to a less extent the adjacent feathers pointed and stiffened. Hargitt recognised 18 species ranging from India, E. Asia and Japan to the Philippines, Celebes and Flores, and one aberrant species with a wide range in Africa.

+ * 55. IYNGIPICUS TEMMINCKI (Malh.).

Temminck's Pygmy Woodpecker.

a. **Picus temmincki** *(1)* Malh., Rev. Zool. 1849, 529; *(2)* Malh., Picidae 1861, I, 155, pl. XXXVI, fig. 3 ♀; *(3)* Brüggem., Abh. Ver. Bremen 1876, V, 49; *(4)* Rosenb., Malay Archip. 1878, 275 (Timincki err.).
b. **Yungipicus temmincki** *(1)* Bp., Consp. Vol. Zygod. 1854, 8; *(2)* Walden, Tr. Z. S. 1872, VIII, 41, 111; *(3)* Salvad., Ann. Mus. Civ. Gen. 1875, VII, 647; *(4)* Meyer, Ibis 1879, 57; *(5)* Platen, Gefied. Welt 1887, 206.
Iyngipicus temmincki *(1)* Hargitt, Ibis 1882, 40; *(2)* Gould, B. Asia VI, pl. XXV, ♂, ♀ (1882); *(3)* Guillem., P. Z. S. 1885, 546; *(4)* Marshall, Die Spechte 1889, 64; *(5)* Hargitt, Cat. B. XVIII, 1890, 335; *(6)* M. & Wg., Abh. Mus. Dresd. 1895, Nr. 8, p. 6; *(7)* iid., ib. 1896, Nr. 1, p. 8; *(8)* Hartert, Nov. Zool. 1896, 159; *(9)* id., ib. 1897, p. 158.

"**Tatangkul kokie**", Manado District, Nat. Coll.
"**Tukang ketjil**", Malay, near Manado, Nat. Coll.
For further synonymy and references see Hargitt 4.
Figures and descriptions. Gould *II*; Malherbe *a II*, *a 1*; Walden *b 1*; Salvadori *b 3*; Hargitt *1, 5*.
Adult male. Mantle, back, and scapulars dusky olive-green, barred with buffy; rump uniform buff; upper tail-coverts and tail darker buff, barred with bistre-brown; wing-coverts and quills darker dusky olive-green, marked with rather large buffy white spots on the outer webs of the coverts near their ends, notched with buffy on both webs[1]) of the quills nearly meeting so as to form bars about 8 mm apart; head above and ear-coverts olive; lores, a supra-orbital stripe, sub-orbital and malar stripe, and a nuchal patch, buffy white; a collar behind ear-coverts, not quite meeting at the nape, scarlet-vermilion; chin and throat slate-grey, the ends of the feathers white; under surface bistre-brown, the margins of the feathers paler and fulvous — whitish on the sides and flanks, — giving a broadly streaked appearance; under wing-coverts white, olive about the metacarpal edge; quills below olive, more broadly barred with white than above (near Manado, [♂] Aug.-Sept. 1892, C 10856).

"Iris brown ["dull crimson", Doherty *9*]; bill greyish brown; tarsus dull olive" (Guillem. *3*).

---

[1]) Hargitt speaks only of the outer webs of the quills as being spotted with white.

Adult female. Similar to the male, but without the scarlet patches on the sides of the occiput (Manado, ♀, Mch. 1871, C 473, 474).

Some specimens present a much more barred appearance on the back than others. This may be a matter of age, the more uniform specimens being probably the younger.

Measurements.

	Wing	Tail	Bill from nostr.	Tarsus
a. (C 10856) [♂] ad. near Manado .	76	30	13.5	13.5
b. (C 5157) [♂] ad. Manado	75	33	12.5	—
c. (C 5158) [♀] ad. Manado .	75	38	—	—
d. (C 474) ♀ ad. Manado . .	79	40	—	13.5
e. (C 473) ♀ ad. Manado	79	30	13	14
f. (Sarasin Coll.) ♀, Tomohon, 5. IV. 94	76	35	13	14
g. (C 14907) ♂, Indrulaman, S. Cel., Oct. 95 (Everett)	79	34	15	14
h. (Sarasin Coll.) ♂, Macassar, 27. VIII. 95 . .	79	35	14	14
i. (Sarasin Coll.) ♀, Macassar, 11. VI. 95 .	85	39	15	14.5

Egg and breeding habits unknown.

Distribution. Celebes — South Peninsula (Wallace *b 2, 4,* P. & F. Sarasin, Everett *8*); Minahassa (Rosenb. *a 4*, Meyer *b 4*, Fischer *a 3*, Guillem. *3*, etc.).

This little Woodpecker has as yet only been recorded from the northern and southern extremities of Celebes. Hargitt makes no mention of differences in the southern birds, such as are found in the other Celebes Woodpecker, *Microstictus fulvus*, in its southern quarters, neither have we been able to detect any, and there can be little doubt that *I. temmincki* will be found to range all over the island. Most writers speak of it as a rare species; it was not found to be so by Meyer near Manado, nor, apparently, by our native collectors in August and September, 1892, when five specimens were obtained in the same neighbourhood. Dr. Platen met with it among the orange and coffee plantations near Rurukan (3000 ft.). Nothing has been recorded of its habits. The allied *I. canicapillus* (Bl.) of Tenasserim is said by Davison (Str. F. VI, 125) to frequent old clearings, moderately thin jungle, groves of trees, etc., but to avoid, as a rule, dense forest or bamboo jungle. *I. gymnophthalmus* of Ceylon and South India is said to live in the highest branches of trees (Legge, B. Ceylon, 187). "Its powers of flight, afforded by its long wings, are considerable".

The nearest known ally of *I. temmincki* is *I. ramsayi* Hargitt of the Sooloo Islands. With one exception the genus *Iyngipicus* is confined to the Oriental Region, but, like *Pelargopsis*, it crosses Wallace's line between Bali and Lombok, a species being found in the latter island, Sumbawa, and Flores. *I. temmincki* and *I. ramsayi* are, unlike the rest, "above olive-brown" — we should add with a strong wash of green in the former — "striped or barred with white"; the remaining Oriental forms are "above black and white, the white uniform or barred", but *I. ramsayi*, in the coloration of its under surface, betrays affinity to *I. aurantiiventris* (Salvadori) of Borneo. *I. temmincki* appears to be one of the most distinct members of the genus. The single species occurring elsewhere outside the Oriental Region is *I. obsoletus*, which ranges, according to Hargitt, from N. E. Africa to Equatorial Africa, westward to Senegambia, a remarkably wide distribution for a genus the species of which are so narrowly located elsewhere. This is an aberrant form, though Edward Hargitt considers it a true *Iyngipicus*. It is very far removed in characters from *I. temmincki*. A few remarks on the distribution of the East Indian Woodpeckers will be found under *Microstictus fulvus*. As is well known to ornithologists, no species of Woodpecker is known east of the Molucca Straits, the family being entirely absent from the Australian region, through occurring elsewhere all over the world — except in Madagascar and the Pacific Islands — from about the Arctic Circle, south to South Patagonia, the Cape, Flores and Celebes (see Marshall, Spechte, 1889, 42, 43).

GENUS MICROSTICTUS[1]) Harg.

A genus of four species restricted to Celebes and the Philippines. In size about equal to a Turtle Dove; exposed culmen not longer than the head; nostrils feathered; tip of upper mandible chisel-edged; third toe longer than the reversed fourth toe; tail nearly as long as the wings, the individual rectrices terminally narrowed, and stiffened like whale-bone; first primary half as long as the second; 4th, 5th and 6th subequal and longest.

+ * 56. MICROSTICTUS FULVUS (Q. G.).

North Celebesian Slate-and-fawn Woodpecker.

a. **Picus fulvus** *(1)* Quoy & Gaim., Voy. Astrol. 1833, text, 228, Atl. pl. 18, fig. 2 (♂); *(2)* Rosenb., Malay Archip. 1878, 275 (lulvus); *(3)* Joest, Das Holontalo 1883, 105.
b. **Hemilophus fulvus** *(1)* Gray, Gen. B. II, 1846, 439; *(II)* Rchb., Handb. Picinae 1854, 385, t. 644, f. 4302—03.
c. **Dryopicus fulvus** *(I)* Malh., Picidae 1861, I, 53, pl. XIV, fig. 1 ♂, 2 ♀.
d. **Campethera fulva** *(1)* Gray, HL. 1870, II, 193; *(2)* Brüggem., Abh. Ver. Bremen 1876, V, 48.
e. **Mulleripicus fulvus** *(1)* Wald., Tr. Z. S. 1872, VIII, 41 pt. (Manado); *(2)* Garrod, P. Z. S. 1873, 630; *(3)* Lenz, J. f. O. 1877, 366; *(4)* Meyer, Ibis 1879, 57, 145; *(5)* Guillem., P. Z. S. 1885, 546.
f. **Alophonerpes fulvus** *(1)* ? Salvad., Ann. Mus. Civ. Gen. 1875, VII, 646; *(2)* W. Blas., Ztschr. ges. Orn. 1886, 87; *(3)* Marshall, Die Spechte 1889, 58 (Borneo err.).
Microstictus fulvus *(1)* Hargitt, Cat. B. XVIII, 1890, 490; *(2)* M. & Wg., Abh. Mus. Dresd. 1895, Nr. 8, p. 6; *(III)* Meyer, Abb. Vogelskel. 1897, pl. CCXXVII.
"**Burong tukan**" (Carpenter Bird). Malay name, Minahassa (Meyer e 4, Guillem. e 5, Nat. Coll.).
"**Rumerkukor**" (Meyer) near Manado.

---

[1] Mr. Hartert (Nov. Zool. 1897) would reinstate the name *Lichtensteinipicus*, intended by Bonaparte for the Philippine form, *M. funebris*, but by the Rules of Nomenclature, Hargitt was justified in rejecting it, since it has, we believe, never been described. Hargitt's *Microstictus* (type, apparently, *M. fulvus* of Celebes) has been properly diagnosed (Cat. B. XVIII, 7, 489).

"**Tatankul**" — near Tanawangko, Meyer, — or "Tatankul Sela", Nat. Coll, Minahassa.
"**Tukan besar**", Malay, Minahassa, Nat. Coll.
"**Bitu-bitua**", Gorontalo, Joest *a 3*.
"**Widi-widua**", Rosenb. *a 2*.
For further synonymy and references see Hargitt *1*.
Figures and descriptions. Quoy & Gaimard *a 1*; Reichb. *b II*; Malherbe *c 1*; W. Blasius *f 2*; Hargitt *1*.

Adult male. General colour above dusky slaty, becoming dark bistre-brown on the upper tail-coverts and tail-feathers, the shafts of which are whity-brown in places, both shafts and webs pure bistre towards the tip; entire front part of head, including forehead, crown, sides of head, and malar region, as far as a line drawn about 5 mm behind the eye, red (yellowish crimson); remaining parts of the head and hind neck slaty like the back, but each feather furnished with a small tip of white. Chin and throat broccoli-brown, speckled with minute white tips to the feathers; remaining under parts buffy clay-colour; tail below broccoli-brown washed with tawny-olive towards the tip; under wing-coverts slaty, touched with clay-colour (Banka Id. off N. Celebes, 16. V. 93 — C 12275). "Iris light yellow; bill black; foot greenish grey" (Guillem. *e 4*).

E. Hargitt remarks that the red immediately behind the eye is developed in very old males only. This appears to be the case in our specimens also, though one or two which must be old, judging from their bills, have no red behind the eye.

Adult female. Similar to the male, but without red on the head, this colour being replaced by dusky slaty grey like the hind head, speckled with smaller white tips (Banka Id. 19. V. 93 — C 12276, Manado, C 5156).

Immature (female). Like the adult female, but with the crown and most of the forehead red; under parts browner (N. Celebes, C 3600). Obs.: Hargitt was evidently quite right in describing a similar specimen as a young female. His opinion is fully confirmed by two immature specimens of the next species from East Celebes; they are casting off the old worn red feathers on the top of the head and are acquiring new ones of slaty with whitish spots, as in the adult female. The young therefore resembles the male more than the female.

Fledging male. Differs from the above plumage (as described *1*) in having the red patch on the forehead and crown more marked and richer in colour; the under parts more of a sandy grey; the tail-shafts lighter and the under surface of the tail greyer (Hargitt).

Measurements (12 specimens). Wing 176—189 mm; tail 150—172; tarsus 30 ca.; bill 28.5 (♀ ad. and ♀? imm.) — 36 (♂ ad.)

Egg and nest. Two, seldom three, eggs in a hole of a dead tree (Meyer *e 4*).

Distribution. Celebes, Northern (and S. E.?) Peninsula: — Lemboh, Banka and Mantohage, — Is. off the Minahassa (Nat. Coll. in Dresd. Mus.); Minahassa (Meyer *e 4*, Guillem. *e 5, etc.*); Gorontalo District (Meyer *e 4*, Joest *a 3*); Togian Islands (Meyer *e 4*); (?) Kandari, S. E. Celebes (Boccari *f 1*).

Touching the occurrence of this form in the S. E. Peninsula, Prof. W. Blasius (Z. ges. Orn. 1885, 239) remarks: "As at this place the southern *Streptocitta*-species: *albicollis* occurs, it would be very remarkable if the *Alophonerpes- (Microstictus)* species were not also the southern *wallacei*." Blasius' opinion is strengthened by the discovery of *M. wallacei* in Central and East Celebes.

Next to nothing has been recorded of the habits of *M. fulvus*: Meyer remarks that it feeds on tree insects, on white ants, caterpillars, etc., as all

Woodpeckers. Lives in pairs. If the male and female lose each other, the male knocks and the female follows the sound".

The genus *Microstictus* is, as Hargitt shows, peculiar to Celebes and the Philippines, being represented in the latter islands by two species, *M. fuliginosus* (Tweedd.) of Mindanao and Samar (Steere) and *M. funebris* (Valenc.) of Luzon, Marinduque and Cagayan. These forms differ widely from *M. fulvus* in coloration, being unicolorous above and below, but they (*M. funebris* at least) seem to possess no structural differences and might almost be described as small melanistic races of *M. fulvus*.

The genus *Microstictus* is most nearly related to *Hemilophus*, the single species of which, *H. pulverulentus*, was found by Hargitt to range from the Himalayas to the Malay Peninsula, Sumatra, Java, and Borneo.

The geographical distribution of the Woodpeckers, as a family, is of high interest. "It is similar", writes Marshall (Spechte, p. 42), "to that of the Cats, Dogs, Martins and Squirrels; the Woodpeckers are wanting, like these, in Madagascar and in by far the greatest part of the Australian Region, where in contrast to the mammals mentioned they occur further in Celebes[1]), as they also, likewise in contrast to these, are found in the West Indies. In two parts of the earth they have attained a remarkable development, a great richness in original, well marked genera and beautiful species, namely in tropical America and in India, a fact which stands in the closest connection with the increased diversity and greater richness of the forests. In America their range extends from Port Famine in the south to beyond the Polar circle in the north, and under the equator from the strand to 14,500 feet above the sea-level in the Andes. They are found in the islands off the coast of Chili and western North America, on most of — perhaps all — the West Indies and Bermudas, but are wanting on Juan Fernandez, Mas a fuero, as also in the Galapagos Islands and in the Bahama Group at least as breeding birds. In the Old World they reach on the West coast from 70° N. to the Cape, but are found in the islands of the Atlantic only in Great Britain, in the Azores and the Canaries. In the eastern half they are met with from about the Polar Circle as far as Celebes and the Philippines, are wanting, however, in the islands east of a line drawn from Cape Navarin along the east coasts of Kamtschatka, the Kurile Islands, the Japanese Archipelago and the Loochoo Islands as far as Cape Engano in Luzon".

As compared with other Picariae in the East Indies the Woodpeckers correspond best as regards their distribution with the *Bucerotidae*; also to some extent with the *Phaenicophaeinae*. All three groups — excepting one genus of

---

[1] Prof. Marshall is, however, in error in excluding Celebes from the range of the Squirrels, four species being known from the island, three of which are peculiar, and none occur farther east; also a species of *Paradoxurus* and one of *Viverra* belong to it. Two other families of mammals, unknown in Madagascar and in the Australian Region — except in Celebes — might be added, viz. the Apes, and the Oxen, though the first are of course confined to the warmer parts of the globe.

Buceros of good flying-powers in marked contrast to its fellows — are absent from the Australian Region; all three, also, are absent from the Sangi Islands, a circumstance which speaks for their recent volcanic upheaval. As in the case of the *Bucerotidae*, Celebes has two Woodpeckers, one rather large, the other a small one, which, moreover, find their nearest allies in the Philippines with Sooloo. The *Bucerotidae* greatly enforce Mr. Wallace's view of the former land-connection of Borneo, Sumatra and Malacca and of the earlier separation of Java from them, and a certain amount of similar proof is brought by the Woodpeckers. As Hargitt shows in his Catalogue the same species of the following Himalayan or tropical genera are peculiar to Malacca, Sumatra and Borneo, whereas Java has its own peculiar species of them, viz.

Malacca, Sumatra and Borneo	Java
Chrysophlegma (2 sp.)[1]	*Chrysophlegma* (2 sp.)
Gauropicoides	absent
Lepocestes	absent
Miglyptes (2 sp.)	*Miglyptes* (1 sp.)
Hemicercus	*Hemicercus*.

On the other hand several other Himalayan genera extending into the Sunda Islands are represented by identical species in Java, Borneo, Sumatra and Malacca, viz. *Tiga*, *Chrysocolaptes*, *Hemilophus* (ranging to the Himalayas in one species), *Thriponax* (also in the Philippines) and *Iyngipicus* (most species of which are local, but one, *I. auritus* Eyt., has the above range extending further to Cochin China). *Micropternus*, an Indo-Chinese form connects the Malay Peninsula, Sumatra and Java by a subspecies of the Indo-Siamese *M. phaeoceps*, while Borneo has its own peculiar species. The wide-spread *Dendrocopus* of Europe, Asia, N. America and southern S. America also has a species in Malacca, Sumatra and Borneo, but is wanting in Java. The two last-named genera are, however, more likely than the others to have extended their range by migration, because they belong to a more northern area — especially *Dendrocopus*. Of particular interest is the reoccurrence of the South and Central American genus *Picumnus* in the Oriental Region from the Himalayas and China south to Sumatra.

To account for the distribution of the Woodpeckers Wallace suggests a Central Asiatic origin, whence they spread to South America by way of North America, and developed into great diversity of form under the favourable conditions of the Neotropical forests (Geogr. Distr. II, 303). Tristram has urged a circumpolar origin. Marshall speaks for their evolution in the New World from the Passerine branch, and their gradual spread into the eastern Palaearctic Region, whence, as in North America, they were driven southwards by the glacial period, to work their way in many cases north again as the cold moderated. This view contains, perhaps, the fullest explanation.

---

[1] Sumatra and Malacca have each a peculiar species.

Marshall by no means makes it clear, however, why Celebes should have been colonised by its two Woodpeckers "at a much later time" than the other Sunda Islands. The two Celebesian species are very distinct from the other members of their genera in the Philippines and the Bornean Province, and they should be regarded as very long separated from the rest. On the other hand, as has been shown, the species of many genera in Java, Borneo and Sumatra are identical with, or but little different from, those of Malacca and S. E. Asia. These, therefore, may be considered either to have more recently invaded these islands by flight, or to have been recently cut off in them by the submergence of land. Though their wings are of good size, Woodpeckers do not appear to attempt long flights, except, perhaps, the *Colaptes* of America, in which it is, as Malherbe says, rapid and prolonged; the keel of the sternum is shallow and the pectoral muscles, judging from the birds' appearance on the wing, are feeble, and the habits of the family only call for flight in launching themselves from one tree to another, or in making their way at a low elevation to the next wood. Their absence in New Guinea and Madagascar, where, as Marshall points out, there can be no lack of suitable insect-food, and in the islands of Polynesia, where indeed insects are very scare, prove them, as might be expected from their flight and habits, to be birds to which a narrow sea-passage is a formidable barrier. On the whole, evidence is in favour of the view that *Microstictus* and *Iyngipicus* reached Celebes at a time when Borneo, Sumatra and Java were united to the Asiatic continent and when Celebes itself and the Philippines were also nearly or quite in touch with the continent. Submergence then put a sufficient barrier between it and Celebes, while new forms from the north subsequently invaded the other countries and gave the Philippines, Borneo and Java an avifauna so much richer in *Picidae* than that of Celebes.

* 57. MICROSTICTUS WALLACEI (Tweedd.).
Wallace's Slate-and-fawn Woodpecker.

a. **Alophonerpes fulvus** *(1)* Cab. & Hein. (nec Q. & G.), Mus. Hein. IV, pt. 2, 1863, 107; *(2)* ? Salvad., Ann. Mus. Civ. Gen. 1875, VII, 646; *(3)* Tristr., Cat. Coll. B. 1889, 102; *(4)* Heine & Rchnw., Nomencl. Mus. Hein. 1890, 216, pt.
b. **Mulleripicus fulvus** *(1)* Wald., Tr. Z. S. 1872, VIII, 41.
c. **Mulleripicus wallacei** *(1)* Tweedd., Ann. & Mag. N. H. 1877, XX, 533; *(2)* id., Collected Orn. Works 562, 667; *(3)* P. & F. Sarasin, Z. Ges. Erdk., Berlin 1895, 327.
d. **Alophonerpes wallacei** *(1)* W. Blas., Ztschr. ges. Orn. 1885, 209, 236, pl. XI, ♂, ♀.
e. **Microstictus wallacii** *(1)* Hargitt, Cat. B. XVIII, 1890, 491.
Microstictus wallacei *(1)* Büttik., Zool. Erg. Weber's Reise in Ost-Ind. 1893, 273; *(2)* M. & Wg., Abh. Mus. Dresden 1896, Nr. 1, p. 8; *(3)* iid., ib. Nr. 2, p. 10; *(4)* Hartert, Nov. Zool. 1896, 159; *(5)* id., ib. 1897, 158.
f. **Lichtensteinipicus wallacei** *(1)* Hart, Nov. Zool. 1897, 164.
"Bantinotto", Tjamba District, Platen *d I*.
"Sumboli", Tonkean, E. Celebes, Nat. Coll.

Figure and descriptions. W. Blasius *d 1*; Tweeddale *c 1*; Hargitt *e 1*.
Adult male. Differs from the adult male of *M. fulvus* in having the red of the head carried much further back, so as to include the occiput and entire sides of head; the tail much darker (Blas. *d 1*, Hargitt *e 1*, ♂, Indrulaman, Oct. 95, Everett — C 14901). Iris light yellow; bill black; feet lead-grey (Platen *d 1*).
Adult female. Differs from the adult female of *M. fulvus* in having the ground-colour of the chin and throat wood-brown, upon which the white tips of the feathers stand out much less distinctly than on the broccoli-brown; tail much darker, "brownish dusky above and below, the shafts nearly black above, their under side dusky, with smoky white bases". Iris, etc. as in male. (W. Blas. *d 1*, Hargitt *e 1*; Tjamba, 24. V. 78 — C 10425).
Young male. With less red on sides of face than in adult male; malar stripe red; lores, orbital region and fore parts of ear-coverts unspotted dusky grey, a few feathers under eye tinged with crimson; ground-colour of chin and throat more dusky, the spots more rounded and very minute; chest and breast of a more dingy buff colour, many of the feathers with dusky shaft-streaks; shafts of tail-feathers browner, and their bases very much paler (ex Hargitt).
Immature [female]. Head above red; in other respects like the adult ♀. (Tonkean, E. Cel., V—VIII. 95 — C 14460 — moulting.)
Measurements (12 specimens). Wing 178—192 mm; tail 145—172; culmen 37—45; rictus 43—52; tarsus 30—33 (W. Blasius *d 1*); bill from nostril c. 30—46 mm.
Distribution. Celebes:— Southern Peninsula (Wallace *c 1*, *e 1*, Platen *d 1*, Weber *1*); Luwu at the head of the Gulf of Boni (Weber *1*), Lembongpangi, S. Centr. Celebes (P. & F. Sarasin *c 3*, *2*); Tonkean, E. Celebes (Nat. Coll. *3*); ? Kendari, S. E. Peninsula (Beccari *a 2*); Dongala, W. Celebes (Doherty *f 1*).
Observation. This Woodpecker is now known from South, Central, West and East Celebes and will very likely prove to be the species of all parts of the island except the North Peninsula, where it is represented by *M. fulvus*. It was probably this species, rather than *M. fulvus*, that Beccari found in S. E. Celebes. There seem to be no differences of colour in *M. wallacei* from different parts of its range, and transitions towards *M. fulvus* are not yet known, so that the two may be regarded as species, though they may turn out later as subspecies of a single form.

ORDER COCCYGES.

Two families are placed in this order or suborder by Shelley (Cat. B. XIX, 209) and Blanford (Faun. Br. Ind., B. III, 1895, 203), viz. the Cuckoos *(Cuculidae)* and Touracos *(Musophagidae)*, an African family. In the former the outer toe is always reversed, and the oil-gland nude; in the Touracos the outer toe is reversible, the oil-gland tufted (Shelley) and "the bill of nearly all the species is curiously serrated or denticulated along the margin" (Newton, D. B. 982).

On the order *Coccyges* Mr. Blanford remarks: "This is distinguished from all other zygodactyle groups, except the Parrots, by possessing the ambiens muscle, a character to which a very high importance was attached by Garrod, and by the deep plantar tendons being arranged as in Gallinaceous birds, and only differing in arrangement from the Passerine plan by being connected by a vinculum; the *flexor longus hallucis* leads to the hallux alone, the *flexor perforans digitorum* serves the other three digits. The palate is desmognathous; basipterygoid processes are wanting". Resemblances to the *Gallinae* and the *Opisthocomus* have been pointed out by Huxley and Garrod; an important point of difference is that the young of the *Gallinae* run about and feed themselves as soon as they are hatched, but the young of the *Coccyges* are hatched naked and helpless, in which they resemble the Pigeons more.

FAMILY CUCULIDAE.

The best external clues to the recognition of a Cuckoo-form are, perhaps, furnished by the zygodactyle foot (the outer toe reversed), the claws usually weak, the bill decurved and usually weak, with the nostrils exposed (except in 2 or 3 genera, *Dasylophus, Lepidogrammus*), the tail-feathers 10 in number (except in *Crotophaga* and *Guira*, which have only 8) and often long and broad. The following anatomical and pterylological characters are noted by anatomical writers: Palate desmognathous; basipterygoid processes absent; caeca present; both carotids present; contour-feathers without any aftershaft; dorsal feather-tract divided between the shoulders, and enclosing a lanceolate naked space on the back; oil-gland present, but nude; young hatched naked, and not passing through a downy stage before acquiring feathers. Shelley recognises 6 subfamilies, of which three, *Cuculinae, Centropodinae,* and *Phoenicophainae* occur in Celebes, and *Scythrops* is entitled to be regarded as a fourth subfamily.

SUBFAMILY CUCULINAE.

The true Cuckoos may be recognised, as Shelley points out, by their long, flat wings which do not fit closely to the body, and the longest primaries overreach the secondaries by more than the length of the tarsus or bill. The tail rarely exceeds the wing in length (*Coccystes, Cacomantis* and *Surniculus* are exceptions). They are parasitic in their habits, at least in most cases.

GENUS HIEROCOCCYX S. Müll.

A genus which hardly differs from *Cuculus* except in having the wing less pointed, the secondaries being about $^2/_3$ to $^3/_4$ wing-length, instead of only about $^1/_2$ as in *Cuculus*, and the tail-feathers crossed with four or five dark bands, instead of having them (in the adult) black with spots of white at intervals against the shafts of the feathers. Six species are recognised by Shelley, ranging from East Siberia and India to Celebes. One of them, *H. sparverioides,* has been observed to build its own nest and brood on its eggs, but it has also been recorded as parasitic, like, so far as is known, its fellow-species.

+ * 58. HIEROCOCCYX CRASSIROSTRIS Tweedd.
Blunt-winged Cuckoo.

Hierococcyx crassirostris *(1)* Wald., Ann. & Mag. N.H. 1872, 305; *(II)* id., Tr. Z. S. 1872, VIII, 116, pl. 13; *(3)* Meyer, J. f. O. 1873, 405; *(4)* Platen, Gefied. Welt 1887, 219; *(5)* Shelley, Cat. B. XIX, 1891, 239; *(6)* M. & Wg., Abh. Mus. Dresd. 1895, Nr. 8, p. 6.

a. **Cuculus crassirostris** *(1)* Brügg., Abh. Ver. Bremen, 1876, V, 59.

"Kokokuk" (meaning simply a Cuckoo), Tondano, Nat. Coll.

Figure and description. Walden *II, 1*; Shelley *5*.

Adult. Head, neck, ear-coverts, and cheeks rather pale slate-grey; upper surface dark rufous brown, the margins of the wings mottled with russet, the rump and upper tail-coverts redder and obscurely tipped with grey like the head; metacarpal joint white; tail rufous brown, brighter than the back, tipped with white and crossed with three or four broad bands of black, the interspaces on the more lateral feathers chiefly white or white mingled with rufous, the middle feathers all rufous brown with a broad terminal band of black and a black spot indicating another band with a small spot of white adjoining; under surface white, thickly barred with slaty black on chin and throat, with purer black across chest, breast, sides and flanks; under tail- and wing-coverts white, the latter with a brown spot here and there at the edge of wing and on the longer series; quills below broccoli-brown, white towards base (near Tondano, Aug.—Sept. 1892 — C 10796).

"Bill black, at base of lower mandible yellow; bare skin round the eyes gold-yellow; feet gold-yellow". Other adult specimens (all marked ♂) for the loan of one of which we are indebted to Mr. Nehrkorn, while the others are in the Sarasin Collection, are like the above. "Iris brown; bill black, below yellow-green; orbits and feet light yellow" (Rurukan, 2. XI. 84 — Platen in Mus. Nehrk. Nr. 931).

Immature. Differs from the adult in having the head above brownish slaty; neck and upper surface chestnut (i. e. lighter and redder), some remains of dark bars on the tertiaries above; below buffy — not pure — white, plentifully barred with blackish brown, except on under wing- and tail-coverts (Minahassa — Nr. 2207). This is like fig. 2 of Lord Walden's plate, but the under surface is more thickly barred. As Captain Shelley points out, the tail in this figure conveys a wrong impression, the two middle feathers being lost; there is no white except at the tip on the middle tail-feather (there is only one remaining) of our specimen, though the shaft is pale at one spot.

Young. A younger specimen in the Nehrkorn Coll. (Nr. 930 — Rurukan, 10. XI. 84: Platen) differs from that described in being still paler rufous brown above, with the mantle indistinctly spotted; below still more thickly barred with black. "Iris brown; bill brown, below yellow-green; orbits and feet light yellow".

Measurements.	Wing	Tail	Tarsus	Bill from nostril
a. (C 10796) ad. Tondano	208	170(c)	24	21
b. (Mus. Nehrk. Nr. 931) ad. Rurukan	203	177	24	—
c. (ib. Nr. 930) juv. Rurukan	198	170	23	19
d. (Nr. 2207) juv. Minahassa	200	175(c)	23	20
e. (Sarasin Coll.) ♂ ad. Kema, 23. X. 93	208	180(c)	25	20
f. (Sarasin Coll.) ♂ vix ad. Loka, S. Cel., 31. XI. 95	202	—	24	19.5
g. (Sarasin Coll.) ♂ ad. Matanna-Tomori, 7. III. 96	216	—	26	20

Distribution. Celebes: — North Celebes (Meyer *1*, Faber in Dresd. Mus., Fischer *a 1*), Rurukan (Platen); near Tondano (Nat. Coll. in Dresd. Mus.), Mount Klabat (Platen *4*), Kema (P. & F. S.); Matanna-Tomori, Central Celebes (id.); Loka, S. Celebes (id.).

This Cuckoo is the most aberrant member of the genus *Hierococcyx*, differing, as Capt. Shelley points out in his key to the species, from the other forms in having the under surface of the quills not distinctly barred, except at the base, a point which also distinguishes it from *Cuculus*. The bill is exceptionally stout and large; the middle tail-feathers display one or two isolated white spots against the shaft, a peculiarity well developed in *Cuculus*, but not seen in *Hierococcyx* except in the present case. The wing of *H. crassirostris* is also less pointed that in *H. sparverioides*, the ends of the secondaries reaching $^5/_7$ of the wing-length as against $^2/_3$ in the latter species, and about $^1/_2$ in *Cuculus*. This form cannot be regarded as a local differentiation of one of the wider-spread members of the genus, it appears to be an independent development of a stock now not existing anywhere else.

The bird was first discovered by Meyer in 1871 and was only known from North Celebes; it would appear to be partial to the hills. Thus, Dr. Platen met with it at Rurukan (3000 ft.) and on the slopes of Mt. Klabat, and we have recently received a specimen from near the high-lying lake of Tondano, while the Sarasins' specimen from near Loka was killed at about 4500 ft. alt.

The tail-feathers of this species are more than usually instructive in their bearing upon the question of changes of coloration provoked by exposure to light. The outermost pair in an adult before us are white barred with black; the white on the next pair, especially on the outer web, is discoloured with rufous brown, giving the appearance that the black had become washy and had "run", soaking into the white, with the result of a rufous brown colour. On the third pair this process is more advanced, especially again on the more exposed outer web, which is now rufous brown, with little isolated pools of white close up to the shaft, showing where the white bars should be, and with the black bars no longer reaching to the external edge of the feather, but blackest near the shaft and then gradually seeming to melt or become diluted until they pass into the rufous brown of the rest of the web; on the inner webs, however, the black and white bands are well separated, particularly towards the base of the feathers. On the fourth pair the same thing is seen still more advanced, the black bars on the outer webs being further reduced, but on the inner webs the black and white colours are well separated. On the fifth or middle pair (Cuckoos have only ten tail-feathers) where both webs are exposed, the black and white colours are converted on both webs into an almost uniform rufous brown, except at the tip of the tail which is white, the broad terminal band of black, a minute pool of white higher up, and some signs of a black spot (the larger half of which is on the inner web), while the shaft partakes of the colours through which it passes; nevertheless, on raising the upper tail-coverts and looking at the base of the feathers, it is seen that here, where they are not exposed, they are white and black, and entirely white at the extreme base.

+ 59. HIEROCOCCYX SPARVERIOIDES (Vig.).
Large Hawk-cuckoo.

a. **Cuculus sparverioides** *(1)* Vigors, P. Z. S. 1831, 173; *(II)* Gould, Cent. Himal. B. 1832, pl. 53; *(3)* David & Oust., Ois. Chine 1877, 63.
Hierococcyx sparverioides *(1)* Bp., Consp. Av. 1850, I, 104; *(2)* Hume, Str. F. III, 1875, 80; *(3)* Hume & Davis., Str. F. VI, 1878, 157; *(4)* Oates, B. Br. Burm. 1883, II, 108; *(5)* A. Müll., J. f. O. 1885, 157; *(6)* Oates, ed. Hume's Nests & Eggs Ind. B. 1890, II, 384; *(7)* Shelley, Cat. B. XIX, 1891, 232; *(8)* Tacz., Faun. Orn. Sib. Orient. II, 1893, 694; *(9)* Rickett, Ibis 1894, 222; *(10)* Bourns & Worces., B. Menage Exped. 1894, 35; *(11)* Blanf., Faun. Br. Ind. B. III, 1895, 211; *(12)* M. & Wg., Abh. Mus. Dresden 1896, Nr. 1, pp. 1, 4; *(13)* Grant, Ibis 1896, 559.
b. **Cuculus strenuus** *(1)* Gld., P. Z. S. 1856, 96; *(II)* id., B. Asia VI, pl. 42 (1862).
c. **Hierococcyx strenuus** *(1)* Cab. & Heine, Mus. Hein. 1862, IV, 28; *(2)* Everett, J. Str. Br. R. A. S. 1889, 171.

For further synonymy and references cf. Shelley 7.
Figures and descriptions. Gould *a II, b II*; David & Oustalet *a 3*; Oates *4*; Shelley *7*; Taczanowski *8*; Blanford *11*; etc.

Young. Above blackish brown, the feathers edged and obscurely barred with tawny-olive, more buff-brown, streaked with blackish brown on hind neck; head above very dark grey, ear-coverts rather lighter, lores whitish; tail grey-brown with three exposed blackish bands and two under the upper tail-coverts, tip brownish white; under parts white, washed with buff, becoming buff-brown on flanks, the whole broadly striped with blackish brown, taking a more sagittate form on the flanks; chin blackish; remiges below dusky, barred with light clay-buff: "foot yellow; eyelids and iris yellow; bill black, below greenish". — Wing 234 mm; tail c. 200; tarsus 25; bill fr. nostr. 16.5. (♀, Gunong Masarang, N. Celebes, 16. IX. 94: P. & F. Sarasin.)

Adult. "Has the lores whitish; crown, nape and sides of head and neck ashy, varying from rather pale to blackish, passing into the colour of the back which with the remainder of the upper parts is rich brown with a purplish gloss; quills barred with white on the inner webs; tail above brown, pale brownish grey below, tipped with white or rufous white, and crossed outside the coverts generally by 3, sometimes by 4, black or dark brown bands, the last the broadest, and the last but one the narrowest and separated from the last by a narrow space; beneath the chin is dark ashy, with a white moustachial stripe on each side; throat white, more or less streaked with ashy and rufous and passing into the more rufous upper breast, which also has ashy shaft-stripes; lower breast, flanks and abdomen white, more or less suffused with rufous and transversely banded with brown; vent, lower tail-coverts and edge of wing white" (Blanford *11*).

Measurements. Wing 216—254 mm; tail 210—235; tarsus 26.8; bill from gape 35.6 (Blanford *11*).

Eggs and breeding habits. According to observations by Miss Cockburn in the Nilghiris *(6)* this Cuckoo builds its own nest of twigs like a Crow's and hatches its eggs. These are three in number, white, with a few touches (sometimes wanting) of light brown. A similar observation by Mr. R. H. Morgan, the nest containing 4 eggs, is recorded

(6). Mr. Hodgson observed the young of this species being fed by other birds, and states that it is parasitic in its breeding-habits *(6).*

Distribution. India, S. E. Siberia, and China, south to Malacca *(7, 8, 11)*; Philippines — Luzon (fide Gould *b 1, 7*), Calamianes (Bourns & Worcester *10*), Palawan (Platen *c 2*), Negros (Whitehead *13*); Labuan (Treacher *c 2*); Java (Diard in Schl., Mus. P.-B. Cuculi 1864, 15); Minahassa, Celebes (P. & F. Sarasin *12*).

A single immature specimen of this Cuckoo, described above, was obtained in North Celebes by the cousins Sarasin in September, 1894. It is probably only a rare winter visitor to the island from China or S. E. Siberia, where David *(a 3)* and Godlewski *(8)* remark upon its presence in summer. Few specimens have as yet been obtained in other parts of the Archipelago.

It is distinguishable from *H. crassirostris* by its having the wing below banded (except towards the tip), as well as more pointed and longer, the bill smaller, the 3 exposed bands on the middle tail-feathers all complete; the banded wing also serves to distinguish it from *H. fugax*, as well as its larger size.

+ 60. HIEROCOCCYX FUGAX (Horsf.).

Horsfield's Hawk-cuckoo.

a. **Cuculus fugax** *(1)* Horsf., Tr. L. S. 1821, XIII, 178 (Java).
b. **Cuculus lathami** *(1)* J. E. Gray, Ill. Ind. Zool. 1834, pl. 34, fig. 2.
Hierococcyx fugax *(1)* Bp., Consp. 1850, I, 104; *(2)* Cab. & Heine, Mus. Heine 1862, IV, 28 (Hieracococcyx); *(3)* Salvad., Cat. Ucc. Borneo 1874, 65 (Hieracococcyx); *(4)* id., Ann. Mus. Civ. Gen. XIV, 1879, 185; *(5)* Hume & Davison, Str. F. VI, 1878, 157; *(6)* Kelham, Ibis 1881, 391; *(7)* Wrdl. Ramsay, ib. 1884, 333; *(8)* A. Müller, J. f. O. 1885, 157; *(9)* Büttik., Notes Leyd. Mus. 1887, 28; *(10)* Sharpe, P. Z. S. 1887, 442; *(11)* id., Ibis 1890, 10; *(11bis)* Everett, J. Str. B. R. A. S. 1889, 171; *(12)* Shelley, Cat. B. XIX, 1891, 236; *(13)* Styan, Ibis 1891, 325, 484; *(14)* Salvad., Ann. Mus. Civ. 1891, XXXII, 45; *(15)* Baldamus, Leben europ. Kuck. 1892, 133; *(16)* Hagen, T. Ned. Aard. Genoots. 1890 (2) VII, 136; *(17)* Bourns & Worces., B. Menage Exped. 1894, 35.
c. **Cuculus hyperythrus** *(1)* Gld., P. Z. S. 1856, 96; *(11)* id., B. Asia 1856, VI, 431 (Shanghai); *(2)* David & Oust., Ois. Chine 1877, 64.
d. **Hierococcyx pectoralis** *(1)* Cab. & Heine, Mus. Hein. 1862, IV, 27 (Philippines); *(2)* Wald., Tr. Z. S. 1875, IX, 161; *(3)* id., P. Z. S. 1878, 946; *(4)* Steere, List Coll. B. Phillip. Is. 1890, 12.
e. **Hierococcyx sparverioides** *(1)* Schrenck (nec Vig.), Reis. Amurlande, Vög. 1859, 257, pl. 10.
f. **Hierococcyx hyperythrus** *(1)* C. & H., Mus. Hein. 1862, IV, 30; *(2)* Wrdlw. Ramsay, Ibis 1886, 157; *(3)* Seebohm, B. Japan. Emp. 1890, 171.
g. **Cuculus asturinus** *(1)* Brügg., Abh. Ver. Bremen 1876, V, 101; *(2)* W. Blas., J. f. O. 1883, 153.
Figures and descriptions. Gould *c II*; J. E. Gray *b I*; Schrenck *e I*; Brüggemann *g 1*; Blasius *g 2*; W. Ramsay *f 2*; Shelley *12*.

Female, not quite fully adult. Above slaty grey, some feathers round neck white, the longest scapulars white barred with ashy; primaries slaty black, tipped with a narrow rim of white; some of the secondaries blackish brown, barred with rufous; upper tail-coverts blackish brown, tipped with rufous brown; tail-feathers rufous brown crossed with 5 black bars, the subterminal one much the broadest; in front of eye white; chin slaty grey; under surface white, chest washed with rufous, breast and sides streaked here and there with brown; under wing-coverts white with a few streaks of brownish; quills below shining dusky, barred on the inner webs — except towards free end of wing — with white, which gradually increases in extent until at their bases the quills are all white. Wing 200 mm; tail 160; tarsus 22; mid. toe without claw 23; bill from nostril 16.

"Bill black, below yellow-green; orbit yellow-green; feet light yellow; iris brown" (Platen in Mus. Nehrkorn, specimen from Rurukan, N. Celebes, 28. I. 85).

Adult. Uniform slaty grey above. .

Young. Upper surface barred all over with rufous and brown (except head, which is getting smoky grey); tail, tail-coverts and secondaries as in the above-described ♀; under surface strongly washed with rufous and marked with broad streaks of rich brown (♀ Mindoro; Platen in Mus. Nehrkorn, Nr. 2485).

Distribution. Lower Amoor (Maack *e 1*); Japan (Seebohm *f 3*); China (Gld. *c 1, c II*, Styan *13*); Philippines (Cabanis *d 1*, Everett *d 3*, Maitland-Heriot *f 2*, Steere *d 4*, B. & W. *17*); North Celebes — Gorontalo (Riedel *g 1*); Rurukan (Platen in Mus. Nehrk.); Borneo (Everett *11bis, 11*, Fischer, Schwaner *3bis, 11*); Java (Horsf. *a 1*); Sumatra (Beccari *4*, Klaesi *9*, Modigl. *14*); Singapore (Kelham *6*); Perak, Malacca (Wray *10*); Salanga (A. Müll. *8*).

In his Catalogue of the Cuckoos *(12)* Capt. Shelley includes under the title *Hierococcyx fugax* three forms, *H. fugax*, *H. hyperythrus*[1]) and *H. nisicolor*, which have been commonly treated by other authors as distinct. According to Major Wardlaw Ramsay *H. hyperythrus* has the "rufous (of the under surface) darker, uniform, or nearly so", inhabits Japan, Amoorland, China, and the Philippines, the young birds being found in Borneo and Malacca, and perhaps Sumatra and Java in winter; *H. nisicolor* has the "rufous paler, mixed with white" and ranges from the Himalayas to Tenasserim, descending perhaps into the Malay Peninsula; *H. fugax* is a species which in maturity retains a plumage like *H. hyperythrus* juv.; its distribution is not definitely stated *(f 2)*. We conclude that a stationary race breeding in the East India Islands is intended; but, if so, Major Ramsay is in error in applying to it the name *H. fugax* Horsf., since that bird has a grey, not a brown, back *(a 1)*. Capt. Shelley finds the type to be an adult of the large series of examples comprehended by him under the title *H. fugax*, a grey-backed species when adult, under which the types of *H. nisicolor* and *hyperythrus* are also included.

Not taking into further consideration *H. nisicolor*, of which Hume & Davison say that "of course it cannot be confounded with *fugax* which has a very much longer bill" *(5)* and the distinctness of which is upheld by Blanford in his recent work, it appears to us highly probable that Major Ramsay's opinion as to the distinctness of *H. hyperythrus*, subspecific at least, from *H. fugax*, the resident form of the East Indies, is correct. Mr. Everett discovered *H. fugax* breeding on Kini Balu: "I found this species laying in the nest of *Culicicapa ceylonensis* on the 29th April. The

---

[1]) The type of *C. hyperythrus* in the British Museum is labelled Manila.

egg was creamy white, slightly spotted with pale yellow-brown and grey, forming a complete zone at the larger end: axis 0.9 inch, diam. 0.65" *(11)*. Major Ramsay appears, however, not to be in the right in holding the resident form to be a brown-backed species when adult, resembling the young Chinese birds; although Dr. Sharpe l. c. does not mention whether Mr. Everett's adult specimens had grey backs, it appears that they had, since there are such specimens from him, and other collectors in Borneo, in the British Museum *(12)*, as shown in Shelley's series. We do not of course venture to separate in any way this stationary form from *H. hyperythrus*, without examining more material than the one or two specimens consulted by us, but desire to call attention to the fact that a certain Cuckoo visits Amoorland, Japan and China in summer, passing over, but not remaining in the Lower Yangtse Basin in migration, as Mr. Styan believes *(13)* — which may, or may not, be identical with one known to breed in Borneo. In any case a new and close comparison would be desirable, as the questions of migration and the differentiation of new forms are interestingly involved in such cases as this.

Only two specimens of this species are as yet known to us from Celebes. The first was named as a new species, *C. asturinus*, by Brüggemann, but Prof. W. Blasius, after a careful examination of the type, came to the conclusion that it is identical with *C. hyperythrus* and *fugax* (see: *g 2*). The second is the female kindly lent to us by Mr. Nehrkorn and described above. It was shot at Rurukan (3000 ft.) in the Minahassa in January 1885 by Dr. Platen, and was most likely a migrant individual from Japan or China.

GENUS CUCULUS L.

In the true Cuckoos the bill is moderate, the nostrils round, in a slightly tubular formation of skin; the wing is long and pointed, the secondaries only about $^1/_2$ the wing-length; the feathers of the rump and upper tail-coverts very thick, the latter about $^2/_5$ the length of the tail; tail graduated; feet delicate, yellow in colour (except in *C. pallidus* (Lath.) and *sonnerati* Lath). Ten species are recognised by Shelley (1891), inhabiting Europe, Africa, Asia, Australia.

61. CUCULUS CANORUS (L.).

Common Cuckoo.

The races of the Common Cuckoo appear to be two or three in number, one of which, a somewhat small Eastern one, occurs in Celebes and other islands of the East Indies probably on migration. Two races of *C. canorus* are :

† 1. The typical Cuculus canorus.

a. **Cuculus canorus** Linn., S. N. 1766, I, 168; Dresser, B. Europe V, 1878, 199, pl. 299; Shelley, Cat. B. XIX, 1891, 245; etc.

188   Birds of Celebes: Cuculidae.

For synonymy see Shelley l. c.
Distribution. Europe, N. Africa, Asia as far east as India, migrating in winter south to the Cape, S. Persia, S. India, as a straggler — Ceylon.
Diagnosis. Larger, wing 203—226 mm (8.0 in.—8.9 in., Legge, B. Ceylon 222); the bars on the under surface usually a very little narrower.

2. Cuculus canorus canoroides (S. Müll.).

b. **Cuculus canorus** *(1)* Horsf., Tr. Z. S. 1821, XIII, 179 (Java); *(1bis)* Wald., Tr. Z. S. 1872, VIII, 115, pt. (Celebes); *(2)* Swinh., P. Z. S. 1871, 395 (China); *(3)* Hartl. & F., P. Z. S. 1872, 100 (Pelew); *(4)* Finsch, J. Mus. Godef. 1875, VIII, 12 (Pelew); *(5)* Hume, Str. F. 1875, III, 78 (Pegu); *(6)* id., Str. F. 1876, IV, 288 (Andamans and India); *(7)* David and Oust., Ois. Chine 1877, 65 (China); *(8)* Finsch, P. Z. S. 1879, 12 (Duke of York Id.); *(9)* Oates, B. Brit. Burmah 1883, 103 (Burmah); *(10)* Styan, Ibis 1887, 230; 1891, 325, 484; *(11)* Seebohm, B. Japan. Emp. 1890, 169 (Japan); *(12)* Shelley, Cat. B. 1891, XIX, 245, pt. (Japan, New Guinea, etc.); *(13)* Baldamus, Leben europ. Kuck. 1892, 17, 18 (? Celebes); *(14)* Seebohm, Ibis 1893, 51 (Loochoo Is.); *(15)* Hose, t. c. 414 (Sarawak); *(16)* Bourns & Worces., B. Menage Exped. 1894, 35.

c. **Cuculus canoroides** *(1)* S. Müll., Verh. Natuurk. Comm. 1839—44, 235 (in part?); *(2)* Gld., Handb. B. Austr. 1865, I, 614; *(3)* Blyth, Ibis 1865, 31, 40; *(4)* Salvad., Orn. Pap. 1880, 1, 328; Agg. 1889, 48, and 1891, 217; *(5)* W. Blas. & Nehrk., Verh. z.-b. Ges. Wien 1882, 417; *(6)* W. Blas., J. f. O. 1883, 115; *(7)* Pleske, Bull. Acad. Petersb. 1884, 116; *(8)* Guillem., P. Z. S. 1885, 565, 624; *(9)* W. Blas., Ornis 1888, 306, 565; *(10)* id., Ibis 1888, 373; *(11)* id., J. f. O. 1890, 145; *(12)* Whitehd., Ibis 1890, 46.

d. **Cuculus optatus** *(1)* Gld., P. Z. S. 1845, 18; *(II)* id., B. Austr. 1848, IV, pl. 84; *(3)* Baldms., Leben europ. Kuck. 1892, 127 (pt.?).

e. **Cuculus canorinus** *(1)* ? Cab. & Heine, Mus. Hein. 1862, IV, 35; *(2)* Salvad., Cat. Ucc. Borneo 1874, 67; *(3)* ? Tacz., J. f. O. 1881, 186; *(4)* ? Dedititus, J. f. O. 1886, 540; *(5)* Everett, J. Str. Br. R. A. S. 1889, 169; *(6)* Baldamus, Leben eur. Kuck. 1892, 126.

f. **Cuculus striatus** *(1)* Swinh. (nec Drapiez), Ibis 1861, 259, 340 (in part); ? *(2)* David & Oust., Ois. Chine 1877, 65.

g. **Cuculus intermedius** (nec Vahl) *(1)* Seeb., B. Japan 1890, 169 partim?; *(2)* M. & Wg., J. f. O. 1894, 241; *(3)* iid., Abh. Mus. Dresd. 1895 Nr. 9, p. 3; *(4)* Hartert, Nov. Zool. 1896, 159, 176, 552.

"**Dandape**" (ad.) or "**Parapaa**", "**Mengantagota**" (juv.) Talaut, Nat. Coll.
Distribution. ? East Siberia *(e 3, e 4)*; China (Swinh. *b 2*, David *b 7*); Japan *(b 11, b 12)*; Burmah (Hume *b 5*, Oates *b 9*); India (Hume *b 6*); Andamans (Hume *b 6*); Philippines (Platen *c 9, c 11*, B. & W. *b 16*); Borneo *(c 4, e 5)*; Great Sangi (Platen *c 9*); Talaut Is., Karkellang and Kabruang (Nat. Coll. *g 2, g 3*); Celebes — North (Meyer *b 1bis*); South (Everett *g 4*); Saleyer, Djampea and Kalao (Everett *g 4*); Bali (Doherty *g 4*); Pelew Is. (Heinsohn *b 3, b 4*); Halmahera, Ternate, Batchian, Morty, Amboina, Waigiou, Aru, New Guinea, Jobi, Duke of York Id., New Britain, N. Australia (Salvad. *c 4*, Shelley *b 12*); Malacca, Singapore, Java, Dahat Id., Timor (Shelley *b 12*).
Diagnosis. Smaller, wing 190 (or less) to 220 mm (specimen in Coll. Salvad. *c 4*); the bars on the under surface usually broader and blacker; wing below with one or two fewer white bars.

Adult. Above slate-grey, paler on sides of face, throat and breast; wings shining dusky brown; tail black, browner on basal part; marked at intervals of about 1.5 cm with small white spots at the side of the shaft, growing larger towards the outer feathers where they are in displaced juxtaposition, tip of tail white; under surface of body buffy white, darkest on under tail-coverts, marked with transverse bars of blackish sepia (except on under tail-coverts which are transverse-spotted) of about 2 mm breadth; quills below barred with white on basal two-thirds, the inner ones uniform white at the base (ad. Karkellang, Talaut, Autumn 1896; Nat. Coll. C 15302).

Young. Upper surface unevenly barred with rufous brown and black, the brown bars also well developed on the tail and as notches on either web of the quills above; below white, washed with cinnamon about the throat and chest, and barred all over with sepia (Karkellang, Sept. 1896, C 15304).

Second year. On moulting the rufous plumage, the bird assumes a dress resembling that of the adult, but browner and darker, all the feathers of the upper parts being edged with whitish (Talaut Is. C 15300, Autumn 1896; C 13080 and C 13787, Nov. 1894 — three specimens in transition-plumage).

Eggs. Elongated oval, a shade narrower at one end; ground-colour pure white, with a slight gloss, very sparsely marked, chiefly towards the larger end, with minute specks and tiny lines of dingy olive-brown and very pale inky purple or purplish grey. 22.6 × 15.2 mm (Oates).

Measurements.	Wing	Tail	Tarsus	Bill from feathers of forehd.
a. (C 1857) Manado (March)	201	170	20	—
b. (C 1860) Manado (March)	195	165	19	16

In summer the Common Cuckoo, *C. canorus*, is spread from the British Islands to East Siberia, China and Japan, varying so little that the majority of ornithologists speak of birds from the East, as from the West, simply as *C. canorus*. South-eastern examples, however, run a little smaller in size and are slightly different in the markings on the under surface. Hume seems to have first recognised the presence in India of two races of the Cuckoo in the small and large individuals which occur there in winter: speaking of a specimen from the Andamans he says: it "is precisely similar to a great number of others that I have obtained in India, and which in common with most other Indian ornithologists I have always called *canorus*. These specimens differ only from others obtained in India, and from European ones, in their slightly smaller size, and possibly a shade slenderer bills" *(b 6)*. The first specimen from Pegu, brought to notice by Hume *(b 5)*, was also undersized, and Oates finds likewise that birds of smaller size (wing 8 in.) than the European Cuckoo, but not otherwise differing, are the rule there *(b 9)*. As in India, they appear, says Mr. Oates, to be merely winter visitors to Burmah, though a few may breed there. In China, as Swinhoe, David, and Styan remark, the Cuckoo is a summer visitor, and it is most probable that birds from there, or from the territories rather further north, are those which pass into Burmah in winter. It should be noticed that neither David & Oustalet in Chinese specimens, Seebohm in Japanese ones, nor Shelley in those of either country, find differences between them and European specimens (*b 7*, *b 11*, *b 12*), though Styan (*b 10*) has since remarked that "the common Cuckoos of Kiukiang are nearly all of a small pale race, with very narrow bars on the under parts". Hume and Oates also treat of them under the name *C. canorus*, and it may be doubted whether an attempt to discuss under a separate trinomial so slight a departure from the typical form is well advised. In the East India Islands, Salvadori and other naturalists carefully hold the eastern form apart under the name *canoroides*, though the Italian ornithologist at the same time calls attention to its very close resemblance to the typical form.

Capt. Shelley (Cat. B. XIX, 252) identifies what Count Salvadori appears to us to call *C. canoroides* with a different species, "*C. intermedius* Vahl"; and Salvadori (Orn. Pap., Agg. 1891, 217) apparently acquiesces in this determination, overlooking the fact that Shelley catalogues several specimens of *C. canorus* from his own especial province, the Moluccas and New Guinea, as well as several of "*C. intermedius*". From Salvadori's remarks we conclude that he was in possession of specimens of the small south-eastern race of the former bird only. "*C. intermedius*", according to Mr. Blanford (Faun. Br. Ind., B. III 1895, 208) — here called *C. saturatus*, — may be distinguished in adult plumage from *C. canorus* by the much darker upper parts, "pure blackish ashy"; the edge of the wing pure white; the size always smaller and the bill a little stronger.

Shelley speaks of the bars as "jet-black"; in *canorus* as "more dusky". The bars on Salvadori's *canoroides* are dusky ("nigricans"), and very little broader than in the *typical canorus*, and the wing 200 mm, in one case 220 mm, as in a good-sized specimen of *C. canorus*.

To the differences pointed out by Oates and Blanford it may be added that "*C. intermedius*" (Shelley, partim) appears to have a very distinct cry. Colonel Tytler (Ibis 1868, 202, under *C. himalayanus*) speaks of it as a peculiar call of "Goog, goog, goog", as heard by him between Simla and Mussurie; Seebohm, who shot the bird in Siberia, as "a guttural and hollow-sounding "hoo" (Ibis 1878, 326); Major Marshall saw the bird (called *C. himalayanus*) as it was uttering "the deep booming notes sounding something between the notes of the Hoopoe and Green Pigeon" (Ibis 1884, 410). Swinhoe's *C. monosyllabicus* (Ibis, 1865, 545), with a cry "like the two notes of *canorus* run into one", is probably the same; but the species spoken of by Major Scully as *C. striatus* (Str. F. 1879, VIII, 254) would appear from its cry rendered by the natives of Nepaul as "Kaifal pakyo" to be *C. micropterus* Gould, whose note according to Oates resembles the word "bho-kutha-kho".

The form, with the monosyllabic call, should, apparently, as Mr. Blanford shows, be called *Cuculus saturatus* Hodgs. It has been supposed to occur in Celebes, but we believe erroneously, the eastern form of *C. canorus*, which visits the East Indies in winter, having been confused with it.

As to *C. canorus* itself, we incline to the opinion that there are two eastern races of it, a more northern one, pale and with narrow bands below, inhabiting Siberia and (wintering in?) China (see Styan, *supra*, and Taczanowski, "Faune Orn. Sib. Orient." 1893, II, 685 — here called *C. canorus borealis* (Pall.)), and a race with broader, blacker bands, *C. canorus canoroides*, the subject of the present article, which visits the East Indies in winter and probably inhabits more southern latitudes than *C. c. borealis* in summer. This view seems to coincide with the opinion of Mr. Hartert, who has most kindly looked into the question for us and, after carefully examining the material at Tring and seeing that in the British Museum, informed us that the birds from South Celebes, Talaut, and the Lesser Sunda Islands "cannot, I think, be the eastern *C. canorus*". They are, we believe, the more south-eastern, *C. c. canoroides*.

The following data apparently show that the Common Cuckoo occurs only as a winter migrant in the East Indies south of the Philippines, though the facts to hand are insufficient and the collecting season varies in different localities.

Japan: Summer visitor (Seebohm *b 11*).
China: Summer visitor (Swinh. *f 1, b 2*, David *b 7*, Styan *b 10*).
India: Winter (Hume *b 6*).
Andamans: Nov. 16th (Hume *b 6*).
Burmah: Aug. to Feb. (Oates *b 9*).
Philippines — Mindanao: 1 in summer 1889 (Platen *c 11*).
Borneo: "Occurs on migration on the coast" (Hose *b 15*).
Talaut Island: 11 specimens, Autumn (Nat. Coll. in Mus. Dresd. & Tring).
Great Sangi: Jan. 25th (Platen *c 11*.).
Celebes: March and April (Meyer and Nat. Coll., Dresd. Mus.).
Ternate: Oct., Nov., Dec. (Bruijn *c 4*, Fischer *c 7*).
Amboina: Nov. (Bruijn *c 4*, Platen *c 5*).
Waigiou: Oct., Nov. (Guillem. *c 8*).
Aru: Dec. (Guillem. *c 8*).
New Guinea: Sept. 18th 1877 (D'Alb. *c 4*).
New Britain: "The eastern representative of our European Cuckoo was repeatedly observed and collected by me from November till January, as well as in New Britain as round about Cape York" (Finsch, *C. canoroides*, Mitth. Orn. Ver. Wien 1884, 92; Vög. Südsee-Inseln 1884, 12).
? North Australia: Jan. (Gould *d II, c 2*).

On the breeding of the *Cuculinae* compare the articles on *Eudynamis melanorhyncha* and *Cuculus saturatus*.

+ 62. ? CUCULUS SATURATUS Hdgs.

Himalayan Cuckoo.

a. **Cuculus intermedius** *(1)* Seeb. (nec Vahl), B. Japan 1890, 169, pt.?; *(2)* Oates, Hume's Nests and Eggs Ind. B. 1890, II, 381; *(3)* Shelley, Cat. B. XIX, 1891, 252, partim; *(4)* Wiglesw., Av. Polyn. 1894, 10, partim; ? *(5)* Büttik., Zool. Erg. Wober's Reise 1893, III, 275; *(6)* Oust., Nouv. Arch. du Mus. 1893, 136.

b. **Cuculus striatus** "Drapiez"[1]) *(1)* Schl., Mus. P.-B. Cuculi 1864, 7, ? partim; *(2)* Swinh., P. Z. S. 1871, 395 ?; *(3)* Hume, Str. F. 1874, 190; *(4)* Oates, B. Br. Burmah 1883, II, 105; *(5)* id., Ibis 1889, 355—359; *(6)* Everett, J. Str. Br. R. A. S. 1888, 170.
c. **Cuculus himalayanus** auct. (nec Vigors, P. Z. S. 1831, 172 — fide Blanford *1, 2*).
d. **Cuculus monosyllabicus** *(1)* Swinh., Ibis 1865, 545.
e. **? Cuculus peninsulae** *(1)* Stejn., Bull. U. S. Nat. Mus. Nr. 29, 1885, 227.
Cuculus saturatus Hodgson, *(1)* Blanf., P. Z. S. 1893, 315—319; *(2)* id., Faun. Br. Ind. B. III, 1895, 207.

Descriptions. Oates *b 5*; Blanf. *2*.

Adult. When similar to *C. canorus*, except that the upper parts are much darker, pure blackish ashy, the lower parts are generally pale buff with the black bands broader and more regular, and the edge of the wing pure white. The size is always smaller, and the bill a little stronger. The young pass through two stages, in neither of which is there a white nuchal spot" (Blanford).

Changing plumage. In the first plumage the back is blackish with white margins, in the second blackish, barred with dark rufous (Oates & Blanf). We have two specimens from Talaut in such a "first" plumage, which are evidently passing directly into the adult dress, the grey feathers of that plumage being intermixed with the others on the head, neck, and throat. We have described these as second-year birds in the former article.

In their rufous stage Mr. Oates distinguishes *C. canorus* as being pale rufous; *C. saturatus* dark rufous, with coarse bars; *C. poliocephalus* chestnut, delicately barred.

Distribution. East Siberia from the Yenesei ? (Seebohm *a 1*) to Kamtschatka ? (Stejneger *e 1*); China (Swinh. *b 2*); Setchuan (Oust. *a 6*); ? Japan (Seebohm *a 1*); India (Oates *b 4*, Blanf. *2*); Pegu and Tenasserim (Oates *b 4*); Andamans and Nicobars (Davison *b 3*, R. G. W. Ram'say *a 3*, Oates *a 2*); ? Sumatra (Shelley *a 3*); ? Java (Shelley *a 3*, Schl. *b 1*); Borneo (Everett *b 6*, Schl. *b 1*); ? Philippines (Shelley *a 3*); ? Pelew Islands (Hartl. & Finsch *a 4*); ? Celebes — Manado (Shelley *a 3*) — Macassar (Wall. *a 3*, Weber *a 5*); ? Batchian, Morty, Flores, Timor, New Guinea, N. Australia (Shelley *a 3*).

We include this form of Cuckoo doubtfully in the Celebes avifauna on the authority of Capt. Shelley, and Mr. Büttikofer; Shelley notes 3 specimens in the British Museum from Manado and one from Macassar, as well as one specimen of the Common Cuckoo, *C. canorus*, also from Manado. These four Manado specimens are from the Tweeddale Collection, presumably the identical four from Meyer mentioned by Lord Walden (Tr. Z. S. 1872, VIII, 115), who remarks that "without the example in full plumage it would have been difficult to say whether the other three did not belong to *C. canoroides* Müller". Shelley now finds them identical with the form he terms "*C. intermedius*", and a specimen obtained by Prof. Weber at Macassar is likewise found to be "*C. intermedius*" by Büttikofer, though what "*C. intermedius*" is we confess we do not know. It seems to be composed of *C. saturatus* and of two races of *C. canorus*. Notwithstanding its near affinity to *C. canorus*, numerous observers — Swinhoe, Seebohm, Marshall, Tytler, Kittlitz, David — agree in

---

[1] Not of Drapiez, according to Seebohm *(a 1)* and Blanford *1*.

finding that *C. saturatus* has a different cry, a note which may be syllabilised as "hoo", resembling somewhat the cry of the Hoopoe.

"Like all the Cuckoos", says Swinhoe *(b 3)*, "these birds in China are only summer visitants", an observation confirmed by David *(b 5)*, and by Godlewsky in the Baikal country, Dauria, Ussuriland and the coasts of the Sea of Japan. So, likewise, according to Seebohm, it — if indeed this species is intended — visits all the Japanese Islands in summer. It is known as a breeding bird in the Himalayas, but in no other part of India as far as Mr. Oates is aware *(2)*; in the winter it visits the low-lying Lucknow division (Reid, Str. F. 1887, X, 451), and Mr. Oates considers that it is probably only a winter visitor to Pegu and Tenasserim *(a 6)*. Davison *(a 4)* did not hear it in the Andamans in December and January, but first noticed its cry on March 14th in the Nicobars; probably, we should think, like *C. canorus*, only the male utters this cry in the breeding season, for there are specimens from the Andamans in the British Museum obtained in December, January and February *(4)*. Here, therefore, also it would appear to be a winter visitant.

Schlegel records the dates of 12 specimens ascribed to *C. striatus* killed in the Moluccas and Timor in the winter months; it appears, therefore, that part of the birds, which visit China, Japan and Siberia in summer, make their way to these islands in winter. Nevertheless some, apparently, remain in the East Indies and even breed there; one of Schlegel's specimens (Nr. 44) is dated "Ternate May 1st 1861", another is a "young one, with red bands, taken from the nest in Java (voyage of S. Müller, Nr. 30)". The following note of Mr. de Bocarmé suggests that it is in the mountains, not in the plains, of Java that the species breeds: "In February and in March this Cuckoo descends in Java from the mountains; it is to be found in the bushes and even in the Rhizophores by the sea-shore. The male is then mute. It is during these two months that this species moults; at a height of more than seven thousand feet and thereabouts, in a European temperature, every part of the forest resounds with the voice of this bird, which is never to be heard but when perched on the top of the highest trees" *(b 2)*.

C. saturatus is parasitic in its habits, but Capt. Hutton, writing in Hume's "Nests and Eggs of Indian Birds" *(2)*, says that "when the young bird is old enough to leave the nest, the foster-parents feed it no longer, and it is then supplied by the old Cuckoo, or at all events by one of the same species. This I have myself repeatedly witnessed . . . . . . At Jeripance, below Mussurie, I have seen the young Cuckoo sitting for hours together on a branch waiting for the return of the adult bird, which continued every now and then to bring supplies of caterpillars wherewith to satisfy the apparently insatiable appetite of the nestling, until at last both would fly off to another spot. To satisfy myself that it was really this Cuckoo that fed the young, I shot one in the very act, and found it to be no other than our summer visitant, *Cuculus intermedius*."

Notwithstanding the careful work of Mr. Oates *(b 7)*, the present Cuckoo is most unsatisfactorily understood owing to the great difficulty of distinguishing the supposed species or subspecies, and a fresh comparison of the Himalayan birds should be made with those of China, Japan and the N.W. parts of the East Indies.

GENUS CHRYSOCOCCYX Boie.

Small Cuckoos of about the size of a Sparrow, easily recognised by the metallic green, bronze, or coppery on the upper parts, and, when barred, on the bars of the under parts. The bill and feet are dark in colour — grey, brown, leaden, or blackish; the wing is rather pointed, the secondaries falling short of the tip by about $1/3$ the wing-length, the first primary is rather short, about $1/2$ the length of the second. The African species — 4 in number — have the under surface of the wing barred and are separated from those of India, the East Indies, Australia, and New Zealand by Capt. Shelley, the latter forms having a single oblique band across the wing below *(Chalcococcyx)*.

† 63. CHRYSOCOCCYX MALAYANUS (Raffl.).

Little Bronze Cuckoo.

a. **Cuculus malayanus** *(1)* Raffles, Tr. L. Soc. 1822, XIII, 286.
Chrysococcyx malayanus *(1)* Horsf. & Moore, Cat. B. Mus. E. I. Co. 1856—58, 706.
b. **Chrysococcyx minutullus** *(1)* Gould, P. Z. S. 1859, 128; *(II)* id., B. Austr. Suppl. 1859, pl. 56.
c. **Lamprococcyx minutullus** *(1)* Cab. & Hein., Mus. Hein. 1862, IV, 15, note; *(2)* Gld., HB. B. Austr. 1865, I, 625.
d. **Chrysococcyx basalis** *(1)* Salvad., Cat. Ucc. Borneo 1874, 62, partim; *(2)* Sharpe, Tr. L. Soc. 1877, (2) I, 320; *(3)* W. Rams., Tweedd. Orn. Works Index 1881, 656; *(4)* ? Steere, List Coll. B. & M. Philipp. Is. 1890, 12.
e. **Chalcococcyx malayanus** *(1)* Wald., P. Z. S. 1878, 945; *(2)* Shelley, Cat. B. XIX, 1891, 298; *(3)* Sharpe, Ibis 1894, 247, 258; *(4)* M. & Wg., Abh. Mus. Dresden 1896, Nr. 1, p. 8; *(5)* Hartert, Nov. Zool. 1896, 159, 572.
f. **Lamprococcyx malayanus** *(1)* Hume, Str. F. 1878, VI, 503; *(2)* North, Pr. L. Soc. N. S. W. (2) IX, 39 (1894).
g. **Chalcites minutullus** *(1)* Rams., Tab. List 1888, 15.
For further synonymy and references cf. Shelley *e 2*.
Figure and descriptions. Gould *b II, c 2*; Shelley *e 2*.

Adult. Above glossy bronze-green, an obscure subterminal band across the middle tail-feathers, bases of the feathers on forehead white; face and under parts white, the feathers of lores, superciliary region and ear-coverts with dark tips, the entire under parts, including the under wing-coverts barred with glossy bronze-brown; tail below with the outermost pair of rectrices crossed with four bands of black and white, next pair hazel on the inner web with three dark bars and a white spot at the tip, outer web brown, third and fourth pairs hazel on the inner web with a terminal white or whitish spot and a subterminal blackish band, passing on to the outer webs which are brown, middle pair below grey-brown, with a dark subterminal band;

remiges below grey-brown, passing on the inner webs where they rest upon the body into pale cinnamon and into white at the base; bill and feet black; "iris reddish, rim of the eyelids cherry-red" (♂, Macassar, 22. June, 1895: P. & F. Sarasin).

Young. Above browner and less glossy than the adult, the edges of some of the feathers pale; below white without bars, becoming greyer on breast and throat, and brown on sides of head; under wing-coverts and much of the inner webs of the remiges buff-white, the former slightly barred, a few sagittate bars on sides, flanks, and under tail-coverts; tail much as in the adult, but hardly a trace of bars (except the subterminal one) on the second feather from the outside (Sosso, S. W. Central Celebes, 11. Aug. 1895: P. & F. Sarasin).

Measurements (3 adults, S. Celebes). Wing 91—94 mm; tail 55—61; tarsus c. 14.5; bill from nostril 10—11.

Eggs. See North *f 2*.

Distribution. Malay Peninsula (Maingay, etc. *e 2*); Sumatra (Raffles *a 1*); Borneo (Motley, etc. *e 2*); Philippines — Mindanao (Everett *e 1*), Negros (Steere *d 2*), ? Basilan (Steere *d 4*), Sooloo Is. (Everett *e 3*); Celebes — Macassar (P. & F. Sarasin *e 4*, Everett *e 5*), Indrulaman (Everett *e 3*), Sosso, S. W. Central Celebes [P. & F. S. *e 4*); Sumbawa (Doherty *e 5*); Flores and Timor (Wallace *e 2*); N. W. and N. Australia (Gould *c 2, e 2*, Ramsay *g 1*); New Guinea and Solomon Islands (fide Shelley *e 2*).

This Cuckoo was first found in Celebes in June and July, 1895, at Macassar by the cousins Sarasin, and again in August at Sosso in South-west Central Celebes — four examples being obtained in all; later in the same year two more were sent to the Tring Museum by Mr. Everett from Macassar and Indrulaman near the Peak of Bonthain. Without making a careful comparison of specimens from other parts of the range of the species, it is not possible to form an opinion as to whether it is migratory or a resident form in Celebes. The distinction of the East Indian and Australasian forms of *Chrysococcyx* is not easy. The present species may generally be known by its small size, the wing, according to Capt. Shelley, not exceeding 97 mm, but equally small measurements may sometimes be found in other species, for instance *C. basalis*. *C. poecilurus* Gray of North Australia, New Guinea and the Moluccas is more thickly banded below, the dark bands being $^2/_3$ the width of the intervening white spaces, the second lateral tail-feather besides the subterminal band displays two spots only to represent the bars on the inner web in *malayanus*; *C. plagosus* (Lath.) has only the subterminal band on the second tail-feather, and the bands below are broader; this species is found by Shelley to inhabit Australia and parts of Papuasia. *C. basalis* (Horsf.) has recently been recorded by Mr. Hartert from South Celebes and shown to differ by the characters given below.

64. CHRYSOCOCCYX BASALIS (Horsf.).

Larger Bronze Cuckoo.

a. **Cuculus basalis** *(1)* Horsf., Tr. L. Soc. 1821, XIII, 179.
Chrysococcyx basalis *(1)* Blyth, J. A. S. B. 1846, XV, 54.

b. **Chrysococcyx lucidus** (nec Gm.), *(1)* Gould, B. Austr. 1848, IV, pl. 89.
c. **Lamprococcyx basalis** *(1)* Gould, HB. B. Austr. 1865, I, 626; *(2)* Salvad., Orn. Pap. I, 1880, 349.
d. **Chalcococcyx basalis** *(1)* Shelley, Cat. B. 1891, XIX, 294; *(2)* Hartert, Nov. Zool. 1896, 159.

For further synonymy and references cf. Salvadori *c 2*; Shelley *d 1*.

Figure and descriptions. Gould *b 1, c 1*; Salvadori *c 2*; Shelley *d 1*.

Diagnosis. "The longer wing (this ♀ has it 97 mm long), both webs of the second rectrix from the outside being rufous for the basal two-thirds, the rather broader and paler bands of the breast, a superciliary whitish line, and a broad dark line from the eye along the sides of the neck distinguish this species without difficulty from *Ch. malayanus*" (Hartert *d 2*).

Distribution. "Australia, Aru Islands, Timor, Flores, Lombock, Java; re-occurring in Malacca" (Shelley *d 1*); South Celebes (Everett *d 2*).

A female specimen of this Cuckoo was obtained by Mr. Everett on Mount Bonthain in 1895. It had not previously been recorded from Celebes. Two of the points of difference between it and *C. malayanus* found by Mr. Hartert do not seem to us to hold good, namely the size is only that of a large example of *C. malayanus*, and a whitish superciliary stripe is found in that bird as well. The geographical distribution of the two birds is very similar, and we have noticed that nearly all the specimens in which the sex has been ascertained is male in *C. malayanus* and female in *C. basalis*!

GENUS CACOMANTIS S. Müll.

Small Cuckoos, about the size of a Lark, with long tails, usually longer than the wing, strongly graduated, the outermost feathers being from about $^1/_2$ to $^2/_3$ the tail-length; wing moderately pointed, secondaries about $^2/_3$ the length of the wing, first primary nearly as long as the secondaries, a broad band of white across the wing seen from underneath. The genus is found in about 10 species from India to Australia and Fiji. *C. passerinus* of India is known to be parasitic.

+ * 65. CACOMANTIS VIRESCENS (Brügg.).

Rufous-bodied Cuckoo.

a. **Cacomantis sepulchralis** *(1)* Wald. (nec. S. Müll.), Tr. Z. S. 1872, VIII, 116; *(2)* Meyer, Ibis 1879, 69.
b. **Cuculus virescens** *(1)* Brügg., Abh. Ver. Bremen 1876, V, 59.
Cacomantis virescens *(1)* Shelley, Cat. B. XIX, 1891, 274; *(II)* Meyer, Vogelskel. 1892, XVIII, p. 47, t. CLXXIII; *(3)* Büttik., Zool. Erg. Weber's Reise in Ost-Ind. 1893, III, 276; *(4)* M. & Wg., Abh. Mus. Dresden 1895, Nr. 8, p. 6; *(5)* iid., ib. 1896, Nr. 1, p. 8; *(6)* iid., ib. Nr. 2, p. 10; *(7)* Hartert, Nov. Zool. 1896, 159; *(8)* id., ib. 1897, 164.

"**Burong-pangil-udjan**" i. e. "**Rain-caller**" (Meyer *a 2*), or "**Burong socangge**" (Nat. Coll. in Mus. Dresd.), Malays of the Minahassa.
"**Embis**" (Meyer), or "**Koko imbantik**" (Nat. Coll.), Natives of the interior of the Minahassa.
"**Sinde**", Tonkean, E. Celebes (Nat. Coll.).
"**Tokulat**", Banggai, ii d.

Figure and descriptions. Walden *a 1*; Brüggemann *b 1*; Shelley *1*; Moyer *c II* (skeleton); M. & Wg. *6*.

Adult. Head and hind neck greenish slate-grey, becoming greener on the upper back, and dark slaty on the rump and upper tail-coverts; wing brownish glossed with bronze-green; metacarpus white, partly concealed; tail black, glossed with purple, the tips of the feathers white, the inner webs of some of the outer ones slightly notched with white; cheeks, chin and throat grey, paler than the head; all the other under parts, including under wing- and tail-coverts, deep cinnamon-rufous; wing below with a broad white band running across the inner webs of the quills near their base (N. Celebes, between Manado and Arakan, C 10876).

Young in first plumage. Very different from the adult: above black, the feathers margined and barred with cinnamon-brownish, taking a more striate form on the head; under parts barred with white and black, the black predominating; under wing-coverts white, with a few black bars and transverse spots; remiges below dusky blackish, with a broad buff-white band across the middle of them; on the outermost quill a whitish spot only; tail black, barred or notched with cinnamon-rufous, the bars mostly not reaching down to the shaft: "iris sepia; feet above greyish, below yellowish; bill above black, below reddish grey" (♂, Kema, 27. Oct. 93, and juv. Tomohon, 30. Apr. 94: P. & F. Sarasin).

Immature in changing plumage. Half in young, half in adult dress; the adult coloration prevails on the under surface, though the rufous is browner and on the breast some of the feathers show vermiculate bars, often obscure; wings as in the young; head, back and tail with adult and young feathering mingled together (E. Celebes, V.—VIII. 95: C 14437).

The young of *C. virescens* is easy to distinguish from that of *C. merulinus* by the under side of the wing, which is notched with rufous-cinnamon all along the inner webs in the latter; the predominating colour of the former is dusky black, of the latter cinnamon.

Measurements.	Wing	Tail	Tarsus	Bill from nostr.
a. (C 10876) ad. near Manado, VIII.—IX. 92 . .	115	138	16	10.5
b. (C 1865) ad. Manado, III. 71 . , .	111	136	14	—
c. (C 1866) ad. Manado, III. 71	118	—	17	—
d. (10875) ad. near Manado, VIII.—IX. 92 . .	111	132	16	—
e. (C 5185) imm. Manado	112	134	15	18
f. (C 5184) imm. Manado	111	124	16	11
g. (C 14695) ad. Banggai, V.—VIII. 95	106	128	13.5	10.5
h. (Sarasin Coll.) ♂, Tomohon, 16. IV. 94 . . .	122	143	—	11
i. (Sarasin Coll.) ♂, Kema, 15. IX. 93 . .	108	127	—	11
j. (Sarasin Coll.) imm., Kema, VIII. 93	107	130	—	—
k. (Sarasin Coll.) ♂, Mapane N. Centr. Cel., 28. II. 95	109	130	—	11
l. (Sarasin Coll.) ♂, Loka, S. Cel., S. IX. 95 . .	114	130	—	11

Skeleton.

Length of cranium	35.4 mm	Length of tarso-metatarsus	15.6 mm
Greatest breadth of do.	15.0 »	Length of digitus III	18.0 »
Length of humerus	23.0 »	Length of sternum	20.0 »
Length of ulna	23.3 »	Greatest breadth of do.	15.0 »
Length of radius	22.0 »	Height of crista sterni	8.0 »
Length of manus	23.0 »	Length of coracoideum	16.0 »
Length of metacarpus	12.0 »	Length of scapula	19.0 »
Length of digitus princ.	11.5 »	Length of clavicle	18.0 »
Length of femur	15.7 »	Length of pelvis	25.0 »
Length of tibia	25.5 »	Greatest breadth of do.	15.0 »
Length of fibula	7.0 »		

Distribution. Celebes: — Minahassa (Wallace *c 1*, Meyer *a 1, a 2*, Fischer *b 1*, etc.); Mapane, N. Centr. Celebes (P. & F. S.); Luwu, S. Centr. Celebes (Weber *3*); Tonkean, E. Celebes (Nat. Coll.); Banggai (Nat. Coll.); S. Celebes (P. & F. S. *6*, Everett *7*); W. Celebes, Tawaya (Doherty *8*).

The nearest ally of this small Cuckoo appears to be *Cacomantis aeruginosus* Salvad. of Buru, Ceram, Goram and Amboina, a form which Count Salvadori, with two specimens from Buru before him, at first identified with it (Ann. Mus. Civ. Gen. 1876, VIII, 373), but afterwards separated (op. cit. 1878, XIII, 458) on the ground that *C. aeruginosus* differs in the rufous chestnut of the under parts being more intense, but less pure and mixed with grey, and extending more towards the chin (Orn. Pap. 1880, I. 336).

Capt. Shelley (1891) again unites the two forms; but, in holding them distinct, we accept Salvadori's more recent view (Orn. Pap., Agg. 1891, 218). It may be pointed out that the Celebes form seems to have a shorter wing and longer tail than *C. aeruginosus*:

C. virescens: wing ca. 114; tail ca. 135.

C. aeruginosus: wing ca. 120; tail ca. 114.

C. merulinus differs in having the upper plumage paler, the grey of the throat carried over on to the upper breast; remainder of body below rufous buff, not cinnamon-rufous, the inner webs of the tail-feathers regularly barred with white, most strongly so on the outer feathers (Shelley); in *C. virescens* the bars are reduced to some inconspicuous notches.

By the grey of the throat being spread on to the breast and by the inner webs of the tail-feathers being barred, or deeply notched, with white, *C. assimilis* Gray of the Moluccas (var. *major* Salvad.) and Papuasia may readily be distinguished.

We have seen a specimen from Mr. Nehrkorn's collection, Nr. 1758, Sooloo Islands (Platen), with a deep rufous throat, breast and under surface, wing 123, tail 150 (ca.) mm; "Iris braun, Schnabel schwarz, Basis, Augenring und Füsse gelblich". This appears to be a new species, unless it is *C. sepulchralis* which Shelley unites with *C. merulinus*. Dr. Sharpe (Ibis 1894, 247, 258) records from Sooloo *C. merulinus* only.

The first notice of the occurence of a *Cacomantis* in Celebes was made by Walden, in reference to three specimens obtained by Meyer in 1871, though in the British Museum there is a specimen of earlier date killed by Wallace at Manado. The cry of the bird, according to Meyer, is tü, tü, tütütü, like a flute. It has recently been sent from various parts of Celebes and from Banggai by other collectors. Feeds on insects.

The nest referred to by Meyer as that of this species must doubtless be that of some other bird, very possibly that of one into whose care its eggs are usually given. The parasitical habits of the genus *Cacomantis* have been established in the case of at least three species — *C. passerinus* (Vahl), *C. flabelliformis* (Lath.) and *C. pallidus* (Lath.) and, apparently, *C. merulinus* (Scop.), cf. Oates, Hume's Nests and Eggs Ind. B. II, 385; North, Nests and Eggs B. Austr. 1889, 243, 244; Baldamus, Leben europ. Kuckucke, 1892, 134—138.

66. CACOMANTIS MERULINUS (Scop.).
Buff-bellied Cuckoo.

a. **Cuculus merulinus** *(1)* Scop., Del. Flor. & Faun. Insubr. 1786, 89.
Cacomantis merulinus *(1)* Cab. & Hein., Mus. Hein. IV, I, 1863, 21; *(2)* Salvad., Cat. Ucc. Borneo 1874, 64; *(3)* Shelley, Cat. B. XIX, 1891, 268 pt.; *(4)* Büttik., Zool. Erg. Weber's Reise in Ost-Ind. 1893, III, 276; *(5)* Hose, Ibis 1893, 414; *(6)* Styan, ib. 433; *(7)* Sharpe, Ibis 1894, 247, 258; *(8)* Grant, ib. 520; *(9)* id., Ibis 1895, 262, 466; *(10)* Blanf., Faun. Br. Ind. B. III, 1895, 218; *(11)* M. & Wg., Abh. Mus. Dresden 1896, Nr. 1, p. 8; *(12)* Grant, Ibis 1896, 474, 560; *(13)* Hartert, Nov. Zool. 1896, 551, 586; *(14)* Kuschel, Orn. Mb. 1895, 156; *(15)* Hartert, Nov. Zool. 1897, 164.
b. **Cacomantis lanceolatus** (S. Müll.), *(1)* Wald., Tr. Z. S. 1872, VIII, 53; *(2)* Meyer, Ibis 1879, 146.
Descriptions. Shelley *3*; Blanford *10*; etc.
Adult. Like the foregoing species, *C. virescens* (Brügg.), but paler above, the grey of the throat spreading over the breast; remaining under parts buff, not cinnamon-rufous; tail-feathers regularly barred with white on the inner webs, whereas in *C. virescens* the bars are reduced to inconspicuous notches (♂, Macassar, 10. July 95: P. & F. Sarasini).

"Iris orange-red; bill red-brown; feet yellow"; wing 104 mm; tail 105 (Platen in Mus. Nehrkorn — Nr. 1759 — Palawan).
Young. Above cinnamon, with broad stripes of dusky on head and neck, taking the form of bars on back and innermost remiges; on the other remiges and the tail-feathers the cinnamon develops into deep notches, extending as bars across the outer tail-feathers; under parts paler cinnamon, striped with dusky on throat and breast, taking the form of sagittate bars lower down and on under wing-coverts; remiges below notched on the inner webs (slightly so on the outer ones) with cinnamon, becoming uniform cinnamon-buff on the basal part of the feathers (♀, Macassar, 19. July 95: P. & F. Sarasin).
Immature (assuming adult dress). Head, neck, and throat chiefly clothed with the grey plumage of the adult, but with cinnamon feathers with dark centre-stripes of the young intermixed; wings and tail barred with cinnamon and dusky as in the young,

but two or three feathers of the adult in the tail; back, breast, and belly with young and adult plumage intermixed, the adult predominating above (♂, Macassar, 10. July 95: P. & F. Sarasin).

Measurements (2 ad., 2 juv. Celebes). Wing 102—106 mm; tail 103—113; tarsus c. 17; bill from nostril c. 12.5.

Egg. Cf. Kuschel *14*: Java.

Distribution. The Indian countries, to Malacca, Java, Bali, and Sumba, S. China and Hainan, south throughout the Philippines to Borneo, and Celebes (see Salvadori *2*, Shelley *3*, Blanford *10*, Styan *6*, Grant *12*, Hartert *13*).

In Celebes — Macassar (Wallace *3*, Weber *4*, P. & F. Sarasin *11*); ? Togian (Meyer *b 2*); W. Celebes, Tawaya (Doherty *15*).

Among the two hundred and more specimens in the British Museum included by Capt. Shelley under the name *C. merulinus*, there is one marked: "♂ ad. Macassar (A. R. Wallace)". This appears to be the specimen which was in the hands of Lord Walden in 1872 *(b 1)*, when he remarked that it appeared to belong to the group of which *C. merulinus* is typical. This determination has recently been confirmed by Mr. Büttikofer, who records two specimens obtained by Prof. Weber at Macassar, and is further proven by four specimens obtained by Drs. P. and F. Sarasin at the same spot. Still further east Shelley records a specimen from Ternate, but this indication is not accepted without query by Count Salvadori (Orn. Pap., Agg. III, 1891, 218).

Shelley remarks that *C. merulinus* appears to consist of three races: viz. "*C. merulinus* (Scop.), a pale bird with the grey of the head and throat sharply defined and with clear regular white bars across the inner webs of the tail-feathers. *C. threnodes* Cab., a rather darker bird with a deeper rufous buff colour on the breast. *C. sepulchralis* (S. Müll.), with the upper parts nearly uniform, the notches on the outer webs of the tail-feathers shaded with rufous and the white bars on their inner webs nearly obsolete". From the last-named form, writes Mr. Büttikofer *(4)*, *C. merulinus* "should certainly be separated. It is distinguishable from *C. sepulchralis* by its smaller size (wing maximum 103, tail maximum 108 mm), further by the very pale ash-grey colour of head, nape, sides of neck, chin, throat and upper chest, which colour extends on to the upper breast in certain individuals, and by the very pale ochre-yellow colour of breast, abdomen and under tail-coverts".

Mr. Hartert is of opinion that *C. threnodes* may also have to be separated again from *C. merulinus*.

GENUS COCCYSTES Glog.

In this genus, the species of which are about the size of a Turtle-dove, the tail is much longer than the wing, and the occipital feathers are lengthened so as to form a crest. Bill moderate, nostril a slightly projecting, oval formation of skin; wing short, secondaries about $^3/_4$ its length, tail strongly graduated, the outermost feathers rather more than $^1/_2$ its length, the upper tail-coverts very long, nearly $^1/_2$ the length of the tail. Shelley recognises 6 species and 2 subspecies, inhabiting Southern Europe, Africa and tropical Asia as far as Celebes and the Philippines. At least two of the species are known to be parasitic in their breeding-habits. As Mr. Blanford remarks: "It is possible that *Coccystes* may prove, when the anatomy and pterylosis are examined, to belong to the *Phoenicophainae*".

+ 67. COCCYSTES COROMANDUS (L.).
Red-winged Crested Cuckoo.

a. **Le Coucou hupé de Coromandel** *(1)* Briss., Orn. 1760, IV, 147, pl. XIa, f. 1.
b. **Cuculus coromandus** *(1)* Linn., S. N. 1766, I, 171.
c. **Le Coucou à collier blanc** *(1)* Levaill., Ois. d'Afr. 1806, V, 213.
d. **Oxylophus coromanus** *(1)* Jerd., Madr. Jrn. 1840, XI, 272.
Coccystes coromandus *(1)* Horsf. & Moore, Cat. B. Mus. E. I. Co. 1854, 693; *(2)* Bulger, Ibis 1869, 159; *(3)* Swinh., P. Z. S. 1871, 394; *(4)* Salvad., Cat. Ucc. Borneo 1874, 67; *(5)* Hume, Str. F. 1875, III, 82; *(6)* Brügg., Abh. Ver. Bremen 1876, V, 465, 531; *(7)* David & Oust., Ois. Chine 1877, 61; *(8)* Hume & Davison, Str. F. 1878, VI, 162; *(9)* Tweedd., P. Z. S. 1878, 946; *(10)* Scully, Str. F. 1879, VIII, 257; *(11)* Legge, B. Ceylon 1880, 249; *(11ᵇⁱˢ)* Vidal, Str. F. 1880, IX, 55; Butler, t. c. 389; *(12)* Müller, J. f. O. 1882, 406; *(13)* W. Blas., J. f. O. 1883, 122, 155; *(14)* Oates, B. Brit. Burmah 1883, II, 117; *(14ᵇⁱˢ)* Davison, Str. F. X, 1883, 360; *(15)* Büttik., Notes Leyd. Mus. 1886, IX, 29; *(16)* Norman, Ibis 1888, 400; *(17)* Hartert, J. f. O. 1889, 370; *(18)* Everett, J. Str. Br. R. A. S. 1889, 173; *(19)* Sharpe, Ibis 1890, 12, 282; *(19ᵇⁱˢ)* Hagen, T. Ned. Aard. Genoots. 1890 (2) VII, 137; *(20)* Shelley, Cat. B. XIX, 1891, 214; *(21)* Oates, Hume's Nests and Eggs Ind. B. 1891, II, 391; *(22)* Styan, Ibis 1891, 317, 325, 484; *(23)* Vorderman, Nat. Tdschr. Ned. Ind. 1892, LI, 380; *(24)* Hose, Ibis 1893, 415; *(25)* Bourns & Worces., B. Menage Exped. 1894, 35; *(26)* Blanf., Faun. Br. Ind. B. 1895, III, 227.
f. **Coccystes coromandus** var. **fuliginiventer** *(1)* Swinh., Ibis 1867, 227.
For further synonymy and references see Shelley 20.
Figures and descriptions. Brisson a *I*; Levaill. c *I*; David & Oust. 7; Scully 10; Legge *11*; Oates *14*; Shelley *20*; Vorderman *23*; Blanford *26*.

Adult. Head above, including nape and ear-coverts, brownish black, a long crest blue-black; on hind neck a white collar; mantle greenish black, passing into glossy bottle-green on the scapulars, inner quills and lesser wing-coverts; the other wing-coverts and quills light hazel, the latter dull brownish at the ends; rump and upper tail-coverts black glossed with blue and green; tail blue-black, the extreme tips of the feathers, chiefly on the outer webs, white; chin, throat and sides of neck, upper breast, and under wing-coverts cinnamon washed with rufous (chestnut-buff or yellowish ferruginous); lower breast and sides white, slightly washed with the colour of the throat; thighs, flanks and abdomen brownish grey, becoming brownish black glossed with blue and tipped with light brown on the under tail-coverts; wing below cinnamon (Borneo, Nr. 7118).

"Iris brown; bill black, at the base somewhat bluish white; orbits grey-blue, feet lead-blue" (Hart. *17*).

Sexes. The male and female closely resemble one another.
Young. The nestling differs from the adult in having most of the feathers of the upper parts tipped with rufous, the collar being shaded with rufous; the tail-feathers broadly tipped with pale sandy buff; the throat buffy white like the chest, and the under tail-coverts rufous buff (Shelley *20*).
Measurements. Wing 155 (♀)—173 mm (♂); tail 229—254; tarsus 25.4; culmen 28—30 (Shelley *20*).
Egg. Very broad oval; fine and glossy. Moderately pale, blue, somewhat greenish, without any spots or specks. 26.6 × 23.2 mm (Oates *21*). Apparently parasitic in breeding habits.
Distribution. India, west to Kumaon, south to Trichinopoly (Norman *16*); Ceylon (Legge *11*); Pegu (Oates *14*); Tenasserim (Feilden *5*); China (Swinh. *f1, 3*, David *7*, Styan *22*); Siam (fide Shelley *20*); Cambodia (Mouhot *20*); Salanga (A. Müll. *12*); Malacca (Davison *8, 20*, Hartert *17*); Penang (Horsf. & Moore *1*); Singapore (Charlton *16*); Sumatra (S. Müll. *4*, Klaesi *15*, Hartert *17*); Bangka (van den Bossche *4*, Büttik. *15*); Java (Horsf., Kuhl, v. Hasselt *4*, Vorderman *23*); Borneo — North, West, South, Central, and North-west (Wallace, Everett, etc. *4, 10, 18, 19, 20*); Philippines — Mindanao (Everett *9, 20*), Siquijor and Palawan (Bourns & Worcester *25*); Celebes — Minahassa (Fischer *6, 13*).

A single example of this Cuckoo was shot on 18th October 1873 by Fischer at a height of 4000 ft. on Mount Lokon not far from Manado in the Minahassa. Only one specimen, as far as we can ascertain, was known from the Philippines prior to 1894, the bird having been shot by Mr. Everett at Zamboanga in Mindanao in March 1878, but Messrs. Bourns & Worcester afterwards obtained the species in Siquijor and Palawan. Everett considers it rare in Borneo, where, up to 1890, he had only met with two specimens; so, too, Dr. Vorderman remarks that it is scarce in Java *(23)*.

There is reason to believe that *Coccystes coromandus* is only a winter visitor to the East India Islands. Mr. Styan *(22)* marks it as a summer visitor to the Lower Yangtse division of China; Capt. Feilden *(5)* speaks of it as "the commonest Cuckoo at Thayetmyo, Tenasserim, ... arriving in the beginning of the rains (April) and the young birds do not leave till October"; in Tipperah and Sikkim further north, as well as in Tenasserim, Mr. Oates *(21)* records two cases of a female ready to lay being shot, but both in Sikkim and Lower Pegu it has been found in December as well *(20)*. Jerdon *(11)* says it is found in Bengal "only during the rains" (the months of our summer). In Ceylon, on the other hand, it is, according to Colonel Legge, a winter visitor, "arriving about October and departing again in April"; also in South India, where it has very rarely been obtained, it may prove to be only a winter visitor. Thus Mr. Vidal's specimen from the South Konkan was killed 2nd January, 1880 *(11bis)*, and one from Kotagherry, Nilghiris (Miss Cockburn) in the British Museum is dated Dec. 5th *(20)*. The following are all the dates we find recorded from the south-eastern parts of its range:

Malacca: July *(20)*.
Salanga; Jan. 5th, 15th, March 7th *(12)*.

Sumatra: 2, sometime between Oct.—Mch. *(15)*.
Java: February *(23)*.
Borneo: Dec. 20th *(4)*, Dec. 29th *(19)*.
Celebes: October 18th *(6)*.
Mindanao: March *(9)*.

Many more observations are, of course, wanted before a decided opinion can be formed that the birds which visit China and Tenasserim in summer are the same individuals as those found in the East Indies in winter. In all probability such is the case. Mr. Hose *(24)* remarks that it "occurs on migration" on the Sarawak coast of Borneo, and, as regards Celebes, we are inclined to regard the species as a rare winter migrant, perhaps only a straggler.

C. coromandus is most likely entirely parasitic in its breeding habits. Thus, Capt. Feilden has shot a young one out of a brood of young Quaker Thrushes (*Alcippe*) and has found an egg, apparently of this Cuckoo, in the nest of that bird. Its food consists of hairy caterpillars, "beetles, grasshoppers, *Mantidae*, and other large insects" (Legge *11*, Hartert *17*).

The nearest ally of this Cuckoo is the *Coccystes glandarius* (L.) of "Southern and Central Europe, as far east as Persia, wintering in Africa" (Shelley), from which *C. coromandus* may be readily distinguished by the white collar across its hind neck and by the absence of large terminal spots of whitish on the wing-coverts. It is of some interest to note that these spots are to be seen to some extent in the immature *C. coromandus*.

GENUS SURNICULUS Less.

A genus of three small Cuckoos of about the size of a Lark, easily recognised by their general plumage of black and resemblance to a Drongo (*Dicrurus*). The tail is about as long or longer than the wing, in two species forked, the outer feathers taking a lateral curl as in adult Drongos; in *S. velutinus* it is square; the outermost rectrix is short, about ²/₃ the length of the tail. Across the base of the wing below is a white band. The genus is found from the Himalayas to the Philippines, Batchian and Java. Davison (Str. F. 1878, VI, 159) observed a young one of *S. lugubris* in Tenasserim being fed by a Drongo; it is supposed, as Blanford says, to deposit its eggs in the nests of these birds.

68. SURNICULUS MUSSCHENBROEKI A.B.M.
Van Musschenbroek's Drongo-cuckoo.

Surniculus musschenbroeki *(1)* Meyer, Rowl. Orn. Misc. 1878, III, 164; *(2)* Salvad., Ann. Mus. Civ. Gen. 1878, XIII, 461; *(3)* id., Orn. Pap. I, 1880, 357; *(4)* Shelley, Cat. B. XIX, 1891, 230; *(5)* Hartert, Nov. Zool. 1896, 159, 165.

Description. Meyer *1*.
Adult. Black, glossed with blue, on the wings with green; under wing-coverts and quills

below shining dusky, a narrow band of white across the basal third of the secondaries (except the inner ones) and of the primaries (except the three or four outer ones), a deep, narrow notch of white on inner web of first primary at half length; tail Drongo-like, the outermost feather short, about ¾ the length of the tail, with about 4 white spots near the shaft, the two terminal ones extending into obscure bars; tibio-tarsal feathers behind white and woolly; one or two of the longest under tail-feathers just tipped with white: bill black; feet (in skin) brown. Wing 136 mm; tail c. 145; tarsus c. 16; bill from nostril 15.5 (type, [♀] ad. Batchian: Meyer — Nr. 1972.

Distribution. Batchian (Meyer *1*); South Celebes — Indrulaman (Everett *5*); North Celebes (Hose, in the British Museum).

The type of this species, which was obtained by one of Meyer's hunters in Batchian, remained the only specimen known for more than twenty years, when the species was rediscovered by Everett on the foot-hills of the Peak of Bonthain, and two males in perfect plumage were sent to the Tring Museum. They differ in no important points from the type.

Capt. Shelley recognises three species of the genus *Surniculus*, viz. *S. lugubris* (Horsf.) ranging from India and Ceylon to Borneo and Java, *S. velutinus* Sharpe known from half a dozen islands of the Philippine group (see Grant, Ibis 1896, 559), and *S. musschenbroeki* from Batchian and, as we now see, N. & S. Celebes. The last is nearly related to *S. lugubris*, which differs by having the under parts glossy black-brown, instead of glossy blue-black, the under tail-coverts are transversely marked with white and the white bars on the outermost tail-feather are well marked. *S. velutinus* has the tail square.

When writing the history of the Cuckoos of Celebes for this work (about three years ago) we remarked with surprise upon *Surniculus* and *Chrysococcyx* as not having yet been found there, but the labours of Mr. Everett and of the Drs. Sarasin have now brought the discovery of both in the south of the island, the latter genus in two species.

A specimen from Mr. Hose, marked "♂, October 1895: Bantik, Celebes", and sent by Dr. Sharpe to the Dresden Museum for comparison with the type, has a much larger white patch on the middle of the occiput and the white bars on the outermost tail-feather less obliterated than in the type. Everett's 2 specimens from S. Celebes differ in much the same way, but it is hardly probable that these are racial characters. The wing of Hose's specimen measures 127 mm.

GENUS EUDYNAMIS Vig. Horsf.

In the Koels, which are about the size of a Turtle-dove, the bill is strong and almost perpendicularly decurved at the tip; the feet are large, the middle toe without the claw being as long as the tarsus; the wing is moderately long, the tail as long as the wing, rounded, the outermost feathers about ³/₄ the length of the longest. The sexes are dissimilar in coloration, the male black, the female spotted or barred with tawny-brown and black. Six species are recognised in the Catalogue of Birds, they range from the Himalayas to Australia.

These Cuckoos are probably all parasitical; the Indian species deposits its eggs in Crows' nests, and in Borneo Whitehead found that the nest of a Mynah was used for this purpose.

+ * 69. EUDYNAMIS MELANORHYNCHA S. Müll.
Black-billed Koel.

Eudynamis melanorhyncha *(1)* S. Müll, Verh. Natuurk. Comm. 1839—44, 176; *(2)* Gray, Gen. B. 1847, II, 464; *(3)* Bp., Consp. Av. 1850, 101; *(4)* Cab. & Heine, Mus. Hein. 1862, IV, 55; *(5)* Finsch, New Guinea 1865, 159 (Celebes tantum); *(6)* Blyth, Ibis 1866, 364; *(7)* Wald., Ibis 1869, 344; *(8)* Gray, HL. 1870, II, 221; *(9)* Wald., Tr. Z. S. 1872, VIII, 53, 112; *(10)* Salvad., Ann. Mus. Civ. Gen. 1875, 650; *(11)* Brügg., Abh. Ver. Bremen 1876, V, 59, 406; *(12)* Lenz, J. f. O. 1877, 371; *(13)* Meyer, Ibis 1879, 69, 146; *(14)* id., Isis 1884, 18; *(15)* Guillem., P. Z. S. 1885, 550; *(16)* W. Blas., Ztschr. ges. Orn. 1886, 96; *(17)* id., Ornis 1888, 566; *(18)* Tristr., Cat. Coll. B. 1889, 82; *(19)* Heine & Rchw., Nomencl. Mus. Hein. 1890, 200; *(20)* Shelley, Cat. B. XIX, 1891, 327; *(21)* M. & Wg., Abh. Mus. Dresd. 1895, Nr. 8, p. 6; *(22)* iid., ib. 1896, Nr. 2, p. 8; *(23)* Hartert. Nov. Zool. 1897, 164; *(XXIV)* Meyer, Abb. v. Vogelskel. 1897, pl. CCXXVIII.

a. † **Eudynamis fascialis** *(1)* Wall., P. Z. S. 1862, 339; *(2)* Blyth, Ibis 1866, 364; *(3)* Wald., Ibis 1869, 345; *(4)* Gray, HL. 1869, II, 221.

b. **Cuculus melanorhynchus** *(1)* Schl., Mus. P.-B. Cuculi 1864, 20.

"Kao", Minahassa, Malays (M. *13*).
"Kororeke" or "Kembaluwan", Minahassa, inland names (M. *13*).
"Kuku buri"; hepatic phase, and
"Konil", mature plumage (Guillem. *15*).
"Koukou werreng", inland near Manado, or
"Koue itam", Nat. Coll. Mus. Dresd.
"Kuow maitem", Tonkean, E. Celebes, iid.
"Totaal mapok", Peling, iid.

Descriptions. S. Müller *1*; Cabanis & Heine *4*; Walden *7, 9*; Salvadori *10*; Brüggemann *11*; Guillemard *15*; Blasius *16*; Shelley *20*.

*A.*¹) **Adult male.** All over glossy blue-black. Bill black; feet black; iris bright red *(15)*, (N. Celebes, C 3583, and others).

B. **Male nearly adult.** Above like the adult, but with some old dull feathers of blackish brown in tail, together with new ones of blue-black; under surface from chest downwards very dark olive with slight greenish reflexions (N. Celebes C 3584).

C. **Younger male.** Above like the adult, but all the quills dusky, except the innermost ones. which are blue-black; under surface from chest downwards wood-brown (or pale

---

¹) The letters *A*, *a*, etc. are used to indicate individuals of opposite sexes of corresponding age, *A* being the fully adult male, *a* the fully adult female. *B* and *b* the next stage of the two sexes, and so on.

cinnamon), closely mottled with sooty, the flanks and under tail-coverts almost entirely of this colour (near Manado, C 10872).

D. **Male with earliest appearances of adult plumage.** Head, hind neck, and some sprouting feathers in the tail blue-black; rest of upper surface very dark glossy green, the quills with blue reflections; a pale stripe from rictus to side of neck; chin and throat blackish, passing into brownish green on chest; rest of under surface sooty brown, varied with cinnamon, especially on the breast and upper abdomen (Manado, C 1844).

Three other [♂] specimens, apparently slightly less developed, have the tail, like the back, very dark green, with greenish or bluish reflections; the cinnamon colour greatly predominating on the under surface and the long rictal stripe of white varied with cinnamon or brown broader and more distinct.

E and e. **First plumage** [♂ and ♀]. Above dark glossy green, uniform, approaching to black on the head; chin and throat dusky; rictal streak of fulvous white passing into the cinnamon-brown of sides of neck; under surface cinnamon, each feather crossed with two or three fine brace-shaped bars of dusky (C 5181 and three others). In one the bars on the under surface are quite absent (C 1843). Tail-feathers of the two others obscurely barred with dark cinnamon (C 263, C 1845). Bill in C 5181 dark horn-brown, under mandible paler; in C 263 and 1845 darker horn; in 1843 black.

Nestling. Plumage uniform above, apparently similar to the above, and below similarly cinnamon with brace-shaped bars (Tomohon, 15. IV. 94: P. & F. Sarasin).

c. **Female with earliest appearances of adult plumage.** Head, neck and mantle bluish black, as in male with earliest appearance of adult dress (see *supra*); back and wing-coverts varied with somewhat obscure bars of rufous brown (cinnamon-hazel); wings composed of old dusky feathers, uniform as in the young, and of new ones (some of them sprouting) crossed with cinnamon-hazel and black bars, the black ones a little the broadest; tail similarly composed of new feathers — some of them growing — barred with cinnamon-hazel and black, one old feather is dusky with the cinnamon markings very obscure; chin and throat black, slightly touched up with whitish brown; rictal streak white; under surface fulvous cinnamon, crossed with narrow brace-shaped blackish bars. Bill blackish horn-colour (near Manado, C 10869).

From the circumstance that the third and fourth quills of one wing and the third and fifth of the other are old feathers, but are nevertheless considerably marked with imperfectly-formed bars of cinnamon in their basal two-thirds, we infer that this specimen may be a second-year bird passing into its third phase of dress. The quills, in what we take to be young specimens in first plumage, are uniform, but perhaps our four young ones are all males. Another [♀] specimen (C 266), apparently of the same age as the above is somewhat more melanistic in character.

b. **Female a little older.** Like the last, but with no remains of a younger dress. Head, neck and mantle bluish black; rest of upper surface evenly barred with cinnamon-hazel and black; chin and throat black; below fulvous cinnamon, finely barred with dusky (near Manado, C 10871).

a. **Adult female.** Like the last, but with the head, neck, chin and throat cinnamon-hazel with dusky edges to most of the feathers, giving a somewhat streaked appearance; the cinnamon-hazel bars of the upper surface much broader than the black ones (♀, Kema, 7. Aug. 93: P. & F. S.). In C 3581 the bill is dark horn-colour, the under mandible paler; in C 10870 almost all black.

Three other specimens afford transitions between the immature stage b and the adult stage a (N. Celebes, C 265, 5183, 264).

Measurements.	Wing	Tail	Tarsus	Bill from nostril
a. (C 1848) ad. [♂] Manado, March 1871	210	210	30	18.5
b. (C 3581) ad. [♀] N. Celebes, 1877	186	192	32	18.5

Distribution. Celebes — Minahassa (Forsten *b 1*, Meyer *9*, Fischer *11 etc.*); Gorontalo (Forsten *b 1*, Meyer *13*); Talissi Island (Guillem. *15*); Banka, Lemboh and Mantehage Islands (Nat. Coll.); Togian Islands (Meyer *13*); Macassar (Wallace *20*); West Celebes (Doherty *23*); E. Celebes, and Peling (Nat. Coll.); Sula Islands (Allen *a 1, 20*).

Brüggemann *(11)* records a specimen as having come from Sangi, but this would appear to be a mistake, since that island has a race of *Eudynamis* of its own, *E. mindanensis sangirensis* W. Blas.

In the foregoing sketch of the several stages of dress passed through by this Cuckoo a conclusion has been arrived at identical with that of Capt. Shelley, but differing radically from those of Lord Walden, Brüggemann, Dr. Guillemard, and Prof. W. Blasius. According to Shelley the adult male "is all over blue-black, the adult female above chestnut evenly barred with olive-black, ... under parts (except chin and upper throat) buff, with numerous narrow wavy black bars"; Brüggemann, on the other hand, states that the sexes are alike both in size and colour, and what Shelley describes as the adult female is held by him, Walden, Blasius and Guillemard to be the first stage of immature dress. That these authorities are in error is shown by the nestling obtained by the Drs. Sarasin, as also, by the specimen described by us as *c* [♀] with earliest appearances of adult plumage, to which attention may be drawn. Here the new and growing feathers are barred with black and rufous brown, while the old worn feathers, which were about to be cast off, are of that more uniform brown appearance belonging to the fledgeling, but supposed by these authors to belong to a more advanced stage. As a great series before us in almost every phase of dress show, the young of both sexes are at first apparently just alike in coloration, but commence to deviate in the second stage, becoming more and more different as they approach maturity. Dissimilarity of the sexes is one of the characteristics of the genus *Eudynamis*, elsewhere, as Indian ornithologists well know.

There is a remarkably close resemblance between the old female of this Celebesian Cuckoo and the young of the wide-spread *Centrococcyx bengalensis* also a species occurring in Celebes. The young dress of *E. melanorhyncha* recalls the uniform coloration of the upper surface of *Coccystes coromandus*.

The only specimen as yet known from the Sula Islands is a male, apparently partially affected by albinism, which was obtained by Mr. Wallace's assistant, Allen, in Sula Besi or Sula Mangoli, and was named as distinct on account of its smaller size. This distinction is not found to hold good by Capt. Shelley, and until more specimens for comparison are forthcoming it should be regarded as identical with *E. melanorhyncha*. Shelley's opinion is confirmed by an example from Peling, resembling the birds of the mainland.

Meyer has described the Black-billed Koel *(13)* as "shy and lively in its actions. Roosts in the darkest spots, in trees, where it can hardly be detected. If danger threatens, or if it hears a particular noise which frightens it, it communicates its alarm from afar to others; and it is no fable that the natives are warned by the bird hours before — if, for instance, a troop of horsemen approaches, or an official with his attendants. The native therefore often makes his preparations according to this bird's behaviour; hearing it in the forest he will always be cautious. But its cry at night he consults as an oracle, and converses with the bird by imitating its cry and interpreting it. If he hears it at night near a house he augurs the death of a man".

"I found mostly nutmegs in its stomach. Before nutmegs were cultivated in the Minahassa, which is only during the last few years [prior to 1871], the bird fed on different fruits, chiefly waringuis, but now nearly altogether on nutmegs, which it swallows whole for the sake of the rind; the nutmeg itself is found uninjured in the crop or stomach; and the bird contributes greatly to the distribution of this spice. It damages the plantations very much. It is said to seek its food at night".

Meyer learnt that it lays its eggs in other birds' nests, but could not ascertain in which, and the cousins Sarasin have recently produced proof of its parasitic habits through the young one described above, which was "brought in a nest which the boy ascribed (in our opinion, however, incorrectly) to *Enodes erythrophrys*". Many interesting facts on the breeding of *Eudynamis honorata* (L.) in India are recorded in Hume's "Nests and Eggs of Indian Birds" (Oates, ed. 1891, II, 392—397); this species always deposits its eggs in Crows' nests, usually that of *Corvus splendens*. So, too, in Burmah *E. malayana* Cab. & Heine lays its eggs, says Mr. Oates, in the nests of Crows; and Dr. Tiraud states that in Cochin China the eggs are also placed in the nests of Mynahs (B. Brit. Burmah II, 120). *E. mindanensis* also makes use of a Mynah's nest (Whitehead, Ibis 1888, 410). It will therefore very probably be found that in Celebes *Corvus enca* is one, perhaps the most usual, foster-parent of *E. melanorhyncha*.

In India the Cuckoo sometimes pays dearly for her imposition: *Corvus culminatus*, writes Mr. Anderson, "is easely duped, while her cunning congener, *C. splendens*, is fully aware of the deception". Colonel Butler records a case of a female specimen being mobbed to death by crows, and of a male (which, as Mr. Anderson says, frequently accompanies the female when she is about to deposit her egg), being so harried by two Crows that he was able to take it by hand. Naturalists, who believe in protective mimicry, may wonder that the males of *Eudynamis* are black corvine-looking birds, while their females, to whom a deceptive resemblance to the Crows might be of value, differ greatly in coloration.

The young are black, as Mr. Whitehead states in the allied *E. mindanensis* (Ibis 1888, 410, Expl. Kini Balu 1893, 145), and Capt. Shelley in *E. honorata*, resembling the adult males instead of, as is much more usual among birds, the female, or of being different from either. Mr. Whitehead to account for it makes the following plausible suggestion: "the 'Phow' *(E. mindanensis)* lays its eggs in the nest of the Yellow-mottled Mynah *(Gracula javanensis)*. The young Cuckoo, being black, does not differ from the young Mynah, and so the deception is carried on until the young bird can take care of itself". Much has been written on the resemblance of the eggs of Cuckoos to those of the birds in whose nests they are placed, and with good reason; but mimicry, if it be so, of the young of its foster-mother by the young Cuckoo brings another phase of the question before us. It is very conceivable that this resemblance is of advantage to the young Cuckoo. The clever *Corvus splendens* of India apparently recognises the imposture and abandons the Cuckoo, as Mr. Hume has observed, when it leaves the nest; nevertheless the philoprogenitive instinct in birds is so irresistible that their discriminative powers often appear scarcely to come into play at all. Those which have hatched the young *Cuculus canorus*, do not think to restrain themselves from attending to his wants; "the actions of his foster-parents", writes Prof. Newton, "become, when he is full grown, almost ludicrous, for they often have to perch between his shoulders to place in his gaping mouth the delicate morsels he is too indolent or too stupid to take from their bill" (Dict. of B. 120). *Corvus splendens*, however, is fully aware that *Eudynamis* is in the habit of imposing her eggs upon her; the species in whose nests *C. canorus* places her eggs do not, perhaps, discover that they have been cheated, but rear the young Cuckoo as a surprisingly fine-grown offspring of their own. Were the Cuckoo the most insignificant of the brood in point of size, as well as being so anomalous in structure — this is the case with *Eudynamis* in the nest of *Corvus* — it would run more risk of neglect, and a special adaptation of plumage might be an advantage to it. The dwarf pig of a litter is, if we are not misinformed, often devoured by the mother.

The origin of the habit of laying their eggs in the nests of other birds probably dates very far back among the Cuckoos; there is reason to suppose that the habit runs through all the genera (except *Coccyzus*) of the subfamily, *Cuculinae*, as defined by Capt. Shelley (Cat. B. XIX, 210) embracing forms differing greatly in coloration and considerably in structure. Thus, Dr. Baldamus (Leben der europ. Kuckucke, 1892) cites cases of parasitism in the genera, *Cuculus*, *Hierococcyx*[1]), *Cacomantis*, *Heteroscenes*[2]), *Lamprococcyx (Chalcococcyx)*, *Eudynamis*, *Urodynamis*[3]), *Scythrops*, *Coccystes*. Prof. Newton, further, speaks of species of *Surniculus*, *Phoenicophaes* and *Zanclostoma* as parasitical (Dict. ot Birds 1893, 125), but we venture to express a doubt in regard to the two last-named genera, which Shelley places in an other subfamily, *Phoenicophaeinae*, inasmuch as *Rhopodytes (Zanclostoma) tristis* and *viridirostris* are known to build their own nests (Hume, Oates ed. t. c. 397, 399).

---

[1]) *H. sparverioides* is apparently only partially parasitical.
[2]) Treated by Shelley as identical with *Cuculus*.
[3]) Perhaps only partially parasitical (cf. Finsch, Mitth. Orn. Ver. Wien 1884, VIII, 126).

The correspondence in habits between *Eudynamis* and *Cuculus* is further displayed by the circumstance that the young Crows, which have the misfortune to find themselves in the same nest with the *Eudynamis*, are sometimes ejected, "probably", as Mr. Hume says, "by the young Cuckoo; I have found the latter in a nest with three young Crows, all freshly hatched, and a week later have found the young Crows "missing" and the young Cuckoo thriving". Mr. Hume was of opinion that the Crows' eggs were not destroyed by the mother Cuckoo; Colonel Butler, on the other hand, says, "when the hen birds lays she often turns some of the Crow's eggs out of the nest, as I have several times examined Crows' nests and found three or four eggs one day, and on examining them a day or two later have found some of the Crows' eggs missing and Coëls' eggs in their place". There is some reason to suppose that the Common Cuckoo, *C. canorus*, sometimes turns out one or more of the foster-parent's eggs (Newton, Dict. 121).

Capt. Hutton's observations on the old *Cuculus intermedius* feeding a young one after leaving the nest have already been quoted from Hume's great work; so, too, as has already been indicated, in the case of the present genus Hume has a similar observation: "I have never seen Crows feeding fully fledged Coëls out of the nest, whereas I have repeatedly watched adult female Coëls feeding young ones of their own species" (Nests and Eggs 1891, II, 393). There is, according to Prof. Newton, no evidence worthy of consideration that the female of *C. canorus* takes any interest "in the future welfare of the egg she has foisted upon her victim, or of its product"; the observations of Mr. Hume and Capt. Hutton, nevertheless, render it certain that two Cuckoos at least, and we suspect all parasitic birds, are not totally devoid of sympathetic instincts for the wants of young members of their own species; though, whether the individuals observed looking after the young Cuckoos were their identical mothers or not, there it no evidence to show.

70. EUDYNAMIS MINDANENSIS (L.).

Philippine Koel.

While the preceding species, *E. melanorhyncha* of Celebes and Sula, is readily distinguishable by its black bill from other members of the genus *Eudynamis*, the distinguishing characters of the remaining species, which are spread out from India and S. China across the Archipelago to New Guinea and Australia, are by no means so strongly pronounced. Especial care, too, is called for in considering the western forms, owing to the circumstance that — in certain parts at least — the species are not perfectly stationary. Thus, the species spoken of as *Eudynamis maculata* by Swinhoe (P. Z. S. 1871, 394), David & Oustalet (Ois. Chine 60) and De La Touche (Ibis 1892, 480) is only a summer visitor to South China; and Mr. Whitehead remarks: "I never heard or shot an adult bird [of *E. mindanensis* in Palawan] after the middle of August, when it no doubt migrates to Borneo and other islands, as most of the birds in Labuan are seen after September during the N.E. Monsoon" (Ibis 1888, 410).

In the Sangi Islands an *Eudynamis* occurs which Prof. W. Blasius has separated as a variety of *E. mindanensis*. It may be doubted whether *E. mindanensis* itself is entitled to more than subspecific separation (i. e. whether individuals of it do not intergrade with *E. honorata* or *E. orientalis*); for the present, however, it appears best, in view of the want of sufficient data and material, to treat it as a species, composed of the following two subspecies:

1. The typical Eudynamis mindanensis.

a. **Cuculus mindanensis** *(1)* Linn., S. N. 1766, I, 69 (ex Brisson).
b. **Eudynamis mindanensis** *(1)* Cab. & Heine, Mus. Hein. 1862, IV, 52; *(2)* Wald., Ibis 1869, 340, ? pt.; *(3)* id., Tr. L. S. 1875, IX, 162; *(4)* Sharpe, ib. 1877, 2nd ser. I, 320, 351; *(5)* Tweedd., P. Z. S. 1877, 543, 691, 823; 1878, 946; 1879, 70; *(6)* W. Blas., Ornis 1888, 306; Ibis 1888, 373; *(7)* Everett, J. Str. Br. R. A. S. 1889, 173; *(8)* Whitehead, Ibis 1890, 46; *(9)* W. Blas., J. f. O. 1890, 138; *(10)* Steere, List Coll. B. Philip. Is. 1890, 12; *(11)* Shelley, Cat. B. XIX, 1891, 321; *(12)* Hartert, J. f. O. 1891, 298; *(13)* Whitehd., Expl. Kini Balu 1893, 145; *(14)* Sharpe, Ibis 1894, 247, 258; *(15)* Bourns & Worces., B. Menage Exp. 1894, 35; *(16)* Grant, Ibis 1895, 115; 1896, 123, 474.
c. **Eudynamis malayana** Sharpe, Ibis 1888, 198 (fide Whitehead).
d. **Eudynamis orientalis** *(1)* Whitehead, Ibis 1888, 409.
Descriptions. Cabanis & Heine *b 1*; Tweeddale *b 5*; Shelley *b 11*.
Adult male. Entirely black, glossed with blue. "Iris bright crimson; bill greenish grey; feet darker greenish grey" (Everett *b 5*).
Adult female. Above dusky, glossed with olive, the head and sides of neck streaked with rufescent brown, the rest of the upper surface spangled (several spots on each feather), and the quills and tail rather narrowly barred, with the same colour or a rather paler shade; chin and throat pale tawny, the bases and margins of the feathers black; remaining under parts paler barred with dusky. The bars on the wings and tail do not quite reach to the shaft. "Iris bright crimson; bill and legs greenish plumbeous" (Everett *b 5*).
Young. "The young of both sexes are black, like the adult male. The only signs of the female plumage in one young female were on the secondaries, which, on the inside of the wing, were slightly barred with brown. Another young female had two or three brown feathers on the back, the wings being dull black" (Whitehead *d 1*).
Distribution. Philippine Islands — Luzon (Everett *b 5*), Mindoro, Mindanao, Basilan, Samar, Marinduque, Guimaras (Everett *b 5*, Steere *b 10*, Meyer *b 3*), Negros (Everett *b 11*), Palawan (Whitehead *d 1, b 8*, Platen *b 6*), Sooloo Islands (Platen *b 9*).

† 2. Eudynamis mindanensis sangirensis (W. Blas.).

e. **Eudynamis niger** *(1)* Brügg., Abh. Ver. Bremen 1876, V, 57.
f. **Eudynamis nigra** *(1)* Fischer, op. cit. p. 538.
g. **Eudynamis sp. nov.?** *(1)* Salvad., Atti Ac. Torino 1878, 1188.
h. **Eudynamis mindanensis** *(1)* Meyer, Isis 1884, 6, 17; *(2)* W. Blas., "Braunschweig. Anz."

11. Jan. 1888, Nr. 9, p. 86; id., Russ' Isis 1888, 78; *(3)* Shelley, Cat. B. XIX, 1891, 321 (Sangi).
k. **Eudynamis mindanensis** var. **sanghirensis** *(1)* W. Blas., Ornis 1888, 566—569.
l. **Eudynamis mindanensis sanghirensis** *(1)* M. & Wg., J. f. O. 1894, 241; *(2)* iid., Abh. Mus. Dresd. 1895, Nr. 9, p. 3.
"**Buago**" ♂, "**Liaga**" ♀ or "**Paparapa**", Talaut, Nat. Coll. in Mus. Dresd.
"**Kuwao maitung**" [♂], "**Kuwao**" [♀] Tagulandang, iid.
Descriptions. Brüggemann *e 1*; Meyer *h 1*; W. Blasius *k 1*.
Adult male. Bill broader and stronger, the ridge of the culmen considerably more rounded; the plumage with greener reflexions than in the *typical E. mindanensis* (W. Blasius *k l*).
Without Philippine specimens for comparison we quote Blasius' diagnosis.

Measurements.	Wing	Tail	Tarsus	Culmen from cere
a. (C 1852) [♂ ad.] Siao	200	192	35	33
b. (C 1853) [♀ ad.] Siao	201	190	30	30.5
c. (C 1851) [♀] Gt. Sangi	199	185	33	34
d. (C 1856) [♀ juv.] Gt. Sangi	183	180	31.5	30.5

Further adults (4 ♂, 4 ♀?) measure: wing 189—202; tail 186—196; bill from nostril 18.5—20.5 mm. The birds were shot early in November (Kabruang) or at the end of October (Salibabu), and several are in moult.
We have since received many more specimens from Talaut and Tagulandang.
Observation. The specimen *d* from Sangi and two others from Talaut are in the same phase of plumage as that described as "*c* [♀]" in our article on the foregoing species; namely they were losing old worn feathers of uniform dusky brown or black in the wings and tail, while new feathers barred with black and light rufous brown were in a growing condition when the birds were killed.
Distribution. Sangi and Talaut Islands — Great Sangi (Fischer *f 1*, Meyer *h 1*, Platen *k 1*), Siao (Meyer *h 1*), Tagulandang and Ruang (Nat. Coll.); Kabruang, Salibabu and Karkellang (Nat. Coll. *l 1* in Dresd. and Tring Mus.).

The bill of this species would appear to get paler with age, that of *E. melanorhyncha* blacker.
This form has never been found in Celebes; Brüggemann's specimens — 3 of which were at first supposed by him to have come from Manado — were obtained, as Fischer afterwards pointed out, in Sangi (presumably Great Sangi).
Attention has been drawn to Mr. Whitehead's remarks (antea p. 209) upon the species in Palawan, where it lays its eggs in the nest of the Yellow-mottled Mynah *(Gracula javanensis)*, and where it is believed by him to migrate to Borneo and other islands during the N.E. monsoon.

SUBFAMILY CENTROPODINAE.

These Cuckoos, or Coucals, vary from medium to large size and are recognisable by the long, nearly straight, hind claw, resembling that of a Lark, and by the shafts of contour-feathers of the head, mantle and breast, which are spinous and thickened. The nostril is a linear slit covered by a sort of operculum of skin; and there is a row of stiff bristles above the eye. The tarsus is naked, except quite at the tibial joint; tail graduated, longer than the wing. The birds seem to be adapted to a life on the ground in jungle and scrub; they are not parasitic in their breeding-habits.

GENUS CENTROCOCCYX Cab. Heine.

In these Coucals the toes are not longer than the tarsus and the hind (first) toe with the claw is longer than the middle toe and claw. The bill is shorter than the head and considerably hooked; the contour-feathers of the head, mantle and breast are dense and close, with thickened pitch-black spinous shafts; wing short and blunt, 4th to 6th quills longest, secondaries shorter by about $1/_6$ the wing-length; tail strongly graduated, longer than the wing by $1/_3$ or $1/_4$, the outermost rectrix about half the length of the tail. The sexes are similar in coloration, but the right testis only of the male is present; it is smaller than the female and broods on the eggs, while only the female has been as yet observed to utter call-notes. The young are quite different from the parents, covered with spines as fledgelings, then assuming a barred plumage.

+ 71. CENTROCOCCYX BENGALENSIS (Gm.).
Lesser Coucal.

a. **Cuculus bengalensis** *(1)* Gm., S. N. 1788, I, 412 (Bengala).
b. **Cuculus javanensis** *(1)* Dumont, Dict. Sc. Nat. 1818, XI, 144.
c. **Centropus affinis** *(1)* Horsf., Tr. Z. S. 1822, XIII, 180 (Java); *(2)* Bernst., J. f. O. 1859, 185; *(3)* id., ib. 1860, 269; *(4)* id., N. T. N. I. XXI, 1860, 27—49, pl. I (anat); *(5)* Sclat., Ibis 1861, 48, note (Java, Malacca, India); *(6)* Tristr., Cat. Coll. B. 1889, 85 (Java, Sumatra).
d. **Centropus lepidus** *(1)* Horsf. l. c. (Java).
e. **Centropus bengalensis** *(1)* Steph., Gen. Zool. 1826, XIV, 213; *(2)* David & Oust., Ois. Chine 1877, 59 (Formosa, Hainan); *(3)* Oates, Str. F. 1877, V, 146 (Pegu); *(4)* Gammie, t. c. 385 (Sikkim); *(5)* Hume & Davis., Str. F. 1878, VI, 171 (Tenasserim); *(6)* Bingham, Str. F. 1880, IX, 169 (Tenasserim); *(7)* Styan, Ibis 1887, 230 (Foochow); *(8)* Shelley, Cat. B. 1891, XIX, 352; *(9)* De La Touche, Ibis 1892, 480 (S. China); *(10)* Styan, ib. 1893, 433 (Hainan); *(11)* De La Touche, Ibis 1895, 336; *(12)* Blanf., Faun. Br. Ind. B. III, 1895, 243.
f. **Centropus dimidiatus** *(1)* Blyth, J. A. S. 1843, XII, 945; *(2)* Swinh., Ibis 1860, 360 (Amoy, Hong-Kong).
g. ? **Centropus rectunguis** *(1)* Strickl., P. Z. S. 1846, 104; *(2)* Schl., Mus. P.-B. Cuculi 1864, 67 pt. (India, Formosa and E. India Islands); *(3)* Shelley, Cat. B. XIX, 1891, 343 (Malacca, Sumatra, Borneo).
h. **Centropus medius** *(1)* Bp., Consp. I, 1850, 108 (Amboina, Java), (ex S. Müll. MS.); *(2)* Wall., P. Z. S. 1863, 23 (Buru, Ceram, Gilolo).
i. **Centropus lignator** *(1)* Swinh., Ibis 1861, 48 (Formosa).
j. **Centrococcyx lepidus** *(1)* Cab. & Heine, Mus. Hein. 1862, IV, 109 (Java); *(2)* Hartert, J. f. O. 1889, 372 (Sumatra); *(3)* Heine & Rchnw., Nomencl. Mus. Hein. 1890, 204 (Java, Borneo).

k. **Centrococcyx affinis** *(1)* Cab. & Heine, Mus. Hein. 1862, IV, 110 (Java); *(2)* Wald., Tr. Z. S. 1872, VIII, 56—60, 112 (Java, Celebes, Flores); *(3)* Meyer, Ibis 1879, 70 (Celebes); *(4)* W. Blas., J. f. O. 1883, 132 (Celebes); *(5)* Meyer, Ibis 1884, 6, 18 (Siao, Sangi, N. Celebes, Halmahera, Ceram); *(6)* Guillem., P. Z. S. 1885, 504 (Sumbawa), 551 (N. Celebes, Limbe); *(7)* Sharpe, Ibis 1888, 198 (Palawan); *(8)* Whitehead, Ibis 1890, 47 (Palawan); *(9)* Heine & Rchw., Nomencl. Hein. 1890, 204 (Java).

l. **Centrococcyx bengalensis** *(1)* Cab. & Hein., Mus. Hein. 1862, IV, 111 (Nepal); *(2)* Wald., Tr. Z. S. 1872, VIII, 59 (India, Burmah); *(3)* Hume, Str. F. 1875, III, 84 (Upper Pegu), 324 (Tenasserim); *(4)* id., Str. F. V, 1877, 28 (Cachar); *(5)* Ball, S. F. VII, 1878, 208 (Ind. Penin.); *(6)* Cripps, t. c. 266 (E. Bengal); *(7)* Hume, S. F. VIII, 1879, 55 (Malacca — Singapore); *(8)* Kelham, Ibis 1881, 395 (Malacca, Singapore); *(9)* Oates, Str. F. X, 1882, 196 (Pegu); *(10)* A. Müll., J. f. O. 1882, 410 (Salanga); *(11)* Oates; B. Brit. Burmah, 1883, II, 127 (India, Burmah, Tenasserim, China, Siam, Cochin China); *(12)* Davison, Str. F. X, 1883, 361 (Wynaard); *(13)* Hume, ib. XI, 1888, 78 (Manipur); *(14)* Heine & Rchw., Nomencl. Hein. 1890, 204 (Nepal); *(15)* Oates, ed. Hume's Nests & Eggs 1891, II, 406 (India); *(16)* Munn, Ibis 1894, 56; *(17)* M. & Wg., J. f. O. 1894, 241; *(18)* iid., Abh. Mus. Dresden 1895, Nr. 8, p. 6; *(19)* iid., ib. Nr. 9, p. 3; *(20)* iid., ib. 1896, Nr. 1, p. 8; *(21)* iid., ib. Nr. 2, p. 11.

m. **Centrococcyx moluccensis** *(1)* Cab. & Heine (ex Bernst. MS.), Mus. Hein. 1862, IV, 113 ("Timor", Tidore or Timor; Ternate); *(2)* Wald., Tr. Z. S. 1872, VIII, 57, 59 (Ternate).

n. ? **Centrococcyx rectunguis** *(1)* Cab. & Heine, Mus. Hein. IV, 114; *(2)* Büttik., Notes Leyd. Mus. 1887, 32 (W. Sumatra); *(3)* Hagen, T. Ned. Aard. Genoots. 1890 (2), VII, 137.

o. **Centropus viridis** *(1)* Swinh. (nec Scop.), P. Z. S. 1863, 266 (S. China, Formosa); *(2)* id., Ibis 1870, 235 (Hainan); *(3)* Guillem., P. Z. S. 1885, 257 (Sula Is.), (fide W. Blas., p. 14).

p. **Centrococcyx javanensis** *(1)* Wald., Tr. Z. S. 1872, VIII, 58, 60 (Java, Malacca, Banjermassing, Celebes); *(2)* id., Ibis 1872, 367 (N. Borneo); *(3)* Salvad., Cat. Ucc. Borneo 1874, 76 (Borneo, etc.); *(4)* id., Ann. Mus. Civ. 1875, VII, 651 (Celebes); *(5)* Sharpe, Ibis 1876, 34 (Borneo); *(6)* Salvad., Ann. Mus. Civ. 1879, XIV, 188 (Sumatra); *(7)* Nicholson, Ibis 1881, 141 (Java); *(8)* 1883, 241 (Sumatra); *(8)* W. Blas., Ztschr. ges. Orn. 1885, 263—270 (S. Celebes); *(9)* id., Ibis 1888, 374 (Palawan); *(10)* id., Oruis 1888, 306 (Palawan), 570 (Sangi); *(11)* Everett, P. Z. S. 1889, 226 (Palawan); *(12)* id., J. Str. Br. R. A. S. 1889, 175 (Borneo, Palawan); *(13)* Sharpe, Ibis 1890, 14 (Labuan), 282 (Himalayas — Celebes); *(14)* W. Blas., J. f. O. 1890, 140 (Sulu Is.); *(15)* Salvad., Ann. Mus. Civ. 1891, (2) XII, 46 (Sumatra); *(16)* Vorderman, N. T. Ned. Ind. 1891, 217 (S. Sumatra); 1892, 383 (Java); *(17)* Hose, Ibis 1893, 416 (Borneo); *(18)* Büttik., Zool. Erg. Weber's Reise Ost-Ind. 1893, III, 276.

q. **Centrococcyx medius** *(1)* Wald., Tr. Z. S. 1872, VIII, 57, 58 (Amboyna, Ceram); *(2)* Salvad., Orn. Pap. 1880, I, 375 (Moluccas), Agg. 1889, 52 and 1891, 220; *(3)* Pleske, Bull. Acad. Petersb. 1884, 117 (Ternate).

r. **Centropus bengalensis** var. **affinis** *(1)* Brügg., Abh. Ver. Bremen 1876, V, 61 (N. Celebes).

s. **Centropus bengalensis** var. **javanensis** *(1)* Brügg. l. c. (N. Celebes).

t. **Centropus moluccensis** *(1)* Rchnw., J. f. O. 1877, 218 (Celebes); *(2)* W. Blas., ib. 1883, 122.

u. **Centropus celebensis** Beddard (nec Q. & G.), P. Z. S. 1885, 183, 184 pterylogr.

v. **Centropus javanensis** *(1)* Sharpe, P. Z. S. 1879, 328 (Borneo); *(2)* id., Ibis 1879, 246 (Borneo).

w. **Centropus javanicus** *(1)* Shelley, Cat. B. XIX, 1891, 354; *(2)* Sharpe, Ibis 1894, 247; *(3)* Grant, t. c. 520; *(4)* Bourns & Worces., B. Menage Exped. 1894, 35; *(5)* Vorderm., N. T. Ned. Ind. 1895, LIV, 329, 333; *(6)* Clarke, Ibis 1895, 476; *(7)*

Hartert, Nov. Zool. 1895, 475; *(8)* id., ib. 1896, 552, 562, 572, 575, 586, 595; *(9)* Grant, Ibis 1896, 474.

x. **Centropus bengalensis lepidus** *(1)* Hartert, Kat. Mus. Senckenb. 1891, 150 (Java); *(2)* id., Ornis 1891, 122.

"**Burong kussu-kussu**" (Bird of the high grass), Manado, Meyer *k 3*; Nat. Coll., Dresd. Mus.
"**Totombarang**", inland name, Minahassa, Meyer *k 3*.
"**Kuluket**", Guillem. *k 6* or "**Koloket**", Nat. Coll., Dresd. M., inland name near Manado.
"**Kalukku**", Maros, S. Celebes, Platen *p 8*.
"**Aroöa**" or "**Aeroöta**", Talaut, Nat. Coll.
"**Karoko**", Balante, E. Celebes, iid.

For further synonymy and references see Salvad. *p 3, q 2*; Cab. & Heine *j 1, k 1, l 1, m 1, n 1*; Shelley *e 8, w 1*.

Descriptions. Hume & Davison *e 5*; David & Oustalet *e 2*; A. Müller *l 10*; Oates *l 11*; Salvadori *q 2*; W. Blasius *p 8*; Shelley *e 8, w 1*; Blanford *e 12*.

Adult male. Head, neck, mantle and all the under parts black, glossed with greenish on the upper parts and chest, more sooty on the lower under parts; the shafts of all the feathers spinous and stiff, jet black; wings cinnamon-rufous, the ends of the primaries, the inner secondaries and scapulars brown with a slight gloss; back sooty; upper tail-coverts and tail black glossed with bronze-green; quills below purer rufous than above; under wing-coverts cinnamon-rufous, slightly varied with brown (N. Celebes, C 3579). "Iris dark brown; bill black; feet slate-grey" (Platen *p 8*).

Nearly adult male. Differs in having the gloss on head, neck and chest blue-black; mantle, like inner quills, deep reddish brown contrasting with the neck, shafts whitish; back and rump black, barred with wood-brown; upper tail-coverts glossed with bluish, tail with bronze-green, terminally fringed with rufous and an imperfect bar of the same on some of the lateral feathers (Maros, S. Celebes, 16. II. 1878, C 12081).

Adult female. Like the adult male, but much larger, the gloss on the head and neck possibly a shade bluer (S. Celebes, Feb., C 12082).

Nearly adult female. Like the nearly adult male, but not quite so far advanced in coloration: many of the feathers in the black plumage of head, neck, and underparts with some part of the shaft white (usually the tip, sometimes the middle); under parts black varied with pale fawn-colour alongside the white portion of the shaft (Nr. 7093).

Younger female. Head and hind-neck black as the last, but most of the feathers with part of the shafts and the adjacent part of feather fulvous white, forming a narrow streak; below pale cinnamon-tawny with pale shafts, barred with dusky on the sides and thighs, and varied with some black feathers barred with tawny; wings and tail still much as in adult; upper tail-coverts barred with tawny brown. (Near Manado, C 10873.) Bill blackish horn-colour, paler at the base.[1]

First plumage [male and female][2]. Head and neck tawny-cinnamon, the sides and bases of the feathers black, giving a streaked appearance; on mantle, back and wings the black

---

[1] This specimen does not uphold Prof. W. Blasius' view (*p 8*, p. 270) that the change of coloration in this species very likely takes place on the head and body — not wings and tail — without moulting, as it is obviously moulting, but the former specimen (Nr. 7093) with the same particoloured appearance below does not show any trace of moulting. We, however, take this simply for the ordinary second plumage. A nestling of *C. sinensis* in the possession of Mr. De La Touche, bought 26th August, 1886, assumed the adult plumage in the following summer; it began to show in patches in the early spring, but unfortunately Mr. De La Touche does not say whether the feathers were moulted or not (Ibis 1892, 480).

[2] Mr. Oates makes the unexpected statement that the adults in winter plumage are clothed like young birds in first plumage *(l 9, l 11)*. We have specimens in the ordinary adult dress from Celebes, Talaut, Sangi, Tagulandang and Ceram dated January, February, March, April, May, June, July, August, October, November.

(here dusky) and tawny-cinnamon take the form of bars of about equal width; tail glossed with green and barred narrowly with tawny-cinnamon; below fawn-colour, almost clear of bars along the middle line of the under surface, more tawny and barred with dusky on the sides, flanks, under wing- and tail-coverts (Minahassa, Aug.—Sept., C 10874, ib. Aug.—Sept. 10795; Gt. Sangi C 1516). Bill light yellow horn-colour, blackish at base of upper mandible.

Nestling about a fortnight old. Like the last-described in all respects, except that the tawny-cinnamon bars on the tail-feathers, which are just making their appearance, are about as broad as the black ones. In the full-grown tail-feathers of the above specimens they are only about $1/_4$ the width of the black bars (Ternate, C 788).

Two unfledged nestlings taken by the Drs. Sarasin in N. Celebes are covered with spines almost comparable to those of a Hedgehog, but from the tip of many of them issues a single, long, white hair or fibre, particularly on the nape and shoulders (Kema, 21. Aug. 1893: P. & F. Sarasin).

Eggs. "White, glossy, somewhat hard-shelled, fatty to the touch, in form elliptical: 32—33 × 23.5 mm" (Nehrkorn MS., 2, collected by Dr. Platen at Rurukan, Minahassa, 11. May 1886).

Nest. "Shaped like an egg, about 10 inches high and 8 inches in diameter. The entrance, 5 by 4, midway between top and bottom. Composed of elephant-grass and the surrounding grasses are bent down and incorporated in the structure. The egg-chamber and sides neatly lined with thatch-grass. The walls about 1 inch thick" (Oates *l 15*).

Number of eggs 2 or 3; Mr. Inglis in Cachar generally found 6.

See, also, Bernstein's admirable description: J. f. O. 1859, 185.

Measurements. The female is usually considerably larger than the male. Prof. W. Blasius *(p 8)* records the following:

Males: length in the flesh 345—370; expanse 410—455 mm,
Females: length in the flesh 390—430; expanse 505—520 (Platen).

Females	Wing	Tail	Tarsus
g. South Celebes (Platen)	174	212	41
i. South Celebes (Platen)	172	227	42
f. South Celebes (Platen)	180	239	43
h. South Celebes (Platen)	176	231	42
k.[1]) South Celebes (Platen)	183	235	42

Males	Wing	Tail	Tarsus
a. South Celebes (Platen)	154	209	39
b. South Celebes (Platen)	159	203	38
e. South Celebes (Platen)	156	201	38
c. South Celebes (Platen)	158	207	39
d. South Celebes (Platen)	157	191	38

Hume records the following measurements taken in the flesh *(e δ)*:

Adult males from various localities, from Johore to Suddhya (Assam): Length 323—330 mm; expanse 404—437; wing 133—140; tail 171—196; tarsus 37—39; bill from gape 28—48; hind toe claw inside 18—24.

---

[1]) The label of this was lost. The letters are those given by Prof. W. Blasius.

Adult females. Length 362—380 mm; expanse 463—476; wing 165—173; tail 213—218; tarsus 41—43; bill from gape 32; hind toe claw inside 23—28.

"And the bills are not only longer, but markedly stouter as a body than those of the males."

From the careful measurements of Hume, Blasius, and Salvadori *(q 2)*, who independently obtain similar results, it is obvious that Mr. Oates's statement, that "the female is of about the same size as the male" *(l 11)*, is not correct; so, too, the measurements of Capt. Shelley *(e 8, w 1)*, which suggest an equal size of the sexes, are misleading.

Variation. It is of interest to note that the males and females of South Celebes measured by Blasius and Platen are respectively much larger than the males and females of Malacca—India measured by Hume; while Salvadori's Moluccan specimens are still larger than those of Blasius.

Distribution. India — Himalayas from Nepal to Sikkim and Assam *(e 8, e 4, l 1)*; Central India (Ball *l 5*, etc., *e 8*); South India — Wynaard (Davison *l 12*); Travancore (Bourdillon *e 8*); Khasia Hills (Chennell *e 8*, Griffith *w 1*); Manipur (Hume *l 13*); Upper Pegu (Oates *l 3*); Burmah (Oates *l 11*); Tenasserim (Davison *e 5*, Bingham *e 6*); South China (Swinh. *o 1*, De La Touche *s 9*); Formosa and Hainan (Swinh. *o 1*, *o 2*, David *e 2*); Siam (fide Oates *l 11*, Tweedd. Coll. *e 8*); Cochin China (fide Oates *l 11*); Salanga (A. Müll. *l 10*); Malacca (Davison *e 8*, Hume *l 7*, Kelham *l 8*); Penang (Brit. Mus. *e 8*); Banguran (Hose *w 7*); Singapore (Davison *e 8*, Hume *l 7*, Kelham *l 8*); Sumatra (Beccari *p 6*, H. O. Forbes *p 7*, ? Klaesi *n 2*, Modigl. *l 15*); Banka (v. d. Bossche *g 2*); Java (Horsf. *c 1*, *d 1*, H. O. Forbes *p 7*, Vorderman *p 16*, etc.); Bali (Doherty *w 8*); Lombok (Vorderman, Doherty *w 5*, *w 8*); Sumbawa (Forsten *g 2*, Guillem. *k 6*); Satonda and Sumba (Doherty *w 8*); Flores (Wall. *e 8*); Timor (Wall. *g 2*, *e 8*); Borneo (Schwaner *g 2*, Mottley Doria & Beccari *p 3*, etc.); Philippine Is. (B. & W. *w 4*, Whitehead *w 9*); Palawan (Whitehead *k 7*, *k 8*, Platen *p 10*); Sooloo Is. (Guillem. *o 3*, Platen *p 14*); Talaut Is. (Nat. Coll.); Great Sangi and Siao (Meyer *k 5*); Tagulandang (Nat. Coll.); Celebes — Lemboh Id. (Guillem. *k 6*); Banka and Menado tua (Nat. Coll., Dresd. M.); Minahassa (Meyer *k 2*, *k 3*, Beccari *p 4*, etc.); Gorontalo (Forsten *g 2*); Balante, E. Cel. (Nat. Coll.); Kandari, S. E. Penin. (Beccari *p 4*); Luwu (Weber *p 18*); Palopo (P. & F. Sarasin); Macassar (Wall. *e 8*, Weber *p 18*); Maros Waterfall (Platen *p 8*, Weber *p 18*); Amboina, Ceram, Buru, Halmahera, Ternate, Tidore, Batchian (Salvad. *q 2*).

The Coucal, or "Lark-heeled Cuckoo", now under consideration has been split up by various authorities — by Cabanis and Heine into 5, by Walden into 6, by Shelley into 2 species, but upon grounds which are very unsatisfactory. Cabanis and Heine laboured under the disadvantage of having insufficient material, each form being represented in only one or two specimens; but two of the species recognised, *C. lepidus* Horsf. and *C. affinis* Horsf., are, as W. Blasius has now satisfactorily proved *(p 8)*, the male and female of the same species. By Shelley two races, *C. bengalensis* and *C. javanicus*, hardly distinguishable at the best, are recorded as occurring together in the same months of the year in the hill-country of N.E. India, in Pegu, and in Formosa. Were the means of distinction employed by Shelley allowed, both could also be recorded from Celebes. What is worse, Shelley is divided in his own mind on the subject of this troublesome species; to wit, Davison's Tenasserim specimens are included under *C. javanicus*, but Hume's remarks relating thereto are planted in the synonymy of *C. bengalensis*; in the sketch of the distribution of *C. bengalensis* the author draws a line at the Burmese countries, though including in the synonymy records by Hume and Davison, Bingham, and Kelham of its plentiful occurrence in Tenasserim, the Malay Peninsula and Singapore; Cabanis and Heine's two specimens of *C. lepidus* from Java are, moreover, included in the synomymy of both the species recognised by Shelley. Authors, on the whole, are generally agreed in uniting specimens from neighbouring localities with which they are acquainted; Schlegel alone has grouped them together as one species, a decision which Büttikofer *(n 2)* rightly declines to undo. Thus Hume *(e 5)* says: "It may be that the Javan bird is distinct, but certainly all those that we have seen from the Malayan Peninsula have been identical with those from various parts of India and Burma". Shelley crosses the bridge from the Malay Peninsula to Java, as Sclater had done thirty years earlier *(c 2)*, and the soundness of which had been again indicated by A. Müller in 1882 *(l 10)*.

In another direction Capt. Shelley agrees with Mr. Oates *(l 11)* and the Abbé David & Dr. Oustalet *(e 2)* as to the identity of the birds from South China, with others of India; though he terms the race *C. javanicus*, and the other authors *C. bengalensis*. In the East Indies birds of the different Great Sunda Islands are very rightly, as it appears to us, united by Shelley as one species with those of the Moluccas, the identity of which with *C. javanensis* had been already indicated by Salvadori *(q 2)* and Meyer *(k 5)*.

With a large series of 22 specimens from Celebes, and further specimens from Tagulandang, Great Sangi, Siao, Talaut, Ceram, Halmahera, Ternate, 4 labelled "Moluccas", 4 or 5 Java, and 1 Sumatra before us, we are able to lend support to this view.

It is certain that, as a species, *C. bengalensis* ranges from the Himalayas and South China throughout the intermediate countries to the Moluccas.

The question next suggests itself: does this species tend to develop any extreme forms of coloration or size in special points of the area over which it is spread, or is it to be regarded as "one harmonious whole"? Before a new species can arise, isolation of some description is necessary; in other words, a group must be prevented by some cause or liking from interbreeding with the rest of the species before its complete separation by colour or structure can take place. An intergrading local race or subspecies will, however, always be likely to come to a head, if we may use the expression, at special centres in the range of a species of uninterrupted distribution, especially if it be stationary. During migration, many birds undoubtedly get lost and do not always find their way back to the place of their birth; such, settling in other localities, may interbreed with the individuals of the species found there and act as a check on the differentiation of local races.

In some parts of its range — perhaps in all — *C. bengalensis* is not a strictly stationary species. Thus, in Cachar, Mr. Inglis says "this Coucal arrives about the beginning of June and departs at the close of the rains; breeds from June till September" *(l 4)*; in Upper Pegu Mr. Oates found it "during 9 months of the year rather an uncommon bird" *(l 3)*; in East Bengal Mr. Cripps "cannot recollect ever having noticed this species in the cold weather", but has observed it from the beginning of May *(l 6)*. Mr. Gammie remarks that it has increased largely of late in Sikkim. It is resident in South China, according to Swinhoe *(o 1)*; resident in Hong-Kong, a straggler to Amoy *(f 2)*; sedentary, according to David and Oustalet, in Hainan and Formosa *(e 2)*; "plentiful" as Kelham says *(l 8)* "at all seasons throughout Perak, Larut, Port Wellesly, Johore, and all the Settlements".

It is not, therefore, surprising that in the East Indies, as the measurements of Hume, W. Blasius and Salvadori tend to show, the species should differ slightly in size from those found from Malacca to India; but there appears to be sufficient stir going on within the species as a whole to prevent the development of any well marked local differences of coloration. In Malacca, Sumatra and Borneo a form, which we believe to be only an occasional variety of this species, has the under wing-coverts black, and is separated by Capt. Shelley as a distinct species, *C. rectunguis*.[1]

What is known about this species is not yet sufficient for deciding whether, and where, trinomials might be judiciously conferred, and this must be left to the future, when the form is better known.

Mr. Gammie gives an excellent description of its habits in Sikkim. "Among the grassy scrub, up to 3500 feet, it is now abundant, where, only a few years ago, it was rarely to be found. In the earlier part of the rainy season its odd, monotonous notes are to be heard in every direction. I am not sure that the male calls, but have shot the female — as I found by dissection — when calling. It has a call of a double series of notes: ¦whoot, whoot, whoot, whoot; then often a pause of four or five seconds, kurook, kurook, kurook, kurook. The whoot is ventriloquistic, sounding as coming from a distance of six or eight yards from the bird. Before calling, it seats itself about five feet from the ground, then you see it draw its neck and body together, slightly puffing out its body-feathers, raising its back and depressing its tail, and for every whoot there is a violent throb of the body as if the bird was in great pain, at the same time the motion of the throat is scarcely perceptible and its bill is closed. Then, as if greatly relieved, it stretches itself out, the feathers fall smooth, and with open mouth and throbbing throat comes the kurook without the slightest attempt at ventriloquism. When searching for the caller one must take no notice

---

[1] It can hardly be *'Centropus rectunguis* Strickl., since that species has "a deep blue tint on the head, neck and breast", and no mention is made by Strickland of black under wing-coverts; but these, and a bright gloss of "green on the head, back and lower breast" are the chief means of distinguishing *C. rectunguis* Shelley.

of the whoot but wait for the kurook. It feeds almost entirely upon grasshoppers and frequents the open, scrubby tracts only. I have never once seen it in larger forest".

In South Celebes in the month of February Dr. Platen found it a quiet and rather shy bird, appearing and disappearing in a mysterious manner before the traveller, as he moved through a hot treeless grass-plain, where the growth is taller than a man. "It flies, on being disturbed or when danger threatens, quickly downwards, then horizontally for a distance, to raise itself as quickly again and cling to another grass-stem, where it looks out for its prey. This species, too, feeds, like our German Cuckoo, without damage to its health, on hairy caterpillars; for on every dissection the walls of the stomach were found covered with brown-black caterpillar hairs. The bird presents the peculiarity that the male is much smaller than the female" *(p 8)*.

Two remarkable discoveries in connection with this species were made and fully described by Bernstein *(c 2, c 3, c 4)*, namely, the male is always to be found hatching the eggs by day (what share, if any, the female took in the work he could not ascertain), and it possesses only the right testicle, the left one being entirely wanting. It should further be borne in mind that the male is the smaller, weaker bird, and that, as Mr. Gammie's observations tend to show, it apparently leaves the female to do the "singing".

The young of this species, and we believe of all *Centrococcyges*, in its first plumage is wonderfully like the adult female of *Eudynamis melanorhyncha*, though it may, of course, be at once distinguished by its nostrils, which are feathered above, by its long Lark-like hind claw, and by the peculiar spinous character of the feathers of its head, neck and body.

This similarity is not kept up between the adults of *C. bengalensis* and the adult male of *E. melanorhyncha*, though both may be said to have developed in a melanistic direction[1]), the latter being entirely black, and *C. bengalensis* and its relations black with rufous wings and back. What is unusual about the case is that the young *Centrococcyx* resembles the adult female of *Eudynamis melanorhyncha* and not the young of that species, which is black, but Mr. Whitehead gives reasons (*antea* p. 209), why the plumage of the young *Eudynamis (E. mindanensis)* may have been specially modified to make it resemble the black young ones of its foster-parents. *Centrococcyx*, not being a parasitical Cuckoo, has no need of such an alteration in its young. This type of plumage is shared by the young of some other *Cuculinae*, such as *Cuculus*, *Cacomantis*, *Hierococcyx*, *Urodynamis*, and it may have a deep phylogenetic significance.

The long Lark-like hind claw of *Centropus* suggests at once that it is a terrestrial bird; and, if we mistake not, the high course grasses in which it lives

---

[1] It is worthy of note that *Centropus* appears to have strong tendencies towards albinism. The Dresden Museum contains two perfect albinos of *C. viridis*, one perfect albino each of *C. ateralbus*, and *C. goliath*, one partial albino of *C. bengalensis*, and several such of *C. ateralbus*.

have to do with the curious spine-like character of the shafts of its contour-feathers. This grass ("Kussu kussu" *Chrysopogon aciculatus* Fr.) grows taller than a man and is so sharp that great care must be used in passing through it or hands and face get badly cut. Similar considerations have led us to examine the outer webs of the primaries, in anticipation that a change might have been wrought by frequent brushing against the stems of the jungle-grass as the bird flew in and out amongst it; but there is nothing remarkable about them, unless it be that the outer webs of the longest primaries are narrowed rather suddenly in their terminal third (more so than in other Cuckoos before us), while the third, fourth, fifth, sixth and seventh — the longest — are all very much of a length, forming a remarkably blunt-tipped wing. The same condition obtains among other birds in other members of the subgenus *Centrococcyx* (*C. eurycercus, C. viridis*), in the subgenus *Centropus* (*C. senegalensis*), and — though less distinctly — in the subgenus *Poliphilus* (*P. phasianus*); but not in *Nesocentor* (*N. menebiki*) and *Pyrrhocentor* (*P. celebensis*), in which the shafts of the contour-feathers are less spiny in character and suggestive of somewhat different habits of life. The interesting point is that in the former subgenera the narrowest part of the outer web of the longest quills is usually some distance from the tip, whereas just before the tip the web becomes a shade broader again. The case should be considered in connection with the racket-feathers of *Prioniturus*.

GENUS PYRRHOCENTOR Cab. Heine.

These Coucals may be distinguished by their having the culmen longer than the head, the hind toe and claw shorter than the middle and reversed fourth toe and claws, the wing very blunt and round, the primaries (6th to 8th longest) very little longer than the secondaries, the tail half as long again as the wing, the contour-feathers of the head, neck, and breast with stiffened shafts, not dense and close, but on the other hand loose and open. The young are not known, immature birds hardly differ from the adults. About 5 species are known, inhabiting the Philippines, Celebes and Kangean.

* 72. PYRRHOCENTOR CELEBENSIS (Q. G.).

Brown Coucal.

Two geographical races of this highly local species have been distinguished: they are:

+ 1. The typical Pyrrhocentor celebensis.

a. **Centropus celebensis** *(1)* Quoy & Gaim., Voy. Astrol. Zool. I, 1830, 230; Atlas Aves 1833, pl. 20; *(2)* Gray, Gen. B. II, 1846, 455; *(3)* Bp., Consp. 1850, I, 108; id., Consp. Vol. Zygod. 1854, 5; *(4)* Finsch, New Guinea 1865, 100; *(5)* Brügg., Abh. Ver. Bremen 1876, V, 60; *(6)* Shelley, Cat. B. XIX, 1891, 365, pt.; *(7)* Hartert, Kat. Senkenb. Mus. 1891, 150.

b. **Centropus bicolor** *(1)* Lesson, Tr. d'Orn. 1831, 137; *(2)* Blyth, J. A. S. B. 1843, XII, 946; 1845, XIV, 203; *(3)* Pucher., Rev. Zool. 1852, 472; *(4)* Hartl., J. f. O. 1855, 421; *(5)* Gray, P. Z. S. 1860, 359 (Gilolo!); *(6)* id., HL. 1870, II, 214; *(7)* Schl., Mus. P.-B. Cuculi 1864, 73; *(9)* Rosenb., Malay. Arch. 1878, 275 (?).

c. **Pyrrhocentor bicolor** *(1)* Cab & Heine, Mus. Hein. 1862, IV, 117; *(2)* Salvad., Orn. Pap. 1880, I, 365; *(3)* Heine & Rchnw., Nomencl. Mus. Hein. 1890, 205.

d. **Pyrrhocentor celebensis** *(1)* Wald., Tr. Z. S. 1872, VIII, 55; *(2)* Salvad., Ann. Mus. Civ. Gen. 1875, VII, 650; *(3)* Meyer, Ibis 1879, 70, 146 pt.; *(4)* W. Blas., J. f. O. 1883, 136; *(5)* Beddard, P. Z. S. 1885, 170, 172, 180, 187; *(6)* Guillem., t. c. 551; *(7)* W. Blas., Ztschr. ges. Orn. 1886, 98; *(VIII)* Meyer, Vogel-Skel. XVIII, 1892, 47, pl. CLXXII; *(9)* Vorderman, N. T. Ned. Ind. 1893, LII, 190, 191; *(10)* M. & Wg., Abh. Mus. Dresden 1895, Nr. 8, p. 6; *(11)* ? Hartert, Nov. Zool. 1896, 160.

e. ? **Pyrrhocentor celebensis celebensis** *(1)* Hartert, Nov. Zool. 1897, p. 160, 164.

"**Koun-Koun**" (Quoy & Gaim. *a 1*) or "**Kung-Kung**" (Nat. Coll., Dresd. Mus.) near Manado.
"**Kuwo**", native Malay name, Meyer *3*.
"**Unguno**", Gorontalo, Joest, Das Holontalo 1883, 106.
"**Ungung-gungo**", Rosenb. *b 9*.

Figures and descriptions. Quoy & Gaimard *a I*, Meyer *d VIII* (skeleton); Pucheran *b 3*; Cabanis & Heine *c 1*; Brüggemann *a 5*; Salvadori *2*; Beddard *d 5* (pteryl. etc.); W. Blasius *d 7*; Shelley *a 6*.

Adult. Head, neck, throat and breast wood-brown, darker above, paler below, more drab-colour on head, passing on the rest of upper surface, wings, and tail into Mars- and walnut-brown; under tail-coverts, flanks and thighs of the same colour passing into wood-brown on the lower breast. Bill black, horn-brown at tip. In some specimens, it is yellowish white at the tip and over part of the lower mandible.

Adult male and female. Alike in coloration, the latter is perhaps a triple larger.

Immature. Differs little from the adult. The head, neck and breast is washed more strongly with a rufous tint, so that the colour here contrasts less strongly with that of the rest of the body, and across the throat a rufous collar is perceptible (Manado, C 5186, Limbotto ♀, C 1814).

Lord Walden remarks: "In *P. celebensis*, the fully adult bird loses the bright yellow-rufous chin-, throat-, neck-, and breast-plumage of the younger bird. These parts become very pale fulvous and contrast with the dark chestnut of the remaining lower region. . . . The young bird is bright rufous throughout" *(1)*.

Measurements.	Wing	Tail	Tarsus	Bill from nostril
a. (C 1827) ♀ ad. Limbotto, July	182	280	50	30
b. (C 1801) ♀ ad. Limbotto, July	179	270	—	26.5
c. (C 1813) ♀ ad. Togian Is., Aug.	183	275	46	—
d. (C 1814) ♀ imm. Limbotto, July	159	224	—	23
e. (Salvad. 2) ♀, Manado, July	190	275	—	—
f. (Sarasin Coll.) ♂, Kema, 20. VIII. 93	183	280	45	28
g. (Sarasin Coll.) ♂, Kema, 13. IX. 93	—	270	—	25
h. (Sarasin Coll.) ♂, Kema, 10. XI. 93	173	250	47	28
i. (Sarasin Coll.) ♂, Tomohon, 8. X. 94	175	295	—	24
j. (C 1810) ♂, Manado, March	173	261	43	25
k. (Salvad. 2) ♂, Manado, July	170	250	—	—

We are also able to compare 32 other specimens, sex not indicated. One displays a white feather on the back, — partial albinism.

Skeleton *(VIII)*.

Length of cranium	70.0 mm	Length of tarso-metatarsus	45.0 mm
Greatest breadth of do.	28.0 »	Length of digitus III	41.0 »
Length of humerus	39.0 »	Length of sternum	27.7 »
Length of ulna	31.5 »	Greatest breadth of do.	25.5 »
Length of radius	28.7 »	Height of crista sterni	7.0 »
Length of manus	33.0 »	Length of coracoideum	28.7 »
Length of metacarpus	17.5 »	Length of scapula	38.0 »
Length of digitus princip.	14.0 »	Length of clavicula	28.0 »
Length of femur	48.0 »	Length of pelvis	50.0 »
Length of tibia	66.0 »	Greatest breadth of do.	29.0 »
Length of fibula	22.8 »		

Distribution. Minahassa (Voy. Astr. *a 1*, Meyer *d 3*, Fischer *a 5*, *etc.*); Gorontalo (Forsten *1*, Meyer *d 3*); Togian Is. (Meyer *d 3*); *(?)* Bonthain Peak, S. Celebes (Doherty fide Hartert *e 1*).

Variation. The female from the Togian Islands in the Dresden Museum corresponds with adults from North Celebes in coloration, but the bill is yellowish white at the end, the under mandible being entirely of this colour except towards the base, where it becomes blackish. The feet are also paler. Several specimens from Celebes itself also offer some amount of pale colour on the bill. Hartert finds his specimens from Bonthain somewhat different.

2. Pyrrhocentor celebensis rufescens M. & Wg.

Abh. Mus. Dresden 1896, Nr. 2, p. 11; Hartert, Nov. Zool. 1897, 160, 164.

"**Kung-kung**", Tonkean, Nat. Coll.

Diagnosis. Differs from the typical form in having the face, neck, throat and breast light cinnamon-rufous, instead of wood-brown; head above clearer brown; mantle and wing-coverts redder brown.

Measurements. Wing 175—185 mm; tail 265—325; tarsus c. 46; bill from nostril 26—31.

Distribution. Eastern and Southern Peninsulas, Celebes; Tonkean (Nat. Coll.), Macassar (Wallace *a 6*, Doherty); Tanette, Mandalli, Maros (Meyer *d 3*); (if the same) Dongala, W. Coast (Doherty).

Observation. Seven specimens of this well-marked race were obtained by native collectors at Tonkean in the eastern part of the E. Peninsula. Meyer *(d 3)* when in Celebes remarked: "the South Celebean specimens appeared to me, when I first saw them, somewhat more brilliantly coloured than those of North Celebes, but afterwards, when I compared the skins in the cabinet, I could find no difference." There are now no specimens from the south in the Dresden Museum, but Mr. Hartert has recently received two from the low country north of Macassar, which "are hardly different from *P. c. rufescens*", and two skins from Dongala on the west coast which "resemble very much the co-types of *P. c. rufescens*", but "are a little paler and less rufous above where they look more like *P. c. celebensis*". A third skin from Dongala was aberrant. It still remains to be proved therefore whether these birds are intermediate: *P. celebensis—rufescens*, or what.

The genus *Pyrrhocentor* was established in 1862 by Cabanis and Heine for this species and *P. unirufus* C. & H. *(c 1)*, with the former for type. To those *P. melanops* (Lesson) of the Philippine Islands, *P. kangeanensis* Vorderm. of the Kangean Islands, and a new species of Prof. W. Blasius (not yet described) from Mindoro may be added. The genus *Pyrrhocentor*, like *Centrococcyx* and some others, is not admitted as distinct in Shelley's Catalogue of the Cuckoos, where all the *Centropinae* are grouped together in the single genus *Centropus*; it appears to us nevertheless to be worthy of distinction as a genus for the following reasons:

Compared with *Centropus bengalensis* and *Centropus senegalensis* we find 1. that the contour-feathers are different in character; in *Pyrrhocentor* the shafts are fine though stiff, the barbs also are long and unconnected, hairlike for the most part of the feather; in *Centropus* and *Centrococcyx* the shafts are spinous, and the barbs have not such a hairlike appearance; 2. The primaries are different, the outer webs being rather suddenly narrowed in their distal half in *Centropus* and *Centrococcyx*, whereas in *Pyrrhocentor* the outer webs are proportionally broader, and the narrowing is slight and proceeds gradually from base to tip. The relative lengths of the primaries taken from specimens in which all the feathers appear to have obtained their normal length are as follows[1]:

	I	II	III	IV	V	VI	VII	VIII	IX	X
Pyrrhocentor celebensis	70	101	122	141	148	153	152	151	149	145
Pyrrhocentor sp. (Mindoro) .	62	87	113	124	134	135	136	138	136	131
Centrococcyx bengalensis (a) . .	74	115	130	135	135	136	133	123	120	111
Centrococcyx bengalensis (b)	60	92	110	117	117	117	113	106	103	97
Centropus senegalensis . .	80	106	125	130	130	130	129	124	119	114

In *Pyrrhocentor*, as the above figures show, the first, second and third quills are relatively shorter than in *Centropus* and *Centrococcyx*; the maximum length is obtained from the sixth to the eighth feather, instead of the fourth to the sixth; the wing is still blunter, the quills seem weaker. The bird has all the appearance of being a very feeble flyer[2]. Consequently it is of the more value in questions concerned with the geographical distribution of birds. 3. The hind claw of *Pyrrhocentor* is much shorter. 4. The young of *Centrococcyx* and *Centropus* is rufous brown above barred with black; we have seen large numbers of *Pyrrhocentor*, but never one differing greatly from the adult. What the nestling is like is as yet unknown; if it is rufous barred with black this phase of plumage would appear to be of very short duration.

The nearest known ally of *P. celebensis* is Vorderman's recently described *P. kangeanensis* of the Kangean Islands, a group lying between Java and Celebes, though much nearer to the former island than to the latter. This form is much larger than *P. celebensis* and has the chestnut colour on the under parts of that

---

[1] Measurements taken from the base of the first primary.
[2] It was not sent to us by our hunters from the islands of Manado tua, Mantehage, Lembeh, and Banka, six or eight miles from the coast of Celebes. In N. Celebes it is very common.

species replaced by an umber tint, and the back and the tail not red-brown, but umber-colour (Vorderman 9).

Meyer found this bird very frequent during his residence from January till July in the Minahassa; he also met with it in South Celebes in September and October where it had previously been found, but not recorded, by Wallace. Platen does not seem to have come across it there, neither did Prof. Weber, the Drs. Sarasin, nor Mr. Everett send any specimens home from the south. Meyer remarks (3) that it makes a nest of brush-wood, like a Pigeon's nest, in trees in the deep forest, and feeds on fruits such as waringui, nutmegs, etc.; but this is much in need of confirmation; observations on its habits are wanting. When von Rosenberg (b 9) says that it is a bird which frequents by preference the high grass and bush and consigns its eggs to other birds he appears to have had *Centrococcyx bengalensis* and *Eudynamis melanorhyncha* in view, as well as this species.

SUBFAMILY PHOENICOPHAINAE.

In the Catalogue of Birds this subfamily is distinguished by its short rounded wing, which fits close to the body, and (from the *Centropodinae*) by the claw of the hind (first) toe being ordinary, not lengthened and Lark-like. In the more typical forms the nostril pierces the horn of the bill unprotected by any formation of skin. The bill is generally large and weak, the tail very long, broad and graduated. The genera seem to stand much nearer to the *Centropodinae* than the *Cuculinae*, or *Scythropinae*.

GENUS PHOENICOPHAES Vieill.

The Malkohas are birds about the size of a Magpie; the bill is large, very high and bloated at the base, compressed at the tip, yellow, green, or particoloured in hue, the nostril small, linear to round in shape, placed low down on the maxilla just above the tomia, the nasal canal running obliquely upwards; a row of stiff bristles over the eye; much of the face naked; feet rather small, the middle toe longer than the fourth by the length of its claw; wing very blunt, primaries overreaching the secondaries only by about half the length of the tarsus; tail $1\frac{1}{2}$ times the length of the wing, or more, the outermost rectrix only about half the tail-length. There are seven species, ranging from Ceylon and Malacca to Celebes, and several of them have been distinguished as distinct genera, but on very slender grounds, depending on the shape of the nasal aperture. They feed principally, as is known from a few forms, on insects or fruit.

Birds of Celebes: Cuculidae.

* 73. PHOENICOPHAES CALORHYNCHUS Temm.
Blue-tailed Malkoha.

In the Southern Peninsula of Celebes this well-marked species has the head of a lighter grey than in the birds of the north and east of the island and has been subspecifically distinguished therefrom. The two known races are:

+ 1. The typical Phoenicophaes calorhynchus.

a. **Phœnicophaus calorhynchus**[1] *(1)* Temm., Pl. Coll. 1825, pl. 349; *(2)* Less., Man. d'Orn. 1828, II, 128; *(3)* Cuv., Règne An. 1829, I, 456; *(4)* S. Müll., Verh. Natuurk. Comm. 1839—44, 234, note; *(5)* Blyth, J. A. S. B. 1845, XIV, 199; *(6)* Gray, Gen. B. 1846, II, 459; *(7)* Bp., Consp. 1850, I, 98; id., Consp. Vol. Zygod. 1854, 5; *(8)* J. & E. Verr., Rev. Zool. 1855, 356; *(9)* Schl., Mus. P.-B. Cuculi 1864, 48; *(10)* Finsch, Neu-Guinea 1865, 159; *(11)* Wall., Malay Archip. 1869, I, 429; *(12)* Gray, HL. 1870, II, 205; *(13)* Wald., Tr. Z. S. 1872, VIII, 52, fig. 5, head; *(14)* Brügg., Abh. Ver. Bremen 1876, V, 57; *(15)* Lenz, J. f. O. 1877, 371; *(16)* Rosenb., Malay. Arch. 1878, 275; *(17)* Meyer, Ibis 1879, 67; *(18)* Guillem., P. Z. S. 1885, 549; *(19)* Hickson, Nat. in N. Celebes 1889, 255; *(20)* Tristr., Cat. Coll. B. 1889, 83; *(XXI)* Meyer, Abb. von Vogelskel. 1897, pl. CCXXIX.

b. **Melias calyorhynchus** *(1)* Less., Tr. d'Orn. 1831, 132.

c. **Zanclostomus calorhenchus** *(1)* Blyth, J. A. S. B. 1842, 1098, 28.

d. **Phœnicophaus melanogaster** *(1)* Blyth (nec Vieill.), Cat. B. Mus. A. S. B. 1849, 75.

e. **Rhamphococcyx calorhynchus** *(1)* Cab. & Heine, Mus. Hein. 1862, IV, 65; *(2)* Sharpe, P. Z. S. 1873, 605, fig. of head; *(3)* Salvad., Ann. Mus. Civ. Gen. 1875, VII, 649; *(4)* Cab. & Rchnw., J. f. O. 1876, 324 (Ceram!); *(5)* Salvad., P. Z. S. 1877, 195; *(6)* id., Orn. Pap. 1880, I, 392; *(6bis)* W. Blas., J. f. O. 1883, 136; *(7)* id., Z. ges. Orn. 1886, 95; *(8)* Hein. & Rchnw., Nomencl. Mus. Hein. 1890, 201; *(9)* Shelley, Cat. B. XIX, 1891, 396; *(10)* Hartert, Kat. V. Mus. Senckenb. 1891, 149; *(11)* M. & Wg., Abh. Mus. Dresden 1893, Nr. 8, p. 6; *(12)* iid., ib. 1896, Nr. 2, p. 11.

"**Wakeke**" or "**Bakeke**" (Meyer *a 17*), "**Burong bakeke**" or **Makekeke**" (Nat. Coll.), near Manado. "**Koko-onde**" or "**Tontonbara**" (Meyer *a 17*), inland name, Minahassa, meaning Foreteller-bird by daytime ("Geloofvogel bij dag" — Dutch).

"**Aluii**", Gorontalo (Rosenb. *a 16*).

"**Djee**", Tonkean and Balante, E. Celebes (Nat. Coll.).

Figures and descriptions. Temminck *a 1*; Meyer *a XXI* (skeleton); Walden *a 13* (head fig.); Sharpe *e 2* (head fig.); Brüggemann *a 14* (juv.); Cabanis & Heine *e 1*; Shelley *e 9*.

Adult (North Celebes). Head above as far as nape dark mouse-grey; neck, mantle, scapulars and wing-coverts (except the greater series) chestnut; chin, throat and breast paler (burnt sienna or hazel); lower back, rump, and upper tail-coverts, abdomen, sides, under wing-coverts, flanks and thighs slaty, paler below, but passing into black on the under tail-coverts; wings and tail glossy dark violet-blue; the greater wing-coverts similarly coloured, but often broadly fringed with dark chestnut where they approximate the chestnut scapulars and median coverts (Manado, C 5177). Bill, tip white, then a black space about 10 mm wide, the rest

---

[1] The name of this species has been variously spelt *Phænicophæus*, *Phoenicophaeus*, *Phænicophaus*, *Phænicophais*, *calyorhynchus*, *calorhynchus*, *callirhynchus*, *calorhenchus*.

yellow, except about the nostril where it is red; under mandible red. "Iris reddish brown; tarsus black" (Guillem. *a 18*).

Sexes. Alike in coloration.

Young. Head above rust-red, the bases of the feathers grey. Rectrices about 40 mm longer than in the adult bird, proportionally narrower and not rounded off broadly, but more gradually pointed. Bill considerably smaller. Upper mandible olive-yellow, without a black space before the tip, but washed out olive-green, the tip itself blackish. Under mandible dirty red, along the ridge and at the tip yellowish (Brüggem. *a 14*).

Measurements.

	Wing	Tail	Tarsus	Culmen
a. (C 1837) ad. Manado — March	181	316	43	46.5
b. (C 3575) ad. N. Celebes	180	321	42	44.5
c. (C 3577) ad. N. Celebes	180	330	42	45
d. (C 3572) ad. N. Celebes	174	325	42	44.5
e. (C 5176) ad. ♂ Manado	185	310	42	44.5
f. (C 5177) ad. ♀ Manado	185	312	41.5	40.5

Nest. "Makes a nest of twigs, like a Pigeon" (M. *a 17*); further observations would be welcome.

Distribution. Celebes — Minahassa (Wallace *10*, Meyer *a 17*, Platen *8*, Guillem. *a 18, etc.*); Gorontalo (Forsten *a 9*, Rosenb. *a 16*, Meyer *a 17*); Togian Islands (Meyer *17*); E. Peninsula, Tonkean and Balante (Nat. Coll.); Ussu, S. E. Cel. (P. & F. Sarasin); Kandari, S. E. Celebes (Beccari *3*).

2. Phoenicophaes calorhynchus meridionalis M. & Wg.

f. **Phœnicophaus calorhynchus** *(1)* Wall., Malay Archip. 1869, I, 340.
g. **Rhamphococcyx calorhynchus** *(1)* W. Blas., Z. ges. Orn. 1883, 262; *(2)* Platen, Gefied. Welt 1887, 218; *(3)* Shelley, Cat. B. XIX, 1891, 396 (Macassar); *(4)* Büttik., Zool. Erg. Weber's Reise 1893, III, 275.
h. **Rhamphococcyx calorhynchus meridionalis** *(1)* M. & Wg., Abh. Mus. Dresd. 1896, Nr. 2, p. 11
i. **Phœnicophaes calorhynchus meridionalis** *(1)* Hartert, Nov. Zool. 1896, 160; id., ib. 1897, 160, 164.

"**Zanissere**", Tjamba Distr., Platen *g 1*.

Diagnosis. Head above as far as the nape pale mouse-grey, instead of dark mouse-grey. "Iris blood-red; feet and orbits black" (Platen *7*).

Observation. We have been able to compare over 70 examples from North Celebes and 11 from the East with 6 from the South, and find that the grey of the head is always darker in birds from the Northern Peninsula, but no other difference is apparent. Not one of these specimens is in the young plumage described by Brüggemann. The sexes are not only alike, but the young bird undoubtedly assumes the adult dress at a very early age.

Measurements.

	Wing	Tail	Tarsus	Culmen
a. (C 12077) ad. ♂ S. Celebes, Tjamba, 4. V. 78	200	345	41	46
b. (C 12080) ad. ♂ S. Celebes, Tjamba, 10. IV. 78	202	313	44	46
c. (C 12075) ad. S. Celebes, Tjamba, 15. VI. 78	174	315	41	41.5
d. (C 12079) ad. ♀ S. Celebes, Tjamba, 10. IV. 78	180	310	44	46
e. (C 12078) ad. ♀ S. Celebes, Tjamba, 7. IV. 78	188	327	42	45
f. (C 12076) ad. ♀ S. Celebes, Tjamba, 19. IV. 78	193	328	43	45

Distribution. South and West Celebes: Macassar (Wallace *f 1, g 3*; Doherty *i 2*), Tjamba Distr. (Platen *g 1, g 2*), Loka, Luwu and Palopo (Weber *g 4*), Indrulaman (Everett *i 1*); W. Celebes, Dongala (Doherty *i 2*).

Specimens from East Celebes are similar to those of the North, but seem to be a shade darker. A specimen from S. E. Celebes in the Sarasin-Collection (Ussu, 20. II. 96) is also dark in coloration, and wants the ivory-white tip to the upper mandible, a condition which we have not seen in any specimen from the rest of the island. A female from West Celebes was found by Hartert to belong to the southern race.

This species has been placed in the special subgenus *Rhamphococcyx* of which it is the only species and peculiar to the Island of Celebes. Its nearest ally known is *Rhinococcyx*, a subgenus represented by one species only, *R. curvirostris* (Shaw and Nodder, of Java. The nostril of *Rhamphococcyx* is an elongated slit parallel to the gape, and the eye is surrounded by a bare space of smooth bluish grey skin; the nostril of *Rhinococcyx* is placed in a groove terminating in a widened, somewhat pear-shaped, orifice, and the bare space, which occupies most of the side of the head is red in colour and rugose in character, looking as through formed of rudimentary feathers, which sprout, but do not come to perfection. *Rhinococcyx* is again, as Dr. Sharpe has pointed out (Tr. Z. S. 1877, 321), closely allied to *Dryococcyx*, a subgenus in which the nostril is a small circular hole, situated in a deep perpendicular groove, which separates the lores from the upper mandible; it is known only by one species from the island of Palawan.

Urococcyx, recently distinguished by Capt. Shelley in the Catalogue of the Cuckoos (pp. 368, 398), was held to have round nostrils not placed in a groove, but Count v. Berlepsch (Nov. Zool. 1895, 72) has shown that in the species found in Borneo the nostril is oval; these forms are very like *Rhinococcyx curvirostris* in coloration, and in Capt. Shelley's system they come between *Rhinococcyx* and *Dryococcyx*, and consist at present of three species, one, *U. erythrognathus*, inhabiting Sumatra, the Natunas and Malacca as far as Tenasserim; one *U. borneensis* (Bp.), Borneo; while the habitat of the other, *U. aeneicauda* (J. and E. Verreaux), was unknown until quite recently, when it was found in Mentawei Id. off the W. coast of Sumatra by Modigliani (Salvadori, Ann. Mus. Civ. Gen. 1894, 590).

Sharpe *(e 2)* remarks on *P. calorhynchus*: "to the true *Phoenicophaes*, it is allied by the shape of the nostrils, but differs in its smooth face and feathered lores". Lord Walden had already called attention to this supposed affinity *(a 13)*, but it appears to us that the very different type of coloration places the typical subgenus *Phoenicophaes*, which is confined to Ceylon, in a position of much more remote relationship to *Rhamphococcyx*, *Urococcyx*, *Rhinococcyx* and *Dryococcyx*, which form a closely related group of subgenera, or, according to Count v. Berlepsch, the well marked species only of one genus, *Phoenicophaes*.

These forms, as indeed the whole of the *Phoenicophaeinae*, are of special interest in questions concerned with the geographical distribution of animals. Among birds which appear to be stationary and settled, it is usual to find new species in localities separated by a narrow reach of sea or other geographical barrier; among the *Phoenicophaeinae* these barriers are often the means of separating forms which some authorities have distinguished as genera. Thus, *Rhamphococcyx* is confined to Celebes, *Rhinococcyx* to Java, *Dryococcyx* to Palawan, *Phoenicophaes* s. str. to Ceylon, *Dasylophus* to Luzon and Marinduque (Steere), *Lepidogrammus* to Luzon, *Hyetornis* to Jamaica, each of which genera Capt. Shelley finds to be represented by a single species; *Coua* with 12 species is peculiar to Madagascar.

The subfamily *Phoenicophaeinae* forms an important link between Celebes and Asia, no member of it being known from the Moluccas or any other part of the Australian Region. Owing to their structure and their habits the *Phoenicophaeinae* appear less likely than any Celebesian genera yet considered to have spread their range by flight over wide stretches of sea. The wing is very short and rounded, the tail very long; the length in the flesh of specimen "a" (antea) of *Phoenicophaes calorhynchus meridionalis* was found by Dr. Platen to be 550 mm, the wing is only 200 mm; Legge shows the wing to be only one-third the total length in the Ceylon species, *P. pyrrhocephalus* (B. Ceylon, 256); and Shelley's measurements prove that the same condition obtains throughout the subfamily, the total length being always more than twice, and often more than three times that of the wing.

It appears that these birds are rarely or never seen on the wing in the sense of taking long flights — at least this is the case with the Indian and East Indian forms. Davison remarks of *Rhopodytes tristis*: "Its flight is weak, and it relies more for its safety on the dense and impenetrable character of the places it prefers to frequent. It has a marvellous capacity for making its way through dense cover" (Str. F. VI, 163); and on the next page he remarks that the habits of *R. sumatranus* are similar. *Urococcyx microrhinus* in Borneo was found by Mottley to conceal itself among the brushwood and when disturbed to take only very short flights (Oates t. c. 125). *Rhinortha chlorophaea*, according to Davison, resembles *Rhopodytes* in all its habits (S. F. VI, 166). Legge speaks of the way in which *Zanclostomus* in Ceylon makes off, threading its way quickly through the most tangled underwood, but in places where it is common it may often be seen flying across roads. *Phoenicophaes* when flushed in the jungle flies up to high branches and quickly gets out of danger, taking short flights from tree to tree. Meyer found that *Phoenicophaes* in Celebes "does not fly away even after being shot at; it sits quiet if a bird by its side falls down; but I always got the impression that it is the fright which rivets it to the spot[1]). It flies quickly, or rather glides or slides through the foliage" *(a 17).*

---

[1]) "Once" — continues Meyer, in a different connection — "some one told me, at Remboken, on the shores of the Tondano lake, that several years before such a bird flew, crying very loudly, over the village,

This gliding motion through the crowns of the trees is compared by Platen to the spectacle afforded at sea by a wandering troop of porpoises: "it is not flying nor hopping, but rather a rhythmical diving and disappearing in and out of the green, and it is impossible to form any approximate notion of the number of individuals" (Gefied. Welt 1887, 218). It will readily be imagined how this gliding mode of travelling is due to the parachute-like action of the great tail; the superior facility for settling afforded by this organ to the common Magpie *(Pica caudata)* is obvious[1]).

From their structure, therefore, these birds appear less capable than usual of sustained flight, and from observations on their habits it is found that they do not make use of their wings for this purpose; consequently there seems to be so much the less reason to suppose that their distribution took place by flight across the straits which separate their habitats.

As Shelley shows, the sexes of all the *Phoenicophaeinae* are similar in coloration, except, perhaps, in *Rhinortha*. *P. calorhynchus* follows the rule, and the young, as is often the case in genera in which the male and female are alike, assumes the adult plumage at a very early age. Young birds, according to Brüggemann, who alone seems to have received specimens, have the tail-feathers narrower and longer than in the adult, and the head rusty instead of grey. In this — the form of the tail-feathers — and in having a line of yellow along the ridge of the under bill they call to mind *Rhinococcyx* of Java.

Facts concerning the food of the Celebesian form are scanty; Meyer found that it feeds on insects; probably, however, its food — as its bill suggests — consists of fruits as well; Legge has found such to compose the chief diet of *Phoenicophaes* and *Zanclostomus* in Ceylon, through insects were discovered in the stomach as well, and Davison remarks that *Rhinortha chlorophaea* feeds "apparently entirely on insects", and a similar fare with the addition of small reptiles is ascribed by Oates to *Rhopodytes tristis* (Oates, B. Brit. Burmah II, 121,122).

SUBFAMILY SCYTHROPINAE.

This subfamily of the Cuckoos contains but a single species, *Scythrops novaehollandiae*, one of the largest Cuckoos, with a body as big as that of a Crow and a much larger bill and tail. Its plumage is Cuculine — brown upper surface, grey head, neck, and breast, and faintly barred under parts; it differs by its enormous bill, which is furnished on the basal half with a shallow groove by the side of the culmen and a second groove lower

---

and that all the inhabitants became frightened as to what might happen"! A distructive fire which took place in the village next day was associated by the natives with this unusual behaviour of the "Foreteller-bird of the daytime".

[1]) De Bocarmé says that *Rhinococcyx* in Java is in the habit of sitting couched upon foliage supported by its tail and half open wings, but not grasping any bough with its foot (Schlegel, Cuculi 49).

down above the nostril. The nostril is oval, cartilaginous above, situated at the base of the maxilla; the wing is long and pointed, longer than the tail, the 3rd quill longest, slightly exceeding the 4th, the secondaries hardly reaching more than half the length of the wing. A partial migrant. Probably parasitical. Feeds on insects, seeds.

GENUS SCYTHROPS Lath.

Description as for the subfamily.

✢ 74. SCYTHROPS NOVAEHOLLANDIAE Lath.
Channel-bill.

Scythrops novae-hollandiae *(1)* Lath., Ind. Orn. 1790, I, 141; *(II)* Temm., Pl. Col. 1824, 290; *(III)* Vieill. & Oud., Gal. Ois. I, 1825, 27, p. 39; *(IV)* Less., Tr. d'Orn. 1831, 128 pl. 23, f. 1; *(V)* Lafresn., Mag. de Zool. 1835, pl. 37 (juv.); *(VI)* Küster, Orn. Atlas 1838, pt. 15, pl. 4; *(6bis)* Sal. Müller, Wieg. Arch. 1846, XII, 116; *(VII)* Gld., B. Austr. 1848, IV, pl. 90; *(8)* Wall., Ibis 1860, 147; *(9)* id., P. Z. S. 1863, 485; *(10)* Schl., Mus. P.-B. Cuculi 1864, 36; *(11)* Gld., Handb. B. Austr. 1865, I, 628; *(XII)* Diggles, Orn. Austr. 1866, pt. IV, pl.; *(13)* Wald., Tr. Z. S. 1872, VIII, 27, 51; *(13bis)* Rosenb., Malay. Arch. 1878, 275; *(14)* Meyer, Ibis 1879, 67; *(15)* Salvad., Orn. Pap. 1880, I, 372, and Agg. 1889, 52; 1891, 220; *(16)* Finsch, Mitth. Orn. Ver. Wien 1884, VIII, 93; *(17)* Meyer, Isis 1884, 18; *(18)* Guillem., P. Z. S. 1885, 549; *(19)* W. Blas., Ztschr. ges. Orn. 1885, 208, 260; *(20)* Ramsay, Pr. L. Soc. N. S. W. 1886, 1094; *(21)* North, ib. 1887, 410; id., ib. 1888, 1780; *(22)* Rams., Tab. List 1888, 15; *(23)* Cox & Hamil., Pr. L. Soc. N. S. W. 1889, 416; *(24)* Tristr., Cat. Coll. B. 1889, 82; *(25)* North, Nests & Eggs B. Austr. 1889, 248, pl. VII, fig. 3 (egg); *(26)* Shelley, Cat. B. XIX, 1891, 330; *(27)* Baldamus, Leben d. Kuckucke 1892, 154; *(28)* Newton, Dict. B. 1893, 84, 125 note; *(XXVIIIbis)* Meyer, Vogelskel. pt. XIX, 60, t. CLXXXV (1894); *(29)* M. & Wg., J. f. O. 1894, 241; *(30)* iid., Abh. Mus. Dresden 1895, Nr. 8, p. 6; *(31)* Hartert. Nov. Zool. 1896, 159, 243; *(32)* Salvad., Ann. Mus. Civ. Gen. 1896, 69.

a. **Australian Channel-Bill** *(I)* Lath., Syn. Suppl. 1801, II, 96, pl. 124; *(II)* id., Gen. Hist. 1822, II, 300, pl. 32.
b. **Scythrops australasiae** *(I)* Shaw, Gen. Zool. 1811, VIII, 2, 378, pl. 50.
c. **Cuculus praesagus** *(1)* Reinw., MS. fide Bp., Consp. 1850, I, 97; *(2)* Reinw., Reiz. Ind. Archip. in 1821, 1858, 592.
d. **Scythrops novae-hollandiae** var. **praesagus** *(1)* Brügg., Abh. Ver. Bremen 1876, V, 56.
"**Laebukua**", Talaut (Nat. Coll.).
"**Ulaäto**", Bone, Gorontalo Distr. (Rosenb. *13bis*).
"**Kapureh**" near Manado ("Purej" — Nat. Coll.).
"**Krok**", Minahassa (Meyer *14*).
"**Uriah**", Tjamba Distr., S. Celebes (Platen *19*).
"**Alo puti**", Bonthain, S. Celebes (Ribbe & Kühn).
For further synonymy and references see Salvad. *15*; Shelley *26*.
Figures and descriptions. Latham *a I*, *a II*; Shaw *b I*; Temminck *II*; Vieillot *III*; Lesson *IV*; Lafresnaye *V*; Küster *VI*; Gould *VII*; Diggles *XII*; Meyer *XXVIIIbis* (skeleton); Salvadori *15*; W. Blasius *19*; North *25* (egg); Shelley *26*.

Adult. Upper surface (except head and neck) dark drab, the feathers of back, upper tail- and wing-coverts broadly terminated with blackish brown, terminal portion of quills blackish brown; tail crossed by a broad subterminal band (70 mm wide) of black, the lateral feathers marked on the inner webs with narrow bars of black, longest on the outermost pair, but generally not reaching to the shaft, the interspaces chiefly filled with pale cinnamon, tip of tail white, broadest (ca. 30 mm) on the lateral feathers; entire head and neck uniform cinereous, a little paler on chin and throat, becoming light grey with a pink wash on the breast, on which are very faint traces of bars; rest of under surface with a wash of buff, the bars becoming more distinct, especially on the flanks and under tail-coverts, where they are light drab-brown (nr. Manado, Aug.—Sept. 1892, C 10868). "Iris cherry-red; skin round eyes cherry-red; bill whity grey, darker at root; feet lead-grey" (Platen *19*).

In another specimen, probably older, marked Celebes — Riedel, Nr. 14009, the bars on the under surface are discernible only on the flanks, thighs and under tail-coverts.

Sexes. They are similar in plumage (Shelley *26*). Prof. W. Blasius *(19)* believes that the male has "a very considerably longer bill, with the point more strongly curved; a larger size in general, the colour of the upper surface somewhat darker, the barring of the under surface scarcely perceptible". Mr. Hartert draws attention to a female which is larger than a male specimen *(31)*.

Young. The cinereous of head, neck and breast of adult is chiefly occupied with feathers of pale cinnamon; back, wing-coverts and tail-coverts broadly terminated with pale cinnamon, usually preceded by a subterminal bar of blackish; ends of wings pale cinnamon and white; the broad white tip of tail washed with the same colour, the lateral tail-feathers mottled with bars of pale cinnamon and black on the outer as well as the inner webs; under surface deeper buff colour, the bars of drab-brown very distinct (N. Celebes — Riedel, Nr. 14010). "Iris light brown; skin round eyes cherry-red; bill whity grey, not darker at the base" (Platen *19*).

A somewhat older immature bird differs from the above in that the cinnamon tips of the feathers are fewer, narrower, and paler, being almost pure white on the quills and tail (Celebes, C 3580).

Egg. Dull white, with faint washed-out pinkish spots and minute dots, also some of a yellowish brown tinge. Size 38×27 mm (North *25*).

Breeding season in Australia. November (North *25*).

Measurements.	Wing	Tail	Tarsus	Bill from nostril (straight)
a. (C 10868) ad. Manado, Aug.—Sep. 92	362	280	45	75
b. (14009) ad. Celebes	378	293	41	73
c. (C 11207) ♀ ad. S. Celebes, 31. March 93	350	281	44	66
d. (C 3580) juv. Celebes	348	275	41	52.5
e. (14010) juv. Celebes	315	257	—	45
e'. (Sarasin Coll.) ♀, Kema, Sept. 9	355	277	41	66
f. (13082) imm. Salibabu Id. 29. X. 93	356	288	—	62
g. (C 7400) ad. Ceram	354	270	—	71
h. (C 7401) ad. Ceram	342	265	—	69
i. (C 7403) ad. Ceram	320	253	—	58
j. (C 7402) ad. Ceram	320	255	—	68
k. (C 7399) ♀ ad. Aru	366	276	—	68
l. (C 10562) ♀ ad. Makisa	345	266	—	55.5
m. (Mus. Nehrk. Nr. 3821) juv. Australia	313	270	—	—

Skeleton.

Length of cranium	118.0 mm	Length of tarso-metatarsus	48.0 mm
Greatest breadth of cranium	44.5 »	Length of digitus I	26.5 »
Length of humerus	91.0 »	Length of digitus II	38.0 »
Length of ulna	89.0 »	Length of digitus III	52.0 »
Length of radius	82.0 »	Length of digitus IV	51.0 »
Length of manus	88.0 »	Length of sternum	60.5 »
Length of metacarpus	43.7 »	Greatest breadth of sternum	51.0 »
Length of digitus I	17.0 »	Height of crista sterni	18.5 »
Length of digitus II	40.5 »	Length of coracoideum	46.5 »
Length of digitus III	13.0 »	Length of scapula	60.0 »
Length of femur	64.5 »	Length of clavicula	47.0 »
Length of tibia	86.4 »	Length of pelvis	83.5 »
Length of fibula	40.0 »	Greatest breadth of pelvis	45.0 »

Distribution. "Universally distributed over the whole continent of Australia" (North 21, 25); Tasmania (North 21); S. E. New Guinea, New Britain, Duke of York, Kei, Aru (Riedel 17), Ceram, Buru, Obi major, Batchian, Ternate (cf. Salvad. 15); Celebes — Minahassa (Forsten 10, Reinw. c 2, Meyer 14, Guillem. 18), Gorontalo (Rosenb. 13bis), Macassar (Wallace 26), Tjamba, S. Celebes (Platen 19); Flores (Allen 9, 15); Talaut — Salibabu Id. (Nat. Coll. 29).

This Cuckoo, the Channel-bill of Latham and of Australian ornithologists, is remarkable at once as being the largest of all the Cuckoos, and for the peculiar structure of its great bill. Prof. Newton (28) remarks that "its systematic position has often been disputed — its large and curiously grooved bill inducing some to refer it to the *Bucerotidae* (Hornbills), while its zygodactyle feet caused others to place it among the *Ramphastidae* (Toucans)". Count Salvadori (15) believes, that it might be made the type of a distinct subfamily constituted by Bonaparte; this we have adopted. The development of its plumage also affords points of interest; the immature bird presents certain analogies to the young of other Cuckoos of less aberrant structure, but the manner in which the cinnamon colour is confined to a single, large, ill-defined spot at the tip of the feathers of the upper surface seems to be peculiar to it.

Gould remarks that in New South Wales it is migratory, arriving in October and departing again in January (VII, 11). Of late years no one in Australia appears to have paid further attention to the subject, except that Mr. North mentions it as a straggler in Cumberland County, N. S. W., and in Tasmania (21). As Count Salvadori remarks, this migration is probably directed towards the equator, and that the species wanders to New Guinea from Australia, and during its migration extends its way to the Kei Islands, Moluccas, Celebes and the islands of the Timor Group. This conclusion does not, however, appear to be entirely correct, at least as regards the Minahassa, Celebes, where, as we are informed by Mr. Cursham, it is present all the year. Meyer found the bird very common in the Minahassa from January till July, and was informed that "during the east monsoon, when it is very dry (May till November), the bird cries much". It is a breeding species there as well as in

Australia, as is shown by von Rosenberg's statement that he once came into possession of an egg now in the Leyden Museum, which fell from the oviduct of a specimen killed by one of his hunters *(13bis)*, and a native told Meyer that he had once taken a young *Scythrops*, together with a young Crow out of a Crow's nest. The bird is often to be seen along with Crows *(14)*.

The plumage of the species in all localities appears to be very similar and stable; in dimensions of wing, bill and tail, however, it is an unusually variable bird. Probably, as Prof. W. Blasius remarks, the bill in the male is larger than in the female.

In Celebes it flies in large flocks, and feeds on fruits, such as the waringui *(14)*. Gould mentions large insects as its food. Latham says that "in the crop and gizzard the seeds of the red gum and peppermint trees have been found ... exuviae of beetles also".

It is probably parasitical, but positive evidence on this point is still wanting. In Gould's Handbook an amusing account of the behaviour of a young *Scythrops* is given. Immediately it was put into the aviary it showed signs of hunger by opening its mouth, and had its wants promptly and carefully attended to by a Laughing Jackass *(Dacelo gigas)*.

ORDER BUCEROTES.

The order contains the single family of the Hornbills, "a very natural and in some respects an isolated group, placed by Prof. Huxley among his *Coccygomorphae*" (Newton), furnished with a bill of large or very large size, supported internally by a bony cellular structure, usually furnished at the base with a peculiarly shaped casque, and having a normal foot — three toes in front and the hallux behind. In size they vary from that of a domestic Pigeon to that of a Turkey; the plumage is usually of a pie-bald character, the tints being abruptly located, and the colours displayed are few, chiefly black, white, and various browns, while blue and red (except on the bare parts of the skin) are absent; in this they differ from the Toucans. The upper eyelid is provided with a row of stiff lashes; the wing is of moderate length, 11 primaries, the outermost one small and often curiously attenuated, the secondaries long; tail long, consisting of 10 feathers. Food: fruit and seeds, "the bigger species also capture and devour a large number of snakes, while the smaller are great destroyers of insects ... They breed in holes of trees, laying large white eggs, and when the hen begins to sit the cock plasters up the entrance[1]) with mud

---

[1]) Sometimes before over the eggs are laid.

or clay, leaving only a small window through which she receives the food he brings her during her voluntary imprisonment" (Newton, D. B. 437).

For anatomical characters see Grant, Cat. B. XVII 1892, 347; Blanford, Faun. Br. Ind., B. III 1895, 140; Gadow, Bronn's Kl. u. Ord. VI, 4, Av. II, p. 234 (1893).

FAMILY BUCEROTIDAE.
Map VII.

The characters as for the order. The family is divided by Elliot and Grant into two subfamilies, *Bucorvinae* (or *Bucoracinae*) and *Bucerotinae*, the former contains only one genus confined to Africa and distinguishable from all other Hornbills by its long tarsus, twice as long as the middle toe and claw, the latter with about 20 genera, belonging to the Indian countries, the East India Archipelago, and Africa, having a short tarsus about equal to or shorter than the middle toe and claw.

GENUS RHABDOTORRHINUS M. & Wg.¹)

The small Celebesian Hornbill is somewhat bigger than a crow, the bill very large, not furnished with a prominent casque, but in the adult the culmen is drawn up into a high sharpe ridge covering more than the basal two-thirds of the bill, with three parallel longitudinal ridges on the side of the maxilla; some transverse ridges occasionally at the base of the lower mandible; the nostril broad oval, situated at the base of the maxilla between the two lowest ridges and partially protected by feathers. The genus consists of a single species, in which the sexes are different, the female being black, the male having a yellow face and throat (changing to white after death with time and exposure of the skin).

† * 75. RHABDOTORRHINUS EXARATUS (Temm.).
Lesser Celebesian Hornbill.

a. **Buceros exarhatus** "Reinw." *(1)* Temm., Pl. Col. 1823, II, 91, pl. 211 (♀); *(2)* Wagl., Syst. Av. Buceros 1827, Nr. 15; *(3)* Less., Man. d'Orn. 1828, II, 107; *(4)* Schl. & Müll., Verh. Natuurk. Comm. 1839—44, Aves 23; *(5)* Reinw., Reis Ind. Archip. in 1821, 1858, 592; *(6)* Gray, HL. 1870, II, 128.
b. **Buceros exarhætus** *(1)* Blyth, J. A. S. B. 1847, XVI, 997.
c. **Buceros exaratus** *(1)* Gray, Gen. B. 1847, II, 400; *(2)* Blyth, Cat. B. Mus. As. Soc. 1849, 44; *(2ᵇⁱˢ)* Gray, P. Z. S. 1860, 356 (Moluccas!); *(3)* Schl., Mus. P.-B. Buceros 1862, 10; *(4)* Finsch, New Guinea 1865, 162; *(V)* Wald., Tr. Z. S. 1872, VIII, 47, pl. V; *(6)* Brügg., Abh. Ver. Bremen 1876, V, 56; *(7)* Lenz, J. f. O. 1877, 370; *(8)* Rosenb., Malay. Archip. 1878, 274; *(9)* Meyer, Ibis 1879, 65; *(9ᵇⁱˢ)* Salvad.,

¹) ῥαβδωτός = striatus, † ῥίς, ῥινός = nasus.

Orn. Pap. 1880, I, 401; *(10)* W. Blas., Ztschr. ges. Orn. 1885, 249; *(11)* Hickson, Nat. in N. Celebes 1889, 20; *(XII)* Meyer, Vogelskel. 1892, pt. XVII, 43, pl. CLXIX.
d. **Hydrocissa exaratus** *(1)* Bp., Csp. 1850, I, 90; *(2)* Dubois, Bull. Mus. Belg. 1884, III, 210.
e. **Anorrhinus exaratus** *(1)* Bp., Consp. Vol. Anisod. 1854, 2; *(2)* W. Blas., J. f. O. 1883, 136.
f. **Hydrocissa exarata** *(I)* Elliot, Monogr. Bucerot. 1878, pp. XIX, XXIII, XXVIII, fig. head (*enarhatus* — err.), pl. 46, text; *(2)* Tristr., Cat. Coll. B. 1889, 95; *(3)* Heine & Rchnw., Nomencl. Mus. Hein. 1890, 169; *(4)* Hartert, Kat. Mus. Senckenb. 1891, 139.
g. **Penelopides exaratus** *(1)* Grant, Cat. B. XVII, 1892, 376; *(2)* Hart., Nov. Zool. 1897, 164.
Rhabdotorrhinus exaratus *(1)* M. & Wg., Abh. Mus. Dresden 1895, Nr. 8, p. 6; *(2)* iid., ib. 1896, Nr. 1, p. 8; *(3)* iid., ib. 1896, Nr. 2, p. 11.

"**Karaka**" (Moyer *c 9*), "**Karkar**" (Nat. Coll.), native Malay name, Minahassa.
"**Karokok**", Alfurous, N. Minahassa (M. *c 9*).
"**Kerek-kerek**", Tondano, Minahassa (M. *c 9*).
"**Hele-hele**" (Rosenb. *c 8*).
"**Tolo-tolo**", Tjamba, S. Celebes (Platen *c 10*).
"**Tahutahu**", Tonkean, E. Celebes (Nat. Coll.).

Figures and descriptions. Temminck *a I*; Walden *c V*; Elliot *f I*; Meyer *c XII* (skeleton); Brüggemann *c 6*; W. Blasius *c 10*; Grant *g 1*.

Old male. Sides of head including forehead, superciliary region, ear-coverts, cheeks, chin and throat pale lemon-yellow (soon fades into white); all other parts black, on back and wings strongly glossed with green; a low casque on bill, reaching to within 20 mm of the tip, rather sharply ridged along the culmen and furnished on the sides with 3 additional ridges and furrows (Manado, C 8119). "Iris red-brown; orbit citron-yellow" (Platen *c 10*).

"Iris crimson, eyelashes black, orbit bright crimson; feet black, soles ochreous; bill pale horn-colour, casque dull ferruginous" (Doherty *g 2*).

Younger male. Like the adult; but the bill much smaller and the casque with its ridges and furrows imperfectly formed. In the old male and female the upper and lower mandibles do not close, except at the basal part and at the tip; in the young bird the edges seem to meet properly all along the bill (Paguatt, Sept., C 8117).

Adult female. Like the male, but the sides of the head, chin and throat, like the rest of the plumage, black; size generally a little smaller (Manado, Nr. 7708). "Iris red-brown; orbit black" (Platen *c 10*).

Measurements. Males — wing 218—252 mm; tail 180—222; rictus 88—102; length in the flesh 480—570; expanse 670—800; from tip of wing to tip of tail 150—170.

Females — wing 205—233 mm; tail 176—203; rictus 83—98; length in the flesh 450—500; expanse 650—760; from tip of wing to tip of tail 140—155 (Blasius and Platen *c 10*).

Skeleton (juv.).

Length of cran. fr. sut. nasofr.	40.0 mm	Length of tarso-metatarsus	45.0 mm
Greatest breadth of cranium	38.7 »	Length of digitus III	44.0 »
Length of humerus	64.0 »	Length of sternum	55.0 »
Length of ulna	88.0 »	Greatest breadth of sternum	34.4 »
Length of radius	81.0 »	Height of crista sterni	10.5 »
Length of manus	59.0 »	Length of coracoideum	40.0 »
Length of metacarpus	34.5 »	Length of scapula	48.0 »
Length of digitus principalis	21.0 »	Length of clavicula	30.5 »
Length of femur	54.5 »	Length of pelvis	55.0 »
Length of tibia	79.5 »	Greatest breadth of pelvis (pub.)	35.0 »
Length of fibula	28.5 »		

Egg. Von Rosenberg, to whom a rotten egg of this species was brought, describes it as rough-shelled, dirty white, and of the size of a Pigeon's egg *(c 8)*.

Nest. Meyer *(c 9)* was informed that "it makes its nest in hollow-trees or between wood, and lays two eggs"; but particulars as to whether the female is walled in with mud by the male are wanting.

Distribution. Celebes — Minahassa (Reinwardt *a 5*, Forsten *c 3*, etc.); Lembeh Id. (Nat. Coll.); Paguatt, Gulf of Tomini (Meyer *c 9*); Mapane (P. & F. Sarasin *2*); Tonkean, E. Celebes (Nat. Coll.); Macassar (Wallace *g 1*); Tjamba (Platen *c 10*); Tawaya, W. Celebes (Doherty *g 2*).

Mr. Elliot *(I)* adds "Malacca" as a locality of this species with Meyer's name as collector, and Dr. Dubois likewise records a specimen, or specimens, in the Brussels Museum as having been obtained by Meyer in Malacca. Meyer never did any collecting in Malacca, and the notice of this locality rests upon mislabelling due to other hands than his.

The genus *Hydrocissa* was, as Mr. Elliot says (Bucerot. pl. XI, text), proposed by Bonaparte *(d 1)* for various species "not nearly related beyond the fact that they belong to the same family". Mr. Elliot and Mr. Grant agree in placing them in three distinct genera, two of which, *Anthracoceros* and *Anorrhinus* were made by Reichenbach a little prior to Bonaparte, the first containing his *Hydrocissa monoceros*, *pica* and *violacea* (synonyms of *A. coronatus*) and *malayana*, while *Hydrocissa galerita* is the type of *Anorrhinus*, a genus which Grant finds to be represented by this species alone. The sixth and last species of Bonaparte's *Hydrocissa*, *H. exarata* of Celebes, is thus alone left in the genus. By Rule 5 of the Stricklandian Code Elliot would be fully justified in allowing the generic name *Hydrocissa* to stand for *exarata*, no type of the genus having been at first specified by Bonaparte, though *monoceros* stands first in the list, were it not that four years later Bonaparte in his Conspectus Volucrum Anisodactylorum, p. 2, distinctly signifies *monoceros* as his type by placing that species (with its synonym, *pica*) alone in the genus *Hydrocissa* and by relegating *exaratus* to the genus *Anorrhinus* Rchb., thus making it unlawful for any subsequent author to transfer the name *Hydrocissa* to any other part of the original genus (see Rules of Zool. Nomencl. 1863 § 5). Grant, therefore, rightly makes *Hydrocissa* a synonym of *Anthracoceros*; but, as we cannot agree with him that *Buceros exaratus* belongs to the Philippine genus, *Penelopides*, we find it necessary to give it a new generic name, *Rhabdotorrhinus*.

R. exaratus, the sole species of this genus, has in Elliot's opinion "no ally in the family, and is remarkable for its crest-like casque, hardly to be distinguished from the maxilla, and is moreover peculiar for the lateral grooves running its entire length. It . . . is probably the sole survivor of a subgroup of this family" *(f 1)*.

Penelopides has a small, smooth, somewhat tubular-looking epithema, and the basal part of the bill is furnished with a thin side-plate on one or both mandibles, which is notched or ribbed, as is sometimes the case in the lower mandible of *R. exaratus*. A broad band of rufous occupies more or less of the middle of the tail in *Penelopides*; in *Rhabdotorrhinus* it is uniform black.

Very little has been recorded of the habits of this species. Meyer *(c 9)* observed: "These birds live in pairs together. Their flight is heavy and slow. They feed on fruits, such as the waringuin *(Urostigma)*; and large flocks are often to be seen together on high fruit-trees. A common bird, not restricted to the north-eastern parts of Celebes. [A MS. note says they were rare until May 14th, and then became very plentiful.] I also found it in Paguatt, in the Gulf of Tomini, in September". The cousins Sarasin got it at Mapane, N. Central Celebes, and the native hunters working for the Dresden Museum killed a specimen with strongly developed ridges and furrows at the base of the lower mandible in East Celebes; Wallace and Platen obtained it also in the Southern Peninsula, but it was not sent by Beccari from S.E. Celebes nor by Meyer from the Togian Islands. No Hornbill has been recorded from the Sangi Islands. Observations on the wanderings of the species in the island are altogether wanting; such is also the case as regards their breeding habits. As Mr. Oates remarks (B. Brit. Burmah II, 91) the breeding female of all species of *Bucerotidae* as far as is known is walled in by the male with mud, etc., a narrow opening only being left, through which the male provides his mate with food. A. D. Bartlett made the surprising discovery on a species in 1869 (P. Z. S. 142; Flower, ib. 150; Murie, ib. 1874, 420—425) that the bird *(B. corrugatus)* has the habit of ejecting a sack like a fig, formed of the epithelial lining of its stomach, filled with undigested fruit. The power of throwing up this sack was supposed by Mr. Bartlett to be possessed only during the breeding season by the male, which thus transfers at a lump the whole contents of his stomach to his imprisoned female; but actual observations in confirmation of this interesting view appear to be still entirely wanting (cf. Newton, D. B. 437). Mr. Whitehead says the food is brought by six or seven other birds (Expl. Kini Balu, 50).

Bingham gives (Str. F. VIII, 459 sq.) interesting observations as to the plastering-up of the female in the nesting cavity by the male before she has begun to lay. The object of this curious process is regarded as a protective one, though it is not exactly known against what foes; the natives of Ceylon attribute it to the bird's fear of monkeys (Horsf. & Moore, Cat. B. Mus. As. Soc. 589; Legge, B. Ceylon 274), as does also Mr. Whitehead in Borneo (Expl. Kini Balu, 1893, 50). In New Guinea and elsewhere, where monkeys are absent, the male also walls in the female. Bernstein (J. f. O. 1861, 116, 117) ascertained from personal observation that the female of *Rhytidoceros plicatus*, the only species found east of Celebes, is similarly plastered up by the male, and is of opinion that *Pteromys* and the larger species of *Sciurus* are more dangerous foes to it than monkeys could be. After sitting for a time, the female loses nearly all her quills and tail-feathers and becomes incapable of flight, and Mr. Gammie remarks on *Aceros nepalensis* (Hodgs.) in Sikkim that she "is said not to leave the nest from the time of her entrance till she comes out with her young ready for flight, a period of about three months" (Oates, Hume's Nests & Eggs III, 79).

GENUS CRANORRHINUS Cab. Heine.

The size of an Eagle; bill very large, narrow and decurved, yellow, with a transversely ribbed plate at its base, a large and high smooth epithema, in shape somewhat like the crest of a Grecian helmet, covering the basal half of the culmen and the crown; the nasal aperture round, situated in the crevice formed at the junction of the casque with the maxilla, and concealed by bristles; throat, cheeks, and periocular skin naked; first primary more than half the length of the wing and not greatly attenuated. Sexes dissimilar. The type of the genus is *C. cassidix*. Three other species included therein by Elliot and Grant are aberrant, and a greater uniformity of generic division might be obtained by restricting this genus to *C. cassidix*.

+ * 76. CRANORRHINUS CASSIDIX (Temm.).
Greater Celebesian Hornbill.

a. **Buceros cassidix** *(1)* Temm., Pl. Col. 1823, II, 66, pl. 210(♂); *(2)* Wagl., Syst. Av. Buceros 1827, Nr. 3; *(3)* Less., Man. d'Orn. 1828, II, 105; *(IV)* Griff., An. Kingd. 1829, II, 434, pl.; *(5)* Less., Tr. d'Orn. 1831, 253; *(VI)* Schl. & Müll., Verh. Natuurk. Comm. 1839—44, Aves, p. 24, pl. 4 (♀); *(7)* Gray, Gen. B. 1847, II, 399; *(8)* Reinw., Reis Ind. Archip. in 1821, 1858, 591; *(9)* Sclat., Ibis 1859, 113; *(10)* Schl., Mus. P.-B. Buceros 1862, 9; *(11)* Finsch, Neu Guinea 1865, 162; *(12)* Wall., Mal. Archip. 1869, I, 364, 429; *(12bis)* Marshall, Schädelhöcker der Vög. (Nied. Archiv Zool. I) 1872, 33, 41, pl. XI, fig. 10; *(13)* Brüggem., Abh. Ver. Bremen 1876, V, 56; *(14)* Rosenb., Mal. Archip. 1878, 273; *(15)* Büttik., Zool. Erg. Weber's Reise in Ost-Ind. 1893, III, 275.
b. **Calao (Cassidix) cassidix** *(1)* Bp., Consp. 1850, I, 90.
c. **Rhyticeros (Cassidix) cassidix** *(1)* Bp., Consp. Vol. Anisod. 1854, 3.
Cranorrhinus cassidix *(1)* Cab. & Hein., Mus. Hein. 1862, IV, 173; *(2)* Wald., Tr. Z. S. 1872, VIII, 47, fig. 1—4; *(3)* Lenz, J. f. O. 1877, 370; *(IV)* Elliot, 1878, pp. XVIII, XXIX, XXXVI, pl. XVI text; *(5)* Meyer, Ibis 1879, 65, 146; *(6)* W. Blas., J. f. O. 1883, 136; *(7)* id., Ztschr. ges. Orn. 1885, 254—260; *(8)* Guillem., P. Z. S. 1885, 54; *(9)* id., Cruise Marchesa 1886, II, 188; *(10)* Platen, Gefied. Welt 1887, 210; *(11)* Hickson, Nat. in N. Celebes 1889, 255; *(12)* Heine & Rchw., Nomencl. Mus. Hein. 1890, 170; *(13)* Hartert, Kat. Senckenb. Mus. 1891, 140; *(14)* Grant, Cat. B. 1892, XVII, 377; *(15)* M. & Wg., Abh. Mus. Dresden 1895, Nr. 8, p. 7; *(16)* iid., ib. 1896, Nr. 2, p. 11; *(XVII)* Meyer, Vogelsk. 1891 pl. 190, 192—4.
d. **Buceros (Cranorrhinus) cassidix** *(1)* Gray, HL. 1870, II, 129; *(2)* Dubois, Bull. Mus. Belg. 1884, III, 192.

"**Alo**", Tondano (Reinw. *a 1, a 10*).
"**Burong-taun**" (Year-bird), Minahassa (Meyer *5*).
"**Uwak**" (Meyer *5* and Nat. Coll.), native name, Lotta, Minahassa.

"**Ngak**", Tonkean, E. Celebes (Nat. Coll.).
"**Ahlo**", Gorontalo (Rosenb. *a 14*).
"**Pankao**", Tulabello, Gorontalo Distr. (Rosenb. *a 14*).
"**Alo**", Tjamba, S. Celebes (Platen 7).

Observation. It is curious, considering the great diversity of language in Celebes, that the native name for this bird is the same in Tondano, Gorontalo and Tjamba, that is from the north to the south.

Figures and descriptions. Temminck *a I*; Griffith *a 4*; Schlegel & Müller *a VI*; Elliot *IV*; Marshall *12bis* (skull in section fig.); Walden *2*; Lenz *3*; W. Blasius *7*; Grant *14*; Meyer *XVII* (skel.).

Adult male. Head, where not covered by the casque, and nape dark chestnut; ear-coverts and entire neck light golden ochraceous; tail white; rest of plumage black, with a green gloss on the upper surface (Gorontalo, C 12173). Iris light brown to red; feet black, soles of the feet grey; claws black; bill yellow, base brownish, with dark brown bands; chin brownish red; round the eyes deep blue; throat light blue, with a dark blue patch in the middle. The casque is smooth in life; and the wavy unevenness in dried specimens is only in consequence of its shrivelling up after death (Meyer *5*).

Dr. Guillemard *(8)* writes: "Iris bright orange; feet black. Length of adult male about 104 cm; weight 5 lb. 3 ozs. Bill orange-yellow; the basal plaques dull red anteriorly; casque dull red; base of throat cobalt-blue, marked with two patches of black and one of dark blue".

The colours of the soft parts in a fine adult male in the Sarasin Collection are indicated as follows: "iris yellow-red; skin of cheeks white and Prussian blue; throat-sack white-blue, across it in the middle a broad Prussian blue band, behind this on either side a black streak; feet black; bill yellow horn-red" (♂, Kema, 19. Aug. 93: P. & F. S.).

Adult female. Like the adult male but smaller, and with the plumage of the head and neck black like the body. Casque yellow.

It is likely that the ribbed plates at the base of the bill are renewed yearly, and that the number of ribs increases with age. A belief that one rib is added to the bill-plates each year has given rise to the name by which the bird is known in the Minahassa, "Burong-taun" or Year-bird (Meyer *5*, Blasius *7*).

Young male. Like the adult female, but with the black plumage of the neck varied with tawny and some of the bases of the black feathers are of this colour; nape varied with chestnut feathers; casque as in female, but lower; bill shorter, base furnished with a thin covering without ribs (Manado, Nr. 8106).

Another specimen is like the last, but has the bill rather larger, and only here and there a tawny feather in the black neck. This may be a female, as some sprouting feathers in the neck are black, none tawny (Lotta, C 12091).

Measurements. Males (3 specimens) — length in the flesh 850—920 mm; expanse 1300—1400 (Platen 7); (9 specimens) wing 400—470 mm; tail 270—328; rictus 175—230 (Blas. 7, Dresden Mus. and Sarasin Coll.).

Females (9 specimens) — length 700—830 mm; expanse 1000—1270 (Platen 7); wing 354—394; tail 234—283; rictus 165—185 (Blas. 7).

Eggs. Unknown.

Skeleton.

Length of cranium	252.0 mm	Length of tarso-metatarsus	63.0 mm
Greatest breadth of cranium	60.3 »	Length of digitus I	52.0 »
Length of humerus	123.0 »	Length of digitus II	63.0 »
Length of ulna	188.0 »	Length of digitus III	76.0 »
Length of radius	166.0 »	Length of digitus IV	73.0 »
Length of manus	112.0 »	Length of sternum	105.0 »
Length of metacarpus	67.0 »	Greatest breadth of sternum	63.5 »
Length of digitus I	21.6 »	Height of crista sterni	25.0 »
Length of digitus II	44.0 »	Length of coracoideum	65.0 »
Length of digitus III	20.0 »	Length of scapula	88.0 »
Length of femur	93.0 »	Length of clavicula	65.0 »
Length of tibia	127.0 »	Length of pelvis	111.0 »
Length of fibula	100.0 »	Greatest breadth of pelvis	64.0 »

Nest. Rosenberg *(a 14)* states that the female is walled-in with mud into the nesting-cavity by the male, but he does not seem to write from personal observation. Meyer *(5)* says that the nest is built on the tops of the highest trees.

A female before us has the quills worn in a remarkable way, as if she had performed incubation under the conditions observed in Africa, India, Java etc.

Distribution. Celebes — Minahassa (Reinwardt *a I, a 8, a 10,* Forsten *a 10,* etc.); Gorontalo (Rosenberg *a 14,* Meyer *5*); Lembeh Id. (Nat. Coll.); Togian Islands (Meyer *5*); Tonkean, E. Celebes (Nat. Coll.); Macassar (Wall. *a 9, a 12, 14*); Tjamba, S. Celebes (Platen *7*), Maros and Luwu (Weber *a 15*).

Cranorrhinus cassidix of Celebes is the type of the genus originally made for it and *C. corrugatus* of Malacca, Sumatra and Borneo by Cabanis & Heine in 1862. Though *C. corrugatus* is retained in the genus by Elliot and by Grant, it is a very different species, both as regards coloration and form of casque. The nearest ally of *C. cassidix* is *C. leucocephalus* of Mindanao and the small island of Camiguin off the north coast of Mindanao, a species differing chiefly from *C. cassidix* in having a narrow band of black at the end of the white tail, the casque laterally corrugated, not smooth, and the ribbed plate on the side of the bill present only on the mandible. It is interesting to note that the tail of *C. cassidix* is sometimes tipped on a few feathers (young ♂ 3rd June, Tjamba, C 12174), or spotted near the tip, with black; in one adult male in the Dresden Museum one of the two middle rectrices has a long streak of black on its terminal third. Prof. W. Blasius, with his usual exactitude of observation, remarks that such spots were not present in the nine adult females from South Celebes examined by him; nor are they to be seen in the females in the Dresden Museum. A female of *Rhytidoceros plicatus* from Ceram has an outer rectrix on the inner web margined with black of irregular width throughout its length (C 12175).

In his great work on the Hornbills Mr. Elliot appears to have attached, if anything, a too great value to the shape of the epithema and not enough to the coloration of the plumage, for the casque appears to be the most variable character in the family and has been modified into all sorts of shapes, whereas the general types of plumage are probably much more stable. The genus most nearly allied to *Cranorrhinus* is unquestionably *Rhytidoceros*, the members of which correspond with *Cranorrhinus cassidix* very closely in general coloration, much better than do *Cranorrhinus corrugatus* and *waldeni*; indeed the latter are as much entitled to generic distinction from *C. cassidix* as are the species of *Rhytidoceros*. *Rhytidoceros* occurs both west and east of Celebes, the four species having the following distribution: *R. undulatus* from Cachar and Pegu down the Malay Peninsula to Sumatra, Borneo and Java; *R. subruficollis* with much the same distribution but not known in Java; *R. narcondami* from the small island of Narcondam in the Bay of Bengal about 170 miles due S. from Cape Negrais; and *R. plicatus* (with its subspecies *ruficollis*) ranging from the Moluccas over New Guinea to the Solomon Islands (Grant, Cat. B. XVII, 1892, 382 sq.).

It is instructive to see how the conclusions deduced from the geographical distribution of the members of one family may be partly undone, partly confirmed, on considering that of another family. The *Bucerotidae*, like the *Phoenicophaeinae*, consist of many narrowly located genera; were the generic distinctions as finely drawn as in the *Phoenicophaeinae* several more generic names would have to be bestowed.

Map VII. Distribution of the Hornbills of the East Indies.

	Nr. of species in genus	Indo-China	Tenasserim	Malacca	Sumatra	Borneo	Java	Philippines	Celebes	Moluccas	Papuasia
1. *Buceros*	2	.	.	†	†	†	†				
rhinoceros		.	.	*	*	*					
sylvestris		.	.	.	.	.		*			
2. *Dichoceros*	1	†	†	†	†						
3. *Hydrocorax*	3	.	.	.	.	.		†			
4. *Anthracoceros*	5	†	†	†	†	†	†	Sooloo			
convexus		.	.	*	*	*		*			
malayanus		.	.	*	*	*					
montani		.	.	.	.	.		Sooloo			
5. *Gymnolaemus*	1							Palawan			
6. *Penelopides*	6					,.		†			
7. *Rhabdotorrhinus*	1							.	†		
8. *Cranorrhinus*	4			†	†	†		†	†		
cassidix									*		
leucocephalus								*			
corrugatus		.	.	*	*	*		*			
waldeni								*			
9. *Rhytidoceros*	4	†	†	†	†	†	†	.	†	†	†
undulatus		*	*	*	*	*	*				
subruficollis		*	*	[*]	*	*					
plicatus										*	*
10. *Anorrhinus*	1	†	†	†	†						
11. *Berenicornis*	1	†	†	†	†						
12. *Rhinoplax*	1	.	†	†	†	†					

The range of the *Phoenicophaeinae* ceases at Celebes, and the Celebesian subgenus *Rhamphococcyx* finds its nearest ally in the Javan subgenus *Rhinococcyx*; the *Bucerotidae* have a species, ranging from the Moluccas to the eastern bounds of Papuasia, whose nearest ally is found in Java, Borneo, Sumatra and the Asiatic countries to the west; while the two species of Celebes show affinity to the Philippine species. On the other hand the Hornbills, like the Coucals, lend very interesting proof to Wallace's view of the recent simultaneous separation of Borneo, Sumatra and Malacca — not Java, Sumatra and Malacca, as might have been expected from the proximity of these islands. (See map VII). In view of the fact that three genera are peculiar to this section of the East Indies, one to Celebes, one to Palawan, two to the other Philippines, and that most species of the East Indies have a very restricted range, it is puzzling to find that *Rhytidoceros* is represented by one species only in all the islands between Halmahera and the Solomons, and is continued in a closely allied species, *R. subruficollis* from Borneo to the Indo-Chinese countries; in other words *Rhytidoceros* has only become differentiated into two species between British Burmah and the confines of Papuasia, while other *Bucerotidae* have become changed into eleven genera and 26 species in half the distance.

An easy way of getting over this difficulty — and, as it would appear, most probably the true way — is on the supposition that, while *Rhytidoceros* has extended its range by flight, most of the other East Indian Hornbills have become located in their several habitats by terrestrial changes, resulting in the production of barriers of sea over which the birds do not venture to fly. On examining the wings of the Hornbills a very curious structure of the primary quills is seen in most species. The first and second are very much attenuated and so weak that they are obviously of no use in flight: it would be of interest to know, indeed, whether they are freely extended in flight, or concealed under the larger quills. All the genera of the East Indian Hornbills are represented in the Dresden Museum, and we find that this modification of the quills is most pronounced in *Anthracoceros*, though nearly as well seen in *Penelopides* and *Rhabdotorrhinus*; it is also found to a greater or less extent in *Buceros*, *Dichoceros*, *Hydrocorax*, *Gymnolaemus*, *Cranorrhinus*, *Rhinoplax*. It is not seen in *Anorrhinus* and *Berenicornis*; but in these two genera the wing is shorter and blunter than in most others, and does not appear adapted for sustained flight; Davison observed that the flight of *Berenicornis* — a contrast to other Hornbills — is noiseless (Str. F. VI, 104). With these two exceptions, the first primary has all the appearance that it is gradually dwindling away; it is probably an effect of disuse, combined with wear and tear of some kind[1]). In *Rhytidoceros plicatus* and *undulatus* the first primary is strong and not noticeably attenuated and may be of some service in flight; in *R. subruficollis* it is attenuated about as much as

---

[1]) As we believe caused by attrition, due to the position of the outermost remex underneath the adjacent remiges in the wing of these clumsy fliers.

in *Cranorrhinus cassidix*. On the whole the wings suggest that *Rhytidoceros* is more given to flying than the other genera. The following actual observations on the flight of the *Bucerotidae* tend to prove this. Prof. Newton in his "Dictionary of Birds" (1893, 435) states that Tickell in his manuscript "Birds of India" (in the library of the Zoological Society of London) "divides the Hornbills of that country into two genera only, *Buceros* and *Aceros*, remarking that the birds of the former fly by alternately flapping their wings and sailing, while those of the latter fly by regular flapping only". *Anthracoceros coronatus* may be taken as typical of the former class, as is shown by Legge's corresponding observation: "when flying it proceeds with rather quick-flapping of the wings, and then sails along with them outstretched, its long tail and motionless primaries giving it a singular aspect... It usually does not take long flights; when it does the momentum of its huge bill and heavy neck are such as to cause it on alighting to topple forward before gaining its equilibrium" (B. Ceylon, 1880, 274). Davison observed the same mode of flight in *Dichoceros bicornis* (Str. F. VI, 99), so, also, Bourdillon in the same species (Str. F. IV, 387), in *Anthracoceros convexus*, here called *Hydrocissa albirostris* (t. c. p. 101), and in *Ptilolaemus tickelli* (p. 104). *Berenicornis* flaps regularly, but "keeps in small parties about the lower trees and undergrowth" (p. 106).

As to *Rhinoplax vigil*, Hartert (J. f. O. 1889, 367, 368) remarks that it is "a bad flier, even worse than *Buceros rhinoceros*". On *Cranorrhinus corrugatus* in E. Sumatra Dr. Hagen (T. Ned. Aardrijk. Genootschap 1890, (2) VII, 139) has published the following observations: it has "like all Hornbills a heavy clumsy flight, accompanied by a peculiar ringing whizzing noise. After about each kilometer of the way (they flew daily from the sea-side to their roosting-trees inland) they rested for some minutes on suitable tall tree-tops. The resting-places are fixed points, and, if they are not scared, the birds may be expected with tolerable certainty every evening at the appointed time". It is thus seen that at least seven of the eleven genera of the *Bucerotidae* of the Indian countries and Great Sunda Islands have poor powers of flight: *Anthracoceros*, *Buceros*, *Dichoceros*, *Ptilolaemus*, *Berenicornis*, *Rhinoplax*, *Cranorrhinus*.

The flight of the wide-spread *Rhytidoceros* appears to be in marked contrast to these.

Davison writes of *Rhytidoceros subruficollis* and *undulatus*: these species "are remarkably strong on the wing, and morning and evening, where they occur, numbers may be seen flying far overhead, sometimes at such a height that they look not bigger than Crows. The strokes of their wings are accompanied by a peculiar metallic or resonant swish[1]) which can be heard at an incredible distance. One is often made aware of these birds flying far overhead by the

---

[1]) Bernstein (J. f. O. 1861, 114) tries to explain the noisy flight by the movement of the air in the large air sacks, an explanation which appears to us very far-fetched; neither does Mr. O. Grant's explanation (C. B. XVII, 347 note) appear satisfactory. In our opinion the sound need not be explained in any other way than the noise caused by the flapping of the wing of many other large birds.

sound of their wings, and on looking up, the birds are seen at such an immense height as to be only just distinguishable" (Str. F. VI, 113).

Our artist, Bruno Geisler, has furnished us with the following valuable notes on *Rhytidoceros*: "During my five-years' travelling and collecting in Ceylon, Java, and German New Guinea I had the opportunity almost daily of observing the flight of Hornbills, especially of *Rhytidoceros* in the last country. Although the bird is not common, its presence is perceived at long distances by its noisy flight. With powerful strokes they pass over the virgin-forest in the morning, usually in pairs, to certain fruit-trees, causing thereby a resounding noise which may be compared to the cutting of half-rotten wood with a big saw. On my frequent boat-trips to the Huon Gulf (New Guinea), when sailing not far from the coast, I was able to watch with ease the extended flight of this bird. Often there were a pair, more rarely a larger number (which then always flew in single file at intervals of about 3 m) flying onward at a height of 100—200 m above the forest, and in spite of the great distance the beating of their wings could still be heard. In like manner, when on the mountains, enjoying the splendid view of the flat coast-land, I have seen the birds flying over wide stretches until they vanished in the distance from my keen sight. On several occasions I had the opportunity of watching the birds come over the fore-lying hills, and then, describing a great bow from a height of about 500 m, they settled upon the trees on the coast. In settling they swoop downwards on outstretched wings with a loud whizzing noise towards the bough in view, always, however, just before reaching it, they make a curve rising up to their seat, when the male often gives utterance to a loud 'go go — gagaga' — as if of satisfaction. In the silent virgin-forest the noise of their flight brings to mind a locomotive-engine in quick motion. That the flight, although it cannot be called laborious, nevertheless taxes the birds' energies very much, may be seen from the circumstance that after a long flight the bird always settles with open bill and perceptibly out of breath.

After the breeding-season an interesting spectacle is afforded by the assembly on the common roosting-tree. After sundown all that have been scattered in the neighbourhood come singly, in pairs, or in groups, some approaching from a height, some bustling out of the thick forest to the tall, free-standing tree. Received with all sorts of ejaculations by those already arrived, the suitable sleeping-places are sought for amid frequent quarrels; and, especially on moonlight nights, the company rarely ceases its somewhat noisy conversation."

From the observations of Davison and Geisler the flight of *Rhytidoceros* appears to be so superior to that of the other Hornbills in the East Indies that the wonder is — not that *Rhytidoceros* ranges from N.W. India to the Solomon Islands without presenting new forms at every zoo-geographical step like other Hornbills, but that the ranges of its species are not even wider than is the case. At all events there is good reason to believe that the anomalous distribution of this Hornbill is a consequence of its fine flying-powers and habit of flying at great heights.

The habits of the Celebesian *Cranorrhinus*, which is rather nearly allied to *Rhytidoceros*, are thus described by Meyer *(5)*: "As one looks down on a forest from a high point, it appears to swim over the green foliage more majestically than any other bird of Celebes. Its flight is heavy, slow, and noisy, and audible from far away. Its cry is very lond, and not immediately to be distinguished from that of the Black Ape of Celebes[1]). They are often seen in pairs together. If the female is shot, the male returns to the spot after having flown away frightened by the shot; and therefore frequently male and female can be procured. On a tree they are very active, jumping from branch to branch; they are fond of fighting and are generally aggressive birds". Rosenberg *(a 14)* once obtained a female owing to this propensity, it being knocked down stunned by another of its sex. "They feed on forest fruits". Layard observed that *Anthracoceros coronatus*, in order to obtain its food when attached to a branch, "resorts to an odd expedient — the coveted morsel is seized in its powerful bill, and the bird throws itself from its perch, twisting and flapping its wings until the fruit is detached; on this the wings are extended, the descent arrested, and the bird regains its footing". A tame specimen had the parrot-like habit of using its bill in regaining its perch (Legge, B. Ceylon 274). In many Hornbills, especially in old examples, the cutting edges of the bill do not meet for a considerable space between the tip and the basal part.

The head of nearly all *Bucerotidae* plays ethnographically a great role throughout the East Indian Archipelago, as Mr. Pleyte has recently shown (Rev. d'Ethn. 1885, 313 and 1886, 464). Wilken (Med. Ned. Zend. gen. 1863, 133) remarked on the present species in the Minahassa, that a head-hunter, who has the intention of carrying out his dreadful work, fixes on his head half the head of a *Buceros*, and, after having been successful, he jauntily sticks an entire head of *Buceros* on his crown for every human head obtained in his expedition (cf. also Meyer's notes in "Negritos" vol. IX Publ. Ethn. Mus. Dresden 1893, p. 9). The same author tells us that the Macassars and Bugis bury the head of this bird unter the chief pole of a new house (Bijdr. taal, land en volkenk. Ned. Ind. 1889, 110).

[1]) Bernstein (J. f. O. 1861, 115) says of the Java species that it roars like a wild beast.

ORDER CORACIAE.

There is little agreement among systematists as to the value of the Bee-eaters, Kingfishers, Rollers, Goatsuckers, and Swifts, as orders or families of the Avian System. We prefer to take the three first groups as families of the *Coraciae*, placing the Goatsuckers and Swifts in a different order, the *Macrochires*, though it is possible that the Rollers (for instance, *Eurystomus*) may have certain affinities with the latter. In the *Coraciae* the bill is strong, often very large or long, the palate desmognathous, the wing short to moderately long; in the *Macrochires* the bill is generally very small, soft, and weak, the gape very deep and wide, the palate aegithognathous, or schizognathous, the wing very long.

FAMILY MEROPIDAE.

In the Bee-eaters the bill is long, thin, sharp, and slightly decurved; the foot has three toes in front and one behind, the tarsus is short, not longer than the hind toe and claw; the tail of moderate length, square, or forked, or with the two middle rectrices prolonged and with the projecting ends attenuated; the secondaries and inner primaries with a heart-shaped tip. The colours are very bright, green being most prevalent, also blue, black, yellow, red, and mixed hues occur. The birds breed in holes in the ground, laying white eggs; their food, consisting of *Hymenoptera* and other insects, is, we believe, generally, if not always, captured on the wing. The family is found in the temperate and tropical parts of the Old World.

For anatomical particulars: see Beddard in Dresser's "Monograph of the Meropidae", Introduction; and Sharpe, Cat. B. XVII 1892, 41, etc.

GENUS MEROPS L.

The typical Bee-eaters have the two middle tail-feathers of the adult lengthened and terminally attenuated, the wing is moderately long and pointed, the first primary very minute, the second the longest. Their powers of flight are great. Green, or yellow-green, is the most characteristic colour, except in two or three African species in which carmine-red predominates. Seventeen species, with a range nearly over the whole of the Old World, are admitted in the Catalogue of Birds, vol. XVII.

✢ 77. MEROPS ORNATUS Lath.
Australian Bee-Eater.
Plate VIII.

Merops ornatus *(1)* Lath., Ind. Orn. Suppl. 1801, p. XXXV; *(II)* Gld., B. Austr. 1840, II, pl. 16; *(3)* Jukes, Voy. "Fly" 1847, 157; *(IV)* Reichb., Hb. sp. Orn. Meropinae 1852, 68, t. 446, f. 2233—34; *(5)* Wall., P. Z. S. 1862, 335, 338; *(6)* Schl., Mus. P.-B. Merops 1863, 4; *(7)* Gld., HB. B. Austr. 1865, I, 117; *(8)* Brügg., Abh. Ver. Bremen 1876, V, 49; *(9)* Macleay, Pr. L. Soc. N. S. W. 1876, I, 37; *(10)* Sharpe, J. Linn. Soc. Zool. 1879, XIV, 686; *(11)* Meyer, Ibis 1879, 57, 145; *(12)* Salvad., Orn. Pap. 1880, I, 401; Agg. 1889, 52; Appendice 1891, 222; *(13)* Meyer, Verh. z.-b. Ges. Wien 1881, 763, 769; *(14)* W. Blas., J. f. O. 1883, 135; *(15)* Meyer, Isis, Dresden 1884, 6, 19; *(16)* Sharpe, Report Voy. "Alert" 1884, 21; *(16 bis)* Pleske, Bull. Ac. Petersb. 1884, 117; *(XVII)* Dresser, Monogr. Merop. 1884—86, 51, pl. XIV; *(18)* Guillem., P. Z. S. 1885, 503, 546, 566; *(19)* Rams., Pr. L. Soc. N. S. W. 1886, 1097; *(20)* id., ib. 1887, 166; *(21)* North, t. c. 441; *(22)* Rams., Tab. List 1888, 3; *(23)* Cox & Hamil., ib. 1889, 401; *(24)* North, t. c. 1024; *(25)* id., Nests & Eggs Austr. B. 1889, 34; *(26)* Meyer, Ibis 1890, 413; *(27)* Büttik., Notes Leyd. Mus. 1891, 210; *(28)* Sharpe, Cat. B. XVII, 1892, 74; *(29)* Meyer, Abh. Mus. Dresd. 1893, Nr. 3, p. 11; *(30)* Newton, Dict. B. 1893, 30; *(31)* M. & Wg., Abh. Mus. Dresden 1895, Nr. 8, p. 7; *(32)* iid., ib. 1896, Nr. 2, p. 11; *(33)* Hartert, Nov. Zool. 1896, 570, 586, 595; *(34)* id., ib. 1897, p. 164.

a. **Guêpier Thouin ou à longs brins** *(I)* Levaill., Hist. Nat. Guêpiers 1807, 26, pl. 4.
"**Korikeri**", near Manado (Nat. Coll. in Mus. Dresd.).
"**Tomonsi bowula**", Poling Id. (iid.).
"**Tumpiliwarata**", Karkellang, Talaut (iid.).

For further synonymy and references cf. Salvadori *12*; Dresser *XVII*; Sharpe *28*.

Figures and descriptions. Gould *II*; Dresser *XVII*; Levaillant *a I*; Reichb. *IV*; Wallace *5*; Salvadori *12*; Meyer *15*; North *21* (egg), *25* (egg); Sharpe *28*.

Adult. Head above, neck, mantle, and upper wing-coverts yellow-green, passing into yellow-chestnut on occiput and nape; and becoming blue-green on the scapular region; lower back and rump turquoise-blue; darker on upper tail-coverts; wings rufous hazel, outer webs of the primaries green, the primaries and secondaries broadly tipped with black, tertiaries green, washed with blue; tail black, washed with green and blue on the exposed webs, the shafts of the two middle feathers produced much beyond the others, with extremely narrow webs which widen slightly at the tip into a sort of spatule; lores, suborbital region and ear-coverts black; below this a malar stripe of turquoise-blue; chin ochre-yellow, more chestnut on upper throat; across lower throat a broad black band; breast and sides yellow-green, passing into light blue on the under tail-coverts; under wing-coverts and most of under side of quills pale rufous, becoming more dusky towards the ends of quills (ad. Tabukan, Gt. Sangi: Meyer — C 2441).

Sexes. Alike in coloration *(28)*.

Young. Differs from the adult in wanting the black band across the throat, in having the general plumage grey-green varied with tips of pale blue, instead of yellow-green, and in having the middle tail-feathers without lengthened spatules; bill shorter (Manado tua, Apr. 93, C 12206; Mantehage, Apr. 93, C 12207; Nat. Coll.).

Observation. A specimen in the Dresden Museum (Nr. 6870) "Moluccas" without a black patch on the throat has, nevertheless, very long spatules 40—50 mm. Another, in which the black throat-bar is very imperfectly formed (C 1855 — Sangi), has spatules of 45 mm length. In others, which are in adult plumage, the spatules vary much in length (C 2213 — Manado = 16 mm; C 2215 — Manado = 33 mm). Specimens with the crown varied with pale blue-tipped feathers amongst the green ones, have spatules of all lengths (in C 12203 — Mantehage = 8 mm; C 2220 — Togian ♀ 1 spat. = 13 mm, the other 25 mm; C 2030 — Manado = 20 mm; C 2218 — Manado = 19 and 21 mm; C 2221 — Manado = 27 mm; C 2212 — Manado = 41 mm; C 1855 — Sangi = 45 and 47 mm; C 2224 — Manado = 72 mm). In several of these cases the two middle rectrices have not attained their full growth, and the spatules rest upon the other rectrices, not reaching to their tip.

It may be seen that the blue feathers in the crown of specimens 2224 (spatule 72 mm), 1855 (spatule 45 mm), 2212 (spatule 41 mm), and others are worn and old, whereas the yellow-green feathers are fresh and perfect, showing that the birds are assuming a green crown. It might be inferred that these are all immature specimens in process of assuming the adult dress, and that the length of the spatules varies greatly in different individuals. After full consideration we find that this view must be rejected; specimens 1855 and 2224 are among those with the very longest bills in a large series, being about 5 mm longer than in young specimens, and it is inconceivable that such a bird as 2224, with one old spatule 72 mm long and a new one sprouting, can be a young bird. The primaries — such as are remaining — are old and very much worn, as if the bird had been engaged in incubation in its nesting-hole in the sand; we take it for a very old female and infer that in the breeding season the crown and under-parts in this sex are strongly washed with blue, much as in the young. Unfortunately, no positive observations on this subject have been, so far as we are aware, yet made.

Measurements. Wing 106—116 mm (Manado, ad. C 2211; Sangi, ad. C 2441 and others); tail, without spatule, 78 mm ca.; bill 26 (juv.)—31 (ad.).

Eggs. 4 or 5 (Gld. 7). Beautiful pearly white; circa 21.6 × 18.3 mm (Australia — North 25); "23 × 20: The eggs of all *Meropidae* are spherical, very glossy and white, when not dirtied by rotting matter in the nest, in which case they often appear clay-yellow, like the eggs of the *Podicipedidae*" (Nehrkorn MS.).

Nest. "The eggs are deposited and the young reared in holes made in the sandy banks of rivers or any similar situation in the forest favourable for the purpose. The entrance is scarcely larger than a mouse-hole, and is continued for a yard in depth, at the end of which is an excavation of sufficient size for the reception of the four or five beautiful pinky-white eggs" (Australia — Gould *II*, 7).

Breeding season. In New South Wales October, November, December (North 25).

Distribution. Throughout Australia as far as known; New Guinea; New Britain; west to Talaut and Sangi; Celebes and Lombok[1].

For exact localities cf. Salvadori *12*, adding Karkellang, Talaut (Nat. Coll. in Dresd. and Tring Mus.); Great Sangi (Fischer *8*, Meyer *15*); Siao (Nat. Coll.); Peling (ii d.); Samao (f. Büttikofer *27*).

Observation. Dr. Sharpe omits Java — as it appears to us with good reason. Borneo is included in its range by Mr. Dresser; but Mr. Everett omits it from his list of Bornean Birds (J. Str. Br. R. As. Soc. 1889).

---

[1] As to specimens from Timor, Lombok etc. see below.

Although there is reason to believe that some few individuals of this Bee-eater remain throughout the year in some or, perhaps, all of the East India Islands in which it occurs, it is in the main a migratory bird, visiting Australia in September, breeding there, and returning northward again in February. Dr. E. P. Ramsay *(19)* speaks of it as more nearly approaching truly migratory species than any other Australian bird, and, though we believe that field observers will still have a great deal to tell on the subject of migration, it is certain no form illustrates the phenomenon of coming and going with the seasons in this region so strikingly as the present conspicuous bird. Gould terms it strictly migratory, arriving in New South Wales in August and departing northward in March *(II)*. In the Mudgee District, N. S. W., Cox & Hamilton write *(23)* that it arrives by September 25th or later, commences to nest at once, and eggs may be taken from November till the end of December; it leaves in February, a few remaining till March. When the "Fly" was in the Torres Straits, Jukes was able to observe the passage of the bird on its journey from Australia to New Guinea, and also its return-journey: "While we were in this neighbourhood (near Mount Egmont Island in Torres Straits) about the end of February, great flocks of the Bee-eater which is common in Australia *(Merops ornatus)* were continually passing to the northward. The white pigeons, also ['*Carpophaga luctuosa*' = *bicolor* or *spilorrhoa*], were going in the same direction in numerous small flocks ... The Bee-eaters go as far to the southward as Sydney during the summer of New South Wales, but we never saw the white pigeons much to the southward of Torres Strait. In September, 1844, they were coming thickly from the northward to Endeavour Strait, and they seem to return in March" *(3)*.

Macleay *(9)* had a similar opportunity during the cruise of the "Chevert" in 1875 of watching the flight of *M. ornatus* across the Torres Straits: "It seems to commence its annual migration southwards as early as August. Throughout the early part of September, I observed, or heard, scattered flocks of from twelve to twenty of them passing the ship at all hours of the day and night, and making direct for the mainland near Cape York. They flew low, and with anything but a steady flight. I imagine their migration is a very slow and painful affair, for it is generally the month of November before they reach their breeding grounds on the Murrumbidgee".

Not all individuals make this transit to Australia, however. In German New Guinea our artist, Bruno Geisler observed *(29)* that individuals are always to be found, but, at the beginning of the year only, flocks of 20 to 30 occur. Sharpe *(10)* records Lawes' discovery of its nesting in South-east New Guinea, where, as in Australia, it lays its eggs in sand; and Mr. North *(21)* describes some eggs procured by Parkinson in New Britain, remarking that they are like Australian ones, but smaller. Dr. Ramsay *(XVII)* has also remarked that "specimens from Port Denison (N. E. Queensland) are somewhat smaller than those from New South Wales"; nevertheless North's subsequently recorded measurements of Australian specimens *(25)* do not bear out his former statement: New Britain 21.6 >< 19.5 mm, a set from Buldery (N. S. W.) average ca. 21.6 >< 18.3 mm. Count Salvadori *(12)* was led by North's remarks doubtfully to refer the New Britain birds to *Merops salvadorii* Meyer, a species, whose validity is still to be proved.

Turning to Celebes and the Moluccas, it appears that this Bee-eater occurs here almost but not quite, entirely in the months when it is absent in Australia — that is, in the Australian winter. Meyer *(11)* writes: "*M. ornatus* Lath. is only numerous during the east monsoon (April—October). Near Manado in May, on the Togian Islands in August". Fischer *(16bis)* says: "this Bee-eater does not occur as a stationary bird, but appears in Ternate in the months March, April and May". Nevertheless Count Salvadori *(12)* was able to record two specimens killed in Halmahera in December, though the remaining 62 of his series from New Guinea and the Moluccas are dated from March to September. It would thus appear that a few remain behind in the islands throughout the year; whether they breed there or not is not known.

Notwithstanding that this species is such a thorough migrant it appears very possible that it is undergoing differentiation into an eastern and a western race. Mr. Dresser has the following remarks on its variations: "Specimens which I have examined vary somewhat in intensity of coloration and especially as regards the amount of blue in the plumage. A specimen in the Tweeddale Collection from Port Albany has the crown, scapulars, and back slightly tinged with turquoise-blue; and one in the British Museum, a female from Dorey, has the abdomen washed with blue and is labelled *Merops caerulescens*. [See our remarks, *antea*, on the probable breeding plumage of the female.] Another example in the British Museum, a female from Lombock, is peculiar in having a tolerably broad band of blue below the black pectoral band . . . I can endorse Mr. Wallace's remarks [5] that specimens from the Sula Islands agree with those from Ternate in having more brown on the head and less blue on the breast than the Timor and Lombock specimens". Meyer *(15)* has since separated specimens from Sumba as a subspecies, *M. ornatus sumbaensis*. Two, apparently second-year birds, have more blue on the throat below the black cross-band than any of over 60 other specimens in the Dresden Museum from Celebes, Sangi, the Moluccas and New Guinea; in two others the amount of blue is inconsiderable. Hartert *(33)* enumerates specimens from Sumbawa, Sumba, and Lombock, some with and some without blue on the throat. This character consequently seems not to be of a positive nature. There can be little doubt that the Sumba birds correspond with those from Timor and Lombok, whose differences struck Mr. Wallace. It would be of much interest to know whether specimens from West Australia have more blue below the black collar than those of New South Wales, or other differences, for if *M. ornatus sumbaensis* be a valid subspecies, there is strong reason to suppose that it will be found to make its way to Australia by a different route from that of the great body which crosses Torres Straits in September and February. We suspect that all the birds of the Lesser Sunda Islands — Bali to Timor — may be identified with *M. ornatus sumbaensis* and that they make their way in due season by way of Timor and Rotti to N.W. Australia across the Timor Sea, which is not without a few small islands and reefs to serve as resting places.

As in *Prioniturus*, the bare shafts of the lengthened middle tail-feathers of adults are not the result of the wear and tear undergone by these feathers between the yearly or half-yearly moult of the species. But it is obvious that the webs of the feathers, and, especially, of the two long central tail-feathers, must get damaged and worn down considerably in a bird which burrows a sort of mouse-hole a yard deep in sand; the tail-feathers in particular will be rubbed and bent about, when the bird turns round in its nesting cavity at the end of its hole. In two or three specimens before us (C 2224, Nr. 13590, C 12205), the spatules seem to have suffered from this cause. In adult specimens, however, the spatules grow out of the shaft perfectly formed; the above mentioned specimen, C 2224, with one old spatule of 72 mm is getting a new one, the bare shaft of which is 30 mm long, while about 10 mm more of bare shaft at the base is still enclosed in the sheath out of which it is growing (see Plate VIII). In birds of the year the tail is square and simple, and it appears that in young specimens of the second year, as in *Prioniturus*, the middle tail-feathers are not greatly longer than the lateral ones, and the projecting part is usually not bare of web, but simply much narrowed and hollowed out a little behind the tip. Those with very long bare spatules are without doubt older birds. The case of the Bee-eaters, like that of *Prioniturus*, where we have discussed the matter more fully, appears to be a remarkable illustration of the inheritance of mutilations, an effect of wear and tear, continually repeated throughout countless generations, being ultimately reproduced in the offspring, the successive ontogenetic changes in which betray the gradual result wrought by the attrition of the sand on its more remote, and less remote, ancestors[1]).

---

[1] It is very possible that the peculiar heart-shaped double tips of the secondaries and of all the primaries (except the longest) owe their formation to the same cause. Has *Merops* the habit of supporting itself on its opened wings when it clings to the steep sand-bank in which it commences to dig its nest? Mr. Bruno Geisler, who observed *Merops* breeding in a river-bank in Java informs us that this is the case. House and Sand Martins also do so and it is of special interest to observe that the same curious quill-formation obtains in them as well. If our supposition is correct, the pressure on the feathers and the oscillation of the body of the pecking bird will force the ends of the webs apart from the shaft on one or both sides of it, forming a little notch, just as is seen if the tip of a perfect quill be tapped gently with the finger or rubbed on blotting paper or other rough substance. It is important to observe that the double-tips of the quills in *Merops* and other birds are arranged in what Mr. Keeler would term "Successional Taxology": — in the inner secondaries the outer web of the feathers has the shorter tip, in the middle of the wing the tips are about equal, in the outer secondaries and primaries the inner web has the shorter tip; in other words, where one tip is more exposed to friction than the other, it is shorter; where they are equally exposed, they are equal. Moreover, in the middle of the wing the secondaries have a very square appearance, as if they had been truncated by some means. We conceive that the wings of *Merops* are much less

The present species, like *M. apiaster* of Europe, deserves its name of Bee-eater. Gould remarks that its food consists of various insects, chiefly *Coleoptera* and *Neuroptera*; Cox & Hamilton observed that it is very destructive to bees. The same writers mention that it will plunge into water for a bath.

78. MEROPS PHILIPPINUS L.
Blue-tailed Bee-eater.
Plate VIII.

a. **Le Grand Guêpier des Philippines** *(1)* Briss., Orn. 1760, IV, 560, pl. 43; *(II)* Daubent., Pl. Enl. VI, pl. 57.

Merops philippinus *(1)* L., S. N. 1767, f. Wald.; *(2)* Blyth, J. A. S. 1846, 369; *(III)* Gld., B. As. 1855, I, pl. 36; *(4)* Schl., Mus. P.-B. Merops 1863, 2; *(5)* Beavan, Ibis 1867, 318; *(6)* Wald., Tr. Z. S. 1872, VIII, 42; *(7)* Meyer, J. f. O. 1873, 405; *(8)* Finsch & Conrad, Verh. z.-b. Ges. Wien 1873, 2, 8 (sep. copy); *(9)* Salvad., Cat. Ucc. Borneo 1874, 89; *(10)* David & Oust., Ois. Chine 1877, 72; *(11)* Hume, S. F. V, 1877, 18; *(12)* Hume & Davis., ib. VI, 1878, 67, 498; *(13)* Ball, ib. VII, 1878, 203; *(14)* Cripps, t. c. 258; *(15)* Tweedd., P. Z. S. 1878, 107, 282, 340, 709; *(16)* Meyer, Ibis 1879, 57; *(17)* Legge, B. Ceylon 1880, 306; *(18)* Vidal, Str. F. IX, 1880, 40; *(19)* Bingh., t. c. 152; *(20)* Butler, t. c. 381; *(21)* Nicholson, Ibis 1881, 143; *(22)* Kelham, t. c. 378; *(23)* Reid, Str. F. X, 1881, 21; *(24)* Davis., t. c. 1883, 350; *(25)* A. Müll., J. f. O. 1882, 396; *(26)* Oates, B. Brit. Burmah 1883, II, 66; *(27)* W. Blas., J. f. O. 1883, 135; *(28)* Vorderman, Ned. Tdschr. Ned. Ind. 1883, 47; *(XXIX)* Dresser, Monogr. Merop. 1884, pp. XVII, 55, pl. XV; *(30)* A. Müll., J. f. O. 1885, 155; *(31)* Guillem., P. Z. S. 1885, 503; *(32)* Hume, Str. F. XI, 1888, 42; *(33)* Hartert, J. f. O. 1889, 364; *(34)* Everett, J. Str. Br. R. A. S. 1889, 164; *(34 bis)* Tristr., Cat. Coll. B. 1889, 97; *(35)* Steere, List B. and Mamm. Philipp. 1890, 9; *(36)* Oates, Hume's Nests and Eggs 1890, III, 63; *(37)* Hartert, J. f. O. 1891, 296; *(37 bis)* Salvad., Anu. Mus. Civ. Gen. 1891, 47; *(38)* Sharpe, Cat. B. XVII, 1892, 71; *(XXXIX)* Meyer, Vogelskel. pt. XVII, 1892, pl. CLXII; *(40)* Mann. Ibis 1894, 60; *(41)* Büttik., Z. Erg. Weber's Reise Ost-Ind. 1893, III, 274; *(42)* Oust., Nouv. Arch. du Mus. 1893, 138; *(43)* Steere, Ibis 1894, 417; *(44)* Bourns & Worces., B. Menage Exped. 1894, 34; *(45)* Vorderm., N. T. Ned. Ind. 1895, LIV, 333; *(46)* Clarke, Ibis 1895, 474; *(47)* Blanford, Faun. Br. Ind., B. III, 1895, 111; *(48)* M. & Wg., Abh. Mus. Dresd. 1896, Nr. 1, p. 8; *(49)* Sharpe, Br. B. 1896, II, 57; *(50)* Jesse, Ibis 1896, 190, 195; *(51)* Hart, Nov. Zool. 1896, 550, 595.

b. **Le Guêpier à queue d'azur ou le Guêpier Daudin** *(1)* Levaill., N. Hist. Guêpiers 1807, 49, pl. 14.

c. **Merops javanicus** *(1)* Horsf., Tr. L. S. 1821, XIII, 171.

d. **Merops daudeni** *(1)* Cuvier; *(2)* Hume, Str. F. II, 1874, 162; *(3)* id., III, 1875, 49.

e. **Merops philippinus** var. **celebensis** *(1)* W. Blas., Ztschr. ges. Orn. 1885, II, 239; 1886, 88,

violently used than is the tail of the Woodpecker, for the rectrices of the latter have the webs tapering off to whalebone-like spines, which are used as a powerful prop to the bird during its pecking and excavating work. At the same time it should be pointed out that a somewhat similar form of quills occurs in some other birds, such as *Carduelis*, where a corresponding cause for their formation cannot, perhaps, be supposed.

254   Birds of Celebes: Meropidae.

"**Burong langir**" (bird of high flight), Minahassa (Meyer *16*).
"**Tguru**", Macassar and Tjamba (Platen *e 1*).
For further synonymy and references see Dresser *XXIX*; Sharpe *38*.
Figures and descriptions. Daubenton *a II*; Levaillant *b I*; Gould *III*; Dresser *XXIX*; Meyer *XXXIX* (skeleton); Legge *17*; Oates *26*; W. Blasius *e 1*; Sharpe *38*; Blanford *47*.
Adult. Above olive-green, scapulars washed with greyish blue; quills — double-tipped — green, broadly tipped with blackish (except the innermost ones) washed with greyish blue on terminal part; rump and upper tail-coverts glaucous blue; tail duller greenish blue, the two middle rectrices much longer than the others, the projecting part much narrowed, but never bare of web, extreme ends slightly widened; above eyes and on forehead at base of bill a streak of turquoise-blue; a submalar streak of the same colour; between them incl. lores, subocular region and ear-coverts black; chin ochre-yellow; throat rufous hazel; upper breast yellowish green, becoming buffy olive-grey on the sides and abdomen; under tail-coverts very pale greyish blue; under wing-coverts and quills towards their bases rufous (♀ Macassar, Platen. 25. I. 78, — C 5367). "Iris blood-red; bill and feet black."
Sexes. Alike (Sharpe *38*, Legge *17*).
Young. The two middle tail-feathers very little longer than the others; all the feathers of the green parts of the upper surface narrowly margined with greyish or bluish; the reddish on the throat much less pronounced; the blue supraloral and malar streaks wanting (Macassar, Oct. 1871, Meyer, C 2261). "Iris dull red, or brownish red" (Legge).

Measurements.

	Wing	Tail without spat.	Bill from nostril
a. (C 5367) ♀ ad. Macassar	130	95	—
b. (C 3511) imm. N. Celebes	128	98	32
c. (C 2263) ♂ imm. Limbotto, N. Cel.	125	99	31
d. (C 3512) imm. N. Celebes	123	95	30
e. (C 3512) imm. Manado	128	90	—
f. (C 2261) ♂ juv. Macassar	125	95	27
g. (Schlüter) ♂ imm. S. Celebes	130	96	34
h. (Schlüter) imm. S. Celebes	126	91	—
Six specimens from the Philippines	121–131	90–100	29–35.5
C 2264 ♂ ad. Singapore, Dec.	132	94	33
C 4853 imm. Calcutta	124	90	32
C 6152 ad. Kupang, Timor	133	91	36

Skeleton.

Length of cranium (fr. sin. nasofr.)	20.5 mm	Length of fibula	14.0 mm
Greatest breadth of do.	17.7 »	Length of tarso-metatarsus	10.0 »
Length of humerus	29.7 »	Length of sternum	34.0 »
Length of ulna	37.7 »	Greatest breadth of do.	17.6 »
Length of radius	35.5 »	Height of crista sterni	10.0 »
Length of manus	30.5 »	Length of pelvis	26.0 »
Length of femur	18.0 »	Greatest breadth of do.	18.0 »
Length of tibia	28.0 »		

Eggs. Usually 4 or 5 (Hume *36*). Spherical, very glossy and white. 23.5 to 24.5 by 20 mm (Nehrkorn MS.). 20.8—24.6 × 17—21.6 (Hume *36*).

Nest. "Breeds in holes in banks. The holes are rarely less than 4 feet deep, and I have known them to extend to 7 feet [often over 7 feet — Bingham *XXIX*]. In diameter they vary from 2—2½ inches" (Hume *36*).

Breeding season. On the Irrawaddy at the end of April (Oates *36*); Nerbudda River, India, by the 1st April (Nuna *36*); Mahanuddy River, India, in May (Blewitt *36*); Lahore, Punjaub. in June (Marshall *36*); Kaukarit River, Tenasserim, in April and May (Bingham *36*).

Distribution. Almost all parts of India (Dresser *XXIX*; Legge *17*); Ceylon (Legge *17*); Andaman Islands (Beavan *5*); Nicobar Islands (Blyth *2*, Davison *d 2*); Burmah (Oates *26*); Tenasserim (Bingham *19, XXIX, etc.*); South China (David *10*); Cochin China (D. & O. *10*; Pierre *38*); Malay Peninsula (Cantor *38*, Davison *38, etc.*); Singapore (Kelham *22, etc.*); Sumatra (Raffles *9*, Hartert *33*, Modigl. *37bis*); Java (Horsf. *9*, Boie *9*, H. O. Forbes *21*, Vorderman *28, etc.*); Bali (Doherty *51*); Lombok (Vorderman, *etc. 9, 45, 51*); Sumbawa (Guillem. *31*); Flores (Wall. *38*); Timor (Wall. *34bis, 38*), Riedel in Dresden Mus.); Celebes — Macassar (Meyer *7, 16*, Conrad *8*, Platen *e 1*), Tjamba (Platen *e 1*), Parc Parc (P. & F. Sarasin *48*), Maros and Luwu (Weber *41*), Limbotto in Gorontalo Distr. (Meyer *16*), Minahassa (v. Musschenbr. & Faber in Dresden Mus.); Borneo (Schwaner, *etc. 4, 9, 34*); Philippine Islands (Cuming *9*, Everett *15, 38*, Steere *etc. 35, 43, 44, 46*).

In addition to these localities Mr. Nicholson *(21)* records two specimens captured at sea by Mr. H. O. Forbes in the Indian Ocean a long way east of the Maldive Islands, viz. one, on Nov. 7th, in 3°26' N. by 17°48' E. (misprint, apparently, for 77°48' E.); the other, on Nov. 9th, in 1°24' N., by 76°43' E.

In his excellent article on this Bee-eater, Mr. Dresser proves from ample quotations from the observations of naturalists that it is in the main a migratory species. The results arrived at may be briefly recapitulated with one or two additional observations as follows:

Oudh and Kumaon: Hot season (May—Sept.), but not in any numbers (Irby).
N. W. Provinces: Arrive November; scarce, June; all gone July, August (Bingham).
East Bengal: Arrive in February, breed in July and August, after which they disappear (Cripps).
Lower Bengal: Chiefly, or only, during the rainy season (summer), (Blyth).
Chota Nagpur: Hot season (Ball).
Central and East India: Cool season.
Calcutta district: Rainy season — June to August; never very plentiful (Munn *40*).
India, W. Coast (Virgola): numbers in January, by April not one to be seen (Bingham).
Ceylon: Arrives in September, leaves in March—April (Legge).
Pegu: Constant resident, but in the rains comparatively few (Oates). Breeds in thousands on the Irrawaddy in April—May (Jordon).
Tenasserim: Partially migratory; appears after the rains and vanishes by the end of May following (Bingham).
South China: Summer visitor (David & Oustalet).
Singapore: Arrives in September; no mention of its occurrence except during the wet season (Kelham).
Celebes: In the Minahassa plentiful only in the dry season during the east monsoon (April—Oct.); rare in the west monsoon (Oct.—April) (Meyer).

From the Philippine Islands Lord Tweeddale *(15)*, Meyer *(16)*, Dr. Sharpe *(38)* and Dr. Steere *(35)* record dates sufficient to show that it occurs there all the year round; in all probability, however, its numbers fluctuate with the season.

This species appears to be much rarer in Celebes than *M. ornatus*. Large numbers of the latter have been sent to the Dresden Museum in two collections formed by native hunters in April—May and August—September, but not one of *M. philippinus* was among them; neither does it seem to have been obtained there by Wallace, Forsten, S. Müller, v. Rosenberg, Fischer, Everett, or Doherty. The Celebes form was separated in 1885 by W. Blasius as var. *celebensis*, on the ground that it differs in wanting a sharp boundary between the red-brown throat on the remaining underparts, and by having a darker, much more olive-brown back and head, and the blue colour of the body much less developed *(e 1)*. Four specimens from South Celebes before us (two of them already discussed l. c. by Prof. W. Blasius) bear out these conclusions; but four others from North Celebes do not, but are like ones from the Philippines and elsewhere, though the North Celebes birds before him were included under var. *celebensis* by Blasius, and an adult female from S. W. Celebes in the Sarasin Collection does not seem to differ from a specimen from Luzon. The validity of the subspecies is not found admissible by Dresser *(XXIX)* or by Sharpe *(38)*; and, in view of the migratory habits of the species as a whole, and the evidence of the specimens before us, we think it hardly possible that the differences pointed out can be of a racial description.

Birds of the year, as already remarked, have the central rectrices but little longer than the lateral ones; but, as Legge says, after arriving in Ceylon in September they quickly acquire the adult tail *(17)*. The narrowed feathers grow out of the sheath perfectly formed from the first, but in this species they never appear to attain to the same length and bareness often found in *M. ornatus*, and, as compared with that species, are in a less advanced state of development. This may have to do with the circumstance that *M. philippinus* makes a larger, though usually much deeper nesting-hole; it is described as being 2 to 2½ in. in diameter, or as large as that of a rat, while that of *M. ornatus* is said by Gould to be like a mouse-hole. "The egg-chamber", says Major Bingham *(XXIX)* "is proportionally larger than that of the smaller species [*M. viridis*], and is, unlike theirs, sometimes lined with a little grass, a few feathers, or the wings of white ants."

Perhaps its greatest known breeding-quarters are the banks of the Irrawaddy. We do not know of any record of its nesting in the East Indies, though it occurs in the Philippines and perhaps in Celebes and elsewhere all the year round.

M. philippinus preys, according to Jerdon on wasps, bees, dragon-flies, bugs, and even on butterflies *(XXIX)*, which it captures on the wing. All observers seem to have been struck by its rapid, soaring, flight, whence its name in North Celebes, "Burong langir" — the bird which flies up very high.

The nearest ally of *M. philippinus* is perhaps *M. ornatus*, which is placed next to it both by Dresser and by Sharpe. Both are migratory and one may be broadly said to represent the other according as they occur north or south of the equator, and it is conceivable that their differentiation into two forms may have been influenced by the circumstance that the breeding seasons in India and Australia are separated by an interval of half a year.

GENUS MEROPOGON Bp.

This genus has the two middle tail-feathers prolonged and attenuated as in *Merops*, but is easily distinguished by its possessing a gorget of broad, lengthy feathers, and by a very different wing-formula: the first quill is about half the length of the longest, the second equal to the sixth, the third, fourth and fifth the longest. It is most nearly allied, apparently, to *Nyctiornis*, which has the tail square, and a curious groove along the ridge of the culmen. *Meropogon* is known only by one species from the Minahassa, North Celebes.

79. MEROPOGON FORSTENI Bp.

Forsten's Bee-eater.

Meropogon forsteni *(1)* Bp., Consp. 1850, I, 164 (ex Temm. MS.); *(2)* Reichb., Hb. spec. Orn., Merops, 1851, 80; *(3)* Bp., Consp. Vol. Anisodact. 1854, 8; *(4)* Wall., Ibis 1860, 142; *(5)* id., Malay Archip. 1869, I, 429; *(6)* Wald., Tr. Z. S. 1872, VIII, 42, 111; *(VII)* Gld., B. Asia I, pl. 39 (1873); *(8)* Meyer, J. f. O. 1873, 405; *(8bis)* Hartlaub in Neumayer's Anleitung 1875, 475; id. ib. 2nd ed. 1888, 395; *(8ter)* Salvad., Ann. Mus. Civ. Gen. 1875, VII, 655; *(9)* Meyer, Ibis 1879, 58; *(X)* id., Vogelskel. pt. 1, 1879, 3, pl. V; *(11)* W. Blas., J. f. O. 1883, 135; *(XII)* Dresser, Monogr. Merop. 1884, 15, pl. IV; *(13)* Platen, Gefied. Welt 1887, 230; *(14)* Sharpe, Cat. B. XVII, 1892, 41, 87; *(15)* Meyer, Vogelskel. 1892, pt. XVII, 34; *(16)* M. & Wg., Abh. Mus. Dresden 1895, Nr. 8, p. 7; *(17)* iid., ib. 1896, Nr. 1, p. 4.

a. **Merops forsteni** *(1)* Schl., Handl. d. Dierk. 1857, 310; *(2)* id., Dierent. 1872 p. 53 fig.; *(3)* id., Mus. P.-B. Merops 1863, 8; *(4)* Finsch, Neu Guinea 1867, 160; *(5)* Meyer, J. f. O. 1871, 231; *(6)* v. Musschenbr., N. T. Ned. Ind. 1876, XXXVI, 382; *(7)* Rosenb., Malay. Archip. 1878, 272.

b. **Pogonomerops forsteni** *(1)* Cab. & Heine, Mus. Hein. 1860, II, 132, note.

c. **Nyctiornis (Meropogon) forsteni** *(1)* Gray, HL. 1869, I, 98.

d. **Nyctiornis forsteni** *(1)* Brügg., Abh. Ver. Bremen 1876, V, 49.

Figures and descriptions. Gould *VII*; Dresser *XII*; Meyer *X, 15* (skeleton); Schl. *a 3*; Walden *6*; Brüggemann *d 1*; Sharpe *14*.

Adult. General colour above, including wings and the two lengthened middle rectrices, parrot-green, becoming dull bluish on the narrowed projecting part of the two middle rectrices, the ends and shafts of the remiges blackish; the lateral rectrices rufous chestnut, the outer webs of the outermost ones and the outer margins of the others green; forehead, crown, cheeks, chin, throat and chest dark hyacinth-blue, the gorget-feathers much elongated, broadened, and pendant; remaining parts of head and neck dull maroon; abdomen, flanks and sides dusky, washed with green;

under tail-coverts hazel margined with green; under wing-coverts silky whitish; basal part of quills below fawn colour. (Minahassa, v. Musschenbr. Nr. 1973). ♂, "Iris brown; bill black; feet blackish grey" (Platen — fide Nehrkorn in lit.).
Sexes. Similar in coloration.
Nest. A burrow in the ground (see below).

Measurements.

	Wing	Tail (lateral rect.)	Tarsus	Bill from nostril
a. (1973) ad. Minahassa	115	107	11.5	34
b. (2208) ad. "Celebes" (Frank)	114	105	11.5	36
c. (C 13896) ♀ ad. Rurukan, 24. III. 94 (P. & F. S.)	117	108	11	36
d. (Sarasin Coll.) ♂ ad. Rurukan, 16. III. 94	117	114	—	37.5
e. (Sarasin Coll.) ♀ Rurukan, 18. III. 94	113	109	—	37.5
f. (Sarasin Coll.) ♀ ad. Rurukan, 16. IV. 95	118	110	—	38

In specimen, b, the tail is square, but the two middle rectrices are wanting[1]).

Skeleton.

Length of cranium	69.0 mm	Length of tibia	26.0 mm
Greatest breadth of cranium	20.5 »	Length of tarso-metatarsus	11.8 »
Length of humerus	31.8 »	Length of sternum	34.4 »
Length of ulna	39.5 »	Greatest breadth of sternum	12.5 »
Length of radius	36.7 »	Height of crista sterni	11.0 »
Length of manus	30.3 »	Length of pelvis	24.0 »
Length of femur	18.6 »	Greatest breadth of pelvis	18.7 »

Distribution. Minahassa, Celebes: Tondano (Forsten *a 3*), Rurukan (P. & F. Sarasin), near Rurukan and Langowan (Meyer *a 5, 9*), Mt. Klabat (Platen *13*), Mt. Masarang (P. & F. S.).

This beautiful Bee-eater, one of the chief treasures of the Celebesian Ornis, was discovered in 1840, by Forsten, who obtained a single male near Tondano at an elevation of 2000 ft. Subsequently, it was sought for in vain by Wallace and von Rosenberg, and doubt was felt as to its Celebesian origin, but in 1871 it was rediscovered by Meyer in the rich virgin forest of the high country near Rurukan, and on the way from Langowan to Panghu, places in the neighbourhood of the Tondano lake. The bird appeared to be not rare, but restricted to certain localities. Since this, specimens have been sent to Europe by Fischer and van Musschenbroek; while Dr. E. Oustalet, as Mr. H. E. Dresser says, purchased several examples from a plumassier, by whom they "would have been cut up for plumes, had not Mr. Oustalet fortunately rescued them from so sad a fate" *(XII)*. Dr. Platen *(13)* describes his encounter with a troop of twenty or thirty specimens of this rare bird near a swampy pool in the virgin forest on the S.W. side of Mount Klabat; they shot past him noiselessly on motionless extended wings, to vanish quickly. and return again, when several fell to his gun. Unfortunately they were moulting. One of Platen's specimens in the Nehrkorn Museum is dated 13. I. 89; thus we know when the species moults. What is known of Forsten's Bee-eater tends to prove that it is a forest-haunting bird of sociable habits, and so far it has been recorded only

---
[1]) By a *lapsus calami* Dr. Sharpe states that the *Meropidae* have "ten rectrices" (Cat. B. XVII, 41). The proper number is twelve.

from a considerable altitude (2000 ft and more), where the several points of its discovery are almost within sight of one another.

The Sarasins paid much attention to this species and succeeded in discovering its breeding habits. They write (in lit. 13. July, 1895): "On 16th April we received nestlings when at Tomohon. The hunter described the nest as follows: On the earth-embankments round the small crater-lake on Mt. Masarang was a small entrance-hole, which led into a passage of over a meter long, this opened at the end into a chamber the size of a cooking-pot, filled with remains of food (ejectamenta), elytra of beetles, insects' wings, etc. He brought a great mass thereof back, of which we send you two glass cylinders full. In this mass he found 2 young ones; one must have been dead already for a long time, for it was full of maggots and yet not turned out of the nest, the other was alive and bored continually into the mass of ejectamenta, in order to hide itself. Remains of egg-shells were also in the nest. We had already often remarked that all examples of *Meropogon* had clay and dirt on the crest-feathers; this manifestly comes from its flying into holes in the ground".

As to the food of the species, shown by the ejectamenta sent by the Drs. Sarasin, Dr. Heller found: "The bulk of the insect-remains is formed of fragments of *Hymenoptera*, and indeed these consist principally of the wings, clypei, single segments of the body and the pollenigerous tarsi of a species of *Apis* (probably *indica* F.), also of the wings and clypei of a species of *Dielis* (perhaps *javanica* Lep.). Besides other *Hymenoptera*-remains, not to be determined at once, remains of beetles are tolerably abundant, for instance, of *Euchlora* (a form almost as big as a Cockchafer), *Macronota sp.*, *Glycyphana regalis* Voll., and others".

Touching the generic characters and nearest affinities of *Meropogon*, Lord Walden remarks (6): "this species has the first primary half the length of the second, which is a little shorter than the third. The third and fourth are longest and equal[1]. The fifth is somewhat shorter than the third and fourth but longer than the second. In the structure of the wing therefore it differs from both *Merops* and *Melittophagus*, but agrees with *Nyctiornis*[2]. The grooved culmen of *Nyctiornis* is not present; but a shallow channel extends from the base of the maxilla, on both sides of the culmen, for two-thirds of its length . . . . The rectrices are truncated as in *Nyctiornis*, but the middle pair are elongated, as in *Merops*, and closely resemble in form and proportion those of *M. philippinus*. The feet are those of the family. The elongated pectoral plumes resemble in character those of *Nyctiornis*. Altogether *M. forsteni* may be regarded as a link uniting *Nyctiornis* to *Merops*, but most nearly allied to *Nyctiornis*".

---

[1] It would be more correct to say the third, fourth and fifth about equal and longest, the first about half their length, the second equal to the sixth. A feather must have been missing in Walden's example.
[2] In *Merops* the first primary is abortive, and the second and third more or less nearly equal and longest. Lord Walden remarks that the *Melittophagus*, which most nearly resembles *M. forsteni* in the graduation of the quills and the form of the rectrices, the middle pair excepted, is *M. bullockoides* Smith.

Comparing the skeleton of *Meropogon forsteni* with that of *Merops philippinus*, Meyer *(15)* points out that: in *M. forsteni* the crista sterni does not rise sharply, there is a fenestra (no incisura) intermedia, occipital protuberance less strongly expressed, processus orbitalis posterior more strongly developed, 7 ribs (in *philippinus* 6).

In the genus *Merops* the nearest known ally of *Meropogon* is probably *Merops breweri* of Gaboon, West Africa, between which species and *Nyctiornis athertoni* of India as far as Siam and Tenasserim, *M. forsteni* is placed by Dresser and by Sharpe; but we cannot agree with Mr. Wallace that *M. breweri* is the "only near ally" of *Meropogon (5)* and Mr. Wallace himself appears to have abandoned this opinion in his later work "Island Life" (1880, 430). Its nearer allies are no doubt *Nyctiornis athertoni* and *amictus*.

The geographical distribution of East Indian *Meropidae* affords an interesting case almost parallel with that of the Hornbills. East of Celebes, as has been shown, there occurs only one Hornbill, a species, ranging from the Moluccas to the Solomons, belonging to a genus of exceptionally good flying-powers, which has in all probability extended its range by flight. In Celebes and the countries to the west occur numerous genera of Hornbills with weaker wings and, consequently, poor powers or indolent habits of flight; and these genera are of narrow geographical range. Similarly, among the *Meropidae* there is found to the east of Celebes only one *Merops*, *M. ornatus*, a species belonging to a genus which, like the wide-spread Hornbill, *Rhytidoceros*, is excellently fitted, by the form of its quills and habits of flight, to making its way on the wing to new territories. Many species of *Merops* are migrants, and *M. ornatus* makes its way annually from Celebes eastward to Australia. It cannot, therefore, be doubted that this species extended its range by flight; it is indeed conceivable, as has been already remarked, that some members of *M. philippinus*, visiting the East India Islands about the equator during the cold season, made their way to the south, instead of returning north, and so became gradually adapted to a different breeding season, thus cutting themselves off from all intercourse with the stock from which they sprang. *Meropogon* and *Nyctiornis* correspond with the ill-flying species of Hornbills, and, like them, extend to Celebes, but no further. Actual observations on their flight are still rather a desideratum, but their wings are much blunter and more rounded than that of that genus, and they appear to be of much quieter habits of flight than *Merops*. Thus Hodgson writes of *N. athertoni*: these birds "seek the deepest recesses of the forest, and there, tranquilly seated on a high tree, watch the casual advent of their prey, and having seized it return directly to their station. They are of dull staid manners, and never quit the deepest recesses of the forest" (Dresser, p. 9). Dr. Hagen refers to *Nyctiornis amicta* in Sumatra as a bird which "sits quiety and motionlessly on a bough above us" (T. Ned. Aard. Genoots. 1890, 2. ser., VII, 141). This should be contrasted for a moment with Dr. Jerdon's account of *M. philippinus*: "the flight of this Bee-eater is very fine and powerful, now dashing onwards with rapid strokes, and a velocity that can beat that of a dragon-fly, having captured which, it flaps along in more measured time, now and then soaring with outspread wings"; — or with Col. Legge's: "the Blue-tailed Bee-eater congregates in large flocks on the wing, dashes to and fro for hours together, ascending to a great height in pursuit of its prey, and keeping up its not unpleasant notes without intermission" (Dresser, pp. 59, 60). Other observers write similarly. The genus *Merops* is consequently as valueless as *Rhytidoceros* for questions connected with geographical distribution; *Nyctiornis* and *Meropogon*, on the other hand, may well be species which by reason of their habits have remained quietly settled in their peculiar habitats since subsidences of land took place and the sea gradually worked its channels. Like as was seen to be the case among the ill-flying Hornbills, Borneo, Sumatra and Malacca have one species of *Nyctiornis*, viz. *N. amictus*, in common; while Celebes has its own peculiar genus, *Meropogon*, a form perhaps of equal value with the Hornbill, *Rhabdotorrhinus*. To make the sequence perfect, Java and the Philippines should also each harbour a corresponding genus; it is, of course, still possible that such will be found there.

FAMILY ALCEDINIDAE.

Perhaps the best means of distinguishing a Kingfisher (except from the Bee-eaters) is furnished by its foot, which is small and feeble with the anterior toes united for the greater part of their length, only the distal phalanx (circa) and the claw being free, the inner toe small, in one genus quite abortive, in others absent. The bill is straight, usually pointed, in a few genera having a terminal hook to the culmen, often very long, never appreciably shorter than the large head, the nostrils linear, exposed. more or less covered above with a membranous operculum, and situated in the side of the maxilla, often overreached by the frontal plumage. The wing is generally short and rounded, but sometimes of a fair length; the tail variable; the legs very short and weak. Blue is the most characteristic colour, occurring as it does — though not in all genera or species — in very pure and brilliant tints; rufous, purple, brown, black, white, and green are also frequent colours, but pure red and especially pure yellow are very rare. The smallest Kingfishers of the genus *Ceyx* are less than a Sparrow in size, the largest, *Dacelo*, as large as a Crow. The family is cosmopolitan in its distribution.

For some anatomical peculiarities see Blanford, Faun. Br. Ind. B. III, 1895, 118; Cunningham, P. Z. S. 1870, 280.

The Kingfishers prey upon fish, insects, reptiles, etc. — some genera exclusively on fish. They have the habit of watching for their prey. They nest in holes, usually in the ground, and lay white eggs.

262   Birds of Celebes: Alcedinidae.

The Celebesian area is very rich in Kingfishers, having 7 genera, two of which, *Ceycopsis* and *Cittura*, are peculiar to it.

GENUS ALCEDO L.

With four toes, the inner one about as long as the hallux; tail very short, shorter than the bill; bill long, straight, rather slender and narrow, higher than broad at base across the nostrils; wing short, the 2nd, 3rd and 4th quills longest, 1st but little shorter, secondaries about $^{1}/_{3}$ their length. Birds somewhat larger than a Sparrow, with the rump and much of the back bright blue, the head barred with blue and black.

Sharpe recognises 10 species, distributed "over the greater part of the Old World, excepting Australia and Polynesia. Absent in the New World". There are three species in Celebes, one probably a migrant.

+ 80. ALCEDO ISPIDA L.
Common Kingfisher.

Alcedo ispida Linn., *(1)* Naum., Vög. Deutschl. 1826, V, 480, pl. 144; *(II)* Gld., B. Eur. 1837, II, pl. 61; *(III)* id., B. Gr. Brit. 1870, II, pl. 10; *(IV)* Sharpe, Monogr. Alcedin. 1870, I, pl. I; *(V)* Fritzsch, Vög. Eur. 1871, pl. XIV, fig. 1; *(6)* Shelley, B. Egypt 1872, 164; *(7)* Hume, Str. F. 1873, I, 168; *(VIII)* Dresser, B. Eur. 1875, V, 113, pl. 290; *(9)* Newton's ed. Yarrell's Brit. B. 1881, II, 443; *(10)* Liebe, J. f. O. 1883, 286; *(11)* Seeb., Hist. Brit. B. 1884, II, 341; *(11bis)* Sharpe, Ibis 1886, 166; *(12)* St. John, Ibis 1889, 157; *(12bis)* Seeb., B. Japan. Emp. 1890, 175; *(13)* Sharpe, Ibis 1891, 110; *(14)* id., Cat. B. XVII, 1892, 141; *(15)* Sclat., Ibis 1892, 573; *(16)* Meyer & Helm, Jb. Orn. Beob. Sachsen VI, 1892, 85; *(17)* Seeb., Ibis 1893, 51; *(18)* Newton, Dict. B. 1893, 485; *(19)* Grant, Ibis 1894, 409; *(20)* Bourns & Worces., B. Menage Exp. 1894, 34.

a. **Alcedo bengalensis** Gm., *(1)* ? S. Müll., Reiz. Ind. Arch. 1858, II, 8; *(II)* Gld., B. Asia 1802, I, pl. 53; *(III)* Sharpe, Monogr. Alcedin. 1870, 11, pl. 2; *(4)* Swinh., P. Z. S. 1871, 347; *(5)* Salvad., Cat. Ucc. Borneo 1874, 92; *(6)* David & Oust., Ois. Chine 1877, 74; *(7)* Meyer, Ibis 1879, 64; *(8)* Legge, B. Ceylon 1880, I, 292; *(9)* Salvad., Orn. Pap. 1880, I, 407; *(10)* Meyer, Isis, Dresden 1884, 6; *(11)* Guillem., P. Z. S. 1885, 504, 547; *(12)* W. Blas., Ornis 1888, 307, 570; *(13)* Everett, J. Str. Br. R. A. S. 1889, 158; *(14)* Sharpe, Ibis 1890, 18, 253; *(15)* Whitehd., t. c. 45; *(15bis)* Oates, Hume's Nests & Eggs 1890, III, 1; *(16)* Styan, Ibis 1891, 325, 483; *(17)* Salvad., Ann. Mus. Civ. Gen. 1891, 47; *(18)* De La Touche, Ibis 1892, 479; *(19)* Hose, ib. 1893, 408; *(20)* Munn, ib. 1894, 56; *(21)* Everett, ib. 1895, 30.

b. **Alcedo ispida bengalensis** *(1)* Tacz., Faun. Orn. Sib. Orient. 1891, I, 194; *(2)* Hartert, Nov. Zool. 1894, 480; *(3)* id., ib. 1896, 550, 571; *(4)* id., ib. 1897, p. 160.

For synonyms and references see Sharpe *14*.

Figures and descriptions. Naumann *I*; Gould *II, III*, a *II*; Sharpe *IV, 14*; Fritzsch *V*; Hume *7*; Dresser *VIII*; Legge a *8*; Salvad. a *9*.

Adult. General colour above greenish blue, greener on the mantle, each feather on head above and hind neck crossed with a subterminal bar of pale blue, wing-coverts tipped with the same colour; a line down middle of back, rump turquoise-blue, becoming more cerulean on the upper tail-coverts; tail duller blue; malar stripe continued to shoulder, greenish blue washed with brighter blue; ear-coverts cinnamon-rufous, paler on sides of neck; chin and throat white, washed with buff, rest of under surface cinnamon-rufous, darkest on the breast (Manado, Celebes, March 1871 — Nr. 6288).

Adult female. Similar to the male, but rather duller and greener, and distinguished by having the basal half of the lower mandible red (Sharpe *14*).

Young. Similar to the adults, but much more dingy in colour, and always recognisable by the ashy colour which overspreads the fore neck and breast, all the feathers being edged with dull ashy (Sharpe *14*).

Measurements. (Nr. 6288) ad. Manado, Celebes, wing 70, tail 31, bill from nostril 32, tarsus 9 mm.

Eggs. 5 to 7; roundish oval, pure white (unblown pinkish). very smooth and glossy; 19—22 ×16.5—18.3 mm (Dresser *VIII*; Hume *a 15bis*).

Nest. Digs a hole in a bank, usually overhanging water, 2 to 3 inches in diameter, and 1½ to 3½ feet deep, a nest of ejected fish-bones gradually formed at the end (Dresser, etc.).

Breeding season. In Europe commences in April (Dresser *VIII*); in India, January or March to June, varying according to locality (Hume *a 15bis*), Calcutta district, July and August (Munn *20*).

Distribution. Nearly the whole of Europe as far as about 60° N.; North Africa (in part only in winter); across Northern Asia as far as the Japanese Islands, throughout China, the Burmese countries and the Indian Peninsula (Meyer & Helm *16*; Sharpe *14*); Malay Peninsula to Borneo, Java, the Philippines (Salvad. *a 9*, B. & W. *20*, Everett *19, a 21*); Talaut Is. — Karkellang (Nat. Coll.); Great Sangi (Hoedt, Platen *a 12*); Siao (Hoedt *a 12*); North Celebes — Manado (Meyer *a 7*), Kema (Guillemard *a 11*); Halmahera; Ternate; Timor (Salvad. *a 9*).

The Common Kingfisher has rufous ear-coverts, thus distinguishing itself from the closely allied *A. moluccana*, in which the ear-coverts are blue like the cheeks and sides of the head. In Flores, Dr. Sharpe *(14)* finds the Common Kingfisher has a little blue before and behind the eye, and separates it as a subspecies, *floresiana*. Hartert says the two forms intergrade.

In uniting in one species, *A. ispida*, the 395 specimens of this form in the British Museum from all parts of Europe, Asia and the East Indies, Dr. Sharpe *(14)* has done a stroke of work for which ornithologists, troubled to know under what name to speak of specimens from this or that locality, will be grateful. There is now perhaps too strong a tendency to regard *A. ispida* as consisting of two races, a larger and typical western, and a smaller eastern *(A. bengalensis)* subspecies; for differences probably exist between specimens from several different localities. Thus, Dr. Sharpe separates the Flores bird *(floresiana)*, Mr. Hume that of Sindh *(sindiana)*, Reichenbach that of the Sunda Islands (var. *sondiaca*), Seebohm that of Japan, China and South Siberia *(bengalensis)*. The Kingfisher is the more puzzling in not being a thorough-going migrant; in the West it is known to visit Corsica, Malta and North Africa in winter; in the East it is a summer migrant in the northern Japanese Islands, a resident in the southern ones (Seebohm *12bis*), a resident about the Lower Yangtse (Styan *a 16*), apparently so in Ceylon *(a 8)*; but, according to Mr. Whitehead it is a winter migrant to Borneo and Palawan, arriving in Palawan on its way south about the middle of September *(a 14, a 15)*, and such we believe it will be found to be in the Sangi Islands and in Celebes, where specimens have been recorded only by Meyer (March) and Guillemard. The time is not yet ripe for splitting up *A. ispida* into subspecies,' even if it is ever desirable to do so.

† 81. ALCEDO MOLUCCANA (Less.).
Blue-eared Kingfisher.

a. **Alcedo ispida** var. **moluccana** *(1)* Lesson, Voy. Coquille I, pt. 1, p. 343 (1826?).
b. **Alcedo ispida** var. *(1)* Lesson, Man. d'Orn. 1828, II, 89.
c. ?**Alcedo bengalensis** *(1)* S. Müll., Verh. Natuurk. Comm. 1839—43, 87, 110; id., Reize Ind. Arch. 1858, II, 8.
d. **Alcedo ispidioides** *(1)* Less., Compl. Buff. 1837, IX, 345; *(2)* Salvad., Ann. Mus. Civ. Gen. 1875, VII, 652; *(3)* id., Orn. Pap. 1880, I, 408; Agg. 1889, 53; *(4)* W. Blas., P. Z. S. 1882, 703; *(5)* id. & Nehrk., Verh. z.-b. Ges. Wien 1882, 418; *(6)* W. Blas., J. f. O. 1883, 136, Nr. 36; *(VII)* Meyer, Vogelskel. pt. VII, 1884, p. 45, pl. LXIII; *(8)* W. Blas., Ztschr. ges. Orn. 1886, 92; *(9)* Grant, P. Z. S. 1888, 192; *(10)* Sharpe, Cat. B. XVII, 1892, 152; *(11)* Meyer, J. f. O. 1892, 258; *(12)* M. & Wg., Abh. Mus. Dresd. 1895, Nr. 8, p. 7; *(13)* iid., ib. 1896, Nr. 2, p. 11; *(14)* Hartert, Nov. Zool. 1896, 175, 244.
e. **Alcedo moluccensis** *(1)* Blyth, J. A. S. B. 1846, XV, 11; *(II)* Sharpe, Monogr. Alcedin. 1870, 21, pl. 4; *(3)* Wald., Tr. Z. S. 1872, VIII, 45; *(4)* Brügg., Abh. Ver. Bremen 1876, V, 55; *(5)* Rosenb., Malay. Archip. 1878, 271; *(6)* Meyer, Ibis 1879, 64; *(7)* E. L. C. Layard, ib. 1880, 299; *(8)* Guillem., P. Z. S. 1885, 547, 566; *(9)* Tristr., Cat. Coll. B. 1889, 90; *(10)* Cab. & Rchw., Nomencl. Mus. Hein. 1890, 163.
f. **Alcedo minor moluccensis** *(1)* Schl., Mus. P.-B. Alcedin. 1863, 8; *(II)* id., Vog. Ned. Ind. Alcedin. 1864, 5, 44, pl. 1, fig. 4, 5; *(3)* id., Rev. Alcedin. 1874, 3.
g. **Alcedo ispida ispidioides** *(1)* Hartert, Nov. Zool. 1896, 571.
h. **Alcedo ispida moluccana** *(1)* Hartert, Nov. Zool. 1897, 160, 163.

"**Kikis wowolean**" (Kingfisher which lives on the river-side), Minahassa, Meyer *e 6*.
"**Radja oudang biru**" or "**Pasa biru**", Minahassa, Nat. Coll. in Mus. Dresd.
"**Sunti**", Tonkean, E. Celebes, iid.
"**Singtobu**", Peling, iid.

For further synonymy and references see Sharpe *d 10*.

Figures and descriptions. Sharpe *e II, d 10*; Schlegel *f II*; Meyer *d VII* (skeleton); Salvadori *d 3*; W. Blasius *d 4, d 8*.

Adult. Similar to *Alcedo ispida*, but with the ear-coverts blue, like the cheeks, not cinnamonrufous; the blue of the upper surface purer, not greenish blue (more campanula-blue instead of turquoise-blue or light cerulean), (Banka Id., N. Celebes, 15. V. 93, C 12258). Iris brown; tarsus bright red; bill black in the male.

Female. The females have the base of the lower mandible orange (Guillem. *e 8*).

Young. Upper surface blacker, the blue bars on the head being narrower, and the scapulars blackish; the blue of the back, rump and upper tail-coverts paler; basal part of mandible pale (N. Celebes, C 3615). "The young bird differs from the adult in having the foreneck and chest obscured by dusky margins to the feathers; the scapulars are blackish slightly washed with blue" (Sharpe *10*). A young male (Dresden Mus. Nr. 6280) is without rufous on the lores, like a specimen (? ♀) from Ceram.

Measurements.

	Wing	Tail	Bill from nostr.	Tarsus
a. (C 12258) ad. [♂] Banka, Celebes 15. V. 93	73	31	31	9.5
b. (C 2316) ♀ Manado, Feb. 71	73	31	32.5	9.5
c. (C 3615) [♂] juv. N. Celebes	74	34	31.5	9
d. (Nr. 6284) ♀ Gorontalo, July 1871	72	31	35.5	9.5
e. (Nr. 6279) — Manado, Mch. 1871	73	33	31	9.5
f. (Nr. 6282) [♂] ad. Manado, Mch. 1871	72	31	34.5	9
g. (Nr. 6281) juv. Manado, Mch. 1871	76	32	28	—
h. (Nr. 6280) ♂ juv. Gorontalo, July 1871	66	28	21	8.5
i. (Nr. 6278) [♀] ad. Gt. Sangi	74	32	33	9.5
j. (C 12614) [♂] ad. Siao, 2. VII. 93	75	31	32.5	9.5
k. (Nr. 6283) ♀ Watubella	75	31	29.5	9.5
m. (C 7379) [♀?] Ceram	74	31	32.5	9
n. (C 9956) ad. ♂ New Britain, 25. V. 86	74	32	32	9

Skeleton.

Length of cranium	61.5 mm	Length of tibia	24.5 mm
Greatest breadth of cranium	17.5 »	Length of tarso-metatarsus	9.0 »
Length of humerus	25.0 »	Length of sternum	24.5 »
Length of ulna	31.0 »	Greatest breadth of sternum	18.0 »
Length of radius	29.0 »	Height of crista sterni	7.0 »
Length of manus	23.0 »	Length of pelvis	22.5 »
Length of femur	16.7 »	Greatest breadth of pelvis	17.0 »

Distribution. Great Sangi (Meyer, Dresden Mus.); Siao (Nat. Coll.); Lembeh and Banka Is., off N. Celebes (Nat. Coll., Dresden Mus.); Celebes: — Minahassa (Wall. *d 10*, Meyer *e 6*, Guillem. *e 8*, etc.), Gorontalo (Forsten *f 1*, Rosenb. *f 3*), South Celebes (Doherty *h 1*), East Celebes (Nat. Coll.), Peling (Nat. Coll.), Sula Islands (Hoedt *f 3*) — Sula Besi (Bernstein *f 3*), Djampea and Kalao (Everett *d 14*); Buru, Amblau, Amboina, Banda, Ceram, Goram, Watubella, Obi, Halmahera, Batchian; Mysol, Salawatti, New Guinea — East Cape, New Britain, Duke of York Id., New Ireland, Solomon Islands (cf. Salvad. *d 3*); Fergusson Id. (Meek *d 14*).

As Dr. Sharpe remarks *(e II)* this bird seems to be the representative species of *Alcedo ispida* in the Moluccas. It will probably be found to be stationary, and *A. ispida* a migrant. Hartert says it is connected with a form of *A. ispida* by intermediate forms.

Very little is known of the habits of *A. moluccana*, but they appear to be the same as those of the common Kingfisher. Mr. Wallace *(e II)* says it frequents the banks of streams and eats small fish; Meyer *(e 6)* found it in Celebes at all times and everywhere near the sea-shore and rivers; Mr. E. L. C. Layard *(e 7)* found it only on the sea-shore in New Britain. The Common Kingfisher of Europe is often to be met with on the sea-shore, where it feeds largely on shrimps and prawns *(e II)*.

82. ALCEDO MENINTING Horsf.
Malayan Kingfisher.

Alcedo meninting *(1)* Horsf., Tr. L. S. 1821, XIII, 172; *(II)* Temm., Pl. Col. 1823, Nr. 239, fig. 2; *(III)* Reichb., Hb. sp. Orn. Alcedin. 1851, 4, pl. 394, fig. 3050–51; *(4)* Schl., Mus. P.-B. Alcedin. 1863, 9; *(V)* id., Vog. Ned. Ind. Alcedin. 1864, 6, 44, pl. 3, fig. 2, 3; *(5bis)* id., Revue Alcedin. 1874, 5; *(6)* Salvad., Cat. Ucc. Borneo 1874, 93; *(7)* id., Ann. Mus. Civ. Gen. 1875, VII, 652; *(8)* Sharpe, P. Z. S. 1879, 320; *(9)* Vorderman, Nat. Tdschr. Ned. Ind. 1882, XLI, 185; *(10)* Oates, Str. F. 1882, X, 188; *(11)* W. Blas., J. f. O. 1883, 115; *(12)* id., Ztschr. ges. Orn. 1886, 94; *(13)* id., Ornis 1888, 307; *(14)* Vorderman, N. T. Ned. Ind. 1891, L, 434; *(15)* id., Notes Leyd. Mus. 1891, 124; *(16)* Hartert, Ornis 1891, 121; *(17)* Sharpe, Ibis 1894, 246, 258; *(18)* Everett, Ibis 1895, 30; *(19)* Vorderm., N. T. Ned. Ind. 1895, LIV, 334; *(19bis)* Blanf., Faun. Br. Ind. B. III, 1895, 125; *(20)* Kuschel, Orn. Mb. 1895, 156; *(21)* M. & Wg., Abh. Mus. Dresd. 1896, Nr. 1, p. 8; *(22)* iid., ib. Nr. 2, p. 11; *(23)* Hartert, Nov. Zool. 1896, 550.

a. **Alcedo asiatica** *(I)* Sws., Zool. Illustr. 1821, 1, pl. 50; *(II)* Sharpe, Monogr. Alcedin. 1870, 25, pl. 5; *(3)* Wald., Tr. Z. S. 1872, VIII, 45; *(4)* Brügg., Abh. Ver. Bremen 1876, V, 55; *(5)* Meyer, Ibis 1879, 64; *(6)* Oates, B. Brit. Burmah 1883, II, 73; *(7)* Guillem., P. Z. S. 1885, 255; *(8)* Everett, J. Str. Br. R. A. S. 1889, 159; *(9)* Whitehead, Ibis 1890, 45; *(10)* Oates ed. Hume's Nests & Eggs 1890, III, 6.

"**Sunti**", E. Celebes, Tonkean, Nat. Coll.
"**Tengkesi mosoni**", Peling, iid.

The following authors more especially discuss local variations of this species:

On the typical Alcedo meninting.

b. **A. meninting** *(1)* Hume, Str. F. 1878, VI, 83, 84; *(2)* Sharpe, Cat. B. XVII, 1892, 157

On Alcedo meninting rufigastra (Wald.).

c. **Alcedo meninting** *(1)* Beavan, Ibis 1867, 319 (Andamans).
d. **Alcedo rufigastra** *(1)* Wald., Ann. & Mag. N. H. 1873, XII, 487 (Andamans); *(2)* id., Ibis 1874, 136 (Andamans).
e. **Alcedo asiatica** *(1)* Hume, Str. F. II, 1874, 174, 494.
f. **Alcedo beavani** *(1)* Hume, Str. F. IV, 1876, 287, 383 (Andamans); *(2)* Sharpe, Cat. B. XVII, 1892, 160 (Andamans).

On Alcedo meninting rufigastra (Wald.),
or a hardly distinguishable continental form, **Alcedo meninting beavani** (Wald.).

g. **Alcedo beavani** *(1)* Wald., Ann. & Mag. N. H. 1874, XIV, 158 (the name has special reference to Beavan's specimen from Maunbhoom); *(2)* id., Ibis 1875, 461 (Burmah); *(3)* Hume, Str. F. VI, 1878, 84, 499 (Tenasserim); *(4)* id., ib. VIII, 1879, 468, note (Sikhim); *(5)* id., ib. IX, 1880, 247 (Cachar); *(6)* id., ib. X, 1882, 188 (Pegu); *(7)* Sharpe, Cat. B. XVII, 1892, 160 (India to Cochin China; Celebes!); *(8)* Gurney, P. Z. S. 1895, 339 (Ceylon); *(9)* Blanf., Faun. Br. Ind. B. III, 1895, 124.

---

¹) Other writers have hitherto identified specimens from Celebes simply as *A. meninting*. Dr. Sharpe accepts their determination and includes Celebes as a locality of the typical form, treated by him as a species for itself. The specimens from Celebes personally examined by him in the British Museum are, however, referred to *A. beavani*!

For further synonyms and references to the species see Sharpe *b 2, g 7*.
Figures and descriptions. Sharpe *a II, b 2, f 2, g 7*; Temminck *II*; Swainson *a I*; Reichb. *III*; Schlegel *V, 4*; Walden *d 1*; Hume *b 1, e 1, f 1, g 3*.
Adult male. Like *A. moluccana*, but smaller; feet smaller and more delicate, inner toe more reduced, claw of inner toe slightly overreaching the base of the terminal joint (claw not counted as such) of middle toe, whereas in *moluccana* it reaches to the middle of it; back, rump, and upper tail-coverts cobalt; the blue of the other parts of upper surface more violet-blue; under surface chesnut cinnamon-rufous (Banggai Id. V.—VIII. 95, C 14728).
Adult female. In some parts of its range — probably in all parts — the under bill of the female is red or rufous.

"The adult female does not differ from the male as is usually asserted; but the blue cheeks and ear-coverts are not assumed so quickly as in the male. Traces of ferruginous are visible in these parts until the bird is aged" (Oates *a 6*, Pegu).

Young. "The young bird differs in having cheeks and ear-coverts ferruginous: in males the blue on these parts is quickly assumed, in females very slowly and some trace of ferruginous in these latter is generally present" (Gorontalo, Riedel — C 11097).

"Young birds able to fly have the bill black with the tip white; the legs pale red and the iris dark brown. One young bird, probably a male, has the whole plumage just as bright as the adult male, and the cheeks and ear-coverts blue" (Oates *10*).

Measurements.	Wing	Tail	Exp. Culmen	Tarsus
a. (W. Blas. *12*) [♂] Gorontalo . .	65	28	36	
b. (W. Blas. *12*) [♀] Gorontalo . .	65	29	39	
c. (Petersb. Mus.) [♂ juv.] Gorontalo . .	61	27	—	8.5
d. (C 11097) [♀ juv.] Gorontalo	60	20	24	8.5
e. (C 14728) ad. Banggai, V.—VIII. 95 (Nat. Coll.)	64	26	39	8.5
f. (C 14593) ad. Peling, V.—VIII. 95 (iid.)	61	26	37	7.5
g. (C 14441) ad. Tonkean, E. Cel., V.—VIII. 95 (iid.) .	64	28	38	8.5
h. (Sarasin Coll.) ♀ Macassar, 15. XII. 94	62	24	36	8.5

Eggs. 4 to 6; very glossy and round, white; average size 19.5 × 17.5 mm (Pegu — Oates *a 6, a 10*); glossy white; rather rounded, 20 × 16.5 mm (Labuan — Low *8*). 21 × 17 mm (Nehrkorn MS.).

Nest. A nest of this species described by Mr. Oates was in the steep bank of a ravine in thick forest. Gallery about one and a half feet long, terminating in a small chamber. Eggs laid on the bare soil *(a 10)*.

Breeding season. Pegu, July (Oates *a 6*).

Distribution of the species. Southern and Central India to the Eastern Himalayas (Sharpe *g 7*, Beavan *g 1*, Hume *g 4, g 5*, etc.); Burmah (Oates *a 6, g 6*); Tenasserim (Davison *g 3*); Andaman Islands (Beavan *c 1*, R. G. W. Ramsay *d 1, e 1*, Davison *e 1, g 7*); Cochin China (Brit. Mus. *g 7*); Malay Peninsula (Wall. *a II*, Davison *b 2*); Salanga (A. Müll. *12*); Penang (Cantor *6*, Wall. *b 2*); Singapore (Davison *g 7*, etc.); Sumatra (Raffl. *6*, Wall. *b 2*, etc.); Banka (Bossche *4*); Billiton (Vorderman *14, 15*); Borneo (S. Müll. etc. *6, a 8*); Palawan (Whitehead *a 9*, Platen *13*); Balabac (Everett *18*); Sooloo Islands (Guillem. *a 7*, Everett *17*); Celebes: — Gorontalo (Riedel *12*, Rosenb. *5bis*), Togian Islands (Meyer *a 5*), Tonkean, E. Celebes (Nat. Coll.), Kandari, S. E. Celebes (Beccari *7*), Macassar (Wallace *a II, 6, g 7*, P. & F. Sarasin), Peling and Banggai (Nat. Coll.); Java (Horsf. *1*, etc.); Lombok (Wall. *a II, b 2*, Vorderman *19*).

This species varies considerably for a Kingfisher in different parts of its range and is split up by Dr. R. B. Sharpe into two species, viz. *A. beavani* Wald., of the Andaman Islands, India and Burmah as far as Bankasoon (South Tenasserim) and Cochin China, occurring again in Celebes; *A. meninting* Horsf. ranging from Bankasoon down the Malay Peninsula to Sumatra, Borneo, Java and Lombok. Mr. Blanford *(19bis, g 9)* follows Dr. Sharpe

These two supposed species intergrade, as far as we can make out, between Burmah and Malacca, and either a system of subspecies should, if possible, have been created, or, as in the case of *A. ispida*, the badly understood local races might have been embraced under one name. The forms, which Dr. Sharpe brings together under the name *beavani*, Mr. Hume would unite as one with *meninting* or separate into two or three; of the South Indian birds he writes: "it must be understood clearly that this is not *beavani* [of the Andamans], but a form lying on the other side of *asiatica*" (Str. F. IV, 383), and later again: "I believe this ought to be separated as a distinct species" (Str. F. X. 1883, 352 note). Specimens from the Andaman Islands run to another extreme of coloration, but in Pegu and Tenasserim an intermediate form of *A. meninting* occurs which "is very close to, and runs into" the typical *A. beavani* *(g 3, g 6)*. Dr. Sharpe finds that the Indian variation repeats itself in Celebes; the fact is curious, but few specimens have been obtained in the island and more material may possibly serve to show that it differs. Three specimens from Borneo and one from Java in the Dresden Museum are more violet in hue than those before us from Celebes, Peling, and Banggai, and the bills of the former seem to be broader — in one case much longer and the wing shorter than in the Celebesian birds; also the reduced toe appears to be smaller. Schlegel remarks on Rosenberg's three specimens from Gorontalo that the colours of the upper surface incline strongly to green *(5 bis)*; the specimens in our hands do not confirm this statement, but show bright blue tints.

On the habits of this bird in Pegu Oates writes *(a 6)*: "This species is restricted to the dense forests, where the ground is broken up by nullahs and ravines. I think it always darts on its prey from a perch and does not hover in the air". In the Andamans its habits appear to differ: "it keeps exclusively (as far as I have observed)", says Davison, "to the salt water creeks, occasionally venturing out to the fishing stakes at the mouths of the creeks. Its voice is weaker, and not nearly so shrill as that of *A. bengalensis* (*A. ispida*); it feeds on small fish, after which it plunges, keeping under water for some considerable time" *(e 1)*.

GENUS PELARGOPSIS Glog.

The members of this genus are about the size of a Turtle-dove; the tail is longer than the bill, rounded; the bill about twice as long as the head, large, the culmen straight, flattened along its ridge; wing moderate, 3rd and 4th quills longest, the 1st a little more than ⅔ their length; inner toe as long or longer than the hallux, longer than the tarsus. About 5 species, with numerous local races, ranging from India to the Philippines, Celebes and Flores.

† * 83. PELARGOPSIS MELANORHYNCHA (Temm.).
Celebesian Stork-billed Kingfisher.

a. **Alcedo melanorhyncha** *(1)* Temm., Pl. Coll. 1826, pl. 391, livr. 66; *(2)* Schl., Handb. d. Dierk. 1857, 306; *(2)* id., Mus. P.-B. Alcedin. 1863, 15; *(IV)* id., Vog. Ned. Ind. Ijsvogels 1864, 10, 47, pl. 2, fig. 1; ? *(5)* id., Rev. Alcedin. 1874, 8.

b. **Dacelo melanorhyncha** *(1)* Less., Tr. d'Orn. 1831, 246 (Java!); *(2)* Finsch, Neu Guinea 1866, 160.

c. **Halcyon melanorhyncha** *(1)* Gray, Gen. B. 1846, I, 79; *(2)* Bp., Consp. 1850, I, 155; *(3)* Cass., Cat. Halc. Philad. Mus. 1852, 10; *(4)* Wall., Ibis 1860, 142; ? *(5)* id., P. Z. S. 1862, 335, 338; *(6)* Gray, HL. 1869, I, 92; *(7)* Rosenb., Malay. Archip. 1878, 271; *(8)* Pelz. u. Lorenz, Ann. k. k. Nat. Hofmus. Wien 1886, I, 258.

d. **Hylcaon melanorhyncha** *(1)* Reichb., Hb. spec. Orn., Alcedin. 1851, 18, pl. 399, f. 3074.

e. **Rhamphalcyon melanorhyncha** *(1)* Bp., Consp. Vol. Anisod. 1854, 10.

Pelargopsis melanorhyncha *(1)* Sharpe, P. Z. S. 1870, 62; *(II)* id., Monogr. Alcedin. 1870, pp. XXVII, 95, pl. 29; *(3)* Wald., Tr. Z. S. 1872, VIII, 45; *(4)* Salvad., Ann. Mus. Civ. Gen. 1875, 13; *(5)* Brügg., Abh. Ver. Bremen 1876, V, 55; *(6)* Meyer, Ibis 1879, 64, 146; *(7)* W. Blas., J. f. O. 1883, 136; *(8)* Guillem., P. Z. S. 1885, 547; *(9)* W. Blas., Ztschr. ges. Orn. 1886, 92; *(10)* Cab. & Rchnw., Nomencl. Mus. Hein. 1890, 165; *(11)* Sharpe, Cat. B. XVII, 1892, 97; *(12)* M. & Wg., Abh. Mus. Dresden 1895, Nr. 8, p. 7; *(13)* Hart., Nov. Zool. 1897, 163; *(XIV)* Meyer, Abb. v. Vogelskel. 1897, pl. II, CCXV.

"**Radja udan puti**" (White king of the Crabs), Minahassa and the islands off the coast, Meyer *6*, Nat. Coll.

"**Bua-Buaa**", Rosenb. *c 7*, ? Gorontalo.

Figures and descriptions. Sharpe *II* (imm.), *1, 11*; Meyer *XIV* (skeleton); Temminck *a I*; Reichb. *d I*; Schlegel *a IV, a 3*.

Adult. Head, neck and body above and below cream-colour, becoming darker, buff, more on abdomen; cheeks, around eye, and forehead drab, the feathers bordered with cream-colour; scapulars, wings and tail drab with green and bluish reflections, changing with the light; inner webs of quills below buffy white, the colour spreading out over the under side of the wing towards the base of the feathers (Banka Id., N. Celebes, 20. V. 93 — C 12254). Iris yellowish brown, feet red-brown; claws and bill black (M. ♂).

Sexes. Alike.

Immature. Like the adult, but entire head above and cheeks pale smoke-grey; darker towards the forehead, with darker centres to the feathers; lores, ear-coverts, middle of mantle, scapulars, wings, tail, and the longer upper tail-coverts brown, much darker than in the adult, with greenish or bluish reflections; back and rump pale blue (!) upon a buff ground; breast and sides of neck barred with faint drab-colour; bill shorter than in the adult, black, horn-colour at extreme tip (near Manado, Aug.—Sept. 1892, C 10865).

270 Birds of Celebes: Alcedinidae.

Measurements.	Wing	Tail	Bill from nostril	Tarsus
a. (C 12254) [♂] ad. Banka I., 20. V. 93	154	101	69	15
b. (C 12255) [♂] ad. Banka I., 18. V. 93	153	—	64	15
c. (Nr. 6245) ♂ ad. Manado	145	87	68	16
d. (Nr. 6244) ad. Manado, Mch. 71	147	91	65	—
e. (Nr. 3809) ad. N. Celebes	151	92	65	15
f. (C 12253) [♀] ad. Manado tua Id., 15. IV. 93	149	87	62.5	15
g. (C 10866) imm. near Manado, Aug.—Sept.	145	87	63.5	15.5
h. (C 10865) imm. near Manado, Aug.—Sept.	152	94	59	16.5

The wing-lengths of 5 additional specimens from Lombeh Id. and the Minahassa fall within the above measurements.

In the young specimen h the tarsus and tibio-tarsal joint is curiously padded below with thick porous, yellow skin. This is not seen in the older birds, and in this specimen it is about to be cast off. Probably the young have the habit of resting on their "heels", and these calosities serve as a protection to them [1]).

Eggs. Unknown. Always two in number (M. *6*).

Distribution. Celebes and the Sula Islands: — Manado tua & Banka Is., off N. Celebes (Nat. Coll. in Dresden Mus.), Lombeh Id. (Guillemard *8*, N. Coll.), Minahassa (Wall. *c 4*, Meyer *6, etc.*), Gorontalo Distr. (Rosenb. *a 5*), Pogoyama, Gulf of Tomini (Guillem. *8*), Paguatt and Posso, Gulf of Tomini (Rosenb. *a 5*), Togian Islands (Meyer *6*); Kandari, S.E. Celebes (Beccari *4*), Tawaya, W. Celebes (Doherty *13*), Sula Besi and Sula Mangoli (Allen *c 5*, Bernst. *a 5* — if identical).

This Kingfisher is, as Dr. Sharpe remarks (*1*), the most distinct species of the genus *Pelargopsis*. The black bill and the absence of blue on the back of the adult bird at once distinguish it from the other members of the genus *Pelargopsis*, about ten in number, which range from Ceylon and India as far as the Philippines and Flores and have all red bills and blue backs and rumps. Nevertheless, the young of *P. melanorhyncha* has, as shown above, a light blue back, which, it would seem, is soon lost — proof that *P. melanorhyncha* is derived from a form with a blue back. The young of other species do not differ appreciably in coloration, as far as is known, from the adults. Hartert (*13*) says that some of his specimens from West Celebes have a small red spot at the base of the maxilla; there is a tendency to redness at the extreme base of the bill in one or two of our specimens from the Minahassa. *P. melanorhyncha* was separated by Reichenbach as a subgenus on the ground of a supposed difference of structure of the bill. Like Dr. Sharpe and others, we do not find that its bill differs in any important particular from those of others of its genus, except in colour. In the Celebes Province *Pelargopsis* has undergone the greatest modification, it has lost the blue on the back and rump and acquired, apparently, its black bill there.

Similarly *Eudynamis melanorhyncha* of Celebes and Sula has a black bill,

---

[1]) See Munn, Ibis 1893, 59, for a similar pad in the young *Pelargopsis gurial*, and Günther, ib. 1890, 411 in *Iynx torquilla*.

though everywhere else the bill of *Eudynamis* is whitish. In Celebes, too, the Lories, *Trichoglossus ornatus* and *Psitteuteles meyeri*, and in Sula *Psitteuteles flavoviridis*, have lost the band of bright colour across the base of the wing below, so that the whole of underside of the quills is uniform dusky shining smoke-grey.

The nearest ally of *P. melanorhyncha* after *P. dichrorhyncha*, appears to us to be *P. gigantea* Wald. — by no means a gigantic member of its genus — of the southern Philippines, a form regarded by Sharpe as a "subspecies" of *P. leucocephala* of Borneo. *P. melanorhyncha* may perhaps be regarded as a somewhat recent immigrant to the Celebes Province — not as one of the ancient types of the country; because 1. its modifications are slight as compared with those of the Bee-eater, *Meropogon*, for instance, with its nearest ally, *Nyctiornis* of India to Borneo, or of the Hornbill, *Rhabdotorrhinus*, with *Penelopides* of the Philippines; 2. members of the genus are known to be birds of good flying powers; 3. a form of *Pelargopsis*, *P. floresiana*, treated by Sharpe as only subspecifically distinct from other forms ranging from India to Java and Borneo, makes nothing of Wallace's line between Bali and Lombock, being found in Flores[1]); 4. the circumstance that *P. melanorhyncha* throws back, when young, to the other members of the genus in having a blue back seems to betoken that the adult plumage is not very ancient.

Touching the last point, a few general remarks on the plumage of the Kingfishers will be found further on under *Ceycopsis fallax*. On the habits of this species Meyer *(6)* notes briefly that it is "on riversides rather rare; always in pairs together. On the 26th of February I shot on the river Tamumpat, near Manado, a female; and the male flew for hours up and down the river, crying for its mate; every half hour it passed my resting-place. It has a very quick arrow-like flight, feeds on large and small fishes, and always lays two eggs".

+ 84. PELARGOPSIS DICHRORHYNCHA M. & Wg.
Red-and-black-billed Kingfisher.
Plate IX.

Pelargopsis dichrorhyncha *(1)* M. & Wg., Abh. Mus. Dresden 1896, Nr. 2, p. 12; *(2)* Hartert, Nov. Zool. 1897, 163.

"Bukaka mawuti", Banggai, Nat. Coll.

Diagnosis. Similar to *P. melanorhyncha*, but with more or less of the culmen and base of the maxilla and most of the under mandible red, the rest black; size somewhat larger.

Measurements. Wing 151—161 mm; tail c. 100; bill from nostril 67—73; tarsus c. 17.

Distribution. Peling and Banggai Islands (Nat. Coll.).

The Stork-billed Kingfisher of Celebes has been recorded from Sula by Wallace (P. Z. S. 1862, 335, 338), Schlegel (Mus. P.-B. Rev. Alced. 1874, 8)

[1]) Dr. Sharpe is of opinion that this species crossed the strait by flight (Monogr. p. XLI).

and Sharpe (Cat. B. 1892 XVII, 97). Mr. Wallace speaks of the base of the bill as red. A comparison of them with *P. dichrorhyncha* is desirable, as Sula birds generally have much more in common with those of Peling than with those of Celebes.

This Kingfisher is easily distinguishable from that of the mainland of Celebes by its dichromatic bill, but the distribution of the two colours red and black thereon is subject to much individual variation; in one or two specimens in the Dresden Museum the entire upper mandible is black except at the base, in another example more than half of it is red; the under mandible seems to be always red for at least the basal half, sometimes the terminal fourth only is black.

The type and co-types, which were obtained by our native hunters in Peling and Banggai between May and August, 1895, are in the Dresden and Tring Museums.

GENUS CEYX Lac.

Small Kingfishers, similar to *Ceycopsis*, but having only three toes, the inner (second) toe being absent.

About 18 species in the Indian and Papuan countries.

† * 85. CEYX[1]) WALLACEI Sharpe.
Sula Three-toed Kingfisher.

a. **Ceyx lepida** *(1)* Wall. (nec Temm.), P. Z. S. 1862, 338; *(2)* Finsch, Neu-Guinea 1866, 161, pt.
Ceyx wallacii *(1)* Sharpe, P. Z. S. 1868, 270; *(II)* id., Monogr. Alcedin. p. 129, pl. 45 (1868); *(3)* Salvad., Orn. Pap. 1880, I, 417; *(4)* Sharpe, Cat. B. XVII, 1892, 182.
b. **Dacelo wallacei** *(1)* Schl., Revue Alcedin. 1874, 34.
c. **Dacelo cajeli** *(1)* Schl., Ned. Tdschr. Dierk. 1866, III, 339, pt. (Soula).
Figure and descriptions. Sharpe *1, II, 4.*
Adult. Above black; head and nape spotted with cobalt, more on the nape, each feather having a mesial stripe of brighter blue; cheeks and wing-coverts streaked with bright cobalt; back very rich shining cobalt; the uppper tail-coverts slightly tinged with ultramarine; scapulars black; quills and tail-feathers blackish, the inner web of the former light rufous towards the base; throat whitish; loral spot and the whole of the under surface deep orange; a patch at the sides of the the uppper breast black; "bill and feet coral-red; iris dark" (A. R. Wallace).
Measurements. Total length 141 mm, culmen 36, wing 63, tail 25, tarsus 7.6 (Sharpe *4*).
Distribution. Sula Islands (Allen *a 1*) — Sula Besi (Bernstein *b 1*), Sula Mangoli (Hoedt *b 1*).

This little three-toed Kingfisher was first obtained by Mr. Wallace's assistant, Allen, in Sula Besi or Mangoli, where further specimens were obtained in 1864 by Bernstein and Hoedt.

---

[1] "Ceyx" — a substantive of masculine gender.

Its nearest ally is *Ceyx cajeli* Wallace of Buru, which differs chiefly in having the cheeks and ear-coverts black, and the back silvery blue, instead of cobalt (Sharpe), and a shorter bill. This form is reported by Schlegel to have been observed in Sula, but this rests no doubt upon an error of identification.

The genus *Ceyx*, as Dr. Sharpe points out, consists of two divisions, "the rufous-backed section and the blue-backed section", which "have distinct ranges, the one being Indo-Malayan, and the other Austro-Malayan" (Monogr. p. XLI). Elsewhere (p. XXIX), the author further divides *Ceyx* into four groups: —

1. The *C. tridactylus*-group with red bills and lilac plumage, corresponding to the rufous-backed section;
2. *C. lepidus*-group with red bills but blue plumage;
3. *C. solitarius* with a black bill, to which *C. gentianus* Tristr. may now be added;
4. *C. philippensis* (= *C. cyanipectus* Lafr.) with a blackish upper and orange lower mandible, to which several recently discovered species may be added, if *C. argentatus* Tweedd. and *flumenicolus* Steere do not constitute a fifth group for themselves.

The present species, *Ceyx wallacei* belongs to the second, or *Ceyx lepidus*-group.

After a comparatively brief study of the genus *Ceyx* under the excellent guidance of Dr. Sharpe's great "Monograph of the Alcedinidae" and his more recent Catalogue of the Kingfishers, it may seem presumptuous to criticise his conclusions; yet it may well be asked why the genus *Ceyx* is put into one subfamily, the *Daceloninae*, and the genus *Alcyone* into another, the *Alcedininae?*

The differences between the extreme members of these two groups, *Dacelo* or *Melidora* and *Alcedo*, are most interesting and important, and in 1892 Dr. Sharpe maintains his subdivisions of 1870; the *Alcedininae* are stated to be distinguished by a "bill long and slender, and perceptibly keeled, habits mainly piscivorous"; the *Daceloninae* have the "bill more or less depressed; culmen rounded or flattened, sometimes even grooved. Habits mainly insectivorous or reptilivorous" (Cat. B. XVII, 93). Yet in the earlier work, the close correspondence of *Alcyone cyanipectus* with *Ceyx philippensis* is commented upon; "Count Salvadori, who has paid much attention to these birds, stipulates for their both being placed in the genus *Alcyone*; but I would rather keep them in the genus *Ceyx*, because we should then have plumage as an additional generic character" (p. VII); later Major Ramsay and Dr. Sharpe decided that *Alcyone cyanipectus* is the male and *Ceyx philippensis* the female of the same species (Ibis 1884, 332, pl. IX) a view which received full confirmation from Messrs. Bourns and Worcester (Ibis 1895, 404), though at one time (Ibis 1895, 112) doubted by Mr. Grant, but afterwards also consented to by him (Ibis 1896, 471). They are now placed by Sharpe in the genus *Ceyx*, by Grant in the genus *Alcyone*. Observations on the habits of *Ceyx* are scanty; they were still scantier at the date of appearance of the 'Monograph', and our leading Alcedinist was induced to conclude that *Ceyx* differed more widely from the *Alcedininae* than now appears to be the case: "*Alcyone* is a fish-eater and partakes of the characteristics of true *Alcedo*; ... *Ceyx*, on the other hand, is a forest-loving genus, living away from the water, feeding on insects, and bearing affinity towards *Halcyon*" (ib. p. VIII). Notes that have since appeared in "Stray Feathers" and elsewhere tend to show that, while *Ceyx* avoids broader waters, it is generally to be found by small streams — sometime, indeed, dried up ones — in deep forest or jungle. Legge (B. Ceylon, 304) says *Ceyx tridactylus* subsists on diminutive fish and small aqueous insects; a writer in Hume's "Nests and Eggs of Indian Birds" (Oates ed. III, 14) says: "as far as I have been able to ascertain, it is entirely a fish-eater, though it may also devour water-insects, small prawns, etc. I have never seen any remains of insects in its stomach". Other observers — De Bocarmé, and Wallace — testify to the insectivorous and crustacivorous habits of species in Borneo and Java (Sharpe, p. 120) while *Ceyx cajeli* — and, therefore, probably its near ally *C. wallacei* — was observed by Wallace to feed on water-insects and small fish (Sharpe, p. 127). From this it appears certain that *Ceyx* must be set down as a semi-piscatorial bird, in habits a link between *Alcedo* and *Halcyon*.

Further, in plumage the genus *Ceyx* is a link between *Alcedo* and certain *Daceloninae*. Dr. Sharpe has called attention to the close resemblance borne by the female *Ceyx cyanipectus* (*Ceyx philippinensis*, Monogr. p. 113) to *Alcedo moluccana*, and remarks that Lesson erroneously referred *Ceyx solitarius* to *Alcedo meninting*. Altogether, the plumage of the blue-backed section of *Ceyx* is distinctly Alcedinine. The red-backed section on the other hand, especially *Ceyx rufidorsus* (= *euerythrus* Sharpe), recalls — save for the absence of the blue rump — the peculiar plumage of the wide-spread *Halcyon coromanda* (*Callialcyon*. Compare, Sharpe, pl. 41 with pl. 57). Thus Sharpe remarks: "the link towards *Halcyon* seems to be in the lilac-backed section of the genus *Ceyx* with the lilac-backed section of the genus *Halcyon*, where the tail is rather shorter than in most of the other members of the genus" (Mon. p. XLVI. In plumage, as in habits, the two sections of the genus *Ceyx* appear to unite the *Alcedininae* and *Daceloninae*.

The question now suggests itself: did *Ceyx* arise from *Halcyon* and *Alcedo* from *Ceyx*, or was the reverse the order of development, or, as might also be thought possible, did both *Alcedo* and *Halcyon* take their origin from *Ceyx*, the one from the blue-, the other from the lilac-backed section of it? Perhaps Dr. Sharpe, following up his conclusions of 1870, and students of the Kingfishers in the future will be able to read as from a book the past history of the family, undoubtedly impressed upon their plumage and structure, but at present hidden from ornithological knowledge. One small token we believe ourselves able to point to, which goes to confirm the conclusions arrived at by Dr. Sharpe from other lines of reasoning — namely, that *Dacelo* and *Melidora* "are the most ancient form of Kingfisher extant", while "*Alcedo*, the most specialized type of the family *Alcedinidae*, ... belongs to a more recent development" (Mon. p. XLIV). This seems to be indicated by the immature plumage of *Ceycopsis fallax* of Celebes, which, as Sharpe says (p. XII), presents "a recognisable link between the Aethiopian *Ispidinae* and the Malayan *Ceyces*", and further on again (p. XLI) "it unites the characters of the two groups of the genus *Ceyx*, which converge from opposite sides upon its flanks, as one may say; for it is red in general plumage, but has a bright blue back". We pass on, therefore, to the consideration of this interesting Celebesian species, the next in our list.

GENUS CEYCOPSIS Salvad.

A genus peculiar to the Celebesian area; small Kingfishers about the size of a Sparrow, recognisable by their having the inner toe much reduced, shorter than the hallux (shorter also than the tarsus and less than half the length of the middle toe); bill red, about half as long again as the head, rather flat, at the base across the nostrils about as broad as it is high; tail small, shorter than the bill; wing rather short, 2nd, 3rd and 4th quills longest, 1st a little shorter. Two closely allied species, inhabiting Celebes and Sangi respectively.

+ * 86. CEYCOPSIS FALLAX (Schl.).
Fallacious three-toed Kingfisher.
Plate X.

a. **Dacelo fallax** *(1)* Schl., Ned. Tdschr. Dierk. 1866, III, 187; *(2)* id., Revue Alcedin. 1874, 32; *(3)* Rosenb., Malay. Archip. 1878, 272, 584.
Ceycopsis fallax *(1)* Salvad., Atti Acc. Sc. Torino 1869, 447; *(II)* Sharpe, Monogr. Alcedin. 1870, pp. XII, XXIX, XLI, XLVI, 135, pl. 48; *(3)* Wald., Tr. Z. S. 1872, VIII, 49, 112; *(4)* Meyer, Ibis 1879, 63, partim; *(5)* W. Blas., J. f. O. 1883, 136; *(6)* Guillem., P. Z. S. 1885, 548; *(7)* id., Cruise "Marchesa" 1886, II, 198; *(8)* W. Blas., Ornis 1888, 572, partim; *(9)* Sharpe, Cat. B. XVII, 1892, 190; *(10)* Hartert, Nov. Zool. 1896, 158; *(11)* id., ib. 1897, 163.
b. **Ceyx fallax** *(1)* Gray, HL. 1871, III, 227; *(2)* Brügg., Abh. Ver. Bremen 1876, V, 55; *(3)* Tristr., Cat. Coll. B. 1889, 92.
"**Radja udang mera ketjil**", Minahassa, Nat. Coll.
Figure and descriptions. Sharpe *II, 9*; Schlegel *a 1, a 2*, Brüggemann *b 2*.
Adult. General colour above dark reddish hazel: head above including nape brownish black spangled with bright ultramarine-blue, each feather having a subterminal spot; on middle wing-coverts a few spots of a rather lighter colour, lores ferruginous; ear-coverts and cheeks also ferruginous, but washed with purple of a magenta tint into which the blue of the head-spots passes on the sides of the crown; lower back and rump whitish turquoise-blue, becoming azure-blue on the upper tail-coverts; quills and tail dusky; chin and throat white; feathers of ear-coverts continued

into a whitish band on sides of neck; remainder of under surface rufous, darkest and more ferruginous on sides of breast, palest and more orange-rufous on the middle parts of under surface, on under wing-coverts, and edge of wing; middle of abdomen lightly washed with lilac; inner edge of quills below towards their base dull rufous vinaceous (near Manado, Aug.—Sept 1892, C 10867). "Iris brown; bill brilliant coral; feet bright coral red" (Guillem. 7, Rosenb. *a 1*).

The lilac tint mentioned in Dr. Sharpe's description *(9)* of male (?) does not extend on to the lower back of this example. In adults according to Schlegel *(a 1)*, the rufous of the under surface passes into whitish on the abdomen. This also does not seem to be the rule; on the other hand the lower breast is washed with lilac in some adults.

Female. According to Mr. Hartert *(11)* the female is of a much darker and less rufous brown above.

Immature. A younger specimen has the blue spots of the head duller and less sharply defined, and of a more magenta tint on towards the nape, under surface and rump and upper tail-coverts without any lilac tint; the wing-coverts browner, some obscure spots of purplish blue on the middle series; bill pale greenish horn-colour, blackish on basal part of ridge of upper mandible (Minahassa, Nr. 1903).

Young. A specimen kindly lent to us by Mr. Pleske is like the last, but with scantier blue tips to the feathers of the head, those on the nape of a paler more lilac purple; breast somewhat darker (N. Celebes — 1889). The young are lighter reddish above, have less blue on the head, and want the lilac wash on the lower breast, the upper mandible is dusky (Lembeh, March 1895, C 14206, 14210). Schlegel, after von Rosenberg, describes the bill of the young as "noirâtre" *(a 1)*; Dr. Guillemard shot a specimen with the bill slate, lower mandible red *(6)*. On referring to Sharpe's recent Catalogue, we find that the bright red bill is a character of the adult in the *Ceyx* genus; that of the young of *Ceyx bournsi* is of a duller red than in the adult, of *C. malamaui* more dusky, whitish horn-colour, of *C. melanurus* horny whitish, of *C. dillwynni* (nestling) blackish horn-colour, of *C. tridactylus* paler, dusky at base of both mandibles.

Eggs and breeding habits. Unknown.

Measurements.	Wing	Tail	Tarsus	Bill from nostril
a. (C 10867) ad. near Manado	56	22	9	28.5
b. (Nr. 1903) imm. Minahassa	58	24.5	9	29
c. (Tring) ad. Lembeh, 3. III. 95 (Nat. hunters)	59	22	—	29
d. (C 14207) ad. Lembeh, 21. II. 95 (Nat. hunters)	60	26	—	29
e. (C 14208) ad. Lembeh, 27. II. 95 (Nat. hunters)	59	—	8.5	—
f. (Tring) ad. Lembeh, 6. III. 95 (Nat. hunters)	54	—	8.5	—
g. (C 14209) ad. Lembeh, 3. III. 95 (Nat. hunters)	57	21.5	9.5	30
h. (Tring) ad. Lembeh, 16. III. 95 (Nat. hunters)	57	—	9.0	30.5
i. (Tring) ad. Lembeh, 4. III. 95 (Nat. hunters)	57	22	—	30
j. (C 14206) juv. Lembeh, 11. III. 95 (Nat. hunters)	58	—	—	27.5
k. (C 14210) juv. Lembeh, 1. III. 95 (Nat. hunters)	58	—	—	27.5

Distribution. Celebes:— Minahassa (v. Duivenbode *a 2*, Meyer *4*, Guillem. 7;); Lembeh Id. (Nat. Coll. in Mus. Dresden and Tring); Gorontalo Distr. (v. Rosenb. *a 2*, Meyer *4*); Dongala and Tawaya, W. Celebes (Doherty *11*); South Celebes, Maros (Meyer *4*), Indrulaman (Everett *10*).

Variation. The specimens from Lemboh vary somewhat individually, some having the bases of the blue-tipped feathers of the head rufous, others most of them dusky, the latter specimens, with one exception *(c)*, have the scapulars duller and browner. These differences are shared by the two immature birds, there being one of each, distinguishable as young by their partially dusky and shorter bills and absence of blue spots on the wing-coverts. The more red or dusky colours are probably differences of a sexual nature.

This little Kingfisher was not discovered until 1863—1864, when von Rosenberg obtained the first specimens in the Province of Gorontalo; it is still a rare species in collections. Meyer remarks *(4)*: "I got this species near Manado, near Gorontalo, and at the waterfall of Maros, in South Celebes; but I did not procure many specimens, perhaps for the reason that it lives in the forest and is a small species. In May a living specimen was in my possession at Manado. The colours of this little species are very delicate. It is the loveliest Kingfisher of Celebes". One of Dr. Guillemard's specimens was, curiously enough, shot on the sea-coast at Wallace Bay, N. Celebes, "on the lonely beach frequented by the Maleos *(Megacephalon maleo)* described by Wallace in his 'Malay Archipelago'" (I, p. 413).

In the foregoing article on *Ceyx wallacei* attention was drawn to Dr. Sharpe's remarks pointing out the interesting position occupied by *Ceycopsis fallax*. In the structure of the foot it is a connecting link between *Ispidina* (with *Myioceyx*) of Africa and Madagascar and *Ceyx* of India and the East Indies; in plumage it is a connecting link between the red-backed section of *Ceyx* of India to the Lesser Sunda Islands, Borneo, and the Philippines, and the blue-backed section of *Ceyx* of Papuasia and the Moluccas to Sula and the Philippines. Further, from the extreme members of the red-backed divisions of *Ceyx* and of *Ispidina* to the blue-backed divisions of the same genera a gradual transition from the plumage of *Callialcyon* to that of *Alcyone* and *Alcedo* is seen. The little *Ceycopsis fallax* occupies a middle position. It appears to us that, when growing into adult plumage, the bird develops towards the *Alcedo*-type, rather than in the opposite direction, since the blue spangles on the head become brighter, bluer, better defined, and extended further back on the hind neck. A future worker on the Kingfishers will no doubt find more such indications; the present one is a hint — if it does not mislead us — that the blue-backed, *Alcedo*-like *Ceyces* are derived from the red-backed ones[1]. That is to say, the *Ceyces*, which are now found between the Philippines and the Solomon Islands, appear to be descended from those now known from India to the Philippines.

It is hardly possible to say between which red-backed and which blue-backed *Ceyx Ceycopsis fallax* most immediately lies; we should place it between the red-backed *Ceyx melanurus* of Luzon and the blue-backed *C. wallacei* of Sula; *Myioceyx* of West Africa has also a good deal in common with it; but other ornithologists may think differently as to its nearest affinities.

---

[1] The young of *Ispidina leucogaster* has most of the crown ferruginous (Sharpe, Cat. 194).

Ceycopsis has four toes — the inner one very minute — Ceyx has three; it may insult the convictions of many anatomists if called upon to admit that a bird with four toes may be intermediate between two sections of a genus with three; still more outrageous seems the assumption that Alcedo with four toes can be descended from Ceyx with three. We do not, of course, suggest that such is the case; but once undoubtedly Ceyx had four toes and the second of Ceycopsis was much larger than it is now.

It is possible that a future worker on the Kingfishers of the genera Alcedo, Alcyone, Ceyx, Ceycopsis, Ispidina and Myioceyx will come to the conclusion that the red and the blue types of plumage in these genera have persisted since all agreed in having four toes; that since then the small inner toe has become completely aborted in both the red and most of the blue forms in the East Indies, except Celebes, where the process as seen in Ceycopsis is not quite complete. In any case Ceycopsis may be regarded as a rather ancient form. We do not know how it got to Celebes, but its ancestors appear to be the red forms of Ceyx of the islands to the west and the Indian countries.

It is a mystery why the second toe in certain Kingfishers has become aborted, and we are at a loss to make a suggestion for the explanation of the fact. Can there be a reason in their habits or their mode of life?

ᵻ * 87. CEYCOPSIS SANGIRENSIS M. & Wg.
Sangi Fallacious Kingfisher.
Plate X.

a. **Ceycopsis fallax** *(1)* Meyer, Ibis 1879, 63 partim; *(2)* id., Isis, Dresden 1884, 6; *(3)* W. Blas., Ornis 1888, 572 partim.

Adult. Similar to *Ceycopsis fallax* of Celebes, but with the blue spots of the head above much larger and continued further down the hind neck, the spots on the sides of occiput almost running into one another, and blue like those of the head, not magenta; the spots on the middle and greater wing-coverts larger, magenta; mantle washed with magenta; bill longer and differing in shape — not so much narrowed in its terminal third or so much broadened at its base; size a little greater (Tabukan, Great Sangi, Meyer in Vienna Museum 1877, Nr. II i).

Immature (with a dusky horn-coloured bill). Just like the adult, but with none of the feathers of the mantle tinted with magenta, the blue on the head and neck a trifle darker and the ear-coverts less strongly washed with magenta (Tabukan, Great Sangi, Meyer — Nr. 6225).

Measurements.	Wing	Tail	Bill from nostril	Tarsus
a. (Vienna Mus.) ad. Gt. Sangi	60¹)	28	33	9
b. (Nr. 6225) imm. Gt. Sangi	62	25.5	31	9

Distribution. Tabukan, Great Sangi (Meyer a 1, a 2), where it appears to be plentiful.

¹) The ends of the quills are frayed, so that 1 or 2 mm should be added to make up their original length.

The most important points of difference in this form of *Ceycopsis* appear to be the differently shaped bill, and the large size and great extent on the sides and back of the nape of the blue spots or bars of the head; on the sides of the nape, in fact, they blend into almost continuous blue. *Ceycopsis* is a peculiarly Celebesian genus, as remarked above.

GENUS HALCYON Sw.

This large genus has been split into eleven, chiefly by Cabanis & Heine and Bonaparte, and it contains indeed a number of forms which might perhaps be allowed subgeneric distinction, but for the present Dr. Sharpe has no doubt done useful work in gathering them together, though it might be puzzling to distinguish all members of the genus by his key.

These Kingfishers are of medium size, from the size of a Lark to that of a Thrush. The tail is longer than the bill; the inner toe, the hallux, and the tarsus are of about equal length with one another. Compared with *Pelargopsis*, *Halcyon* may be recognised by its flatter bill and rounded culmen; measured at a level with the anterior end of the nostril the bill is as broad as it is deep; in *Pelargopsis* it is very much narrower than deep; on the other side of *Halcyon* may be placed *Dacelo* and *Sauromarptis* of the Australian Region, in these the bill is very broad, the length from the nostril being only 2 to $2^1/_2$ times its width, as against 3—4 times in *Halcyon*, or sometimes over 4 times in *Pelargopsis*.

Mr. Blanford (Faun. Br. Ind. B. III, 1895, 119) recognises *Halcyon* (containing *H. pileata*), *Callialcyon* (*H. coromanda*), and *Sauropatis* (*H. chloris*, also *H. sancta*) as distinct genera; in the first-named the primaries are white at the base and the bill is red; the bill is also red in *Callialcyon*, but in *Sauropatis* it is black on the upper mandible and on the terminal part of the lower. In *Sauropatis* it should be added that the first primary is nearly as long as the second and third and about equal to the fourth; in *Callialcyon*, as in typical *Halcyon*, it is shorter — the tip about half way between the secondaries and the wing-tip. Sharpe (1892) recognises 53 species of *Halcyon*, inhabiting Africa, the warmer parts of Asia to Australia.

88. HALCYON COROMANDA (Lath.).
Ruddy Kingfisher.

This species ranges from the Himalayas, East China, and Japan, south to the Philippines, Celebes, Java, and the Andaman and Nicobar Islands[1]. It varies locally. The Celebes form has received a name, and it may, apparently, be allowed to stand as a subspecies without causing confusion.

---

[1] It is absent, according to Blanford, in the Indian Peninsula.

✱ **Halcyon coromanda rufa** (Wall.).

a. **Halcyon rufa** *(1)* Wall., P. Z. S. 1862, 338.
b. **Dacelo coromandeliana** *(1)* Schl., V. Ned. Ind. Ijsvogels 1864, 24, 56 (Celebes, Sula).
c. **Dacelo rufa** *(1)* Finsch, New Guinea 1866, 160.
d. **Halcyon coromanda** *(1)* Sharpe, Monogr. Alcedin. 1870, 155 (Celebes), pl. 57, front figure (type of *H. rufa*); *(2)* Guillem., P. Z. S. 1885, 548; *(3)* Hickson, Nat. in N. Celebes 1889, 90; *(4)* M. & Wg., Abh. Mus. Dresd. 1895, Nr. 9, p. 3 (Talaut).
e. **Callialcyon**[1] **rufa** *(1)* Wald., Tr. Z. S. 1872, VIII, 44; *(2)* Salvad., Ucc. Borneo 1874, 102; *(3)* id., Ann. Mus. Civ. Gen. 1875, VII, 653; *(4)* Meyer, Ibis 1879, 62, 146; *(5)* W. Blas., J. f. O. 1883, 136; *(6)* Meyer, Isis, Dresden 1884, 6; *(7)* W. Blas., Ztschr. ges. Orn. 1885, 246; *(8)* id., ib. 1886, 90; *(9)* id., Ornis 1888, 572.
f. **Dacelo coromanda** *(1)* Schl., Revue Alcedin. 1874, 17 (Celebes, Sula, Sangi).
g. **Halcyon coromanda var. rufa** *(1)* Brügg., Abh. Ver. Bremen 1876, V, 54.
h. **Halcyon coromanda rufa** *(1)* Stejn., Pr. U. S. Nat. Mus. 1887, X, 403; *(II)* Meyer, Abb. v. Vogelskel. 1897, II, pl. CCXVI.
i. **Halcyon**[2] **rufus** (subsp. Celebos) *(1)* Sharpe, Cat. B. XVII, 1892, 221.
j. **Callialcyon coromanda rufa** *(1)* M. & Wg., Abh. Mus. Dresd. 1895, Nr. 8, p. 7; *(2)* iid., ib. 1896, Nr. 2, p. 12.

"Radja udan mera" (Red King of the Crabs), North Celebes, Meyer e 4.
"Radja udan mera besar", near Manado, Nat. Coll.
"Bengkah mahamu", Siao, Nat. Coll.
"Sumpotito", Tjamba, S. Celebes, Platen e 7.
"Sangkul", Peling, Nat. Coll.
"Pisawato", Karkellang, Talaut, ii d.

Figures and descriptions. Sharpe *d I, i 1*; Meyer *h II* (skel.); Wallace *a 1*; W. Blas. *e 7*.

Adult [♂]. General colour rufous, darker above, chestnut-red on the wings and tail; from crown to scapulars, wing-coverts, upper flanks, upper tail-coverts, middle tail-feathers and outer webs of the others, cheeks, ear-coverts and sides of neck washed with magenta, especially strongly on mantle and scapulars; lower back and rump pale silvery blue; under surface orange-rufous, much darker and slightly washed with magenta across the lower throat, forming a sort of ill-defined collar (near Manado, Aug.—Sep. — C 10862). Iris dark-brown; bill, feet, and claws red (M. *e 4*).

Sexes. Similar.

Immature. Like the adult, but with a weaker magenta wash on the upper surface, seen only on hind head and nape, lower mantle, scapulars, upper wing- and tail-coverts; the blue of the rump darker, azure; the feathers of the cheeks, throat and breast terminally fringed with brown, forming wavy cross-bars; bill red (near Manado, Aug. till Sept. — C 10863).

---

[1] The genus *Callialcyon* was made by Bonaparte in 1850, Consp. Avium, 156. The type of *Entomothera* of Horsfield (1820), mentioned as *Halcyon coromanda* by Sharpe, is not specified in the original text (Tr. L. S. XIII, 173), and *Ceyx tridactylus*, as the first species of the genus, has the best claim to the name. *Callialcyon* differs notably from *Halcyon* in arrangement of primaries, coloration, shortness of tail.

[2] The genus *Halcyon* was made by Swainson in 1820 (Zool. Illustr. pl. 27; Classif. B. 1837, II, 335). Since writing his "Monograph of the Kingfishers" between 1868—70, it appears that Dr. Sharpe has made a mythological discovery. Swainson conceived Halcyon to be one of the forms of Alcyone's name, Aeolus's daughter, who was changed into a Kingfisher, though Canon Tristram doubts it (Ibis 1893, 213). From his recent Catalogue of the Kingfishers it appears that Dr. Sharpe has found Halcyon to be a person of the male sex, presumably a lover of Alcyone's, though Alcyone has hitherto always been accounted the faithful wife of Ceyx.

Young. Two young specimens — perhaps a fortnight out of the nest, with the hard white points for breaking the egg still left on the tip of the upper bill, are like the adults; in one specimen (? ♂ — C 12583) the magenta wash is more strongly developed than in many adult birds, but a slight shade darker; in the other (? ♀ — C 12581) this colour is less strong than in its fellow of the same age. The blue of the rump a trifle darker in the latter, but both are intermediate between adults with the darkest and palest rumps. Feathers of under surface from chin to middle of abdomen terminally fringed with dusky brown, more broadly than in the older immature bird; bill whitish yellow, in the second specimen dusky horn-colour in places. The shape of the maxilla is noteworthy from the manner in which it overlaps at the tip and is bent down, and the mandible also is peculiar from the keel-shaped prominence along the middle. This calls to mind somewhat the shape of the bill in *Melidora*. The adult bird has no hook and no noteworthy keel. Feet and claws pale brown (Siao — Nat. Coll.).

Observation. In Museum-specimens the bright red of the bill fades in course of years into ochraceous buff, or some still paler tint.

Measurements.	Wing	Tail	Tarsus	Bill from nostril
a. (Nr. 6179) ad. Manado	114	75	16	49
b. (Nr. 1906) ad. Minahassa	114	71	17	49
c. (Nr. 6181) ad. Manado	118	—	16	49
d. (C 10862) ad. near Manado, VIII.—IX. 92	116	70	16	47.5
e. (C 10864) ad. near Manado, VIII.—IX. 92	115	—	16.5	49.5
f. (C 10863) imm. near Manado, VIII.—IX. 92	114	—	16.5	49
g. (C 10507) ♂ ad. Siao	118	72	16	46
h. (C 12582) ad. Siao, 28. VI. 93	117	74	15	47
i. (C 12581) juv. Siao, 9. VII. 93	91	24	15	21
j. (C 12583) juv. Siao, 23. VI. 93	97	29	16	21
k. (Nr. 6180) ad. Gt. Sangi	115	72	15.5	46
l. (C 13868) ad. Minahassa, 17. XII. 94	114	71	18	48
m. (C 14201) ad. Lembeh Id., 26. II. 95	121	71	17	48
n. (C 14202) vix ad. Lembeh Id., 12. III. 95	126	69	17	43
o. (Sarasin Coll.) ♂ imm. Minahassa, 17. II. 94	126	71	17	49
p. (Sarasin Coll.) ♂ ad. Macassar, XII. 95	121	69	—	50
q. (Sarasin Coll.) ♂ ad. Macassar, XII. 95	118	70	16.5	51
r. (C 14590) ad. Peling Id., V.—VIII. 95	110	68	15	—
s. (C 14591) vix ad. Peling Id., V.—VIII. 95	114	70	16	45.5
t. (C 14592) ad. Peling Id., V.—VIII. 95	115	68	—	48
u. (C 13780) vix ad. Karkellang, Talaut, XI. 94	127	75	16.5	45

Variation. In Celebes the young on leaving the nest has the rich magenta hue of the upper parts as strongly developed as in the adult; the tarsus (as is usual in birds) attains its full length at a much earlier age than the bill, wing and tail.

Immature specimens, as we take them, which are only recognisable by the impure red of the bill — a certain amount of duskiness being present at its base — lose the magenta tint of the young almost completely (it is most pronounced on the tail) and are simple cinnamon-rufous above.

In the adult bird the magenta tint is again developed and is very rich.

The blue tint on the rump varies independently of age in specimens from the same locality killed at the same date; silvery- or pearl-blue is the commonest colour, but it may be found as deep as azure.

We are not prepared to maintain that the bird varies locally in any part of the province specially treated of in this work. More material from Talaut is, of course, desirable.

Three adults from Borneo of the form conceived by Sharpe as *C. coromanda* (Lath.) are a little smaller than those of Celebes, having wing 105—110 mm; bill 42.5—45. One from Palawan. ♀ scarcely adult, is large, wing 124 mm; bill 48; tarsus 17 (Nr. 12533).

The Bornean race is distinguishable from that of Celebes by the rather darker magenta of the upper surface, and by its slightly smaller size; our immature Palawan bird has no magenta wash above, the under surface ochraceous buff instead of orange-rufous.

Egg of the species. White, 28—29 × 26—27 mm (Nehrkorn MS.).
Egg and breeding habits in Celebes unknown.

Distribution. Celebes, Sula, Sangi and Talaut Islands: — Minahassa (Rosenb. *f 1*, Meyer *e 4*, Guillem. *d 2*, etc.); Lembeh Id. (Nat. Coll.); Gorontalo Distr. (Ros. *f 1*, Meyer *e 4*); Togian Islands (Meyer *e 4*); Macassar (Wallace *e 1*, *i 1*, P. & F. Sarasin); Tjamba, S. Celebes (Platen *e 7*); Peling (Nat. Coll.); Sula Besi and Mangoli (Allen *a 1*, Bernst. & Hoedt *f 1*); Talissi Id. (Hickson *d 3*); Siao (v. Duivenb. *f 1*, Nat. Coll.); Great Sangi (Hoedt *f 1*, Platen *e 9*, Meyer); Karkellang, Talaut (Nat. Coll.).

Without sufficient material from other parts of the range of the species we are obliged to confine our studies to the subspecies of the Celebesian Province. On the variation of the species Dr. Sharpe *(i 1)* makes the following instructive note: "The ordinary form of *H. coromandus* from the Himalayas, Manchuria and Japan is rather pale. The insular forms are darker and richer in colour, especially the one from the Andamans. Specimens from Borneo are rather smaller; but the only race deserving of subspecific separation seems to me to be the bird from Celebes, which Mr. Wallace called *Halcyon rufa*".

The beautiful Kingfisher, *C. coromanda*, appears to be on the whole a stationary species, though probably migratory in the most northern parts of its range. Seebohm writes (B. Japan. Emp. 173): "it is said to be only a summer visitor to Yezzo, but to be a resident in the other islands of the Japanese group". Sharpe *(i 1)* records 16 specimens from Sikkim dated from March to November, but not in the cold season. It appears to be no common bird on the continent of Asia; in India it must be extremely scarce, judging from the remarks of Jerdon quoted by Sharpe *(d 1)* and from the circumstance that the volumes of "Stray Feathers" contain no mention of its occurrence, except in Sikkim; in 1874 (II, 494) Mr. Hume was able to compare "some forty Sikkim and Tenasserim birds" with others from the Andamans. Blanford (Faun. Br. Ind. B. III 1895, 135) records it from "the Lower Himalayas up to about 5000 feet, in Eastern Nepal, Sikkim and farther East". Oates says it is one of the rarest Kingfishers in British Burmah; Davison found it "by no means a common species anywhere" in Tenasserim. From China until the last few years no specimen had been recorded; in "Ibis" 1889, 446; 1891, 484, Mr. Styan was able to refer to a single specimen killed at the mouth of the Yang-tse, and to another from Manchuria now in the British Museum.

Meyer writes that in Celebes it is "generally found in bamboo brushes near rivers, generally several together. It is not a rare bird, but is not to be procured without great patience. In the stomach I found fishes, ants, etc." Davison (S. F. VI, 75) saw it most plentifully in Tenasserim on the coast, though it was also to be met with by inland creeks; in the Andamans Hume (S. F. II, 170) remarked that it affected the gloom of the mangrove swamps, and never visited the clearings or the open coast; in Labuan Whitehead speaks of it as a swamp-loving species, frequenting the beds of Nipa palms near the coast (Ibis, 1890, 20).

Callialcyon coromanda is most likely an ancient form. We agree with Dr. Sharpe in regarding it as the nearest existing *Halcyon*-form having affinity to the red-backed group of *Ceyx*, and it is of some interest to observe that the distribution of that group and of *C. coromanda* is somewhat similar; namely, neither pass into the Australian Region; but *Callialcyon* occurs also in China and Japan, where *Ceyx* does not. The close similarity of the young *Callialcyon* to the adult must not be cited as a token of antiquity; though it is hardly reasonable to regard the slightly differing local races of *Callialcyon* as very ancient, the young nevertheless appear to resemble their parents more closely than they do one another. In other words, the influence of the parents is superimposed upon the ancestral qualities, and obscures them — to what extent we do not know. The young *Callialcyon* described by Sharpe (i 1) appears to correspond with a pale race, while our two from Siao seem by comparison as if saturated with a deeper rufous tint below and with magenta above. All, however, agree in having the under parts crossed with wavy lines of dusky brown.

In view of its slight differentiation as a subspecies, it appears hardly possible to regard *C. coromanda rufa* as other than a recent immigrant to the Celebesian Province from some part of the Sunda Islands or Philippines. The next species, *Halcyon pileata*, which has the same distribution, is a much more thorough-going migrant than *C. coromanda* now is. We conceive that this species may have relinquished the habit of migration in comparatively recent times.

+ 89. HALCYON PILEATA (Bodd.).
Black-capped Kingfisher.

a. **Alcedo pileata** *(1)* Bodd., Tabl. Pl. Enl. 1783, p. 41; *(2)* v. Musschenbr., N. T. Ned. Ind. 1876 XXXVI, 377.

b. **Halcyon atricapilla** (Gm. 1788), *(1)* Gould, B. Asia 1860, I, pl. 45; *(2)* Jerd., B. Ind. 1862, 226; *(3)* Hume; Str. F. 1874, II, 168; 1876, IV, 287.

Halcyon pileata *(1)* Sharpe, Monogr. Alcedin. 1868, 169, pl. 62; *(2)* Armstr., Str. F. 1876, IV, 306; *(3)* Hume & Davison, Str. F. 1878, VI, 75, 499; *(4)* Legge, B. Ceylon 1880, 301; *(5)* Vidal, Str. F. 1880, IX, 49; *(6)* Bingh., t. c. 154; *(7)* Butler, t. c. 382; *(8)* Kelham, Ibis 1881, 380; *(9)* Oates, B. Brit. Burmah 1883, II, 83; *(10)* Guillem., P. Z. S. 1885, 548; *(11)* Büttik., Notes Leyd. Mus. 1887, 38; *(12)* Taczan., P. Z. S, 1888, 462; *(12bis)* Hartert, J. f. O. 1889, 365; *(13)* Everett, J. Str. Br. R. A. S. 1889, 160; *(14)* Whitehead, Ibis 1890, 45; *(15)* Steere, List B. Philipp. B. 1890, 11; *(15bis)* Hagen, T. Ned. Aard. Genoots. 1890 (2) VII, 144; *(16)* Styan, Ibis 1891, 317, 325, 483; *(17)* Stejn., Pr. U. S. Nat. Mus. 1891, XIV, 495; *(18)* De La Touche, Ibis 1892, 479; *(19)* Hose, Ibis 1893, 409; *(20)* Styan, t. c. 433; *(21)* Hart., Nov. Zool. 1894, 480; *(22)* Bourns & Worces., B. Menage Exped. 1894, 34; *(23)* Everett, Ibis 1895, 30; *(24)* Blanf., Faun. Br. Ind. B. III, 1895, 133.

c. **Entomobia**[1]) **pileata** *(1)* Cab. & Heine, Mus. Hein. 1860, II, 155; *(2)* Salvad., Cat. Ucc. Borneo 1874, 102; *(3)* Wald., Tr. Z. S. 1875, IX, 154; *(4)* David & Oust., Ois. Chine, 1877, 75; *(5)* Meyer, Ibis 1879, 61; *(6)* Tacz., J. f. O. 1881, 180; *(7)* W. Blas., ib. 1883, 124, 147; *(8)* id., Ornis 1888, 307; *(9)* Vorderman, N. Tdschr. Ned. Ind. 1889, XLIX, 423; *(10)* Salvad., Ann. Mus. Civ. Gen. 1891, 48; *(11)* Tacz., Faun. Orn. Sib. Orient. 1891, I, 192.

d. **Dacelo pileata** *(1)* Schl., Mus. P.-B. Alcedin. 1863, 27; *(II)* id., Vog. Ned. Ind. Ijsv. 1864, 22, 54, pl. IX, fig. 2; *(3)* id., Revue Alcedin. 1874, 18.

e. **Halcyon pileatus** *(1)* Sharpe, Cat. B. XVIII, 1892, 229.

For further synonymy and references see Sharpe *e 1*.

Figures and descriptions. Gould *b I*; Sharpe *I, e 1*; Schlegel *d II*; Hume *3*; Legge *4*; David & Oust. *c 4*; Oates *9*; Taczanowski *c II*; Blanford *24*.

Adult. Above bright French-blue; entire head, including malar region and nape, and wing-coverts black; chin, throat and a broad collar round the neck white, passing into cinnamon-rufous on lower breast, abdomen, and under tail- and wing-coverts; remiges and tail below black, the primaries for their basal half or more white, forming a broad patch.

Wing 124 mm; tail 80; tarsus 14, bill from nostril 53. (Main, Minahassa, 16. Feb. 1894: Nat. Coll. — C 13241).

"Bill deep red; mouth pale red; eyelids pinkish plumbeous, covered with white feathers, except on the edges where they are black; foot dark red, brownish in front of the tarsus; claws dark horn-colour; iris dark brown" (Oates *9*). Length 280 mm; culmen 61; wing 127; tail 86; tarsus 14 (Sharpe *I, e 1*).

Adult female. Does not differ in colour from the male. Length 280 mm; culmen 61; wing 130; tail 76; tarsus 13 (Sharpe ib.).

Young. Birds of the year have the black of the upper parts and the blue of the back and rump less pure, and the sides of the chest and breast, as also the feathers of the hind-neck collar, marked with crescentic tippings of blackish brown; but in some examples the latter part is striated with brown instead of barred. These crescentic markings appear to remain until the bird is fully aged, as they are present in many specimens which have the upper surface in beautiful adult feather (Legge *4*). Bill shorter, yellow-brown in colour (Styan *16*). Irides bluish grey; legs and feet dark red-brown; soles pale red; bill dark brown, except at extreme tip of lower mandible for about 0.25 in. from tip, the sides and angle of gonys, the gape and one-third of upper mandible from tip gradually coming to a point on ridge of culmen, which

---

[1]) *Entomobia* — nomen nudum.

is of a very pale yellowish orange. Length 190 mm, wing 89; tarsus 16; bill 44.5 (Hume & Davis. *3*).

Eggs. 6; a little more oblong than those of *Alcedo ispida*, shell less smooth and less glossy, somewhat rough in certain places; colour pure and transparent white. Average size: 33.6×28.5 mm (Taczanowski *12*).

Nest. A single nest of this species found by Mr. Kalinowski in Corea was formed in the perpendicular sandy wall of a ravine, at a height of 4 meters. Hole like that of *A. ispida*, entrance elliptical, tunnel a meter deep, nearly horizontal and curved, widened and deepened at the bottom, lined with a thick bed of the bones of frogs and lizards, intermingled with the remains of large insects. The nest appeared to have done service for several years (Taczan. *12*).

Abbé David writes *(c 4)*: "Je l'ai trouvé nichant sur un grand arbre des montagnes de Pekin" [?].

Breeding season. Corea, summer (Taczan. *12*); Yangtse, young birds leave the nest in July (Styan *16*).

Distribution. Bengal and the Indian Peninsula (Jerdon *b 2*, Vidal *5*, Butler *7*); Ceylon (Layard *4*); Burmah (Beavan *c 2*, Oates *9*); Tenasserim (Briggs *c 2*, Davison *3*, etc.); Andaman Is. (Beavan *c 2*, Hume *3*, etc.); Nicobar Is. (Hume *3*); Malay Peninsula (Cantor *c 2*, Kelham *8*, etc.); Singapore (Kelham *8*, Davison *e 1*); Sumatra (Kreling *c 2*, Klaesi *11*, Hartert *12*^{bis}, Modigl. *c 10*); Borneo (Wall. *c 2*, Doria & Beccari *c 2*, Everett, etc. *13*); Bunguran, Natuna (Everett *21*); Celebes, Minahassa (v. Musschenbroek *a 2*, *c 5*, *c 7*, Guillem. *10*, Nat. Coll.); Philippines — Balabac and Basilan (Steere *15*), Tawi Tawi (Bourns & Worcester *22*), Balabac (Everett *23*), Palawan (Whitehead *14*, Platen *c 8*); Cambodia (Mouhot *e 1*); Siam (Schomburgk *c 2*, Conrad *e 1*); Cochin China (Day *c 2*, Oustalet *c 4*); Hainan (Styan *20*); China (David *c 4*, Styan *16*, De La Touche *18*); Corea (Kalinowski *12*); Askold Id. (Taczan. *c 6*); Japan — Nipon (Stejneger *17*).

This species, the Black-capped Kingfisher of Indian naturalists, has hitherto been recorded only twice from Celebes, first by van Musschenbroek who states *(a 2)* that one was shot by his son between Manado and Tanahwangko in November, which was sent to the Leyden Museum *(c 5, c 7)*, and again in 1885 it was found near Kema, 20 miles from Manado, by a native hunter left there to collect by Dr. Guillemard, by whom two immature and one adult birds were obtained. Dr. Guillemard has overlooked van Musschenbroek's notice, which is thus confirmed. The Dresden Museum has recently received two further specimens from Celebes obtained the first at Likoupang, 16. Febr. 1894, the other at Manado, 24. March, 1895. This Kingfisher, like *Alcedo ispida*, is a migratory species, and in all probability the Celebesian specimens obtained were winter migrants from East Asia. In proof of this statement we tabulate the following observations:

Corea: "rather common in summer, nests, and leaves the country for the winter" — Taczanowski *12*.

China: "disappears at the end of summer, retiring to Cochin China" — David & Oustalet *c 4*. "Summer visitant to the Lower Yangtse Basin". "Comes annually to breed in the Yangtse Valley" — Styan *16*.

"Not uncommon at Foochow in spring and from the end of August to the beginning of October. It occurs also sparingly in winter" — De La Touche *18*.

Tenasserim: the numbers vary according to the season. "In January and February excessively numerous along the higher portions of the Pakchan; on going over the same part of the river in May, not a single bird was to be seen" — Davison *3*.

Sumatra: during the rainy season (October — January) a very common bird which one hardly ever sees at other seasons — Hagen *15 bis*.

Palawan: "a winter migrant, arriving about 23rd Sept." — Whitehead *14*.

In addition to these observations attention should be drawn to the numerous dates of the specimens in the British Museum recorded by Dr. Sharpe *(e 1)* viz: India, 5 specimens, December and January; Burma 8, December, January; Tenasserim, 45, September to April; Malay Peninsula, 25 (some without dates), October to March, but one also in August; S. E. China, 3, January. Only one specimen was known from Ceylon at the time of publication of Legge's great work, "which must have been a straggler driven to the coasts of Ceylon by the northerly winds of December". So, too, only a single specimen is as yet known from Japan, a recent addition to the avifauna of that country *(17)*. The migratory habits of the species serve to explain these occurrences.

It is an interesting question why this species is migratory, while *Callialcyon coromanda*, which has very much the same range, is stationary except in the northern parts of its range. The latter species, as Dr. Sharpe has shown, now appears to be undergoing subdivision into many species or subspecies within itself. At one time, we may suppose, it was a migrant like *H. pileata*, but has gradually become almost completely stationary. *H. pileata* is perhaps beginning to settle down in the same way, as its occurrence in South China in summer and winter and in the Malay Peninsula in winter and summer tends to show. The successful multiplication of the southern residents and their gradual spreading would prove a destructive agent to wandering habits on the part of others of the species.

The long Northern Peninsula of Celebes presents a remarkable boundary to species migrating from East Asia to warmer quarters for the winter. This has already been noticed in the case of several species. It is unusual for migrant Asiatic species to pass beyond this boundary into the Moluccas. Nor is this to be wandered at. The Celebes Sea is shut in like a basin with the long wall of North Celebes for its southern border, and the Sangi chain of islands between Mindanao and the Minahassa on the east. Kingfishers, being heavy birds of strong but nervous flight, heading their way straight, apparently, like a bolt towards a given mark, will be little likely to wander away from the path thus made geographically easy for them. This may explain why *Callialcyon coromonda* is not found east of the Celebesian Province.

The present Kingfisher appears to be omnivorous in its food. David never saw it fishing in China, but found it to be especially partial to beetles (*Cantharidae* and *Mylabridae*); the nest already noticed, found by Kalinowski in Corea, was bedded with the bones of frogs and lizards; in the Malay Peninsula Kelham, who has seen it dart down with a splash into the water and catch a frog, found that it fed there upon frogs, small fishes, and crabs *(S)*.

+ 90. HALCYON SANCTA Vig. Horsf.
Sacred Kingfisher.

Halcyon sancta *(1)* Vig. & Horsf., Tr. L. S. 1826, XV, 206; *(II)* Gld., B. Austr. 1840, II, pl. 21; *(III)* Diggles, Orn. Austr. 1869, pl. 2, fig. 2; *(IV)* Sharpe, Monogr. Alcedin. 1870, 239, pl. 91; *(5)* Finsch, Journ. Mus. Godef. 1875, VIII, 50; *(6)* Brüggem., Abh. Ver. Bremen 1876, V, 54; *(7)* E. L. & L. C. Layard, Ibis 1882, 503, 543; *(8)* Finsch, Vög. der Südsee 1884, 8, 24 (Mitth. Orn. Ver. Wien 1884, 55, 95); *(9)* Rams., Pr. L. S. N. S. W. 1886, 2. ser. I, 1086; *(10)* id., Tab. List Austr. B. 1888, 3; *(11)* Cox & Hamil., ib. 1889, IV, 401; *(12)* North, Nests & Eggs B. Austr. 1889, 37; *(13)* Everett, J. Str. Br. R. A. S. 1889, 161; *(14)* Wiglesw., Aves Polyn. 1891, 13; *(15)* Sharpe, Cat. B. XVII, 1892, 267; *(16)* M. & Wg., J. f. O. 1894, 242 partim; *(17)* Madarasz, Aquila 1894, 99; *(18)* M. & Wg., Abh. Mus. Dresden 1896, Nr. 2, p. 12; *(19)* Hart., Nov. Zool. 1896, 551, 571, 595.

a. **Todirhamphus sanctus** *(1)* Bp., Consp. 1850, I, 156; *(II)* Reichb., Hb., Alcedin. 1851, 33, t. 418 fig. 3131—33; *(3)* Wall., P. Z. S. 1862, 338; *(4)* Gld., Hb. B. Austr. 1865, I, 128.

b. **Sauropatis sancta** *(1)* Cab. & Heine, Mus. Hein. 1860, II, 158; *(2)* Wald., Tr. Z. S. 1872, VIII, 44; *(3)* Salvad., Cat. Ucc. Borneo 1874, 104; *(4)* Meyer, Ibis 1879, 62, 145; *(5)* Salvad., Orn. Pap. 1880, I, 476; *(6)* Meyer, Isis, Dresden 1884, 6; *(8)* Guillem., P. Z. S. 1885, 548, 568, 627; *(9)* W. Blas., Ornis 1888, 575; *(10)* Salvad., Orn. Pap. Agg. 1889, I, 58; *(11)* Vorderman, N. Tdschr. Ned. Ind. 1891, 440; *(12)* id., Notes Leyden Mus. 1891, 124.

c. **Dacelo sancta** *(1)* Schl., Mus. P.-B. Alcedin. 1863, 35; *(II)* id., Vog. Ned. Ind. Alcedin. 1864, 27, 59, pl. X, f. 1; *(3)* id., Revue Alcedin. 1874, 26.

"**Dadubatang bahewa**", Karkellang, Talaut.
+"**Saika anej**", Kabruang, Talaut.
"**Bengka budia**", Great Sangi.
"**Sisakomang kadio**", Siao.
"**Bengka**", Tagulandang and Biarro.
"**Kiskis posiposi**", Manado tua.
"**Tengkesi ise ise**", Tonkean, E. Celebes.
"**Tengkesi mosoni**", Banggai.

The above native names have all been communicated to us by our native hunters there.

For further synonymy and references see Salvadori *b 5, b 10*; Sharpe *15*.
Figures and descriptions. Gould *II*; Reichb. *a II*; Schlegel *c II*; Diggles *III*; Sharpe *IV, 15*; Salvad. *b 5*; Ramsay *9*; North *12* (egg).

Adult. Head above, mantle and scapulars bluish sage-green, shading off on the sides of the occiput into verditer-blue, the colour of the back, rump and upper tail-coverts; wing-coverts, quills and tail duller blue, inner webs and ends of quills

blackish; lores orange-rufous; under surface and a broad collar around hind
neck white, pure on chin and throat, washed with orange-rufous on the other parts,
most intensely on the collar, flanks and abdomen; malar stripe, ear-coverts,
continuously with a broad line around neck bordering the blue-green of
the head, blackish; upper edge of mantle blackish; on sides of neck and breast
some faint dusky bars (probably remains of immature plumage) (Siao, 15. VI. 93
— C 12612). Iris brown; tarsus dull green; bill black, basal half of lower mandible
white (Guillem. *b 8*).

Young. Like the adult, but duller in colour; breast and collar barred with dusky; wing-
coverts margined with pale fawn-colour; lores and margins of feathers of forehead pale cinnamon (Great Sangi, 22. VII. 93 — C 12708).

Measurements.

	Wing	Tail	Tarsus	Bill from nostril
a. (Nr. 8177) ♂ ad. Kalindong, N. E. Celebes, April 1871	90	56	11.5	32
b. (C 12612) ad. Siao, 15. VI. 93	86	55	11.5	33.5
c. (C 12613) juv. Siao, 3. VII. 93	89	55	—	—
d. (C 12708) juv. Gr. Sangi, 22. VII. 93	93	60	12.5	33.5
e. (Nr. 8174) imm. Gr. Sangi	89	56	—	32.5
f. (Mus. Berol.) ♂ ad. Togian Is. Aug. 71	90	61	12.5	34.5
g. (Mus. Berol.) ♀ ad. Togian Is. Aug. 71	89	55	—	31
h. (C 13090) imm. Kabruang, 2. XI. 93	96	59	—	38.5

Other specimens are dated: 3 Siao — June till July; 1 Sangi — July (Nat.
Coll.); Tagulandang and Biarro — September; Manado tua — May; Talaut —
autumn; E. Celebes and Banggai — May to August.

Eggs. Usually 5; pearly white; average of five measured 26.2×22.1 mm (North *12*),
25×21.5 mm (Nehrkorn MS.).

Nest. In Australia — eggs deposited on the decaying wood in a hollow branch, or hole of
a tree, usually an *Eucalyptus* (North *12*), the hollow spouts of the gum — and boles
of the apple-trees *(Angophorae)* generally selected (Gld. *II, a 4*); usually a hole 12
to 18 inches long dug in the arboreal nest of white Ants (Ramsay *IV*); in New
Caledonia — nests in holes, sometimes dug by itself in banks, or in hollow trees,
often at a considerable altitude from the ground (Layard *7*).

Breeding season. Australia — commences in October and lasts till December (Gld. *a 4*),
commences in September (North *12*), eggs to be had from the beginning of October
to November 20th in the Mudgee District, N. S. W. (Cox & Hamilton *11*); New
Caledonia — nesting season from November to January (Layard *7*).

Distribution. All parts of Australia and Tasmania (E. P. Ramsay *10*, etc.); Islands of Torres
Straits *(b 5)*; Norfolk Island (Gray *b 5*, Lay. 7)[1]; New Caledonia, Loyalty Islands,
and New Hebrides (Lay. etc. *14*); ? Tonga Islands — Vavao (Verr. *c 3, 14*); Pelew
Islands (Semper *5*); Papuasia from the Solomon Is. and New Ireland to New
Guinea, Waigiou, Aru, Kei (Salvad. *b 5*); the Moluccas from Banda, Ceram and
Buru to Halmahera (Salvad. *b 5*); Sula Islands (Allen *a 3*, Hoedt *c 3*); Togian
Islands (Meyer *b 4*); Celebes — Macassar (Wall. *b 2, 15*); Gorontalo (Rosenb. *c 3*);
Minahassa (Meyer *b 4*, Guillem. *b 8*); Manado tua, Biarro and Tagulandang (Nat.
Coll.); Sangi Is. — Siao (v. Duivenb. *c 3, b 9*, Nat. Coll. in Dresd. Mus.); Great
Sangi (Platen *b 9*, Nat. Coll. in Dresd. Mus.); Karkellang and Kabruang — Talaut

---

[1] According to Sharpe the species inhabiting Lord Howe's Island and Norfolk Island is *H. vagans*.

Is. (Nat. Coll. in Dresd. and Tring Mus.); South Borneo *(b 3, 13)*; S. E. Sumatra (E. C. Buxton *15*); Banka (Vosmaer *c 3*); Billiton (Vorderman *b 11, b 12*); Walter *15*); Mendanao, near Billiton — NB. not Mindanao (Vorderman *b 12*); Java, Lombok, Timor, Timorlaut (Salvad. *b 5, b 10*); Bali and Sumbawa (Doherty *19*).

This Kingfisher is, very probably — if not a migrant to Celebes — increased in numbers by migration to the country during the Australian winter, though it occurs in the Province throughout the year. Few specimens have been obtained on the mainland of the island; in 1874 there was only one example from Celebes in the Leyden Museum, viz. from Gorontalo, and we can only find notice of one specimen from South Celebes, shot by Mr. Wallace near Macassar. Meyer met with it only at Kalinaong in the N. E. Minahassa and gained the impression that it is a rare bird. The only other ornithologist who has recorded it from Celebes appears to be Dr. Guillemard, who obtained a female specimen at Manado, but we have several from our native hunters labelled Tonkean, E. Celebes. It seems to be more common on the smaller islands. In June and July our hunters seem to have found it rather plentiful in Siao, but less so in Great Sangi in July, judging from the proportion of specimens recently sent to the Dresden Museum. Prof. W. Blasius records four young specimens killed by Dr. Platen's hunters on Great Sangi in July and August, and we have examined a number of specimens from Talaut, Tagulandang, Banggai, etc.

In Australia it is known to be a migrant form, though here, as in the tropical parts of its range, a few seem always to be met with. Thus Gould writes *(a 4)*: "It is a summer resident in New South Wales and throughout the southern portion of the continent, retiring northwards after the breeding-season. It begins to disappear in December, and by the end of January few are to be seen: solitary individuals may however be met with even in the depth of winter. They return again in spring, commencing in August, and by the middle of September are plentifully dispersed over all parts of the country". In the Mudgee District Cox & Hamilton observe that it arrives early in September, and it has been noticed there as late as March 1st. Dr. Sharpe records one from North Australia in July, where we suppose it is always present.

In the East Indies — except the S. E. parts — the great majority of properly dated specimens in collections have been obtained when the species is away from the southern parts of Australia, i. e. between March and the end of September, though here and there individuals killed at other times of the year are recorded, showing that the migration to the South east and to Australia is not thorough.

Thus Schlegel records a specimen from Sula, 27th November, and Count Salvadori one from Halmahera, December, and specimens were obtained by our hunters in Talaut in November.

This bird, like others, sometimes goes astray in its migrations. During

Dr. Finsch's cruise in the South Seas a specimen flew on board in 15° S. by 157° E. "The nearest land 300 sea-miles to the north-west was South-east Id. of the Louisiade Group, to the north-east, circa 330 sea-miles, Rennell Id. of the Solomon Group, due west we lay 690 sea-miles from the coast of Australia and east 570 miles from Esperitu Santo in the New Hebrides. This most remarkable and certainly very rare case of flying astray on the part of a bird fitted apparently with such moderate powers of flight calls for so much the more wonder, as we had had very unsettled weather on the preceding days, with heavy storm and squalls from north-west to north-east in all directions of the compass" (8).

In view, perhaps, of the migratory habits of the species Count Salvadori has accepted the localities Tonga Islands and Pelew Islands without query; we question the first, since it rests only upon the evidence of that unreliable authority, Verreaux, and, as to the second, Finsch is of opinion that Semper's label recording a specimen from the Pelew Islands may be erroneous, since the species was not obtained there by Kubary during his 1½ years' collecting in the islands (5), but, after having made inquiries, we have found out that Semper collected birds only on the Pelew Islands and not on the Philippines and that there is no reason whatever to doubt that locality. The birds in question may have been stragglers to these localities.

Notwithstanding the amount of collecting done of late years in North Borneo by Everett, Whitehead and others, this species has only been obtained in the south of the island. We infer that the bird reaches this part of Borneo from the chain of islands running from Sumatra to Timor and Timorlaut. In West Australia birds apparently differ from those of East Australia, but in what way we cannot say, for Gould (a 4) says the former are "a trifle larger in all their measurements; but otherwise present no differences of sufficient importance to warrant their being considered as distinct"; whereas Dr. E. P. Ramsay (9) says they are "slightly smaller and of a clearer blue on the back than our N. S. W. specimens, with a narrow, well-defined white collar and nuchal spot. Wing 93 mm, bill from nostril 38, total length 178". From these measurements it would appear that the word "smaller" may be a *lapsus calami*, "larger" being intended. As in the case of *Merops ornatus*, we suspect — should differences be proven — that specimens from Sumatra to Timor and S. Borneo will be found to have more in common with West Australian than with East Australian, Papuan, Moluccan and Sangi birds, and that the former migrate across the Timor Sea, the latter across Torres Straits; but this opinion we utter with all reserve.

The haunts of this Kingfisher are very varied: "the most thickly wooded brushes, the mangrove forests which border, in many parts, the armlets of the sea, and the more open and thinly timbered plains of the interior, often in the most dry and arid situations far distant from water" (Gould *a 4*).

Its food is similarly various; Gould speaks of *mantes*, grasshoppers, caterpillars, lizards, and very small snakes; E. P. Ramsay *(IV)*, and Cox & Hamilton *(11)*, unlike Gould, have seen it plunge into the water also and take fish.

A very near ally of this species is *H. vagans* of New Zealand, which Dr. Sharpe speaks of as "a large and richly coloured island race of *H. sanctus*". Another closely related form is *H. chloris*, the next species on our list, from which, as Salvadori remarks, it is easily distinguishable by its much smaller dimensions, and by the fulvous colour of the cervical collar, and of the under parts *(b 5)*.

91. HALCYON CHLORIS (Bodd.).
White-collared Kingfisher.

This plentiful species is distributed from the coasts of the Red Sea and India south and east as far as the Pelew Islands, the Solomon Islands, the New Britain Group, and perhaps even the Fiji Islands. According to Dr. R. B. Sharpe's recent researches (Cat. B. XVII, 1892, 272—283), a tendency to become differentiated into local forms finds expression in many different localities within this vast range, yet the differences are such that — locality unknown — it is certain that an adult specimen could not be determined as this or that form by means of a description alone, to say nothing of younger specimens. Moreover, "here and there", as Dr. Sharpe says, "is to be found a specimen which seems to be intermediate between the races; and as several of the latter appear to be migratory, it is quite possible that some of them hybridize with the resident birds of the countries which they visit". Instead, therefore, of treating these supposed races as being several distinct species and subspecies, as Sharpe has done, we cannot help thinking that the whole should have been viewed as one species, *H. chloris*, which perhaps branches out into numerous subspecies. Working ornithologists are well aware that it is much easier to make a new species than to "kill" an old one — especially so long as the author of it is living — and most inconvenient in this way are species which are entitled only to rank as possible subspecies Naturalists either entirely admit their validity as species or entirely ignore them, and the result is confusion and mutual disrespect for the other's judgment. Would the former only consent to attach a trinomial or some other token to indicate that the variation in the species comes to a particular head there, this stumbling-block would be done away; it would then become understood that he who employs the simple binomial speaks broadly of the species as a whole, while he who employs some special additional sign refers thereby to some particular section of the species. That the best way to do this latter is to employ the tri-, quadri-, quinqui-, etc.-nomials suggested by American ornithologists we do not believe; when the need for it becomes sufficiently pressing, it is to be hoped that some much briefer, simpler method will be found.

We have paid great attention to this species and have before us at this moment a series of 80 skins from the Celebesian area alone, not to speak of a great number of others which have passed through our hands in the last few years, yet we do not feel able to say anything about it, except that it varies individually in the same localities to a very great extent, both in size, in the size and form of the bill, in the different shades of blue and green of the upper surface, and in the presence or absence of bars on the under surface; moreover, as soon as the bird can fly, it wears the adult dress and even experienced ornithologists might take such, before they are full grown, for a smaller species or subspecies.

+ The typical Halcyon chloris.

a. **Alcedo chloris** *(1)* Bodd., Tabl. Pl. Enl. 1873, p. 49; *(2)* Rosenb., Malay. Archip. 1878, 272.

b. **Alcedo collaris** (Scop.), *(1)* Kittl., Kupfert. Vög. 1833, I, 10, pl. 14, fig. 1.

c. **Halcyon collaris** *(1)* Swains., Zool. Illustr. 1820, I, pl. 27.

d. **Halcyon chloris** *(1)* Gray, Gen. B. 1846, I, 79; *(II)* Sharpe, Monogr. Alcedin. 1870, 229, pl. 87; *(3)* Hume & Davison, Str. F. 1873, I, 451; *(4)* Brüggem., Abh. Ver. Bremen 1876, V, 53; *(5)* Lenz, J. f. O. 1877, 367; *(6)* Sharpe, P. Z. S. 1879, 332; *(7)* Everett, J. Str. Br. R. A. S. 1889, 161; *(8)* Whitehead, Ibis 1890, 45; *(9)* Steere, List Coll. B. Philipp. Is. 1890, 11; *(X)* Sharpe, Cat. B. XVII, 1892, 273, pl. VII, fig. 3; *(11)* Bütt., Zool. Erg. Weber's Reise in Ost-Ind. 1893, III, 275; *(12)* M. & Wg., J. f. O. 1894, 242; *(13)* Studer, Gazelle-Reise 1889, III, 198; *(14)* Hartert, Ornis 1891, 121; *(15)* Bourns & Worces, B. Menage Exp. 1894, 34; *(16)* Vorderm., N. T. Ned. Ind. 1895, LIV, 335; *(17)* Hart., Nov. Zool. 1895, 474; *(18)* M. & Wg., Abh. Mus. Dresden 1895, Nr. 8, p. 7; *(19)* iid., ib. Nr. 9, p. 3; *(20)* iid., ib. 1896, Nr. 1, p. 8; *(21)* iid., ib. Nr. 2, p. 12; *(22)* Büttik., Notes Leyden Mus. 1896, XVIII, 172; *(23)* Hart., Nov. Zool. 1896, 158, 175, 562, 571, 586, 595; *(24)* id., ib. 1897, 158; *(XXV)* Meyer, Vogelskel. 1807, pl. CCXVI.

e. **Halcyon forsteni** *(1)* Bp., Consp. 1850, 157 (ex Temm. MS.); *(2)* Gray, HL. 1869, I, 93; *(III)* Sharpe, Monogr. Alcedin. 1870, 235, pl. 89; *(4)* id., Cat. B. XVII, 1892, 279.

f. **Todiramphus forsteni** *(1)* Reichb., Hb. spec. Orn., Alcedin. 1851, 30.

g. **Cyanalcyon forsteni** *(1)* Bp., Consp. Vol. Anisod. 1854, 9, Nr. 313.

h. **Dacelo forsteni** *(1)* Schl., Mus. P.-B. Alcedin. 1863, 37; *(II)* id., Vog. Ned. Ind. Alced. 1864, 29, 60, pl. XI, f. 1; *(3)* Finsch, Neu Guinea 1865, 161.

i. **Sauropatis forsteni** *(1)* Wald., Tr. Z. S. 1872, VIII, 44; *(2)* Meyer, Ibis 1879, 62.

j. **Todiramphus collaris** *(1)* Wall., P. Z. S. 1862, 348.

k. **Sauropatis chloris** *(1)* Cab. & Heine, Mus. Hein. 1860, II, 160; *(2)* Wald., Tr. Z. S. 1872, VIII, 44; *(3)* Salvad., Cat. Ucc. Borneo 1874, 103; *(4)* id., Ann. Mus. Civ. Gen. 1875, VII, 653; 1876, IX, 53; *(5)* Meyer, Ibis 1879, 61, 146; *(6)* Salvad., Orn. Pap. 1880, I, 470; *(7)* W. Blas., J. f. O. 1883, 135; *(8)* Kutter, ib. 1882, 171; 1883, 301; *(9)* Meyer, Isis, Dresden 1884, 6, 19; *(10)* Guillem., P. Z. S. 1885, 256, 504, 547, 568; *(11)* W. Blas., Ztschr. ges. Orn. 1885, 244; 1886, 90; *(12)* id., Ornis 1888, 307, 573; *(13)* Salvad., Orn. Pap., Agg. 1889, I, 57; *(14)* Hickson, Nat. in N. Celebes 1889, 90.

l. **Dacelo chloris** *(1)* Schl., Mus. P.-B. Alcedin. 1863, 32; *(II)* id., Vog. Ned. Ind. Alcedin. 1864, 26, 58, pl. X, f. 3, 4; *(3)* id., Rev. Acc. 1874, 21.

m. **Halcyon meyeri** subsp. *(1)* Sharpe, Cat. B. XVII, 1893, 282.
n. **Halcyon sancta** (nec Vig. & Horsf.) *(1)* M. & Wg., J. f. O. 1894, 242, pt.; *(2)* iid., Abh. Mus. Dresden 1895, Nr. 9, p. 3.
"**Saika**", Talaut, Nat. Coll.
"**Bengka ngake**", Great Sangi, iid.
"**Sisakomang**", Siao, iid.
"**Sisakomang koke**", Siao, immature, iid.
"**Radja udan biru**", (Blue King of the Crabs), Malay name, Minahassa, Meyer *k 5*.
"**Kikiskatanaän**", Alfurous, Minahassa, M. *k 5*.
"**Kiskis**", Minahassa and the islands off the coast, Nat. Coll.
"**Kiss-kiss**", Talissi, Hickson *k 14*.
"**Doë**", Gorontalo?, v. Rosenberg *a 2*.
"**Tjiki**", Tjamba, Macassar, and Maros, S. Celebes, Platen *k 11*.

For further synonymy and references compare Salvadori *k 6*, *k 13* and Sharpe *d X*.

Figures and descriptions. Kittlitz *b 1*; Swainson *o 1*; Sharpe *d II, d X, e III, e 4*; Meyer *d XXV* (skeleton); Schlegel *l II*; Hume *d 3*; Brüggemann *d 4*; Salvadori *k 6*; W. Blasius *k 11*; Schlegel *h II (H. forsteni)*.

Fürbringer (Untersuchungen 1888, pl. III, fg. 250, pl. IV, fg. 37) has figured some parts of the skeleton.

Adult. General colour above greenish blue, becoming more cerulean on sides of head and nape, wing-coverts, quills, and tail; lores fulvous white; subloral region, ear-coverts, continuously with a narrow band round neck, black, on the ear-coverts washed with blue; entire under surface, a broad collar and a very small nuchal mark above the black band white (N. Celebes, C 3618).

The white nuchal spot is not noticeable in most specimens.

Iris brownish; bill black, below reddish white (M. *f 5*). Iris brown; bill black, basal half of under bill yellow-white; feet black (Platen *f 11*).

Sexes. Similar in colour (Sharpe *d X*).

Young. Similar to the adult in coloration, recognisable as young by the small bill — in this specimen further by the wing and tail not yet full grown. The bill is short (from nostril 29 mm), the extreme tip white, and decurved, it is also less broad, width where the first submalar feathers sprout 15 mm, as against 17.5 in an adult. This specimen is of a strong green tint above and slightly barred below; those described by Brüggemann below are of a fine blue colour above, and not barred below (♂ juv. Macassar, 11. IX. 95: P. & F. Sarasin).

Two specimens. "Bill almost as broad as in the adult, but only 28 mm long, dark horn-grey, the tips of both mandibles white. Upper mandible strongly hooked, also the tip of the under mandible somewhat turned downwards. Feathers of the head still entirely enclosed in the silver-white sheaths. So too on the back, the wings, and under surface many sheaths are still noticeable. Wing 79 mm. Above fine blue. Below pure white, in one specimen with slight traces of blackish borders on the feathers of the breast" (Brüggem. *d 4*).

Brüggemann rightly remarks that the form of bill in the young bird is of special interest and fully confirms the views to which Sharpe has given utterance in his excellent review of the genetic affinities of the Kingfishers.

The young vary in tint from bright China-blue to verdigris-green, just as do the adults. We believe a white nuchal spot is always present, though it is often completely concealed and probably often not seen in life. There is often nothing to distinguish the young but their smaller bill, and just this point is likely to lead them to be taken for a smaller race. For a long time we took a small series of specimens from Talaut for a race of the small *H. sancta (n 1, n 2)*, then with an increased series of about 25 specimens we labelled it with a ? "small race of *chloris*"; finally, having studied the young and found every possibly transition between specimens with bills of 30 to adults with bills 47 mm in length, it appears obvious that the small specimens are simply young ones which have not yet attained their full size.

Measurements.	Wing	Tail	Bill from nostril	Tarsus
North Celebes, 10 adults	105-114	66 c.	33.5-40.5	13 c.
North Celebes, 2 imm.	104, 105	65, 68	33.5-38	—
S. Cel., Tanette, 1 ad., Sept., Maros, ♀ ad., Nov. 1871	109-114	62-65	35, 35	—
Manado tua and Mantehage Is., 4 imm., April 1893	108-112	65 c.	33.5-41	—
Banka Id. off the Minahassa, 2 ad., May 1893 . .	105, 106	63, 70	37.5-42.5	—
Siao, 1 ad., June 1893	114	68	40	—
Siao, 2 imm., June 1893	104, 105	62, 64	34, 36	—
Great Sangi, 3 ad., July 1893	104-111	68-69	37-41	—
Togian Is., ♂ ad., ♀ ad., August 1871	103, 106	65, 65	37, 37	—
Talaut Is., Kabruang, 5 ad., Salibabu, 2 ad., Oct.-Nov. 93	110-118	68-75	43-46	—
Talaut Is., Kabruang, 4 imm., Nov. 1893	108-113	66-73	38-43	—
East Celebes, Peling and Banggai — 33 examples	102-119	—	36-47	—
Young and immature specimens.				
Karkellang, 12, Kabruang, 1, late autumn 93, 94, 96	90-104	54-65	30-38	13 c.
E. Celebes and Peling, 5 juv., May-August 1896 .	101-105	—	34-37	—

Eggs. Mindanao: 4 in number; 28.7—31.2 × 23.1—24.0 mm; pure white, smooth, glossy, short oval, or somewhat ovate, fine-grained with pores only slightly indicated (Kutter *k 8*); Labuan: 3, pure white; 29.2—30.5 × 24.1—25.4 (Sharpe *d 6*).

Nest. In holes in trees, or in a termites' nest (Kutter *k 8*). The Sarasin Collection contains a fragment of the rotten stem of a tree with a hole of about 45 mm diam. through it, apparently pecked out by the bird, and the lining of the nest, consisting only of a small quantity of vegetable fibres and grasses (Tomohon, orig. label lost).

Breeding season. Mindanao, March and April (Kutter); Labuan, nest found March 22nd (Low *d 6*).

Distribution. In the Celebes Province: — Talaut Is. (Nat. Coll. Dresden & Tring Museums) Great Sangi (Bruijn *k 4*, Platen *k 12*, Nat. Coll.); Siao (Hoedt *l 3*, v. Duivenb. *l 3*: Nat. Coll.); Tagulandang, Ruang (Nat. Coll.); Talissi (Hicks. *k 14*); Banka, Mantehage, Manado tua and Lembeh (Nat. Coll.); Minahassa (Meyer *k 5*, etc.); Gorontalo (Rosenb. *l 3*, etc.); Togian Is. (Meyer *k 5, m 1*); E. Celebes (Nat. Coll.); Kendari, S. E. Peninsula (Beccari *k 4*); South Peninsula (Wall. *k 2*, Platen *k 11*, Meyer *k 5*, Weber *d 11*); Peling and Banggai (Nat. Coll.); Sula Islands — Sula Besi (Wall. *j 1*, Bernst. *l 3*).

As is commonly supposed to be the case with excessively numerous species, this Kingfisher is subject to very considerable individual variation — for a Kingfisher indeed, in which family the species are very stable, the individual variation of *H. chloris* may be termed great. A couple of hundred specimens or more of this species from the Celebesian area have passed through our hands in the last few years and it has on several occasions received close attention from us. The result is, as already indicated, that we are unable to point to a single local race or subspecies; if local variation exists, it is entirely swamped by the individual variation, and the racial development could only be ascertained by long investigation on hundreds of specimens and by so determining the average individual of this or that spot. Some of the more prominent points of variation in the Celebesian area are: the bill-length from 30 mm (young—but not appearing so) to 47 mm, its width noticeably different, the culmen sometimes slightly recurved; the white nuchal spot, sometimes large, sometimes present only on the basal part of the feathers and concealed, the tips being blue; the colour of the ear-coverts — blackish, or strongly washed with blue or green; the colour of the upper surface varying, as already pointed out, from bright China-blue to verdigris-green. At one time we thought the Talaut race ran larger in size and in size of bill than in Celebes, a larger series has, however, shown that this is not the case; at another time we thought that Talaut possessed two races — a large and a small one, but later investigations simply go to prove that the small one is young and not full-grown. Dr. Sharpe has broken up *H. chloris* into many species and subspecies: were his views to be adopted, not only *H. chloris*, but also *H. armstrongi* Sh., *H. forsteni* Bp., *H. solomonis* Salvad., *H. humii* Sh., *H. meyeri* Sh. and perhaps others would apparently have to be admitted into this work — all from the Celebesian region. *H. humii* we hold simply for young, *H. armstrongi*, *H. solomonis*, and *H. forsteni* for individual variations.

In some of the Moluccas Dr. Sharpe finds that specimens with black ear-coverts joining a very broad black nape-band are prevalent, but does not find it possible to draw a line of division between them and others from the same or neighbouring islands. Most of the Moluccan examples in the Dresden Museum have a slight wash of blue on the ear-coverts: on the contrary those of Timorlaut are without colour on this part, being simply black.

In Malacca, Sumatra, and Borneo there can be no doubt that *the typical chloris* intergrades with *H. armstrongi* of the Burmese countries; Sharpe records both forms from the two latter localities, and there is no reason to believe that they separate themselves at a given season by migration.

Again, in Acheen, Sumatra, Dr. Sharpe records both *the typical chloris* and *H. humii* of the Malay Peninsula and Siam at the same date.

Count Salvadori questions the identity of specimens from the Pelew Islands with *H. chloris*; a specimen from there in the Dresden Museum is remarkably green on the head, mantle and scapulars, but not more so than one from Timorlaut, where much bluer specimens also occur.

We unite *Halcyon forsteni* with this form of *H. chloris*, believing, like Count Salvadori *(k G)*, that it is only a melanotic variety of it. The feathers of the under surface of the type of *forsteni* which we have seen, are muddled with greenish black, not regularly barred as shown in Sharpe's *(e III)* figure of it. The head, back, and scapulars of this specimen are very green. It is worthy of note that those specimens from the Celebes Province which are greenest above most usually have the breast barred with dusky, while those of the bluest upper surface are generally pure white below.

Sharpe has separated the two specimens from the Togian Islands in the British Museum as *H. meyeri*, a subspecies of *H. humii*, on the ground of their possessing black ear-coverts and nuchal collar, a wash of fulvous on the sides of the body, and the bill upturned at the tip in a very conspicuous manner. After examining the two specimens from Togian (Meyer) in the Berlin Museum, kindly forwarded to us for the purpose by Prof. Reichenow, we find ourselves unable to admit the racial distinction of Togian birds; the bill in these two examples does not incline upward more than in many specimens from the mainland of Celebes, the ear-coverts of the female are blue, of the male washed with blue, the sides are indeed slightly fulvous, but this is also seen in some Celebesian specimens. The blue of the upper surface does not differ from that of some specimens from Celebes. Whether the greener-backed birds also occur in Togian, as they do in Great Sangi, Talaut and elsewhere, remains to be ascertained. *H. chloris* in the East Indies may be compared to the Sparrow in Europe, so far as variation is concerned.

GENUS MONACHALCYON Rchb.

These Kingfishers are of rather large size, and may be recognised by their having the tarsus as long as the middle toe without the claw; the 4th and 5th primaries are longest, the 1st shorter than the secondaries; the bill is red, conical, deeper than wide across the nostril, the length from the nostril less than 3 times the width; the tail is long, about 3 times the length of the bill from the nostril, rounded. The typical forms of Celebes are forest-haunting birds. Four species, one in Flores.

* 92. MONACHALCYON MONACHUS (Bp.).
Blue-cowled Kingfisher.

The typical form of this species inhabits the Northern Peninsula of Celebes and has a China-blue head. In the Eastern Peninsula a form with a black head occurs. From Western Celebes Mr. Hartert has described a bird which is treated by him as a subspecies of the typical form, but further proof is required before the very different black-headed *M. capucinus* can be brought in as a third subspecies. The young of the typical *monachus* is known to be like the adult; the young of *M. capucinus* is not known.

+ 1. The typical **Monachalcyon monachus**.

a. [Dacelo princeps Forsten in lit.].
b. **Halcyon monachus** [*(1)* Gray, Gen. B. I, 1846, 79, pt. (descr. null.)]; *(2)* Bp., Consp. 1850, I, 154; *(3)* Cass., Cat. Halc. Philad. Mus. 1852, 8; *(4)* Wall., Ibis 1860, 142; *(5)* Brüggem., Abh. Ver. Bremen 1876, V, 50.
c. **Dacelo monachus** [*(1)* Gray, List Fissirostr. Br. Mus. 1848, 53, pt. (descr. null.)]; *(2)* Blyth, Cat. B. Mus. As. Soc. 1849, 46.
d. **Monachalcyon princeps** *(1)* Wald., Tr. Z. S. 1872, VIII, 43; *(2)* Salvad., Ann. Mus. Civ. Gen. 1875, VII, 653, pt.; *(3)* Meyer, Ibis 1879, 60; *(4)* Salvad., Orn. Pap. 1880, I, 503; *(5)* W. Blas., J. f. O. 1883, 135.
e. **Halcyon princeps** *(1)* Cass., Cat. Halc. Philad. Mus. 1852, 8.
f. **Paralcyon princeps** *(1)* Bp., Consp. Vol. Anisod. 1854, 9.
g. **Dacelo princeps** pt. *(1)* Schl., Mus. P.-B. Alced. 1863, 24, pt.; *(II)* id., Vog. Ned. Ind. Alced. 1864, 20, 53, pl. 7, fig. 1, 2 (only); *(3)* Gray, Hl. 1869, I, 89; *(4)* Schl., Revue Alced. 1874, 16.
Monachalcyon monachus *(I)* Sharpe, Monogr. Alced. 1870, 255, pl. 98, figs. 1, 2 (only); *(2)* Lenz, J. f. O. 1877, 367; *(3)* Reichnw., J. f. O. 1883, 122; *(4)* Guill., P. Z. S. 1885, 547, pt.; *(5)* Tristr., Cat. Coll. B. 1889, 94; *(6)* Heine & Rchnw., Nomencl. Mus. Hein. 1890, 168; *(7)* Sharpe, Cat. B. XVII, 1892, 294; *(8)* M. & Wg., Abh. Mus. Dresden 1896, Nr. 2, 13; *(9)* Hart., Nov. Zool. 1896, 562.
h. **Actenoides hombroni** *(1)* Reichnw. (nec Bp.)., J. f. O. 1877, 218.
i. **Monachalcyon monachus monachus** *(1)* Hart., Nov. Zool. 1897, 163.
"**Radja udan kapala biru**" (King of the crabs with blue head), Minahassa, Malay, Meyer *d 3*, Nat. Coll.
"**Kikis tambo**", Minahassa, Alfurous, Meyer *d 3*.
Figures and descriptions. Sharpe *I, 7*; Schlegel *g II, g 1*; Brüggemann *b 5*.

Adult male. General colour above green-olive, tail, upper tail-coverts and outer webs of primaries washed with verditer-blue; entire head, including ear-coverts and malar region, dark China-blue, paler on sides; chin and throat white; remaining under parts orange-cinnamon-rufous, continuous with a narrow collar round neck darker and redder, scarcely meeting behind; under wing-coverts pale buffy cinnamon. Bill red (near Manado, Aug.—Sept., C 10860, March, Nr. 6148).

Young (male). Two specimens, about a fortnight out of the nest, are just like the adult male, but with the breast crossed with wavy lines of dusky; the lores and forehead touched with cinnamon; bill short, tip a little bent down, horn-brown, yellowish in places, especially towards tip (Manado tua Id., 8. IV. 93, C 12257, Lotta, Minahassa, 14. IV. 93, C 12256).

Female. Like the adult male, but with only the upper part of head blue — malar region, ear-coverts, sub- and supra-orbital region, lores, and bases of feathers of forehead, cinnamon-rufous like the under surface, but darker except above the eye; chin and throat wood-brown (Manado, March, C 2286). Three other specimens, marked female (C 2285, Nr. 6152, 6151) have a greater or less amount of greyish blue on the malar region, ear-coverts and under the eye.

Observation. From the circumstance that the specimen (Nr. 6151) with most blue on the sides has a somewhat shorter and more immature-looking bill, we believe that the young female will be found to resemble the male, having the entire sides of the head blue, but that in the adult female these parts gradually become cinnamon-rufous, whereas the old male hardly differs from the young. Brüggemann (*b 5*)

Birds of Celebes: Alcedinidae.

was certainly in error in stating that this species appears in two different dresses according to the season, the first dress described being that of the adult male, the second of the female. We have specimens in both dresses killed in March and again six months later in August—September, and further properly sexed examples described or remarked upon by Schlegel *(g 1)*, Salvadori *(d 2)*, and Sharpe *(7)*, prove that the differences are of a sexual nature.

Measurements.	Wing	Tail	Bill from nostril	Tarsus
a. (Nr. 6148) ♂ ad. Manado, Mch. 1871 (M.) . . .	146	113	37	—
b. (C 10860) [♂] ad. Manado; Aug., Sept. 1892 (N. C.)	149	119	38.5	20
c. (C 2282) ♂ vix ad. Manado, Mch. 1871 (M.) . . .	141	—	36	—
d. (Nr. 6153) [♀] ad. Manado	146	—	38.5	21
e. (C 2286) ♀ ad. Manado, Mch. 1871 (M.)	142	121	39	21
f. (C 2285) ♀ ad. Manado, Mch. 1871 (M.) . . .	150	117	40	21
g. (Nr. 6152) ♀ ad. Manado, Mch. 1871 (M.) . .	144	113	39.5	—
h. (Nr. 6151) ♀ vix ad. Manado, Mch. 1871 (M.) . .	144	116	37	—
i. (Nr. 6150) ♀ ad. Manado . . . , ...	147	—	39	—
j. (C 10859) [♀] vix ad. Manado, Aug., Sept. 1892 (N. C.)	143	114	37	—
k. (C 12256) juv. Lotta, Msa., 14. IV. 93 (N. C.) . . .	122	58	28	20
l. (C 12257) juv. Manado tua Id., 8. IV. 93 (N. C.) . .	120	59	26	19
m. (C 14192) [♂] ad. Lembeh Id., Feb. 1895 (N. C.) .	146	110	40	21
n. (C 14193) [♂] ad. Lembeh Id., Mch. 1895 (N. C.) . .	141	110	35	—

Observation. In old examples the bill appears to be thinner as well as longer, being a little more narrowed at the sides near the base and the ridge of the culmen sharper.

Eggs. 3; white and transparent; 30 × 25 mm (Meyer *d 3*).

Nest. In a white ants' nest the size of a gourd. Half of the ants' nest destroyed by the bird, cavity for the eggs six inches in diameter, entrance two inches in diameter and nine inches long. The nest was still partly inhabited by the ants (Meyer *d 3*). Two other eggs of this species were found by Meyer lying on the ground and dirty.

Breeding season. The above nest was found in the Minahassa, March 17th. The young fly — as shown by the two specimens in the Dresden Museum — at the end of March or in April.

Distribution. Celebes, Northern Peninsula: Minahassa (Forsten *g 1*, Wallace *b 4*, I, Rosenb. *g 4*, Meyer *d 3*, etc.); Gorontalo Province (Rosenb. *g 4*); Manado tua Id. and Lembeh Id. (Nat. Coll. in Dresden Mus.).

† 2. Monachalcyon monachus intermedius Hart.

Nov. Zool. 1897, 163.

Diagnosis. Differs conspicuously from the typical form of N. Celebes in having the head of a much deeper blue, and with a distinct, though faint, greenish tinge; the tail a little less washed with blue; the breast and abdomen a shade lighter than in the most males of the typical form, the bill apparently a little thicker. Wing about 142 mm; tail about 127; beak 50 (Hartert).

Distribution. Western Celebes — Tawaya (Doherty).

Observation. A single specimen, a male, was obtained by Mr. Doherty at the above spot. Mr. Hartert had the great kindness to bring this specimen, together with the only known female of *M. capucinus*, to Dresden at the meeting of the German Ornithologists' Union 1897, and he agrees with us that, while *M. capucinus* may be allowed to stand as a good species, easily distinguishable by its greenish black head, the form *intermedius* should rank only as a race of *M. monachus* typical.

As Wallace and Meyer remark *(b 4, d 3) M. monachus* lives, not near river-banks, but in the forest, and Wallace noted that it is insectivorous in its habits, feeding on *coleoptera*, *gryllae*, etc. *(I)*. The bird appears to moult about August, as shown by a specimen in the Dresden Museum (C 10860).

Touching its relationship with other genera Sharpe considers (Monogr. p. XLV) that *Monachalcyon* "shows no direct affinity to any existing genus, and the only place I can assign to it is in the vicinity of *Tanysiptera* ... It is very probably derived from the same parent-stock, and, being isolated in the island of Celebes, has been modified into its present form". The author also remarks (p. XVIII) that in form of bill *Monachalcyon* seems intermediate between *Tanysiptera* and *Halcyon*, being in general plumage not far removed from the cinnamon group of *Halcyones*. We differ in so far from Dr. Sharpe as to think that *Monachalcyon* may be described as a small-billed and long-tailed *Halcyon*, and that its nearest existing allies are the long-tailed *Halcyones* of the Philippines, *H. hombroni* of Mindanao and *H. lindsayi* and *moseleyi* of Luzon and Negros. The first-named species has indeed been confounded with *M. monachus* by so experienced an ornithologist as Prof. Reichenow *(3)*. In his recent Catalogue of the Kingfishers (Cat. B. XVII, 1892, 296) Dr. Sharpe includes *Caridonax fulgidus* of Flores and Lombok in the genus *Monachalcyon*, and, like Hartert, we acquiesce in this view, though this species has a rounded culmen (in *M. monachus* the ridge is rather sharp), and the tail much more graduated, the outermost feather being only $^{3}/_{7}$ the length of the tail instead of about $^{5}/_{6}$ as in *M. monachus*, and it is said to haunt low woods and thickets, while *M. monachus* is a forest bird. It differs moreover so greatly in coloration that it may prove to be less nearly related to *Monachalcyon* than are the nearest members of *Halcyon*, *H. hombroni* and its allies.

† * 93. MONACHALCYON CAPUCINUS M. & Wg.

Black-cowled Kingfisher.

Plate IX.

Monachalcyon capucinus *(1)* M. & Wg., Abh. und Ber. Mus. Dresden 1896, Nr. 2, p. 12; *(2)* Hart., Nov. Zool. 1897, 160.

a. **Monachalcyon monachus capucinus** *(1)* Hart., Nov. Zool. 1897, 163.

"Bukaka daka daka", Tonkean, E. Celebes (Nat. Coll.).

Diagnosis. Differs from *M. monachus* of North Celebes by having the head and face black (instead of China-blue), the tail olive-green (not washed with blue), the remiges dusky olive-green (not washed with blue), the remaining upper parts clearer olive-green, the

whitish of the chin and throat not extending so far down towards the jugulum (type [♂] Tonkean, East Celebes, May—Aug. 1896: Nat. Coll. — C 14761).
Wing 150 mm; tail 115; bill from nostril 41; tarsus 21.

Female. Like the male, but the superciliary region, face and ear-coverts ferruginous, below the eye blackish and at the base of bill varied with blackish; back a shade yellower olive; "eye brownish grey, bill orange [red], feet orange-brown"; wing 146 mm; tail 120; tarsus 22; bill from nostril 42 (♀, Macassar, July 1896: W. Doherty in the Tring Museum).

Distribution. Eastern and Southern Peninsulas of Celebes: Tonkean (Nat. Coll.), Macassar (Doherty).

At the time of writing only two specimens of this species are known, the type, a male in the Dresden Museum, and a female from Macassar in the Tring Museum most kindly lent to us for comparison by Mr. Hartert. Possibly the head of the latter is not quite so deep black as in the male, the middles of the feathers having a stronger sea-green tint and being more powdery-looking. The feathers appear to be fresher than in the type, the more powdery look is probably transient and the head might get blacker with wear. The back is a shade yellower olive. These differences are so small that they will probably be more than bridged over by individual variation. There is a gloss of blue-green on the head of the male, hardly perceptible except on the cheeks and ear-coverts, and the concealed middles of the feathers are greyish — not clear black. It would be a matter of much interest if this well-marked form should prove in other parts of Celebes to pass into the typical *M. monachus* by imperceptible gradations. Mr. Hartert, as shown above, has described a form from Tawaya, West Celebes, which he considers intermediate (cf. *supra*). It differs from *M. capucinus* by having a deep blue, or deep greenish blue, cowl, and a wash of blue on the tail.

The Eastern Peninsula of Celebes is separated from the Northern Peninsula only by the 40 miles or so of width of the Gulf of Tomini, but by land the distance is much greater than from the Southern Peninsula. It can therefore call for no great surprise, if the fauna of the Eastern Peninsula proves to have more in common with the Southern than with the Northern Peninsula. Besides *M. capucinus*, other characteristic southern forms which have been found in the Eastern Peninsula are *Microstictus wallacei* and *Pyrrhocentor celebensis rufescens*. The specimens of *Phoenicophaes* are more like the northern individuals; some other birds are not yet known from other quarters of Celebes.

+ 94. MONACHALCYON PRINCEPS Rchb.[1]
Juvenile Cowled Kingfisher.

a. [Dacelo cyanocephala Forsten in lit.]
b. [Dacelo monachus Temm. MS. Leyd. Mus.]

[1] For a long time this species was held by Schlegel, Sharpe, and other ornithologists to be the young of *M. monachus*. It was recognised as distinct by Temminck and Forsten, but the names chosen

c. [**Halcyon monachus** pt. *(1)* Gray, Gen. B. I, 1846, 79 (descr. null.).]
Monachalcyon princeps *(1)* Reichb., Handb. Alced. 1851, 38, t. 425, fig. 3157; *(2)* Salvad., Ann. Mus. Civ. Gen. 1875, VII, 653 part. (descr. of young); *(3)* M. & Wg., Abh. Mus. Dresden 1895, Nr. 8, p. 7.
d. **Dacelo princeps** pt. *(1)* Schl., Mus. P.-B. Alced. 1863, 24; *(II)* id., Vog. Ned. Ind. Alced. 1864, 20, 53, pl. 7, fig. 3; *(3)* id., Revue Alced. 1874, 16.
e. **Monachalcyon monachus** pt. *(1)* Sharpe, Monogr. Alced. 1870, 255, pl. 98, fig. 3; *(2)* Guillem., P. Z. S. 1885, 547.
f. **Halcyon cyanocephala** *(1)* Brügg., Abh. Ver. Bremen 1876, V, 51.
g. **Monachalcyon cyanocephalus** *(1)* Meyer, Ibis 1879, 60; *(2)* Sharpe, Cat. B. XVII, 1892, 295; *(3)* Hart., Nov. Zool. 1897, 163.
h. **Monachalcyon cyanocephala** *(1)* W. Blas., J. f. O. 1883, 135.
Figures and descriptions. Reichb. *I*; Schl. *d II, d 1*; Sharpe *e 1, g 2*; Brüggem. *f 1*.
Adult male. General colour above dark bistre-brown, the mantle, wing-coverts, scapulars and upper tail-coverts terminally bordered with cinnamon; entire head and nape, ear-coverts, and malar region, cyanine-blue; lores cinnamon, the forehead also slightly touched with cinnamon; neck cinnamon, the feathers crossed with a subterminal bar of dusky, the colour of their basal parts; under parts white, washed with cinnamon on breast, sides and under tail-coverts, and crossed from the upper chest downwards with somewhat broad brace-shaped bars of brown, broadest on the sides; under wing-coverts and inner webs of quills below towards their basal part cinnamon, the former marked with a few brown bars ([♂] Gorontalo, Nr. 8186). Bill very pale horn, darker about the middle of the upper mandible.
Immature [male]. Like the adult male, but the under surface and under wing-coverts more broadly and plentifully barred with brown; the cinnamon margins to the feathers of the upper surface of a more rufous cinnamon tint and preceded by a subterminal dusky line passing into the brown of the back (Gorontalo, Nr. 8185).
Adult female. Differs from the form described as the adult male in having a stripe of pale cinnamon stretching from the lores above the eye to the ear-coverts and a similar stripe from the rictus below the eye to the ear-coverts, with a blue stripe between them; the under surface more plentifully barred. "Iris brown; bill horn-yellow-brown; feet brown" (Rurukan, ♀, 5. XI. 84: Platen in Mus. Nehrkorn, Nr. 916).

An adult female in the Sarasin Collection, killed in the breeding-season, had the "bill yellow, above greyish, also below at the sides; iris dark brown; foot yellow-grey" (♀ Rurukan, 26. III. 94, P. & F. S.).
Young female. With cinnamon superciliary and malar stripes as in the adult female, the malar stripe more extended and broader, the ear-coverts black, the bars on the breast narrower and more irregular. Bill shorter, blackish in colour, tip whitish (♀, Tomohon, 7. VI. 94: P. & F. Sarasin). See, also, Brüggemann *f 1*.
Young male (just about able to fly). Like the adult male — without the superciliary and malar stripes of the female; a little cinnamon on the lores; bill much smaller and shorter than in the adult and with a slight hook at the tip, terminal third yellowish white, basal part blackish horn (♀, Tomohon, 27. III. 94: P. & F. S.).

by these authors were never published. Its distinctness was again pointed out by Dr. Brüggemann in 1876 (*f 1*), who gave the present species Forsten's MS. name *cyanocephala*, but in 1851 Reichenbach had already figured and described it as *M. princeps* from a specimen in the Dresden Museum. It appears that in calling this form *princeps* and the foregoing species *monachus* Reichenbach and Bonaparte misapplied Forsten's and Temminck's unpublished names, but this, though a little to be regretted, does not call for further comment.

Birds of Celebes: Alcedinidae.

Measurements.	Wing	Tail	Bill from nostril	Tarsus
a. (Nr. 3688) ad. [♂] Celebes (type of *M. princeps*)	114	88	—	18
b. (Nr. 8186) ad. [♂] Gorontalo	120	96	34	17
c. (Nr. 8185) vix ad. [♂] Gorontalo	115	90	33	18
d. (Nr. 3687) ad. [♂] Manado	114	85	34.5	—
e. (Sarasin Coll.) ♀ ad. Rurukan, 26. III. 94	117	85	31	18

Eggs. "The two eggs of this species sent to me by Dr. Platen from Rurukan in the Minahassa are, like those of all *Alcedidae*, white, yet without gloss; the shell very delicate. They measure 33 × 25.5 mm" (Nehrkorn MS.).

Distribution. Celebes, Northern Peninsula: Minahassa (Forsten *d 1*, *d II*, v. Rosenberg *d 3*, *f 1*, etc.); Gorontalo District (Riedel in Dresd. Mus.).

As in the case of *M. monachus* the two different dresses, which we have described as of a sexual character, have been treated by Brüggemann *(f 1)* as seasonal dresses. Specimens, the sex of which have been ascertained, are very scarce, but through the kindness of Mr. Nehrkorn we have been able to describe an adult female with two cinnamon stripes on the sides of the head. The specimen figured by Schlegel *(d II)* has also two stripes on the sides of the head and is marked ♀ juv. The Drs. P. & F. Sarasin, when in Celebes, paid much attention to this question, and sent home an adult female ready to breed, a quite young female, and a male just out of the nest. They and the two above prove that the male has a complete blue cowl, covering the head, face, ear-coverts and nape, while the female has a cinnamon superciliary and a similar malar stripe, and these sexual differences — as is known to be the case in some other Kingfishers — exist from the very first. The example described by Dr. Sharpe in the Catalogue of Birds should be an adult female.

The plumage of this species has a curiously immature appearance which led to its being treated of as the young of *Monachalcyon monachus*, until its distinctness was pointed out by Brüggemann in 1876 *(f 1)*. Chief among its differences are the relation of the primaries to one another; in the present species the third is nearly as long as the fourth and fifth and longer than the sixth, in *M. monachus* the sixth is equal to, or longer than, the third. *M. princeps*, too, is altogether a much smaller species.

The plumage of *Monachalcyon princeps* appears to be ancient in character. Dr. Sharpe, speaking of it as the young of the foregoing species, remarks (Monogr. 1870, p. XVIII, XLV) that it very much resembles the young of *Tanysiptera* (compare Monograph, plate 98, *M. monachus* — fig. on the left, with plate 101, *Tanysiptera doris* and plate 106, *T. hydrocharis*). The adult *Tanysiptera* has of course made a vast departure from this plumage. Again *M. princeps* has much in common with *Halcyon lindsayi* of Luzon and *H. hombroni* of Mindanao, though, while *M. monachus* most closely resembles *H. hombroni*, the present species bears most likeness to *H. lindsayi*.

GENUS CITTURA Kaup.

This genus is peculiar to the Celebes Province. Size medium; the bill very broad and flat, much broader than deep, in length from the nostril twice its width or less (in *C. sangirensis* sometimes a little more), in colour red when adult; the tail is long, about 3 times the length of the bill, strongly graduated, the outermost feathers little more than half the length of the tail; the tarsus is scarcely shorter than the middle toe without the claw; the 1st primary is about equal to the secondaries, the 3rd, 4th and 5th the longest. Two species. Sexes slightly different.

✢ * 95. CITTURA CYANOTIS (Temm.).

Broad-billed Kingfisher.

a. **Dacelo cyanotis** *(1)* Temm., Pl. Col. pl. 262, 1824 (Sumatra!); *(2)* Less., Tr. d'Orn. 1831, 248 (Sumatra!); *(3)* Gray, Gen. B. 1846, I, 78; *(4)* id., List Fissirostr. Brit. Mus. 1848, 52; *(5)* Bp., Consp. 1850, I, 154; *(6)* id., Consp. Vol. Anisod. 1854, 9, Nr. 113; *(7)* Wall., Ibis 1860, 142; *(8)* Schl., Mus. P.-B. 1863, 22; *(IX)* id., Vog. Ned. Indië, Alced. 1864, 18, 51, pl. 6, fig. 1, 2; *(10)* Finsch, Neu Guinea 1866, 160; *(11)* Gray, HL. 1869, I, 89; *(12)* Schl., Rev. Alcedin. 1874, 14; *(13)* Rosenb., Malay. Archip. 1878, 271.
Cittura cyanotis *(1)* Kaup, Verh. nat. hist. Ver. Hessen 1848, 68 (Familie d. Eisvögel 8); *(II)* Reichb., Hb. spec. Orn. Alcedin. 1851, 38, t. 429, f. 3170; *(III)* Sharpe, Monogr. Alcedin. 1868, pp. XX, XXXVII, 301, pl. 119; *(4)* Wall., Malay Archip. 1869, I, 413; *(5)* Wald., Tr. Z. S. 1872, VIII, 44; *(6)* Salvad., Ann. Mus. Civ. 1875, VII, 654; *(7)* Brüggem., Abh. Ver. Bremen 1876, V, 54; *(8)* Lenz, J. f. O. 1877, 368; *(IX)* Rowley, Orn. Misc. 1878, III, 131—143, pl. XCIX; *(10)* Meyer, l. c. remarks; *(11)* id., Ibis 1879, 63; *(12)* W. Blas., J. f. O. 1883, 136; *(13)* Meyer, Isis, Dresden 1884, 19; *(14)* Guillem., P. Z. S. 1885, 548; *(15)* W. Blas., Ztschr. ges. Orn. 1886, III, 90; *(16)* Tristr., Cat. Coll. B. 1889, 95; *(17)* Heine & Rchw., Nomencl. Mus. Hein. 1890, 168; *(18)* Sharpe, Cat. B. XVII, 1692, 292; *(19)* M. & Wg., Abh. Mus. Dresden 1895, Nr. 8, p. 7; *(20)* iid., ib. 1896, Nr. 2, p. 13; *(21)* Hart., Nov. Zool. 1897, 163.

"**Kikis talun**", Minahassa, Alfurous name, Meyer *10, 11*.
"**Radja udang utan**", Malay near Manado, Nat. Coll.
"**Bulu bebek**", Minahassa, Guillem. *14*.
"**Bukaka memejang**", Tonkean, E. Celebes, Nat. Coll.
Figures and descriptions. Temminck *a I*; Schlegel *a IX*; Sharpe *III* (♂), *18*; Rowley *IX* (♀); Reichb., *II*; Salvad. *6*; Brüggemann *7*; Lenz *8*; Meyer *10*; W. Blasius *15*.
Adult male. General colour above raw umber washed with olive, becoming buff on the outer webs of the scapulars, hazel-chestnut on the rump and tail, and washed with orange-rufous on the head above and nape; a narrow line above and below the eye, continuous with a broad stripe on the sides of the occiput, dark marine-blue; the tips of the feathers of the supraorbital region, continued above the blue stripe, ochraceous buff, the bases dark blue, dusky black at the extreme base; wing-coverts dark marine-blue, where concealed by the scapulars black, quills dusky, the inner

ones washed with pale russet; entire under surface pinkish and vinaceous buff, becoming ochraceous buff on the cheeks and ear-coverts, where some of the feathers are particoloured, being marine-blue on the upper web and helping to form the blue stripe on the sides of the occiput; metacarpal edge pinkish buff; under wing-coverts of the same colour, but blackish near the metacarpal edge (Manado, C 2293). Iris rosy red; bill and feet dark red; claws blackish brown (Meyer *11*).

Adult female. Like the male, but with the wing-coverts, superciliary stripe and parietal patch black or blue-black instead of blue, the superciliary line of feathers, continued above the parietal patch, tipped with silver-white or pearl-blue (Manado, Nr. 6199, C 2299; ♀ ad. Pinogo, Bone Valley, 9. Jan. 1894: P. & F. Sarasin).

Young male. Resembles the adult male in coloration. Bill black (Meyer *10, 11*).

Young female. Resembles the adult female; wing-coverts more wood-brown, the inner ones black; bill black, shorter than in adult (Manado, Nr. 6198). "Iris rose-colour; feet red-brown; bill red-brown, in part black" (Kema, 16. Aug. 93: P. & F. S.).

Measurements.	Wing	Tail	Bill from nostril	Tarsus
a. (C 2293) [♂] ad. Manado	101	95	28	14.5
b. (C 2292) [♂] ad. Manado	100	98	—	14.5
c. (Nr. 6196) [♂] ad. Manado	98	93	28	—
d. (Nr. 6197) [♂] imm. Manado	100	94	24	14.5
e. (C 2304) [♀] ad. Manado	100	95	28	15.5
f. (C 2299) [♀] ad. Manado	100	92	27	14.5
g. (C 2301) [♀] ad. Manado	103	96	28	—
h. (C 2298) [♀] ad. Manado	100	—	28.5	—
i. (Nr. 6199) [♀] ad. Manado	103	88	27.5	—
j. (Nr. 6198) [♀] juv. Manado	96	—	21 ca.	—
k. (C 14198) [♀] ad. Lembeh Id. 28. II. 95 (Nat. Coll.)	99	97	31	15
l. (C 14199) [♀] ad. Lembeh Id. 2. III. 95 (Nat. Coll.)	104	98	28	—
m. (C 14200) [♂] ad. Lembeh Id. 26. II. 95 (Nat. Coll.)	99	94	29	—
n. (Sarasin Coll.) ♀ ad. Pinogo, 9. I. 94	99	88	28	—
o. (Sarasin Coll.) ♀ vix ad. Ussu, S. E. Celebes 20.II.96	96	85	28	14
p. q. r. (14444—46) [♀ ♀, ♂ juv.] Tonkean, E. Cel. (N. C.)	100-105	92-101	—	—

Observation. Of 19 other specimens recently sent to the Dresden Museum from the Manado District by our native hunters, 6 had the superciliary stripe of light pearl-blue-tipped feathers, and black lesser and middle wing-coverts, the remaining 13 had the superciliary tips ochraceous and the ear-coverts and wing-coverts deep blue, the difference — a sexual one — being well-marked, as in those now before us.

Eggs and breeding habits unknown.

Distribution. Celebes — Northern Peninsula: Minahassa (Forsten *a 8*, Wallace *a 7*, v. Rosenberg *a 12*, Meyer *10, 11*, etc.), Gorontalo Distr. (Rosenb. *a 12*, P. & F. S. *19*); E. Celebes (N.C.); Ussu, S. E. Celebes (P. & F. Sarasin); Tawaya, W. Celebes (Doherty *21*).

This Kingfisher was for a long time known only from the northern arm of Celebes, though further explorations in the island have now shown that it has a wider range.

It has recently been found in West Celebes by Doherty, at Tonkean, East Celebes, by our native hunters, and in S. E. Central Celebes by the Drs. Sarasin. Our two females from East Celebes and the Sarasins' female from the South-east have the silvery tips of the long superciliary stripe longer than in females from the north, and seem to present a slight racial departure.

As to the habits of this species: "It lives", says Meyer *(10, 11)*, "like *Monachalcyon princeps*, only in the forest, not on river-sides; and it is not at all a rare bird, according to my experience. It likes to sit dreaming alone on branches of trees. Its cry is, five or six times one after another, kebekek. In the stomach I found insects, crabs, worms, etc".

"Male and female are easily to be distinguished, viz. from the colour of the wing-coverts and the sides of the head, which is blue in the male, black or bluish black in the female; the male has no white superciliary spots. Even the young ones, which were alive in my possession, show this difference". Among the Kingfishers the same condition obtains in *Monachalcyon*, possibly in some species of *Halcyon* in Western Polynesia (Wg., Aves Polyn. 16), and to some extent in some species of *Ceryle*.

The only other species of the genus *Cittura* is *C. sangirensis* Sharpe of the Sangi Islands, an important link between Celebes and those islands, distinguishable by its larger size, black frontal band and malar region at the gape and by its mauve ear-coverts, sides of neck, and chest. Both sexes of this species, too, resemble the female of *C. cyanotis* in having a pearl-blue or silvery line of feather-tips over the eye, but the male has blue wing-coverts, the female black ones, as in the sexes of *C. cyanotis*. Perhaps, as Meyer has remarked *(10)*, allied species are still to be discovered somewhere in Borneo or the neighbourhood.

Of the two, *C. sangirensis* may perhaps be the one more likely to have varied in a lesser degree from the ancestral type. The female of *C. cyanotis*, which may probably be regarded as of a more ancient type than the male, partakes of the coloration of *C. sangirensis* not only in the points mentioned, but in having usually a wash of phlox-purple about the ear-coverts and throat, and it has been confused with *C. sangirensis* by Lenz *(8)* and Guillemard *(14)*. Mr. Keeler, too, believes that the colour of the basal part of feathers of birds in general have a phylogenetic value; if this be so, *C. cyanotis* should be sprung from a bird with a black or dark crown and malar region, but a pale chin (as it still has). *C. sangirensis* has the forehead, a broad eyebrow, occipital side-patch, malar and subocular region black; in this respect, therefore, it is possible that it should be regarded as the more ancestral form.

96. CITTURA SANGIRENSIS Sharpe.
Sangi Broad-billed Kingfisher.

Cittura sanghirensis *(1)* Sharpe, P. Z. S. 1868, 270, pl. 27 ♂; *(II)* id., Monogr. Alcedin. 1868, 299, pl. 118 ♂; *(3)* Newton, Ibis 1869, 215; *(4)* Salvad., Ann. Mus. Civ. Gen. 1876, IX, 53; *(V)* Rowley, Orn. Misc. 1878, III, 132, pl. (♀); *(6)* Meyer, ib. 136—140, remarks; *(7)* id., Ibis 1879, 63; *(VIII)* id., Vogelskel. 1882, 21, pl. XXVI; *(9)* id., Isis, Dresden 1884, Abh. I, 6, 19; *(10)* W. Blas., Ztschr. ges. Orn. 1886, 91; *(11)* id., Ornis 1888, 576—578; *(12)* Tristr., Cat. Coll. B. 1889, 95; *(13)* Sharpe, Cat. B. XVII, 1892, 293.

a. **Dacelo sanghirensis** *(1)* Finsch, pt. in Sharpe *II*; *(2)* Gray, HL.·1869, I, 89; *(3)* Schl., Mus. P.-B. Rev. Alced. 1874, 14.

b. **Cittura cyanotis** pt. *(1)* Lenz, J. f. O. 1877, 368.

"**Bengka kehu**", Great Sangi, Nat. Coll.

Figures and descriptions. Sharpe *I, II, 13*; Rowley *V*; Meyer *VIII* (skeleton), 6; W. Blasius *11*.

Adult male. Head above russet, washed with orange-rufous, passing into raw umber on hind neck and middle of back; outer webs of scapulars light cinnamon, passing into the colour of the back; upper tail-coverts and tail dark cinnamon-rufous; a frontal band, a patch at base of lower bill, continuous with the sub- and supra-orbital region black, becoming very dark marine-blue on the sides of the occiput; above this a long superciliary line of pearly white-tipped feathers; wings dusky black; wing-coverts very dark marine-blue; chin and upper throat white; ear-coverts, sides of neck, and a broad wash across the upper chest light phlox-purple; remainder of under surface buff, washed with cinnamon-rufous on lower flanks and tail-coverts; edge of wing and under wing-coverts white, but black next to the edge of wing (Gr. Sangi — Nr. 6193). Bill sealing-wax red; iris light red; feet red-brown (Platen *11*).

Adult female. Like the male, but the parietal stripe and wing-coverts black, with little or no blue colouring in it (Blas. *11*; Gr. Sangi — C 12703, 12704).

Young male. Like the adult male; bill shorter, red-brown: — Platen (Gr. Sangi, Nr. 6192), with very small pale tips to the wing-coverts (Blas. *11*).

Young female. Like the adult female; bill shorter, a little hooked at the tip, red-brown: — Platen (Blas. *11*; Gr. Sangi, Nr. 6195).

Iris and feet of young as in adult (Platen *11*).

Observation. A specimen from Siao before us does not differ from those of Great Sangi.

Measurements.	Wing	Tail	Bill from nostril	Tarsus
a. (Nr. 6194) [♂] ad. Gr. Sangi	113	101	35	16.5
b. (Nr. 6193) [♂] ad. Gr. Sangi	108	98	34	16
c. (Nr. 6191) [♂] vix ad. Siao	110	98	34.5	16
d. (Nr. 6192) [♂] juv. Gr. Sangi	108	100	29.5	16.5
e. (C 12703) [♀] ad. Gr. Sangi, 21. VII. 93 . .	114	100	35	17
f. (C 12704) [♀] ad. Gr. Sangi, 13. VII. 93[1] . .	107	94	35	—
g. (Nr. 6195) [♀] juv. Gr. Sangi	103	55	25	15.5

Thirteen adults measured by Prof. Blasius *(11)* vary as follows: wing 106—117; tail 97—109; culmen 39.5—45; tarsus 16—18 mm; length in the flesh (Platen) 260—270.

---

[1] This specimen is moulting. A number of other species appear to moult in July in the Sangi Islands.

Skeleton.

Length of cranium	. . . 78.0 mm	Length of tibia	. . . . . 38.0 mm
Breadth of cranium	. . 25.6 & 27.5 »	Length of tarso-metatarsus	. 16.2 »
Length of humerus	. . . 93.6 »	Length of sternum	. . . 25.5 »
Length of ulna	. . . 44.0 »	Greatest breadth of sternum	. 21.0 »
Length of radius	. . . 40.5 »	Height of crista sterni	. . 7.8 »
Length of manus	. . . 29.4 »	Length of pelvis	. . . . 29.0 »
Length of femur	. . . 22.6 »	Greatest breadth of pelvis	. 23.0 »

Egg and breeding habits unknown.

Distribution. Sangi Islands — Great Sangi (Hoedt *a 3*, v. Duivenb. *a 3*, Meyer *6*, *9*, Bruijn *4*, Platen *11*, Nat. Coll. Mus. Dresd.); Siao (Meyer *6*, *9*).

This species appears to be fairly abundant in Great Sangi, but scarce in Siao, where it has only been obtained as yet by Meyer's hunters. It is a handsomer species than its near and only known ally, *Cittura cyanotis* of Celebes, from which it may be easily distinguished by its black frontal band, black malar region, light phlox-purple ear-coverts and chest, and larger size. Its coloration, as pointed out in our article on that species, appears to be of a more ancestral type.

Cittura ranks low as a form of Kingfisher, and the bill of the young is of interest, partaking as it does more of the *Melidora*-type than does that of the adult *Cittura*, and it lends further confirmation therefore to Dr. Sharpe's views. The bill of the young *Cittura* is dark in colour, a little hooked at the tip, and very flat towards the base, the nostrils being situated at the inner edge of a rather flat ledge, with the ridge of the culmen standing up high between them; the bill of the adult is red, without a hook, and fairly round at the sides, the ridge of the culmen less raised. In his notable "Monograph" Sharpe writes, p. XLIV: "Turning . . . to *Melidora* we seek the links which may still be left us on the globe whereby to connect forms apparently so different, and at once seize upon *Cittura* as the nearest approach to this extreme form; for here is also seen the grooved bill, although the maxillary hook is absent". The probable significance of the little maxillary hook in the young *Cittura*, and of the dark colour of its bill, that of *Melidora* being black — thus become apparent; but the shape of the bill about the nostrils does not seem to point to a form like *Melidora*, but more likely to a lost ancestral form of both. *Melidora* is confined to New Guinea, *Dacelo* to Australia and New Guinea, and the insectivorous Kingfishers in general "have their greatest development in the Austro-Malayan subregion, while the piscivorous Kingfishers are found all over the globe, except Oceania" (Sharpe, p. XLVII). In considering the geographical distribution of the Kingfishers it may be of importance to observe that the countries which have preserved for us what appear to be the most ancient Kingfishers are tenanted by ancient faunal and floral types in general. Up to the present palaeontology has brought no evidence of the existence of the lower forms of true Kingfishers in other parts of the globe; as Prof. Newton remarks (Dict. B. 1893, 489), "the only fossil referred to the neighbourhood of the Family is the *Halcyornis toliapicus* of Sir R. Owen (Br. Foss. Mamm. and Birds, p. 554) from the Eocene of Sheppey — the very specimen said to have been previously placed by König (Icon. foss. sectiles, fig. 153) in the genus *Larus*". Mr. Lydekker (D. B. 281) also considers this fossil to be Larine in character. Prof. Zittel, besides, mentions (Hdb. d. Palaeont. 1890, III, 852) uncertain remnants of *Alcedo* in the Eocene of Paris and of *Ceryle* in the bone-caves of Brazil. Under these circumstances there is no telling what palaeontology will have to say on the subject in the future, and it may be hazardous to form an opinion from our present very scanty knowledge.

FAMILY CORACIIDAE.

In external appearance the Rollers are very like a Daw or Jay. The best means of distinguishing them at sight from such corvine Passeres is furnished by the nostril, which in the *Coraciidae* is elongated and covered above by an operculum of skin, on to which the plumage of the forehead extends, while in corvine birds the nostril is a roundish hole screened by long bristles. Anatomically, of course, there are important differences: the Rollers have both carotid arteries, the skull is desmognathous, the deep flexor tendons of the foot become fused before branching off to the four toes (see Beddard in Dresser's Mon. Corac.). The Broad-billed Rollers *(Eurystomus)* are again very like the Broad-bills *(Eurylaemi)*, but the form of the nostril is sufficient for their distinction.

According to Gadow, the Rollers are distinguishable from the other members of the order *Coraciae (Momotidae, Alcedinidae, Meropidae,* and *Upupidae)* by having 14 cervical vertebrae, the spina interna being wanting; from the Kingfishers and Bee-eaters the Rollers differ in the form of the bill and foot, the bill being decurved, stout, and with a pendant tip or hook, the foot having the anterior toes free, or free from the basal joint, instead of having the third and fourth toes united down to the penultimate joint. In *Merops* the carotid-arrangement is different — the left one only being present.

The *Coraciidae* breed in holes in trees and lay white eggs.

GENUS CORACIAS L.

Bill nearly as long as the head, black or blackish, decurved, the tip overlapping the under mandible, compressed, much deeper than broad, nostril elongated, with a membrane above covered with the plumage of the forehead; some stout bristles above the gape; tarsus about as long as the hind toe and claw; wings moderate, 2nd, 3rd, and 4th quills longest; tail moderately long, square (in some species with the outermost rectrices lengthened so as to form a deep fork). In size these birds are about equal to a Jay. The chief colours are blue and green. Dresser recognises twelve species inhabiting Africa, Europe and Southern Asia, the genus being absent in America and the Australian Region and, curiously enough, absent in the Great Sunda Islands, excepting Celebes.

+ * 97. CORACIAS TEMMINCKI (Vieill.).
Celebesian Roller.

a. **Le Rollier Temminck** *(1)* Levaill., Ois. Parad. Rolliers.1806, III, Suppl. p. 46, fig. D.
b. **Galgulus temmincki** *(1)* Vieill., N. Dict. 1819, XXIX, 435; *(2)* id., Enc. Méth. 1823, 869.
Coracias temmincki *(1)* Wagler, Syst. Av. 1827, 215; *(2)* Gray, Gen. B. I, 1846, 62; *(3)* id., List Coraciae Br. Mus. 1848, 33; *(4)* Bp., Consp. 1850, I, 167; *(V)* Rchb., Hb. Merop. 1851, 51, t. 434, fig. 3187; *(6)* Gray, P. Z. S. 1858, 189 (N. Guin.); *(7)* id., Cat. B. New Guin. 1859, 189; *(8)* Cab. & Hein., Mus. Hein. 1860, II, 118; *(9)* Gray, P. Z. S. 1861, 433; *(10)* Wall., Ibis 1864, 41; *(11)* Blyth, ib. 1866, 345, 346; *(12)* Finsch, Neu Guinea 1866, 160; *(13)* Schl., Mus. P.-B. Coraces 1867, 138; *(XIV)* Gld., B. Asia pl. 56 (1869); *(15)* Gray, HL. 1869, I, 75; *(16)* Wall., Malay Archip. 1869, I, 337, 429; *(17)* Sharpe, Ibis 1871, 184; *(18)* Wald., Tr. Z. S. 1872, VIII, 43; *(19)* Salvad., Ann. Mus. Civ. Gen. 1875, VII, 655; *(20)* Brügg., Abh. Ver. Bremen 1876, V, 49; *(21)* Rosenb., Malay. Archip. 1878, 271; *(22)* Meyer, Ibis 1879, 59; *(23)* Salvad., Orn. Pap. 1880, I, 512; *(24)* Wall., Island Life 1880, 433; *(25)* W. Blas., J. f. O. 1883, 135; *(26)* id., Ztschr. ges. Orn. 1885, 242; 1886, III, 88; *(27)* Guillem., P. Z. S. 1885, 546; *(28)* Platen, Gefied. Welt 1887, 205; *(29)* Tristr., Cat. Coll. B. 1889, 98; *(30)* Heine & Rchw., Nomencl. Mus. Hein. 1890, 157; *(31)* Sharpe, Cat. B. XVII, 1892, 26; *(32)* Büttik., Zool. Erg. Weber's Reise in Ost-Ind. 1893, 274; *(XXXIII)* Dresser, Mon. Corac. 1893, 49, pl. XIII; *(34)* Newton, Dict. B. p. 794 (1894); *(35)* M. & Wg., Abh. Mus.. Dresden 1895, Nr. 8, p. 7; *(36)* iid., ib. 1896, Nr. 1, p. 5; *(37)* iid., ib. 1896, Nr. 2, p. 13; *(38)* Hart, Nov. Zool. 1896, 158; *(XXXIX)* Meyer, Abb. v. Vogelskel. II, 1897, pl. CCXIV.
c. **Rollier d'Urville** *(1)* Quoy & Gaim., Voy. Astrol., 1830, Atl. pl. 16.
d. **Coracias papuensis** *(1)* Quoy & Gaim., op. cit. text, 1830, 220; *(2)* Sclat., Pr. Linn. Soc. 1858, 155.
e. **Coracias pileatus** *(1)* Bp., Consp. Vol. Anisod. 1854, 7, Nr. 210 (ex Reinw. MS.); *(2)* Reinw., Reis Ind. Archip. in 1821, 1858, 592.
f. **Eurystomus pileatus** *(1)* Gray, P. Z. S. 1860, 346 (Moluccas).

"**Tokkakak**", ("**Tonkaka**"—Guillem. 27), Manado District, Mantehage, Lembeh and Banka Is., Nat. Coll. in Dr. Mus.
"**Kapala biru**" (Blue head), Malay, Minahassa, Meyer 22.
"**Fateh rokos**", Minahassa, Meyer 22.
"**Lungun-geü**", ?Gorontalo Distr., Rosenb. 21.
"**Patjujung Dapo**", Tjamba Distr., S. Celebes, Platen 26.
"**Djohak**", Tonkean, E. Celebes, Nat. Coll. in Dr. Mus.
Figures and descriptions. Dresser *XXXIII*; Gould *XIV*; Meyer *XXXIX* (skel.); Levaillant *a I*; Reichb. *V*; Quoy & Gaimard *c I*; Brüggem. *20*; W. Blas. *26*; Sharpe *31*.

Adult. Mantle, back, scapulars, and innermost quills sago-green, in certain lights olive-green, and touched on the borders of the feathers with verditer-blue; entire head above, including supra-orbital region, and upper tail-coverts Nile-blue, with a strong turquoise-blue gloss; wings, tail, and rump hyacinth-blue glossed with purple, the ends of quills and tail-feathers more dusky, the concealed inner webs — when seen held vertically with the light falling on them — bright dark bronze; ear-coverts and subocular region brownish bronze; under parts, continuous with a broad collar around hind neck, dusky blue, washed on the under parts with mauve-purple, and becoming violet-blue on the under tail-coverts, the hackle-feathers on chin touched with pale-blue, on throat with light violet; under wing-coverts, and most inner webs of quills below, and tail below, hyacinth-blue glossed with purple, the outer webs, the adjacent part of the inner webs, and the ends of the quills blackish, or dusky bronze, when the bird is held in an inclined position, bill lowest, with the light falling on it. (Near Manado, Aug.-Sept. 1892, C 10857). Iris brown; bill black; tarsus dirty brownish black (Guillem. 27).

Observation. The coloration of the wings of this species is of much interest. The wings appear to be to some extent metallic; at all events the purplish blue is soon more widely spread on their under surface in certain lights. On the upper surface they are purplish blue, except on a narrow portion of the inner webs, probably always covered by the superjacent feathers where they are bronze or blackish, according to the light; on the under surface the converse is seen — where they are blue above they tend to bronze below, where they are bronze above they are blue below. The lines of demarkation of the tints, however, do not correspond above and below, nor are the colours on many quills — especially on the inner ones — sharply separated, but tend to blend with one another.

Sexes. Alike, but the female less brilliantly coloured (Meyer 32 — Limbotto, July, C 2191).

Young. Similar to the adults; bill much shorter, but otherwise like; the collar of dusky blue round the hind neck much narrower; the pale blue of the crown greyer and slightly washed with brown, the tips of the feathers brownish giving an indistinct barred appear-ance, the pale blue hackle-feathers of the chin extending further down on to the throat (N. Celebes, C 3597; near Manado, Aug.—Sept. 1892, C 10858). Iris light hazel; bill black; feet yellow-brown (Platen 26 — ♂ juv.). See, also, W. Blasius 26, description ♂ juv.

Measurements (8 specimens). Wing 175 juv.—190; tail 127—144; bill from nostril 23.5 juv. —31; tarsus 22 juv., 23 ad.

Eggs and Nest. Unknown.

Breeding season. One of the young specimens in the Dresden Museum, killed Aug. or Sept., appears to be about six weeks old; hence the eggs would have been laid in June or July. Prof. W. Blasius's *(26)* young specimens from South Celebes, killed May 13th, must have been the product of an egg laid at least as early as March.

Distribution. Celebes: — Minahassa (Reinwardt e 2, Forsten 13, v. Rosenberg 13, etc.); Banka, Lembeh and Mantehago Is. off the Minahassa (Nat. Coll., Mus. Dresd.); Gorontalo District (Rosenb. 13, Meyer 22); Tonkean, E. Celebes (Nat. Coll. in Dresd. Mus.); Kandari, S. E. Peninsula (Beccari 19); near Macassar (Wallace 16, 31, Everett 38); Tjamba, S. Celebes (Platen 26); Kadjang, S. Celebes (Weber 32); Bulokomba, S. Celebes (Everett 38).

The habitat of the species was recorded by Quoy & Gaimard *(d 1)* as New Guinea, which for a long time remained an uncertain locality for it, but

Salvadori *(23)* considers it indubitable, as we do too, that no species of the genus *Coracias* occurs in New Guinea and that the indication of Quoy & Gaimard is, therefore, wrong. The species seems to be strictly confined to Celebes.

In North Celebes this is a common bird. It follows the hoe of the fieldlabourer, says Platen *(28)*, examining the freshly broken ground for booty, always pursued and attacked by the greedy Crow, *C. enca*. Meyer *(22)* remarks that it "usually flies singly; but after feeding, several play together. They frequently sit on dead twigs and look out for grasshoppers and other insects; then suddenly rushing upon their prey they return to their perch. Cry tschirrr". In South Celebes Wallace found it a rare bird; Platen and Weber, too, obtained each but a single specimen. Wallace *(16)* observes that "it has a most discordant voice, and generally goes in pairs, flying from tree to tree, and exhibiting while at rest that all-in-a-heap appearance and jerking motion of the head and tail which are so characteristic of the great Fissirostral group to which it belongs. From this habit alone the kingfishers, bee-eaters, rollers, trogons, and South American puff-birds, might be grouped together by a person who had observed them in a state of nature, but who had never had an opportunity of examining their form and structure in detail".

Later in the same work *(16)* Mr. Wallace remarks that "the Celebes Roller is an interesting example of one species of a genus being cut off from the rest. There are species of *Coracias* in Europe, Asia, and Africa, but none in the Malay Peninsula, Sumatra, Java, or Borneo. The present species seems therefore quite out of place; and what is still more curious is the fact, that it is not at all like any of the Asiatic species, but seems more to resemble those of Africa". These views appear also to be those of no less an ornithologist than Dr. Sharpe who places *(31) C. temmincki* between *C. olivaceiceps* Sharpe of South Africa and *C. cyanogaster* Cuv. of Senegambia. Mr. Dresser *(XXXIII)* does not enter into the question of its affinities, but places it last in the genus next to *C. cyanogaster*. In our opinion *C. temmincki* has nothing to do with the latter species, which was placed with other African forms in a different genus by Bonaparte on account of its Swallow-tail, and it is very questionable whether *C. olivaceiceps* can be regarded as a nearer ally than *C. affinis* McClell. of India. In a later work *(24)* Mr. Wallace cites *C. temmincki* as a Himalayan form in Celebes, a designation which has at least an equal right to acceptance with his earlier view of the African affinities of the species.

In that the range of the genus is bounded to the east by the Moluccan Straits, *Coracias* conforms to the general rule found among Asiatic forms in Celebes. Its absence in Borneo, Java, Sumatra and Malacca is an enigma. It may be suggested that the genus was once represented in these countries, and has become extinct there; or it may be thought possible that an ancestral form, a migrant like *C. garrulus*, accidentally straggled to Celebes, and settled in Africa, varying in the same direction in both places; or half a dozen other hypothetical explanations may be suggested, which would, however, not bring the problem to a solution in the present deficient state of our knowledge.

GENUS EURYSTOMUS Vieill.

Differs from *Coracias* by its shorter, broader bill, broader than deep across the nostril, not black in colour, but yellow or chiefly red (in the young duskier), the nostril a long, linear slit; no rictal bristles; the toes united for their basal joint; wing rather long and pointed, the 2nd quill a little longer than the 1st and 3rd. About seven or eight species, ranging from India to Australia and New Zealand; also Africa, and Madagascar.

98. EURYSTOMUS ORIENTALIS (L.).
Indo-Australian Broad-billed Roller.

As a species this bird ranges from the Himalayas and Amoorland in the north to Australia and, occasionally, to New Zealand in the south. It varies, of course, individually, like everything else that lives, and also to some extent racially, but both questions are very improperly understood. The two chief authorities of the present day on the Rollers, Dr. Sharpe and Mr. Dresser, are not in unison on the racial question: Dr. Sharpe (P. Z. S. 1890, 551; Cat. B. XVII 1892, 36, 38) would have it that there are three species and one subspecies; Mr. Dresser (Ibis 1891, 91—102, Mon. Corac. 1893, 70, 76) maintains the more generally accepted view of two species — a northern and a southern one. We prefer to avoid this troublesome question with the remark that Dr. Sharpe is probably wrong[1]) and Mr. Dresser probably not entirely

---

[1]) We have paid considerable attention to Dr. Sharpe's views. Treating his three species and one subspecies as four races of *E. orientalis*, the following should be their geographical ranges:

1. **The typical E. orientalis** (Linn.), the stationary resident form of the East Indies from Cachar and the Burmese countries to the Great Sunda Islands (Sharpe; as far as Talaut (Nat. Coll. *ic*), Sangi, Celebes (Mus. Dresd.), Peling (Nat. Coll.) and Halmahera (Salvadori).

2. **E. orientalis calonyx** (R. B. Sharpe), a more northern, migratory form, ranging from the Himalayas, where some examples at least appear to be stationary in places, though in Terai below Kumaon it is said by Mr. Thompson to arrive in April, breed in May, and leave in July and August (P. Z. S. 1890, 552), to "Amoor-land, Manchuria, and Northern China in summer, apparently wintering in Tenasserim and the Malayan Peninsula and Borneo" (Sharpe), Labuan, Java, Keeling Island, Ceylon, Sangi (Dresser, Ibis 1891, 101), Celebes (Dresser, and in the Dresden Museum).

3. **E. orientalis laetior** (R. B. Sharpe), "Southern India and Ceylon" (Sharpe), apparently a resident in Ceylon but very rare (Legge, B. Ceylon, 286), believed by Bourdillon (Str. F. 1876, IV, 382) to be a visitor only to Travancore, S. India.

4. **E. orientalis australis** (Swains.), apparently a more or less regular migrant, ranging from Celebes (Sharpe), Sula (Sharpe) and Buru (Salvad. Orn. Pap. I, 1880, 504) to Papuasia, Australia — where, in the Victoria basin, "it arrives early in spring, and, having brought forth its progeny, retires northwards on the approach of winter" (Gould, Handbook B. Austr. I, 120; in N. S. W. earliest arrival observed by Caley (Tr. Z. S. XV, 202) on October 3rd, 1809, disappearing early in February. Most plentiful about Christmas. — "Occasionally reaching to New Zealand" (Sharpe), four examples being known to Buller, stragglers killed recently on the West coast at the time the species visits Australia (B. New Zealand, 2nd. ed. 1888, I, 118).

right, and we think that other writers in a similar position to ours would also do well, in order not to involve themselves and others in more perplexity, to disregard the variations of this species and to speak of it simply as *E. orientalis*, until some one has thoroughly taken the whole matter in hand and shown what these variations really mean. The following references bear upon the occurrence of *E. orientalis* in the Celebesian Province.

Eurystomus orientalis (L.).

a. **Eurystomus pacificus** *(1)* Wall., P. Z. S. 1862, 339; *(2)* Legge, B. Ceylon, 1880, 285; *(III)* Dress., Mon. Corac. 1893, 75, pl. XIX.

Eurystomus orientalis *(1)* Schl., Mus. P.-B. Coraces, 1867, 139, 140; *(2)* Wald., Tr. Z. S. 1872, VIII, 43; *(3)* Brügg., Abh. Ver. Bremen 1876, V, 49; *(4)* Salvad., Ann. Mus. Civ. Gen. 1876, IX, 53; *(5)* Rosenb., Malay. Archip. 1878, 271; *(6)* Meyer, Ibis 1879, 60; *(7)* Salvad., Orn. Pap. 1880, I, 508; *(8)* W. Blas., J. f. O. 1883, 135; *(9)* Meyer, Isis, Dresden 1884, 6; *(10)* Guillem., P. Z. S. 1885, 546; *(11)* W. Blas., Ztschr. ges. Orn. 1886, 89; *(12)* id., Ornis 1888, 570; *(13)* Hickson, Nat. in N. Celebes 1889, 90; *(14)* Dresser, Ibis 1891, 101; *(15)* Büttik., Zool. Erg. Weber's Reise in Ost.-Ind. 1893, 274; *(16)* M. & Wg., J. f. O. 1894, 242; *(XVII)*

Two other forms, *E. crassirostris* Scl. of New Guinea and the New Britain Group, and *E. solomonensis* Sharpe of the Solomon Islands, are very nearly related to, but apparently do not intergrade with, the races of *E. orientalis*. *E. orientalis australis* and *E. crassirostris* occur in New Guinea together.

We give a key to Sharpe's supposed subspecies of *E. orientalis*, based upon the keys of Sharpe and Salvadori, and specimens before us:

  a. Paler = *E. orientalis australis*.
  b. Brighter, head more dusky.
    b' Terminal half of tail black.
    b" Head darker; blue colour of underparts more intense = *E. orientalis laetior*.
    c" Colours less intense = *the typical E. orientalis*.
    c' Terminal half of tail distinctly shaded with purplish blue = *E. orientalis calonyx*.

It should be observed that hard and fast lines cannot be drawn between subspecies — by definition they intergrade — and the above key is specially applicable only to the individuals from certain geographical areas in which the subspecies may be said to come to a head. Thus specimens from North China, where they are found in summer, appear from Sharpe's remarks (Ibis 1893, 562) to be always determinable as *E. orientalis calonyx*: at this time of the year the race which stays nearer the equator in the Sunda Islands and breeds there, is supposed by him to be *the typical E. orientalis* — at least nine specimens sent by Mr. Everett to Dr. Sharpe — date not stated — all proved to be of this race, but we have elsewhere recorded *calonyx*, as well as *orientalis* and intermediate specimens, from Peling in summer (Abh. Mus. Dr. 1896, Nr. 2, p. 13). Those which migrate to Australia in summer are of the race, *E. orientalis australis*. In Celebes, and doubtless in many other quarters, such as Burmah, S. India, the Philippines, Halmahera, a confusion of forms is found — at all events we cannot detect any well-defined races there. It will have been noted that three (!) of Sharpe's four subspecies of *E. orientalis* have been recorded from Celebes; moreover many of the 40 specimens in the Dresden Museum from there, Peling, Talaut and Sangi belong neither to *the typical orientalis* nor to *orientalis calonyx*, but are intermediate between the two, and such was the case with two of Mr. Dresser's five specimens from Celebes. We understand Sharpe to be of opinion that these specimens of Dresser's are intermediate between *E. orientalis* and *australis* (Cat. B. XVII, 34, footnote).

Among other things it should be borne in mind that the adults found in North China in summer are in full breeding plumage; probably they are duller in colour in winter and more like *the typical orientalis*. That this is so is shown by a specimen in the Dresden Museum from Great Sangi (C 638); it is moulting, and the new secondaries are washed with dark purplish on the external edges to their ends, while two or three old feathers are dull dusky, and the blue is not seen for the terminal inch of their outer webs, but only on the more basal part. The purple-blue is also just perceptible on the outer edges of the tail to its very tip, though indeed it would never be so conspicuous as in Sharpe's figure of *calonyx*.

Dresser, Mon. Corac. 1893, 67, pls. XVII, XVIII; *(18)* M. & Wg., Abh. Mus. Dresden 1895, Nr. 8, p. 8; *(19)* iid., ib. 1895, Nr. 9, p. 4; *(20)* iid., ib. 1896, Nr. 2, p. 13; *(21)* Hart., Nov. Zool. 1896, 176; *(22)* id., ib. 1897, 159.

b. **Eurystomus australis** *(1)* Sharpe, Cat. B. XVII, 1892, 36.

"**Hendingo-Opo**", Gorontalo?, Rosenb. *5.*
"**Tjetje**", Malay name, N. Celebes, Meyer *6.*
"**Kokotaka**", Talissi Id., Hickson¹) *13.*
"**Poopoopung**", Siao, Nat. Coll. Mus. Dr.
"**Atera**", Talaut, iid.

For full synonymy and references see Salvad., Orn. Pap. 1880, I, 503 *(E. pacificus)* and 508 *(E. orientalis)*, Agg. 1889, 60; Sharpe, Cat. B. XVII, 1892, 33, 36 *(E. laetior, E. australis)*, 38 *(E. calonyx)*; Dresser *XVII, a III (E. orientalis)*.

Figures and descriptions. Dresser *XVII, a III;* Gould, B. Austr. II, pl. 17 *(E. pacificus,;* Sharpe l. c. pl. II, fig. 1 *(E. orientalis)*, fig. 2 *(E. calonyx)*; Legge *a 2*; Salvad., ll. cc.; Meyer, Mitth. Z. Mus. Dresden 1875, I, 19; W. Blas. *II*; etc.

Skeleton. Eyton, Blanchard and Fürbringer have figured the skeleton or parts of it.

Adult. General colour above brownish French-green, becoming dark olivaceous seal-brown no top and sides of head, chin, and sides of neck, washed with bluish on the scapulars, innermost quills and upper tail-coverts; around orbital region and forehead blackish; wing-coverts strongly washed with cerulean-blue; quills black washed on their outer webs with cyanine-blue (except the innermost ones which are concolourous with the scapulars), the primaries crossed about their middle length by a broad band of pale blue, not seen on the outer web of the first primary; tail black, washed with cyanine-blue, most strongly towards its basal part; on throat an ill-defined patch of French-blue washed with mauve; under-parts, from breast downwards, light sea-green, the blue tint in it being strongest on the abdomen; quills below (except for the band of pearl-blue) black, the inner webs hyacinth-blue for the most part (Great Sangi, C 838). "Iris light brown; bill sealing-wax red, tip black; feet brown-red" (Platen *12*).

Sexes. Similar.

Young. Differs from the adult in being duller in colour and in having a black bill, and in wanting the bright blue patch on the throat, which is greenish, a little duller than the abdomen (Sharpe).

Measurements (8 adult specimens: Celebes and Great Sangi — 5 dated April, July). Wing 180—194; tail 96—105; tarsus 17; bill from nostril 16.5—18 mm.

Eggs. India. Very broad oval, pure white and faintly glossy; 34—36 × 29—29.5 mm (Oates, Hume's Nests and Eggs Ind. B. III, 37); Australia *(E. australis)*, two or three in number, dull white, rather glossy, sometimes variable in form: 36.8 × 26.7 oblong, 34 × 27.9 roundish (North, Nests and Eggs Austr. B. 35 pl. XIV fig. 1); three, sometimes four eggs, beautiful pearly white, considerably pointed at the smaller end (Gould.).

Nest. Breeds in holes in trees: no nest, eggs laid on a few chips of rotten wood — India (Bourdillon in Oates, l. c.); no nest, eggs laid on the dust formed by the decayed wood — Australia (Ramsay in North l. c.).

Breeding season. In South India nest found March 17th and April 20th, in the Himalayas, May (Oates); in Australia, October (North); from September to December (Gld., Handb. B. Austr. I, 120).

---

¹) Dr. Hickson remarks: "my boys called this bird 'Kokotaka', but Meyer says that the native name is 'Tjetje'." In N. Celebes 'Tokkakak' (= 'Kokotaka'?) is the name for *Coracias temmincki*.

Distribution of the species. From the Himalayas and Amoor-land southwards to Ceylon, the islands of the Bay of Bengal, throughout the East Indies and Papuasia — except the New Britain and Solomon Groups — to Victoria and South Australia, occasionally wandering to New Zealand.

On the habits of *E. orientalis*, the Broad-billed Roller, in Celebes Meyer writes: "This bird sleeps in the morning, and searches for food at midday; in the evening it flies after beetles. It is to be seen near river-banks, where it sits a long time quietly on a branch of a tree over the water, and can easily be shot, not being a shy bird". Similarly its mode of life is described by Morgan in Malabar (Str. F. 1874, II, 531), by Oates and by Davison in Burmah and Tenasserim (B. Brit. Burm. II, 71; Str. F. VI, 72), by Whitehead in Borneo (Ibis 1890, 21), by Gould in Australia. Hartert (J. f O. 1889, 364), when in Assam, watched the bird performing plays of flight above high forest during the hot hours of midday after the manner of the Roller. On the whole, however, the bird may be put down as of a sedentary disposition and almost semi-crepuscular in habits. But that it has fine powers of flight is shown indirectly by the shape of its wings, and directly by its occasionally straggling to New Zealand, a distance of upwards of 1000 miles from the nearest point of Australia. It preys upon insects — chiefly, though probably not solely, beetles — caught on the wing or on the ground; a tame specimen kept by Blyth acquired a great liking for plantains (B. Ceylon, 286). It is, as Gould remarks, a very bold bird at all times, particularly so during the breeding season; as is too often the case, however, boldness is accompanied by cruelty, and it is said, writes Gould, to take young Parrots from their holes and kill them, an observation confirmed by E. P. Ramsay, who remarks that "they not unfrequently fight with, and dispossess the *Dacelo gigas*, and I have seen them take the young of the bird and throw them out of the nest" (Pr. L. Soc. N. S. Wales VII, 46). Its note is described by Meyer as kiak, kiak, by Whitehead as kick, kick, "sounding somewhat like the noise made by coachmen to horses" (l. c.); it is also said to have a single, full, deep-toned whistle (Legge, l. c.).

The plumage in the genus *Eurystomus* is remarkably similar to that of *Coracias*. The two genera are remotely related, hence the type of plumage is probably very ancient. It is interesting in this connection to observe that the sexes are alike and that the young hardly differ from them.

Eurystomus orientalis appears to be a species in process of undergoing differentiation into numerous smaller ones, a process which is not yet complete in most quarters. As the bird is migratory both in its northern territories and in Australia there can be no doubt that its distribution has taken place by flight. The genus *Eurystomus* also occurs in Madagascar and tropical Africa, being represented there by three species nearly related to one another, but very distinct from *E. orientalis*; they are placed in a different genus, *Cornopio* by Cabanis & Heine (Mus. Hein. II, 119) and Heine & Reichenow (Nomencl. Mus. Hein.

157). It is therefore most likely that these African and Madagascar forms and the Indo-Australian *E. orientalis* are sprung from ancestors belonging to some part of Asia, Africa, or Madagascar — at all events not to Australia. *E. orientalis*, we may suppose, reached the East Indies by migration in winter from its northern haunts. Pursuing their way southwards, instead of northwards, some of them have become adapted to the different breeding-season and different period of migration of birds in Australia, and under this separation from the northern birds have begun to develop racial differences, such as we find in *the typical Falco peregrinus* and the Australian *F. peregrinus melanogenys*, and much more strongly pronounced in the Australian migrant *Merops ornatus* and the Indian migrant *M. philippinus* or in the Australian migrant *Halcyon sancta* and the East Indian resident *Halcyon chloris*. *Eurystomus crassirostris* of the New Britain Group and New Guinea and its surrounding islands, and *E. solomonensis* do not appear to intergrade with *E. orientalis* which occurs in some of the same localities as the former, this being the resident form, the other, perhaps, solely or mainly a migrant. In the Great Sunda Islands, and the Burmese countries it is the opinion of Dr. Sharpe, that a resident form, distinguishable from the migrants from the north is coming into existence, but in some localities — among them Celebes — it appears that resident birds and migrants from north and south occur together and that a certain amount of interbreeding takes place, rendering exact nomenclature impossible.

We may pause for a moment to think what theories on former physical geography a naturalist in the distant future might build up, when these races of *E. orientalis* may be supposed to have become resident and well differentiated species in India, the East Indies, Australia and New Zealand!

ORDER MACROCHIRES.

The Goatsuckers and Swifts, which catch their insect-prey on the wing (except *Steatornis* which feeds on fruit) and possess an enormous gape reaching to beneath the eyes or further, but a minute and weak bill, very long wings, small feet and tarsi.

The tail-feathers are 10 in number; there are also 10 primaries in both (see, also, Hartert, Cat. B. XVI; Gadow in Bronn's Kl. & Ord.; Blanford, Fauna Br. Ind., Birds III, 162).

FAMILY CAPRIMULGIDAE.

The Goatsuckers are crepuscular or nocturnal in habits, wearing a nocturnal plumage — soft, with the pattern broken up and vermiculate, without bright pure tints, such as pure blue, yellow or red, but chiefly brown, grey, and dusky colours; often with "recognition marks" of white on the throat and tail. The bill is small and soft, the nasal aperture membranous and tubular, the gape very wide, reaching below the eye. Five subfamilies of the *Caprimulgidae* may be distinguished; their differences are pointed out below (p. 325) under *Lyncornis macropterus*.

GENUS CAPRIMULGUS L.

The true Goatsuckers may be recognised by their having a row of strong bristles above the gape directed forwards; no ear-tufts, a rounded tail with white patches on the outer rectrices of the male (usually absent in the female). Outer toe with four phalanges, claw of middle toe pectinated. The genus is of almost cosmopolitan distribution.

99. CAPRIMULGUS MACRURUS Horsf.

Horsfield's Nightjar.

Mr. Hartert's careful researches on the Goatsuckers offer proof that *C. macrurus* as a species should be viewed as made up of two subspecies, i. e. *the typical macrurus* ranging from Malacca to southern Australia (Ramsay) and *C. macrurus albonotatus* (Tick.) of N. India; while the territories from the foot of the Himalayas to Tenasserim produce intermediate forms. These latter might be distinguished as *C. macrurus — albonotatus.*

←1. The typical Caprimulgus macrurus.

a. **Caprimulgus macrurus** *(1)* Horsf., Tr. Linn. Soc. 1821, XIII, 142; *(II)* Gld., B. Austr. 1848, II, pl. 9; *(3)* id., Handb. B. Austr. 1865, I, 100; *(4)* Salvad., Cat. Ucc. Borneo 1874, 117; *(5)* Brügg., Abh. Ver. Bremen 1876, V, 464; *(6)* Salvad., Orn. Pap. 1880, I, 528 (pt. — non ex India); *(7)* W. Blas., J. f. O. 1883, 115, 122, 136; *(8)* Ramsay, Tab. List 1888, 2; *(9)* North, Nests and Eggs Austr. B. 1889, 29; *(10)* Salvad., Agg. Orn. Pap. 1889, I, 62; *(11)* Sharpe, Ibis 1890, 22, 283, 288; *(12)* Everett, J. Str. Br. R. A. S. 1889, 153; *(12bis)* Hagen, T. Ned. Genoots. 1890, (2) VII, 142; *(13)* Hartert, Kat. Vogels. Senckenb. Mus. 1891, 120; *(14)* id., Cat. B. XVI, 1892, 537 (pt. non ex Tenasserim, Burma); *(15)* id., Ibis 1892, 281—283; *(16)* Hose, ib. 1893, 406; *(17)* Mad., Aquila 1894, 97; *(18)* Bourns & Worces., B. Menage Exp. 1894, 34; *(19)* Blanf., Faun. Br. Ind. B. III, 1895, 188, pt.; *(20)* Hart., Nov. Zool. 1896, 175, 562, 593; *(21)* id. Tierreich, 1897, I, 54 *(typicus).*
b. **Caprimulgus salvadorii** *(1)* Sharpe, P. Z. S. 1875, 99, pl. XXII, fig. 1.
For further synonymy and references see Salvadori *a 6, a 10*; Hartert *a 14.*
Figures and descriptions. Gould *a II, a 3*; Sharpe *b I*; Salvadori *a 6*; North *a 9* (egg); Hartert *a 14.*
Adult male. Upper surface pale brown, finely vermiculated with greyish, especially on the top of the head; centre of crown longitudinally streaked with brownish black; an indistinct fulvous band across the hind neck; back and rump marked with deep

brown; scapulars and wing-coverts with more or less bright buff patches, the
former mostly velvety brownish black; first primary deep brown, a large spot on
the inner web and a broad white patch across both webs of the next three primaries, often an indication of a white spot on the fifth quill; secondaries deep
brown, with narrow interrupted rufous bars, the two outer pairs of rectrices
largely tipped with white, these tips varying in extent, being generally about 2 inches
(50 mm) in length; chin and sides of throat rufous brown, finely barred with
blackish brown; throat with a very large white spot, bordered at the lower part
with deep black; abdomen rufous buff, barred with brown. Total length about
292 mm; wing 180—198 (about 190 on an average); tail 142—160; tarsus 17.8,
feathered in front for almost its whole length (Hartert *a 14*). Iris blackish brown;
bill black; feet and claws reddish brown (Gould *3*).

Adult female. Differs from the adult male in having the outer web of the first primary
spotted with rufous, in having pale rufous marks on the primaries instead of white
ones, in the tips to the outer rectrices being less in extent and tinged with buff or
rufous, speckled with brown on the tip of the outer web (Hartert *a 14*).

Young. The markings less developed, the young male has the white patches on the primaries
and rectrices tinged with rufous and less in extent. The nestling is covered with
buffy down (Hartert *a 14*).

Distribution. Almost the whole of Australia (Ramsay *8*); New Britain; Aru; New Guinea;
Waigiou; Halmahera; Obi; Ceram; Buru; Celebes — Gorontalo; Timorlaut; Timor;
Lombok; Java; Borneo; Palawan; ? Philippines; Sumatra; Malacca (Salvad. *a 6*,
a 10, Brüggem. *a 5*, Everett *a 12*, Hartert *a 14*, *a 15*); Saleyer and Djampea
(Everett *a 20*); Sumbawa (Doherty *a 20*).

2. Caprimulgus macrurus albonotatus (Tick.).

c. **Caprimulgus albonotatus** *(1)* Tick., J. A. S. B. 1842, XI, 580; *(2)* Jerd., B. Ind. 1862,
I, 194; *(3)* Hume, Str. F. 1875, III, 45; *(4)* id., ib. 1878, VI, 58; *(5)* Hartert,
Cat. B. XVI, 1892, 540 (subsp.) partim; *(6)* Blanford, Fauna Br. Ind. B. III, 1895,
188, pt.

d. **Caprimulgus macrurus albonotatus** *(1)* Hartert, Ibis 1892, 282; *(2)* id. Tierreich, 1897, I,54.

Diagnosis. Differs from the *typical macrurus* in being altogether a larger bird (wing 229 mm,
against 197—203 in *macrurus*), both above and below altogether a lighter coloured
and more buffy bird, with broader white or buffy or creamy white margins to the
scapulars and wing feathers, and with the whole lower parts comparatively uniform;
whereas in *macrurus* the breast is much darker and contrasts strongly with the much
paler abdomen (Hume *c 4*).

Variation. Specimens from the plains of North-western India are very light coloured; the
markings paler brown (than in the *typical macrurus*), the scapulars less brilliantly
marked, the dark spots on the crown less numerous, narrower, and more confined to
the middle of the crown; wings very long; lower parts very strongly tinged with pale
sandy rufous. Wing 203—218 mm; tail 165—178 [as against 180—198 and 142—160
respectively in the *typical macrucus*], (Hartert *c 5*).

Distribution. India throughout the Himalayas at low elevations, in the North-west Provinces,
Bengal, Chutia Nagpur, and Raipur, and in Burma (Blanford).

3. Caprimulgus macrurus — albonotatus.

e. **Caprimulgus nipalensis** [Hodgs., Icon. ined. in Mus. Brit.], *(1)* Hartert, Ibis 1892, 283.
f. **Caprimulgus macrurus** *(1)* Hume, Str. F. 1875, VI, 58; *(2)* Oates, B. Brit. Burmah 1883,

II, 20; *(3)* id., ed. Hume's Nests & Egg. Ind. B. 1890, III, 45; *(4)* Hartert, Cat.
B. XVI, 1892, 537, pt.; *(5)* id., Ibis 1892, 282, 283 (intermediate forms).
g. **Caprimulgus albonotatus** *(1)* Oates, B. Brit. Burmah 1883, II, 19, ? pt.; *(2)* Hartert,
Cat. B. XVI, 1892, 541 (ex Mont. Himalaya).
Distribution. Himalayas, mountains of Burmah and Assam, Tenasserim, Siam and China.
Variation. On comparing good series of birds from Tenasserim, Burmah, and Assam it
has been found that in the latter localities many intermediate phases between the
typical macrurus and *macrurus albonotatus* occur, so that it is impossible to draw
any line between the two forms. Specimens from the foot of the Himalayas also
may be regarded as intermediate between the true *albonotatus* and *macrurus* (after
Hartert *f 5*).

"Specimens from Borneo", says Mr. Hartert, "are very dark and small: ...
specimens from Waigiou and Aru are perfectly similar to those from North Borneo;
they belong to the very dark blackish island-form".

Egg (of the species). Australia — 2 in number, light rich cream-colour, fading to whitish
after being emptied, clouded all over with fleecy markings of pale slaty lilac, which
appear as if beneath the surface of the shell; 28 × 21 mm (Ramsay *a 9*); West
Java — 2, elliptical, faint white tending to yellowish, marked with scattered spots,
mostly not large, of reddish grey-brown and ash-grey, the latter in particular numerous
towards the blunt end, though not forming an appreciable circlet; 28—30 × 21—22
(Bernstein, J. f. O. 1859, 182); Tenasserim — 2 in number; somewhat cylindrical
ovals; shell very fine and smooth, excessively close-grained, very thin; delicate creamy
pink, everywhere rather thinly spotted, streaked, clouded, and marbled with very
pale, somewhat brownish purple, and very pale subsurface-looking inky grey; (ten
specimens) 29—33 × 20—23 mm (Hume *f 3*).

Nest. None — the eggs are laid upon the bare ground: Australia (Rams. *a 9*); Tenasserim
(Bingham *f 3*); Java — a slight depression, upon a few bamboo-leaves (Bernstein l. c.).

Breeding season. Tenasserim — March, April (Bingham *f 3*).

The right of *C. macrurus* to be included in the Celebes avifauna rests upon
two specimens, obtained by von Rosenberg or Riedel in Gorontalo. These
latter were determined by Brüggemann, who pointed out that Lord Walden's
specimen of a *Caprimulgus* obtained by Meyer in Celebes appeared to be
identical with them. In this Brüggemann was in error; Meyer's specimen
has since been made the type of a distinct species, *Caprimulgus celebensis* Grant.
Mr. Everett obtained *C. macrurus* in Saleyer and Djampea. A nestling from
the latter island shows that the species breeds there.

It is probable that *C. macrurus* is not strictly stationary; the Nightjars feed
upon insects, which become scarcer at certain seasons, when local movements
on the part of the birds will probably take place. Such, in particular, should
be the case in the southern parts of Australia. Like the Owls, the Celebes
Nightjars appear to have nothing to say on the question of the former distribution of the land and water round about; like them they appear from their
nocturnal habits to be specially liable to be carried across straits of the sea by
winds, and if in addition the species has migratory tendencies its range will be
likely to become all the wider. Whereas, for instance, one finds two distinct

species of the Woodpecker *Microstictus*, one occurring in North, the other in South Celebes, the two subspecies of *Caprimulgus macrurus*, which appear to be of about equal value, are found one in India, the other in the East Indies and Australia.

⊥ * 100. CAPRIMULGUS CELEBENSIS Grant.
Celebesian Nightjar.
Plate XI.

a. **Caprimulgus** sp. *(1)* Wald., Tr. Z. S. 1872, VIII, 115.
b. **Caprimulgus manilensis** *(1)* Hartert, Cat. B. XVI, 1892, 544, partim.
Caprimulgus celebensis *(1)* Grant, Ibis 1894, 519; *(2)* Hartert, ib. 1896, 371; *(3)* id. Tierreich 1897, I, 53.
"Trio", Lemboh (Nat. Coll.).

Diagnosis. This species, with its Philippine ally *C. manilensis*, differs, according to Hartert, from *C. macrurus* chiefly in the smaller extent of the white spots on the primaries and of the white tips of the lateral rectrices, especially in the latter (white tips of two outer pairs about 25 mm broad, in *macrurus* about 50 mm). Abdomen brownish buff, barred with dark brown, but not so regularly as in *C. macrurus*, the broader buff tips of the feathers producing a more spotted appearance.

Measurements. Total length 305 mm; wing 180; tail 147; tarsus 20; mid. toe with claw 26.7; culmen 21.6; longest rictal bristle 35.6 (Grant *1*).

Distribution. Celebes (Meyer *a 1, b 1, 1*); Lemboh Id. (Nat. Coll. in Dresd. Mus.).

A single specimen of this species now in the British Museum was obtained by Meyer in Celebes — probably in the Minahassa — in 1870—1871. It was not immediately determined by Lord Walden, and Brüggemann having received the foregoing *C. macrurus* from Gorontalo expressed the opinion that Meyer's example would prove to be the same, but he apparently overlooked Walden's determination of the specimen as *manilensis* in a subsequent work (*a 1*). Walden's decision was confirmed by Mr. Hartert, but Mr. Ogilvie Grant who has recently had occasion to compare the Celebesian specimen with Philippine ones, remarks: "How the bird from Celebes came to be identified with *C. manillensis*, G. R. Gray, by both Lord Tweeddale and Mr. Hartert I am at a loss to understand, for two more totally distinct species of Goatsucker can hardly be imagined. In the Celebes bird the rictal bristles are much longer and stouter than in any other species of *Caprimulgus*, and extend far beyond the end of the culmen, the longest ones being about once and a half the length of the culmen from its base, and more than twice as long as the exposed part of the culmen. This character alone is sufficient to distinguish the Celebes bird at a glance, for in *C. manillensis* the bristles are very much finer and are very little longer than the culmen. The example from Celebes (apparently not quite adult, the primaries being tipped with buff) is most nearly allied to *C. andamanicus* Hume, the two outer tail-feathers being rather narrowly tipped on both webs with white, the white extremity being less than an inch in length;

the white spots on .the inner webs of the primaries are small, but clearly defined".

A second specimen was killed on Lembeh Island in March, 1895, and sent to the Dresden Museum; we give a figure of it. We find Mr. W. R. Ogilvie Grant's statement as to its distinction from *C. manilensis* couched in too energetic language; instead of — as his words might lead the reader to suppose — belonging almost to two different genera, the two forms seem to us to be so closely allied as almost to endanger the specific distinction of *C. celebensis*, and to awake a doubt as to whether it may not ultimately have to be reduced to the rank of a subspecies.

101. CAPRIMULGUS AFFINIS Horsf.

Allied Nightjar.

Caprimulgus affinis *(1)* Horsf., Tr. L. S. XIII, 1821, 142; *(2)* Sclat., P. Z. S. 1863, 212; *(3)* Walden, Tr. Z. S. 1872, VIII, 114; *(4)* Salvad., Cat. Ucc. Borneo 1874, 115; *(5)* Tweedd., P. Z. S. 1877, 691; *(6)* Vorderman, Nat. Tdschr. Ned. Ind. 1883, XLII, 54; *(7)* Guillem., P. Z. S. 1885, 504; *(8)* Tristr., Cat. Coll. B. 1889, 112; *(9)* Everett, J. Str. Br. R. A. S. 1889, 153; *(10)* Vorderman, Notes Leyden Mus. 1891, 25; *(11)* id., N. T. Ned. Ind. 1891, L, 448; *(12)* Hartert, Ibis 1892, 280; *(13)* id., Cat. B. XVI, 1892, 549; *(14)* Büttik., Zool. Erg. Weber's Reise Ost-Ind. 1893, III, 292; *(15)* Vorderm., N. T. Ned. Ind. 1895, LIV, 336; *(16)* Hart., Nov. Zool. 1896, 158, 549, 570, 595; *(17)* id., ib. 1897, 163; *(18)* id., Tierr. 1897, I, 50.

a. **Engoulevent des Roseaux** *(1)* Hombr. & Jacq., Voy. Pôle Sud 1846, pl. 21, fig. 2.
b. **Caprimulgus arundinaceus** *(1)* Bp., Consp. 1850, I, 60; *(2)* Jacq. & Pucher., Voy. Pôle Sud, Texte 1853, 93; *(3)* Salvad., Cat. Ucc. Borneo 1874, 116.
c. **Caprimulgus faberi** *(1)* Meyer, Isis, Dresden 1884, 20 (fide Hartert).
For further synonymy and references see Hartert *13*.
Descriptions. Hombron & Jacquinot *a 1*; Walden *3*; Vorderman *6*; Hartert *13*; Meyer *e l*.

Adult [male]. Ground-colour above, including upper surface of tail, broccoli-brown, thickly vermiculated, sprinkled, and spotted with sepia, clove-brown, and black, the crown, nape and mantle being most strongly marked — not streaked — with black, the scapulars marked with broken spots of light cinnamon; first four primaries blackish brown, outer webs paler, across their middle length a broad white band, only seen on the inner web of the first quill, on both webs of the other three; remaining quills (except the innermost ones which share the colour of the back) and greater wing-coverts brownish black barred with tawny-cinnamon, the bars being somewhat broken up with black; outermost tail-feather white, base of inner web barred with black and pale tawny-cinnamon; the next, white, mottled with bars of black and light tawny-cinnamon towards its base, most extensively on the inner web; general colour of under surface pale cinnamon, closely sprinkled and barred with broken markings of dark sepia on throat and breast, and rather regularly marked with narrow bars of sepia on the remaining under parts; under tail-coverts uniform cinnamon-whitish; on each side of throat a white patch (Java, Nr. 8898).

Adult female. Differs from the male in having no white whatever on the tail, the rectrices being banded with pale rufous grey and dark brown, paler on the tips (Hartert *13*).

Variation. The spots on the throat vary in size in both sexes and are sometimes almost confluent, often not pure white but buff, this latter colour being apparently a sign of immaturity, as also is the buff colour of the white patches on the primaries (Hartert *13*).

Measurements.

	Wing	Tail	Tarsus	Bill from nostril
a. (Nr. 8898) Java	167	98	18	7
b. (Nr. 13732) Sumatra (type of *C. faberi*)	166-67	107	—	—

Egg and breeding habits. Unknown[1].

Distribution. Sumatra (Raffles *13*, Faber *c 1*); Java (Horsf. *1*, Vorderman *6*); Billiton (Vorderman *10, 11*); Banjermasin, S. Borneo (Hombron *b 2*, Mottley *2, 4, 9*); Celebes (Meyer *3, 13*, Everett *16*, Doherty *17*); Lombok (Wall. *3, 4, 13*, Doherty *16*); Sumbawa (Guillem. *7*, Doherty *16*); Flores (Weber *14*); Timor (Wall. *13*).

A single specimen only of this species obtained by Meyer was the sole proof of its occurrence in Celebes till September, 1895, when a second was killed at Macassar by Everett and determined by Hartert (*16*) as a perfectly adult male. Later two further specimens were obtained by Doherty in West Celebes. Meyer's example is now in the British Museum and is marked by Hartert as immature.

The nearest ally of *C. affinis* is *C. griseatus* (Wald.) of Luzon, a rare bird in collections, which will probably prove to be only subspecifically distinct from *C. affinis* when the intermediate localities have provided ornithologists with specimens. Walden (Tr. Z. S. 1875, IX, 326) speaks of this form as intermediate in dimensions between *C. affinis* and *C. monticola* of India and South China to Tenasserim, but not yet known, apparently, from the intermediate Malay Peninsula. *C. monticola* is, however so much bigger than the other two that it is hardly likely that it will be found to intergrade with them.

GENUS LYNCORNIS J. Gd.

Distinguishable from *Caprimulgus* by having no rictal bristles, no white on the tail-feathers; the parietal plumage lengthened so as to form "ear-tufts". Outer toe with four phalanges, claw of middle toe pectinated. Hartert (1897) recognises 6 species ranging from N. E. India to New Guinea.

102. LYNCORNIS MACROPTERUS Bp.
Celebesian Eared Nightjar.
Plate XI.

Lyncornis macropterus [Temm., Mus. Leyd.], *(1)* Bp., Consp. 1850, I, 62; *(2)* Wall., Ibis 1860, 141; *(3)* Gray, Hl. 1869, I, 605; *(4)* Walden, Tr. Z. S. 1872, VIII, 47,

---

[1] Dr. Kutter (J. f. O. 1882, 175) says that the egg of *C. griseatus* of the Philippines is very similar to that of *C. affinis* said to be described by Baron König-Warthausen under *C. bisignatus* (J. f. O. 1868, 373).

112; *(5)* Rosenb., Malay. Archip. 1878, 271; *(6)* W. Blas., J. f. O. 1883, 136; *(7)* Guillem., P. Z. S. 1885, 549; *(8)* Tristr., Cat. Coll. B. 1889, 114; *(9)* Hartert, Cat. B. XVI, 1892, 605; *(10)* M. & Wg., Abh. M. Dr. 1895, Nr. 8, p. 8; *(11)* iid., ib. 1896, Nr. 1, p. 5; *(XII)* Meyer, Vogelsk. 1897, pl. CCXIII; *(13)* Hart., Tierr. 1897, I, 25.
a. **Caprimulgus macropterus** *(1)* Schl., Handl. d. Dierk. 1857, I, 223; *(2)* Finsch, Neu-Guinea 1866, 162.
"**Tulie**", ? Gorontalo, Rosenb. *5*.
"**Trio**", Gorontalo Distr. (or Minahassa), Guillem. *7*.
"**Kokotaka**", Minahassa, Nat. Coll.
"**Burong malas**", Malay, iid.
"**Burong malam**" (Evening-bird), Malay, Manado, Meyer MS.
Figure and descriptions. Meyer *XII* (skel.); Bonaparte *1*; Hartert *9*.
Adult [nearly]. General colour above brownish black, more dusky on the wings, the whole varied with cinnamon and light tawny-olive; entire head above very finely vermiculated with black and russet, marked with large scattered shield-shaped spots of sooty black narrowly bordered with russet; behind the head the feathers broadly fringed with tawny-olive, forming a collar; hind neck, mantle and middle of back sooty black, somewhat vermiculated with cinnamon on the mantle and back; rump, upper tail-coverts and tail more strongly speckled and vermiculated with cinnamon, five or six narrow broken bars of which colour are discernible on the tail; on the scapulars and inner quills the cinnamon tint prevails over the black, scapulars marked with some large shield-shaped spots of black; quills dusky, speckled with tawny-olive in the form of ill-defined bars; upper wing-coverts speckled with russet; superciliary region, ear-coverts, malar region and chin, black, barred with russet; across the throat a broad white band; upper breast and sides of neck black, the feathers fringed with russet; below this a broad band of almost clear cinnamon; remaining under parts cinnamon, brokenly barred and speckled with black, the bars being well formed on the flanks; under wing-coverts barred with blackish and cinnamon (near Tondano, Aug.—Sept. 1892, Nat. Coll. — C 10794). Iris brown; bill reddish brown; foot clear brown (Guillem. *7*).
Variation. The species varies considerably; in six other specimens before us the cinnamon bars on the tail are much broader and better defined than in that described, and the upper surface paler — more or less. One specimen (C 12868) is almost entirely without spots on the crown; this is perhaps the oldest. The white patch on the throat also varies in extent.
Sexes. Apparently similar.
Measurements (7 specimens). Wing 252—262 mm; tail 151—161; bill from nostril 7—8; tarsus 17.5 ca.
Egg. "On 4th April, 1895, a hunter brought us in Tomohon a female and a completely smashed egg, which he had found lying simply on the ground. The fragments showed a reddish white colour with rather dark red-brown spots" (P. & F. Sarasin *11*).
Distribution. Celebes, Northern Peninsula: — Minahassa (Wallace *2*, *9*, v. Musschenbroek, etc.), Gorontalo Distr. (Riedel, Mus. Dresd., Guillemard *7*).

Mr. Wallace speaks of this species as abundant about Manado, where it appears soon after sunset, chasing insects with rapid evolutions. Guillemard found it especially plentiful near Likoupang on the north coast, and v. Rosenberg, whose notes have special reference to Gorontalo, mentions it as very common, particularly by river-banks. Nevertheless it has escaped many collectors, and has not been recorded from any part of Celebes but the Northern Peninsula. It is closely related to the Philippine species of *Lyncornis*, especially *L. mindanensis*, its nearest neighbour, which, however, as Mr. Hartert shows, differs in having the abdomen much more evenly barred. Should *L. macropterus* prove to be confined to North Celebes, it might be reasonably inferred that the species is a recent immigrant from Mindanao.

The genus *Lyncornis* is known to range from Burmah to Java, Celebes and the Philippines, occurring again in Salawatti and New Guinea.

Although the Goatsuckers of Celebes appear to have nothing important to say on the past conditions of this island, the geographical distribution of the Goatsuckers of the world in general affords several points of interest. In 1866 (P. Z. S. 127) Sclater divided the *Caprimulgidae* into 3 subfamilies:

1. *Podarginae*, most typical of Australia, in which the genus *Nyctibius* of South America to Jamaica and Mexico was included;
2. *Steatornithinae*, the frugivorous Cave-goatsucker of S. America from Guiana to Peru;
3. *Caprimulginae*, the typical Goatsuckers, of which S. America is by far the richest in genera, after which comes Africa.

Later, Garrod's anatomical researches led him to the conclusion that *Steatornis* should be separated as a family for itself (P. Z. S. 1873, 534; Coll. Papers, 186). In 1885 (P. Z. S. 153), Beddard divided the *Caprimulgidae* into four subfamilies: 1. *Steatornis*; 2. *Podargus* and *Batrachostomus*; 3. *Aegotheles*; 4. *Caprimulgus*, *Chordeiles*, and *Nyctidromus*. This anatomist differs in so far from Garrod as to conclude that "*Podargus*, *Batrachostomus* and *Aegotheles* are much nearer to *Steatornis* than to *Caprimulgus*, but should be placed in an intermediate position". Recently Hartert has separated these groups of forms into three different families with subfamilies as follows:

1. Caprimulgidae.
    a. *Caprimulginae*, chiefly South American, 11 peculiar genera being found in America and Jamaica, 3 in Africa, 1 in the Oriental Region and New Guinea, 1 in the Australian Region, while 1 *(Caprimulgus)* is common to all, though sparsely represented in the direction of Australia.
    b. *Nyctibiinae*, tropical parts of S. America to Jamaica and Mexico. (Sclater has placed this subfamily in Hartert's *Podargidae*.)
2. Podargidae.
    a. *Podarginae*, two genera, the one confined to Australia and Papuasia, the other to the Oriental Region from the Himalayas to Java and the Philippines.
    b. *Aegothelinae*, confined to Australia and Papuasia, with one species in New Caledonia.
3. Steatornithidae.
    South America from Guiana to Peru.

Although we should prefer to recognise one family of five subfamilies, rather than three families with two subdivisions as Mr. Hartert does, — for the *Nyctibiinae* connect the *Caprimulgidae* with *Podargidae*, from which again the

Steatornithidae, according to Mr. Beddard, are not far removed — it is nevertheless well to be made aware of the differences which separate the several groups. Some of these differences may be recapitulated as follows:

	Caprimulginae	Nyctibiinae	Podarginae	Aegothelinae	Steatornithinae
Phal. in outer toe	Four	Five	Five	Five	Five
Claw of middle toe	Pectinated	Not pectinat.	Not pectinated	Not pectinated	Not pectinat.
Oil-gland . . .	Pres. (small)	Present?	Absent	Absent	Present
Powd.-dwn. patches	Absent	Present	—	—	—
Syrinx . . .	Tracheo-bronchial	—	Approximates bronch. structure	Tracheo-bronchial	Bronchial
Stern., nr. of incis.	One pair	Two pairs	Two pairs	Two pairs?	One shallow pair
Palate . .	Aegithognathous[1]	Schizognath.	Desmognathous	Desmognath.?	Desmognath.
Biceps-slip . . .	Present	—	Absent	Absent	Absent
Fem.-caud. muscle	Present	—	Present	Present	Absent
Caeca	Present	—	Present	Absent	Present
Food	Insectivorous	Insectivorous	Insectivorous	Insectivorous	Frugivorous
Eggs . . .	Coloured	—	Pure white	Do., or palely striated	Pure white
Nest . . .	None, eggs laid on ground	—	Flat, of small sticks on horiz. branches	In hollow trees	In caves

In external appearance the *Podarginae* are very like the *Nyctibiinae*; structurally they differ in at least one very important point, i. e. the structure of the palate. Nevertheless in this respect also the two subfamilies approach one another more nearly than the other *Caprimulgidae*: "we have almost a transition to the Desmognathous structure in *Nyctibius*" (Huxley, P. Z. S. 1867, 456).

It is remarkable that Australia, the part of the world almost the most remote from South America, should furnish the link, *Podargus*, between the South American forms *Nyctibius* and *Steatornis*, which would otherwise be best separated as families for themselves. The distribution of the *Caprimulgidae* may be compared with that of the *Struthiones*. But it seems to have taken place at two distinct epochs; — an earlier one when the dispersal of the main groups took place, and a later one which determined that of the subfamily, *Caprimulginae*. The latter group from its great richness in genera, may be supposed to have been longer in America than elsewhere, the various structural modifications being a matter of time. The Palaearctic Region possesses only the cosmopolitan genus *Caprimulgus*, which most likely entered it from America in comparatively recent times; while *Lyncornis* and *Eurostopodus* and the peculiar *Caprimulginae* of Africa seem to have proceeded south and become isolated after an exodus from America at an earlier period.

[1] cf. Hartert, Ibis 1896, 370.

The *Caprimulgidae* are again of interest, as Mr. Hartert shows, from the high development of protective coloration seen in them. "The colours of the *Caprimulgidae* ... vary much according to the surroundings, and especially in relation to the soil. So, as a rule, we find yellow, buff, and isabelline-coloured species in sandy deserts and desert-like localities, darker species in more wooded countries, and richly coloured ones in tropical forests. But even the same species varies much, according to the soil that it frequents, and therefore several species readily form more or less well-defined local races, often well worthy of subspecific rank. In many cases, of course, as usual, we do not understand the reasons why a certain form is differentiated, because we have so little knowledge of the influence of climate, the amount of rainfall, the surroundings, and the food" (Ibis 1892, 275).

The need for protective coloration in so defenceless a bird as the Goatsucker which rests on the bough of a tree, or on the ground, in the day-time and seeks its insect-food at night, is manifest. Its curious habit, also, of sitting along the bough, instead of across it like other birds, is no doubt a protective measure. Its plumage of dark brown and grey varied with all sorts of patches, spots and vermiculations is admirably adapted to conceal it when lying lengthways on a bough covered with lichens, moss, brown bark and the moving shadows of leaves. So long as it remains thus at rest it is an inconspicuous object; the white outer tail-feathers, wing-band and throat-patch of many species of the *Caprimulgidae* are then hidden by overlying feathers or by its resting-place below, but so soon as it takes to wing these white patches must at once become striking marks. They exemplify particularly well the "recognition markings" of Wallace.

It is a curious fact that the Owls, also nocturnal birds, are on the whole much like the Goatsuckers in coloration. In their case, too, protective coloration may be reasonably urged to account for the fact, though they are not at all defenceless birds like the Goatsuckers. It is well known how a troop of Sparrows delight in mobbing an Owl in the day-time, as indeed also an escaped Parrot, Canary, or such like, and for the destruction of Birds of Prey and Crows one of the most irresistible lures is an Eagle Owl, which, when seen fastened in the open, will have a bad time from the beaks and claws of the diurnal Birds of Prey and Crows, if these are not shot down as they come by the hunter under cover close by. The plumage of the Owls, likewise resembles that of the Goatsuckers in being soft, and their flight is likewise noiseless, a fact explained in the case of the Owls by the need they have of approaching their prey without being heard. In the case of the Goatsuckers this explanation does not suffice — nor indeed does it in the case of the insectivorous Owls. The Goatsuckers, as a rule, like their allies the Swifts, catch their insect-prey on the wing, but the flight of the Swifts is rushing and loud.

Quiet flight therefore cannot be of any vital importance; a wide mouth, speed, and dexterity in turning seem to be the only essentials. It may also be added that insects are much less sensitive to sound than birds and mammals, or have quite different perceptions of it. The cause may be quite different. The manifest influence of light, shelter, and exposure on the plumage will be found referred to many times in earlier and later pages *(Loriculus, Hierococcyx crassirostris, Graucalus, Coracias)*; the converse of sunlight (darkness) must of course also have effect, and in the soft glossless plumage of the Owls and Goatsuckers it may be that we see the influence of the want of light upon the feathers of birds. Night falls upon other birds also, but these are then sleeping and their vital energies are less active or otherwise employed.

FAMILY CYPSELIDAE.

The Swifts are diurnal birds of aerial habits, taking and devouring their insect-food on the wing. The bill is small, the nostrils membranous, the gape deep and wide, extending to below the eyes, the palate aegithognathous; the secondaries are much reduced, the primaries very long, 3—4 times the length of the secondaries or more; there are said to be no median wing-coverts; the legs are very short. The keel of the sternum is very high, the posterior margin unnotched. Three subfamilies have been distinguished — all occurring in the Celebesian Province: the *Cypselinae* with the three anterior toes consisting each of three phalanges, the tarsus feathered; the *Chaeturinae* with the tarsus bare and the toes normal, i. e. the middle one with 4, the outer with 5 phalanges; the *Macropteryginae* with fenestra in the hinder margin of the sternum, and the tail extending about to the tips of the wings and deeply forked.

GENUS CYPSELUS[1] Ill.

The typical Swifts may be recognised by the toes which are naked and all directed forwards. Hartert (1892) recognised 16 species, of cosmopolitan distribution. Many of them migratory. They usually breed in holes in buildings or rocks, not forming a nest, laying white eggs.

103. CYPSELUS PACIFICUS (Lath.).
Australian Swift.

a. **Hirundo pacifica** *(1)* Lath., Ind. Orn. Suppl. 1801, 58.
b. **Cypselus australis** *(1)* Gld., P. Z. S. 1839, 141; *(II)* id., B. Austr. 1848, II, pl. 11; *(III)* Diggles, B. Austr. 1877, p. 20, pl. 20.

---

[1] The generic name *Micropus* of Meyer & Wolf (Taschenb. 1810, 280) antedates *Cypselus* of Illiger (Prodr. 1811, 230) by a year. By rule 10 of the Stricklandian Code the name *Micropus* is to be rejected.

328   Birds of Celebes: Cypselidae.

c. **Cypselus vittatus** *(1)* Jard. & Selb., Ill. Orn. 1843, IV, pl. 49; *(2)* Swinh., Ibis 1860, 48, 429; 1861, 254, 328; *(3)* id., ib. 1863, 253; *(4)* id., P. Z. S. 1863, 263; *(5)* Blyth, Ibis 1863, 263.

Cypselus pacificus *(1)* Blyth, J. A. S. B. 1843, XIV, 212, 548; *(2)* Scl., P. Z. S. 1865, 599; *(3)* Gld., HL. B. Austr. 1865, I, 105; *(4)* Swinh., Ibis 1870, 89; *(5)* Blyth, ib. 161; *(6)* Jerd., ib. 1871, 355; *(7)* Tacz., J. f. O. 1872, 351; *(8)* Swinh., Ibis 1874, 435; *(9)* Salvad., Ucc. Borneo 1874, 119; *(10)* Hume, Str. F. 1875, 14, 43; *(11)* Swinh., Ibis 1876, 331; *(12)* Hume & Davis., Str. F. 1878, VI, 48; *(13)* Salvad., Orn. Pap. 1880, I, 534; *(14)* Hume, Str. F. 1880, IX, 246; 1882, X, 185; *(15)* Oates, B. Brit. Burmah 1883, II, 1; *(16)* Ramsay, Tab. List 1888, 3; *(17)* Seeb., B. Japan 1890, 177; *(18)* Tacz., Faun. Orn. Sib. Orient. 1891, I, 168; *(19)* Pleske, Mél. Biol. Ac. Petersb. 1892, XII, pt. 2, 294; *(20)* Oust., Nouv. Arch. Mus. 1893, 138; *(21)* id., ib. 1894, 112; *(22)* Styan, Ibis 1894, 334; *(23)* Blanf., Faun. Br. Ind. B. 1895, III, 167.

d. **Micropus pacificus** *(1)* Stejn., Bull. U. S. Nat. Mus. Nr. 20, 1885, p. 321; *(2)* Hartert, Kat. V. Mus. Senckenb. 1891, 119; *(3)* id., Cat. B. XVI, 1892, 448..

e. **Apus pacificus** *(1)* Hartert, Tierreich 1897, I, 86.

For further synonymy and references cf. Salvad., *13*; Hartert *d 3*.

Figures and descriptions. Gould *b II*; Jardine & Selby *c I*; Diggles *b III*; Hume *10*; Salvadori *13*; Oates *15*; Hartert *d 3*; Taczanowski *18*; Blanford *23*.

Adult. Seal-brown, tail darker; across the uropygium and flanks a broad white band with dark shaft streaks, the feathers of the upper surface narrowly fringed with white; chin and throat white with dark shafts streaks; remaining under parts brown, paler than the upper surface, each feather broadly fringed with white (Ussuri, ♀, 25. V. 1884 — C 10426). Irides brown; eyelids pinkish grey; bill black; inside of mouth fleshy; feet pinkish; claws dark horny (Oates *10*).

Measurements.

	Wing	Tail	Tarsus	Bill from gape
a. (C 10426) ♀ ad. Ussuri, Manchuria	190	83	11	18
b. (C 10509) ad. Cape York, Australia	181	76	—	19 c.
c. (Nr. 2572) juv.? Siao, Sangi	167	72	—	19 c.

Eggs. China: 2 in number; pinkish white until blown, then becoming unpolished white. Average 20.5 × 17.8 mm (Swinh. *8*). 20 × 16 mm (Nehrk. MS.). See, also, Tacz. *18*.

Nest. In holes and crannies of the rocks; a shallow saucer, nearly 4 inches in greatest breadth, of refuse-straw and a few bits of catkins and feathers, all strongly agglutinated with a gelatinous matter, doubtless the bird's saliva. Another nest, perhaps the accumulation of six years, consisted of six nests one placed above another and strongly glued to it (Swinh. *8*).

Breeding season. China, breeding in numbers, June 22nd.

Distribution. Southern Siberia from the Yenesei to Japan (Seebohm *17*); Tibet (Bonv. & Pr. d'Orl. *20*); N. W. India from Upper Assam, Cachar and the Khasia Hills (Jerdon *6*, *15*) east to China (Swinh. *c 2*) and Formosa (Swinh. *c 3*), south to Burmah (Oates *15*); Tenasserim (Davison *12*); Malay Peninsula (Blyth etc. *9*); Sangi Islands — Siao (Meyer in Dresd. Mus.); New Guinea (Salvad. *13*); as far as Victoria and South Australia (Ramsay *16*).

---

having been previously used by Linnaeus for a plant-genus (Syst. Nat. 1766, II, 580). Zoologists are divided in opinion as to the admissibility of such names into zoology, and, until a general agreement as to nomenclature has been arrived at, we prefer not to allow *Micropus* to replace the familiar name *Cypselus*.

There is in the Dresden Museum a single specimen, killed by Meyer's hunters in Siao, which we somewhat doubtfully identify as a young one of this species. The only other known species it could belong to is *C. leuconyx* Blyth, "as yet only known from the Himalayas and rocky hills of Central India, but", says Mr. Hartert, "further investigation will probably add more to the knowledge of its distribution". Of this form Hume *(10)* remarks it is altogether a much smaller bird: wing 150 to 158 as a maximum; the under surface with much narrower and less marked white fringes than in *pacificus*; the whole of the feet (not the claws, as has been erroneously stated) very pale-coloured, almost albescent in some specimens. In *C. pacificus* Hume finds the wing varies from 178 to 190. The specimen from Siao — wing 167 mm — is thus intermediate between *C. leuconyx* and *C. pacificus* in size, but, since *leuconyx* is said not to exceed 158 mm in the wing, the smallness of the Siao specimen seems only to be explained on the ground that it is a young specimen of the larger species. The white fringes on the under surface are hardly noticeable, but the ends of most of the feathers are worn off.

Swinhoe found this species to be only a summer visitor to China, arriving in the spring and going south in the winter. Taczanowski records its migrations in S. E. Siberia. Its range extends to South Australia, where, as Mr. Hartert remarks, it is probably a winter visitor. Gould also 50 years ago believed that such would prove to be the case.

Australian ornithologists, so far as we have been able to ascertain, have not made any further observations on this interesting subject, except that Cox and Hamilton record it as seen among flocks of *Chaetura caudacuta*, a Swift with a similar range, observed in New South Wales from December to April and in July and August (P.L.S., N.S.W. 1889, 2nd ser., IV, 399).

GENUS CHAETURA Steph.

The Pin-tailed Swifts are easily distinguished by the shafts of the rectrices which project beyond the webs as sharp, strong spines. The toes have the normal number of phalanges, the tarsus is naked.

About 30 species, of almost cosmopolitan distribution.

⊹ * 104. CHAETURA CELEBENSIS (Scl.).
Steel-blue Pin-tailed Swift.
Plate XII.

a. **Chætura gigantea** var. **celebensis** *(1)* Sclat., P. Z. S. 1865, 608; J. f. O. 1867, 130.
b. **Hirundinapus giganteus** *(1)* Wald. (nec Temm.), Tr. Z. S. 1872, VIII, 46.
c. **Chætura gigantea** *(1)* Rosenb., Malay. Archip. 1878, 271.
d. **Hirundinapus celebensis** *(1)* Salvad., Ann. Mus. Civ. Gen. 1878, XII, 320; *(2)* W. Blas. J. f. O. 1883, 114.

Chætura celebensis *(1)* Hartert, Cat. B. XVI, 1892, 476; *(2)* W. E. Clarke, Ibis 1894, 533; *(3)* id., Ibis 1895, 474; *(4)* M. & Wg., Abh. Mus. Dresden 1896, Nr. 1, p. 5.

Adult. Deep steel-black, back and rump with purple gloss; two white spots on the sides of the forehead; sides of the abdomen and under tail-coverts white. In size similar to *C. gigantea*; — total length nearly 229 mm, wing 203, tail 71, tarsus 16.5 (Hartert *1*).

Female. A specimen in the Sarasin Collection marked "♀ juv." (but we cannot see any signs of immaturity) answers to Mr. Hartert's description of the species, except that the lores are reddish brown, not white: wing 208 mm; tail 63; tarsus 16; bill from nostril c. 6 (♀, Tomohon, 29. III. 95: P. & F. S.).

Distribution. N. Celebes — Manado (Leyden Museum, *a 1, d 1*), Tomohon (P. & F. Sarasin); Philippines — Negros (Keay *2*).

Only two specimens of this species were known for a space of thirty years. The name of the collector has not been recorded, but they are stated by Sclater and Hartert to have come from Manado. The latter remarks that they belong to a very distinct species. Next it was recorded from Negros by Mr. W. Eagle Clarke, and in March 1895 a third Celebesian specimen was obtained by the Sarasins. The nearest ally of *C. celebensis* is *Chaetura indica* Hume of India and Ceylon, occurring also in the Andamans in summer as well as in winter (Hume, Str. F. II, 156). This has the white spots on the lores, but differs from *C. celebensis* in the smoky brown — not steel-blue — colour of its under surface. *C. gigantea*, which ranges from the Malay Peninsula to Java, Borneo and Palawan (Hartert), is similar to *indica*, but wants the white spots on the lores. It is strange that birds of such grand flying-powers as these Swifts should seem to have such restricted ranges, but the difficulty of shooting them is no doubt very great.

Chaetura caudacuta (Lath.), however, ranges from Kamtschatka to Tasmania, and has occurred as a straggler in Great Britain. This species also must needs pass over Celebes on its way south to Australia, which it visits during the winter of the North. It has also been observed in Australia in the southern winter months, July and August (Cox & Hamilton, P.L.S., N.S.W. 1889, 399).

Chaetura is interesting for its curious spined tail. Hartert remarks that in *C. gigantea* and in *C. indica* (and, of course, in its ally *C. celebensis*) "the rectrices are very acute and run into a very long spine, whereas in *C. caudacuta*¹) and *C. nudipes* they are rounded and the spine is only half as long".

C. caudacuta may be at once distinguished from *C. celebensis* by its white throat. Another species with a white throat is *C. picina* Tweedd., for a long time known only by one specimen from Mindanao.

---

¹) The secondaries in Gould's plate of this species in the "B. Austr." are drawn wrong — much too long.

GENUS COLLOCALIA G. R. Gray.

The Edible-nest Swifts are birds of small to almost the smallest size (not regarding the long wing), furnished with a square tail of ten normally shaped feathers, bare tarsus and toes with the ordinary number of phalanges. They are cave-dwellers. Some of the species form their nests entirely of their own glutinous saliva, others make use of various materials, but probably use saliva for securing them and attaching the nest. Two white eggs. The number of species is uncertain; in 1897 Hartert recognised 14 and some subspecies. They range from India to Australia and occur on some of the islands of the Indian and Pacific Oceans.

† 105. COLLOCALIA FUCIPHAGA (Thunb.).

Indian Edible-nest Swiftlet.

Mr. Hartert (1892) separates as a subspecies *C. fuciphaga brevirostris* (McClell.) of the Himalayas. The present article relates only to the more typical forms.

a. **Hirundo fuciphaga** *(1)* Thunb., Act. Holm. 1772, XXXIII, 151, pl. IV.
b. **Hirundo vanicorensis** *(1)* Quoy et Gaim., Voy. Astrol., Zool. 1830, I, 206, pl. XII, f. 3.
c. **Collocalia unicolor** (Jerd.), *(1)* Hume, Str. F. 1876, IV, 374; *(2)* Oates, ed. Hume's Nests & Eggs Ind. B. 1890, III, 28.
d. **Collocalia nidifica** *(1)* Gray, Gen. B. I, 1845, 55; *(2)* Bernst., J. f. O. 1859, 112—118; *(3)* Rosenb., Malay. Archip. 1878, 271; *(4)* Salvad., Atti Ac. Sc. Tor. 1880, XV, 344.
Collocalia fuciphaga *(1)* Bp., C. R. 1855, XLI, 977; *(2)* Wald., Tr. Z. S. 1872, VIII, 46; *(3)* Salvad., Cat. Ucc. Borneo 1874, 120; *(4)* id., Orn. Pap. 1880, I, 544; *(V)* Meyer, Vogelskel. pt. V, 1883, 36, pl. 46, fig. 2; *(6)* Pryer, P. Z. S. 1884, 532—538; *(7)* Finsch & Meyer, Ztschr. ges. Orn. 1886, 14; *(8)* Pryer, Auk 1888, 335; *(9)* Lucas, Auk 1889, 9, 11; *(10)* Salvad., Orn. Pap. Agg. 1889, I, 63, III, 224; *(11)* Everett, J. Str. Br. R. A. S. 1889, 152; *(12)* Wiglesw., Av. Polyn. 1891, 17; *(13)* Vorderman, Nat. T. N. Ind. 1891, L, 450; *(14)* id., Notes Leyd. Mus. 1891, 125; *(15)* Hart., Cat. B. XVI, 1892, 498; *(16)* Brns. & Worc., B. Menage Exp. 1894, 35; *(17)* Hart., N. Z. 1895, 472; *(18)* Grant, Ibis 1895, 461; *(19)* Blanf., F. Br. Ind. B. III, 1895, 176; *(20)* Hart., Ibis 1896, 369; *(21)* id., Tierr. 1897, I, 67 pt.
e. **Collocalia vanicorensis** *(1)* Gray, B. Trop. Is. 1859, 4; *(2)* Wiglesw., Av. Polyn. 1891, 18.
f. **Collocalia fusca** err. *(1)* Hickson, Nat. in N. Celebes 1889, 49, 91.

For further synonymy and references see Salvadori *4, 10*; Hartert *15*; Wiglesw. *e 2*.
Figures and descriptions. Quoy & Gaim. *b I*; Meyer *V* (skeleton); Salvadori *4*; Hartert *15*; Blanford *19*.
Adult. Upper surface dark sooty brown with very little gloss; head, wings and tail darker and more glossy; feathers in front of the eye whitish with dark brown tips; lower surface brownish grey with darker shaft-stripes; under wing-coverts blackish brown. Total length about 114 mm; bill at base 3.8; wing 112—119; tail 56 (Hartert *15*).
Young. Similar to the adult (Hartert).

Skeleton.

Length of cranium	23.0 mm	Length of tibia	19.0 mm
Greatest breadth of cranium	12.5 »	Length of tarso-metatarsus	9.5 »
Length of humerus	8.4 »	Length of sternum	16.0 »
Length of ulna	12.0 »	Greatest breadth of sternum	11.0 »
Length of radius	11.0 »	Height of crista sterni	9.0 »
Length of manus	27.2 »	Length of pelvis	16.5 »
Length of femur	12.5 »	Greatest breadth of pelvis	14.5 »

Eggs. Nilghiris, S. India — usually 2 in number, dull, almost wholly glossless white, as a rule slender elongated ovals, almost cylindrical, sometimes absolutely cylindrical; at times slightly pyriform: 20—22.8 × 13.5 — 14.7 mm, average size 21 × 13.7 (Hume *c 2*); 20—22 × 13.5 mm (Nehrkorn MS.). The eggs are similarly described by other writers.

Nest. Composed of the saliva of the bird — in India mixed with moss and feathers (Hume *c 2*), in Borneo mixed, apparently, with the substance of a gum-like Alga (Pryer *6, 8*). Breeds in caves, usually in vast colonies.

Distribution. From India south-east to New Caledonia and the New Hebrides, and from Ceylon and the Andamans east to the Marianne and Caroline Islands (see Salvad. *4, 10;* Wg. *12, e 2*). Recorded from Celebes: Macassar (Wallace) by Walden *(2)* and Hartert *(15)*.

The members of the genus *Collocalia* are, like all Swifts, birds of great flying-powers, and, as is usually the case with such they do not conform to the geographical bounds which may be drawn for ill-flying species or for species sedentary for other reasons. Nevertheless it is found that several species have very restricted ranges, while others, *C. francica* and *C. esculenta*, are, like the present species, widely distributed. The *Collocaliae* are, apparently, restricted to districts where there are rocks and caves suitable for them to roost and breed in; these they haunt in common with bats. From the thick deposits of guano on the floors of some of these caves it is evident that they have been in use for a long period; in a cave in Ceylon the guano-deposit was, as Colonel Legge was informed, 30 feet thick, and Pryer mentions *(6)* that a pole thrust down 18 feet into the deposit on the floor of the great cave "Simud Putech" of North Borneo did not reach the bottom. This cave is described by Daly (P. Z. S. 1888, 110), as being from 850 to 900 feet in depth, coated with a layer of guano of from 5 to 15 feet deep. The birds are remarkable for their adhesiveness to any selected roosting or nesting spot; in certain parts the species are migratory visiting their breeding-caves and leaving them regularly; and Davison describes a case (Str. F. II, 159) where "a large number of these birds had taken up their sleeping quarters against the roof of a shed on Viper Island, Port Blair, occupying about a square yard of the surface; this place they continued to occupy till the shed was destroyed, when, of course, they disappeared; but after a time another shed was built exactly on the same site, and as soon as the roofing was completed back came all the Collocalias and re-occupied the same spot on the roof of the new shed as they had occupied in the old". The same excellent observer describes the attempts of a pair, continued for nearly a week, to fix scraps of moss, the commencement of their nest, on the painted ceiling of a room, "but failed to get a single piece to stick, and so at last gave it up as a bad job".

The nests are chiefly formed of the saliva of the birds more or less intermixed with feathers and sometimes with moss. The high value (as an aphrodisiacum) set upon them by the Chinese for soup is well known, and it is said that nests of the purest quality — the white ones — are worth their weight in silver[1]). According to Mr. Daly, the white nests of *C. fuciphaga* are those which are gathered before the bird has commenced to lay any eggs, and which are composed of clear transparent mucilaginous matter, with very few feathers mixed up with them. The next quality is the red or grey nest, partly mixed with feathers and in which eggs are sometimes found; the least valuable are the black nests, which are much mixed with feathers and sometimes contain fledglings. The black nests have been supposed to belong to different species or varieties, but Mr. Daly says these nests are those which have been "overlooked at the previous gathering and have darkened or deteriorated from exposure to water and to the atmosphere of the caves. The partial decomposition of the mucous matter renders them the least valuable".

Bernstein *(d 2)* has called attention to the enormous development of the salivary glands in these birds, especially the glandulae sublinguales. Their great size is attained only during the time that the birds are building their nests, "after this, even during the laying of the eggs, they atrophy and appear but little bigger than those of other birds. At that time (the building time) they appear, when the bill of the bird is opened, like two large pads at the side of the tongue. They secrete in quantities a thick, adhesive slime ..., resembling gum-arabic ... which can be drawn out of the mouth in somewhat long threads. If the end of such a thread of slime is placed on the point of a bit of stick and the latter is then turned slowly on its axis, in this manner the whole mass of saliva for the moment present can be drawn out of the mouth and even out of the orifices of the glands". As Bernstein found from some tame specimens the supply of saliva stands in direct relation to the quantity of food supplied: "it was very small, when the birds had gone hungry a few hours". Bernstein came to the conclusion that the glue-like substance of the nest was formed solely by these secretions of the bird, and believed that the feathers seen in the nest are such as have been caught and stuck by this adhesive matter when half dry. Nevertheless, chemical analysis ultimately proved the presence also, in addition to the bird's saliva, of a gumlike substance from an Alga in the nests of *C. fuciphaga* obtained by Pryer in North Borneo *(8)*. That *C. fuciphaga* should employ such foreign substances sometimes or always is not surprising, since some other *Collocaliae* are known to make use of moss, lichens, etc., as does *C. fuciphaga* itself in India *(c 2)*.

---

[1]) In 1896 there were imported into China 3,600 000 edible birds-nests.

⊹ 106. COLLOCALIA ESCULENTA (L.).
White-bellied Swiftlet.

a. **Hirundo esculenta** *(1)* Linn., S. N. 1766, XII, 343.
Collocalia esculenta *(1)* Wald., Tr. Z. S. 1872, VIII, 46; *(2)* Meyer, Ibis 1879, 65; *(3)* Salvad., Orn. Pap. 1880, I, 540; *(4)* Guillem., P. Z. S. 1885, 549; *(5)* Hickson, Nat. in N. Celebes 1889, 49, 91; *(6)* Salvad., Orn. Pap. Agg. 1889, 63; *(7)* Hartert, Cat. B. XVI, 1892, 509; *(8)* Tristr., Ibis 1892, 296; *(9)* Büttik., Notes Leyden Mus. 1892, 194; *(10)* M. & Wg., Abh. Mus. Dresden 1895, Nr. 8, p. 8; *(11)* iid., ib. 1896, Nr. 2, p. 14; *(12)* Salvad., Ann. Mus. Civ. Gen. 1896, XXXVI, 73; *(13)* Hartert, Nov. Zool. 1896, 158, 175, 243, 570; *(14)* id., Tierreich 1897, I, 70.
b. **Collocalia hypoleuca** *(1)* Gray, P. Z. S. 1858, 170, 189; *(II)* id., Cruise "Ouraçon", B. 1873, 356, pl. II, fig. 1; *(3)* Rosenb., Malay. Archip. 1878, 271; *(4)* Salvad., Atti Ac. Sc. Tor. 1880, 346.
c. **Collocalia viridinitens** *(1)* Gray. Ann. and Mag. N. H. 1866, XVII, 120.
d. **Collocalia sp.** *(1)* Büttik., Notes Leyd. Mus. 1891, 210.
"Tetekek", Peling, "Tepede", Banggai (Nat. Coll.).
For further synonymy and references see Salvad. *3, 6*, Hartert *7*.

Adult. Above glossy dark bluish green, tail and outer webs of quills bluer, inner webs of latter dusky, rectrices, except the two middle ones, with a large patch of white on basal half of the inner webs; bases of loral plumes white; chin and sides of face dark mouse-grey; throat, breast and sides mouse-grey with white margins, these margins increasing in width on the lower parts, so that the abdomen is pure white; under tail-coverts — the longest uniform blue-black, the shorter ones blackish green, broadly bordered with white, the shortest uniform white; under wing-coverts blackish, fringed with white (Minahassa — Nr. 821). Bill and feet dark brown (Platen, Mus. Nehrkorn).

Sexes. Alike.

Young. Not so glossy above, but otherwise similar to the adult (Hartert *7*).

Measurements.	Wing	Tail	Bill from gape	Tarsus
a. (Nr. 821) ad. Minahassa (Faber)	94	44	8.5	—
b. (Nr. 820) ad. Minahassa (Faber)	85	35	8	—
c. (Sarasin Coll.) ♂ ad. Tomohon, 12. X. 94	97	40	—	7.5

A series of 17 from Peling and Banggai were very small; wing 69—90 mm, average 80.

Eggs. Amboina. Elliptic and white, like the eggs of all *Cypselidae* and *Trochilidae*. Size 17.5×11 mm. They are similar to a series of Humming-birds eggs, the shells of which, however, are somewhat more delicate (Nehrkorn MS.). Always 2 in a clutch (Hartert).

Nest. A nest in the Sarasin Collection "is built entirely of lichens and was stuck fast on a rock with saliva. In form it resembles perfectly the ordinary edible nests. Of these we saw many, as we were returning along the coast from Malibagu back to Kema. As is known the harvesting of these nests is a monopoly of the king of Bolang Mongondo" (P. & F. Sarasin in litt.). By Mr. Everett "on November 2nd a colony was found breeding in a cave near Indrulaman. The nests were not edible, but consisted of moss, rootlets, and little twigs, agglutinated and fixed to the walls of the cave with saliva" (Hartert *13*).

Breeding season. Dr. Hickson *(5)* found this species and *C. fuciphaga* breeding in countless numbers in two caves at Tanjong Aros, Talissi Island, North Celebes on August 28th. In Amboina Dr. Platen collected quantities of eggs on July 23rd, 1881.

Distribution. Celebes — Talissi Island (Guillem. *4*, Hickson *5*), Strait of Lembeh (Meyer *2*), Minahassa (P. & F. Sarasin), Macassar (Wallace *1, 7*), Maros (Guillem. *4*), Bonthain (Everett *13*); Peling and Banggai (Nat. Coll.); Moluccas and Papuasia as far as the Solomon Islands (Salvad. *4, 6*) to Cape York, Australia (Hartert *7*).

This species has a more southern range than *C. fuciphaga*, from which it may be distinguished by its white abdomen and somewhat smaller size.

As is proved by the thickness of the guano deposits (see preceding article) these birds have evidently tenanted certain caves for enormous periods, and it is not surprising that this species seems to vary locally or even "troglodytically"; for instance the specimens from Peling and Banggai examined by us were very small.

+ 107. COLLOCALIA FRANCICA (Gm.).
Little Grey-rumped Swiftlet.

a. **Hirundo francica** *(1)* Gm., S. N. 1788, I, 1017.
Collocalia francica *(1)* Gray, List B. Br. Mus. 1848, II, Fissirostr. 21; *(II)* M.-Ed. & Gr., Ois. Madag. 1879, 198 pl. 72—5; *(3)* Hart., Cat. B. XVI, 1892, 503; *(4)* Blanf., F. Br. Ind. B. III, 1895, 178; *(5)* Hart., N. Z. 1896, 175; *(6)* id., Tierr. 1897, I, 68.
b. **Macropteryx spodiopygia** *(1)* Peale, U. S. Expl. Exped. 1848, 170, pl. 49.
c. **Collocalia spodiopygia** *(1)* Cass., U. S. Expl. Exped. 1858, 184, pl. 12, f. 3; *(2)* Finsch & Hartl., Orn. Centralpol. 1867, 48; *(3)* Salvad., Orn. Pap. 1880, I, 546; *(4)* id., Agg. O. P. 1889, 63; *(5)* Wiglesw., Av. Polyn. 1891, 18.
d. **Collocalia terrae-reginae** *(1)* Gould, B. New Guinea 1875, IV, pl. 38.
e. **Collocalia infuscata** *(1)* Salvad., Atti Ac. Sc. Tor. 1880, XV, 348.

For further references cf. Finsch & Hartlaub *c 2*; Milne-Edwards & Grandidier *II*; Salvadori *c 3, c 4*; Wiglesworth *c 5*; Hartert *3*.

Figures and descriptions. Gould *d 1*; Peale *b 1*; Cassin *c 1*; M.-E. & G. *II*; F. & H. *c 2*; Salvadori *c 3*; Hartert *3*; Blanford *4*.

Diagnosis. Distinguishable from *C. fuciphaga* by its having a grey or dusky white band across the rump, from *C. esculenta* by its having no white on the abdomen, and by its larger size.

Measurements. Wing 111—114 mm; tail 53 (Hartert *3*).

Skeleton see Milne-Edwards & Grandidier *II*.

Eggs and nest cf. Finsch & Hartlaub *c 2*; Hume, Nests and Eggs Ind. B. Oates ed. III, 35.

Distribution. Burmah, through the Philippines and the East Indian Islands to North Australia, east as far as Fiji and Samoa, west to Mauritius and Bourbon (cf. Hartert *3*). — In the Celebesian area: Djampea Island (Everett *5*).

Quite recently Mr. Hartert was able to record two fully feathered nestlings and a number of nests and eggs from Djampea between Celebes and Flores, which he ascribes to this species. It has not as yet been recorded from the mainland of Celebes. The Dresden Museum recently got specimens from Mindanao. The species is a puzzling one, and varies in a way that is likely to cause little satisfaction to trinomialists.

GENUS MACROPTERYX Sw.

The subfamily *Macropteryginae* consists only of the present genus. Its appearance is perfectly Cypseline, but it is easily recognisable by its long forked tail, the outermost rectrices reaching as far as the ends of the wings; the toes are ordinary, the tarsus bare and very short, about as long as the hind toe without the claw. The superciliary plumes are long, forming tufts. There are two fenestrae in the sternum posteriorly. Its nidification is very peculiar, resembling most that of the Goatsucker, *Batrachostomus* (see below). Five species have been recognised by Hartert, ranging from India to the Solomon Islands.

† * 108. MACROPTERYX WALLACEI (J. Gd.).
Celebesian Crested Swift.

a. **Dendrochelidon wallacei** *(1)* Gld., P. Z. S. 1859, 100; *(II)* id., B. Asia I, pl. 23 (1859); *(3)* Wall., P. Z. S. 1862, 339; *(4)* Sclat., P. Z. S. 1865, 616; J. f. O. 1867, 140; *(5)* Gray, HL. 1869, I, 65; *(6)* Rosenb., Malay. Archip. 1878, 271; *(7)* W. Blas., J. f. O. 1883, 125; *(8)* Heine & Rchnw., Nomencl. Mus. Hein. 1890, 190.

Macropteryx wallacei *(1)* Finsch, Neu Guinea 1866, 162; *(2)* Walden, Tr. Z. S. 1872, VIII, 45; *(3)* Meyer, Ibis 1879, 65, 146; *(4)* W. Blas., J. f. O. 1883, 136; *(5)* Guillem., P. Z. S. 1885, 548; *(6)* W. Blas., Ztschr. ges. Orn. 1855, 246; *(7)* id., ib. 1886, 95; *(8)* Hickson, Nat. in N. Celebes 1889, 91; *(9)* Tristr., Cat. Coll. B. 1889, 112; *(10)* Hartert, Cat. B. XVI, 1892, 515; *(11)* Büttik., Zool. Erg. Weber's Reise Ost.-Ind. 1893, III, 275; *(12)* M. & Wg., Abh. Mus. Dresden 1895, [Nr. 8, p. 8; *(13)* iid., ib. 1896, Nr. 1, pp. 5, 9; *(14)* iid., ib. 1896, Nr. 2, p. 13; *(15)* Hart., Nov. Zool. 1896, 175.

b. **Dendrochelidon klecho** var. **wallacei** *(1)* Brügg., Abh. Ver. Bremen 1876, V, 55.

"**Peapatta**", Gorontalo, Ros. *a 6.* "**Burong padang**" (Knife-bird), Malay, Manado, Meyer *3.*
"**Pavas**", Manado, Guill. *5.* "**Pavas**", Talissi, Hicks. *8.* "**Pisok besar**", Banka, Nat. Coll.
"**Rewata**", Tjamba, S. Cel., Platen *6.* "**Diding**", Tonkean, E. Cel., Nat. Coll.
"**Papingis**", Peling; "**Tepede babasar**", Banggai, Nat. Coll.

Figure and descriptions. Gould a *II*, a *1*; Guill. *5* (young); W. Blas. *6*; Hart *10*; Bütt. *11.*

Adult male. Head above, wing-coverts, exposed webs of quills, and tail, shining blackish blue, varied on crown with bronze-green, and passing into bronze-green with blackish blue margins on most of the feathers on hind neck, mantle, and scapulars, which again shades off into the dark grey of the back, rump, and upper tail-coverts; the longest tertiary coverts lighter grey; ear-coverts dull dark chestnut, face black glossed with green; a very narrow superciliary stripe of grey; under parts from chin downwards dark grey, becoming whitish tinged with buff and fulvous on abdomen and shaded with greenish on sides of breast; under wing-coverts dark brown, washed with bluish at metacarpal edge (Minahassa [♂] — C 3499). Iris brown; feet grey; claws blackish; bill black (Meyer *3*).

Adult female. Like male but the ear-coverts greenish steel-blue, not chestnut (Hartert *10*).

Young. The under surface marked irregularly with buff feathers, which are white at the base and tipped with brownish black; the tertials are white, and the primaries and secondaries strongly tipped with that colour (Guillem. *5*).

Changing plumage. Young birds assuming the adult dress have a curious appearance. The feathers of immaturity are pale brown, tipped with white, with subterminal black bars, those of the upper parts are dull blackish brown with buff-brown edges; these are mixed with the grey or glossy blackish feathers of maturity (♂ juv. Palopo, 22. I. 95: Sar.).

Variation. A series of specimens from E. Celebes, Peling and Banggai prove to be more bronzy on the head and wing-coverts than others from North Celebes, which show stronger tints of steel-blue; still two examples from the latter locality do not differ from the specimens in question.

Measurements. Wing 171 (juv.)—190 (adult); tail 92 (juv.)—124 (adult); rictus 19.5—20.9; tarsus 8.0 circa (W. Blasius *6*, and specimens Dr. Mus.).

Eggs and nest. Unknown.

Distribution. Celebes Province — Southern Peninsula: Macassar (Wallace *a 1, 10*), Tjamba (Platen *6*), Luwu and Boni (Weber *11*), Palopo (P. & F. Sarasin); Northern Peninsula: Minahassa (Wallace *10*, Meyer *3*, etc.), Talissi Id. (Hickson *8*), Banka Id. (Nat. Coll. in Dresd. Mus.), Togian Islands (Meyer *3*); E. Peninsula, Tonkean (Nat. Coll.); Peling and Banggai (iid.); Sula Islands (Allen *a 3, 10*).

This species is most nearly related to *M. longipennis* (Rafin.) = *klecho* Horsf., which ranges from Tenasserim to Java and Borneo. *M. wallacei* differs by its larger size (wing about 178 as against 165 mm), the bluer — less green — tint of the upper parts, the lighter grey of the under surface, and in the male adult the darker chestnut of the earcoverts (Büttik. *11*, Hartert *10*).

The genus *Macropteryx* is separated as a family, *Dendrochelonidae*, by Lucas (Auk 1889, 8), and as a subfamily of the *Cypselidae* by Hartert, who regards it as an approach to the *Caprimulgidae*. Among the Swifts it seems to be related to the *Chaeturinae*, also marked off as a subfamily in the Catalogue of Birds. In the typical Swifts, *Cypselinae*, the front toes contain each only three plalanges: in *Chaetura* and *Macropteryx* the normal numbers — five in the outer toe, four in the middle toe — are seen. In their peculiar nesting-habits, the *Macropteryginae* offer, as Hartert remarks, a striking correspondence with the curious nidification of the Goatsucker, *Batrachostomus*. Bernstein seems to have been the first to discover the nidification of *Macropteryx*, and his admirable observations (J. f. O. 1859, 184) on *M. klecho* in Java remain unsurpassed in fullness and interest. "It chooses", he says, "a free-standing branch, high in the crown of a tree, to build its nest on. If the choice of such a spot is remarkable for a member of the family of the *Cypselidae*, the relations in size between the bird, nest, and egg are far more remarkable. The nests, by its more or less semicircular form and the manner in which the component materials are bound together, calls to mind to some extent the nests of the *Collocaliae*; it is however much smaller and flatter than these. The nests measured by me were in depth about 10 mm by 30—40 mm broad against 50 mm in those of the much smaller *C. nidifica* (*esculenta* Horsf.). The nest is always found on a horizontal branch, usually not more than an inch thick, which at the same time forms the hinder wall of the nest; here it is fixed so as to form at the side thereof a somewhat flat, longish, semicircular saucer, just large enough to contain the single egg. The walls of the nest are extremely thin, and delicate, scarcely thicker than parchment. They consist of feathers, some few bits of tree-lichens and small pieces of bark, which materials are glued together with a sticky vehicle, without doubt, as in the *Collocaliae*, the saliva of the bird; occasionally in these birds also the salivary glands swell remarkably at the time of reproduction. The smallness and fragility of the nest does not allow of the brooding bird sitting thereon. On the contrary, as I have repeatedly observed, it sits on the bough and covers the nest and the egg in it with the belly alone. The latter, with a length of 25 mm and a maximum breadth of 19 mm, answers perfectly to the size of the bird. It it of regular, perfectly oval form, so that it is not possible to make out with certainty a sharp or a blunt end. Its colour is a very pale sea-blue, which colour gets paler after blowing the egg, and then white, with faint tinges of blue. According to my observations the bird rears two broods a year, one immediately after the other, but rarely makes use of one and the same nest for this purpose. With such a low rate of reproduction, it can cause no surprise that this bird, though probably occurring all over Java, is nowhere very plentiful". The nest of *Macropteryx coronatus* (Tick.) is very similarly described by Hume (Oates ed. Hume's Nest and Eggs Ind. B. III, 36). The nest of *Batrachostomus hodgsoni* (Gray) is described in the same work (p. 40) as a circular pad barely $3\frac{1}{2}$ inches in diameter, about $\frac{3}{4}$ inch thick. This is placed on the upper surface of a horizontal bough, and receives one white egg. In this case, apparently, there would be nothing to prevent monkeys and other marauders from helping themselves to the white egg, but that *Batrachostomus* is a nocturnal bird and doubtless is sitting upon the nest all through the day.

The genus *Macropteryx*, is most strongly represented in the Oriental Region, but ranges from India to the Solomon Islands.

Macropteryx mystacea (Less.). Gray (HL. 1869, I, 65) mentions Celebes as a locality for this species, which ranges from Halmahera to the Solomons, and Rosenberg (Mal. Arch. 271) repeats this indication, doubtless derived from Gray (see W. Blasius, J. f. O. 1883, 125). There appears to be no proof of this species having ever been found in the Province.

Macropteryx comata (Temm.). As Salvadori points out (Cat. Ucc. Borneo, 1874, 124), Cassin indicates (Cat. Hirund. Mus. Philad. 1853, 15) individuals of this species from Celebes and Timor. The localities are repeated by Gray (l. c.), by von Rosenberg (l. c.) "Celebes", and by Hartert (Cat. B. XVI, 518); but in the case at least of Celebes no evidence of their correctness has been produced.

ORDER PASSERES.

To this order belong pre-eminently the Song-birds, the males of the majority of which have the habit in the breeding season — in many cases at other times of the year also — of uttering a sequence of notes almost always pleasing to the human ear (and doubtless still more so to the singer and his partner), though a few, such as the Crows, have no song properly so called, while some birds belonging to other orders, such as, for instance, the common Fowl and the Redshank, have a very appreciable form of song. The *Passeres* number some 6000 species, ranging in size from the most minute to the Raven and Lyre-bird. The great majority of them feed upon insects and other invertebrate organisms, seeds and fruits, some few occasionally on carrion, and occasionally *(Laniidae, Corvidae)* on small birds and other small vertebrates. As architects they are the most skilful nest-builders in the bird-world, and those which breed in holes as a rule (if not always) form a comfortable lining thereto, wherein they differ from certain "Picariae", such as *Merops*, which do not line their nesting-holes. The young are hatched blind and naked.

There is no difficulty in distinguishing the *Passeres* from other groups of birds, except from certain Picarian forms, such as the Swifts, which approach the Swallows in external appearance and habits, and the Rollers, which are of a corvine appearance, or the Broad-billed Rollers, which resemble the *Eurylaemidae*. For the distinction of the *Passeres*, avian anatomists attach most importance to the toes and their flexor tendons, to the palatal bones, and to the syrinx.

The toes are normal, that is, there are three in front and the hallux behind, and the hallux is served by a free flexor tendon, not united to the flexor tendon of the three anterior toes, except in the *Eurylaemidae*, which are now commonly allowed to rank as a distinct suborder, or order.

The palate is aegithognathous (as it is indeed also in certain Picarian forms). "In the true aegithognathous structure the vomer is broad, abruptly truncated in front and deeply cleft behind, so as to embrace the rostrum of the sphenoid; the palatals have produced postero-external angles, the maxillo-palatals are slender at their origin, and extend obliquely inwards and backwards over the palatals, ending beneath the vomer in expanded extremities, not united either with one another or with the vomer" (Newton, D. B. p. 2).

The syrinx has received great attention especially from Garrod and Gadow. The former used it as the diagnostic for forming two main groups of the *Passeres*, viz.

Acromyodi: "in which the intrinsic muscles of the syrinx are fixed to the ends of the bronchial semi-rings";
Mesomyodi: "in which the intrinsic muscles of the syrinx are fixed to the middle of the bronchial semi-rings" (Oates, Faun. Br. Ind. B. I. p. 5).
Gadow makes a somewhat different division:
Anisomyodae: in which the muscles of the syrinx are not bilaterally symmetrical, but either entirely lateral, or purely dorsal, or ventral;
Diacromyodae: in which the muscles of the syrinx are bilaterally symmetrical, being inserted in the dorsal and ventral ends of the bronchial rings (Bronn's Kl. & Ord. 1893, VI, pt. 4, II, p. 272).

FAMILY PITTIDAE.

The Pittas are *Passeres* of terrestrial habits and may be easily recognised by their long tarsi, short tails (about as long as the tarsus or a little longer), and brilliant coloration; glossy green, deep velvety-black, turquoise and other blues, and pure red being common tints.

The formation of the syrinx led Garrod to refer the Pittas to his *Mesomyodian Passeres*. As Prof. Newton remarks, "this in itself was an unexpected determination, for all the other birds of the group, as then known, inhabit the New World, where no Pittas occur". Dr. Gadow makes the Pittas a family of the *Clamatores*-section of his *Anisomyodian Passeres*, the other families of this group being American and one small one of three species, the *Xenicidae*, New Zealandian. In the opinion of Gadow the *Pittidae* form a transition from the *Clamatores* to the *Subclamatores* (consisting of the single family *Eurylaemidae*).

The bill is about as long as the cranium, without rictal bristles, the nostril oval; there are 10 primaries, the 1st large, about as long as the secondaries; 12 rectrices. The sexes are similar in coloration.

GENUS PITTA Vieill.

The family *Pittidae* consists of about 50 species, which fall into a number of sections, but it is doubtful if more than two of them should be separated generically, viz. *Anthocincla* which has long "ear-tufts" and a longer more compressed bill, found in a Burmese species, and *Melampitta* of New Guinea, which is entirely black and has the frontal plumes short, erect, and the tail longer (cf. Sclater, Cat. B. XIV, 1888, 412; Oates, Faun. Br. Ind. B. II, 1890, 387).

+ * 109. PITTA CELEBENSIS Müll. Schl.
Celebes Red Pitta.

Pitta celebensis *(1)* M. & S., Verh. Natuurk. Comm. 1839—44, Aves, Pitta, p. 18 (ex Forsten in litt.); *(2)* Gray, Gen. B. I, 213 (1846); *(III)* Westerm., Bijdr. t. d. Dierk. 1854,

I, 46, Pitta, pl. 3; *(4)* Schl., Handl. Dierk. 1857, 254; *(5)* Wall., Ibis 1860, 142; *(VI)* Schl., Vog. Ned. Ind. Pitta 1863, 17, 34, pl. 4, fig. 4; *(7)* id., Mus. P.-B. Pitta 1863, 6; *(8)* Wall., Ibis 1864, 105; *(9)* Finsch, Neu Guinea 1865, 167; *(10)* Gray, HL. 1869, I, 296, Nr. 4377; *(10bis)* Wall., Malay Archip. 1869, I, 366; *(11)* Schl., Revue Pitta 1874, 10, pt.; *(12)* Salvad., Ann. Mus. Civ. 1875, VII, 663; *(13)* Brüggem., Abh. Ver. Bremen 1876, V, 64; *(14)* Meyer, Rowley's Orn. Misc. 1877, II, 327; *(15)* Sclat., P. Z. S. 1877, 99; *(XVI)* Gld., B. New Guinea IV, pl. 34 (1878); *(17)* Rosenb., Malay. Archip. 1878, 272; *(18)* Guillem., P. Z. S. 1885, 552; *(19)* W. Blas., Ornis 1888, 102; *(20)* Sclat., Cat. B. XIV, 1888, 436; *(21)* Tristr., Cat. Coll. B. 1889, 119; *(22)* Hartert, Kat. Mus. Senckenb. 1891, 107; *(23)* Whitehead, Ibis 1893, 505; *(XXIV)* Elliot, Monogr. Pitt. pt. III, 1894; *(25)* M. & Wg., Abh. Mus. Dresd. 1895, Nr. 8, p. 13; *(26)* iid., ib. 1896, Nr. 2, p. 18; *(27)* Hartert, Nov. Zool. 1897, 163.

a. **Brachyurus celebensis** *(1)* Bp., Consp. 1850, I, 255; *(II)* Elliot, Mon. Pitt. 1863, pl. XVII; *(3)* id., Ibis 1870, 418.

b. **Erythropitta celebensis** *(1)* Bp., Consp. Vol. Anisod. 1854, 7; *(2)* Wald., Tr. Z. S. 1872, VIII, 62; *(3)* Meyer, Ibis 1879, 126; *(4)* Platen, Gefied. Welt 1887, 218; *(V)* Meyer, Vogelskel. I, 1892, 33, t. CLXI.

"Mopo" (= Grandfather), Minahassa, Meyer *b 3*.
"Mupu sava merah", Minahassa, Guillem. *18*.
"Tenge-tenge", [? Gorontalo], Rosenb. *17*.
"Tagongong", Tonkean, E. Celebes, Nat. Coll.

Figures and descriptions. Gould *XVI* [immature[1])]; Westermann *III*; Schlegel *VI*; Elliot *a II, XXIV, a 3*; Meyer *b V* (skeleton); Müller & Schlegel *1*; Sclater *20*.

Adult. Head above chestnut with a vertical stripe from forehead to occiput glaucous blue; mantle, back, scapulars and lesser wing-coverts dull French-green, changing with the light; remaining upper-parts dull blue, two or three wing-coverts near the carpal edge partly white forming a concealed spot; bastard-wing and primaries black, the latter greyish at their distal ends, a white speculum formed by the second, third and fourth quills; lores, ear-coverts, chin and upper throat rufous isabella-colour, touched with blue on the ear-coverts and above the eye; jugular collar black, passing narrowly round hind neck; a broad pectoral collar below the last cerulean-blue, passing narrowly round hind neck, on sides of chest and body greenish, broadly bordered on the breast with a black band; remaining under-parts deep scarlet; under wing-coverts drab, the axillaries tipped with white (ad. near Manado, Aug.—Sept. 1892: Nat. Coll. — C 10886, ♂ ad. Rurukan, 21. VII. 84, Platen in Mus. Nohrk., Nr. 947).

Iris greyish brown (M. *b 3*) or pale olive (Wall. *8*) or ruddy brown (Guill. *18*); bill black, feet greyish black, dusky lead-colour, pinkish slate or grey (*b 2, 8, 18*, Platen).

Young in changing plumage. The vertical stripe buff, a patch on the jugulum buff-white bordered below with black; remaining under-parts isabella-colour, washed with darker brown on the breast, sides and ear-coverts, paler and washed with scarlet on the abdomen; a few blue feathers of the pectoral collar present and others of deep scarlet on the abdomen; upper surface much as in the adult; bill short, most of under bill and terminal part of upper yellowish white (Mantehage Id., 22. IV. 93: Nat. Coll. — C 12199).

---

[1]) The white spot on the jugulum is a sign of immaturity, though it may also be common in adult females.

Measurements. Adults — wing 103—109 mm; tail 41; tarsus 38 c.; bill from nostr. 14.5—16; immature — wing 101; tarsus 36; bill fr. nostr. 12.5.

Skeleton.

Length of cranium fr. sin. nasofr.	25.9 mm	Length of fibula	23.3 mm
Greatest breadth of cranium .	18.5 »	Length of tarso-metatarsus .	38.0 »
Length of humerus	30.2 »	Length of sternum	33.5 »
Length of ulna	35.6 »	Greatest breadth of sternum .	20.0 »
Length of radius	32.8 »	Height of crista sterni . . .	11.0 »
Length of manus	31.0 »	Length of pelvis	33.0 »
Length of femur	28.5 »	Greatest breadth of pelvis .	20.5 »
Length of tibia	49.0 »		

Distribution. Celebes — Minahassa (Forsten *1*, Wallace *5, 20*, etc.); Bolang-Mongondo (P. & F. Sarasin); Gorontalo (Rosenb. *11*); Mantehage Id. (Nat. Coll.); Togian Islands (Meyer *b 3*); Tonkean, E. Celebes (Nat. Coll.); Tawaya, W. Celebes (Doherty *27*); Maros, S. Celebes (Wallace *10bis*); Segeri, S. Celebes (Meyer *b 3*).

The Red Pitta of Celebes seems to be a somewhat rare bird in the south of the island. Writing on its habits Meyer has remarked *(b 3)*: "Although not so difficult to procure as the Black-headed, it is nevertheless a bird which it is not easy to get a shot at, being very quiet in the day-time and seldom calling except in the morning and evening its 'tüüüü tschui'. In the evening the cry 'oppo' (origin of the native name) is heard, with which the male and female call one another, the notes sounding melancholy and protracted. 'Oppo' means, in the language of the country, 'grandfather'; and the natives tell the tale, that once a child, which had gone with its grandfather into the forest, got astray and was transformed into a bird, which now always calls for its grandfather".

"*Pitta celebensis* only runs on the ground, and is very shy and watchful; it glides noiselessly through the leaves; and as its back is green it can only with difficulty be detected. To approach it one must creep through the densest brushes; and without imitating the call of the bird its pursuit would be in vain. But if the hunter imitates the cry he can draw the bird almost to the muzzle of his gun. During the day-time they go singly, in the evening in pairs together. The nest is to be found in brushes near small pools.

The bird digs a hole in the slope of the river-bank, and builds its nest therein of wood and leaves, lined with cotton or hairy plant-materials (for instance, from *Arenga saccharifera* Lab.). It lays two eggs. . . . This *Pitta* feeds on beetles, small caterpillars, etc."

Dr. Platen and v. Rosenberg likewise mention the quiet, concealed habits of this bird. The latter describes its nest, or that of *P. forsteni*, as being always placed upon the ground and formed of leaves and moss, lined with fine grass-stalks.

Pitta celebensis is one of a group in which Mr. Whitehead has recently *(23)* enumerated 12 species, and *P. inspeculata* M. & Mg. should now be added: "a very compact group, all the species having brown or black-and-brown throats, beneath which is a pectoral band of shining greyish blue, sometimes flanked with green, and divided by a black band or otherwise from the bright scarlet breast".

The group is distributed as follows:

Habitat	P. kochi	P. erythrogaster	P. propinqua	P. inspeculata	P. coerulitorques	P. palliceps	P. celebensis	P. rubrinucha	P. rufiventris	P. cyanonota	P. maculata	P. finschi	P. loriae	Authorities
Luzon	*	*												Sclater, R. G. W. Ramsay
Tawi Tawi		*												Bourns & Worcester
Tablas, Sibuyan		*												Bourns & Worcester
Calamianes		*												Bourns & Worcester
Panay, Guimaras		*												Bourns & Worcester
Masbate		*												Bourns & Worcester
Siquijor		*												Bourns & Worcester
Mindoro		*												Steere
Palawan		*	*											Sharpe, W. Blas., Everett
Balabac		*												Everett
Zamboanga		*												Sclater
Guimaras		*												Steere
Marinduque		*												Steere
Samar			*											Steere
Basilan		*												Sclater, R. G. W. Ramsay
Mindanao		*	*											Sharpe, Steere, M. & Wg.
Sooloo		*												Sharpe, W. Blasius
Talaut				*										M. & Wg.
Great Sangi					*									Sclater, M. & Wg.
Siao					*									Brügg., Schleg., M. & Wg.
Celebes						*								M. & Wg., Schlegel
Buru							*							Salvadori, Sclater
Batchian								*						Salvadori, Sclater
Ternate									*					Salvadori, Sclater
Halmahera								*						Salvadori, Sclater
Guebeh									*					Salvadori, Sclater
Dammar								*						Salvadori, Sclater
Obi								*						Salvadori, Sclater
Papuan Islands											*			Salvadori, Sclater
S. E. New Guinea												*		E. P. Ramsay, Salvadori
Su-a-u Island[1]													*	Salvadori
Cape York											*			Salvadori

It will be observed that the distribution of the Philippine species appears to be somewhat mixed up as regards Luzon, Mindanao and Palawan, and a re-examination of specimens with special regard to age and sex might prove instructive. *P. erythrogaster* is very possibly the female or young male of *propinqua*.

---

[1] An island near South Cape, New Guinea.

P. celebensis is most like its nearest neighbour *P. palliceps* Brügg. of Siao, which has a much paler chestnut head, the black band below the blue pectoral collar absent, or nearly so, and the bill longer; *P. rubrinucha* Wall. of Buru has no black band below the blue, and also a scarlet patch on the nape. These three species agree in having a broad pale blue vertical stripe, but this often appears in specimens of *P. mackloti* and we find signs of it in most individuals of *P. inspeculata* and in *P. cyanonota*. Whether these signs are the last remnants of a character in process of undergoing obliteration, or whether they indicate a new character coming into existence — a character most advanced in development in the Celebes, Buru and Siao birds — may possibly be ascertained some day. In *Pitta inspeculata* we are able to point to a character which is undoubtedly becoming obliterated.

+ * 110. PITTA PALLICEPS Brügg.
Siao Red Pitta.

a. **Pitta celebensis** pt. *(1)* Schl., Mus. P,-B. Rev. Pitt. 1874, 10 (Siao).
Pitta palliceps *(1)* Brüggem., Abh. Ver. Bremen 1876, V, 64, t. II, f. 7, 8, 9 (bill); *(2)* Salvad., Ann. Mus. Civ. Gen. 1876, IX, 54; *(3)* Meyer, Rowley's Orn. Misc. 1877, II, 327, 328; *(4)* id., Isis, Dresden 1884, 6; *(5)* Sclat., Cat. B. XIV, 1888, 436; *(6)* W. Blas., Ornis 1888, 602, 637; *(7)* Whitehead, Ibis 1893, 505; *(VIII)* Elliot, Mon. Pitt. pt. V, 1895, pl.
"Lihange mahamu", Siao, Nat. Coll.
Figure and descriptions. Elliott *VIII*; Brüggemann *1*; Sclater *3*.
Adult. Like *Pitta celebensis*, but the head and neck much paler, more yellow-chestnut, not bordered below with black across the hind neck, the vertical stripe apparently less developed; the secondaries green like the back, this colour extending half over the rump; the blue pectoral band not fringed below with black feathers, though the bases of these are black and a dark border may be seen; bill somewhat long and thin (Siao, 8. VII. 93: Nat. Coll. — C 12619).
Young in first plumage. Young-Robin-like *(Erythaca)*: above greyish olive-brown, with a greenish wash on the back and wings, tail and the outer edges of some of the quills greyish blue; lesser wing-coverts blackish; under-parts, including sides of neck and breast with pale buffish centres to the feathers, creating a mottled appearance; lower abdomen and under tail-coverts buff; on jugulum a buff cross-patch bordered below with blackish; the two white specula near the carpal edge and on the quills well developed; both mandibles yellowish white at tip (juv. Siao, 28. VII. 93: Nat. Coll. — C 12621).
Observation. In the type specimen Brüggemann makes no mention of green secondaries, and these, therefore, do not perhaps present a stable point of difference between this form and *celebensis*.

Measurements.	Wing	Tail	Tarsus	Bill from nostrils
a. (C 12619) ad. Siao	103	39	39	16
b. (C 12621) juv. Siao	94	34	39	12
c. (C 13527) ad. Tagulandang, Aug. 94	109	40	39	15
d. (C 13528) ad. Tagulandang, Aug. 94	107	40	41	16

Distribution. Siao (Hoedt *a 1*, v. Duivenb. *a 1*, Fischer *1*, Nat. Coll. in Dresden Mus.); Tagulandang (Nat. Coll.).

This form of Red Pitta is confined, so far as is known, to Siao and Tagulandang, two of the southern islands of the Sangi Group. The habitat of the type, originally indicated as "Sangir", has now been fairly satisfactorily settled as being Siao. Only 12 specimens in the Leyden, Darmstadt and Dresden Museums are on record, the majority being in immature plumage. *P. palliceps*, as Meyer has elsewhere *(3)* remarked, is undoubtedly only an insular variation of *P. celebensis*; the Great Sangi bird is somewhat further removed and resembles rather more closely *P. erythrogaster* of the Philippines; nevertheless *P. palliceps*, in virtue of its pale head, absence of all signs of a black ring round the hind neck, and of its similar blue pectoral collar and throat, is undoubtedly an approach towards *P. caeruleitorques* of Great Sangi. In fact the four forms afford regular transitions, just as do their respective habitats, and it is highly probable that Siao and Sangi served as stepping-stones and halting-places and became colonised by Red Pittas from Celebes or from Mindanao. In which direction this emigration took place may possibly be understood when the meaning of the disappearing (or growing) white speculum and vertical stripe of blue among other characters has been correctly interpreted.

As Prof. W. Blasius remarks *P. palliceps* is the only known species peculiar to Siao, unless indeed *Scops siaoensis* should prove to be distinct from *S. manadensis*, but further research will no doubt bring to light some peculiar forms.

✳ 111. PITTA CAERULEITORQUES Salvad.

Sangi Red Pitta.

Pitta caeruleitorques *(1)* Salvad., Ann. Mus. Civ. Gen. 1876, IX, 53; *(II)* Rowley, Orn. Misc. 1877, II, 324, pl. LXIV; *(3)* Meyer, t. c. p. 327, 328; *(4)* Salvad., Atti Ac. Sc. Tor. 1878, XIII, 1187; *(V)* Gld., B. New Guinea, pl. 32 (1878); *(6)* Meyer, Isis, Dresden 1884, 6; *(7)* W. Blas., Ornis 1888, 601; *(8)* Sclat., Cat. B. XIV, 1888, 433; *(9)* Whitehead, Ibis 1893, 505; *(10)* M. & Wg., J. f. O. 1894, 246; *(XI)* Elliot, Mon. Pitt., pt. IV, 1895, pl.

Figures and descriptions. Rowley *II*; Gould *V*; Elliot *XI*; Salvadori *1*; W. Blasius 7; Sclater *8*.

Adult. Like *P. celebensis* ad., but the head above and neck much paler yellowish chestnut — almost ferruginous, and without any blue vertical streak, the lower neck bordered against the mantle with a fringe of China-blue, like that of the pectoral band, without any intermediate black wing; the green parts of the upper surface somewhat greyer; throat and face darker and more russet; under parts similar, but the black band below the pectoral collar narrower (adult, Great Sangi: Meyer — Nr. 1910). "Iris brown; bill black; feet blue-grey" (Plate n 7).

Another specimen (1909) has the upper surface slightly bluer than in that taken for comparison.

Measurements (2 ex.). Wing 98, 101 mm; tail 36 c.; tarsus 38; bill from nostril 15.5, 16.
Distribution. Great Sangi (Meyer *II*, Bruijn *1, 4*, Plate n 7).

Pitta caeruleitorques is intermediate between *P. palliceps* of Siao and *P. erythrogaster* of Mindanao and other Philippines, the former differing by its vertical stripe of pale blue, its greener back and absence of a blue collar above the mantle and of a black bar below the pectoral band, while *P. erythrogaster* has the pectoral band green, with only a small amount of blue in the middle of it and no black bar below, the blue and green tints of the upper surface much brighter, and the head above not uniform, but showing two dusky streaks on the sides of the crown. When Count Salvadori speaks of *P. caeruleitorques* and *erythrogaster* as the only two Red Pittas with a blue, cervical collar, he is not strictly correct, since this character is sometimes well developed in *P. celebensis*. It occurs again in the since described *P. propinqua* (if that species be really distinct from *P. erythrogaster*) and in *P. inspeculata*.

Prof. W. Blasius points out that in the specimen obtained by Dr. Platen the speculum on the primaries is developed only on the 3rd and 4th quills. This is also the case in the two adult examples in the Dresden Museum (and indeed the white is only seen on the inner web of the 3rd quill, but on both webs of the 4th), but an immature specimen has white on the inner web of the 2nd, on both webs of the 3rd and 4th and a trace on the 5th. In *P. celebensis*, *palliceps*, *erythrogaster* and *propinqua* it is seen on the inner web of the 2nd and on both webs of the 3rd and 4th, so that it appears certain that the speculum in *P. caeruleitorques* has undergone reduction. In the next article this matter is examined a little more fully.

+* 112. PITTA INSPECULATA M. & Wg.
Talaut Red Pitta.

Pitta inspeculata *(1)* M. & Wg., J. f. O. 1894, 245, pl. III; *(2)* iid., Abh. Mus. Dresd. 1895, Nr. 9, p. 6; *(III)* Elliot, Mon. Pitt., pt. V, 1895, pl.

"**Angkaruii**" or "**Antarawuwung**", Kabruang, Salibabu and Karkellang, Nat. Coll.

Figures and descriptions. M. & Wg. *I*; Elliot *III.*

Adult. Head and neck ferruginous chestnut, rather darker on the sides of crown and of forehead, washed with bluish grey on middle of forehead, orbital region and cheeks; upper surface China-blue, a brighter fringe bordering the chestnut of the hind neck; bastard-wing and primaries black, the inner primaries washed externally with blue, no speculum; chin and throat dusky brown, forming into black upon the jugulum; pectoral collar China-blue all round, continuous with the blue of the mantle; remaining under parts deep scarlet; the sides bordering the wing blue; under wing-coverts slaty black washed with blue; longest under tail-coverts blue; quills and tail below shining black. (Type — ad. Kabruang, 7. XI. 93: Nat. Coll. — C 13174.)

The type is without any sign of a white speculum on either the primaries or against the metacarpal edge, though in a few other fully coloured specimens a small white spot or spots may be found on searching for such on the fourth primaries and near the metacarpal edge; a narrow black bar below the pectoral blue is also sometimes present.

Immature. Like the adult, but the throat and jugulum black varied with a few white or pale fawn-feathers, chin mostly cinnamon; a few pale fawn feathers on the abdomen; shoulder greenish tawny-olive; mantle varied with greenish olive; a small speculum formed by a white spot on the inner web of the 3rd quill and on the outer web of the 4th, just present also on the inner web. (Imm. co-type, Kabruang, 7. XI. 94; Nat. Coll. — C 13175.)

Distribution. Talaut — Karkellang, Kabruang and Salibabu (Nat. Coll. in Dresd. and Tring Museums).

The type and co-types of this species were obtained by native hunters sent out on behalf of the Dresden Museum; the species must be rather plentiful, six specimens being obtained in Kabruang and five in Salibabu. This was at the end of October and beginning of November, when the birds were moulting. Two large series were obtained in 1894 and 1896 in the larger island of Karkellang.

P. inspeculata is the only known Red Pitta in which the two spocula on the wing are absent. From *Pitta celebensis* it differs by its back being blue, not green, by its wanting the blue vertical stripe, and rather broad black bar on the breast. It is most like *P. cyanonota* of Ternate and Guebeh in colour, but has the head darker chestnut, the blue of the back more intense, a cervical collar of brighter China-blue, not seen in *cyanonota*, the throat darker and the jugulum black; there is an entire absence or slight indication only of a black band below the blue pectoral collar. In respect of the blue cervical collar it agrees with *P. caeruleitorques* of Sangi and *P. erythrogaster* and *propinqua* of the Philippines, but is easily recognisable from the two last by its entirely blue upper surface and pectoral collar, and from *caeruleitorques* by its darker head, blue upper surface, blue sides, as well as middle, of the pectoral collar, and by the reduction or absence of the black band below it.

Pitta inspeculata is perhaps the most interesting member of the Red Pittas in virtue of the disappearance of the two white specula — on the primaries and against the carpal edge. An examination of the speculum on the primaries of the Red Pittas brings to light some interesting facts. We give a table showing the variation of the quill-specula of all the Red Pittas in the Dresden Museum at the time of writing (July 1894):

Species	Specimen	Distribution	2nd P. out./inn. web	3rd P. out. web	3rd P. inn. web	4th P. out./inn. web	5th P. out./inn. web	Remarks
1. *P. mackloti* ♂ ad.	Nr. 13306	N. Guinea		*	*	*	*	
2. *P. mackloti* ♀ ad.	Nr. 1890	N. Guinea		*	*	*	*	
3. *P. mackloti* ♀ ad.	Nr. 1891	N. Guinea		*	*	*	.	Just seen on 5th inner.
4. *P. mackloti* ♂ ad.	Nr. 1893	Jobi		*	*	*	*	Just seen on 3rd outer.
5. *P. mackloti* ♂ ad.	Nr. 1896	N. Guinea		*	*	*	*	Indicated on 6th outer.
6. *P. mackloti* ♂ ad.	C 1417	N. Guinea		*	*	*	.	Just seen on 5th.
7. *P. mackloti* ♂ ad.	C 1431	Jobi		*	*	*	.	
8. *P. mackloti* ♀ ad.	C 1426	N. Guinea		*	*	*	*	Large.
9. *P. mackloti* ♀ ad.	C 1414	N. Guinea		*	*	*	.	

348   Birds of Celebes: Pittidae.

Species	Specimen	Distribution	2nd P. out/inn. web	2nd P. out/inn. web	3rd P. out/inn. web	3rd P. out/inn. web	4th P. out/inn. web	4th P. out/inn. web	5th P. out/inn. web	5th P. out/inn. web	Remarks
10. *P. mackloti* ad.	C 1432	Jobi	.	.	.	*	*	*	*	*	Large.
11. *P. mackloti* ad.	C 1435	Jobi	.	.	.	*	*	*	*		
12. *P. mackloti* ♂ ad.	C 1423	N. Guinea	.	*	*	*	*	*	*	.	On 3rd & 5th out. small.
13. *P. mackloti* ♀ ad.	C 1416	N. Guinea	.	.	.	*	*	*	*	.	On 5th inner slightly.
14. *P. mackloti* ♂ ad.	C 7105	Aru	.	.	.	.	*	*	*	.	Moulting.
15. *P. mackloti* ♂ imm.	C 1425	N. Guinea	.	.	.	.	*	*	*	.	
16. *P. mackloti* ♂ imm.	C 1428	N. Guinea	.	*	*	*	*	*	*	*	
17. *P. mackloti* ♀ imm.	C 9962	N. Britain	.	*	.	*	*	*	.	.	
18. *P. mackloti* ♂ imm.	Nr.1897	N. Guinea	.	.	.	*	*	*	*	*	
19. *P. mackloti* ♀ imm.	Nr.1895	N. Guinea	.	.	.	*	*	*	*	*	Small on 5th.
20. *P. mackloti* ♀ imm.	Nr.1892	N. Guinea	.	.	.	*	*	*	*	*	Small on 5th.
21. *P. mackloti* ♂ imm.	Nr.1894	N. Guinea	.	.	.	*	*	*	*	*	
22. *P. rufiventris* ♂ ad.	Nr.13304	Batjan	.	.	.	.	*	*	*	.	
23. *P. rufiventris* ad.	Nr.6990	Moluccas	.	.	.	*	*	*	*	*	On 5th bilat. unequal.
24. *P. rufiventris* ad.	Nr.1888	Halmahera	.	*	.	*	*	*	*	*	Small on 2nd and 5th inn.
25. *Pitta* juv.	—	"Moluccas"	.	*	*	*	.	.	.	.	
26. *P. cyanonota* ad.	Nr.1889	Ternate	.	.	.	*	*	*	.	.	Moult., formula doubtf.
27. *P. celebensis* ad.	C 10886	Celebes	.	*	*	*	*	*	.	.	Large.
28. *P. celebensis* ad.	C 10879	Celebes	.	*	*	*	*	*	.	.	
29. *P. celebensis* ad.	Nr.1908	Celebes	.	*	*	*	*	*	.	.	Very small on 5th.
30. *P. celebensis* ad.	Nr.1907	Celebes	.	*	*	*	*	*	*	.	Large.
31. *P. celebensis* ♀ ad.	P.&F.S.	Celebes	.	*	*	*	*	*	.	.	Indicated on 5th outer.
32. *P. celebensis* imm.	C 12190	Celebes	.	*	*	*	*	*	.	.	Indicated on 5th outer.
33. *P. palliceps* ad.	C 12619	Siao	.	*	*	*	*	*	.	.	
34. *P. palliceps* juv.	C 12621	Siao	.	*	*	*	*	*	*	.	
35. *P. erythrogaster* ad.	Nr.6988	Philippines	.	*	*	*	*	*	*	*	Just seen on 5th inner.
36. *P. erythrogaster* ad.	Nr.3243	Luzon	.	.	*	*	*	*	.	.	Just seen on 2nd outer.
37. *P. erythrogaster* ♀ ad.	Nr.13503	Mindanao	.	.	*	*	*	*	.	.	
38. *P. propinqua* ♂ ad	Nr.12845	Palawan	.	*	*	*	*	*	.	.	
39. *P. caeruleitorques* ad.	Nr.1909	GreatSangi	.	.	.	*	*	*	.	.	
40. *P. caeruleitorques* ad.	Nr.1910	GreatSangi	.	.	.	*	*	*	.	.	
41. *P. caeruleitorques* ♀ imm.	C 4185	GreatSangi	.	*	*	*	*	*	.	.	On 5th just indicated.
42. *P. inspeculata* ad.	C 13174	Talaut	.	.	.	*	*	.	.	.	
43. *P. inspeculata* ad.	C 13119	Talaut	.	.	.	.	.	.	.	.	Spec. quite abs., moult.
44. *P. inspeculata* ad.	C 13116	Talaut	.	.	.	.	.	.	.	.	Spec. quite abs., moult.
45. *P. inspeculata* ad.	C 13120	Talaut	.	.	.	.	.	*	*	.	A small spot.
46. *P. inspeculata* ad.	C 13118	Talaut	.	.	.	.	*	.	.	.	A spot on left wing only.
47. *P. inspeculata* ad.	C 13117	Talaut	.	.	.	.	.	*	.	.	A minute spot.
48. *P. inspeculata* imm.	C 13175	Talaut	.	.	.	.	*	*	*	.	Very small on 4th inner.

Further specimens subsequently received from Talaut prove that the white speculum is not always better developed in young individuals.

The above tables show first that the quill-speculum in the Red Pittas is very variable, and indeed it is impossible to indicate all the slight differences in size and shape of the spots in specimens from the same locality and on the right or left wing of the same individual; secondly that Papuan, and, apparently, also Moluccan, specimens generally have none of the speculum on the second quill, but on the 3rd, 4th and 5th; thirdly that Celebesian and Philippine specimens generally have none of the speculum on the 5th quill, but on the 2nd, 3rd and 4th. As several exceptions to this rule are shown above, a few Papuan specimens having the speculum commencing on the second quill, as in Celebes birds, and a few Celebes birds having it carried on as far as the fifth quill, as in Papuan ones, we must, apparently, infer that the speculum, was originally, larger than at present and embraced a part of the 2nd, 3rd, 4th, 5th and, judging from Nr. 5, 6th quills; and that on the 6th it is now almost entirely lost, while Papuan birds have further lost it almost for good on the 2nd, and Celebes and Philippine ones on the 5th. It appears likely enough that at one time the white was more extended, for the speculum appears almost throughout the varied genus *Pitta*, and in some forms, such as *P. sangirana* and *P. cyanoptera* (members of different groups), the white occupies the whole of the ten primaries except the basal portion and ends of the feathers and inner web of the tenth. In other species, such as the beautiful *P. granatina* and *ussheri* of Borneo and *P. forsteni* of Celebes, the speculum has completely disappeared.

Some species, as Mr. Whitehead remarks, have increased the amount of white on the primaries, at least in *P. atricapilla*, *muelleri* and *sangirana* it is more extended in the adult than in the young.

The adult *P. caeruleitorques* of Sangi seems to have lost it on both the 2nd and 5th, while in the adult *P. inspeculata* of Talaut it is found only as a small spot on the fourth quill or is entirely wanting on all of them. In one immature *P. inspeculata* a moderate speculum still makes its appearance, and in the immature *palliceps* and *caeruleitorques* it seems to be a good deal more extended than in the adult; this is not the case in immature Papuan specimens. — It seems certain that there once existed in the Red Pittas and in many of their fellows, and, apparently, still exists, a tendency to lose the white speculum: but what causes set the process of obliteration at work must be a matter of conjecture. We believe, however, from the existence of volcanoes there, that Sangi and Talaut are not fragments of a former continent, but have been upheaved from the sea, and therefore colonised by flight, or through the agency of winds or sea-currents; and when *Pitta* got there, it is perhaps more likely that it would vary than if it had stayed at home, and the obliterating process of the speculum, if nearly latent before, might be brought to greater activity.

It is possible that a melanising influence exists in Talaut, but this is very doubtful. The Talaut *Oriolus* is melanotic, and the *Dicaeum* is darker than its nearest allies. On the other hand the black borders on the wing of the *Eos* are narrower than in Sangi.

Pitta cyanonota Temm. Two specimens of this species of Ternate and Guebeh purporting to have come from Celebes are in the British Museum, but the habitat is very rightly queried by Sclater (Cat. B. XIV, 435). No collectors' names are mentioned, and there can be little doubt that the labels are erroneous.

Pitta rufiventris Cab. & Heine. The same remark applies to a young specimen of this bird labelled "Celebes" in the same collection (l. c. p. 434).

+ * 113. PITTA FORSTENI Bp.
Celebes Green Pitta.

a. **Pitta melanocephala** *(1)* Müll. & Schl. (nec Wagl.), Verh. Natuurk. Comm. 1839—44, Aves, Pitta, p. 19 (ex Forsten MS.); *(2)* Gray, Gen. B. I, 214 (1846); *(III)* Westerm., Bijdr. t. d. Dierk. 1854, I, 46, Pitta, pl. II; *(4)* Schl., Handl. Dierk. 1857, 254; *(V)* id., Vog. Ned. Ind. 1863, 5, 30, pl. II, f. 1; *(6)* id., Mus. P.-B. Pitta 1863, 4; *(7)* Finsch, Neu-Guinea 1865, 167; *(8)* Schl., Rev. Pitt. 1874, 9; *(9)* Rosenb., Malay. Archip. 1878, 272; *(10)* W. Blas., Ornis 1888, 598.
b. **Brachyurus forsteni** *(1)* Bp., Consp. 1850, I, 256; *(II)* Elliot, Mon. Pitt. 1863, pl. XXIV; *(3)* id., Ibis 1870, 419.
c. **Melanopitta forsteni** *(1)* Bp., Consp. Vol. Anisod. 1854, 7; *(2)* Wald., Tr. Z. S. 1872, VIII, 62; *(3)* Meyer, Ibis 1879, 126; *(4)* W. Blas., J. f. O. 1883, 132; *(5)* Platen, Gefied. Welt 1887, 218.
Pitta forsteni *(1)* Wall., Ibis 1864, 106; *(2)* Gray, HL. 1869, I, 295, Nr. 4363; *(3)* Wald., Tr. Z. S. 1875, IX, 189; *(4)* Rowley, Orn. Misc. II, 1877, 331, note; *(V)* Gould, B. New Guinea IV, pl. 30 (1879); *(6)* Sclat., Cat. B. XIV, 1888, 442; *(7)* Whitehead, Ibis 1893, 490, 499; *(VIII)* Elliot, Mon. Pitt. 1895, pt. V, pl.; *(9)* M. & Wg., Abh. Mus. Dresd. 1895, Nr. 8, p. 13.
"**Mopo idiu**" (Green Grandfather), Minahassa, Meyer *c 3*.
Figures and descriptions. Gould *V*; Elliot *b II, b 3, VIII*; Westermann *a III*; Schlegel *a V*; Sclater *5*; Whitehead *7*.
Adult. Entire head all round, neck and upper throat glossy velvety-black; upper surface glossy green, showing bright yellow, malachite or dark olive tints according to the light; sides of neck, jugulum, breast and sides of body bluer green, similarly variable; primaries, bastard-wing and concealed portion of secondaries black; exposed portion of secondaries and innermost primaries, greatest wing-coverts and tail green, darker than the back; lesser wing-coverts and upper tail-coverts metallic silver-blue; abdomen and under tail-coverts deep scarlet; a patch of black separating the abdomen from the breast; entire under side of wing and tail shining black. (♂ ad. Rurukan, 17. I. 83: Platen in Mus. Nehrkorn, Nr. 946; ♀ ad. on the Gunung Sudara, Minahassa, 15. X. 93: P. & F. Sarasin Nr. 105). "Iris dark; bill black; legs grey" (Sarasins, MS.).
Measurements (6 ex. in the Sarasin Collection, and 4 Dresden Mus.). Wing 116—126 mm; tail c. 40—47; tarsus c. 39—44; bill from nostril c. 16—17.
Nestlings. Two in the Sarasin Collection are partially covered with pin-feathers, those along the sides of the body pale fulvous, the rest lead-colour. The arrangement of the ten primaries and twelve rectrices is well seen. "In stomach: remains of insects" (4. June, 1894: P. & F. S.). The mother and the nest accompanying.
Eggs. "Dr. Platen found on May 6th, 1886, a nest with 2 eggs near Rurukan in the Minahassa. They differ from one another:

a, has dark brown surface spots on a white ground;
b, light brown ones on a like ground.
Both eggs measure 30 × 23.5 mm. In the delicacy of the markings they differ little from other *Pitta*-eggs" (Nehrkorn, MS.).

Nest. The nest obtained by the Sarasins with the above two nestlings is a most curious object. In shape a cylinder, closed at one end, or a large pocket, about 22 cm long by about 12 in external diameter. "According to the statement of the finder, the nest was situated on the ground; it represents a moss cylinder, open in front, and shut off by a tree-stem behind" (P. & F. S.). The side which seems to have lain on the ground is composed almost entirely of dark vegetable-fibres, the upper surface and sides chiefly of moss; the dark fibres form the internal structure.

Distribution. North Celebes: Minahassa (Forsten *a 1, a 6*, Meyer *c 3*, etc.), Gorontalo Distr. (Riedel in Dresd. Mus.).

The Green Pitta of Celebes is a rare bird in collections, and is known only from the Northern Peninsula of the island. It was not obtained by Wallace or von Rosenberg; Meyer found it a difficult bird to shoot and met with it only near Manado in December, 1870, and once again later, and Platen observed both it and *P. celebensis* beneath the underwood in the forest near Rurukan. In the stomach the Sarasins found insects.

P. forsteni is well distinguished from *P. sangirana* of Great Sangi, *P. muelleri* of Borneo and *P. atricapilla* of the Philippines by its primaries being entirely black like the rest of the wing, instead of, as in these three species, white with the basal part and distal ends black. In respect of the wing *P. forsteni* more nearly resembles *P. novaeguineae* M. & S. and its allies of the Papuan group, but these birds, having the black of the throat carried further down on to the jugulum, the sides of the abdomen blue or washed with blue, and a different green on the breast, seem to be rather further removed. In *P. mafoorana* Schl., as in *P. forsteni*, no white speculum is seen; in *P. novaeguineae* and *rosenbergi* it is very small. The variation of the quill-speculum is much more striking than in the Red Pittas, it seems to be undergoing, or has already undergone, obliteration east of Celebes, but to be on the increase north and west. The group, numbering 11 or 12 species (see Sclater *6*, Whitehead *7*), ranges from the Himalayas to New Guinea and North Australia.

⊣ * 114. PITTA SANGIRANA (Schl.).

Sangi Green Pitta.

a. **Pitta atricapilla sanghirana** *(1)* Schl., Ned. Tdschr. Dierk. III, 1866, 190.
Pitta sanghirana *(1)* Elliot, Ibis 1870, 411; *(2)* Salvad., Ann. Mus. Civ. Gen. IX, 1876, 54; *(III)* Rowley, Orn. Misc. II, 1877, 329, pl. LXV; *(4)* Meyer, t. c. 330; *(5)* W. Blas., Ornis 1888, 597; *(6)* Sclat., Cat. B. XIV, 1888, 440; *(7)* Whitehead, Ibis 1893, 499.
b. **Pitta atricapilla** pt. *(1)* Schl., Rev. Pitt. 1874, 5 (Sangi).
"**Kopau**", Great Sangi, Nat. Coll.
For further synonymy cf. W. Blasius *5*.

Adult. Similar to *P. forsteni*, but the green above and below darker, the blue on the wing-coverts, upper tail-coverts and rump (where it is more extended) darker, upper abdomen black, the lower abdomen and under tail-coverts scarlet, primaries white, tipped about 20 mm broad on the 1st, about 2 mm broad on the inner ones — with black, and with the bases of the feathers black (ad. Great Sangi, 31. VII. 93: Nat. Coll. — Nr. 12700). Iris brown; bill black; feet blue-grey or grey-reddish (Platen 5).

Immature. Head above and under-parts rufous isabella-colour, almost buff on chin and throat, sides of the body green, breast varied with green; neck and nape blackish; back and mantle green, nearly as in adult; wings chiefly broccoli-brown, no silver-blue on the lesser coverts; the white on the primaries much less extended than in the adult, the first quill black, the white commencing on the inner web of the second (Gt. Sangi: Meyer — Nr. 1916).

Nestling. Much more dusky than the immature bird: head all round neck and chin black, throat varied with buff; upper surface dusky bottle-green; uropygium bluish; upper tail-coverts brighter blue; the white on the quills distributed as in the immature bird, rest of the quills and tail black, under-parts brownish, paler on abdomen, bill black, yellower towards the gape, and the tips of both mandibles yellowish: wing 92 mm; tail 25; tarsus 40; bill from nostril 11 (Great Sangi, 16. VII. 93: Nat. Coll. — C 12702).

Measurements (23 adults: W. Blasius 5). Wing 102—111 mm; tail 34—45; culmen 21—24.5; tarsus 39—43.

Distribution. Great Sangi (Rosenberg *a 1, b 1*, Hoedt and v. Duivenbode *b 1*, etc.).

A fine series of 25 examples of this species obtained by Dr. Platen and his hunters passed into the hands of Prof. W. Blasius in 1888, who, after pointing out that the white on the quills is least extensive in young birds and very variable in extent in fully coloured individuals, remarks that *P. atricapilla* Less. of the Philippines, one of its nearest allies, is smaller and has a very much lighter, almost yellowish green tint on the back and under-parts. *P. muelleri* (Bp.) of Sumatra and Borneo is lighter green above, lighter and bluer green below, has the lesser wing-coverts and the tail-coverts slightly more silvery and paler blue, the black on the abdomen less extended, and the bill, apparently, smaller. Whether these differences are bridged over by individual variation we have insufficient material to judge; if such is the case, then *sangirana*, *muelleri* and the *typical atricapilla* should be reduced to the rank of subspecies.

Unlike *P. forsteni* in Celebes the present species seems to be rather plentiful in Great Sangi, and it is curious that 20 of Dr. Platen's 25 examples were males and only 3 adult females. Though very difficult to approach under ordinary circumstances, it is well known to East Indian hunters that by imitating the call-note the bird can be drawn almost to the muzzle of the gun. Perhaps, in the case of our species, the male bird answers better to the call of the hunter.

115. PITTA CYANOPTERA Temm.

Chinese Blue-winged Pitta.

Pitta cyanoptera *(1)* Temm., Pl. Col. pl. 218 (1823); *(II)* Schl., Vog. Ned. Ind. Pitta 1863, 9, 32, pl. 4, fig. 1; *(3)* Salvad., Cat. Ucc. Borneo 1874, 235; *(4)* W. Blas., "Braun-

schweig. Anzeigen" (newspaper!), 3. März 1886; *(5)* Sclat., Cat. B. XIV, 1888, 420; *(6)* Sharpe, Ibis 1889, 442; *(7)* Whitehead, l. c. notes; *(8)* Everett, J. Str. Br. R. A. Soc. 1889, 147; *(9)* Sharpe, ib. 1890, 281; *(10)* Oates, ed. Hume's Nests & Eggs Ind. B. 1890, II, 283; *(11)* Oates, Faun. Brit. Ind. B. II, 1890, 392; *(12)* Hose, Ibis 1893, 403; *(13)* Whitehead, t. c. 494; *(14)* Sharpe, Ibis 1894, 421, 544.
a. **Brachyurus cyanopterus** *(1)* Elliot, Mon. Pitt. 1863, pl. IV.
b. **Pitta moluccensis** *(1)* Oates, Str. F. V, 1877, 149; *(2)* David & Oust., Ois. Chine 1877, 144; *(3)* Hume & Davis., Str. F. VI, 1878, 240; *(4)* Oates, B. Brit. Burmah 1883, I, 415.
For synonymy and further references see Salvadori *3*; Sclater *5*.
Figures and descriptions. Temminck *I*; Schlegel *II*; Elliot *a I*; David & Oustalet *b 4, 11*; Sclater *5*.
Adult. Head above isabella-colour, with a black vertical stripe; face, sides of head, and neck black; back, scapulars and inner quills olive-green variable in different lights; upper wing- and tail-coverts glossy intense campanula-blue; bastard wing and secondaries black, the latter edged with bluish; primaries white, their bases and the ends of the outer ones black; tail black, tipped with bluish; throat white; under parts cinnamon-buff; middle of abdomen scarlet, paler on under tail-coverts; under wing-coverts black (Baram, N. W. Borneo, Nr. 13305). Iris black; bill dark brown; legs pale pink (Whitehead *7*).
Sexes. Said to be similar.
Young. They have the coronal streak broader, and the feathers of the crown are narrowly margined with black; wing-coverts dull blue, the other parts less bright than in the adult (Oates *11*).
Eggs. "Both the eggs in my collection were collected by Oates in Pegu. They belong to two different clutches, of which the two fellow specimens were in the former Seebohm collection (now British Museum). The larger example measures 30 × 22.5 mm, has a white ground, with washy lilac-grey subjacent, and a few small dark brown superjacent spots. The eggs have resemblance with those of *Chibia hottentotta, brevirostris* and *laemosticta.* Nr. 2 is considerably smaller and measures 26 × 21 mm. The large brown surface spots are vermiform and evenly distributed over the whole egg. Dates: Nr. 1, June 27th, 1877; Nr. 2, June 20th, 1878" (Nehrkorn MS.). Hume *(b 1)* describes the eggs as "far more thickly marked and richly coloured than those of any other Ground Thrushes with which I am acquainted". Davison found as many as 6 in a nest; Oates says 4—6.
Nest. A huge structure of sticks, leaves and roots, bound together with earth, placed on the ground in an open place or against the root of a tree; globular in shape, the entrance at one side close to the ground (Oates *b 4*).
Distribution. Arakan (Blyth *3*, Oates *b 4*); Pegu (Oates *b 4*); Tenasserim (Beavan *3*, Davison *b 3*); Corea (David *b 2*); South China (Swinhoe *3*, David *b 2*); Siam and Cambodia (Brit. Mus. *5*); Malay Peninsula (Moore, Wallace *3*, etc.); Sumatra (Raffles etc. *3*); Borneo (Whitehead, etc. *8*); Celebes — Minahassa (Platen *4*); ?Java (Temm. *3*, Elliot *3*); ?Luzon (Mus. Turin *3*).

This *Pitta* is a migrant; there is, therefore, nothing very surprising in the discovery of an example at Rurukan in the Minahassa by Dr. Platen in 1884—85. Burmah and Tenasserim are known as its breeding-grounds, from where it disappears, except from S. Tenasserim, in the winter or dry season (cf. Davison *b 3*, Oates *b 4, 11*, Hume *b 1*). On Tega Island, Borneo, Mr. John Whitehead met with it in numbers during April 1886, but in the following year Mr. Everett sent a collector to that island, where he procured numbers of *Pitta muelleri*, and no *P. cyanoptera*. When Mr. Whitehead visited the island there were no *P. muelleri* to be seen. This species *(cyanoptera)* is fairly plentiful in Labuan in some seasons; at other times it is not to be met with *(7)*. Mr. Whitehead remarks that this species "takes flight more often when alarmed than *P. ussheri* or any other species of this genus that I have met with", as might indeed be expected in a form given to making especially long journeys on the wing. Its absence or rarity in the Philippines is instructive; its route of migration, like that of *Lanius tigrinus*, seems to be a southern one. In some quarters it seems to occur all the year round; according to Mr. Hose *(12)*, this is the case in the Baram District of Borneo, and, according to Mr. Oates, in Southern Tenasserim.

Davison *(b 3)* makes the following remarks on the habits of this bird: "This species is fond of perching on trees; you may continually see them high up upon high trees calling vociferously. They are not at all wild or shy birds; they feed freely on ants and their larvae, all insects, grubs, and land shells. I never noticed this or any of its congeners coming to the water to drink. This and the closely-allied *P. megarhyncha* seem to frequent most commonly thin tree jungle, where there is not much underwood, and the mangrove swamps, but they also occur abundantly in gardens and plantations. They both have a fine clear double note, which may constantly be heard in the morning and evening wherever they occur. They are decidedly noisy and often call all day, and on moonlight nights a great part of the night also".

P. megarhyncha Schl., which this excellent field naturalist cites as a close ally of the present species, is recognisable, as Hume shows (Str. F. VI, 242), by its much longer, slenderer and excessively straight-culmened bill, the duller and darker brown of the head, the absence of the black vertical stripe, the narrower black collar, etc. *P. nympha* T. & S. of China and Formosa, *P. brachyura* (L.) of India and Ceylon to Tenasserim and *P. irena* of Timor, Sula, Boano near Ceram and Ternate are distinguishable at a glance by their having the blue of the wings confined to the lesser coverts, and turquoise, instead of campanula-blue, and the white on the primaries restricted to a speculum.

+ 116. PITTA IRENA Temm.
Buff-browed Pitta.

a. "**Brève irène**" *(1)* Temm., Pl. Col. 591 (1836).
b. **Pitta crassirostris** *(1)* Wall., P. Z. S. 1862, 188, 335, 339 (Sula); *(2)* id., Ibis 1864, 104, 106; *(3)* Finsch, Neu Guinea 1865, 167; *(4)* Gray, HL. 1869, I, 295, Nr. 4348; *(5)* Sclat., Cat. B. XIV, 1888, 427.
c. **Pitta brachyura** *(1)* Schl. (nec Linn.), Vog. Ned. Ind. Pitt. 1863, 13, 33, pl. 3, fig. 2, 3; *(2)* id., Mus. P.-B. Pitt. 1863, 11; *(3)* id., Rev. Pitt. 1874, 14.

Pitta irena Temm., *(1)* Salvad., Orn. Pap. II, 1881, 391; *(2)* Sclat., Cat. B. XIV, 1888, 427; *(3)* Salvad., Agg. Orn. Pap. 133 (1890); *(4)* Whitehd., Ibis 1893, 407.
"**Lihange**", Tagulandang, Nat. Coll.
For further synonymy and references cf. Salvadori *1, 3*.
Figures and descriptions. Temminck *a 1*; Schlegel *c 1*; Wallace *b 1*; Salvadori *1*, Sclater. *b 5, 2*, Whitehead *4*.
Adult. Entire head, hind neck, chin and middle of throat black; superciliary stripe from forehead to nape deep buff; upper surface changeable olive-green; lesser wing-coverts and tail-coverts turquoise-blue; primaries black, tips greenish, a white speculum formed on the 2nd to 6th quills; secondaries edged externally with bluish green; tail black, tipped with green; under-parts deep buff, darker on the breast and sides; abdomen dark red, the bases of the feathers black; crissum and under tail-coverts more scarlet: wing 117 mm; tail 41; tarsus 39; bill from nostril 15.5 (ad. Tagulandang Id., Aug. 1894, C 13526). Bill black, base of lower mandible horny, feet pale horn or flesh-colour, iris dark (Sula — Wallace *b 2*).
Distribution. Timor (S. Müller *c I*); Tagulandang Id. between Celebes and Sangi (Nat. Coll.); Sula Besi and Sula Mangoli (Allen *b 1*, *[b 5*, Hoedt *c 3*); Ternate (Bernstein *c 2*, Bruijn *1*); Boano off Ceram (Hoedt *c 3, 1*). (S. Müller *c I* states that he met with it also in Samao near Timor).

This species presents another of those cases in which two leaders in ornithology handle the same subject in different manners. The Sula bird is said to have a thicker, more compressed bill than that of Timor (we do not find that Timor birds are otherwise smaller, judging from one before us), and Mr. Wallace's name *crassirostris* is allowed to stand for the Sula form by Mr. Sclater, while Count Salvadori finds no sufficient grounds for its separation as a species. It appears likely enough that the peculiarities of the Sula bird are not great enough to warrant its position as a distinct species, but that occasional individual variation connects it with Ternate and other examples. The occurrence of this species on Tagulandang is comparable to the presence of *Columba albigularis* there.

+ * 117. PITTA VIRGINALIS Hart.
Djampea Buff-browed Pitta.

Pitta virginalis *(1)* Hart, Nov. Zool. 1896, 174, 182.
Description. Hartert, *l. c.*
Diagnosis. Differs from *Pitta irena* by having the deep buff superciliary stripe about twice as broad; the chin black for about 10 mm only as against about 25 mm in *P. irena*; size a little smaller.
Measurements. Wing 103—109 mm; tail 38—39; tarsus c. 38; bill from nostril c. 15 (Hartert ♂, and ♀ Djampea, Dec. 1895: Everett — C 14864).
Distribution. Djampea (Everett).

This was the only *Pitta* found by Mr. Everett on Djampea Island; none were sent from Kalao or Saleyer, where they certainly also occur. It belongs to the same group as *P. irena* of Timor and the Moluccas to Sula and Tagulandang, *P. concinna* of Lombok and Flores, *P. maria* of Sumba, and *P. vigorsi* of Banda and Timorlaut. *P. irena*, *concinna* and *maria* are distinguishable by having much more black on the chin and throat, *P. vigorsi* by having no black here, and by other points (cf. Hartert, Nov. Zool. 1896, 175, 585).

FAMILY HIRUNDINIDAE.

The Swallows are birds of aerial habits, taking their insect-food on the wing. For the rest, they may be best recognised by the small, flat, triangular bill, and by their possessing only nine primaries. They belong to Gadow's *Diacromyodine* group of *Passeres* in virtue of the arrangement of the muscles of the syrinx.

The family belongs both to the Old and New Worlds. There are only a few species in Australia and Papuasia, and these seem to have got there by flight. The genus *Hirundo* only is represented in Celebes.

GENUS HIRUNDO L.

The true Swallows, with a flat, triangular bill; the tail forked to nearly square; the wing long, the first and second quills the longest and the secondaries about half their length; the secondaries with double, or heart-shaped, tips; the tarsus small and bare, about as long as the hind toe and claw; the plumage with a metallic gloss on the upper surface. "They all construct nests of mud lined with feathers, some making their nests cup-shaped, while others add a long tubular entrance. The eggs in some species are speckled, in others white without any marks" (Oates). By their nidification, as well as by their other habits, they show a certain amount of correspondence with some Swifts, for instance, *Collocalia*, differing herein from other *Passeres*.

Two species in Celebes, one a migrant, and both widely distributed elsewhere, and doubtless birds which have extended their range by flight over sea.

118. HIRUNDO RUSTICA L.
Common Swallow.

The Common Swallow is almost cosmopolitan in its range, extending north to about the polar circle, south in winter to the Cape, North Australia, and South Brazil; absent in the Pacific Islands[1]), and New Zealand, and apparently in Madagascar. Within this area it varies considerably and has been divided into a number of species. Sharpe recognises five races or subspecies, other

---

[1]) Except the Pelew Islands, from where there is a specimen in the British Museum, fide Sharpe, Cat. B. X, 137, a record overlooked by Wiglesworth in "Aves Polynesiae". Schmeltz (Ethn. Abth. Mus. Godef. 1881, 391) speaks of *Hirundo rustica* and *Chelidon urbica* as occurring at Yap in the Carolines, but no reference to actual specimens is made, though they may have been in the Museum Godeffroy.

writers make other subdivisions. What is necessary to know is where the extremes of variation in particular directions are found; the intermediate forms, which probably compose the larger half of the species can then be indicated by hyphens connecting the names of the races between which they lie. Thus, in Japan Dr. Sharpe seems to have discovered the summer head-quarters of one extreme, in Europe another extreme is found, i. e. the *typical H. rustica* with the black breast-band unbroken: "there is never any difficulty in recognizing birds from Japan as unmistakeable *H. gutturalis* with the breast-band completely divided [by the chestnut of the throat breaking through it]; and I have never seen an intermediate or doubtful specimen from the Japanese Islands". Celebes and the other East Indies are without doubt the winter habitat of the true *H. rustica gutturalis*, and one of the Celebes specimens before us may be thus identified. Further, Sharpe records the *typical rustica* from Malacca, the Philippines and Batchian — a record overlooked by Count Salvadori — and from Palawan and Borneo (Ibis 1888, 200; 1893, 561) and with Mr. Wyatt later from Celebes, Halmahera, Buru and Amboina (Mon. Hirund. pt. XVII, 1893); other specimens from Celebes, Java, Luzon, etc., after the finding of these authors, do not admit of such positive identification, but are intermediate between the typical and the Japanese races like three others from Celebes and elsewhere before us. These are best indicated as *H. rustica — gutturalis*, a less clumsy method not being possible with the present system of nomenclature. The following literature relates to the occurrence in the Celebes Province of:

1. **Hirundo rustica gutturalis** (Scop.) and **H. rustica — gutturalis**.

a. **Hirundo gutturalis** (Scop.), *(1)* Wald., Tr. Z. S. 1872, VIII, 65; *(2)* Meyer, J. f. O. 1873, 405; *(3)* Salvad., Ann. Mus. Civ. Gen. 1876, IX, 55; *(4)* Meyer, Ibis 1879, 128; *(5)* Salvad., Orn. Pap. 1881, II, 1; *(6)* Meyer, Isis, Dresden 1884, 6, 22; *(7)* Sharpe, Cat. B. X, 1885, 134; *(8)* W. Blas., Ztschr. ges. Orn. 1886, 108; *(9)* id., Ornis 1888, 580; *(X)* Sharpe & Wyatt, Mon. Hirund. pt. XVII pl. (1893).
b. **Hirundo rustica** *(1)* Rosenb., Malay. Archip. 1878, 271; *(II)* Sharpe & Wyatt, Mon. Hirund. pt. XVII pls. (1893); *(3)* M. & Wg., Abh. Mus. Dresd. 1895, Nr. 8, p. 8.
Figures and descriptions. Sharpe & Wyatt *a X, b II*; Salvadori *5*; Sharpe *7*; W. Blasius *8*.
Adult male *(H. rustica gutturalis)*. Above glossy steel blue-black; wings and tail dusky, the rectrices, except the two middle ones, furnished with a large spot of white on the inner webs 5—10 mm from the tip, the outermost rectrices much lengthened and narrow, forming a deep fork; forehead, chin, and throat chestnut, on sides of neck a blue-black collar, broken through on the upper chest by the colour of the throat; under surface light pinkish buff (Macassar, ♂, Jan. 1873 — C 510). Wing 114 mm; tail (middle rectrices) 42; tail (outer rectrices) 93; tarsus 10, rictus 15.
Sexes. They are similar (Sharpe).
Immature. Like the adult, but the head above dusky brown; chin, throat, and forehead at base of upper bill cinnamon; a broad pectoral collar dusky glossed with blue-black (Macassar, ♂, Jan. 1873 — C 571: apparently, *H. rustica—gutturalis*).
Eggs. North India and Afghanistan — *H. rustica* — 3—6 in number; white to pale salmon-

pink, spotted and speckled with brownish red and inky purple: size 17.8—21.3×12.7
—14 mm (Oates ed. Hume's Nests and Eggs Ind. B. 1890, II, 184). The eggs of
H. gutturalis do not differ from those of *rustica*, are only a little smaller (Nehrkorn
MS.).

Nest. An open, saucer-shaped, or half-saucer-shaped, structure of mud, fixed upon or against
a beam or such like under cover, lined with feathers, etc. It is doubtful whether
this species breeds in Celebes.

Distribution in the Celebes Province. Great Sangi (Bruijn's Coll. *a 3, a 9*); Minahassa
(Meyer *a 4*, Riedel *a 8*); Togian Islands (Meyer *a 4*); Southern Peninsula of Celebes
(Meyer *a 4*).

H. gutturalis is recorded by Salvadori from most of the East India Islands
as far as the north coast of Australia, by Sharpe *(a 7)* from the Pelew Islands,
by Seebohm (Ibis 1890, 102) from the Bonin Islands, where two specimens
were observed for one day only, one being shot.

Taczanowski (Faun. Orn. Sib. Orient. 1891, I, 173) records this race as
far north as Kamtschatka. Hartert has recently noted its occurrence in the
Natuna Islands and Sumba (Novitates Zool. 1894, 480; 1895, 471; 1896, 585),
Bourns and Worcester in several additional islands of the Philippines.

Specimens captured at sea near the East Indies are recorded by Sharpe
(a 7), by Finsch & Conrad (Verh. z.-b. Ges. Wien, 1873, p. 1 sep. copy),
by the "Challenger" (see *a X*), and there is a specimen in the Dresden Museum
taken by Meyer in the China Sea more than 100 miles west of Borneo, end
of April, 1872. Though in the main a migrant, it is possible that some individuals of this species remain to breed in the East Indies. It was obtained by
Meyer on the Togian Islands in August, and it has been recorded from Amboina in May, Tifore in August, New Guinea in June *(a 5, a 7)*.

Only two Swallows are known from Celebes, the present species and *H.
iavanica*, which latter may be distinguished by the absence of the black pectoral
collar and by its brownish grey — not white and buffy — under surface; also it
does not acquire the long forked tail of *H. rustica*, having the outer rectrices
only a little (in the adult about 10 mm) longer than the middle ones.

119. HIRUNDO JAVANICA Sparrm.
Java Swallow.

Hirundo javanica *(1)* Sparrm., Mus. Carls. 1789, II, pl. 100; *(II)* Temm., Pl. Col. 1823,
pl. 83; *(2bis)* Bernst., J. f. O. 1859, 267; *(3)* Wall., Ibis 1860, 147; *(4)* id., P. Z. S.
1862, 340; *(5)* Wald., Tr. Z. S. 1872, VIII, 66; *(6)* Salvad., Cat. Ucc. Borneo 1874,
126; *(7)* Hume, Str. F. 1876, IV, 374; *(8)* Hume & Davison, ib. 1878, VI, 43;
(9) Meyer, Ibis 1879, 128, 146; *(10)* Hume, Str. F. 1879, VIII, 47, 1880, IX, 120;
(11) Legge, B. Ceylon 1880, 597; *(12)* Salvad., Orn. Pap. 1881, II, 3; *(13)* Davison,
Str. F. 1883, X, 345; *(14)* Meyer, Isis, Dresden 1884, 6, 22; *(15)* Pleske, Bull.
Ac. Petersb. 1884, 121; *(16)* Guillem., P. Z. S. 1885, 261, 419, 552; *(17)* Sharpe,
Cat. B. X, 1885, 142; *(18)* Ramsay, Tab. List 1888, 2; *(19)* W. Blas., Ornis 1888,
308, 580; *(20)* Everett, J. Str. Br. R. A. S. 1889, 134; *(21)* Hartert, J. f. O. 1889.

354; *(22)* Salvad., Orn. Pap. Agg. 1890, 69; *(23)* Steere, List. Coll. B. & M. Philipp. Is. 1890, 16; *(24)* Whitehead, Ibis 1890, 49; *(25)* Sharpe, t. c. 280; *(26)* Oates, ed. Hume's Nests & Eggs Ind. B. 1890, II, 186; *(26bis)* Oates, Faun. Br. Ind. B. 1890, II, 279; *(27)* Vorderman, Nat. Tdschr. Ned. Ind. 1891, L, 451; *(28)* id., Notes Leyden Mus. 1891, 125; *(29)* Hartert, J. f. O. 1891, 294; *(30)* Salvad., Ann. Mus. Civ. Gen. 1891, (2) XII, 49; *(XXXbis)* Sharpe & Wyatt, Mon. Hirund. pt. XV, pl. (1892); *(31)* Hose, Ibis 1893, 390; *(32)* Büttik., Zool. Erg. Weber's Reise Ost-Ind. 1893, III, 277; *(33)* Sharpe, Ibis 1894, 256, 259; *(34)* Bourns & Worces., B. Menage Exp. 1894, 42; *(35)* Vorderm., N. T. Ned. Ind. 1895, LIV, 336; *(36)* Hartert, Nov. Zool. 1895, 471; *(37)* Grant, Ibis 1895, 455; *(38)* M. & Wg., Abh. Mus. Dresd. 1895, Nr. 8, p. 8; *(39)* iid., ib. 1896, Nr. 1, p. 9; *(40)* iid., ib. 1896, Nr. 2, p. 14; *(41)* Hart., Nov. Zool. 1896, 595; *(42)* Büttik., Notes Leyden Mus. XVIII, 1896, 174; *(XLIII)* Meyer, Abh. Vogelskel. 1897, II, pl. CCXIII.

a. **Hirundo frontalis** *(1)* Quoy & Gaim., Voy. Astrol. Zool. 1830, I, 204, pl. 12, f. 1.
b. **Hirundo domicola** *(1)* Jerd., Madr. Journ. 1844, XIII, 173; *(2)* id., B. India 1862, I, 158.
c. **Hypurolepsis domicola** *(1)* Gld., B. Asia, I, pl. 32 (1868); *(2)* Hume, Str. F. 1874, II, 155.
d. **Hipurolepsis javanica** *(1)* Oates, B. Brit. Burmah 1883, I, 308.

"**Pisok**", near Manado, Manado tua, Lembeh, and Banka Islands, Nat. Coll.
"**Laloa**", Tonkean, E. Celebes, iid.

For further synonymy and references see Salvadori *12* and *22*; Sharpe *17*; Sharpe & Wyatt *XXX*.

Figures and descriptions. Gould *c I*; Sharpe & Wyatt *XXXbis*; Sparrm. *I*; Temminck *II* (fig. pess.); Quoy & Gaimard *a I*; Meyer *XLIII* (skel.); Legge *11*; Salvadori *12*; Sharpe *17*; Oates *d 1*.

Adult. Above black, glossed with greenish blue, wings and tail dusky; forehead, cheeks, chin, throat and upper chest hazel, darker on the forehead; under surface smoke-grey with darker shaft streaks, under wing-coverts, flanks and under tail-coverts browner, the terminal part of the last black, tipped with whitish; quills below hair-brown, the shafts pale (Macassar, ♀, Jan. 1873 — C 512). Iris brown; bill and feet black (Guillem. *16*).

Sexes. They are alike.

In some specimens the under surface is browner and darker, and the forehead and throat darker, being chestnut; this appears to be a sign of age or the breeding plumage.

Young. Like the adult, but the forehead dusky brown with paler margins to the feathers — not chestnut; head above dusky brown; chin, throat and upper chest very pale hazel; middle of abdomen whitish (N. Bohol ♀, Oct. 1877 — C 5403, Banka Id., Celebes, 12. V. 93 — C 12283).

Measurements.	Wing	Tail	Bill from nostril	Tarsus
a. (C 512) ♂ Macassar	104	48	6	10
b. (C 12282) ad. Manado tua, 9. IV. 93	103	45	6	—
c. (C 12283) juv. Banka Id.	103	44	5.5	9.5
d. (C 1854) imm. Great Sangi	106	48	5.5	—
e. (C 6818) ♂ ad. Buru, 27. III. 82 (Riedel)	106	48	6.5	—
f. (C 15696) ad. Timorlaut, May 1883	98	46	6	—
g. (C 5403) ♂ juv. Bohol, Philippines	101	41	6	—

Others from East Celebes and elsewhere do not differ in size; in two the wing is 110, 111 mm, but about 103 seems to be the average.

Egg. India: closely resembles that of *H. rustica*, but is decidedly smaller; moderately broad oval, slightly compressed towards one end, ground-colour pink-white, very finely speckled and spotted, thinly at the small, more densely at the large end, with different shades of dull purple and brownish red: 2 to 5 to a sitting: 16.3—19.6 × 12.2—14.5 mm (Hume *26*). "Examples in my collection from Celebes and Borneo resemble those of our *rustica*, are, however, smaller and measure 17.5 × 13 mm" (Nehrkorn MS.). So, also, two specimens in the Sarasin Coll. (8. Sept.).

Nest. A deep half-saucer. Of mud containing many fresh-water snail-shells, well lined with fine black hair-like vegetable fibres, a few feathers, a small piece of cotton (Kema, N. Celebes, specimen collected by Drs. Sarasin). In India various writers *(26)* mention a thick bed of feathers only as forming the lining of the nest; Mr. Theobald *(26)* speaks also of vegetable down, and Col. Legge *(11)* of "feathers, threads, small pieces of rag", and other things it may chance to pick up. Its habits of nidification, therefore, appear to differ somewhat between India and Celebes, as is not surprising in a case where migration and intermixture of individuals do not appear to be the rule.

Breeding season. North Celebes — July (Sarasin MS.); South India — 2 broods in succession from February to April (Davison *26*); April to June (Wait *26*); Ceylon — April to June (Legge *11*); Tenasserim — lays in the second week of April (Theobald *26*).

Distribution. South India and Ceylon (Legge *11*); Andamans (Hume *c 2*); Tenasserim (Davison *8*, Theobald *26*); Malay Peninsula (Hume *10*, etc.); Sumatra (Wallace *17*, Hartert *24*, Modigl. *30*); Billiton and Mendanau (Vorderman *27*, *28*); Borneo (Wallace, Mottley, etc. *20*); Palawan (Everett *20*, Whitehead *24*); Philippines — Mindoro, Samar, Negros, Basilan, Leyte, Bohol, Dinagat, Cebu, Mindanao (Steere, Platen, etc. *29*), Sooloo (Guillem. *16*); Great Sangi (Meyer *14*, Platen *19*); Lembeh, Manado tua and Banka Is. (Nat. Coll. in Dresd. Mus.); Celebes — North Peninsula (Wallace *3*, Guillem. *16*, Sarasin); E. Peninsula (N.C.); South Peninsula (Wallace *5*, Meyer *9*, Weber *32*); Togian Islands (Meyer *9*); Sula Islands (Allen *3*, *17*), Natuna Is. (Hose *36*); Java, Lombok, Timor (Salvad. *12*, Sharpe *17*); Timorlaut (Riedel, Mus. Dresd.); Ternate, Morty, Batchian, Buru, Amboina, Matabello (rect. Watubella), Kei, Salawatti, New Guinea, Jobi, Aru, Yule Id., Duke of York Id. (Salvad. *12*); Islands of Torres Straits (Sharpe *17*); Cape York, Australia (Ramsay *18*).

The Java Swallow seems to be a stationary species. Davison *(13)* speaks of it as a resident and very common about the Nilghiris, Bourdillon *(7)* as "a resident travelling but little" in the Travancore Hills. Hume believes it to be a migrant to the Andamans, but it is known in literature as a breeding species in South India, Ceylon, Tenasserim, Billiton, Borneo, Java, Celebes, Duke of York, and doubtless elsewhere. Sharpe *(17)* says that "specimens from Travancore are much duller beneath than any other birds yet examined, and those from South-eastern New Guinea and Torres Straits are palest below . . . but light-bellied examples are found in Borneo and other localities". These may be incipient subspecific distinctions, but the bird appears to be darker in its mature dress. One from Timorlaut in the Dresden Museum is especially dark below.

H. javanica is easily distinguishable from *H. rustica* by its wanting, when adult, the long forked tail of that species, by the absence of the black pectoral collar and by the brown tint of its under surface.

The nearest relative of *H. javanica* is *H. neoxena* of Australia. This form is larger and has a longer tail. It is a migratory bird in Tasmania and New South Wales, as Gould observed (Handbook I, 108), according to whom it was also found in New Guinea by Wallace. This record is of course in the highest degree probable, though no confirmation of it has been found since.

The Java Swallow is a familiar bird in the Minahassa. In sending the above-described nest the Drs. Sarasin write: "This comes from our house at Kema, nearly every room of which was at first tenanted by a pair of Swallows. Where it could be managed we left the pretty creatures in peace, but in the library we were obliged to remove the nest. The birds, however, are hardly to be driven away; they were always making renewed attempts to take possession of the old spot. It seems, besides, that every pair claims a room or a closet for itself and will not put up with the presence of another. The song of these birds is very melodious and agreeable, albeit gentle; it calls to mind the soft babbling of a little brook".

Bernstein *(2)* and Davison *(26)* write similarly on the adhesiveness of these birds to their old nesting-spot; the latter says: — "about a week after the first brood have flown the old birds begin to remove the topmost feathers of the nest, replacing them by fresh ones. Three eggs are then again laid, and a second brood reared. After this brood have flown, the old birds still continue to occupy the nest at night, or, more correctly, to occupy the edge of the nest, for they do not get into it, but merely sit close together on its edge. The same nest is occupied the following year, the upper feathers being removed and replaced by fresh ones. Should the nest have been destroyed a fresh one is built on the same site".

The wonderful attachment of birds for house and home should always be remembered in connection with the vast return wave of migratory birds to their summer nesting-haunts, a wave which cannot be compared with one of the sea driven blindly before the wind, but a living wave wherein, we may suppose, that each of the myriad old and, perhaps, young individuals composing it, is striving towards a particular spot, hundreds or, may be, thousands of miles away. Here, as we know, they are generally to be found in due season.

In his "Neu Guinea" 1866, 162, Dr. Finsch marks *Hirundo nigricans* Vieill. as occurring in Celebes, but no confirmation of this statement has come to hand. *Petrochelidon nigricans* ranges from New Guinea and Kei to Australia, a subspecies also occurring in Timor and Flores.

FAMILY MUSCICAPIDAE.

The Flycatchers form a large family of small birds, varying in size from that of a Wren to that of a Lark, of which Sharpe as long ago as 1879 recognised 69 genera, while Newton (1893) is prepared to admit some 60. Many of these afford such near approaches to other families, viz. to the *Turdidae*, *Laniidae*, *Sylviidae* and *Campophagidae*, that their discrimination is often a matter of great difficulty, or is even impossible. They may be distinguished from the *Campophagidae* by not having a dense Cuckoo- or Pigeon-like plumage on the rump, and the nostrils not hidden, though some bristles from the forehead project over them; compared with the *Sylviidae* the Flycatchers have, as Seebohm and Oates point out, a mottled plumage when young, whereas the nestling Warbler is like its parents, but more brightly coloured; the young *Turdidae* (sometimes the adults) have a mottled or squamose plumage of the type of the young Flycatcher, but the Thrushes seek most of their food on the ground and their toes and tarsi are longer and stronger, and the nostril is exposed, not overlapped by any hairs from the forehead (Oates); the typical *Laniidae* may be recognised by their strong bills, with a hook with a notch and a "tooth" behind it, but the prominence of this character fades away in other forms, and there seems to be no perfect criterion for distinguishing them from the Flycatchers.

In the *Muscicapidae* the bill is generally broad and weak, furnished with a small notch near the tip, the gape fenced with bristles, a few projecting from the forehead over the nostrils; the first primary varies from very minute to about half the length of the wing, 3rd—5th the longest; tail of 12 feathers, rounded or square, the middle feathers sometimes lengthened; tarsus and toes rather small and weak. They feed on insects taken on the wing, and nest in holes, or form cup-shaped nests in the open.

The *Muscicapidae* are absent in the New World. The Ethiopian and Australian Regions are richest in genera, though nearly equalled by the Oriental Region, the proportions being 22, 21, 20 in the Catalogue of Birds (Sharpe, vol. IV), but a large proportion of those occurring in the Oriental Region are found in other regions also. In this family Celebes might have been expected to display strong Australian affinities; it has, however, none, since the two Australian genera, *Rhipidura* and *Gerygone*, occurring in Celebes pass on much further into the Oriental Region, and for this and other reasons their distribution can in most cases only be accounted for on the supposition of flight across sea-channels. *Monarcha* and *Myiagra* are links between Djampea, the Sangi Islands, etc. and the Australian Region, though the species, or subspecies have most likely got there by flight. *Zeocephus talautensis* has its nearest affinities in the Philippines, to which the genus was hitherto believed to be restricted. The remaining genera of Flycatchers in Celebes — not counting a migratory *Musci-*

capa — are typically Oriental, and do not occur in the Australian Region, though in some of the Lesser Sunda Islands. They are: *Siphia, Stoparola, Hypothymis, Muscicapula*, and *Culicicapa*; these are all absent on the eastern side of the Molucca Straits. The Long-tailed Flycatchers, *Terpsiphone*, of the Ethiopian and Indian Regions as far as the Lesser Sunda Islands, have not yet been found in Celebes.

GENUS MUSCICAPA L. after Briss.

The typical Flycatchers are small birds of plain plumage — chiefly brown above and streaked with brown below, or black and white; the wing is long, much longer than the tail, the secondaries about $^2/_5$ the wing-length, the first quill very minute, the second one long, as long or longer than the fifth; bill rather small, with scanty rictal and frontal bristles; tarsus shorter than the middle toe and claw, blackish in colour. The more typical species are Palaearctic and migratory.

120. MUSCICAPA GRISEOSTICTA (Swinh.).
Chinese Flycatcher.

a. Muscicapa hypogrammica (Gray), *(1)* Finsch, Neu Guinea 1865, 168; *(2)* W. Blasius, J. f. O. 1883, 115.
b. Butalis hypogrammica *(1)* Wald., Tr. Z. S. 1872, VIII, 66.
c. Butalis griseosticta (Swinh.), *(1)* David & Oust., Ois. Chine 1877, 122.
Muscicapa griseosticta *(1)* Sharpe, Cat. B. IV, 1879, 153; *(2)* Salvad., Orn. Pap. 1881, II, 80; *(3)* W. Blas., P. Z. S. 1882, 706; *(5)* Pleske, Bull. Ac. Petersb. 1884, 123; *(6)* Guillem., P. Z. S. 1885, 632; *(7)* W. Blas., Ornis 1888, 311; *(8)* Everett, J.Str.Br.R.A.S. 1889, 127; *(9)* Salvad., Agg. Orn. Pap. 1890, 81; *(10)* Whitehead, Ibis 1890, 49; *(11)* Styan, Ibis 1891, 322, 349; *(12)* Hartert, J. f. O. 1891, 294; *(13)* De La Touche, Ibis 1892, 408, 424; *(14)* Bourns & Worces., B. Menage Exp. 1894, 40; *(15)* Grant, Ibis 1895, 441; *(16)* M. & Wg., Abh. Mus. Dresd. 1895, Nr. 8, p. 8; *(17)* iid., ib. 1895, Nr. 9, p. 4; *(18)* iid., ib. 1896, Nr. 1, p. 9; *(19)* Hart., Nov. Zool. 1896, 156; *(20)* Grant, Ibis 1896, 540.
d. Muscicapa manillensis (Bp.) *(1)* Sharpe, Ibis 1888, 200.
"Monotaroda" or "Manudorio", Talaut, Nat. Coll.
"Burong pohon", Minahassa, iid.

For further synonymy and references see Sharpe *1*; Salvadori *2*.
Descriptions. Sharpe *1*; Salvadori *2*; W. Blasius *3*; David & Oust. c *1*.
Adult male. Above grey-brown (almost hair-brown: Ridgway), the feathers of the head above with darker middles, the wing-coverts with paler edgings, the inner secondaries with whitish edgings; remiges and terminal part of tail blackish brown; under parts white, streaked with grey-brown on the breast, sides, jugulum and sides of throat; supraloral region and around the eye whitish, in front of eye duskier; under wing-coverts dark fawn: "iris dark sepia; bill black, base of under mandible yellow; legs and feet black". P. & F. S.
Measurements. Wing 83—87 mm; tail c. 50; tarsus 14; middle toe and claw c. 15.5; bill

from nostril 8 (♂, Mt. Loka, N. Celebes, c. 1400 m, 8. X. 95: Sarasin Coll.; and others).

Female. Like the male in coloration.

Young (moulting). Differs from the adult in having a few (remaining) feathers on the scapulars each with a single large white subterminal spot, some of the upper tail-coverts tipped with white, the greater wing-coverts broadly edged with white: "iris dark sepia; legs black-brown; feet below yellowish; bill black, under mandible at the base yellow" (♂, Rurukan, N. Cel., 16. X. 94: Sarasin Coll.).

Nest and eggs. Unrecorded.

Distribution. China; Formosa; Philippine Is. (Salvad. 2, Bourns & Worcester *14*, Whitehead *d 1, 10*, Platen *7*, Everett *15, 20*); Talaut Islands (Nat. Coll. *17*); Celebes *(a 1)*; — N. Peninsula (P. & F. Sarasin *16*, Nat. Coll., Mus. Drosd.); S. Peninsula (P. & F. Sarasin *18*, Everett *19*); Morty, Halmahera, Batchian, Tidore, Amboina, Ceram, New Guinea (Salvad. *2*); Ternate (Fischer *5*); Waigiou and Mysol (Guillem. *6*).

A migratory Flycatcher — apparently the eastern representative of the European *M. grisola* — the breeding-grounds of which seem to be North China, the winter quarters the chain of East India Islands which form a coast-line to the Pacific. Abbé David remarks that it is very abundant in summer in all China, passing Pekin twice a year — in May and June and in August and September, and, as Mr. Styan observed, it passes through the Lower Yangtse Basin in May and August. It nests, therefore, apparently north of Pekin. Its descent into its winter quarters is probably made by way of Formosa and the intermediate small islands to the Philippines. Mr. Whitehead observed it to be "a winter migrant to Palawan, arriving about the 10th September", and we find no dates to show that any remain in the East Indies in summer[1]).

For a long time its right to be included in the Celebes list rested only upon the fact that the island is marked as a locality for it in Finsch's "Neu-Guinea", 168; whence Gray's similar indication seems to be drawn (Handl. I, 321); but from general reasons it was certain to occur there. Positive evidence was furnished by the Sarasins, who got seven specimens in the hill-country of the Minahassa in September; October and November, 1894, and one from Mt. Bonthain in October, 1895; Mr. Everett also got it on the foot-hills of Mt. Bonthain in 1895 sometime after September 28th, and a specimen was killed in March, 1895, and sent from the Minahassa by our native collectors. One specimen was obtained by the same in the Talaut Islands in November, 1894, but in the autumn of 1896 it seems to have visited these islands in greater force, 14 specimens having been then shot and sent to the Dresden Museum. Some of these are now at Tring.

---

[1]) Dr. Fischer's note, therefore, that it occurs all the year round on the Island of Ternate *(5)*, must be regarded as erroneous and misleading, and for similar reasons some of his observations on other species fall to the ground.

GENUS MUSCICAPULA Blyth.

These small Flycatchers differ from *Muscicapa* by their shorter, blunter wing, the second primary being shorter than the fifth, and the secondary quills relatively much longer than in that genus, being about ⁴/₅ the length of the wing; and the sexes are different in coloration. The species *M. hyperythra* stands perhaps nearer to the genus *Siphia* than it does to *M. westermanni*. The genus is Oriental.

121. MUSCICAPULA WESTERMANNI Sharpe.
Little Malay Pied Flycatcher.

Muscicapula westermanni *(1)* Sharpe, P. Z. S. 1888, 270; *(2)* id., Ibis 1888, 385; *(3)* id., Ibis 1889, 196; *(4)* Everett, J. Str. Br. R. A. S. 1889, 128; *(5)* Sharpe, Ibis 1890, 276, 286, 291; *(6)* Vorderm., N. T. Ned. Ind. LI, 1891, 389; *(7)* Grant, Ibis 1894, 506; *(8)* id., Ibis 1895, 442; *(9)* id., Ibis 1896, 464, 540; *(10)* M. & Wg., Abh. Mus. Dresden 1896, Nr. 1, p. 9; *(11)* Hart., Nov. Zool. 1896, 156, 541, 548, 561, 569, 595; *(12)* id., ib. 1897, 158.

Descriptions. Sharpe *1* (♀); Vorderman *6* (♂).

Adult male. Upper parts, face, and ear-coverts glossy black; a broad superciliary stripe extending to the sides of the nape white; inner greater wing-coverts and a broad outer edging on the three inner remiges white; the five lateral pairs of rectrices white at their base for about ¹/₂ their length in the outermost, increasing to about ³/₄ in the fifth pair, middle pair white at base only; chin, throat, and under parts white; thighs blackish; remiges below dusky, the inner edges, where they rest upon the body, whitish: "iris grey; feet blackish; bill black" — Doherty *12* (♂ ad. Erelompoa, N. W. from Loka, S. Celebes, c. 1300 m, 3. XI. 95: P. & F. Sarasin).

Female. Above dark bluish grey, with a slight tinge of brown on the head, stronger on the lower back and rump, inclining to russet on the upper tail-coverts; wings dusky with bistro-brown edgings to the feathers (paler on the greater coverts), lesser coverts bistre; tail-feathers brown, externally rufous-brown; lores and checks whitish, tinged with buff, on the ear-coverts passing into the grey of the upper parts; chin, throat, and under parts greyish white; thighs brown: "iris dark" (♀, Lompobatang, S. Cel., c. 2400 m, 6. XI. 95: P. & F. Sarasin).

Young (male). Remiges and tail black and white as in the adult male; remaining upper parts dull tawny, with black edges and bases to the feathers; below greyish white, the feathers of the breast and throat (faintly) barred with dusky (Bonthain Peak, S. Celebes, c. 1300 m: Sarasin Coll.).

Measurements (2 ♂♂, 1 ♀ ad. — South Cel.). Wing 54—55 mm; tail c. 40; tarsus c. 15; middle toe and claw c. 13; bill from nostril c. 7.

Nest and Eggs. "The nest was placed in a creeper in the big forest, at about 40 foot from the ground; it was quite a small pile of moss, deep, and lined with fine white roots, a very pretty bit of work, and contained one small fawn-coloured egg. They would most probably have laid two eggs, after the manner of most species in these latitudes" (Whitehead *3*).

Distribution. Tenasserim (fide Grant *7*); Perak (Wray *1*); Borneo (Whitehead *2, 3, 4, 5*); Philippines — Luzon and Negros (Whitehead *7, 8, 9*); S. Celebes — Bonthain

Mountains (P. & F. Sarasin *10*, Everett *11*, Doherty *12*); Java (Vorderman *6*, Doherty *11*); Bali (Doherty *11*); Lombok (Doherty and Everett *11*); Sumbawa (Doherty *11*).

Since its discovery in Perak in 1888 much has been learnt about this little species, and it is now known to range from Tenasserim to the Philippines, Celebes, and Sumbawa. It seems to inhabit only the high mountain-regions. In Java Doherty got it at 9000—10000 feet; in Borneo, according to Whitehead, it ranges from 4000—9000 feet; in Celebes it has as yet been found only on the lofty Peak of Bonthain and the mountains abutting on it. It was found here by the Sarasins and by Everett at about the same time, and that it breeds here in the latter half of the year is shown by quite young specimens in both collections.

It bears, sex for sex, some resemblance to *Lalage* which may be regarded as a distant relative of very large size. From *M. hyperythra*, its fellow-inhabitant of the hills of these parts, it differs not only by its coloration, but by its smaller feet and claws, which are blackish in colour, by its smaller first primary, and flatter bill. It is closely allied to *M. melanoleuca* (Hodgs.) of the Himalayas and E. India, the female of which, according to Sharpe and Grant, is much browner above, but the males appear to be exactly similar.

† 122. MUSCICAPULA HYPERYTHRA (Blyth).
Rufous-breasted Blue Flycatcher.
Plate XIII.

a. Dimorpha superciliaris (nec Jerd.), *(1)* Blyth, J. A. S. B. 1842, XI, 190.
b. Muscicapa hyperythra (1) Blyth, J. A. S. B. 1842, XI, 885.
Muscicapula hyperythra *(1)* Sharpe, Cat. B. IV, 1879, 206; *(2)* id., Ibis 1888, 385, *(3)* Everett, J. Str. Br. R. A. S. 1889, 127; *(4)* Sharpe, Ibis 1890, 276, 291; *(5)* id., Ibis 1893, 551; *(6)* Hose, t. c. 396; *(7)* Grant, Ibis 1894, 505; *(8)* M. & Wg., Abh. Mus. Dresd. 1895, Nr. 8, p. 9; *(9)* Hart., Nov. Zool. 1896, 156, 548, 561, 569, 595; *(10)* id., ib. 1897, 158.
c. Cyornis hyperythrus (1) Oates ed. Hume's Nests & Eggs Ind. 1890, II, 2; *(2)* id., Faun. Br. Ind. B. 1890, II, 15.
For further synonymy and references see Sharpe *1*, Oates *c 2*.
Descriptions. Sharpe *1*, Oates *c 2*.
Adult male. Above dark slaty blue; alula, remiges and tail blackish; supraloral stripe extending above the eye white; lores and base of forehead, malar region, angle of chin, and ear-coverts blackish, washed with slaty blue; throat and breast orange-rufous, paling into whitish on abdomen and under tail-coverts; thighs olive, washed with slaty blue; sides washed with olive and rufous; under wing-coverts whitish, edge of wing slaty blue (♂ ad., Masarang hills, N. Cel., 16. VII. 94: Sarasin Coll.).

"Legs and feet very pale silvery to fleshy pink, the terminal joints of the toes and the claws slightly brownish; bill black; iris deep brown" — Hume.

Male, scarcely adult. Like the adult male, but somewhat greyer above; the wings in changing plumage, for the most part resembling those of the female; the abdomen less clear whitish (Lompobatang, S. Cel. c. 2500 m; Oct. 1895: Sarasin Coll.).

Female. Above olive, tinged with grey, tail and wings externally bistre-brown; lores, base of forehead, around the eye, and throat isabelline, inclining to orange-rufous on the breast and under wing-coverts; abdomen whitish, olivaceous on the sides, flanks, and thighs: "bill black; iris dark brown; legs and feet reddish grey" ($\bigcirc$, summit-region of Mt. Klabat, N. Cel., end Sept. 1893: Sarasin Coll.).

Young. Above warm dark brown, blackish on head, the feathers everywhere with broad mesial streaks of tawny; wings and tail blackish brown with warm brown edgings; below tawny, whitish on throat and abdomen and under tail-coverts, the feathers on sides of throat, breast, sides and flanks margined with dusky: "bill above black-brown, tip yellow, below yellow and brown; iris dark brown; legs reddish yellow; feet yellow" ($\male$, summit-region of Mt. Klabat, 25. Sept. 1893: Sarasin Coll.).

Measurements.	Wing	Tail	Tarsus	Bill fr. nostr.
a. (Sarasin Coll.) ♂ ad. Mt. Masarang, N. Cel., 16. VII. 94	60	40	19	7
b. (Saras. C.) ♂ ad. Mt. Lokon, N. Cel., c. 1200 m, 1. VII. 94	60	38	18.5	—
c. (Sarasin Coll.) ♀, Mt. Klabat, end Sept. 93. . . .	57	34	18	6
d. (Sar. C.) ♂ vix ad. Lompobatang, S. Cel. c. 2500 m, Oct. 95	66	45	20	7

The female is evidently smaller than the male. The specimen from S. Celebes is very large. Oates' measurements of Indian specimens are equal to those of N. Celebes.

Nest and eggs. Nest a deep cup of moss and moss-roots, placed under the roots, etc. of a tree: eggs 4 or 3, pale greyish or brownish white, finely freckled and mottled, chiefly at the large end, with dingy brownish red: 17.3×11.2 mm (Hodgson c 1).

Distribution. India (Oates etc. c 2); Malay Peninsula (fide Sharpe 1); Sumatra (Wallace 1); Borneo (Whitehead 2, 3, Hose 6); W. Java (Wallace 1); Bali (Doherty 9); Lombok (Doherty and Everett 9); Sumbawa (Doherty 9); Celebes — North (P. & F. Sarasin 8), South (Everett 9, P. & F. S., Doherty 10).

This little Flycatcher is very like a *Siphia*, sex for sex. The white supra-loral stripe of the male is its most striking distinguishing character: it is much smaller than *Siphia* and has also a smaller bill. The credit of its discovery in Celebes belongs to the Sarasins, who got it at high elevations in the Minahassa in 1893—94. It was also found by Everett and Doherty and — in one young specimen — by the Sarasins on the lofty Bonthain Range in the south of the island. In India, Borneo, Bali, Lombok, and Sumbawa, it is known only from the mountains, or from mountainous countries; and it is evident that it is purely a hill-species. It is probably most nearly allied to *M. hodgsoni* (Verr.) of Indo-China, the male of which wants the white eyebrow.

GENUS SIPHIA Hdgs.

The Celebesian species of this genus are, when adult, blue or olive above, and chiefly orange-rufous below; some forms from other parts have the rufous confined to the region of the throat. The wing is longer than the tail, the secondaries about $^4/_5$ as long as the primaries, the second primary hardly exceeding them; the rictal bristles are well developed, reaching to within the terminal third of the culmen, the bill from the nostril about half as long as the tarsus, the tarsus about as long as the middle toe and claw. The sexes generally differ somewhat in coloration; the young are mottled. The genus ranges (cf. Sharpe) from India to Timor and the Philippines.

† 123. SIPHIA BANYUMAS (Horsf.).
Blue-and-rufous Flycatcher.
Plate XIV.

It has been shown by Mr. Hartert *(f 1)* that this form differs racially in Borneo and in Celebes, the females in the former country having white lores and the males showing some less striking divergences from Celebes birds. Hartert identifies the Borneo birds with those of Java; unfortunately the female of the latter seems never to have been described, and it is not yet certain that the Celebes birds are distinct from, and those of Borneo identical with, those of Java. Probably neither are quite the same as the latter, but, until this is known to be the case, it may be preferable not to commence to break up the species.

a. **Muscicapa banyumas** *(1)* Horsf., Tr. L. S. 1821, XIII, 146; *(II)* id., Zool. Researches in Java 1824, pl. 38.
b. **Muscicapa cantatrix** *(1)* Temm., Pl. Col. 1823, pl. 226 (♂ only, fide Sharpe).
c. **Niltava banyumas** *(1)* Gray, Gen. B. 1846, I, 264; *(2)* Brügg., Abh. Ver. Bremen 1876, V, 68.
d. **Cyornis banyumas** *(1)* Bernst., J. f. O. 1859, 265 (Nat. Tdschr. Ned. Ind. 1860, XXII, 19); *(2)* Wald., Tr. Z. S. 1872, VIII, 117 (?pt. not Borneo); *(3)* Salvad., Cat. Ucc. Borneo 1874, 130 (?pt.); *(4)* W. Blas., J. f. O. 1883, 137; *(5)* Vorderman, Nat. Tdschr. Ned. Ind. 1886, XLV, 362; *(6)* Tristr., Cat. Coll. B. 1889, 202 (?pt.).
Siphia banyumas *(1)* Sharpe, Cat. B. IV, 1879, 449 (?pt.); *(2)* Nicholson, Ibis 1882, 68; *?(3)* Everett, P. Z. S. 1889, 226; *?(4)* id., J. Str. Br. R. A. S. 1889, 132; *?(5)* Sharpe, Ibis 1890, 276; *(6)* Hartert, Kat. Senckenb. Mus. 1891, 96 (?pt.); *(6bis)* id., Ornis 1891, 120; *? (7)* Hose, Ibis 1893, 398; *(8)* Büttik., Zool. Erg. Weber's Reise in Ost-Ind. 1893, III, 277, 286; *(9)* M. & Wg., Abh. Mus. Dresd. 1895, Nr. 8, p. 9; *(10)* iid., ib. 1896, Nr. 1, pp. 5, 9.
e. **Siphia rufigula** (err.), *(1)* Meyer, Ibis 1879, 128.
f. **Siphia omissa** *(1)* Hartert, Nov. Zool. 1896, 71, 157, 171, 172; *(2)* id., ib. 1897, 158.
"Siongsiong", near Tondano, Nat. Coll.

For further synonymy and references compare Hartert *f 1*, and see Salvadori *d 3*; Sharpe *1*.
Figures and descriptions. Horsfield *a II*; Temminck *b 1*; Sharpe *1*; Vorderman *d 5*.
Adult male. Above, including tail and exposed edges of quills, China-blue on a dark ground, duskier on the crown; forehead, superciliary region, and lesser wing-coverts much brighter blue; lores, chin, malar region, sides of throat, and ear-coverts slaty black, washed with blue posteriorly; throat and under surface dark orange-rufous, paler on under wing-coverts and middle of abdomen; wings and tail below shining dusky, paler on the inner webs of the quills, washed with buff on the inner-

most ones (♂, Tomohon, N. Cel. 6. IV. 94, Sarasin Coll.). "Iris grey-brown; bill black; feet violet-grey" (Platen in Mus. Nehrkorn, Rurukan, ♀, 21. IV. 85).

Adult female. Above and below much as in the male, but the blue above duller, with a wash of olive on the head and neck; differs chiefly by not having the lores, orbital region, chin, and submalar region black or blue-black, but orange-rufous, like the under parts, only paler; ear-coverts olive-brown, with pale shafts, slightly washed with blue posteriorly (♀, Loka, S. Cel., 11. X. 95: Sarasin Coll.).

The colour of the lores is whiter (i. e. very little tinged with rufous) in some specimens than in others, it generally extends narrowly across where forehead and bill meet, but not so in one example (or two) of those consulted for this article.

Young. Above dark brown, each feather with a spot of tawny-cinnamon, narrowly bordered with black; on the head these spots take a more striate character; the rufous feathers of the under-surface fringed with black, giving a streaked appearance (Tondano, Aug.—Sept. 1892 — Nat. Coll.: C 10812).

Measurements.	Wing	Tail	Tarsus	Bill from nostr.
a. (Sarasin Coll.) ♂ ad., Tomohon, N. Cel., 6. IV. 94 .	74	54	19.5	9
b. (Sarasin Coll.) ♂ ad., Lake Posso, Centr.Cel., 12. II. 95	74	55	17.5	9.5
c. (Sarasin Coll.) ♂ ad., Loka, S. Cel., 6. X. 95 . . .	78	62	19	11
d. (C 10808) [♂] ad., Tondano, N. Cel., VIII.—IX. 92 .	75	56	17.5	9
e. (C 10809) [♂] ad., Tondano, N. Cel., VIII.—IX. 92 .	75	57	—	9.5
f. (Sarasin Coll.) ♀, Tomohon, N. Cel., 6. IV. 94 .	73	53	19	9.5
g. (Sarasin Coll.) ♀, Tomohon, N. Cel., 21. XI. 94 .	72	55	18	9
h. (Sarasin Coll.) ♀, Rurukan, N. Cel., 24. XI. 94 . .	73	55	18.5	10
i. (Sarasin Coll.) ♀, Loka, S. Cel., 11. X. 95 . . .	72	55	18	9.5
j. (C 10811) [♀], Tondano, N. Cel., VIII.—IX. 92 .	73	—	—	9.5
k. (C 10810) [♀], Tondano, N. Cel., VIII.—IX. 92 . .	74	55	—	10
l. (C 14916) ♂ ad., Saleyer, Nov. 95 (Everett).	74	57	19.5	9.5
m.(C 11065) [♂] ad., Java (v. Schierbrand) . .	76	63	18	—
n. (C 11066) [♂] ad., Java (v. Schierbrand) . .	77	61	18	10

Observation. Lord Walden's statement that Celebes examples are undistinguishable from those of Java is confirmed by the above series, so far at least as the males are concerned, except that the Javan birds seem to have less black on the chin and sides of the throat. They are not darker (but, if anything, paler) blue above, they do not show a tiny white spot on the throat, nor is there a different distribution of the colour on the throat — differences observed by Mr. Hartert in the *Siphia* from Borneo. The female of *S. banyumas* is stated by Sharpe and Hartert to have white lores, but this statement relates to the female from Borneo, and we are not aware that this is the case in the typical form from Java. The male *l* from Saleyer is paler below than those of the mainland; black of chin narrow.

Eggs. Java: 2 in number; dirty white, sometimes partaking of a yellowish, sometimes of a greenish tint, slightly glossy; sprinkled with dirty red spots which blend to some extent with the ground-colour, most plentiful at the large end which appears in consequence as if marbled with rusty; size 21 × 15 mm (Bernstein c I).

Nest. Composed chiefly of the horse-hair-like threads of the Areng palm, some few roots, a little moss, bits of leaf; well built, in shape a half-sphere. Twice found among the ferns and other parasites on the stem of an Areng palm, once in the crevice formed by the boughs of a very mossy tree (Bernstein c I).

Distribution. Java (Horsfield *a 1*, Bernstein *d 1*, etc.); Salcyer (Weber *8*, Everett *f 1*); Celebes: — South Peninsula (Weber *8*, P. & F. Sarasin *10*, Everett *f 1*), Central Celebes (P. & F. S. *10*), W. Celebes (Doherty *f 2*), N. Peninsula (Meyer *d 2*, Fischer *c 2*, etc.); ?Palawan (Everett *3*, *4*); ?Borneo (Doria & Beccari, etc. *d 3*, *4*); ?Labuan (Low *4*); ?Sumatra (Raffles *d 3*); ?Penang (Wallace *1*).

This Blue-backed Flycatcher appears to be a species which has recently extended its range, and in Celebes it may probably be regarded as a slightly modified colonist from Java.

It is related to *S. philippinensis* Sharpe of the Philippine Islands, a species which Lord Tweeddale, while noticing certain differences, did not venture to separate. In this form, as Dr. Sharpe shows (Tr. Linn. Soc. 1889, I, 324), the abdomen and under tail-coverts are white, instead of orange-rufous. In the allied *S. djampeana* the sexes are said by Mr. Hartert to be much alike and the ♀ hardly to differ from the ♂ of the Celebes birds except by the slightly larger bill and darker, almost entirely black, ear-coverts and malar region. *Siphia rufigula* (Wall.) of Celebes is also somewhat similar, but may easily be distinguished by its white abdomen and under tail-coverts and the absence of black on its chin. In some of his excellent field-notes on Javan birds Bernstein *(d 1)* writes as follows on the habits of *Siphia banyumas*: "It inhabits by preference the groves and shrubs around villages at some elevation not far from the hill-forests, as also the coffee plantations and the forests themselves, though it is seen far less plentifully in their depths than along their borders. On the plains on the contrary it belongs to the birds of rarer occurrence. Sitting upon a prominent bough it watches attentively for insects, which it catches very cleverly on the wing, and then, turning back to the perch it had left or to another bough, devours them. Its song is rather simple, and when Temminck *(b 1)* describes it as excellent he is not well informed". In Celebes it also seems to be a hill species; Platen sent it to Mr. Nehrkorn from Rurukan, 3000 feet; the Sarasins got it in the same neighbourhood, also at Lake Posso; they, Weber and Everett about Mt. Bonthain, Doherty at 4000 and 6000 ft. in W. Celebes. Our native collectors recently obtained a number of specimens near Tondano, 2000 ft. No specimens have as yet been recorded from the lowlands of the island. Hose obtained the Bornean race only at a height of 2000 feet on Mount Dulit, whereas *Siphia nigrogularis* "is the usual low-country form of Blue Flycatcher" in Borneo. *Siphia rufigula* of Celebes has, on the other hand, apparently been obtained only in the low country of Celebes.

The curious mottled plumage of the young corresponds with that of the young *Muscicapa grisola* of Europe. From this type the adult *Muscicapa* has departed less widely than the adult *Siphia*.

124. SIPHIA DJAMPEANA Hart.

Djampea Blue-and-rufous Flycatcher.

Plate XIV.

Siphia djampeana Hartert, Nov. Zool. 1896, 172.

Male. Differs from *S. banyumas* from Java, Celebes and Saleyer by having the upper throat white where it meets the black of the chin and submalar region, the black here almost untinged with blue and broader than in *S. banyumas* (♂ ad. Djampea, Dec. 1895: Everett, C. 14868).

Female. Like the male, but the white on the throat almost completely absent (♀ vix ad. Djampea, Dec. 95: Everett, C 14869).

Measurements. Wing 78—81 mm; tail 67—69; tarsus 18—19; culmen 17 (Hartert).

Distribution. Djampea Island between Celebes and Flores (Everett).

This is one of the numerous additions made by Mr. Everett to the avifauna of the Celebesian Province in the latter months of 1895. A good series of specimens was obtained. The most curious point about the bird seems to be that the female is much like the male, or still more like the male of *Siphia banyumas*, from which it is not easily distinguished. The female of *S. banyumas* is a simpler bird. Altogether *S. djampeana* seems to present a more advanced stage of evolution.

125. SIPHIA KALAOENSIS Hart.

Kalao Blue-and-rufous Flycatcher.

Plate XIV.

Siphia kalaoensis Hartert, Nov. Zool. 1896, 172.

Male. Above like *S. djampeana* and *banyumas*, from which it differs by having the breast and middle of the throat white. The abdomen is pale orange-rufous, whiter on the under tail-coverts. The black on the chin and sides of throat broad, and slightly tinged with blue (♂, Kalao, Dec. 1895: Everett, C 14899).

Female. "Like the male, except that the breast is strongly washed with orange-rufous and that the under tail-coverts are coloured like the abdomen. The female is, therefore, practically indistinguishable from the male of *S. djampeana*, but the breast is paler" (Hartert).

Measurements. Wing 76—78 mm; tail 65—68; tarsus 19—20; culmen 16—17. The female is a little smaller, wing 72 mm (Hartert). Mr. Hartert evidently measures the tail from its extreme base; we take it from the oil-gland, which gives a result of about 8 mm less for this species.

Distribution. Kalao Island between Celebes and Flores (Everett).

This is an interesting species. Mr. Hartert draws attention to the increase in white on the under parts of the Flycatchers from Celebes and Saleyer as they range to Djampea and Kalao. It is greatest in the male of Kalao, the female of the Kalao race is like the male of the Djampea race, the female of the

Djampea race like the male of that of Saleyer and the mainland of Celebes. The females of these birds seem to show a lower organisation in this genus, the males a more advanced stage; if this be so, *S. banyumas* is obviously indicated as the ancestral form of *S. djampeana* and the latter of *kalaoensis*. See, pp. 160—169, *Loriculus*.

+ *126. SIPHIA RUFIGULA (Wall.).
Lowland Blue-and-rufous Flycatcher.

a. **Cyornis rufigula** *(1)* Wall., P. Z. S. 1865, 476; *(II)* Wald., Tr. Z. S. 1872, VIII, 66, pl. VII, fig. 3.
Siphia rufigula *(1)* Blyth, Ibis 1866, 372; *(2)* Sharpe, Cat. B. IV, 1879, 454; *(3)* M. & Wg., Abh. Mus. Dresd. 1896, Nr. 1, p. 9; *(4)* Hart., Nov. Zool. 1897, 159.
b. **Niltava rufigula** *(1)* Gray, HL. 1869, I, 326, Nr. 4887; *(2)* Rosenb., Malay. Archip. 1878, 273.
Figure and descriptions. Walden *a II*; Wallace *a I*; Sharpe *2*.
Adult male. Dark ashy blue; throat and breast bright rufous, becoming pale on the belly and pure white on the under tail-coverts; front of the eye and ear-coverts blackish; under wing-coverts rufescent white; quills and tail-feathers dusky, ashy-margined; iris dark; bill black; foot nearly white (Wall. *a 1*).
Total length 132 mm; wing 69; tail 57; tarsus 16.5; culmen 11.4 (Sh. *2*).
Female. Very different from the male: brown above, and without the black face. Upper parts grey-olive, greyest on head, brightening into brownish chestnut on the upper tail-coverts and outer edges of the rectrices; wing-coverts and quills dusky with warmer brown tips and edges; lores, orbital region, chin, throat, and breast orango-rufous, paler on chin and throat; cheeks and ear-coverts darker, passing posteriorly into the grey-olive of the nape; abdomen and under tail-coverts white; sides and thighs olivaceous; under wing-coverts brown-buff: bill dark brown; legs, feet, and claws yellowish in the skin. Wing 60 mm; tail 40; tarsus 17; bill from nostril 8 (♀, Mapane, Centr. Cel. 28. II. 95: Sarasin Coll.).
Mr. Hartert was so good as to compare this specimen with those at Tring and to confirm its identity.
Distribution. Celebes: Minahassa — Manado (Wallace *a 1, 2*); Mapane, southern shore of the Gulf of Tomini (P. & F. Sarasin *3*); Macassar (Doherty *4*).

For thirty years a single male specimen of this species in the British Museum was the only one definitely known, then a female was obtained in Central Celebes by the cousins Sarasin, and latterly a male and a female in the Southern Peninsula by Doherty. About the same time a pair and a male of a bird much resembling the female of *S. rufigula* were sent by Everett and Doherty from the Peak of Bonthain to the Tring Museum, and Mr. Hartert soon became aware that he had to do with a lowland and a mountain species. The differences of *S. bonthaina* from the female of *S. rufigula* are recapitulated from Hartert's notes in the next article; the male of *S. rufigula* looks quite different by reason of its blackish lores and ear-coverts and dark ashy blue upper parts.

It appears to be nearly related to *S. philippinensis* Sh., a considerably larger form: wing 79 as against 69 mm (Sharpe). *S. banyumas* of Celebes may be distinguished by its entire under surface of orange-rufous and by its black chin and sides of throat.

Among the *Muscicapidae* there is no species connecting the Australian Region with Celebes, since Australian forms which reach Celebes pass on into the other Great Sunda Islands, the Philippines, or further. This is of interest from the fact that the Australian Region rivals Africa in the question as to which area possesses the greater number of peculiar *Muscicapidine* genera. Celebes viewed as Australasian ground might be expected to have a share of the peculiar Australian Flycatchers. On the other hand several Oriental genera connect Celebes with the countries to the west, without passing into the Moluccas on the east. *Siphia* is one of these, its range being from the Himalayas and East China to Celebes, Java, and Timor.

┼* 127. SIPHIA BONTHAINA Hart.
Mountain Flycatcher.

Siphia bonthaina *(1)* Hart., Nov. Zool. 1896, 157; *(2)* id., ib. 1897, 158.

Male. "Above olive, quills margined with rufous-brown on the outer webs and with light brown on the inner webs. Tail deep chestnut, more brownish on the tip; upper tail-coverts of the same colour. A large spot over the lores; from the base of the bill to the middle of the eye pale ochraceous. Chin, throat, and breast light ochraceous. Abdomen white, bases of feathers slate-colour. Under tail-coverts white with an ochraceous shade; under wing-coverts very pale brownish. L. t. about 110 mm; wing 65; tail 47" (Hartert *1*).

"Iris deep chestnut; feet slaty grey, soles pale reddish; bill black" (Doherty *2*).

Female. "Has the wing only 61 mm, the tail 45 mm, and the chin, throat, and breast are very much paler than in the male" (Hartert *1*).

Distribution. Mountains of Celebes: Bonthain, c. 6000 ft. (Everett, Doherty).

At the time of writing, the only specimens of this species known are three — two males and a female in the Tring Museum. They bear much resemblance to the female of the lowland *S. rufigula*, and may probably be regarded as representing a lowly organised species in respect of coloration. Mr. Hartert, who has carefully established the validity of this form, writes in lit.: "The upper side of *S. bonthaina* is olive (without any greyish tint in it), while it is olive-grey in the female of *rufigula*. The tail of *S. bonthaina* is chestnut throughout, while it is deep brown with chestnut outer margins in *S. rufigula* ♀". He adds *(2)*: "The wing of *S. rufigula* is decidedly shorter, measuring only 57 mm[1])". The tarsus of *S. rufigula* is much shorter 16 mm (19 to 20 in *S. bonthaina*).

---

[1]) In the Sarasins' example 60 mm.

GENUS STOPAROLA Blyth.

These Flycatchers are distinguishable by their grey-blue or verditer coloration, the males differing from the females by their black lores. The wing formula is very like that of *Siphia*, from which it differs chiefly by the bill, which is shorter, though equally broad at the base, and the tarsus is also relatively a little shorter. The plumage of the rump and flanks seems to be of a softer and thicker character. The male has a sweet song (Legge, Sarasins). The genus is found from Afghanistan to the Philippines, Celebes, and Java.

+*128. STOPAROLA SEPTENTRIONALIS Bütt.

North Celebesian Blue Flycatcher.

Plate XV.

Stoparola septentrionalis *(1)* Büttik., Notes Leyd. Mus. 1893, 169; *(2)* M. & Wg., Abh. Mus. Dresden 1895, Nr. 8, p. 9; *(3)* iid., ib. 1896, Nr. 1, p. 5.

Description. Büttikofer *l.*

Male. Verditer-blue, paling into whitish on the abdomen and white on the under tail-coverts. Forehead and superciliary stripe, and upper throat bright pale blue; lores jet-black; angle of chin at the gonys blackish; remiges and tail dusky, washed externally with the verditer-blue of the back; wing and tail below dusky, the under wing-coverts (except the outer ones) and the inner edges of the remiges white: bill black, legs and feet dark in skin (♂, Mt. Lokon, N. Cel., 1. VII. 94: Sarasin Coll.).

Female. Like the male, but without the black lores (though the feathers here have black bases), the bright superciliary stripe hardly at all pronounced, and there is less bright light blue on the forehead and upper throat; the blue on the upper surface, throat, and breast a shade darker and duller (♀, Tomohon, 4. VI. 94: Sarasin Coll.).

Nestling. Remiges and rectrices (half-grown) as in the adult; head and neck blackish, each feather with a tawny subterminal spot; remaining upper parts dull bluish grey with duller tawny spots; under parts whitish washed with tawny, brighter tawny on breast, all the feathers with U-shaped margins of blackish: bill yellowish, browner above (♂ ?, Tomohon, 13. XI. 94: Sarasin Coll.).

Measurements.

	Wing	Tail	Tarsus	Bill from nostril
a. (Sarasin Coll.) ♂ ad. Tomohon, 3. V. 94	70	53	16	8
b. (C 13898) ♂ ad. Tomohon, 6. V. 94 (P. & F. S.)	71	51	16	8
c. (Sarasin Coll.) ♂ ad. Mt. Lokon, 1. VII. 94	74	55	16	—
d. (Sarasin Coll.) ♂ ad. Mt. Masarang, 25. IV. 94	73	54	—	7.5
e. (Sarasin Coll.) ♀ Tomohon, 8. IV. 94	71	52	15.5	8
f. (Sarasin Coll.) ♀ Tomohon, 4. VI. 94	70	50	—	7.5
g. (Sarasin Coll.) ♀ Tomohon, 18. X. 94	69	48	—	7.5

Nest. Of small roots, moss, and fibres, lined with hair-like vegetable fibres; stated by the finder to have been situated in grass: in shape a flat hemisphere, the cup 45 mm diam. by 35 deep, the walls about 35 thick (see plate: Tomohon, 8. June 1894: Sarasin Coll.).

Distribution. North Celebes — Minahassa, Tondano (von Rosenberg *I*), Manado District (v. Duivenbode *I*), Tomohon, Mt. Masarang and Mt. Lokon (P. & F. Sarasin).

Mr. Büttikofer remarks that as to colour this species may be best compared with *S. melanops* (Vig.) of India, from which it differs in having the inner web of the tail-feathers black instead of blue, and in its size being much smaller. A note of von Rosenberg's says it is "frequently seen in brushwood and low trees, living upon insects".

With this species and the next was made the interesting addition of another Oriental Flycatcher-genus to the avifauna of Celebes, a *Muscicapidine* form not known in the Australian Region. The genus ranges from the Himalayas as far as Afghanistan, east to South China, the Philippines, Celebes, and Java.

The Sarasins have added much to what is known of this species. They obtained a nice series of specimens, with the sex properly ascertained, showing the difference between the male and the female; also a nest containing the two nestlings shown in the plate. Like its near relation, *S. meridionalis* Büttik., of South Celebes, it appears to be an inhabitant of the hill-country.

✦ * 129. STOPAROLA MERIDIONALIS Bütt.

South Celebesian Blue Flycatcher.

Stoparola meridionalis *(1)* Büttik., Notes Leyden Mus. 1893, XV, 170; *(2)* id., Zool. Erg. Weber's Reise in Ost-Ind. 1893, III, 278; *(3)* Hart., Nov. Zool. 1896, 158; *(4)* id., ib. 1897, 158.

Male. Similar to *S. septentrionalis*, but larger, of a darker and duller blue, especially on the throat and breast; the bright light blue on the forehead, superciliary region and upper throat less extensive: "iris deep chestnut; feet black; beak black" (Doherty *4*); (♂, c. 6000 ft. on Lompobatang, S. Cel., 14. X. 95, Sarasin Coll.).

Female. Similar to the male, but the lores not black, being blackish with blue tips to the feathers; size a little smaller (♀, Bonthain Peak, 6000 ft., Oct. 95: Everett — C 14887).

Young. Above tawny, with black edges to the feathers and blackish bases; wings and tail blackish washed with verditer-blue, as in the adult, wing-coverts tipped with tawny; under parts tawny on breast, paler on throat and abdomen, with marginal U-shaped bars of blackish, almost completely absent on under tail-coverts; under wing-coverts buff-white (♂ juv., 1200 m, near Loka, S. Cel., 13. X. 95; Sarasin Coll.).

Measurements.	Wing	Tail	Tarsus	Bill from nostril
a. (Sarasin Coll.) ♂, Lompobatang, 14. X. 95	79	66	17	8
b. (Sarasin Coll.) ♂, Loka, 11. X. 95	81	60	17	8
c. (C 14886) ♂, Bonthain Peak, X. 95 (Everett)	81	62	17	—
d. (C 14887) ♀, Bonthain Peak, X. 95 (Everett)	76	55	17	8
e. (Sarasin Coll.) ♂ juv. near Loka, 13. X. 95	77	56	—	7

Distribution. South Celebes: Macassar District (Teijsmann *1*); Mt. Bonthain and its neighbourhood (Weber *2*, Everett *3*, P. & F. Sarasin, Doherty *4*).

This species is as yet definitely known only from high elevations, 4000 to 10000 feet, on the Bonthain mass of mountains. The Drs. Sarasin write:

"We heard the bird singing on the topmost point of Lompobatang, on the tallest tree in South Celebes. The song is very melodious". It and *S. septentrionalis* are evidently only geographical races.

GENUS HYPOTHYMIS Boie.

The Blue Flycatchers may be distinguished from all the foregoing by their having the tail as long as the wings; the wing is more rounded, the second primary shorter than the secondaries, which are about $^{7}/_{8}$ the length of the wing, the 4th, 5th and 6th primaries the longest. The bill is moderately large, the culmen from the frontal suture almost as long as the cranium; the rictal bristles large and strong, the longest reaching nearly as far as the tip of the bill. The genus is found from the Himalayas as far as the Philippines, Celebes, Sula, and the Lesser Sunda Islands.

✶ 130. HYPOTHYMIS PUELLA (Wall.).

Long-tailed Blue Flycatcher.

a. **Muscicapa coerulea** *(1)* S. Müll. (nec Gm.), Verh. Natuurk. Comm. 1839—43, 91; *(2)* id., Reiz. Ind. Archip. 1858, II, 15.
b. **Myiagra puella** *(1)* Wall., P. Z. S. 1862, 340 (ex Gray MS.); *(2)* Gray, HL. 1872, I, 328, Nrs. 4931, 4932; *(3)* Brügg., Abh. Ver. Bremen 1876, V, 68; *(4)* Rosenb., Malay. Archip. 1878, 273.
Hypothymis puella *(1)* Wald., Tr. Z. S. 1872, VIII, 66, pl. VII, fig. 2; *(2)* Salvad., Ann. Mus. Civ. Gen. 1875, VII, 656; *(3)* Lenz, J. f. O. 1877, 373; *(4)* Sharpe, Cat. B. IV, 1879, 277; *(5)* Meyer, Ibis 1879, 128, 146; *(6)* W. Blas., J. f. O. 1883, 117, 137; *(7)* Meyer, Isis, Dresden 1884, 25; *(8)* Blas., Ztschr. ges. Orn. 1885, 278; 1886, 112; *(9)* Guillem., P. Z. S. 1885, 553; *(10)* Tristr., Cat. Coll. B. 1889, 197 (Sula Is.); *(11)* Heine & Rchnw., Nomencl. Mus. Hein. 1890, 35; *(12)* Büttik., Zool. Erg. Weber's Reise Ost-Ind. 1893, III, 278; *(XIIbis)* Meyer, Vogelskel. I, 1894, t. CCIII; *(13)* M. & Wg., Abh. Mus. Dresd. 1895, Nr. 8, p. 9; *(14)* iid., ib. 1896, Nr. 1, p. 9; *(15)* iid., ib. 1896, Nr. 2, p. 14; *(16)* Hart., Nov. Zool. 1896, 157; *(17)* id., ib. 1897, 157, 162.
c. **Myiagra azurea** *(1)* Pelz. (nec Bodd.), Verh. z.-b. Ges. Wien 1873, April (fide Blas. 6).
"**Rui**" (Meyer 5) or "**Ruirui**" (Nat. Coll.), Minahassa — a name taken from its call-note.
"**Ting-kuikui**", Tonkean, E. Celebes (N.C.).
"**Tangkui**", Banggai (N.C.).
Figures and descriptions. Walden *1*; Meyer *XIIbis* (skeleton); Wallace *b 1*; Brüggemann *b 3*; Sharpe *4*.
Adult male. Campanula-blue; brightest on the head, lores, chin and lesser wing-coverts, duller and greyer on lower back and breast, passing almost into olive-grey on the abdomen, flanks and under tail-coverts; exposed webs of wings and tail blue like the back, below dusky olive-grey; on forehead and chin at base of bill a narrow edge of black; under wing-coverts whitish (Kema, ♂, Febr. 1873: C 743, Meyer).
Iris blue (Meyer 5, Platen 8) — eyelids blue, iris dark (Wall. *b 1*) — iris brown (Guillem. *9* — the differences in the statements of the collectors may be of a

sexual nature); feet blue, like the belly; claws black; bill blue, like the head; even the bones are blue (M. δ).

Female. Like the male, but somewhat greyer and duller (Manado — C 755).
Nestling. Grey; belly and breast white, head grey (Meyer δ).
Measurements (23 specimens — Manado, Togian Is., Banka, Manado tua, Mantehage). Wing 67—77 (Manado, C 742 and C 744); tail 74—84: tarsus 16—16.5; bill from nostril 8.7—9.5. Average length of wing 72—73 mm.

Skeleton.

Length of cranium	35.0 mm	Length of tarso-metatarsus	17.5 mm
Greatest breadth of cranium	15.0 »	Length of digitus I	12.0 »
Length of humerus	18.0 »	Length of digitus II	10.0 »
Length of ulna	22.0 »	Length of digitus III	15.0 »
Length of radius	20.0 »	Length of digitus IV	12.0 »
Length of manus	18.0 »	Length of sternum	19.0 »
Length of metacarpus	10.0 »	Greatest breadth of sternum	12.5 »
Length of digitus I	3.0 »	Height of crista sterni	6.4 »
Length of digitus II	8.5 »	Length of coracoideum	16.0 »
Length of digitus III	2.5 »	Length of scapula	20.0 »
Length of femur	15.5 »	Length of clavicula	15.0 »
Length of tibia	24.0 »	Length of pelvis	12.0 »
Length of fibula	9.0 »	Greatest breadth of pelvis	15.0 »

Eggs. Unrecorded.
Nest. Cup-shaped, the size of a Chaffinch's, rather loosely built of moss and coarse straw-like strips of dead grasses or wood, ornamented externally with a few large seeds, lined with fine root-fibres. Placed on the twigs of a tree. Height 50, breadth 65, breadth of pocket 52, depth of pocket 38 mm (Manado, Nr. 83: Meyer).
Breeding season. This nest was taken in March, 1871, and contained two nestlings. The number of eggs is, therefore, probably only two, which were laid in this case in February.
Distribution. Celebes: — Minahassa (Wall. 4, M. 5, etc.); Manado tua, Mantehage, Lembeh, Banka off the Minahassa (Nat. Coll. in Dresd. Mus.); Hill-forests between the Minahassa and Mongondo (P. & F. Sarasin); Gorontalo Distr. (Meyer δ); Togian Islands (Meyer δ); Kandari, S. E. Celebes (Beccari 2); W. Celebes (Doherty 17); Contr. Celebes, Mapane (P. & F. S. 14); E. Celebes (N. C.); small island off Buton Id. (S. Müller a 1); S. Peninsula (Wallace 4, Platen 8, etc.); Palopo at head of Gulf of Boni (Weber 12); Peling and Banggai Id. (N. C.); Sula Islands (Allen b 1, 4, 11).

The common Blue Flycatcher of Celebes and Sula is a very distinct form most nearly allied to *H. rowleyi* (Meyer) of Great Sangi. *H. rowleyi* differs in having the upper parts much darker blue and the under surface uniform bluish cinereous, and in its much larger size. *H. occipitalis* (Vig.), the range of which is given by Sharpe (Cat. B. IV, 276; Ibis 1890, 276) as from Tenasserim to Flores, Borneo, and the Philippines, is easily distinguishable by its black pectoral collar and black nuchal patch; this form appears to be only subspecifically distinct from *H. azurea* (Bodd.) of Burmah and India. From its well marked differences *H. puella* might be regarded as a rather ancient inhabitant of Celebes.

The genus *Hypothymis* forms another connecting-band between Celebes and the Oriental Region, no Flycatcher of this genus being known in Australasia, not counting the Lesser Sunda Islands. *H. puella* is further of interest as an inhabitant of both Celebes and Sula. The type of the species is from Sula, and the birds from these islands are said by Sharpe to be "a little deeper azure-blue than the Celebes birds", but neither he nor Wallace find them separable. Neither are there any apparent differences in those from Peling and Banggai, nor are any differences to be seen in specimens from the islands off the coast of the Minahassa and in the Gulf of Tomini.

+ * 131. HYPOTHYMIS ROWLEYI (A. B. M.).
Sangi Blue Flycatcher.

a. **Zeocephus rowleyi** *(1)* Meyer, Rowley's Ornithol. Miscellan. 1878, III, 163; *(2)* id., Isis, Dresden 1884, 6.
Hypothymis rowleyi *(1)* Sharpe, Cat. B. IV, 1879, 278; *(II)* Gould, B. New Guinea II, pl. 20 (1882); *(3)* W. Blas., Ornis 1888, 581.
Figure and descriptions. Gould *II*; Meyer *a 1*; Sharpe *1*.
Adult. Above dusky China-blue, brightest on supra-loral region, lesser wing-coverts and mantle; the quills washed externally with greyer blue; under parts from chin downwards bluish French-grey, darker on the sides of the breast, paler and washed with buff about the anal region; quills and tail below dusky smoke-grey, the inner webs of the quills whitish. Bill — in dry specimen — blackish, under mandible paler; feet and claws greyish. Wing 96 mm; tail 92; tarsus 21; bill from nostril 9.5 (Tabukan, Great Sangi — Nr. 2956, type: Meyer).
Distribution. Great Sangi (Meyer).
Remarks. The type of this Flycatcher in the Dresden Museum remains up to the present the only specimen on record. It is much larger than *H. puella* of Celebes: wing 96 mm as against 77 maximum; the upper surface is of a much darker blue, and this colour is not carried on to the chin, throat and chest, which are uniform with the rest of the under surface. It differs, moreover, in the form of the bill from *H. puella*, the nostrils being more deeply sunk in larger cavities, between which the ridge of the culmen stands up more strongly. In this respect, it appears to be a connecting link between *Hypothymis* and *Zeocephus cyanescens* Sharpe of Palawan, a species a good deal similar in coloration; but distinguishable by its black lores and black line along the forehead and chin at the base of the bill, by its larger bill and longer tail.

The type of *H. rowleyi* was found by one of Meyer's hunters near the village of Tabukan on Great Sangi. Up to the present no Blue Flycatcher has been discovered on Siao or the other Sangi Islands, and it is probable that the nearest relations of *H. rowleyi* may still be found — there, or elsewhere. At present its nearest affinities appear to be with *H. puella* of Celebes and Sula, while *Stoparola panayensis* also deserves consideration as an ally.

Hypothymis manadensis (Q. G.). This species, as Oustalet has shown (Bull. Soc. Philom. Paris, Dec. 1877), is from New Guinea — not Celebes. It is the same as *Monarcha dichrous* Gray (cf. Sharpe, Cat. B. IV, 273, 421; Salvadori, Orn. Pap. II, 29).

GENUS RHIPIDURA Vig. Horsf.

The tail of the Fan-tailed Flycatchers is much longer than the wing, graduated, the feathers broad; the 4th and 5th remiges are longest, the second about as long as the secondaries; the bill is moderate, thickly beset with rictal and frontal bristles, reaching nearly to the tip of the bill; the tarsus is (always?) longer than the middle toe and claw. Büttikofer (1893) recognises 75 species, ranging from Australia and New Zealand as far as the Himalayas and many islands of Polynesia, as well as throughout the East India Archipelago.

+ * 132. RHIPIDURA CELEBENSIS Bütt.
Southern Fan-tailed Flycatcher

Rhipidura celebensis *(1)* Büttik., Notes Leyden Mus. 1893, XV, 79; *(2)* Hart., Nov. Zool. 1896, 167, 173, 182, 585.

Adult [male]. Crown, hind neck, fore part of mantle, lesser wing-coverts, and thighs dark earthy brown; quills and greater wing-coverts sepia-brown, edged with the colour of the crown; remaining upper surface, including forehead, rufous; tail blackish, exposed webs iron-grey, basal part — except on outermost pair — red, tips white, broadest on outermost pair, on the outer webs of which the white runs some distance towards the base; lores, sides of face, throat and chest black, the chest bordered below with white-tipped feathers; malar streak from chin to sides of chest white; chin white; remaining under-parts white, tinged with fulvous on flanks, vent, and under tail-coverts; inner edge of quills below ashy whitish. Bill dark brown, white at base of lower mandible, feet dark brown. Wing 66 mm; middle tail-feathers 85, outermost 62; tarsus 20; culmen 13 (ex Büttik.). "Iris dark brown; bill dark brown; mandible ochreous dark brown towards the tip; legs pale grey" (Everett *2*).

Distribution. ? South Peninsula, Celebes — Macassar (Teijsmann); Djampea and Kalao (Everett *2*).

This Flycatcher was described from a single specimen in the Leyden Museum. Mr. Büttikofer remarks that it is "very closely allied to *R. semicollaris* M. & S. from the Timor group, but easily distinguished by the darker tinge of the earthy brown parts of the upper surface and the broader black band across the lower throat and chest". In *R. teijsmanni* of South Celebes the tail is cinnamon-red, with the terminal third sepia-brown, and across the chest there is only a narrow bar of black, not bordered with white below.

The nearest ally of *R. celebensis* at present known is *R. sumbensis* Hart., to which Mr. Hartert allows only subspecific rank, though he has not shown that the two forms intergrade. In three skins of the latter no differences of colour could be detected by Mr. Hartert, but they were not well prepared specimens; they proved to be considerably larger in size. Mr. Hartert queries the label "Macassar" indicated as the locality of Mr. Büttikofer's type, and it is indeed strange, if the bird occurs there, that it should have escaped Wallace,

Meyer, Platen, Weber, the Sarasins, Everett, Doherty etc. Teijsmann visited Saleyer, but we do not know that he was ever on Djampea or Kalao.
Mr. Hartert's male from Kalao was found to be identical with Teijsmann's specimens by Mr. Büttikofer.

+ *133. RHIPIDURA TEIJSMANNI Bütt.

Teijsman's Fan-tailed Flycatcher.

Rhipidura teysmanni *(1)* Büttik., Notes Leyden Mus. 1893, XV, 80; *(2)* id., Zool. Erg. Weber's Reise in Ost-Ind. 1893, III, 278; *(3)* Hart., Nov. Zool. 1896, 157; *(4)* id., ib. 1897, 158.

Description. Büttikofer *1*.

Adult. Crown, sides of head, neck and upper part of mantle olive-brown; forehead, back, rump, flanks, thighs, upper and under tail-coverts and basal two-thirds of all the tail-feathers cinnamon-red; terminal third of tail sepia-brown, fringed towards the tip with cinnamon-red, and tipped with ashy fulvous, most broadly on the outermost feathers; upper wing-coverts sepia-brown, edged with olive-brown; primaries sepia-brown, secondaries olive-brown, the exposed webs of all, except first primary, fringed with cinnamon; chin and upper throat pure white; a rather narrow black bar across the chest; under parts cinnamon; pale whitish fulvous on middle of breast and abdomen; under wing-coverts fulvous; quills below very broadly edged on the inside with vinaceous. Bill blackish, whitish at base; feet pale brown. Wing 69 mm; middle tail-feathers 80, outermost 60; tarsus 19; bill from front 12 (ex Büttik. *1*). "Iris very dark chestnut-brown; feet pale purplish; beak blackish, pale at base of mandible" (Doherty *4*).

An adult male differs from Mr. Büttikofer's description by having the breast below the black jugular collar greyish olive, in the middle inclining to buff, and only about the terminal fourth of the tail is dull sepia or blackish (♂, Loka, S. Cel., 6. X. 95: P. & F. Sarasin).

Young. More suffused with cinnamon-red than the adult, especially on the outer edges of the wings and on the flanks. Black jugular collar absent; the cinnamon-red loral patch nearly absent; under parts darker cinnamon, stained with brown on the breast; chin and throat greyish white (♂ Loka, 10. XI. 95: P. & F. S.).

Distribution. South Peninsula of Celebes — (?) Macassar (Teijsmann *1*), Mt. Bonthain (Weber *2*, Everett *3*, P. & F. Sarasin, Doherty *4*).

This Flycatcher was recently described by Mr. Büttikofer after a single specimen obtained by Teijsmann and labelled at Macassar (but in all probability from the mountains); and two others killed at Loka, 4000 feet, were subsequently sent to the Leyden Museum by Prof. Weber. In the Loka neighbourhood it was found by the Sarasins, Everett and Doherty. It appears to be a very distinct species. Mr. Büttikofer remarks that "*R. rufifrons* from Australia may be considered its nearest ally. From this latter as well as from the other species of the group [with forehead, back, and base of tail cinnamon-red], it differs principally in the red of the basal part of the tail being much more widely distributed, fully occupying the basal two-thirds and being as plainly visible on the under surface as on the upper, while in all the other species the tail, when closed, hardly shows any red region beyond the under tail-coverts".

In his careful description Mr. Büttikofer remarks: "The shafts of the tail-feathers have the color of the accompanying parts of their webs, but the red of the basal part runs, though not very far, into the black terminal third". This seems at first sight to show that the red pigment is on the increase and is supplanting the black. Were the black, on the other hand, to encroach into the red it might be inferred with some degree of reason that the extent of red on the tail had once been greater than at present. But it may be that a contradiction to this is presented by the young, which is more suffused with cinnamon-red, and this colour is seen all along the shaft to the tip of the tail and dimly against the blackish web alongside the shaft for its terminal fourth. The young plumage of birds is often supposed to be more ancient than the adult plumage; if so, the tail of *R. teijsmanni* is increasing in blackness on its distal end, and the cinnamon is receding towards the base. But such questions are hardly for the present generation.

The genus *Rhipidura* is unquestionably Australasian in type. Of Mr. Büttikofer's 75 species 62 belong to the Australian Region, including the Lesser Sunda Islands. The remaining range from the Philippines, Celebes and the other Great Sunda Islands to the Himalayas. The Papuan Islands — 24 species, and the Lesser Sunda Islands — 10 species — are by far the richest in forms, and it is here, possibly, that the genus originated. We infer that its distribution took place mainly, if not entirely by flight, from the fact that numerous species are found in the volcanic islands of Central and North-west Polynesia, and that of two Indian species Mr. Büttikofer places one between species of New Guinea and Fiji, the other between species of New Guinea and the Solomon Islands. Consequently the genus cannot be taken into consideration on the question of the former distribution of land and water in the East Indies, nor can the two Celebesian species be regarded as trustworthy links between Celebes, Timor and Australia, though like *Cacatua sulphurea* and *Circus assimilis* they probably reached Celebes from those countries.

On *R. teijsmanni* the Sarasins write: "It is very plentiful on the whole Peak of Bonthain up to great elevations, 1500 m and more, and indeed a characteristic bird. It flies in pairs and quietly allows itself to be watched. The tail is carried spread out, like a fan".

Rhipidura lenzi W: Blas. The habitat of this species was originally indicated by Lenz (J. f. O. 1877, 374) as North Celebes, and in 1883 it was found to be new and was described by W. Blasius (J. f. O. 1883, 145). The correctness of the habitat was questioned by Meyer (Isis 1884, 26), and the same year the species was recorded by H. O. Forbes from Amboina (P. Z. S. 1884, 431), and later again mentioned from the same island by Büttikofer (Notes Leyden Mus. 1893, XV, 92). These two Amboina specimens are somewhat smaller than W. Blasius's type, and it may be doubted where that bird really came from, though Amboina or the neighbourhood are the most probable localities (cf. also Salvad., Agg. Orn. Pap. 1890, 77).

GENUS ZEOCEPHUS Bp.

A Philippine genus of Flycatchers, now known also from Talaut. The bill is very large, the culmen from the cranial suture about as long as the cranium, and longer than the tarsus, the rictal bristles large; wing rather long, the second quill about as long as the secondaries: the tail long, as long or longer than the wings, the two middle feathers more or less produced. The Talaut bird and one of the Philippine species are of a cinnamon-rufous colour, another form is greyish blue. The feathers of the head are rather short and velvety.

✦ * 134. ZEOCEPHUS TALAUTENSIS M. & Wg.

Talaut Ferruginous Flycatcher.

Plate XVI.

Zeocephus talautensis *(1)* M. & Wg., J. f. O. 1894, 243; *(2)* iid., Abh. Mus. Dresd. 1895, Nr. 9, p. 4.

"Tabahee", Talaut, Nat. Coll.

Adult. Above and below deep orange-ferruginous, somewhat the darkest on the under surface; tail duller; chest, and ear-coverts tinged with chestnut; inner webs of quills dusky for about their terminal half. Bill and feet in the skin leaden black (type, Salibabu, "♂" 28. Oct. 1893, Nat. Coll. — C 13162).

Younger. Like the adult, but below much paler; the lower breast and abdomen whitish; the head above washed with brown; tail-feathers darker, passing into dusky for about the terminal 10 mm of the 3 middle pairs, shafts dark brown. Under bill greyish horncolour (Kabruang, 11. XI. 93 — C 13160).

Measurements. Wing 88—95 mm; tail 82—95; bill from nostril 14.5—15.5; tarsus 17 mm c.

Distribution. Talaut Is. — Kabruang, Salibabu and Karkellang (Nat. Coll. in Dresd. and Tring Museums).

Numerous examples of this Flycatcher were obtained by our native collectors in Talaut on three expeditions in 1893, 1894 and 1896. They were killed in the autumn or late autumn, and at this time, when the rainy season is commencing, the birds moult. *Zeocephus* has hitherto been known only as a Philippine genus, and *Z. talautensis* speaks for the Philippine character of its habitat, also shown by other species; though again other Talaut species have their nearest affinities in Sangi and elsewhere.

The present species may be distinguished from *Z. rufus* (Gray) of the Philippines — Luzon, Panay, Negros, Marinduque, Mindoro, Mindanao, Basilan,

Sooloo, Tablas, Romblon, Sibuyan, Cebu (Steere, Platen, Everett, Bourns & Worcester, etc.) — by its having, when adult, the two middle tail-feathers produced only about 5 mm beyond the others, instead of about 35 mm, and by its brighter and more orange plumage. The paler immature bird, which we have received from Salibabu as well as from Kabruang, resembles the young of *Z. rufus* (separated as *Z. cinnamomeus* by Sharpe), but differs, as shown *(1)*, in certain points.

GENUS MONARCHA Vig. Horsf.

Structurally similar to *Zeocephus*, but the tail much shorter than the wing, and the middle pair of rectrices not lengthened, the bill somewhat smaller, the culmen from the suture about as long as the tarsus and barely as long as the cranium. The genus is found from Australia as far as Talaut, Sangi, Sula, Djampea, and Timor.

+* 135. MONARCHA COMMUTATUS Brügg.
Sangi Grey-and-rufous Flycatcher.
Plate XVI.

a. **Monarcha commutata** *(1)* Brüggem., Abh. Ver. Bremen 1876, V. 68.
b. **Monarcha inornatus** partim *(1)* Salvad., Orn. Pap. 1881, II, 14 (Celebes).
Monarcha commutatus *(1)* W. Blas., J. f. O. 1883, 120, 156, 161; *(2)* Meyer, Isis, Dresden 1884, 6, 22; *(3)* id., Ztschr. ges. Orn. 1886, 24; *(4)* W. Blas., Ornis 1888, 580, 641; *(5)* M. & Wg., Abh. Mus. Dresd. 1896, Nr. 2, p. 14.
"Tarumisi kaooraneng", Siao, Nat. Coll.
Descriptions. W. Blasius *1*; Meyer *2*; Brüggemann *a 1.*
Adult. Upper surface, throat and upper chest slate-grey, slightly washed with olive-grey; remaining under parts dark orange-rufous; forehead and chin at the immediate base of bill blackish; wings and tail dusky, the exposed parts of the feathers washed with the colour of the upper surface. Bill lead-colour with patches of silvery, tips whitish (Siao, 25. VI. 93: Nat. Coll. — C 12623).
Young. Wings and tail fulvescent; front and chin less black; fore-neck below sensibly tinted with cinnamon (Meyer *2* — Great Sangi). It is not impossible that this may be racially distinct from the Siao form.

Measurements.	Wing	Tail	Tarsus	Bill from nostril
a. (Darmst. Mus. type ex *2*) ad. N. Celebes?	86	79	17—18	—
b. (Dresden Mus.) ad. Siao	87	80	18	—
c. (C 12623) ad. Siao, 25. VI. 93	82	72	18	13.5
d. (C 12624) ad. Siao, 27. VI. 93	—	73	18.5	13
e. (C 12622) ad. Siao, 18. VI. 93	87	74	18.5	13

Distribution. Sangi Islands: Siao (Meyer *2*, Nat. Coll. in Dresden and Tring Museums), Great Sangi (Meyer *2*); ? Celebes — near Manado (Fischer *a 1, 1*).

The type *M. commutatus* was indicated to have come from Manado, but it now appears most certain that the specimen was mislabelled and it probably came from Siao. The locality Great Sangi rests at present only upon a young specimen in the Dresden Museum, which, as Meyer has already said *(2)*, may possibly prove to be not perfectly identical with the Siao birds. Siao, therefore, is as yet the only positively ascertained locality for the species or race. Here it seems to be not uncommon, as our native collectors were able to send us the full complement of five specimens asked for. In Great Sangi, the bird was obtained neither by our native collectors nor by Dr. Platen.

As compared with adult specimens of *M. inornatus* — one from Aru, two from (?) the Moluccas, four from Peling and Banggai, two from Djampea, and a large series from Talaut — *M. commutatus* is distinguishable by the somewhat darker grey of its plumage and its blacker edge of forehead and upper corner of chin. Prof. W. Blasius rightly remarks that it seems to be a darker and larger variety of *inornatus*; individuals appear, however, to vary a good deal in size. It is difficult to know how to treat of such a form as this; it might be preferable to view it as a subspecies of *M. inornatus*, but that species is involved in much obscurity at present as regards its local variations, and its treatment as species and subspecies must be left to the future.

The genus *Monarcha* is typical of the Australian and what Wallace terms the Austro-Malayan subregions. The occurrence of a form of *M. inornatus* in the Sangi Islands is a link between them and the Moluccas and Papuasia, but the wide distribution of this species tends to show that is has spread its range recently by flight, and is has, therefore, nothing reliable to say for a former land-connection between Sangi and the Moluccas.

The plumage of this bird has a curious resemblance to that of the adult male *Monticola solitaria*.

✢ 136. MONARCHA INORNATUS (Garn.).

Grey-and-rufous Flycatcher.

a. **Muscicapa inornata** *(1)* Garnot, Voy. Coquille, Zool. Atl. 1826, pl. 16, fig. 2.
b. **Drymophila cinerascens** *(1)* Temm., Pl. col. 1827, pl. 430, fig. 2.
c. **Monarcha cinerascens** *(1)* Wall., P. Z. S. 1862, 335, 341.
Monarcha inornatus *(1)* Sharpe, Cat. B. IV, 1879, 431; *(2)* Salvad., Orn. Pap. II, 1881, 14; Agg. 1889, 71; *(3)* Tristr., Cat. Coll. B. 1889, 201; *(4)* M. & Wg., J. f. O. 1894, 244; *(5)* iid., Abh. Mus. Dresd. 1895, Nr. 9, p. 4; *(6)* iid., ib. 1896, Nr. 2, p. 14; *(7)* Hart., Nov. Zool. 1896, 173, 241.
"**Tabaheo mawora**", Talaut, Nat. Coll.
"**Tangkuis**", Peling; "**Tangkuisi**", Banggai (Nat. Coll.).
For full synonymy see Salvadori 2.
Figures and descriptions. Garnot *a 1*; Temminck *b 1*; Sharpe *1*, Salvadori *2*.
Adult. Like *Monarcha commutatus* (see *antea*), but the grey of the upper surface, throat, and chest, paler — the last-named parts olive-grey as against almost mouse-grey.

Distribution. Talaut Is. — Karkellang and Kabruang (Nat. Coll.); Peling and Banggai (Nat. Coll.); Sula Islands (Allen c I, 1, 3); Djampea (Everett 7); Timor (Wallace 1) and the Moluccas (Salvadori 2) to Aru, New Guinea, the Admiralty Islands and Duke of York Island (Salvad. 2).

Observations. A specimen from the Sula Islands in the British Museum and another from the same locality in the Tristram Collection are identified with this species by Dr. Sharpe and Canon Tristram.

Latterly a score of specimens collected in Talaut in the autumn of 1893, 1894, and 1896 have been sent to the Dresden Museum, and curiously enough they do not agree with the Sangi race, *M. commutatus*, but are like the Eastern form.

A few specimens from Peling and Banggai are more suffused with chestnut-rufous below and paler grey on the head and face than those of Talaut, Djampea, Aru, and others labelled "? Moluccas".

Two from Djampea are darker grey on the throat and breast than one from Aru.

Salvadori includes Meyer's *M. geelvinkianus* and *fuscescens* in the synonymy of *M. inornatus*, a species which might some day perhaps be split up into 20 geographical races with three or more names attached to each — if any good purpose were served by so doing. Individual variation in this species seems to be confined within somewhat narrow limits.

It may safely be assumed to have spread its range by flight. Its absence on the mainland of Celebes is very curious.

ɣ * 137. MONARCHA EVERETTI Hart.
Djampea Black-and-white Flycatcher.
Plate XVII.

Monarcha everetti (1) Hart., Nov. Zool. 1896, 173, 182.

Adult male. Upper surface, including face, throat and jugulum black, glossed with steel-blue; rump and upper tail-coverts, body below, under wing-coverts and remiges where they rest upon the body white; thighs black, with some white tips to the feathers; tail black, white at the concealed base, the lateral feathers white for the terminal half, the white decreasing to a tip of a few mm in the 4th pair and disappearing on the middle pair (♂, Dec. 1895, Djampea Id.: Everett — C 14871).

"Iris dark brown; bill and legs light blue; claws dark grey" — Everett 1.

Immature [male]. Mantle cinereous grey (H. 1).

Measurements. "Length about 14 cm; wing 66—69 mm; tail about 70—72; tarsus 19; culmen 16—17" (Hartert 1).

Female. Entirely different from the male: "Above cinereous grey, slightly washed with brown. Lores whitish. A spot behind the eye pale whitish grey. Wings dark brown, inner webs white towards the base. No white on rump and upper tail-coverts; tail as in the male. Under surface whitish, washed with pale orange-rufous, especially on the breast; abdomen almost white. Thighs pale brownish, under wing-coverts and axillaries dirty white. Iris chocolate; bill pale lead-blue, black at apex; legs dark slate-blue; claws blackish" (Hartert).

Distribution. Djampea Island between Celebes and Flores.

This, as Mr. Hartert points out, is a very distinct species, belonging to Gould's genus *Piezorhynchus*, "if that genus can be separated from *Monarcha*". He adds that in this genus "the female always differs from the male"; but it would seem that the word "generally" should have been used here. Sharpe (1879) admits the genus *Piezorhynchus*, with 20 species in the Australian Region, said to be distinguishable from *Monarcha* by the velvety character of the plumage of the head; Salvadori (1881) unites it with *Monarcha*.

GENUS MYIAGRA Vig. Horsf.

Bill moderately long, very broad and flat, across the nostrils about twice as broad as deep, rictal bristles well developed; wings longer than tail, the second primary about equal to the secondaries in length; feet small, the middle toe and claw shorter than the tarsus. The genus is found from Australia to the Lesser Sunda Islands and the small islands to the south of Celebes, the Moluccas, and many groups of Polynesia.

138. MYIAGRA RUFIGULA Wall.

Timor Broad-billed Flycatcher.

Myiagra rufigula *(1)* Wall., P. Z. S. 1863, 485, 491; *(2)* Pelz., J. f. O. 1875, 51; *(3)* Sharpe, Cat. B. 1879, IV, 382; *(4)* Sclat., P. Z. S. 1883, 55; *(5)* Büttik., Notes Leyden Mus. 1892, XIV, 197; *(7)* Hart., Nov. Zool. 1896, 171, 585.

Adult male. Above dark lead-grey; entire head above, nape, and ear-coverts blackish leaden, with a greenish gloss, on forehead next the bill paler; chin, throat, and chest ferruginous; remaining under parts white, washed with buff, more strongly at the sides and on the under wing-coverts; remiges and tail blackish, the latter bordered above with the lead-grey of the back, the remiges with browner grey; paler below, especially where they rest upon the body (♂, Bonerate Id., 20. III. 96 P. & F. Sarasin).

Female. Differs from the male in having the head, nape, and ear-coverts similarly glossed with greenish, but less dark leaden — very little darker than the back, — the ferruginous of the chin, throat, and chest lighter. The outermost tail-feather whitish along the outer web, and it and the next rectrix incline to brownish white distally. "Iris dark brown; bill black, mandible pale blue, with black tip; legs and claws greyish black" — Everett 7 (♀, Bonerate, 20. III. 96: Sarasin Coll.).

Measurements (4 specimens). Wing 66—69 mm; tail 63—66; tarsus c. 17.5; bill from nostril c. 8.5.

Distribution. Timor and Samao (Wallace *1, 3*); Sumba (ten Kate *5*, Doherty *7*); Djampea and Kalao (Everett *7*); Bonerate (P. & F. Sarasin).

This Broad-billed Flycatcher was discovered first by Everett in December, 1895, and then by the Sarasins in March, 1896, in the above mentioned small islands between Celebes and Flores, which are included in the Celebes Province in this work. The type came from Samao.

The genus *Myiagra* is an Australasian type, most plentifully represented in Papuasia and the Melanesian Islands. The present species is probably most nearly related to *M. fulviventris* Sclat. of Timorlaut, which differs by its fulvous belly and under wing-coverts. *M. albiventris* (Peale) of Samoa has, as Hartert points out, black lores, a larger bill and less rufous below, so distinguishing itself from this Tenimber form.

GENUS CULICICAPA Swinh.

These little Flycatchers have the bill as broad and flat as *Myiagra*, but more pointed and triangular, the rictal bristles are large, nearly as long as the bill. The tail is square, shorter than the wing, the quill-formula much as in *Myiagra*; the middle toe and claw slightly shorter than the tarsus. They are well characterized by their principal colours of yellow and yellow-olive, and the sexes are similar. Only two species are known, with a range from the Himalayas to the Philippines and Celebes.

139. CULICICAPA HELIANTHEA (Wall.).
Wallace's Yellow Flycatcher.

a. **Muscicapa helianthea** *(1)* Wall., P. Z. S. 1865, 476; *(2)* Gray, HL. 1869, I, 321; *(3)* Rosenb., Malay. Archip. 273.
b. **Myialestes helianthea** *(1)* Wald., Tr. Z. S. 1872, VIII, 66, pl. 7, fig. 1; *(2)* Brüggem., Abh. Ver. Bremen 1876, V, 68; *(3)* Wallace, Island Life 1880, 433; *(4)* Guillem., P. Z. S. 1885, 552.
Culicicapa helianthea *(1)* Salvad., Cat. Ucc. Borneo 1874, 135; *(2)* Sharpe, Cat. B. IV, 1879, 370; *(3)* M. & Wg., Abh. Mus. Dresd. 1895, Nr. 8, p. 9; *(4)* iid., ib. 1896, Nr. 2, p. 14; *(5)* Grant, Ibis 1896, 543; *(6)* Hart., Nov. Zool. 1896, 157, 171; *(7)* id., ib. 1897, 158; *(8)* Grant, Ibis 1897, 227.
c. **Xantholestes helianthea** *(1)* Sharpe, Tr. Linn. Soc. 1877, (2) I, 327.
d. **Xantholestes panayensis** *(1)* Sharpe, Tr. L. Soc. 1877, (2) I, 327.
e. **Culicicapa panayensis** *(1)* Sharpe, Cat. B. IV, 1879, 371; *(2)* Steere, List Coll. B. & M. Philipp. 1890, 16; *(3)* Bourns & Worces., B. Menage Exped. 1894, 41; *(4)* Grant, Ibis 1894, 506; *(5)* id., ib. 1895, 443.

"Koko intiwoho", Manado Distr., Nat. Coll.
"Tangkui mosoni", Banggai, Nat. Coll.

Figure and descriptions. Walden *b 1*; Wallace *a 1*; Brüggemann *b 2*; Sharpe *2*.
Adult. Above olive-green-yellow; darkest on the crown, becoming bright yellow on the rump; quills and tail-feathers blackish, the outer webs fringed with greenish ochre; lores yellow; ear-coverts yellower than the crown; chin, throat and remaining under parts lemon-yellow, purest and lightest on chin, throat, under wing-coverts and flanks, dirtier on the breast and sides (Lotta near Manado, 8. V. 93: Nat. Coll. — C 12279).

"Iris dark brown; maxilla dark sepia, mandible ochraceous orange; legs and claws light sepia, soles of feet yellow" — Everett *6*.

Sexes. Similar in coloration (Brüggem. *b 2*, Hartert *6*).

Measurements (7 examples). Wing 58 (♀) — 63 mm; tail c. 50; tarsus c. 13; bill from nostril c. 6.5.

Distribution. Celebesian and Philippine areas: Celebes — Minahassa (Wall. *a 1, 2*, Guill., P. & F. Sarasin, Nat. Coll.); South Peninsula (Everett *6*, Doherty *7*); Banggai Island (Nat. Coll.); Saleyer Island (Everett *6*). Philippines — Luzon, Panay, Negros, Tawi-Tawi, Tablas, Romblon, Sibuyan, Guimaras, Masbate, Siquijor, Palawan (Moseley, Steere, Bourns & Worcester, Whitehead *d 1, e 2, e 3, e 4, e 5, 5, 8*).

Until the last year or two this little Flycatcher was known only from the northern province of Celebes, the Minahassa; but it has now been discovered south as far as Saleyer and east as far as Banggai, and Mr. Ogilvie Grant has found the bird in the Philippines to be identical, so adding a broad area to its range in the north.

The wide-spread *C. ceylonensis* (Swains.) ranging from the Himalayas to Ceylon, Java, and Borneo differs in having the entire head and throat ashy grey, and *C. helianthea* was separated from it nominally by Dr. Sharpe as a different genus (*c 1*), but no generic characters were pointed out, and later the author reunited them with *Culicicapa*.

Mr. Wallace mentions *C. helianthea* as a Himalayan type in Celebes; the genus, however, is as much Sundan as Himalayan. To the east *Culicicapa* is not known in the Australian Region, and, like nearly all the other Flycatchers of Celebes, it seems to suggest the probability of the former connection of the island with Asia, and separation by the sea from Australasia.

GENUS GERYGONE J. Gd.

The position of this genus seems to be between the *Muscicapidae* and *Sylviidae*. The sexes are similar in coloration and the young are probably very like them, though at least one species *(G. brunneipectus)* has brownish edgings to the feathers of the breast. The tarsus is rather large, much longer than the middle toe, by which it approaches the Warblers; the bill is more Flycatcher-like, of moderate size, about $1\frac{1}{2}$ times as broad across the nostril as deep, the nasal area is membranous, the aperture long oval apparently, the rictal bristles few and rather small; wing longer than tail, the 3rd, 4th and 5th quills the longest, the second equal to, or a little longer than the secondaries. The genus is pre-eminently Australasian, but occurs west as far as Borneo.

+ * 140. GERYGONE FLAVEOLA Cab.
Drab-and-yellow Flycatcher.

Gerygone flaveola *(1)* Cab., J. f. O. 1873, 157; *(2)* Meyer, t. c. 404; *(3)* Salvad., Ann. Mus. Civ. Gen. 1875, VII, 665; *(4)* Sharpe, Notes Leyden Mus. 1878, I, 29, part.; *(V)* id., Cat. B. IV, 1879, 214, part., pl. V, fig. 2; *(6)* W. Blas., J. f. O. 1883, 117, 125; *(7)* Guillem., P. Z. S. 1885, 263, pt., 414, pt.; *(8)* Salvad., Ann. Mus. Civ.

Gen. 1891, XXXII, 53; *(9)* Davison, Ibis 1892, 100; *(10)* Sharpe, Ibis 1893, 561; *(11)* Büttik., Notes Leyden Mus. 1893, XV, 174, 175; *(12)* id., Zool. Erg. Weber's Reise Ost-Ind. 1893, III, 278; *(13)* M. & Wg., Abh. Mus. Dresden 1896, Nr. 1, p. 10; *(14)* Hart., Nov. Zool. 1896, 157; 171; *(15)* id., Nov. Zool. 1897, 158, 162.

Figure and descriptions. Sharpe *V*, Cabanis *1*.

Adult. Above drab, washed with olive, head greyer, edges of quills paler; lores dusky whitish; ear-coverts drab, washed with yellow; under surface, including submalar region, sulphur-yellow, darker and brighter on throat, paling down into yellowish white on the under tail-coverts; under wing-coverts almost pure white; inner edges of quills whitish; inner edges of lateral pairs of rectrices whitish near tip: "Iris crimson or red-brown; beak black, in one, evidently younger bird, the base of the mandible is pale" (Doherty *15*); feet blackish *(15)*. (Macassar, ♂, January, 1873: Meyer — C 491.)

Sexes. Similar (♀ ad. C 448).

Measurements.

	Wing	Tail	Tarsus	Bill from nostril
a. (C 491) ♂ ad. Macassar, I. 1873 . .	56	40	18.5	6.5
b. (C 449) ♂ ad. Macassar, I. 1873 .	54	38	17.5	6.5
c. (C 448) ♀ ad. Macassar, I. 1873	54	39	17.5	7.0
d. (Sarasin Coll.) ♂ ad. Loka, S. Cel., 21. X. 95 .	52	37	17	6.5
e. (Sarasin Coll.) ♂ ad. Macassar, 19. VII. 95	52	38	16.5	6.5
f. (Sarasin Coll.) ♂ ad. Macassar, 16. VII. 95 .	52	37	—	7
g. (Sarasin Coll.) — Lake Posso, Feb. 95 .	54	40	—	7

Mr. Hartert *(15)* remarks that the wings of his two examples from Saleyer Island are only 49—50 mm long, and they are slightly paler, "though this latter character is probably due to their being in worn plumage".

Distribution. Celebes — Macassar (Meyer *1*, in Berlin and Dresd. Mus., P. & F. Sarasin); Bonthain Mountains (P. & F. Sarasin, Everett *14*, Doherty *15*); Luwu at the head of the Gulf of Boni (Weber *12*); Kandari, S. E. Celebes (Beccari *3*); Lake Posso, Central Celebes (P. & F. Sarasin); Dongala and Tawaya, W. Celebes (Doherty *15*).

This was one of Meyer's discoveries in South Celebes. No specimens of this species have as yet been obtained in North Celebes, and the bird is somewhat rare in collections. It is most nearly allied to *G. salvadorii* Büttik. of Borneo, or perhaps to the Sooloo form, in which Dr. Guillemard points out some distinguishing characters. The Borneo form was at first united by Dr. Sharpe with *G. flaveola*, but, in separating it as *G. salvadorii*, Mr. Büttikofer *(11)* calls attention to its smaller size, the sides of the breast olive-brown instead of yellow, its wanting the whitish lores and the yellowish wash on the ear-coverts, the upper surface a shade darker, the white near the ends of the inner webs of the tail-feathers forming much larger spots.

Another ally is Salvadori's *Gerygone modiglianii* of Sumatra *(8)*, which Sharpe considers identical with Davison's *G. pectoralis* of the Malay Peninsula (Ibis 1892, 99; 1893, 119, 561) and which, in his opinion, is questionably distinct from the Bornean form of Büttikofer. From *G. flaveola*, as Salvadori points out *(8)*, *G. modiglianii* differs in having a distinct dusky collar on the sides of the neck, the upper surface darker, the yellow below a little clearer, a distinct subterminal black band on the tail. *G. sulfurea* Wall. of Solor near Flores differs by its smaller size, a broad median black band across the tail-feathers, with a large white spot at the tip of the inner web (Sharpe).

The genus *Gerygone*, as may be seen from Count Salvadori's great work, is absent, so far as is known, from the Moluccas. The genus is divided by Sharpe into two, *Gerygone* and *Pseudogerygone*, the latter with the secondaries of somewhat increased length — an unsatisfactory character, we think, whereon to base a genus. Taking the two together as one genus, *Gerygone* ranges from the Malay Peninsula through the Sunda Islands and Philippines, to Australia, New Guinea, New Caledonia, Norfolk Island, New Zealand and the Chatham Islands, as shown by Sharpe. From their richness in species it may be supposed to have arisen in New Guinea or Australia, whence it seems to have spread its range by flight.

Gerygone is placed among the *Muscicapidae* by Sharpe, who rightly remarks that it is one of those genera which evince a great likeness to the Warblers. It is in fact almost, if not quite, impossible to draw a line between the Flycatchers and Warblers. Gould, who left the family to which it belongs doubtful, says that "their food consists of insects of the most diminutive size, such as aphides, gnats and mosquitos. They mostly frequent the thick umbrageous woods, where they flit about under the canopy of the dense foliage, or sally forth into the open glade like true Flycatchers" (Hb. B. Austr. 1865, I, 265).

GENUS PRATINCOLA K. L. Koch.

The Chats are treated as a family, *Saxicolidae*, by many authors; by Oates (Faun. Br. Ind. B. II, 1890, 57) as a subfamily of the *Turdidae*. We do not know how *Pratincola* can be separated from the Flycatchers, all that can be said for it is that its bill is less wide and its feet larger than in most of the *Muscicapidae*. Bill shorter than cranium, across the nostril about as wide as deep; nostril roundish, the feathers of the forehead impinging to its base; rictal bristles moderate, two or three frontal bristles reaching over the nostril; chin-feathers with hairy ends; 2nd primary about as long as the secondaries; tip of wing formed by the 3rd—6th primaries; tail rounded, shorter than wing; tarsus longer than middle toe and claw. Sexes dissimilar. Range: Europe, Africa, Asia, as far as Celebes and Flores.

← 141. PRATINCOLA CAPRATA (L.).
Indian Pied Bush-chat.

a. **Motacilla caprata** *(1)* Linn., S. N. 1766, I, 335.
b. **Saxicola caprata** *(1)* S. Müll., Verh. Natuurk. Comm. 1839—44, 87; *(2)* id., Reizen Ind. Archip. 1858, II, 8.

Pratincola caprata *(1)* Wald., Tr. Z. S. 1872, VIII, 63; *(2)* Salvad., Ucc. Borneo 1874, 252; *(3)* Wald., Tr. Z. S. 1875, IX, 193; *(4)* Cab. & Rchw., J. f. O. 1876, 319; *(5)* Salvad., P. Z. S. 1877, 194; *(6)* Tweedd., P. Z. S. 1877, 696, 761; 1878, 710; *(7)* Sharpe, Cat. B. IV, 1879, 195; *(8)* Legge, B. Ceylon 1880, 431; *(9)* Salvad., Orn. Pap. II, 1881, 420; *(10)* Scully, Ibis 1881, 440; *(10bis)* W. Rams., Tweedd. Orn. Works 1881, App. 657; *(11)* Vorderm., N. Tdschr. Ned. Ind. 1882, XLII, 65; *(12)* Davison, Str. F. X, 1882, 307; 1889, 389; *(13)* W. Blas., Verh. z.-b. Ges. Wien 1883, 75; *(13bis)* Oates, B. Br. Burmah 1883, I, 281; *(14)* Marshall, Ibis 1884, 415; *(15)* C. Swinh. & Barnes, Ibis 1885, 124; *(16)* W. Blas., Ztschr. ges. Orn. 1885, 277; *(17)* Guillem., P. Z. S. 1885, 506; *(18)* Dresser, Ibis 1889, 86; *(20)* St. John, t. c. 163; *(21)* Radde & Walter, Ornis 1889, 61, 168, 172, 221, 256, 259, 266; *(22)* Tristr., Cat. Coll. B. 1889, 143; *(23)* Dresser, Ibis 1890, 342; *(24)* Oates, Faun. Brit. Ind. B. II, 1890, 59; *(25)* id., ed. Hume's Nests & Eggs Ind. B. II, 1890, 41; *(26)* Steere, List Coll. B. & M. Philipp. Is. 1890, 16; *(27)* Büttik., Notes Leyden Mus. 1891, XIII, 211; *(28)* Hartert, Kat. Mus. Senckenb. Vög. 1891, 3; *(29)* id., J. f. O. 1891, 201; *(30)* Büttik., Notes Leyd. Mus. 1892, 197; *(31)* id., Zool. Erg. Weber's Reise Ost-Ind. 1893, III, 277, 293; *(32)* Grant, Ibis 1894, 505; *(33)* Bourns & Worces., B. Menage Exp. 1894, 39; *(XXXIV)* Dresser, B. Eur. Suppl. I, 1895, 33, pl. 641; *(35)* Grant, Ibis 1895, 441; *(36)* M. & Wg., Abh. Mus. Dresd. 1896, Nr. 1, p. 12; *(37)* Hart., Nov. Zool. 1896, 150, 166, 555, 580, 593; *(38)* id., ib. 1897, 155, 161.

"**Tjingan**", Tjamba Distr., Platen *16*.

For further synonymy and references see Sharpe 7.

Figure and descriptions. Dresser *XXXIV*; Sharpe 7, Vorderman *11*; Oates *13bis*, *24*.

Adult male. Black; wing-coverts nearest the body for about two-thirds across the wing, rump, upper and under tail-coverts and anal region white. "Iris brown; bill and feet black" (♂, Tjamba Distr., S. Celebes, 30. IV. 78: Platen — C 5376). Wing 67 mm; tail 55; tarsus 22; bill from nostril 8.

Female. Hair-brown, paler below, almost wood-brown on abdomen, lores, chin, and malar region; the middles of the feathers darker; lower rump, upper tail-coverts and under tail-coverts white, tinged with buff; tail black; wing-coverts blackish, with pale brown edges, some of the concealed inner greater and middle coverts mostly white; remiges brown, edged with wood-brown; under wing-coverts whitish brown; bill blackish; legs and feet black (♀, Lake Posso, Centr. Cel., 15. II. 95: Sarasin Coll.).

Young. Mottled all over like a young Robin: above fulvescent edged with darker brown; head blackish brown, streaked down the centre of the feathers with deep fulvous; wing-coverts and quills edged with fawn-colour; under-parts fulvous, slightly varied with dark edges to the feathers of the breast (Sharpe 7).

Eggs. "Dr. Platen obtained eggs of this bird near Rurukan in the Minahassa, which are somewhat paler than those of our *rubicola*. An egg in my collection collected by Oates in Pegu, 20. IV. 81, corresponds with them, is only a trifle smaller, but similarly coloured" (Nehrkorn MS.).

Usually four in number, often three, occasionally five. Rather broad ovals, somewhat pointed; delicate pale bluish green, speckled, mottled and streaked with brownish red, densest at the large end. They vary enormously in size, viz: 15.2—19.5 × 11.2 —16.3 mm; average (50 specimens) 17 × 14 mm (India, Hume *25*).

Nest. Usually a shallow, somewhat saucer-shaped pad, composed of soft grass, fine roots, and lined with the same, hairs or other soft material; generally placed on the ground in a hole or impression *(25)*.

Distribution. Transcaspia (Radde *21, 18, 23*); S. E. Persia (Blanford *7*); Afghanistan (St. John *20*); Baluchistan (Blanford *7*); Himalayas and India (Hume etc. *24, 26*); Arrakan (Blyth); Burmah (Oates *13bis*); Tenasserim (W. Ramsay and Davison *13bis, 24*); Java (Horsfield & Wallace *7*, Vorderman *11*); Lombok (Wallace *7*, etc. *37*); Sumbawa (Guillemard *17*); Sumba (ten Kate *30*, Doherty *37*); Flores (Wallace *7*, Weber *31*); Timor (Wallace *7*, ten Kate *27*); Samao (ten Kate *27*); Saleyer Island (Everett *37*); Celebes — S. Peninsula (S. Müller *b 1, b 2*, Wallace *7*, Weber *31*, Platen *16*, etc.), West Celebes (Doherty *38*), Central Celebes — Lake Posso (P. & F. Sarasin *36*), N. Peninsula — Rurukan (Platen fide Nehrkorn); ?Borneo (Brunsw. Mus. *13*); Philippines — Luzon, Panay, Cebu, Negros, Bohol, Masbate, Siquijor *(10bis, 26, 33, 29)*; ?Palawan (Hartert *29*).

Ceylon and Southern India are inhabited by a Stonechat, *P. bicolor* Sykes, which seems to be only a subspecies of *P. caprata*. According to Legge, the typical form differs from *P. bicolor* by its smaller size, the smallness of its bill, the more glossy and intense hue of the black of its upper surface and breast, "and there is generally, more especially in Malay specimens, more white on the rump" *(8)*. The Indian female is described by Oates as having the upper tail-coverts ferruginous; they are almost white in Celebes. *P. caprata* was originally described from Luzon. This bird seems to be a resident in the East India Islands, and one that has established itself there in recent times. Platen got its eggs in North Celebes, Everett in the South. It ranges down the chain of islands from Java to Timor, but its occurrence in Borneo still remains to be fully established, for the two specimens from Verreaux in the Brunswick Museum do not afford sufficient proof. It is a rare species in Celebes and the eggs in the Nehrkorn Collection are the only evidence of the presence of the bird in the north of the island, therefore it requires specimens of the bird itself to establish its occurrence positively; in the south it seems to be rather more plentiful. Five or six of the Philippine Islands have produced examples, Luzon the largest number. It (or a form of it) is, as Mr. Oates remarks *(24)*, a resident species throughout the whole of India and Burmah, except the southernmost part of the Peninsula (where *P. bicolor* replaces it) and portions of Tenasserim. In Turkomenia, however, Drs. Radde & Walter record it as a summer visitant. It has been stated to occur in New Guinea, in virtue of a specimen in spirit obtained during the voyage of the "Gazelle" *(4)*, but Count Salvadori *(5, 9)* is, we should think, probably quite right in regarding the label as erroneous.

The genus *Pratincola*, according to Sharpe, ranges, as remarked above, over the whole of Europe, Africa and Asia extending into the East Indies as far as Celebes and Flores. *P. caprata* is a very distinct species, the bird most like it being, perhaps, *P. albofasciata* (Rüpp.) of Abyssinia, which has a white patch on the sides of the neck.

ORDER ACCIPITRES.

The diurnal Birds of Prey: Hawks, Eagles, Falcons, the Osprey, Vultures; distinguishable by the rigid, hooked, and sharply pointed bill; powerful, hooked claws; three toes in front and the fourth behind (except in the Osprey, in which the outer toe is reversible); usually of great powers of flight; 11 primaries; plumage often varied in coloration, but of sober tints: brown, black, white, grey, rufous, and purplish being found, but pure yellow, blue, red, bright green, and metallic tints are wanting. The *Accipitres* usually build a nest of sticks; the eggs are white or whitish in ground-colour, in many genera very handsomely varied with markings of rufous or brown; the young are hatched helpless and clothed with down.

FAMILY FALCONIDAE.

Containing all the Birds of Prey, except the Osprey (which is distinguished from them by its having the outer toe reversible and pterylologically by its having no aftershaft to the feathers) and the Vultures (which have the head and neck bare, or clothed in down).

GENUS SPILORNIS G. R. Gray.

Birds of Prey of medium size (about as large as a Raven), stout and compact in form, wings moderate; bill not denticulate; a broad nuchal crest; tarsi naked (except the upper fourth anteriorly), reticulated with hexagonal scales; toes short; under parts marked with transverse spots or bars of white; food: chiefly reptiles and amphibians; number of eggs laid: one or two. The genus contains about 12 species of a local and stationary character, distributed from the Himalayas and South China to the Andaman Islands, Celebes and Sula.

* 1. **SPILORNIS RUFIPECTUS** J. Gd.[1]

Russet-breasted Serpent-harrier.

Under this specific name we include two well-pronounced geographical races, *1. the typical Spilornis rufipectus* of the mainland of Celebes, *2. Spilornis rufipectus*

---

[1] The abbreviation of the author's name at the head of each article is taken from the Berlin "Liste der Autoren" (1896), though the form adopted does not always meet with our approval.

sulaensis of the Sula group, and *3.* the individuals inhabiting the Peling group, which are intermediate between these two races. For the treatment of this, and of similar cases, the following method of nomenclature may be adopted without prejudice to ornithology.

1. The typical Spilornis rufipectus.

a. **Spilornis rufipectus** *(1)* Gould, P. Z. S. 1857, 222; *(II)* id., B. Asia, I pl. IX (1860); *(3)* Wall., P. Z. S. 1862, 338, pt.; *(4)* id., Ibis 1868, 16, 21; *(5)* Wald., Tr. Z. S. 1872, VIII, 35; *(6)* Sharpe, Cat. B. I, 1874, 291; *(7)* Salvad., Ann. Mus. Civ. Gen. 1875, VII, 643; *(8)* Brügg., Abh. Ver. Bremen 1876, V, 46; *(9)* Gurney, Ibis 1878, 96, 102; *(10)* W. Blas., J. f. O. 1883, 135; *(11)* Gurney, List Diurn. B. of Prey 1884, 17; *(12)* W. Blas., J. f. O. 1885, 403; *(13)* id., Ztschr. ges. Orn. 1885, 222; *(14)* Guillem., P. Z. S. 1885, 544; *(15)* Hickson, Nat. in N. Celebes 1889, 89; *(XVI)* Meyer, Vogelskel. II 1892, 27, t. CLVII; *(17)* Bütt., Zool. Erg. Webers Reise 1893 III, 271; *(18)* Sharpe, Ibis 1893, 552; *(19)* M. & Wg., Abh. Mus. Dresd. 1895 no. 8, p. 3; *(20)* iid., ib. 1896 no. 1, p. 7; *(21)* iid., ib. 1896, no. 2, p. 7; *(22)* Hartert, Nov. Zool. 1896, 161; *(23)* id., ib. 1897, 150.

b. **Circaëtus bacha celebensis** *(1)* Schl., Mus. P.-B. Buteones 1862, 27; *(2)* Rosenb., Malay. Archip. 1878, 271.

c. **Circaëtus rufipectus** *(1)* Schl., Valkv. Ned. Ind. 1866, 37, 72, pl. 2³, figs. 1—3; *(2)* Gray, HL. 1869, I, 15; *(3)* Schl., Rev. Accip. 1873, 114.

"Kokodschi", Tjamba, S. Celebes, Platen *a 13.*
"Berna" (albescent young), Tjamba, Platen *a 13.*
"Buliëa-mohengo", Gorontalo Distr., N. Celebes, v. Rosenberg *b 2.*
"Kiokkiok", near Manado, Nat. Coll. in Dresd. Mus.
"Boina", Balante, E. Celebes, Nat. Coll.
"Sikep utang besar", Lembeh Id., Nat. Coll.

Figures and descriptions. Gould *II. 1*; Schlegel *c I*; Meyer *XVI* (skeleton); Sharpe *6.*

Diagnosis of race. Wing relatively longer than in *S. rufipectus sulaensis* (see. table of measurements), remiges below greyish white, broadly tipped and barred with blackish. These bars coalesce on the secondaries and base of primaries, enclosing spots of white, mottled with brown.

Old female. Upper surface dark brown, glossed with purple; sides and top of head, crest, and throat black, ear-coverts washed with grey; hind neck dusky, the margins of the feathers here and there, and on the crest, fulvous brown; secondaries and some of the upper tail-coverts tipped with white; carpal edge spotted with white; tail above pale brown, tipped with whitish, and crossed with four broad black bands — the basal one rather indistinct; breast mummy-brown; remaining under-parts — including under wing- and tail-coverts — darker, the lower breast spotted, the sides, abdomen, thighs and under tail-coverts closely barred with white; under side of wing broadly barred and spotted with white (Manado, Nr. 6682).

Younger female. Like the above, but the brown of the upper plumage paler and duller without so much purple gloss; hind neck pale brown, without (or with only a few) yellow-brown margins here and on the crest, the black feathers of which are more or less broken up with yellowish white; tail crossed with three black bands, the basal one indistinct; breast much paler and more of a dark wood-brown tone (S. Celebes: Platen — C 10707).

Male. Similar to the female, but the white bars of the under parts more sharply defined and extending further towards the breast (whereas in the female they form sooner into disconnected spots); the brown bars on the under tail-coverts narrower (♂ vix ad. S. Celebes — Platen, C 10708; 3 ♂♂ ad. & vix ad. Sarasin Coll., N. & S. Celebes).

"Iris gold-yellow; periocular skin and cere green-yellow; bill blue-black; feet lemon- (or gold-) yellow" in both sexes (Platen *a 13*).

Female in albescent immature plumage. Head, crest and neck fulvous white with dark brown shaft-streaks; the upper parts display a varied plumage of sepia and fulvous brown, the feathers in general having dark centres and pale bases and margins; primaries and secondaries tipped with white; tail pale brown above, white below, and crossed with 5 to 6 indistinct dark bands, tip white; whole of under surface buffy white, streaked from the breast downwards with dark brown, which often spreads out in a washy manner in lighter brown over much of the feather. In this specimen the cross-barred feathers of maturity are sprouting at the flanks (S. Celebes — Nr. 6683).

A male in albescent plumage recently obtained by the Drs. Sarasin at Kema, Aug. 5th, 1893, corresponds with the above description of the female; ear-coverts and subocular region black; under surface purer white, with fewer and smaller brown streaks, here and on the upper surface not showing a general wash of rufous apparent on comparison in the other specimen. "Iris yellow; legs and feet yellowish grey; bill black, at the base blue" (Sarasin). A second male is much more rufo-fulvous in general tint than the other (Macassar, 12. IX. 95: Sarasin Coll.).

A specimen in the Leyden Museum (N. Celebes — Faber, 1883) is half in albescent plumage, half in adult. Wings and tail as in albescent specimens; under wing-coverts white, some tipped with rufous brown; back, breast, abdomen and thighs much as in ♀ ad.

First plumage. The full-fledged young of this species is not known, but in the cases of *Spilornis bacha* (Java), *S. cheela* (India) and *S. spilogaster* (Ceylon), young birds of each in the first stage of dress have been described or figured (Schlegel, Valkv. N. I. pl. 22, f. 3; Sharpe, Cat. B. I, 287; Legge, B. Ceylon 1880, 62; Bornstein, J. f. O. 1860, 425), from which it will be seen that the first plumage often — perhaps always — much resembles that of full maturity. Whether young birds always assume this mature-looking dress on first leaving the nest, and then lose it and put on the immature albescent plumage, and finally recover the adult type of coloration, or, whether the members of the genus are dimorphous when young — both mature-plumaged and albescent individuals existing from the nest, — are questions upon which opinion is divided, and facts, unfortunately, are as yet insufficient to allow of their being answered. The albescent type of immature plumage probably occurs in all species of the genus. Specimens in this dress are figured by Schlegel in the cases of *S. bacha*, *S. rufipectus* and *S. sulaensis* (*b I*; *c I*; Schl., Valkv. pl. 22, f. 3); similar immature birds of *S. davisoni* Hume, *S. rutherfordi* Swinh. and *S. cheela* Lath. have been described (Hume, Str. F. II, 148; Bingham, ib. IX, 144; Oates, B. Brit. Burmah, II, 194; Sharpe, Cat. B. I, 287), and there is a ♀ specimen of *S. holospilus* Vigors from Mindanao in this plumage in the Dresden Museum (Nr. 13822). Gurney considered this to be the second plumage *(8)*, Schlegel and Colonel Legge express the opinion that it is a more or less frequent variation of dress assumed in the nest itself. Longitudinal streaks on the feathers of birds appear to represent a more original, less highly differentiated plumage than do cross-bars; this is shown by the fact that almost all birds of prey, which when adult acquire a cross-barred under-side, have this region streaked or drop-marked when immature.

If Gurney's view, that the pale, streaked specimens of *Spilornis* are in the second plumage, be correct, the curious case would be seen of a species regularly reverting from a higher stage of dress to a lower one and, subsequently, re-acquiring the more highly differentiated coloration.

As pointing to the probability that both albescent, streaked individuals and also dark, spotted ones exist from the nest in the case of *Spilornis*, Colonel Legge points out that of the Booted Eagle *(Nisaetus pennatus)* both dark and light young ones have been taken out of the same nest; but the case is not strictly a parallel to that of *Spilornis*, inasmuch as *N. pennatus* has two different phases of adult dress, a light and a dark one, and, when dimorphous pairs of young ones have been found, they are said to be sprung from a light male and dark female, or vice-versa (B. Ceylon, 62).

The albescent plumage is found in both sexes.

Skeleton. Length of cranium . . . . . 77.5 mm | Length of fibula . . . . 84.0 mm
Greatest breadth of cranium 42.0 » | » » tarso-metatarsus . . . 76.0 »
Length of humerus . . . . . . 97.0 » | » » sternum . . . . 61.0 »
» » ulna . . . . 110.0 » | Greatest breadth of sternum . . 35.5 »
» » radius . . . . . . 105.0 » | Height of crista sterni . . . . . 13.5 »
» » manus . . . . . . 87.0 » | Length of pelvis . . . . 72.0 »
» » femur . . . . . . 65.0 » | Greatest breadth of pelvis . . . 31.5 »
» » tibia . . . . 105.5 » |
(Siao, Sangi in Mus. Berol. XVII.)

Nidification. Unknown.

Distribution. Celebes. South Peninsula (Wallace *a 1*, Guillemard *a 14*, Platen *a 13*, Weber *a 17*, etc.); Central Celebes — Luwu Distr. (Weber *a 17*, P. & F. Sarasin *a 20*); S. E. Peninsula — Kendari (Beccari *a 7*); E. Peninsula (Nat. Coll. *a 21*); N. Peninsula (Forster *b 1*, v. Duivenbode *c 3*, etc.); Talissi Id. (Hickson *a 15*); Lembeh Id. (Nat. Coll. in Dresd. Mus.); Siao — known only from skeleton (Meyer *a XVI*).

2. Spilornis rufipectus sulaensis (Schl.).

d. **Circaetus sulaensis** *(1)* Schl., Valkv. Ned. Ind. 1866, 38, 72, pl. 23, figs. 4—6; *(2)* Gray, HL. 1869 I, 15.

e. **Spilornis sulaensis** *(1)* Wall. Ibis 1868, 16; *(2)* Sharpe, Cat. B. 1874, I, 292; *(3)* Gurney, Ibis 1878, 102; *(4)* id. Diurn. B. of Prey 1884, 17; *(5)* Sharpe, Ibis 1893, 552.

f. **Circaetus rufipectus sulaensis** *(1)* Schl., Mus. P.-B. Rev. Accip. 1873, 114.

Figures and descriptions. Schlegel *d I*; Sharpe *e 2*.

Diagnosis. Wing relatively shorter than in the *typical S. rufipectus* (see table); under-side of quills greyish white, passing into blackish at the distal ends, and crossed by three or four well-marked bars of blackish, much narrower than those of the typical form. These bars do not coalesce on the basal half of the quills in the same manner as in that form, but pass separately across the wing.

Distribution. Sula Islands, Sula Besi and Sula Mangoli (Allen *d 1*, Bernstein & Hoedt *e I*, *e 1*).

In the Leyden Museum are seven specimens — 3 ♂ ad., 3 ♀ ad. and 1 ♀ juv. albescent — from Sula. The males have the breast paler; on the lower breast and abdomen the white bars are broader and the brown ones narrower, and the barring on the lower breast is better defined than in the other sex; under tail-coverts white (in one specimen slightly barred towards the tip). The females of Sula have the under tail-coverts barred and in regard to the barring of the under surface would appear to resemble the males of Celebes, but there are only one or two males with the sex satisfactorily ascertained in the Leyden Museum for comparison. The female of Celebes is more spotted below.

3. Spilornis rufipectus < sulaensis.[1]

g. **Spilornis sulaensis** *(1)* M. & Wg., Abh. Mus. Dresden 1896 no. 2, p. 7.

"**Alaji Kabat**", Peling; "**Alaji**", Banggai, Nat. Coll.

Diagnosis. Intermediate between the *typical S. rufipectus* and *S. rufipectus sulaensis*, but on the whole more like the latter race.

Distribution. Peling group between Sula and E. Celebes: — Peling and Banggai (Nat. Coll.).

Measurements.	Wing	Tail	Tarsus	Culmen from cere
(Nr. 6682) [♀] North Celebes	370-5	250	79	28
(Nr. 2199) [♀] North Celebes	340-4	250	74	26
(*C* 846) [♀] N. Celebes, March 1871 (Meyer)	360	240	75	—
(*C* 10844) [♀] N. Celebes, Aug.—Sept. 92 (Nat. Coll.)	344	230	74	27
(*C* 10845) [♀] N. Celebes, Aug.—Sept. 92 (Nat. Coll.)	353	236	74	27.5
(Sarasin Coll.) ♂, N. Celebes, 20. Oct. 93 (P. & F. S.)	340	230	78	25.5
(Sarasin Coll.) ♂, N. Celebes, 6. Oct. 94 (P. & F. S.)	342	242	78	27
(Sarasin Coll.) ♂ juv., N. Celebes, 5. Aug. 93 (P. & F. S.)	337	230	73	27
(*C* 14247) [♂] Lembeh Id. March 95 (Nat. Coll.)	337	218	73	25
(*C* 10708) ♂, South Celebes, 6. May 78 (Platen)	330	215	75	28
(*C* 10707) ♂, South Celebes, 12. March 78 (Platen)	325	220	73	27.5
(Schlüter Coll.) ♀, South Celebes (Platen)	344	229	72	—
(Sarasin Coll.) ♂, South Celebes, 21. Sept. 95 (P. & F. S.)	322	223	74	29
(Sarasin Coll.) ♂ juv., S. Celebes, 12. Sept. 95 (P. & F. S.)	332	232	70	26
(Sarasin Coll.) ♀, Centr. C., Palopo, 21. Jan. 95 (P. & F. S.)	328	—	72	28
(*C* 14295) [♀], East Celebes, May-Aug. 95 (Nat. Coll.)	363	236	78	—
(Norwich Mus.) ad. Macassar (Wallace)	345	—	71	—
(Norwich Mus.) ♂ ad. Macassar (Wallace) Communicated	345	—	71	—
(Norwich Mus.) ♂ ad. Macassar (Wallace)	338	—	69	—
(Norwich Mus.) ad. Macassar (Wallace) by Mr.	345	—	76	—
(Norwich Mus.) imm. N. Celebes (Meyer) Gurney:	345	—	71	—
(Norwich Mus.) vix ad. N. Celebes (Whitely)	355	—	76	—
5 adults, Peling Id., V—VIII. 95 (Nat. Coll.)	312-38	—	—	—
4 adults, 2 juv. Banggai Id., VI. 95 (Nat. Coll.)	305-25	—	—	—
(Norwich Mus.) ♂, Sula Islands	314	221	73	27
(Norwich Mus.) ♀ juv., Sula Islands	305	220	72	26

[1] As will be found more fully explained under the *heading: Haliastur indus*, a long connecting hyphen (*e. g. S. rufipectus — sulaensis*) is used in this work to indicate the connecting forms between subspecies; or, where it may be safely said that such individuals have stronger leanings to one subspecies than to the other, the sign > or <, respectively (*S. rufipectus > sulaensis*) indicating that the affinities are more with the typical form, or (*S. rufipectus < sulaensis*) with the latter form.

The fine series of this Serpent-harrier before us displays great variation in tint; on the breast, for instance, from pale russet to dark Vandyck-brown. The darkest examples are probably the oldest. The specimens from North Celebes are, with one exception, darker than those of the southern peninsula, and Mr. Gurney writes that the specimen from Kema (N. Celebes) in the Norwich Museum "is decidedly darker than our four from Macassar, especially on the breast". Two in the Leyden Museum from Pare-Pare near Macassar do not appear to differ from those of the north, and Mr. Hartert remarks *(a 22)* of some fresh specimens from the southern peninsula in the Tring Museum that the breast is much paler in some examples, darker in others. None of the southern examples before us appear to be old birds. The birds from Peling and Banggai vary in the same way, — one is Vandyck-brown on the breast, most of the others much paler.

Outside the Celebes Province a very near ally of *S. rufipectus* is found in Dr. Sharpe's recently described *Spilornis raja* (Bull. B. O. C. 1893, no. X; Ibis 1893, 552, 569), a specimen of which was sent by Mr. Edward Bartlett from Kuching, Sarawak. This is said to be most like the Sula race, but differs in having the white bars of the breast, abdomen and axillaries strikingly narrower. From Sharpe's measurements it would appear to have the wing of the Sula form (viz. 309 mm), but a shorter tail (178 mm).

The food of Serpent-harriers consists chiefly of frogs, snakes and other Reptilia. The Indian species lay one or two eggs; most usually only one.

The genus *Spilornis* is an important one in questions of geographical distribution. Owing to the small number of eggs laid, the species do not suffer from overcrowding; and there appear to be no causes for its ever shifting its quarters. The genus is a purely Indian one, and is not found further east than the Sula Islands; and the close connection of these islands with Celebes is shown by the fact of their possessing the same species, which also — if identical — occurs on Siao in the Sangi Islands.

Attention might here be called to the close similarity of the plumage of *Circus assimilis* Jard. & Selby to certain species of this genus, especially *Spilornis holospilus* Vig. of the Philippines.[1]) Mimicry is here out of the question, as *C. assimilis* does not occur in the Philippines, and the similarity must be taken as pointing to kinship of the two genera (see below under *Circus assimilis* and *Pernis celebensis*).

GENUS CIRCUS Lac.

The Harriers are of slender form, with very long wings, tail, and tarsus, and somewhat short and not powerful toes. A more or less well developed facial ruff extending from behind the ear-coverts across the throat:

[1]) Cf. also *Spilornis panayensis* Steere (List B. Philip. Is. 1890 p. 7).

bill somewhat weak, with a blunt festoon: tarsi naked (except at the top anteriorly), clad in front with transverse shields, elsewhere with small reticulate scales. Food: amphibians, reptiles, small mammals, etc. Eggs 2—4 in number. About 18 species, migratory and stationary, distributed over the greater part of the world.

2. CIRCUS ASSIMILIS Jard. Selby.

Allied Harrier.

Circus assimilis *(1)* Jard. & Selby, Ill. Orn. 1826, I, pl. 51 type examd.[1]; *(2)* Schl., Mus. P.-B. Circi, 1862, 9; *(III)* id., Valkv. 1866, 29, 66, pl. 20, f. 2, 3; *(4)* Walden, Tr. Z. S. 1872, VIII, 37; *(5)* Sharpe, Cat. B. 1874, I, 63; *(6)* Gurney, Ibis, 1875, 225; *(7)* id., Diurn. B. of Prey, 1884, 23; *(8)* W. Blas., Ztschr. Ges. Orn. 1885, 205, 234; *(9)* North, Nests & Eggs B. Austr. 1889, 1, pl. II, f. 4 (egg); *(10)* Büttik., Webers Reise in Ost-Ind. 1893 III, 272; *(11)* M. & Wg., Abh. Mus. Dresden 1896 no. 1, p. 7; *(12)* Hartert, Nov. Zool. 1896, 163.

a. Circus jardinii Gld., P. Z. S. 1837, 141; *(1)* id., B. Austr. 1848, I, pl. 27; *(2)* S. Müll., Reizen Ind. Arch. 1857, II, 8; *(3)* Gld., Handb. B. Austr. 1865, I, 60; *(4)* Schl., Rev. Acc. 1873, 50.

b. Spilocircus jardinii *(1)* Kaup, Isis, 1847, 102.

c. Strigiceps jardinii *(1)* Bp., Consp. 1850, I, 34.

"Bokan buri", S. Celebes, Platen 8.

For further references see Sharpe 5.

Figures and descriptions. Jardine & Selby *1*; Gould *a 1, a 3*; Schlegel *III*; North *9* (egg); Kaup *b 1*; Sharpe *5*; W. Blas. *8*.

Male, nearly adult. General colour above brownish ash, darker on head; forehead, ear-coverts and crown with rufous margins to the feathers; secondaries pure ashy, banded with dark brown — indistinctly on the inner web; wing-coverts, scapulars and upper tail-coverts marked with short bars or large spots of white, which are more indistinct and ashy on exposed parts of the plumage; shoulder rufous; tail above ashy, below white, crossed with seven bars of blackish and terminally margined with white; under surface — including under wing- and tail-coverts and thighs — cinnamon-rufous, lighter on the thighs and abdomen, and spangled all over with white spots arranged two and two at short intervals on the opposite webs of the feathers. "Iris sulphur-yellow; cere and bill bluish grey (cere pale yellow — Wallace); tip of bill black; feet citron-yellow" (Platen). Nr. 6735, Tjamba, May).

Old. Crown of head, cheeks and ear-coverts tawny-rufous, with blackish mesial streaks to the feathers (♂, Lake Posso, 14. Feb, 95, P. & F. Sarasin).

Female. Like the male, but larger.

Young. Above brown with fulvous margins to the feathers; upper tail-coverts white washed with rufous and having dark brown centres; tail sepia-brown tipped with

---

[1] The type of *Circus assimilis J. & S.* in the British Museum is immature and not normal, differing from all other specimens there of this species in the coloration of the wings and tail. The tail is nearly uniform brownish ashy with a rufous wash at its sides, marked with 3 or 4 imperfect bars of brown towards the base, followed by a clear space, with an imperfect terminal bar. Upper tail-coverts white, a few of the longer ones with a bar of brown towards the tip.

tawny-buff and crossed with six bands of black; below pale tawny-buff, lightest on
abdomen and thighs, and streaked with dark brown on breast and under tail-coverts
(ex Sharpe).

Eggs. 2 or 3, white, with a bluish green tinge on the inner surface; 51—52 × 38—39 mm
(Australia — A. J. North *9*). Uniform white, a little smaller than those of *C.
rufus* of Europe: 49—51 × 39—39 mm: Nehrkorn, MS.).

Nest. Flat, of small sticks and twigs, lined with green leaves; usually placed among the thick
branches of a low tree (Australia).

Breeding-time in Australia. Sept.—Nov.

Measurements.	Wing	Tail	Tarsus	Culmen from cere
a. (Sarasin Coll.) ♂ ad. Lake Posso, Central Celebes, 14. II. 95 (P. & F. S.)	375	235	84	20
b. (C 10709) ♂ vix ad. S. Celebes, 16. IV. 78 (Platen)	390	260	93	19
c. (C 10710) ♀ vix ad. S. Celebes, 25. V. 78 (Platen)	425	275	98	20.5
d. (Nr. 6735) ♂, S. Celebes, 6. V. 78 (Platen)	385	250	93	—
e. (C 11293) [♀] New South Wales	435	280	101	21
f. (C 11092) [♀] juv. Australia	437	280	107	22

Distribution. N. Celebes—Minahassa (Riedel *b 4*), Gorontalo District (Forsten *2*, Rosenberg
b 4), Central Celebes (P. & F. Sarasin *11*), S. Celebes (S. Müller *2, b 2*, Weber *10*,
Platen *8*, Everett *12*); Australia—apparently throughout (Ramsay *9*); Tasmania
(Norwich Mus. *6*).

This well-marked Harrier is remarkable for its distribution, occurring, as
it does, in Celebes and Australia and, so far as is yet known, on none of the
intervening islands. To the north and west of Celebes, the Chinese *Circus spilonotus* Kaup is found in the Philippines and Borneo, and, according to Mr. Everett and Mr. Whitehead (J. Straits Br. R. A. S. 1889, 180; Ibis, 1890, 43),
this species is "a regular winter migrant" to Borneo and Palawan. East of
Celebes, a Harrier (*C. spilothorax*, Salvad. & D'Alb.) has been discovered at
Yule Island in the Papuan Gulf, South New Guinea, corresponding closely in
coloration with *C. maillardi* Verr. (*C. macrocelis* Newton) of Madagascar and
Réunion and with *C. wolfi* Gurney and *C. gouldi* Bp. of New Caledonia and
Australia, respectively (Salvad., Orn. Pap. I, 71); but neither this species, nor
C. spilonotus, have anything to do with *C. assimilis*.

The similarity of the adult plumage of this Harrier to that of certain
species of *Spilornis* is worthy of notice. The type of coloration may be
ancient. It should be remarked that Gurney, whose arrangement of the
Falconidae we follow, places the subfamily *Circinae* much nearer to the *Circaetinae* (containing *Spilornis*) than has been the custom with most other authors (*7*).

C. assimilis is, we believe, a stationary species in Celebes, though we have
only been able to collect the following few dates of occurrence there ranging
from February to November:

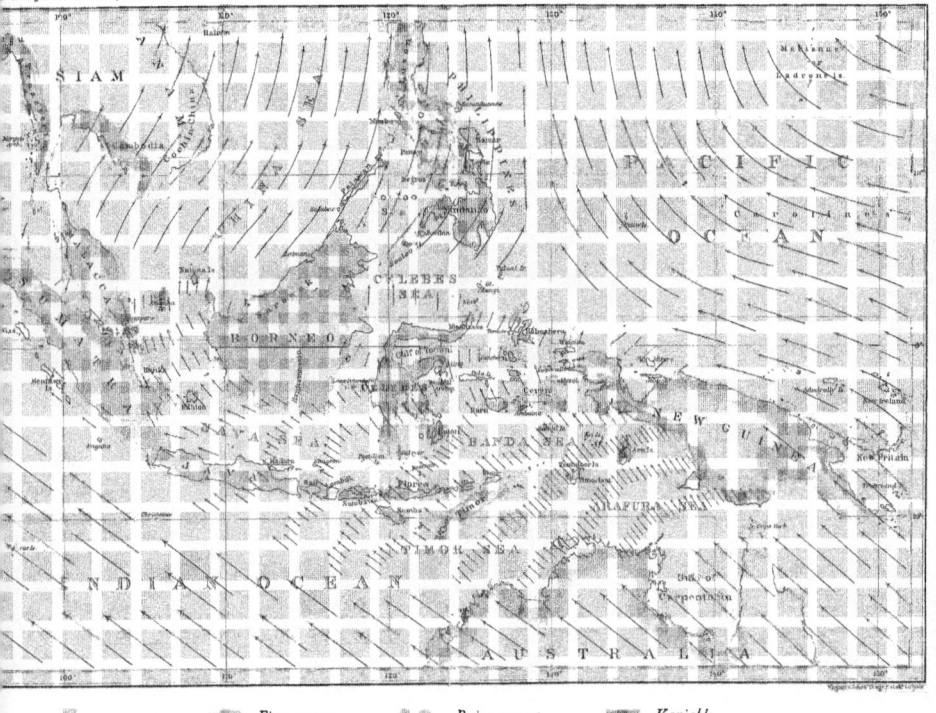

Winds and Rains: April - September

Winds and Rains: October – March

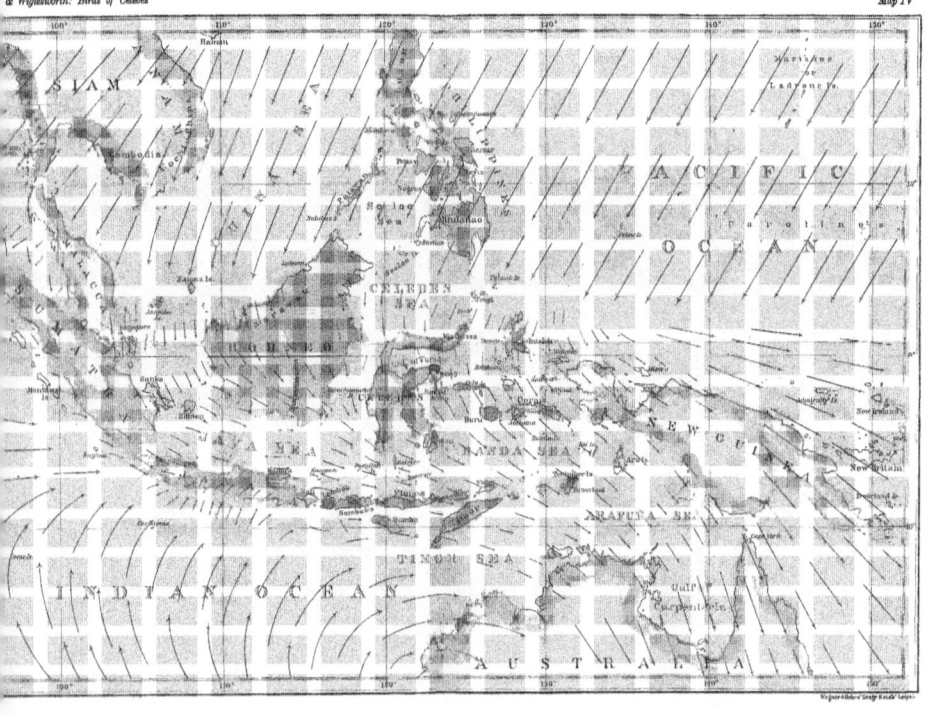

Fine season   Rainy season   Variable

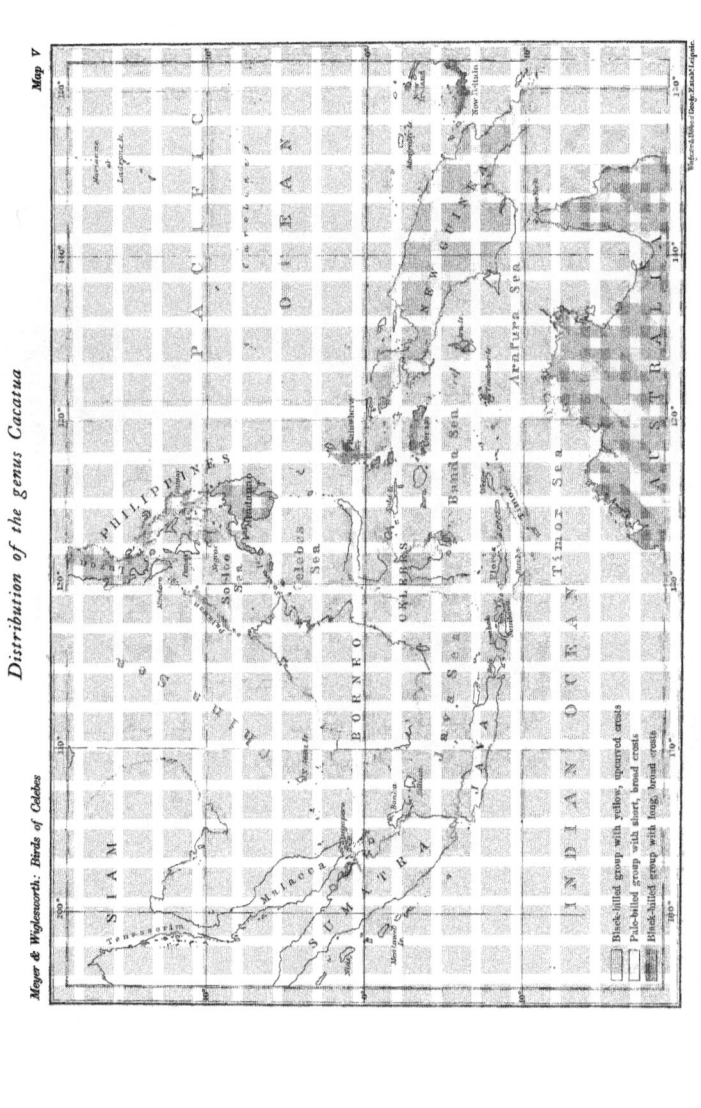

Meyer & Wiglesworth: Birds of Celebes

Distribution of the genus *Cacatua*

Map V

Meyer & Wiglesworth: Birds of Celebes

Distribution of the genus Loriculus

Map VI

Females with red crowns and blue cheeks
Females without red on the crown or blue on the cheeks
Loriculus galgulus
Loriculus vernalis

1. Loriculus vernalis. (2. L. indicus, not on map). 3. L. gulgulus. 4. L. pusillus. 5. L. flosculus. 6. L. philippensis. 7. L. regulus. 8. L. worcesteri. 9. L. chrysonotus. 10. L. bonapartei. 11. L. siquijorensis. 12. L. apicalis. 13. L. bonapartei. 14. L. catamene. 15. L. exilis. 16. L. stigmatus. 17. L. quadricolor. 18. L. sclateri. 19. L. sclateri ruber. 20. L. amabilis. 21. L. aurantiifrons. 22. L. aurantiifrons meeki. 23. L. tener.

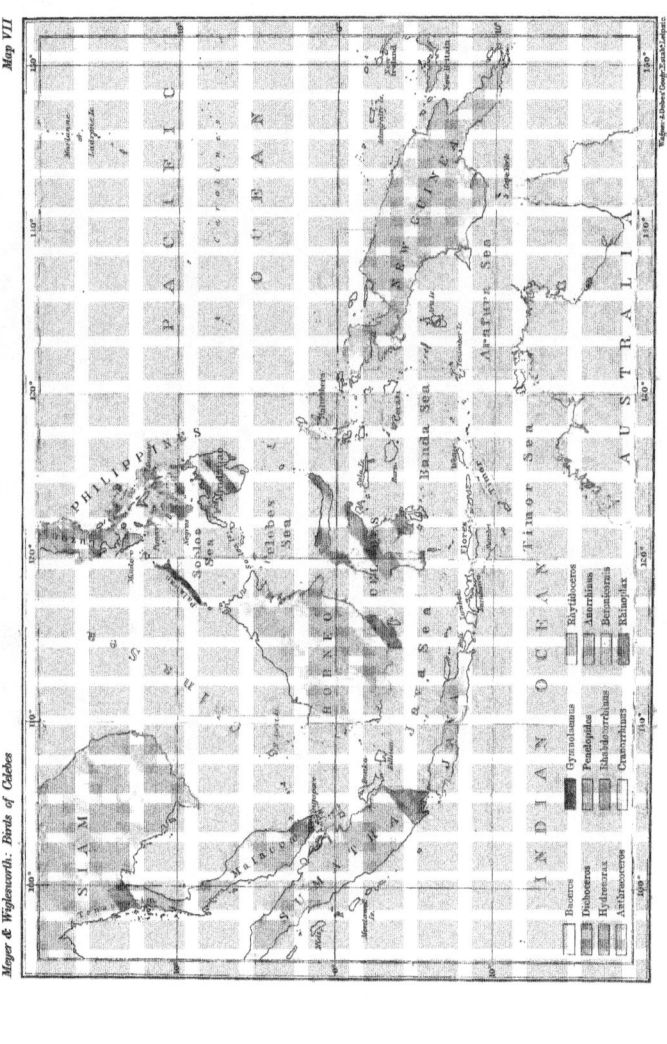

Distribution of the Bucerotidae.

Spilospizias trinotatus (Bp.)
imm.

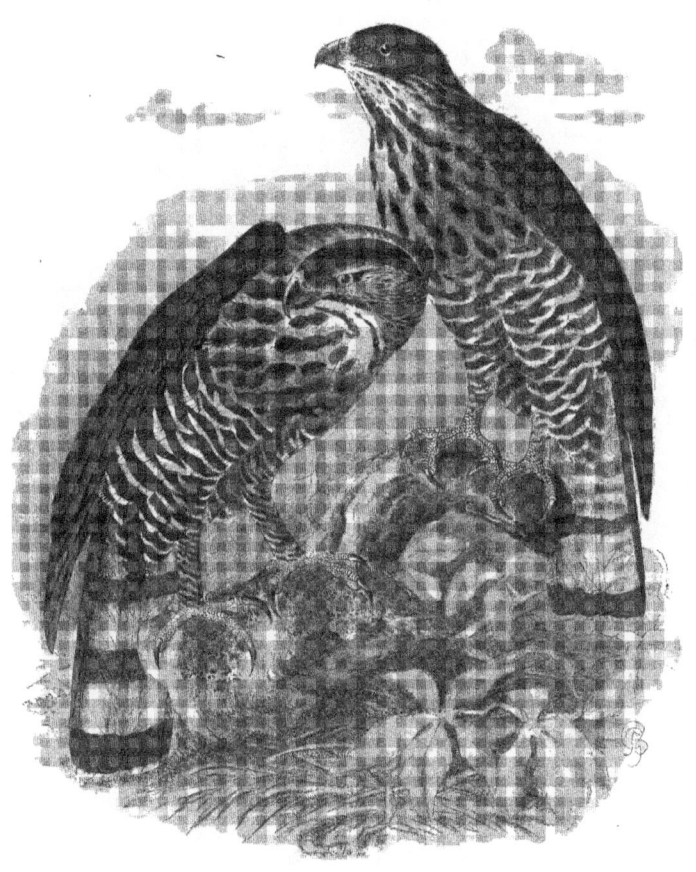

Spizaetus lanceolatus Temm. & Schl.
ad.

Pernis celebensis (Wall.)
ad.

Spizaetus lanceolatus Temm. & Schl.
jur.

Pernis celebensis (Wall.)
juv.

Ninox ochracea (Schl.)
ad. et juv.

Prioniturus platurus (Vieill.) . *1. mas juv., 2. mas imm., 3. mas imm., 4. fem. ad. moulting, 5. mas ad.*

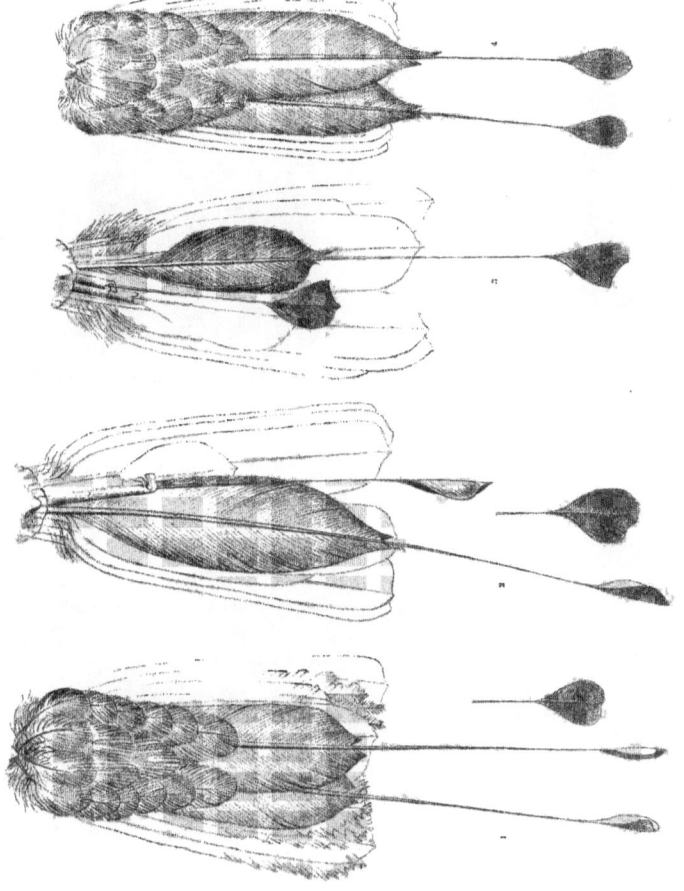

1.–3. *Prioniturus flavicans* Cass. 1. fem. ad, 2.–3. mas moulting   4. *Prioniturus platurus* (Vieill.), mas ad.

Aprosmictus sulaensis Rchw.

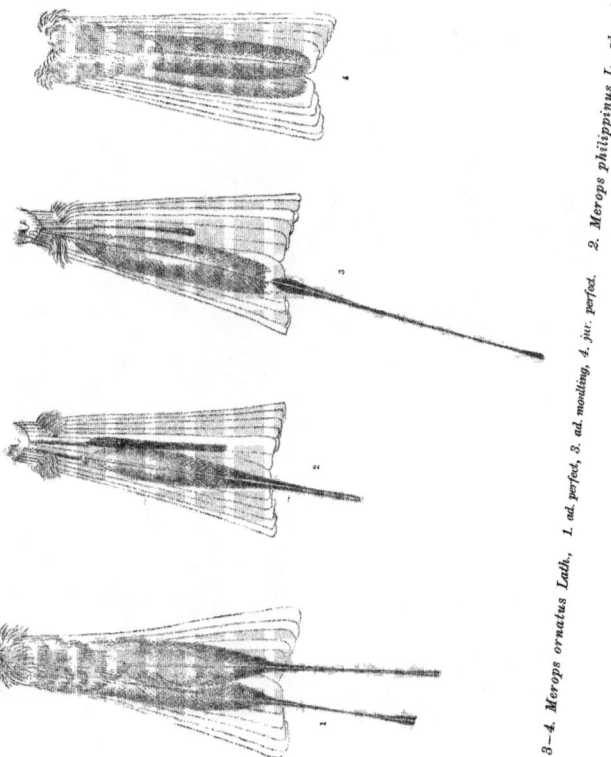

1, 3—4. *Merops ornatus* Lath., 1. ad. perfect, 3. ad. moulting, 4. juv. perfect. 2. *Merops philippinus* L., ad. moulting.

Pelargopsis dichrorhyncha M.&Wg.
Monachalcyon capucinus M.&Wg.

1. Ceycopsis fallax (Schl.)   2.–3. C. sangirensis M. & Wg., *2. ad., 3. juv.*

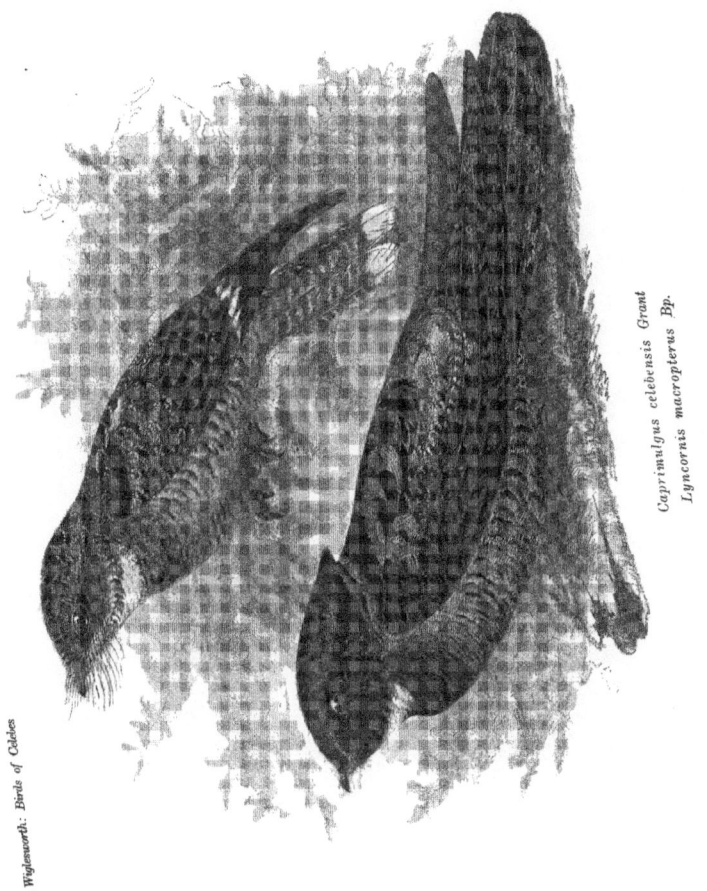

Meyer & Wiglesworth: Birds of Celebes

Caprimulgus celebensis Grant
Lyncornis macropterus Bp.

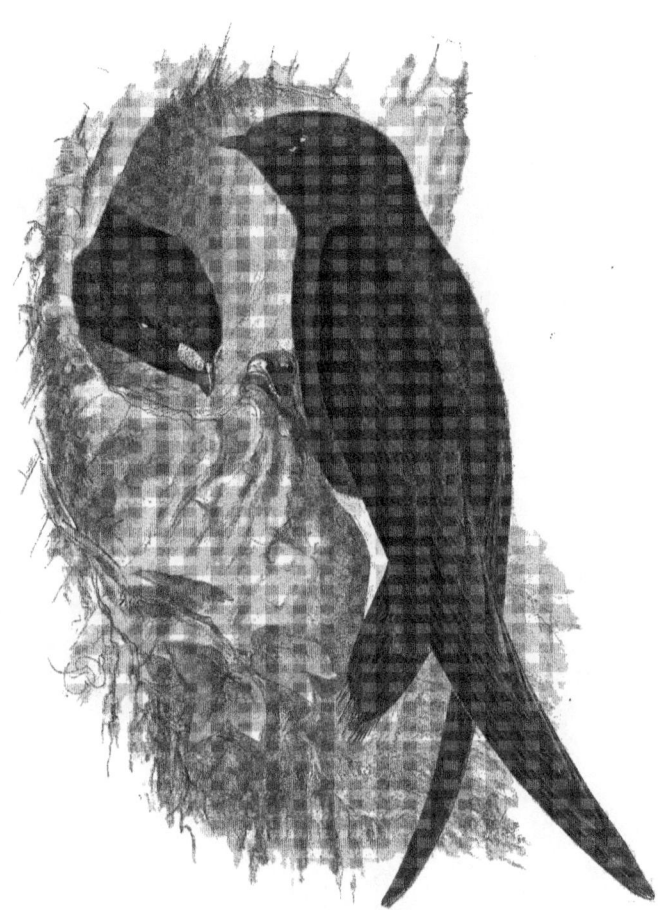

Chaetura celebensis (Sclat.)
fem. juv. & mas.

Muscicapula hyperythra (Blyth)
mas, fem. et juv.

1. Siphia banyumas (Horsf.)
2. Siphia djampeana Hart.
3. Siphia kalaoensis Hart.

Stoparola septentrionalis Bütt.
mas, fem. et juv.

1. Monarcha commutatus Brügg. 2.–3. Zeocephus talautensis M.&Wg., 3. ad. et 2. juv.

Monarcha everetti Hart.
Pachycephala teijsmanni Bütt.
Pachycephala everetti Hart.

PUBLISHED BY R. FRIEDLÄNDER & SOHN, BERLIN.

A. B. MEYER
Abbildungen von Vogelskeletten

herausgegeben mit Unterstützung der Generaldirection
der Königlichen Sammlungen für Kunst und Wissenschaft in Dresden.
2 Bände mit 242 Tafeln in Lichtdruck.
1879—97. Gross-4.
Preis 360 Mark (18 £).

Neuer Beitrag
zur Kenntniss der Vogelfauna von Kaiser Wilhelmsland
besonders vom Huongolfe

nebst Bemerkungen über andere papuanische Vögel und einer Liste aller bisher
von Kaiser Wilhelmsland registrirten.
MIT 1 KARTE (VOM HUONGOLFE) UND 1 TAFEL IN LICHTDRUCK.
1893. Gross-4.
Preis 8 Mark (8 sh.).

ZWEI NEUE PARADIESVÖGEL
(Pteridophora alberti und Parotia carolae),
MIT 2 COLORIRTEN TAFELN.
1895. Gross-4.
Preis 8 Mark (8 sh.).

L. W. WIGLESWORTH
AVES POLYNESIAE.
A Catalogue of Birds of the Polynesian Subregion (not including the Sandwich Islands).
1892. Royal-4.
Price 14 Mark (14 sh.).

L. TACZANOWSKI
ORNITHOLOGIE DU PÉROU.
4 volumes.
1884—86. grand in 8.
Prix 59 Mark (2 £ 19 sh.).

Printed by Breitkopf & Härtel, Leipzig.

www.ingramcontent.com/pod-product-compliance
Lightning Source LLC
Chambersburg PA
CBHW021227300426
44111CB00007B/451